(Continued on inside back cover)

Encyclopedia of Pest Management

Encyclopedia of Pest Management

edited by

David Pimentel

Cornell University
Ithaca, New York, U.S.A.

MARCEL DEKKER

MARCEL DEKKER, INC. NEW YORK • BASEL

ISBN: on-line 0-8247-0517-3
print 0-8247-0632-3

This book is printed on acid-free paper.

Headquarters
Marcel Dekker, Inc.
270 Madison Avenue, New York, NY 10016
tel: 212-696-9000; fax: 212-685-4540

Eastern Hemisphere Distribution
Marcel Dekker AG
Hutgasse 4, Postfach 812, CH-4001 Basel, Switzerland
tel: 41-61-261-8482; fax: 41-61-261-8896

World Wide Web
http://www.dekker.com

The publisher offers discounts on this book when ordered in bulk quantities. For more information, write to Special Sales/Professional Marketing at the headquarters address above.

Current printing (last digit):

10 9 8 7 6 5 4 3 2 1

PRINTED IN THE UNITED STATES OF AMERICA

Encyclopedia of Pest Management

David Pimentel, Editor
Cornell University

Editorial Advisory Board

Matt Liebman *Department of Agronomy, Iowa State University, Iowa, U.S.A.*

Keith Moody *International Rice Research Institute (retired), Makawao, Hawaii, U.S.A.*

Ida Nyoman Oka *Bali, Indonesia*

Larry P. Pedigo *Iowa State University, Ames, Iowa, U.S.A.*

H.F. van Emden *School of Plant Sciences, University of Reading, United Kingdom*

Frank G. Zalom *Statewide IPM Project, University of California, Davis, California, U.S.A.*

List of Contributors

John All / *University of Georgia, Athens, Georgia, U.S.A.*

James A. Allen / *Paul Smith's College, Paul Smiths, New York, U.S.A.*

Stanley Anderson / *University of Wyoming, Laramie, Wyoming, U.S.A.*

Randy Anderson / *USDA-ARS, Brookings, South Dakota, U.S.A.*

Jude Andreasen / *U.S. Environmental Protection Agency, Washington, D.C., U.S.A.*

Alicia Armentia / *Hospital Rio Hortega, Valladolid, Spain*

Salvatore Arpaia / *Metapontum Agrobios, Metaponto, MT, Italy*

Gavin Ash / *Charles Sturt University, Wagga Wagga, New South Wales, Australia*

William W. Au / *The University of Texas, Galveston, Texas, U.S.A.*

Miriam Austerweil / *Institute of Agricultural Engineering, Bet Dagan, Israel*

M.L. Avery / *National Wildlife Research Center, Gainesville, Florida, U.S.A.*

Karen L. Bailey / *Agriculture and Agri-Food Canada, Saskatoon Research Centre, Saskatchewan, Canada*

Piara S. Bains / *Alberta Agriculture Food and Rural Development, Edmonton, Alberta, Canada*

R.A. Balikai / *University of Agricultural Sciences, Dharwad, Karnataka, India*

Barbara I.P. Barratt / *AgResearch, Ltd., Invermay Agriculture Center, Mosgiel, New Zealand*

Johann Baumgärtner / *International Centre of Insect Physiology and Ecology (ICIPE), Nairobi, Kenya*

Lars Bergström / *Swedish University of Agricultural Sciences, Uppsala, Sweden*

Alan A. Berryman / *Washington State University, Pullman, Washington, U.S.A.*

P.S. Bisen / *Institute of Microbiology and Biotechnology, Barkatullah University, Bhopal, Madhya Pradesh, India*

William W. Bockus / *Kansas State University, Manhattan, Kansas, U.S.A.*

Michael J. Bodenchuk / *U.S. Department of Agriculture, Salt Lake City, Utah, U.S.A.*

Gilles Boiteau / *Agriculture and Agri-Food Canada, Fredericton, New Brunswick, Canada*

Mary Bomford / *Bureau of Rural Sciences, Canberra, Australia*

John H. Borden / *Simon Fraser University, Burnaby, British Columbia, Canada*

Donald Boulter / *University of Durham, Durham, United Kingdom*

Céline Boutin / *Canadian Wildlife Service, Hull, Québec, Canada*

Susan M. Boyetchko / *Agriculture and Agri-Food Canada, Saskatoon Research Centre, Saskatchewan, Canada*

Bent Bromand / *Danish Institute of Agricultural Sciences, Slagelse, Denmark*

William L. Bruckart III / *USDA-ARS, Foreign Disease–Weed Science Research Unit, Ft. Detrick, Maryland, U.S.A.*

Carl A. Bruice / *John Taylor Fertilizers, Wilbur-Ellis Company, Sacramento, California, U.S.A.*

Peter A. Burnett / *Agriculture and Agri-Food Canada, Lethbridge, Alberta, Canada*

Frederick H. Buttel / *University of Wisconsin, Madison, Wisconsin, U.S.A.*

Dongfeng Cao / *Cornell University, Ithaca, New York, U.S.A.*

Kitty F. Cardwell / *International Institute of Tropical Agriculture, Cotonou, Bénin*

Steven J. Castle / *USDA-ARS, Phoenix, Arizona, U.S.A.*

Chris P. Chanway / *University of British Columbia, Vancouver, British Columbia, Canada*

Baik Kee Cho / *The Catholic University of Korea, Seoul, South Korea*

Sharon A. Clay / *South Dakota State University, Brookings, South Dakota, U.S.A.*

Matthew Colloff / *CSIRO Entomology, Canberra, Australia*

Mary L. Cornelius / *Southern Regional Research Center, New Orleans, Louisiana, U.S.A.*

Joseph D. Cornell / *State University of New York, Syracuse, New York, U.S.A.*

Robert M. Corrigan / *RMC Pest Management Consulting, Richmond, Indiana, U.S.A.*

Ian P. Craig / *AEMS Pty, Ltd., Toowoomba, Queensland, Australia*

Thomas W. Culliney / *Hawaii Department of Agriculture, Honolulu, Hawaii, U.S.A.*

Paul D. Curtis / *Cornell University, Ithaca, New York, U.S.A.*

Christopher F. Curtis / *London School of Hygiene & Tropical Medicine, London, United Kingdom*

Kent M. Daane / *University of California, Berkeley, California, U.S.A.*

Abhaya M. Dandekar / *University of California, Davis, California, U.S.A.*

Joanna Davies / *IACR-Long Ashton Research Station, Bristol, United Kingdom*

Franck E. Dayan / *United States Department of Agriculture, University, Mississippi, U.S.A.*

J. Renato de Freitas / *University of Saskatchewan, Saskatoon, Saskatchewan, Canada*

Victor de Vlaming / *Water Resources Control Board, Sacramento, California, U.S.A.*

Kathleen Delate / *Iowa State University, Ames, Iowa, U.S.A.*

Doug Derksen / *Agriculture and Agri-Food Canada, Brandon, Manitoba, Canada*

Malcolm D. Devine / *Aventis CropScience Canada Co., Saskatoon, Canada*

S.P. Dhua / *Regional Network on Safe Pesticide Production and Information for Asia and the Pacific (RENPAP) UNDP/UNIDO, New Delhi, India*

Jan Dich / *Karolinska Institutet and Karolinska University Hospital, Stockholm, Sweden*

Christina DiFonzo / *Michigan State University, East Lansing, Michigan, U.S.A.*

Barbara Dinham / *Pesticide Action Network UK, London, United Kingdom*

Joseph M. DiTomaso / *University of California, Davis, California, U.S.A.*

Gary J. Dorr / *University of Queensland, Gatton, Queensland, Australia*

Darna L. Dufour / *University of Colorado, Boulder, Colorado, U.S.A.*

Stephen O. Duke / *United States Department of Agriculture, University, Mississippi, U.S.A.*

Jim M. Dunwell / *The University of Reading, Reading, United Kingdom*

Etienne Duveiller / *CIMMYT, Kathmandu, Nepal*

Charles Eason / *CENTOX Centre for Environmental Toxicology, Lincoln, New Zealand*

Clive A. Edwards / *The Ohio State University, Columbus, Ohio, U.S.A.*

Sanford D. Eigenbrode / *University of Idaho, Moscow, Idaho, U.S.A.*

George Ekstrom / *National (Swedish) Chemicals Inspectorate, Solna, Sweden*

Matthew Escobar / *University of California, Davis, California, U.S.A.*

Ana María Evangelista de Duffard / *Universidad Nacional de Rosario, Rosario, Argentina*

Brian A. Federici / *University of California, Riverside, California, U.S.A.*

Gillian Ferguson / *Ontario Ministry of Agriculture, Food, and Rural Affairs, Harrow, Ontario, Canada*

Paul Fields / *Cereal Research Centre, Agriculture and Agri-Food Canada, Winnipeg, Manitoba, Canada*

Maria R. Finckh / *University of Kassel, Witzenhausen, Germany*

Nancy Fitz / *U.S. Environmental Protection Agency, Washington, D.C., U.S.A.*

Shelby J. Fleischer / *Pennsylvania State University, University Park, Pennsylvania, U.S.A.*

W.J. Florkowski / *University of Georgia, Griffin, Georgia, U.S.A.*

Brenda Frick / *University of Saskatchewan, Saskatoon, Saskatchewan, Canada*

Monika Frielinghaus / *Institute of Soil Landscape Research, ZALF Müncheberg, Germany*

Joe Funderburk / *University of Florida, Quincy, Florida, U.S.A.*

Ján Gallo / *Slovak Agricultural University, Nitra, Slovakia*

Abraham Gamliel / *Institute of Agricultural Engineering, Bet Dagan, Israel*

Angharad M.R. Gatehouse / *University of Newcastle, Newcastle upon Tyne, United Kingdom*

Cesare Gessler / *Institute of Plant Sciences—Pathology, Swiss Federal Institute of Technology, Zürich, Switzerland*

Andrew Gilbert / *Ministry of Agriculture, Fisheries, Food, Sand Hutton, York, United Kingdom*

Bruce D. Gossen / *Agriculture and Agri-Food Canada, Saskatoon Research Centre, Saskatoon, Saskatchewan, Canada*

Ragini Gothalwal / *Institute of Microbiology and Biotechnology, Barkatullah University, Bhopal, Madhya Pradesh, India*

Tetsuo Gotoh / *Ibaraki University, Ibaraki, Japan*

Simon Gowen / *The University of Reading, Earley Gate, Reading, United Kingdom*

Simon R. Gowen / *The University of Reading, Earley Gate, Reading, United Kingdom*

Michael E. Gray / *University of Illinois, Urbana, Illinois, U.S.A.*

Fred A. Gray / *University of Wyoming, Laramie, Wyoming, U.S.A.*

Geoff Gurr / *University of Sydney, Orange, New South Wales, Australia*

Andrew Paul Gutierrez / *University of California, Berkeley, California, U.S.A.*

N.G.M. Hague / *The University of Reading, Earley Gate, Reading, United Kingdom*

Ann E. Hajek / *Cornell University, Ithaca, New York, U.S.A.*

Charles A.S. Hall / *State University of New York, Syracuse, New York, U.S.A.*

Franklin R. Hall / *Ohio State University, Wooster, Ohio, U.S.A.*

Guy J. Hallman / *Agricultural Research Service, Weslaco, Texas, U.S.A.*

Denis Hamilton / *Animal and Plant Health Service, Brisbane, Queensland, Australia*

Ron B. Hammond / *Ohio State University, Wooster, Ohio, U.S.A.*

Elizabeth Harausz / *Cornell University, Ithaca, New York, U.S.A.*

Peter Harris / *Agriculture and Agri-Food Canada, Lethbridge, Alberta, Canada*

Kelsey A. Hart / *Cornell University, Ithaca, New York, U.S.A.*

Leon G. Higley / *University of Nebraska, Lincoln, Nebraska, U.S.A.*

Paul D. Hildebrand / *Agriculture and Agri-Food Canada, Kentville, Nova Scotia, Canada*

Vaughan A. Hilder / *University of Durham, Durham, United Kingdom*

Nancy C. Hinkle / *University of California, Riverside, California, U.S.A.*

Heikki Hokkanen / *University of Helsinki, Helsinki, Finland*

John M. Holland / *The Game Conservancy Trust, Fordingbridge, Hampshire, United Kingdom*

Xiao-Yue Hong / *Nanjing Agricultural University, Nanajing, Jiangsu, China*

David J. Horn / *Ohio State University, Columbus, Ohio, U.S.A.*

Marjorie A. Hoy / *University of Florida, Gainesville, Florida, U.S.A.*

Hei-Ti Hsu / *Beltsville Agricultural Research Center, Beltsville, Maryland, U.S.A.*

Ruguo Huang / *AMCO Produce, Inc., Leamington, Ontario, Canada*

Richard S. Hunt / *Natural Resources Canada, Canadian Forest Service, Victoria, British Columbia, Canada*

Mark D. Hunter / *University of Georgia, Athens, Georgia, U.S.A.*

William R. Jarvis / *Agri-Food and Agriculture Canada, Harrow, Ontario, Canada*

Janice King Jensen / *U.S. Environmental Protection Agency, Washington, D.C., U.S.A.*

Ike Jeon / *Kansas State University, Manhattan, Kansas, U.S.A.*

Jørgen B. Jespersen / *Danish Pest Infestation Laboratory, Lyngby, Denmark*

Judy Johnson / *Horticultural Crops Research Laboratory, Fresno, California, U.S.A.*

Adrian Johnston / *Potash and Phosphate Institute of Canada, Saskatoon, Saskatchewan, Canada*

Carl J. Jones / *University of Tennessee, Knoxville, Tennessee, U.S.A.*

Davy Jones / *University of Kentucky, Lexington, Kentucky, U.S.A.*

Vincent P. Jones / *Washington State University, Wenatchee, Washington, U.S.A.*

Adel A. Kader / *University of California, Davis, California, U.S.A.*

Inger Källander / *Swedish Ecological Farmers Association, Katrineholm, Sweden*

Michael A. Kamrin / *Michigan State University, East Lansing, Michigan, U.S.A.*

Rory Karhu / *University of Wyoming, Laramie, Wyoming, U.S.A.*

Marianne Karpenstein-Machan / *University Kassel, Witzenhausen, Germany*

Phillip E. Kaufman / *Cornell University, Ithaca, New York, U.S.A.*

Lawrence M. Kawchuk / *Agriculture and Agri-Food Canada, Lethbridge, Alberta, Canada*

George G. Kennedy / *North Carolina State University, Raleigh, North Carolina, U.S.A.*

Matthias Kern / *Deutsche Gesellschaft für Technische Zusammenarbeit (GTZ) GmbH, Bonn, Germany*

Peter Kerr / *Pest Animal Control Cooperative Research Center, CSIRO Sustainable Ecosystems, Canberra, Australian Capital Territory, Australia*

Bhupinder P.S. Khambay / *IACR-Rothamsted, Harpenden, Hertfordshire, United Kingdom*

Zeyaur R. Khan / *International Centre of Insect Physiology and Ecology (ICIPE), Nairobi, Kenya*

Mohamed Khelifi / *Université Laval, Québec, Canada*

A.E. Kiszewski / *Harvard School of Public Health, Boston, Massachusetts, U.S.A.*

David W. Koch / *University of Wyoming, Laramie, Wyoming, U.S.A.*

Marcos Kogan / *Oregon State University, Corvallis, Oregon, U.S.A.*

Birgitta Kolmodin-Hedman / *Karolinska Institute, Stockholm, Sweden*

Zlatko Korunic / *Diatom Research and Consulting, Guelph, Ontario, Canada*

E.S. Krafsur / *Iowa State University, Ames, Iowa, U.S.A.*

David P. Kreutzweiser / *Canadian Forest Service, Sault Ste. Marie, Ontario, Canada*

Sheldon Krimsky / *Tufts University, Medford, Massachusetts, U.S.A.*

Lori Lach / *Cornell University, Ithaca, New York, U.S.A.*

Andrew Landers / *Cornell University, Ithaca, New York, U.S.A.*

Douglas A. Landis / *Michigan State University, East Lansing, Michigan, U.S.A.*

Tomaz Langenbach / *Universidade Federal do Rio de Janeirollha do Fundão, Rio de Janeiro, Brazil*

Wayne Thomas Lanini / *University of California, Davis, California, U.S.A.*

Stephen Lefko / *Monsanto Company, DeKalb, Illinois, U.S.A.*

Anne Légère / *Agriculture and Agri-Food Canada, Sainte-Foy, Quebec City, Canada*

David E. Legg / *University of Wyoming, Laramie, Wyoming, U.S.A.*

Hugh Lehman / *University of Guelph, Guelph, Ontario, Canada*

Gilles D. Leroux / *Université Laval, Québec, Ontario, Canada*

Hermann Levinson / *Max-Planck-Institute of Behaviour Physiology, Seewiesen, Germany*

Anna Levinson / *Max-Planck-Institute of Behaviour Physiology, Seewiesen, Germany*

Gavin Lewis / *JSC International Ltd., Harrogate, Yorkshire, United Kingdom*

Leslie C. Lewis / *United States Department of Agriculture, Ames, Iowa, U.S.A.*

W.J. Lewis / *Crop Protection and Management Research Unit, USDA-ARS, Tifton, Georgia, U.S.A.*

Shao Lin / *State University of New York at Albany, Rensselaer, New York, U.S.A.*

John E. Lloyd / *University of Wyoming, Laramie, Wyoming, U.S.A.*

Jeffrey A. Lockwood / *University of Wyoming, Laramie, Wyoming, U.S.A.*

Chris J. Lomer / *Royal Veterinary and Agricultural University (KVL) Copenhagen, Denmark*

Kathrine Hauge Madsen / *The Royal Veterinary and Agricultural University, Frederiksberg, Denmark*

Ken I. Mallett / *Natural Resources Canada, Canadian Forest Service, Edmonton, Alberta, Canada*

Barbara Manachini / *University of Milan, Milan, Italy*

Rex E. Marsh / *University of California, Davis, California, U.S.A.*

James J. Matee / *Tropical Pesticides Research Institute, Arusha, Tanzania*

G.A. Matthews / *Imperial College of Science, Technology, and Medicine, Berkshire, United Kingdom*

Thomas J. Mbise / *Tropical Pesticides Research Institute, Arusha, Tanzania*

William John Hannan McBride / *Cairns Base Hospital, Queensland, Australia*

Ann McCampbell / *Multiple Chemical Sensitivities Task Force of New Mexico, Santa Fe, New Mexico, U.S.A.*

Mary Ruth McDonald / *University of Guelph, Guelph, Ontario, Canada*

Robert J. McGovern / *University of Florida, Bradenton, Florida, U.S.A.*

Crawford McNair / *Simon Fraser University, Burnaby, British Columbia, Canada*

Robert McSorley / *University of Florida, Gainesville, Florida, U.S.A.*

Peter Messenger / *Aventis CropScience France, S.A., Villefranche-sur-Saône, France*

Thomas A. Miller / *University of California, Riverside, California, U.S.A.*

J. David Miller / *Carleton University, Ottawa, Ontario, Canada*

Lowell A. Miller / *USDA/National Wildlife Research Center, Fort Collins, Colorado, U.S.A.*

Nicholas J. Mills / *University of California, Berkeley, California, U.S.A.*

Pierre Mineau / *Canadian Wildlife Service, Ottawa, Ontario, Canada*

Mohyuddin Mirza / *Alberta Agriculture Food and Rural Development, Edmonton, Alberta, Canada*

Carl J. Mitchell / *Centers for Disease Control and Prevention, Fort Collins, Colorado, U.S.A.*

S. Mohan / *Tamil Nedu Agricultural University, Coimbatore, India*

David Moore / *CABI Bioscience UK Centre (Ascot), Berkshire, United Kingdom*

Lindsay Moose / *U.S. Environmental Protection Agency, Washington, D.C., U.S.A.*

Charles J. Muangirwa / *Tropical Pesticides Research Institute, Arusha, Tanzania*

Graeme Murphy / *Ontario Ministry of Agriculture, Food, and Rural Affairs, Vineland Station, Ontario, Canada*

Deborah A. Neher / *University of Toledo, Toledo, Ohio, U.S.A.*

Phillip Nolte / *University of Idaho, Idaho Falls, Idaho, U.S.A.*

Patrick S. O'Connor-Marer / *University of California, Davis, California, U.S.A.*

Ida Nyoman Oka / *Amlapura (Karangasem), Bali, Indonesia*

T.O. Olagbemiro / *IACR-Rothamsted, Hertfordshire, United Kingdom*

Michael T. Olexa / *University of Florida, Gainesville, Florida, U.S.A.*

Dawn M. Olson / *Crop Protection and Management Research Unit, USDA-ARS, Tifton, Georgia, U.S.A.*

Philip Oduor Owino / *Kenyatta University, Nairobi, Kenya*

Margareta Palmborg / *Swedish Poisons Information Centre, Stockholm, Sweden*

Maurizio G. Paoletti / *Padova University, Padova, Italy*

Timothy Paulitz / *United States Department of Agriculture, Washington State University, Pullman, Washington, U.S.A.*

Elke Pawelzik / *Georg-August-University Goettingen, Goettingen, Germany*

Larry P. Pedigo / *Iowa State University, Ames, Iowa, U.S.A.*

Meir Paul Pener / *The Hebrew University of Jerusalem, Jerusalem, Israel*

John H. Perkins / *The Evergreen State College, Olympia, Washington, U.S.A.*

Rajinder Peshin / *Sher-e-Kashmir University of Agricultural Sciences & Technology, Jammu, India*

Robert K.D. Peterson / *Dow AgroSciences, Indianapolis, Indiana, U.S.A.*

P. Larry Phelan / *The Ohio State University, Wooster, Ohio, U.S.A.*

John A. Pickett / *IACR-Rothamsted, Hertfordshire, United Kingdom*

David Pimentel / *Cornell University, Ithaca, New York, U.S.A.*

Richard J. Pollack / *Harvard School of Public Health, Boston, Massachusetts, U.S.A.*

Nick Price / *Central Science Laboratory, Ministry of Agriculture, Fisheries, and Food, Sand Hutton, York, United Kingdom*

Zamir K. Punja / *Simon Fraser University, Burnaby, British Columbia, Canada*

Serge Quilici / *CIRAD-FLHOR Réuion, Saint-Pierre Cedex, France*

Jorge Rabinovich / *Universidad de Buenos Aires, Buenos Aires, Argentina*

Glen C. Rains / *University of Georgia, Tifton, Georgia, U.S.A.*

B. Rajendran / *Tamil Nadu Agricultural University, Cuddalore, India*

Somiahnadar Rajendran / *Central Food Technological Research Institute, Mysore, India*

Uri Regev / *Ben-Gurion University, Beer-Sheva, Israel*

E.D. Richter / *Hebrew University-Hadassah School of Public Health and Community Medicine, Jerusalem, Israel*

Les Robertson / *Sugar Research and Development Corporation, Innisfail, Queensland, Australia*

Geraldo Stachetti Rodrigues / *Embrapa Environment, Jaguariúna, Sao Paolo, Brazil*

Wendell L. Roelofs / *Cornell University, Geneva, New York, U.S.A.*

Jennifer Ruesink / *University of Washington, Seattle, Washington, U.S.A.*

Gunnar Rundgren / *Grolink, Höje, Sweden*

Victor O. Sadras / *CSIRO Land & Water, Glen Osmond, South Australia, Australia*

Peter Sandøe / *The Royal Veterinary and Agricultural University, Frederiksberg, Denmark*

Chris Sansone / *Texas Agricultural Extension Service, San Angelo, Texas, U.S.A.*

Peter J. Savarie / *National Wildlife Research Center, Fort Collins, Colorado, U.S.A.*

David S. Seigler / *University of Illinois, Urbana, Illinois, U.S.A.*

Cetin Sengonca / *University of Bonn, Bonn, Germany*

Mike W. Service / *Liverpool School of Tropical Medicine, Liverpool, England*

Alexei Sharov / *Virginia Polytechnic Institute and State University, Blacksburg, Virginia, U.S.A.*

Roland Sigvald / *Swedish University of Agricultural Sciences, Uppsala, Sweden*

Grant R. Singleton / *CSIRO Sustainable Ecosystems, Australian Capital Territory, Canberra, Australia*

Edward H. Smith / *Cornell University, Ithaca, New York, U.S.A.*

Daniel E. Sonenshine / *Old Dominion University, Norfolk, Virginia, U.S.A.*

David B. South / *Auburn University, Auburn, Alabama, U.S.A.*

Bernhard Speiser / *Research Institute of Organic Agriculture (FiBL), Frick, Switzerland*

Andrew Spielman / *Harvard School of Public Health, Boston, Massachusetts, U.S.A.*

James J. Stapleton / *University of California, Parlier, California, U.S.A.*

Tove Steenberg / *Danish Pest Infestation Laboratory, Lyngby, Denmark*

J.H. Stephens / *MicroBio RhizoGen. Corp., Saskatoon, Saskatchewan, Canada*

Mark M. Stevens / *Yanco Agricultural Institute, Yanco, New South Wales, Australia*

Jamie Stevens / *University of Exeter, Exeter, United Kingdom*

F. Craig Stevenson / *Saskatoon, Saskatchewan, Canada*

John H. Stevenson / *Harpenden, Hertfordshire, United Kingdom*

Graham Stirling / *Biological Crop Protection Pty. Ltd., Moggill, Queensland, Australia*

Praful Suchak / *Suchak's Consultancy Services, Mumbai, India*

Charles G. Summers / *University of California, Davis, California, U.S.A.*

D. Swarup / *Indian Veterinary Research Institute, Izatnagar, UP, India*

Maurice J. Tauber / *Cornell University, Ithaca, New York, U.S.A.*

John R. Teasdale / *USDA-ARS, Beltsville, Maryland, U.S.A.*

Jalpa P. Tewari / *University of Alberta, Edmonton, Alberta, Canada*

Fred Thomas / *CERUS Consulting, Chico, California, U.S.A.*

H. David Thurston / *Cornell University, Ithaca, New York, U.S.A.*

Robert M. Timm / *University of California, Hopland, California, U.S.A.*

James A. Traquair / *Agriculture and Agri-Food Canada, London, Ontario, Canada*

James H. Tumlinson / *Center for Medical, Agricultural, and Veterinary Entomology, USDA-ARS, Gainesville, Florida, U.S.A.*

Michael A. Valenti / *Delaware Forest Service, Dover, Delaware, U.S.A.*

Helmut van Emden / *The University of Reading, Reading, Berkshire, United Kingdom*

Robert Verkerk / *Imperial College of Science, Technology, and Medicine, Berkshire, United Kingdom*

Heike Vibrans / *Colegio de Postgraduados en Ciencias Agrícolas, Montecillo, Mexico*

Clément Vigneault / *Agriculture and Agri-Food Canada, Saint-Jean-sur-Richelieu, Québec, Ontario, Canada*

Paul Vincelli / *University of Kentucky, Lexington, Kentucky, U.S.A.*

Charles Vincent / *Agriculture and Agri-Food Canada, Saint-Jean-sur-Richelieu, Quebec, Canada*

Richard Wall / *University of Bristol, Bristol, United Kingdom*

F.L. Walley / *University of Saskatchewan, Saskatoon, Saskatchewan, Canada*

Douglas Walsh / *Washington State University, Prosser, Washington, U.S.A.*

W.S. Washington / *Institute for Horticultural Development, DNRE, Melbourne, Victoria, Australia*

Alan K. Watson / *International Rice Research Institute, Makati City, Philippines*

D.W. Watson / *North Carolina State University, Raleigh, North Carolina, U.S.A.*

Michael J. Weiss / *University of Idaho, Moscow, Idaho, U.S.A.*

Gerald Wilde / *Kansas State University, Manhattan, Kansas, U.S.A.*

H. Robinson William / *Urban Pest Control Research and Consulting, Christiansburg, Virginia, U.S.A.*

Ingrid H. Williams / *IACR-Rothamsted, Harpenden, Hertfordshire, United Kingdom*

Jeffrey P. Wilson / *University of Georgia Coastal Plain Experiment Station, USDA-ARS, Tifton, Georgia, U.S.A.*

Nicholas Woods / *University of Queensland, Gatton, Queensland, Australia*

Steve Wratten / *Lincoln University, Canterbury, New Zealand*

Amor Yahyaoui / *ICARDA, Aleppo, Syria*

Kohji Yamamura / *National Institute for Agro-Environmental Sciences, Tsukuba, Japan*

Frank G. Zalom / *University of California, Davis, California, U.S.A.*

Kun Yan Zhu / *Kansas State University, Manhattan, Kansas, U.S.A.*

Preface

The Encyclopedia of *Pest Management* focuses on the identification and management of diverse pest species that damage and/or destroy food and livestock products as well as forest products. Throughout the world we are faced with rapid growth in the human population; ensuring an adequate food supply is a prime concern for everyone.

Of major importance is the growing threat to the security of the human food supply. Signaling the seriousness of the human population explosion to our food security is the recent World Health Organization report that indicates that more than 3 billion people are now malnourished. This is the largest number and proportion of malnourished ever reported in history. Malnourishment is a serious disease itself but it also increases human susceptibility to other debilitation diseases like malaria, diarrhea, and AIDS. Sick and diseased people find it difficult to work and even enjoy the other daily activities of their lives.

More stringent efforts are needed to conserve and protect the basic environmental resources that sustain the food system. These resources include fertile land, water, energy, as well as diverse biological resources. Consider that more than 99.7% of the world food is produced on the terrestrial ecosystem while less than 0.3% of the food is produced in the oceans and other aquatic ecosystems. Looking to the future, more food will have to come from the land and less from the oceans. Urgently needed are safe, successful control methods for the destructive pests that ravage and destroy the food and the other resources that sustain a productive agricultural system.

Worldwide, more than 40% of world food production is lost from the food supply because crops are destroyed by insects, diseases, weeds, and some vertebrate animals. This tremendous kiss if needed food occurs despite the application of about 3 billion kilograms of pesticides plus other pest controls now being used in world agriculture. Once the remaining food is harvested, other insect microbe, and vertebrate pests destroy an additional 20% during storage and transport. As a result, more than half of all food produced is lost to pests, despite human efforts to protect the food resources. Clearly, everything possible must be done to reduce the loss of food to pests, both pre-harvest and post-harvest. Renewed efforts to more effectively protect food crops, as well as livestock, and forest products must become high priority.

The scientific leads in pest management throughout the world have contributed to this encyclopedia. All articles were peer reviewed to reinforce the accuracy and objectivity of each article. The articles assess the benefits and risks of various pest management technologies, including pesticides as well as non-chemical controls, with every effort made to use quantitative data. In addition, the environmental and public health impacts of pest control are discussed. We anticipate frequent updates as new information on pest management becomes available.

The editor is grateful to the specialists and colleagues throughout the world who are experts in the field of pest management and control. They have provided valuable advice and assistance concerning specific pests and management practices. In particular the Advisory Board members and Maria Kelly, Encyclopedia Editor and Supervisor, gave the editor tremendous support, guidance, and assistance in the development and production of this volume.

David Pimentel
Editor

Contents

ACARICIDES

Douglas Walsh
Washington State University, Prosser, Washington, U.S.A.

A

INTRODUCTION

Acaricides are pesticides applied to suppress populations of pest mites. Their use has increased substantially over the past half century as mite communities have grown. Affected by both naturally occuring and synthetic sources, mite infestations can be controlled by the application of various agaricides. This entry explores both the adaptability of mites and the effects of agaricides.

DISCUSSION

Advances in production agriculture have intensified crop damage from mite infestation. Van de Vrie et al. (1) observed that outbreaks of mite populations were uncommon historically in agroecosytems where productivity languished far below the levels achieved in modern production agriculture. Spider mite populations stayed below observable levels due to natural regulation by predators, disease, and poor nutrition from low-quality host plants. Van de Vrie et al. went on to observe that mite populations often experienced outbreaks in agroecosytems where production levels were bolstered by the use of synthetic inputs including fertilizers and pesticides. When crop production is optimized (i.e., not limited by water, nutrients, competition from weeds, or predatory mites and insects) the plants in production become an excellent food source for mite pests. Under these conditions, the developmental rate, fecundity, and lifespan of mites are increased and contribute to population outbreaks.

A number of mite species can achieve pest status at high population abundance. Spider mites develop through several stages: egg, six-legged larva, eight-legged protonymph, deutonymph, and adult. Males typically reach maturity before females, and will position themselves near developing quiescent females. When an adult female emerges, copulation will often occur immediately. Under optimal conditions, most mite species can develop from egg to adult in 6 to 10 days. Egg laying can begin as soon as 1 or 2 days after maturing to adults. Most spider mite species overwinter as mated adult females. A notable exception is the European red mite, which overwinters as eggs.

The feeding of such small pests causes a huge drain on crops. At the microscopic level, significant quantities (relative to mite size) of plant juices pass through the digestive tract of spider mites as they feed on leaf tissues. McEnroe (2) estimated this volume at 1.2×10^{-2} ml/mite/h — roughly 50% of the mass of an adult female spider mite. Leisering (3) calculated the number of photosynthetically active leaf cells that are punctured and emptied per mite at 100 per min. In gut content studies of two-spotted mites, Mothes and Seitz (4) observed only thylakoid granules inside the digestive tract following feeding. The thylakoid grana on which spider mites feed are key photosynthetic engines in plant cells. The grana consist of 45–50% protein, 50–55% lipid, and minute amounts of RNA and DNA (5). Water and other low-density plant cell contents are directly excreted (2).

At the macroscopic level, damage from mite feeding can cause leaf bronzing, stippling, or scorching. For most horticultural crops, economic loss is caused by a drop in yield or quality due to reduction in photosynthesis.

Spider mite outbreaks are promoted by hot, dry weather. Water stress, wind, and dust all contribute to the potential for mite outbreaks. When mite outbreaks occur, acaricide treatments are often used for suppression.

Acaricides Types and Effects

Smothering agents: Solutions containing petroleum-based horticultural oils, vegetable oils, or agricultural soaps are applied to many crops. Application of these types of products kills spider mites through suffocation. Unfortunately oils and soaps can prove phytotoxic to crop plants.

Organochlorines: Endosulfan and dicofol are two organochlorine miticides registered for use on many crops. Unlike many other organochlorine pesticides, endosulfan and dicofol are relatively nonpersistent in the environment. Organochlorine acaricides interfere the transmission of nerve impulses and disrupt the nervous system of pest mites. Organochlorine acaricides are more effective at killing mites at warmer temperatures. Overuse of organochlorine acaricides in commercial situations has resulted in tolerance in many pest mite populations.

Organophosphates and carbamates: Many organophosphate and carbamate pesticides have acaricidal activity. Studies have demonstrated significant mite control with applications of parathion, TEPP, and aldicarb. Spider mites are listed as target pests on many organophosphate and carbamate products. However, many mite populations following long-term exposure have developed resistance to the toxic effects of organophosphates (6).

Organotins: Miticides in this category were synthesized in the 1960s and 1970s and registered for commercial use in the 1970s. They have been used extensively to quickly knock down spider mite populations through contact activity. Fenbutatin-oxide has been used extensively since 1970s. Cyhexatin was used extensively in the 1970s and 1980s but regulatory actions now limit its use. Efficacy of the organotin acaricides is improved with warmer weather. Overuse of cyhexatin during the mid-1980s led to the development of resistance. However, populations of spider mite pests can regain susceptibility to organotins following a period of nonexposure (7).

Propargite: This acaricide has been used since the 1960s. It provides effective suppression of pest mites on many crops. Regulatory constraints have resulted in the cancellation of a number of uses. Identification of propargite as a dermal irritant has led to substantial increases in the time required following application before reentry is permitted into the treated site.

Ovicides: Clofentazine and hexythiazox are selective carboxamide ovicidal acaricides. Spider mite eggs exposed to either compound fail to hatch. These acaricides are selective and aid in the conservation of populations of beneficial arthropods. These acaricides are typically used relatively early in the production season before mite populations reach outbreak conditions.

Antimetabolites: A number of new miticidal compounds have been developed in the past 15 years; e.g., avermectins, pyridazinones, carbazates, and pyroles. Pest mortality results from disruption of metabolic pathways typically within the mitochondria of nerve cells of spider mites (8). Avermectins and related compounds are fermentation products derived from mycelial extracts of *Streptomyces* species (9). Avermectins are locally systemic (translaminar) in plant tissues (10). A number of products have been commercialized in recent years. Pyridaben is a pyridazinone recently registered ornamentals and some tree crops. Bifenazate is a carbazate acaricide with a new mode of action is not clearly understood but proven toxicologically safe in mammalian studies. Bifenazate is registered on ornamentals; food uses are pending. Chlorfenapyr is a synthetic pyrole that has been commercially available on cotton; other uses are pending.

Synthetic pyrethroids: Fenpropathrin and bifenthrin are two synthetic pyrethroid insecticides registered for control of spider mites. Spider mites have a well-documented history of rapidly developing resistance to pyrethroid insecticides, and resurgence of spider mite populations following pyrethroid application is typical.

CONCLUSION

Mite pests can prove difficult to control with acaricides due to their potential for high population abundance, small size, and propensity to live on the bottom surfaces of leaves or within the folds of plant tissues. Good acaricide spray coverage is essential for mite control, particularly for acaricides that kill on contact with the pest mite.

Following repeated exposure, spider mite populations have a history of rapidly developing resistance to acaricides. Alternating acaricides that have different modes of action reduces the potential for development of resistance to acaricides within specific modes of activity. Other techniques to discourage resistance development include spraying only when necessary and treating only the infested portions of the crop. Organophosphate, carbamate, and pyrethroid insecticide applications can induce spider mite outbreaks. If possible, avoid early-season insecticide application or apply insecticides that are less disruptive to beneficial arthropods. Careful selection and use of insecticides can potentially reduce the number of miticide applications required later in the season.

REFERENCES

1. Van de Vrie, M.; McMurtry, J.A.; Huffaker, C.B. Ecology of mites and their natural enemies. A Review. III Biology, Ecology, Pest Status, and Host Plant Relations of Tetranychids. Hilgardia **1972**, *41*, 345–432.
2. McEnroe, W.D. The role of the digestive system in the water balance of the two-spotted spider mite. Adv. Acarol. **1963**, *1*, 225–231.
3. Leisering, R.O.B. Beitrag zum phytopatologischen wirkungsmeechanismus von *Tetranychus urticae*. Pflanzenschutz **1960**, *67*, 525–542.
4. Mothes, U.; Seitz, K.A. Functional microscopic anatomy of the digestive system of *Tetranychus urticae* (Acari: Tetranychidae). Acarologia **1981**, *22*, 257–270.
5. Noggle, G.R. The Organization of Plants. In *Introductory Plant Physiology*; Noggle, G.R., Fritz, G.J., Eds.; Prentice Hall: Englewood Cliffs, NJ, 1983; 9–38.
6. Smissaeret, H.R.; Voerman, S.; Oostenbrugge; Reenooy, N. Acetylcholinesterases of organophosphate-suceptible and resistant spider mites *Tetranychus urticae*. J. Agr. Food Chem. **1970**, *18*, 66–75.
7. Hoy, M.A.; Conley, J.; Robinson, W. Cyhexatin and fenbutatin-oxide resistance in Pacific spider mite *(Acari: Tetranychidae)* stability and mod of inheritance. J. Econ. Entomol. **1988**, *81*, 57–64.
8. Hollingsworth, R.M.; Ahammadsahib, K.I.; Gadelhak, G.; McLaughlin, J.L. New inhibitors of complex I of the mitochodrial electron transport chain with activity as pesticides. Transactions **1994**, *22*, 230–233.
9. Burg, R.W.; Stapley, E.O. Isolation and characterization of the producing organism. In *Ivermectin and Abamectin*; Campbell, W.C., Ed.; Springer-Verlag: New York, 1989; 24–32.
10. Walsh, D.B.; Zalom, F.G.; Shaw, D.V.; Welch, N.C. Effect of strawberry plant physiological status on the translaminar activity of avermectin B1 and its efficacy on the two-spotted spider mite *Tetranychus urticae* Koch *(Acari: Tetranychidae)*. J. Econ. Entomol. **1996**, *89* (5), 1250–1253.

ACUTE HUMAN PESTICIDE POISONINGS

A

E.D. Richter
Hebrew University–Hadassah School of Public Health and Community Medicine, Jerusalem, Israel

INTRODUCTION

Worldwide, pesticides are responsible for an estimated 220,000 deaths and 26 million cases of acute poisoning annually, mostly in agricultural workers and rural communities (1). Children are particularly vulnerable to toxic effects of pesticides, including acute poisoning, as are adults with poor nutrition, infection, and substance abuse (2–4).

In the United States, some 25% of 80,000 children and 72% of 34,000 adults exposed to common pesticide products are reported to develop symptoms of poisoning. EPA estimates that approximately 250–500 diagnosed cases occur per 100,000 agricultural workers, (including pesticide handlers). This estimate implies there are some 6250 to 12,500 cases per year in a total workforce of 2.5 million farmers, but the actual numbers are believed to be twice as high (3). In California, the state with the largest agricultural economy in the United States, there was an average of 665 cases of reported occupational poisoning from among 600,000 workers in years 1991–1996 (4). In Nicaragua, interviews of field workers indicate that some two-thirds of all cases of pesticide poisoning are not reported (5, 6).

Surveillance systems for acute poisoning from pesticides, especially in developing countries, reflect the poor status of reporting of all work-related illness. Underreporting of episodes of poisoning occurs because victims often do not have access to treatment. If they are seen, they may be misdiagnosed, and if diagnosed and treated, are not reported. To make matters worse, many farm workers fear costs of medical care, job loss, and retaliation, or are not trained to recognize symptoms.

AGRICULTURE

Suicide attempts account for up to two-thirds of all acute poisonings in developing countries, and result from the widespread availability of pesticides in rural environments. Acute pesticide poisonings are more severe and frequent in the agricultural work sector than in the general community, because usage and concentrations are much greater (7–10). Poisonings also occur from exposure during mishaps with batch concentrations during manufacture (11).

The risks are related to the toxicity and quantity of the agents used, mode of application, persons exposed during application, length and frequency of use and contact, mode of field management and harvesting of treated crops, and the adequacy of measures to protect field workers. Risks for acute poisoning increase especially with use of pesticides with an LD50 of less than 50 mg/kg body weight, but higher LD 50s are not guarantors of absence of risk, and certainly not so from swallowing. Illness in the general community may result from misuse and lapses in practice in vector control. *Observations suggest skepticism concerning the feasibility and efficacy of protective clothing and gear to prevent acute poisoning, especially in tropical climates.*

SOURCES AND ROUTES OF EXPOSURE

Ingestion—intentional or accidental—and skin absorption from spills of concentrates and contamination of clothing or repeated exposure to leaf residues can lead to acute poisoning and life-threatening emergencies (3). Skin abrasions increase absorption. Drift from pesticide spraying (44%) and field residues (33%), notably in cotton and grapes, account for some three-quarters of all cases of reported acute illness in California, mostly when farmworkers are doing fieldwork, such as picking, field packing, weeding, and irrigating. Ratios for illness/use (per kg applied) were highest with applications of restricted organophosphates to fruit trees, followed by vegetables and melons. Air applications resulted in higher ratios for illness per kg of active agent applied compared to ground applications (4). Headache, fatigue, dizziness, nausea, problems with digestion, diarrhea, and breathing as well as insomnia have been associated with proven exposure to adults and children to drift of organophosphates from aerial spraying, as well as from ingestion of vegetables

Table 1 Pesticide poisoning: Screening questions

For adults:
Are pesticides used at home or work?
Were the fields wet when you were picking?
Has there been spraying while you were working or in the last few days?
Did you get sick during or after working in fields?
For children (questions to parent or guardian):
Any containers with pesticides stored in home, nursery, or yard?
Any exposure to pesticides, solvents, fumes, or gases?
For both adults and children:
Are there others with similar complaints/illnesses?[a]

[a]Exploits sentinel case, leads to investigation of high-risk situations, and prevents future poisonings.

with high residue levels. In low-flying aerial spray pilots, these effects increase risks for crashes (12).

RECOGNITION OF ACUTE POISONING

Table 1 presents a checklist of questions to increase the index of suspicion and diagnostic accuracy. Acute poisonings may be missed or misdiagnosed. In children, acute episodes of organophosphate poisoning may be misdiagnosed at first as: aspiration pneumonia or pneumonitis, coma or encephalopathy or head trauma, diabetic ketoacidosis, intracranial aneurysm, respiratory infections, seizure disorders, or microbial diarrhea (3).

In adults, recognition is based mainly on a history of circumstances of exposure of the presenting symptoms. Vomiting or spitting out of the toxic substance and strong odors usually follow ingestion. Heat-related illnesses, including heat stroke, heat cramps, and dehydration in hot climates, complicate diagnosis and management of acute poisoning among agricultural field workers, since these conditions mimic or aggravate many of the signs and symptoms of from pesticide poisoning.

GUIDELINES FOR MANAGEMENT OF ACUTE POISONINGS

EPA has recommended (3) that for all pesticides, the standard procedures are: 1) removal, bagging, and isolation of contaminated clothing, and rapid decontamination of skin and hair (including skin folds and areas under fingernails) with soap, water, and showers, shampoos, and flush irrigation of the eyes; 2) airway protection, using suction or intubation if the patient is unconscious, and if necessary, providing oxygen; 3) gastrointestinal decontamination in the case of ingestion; and 4) management of seizures [Diazepam (Valium) and barbiturates]. Adequate oxygen levels in tissues are especially important before giving atropine to victims of acute organophosphate poisoning. But in the case of paraquat and diquat, oxygen is contraindicated because these agents enhance the toxicity of oxygen to the lung tissue.

Antidotes, intubation, and gastrointestinal decontamination (including activated charcoal) have their complications and may be used too late to provide benefit.

Anticholinesterases: Organophosphates and Carbamates

These agents are still the most widely used insecticides and include the some of the most toxic agents in use. These agents probably still account for the largest number of acute life-threatening episodes worldwide. They are also the only group for which there are highly specific antidotes: atropine and the oximes. With certain organothiophosphates, notably parathion, the P = S group undergoes conversion to P = O, a more labile but much more toxic intermediate exposure. This conversion occurs in situ to residue on leaves under light and heat, and in vivo, and in creases risk for toxic exposures for days following application.

The acute life-threatening effects of organophosphates and carbamates result from inhibition of acetylcholinesterase, the enzyme that breaks down the neurotransmitter acetylcholine at the neuromuscular junction from nerve fibers to skeletal and smooth muscles, glandular cells, and the ganglia of the autonomic and central nervous system. The signs and symptoms of acute poisoning from anticholinesterases result from accumulation of acetylcholine at these sites and excess stimulation of smooth muscles and ganglia (muscarinic effects), skeletal muscle (nicotinic effects), and the central nervous system. The signs, symptoms, and severity vary with the specific organophosphate or carbamate, and are influenced by toxicity of the agent, dose, route of absorption, degree of solubility in fat tissue, genetic determinants of susceptibility (13), and general health status of the individual. The signs and symptoms usually are headache, dizziness, weakness, blurred vision, confusion, small pupils, bronchorrhea, bronchospasm, cough, pulmonary edema, nausea and vomiting, diarrhea, abdominal pain, sweating, salivation, muscle twitching, and slow heart rate, which can progress to sinus arrest or various arrhthymias. Slow heart rate, muscle twitching, seizures, and lethargy are especially common in children. The most common cause of death is respiratory failure with cardiac complications.

Rapid diagnostic tests for acute poisoning can measure inhibition of cholinesterase in red blood cells or serum, although in the case of the *N*-methyl carbamates, blood cholinesterase levels undergo in vitro reactivation of the carbamylated enzyme, and therefore may provide false ''normal'' results. Odor from contaminated clothes or exhaled breath is a helpful diagnostic sign.

Emergency treatment requires oxygenation via an adequate clear airway (see above), and giving atropine sulfate—intravenously or intramuscularly—until the victim recovers from the acute episode. Atropine blocks receptor sites for muscarinic, but not nicotinic effects from excessive levels of acetylcholine, so twitching, muscle weakness, and respiratory depression may not be relieved. Oximes (Pralidoxime–Protopam, 2-PAM) reactivate cholinesterase by reversing the bond between cholinesterase and organophosphates, and relieve the nicotinic as well as the muscarinic effects. These agents may complicate intoxications by carbamates.

Relapses and complications may occur after treatment of the acute episode, and may include arrhythmias and paralysis of facial, neck, and proximal limb muscles. Delayed neuropathy occurs from inhibition of neuropathy target esterase of peripheral or central nerve fibers, which produces weakness and parasthesias. There also are residual neurobehavioral impairments of mood state, behavior, memory, motor performance, and cognitive function.

Organochlorine Insecticides

These agents are absorbed via gut, lung, and skin and rapidly deposited in fat tissue dechlorinated and oxygenated, conjugated, and excreted by the biliary route. Tissue residues are detectable in blood, milk, and fat, but acute poisoning is not common. Early manifestations of poisoning are sensory hyperesthesia and paresthesia of the face and extremities, followed by headache, dizziness, nausea, vomiting, incoordination, tremor and mental confusion, and then myoclonic jerking movements and tonic-clonic convulsions, coma and respiratory depression. Seizures may appear as late as 48 h after exposure, often without warning, may recur, and are extremely difficult to control. Dieldrin, aldrin, and chlordane are more apt to cause delayed or recurrent seizures than are lindane and toxaphene, which are more rapidly biotransformed and excreted. Seizures may follow use of lindane to treat scabies, in which skin is abraded. The treatment requires decontamination (possibly with the use of activated charcoal), anticonvulsants (benzoiazepines and phenobarbital), oxygen, and cardiopulmonary monitoring.

Arsenicals and Mercurials

Arsenicals are restricted or rarely used, but may reenter the exposure chain in unexpected settings. Arsenicals may produce life-threatening episodes from ingestion. The central nervous system, blood vessels, kidneys, and liver are the target organs. In acute episodes, presenting signs are a garlic odor of the breath and feces, metallic taste in mouth, and gastrointestinal complaints. The mercurials produce progressive, chronic, debilitating, and irreversible neurologic impairments.

Herbicides: Paraquat and Diquat

Misuse and accidental or intentional ingestion of these herbicides results in burning pain in the mouth, throat, chest, and upper abdomen, and pulmonary edema, pancreatitis, and kidney and neurologic damage. Most deaths result from ingestion with suicidal intent. The lung is the primary target organ for paraquat, but not diquat, even though toxicity from inhalation is rare, and death results from generation of free oxygen radicals producing pulmonary edema, asphyxia, tissue anoxia, and subsequent fibrosis. Treatment is based on immediate gastrointestinal decontamination with bentonite, Fuller's earth, or activated charcoal. Other herbicides have low mammalian toxicities.

Fumigants

Methyl bromide, still widely used, is severely irritating to the lower respiratory tract and is neurotoxic. Respiratory distress may appear 4–12 h after exposure, and can cause headache, dizziness, nausea, vomiting, tremor, slurred speech, and ataxia. Skin contact causes a chemical burn.

Pyrethroids, Other Biologicals, Insecticides of Biological Origin (e.g., Bacillus Thuringiensis), Acaricides, and Repellents

These agents include the pyrethroids, alkyl phthalates, benzyl benzoate, boric acid and the borates, chlordimeform, chlorobenzilate, cyhexatin, diethyltolbutamide, the fluorides, and nicotine and generally are considered to carry little risk for acute poisoning.

CONCLUSION

Data on manufacture, trade, and use (especially of agents that have been banned or restricted for use) enable assessment of potential for risks for acute poisoning. The incidence of acute pesticide poisoning will fall wherever

Table 2 Prevention of poisoning: Recommendations of pesticide action network

Goal: Phase out most toxic pesticides and promote healthy and sustainable alternatives
Enact regulations to reduce farmworker exposure
• Ban aerial spraying
• Prohibit backpack spraying for restricted pesticides
• Expand buffer zones, posting and notification requirements
Strengthen enforcement and penalties for violators of existing regulations
• Improve reporting of poisonings and training
• Ban incentive bonuses for reporting no injuries or illnesses
Improve farmworker access to medical treatment
• E.g., migrant clinics
• Provide insurance coverage
Ensure farmworker and public right-to-know
• Posted reentry intervals
• Provide literacy-independent instructions
• Establish databases on:
• Pesticide uses
• Violations
• Pesticide illnesses user or grower

(From Ref. 4.)

governments and stakeholders promote reduced use of pesticides and ban, or severely restrict, the use of agents with LD50s < 50 mg/kg body weight (see Table 2). Data on increasing trends in crop yield following large reductions in use of pesticides and shifts to less toxic substitutes refute the claim that reduced use undermines agricultural productivity (12). Standards for packaging resistant to tampering are mandatory, as are right-to-know, expanded and improved training for stakeholders in the recognition and management of pesticide poisoning, and the use of preventive strategies, including protective gear. Outreach of poison control programs to high-risk settings is essential, not only for diagnosis and treatment, but for epidemiologic investigation of sentinel events. Epidemiologic surveillance evaluates the efficacy of such measures.

REFERENCES

1. *WHO. Our Planet, Our Health.*, Report of the WHO Commission on Health and Environment, Geneva, 1992; World Health Organization.
2. Dinham, B. *The Pesticide Hazard: A Global Health and Environmental Audit*; The Pesticide Trust: London, 1993.
3. *EPA. Pesticides and National Strategies for Health Care Providers*, Workshop Proceedings, April 22–23, 1998; www.epa.gov.
4. Reeves, M.; Schafer, K.; Hallward, K.; Katten, A. In *Fields of Poison, California Farmworkers and Pesticides, 1999;* Pesticide Action Network North America Regional Center (PANNA). http:/www.panna.org.fields.
5. Wesseling, C.; McConnell, R.; Partanen, T.; Hogstedt, C. Agricultural pesticide use in developing countries: health effects and research needs. Int. J. Health. Serv. **1997**, *27* (2), 273–308.
6. Keifer, M.; McConnell, R.; Pacheco, A.F.; Daniel, W.; Rosenstock, L. Estimating underreported pesticide poisonings in Nicaragua. Amer. J. Industrial Med. **1996**, *30* (2), 194–201.
7. Agarwal, S.B.A. Clinical, biochemical, neurobehavioral and sociological study of 190 patients admitted to hospital as a result of acute organophosphate poisoning. Environ. Res. **1993**, *62* (1), 63–70.
8. Van der Hoek, W.; Konradsen, F.; Athukorla, K.; Wanigadewa, T. Pesticide poisoning: a major health problem in Sri Lanka. Soc. Sci. Med. **1998**, *46* (4/5), 495–504.
9. Shin, D.C.; Kim, H.J.; Jung, S.H.; Park, C.Y.; Lee, S.Y.; Kim, C.B. Pesticide poisoning and its related factors among Korean farmers. Med. Lav. **1998**, *89* (Suppl 2), S129–S135.
10. Kishi, M.; Hirschhorn, N.N.; Djajadisastra, M.; Satterlee, L.N.; Strowman, S.; Dilts, R. Relationship of pesticide spraying to signs and symptoms in Indonesian farmers. Scand. J. Work Environ. Health **1995**, *21* (2), 124–133.
11. Huang, J.; Zhang, S.; Ding, M.; Zhou, A.; Zhang, J.; Zhang, J. Acute effects of carbofuran in workers of two pesticide plants. Med. Lav. **1998**, *89* (Suppl 2), S105–S111.
12. Richter, E.D.; Gasteyer, S.; El Haj, S.; Safi, J.; Jaqhabir, M.; Safi, J. Agricultural sustainability, pesticide exposures and health risks: Israel, The Palestinian National Authority and Jordan. Ann. NY Acad. Sciences **1997**, *837*, 269–291.
13. Furlong, C.E.; Li, W.F.; Richter, R.J.; Shih, D.M.; Lusis, A.J.; Alleva, A.J.; Costa, L. Genetic and temporal determinants of pesticide sensitivity: role of Paraoxonase (PON1). Neurotoxicology **2000**, *21* (1/2), 91–100.

AERIAL APPLICATION

Ian P. Craig
AEMS Pty, Ltd., Toowoomba, Queensland, Australia

Nicholas Woods
Gary J. Dorr
University of Queensland, Gatton, Queensland, Australia

INTRODUCTION

At the beginning of the 1980s it was estimated that there were approximately 26,000 aircraft worldwide involved with the application of pesticides, with the numbers roughly equally divided between the United States of America, the USSR, and the rest of the world (1). This figure has probably decreased in recent years, primarily due to the decline in communal farming in the USSR and Eastern Europe, but also due to improvements in ground application equipment and public concern over spray drift in some European countries. However, aircraft are still used extensively in public health and are still required for the economic production of many crops such as cotton, rice, fruit and vegetables, sugar cane, and forestry. In 1998 there were 2600 registered aerial operators in the United States of America with approximately 6000 aircraft. According to recent United States Department of Agriculture figures, out of 125 million hectares harvested in the United States in 1998, approximately 25% of the land area receiving crop protection products was sprayed using aircraft.

The use of aircraft as a means of agrochemical application has developed due to the greater speed, better timing, and efficiency of application offered by an airborne platform. Aircraft are able to apply agricultural products rapidly over large areas within narrow optimum application windows, often essential for control of the pest. Aircraft can be used over impenetrable canopies and wet/irrigated areas, and problems of soil compaction disease spread associated with ground vehicles are entirely avoided. However, disadvantages include the high risk of flying close to the ground and the fact that aircraft (especially fixed wing) attract attention to the process of pesticide application and cause public concern. Aircraft also release spray at greater height and speed, which may increase drift compared to ground-based equipment.

AERIAL SPRAY METHODS

Ultra-Low Volume (ULV) Spraying

ULV pesticides formulated in low-volatile oil-based carriers are applied "straight from the can" at total application rates of 1–10 L/ha. This low rate of carrier is achieved by generating small droplets with a volume median diameter (VMD) of approximately 50–100 μm, usually using rotary cage type atomizers (Fig. 1). Such droplet sizes allow large numbers of droplets to be generated resulting in high droplet coverage (expressed in terms of droplet number per square cm) and high efficacy. This technology is particularly suited to the control of airborne locusts and in forestry where there are vast areas of tall canopy to be penetrated. ULV application has also been successfully utilized in the production of cotton in Africa, Asia, and Australia. Because the droplet size used is small, deposition within the first swath may be as low as 25% and it usually takes several swaths for the deposit on the ground to build to its full value. The sequential swaths once overlapped produce a uniform deposit across the crop. The process is highly efficient and cost effective, mainly due to the fact that less ferry time is required. However ULV application can have significantly higher drift potential than other aerial application methods.

Low Volume (LV) Spraying

Most aerial application is carried out using a water-based carrier to dilute the pesticide product, the most common being an emulsifiable concentrate (EC) formulation. Droplet size may range from 100 to 250 m and bulk application rates may range from 10 to 100 L/ha. Nozzles are usually hydraulic with flat fan or anvil types being the most common.

Fig. 1 In-flight photograph of Micronair® rotary cage atomizers.

Large Droplet Placement (LDP) Spraying

LDP spraying is defined as water-based spraying with a droplet VMD greater than 250 μm. Bulk application rates are generally 30–100 L/ha. Large droplets have high sedimentation velocities and are not greatly influenced by vertical air movement and turbulence. LDP methods are therefore used to reduce spray drift when spraying has to be undertaken close to susceptible areas.

AIRSTREAM EFFECTS ON DROPLET SIZE

By simulating aircraft airspeeds in a windtunnel, the effect of airspeed on the droplet size produced by aircraft spray delivery systems has been evaluated (2, 3). With ULV sprays, the droplets are small and resist shattering due to airstream effects upon release from the aircraft. However, with LDP water-based sprays, the larger droplets are more vulnerable. To minimize breakup of these droplets, hydraulic nozzles can be oriented so that the spray is emitted parallel to the airstream. The effect of airstream velocity on the spray, however, is still very important. Increasing airspeed from 100 to 130 knots can reduce droplet VMD by 100 μm and may increase the volume percentage of the spray less than 100 μm (usually less than 5%) by two to three times.

WINGTIP VORTICES

Since the lift generated by a wing produces an area of high pressure on the lower surface of the wing and an area of low pressure on the upper surface of the wing, there is a tendency for air to spill over at the tips in the direction from high to low pressure. This causes tip vortices to be formed and these give rise to vortex or lift-dependent drag. It is these vortices, necessary for lift, which give rise to spreading of the spray and the subsequent swath width of the aircraft. However, if the boom length is as long as the wingspan of the aircraft, then excessive amounts of spray become incorporated in the vortices therefore increasing drift. To minimize incorporation of spray into wingtip vortices, boom length should not exceed 65% of the aircraft wingspan (Fig. 2)

AIRCRAFT SPRAY DEPOSITION

Aircraft spray deposition can now be accurately predicted using computer models that are based on diffusion theory associated with a sedimenting spray cloud. Equations that estimate the downwind distribution of droplets dispersing and settling from a line source (4, 5) are used as the basis of the far wake component of other more complex models, for example AgDISP (6), FSCBG, and AgDRIFT (7), which compute a complex source incorporating aircraft flow field effects. The outputs from these models have been extensively validated against experimental data (8–10).

Averaged over a wide range of conditions and droplet sizes, say from 70 to 250 μm VMD, it appears that roughly two-thirds of the spray released from aircraft is usefully deposited on crop surfaces (10). This may be due to the fact that similar proportions of the spray are within an optimum biological size range of say 50–300 μm (Fig. 3). Of the remaining third, some is lost to the ground, and some is lost to the atmosphere, generally in proportions relating to the droplet size of the spray. A certain percentage of the spray forms a deposition tail extending

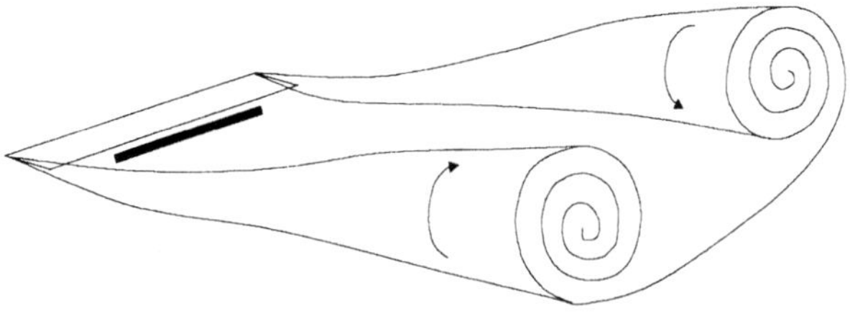

Fig. 2 Wingtip vortices; boom length should not exceed 65% of the wingspan of the aircraft.

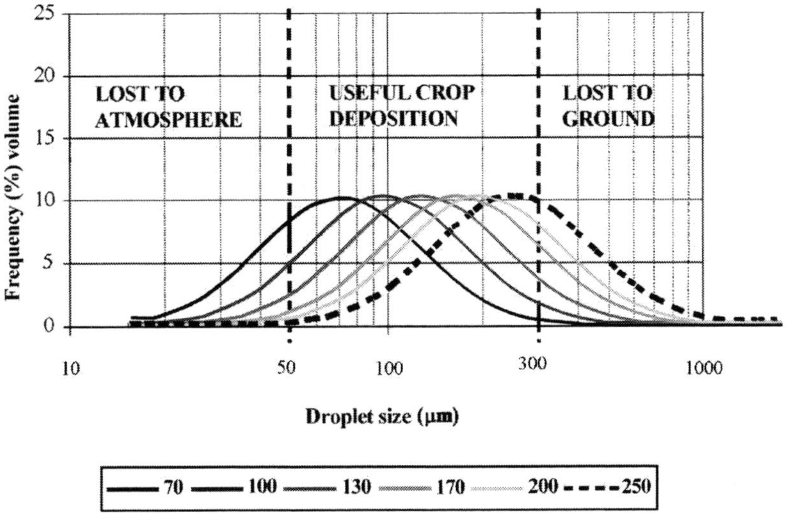

Fig. 3 Schematic droplet size distributions representing the range between ULV (70 μm VMD) and LDP (250 μm VMD). Droplets between about 50 μm and 300 μm constitute useful crop deposition.

downwind from the sprayed field (Fig. 4). The thickness of this tail depends primarily on droplet size, spray release height and atmospheric stability (11). If a 500 m-wide field is sprayed with a release height of 3 m under neutral atmospheric conditions, off target deposition 500 m downwind from the field is typically 1%, 0.3% and 0.2% of the field applied rate for ULV (70 μm), LV (170 μm) and LDP (250 μm) applications respectively (Fig. 4). Computer predictions indicate that for off target deposition to fall to less than 0.1% of the field applied rate, downwind buffer distances of approximately 2 km, 1 km, and 750 m are required for ULV, LV, and LDP applications respectively. Using spray drift prediction models (7), specific buffer distances may be determined for a particular chemical and a particular sensitive target and will vary according to toxicity of the chemical, field source width, application parameters, and meteorology.

THE FUTURE

The trend in aerial application is now toward water-based LDP application to reduce spray drift. The use of helicopters may increase due to their lower flying speed and ability to deliver larger droplets. Helicopter application is generally more expensive than fixed wing, but with the

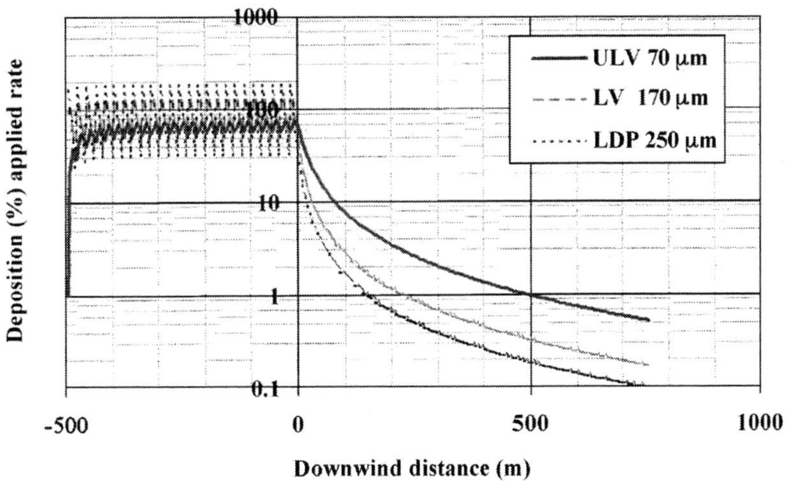

Fig. 4 Gaussian Diffusion Model prediction of typical aerial ULV, LV, and LDP applications showing the overlapped deposit across a field and downwind deposition profiles.

trend toward larger volumes applied, ferry costs may be considerably reduced by field side loading. As with fixed wing aircraft, nozzle type, positioning, and orientation are still important factors that influence drift. However, the boom is further away from the wing with helicopters compared to fixed wing aircraft and therefore the effect of the wingtip vortices on the spray may be less pronounced. Additionally, it is possible for helicopters to fly closer to the ground due to lower flying speed and greater visibility. These factors tend to indicate that less drift may be possible with helicopter applications.

The driftable fraction of the spray can be significantly reduced with LDP application technology, but larger droplets tend to be less efficacious, particularly when applied as insecticides, and can lead to high ground deposition and wastage of chemical. The aerial application industry requires a nozzle that can generate droplets approximately 250 μm in diameter (VMD) without the production of significant fines (droplets less than 100 μm). Computer predictions (12) have indicated that if this could be achieved, LDP application could be optimized leading to a significant reduction in buffer distances required for aerial application.

REFERENCES

1. Spillman, J.J. A view of agricultural aviation worldwide. Aerogram—the newspaper of the college of aeronautics. Cranfield **1988**, *5*, 2–5.
2. Kirk, I.K. In *Application Parameters for CP Nozzles*, ASAE Meeting, Nevada, 1997, Paper No. AA97-006.
3. Woods, N.; Craig, I.P.; Dorr, G. In *Droplet Size Analysis of Aircraft Nozzle Systems Applying Oil and Water Based Formulations of Endosulfan Insecticide*, Eight International Conference on Liquid Atomization and Spray Systems, Pasadena, CA, 2000.
4. Bache, D.H.; Sayer, W.J.D. Transport of Aerial Spray 1. A Model of Aerial Dispersion. Agric. Met. **1975**, *15*, 257–271.
5. Lawson, T.J.; Spillman, J.J. Particle Transmission and Distribution in Relation to the Crop. In *Aerial Application of Pesticides*; Cranfield Institute of Technology: Bedford, UK, 1978; Shortcourse Notes, September 1978.
6. Bilanin, A.J.; Teske, M.E.; Barry, J.W.; Ekblad, R.B. AGDISP: The aircraft dispersion model, code development and experimental validation. Trans. ASAE **1989**, *32*, 327–334.
7. Teske, M.E.; Bowers, J.F.; Rafferty, J.E.; Barry, J.W. FSCBG: An aerial spray dispersion model for predicting the fate of released material behind aircraft. Environ. Toxicol. and Chem. **1993**, *12*, 453–464.
8. Akesson, N.B.; Yates, W.E. Physical Parameters Effecting Aircraft Spray Application. In *Chemical and Biological Controls in Forestry*; Garner, W.Y., Harvey, J., Eds.; ser. 238, American Chemical Society: Washington, DC, 1984.
9. Dorr, G.; Woods, N.; Craig, I.P. In *Buffer Zones for Reducing Drift from the Application of Pesticides*, SEAg International Conference on Engineering in Agriculture, Perth, Paper No. 8.
10. Woods, N.; Craig, I.P.; Dorr, G. Spray Drift of Pesticides Arising from Aerial Application in Cotton. J. of Environ. Qual., *in press*.
11. Bird, S.L.; Esterly, D.M.; Perry, S.G. Atmospheric pollutants and trace gases—off target deposition of pesticides from agricultural aerial spray applications. J. of Environ. Qual. **1996**, *25*, 1095–1104.
12. Craig, I.P.; Woods, N.; Dorr, G. A simple guide to predicting aircraft spray drift. Crop Prot. **1998**, *17* (6), 475–492.
13. Quantick, H.R. *Aviation in Crop Protection, Pollution and Insect Control*; Collins UK, 1985.

AIRBLAST SPRAYERS

Andrew Landers
Cornell University, Ithaca, New York, U.S.A.

INTRODUCTION

The application of pesticides to fruit trees and bush crops has been a challenge for many years due to variations in structure and canopy of the target crop. Airblast sprayers are used to apply pesticides to plantation crops as diverse as grapes, apples, pecans, and citrus trees. They comprise a tank containing liquid pesticide, a pump, and manifold fitted with nozzles. A fan is used to create a high volume of air into which pesticide is dispensed and projected as a large radial plume towards the target. The air blast from the fan moves the pesticide droplets into the canopy, displacing the air within the canopy and depositing pesticide onto the target surfaces.

TRADITIONAL DESIGNS

Traditional airblast sprayers direct the air from a single, axial flow fan, mounted directly behind the sprayer in an upward and outward direction. Axial fans are designed to move large volumes of air at low pressures through a shroud or former which turns the air 90° through an outlet; nozzles are situated radially in the outlet, see Fig. 1. In a review of fan design parameters, generally an increase in fan diameter rather than fan speed is a more efficient way of increasing airflow rate (1). To accommodate varying crop canopies, e.g., as the season progresses, many modern sprayers are fitted with adjustable pitch propellors to provide variable airflows. An adjustable deflector plate should be fitted at the top and base of the air outlet to direct the air towards, and confine it to the target canopy.

Fan size must be matched to canopy size, for example, large diameter fans are required to provide enough air (volume and speed) for citrus and pecan trees compared to smaller fans for single row grape canopies. As the trend towards planting dwarf and semidwarf apple trees continues, so the size of fan needs to alter if drift is to be minimized. Similarly, as canopy size has reduced and sprayer design improved, so pesticide application rate has decreased, for example, from thousands of litres per hectare down to less than a hundred. When changing to high density orchards it makes sense to determine whether the same amount of material per hectare should be applied as in standard plantings (2). Spray deposition studies have shown significant changes in spray capture efficiency due to canopy volume and leaf area.

In many cases, hydraulic pressure, created by a high pressure pump, delivers pesticide to a swirl plate and cone or hollow cone nozzles. The pesticide breaks up into a fine spray, with a volume median diameter (VMD) < 150 μm. The size of the swirl plate and nozzle determines the droplet size.

Alternatively a low pressure pump may deliver liquid to an air-shear nozzle where liquid is sheared into fine droplets by an air stream passing over a shear plate. Droplet size is determined by a balance between liquid/air at the shear nozzle. For the same liquid flow, the higher the velocity of air, the smaller the droplets and vice versa. Air shear nozzles tend to provide finer atomization than hydraulic nozzles. A fine spray is needed to achieve good coverage but losses can increase due to drift and evaporation.

Depending on machine design, airblast sprayers will deliver pesticides to one or two sides of a canopy, or in the case of high output sprayers, two or three rows of trees or vines. Coverage will depend upon many factors including canopy density, airflow (velocity and speed), nozzle configuration, droplet size, leaf surface, and weather.

It is imperative that air-assisted sprayers are matched to target morphologies (3). For example, some orchard sprayers used today are based on specifications derived many years earlier. It is not uncommon to see these machines treating hedgerow plantings with spray issuing over 180° arc, so that, despite advice to the contrary, some is pushed wastefully and dangerously towards the sky.

MODERN DESIGNS

Cross Flow Fans

Cross flow fans comprise a long impellor that immediately turns the air through 90°. They are less efficient than axial

Fig. 1 An engine-driven airblast sprayer suitable for very large trees.

fans and operate at lower pressures. The long slit outlet produces an even air movement and the design allows easy adjustment over the target canopy. The vertical curtain of air moves horizontally towards the canopy.

Canopy Sensors

Drift problems increase when a space occurs within the row. Air blast sprayers can be fitted with optical sensors that detect the presence of target canopies. The systems map an image of the canopy profile, sensing tree shape and size. Sensors, operating with either lasers or ultrasonics, are linked to an on-board computer that automatically switches on/off electric solenoid valves to operate individual nozzles. The advantages include reduced drift and ground deposition, reduced pesticide use, and improved logistics. Losses are directly correlated to the portion of gaps in the fruit wall whilst spraying apple trees, a three-year trial showed comparable efficacy to a conventional airblast sprayer with pesticide savings of about 25–30% and a reduction in drift by about 50% (4).

Directed Air Jets

A number of sprayers direct the airblast to the target via adjustable outlets mounted above and below the canopy. Each outlet incorporates a hydraulic nozzle surrounded in a stream of air. The outlets may be mounted on either side of the sprayer or fitted to a frame allowing two to three rows to be sprayed (see Fig. 2). Outlet angle is adjustable, depending upon growth stage, shape, and density of the canopy, allowing excellent penetration with minimum drift.

Tunnel Sprayers

Tunnel sprayers straddle the canopy and comprise a large tunnel with vertical hydraulic spray booms. The spray is directed through the canopy and any excess spray is captured by the tunnel walls, and recirculated into the spray tank, allowing the sprayer to travel greater distances per tank fill (see Fig. 3). Recycling reduces the amount of pesticide used by an average of 35% in vineyards. The reduction in pesticide use is greatest in the early prebloom season compared to late season when the canopy is in full foliage. A reduction in drift of 90% was attained when compared to airblast sprayers in vineyards (5). Soil contamination is also greatly reduced. Tunnel sprayers working in orchards usually require vertical crossflow fans to improve deposition.

Propeller Fans

A number of sprayers use small propeller fans to direct high volume, low velocity air into the canopy. Pesticide is fed into a high-speed spinning cage, centrifugal forces spread the liquid and throw it from the periphery (6). Fitting the units to a boom and driving the rotary cages with hydraulic motors provides the centrifugal force. The majority of droplets, (95% approximately), are the same

Fig. 2 A directed air jet sprayer in a vineyard.

Fig. 3 A tunnel sprayer for use in orchards or vineyards.

size, depending upon flow rate and cage speed. Advantages include less water, resulting in better timeliness, and a more targeted spray. Research in North America and Europe shows that small droplets on target are effective at controlling diseases and insects.

DRIFT

Pesticide drift is an important and costly problem facing applicators. Drift results in damage to susceptible off-target crops, environmental contamination to watercourses, and a lower than intended rate to the target crop, thus reducing the effectiveness of the pesticide. Pesticide drift also affects neighboring properties, often leading to concern and debate. Throughout the world legislation is being directed at applicators in an attempt to reduce drift. Research in the USA shows that beyond 30 m from the edge of a target orchard or vineyard, ground deposits are usually less than 0.3% of the applied amount (7). Beyond 30 m downwind of the same application zone, airborne deposits are less than 3% of the applied amount.

There are two types of drift: airborne drift, often very noticeable, and vapor drift. The amount of vapor drift will depend upon atmospheric conditions such as humidity, temperature, and the product being applied, and can occur days after an application is made.

Drift Reduction

Drift is influenced by many interrelated factors including droplet size, nozzle type and size, sprayer design, weather conditions and, last but not least, the operator.

In the past, target canopies were drenched with high volumes and coarse droplets resulting in trees or vines dripping with excess pesticide. The belief that too much is better than too little is misplaced. Dripping canopies lead to environmental pollution such as soil contamination, and an excessive number of tank loads per hectare results in poor time management.

Lower volumes may result in smaller droplets, although there is a limit to droplet size because of concerns about drift. Droplets under 150 μm generally pose the greatest hazard; droplets less than 50 μm have insufficient momentum to impact on the target as they remain suspended in the air indefinitely or until they evaporate. Traditional air blast sprayers give the greatest cause for concern as they produce many small droplets which are often off-target.

Modern nozzle technology such as air inclusion nozzles produce larger droplets than conventional cone nozzles. Large droplets normally roll off the leaf but air inclusion nozzles create air bubbles within the larger droplets which then collapse on contact with the leaf, dissipating the energy and dispersing the liquid. Recent research in North America and Germany has shown promising results using air inclusion nozzles with air blast sprayers in trees and bushes, although further trials are necessary in apple orchards.

If drift is to be reduced, a management strategy that includes operator training, the correct use of a well-calibrated sprayer, 50 m buffer zones, alertness to changes in weather conditions, and a well-planned spraying regime must be implemented (8). ''Increasing'' the ease of flexibility of the sprayer as well as operator skills in adjusting sprayer operations to target and risk characteristics remains a significant opportunity not yet achieved. Applying the correct product to the correct target at the correct time with the correct equipment is the key to good application.

REFERENCES

1. Matthews, G.A. Tractor Mounted or Trailed Air-assisted Sprayers. In *Pesticide Application Methods,* 2nd Ed.; Longman Scientific & Technical: Harlow, UK, 1992; 224–228.
2. Hall, F.R. Application to Plantation Crops. In *Application Technology for Crop Protection*; Matthews, G.A., Hislop, E.C., Eds.; CAB International: Wallingford, UK, 1993; 187–213.
3. Hislop, E.C. *Review of Air-assisted Spraying in Crop Protection*, Proceedings of a Symposium Organized by the British Crop Protection Council, Swansea, January 7–9, 1991; Lavers, A., Herrington, P., Southcombe, E.S.E., Eds.; BCPC: Farnham, UK, 1991; 3–14, University College, Monograph No. 46.
4. Koch, H.; Weisser, P. Sensor Equipped Orchard Spraying–Efficacy, Savings and Drift Reduction. In *Aspects of Applied Biology 57*; The Association of Applied Biologists: Wellesbourne, UK, 2000; 357–362.
5. Ganzelmeier, H.; Rautmann, D.; Drift, D. Drift Reducing Sprayers and Sprayer Testing. In *Aspects of Applied Biology 57*; The Association of Applied Biologists: Wellesbourne, UK, 2000; 1–10.
6. Landers, A.J. Drift Reduction in the Vineyards of New York and Pennsylvania. In *Aspects of Applied Biology 57*; The Association of Applied Biologists: Wellesbourne, UK, 2000; 67–73.
7. Fox, R.D.; Derksen, R.C.; Brazee, R.D. In *Airblast/Air-assisted Application Equipment and Drift*, North American Conference on Pesticide Drift Management, Portland, Maine, March–April, 29–1, 1998; Buckley, D., Ed.; University of Maine Cooperative Extension: Orono, ME, 1998; 108–129.
8. Landers, A.J. Spray drift management. Am. Fruit Grower **1999**, *119* (2), 36–38.

ALLELOCHEMICS

John H. Borden
Simon Fraser University, Burnaby, British Columbia, Canada

INTRODUCTION

Allelochemics comprise a vast number of known compounds, and hundreds more yet to be discovered, that mediate behavioral or physiological interactions between organisms of different species. They may benefit the emitter, the receiver, or both. Most are still mainly of scientific interest, but a few (principally attractants or repellents for insect pests) have been incorporated into operational pest management protocols.

TERMINOLOGY

To clarify in part the emerging maze of newly discovered message-bearing chemicals, or *semiochemicals* (Gk. *semeion*, sign or signal), the term *allelochemic* (Gk. *allelon*, of each other) was coined in 1970 to embrace any semiochemical with interspecific activity (1). Thus allelochemics are distinguished from *pheromones* (Gk. *pherum*, to carry; *horman*, to excite) that convey a message between organisms of the same species. Three categories of allelochemics are commonly recognized.

Kairomones (Gk. *kairos*, opportunistic) are allelochemics that provide an adaptive advantage to the perceiver. In most cases there is no benefit, or even harm, to the emitter, for example, the attraction of predators to the odor of their prey. The evolution of such chemicals as true biological signals would be disadaptive, and therefore unlikely. Therefore some semantic purists remove all evolutionary implications pertaining to kairomones by referring to them as infochemicals.

Allomones (Gk. *allos*, other) are allelochemics that convey an adaptive advantage to the emitter. The repellent odor of an alarmed skunk is often used as an example. However, in an evolutionary sense it may also be adaptive for the receiver to be able to detect and avoid the skunk's odor, for example, for a predator not to be "tagged" with an aroma that warns potential prey of its presence. Therefore, a skunk's odor is more aptly termed a *synomone* (Gk. *syn*, with), an allelochemic that conveys a mutual advantage to both the emitter and the receiver.

Most allelochemics have a *releaser* effect, in which behavioral responses are evoked. However, they may also have a *primer* effect, in which a physiological or biochemical function is stimulated or inhibited.

NATURAL OCCURRENCE

Table 1 provides a small window on the thousands of allelochemic interactions that occur in nature. The compounds that mediate these interactions are equally diverse (Fig. 1). Very few interactions are mediated by a single compound; most involve relatively simple blends; and some, for example, floral fragrances, comprise dozens of compounds in a single blend. Although Table 1 provides examples of allelochemic interactions among terrestrial plants, arthropods, and vertebrates, many are also found among aquatic organisms, and examples occur in all Kingdoms and Phyla.

Kairomonal interactions include the attraction of many species of phytophagous insects to their host plants, entomophagous insects to their prey or to insects that they parasitize, and blood-feeding diptera to their vertebrate hosts (1–3). They also include the avoidance by prey species of odors associated with predators, a phenomenon found in five animal phyla, but curiously not yet among terrestrial insects (4). Sometimes more than one type of semiochemical may be involved; for example, the aggregation of bark beetles necessary to mass attack and kill a tree is mediated by a blend of aggregation pheromones synergized by host tree kairomones.

Allomonal interactions may employ "trickery" (5), for example, bolas spiders that emit moth sex pheromones that lure mate-seeking male moths to their death, and many species of myrmecophiles (ant lovers) that gain access to ant nests by chemically mimicking the cuticular recognition compounds of ants on which they prey. The most well-known allomones are released by or are contained in plants, and have a primer effect. Allelopathic allomones are often leached from the leaves (or other parts) of plants of one species, and inhibit the germination of seeds or growth of plants in other species that could be potential

A

Table 1 Examples of natural occurrence and function of allelochemics[a]

Type of allelochemic and source	Example
Kairomone:	
Plants	Attraction of Colorado potato beetles, *Leptinotarsa decemlineata*, to 6-carbon leaf volatiles, e.g., (*E*)-2-hexen-1-ol-(**2**), from solanaceous plants
	Attraction of ambrosia beetles, *Trypodendron lineatum, Gnathotrichus sulcatus* and *G. retusus*, to ethanol (**1**) from moribund coniferous trees, logs, and stumps
	Attraction and stimulation of oviposition by onion maggots, *Delia antigua*, in response to mono- and disulfides, e.g., dipropyl disulfide (**3**) from onions, *Allium cepa*
	Employment of host tree kairomones, e.g., α-pinene (**8**) from conifers and α-cubene (**10**) from elms as synergists of aggregation pheromones that mediate mass attack of trees by bark beetles e.g., the Douglas-fir beetle, *Dendroctonus pseudotsugae*, and the smaller European elm bark beetle, *Scolytus multistriatus*, respectively
Insects	Stimulation in a parasitic chalcidoid wasp, *Trichogramma evanescens*, of searching for and oviposition in corn earworm eggs, *Helicoverpa zea*, by tricosane (**15**) in moth scales adhering to newly laid eggs
	Attraction of predaceous clerid beetles, *Thanasimus* and *Enoclerus* spp. to aggregation pheromones, e.g., ipsenol (**19**), ipsdienol (**20**), and frontalin (**22**) of their bark beetle hosts
Vertebrates	Attraction of blood-feeding mosquitoes to CO_2 (**23**) exhaled by mammals
	Attraction of tsetse flies, *Glossina* spp., to volatiles, e.g., acetone (**24**) and 1-octen-3-ol (**25**) from bovine animals on which they feed
	Avoidance by voles, *Microtus* spp., of volatile chemicals in the urine of mustellid predators, e.g., 2-propylthiotane (**26**) and 3-propyl-1,2-dithiolane (**27**) from short-tailed weasels, *Mustella erminea*
Allomone:	
Plants	Inhibition of germination of growth of one species of plant by allelopathic chemicals produced in the leaves, roots, or other tissues of another plant, e.g., by juglone (**7**) leaching from the leaves of black walnut trees, *Juglans nigra*
	Disruption of growth, metamorphosis, or reproduction of insect herbivores by producing insect hormones or hormone analogues, e.g., juvabione (**11**), a juvenile hormone mimic in true firs, *Abies* spp.
Spiders	Emission of moth sex pheromones, e.g., (*Z*)-9-tetradecenyl acetate (**18**) by bolas spiders, *Mastophora* spp., to attract male moths as prey
Insects	Mimicking the cuticular recognition compounds, e.g., 11-methyl penta-cosane (**16**) of larval ants by caterpillars of lycenid butterflies, syrphid fly larvae, and scarab beetle larvae, thereby gaining entry into and acceptance within ant nests, where they prey on ant brood
Synomone:	
Plants	Avoidance of nonhost plants by insects in response to volatiles emitted by the plants, e.g., repellency of coniferophagous bark beetles, e.g., the mountain pine beetle, *Dendroctonus ponderosae*, to conophthorin (**13**) (also a repellent pheromone of cone and twig beetles) in the bark of birches, *Betula* spp.
	Repellency of black bean aphids, *Aphis fabae*, to methyl salicylate (**4**) in the volatiles of nonhost plants
	Tritophic interaction in which corn plants, *Zea mays*, respond to volicitin (**14**) (a kairomone) in the saliva of beet armyworm caterpillars, *Spodoptera exigua*, feeding on them by producing specific blends of volatiles (synomones), e.g., (*E*)-4, 8-dimethyl-1,3,7-nonatriene (**9**) that attract females of a parasitic wasp, *Cotesia marginiventris*, which in turn oviposit on the feeding caterpillars
	Attraction of honey bees, *Apis mellifera*, to multicomponent blends of floral volatiles, e.g., geraniol (**12**), of many species of angiosperm plants, with mutual benefit of pollination to the plant and a pollen and nectar source to the bees
Insects	Antagonists in the blends of moth sex pheromones that repel males of related species, ensuring reproductive isolation even though the major pheromone components are attractive to males of both species, e.g., (*Z*)-9-tetradecanal (**17**), a pheromone component of threelined leafrollers, *Pandemis limitata*, that inhibits response of obliquebanded leafroller males, *Choristoneura rosaceana*, to threelined leafroller females
	Mutual repellency between aggregation pheromones of two species of bark beetles, e.g., (*R*)-(−)-ipsdienol (**20**) produced by pine engravers, *Ips pini*, and (*S*)-(−)-ipsenol (**19**) produced by California fivespined ips, *I. paraconfusus*, reserving the host phloem resource for the first-arriving species, and avoiding interspecific exploitative competition

[a]Numbers in parentheses correspond with numbered compounds in Fig. 1.

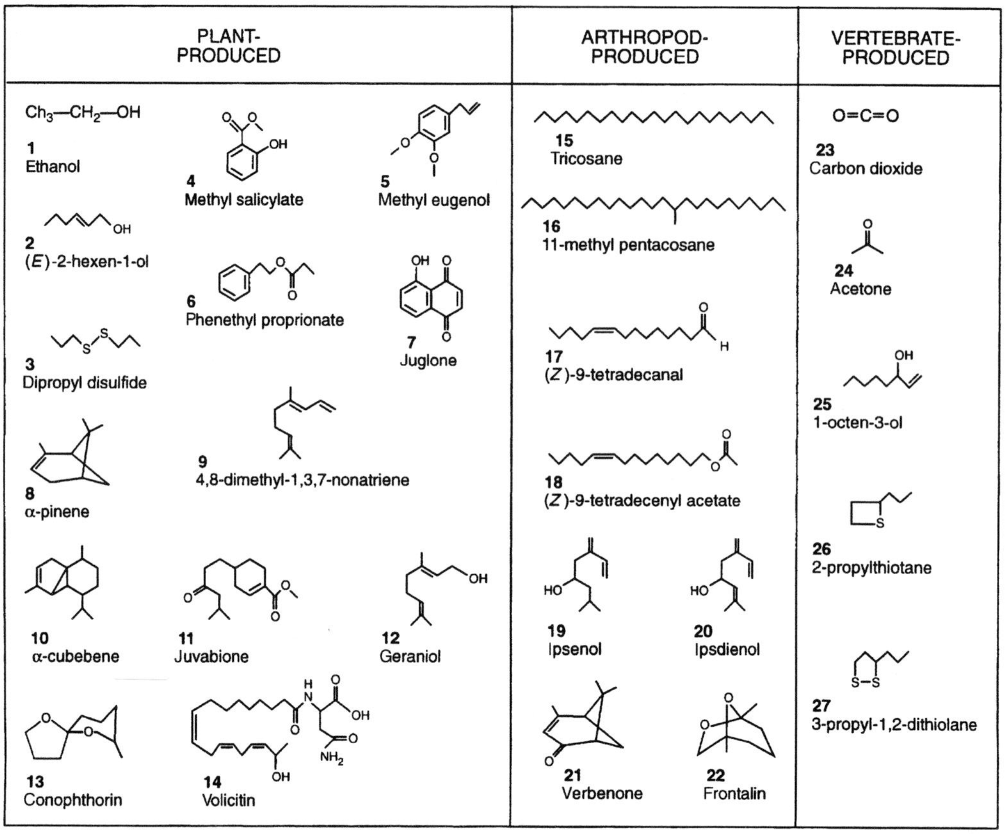

Fig. 1 Structural formulae of compounds given in Table 1 exemplifying some of the chemical diversity among allelochemics as follows: primary alcohol (**1, 2**), secondary alcohol (**25**), disulfide (**3**), aromatic ester (**6**), unsaturated ester (**18**), monoterpene (**8**), sesquiterpene (**9, 10**), sesquiterpenoid (**11**), terpene alcohol (**12, 19, 20**), terpene ketone (**21**), spiroacetal (**13**), fatty acid derivative conjugated to an amino acid (**14**), straight chain hydrocarbon (**15**), branched hydrocarbon (**16**), unsaturated aldehyde (**17**), bicyclic ketal (**22**), atmospheric gas (**23**), ketone (**24**), thiotane (**26**), and thiolane (**27**).

competitors (6). Some plants may also produce defensive allomones against insect herbivores, for example, hormones or analogues of hormones that disrupt the growth and metamorphosis of their insect enemies.

Among the many examples of synomones are repellents that ensure reproductive isolation between closely related species of insects, or mitigate against the occurrence of interspecific exploitative competition for a limited host resource. Often synomones are the same compounds as one or more of the components that convey a pheromonal message, for example, to attract mates or to aggregate on or near a food source. There is increasing evidence that host-seeking phytophagous insects not only use kairomones to find their host plants, but also use synomones to avoid nonhost plants on which their fitness would be greatly reduced. Synomonal floral scents provide a mutual benefit to flowering plants that gain from pollination by insects that in turn are attracted to a nutritious nectar or pollen source.

Another type of allelochemic interaction involves three trophic levels and the action of both kairomonal and synomonal stimuli (3). In one remarkable example of this type of tritrophic interaction, corn plants being fed on by beet armyworm caterpillars are exposed to minute amounts of a kairomone called volicitin in the insect's saliva. Volicitin has a primer effect, eliciting the plant to synthesize a specific blend of volatile synomonal compounds that attract females of a parasitic wasp. The wasp oviposits in the beet armyworm larvae, benefiting by finding its host, and in turn providing an advantage to the plant by parasitizing and killing the caterpillar.

PRACTICAL APPLICATIONS

Knowledge about the natural occurrence and role of allelochemics opens up a huge, but relatively untapped, potential for exploiting them as pest management tools (2). In some cases the knowledge itself is important. Plant breeders may seek varieties of plants that contain or release chemicals that deter feeding or development by phytophagous insects. Plants with allelopathic characteristics, for example, *Eucalyptus* spp., may be useful in landscaping to reduce weed problems. Species or varieties rich in attractive kairomones may be used as trap crops for various insect pests. In the production of transgenic agricultural crops it is critical not to loose the capacity for tritrophic interaction that will ensure parasitism of herbivorous insects, lest the genetically modified plants be more vulnerable to insect pests than unmodified plants.

In other cases, the capacity to use allelochemics as pest management tools may be demonstrated, but technological, economic, or social limitations may prevent their use. Allelopathic allomones from plants are under consideration for development as a new class of biodegradable herbicides (7). But to date none can compete with conventional chemical herbicides with regard to ease of synthesis, capacity for formulation, efficacy, and/or safety. If used widely, some allelopathogens may pose an unacceptable threat to environmental or human health. Recent investigations show considerable promise for using nonhost volatiles to "disguise" herbivorous host plants or trees as nonhosts, but commercial formulations have not yet appeared on the market, in part because of the challenge and expense of registering those allelochemic products as pesticides. Similar problems beset the development and use of predator volatiles to protect plants from damage by herbivorous vertebrates such as deer and voles.

Despite the above limitations, a few kairomones have found widespread commercial use (2, 8). Among them is methyl engenol, which is used worldwide for capturing tephritid fruit flies, both for detection of unwanted introductions and for direct suppression of populations in a lure and kill tactic employing an insecticide-laced substrate baited with methyl engenol. A lure and kill tactic is also used effectively for control of tsetse flies that are drawn by acetone and 1-octen-3-ol baits to insecticide-treated "target" traps that simulate the silhouette of a large vertebrate. Other applications combine kairomones with pheromones. A combination of phenethyl proprionate and methyl engenol with the sex pheromone of the Japanese beetle is used in many thousands of traps in the United States each year. Similarly the kairomones ethanol and α-pinene have been used since 1981 in combination with aggregation pheromones in commercial mass trapping programs for three species of ambrosia beetles in British Columbia.

Allelochemics may also find use in the future in the application of "push–pull" tactics, in which one repellent volatile treatment is used to protect a plant or group of plants from attack by insects (push), and another attractive treatment is used in baited traps or trap plants to pull the insects away. One outstanding example of a successful push–pull application saved a rare stand of endangered Torrey pines in California from being killed by the California five spined ips. Two repellent synomones, verbenone, produced by western pine beetles, and (−)-ipsdienol, produced by pine engravers, were deployed inside the uninfested portion of the stand, and traps baited with attractive aggregation pheromone were arrayed in an adjacent area of beetle-killed pines. Over 86 weeks beginning in May 1999, 330717 beetles were captured. The program ran until September 1993, but not a single Torrey pine died after August 1992.

CONCLUSION

Despite many studies, most natural allelochemic interactions are yet to be discovered. The adoption of allelochemics as pest management tools has been limited. However, there is great potential for judicious selection and commercial development of allelochemics, particularly in integrated pest management programs that will combine a number of alternative ecologically based tactics with the reduced use of conventional chemical pesticides.

[See also the articles Toxins in Plants; Mechanisms of Resistance: Antibiotics, Antixenosis; Antagonistic Plants; Trap Crops; Insect Growth Regulators; Repellents; Chemical Sex Attractants; Ovipositional Disruption Employing Semiochemicals; Trapping Pest Populations; Natural Pesticides in this encyclopedia.]

REFERENCES

1. Whittaker, R.; Feeney, P. Allelochemics: chemical interactions between species. Science **1971**, *171*, 757–770.
2. Metcalf, R.; Metcalf, E. *Plant Kairomones in Insect Ecology and Control*; Chapman & Hall: New York, 1992; 168.
3. Vet, L.; Dicke, M. Ecology of infochemical use by natural enemies in a tritrophic context. Annu. Rev. Entomol. **1992**, *37*, 141–172.
4. Kats, L.; Dill, L. The scent of death: chemosensory assessment of predation risk by prey animals. Ecoscience **1998**, *5*, 361–394.
5. Stowe, M.; Turlings, T.; Loughrin, J.; Lewis, W.; Tumlinson, J. The Chemistry of Evesdropping, Alarm, and Deceit. In *Chemical Ecology: The Chemistry of Biotic Interaction*; Eisner, T., Meinwald, J., Eds.; National Academy Press: Washington, DC, 1995; 51–65.
6. Zindahl, R. *Fundamentals of Weed Science*; Academic Press: New York, 1993; 135–146.
7. Cutler, G. Allelopathy for Weed Suppression. In *Pest Management: Biologically Based Technologies*; Lumsden, R., Vaughn, J., Eds.; American Chemical Society: Washington, DC, 1993; 290–302.
8. Borden, J. Disruption of Semiochemical-Mediated Aggregation in Bark Beetles. In *Insect Pheromone Research: New Directions*; Cardé, R., Minks, A., Eds.; Chapman & Hall: New York, 1997; 421–438.

ALLERGENIC REACTIONS TO PESTICIDES AND PESTS

Alicia Armentia
Hospital Rio Hortega, Valladolid, Spain

INTRODUCTION

Given the ubiquity of arthropods in stored vegetable products, it comes as a bit of a surprise that there are so few reports in the medical literature on allergy to arthropods as inhalants, compared to the many studies dedicated to other allergens. It would seem important, therefore, to evaluate its prevalence in a more numerous series of patients and to study not only the distribution of arthropods and their relationship to clinical symptoms but also the nature of their capacity as occupational allergens. The clinical importance of pests and pesticides as causes of hypersensitivity reactions will be the object of this entry.

ALLERGIC DISEASE IN CEREAL WORKERS BY STORED GRAIN PESTS

Workers occupationally exposed to grain dust have a high prevalence of respiratory symptoms (1–4). Grain dust contains biologic contaminants, including mold spores, pollens, and insects. The most prominent allergenic components involved in cereal allergy are salt soluble proteins, and recently several allergens associated with flour allergy have been purified. All of them exhibit inhibitory activities against arthropods' alpha-amylases and trypsin (4). Nevertheless, the pathogenesis of respiratory symptoms in grain workers and bakers remains obscure when sensitization to cereal flour has not been proved.

Allergy to storage mites in the form of bronchial asthma and allergic rhinoconjunctivitis has been reported among farmers, bakers, and grain elevator workers. The most common storage mites involved belong to the family Glycyphagidae (genera *Lepidoglyphus*, *Glycyphagus*, and *Blomia*), Acaridae (genera *Tyrophagus*, *Acarus*, *Suidasia*, and *Aleuroglyphus* and Chortoglyphydae, genus *Chortoglyphus*) (1). Recent publications based on the analysis of farm dust and clinical parameters have pointed out that *Lepidoglyphus destructor* is the most important storage mite causing symptoms involving both upper and lower respiratory tract among farmers in Europe (5). Campbell et al. have sampled barn air from barns in United States and detected abundant *Lepidoglyphus destructor* derived aeroallergens. Our data regarding mite infestation in samples of cereal flour and clinical data are consistent with those of these authors (3).

Numerous reports of occupational allergy have been attributed to members of the Coleoptera order. *Tenebrio molitor* is a member of the coleoptera insect order, which includes beetles and weevils. Mealworms are larvae of *Tenebrio molitor* and normally infest stored grain. In addition, they are widely used as fishing bait. Bernstein et al. demonstrated that inhaled particulates from *Tenebrio* exoskeletons are potent sensitizers and elicit Ig-E mediated occupational asthma. Schroeckenstein et al. described a subject with occupational allergy to *T. molitor* and by means of immunoblotting techniques, studied the allergens involved in this patient's immunologic response and confirmed the fact that beetles of the Tenebroid family are potentially significant allergens for workers exposed to grains or grain products. The RAST inhibition showed immunologic cross-reactivity between *T. molitor* and *Alphitobious diaperinus*, as well as a slight cross-reactivity with blowfly (6). On the other hand, a barley flour inhibitor of *Tenebrio molitor* alpha-amylase is a major allergen associated with baker's asthma disease (Fig. 1).

Cockroaches are an important source of allergens. Several studies indicate that a high percentage of patients with asthma have significant skin test reactivity and serum IgE antibodies to cockroach allergens (7). Nevertheless, there is the possibility that sensitization to cockroaches may be underestimated because of the limited number of german cockroach extract that are currently marketed for skin testing. Lehrer et al. demonstrated that commercial cockroach extracts vary in allergenic activity and that all house dust extracts tested contain cockroach allergens.

In 1997, we performed an epidemiological analysis of the sensitization to these stored-grain pests on 4379 patients residing in an area of cereal industries (3). Of these patients, 1395 were cereal workers and 2984 did not work

Fig. 1 *Tenebrio molitor* and mealworms.

with cereal. The questionaire consisted of questions relating to rhinoconjunctivitis, coughing, wheezing and shortness of breath, allergic and occupational precipitants, contact with cereal, animal fodder, observation of insects in flour or cereals by the patient, length of exposure, and other epidemiological data. Specific IgE antibodies to the extracts were demonstrated by prick tests and RAST. Association between respiratory symptoms and occupational exposure was confirmed by challenge tests (specific and methacholine). The prevalence of mite sensitization in the total sample studied (4379) was 18.96% (SEM 0.58, 95% CI 16.93–19.19). The prevalence of sensitization to storage mites among mite-sensitive patients was 11.88% (SEM 1.15, 95% CI 9.63–14.3). Among the cereal workers, the most frequent sensitization was that to *D. pteronyssinus* (58%), followed by *D. farinae* (48%), *Lepidoglyphus destructor* and *Tyrophagus putrescentiae* (38%), *Blomia kulagini* (34%), and *Acarus siro* and *Chortoglyphus arcuatus* (24%). In addition to this, 22% of these patients presented negative prick tests and RAST for *Dermatophagoides* species with positive test to storage mites. Of the workers, 50% were sensitized to *Tenebrio molitor* (SEM 0.7, CI 95% 36–64) and 36% to *Blatta orientalis* (SEM 0.67 , CI 95% 23–49). The identification of mites, tenebroids, and cokroaches in dust samples yielded useful data for the diagnosis of our patients. These results, together with the aforementioned studies confirm the fact that beetles of the Tenebrionidae family are potentially significant allergens for workers exposed to grains or grain products. Reasons for the lack of interest in arthropod allergy would include the difficulty of collecting live arthropods for preparing extracts useful for diagnosis. The use of biologically standardized extracts of stored grain pests should improve the accuracy of prick testing in the diagnosis of allergic respiratory disease. In fact, we found skin prick testing the most sensitive diagnostic criteria.

A

OCCUPATIONAL ASTHMA BY STORAGE MITES CONTAMINATING FOODS

Only few reports documented allergic symptoms realated to the manipulation of certain foods. The histamine-releasing power of certain foods, such as cheese and butcher's products has been documented, as well as the importance that the ingestion of these products could have in sensitive patients. Nevertheless, allergy to cheese mites has been rarely reported, although it is known that the fungi of its rind are able to produce asthma and pneumonitis by hypersensitivity.

Recently, we have reported four cases of occupational allergy caused by the manipulation of different foods: cheese, ''chorizo'' (spanish cured sausage), ham, and garlic (8). We have not found previous studies about the allergenic power of chorizo mites, although chorizo also contains additives (talc, nitrites) and other histamine-releasing products. On the other hand, only one published study exists about occupational allergy caused by the manipulation of ham. Garlic is considered to be one of the most common causes of dermatitis of the fingertips in suppliers of foodstuffs. However, its association with asthma or rhinitis has rarely been reported. In 1940, Henson reported a case of occupational asthma induced by garlic dust in a sausage maker. We have found six cases, and have suggested an etiopathogenic mechanism of IgE-mediated hypersensitivity. Challenge and immunologic tests showed a relevant association with sensitization to *Tyrophagus* and a superinfestation by this mite in different samples of garlic dust.

In summary, there are few studies in which the manipulation of food sources contaminated with mites could be the cause of occupational asthma. Nevertheless cases of anaphylaxis developed after eating food contaminated by storage mites has been recently reported. Matsumoto (1996) made the interesting hypothesis that years ago, many mites contaminated stored foods, and possibily, the routine ingestion of the mites through ordinary meals might prevent systemic allergic reaction, since anaphylaxis in response to storage mites has not been reported before. They argued that this protective factor has been lost by the sanitary controls, making mite-related asthma more common.

We conclude that in asthma induced by inhalation of food dusts, possible common contaminating allergens should be checked for, specifically, storage mites.

ALLERGENIC REACTIONS TO PESTICIDES

Pesticides can produce severe pulmonary fibrosis when ingested (paraquat poisoning); cholinergic crises with excessive bronchial, salivary, ocular and intestinal secretions, sweating, miosis, severe bronchospasm, bradycadia, and paralysis (organophospates); chest tightness and muscle weakness due to cholinesterase inhibitors (carbamates); malaise, headache, dyspnea, sweating, thirst, hypothermia, respiratory failure, and death (Dinitrophenols). All these effects can be explained by pharmacologic and toxix mechanisms, but as far as we know, a specific IgE-mediated mechanism has not have been found yet although various pesticides could increased histamine release and exacerbate allergic diseases. Delayed hypersensitivity cutaneous reactions have been frequently reported: contact urticaria (chlorothalonil, permethrin), photocontact dermatitis (fenitrothion), contact pemphigus (dihydrodiphenyltricholrethane), and toxic epidermal necrolysis (organophosphates).

We have suggested that certain food, cocoa, and cola drinks could contain cereal in a masked way and could cause severe allergic reactions (9). The recent syndromes due to toxicity in cola drinks in Belgium and in a baby cereal formula in an Andean area resemble that reported in toxicity in cereals contaminated with PCB dioxins. In fact, they are similar to those observed in patients in Yusho, Japan (1968) and Yu-Cheng, Taiwan (1978) who ingested PCB-contaminated rice oil, and in people from Spain who ingested contaminated and adulterated rapeseed oil (1981).

REFERENCES

1. Blainey, A.D.; Topping, M.D.; Ollier, S.; Davies, R.J. Allergic respiratory disease in grain workers: the role of storage mites. J. Allergy Clin. Immunol. **1989**, *84*, 296–303.
2. Armentia, A.; Tapias, J.; Barber, D.; Martin, J.; De la Fuente, R.; Sanchez, P.; Salcedo, G.; Carreira, J. Sensitization to the storage mite *Lepidoglyphus destructor* in wheat flour respiratory allergy. Ann. Allergy **1992**, *68*, 398–430.
3. Armentia, A.; Martinez, A.; Castrodeza, R.; Martínez, J.; Jimeno, A.; Méndez, J.; Stolle, R. Occupational Allergic Disease in Cereal Workers by Stored Grain Pests. J. of Asthma **1997**, *35*, 369–378.
4. Armentia, A.; Sánchez, R.; Gómez, L.; Barber, D.; Salcedo, G. In vivo allergenic activities of eleven purified members of a major allergen family from wheat and barley flour. Clin. Exp. Allergy **1993**, *23*, 410–415.
5. Johansson, E.; Borga, A.; Johansson, O. Van Hage-Hamsten immunoblot multi-allergen inhibition studies of allergenic cross-reactivity of the dust mites *Lepidoglyphus destructor* and *Dermatophagoides pteronyssinus*. Clin. Exp. Allergy **1991**, *21*, 511–518.
6. Schroeckenstein, D.C.; Meier-Davis, S.; D.M.V.; Bush, R.K. Occupational sensitivity to *Tenebrio molitor* linnaeus (Yellow Mealworm). J. Allergy Clin. Immunol. **1990**, *86*, 182–188.
7. Schou, C.; Lind, P.; Fernandez, Caldas E.; Lockey, R.; Lowenstein, H. Identification and purification of an important cross-reactive allergen from American (*Periplaneta americana*), and German (*Blattella germanica*) cockroach. J. Allergy Clin. Immunol. **1991**, *86*, 935–946.
8. Armentia, A.; Fernández, A.; Pérez, C.; De la Fuente, R. Occupational allergy to mites in salty ham, chorizo and cheese. Alergol. Immunopathol. **1994**, *4*, 152–154.
9. Armentia, A.; Martín, J.M.; Arranz, M.L.; Martín, F.J.; Bañuelos, M.C.; Navas, C. Asthma after ingestion of cereal allergens from cocoa and cola drinks. J. Invest. Allergol. Clin. Immunol. **1999**, *9* (Suppl. 1).

ANTAGONISTIC PLANTS

A

Philip Oduor Owino
Kenyatta University, Nairobi, Kenya

INTRODUCTION

Plant parasitic nematodes cause significant crop losses in Africa and other parts of the world. Infected plants suffer from water deficiency and low yields, and have necrotic and/or galled roots. Control of nematodes has been mainly through the use of chemicals and host resistance. However, the existence of physiological races in the pathogen's population has complicated efforts to breed for resistant cultivars. Chemical control is effective but difficult to sustain for long-term benefits. The high cost of nematicides and their toxic effects also make them less attractive. Some nematicides such as Nemagon and Fumazon have now been banned from the world market and this has placed severe constraints on strategies for nematode control. Interest in developing alternative control measures that are safe and economically attractive has now intensified worldwide. The use of antagonistic plants is viewed as a viable nematode management option (1–3).

THE NATURE OF ANTAGONISTIC PLANTS

Antagonistic plants are defined as plants that produce chemicals in their roots that are toxic and/or repellant to phytonematodes in the soil ecosytem (4). These plants include *Tagetes erecta* L; *Tagetes patula* L; *Datura stramonium* L; *Ricinus communis* L; and *Asparagus officinalis*. Fresh roots of asparagus produce asparaguric acid glycoside that is toxic, even when diluted, to most plant parasitic nematodes. Root exudates from *Tagetes*, *Datura*, and *Ricinus* spp. induces premature hatching of nematode eggs, blocks the processes of mitosis and meoisis, and reduces galling intensity on roots of susceptible plants. This has been attributed, in part, to the toxic effects of the alkaloids terthienyl, hyosine, and ricinine present in *Tagetes, Datura*, and *Ricinis* spp., respectively. These compounds may also disrupt female taxis to roots or male taxis to female (4). Other plants with antagonistic properties include some crucifers and citrus. Root diffusates from crucifers reduce the pathogenicity of nematodes on potato, while a compound in citrus roots is toxic to *Tylechulus semipenetrans* (1).

Antagonistic Plants in Cultural Pest Control

Antagonistic plants may have a great nematode-control potential in agriculture if properly utilized in crop rotation and intercropping systems (2). For example, intercropping food crops with nematicidal plants is now a nematode management strategy in Tanzania, India, and Zimbabwe and has also been recommended for Pakistan (5). Field trials with *T. minuta*, *D. stramonium*, and *R. communis* are promising (256). These plants reduce galling intensity and enhance tomato performance significantly. In India, a rotation of *D. stramonium*, maize, tomato, and pepper reduced the population of root-knot nematodes by 30% but the level of nutrient depletion by the antagonistic plant was 15% (4). Integration of these plants with the biological agent, *Paecilomyces lilacinus* Thom (Sam), gave better results in Kenya (378). Tomato plants grown in soils planted with the various antagonistic plants in combination with *P. lilacinus* develop significantly heavy shoots and roots and relatively fever root galls compared to controls (Table 1). Cases where antagonistic plants are used in crop rotation or intercropping systems are now increasing (4). For instance, in Indonesia, *Tagetes* sp. *Crotalaria usaramoensis*, corn, and sweet potato (*Ipomea batatas*) are used to reduce *Meloidogyne* spp. density in the soil. For cereal-based cropping systems, the following crop sequences for root-knot nematode control are recommended in the Philippines: rice–mung bean (*Phaseolus aureus*)–corn–cabbage–rice, rice–tobacco (*Nicotiana tobacum*)–rice and rice–tobacco and *Tagetes* spp. There is also considerably less galling by *Meloidogyne* spp. on potato (*Solanum tuberosum*) roots intercropped with onion (*Allium* spp.), corn, and marigold compared with galling found on potatoes alone. Although antagonistic plants are gaining popularity as pest management tools, their benefits and risks must be understood

Table 1 Effect of tomato intercropping with *Datura stramonium, Ricinus communis,* and *Tagetes minuta,* and soil treatment with Captafol and Aldicarb, on gall index, number of juveniles, tomato growth and fungal parasitism of *Meloidogyne javanica* eggs by the fungus *Paecilomyces lilacinus*, 50 days after inoculation

Soil treatment[a]	Egg parasitism (%)	No. of juveniles/ 300 cm^3 soil	Shoot dry weight (g)	Shoot height (cm)	Gall[b] index (0–4)
Ne "only" untreated	1.0de	670a	1.5f	26.6f	4.0a
Soil "only" untreated	0.0e	0d	3.5b	43.1ab	0.0e
F+Ne	23.2c	660a	1.6f	30.1e	3.0ab
F+Cap+Ne	1.3de	635a	1.9f	31.2e	3.9a
F+Ald+Ne	27.6b	12de	4.6a	45.3a	1.4d
F+Tag+Ne	29.8ab	161c	2.4e	38.4b	2.1c
F+Dat+Ne	28.3ab	173c	2.5de	36.4cd	2.2c
F+Ric+Ne	30.9a	210c	2.9cd	35.4cde	2.4c
Tag+Ne	2.1de	209c	2.8d	33.4cde	2.9b
Dat+Ne	3.0d	183c	3.0cd	36.6c	2.8b
Ric+Ne	2.8de	204c	3.2bc	37.7bc	3.0ab
Ald+Ne	1.2de	14d	4.5a	46.1a	3.8a
Cap+Ne	0.0e	460b	1.9f	36.1d	4.0a

Numbers are means of 10 replications. Means followed by different letters within a column are significantly different ($P = 0.05$) according to Duncan's Multiple Range Test.

[a]Ne = nematode; F = P. *lilacinus*; Cap = Captafol: Ald = Aldicarb: Tag = *T. minuta*; Ric = *R. communis*.

[b]Gall index was based on a 0–4 rating scale, where 0 = no galls and 4 = 76–100% of the root system galled.

(From Ref. 7.)

thoroughly before one can exploit their potential in pest control.

Benefits and Risks of Antagonistic Plants

Benefits and risks associated with the utilization of antagonistic plants in agriculture are varied. Phytonematoxic plants such as *R. communis*, *D. stramonium*, *Tagetes* spp. *Crotalaria* spp. *A. najus*, and *Datura metel* L. are traditionally gaining popularity due to their medicinal significance (9). The flowers of *D. metel* are used against asthma, while *Crotalaria* spp. is a nitrogen-fixing legume. Castor oil from *R. communis* is used for making soaps and waxes: rinolecic acid from castor seeds is a valuable laxative (9). Despite these attributes, antagonistic plants may pose a serious threat to food production if not well utilized. They may compete with economically important crops for space and nutrients. In addition they are slow in action, an attribute

Table 2 Effect of soil treatment with Aldicarb, *Tagetes minuta, Datura metel,* and *Datura stramonium* on root-knot nematodes in tomatoes (greenhouse test)

Soil treatment[a]	Shoot height (cm)	Shoot dry weight (g)	Gall index (0–5)	Galls ($no.g^{-1}$) root weight	Nematodes, no. $(300\ ml)^{-1}$ soil
Soil + Ne[b], untreated (control)	42.4c[c]	1.3e	4.4a	510.0a	564.1a
Soil only, untreated	97.8a	4.9b	0.0c	0.0d	0.0e
Soil + Ne + Aldicarb	116.3a	6.1a	1.0c	23.3d	18.4d
Soil + Ne + *D. metel*	73.6d	3.7c	2.1b	77.4c	176.3c
Soil + Ne + *T. minuta*	65.4b	2.6d	2.0b	134.9b	310.4b
Soil + Ne + *D. stramonium*	73.4b	3.1c	2.2b	88.4c	170.4c
Soil + Ne + *R. communis*	69.0b	3.3c	2.4b	90.0c	173.0c

[a]Autoclaved soil used

[b]Ne, nematode eggs added to soil

[c]Means followed by the same letter within each column are not significantly different at the 5% level (Duncan's Multiple Range Test).

(From Oduor-Owino, P. Effects of Aldicarb and Selected Medicinal Plants of Kenya on Tomato Growth and Root-Knot Severity, unpublished data, 1992.)

Table 3 Effect of soil treatment with Aldicarb, *Tagetes minuta, Datura metel,* and *Datura stramonium* on infection of tomato by root-knot nematodes (field test)

Treatment	Shoot height (cm)	Shoot dry weight (g)	Fruit yield (g)	Galls (no. g^{-1} root weight)	Nematodes, no. $(300\ ml)^{-1}$ soil
Untreated (control)	80.3d[a]	40.5d	380.3e	69.1a	150.4a
Aldicarb	187.3a	135.1a	3800.4a	4.50d	6.4d
D. metel	157.1b	89.3b	2590.1b	6.4c	17.3bc
T. minuta	107.1c	45.1d	761.1c	11.4b	21.1b
D. stramonium	150.1b	69.4c	2030.4b	9.6b	18.4c

[a]Means followed by the same letter within each column are not significantly different at the 5% level (Duncan's Multiple Range Test).
(From Ref. 2.)

that makes them less attractive for use in a commercial setting. Because of this scenario, it is important that scientific disciplines work together in order to develop a viable pest control system. What is good for the nematologist may not be good for either the agronomist or economist.

FUTURE CONCERNS[a]

There is increasing internal awareness of the value of natural plants and their products in the development of new drugs and formulation of materials that can be used for pest control. Since some of the antagonistic plants can also be used to treat human ailments (9), they may attract intensive scientific evaluation, recognition, and funding. However, more work should be done to reexamine the future of antagonistic plants in nematode control and in the drug industry. Efficacy of these plants against nematodes and their utilization in the pharmaceutical industry will depend highly on the concentrations of the active ingredients in their tissues (10). It will also depend on whether they can stimulate activity of most biocontrol fungi and plant growth consistently. They have so far enhanced tomato growth in the greenhouse and in the field significantly (27) (Tables 2 and 3), but more trials are needed in order to understand the relationship between antagonistic plants, natural enemies, and crop performance.

REFERENCES

1. Caswell, E.P.; Tan, C.S.; De Frank, J.; Apt, W.J. The influence of root exudates of *Chloris gayana* and *Tagetes patula* on *Rotylenchulus reniformis*. Revue de Nematologie **1991**, *14* (2), 581–587.
2. Oduor-Owino, P. Effects of Aldicarb, *Datura stramonium, Datura metel* and *Tagetes minuta* on the pathogenicity of root-knot nematodes in Kenya crop protection. Organic Soil Amendment **1993**, *12* (4), 315–317.
3. Oduor-Owino, P.; Sikora, R.A.; Waudo, S.W.; Schuster, R.P. Effects of aldicarb and mixed cropping with *Datura stramonium, Ricinus communis* and *Tagetes minuta* on the biological control and integrated management of *Meloidogyne javanica.* Nematologica **1996**, *42*, 127–130.
4. Yeates, G.W. How plants affect nematodes. Advances in Ecological Research **1987**, *17* (2), 61–137.
5. Oduor-Owino, P.; Waudo, S.W. Effects of antagonistic plants and chicken manure on the biological control and fungal parasitism of root-knot nematode eggs in naturally infested field soil. Pakistan Journal of Nematology **1995**, *13* (2), 109–117.
6. Oduor-Owino, P.; Waudo, S.W. Comparative efficacy of nematicides and nematicidal plants on root-knot nematodes. Trop. Agric. **1994**, *71* (4), 272–274.
7. Oduor-Owino, P. *Fungal Parasitism of Root-knot Nematode Eggs and Effects of Organic Matter, Selected Agrochemicals and Intercropping on the Biological Control of Meloidogyne javanica on Tomato*; Ph.D. Thesis, Kenyatta University: Nairobi, Kenya, 1996; 132.
8. Oduor-Owino, P.; Sikora, R.A.; Waudo, S.W.; Schuster, R.P. Tomato growth and fungal parasitism of *Meloidogyne javanica* eggs as affected by nematicides, time of harvest and intercropping. East African Agricultural and Forestry Journal **1995**, *61* (1), 23–30.
9. Oduor-Owino, P.; Waudo, S.W. Medicinal plants of Kenya: effects on *Meloigogyne incognita* and the growth of okra. Afro-Asian Journal of Nematology **1992**, *2* (1), 64–66.
10. Oduor-Owino, P. Effects of marigold leaf extract and captafol on fungal parasitism of root-knot nematode eggs-kenyan isolates. Nematologia Mediteranea **1992**, *20*, 211–213.
11. Oduor-Owino, P.; Waudo, S.W.; Sikora, R.A. Biological control of *Meloidogyne javanica* in Kenya: effect of plant residues, benomyl, and decomposition products of mustard (*Brassica campestris).* Nematologica **1993**, *39*, 127–134.

[a]See also *Biological Control of Nematodes*, pages 61–63; *Risks of Biological Control*, pages 720–722; *Toxins in Plants*, pages 840–842; *Pest–Host Plant Relationships*, pages 593–594; *Intercropping for Pest Management*, pages 423–425.

A

AQUATIC WEED CONTROL WITH HERBICIDES

Joanna Davies
IACR-Long Ashton Research Station, Bristol, United Kingdom

INTRODUCTION

Without appropriate management many aquatic plants, particularly exotic or non-native species, can develop into troublesome weeds that cause numerous and varied problems. Aquatic weeds can obstruct water flow in rivers and drainage channels, thus impeding drainage of surrounding land and causing flooding. The presence of excessive aquatic vegetation can also accelerate channel deterioration by increasing the accumulation of silt, interfere with navigation and recreational uses, and disrupt domestic and industrial water supplies. Furthermore, aquatic weeds can present a health hazard to wildlife and humans. Reduced flow can encourage mosquitoes and increase the risk of water-borne diseases such as malaria and bilharzia (schistosomiasis). In addition, toxins produced by blue-green algae have caused allergic reactions in humans and the death of wild animals and farm livestock in many countries. In recent years, aquatic weed growth has been exacerbated by increased nutrient input from agricultural fertilizers, sewage effluents, and industrial pollutants.

USE OF HERBICIDES FOR AQUATIC WEED CONTROL

Management options for aquatic weeds include mechanical, biological, environmental, and chemical control measures. However, unlike terrestrial weeds where control measures are largely determined by the nature of the weed problem, options for aquatic weeds are predominantly influenced by the functions of the waterbody. Chemical control is often limited where water is required for domestic drinking water supply or for the irrigation of crops or livestock. The use of herbicides under these circumstances is not necessarily prohibited but requires strict adherence to label recommendations and observation of recommended irrigation intervals. Despite these restrictions, chemical control is generally preferred where mechanical control is prohibited due to excessive cost, scale of problem, limited accessibility, difficulties with the disposal of cut vegetation, and in situations where mechanical methods are likely to exacerbate the weed problem by dispersing vegetative fragments. Under these circumstances, many aquatic weed species may be effectively controlled with herbicides.

CLASSIFICATION OF AQUATIC WEEDS

Aquatic vegetation can be classified into four categories according to growth habit. For the purposes of weed management, plants are classified as target (weed) or nontarget species, depending on local site conditions.

Emergent plants have stems and leaves that protrude above the water surface. This category includes reeds and rushes and marginal plants that extend into the water.

Floating plants have leaves that float on the water surface. This category includes free-floating plants, such as the duckweeds (*Lemna* species), and species that are rooted in the sediment, such as the water lilies (*Nuphar, Nymphaea*, and *Nymphoides* species).

Submerged plants form the majority of their leaves below the water surface although some species may produce some floating leaves when flowering (*Cabomba* species). They are often rooted in sediment as in the case of the water-milfoils (*Myriophyllum* species) but may be free-floating beneath the water surface as in the case of ivy-leaved duckweed (*Lemna trisulca*) and the hornworts (*Ceratophyllum* species).

Algae are submerged unicellular or multicellular filamentous plants that reproduce rapidly in favorable conditions to form scum or mats on the water surface.

It should be noted that several aquatic plant species produce growth forms that fall into more than one category. For example, arrowhead (*Sagittaria* species) produces emergent leaves in shallow water and submerged leaves in deeper flowing water, while other species produce submerged leaves during spring but emergent or floating leaves later in the season.

HERBICIDES FOR AQUATIC WEED CONTROL

Of the 200-plus herbicides registered in the United Kingdom and United States, only nine active ingredients are available for use in aquatic situations. Research into the development of aquatic herbicides is limited by their small market value and the high cost of developing a product to meet the stringent toxicological and technical challenges presented by the aquatic environment. In particular, the complexity of aquatic ecosystems, combined with the multifunctional nature of many water bodies, dictates that herbicides must meet extremely rigid environmental and toxicological criteria to satisfy registration requirements. Furthermore, unlike terrestrial crop protection products, aquatic herbicides, particularly those used to control submerged vegetation, must be absorbed rapidly from dilute and often flowing aqueous solution. The few products currently meeting these criteria are listed in Table 1. These compounds cause plant death by inhibiting tissue development, photosynthesis, respiration, or the biosynthesis of amino acids or cellulose. They may be classified as systemic or contact compounds according to their absorption characteristics. Systemic herbicides require translocation to their site of action within the plant. Consequently, they are relatively slow to act but generally provide effective long-term control for those species that tend to regrow from underground root systems or vegetative fragments. In contrast, contact herbicides, such as diquat, rapidly kill plant tissue on contact, thus restricting translocation of the herbicide within the plant. These herbicides generally provide only short-term control as regrowth often occurs from unaffected plant organs. In these cases, sequential applications may be necessary to maintain control over the growing season.

Herbicide Application

Herbicide applications to water are generally made using hand-operated knapsack sprayers operated from bank or boat, or spray booms mounted to boats, tractors, helicopters, or planes. Much of this equipment is modified from conventional agricultural sprayers and nozzles, although the injection of herbicides into deep water or onto channel beds may require the use of specifically designed weighted, trailing hoses fitted to boat-mounted spray-booms.

Herbicide Selectivity

Due to the small number and limited selectivity of products available for aquatic weed control, additional strategies are required to enhance the performance of herbicides in the aquatic environment (Table 2). Lack of selectivity is a particular problem but may be improved by manipulating herbicide placement, application rates, and application timing. Selective placement of the herbicide by using hand-held knapsack sprayers or weed wipers to direct spray onto target plants is particularly successful for the selective control of emergent or floating-leaved vegetation. Similarly, the contact activity of herbicides such as diquat can be exploited to improve selectivity. For example, diquat can be used to selectively control water hyacinth (*Eichhornia crassipes*), which is unable to recover after treatment, in bulrushes (*Scirpus* species), which are able to regenerate from unexposed rhizomes below the water surface.

Table 1 Active ingredients registered for aquatic weed control in the United Kingdom and United States

Herbicide	Absorption characteristics	Mode of action
2,4-D amine	Systemic	Plant hormone mimic
Copper[a]	Contact	Photosynthetic inhibitor
Dichlobenil[b]	Systemic	Cellulose biosynthesis inhibitor
Diquat	Contact	Photosynthetic inhibitor
Diquat alginate[b]	Contact	Photosynthetic inhibitor
Endothall[a]	Contact	Respiratory inhibitor
Fluridone[a]	Systemic	Photosynthetic inhibitor
Glyphosate	Systemic	Aromatic amino acid biosynthesis inhibitor
Terbutryn[b]	Systemic	Photosynthetic inhibitor

[a]Not registered for use in the United Kingdom.
[b]Not registered for use in the United States.

Table 2 Selectivity of herbicides available for aquatic weed control

Vegetation type	2,4-D amine	Copper	Dich lobenil	Diquat	Diquat alginate	Endothall	Fluridone	Glyp hosate	Terb utryn
Emergent reeds and sedges							×	×	
Floating and rooted	×		×	×	×		×	×	
Floating and not rooted	×			×			×	×	
Submerged	×		×	×	×	×	×		×
Submerged in flowing water					×				
Algae		×		×	×	×			×

Alternatively, selectivity may be achieved by adjusting herbicide application rates. For example, water hyacinth and spatterdock (*Nuphar luteum*) can be controlled by granular formulations of 2,4-D. However, selective control of water hyacinth in spatterdock can be achieved by using a lower rate of 2,4-D specifically recommended for water hyacinth.

Selectivity can also be improved by manipulating the timing of applications to coincide with periods of maximum accessibility in target plants or least susceptibility in nontarget plants. For example, those species that produce submerged leaves in the spring followed by emergent leaves in the summer are often more effectively controlled by applying a foliar-acting herbicide later in the season when vegetation has emerged above the water surface. Similarly, some aquatic species are better controlled when plants are translocating photosynthate downwards to their overwintering storage organs later in the growing season. At this time, foliar-acting herbicides, such as glyphosate, are more readily translocated to the root system and give better long-term control of species such as *Scirpus lacustris* and *Typha latifolia*. These species are less susceptible during the early growing season. In contrast, *Glyceria maxima* and *Phragmites australis* are susceptible to glyphosate applied at any time. Thus *Glyceria* and *Phragmites* can be eradicated over time from stands of *Scirpus* and *Typha* by early applications.

FACTORS AFFECTING THE PERFORMANCE OF AQUATIC HERBICIDES

There are several factors specific to the aquatic environment that affect the performance of aquatic herbicides. These include water flow, water pH, turbidity, and water hardness.

When applying herbicides to flowing water, special techniques are needed to ensure that the herbicide remains in contact with plant tissue long enough to allow absorption of effective concentrations. Methods to prolong contact times include the use of formulations such as viscous alginate gels, slow release pellets, or invert emulsions and polymers to aid sinking of the herbicide. Contact times may also be improved by making sequential applications using injection equipment or trailing hoses to deliver the herbicide closer to the target plants. In all cases, applications to flowing water bodies should be made against the direction of flow to achieve uniform concentrations and avoid build-up of herbicide downstream.

The activity of aquatic herbicides may be reduced in turbid water where the presence of suspended sediments or particulate material may bind herbicide molecules to produce inert complexes. This is a particular problem for diquat and glyphosate, but can be minimized by using clean water to prepare spray solutions and avoiding the disturbance of sediment when using boats and trailing booms.

Water pH affects the performance of aquatic herbicides by influencing their solubility and absorption by plant tissue. In particular, the activity of 2,4-D amine is increased in acidic water. Although the use of formulations including buffering agents can improve dispersion and solubility of herbicides when preparing spray solutions, the pH of receiving waterbodies cannot be manipulated. However, it may be beneficial to be aware of pH so that application rates of herbicides, such as 2,4-D, can be lowered when applying to acidic water.

Water hardness also influences the performance of herbicides, such as glyphosate, that tend to form inert precipitates with calcium carbonate. As glyphosate is not used to control submerged vegetation, its activity is unlikely to be reduced by the hardness of the receiving waterbody but may be affected if hard water is used to prepare spray solutions. This problem may be minimized by using soft or distilled water to prepare solutions.

PROBLEMS ARISING FROM THE USE OF AQUATIC HERBICIDES

The control of submerged vegetation with herbicides can create large quantities of decaying weeds that cause deoxygenation of the water due to a high bacterial biological oxygen demand. This may lead to the death of fish and other aquatic organisms particularly during summer months when deoxygenation is more rapid due to lower dissolved oxygen levels and increased rates of decomposition caused by higher water temperatures. De-oxygenation can largely be avoided by restricting applications to the early growing season. Where later treatments are essential, applications should be restricted to discrete localized areas of a water body, or slow-release formulations should be used to avoid a sudden build-up of decaying weed.

BIBLIOGRAPHY

Guidelines For The Use Of Herbicides On Weeds In Or Near Watercourses And Lakes; Ministry of Agriculture, Fisheries and Food: London, UK, 1995; 51.

Hoyer, M.V.; Canfield, D.E. *Aquatic Plant Management In Lakes And Reservoirs*; North American Lake Management Society, Madison: Wisconsin, US, 1997; 103.

AREAWIDE PEST MANAGEMENT

Marcos Kogan
Oregon State University, Corvallis, Oregon, U.S.A.

INTRODUCTION

The expression "areawide suppression" of key pests appeared with certain frequency in the entomological literature of the 1960s and 1970s, often equated with eradication or the attempt at total elimination of a pest species over large areas. In a broader sense, however, any attempt at controlling a pest within a region, beyond single field boundaries, was conceived as an areawide program. Examples included the mandatory destruction of cotton stalks to suppress overwintering populations of pink bollworm and boll weevil and other cotton-centered programs, programs for control of citrus pests, and programs for insects of medical or veterinary importance (1–4). In the early 1980s, the areawide qualifier was used in connection with pest management. Areawide pest management was considered an approach for controlling crop pests in a concerted way over a natural agroecological region, for example, the Mississippi Delta, the intermountain valleys of the Pacific Northwest, or the Midwestern Great Plains (5). A synthesis of the concepts of areawide pest suppression and integrated pest management (IPM) resulted in a new strategy for agricultural pest control, areawide IPM (6, 7). The essential requisite for implementation of an areawide IPM program is the availability of tactics that are most effective if implemented regionally (areawide) rather than locally. Examples were mass release of parasitoids or predators, destruction of noncrop alternate hosts and refugia, releases of sterile insects, mating disruption by means of pheromone saturation, or combinations of those tactics (7).

THE PRINCIPLES

Areawide suppression of key pests rests on principles that differ substantially from the tenets of integrated pest management (IPM) because of the focus on the key pest and the approach for its management (1, 5). A key pest is a serious, perennially occurring, persistent species that dominates control practices because, in the absence of deliberate control, the pest population will remain above the economic injury level (EIL). The codling moth, *Cydia pomonella*, is the prototypic key pest for pome fruit crops in the western U.S. A single worm in an apple is a loss, so EILs are very low because the crop value is high (7).

Areawide Suppression of Key Pests

The following is a set of elements that typify an areawide suppression program (6).

1. The program objective is to suppress the key pest to low residual levels (if it cannot be eradicated).
2. The program is implemented over large geographical regions.
3. The program is centrally coordinated and commonly financed by federal or state agencies. In the U.S. this role is performed mainly by the U.S. Department of Agriculture.
4. Program managers, usually representing the coordinating agencies, are the principal decision makers.
5. The program may involve a mandatory component to ensure full participation by growers.

Integrated Pest Management

IPM has been the dominant paradigm in crop protection during the past 30 years. The following are some of the main characteristics of an IPM program (8).

1. IPM programs are targeted against the pest complex within the crop community. Key pests often command the most attention and are the deciding factor in designing an IPM program, but the key pest is treated within the context of the entire biotic community.
2. Most IPM programs are implemented within single fields, whole farms, or any larger spatial scale, depending on the level of program integration. Most successful IPM programs have used individual fields as the base unit for decision making.

3. IPM programs are coordinated by individual growers or cooperatively by growers, consultants, and Extension personnel, depending on spatial scale.
4. Growers are the principal decision makers and the concepts of economic injury levels and economic thresholds are the bases for the decision support system.
5. IPM programs aim at keeping populations of all pests, key and secondary, below economic injury levels.
6. As growers are the principal decision makers, IPM is based totally on voluntary participation by growers.

Areawide IPM

Areawide IPM represents a synthesis between IPM and areawide suppression of key pests (Fig. 1).

1. Areawide IPM is targeted against a key pest whose control opens new IPM options for the secondary pests. The key pest is controlled through tactics whose success depends on regional adoption.
2. Management of the key pest is achieved within regionally coordinated programs. All other pests or potential pests in the crop community are monitored and kept below EILs on a field-by-field basis, just as in classical IPM, using site specific decision making tools and remedial or preventive control tactics best suited to the specific site and risk tolerance of the grower.
3. Areawide IPM may be implemented over meso or macro scales, but it must be regionally focused.
4. Implementation of an areawide IPM program must be centrally coordinated, but management and coordination must be achieved with full grower participation.
5. At its initial phases of implementation, areawide IPM may provide financial incentives for growers to participate if the selected management technique is either more expensive or perceived as riskier than the standard pesticide-based management program. Fig. 1 summarizes the main characteristics of the three systems.

EXAMPLES OF AREAWIDE IPM

Table 1 provides a list and characteristics of some exemplary areawide IPM programs being implemented around the world. To better illustrate the approach, a more detailed analysis is included of the first program implemented under this new paradigm—the areawide management of the codling moth on apple and pear orchards in the Pacific Northwest of the U.S.

A Paradigm for Areawide IPM: The Codling Moth Program

The codling moth is the key arthropod pest of apples and pears in the major producing areas of California, Oregon, and Washington. In the early 1990s three or four sprays of the broad-spectrum organophosphate insecticide azinphos-methyl (Guthion®) were used in the region to control the pest. Signs of up to sevenfold resistance to the insecticide were detected first in California populations,

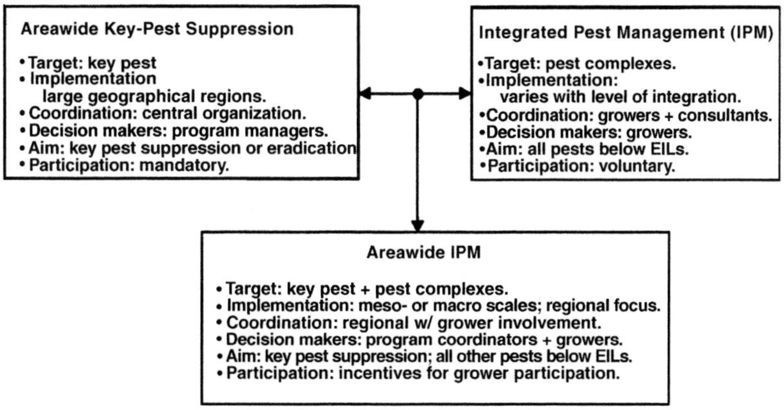

Fig. 1 Areawide IPM as the synthesis between the large-scale approach to suppress key pests (areawide key pest suppression) and the classical integrated pest management approach.

Table 1 Examples of recent or ongoing areawide IPM and a few eradication programs around the world

Crop/cropping system	Region	Year of s program	Target pest(s)	Objective	Main tactic(s)	Refs.
Corn	Various states, USA	1996-date	Corn rootworms	IPM	Pesticide baits	1, 9
Cotton	San Joaquin Valley, CA	1968-date	Pink bollworm	Eradication/ IPM	SIR; MD; crop residue destruction; trapping	3
	Mississippi Delta, USA	1980s-'90s	Cotton bollworm; tobacco budworm	IPM	***NPV*** sprays	10
	Various states, USA	1977	Boll weevil	Eradication	SIR; crop residue destruction; pesticides targeted to early and late populations.	11
Fruit/ vegetable crops	Chile, Japan (successful); other countries in progress		Tephritid fruit flies	Eradication	Pesticide baits; SIT	1, 12
Melons; vegetable crops	Arava, Israel	1990s	Aphid and whitefly transmitted viruses; fruit flies; spider mites	IPM	Sanitation; destruction of crop residues; control of weed reservoirs; crop-free periods; mass releases of predaceous mites	13
Cotton/potato/ vegetables cropping system	Bet Shean Valley, Israel	1994	*Spodoptera littoralis*; whiteflies; bollworms; potato tuberworm	IPM	MD; IGRs; IRM	13
Pome fruit crops	CA, OR, WA, USA	1995-1999	Codling moth	IPM	MD	1, 7, 9, 14
Soybean	Parana, Brazil	1993-date	Soybean caterpillar; southern green stink bug	IPM	*Baculovirus*; mass releases stink bug parasitoids; low rates pesticides	15
Range land	Western USA	1997	Leafy spurge, *Euphorbia esula*	IPM	Biocontrol	9

Acronyms: MD = mating disruption with pheromone saturation; SIT = sterile insect technique; IRM = integrated resistance management program; IGR = insect growth regulators.

and later, at lower levels, in the other states. Pilot tests with commercially produced codling moth pheromone dispensers showed that control was adequate. Mating confusion was obtained at rates of about 1000 dispensers per hectare and Guthion applications were drastically reduced or eliminated with no increase in the amount of fruit injury by codling moth larvae. Research demonstrated that, for optimal mating disruption results, it was necessary to maintain a minimum perimeter-to-area ratio to reduce reinfestations from nontreated orchards. An areawide approach was key to the success of the program.

In 1995 researchers and extension pome fruit specialists from three western universities (University of California, Berkeley; Oregon State University; and Washington State University at Wenatchee) and the Agricultural Research Service (ARS) Agricultural Research Laboratory, Wapato, Washington, developed a program for the suppression of the codling moth based mainly on mating disruption through the timely dispensation of slow-release, plastic, pheromone impregnated "ties." It was anticipated that replacement of nonselective, neurotoxic insecticides by the ecologically friendly pheromone saturation technology should relax pressure on natural enemies of secondary pests [mainly spider mites (*Tetranychus* spp.) and pear psylla (*Cacopsyla pyricola*)] thus further reducing the need for pesticides within the orchard and improving the habitat for naturally occurring or introduced natural enemies.

A pilot test of the program was initiated on five sites, one each in California and Oregon, and three in Washington, for a combined area of about 1440 ha. The pilot test areas were well-defined ecological units, such as small intermountain valleys, or semi-isolated sectors in major production valleys, ranging from 180 to 440 ha of contiguous orchard blocks. Mating disruption dispensing protocols and monitoring systems and management protocols for secondary pests were established. Participating growers received a financial subsidy, because cost of the mating disruption, at the time, was about twice the cost of a Guthion-based program (1, 7, 9, 14).

A public/private sector partnership was established early in the development of the program to assure full expression of the growers' concerns, to benefit from their experience, to give them ownership of the program, and to provide participation in the decision making process from its inception. An organizational structure was developed to coordinate all components of the program with site-specific coordination performed by growers and technical personnel from the participating public institutions.

Within five years, the program was expanded to 22 sites, for a total of 8900 ha. Guthion use was reduced by an average of 75%, and secondary pests, particularly pear psylla, spider mites, and the western tentiform leafminer (*Phyllonorycter elmaella*), were generally kept under adequate natural control. Control of lepidopterous leafrollers (Lepidoptera: Tortricidae) was achieved with soft pesticides, such as *Bacillus thuringiensis* and growth regulators. When the additional costs of control of secondary pests were factored in, mating disruption became competitive with chemical control of the codling moth, and growers started adopting the technology even in the absence of subsidies. By 1998 it was estimated that about 16,000 ha of apple orchards in Washington used the mating disruption technique. In addition to the overall economic and ecological benefits of the program, there were additional social benefits related to orchard labor productivity and welfare (8, 13).

EXTENDING THE AREAWIDE IPM PARADIGM

To what extent is the codling moth areawide IPM program paradigm applicable for other pests on other crops? The main factors to help identify a candidate for an areawide IPM approach are: a) characteristics of the key pest(s); b) the set of management technologies available to suppress the key pest(s); and, possibly, c) the extent that management of the secondary pests on the crop is linked to the suppression of the key pest(s). The following generalizations seem appropriate for assessing the feasibility of adopting the areawide IPM approach (7).

Candidates for areawide IPM are key pests of high value crops. The codling moth is the foremost key pest on apples and pears in the Pacific Northwest. In some other crops, several species may share that honor, in which case it may be necessary to determine which can be most effectively suppressed using the areawide approach with the others being the target of site-specific IPM. In most regions of the world, whiteflies on cotton are one of several key pests; other key pests are the boll weevil, *Anthonomus grandis*, and the tobacco budworm, *Heliothis virescens*.

Control of the key pest is possible with soft management methods. In the absence of a specific technology, it may be necessary to design a multitactical approach including biological control, host plant resistance, and cultural methods, to achieve adequate population regulation.

Suppression of the key pest releases natural control of secondary pests. The importance of the availability and choice of a soft technology as the pivotal tactic in an areawide IPM program is the opportunity that this technology opens to promote natural control or applied biological controls of secondary pests.

There is a core of innovative growers ready to adopt the new technologies. The best promoters of new technologies are peers. It is essential to promote grower input early in the planning of the areawide IPM program. As growers feel that they have ownership of the program, they will be favorably inclined to participate.

Planning and implementation of an areawide IPM strategy for polyphagous key pests, such as Heliothini or whiteflies, are likely to pose far greater challenges than those encountered with the oligophagous codling moth. Besides the most obvious problems of the inherent biological variability of the pest complex, the multiplicity of host crop associations, and the lack of a unified and uniformly effective soft technology for all affected crops, there are sharp sociological differences to overcome in bringing about the cooperation of a diverse grower community. Given, however, the nearly universal nature of the problems caused by these pest complexes, it is certainly essential to adopt a strategic plan that takes into account an areawide approach. It may, however, be necessary to emphasize the regionally specific IPM aspects of the program rather than strive for a universally acceptable unified strategy, as exemplified by the codling moth program.

REFERENCES

1. Tan, K.H. Joint Proceedings of the International Conference on Area-Wide Control of Insect Pests (May 28–June 2, 1998) and the Fifth International Symposium on Fruit Flies of Economic Importance (June 5, 1998), Penang, Malaysia. In *Area-Wide Control of Fruit Flies and Other Insect Pests*; Penerbit Universiti Sains Malaysia: Pulau Pinang, Malaysia, 782.
2. Bottrell, D.; Rummel, D.R. Response of *Heliothis* populations to insecticides applied in an area-wide reproduction diapause boll weevil suppression program. J. Econ. Entomol. **1978**, *7*, 87–92.
3. Huber, R.T.; Moore, L.; Hoffmann, M.P. Feasibility study of area-wide pheromone trapping of male pink bollworm *Pectinophora gossypiella* moths in a cotton insect pest management program. J. Econ. Entomol. **1979**, *72*, 222–227.
4. Stinner, R.E. Biological monitoring essentials in studying wide-area moth movement: example of *Heliothis zea* in North Carolina. In *Movement of Highly Mobile Insects, Concepts and Methodology in Research*; Rabb, R.L., Kennedy, G.G., Eds.; Proceedings of a Conference, Raleigh, NC, University Graphics, North Carolina State Univ.: Raleigh, NC, 1979; 199–211.
5. Knipling, E. F. The Basic Principles of Insect Population Suppression and Management. In *Agriculture Handbook*; USDA: Washington, DC, 1979; 512, 659.
6. Knipling, E.F. In *Areawide Pest Suppression and Other Innovative Concepts to Cope with Our More Important Insect Pest Problems*, Minutes Annual Meetings: National Plant Board, Sacramento, CA, 1980; 68–97.
7. Kogan, M. Areawide Management of Major Pests: Is the Concept Applicable to the Bemisia Complex?. In *Bemisia, 1995: Taxonomy, Biology, Damage, Control and Management*; Gerling, D., Mayer, R.T., Eds.; Intercept: Andover, UK, 1996; 643–657.
8. Kogan, M. Integrated pest management: historical perspectives and contemporary developments. Annu. Rev. Entomol. **1998**, *43*, 247–270.
9. Chandler, L.D.; Faust, R.M. Overview of areawide management of insects. J. Agric. Entomol. **1998**, *15*, 319–325.
10. Luttrell, R.G. Cotton pest management, Part 2. A U.S. Perspective. Annu. Rev. Entomol. **1994**, 527–542.
11. Rummel, D.R. An area-wide boll weevil (*Anthonomus grandis*) suppression program: organization, operation and economic impact. Misc. Pubs. Texas Agric. Exp. Stn. **1976**, *1276*, 152–159.
12. Aluja, M.; Liedo, P. *Fruit Flies, Biology and Management*; Springer-Verlag: New York, 1993; 492.
13. Ausher, R. Implementation of integrated pest management in Israel. Phytoparasitica **1997**, *25*, 119–141.
14. Calkins, C.O. Review of the codling moth areawide suppression program in the Western United States. J. Agric. Entomol. **1998**, *15*, 327–333.
15. Correia-Ferreira, B.S.; Domit, L.A.; Morales, L.; Guimãres, R.C. Integrated soybean pest management in micro river basins in Brazil. IPM Rev. **2000**, *5*, 75–80.

ARTHROPOD HOST PLANT RESISTANT CROPS

Gerald Wilde
Kansas State University, Manhattan, Kansas, U.S.A.

INTRODUCTION

Resistance of plants to pest attack is defined as the relative amount of heritable qualities possessed by the plant that influence the ultimate amount of damage caused by the pest. The use of plant resistance to manage arthropod pest populations provides an ideal approach to integrated pest management because it is biologically and economically sound, environmentally friendly, and generally compatible with other management tactics or strategies. The cultivar forms the foundation on which all pest management programs and tactics are applied, and its effects are specific, cumulative, and persistent.

Because of the many advantages plant resistance offers, virtually every cultivated crop has been evaluated for this trait and one or more resistant sources have been identified. The challenge has been to incorporate these resistant sources into agronomically adapted and consumer acceptable, high-yielding cultivars. In addition to traditional breeding methods, the use of modern breeding techniques and genetic transformation of crops has opened the door to other ways of identifying, incorporating, and employing pest-resistance genes to effectively and economically manage arthropod pest populations. The use of resistant cultivars contributes significant economic and social benefits and sustainable agricultural systems to the world's farmers. The positive effects of resistant cultivars have been demonstrated repeatedly in crops as diverse as wheat, alfalfa, grape, sorghum, maize, rice, apple, and cotton.

PERCENTAGE OF CROPS THAT HAVE SOME DEGREE OF PEST RESISTANCE

Plant resistance has been employed to a greater or lesser degree in practically all of the major food, feed, and fiber crops. Table 1 lists a number of major crops grown in the world and the number of pests for which resistance has been employed to at least some extent in the field. Hectarage planted to resistant cultivars varies for each pest and crop and over time as new varieties and hybrids (both susceptible and resistant) are grown and, in some instances, as new pest biotypes (pest populations that are capable of damaging previously resistant sources) develop. For example, most of the modern rice varieties and hybrids grown in China, India, and other countries are resistant to one or more major pests. Resistant American grape rootstocks have been used extensively over the world to control *Phylloxera vittifolae* (Fitch). A large percentage of the alfalfa planted in the United States is comprised of varieties resistant to aphid species. Sorghum hybrids with resistance to the greenbug have occupied up to 80% of the hectarage in the United States. Significant hectarages of wheat and barley in the United States, Canada, and North and South Africa have resistance to at least one pest. Most commercial soybean varieties are resistant to the potato leafhopper. Several cotton varieties carrying genes for resistance to jassids (*Empoasca* sp) are grown widely in Africa, India, and the Philippines. In the United States, more than 65% of commercial maize hybrids have some resistance to corn leaf aphid, > 90% have some resistance to first generation European corn borer, and > 75% have some resistance to second generation corn borer.

However, many more resistance genes have been identified in all crops than have been used in modern commercial varieties and hybrids, because incorporating them into high yielding cultivars acceptable to growers has been difficult. Recently, transgenic crops have been utilized to combat major insect pests. Hybrids or varieties with insect-resistance genes have been developed in cotton, maize, and potato. An estimated 6.7 million hectares of transgenic corn resistant to the European corn borer, 2.5 million hectares of transgenic cotton resistant to several pests, and 20,000 hectares of transgenic potato resistant to Colorado potato beetle were grown in the world in 1998. The hectares planted to transgenic crops are likely to increase as additional countries register these products and this technology is used on additional crops. For example,

Table 1 List of some major crops grown in the world and number of arthropod pests for which resistant cultivars have been used in the field by growers for pest management

Crop	No. of pests
Alfalfa	6
Apple	1
Asparagus	1
Barley	3
Bean	1
Cassava	2
Chickpea	0
Cotton	6
Grape	1
Lettuce	1
Maize	10
Millet	1
Oat	1
Pea	1
Peanut	4
Potato	1
Raspberry	1
Rice	14
Rye	1
Sorghum	6
Soybean	1
St. Augustine grass	1
Sugar beet	1
Sugarcane	3
Sunflower	1
Sweet clover	1
Sweet potato	1
Wheat	7

specific biotechnology applications are being field tested for rice and wheat, which together occupy 400 million hectares globally.

EFFECT OF PLANT RESISTANCE ON PEST POPULATIONS

The growing of pest-resistant cultivars can be used as a major control tactic or adjunct to other measures. Historically, the use of resistant cultivars combined with other tactics has resulted in a reduction of many pest species to subeconomic levels. Even small increases in resistance enhance the effectiveness of cultural, biological, and insecticidal controls. The extent to which growing resistant plants affects pest populations is dependent upon the level of resistance expressed, the mechanisms of resistance involved, and the number of hectares grown. The growing of resistant wheat on 50% of the hectarage in Kansas has been shown to reduce Hessian fly populations to extremely low levels. Resistance in wheat to wheat curl mite (ca. 25% of the hectarage) was effective in limiting the spread of wheat streak mosiac virus, which the mite transmits. The incorporation of leaf and stem pubescence into most commercial soybean varieties has resulted in population suppression of the potato leafhopper over the past 60 years. As the hectarage of sorghum resistant to the greenbug increased to > 50%, the area of sorghum treated with insecticide was reduced by 50%. Tenfold reductions in pest populations have been observed where insect-resistant rice cultivars have been grown widely.

ECONOMIC AND SOCIAL BENEFITS

Assessing the economic benefits of plant resistance is difficult in the context of integrated pest management programs and is likely to be underestimated frequently and substantially. Even determining the obvious advantages (yield benefits and reduced production costs) may be difficult over a large area where pest populations vary from locality to locality and year to year. Other environmental benefits, such as cleaner water and food, reduced risks to farmers, more flexibility in planting and cropping systems, reduced disease transmission, and reduced secondary pest outbreaks, also are difficult to quantify. Nevertheless, some specific estimates are available. In the United States alone the estimated valued of using arthropod-resistant alfalfa, barley, corn, sorghum, and wheat cultivars is more than $1.4 billion each year. The net economic benefit of greenbug resistance in U.S. sorghum production is estimated at close to $400 million annually. The global economic value of arthropod-resistant wheat has been estimated at $250 million annually. The value of resistance to aphids in alfalfa in the major alfalfa-producing states of the United States is estimated at more than $100 million annually. Breeding for pest resistance in rice has been estimated to be responsible for one-third of recent yield increases and $1 billion of additional annual income to rice producers. The net return of insect-resistant Bt maize in the United States and Canada has been estimated in some studies at $42.00–$67.30 per hectare, but other studies have indicated less of an economic return. The average net economic return of insect-resistant Bt cotton in 1997 was $133 per hectare.

BIBLIOGRAPHY

Antle, J.M.; Pingali, P.L. Pesticides, productivity and farmer health: A philippine case study. Am. J. Agric. **1994**, *76*, 418–430.

Global Plant Genetic Resources for Insect-Resistant Crops; Clement, S.L., Quisinberry, S.S., Eds.; CRC Press: New York, 295.

Harvey, T.L.; Martin, T.J.; Seifers, D.L. Importance of plant resistance to insect and mite vectors in controlling virus diseases of plants: resistance to the wheat curl mite (Acari: Eriophyidae). J. Agric. Entomol. **1994**, *11*, 271–277.

Hyde, J.; Martin, M.A.; Preckel, P.V.; Edwards, L.R. The economics of Bt corn: valuing protection from the European corn borer. Rev. Agric. Econ. **1999**, *21*, 442–454.

James, C. *Global Review of Commercialized Transgenic Crops*; ISAAA Briefs No. 8, ISAA: Ithaca, NY, 1998, 43.

In *Insect Resistant Maize: Recent Advances and Utilization*, Proceedings of an International Symposium, International Maize and Wheat Improvement Center (CIMMYT), Nov 27–Dec 3, 1994; Mihm, J.A., Ed.; CIMMYT: Mexico, D.F., 1997; 302.

Painter, R.H. *Insect Resistance in Crop Plants*; University of Kansas Press: Lawrence, KS, 1968, 520.

Smith, C.M. *Plant Resistance to Insects. A Fundamental Approach*; John Wiley & Sons: New York, 1989, 286.

Smith, C.M.; Quisinberry, S.S. Value and use of plant resistance to insects in integrated pest management. J. Agric. Entomol. **1994**, *11*, 189–190.

van Emden, H.F. Host-Plant Resistance to Insect Pests. In *Techniques for Reducing Pesticides: Environmental and Economic Benefits*; Pimentel, D., Ed.; John Wiley: Chichester, England, 1997, 124–132.

In *Economic, Environmental, and Social Benefits of Resistance in Field Crops*, Proceedings, Thomas Say Publications in Entomology, Wiseman, B.R., Webster, J.A., Eds.; Entomological Society of America: Lanham, MD, 1999, 189.

Contribution No. 00-252-B of the Kansas Agricultural Experiment Station.

A

AUGMENTATIVE CONTROLS

Kent M. Daane
Nicholas J. Mills
University of California, Berkeley, California, U.S.A.

Maurice J. Tauber
Cornell University, Ithaca, New York, U.S.A.

INTRODUCTION

Augmentative biological control is the periodic release of mass-produced or mass-collected natural enemies (1, 2). It is used when resident natural enemies occur too late in time or too low in number to provide adequate pest suppression. Augmentation includes *inoculation*—"seeding" natural enemies in the release area, with pest suppression resulting from their offspring in subsequent generations; and *inundation*—mass-releasing natural enemies to overwhelm the pest population and provide immediate control. Whenever feasible, early-season, inoculative release is preferred because it requires fewer natural enemies and provides control over a longer period. Here, we address augmentation of insect and mite predators and insect parasitoids.

The effectiveness of augmentation programs varies depending on the natural enemy species released, targeted pest, and release environment (2). For example, open fields, row crops, and orchards present a more difficult environment for successful natural enemy release than protected environments, such as greenhouses. Released natural enemies may disperse from the target site, perform poorly at ambient temperatures, or fall prey to resident predators. Requirements for effective augmentation include an ability a) to rear or collect predictable quantities of natural enemies of high quality, b) to store, transport, and release natural enemies effectively, and c) to understand the compatibility of released natural enemies with the target pest(s) and other management practices (2).

GREENHOUSE PESTS

Worldwide, greenhouses cover ~ 300,000 ha (~ 65% vegetables and 35% ornamentals). There are many successful augmentation programs in these environmentally controlled, protected environments, especially in Europe where the practice is commonplace (2, 3).

Predaceous mites (*Mesoseiulus longipes, Phytoseiulus persimilis*, and *Neoseiulus californicus*) can be used to control phytophagous mites on vegetables and ornamentals. Weekly inundative (high pest pressure) or 1–2 inoculative (low pest pressure) releases of 10,000–25,000 predaceous mites per ha provide control. Information on mite life-history traits has improved release effectiveness by matching predator to prey species and habitat. For example, *Phytoseiulus* works well at high humidities, high pest densities, and on low-lying, dense plants; while *Neoseiulus* performs well at low humidities, low pest densities, and on tall-growing plants.

The greenhouse whitefly, *Trialeurodes vaporariorum*, is controlled on vegetables using inoculative releases of both the pest and its parasitoid, *Encarsia formosa* (2, 3). This "pest-in-first" approach utilizes an initial introduction of 1 whitefly per 10 plants followed 10 days later with three successive releases of 4 parasitoids per plant. A more common approach uses 50–90 "nursery" plants per ha, which support both whiteflies and parasitoids. Although these inoculative releases are preferred, 6 (or more) weekly inundative releases of 35,000 parasitoids per ha also provide control. For the silverleaf whitefly, *Bemisia argentifolia*, on ornamentals, inoculative release of both a parasitoid, *Eretmocerus californicus*, and a beetle, *Delphastus pusillus*, can provide control.

For aphid pests, inoculative releases of a cecidomyiid midge, *Aphidoletes aphidimyza*, can offer season-long pest suppression. *Aphidoletes* has several features that favor its use: a) wide commercial availability; b) readily reared, stored, and transported; c) broad array of aphid species as prey; and d) numerical and/or functional responses to increasing prey density (4). The midge is released at 5000–10,000 per ha, and can sustain populations if there is loose soil in which the midge can pupate. Inundative releases of green lacewings (*Chrysoperla carnea* and *Chrysoperla rufilabris*), are also used (5). Lacewings are generalist predators that feed on many soft-bodied insects and mites. They are most commonly released as eggs, mixed in a medium (e.g., corn grit), at 10,000–100,000 eggs per ha. Release of lacewing larvae is more effective; however, there is substantial increase in the average price for "prefed" larvae ($0.0305) compared with eggs ($0.0054). Commercially available aphid parasitoids include *Aphidius colemani* and *Aphidius matricariae*, which have been released weekly at 1000–5000 per ha to suppress aphids on sweet peppers.

Other "specialized" parasitoids include *Dacnusa sibirica* and *Diglyphus isae*, which are used in season-long inoculative releases (1 per m^2 every 3 weeks) to control the leafminer, *Liriomyza trifolii*, on chrysanthemums. However, glasshouses often have multiple pests (e.g., leafminer, mealybug, scale, and thrips) and, under these circumstances, "generalist" predators are often favored because they feed on a wide range of prey species (2, 3). These generalist predators include green lacewings, anthocorids (*Orius* spp.) and mirids (*Macrolophus caliginosus* and *Deraeocoris brevis*). For example, mealybug pests of greenhouse and interiorscape ornamentals are suppressed by inoculative release of the mealybug destroyer, *Cryptolaemus montrouzieri*, combined with inundative releases of green lacewings. In Canadian glasshouse vegetables, the western flower thrips, *Frankliniella occidentalis*, is controlled by release of predatory mites (10 *Neoseiulus cucumeris* per plant, followed 2 weeks later by 1 *Iphiseius degenerans* per plant) after flowers set (both predators fed on pollen) (2). Later in the season, *Orius insidrosus* is released (1 per plant) if thrips or mite populations begin to increase.

STABLE AND POULTRY FARM PESTS

House flies, *Musca domestica*, and stable flies, *Stomoxys calcitrans*, are important pests of dairy and poultry farms. Fly control is based on a manure removal management program; however, inundative releases of parasitoids (*Nasonia vitripennis*, *Muscidifurax* and *Spalangia* species) have a significant role. Weekly releases of 100–200 parasitized-fly-pupae per cow or 500–1000 pupae per calf on dairy farms, and 10 pupae per hen on poultry farms are needed throughout the fly season to provide control.

STORED PRODUCT PESTS

Pests of grain storage bins include moth and beetle larvae. It is noteworthy that grain storage bins, like greenhouses, are enclosed systems. Both pests can be suppressed with inoculative release of the warehouse pirate bug, *Xylocoris flavipes*; a braconid wasp, *Bracon hebetor*, is also available for moth pests.

ORCHARD AND VINEYARD PESTS

While there has been greater success releasing parasitoids, there has been exciting work with predator releases into orchard and vineyard crops (2). This includes mass-produced, insecticide-resistant mites (*Galandromus occidentalis* and *Typhlodromus pyri*) to control spider mites. This method was first used in California almonds; however, the natural occurrence of insecticide-resistant mites reduced the need for inoculation programs. Currently, cost-subsidized programs utilizing insecticide-resistant mites are in use in Canadian and New Zealand orchards.

One of the most successful augmentative release programs has been against California red scale, *Aonidiella aurantii* (2). Beginning in 1956, mass production and inoculative releases of *Aphytis melinus* by the Fillmore Citrus Protection District has suppressed *Aonidiella* populations (additional releases are used when scale control is disrupted by ants, dust, or spray drift). A similar program is being used in California's Central Valley, with releases of 120,000–250,000 *Aphytis* per ha.

Egg parasitoids (*Trichogramma* spp.) are commonly released against moth pests (2). Codling moth (*Cydia pomonella*) damage in pears and walnuts in California can be reduced with four weekly inundative releases of 200,000 *Trichogramma platneri* per ha during each generation of the pest. A forest pest, spruce budworm, *Choristoneura fumiferana*, can be managed with inundative releases of 1,200,000 *Trichogramma minutum* per ha; however, this program is not commercially practiced.

ROW CROP PESTS

One of the first commercially successful uses of augmentation was against spider mites (*Tetranychus* spp.) on strawberries and cotton (2). Inoculative release of the predatory mite *Phytoseiulus persimilis* in strawberries began in the 1960s when available miticides were no longer effective. Inundative releases of 75,000 *Phytoseiulus* per ha, spread over the entire field, suppress spider mite densities. Inoculative releases using early-season seeding of scattered "nursery" plants (5 *Phytoseiulus* per plant) can also be used, although this practice is not as common.

The boll weevil, *Anthonomus grandis*, in cotton can be successfully controlled with inoculative releases of 2500

Catolaccus grandis per ha. Similarly, early-season inoculative releases of *Pediobius foveolatus* against Mexican bean beetle, *Epilachna varivestis*, are effective in beans in the eastern United States. For bollworms (*Helicoverpa* spp.) in cotton and tomatoes, inundative release of 750,000 green lacewings per ha controls the pests; however, such high release rates are not economically competitive with other control programs (5). The level of pest suppression after lacewing releases can vary greatly; recent studies underscore that release methodology, improper matching of lacewing species with the targeted pest and/or crop, and intraguild predation must be better understood to maximize release effectiveness.

One of the most effective use of inundative biological control in row crops has been the release of *Trichogramma brassicae* and *T. nubilale* against the European corn borer, *Ostrinia nubilale*, in both Europe and North America (2). *Trichogramma* species have also been used against *Heliothis* species in cotton and *Diatraea saccharalis* in sugarcane. Many countries also use inundative releases of the larval parasitoids, *Cotesia flavipes, Metagonistylum minense*, and *Paratheresia claripalpis* for successful control of sugarcane borer.

HOME GARDEN AND LANDSCAPE PESTS

The convergent lady beetle, *Hippodamia convergens*, is one of the most commercially available insect predators (2). It is popular, in part, because it is quite visible and because it can feed on a range of small insects (aphids are their preferred prey). This beetle is not mass reared but collected at overwintering aggregation sites and then stored at low temperatures until release. While popular, most studies suggest that adult beetles may disperse from the release area within days after liberation. The Chinese mantid, *Tenodera sinensis*, is also released against a variety of garden pests, and its effectiveness, too, is unknown.

SUMMARY

The market for biologically based pest controls is near $350 million, or 1/20 of the chemical insecticide market. Of this, only about 10% of the products are arthropod natural enemies (6), the remainder being products based on nematodes, bacteria, fungi, botanicals, or behavior-modifying chemicals. Nevertheless, there is a projected increase in the use of augmentation, driven by consumers' desire for pesticide-free produce, loss of current pesticides, and improvements in augmentation programs. To meet these needs, researchers and the insectary industry are working to develop more efficient programs and better consumer guidelines.

REFERENCES

1. Parrella, M.P.; Heinz, K.M.; Nunney, L. Biological control through augmentative releases of natural enemies: a strategy whose time has come. Am. Entomol. **1992**, *38*, 172–179.
2. *Mass-reared Natural Enemies: Application, Regulation, and Needs*; Hoffmann, M.P., Inscoe, M.N., Glenister, C.S., Ridgway, R.L., Eds.; Thomas Say Publications in Entomology, Entomological Society of America: Lanham, MD, 1998; 332.
3. van Lenteren, J.C. A greenhouse without pesticides: factor fantasy. Crop Protection **2000**, *19*, 375–384.
4. *Biological Pest Control, The Glasshouse Experience*; Scopes, N., Hussey, N.W., Eds.; Cornell University Press: New York, 1985; 240.
5. Tauber, M.J.; Tauber, C.A.; Daane, K.M.; Hagen, K.S. New tricks for old predators: implementing biological control with *Chrysoperla*. Am. Entomol. **2000**, *46* (1), 26–38.
6. Hunter, C.D. Suppliers of Beneficial Organisms in North America, California Environmental Protection Agency, Department of Pesticide Regulation: Sacramento, CA, 1997. www.cdpr.ca.gov/docs/ipminov/bensuppl.htm (accessed February 20, 2001).

AUXILIARIES: DEODORANTS, DILUENTS, SOLVENTS, STICKERS, SURFACTANTS, AND SYNERGISTS

Peter Messenger
Aventis CropScience France S.A., Villefranche-sur-Saône, France

INTRODUCTION

Auxiliaries (synonyms: additives, inerts, formulants) are an essential part of an agrochemical product composition. They are present in order to facilitate the application of the active ingredient, and to ensure adequate shelf-life of the commercial product (minimum 2 years). The criteria for selection include the need to be user and environment friendly, and, while auxiliaries are considered to be "inert" in comparison with the active ingredient, the biological efficacy of the product may be optimized by the formulation constituents (for example, surfactants to provide wetting and penetration, and polymers to improve rainfastness).

It is the role of the formulation chemist to select the best combination from the many commercial auxiliaries available in order to achieve the objectives given above. The choice is influenced not only by efficacy/quality considerations, but also by economic, toxicological, and legal constraints. Ignoring the use of water in agrochemical formulations, it can be estimated that for every tonne of active ingredient applied to the target organism, between 0.1 and 0.5 ton of inert auxiliaries are also applied, hence the close scrutiny that these additives are being shown by regulatory authorities in developed countries.

THE ROLE OF AUXILIARIES IN AGROCHEMICAL FORMULATIONS

For an agrochemical formulation to achieve the necessary standards for registration/sale, it must:

- Be chemically and physically stable for the guaranteed shelf life (minimum 2 years)
- Provide the required biological activity on the target organism, without any undesired ill-effect to nontarget organisms (user, crop, wildlife, etc.)
- Conform with all existing regulatory requirements in the country of use
- Be compatible with the commercial packaging for the guaranteed shelf life (no leaks, no smells, no deformation of the pack, etc.)
- Possess the necessary dispersion/application characteristics compatible with the intended application system

Almost without exception, the active ingredient alone cannot meet all these criteria without being blended and processed with a carefully chosen set of auxiliaries to produce a commercial composition corresponding to all marketing and legal requirements. Each auxiliary has a clearly defined role to play in achieving the formulation standards required (Table 1).

This table can be further simplified into three principal functions:

1. Auxiliaries that impart the correct physico-chemical and application properties (e.g., antifreeze, diluent, dispersing agent, stabilizer)
2. Auxiliaries that enhance bioavailability (e.g., sticker, wetting agent, solvent)
3. Auxiliaries present for other marketing and regulatory reasons (e.g., deodorant, dye)

The use of auxiliaries is not only essential for application and stability of the commercial composition, but also permits an optimization of the dose rate, which is a positive step toward reducing worker exposure and environmental risks. Excluding solvents and diluents, the overall concentration of auxiliaries used in the formulation is generally much lower than the concentration of the active ingredient.

Surfactants

Without a doubt, surfactants are the most important and versatile group of additives in agrochemical formulations, and are characterized by having a hydrophilic and a lipophilic group in the same molecule (1). They are essential for product quality as they ensure chemical compatibility with the active ingredient, and provide physical

Table 1 List of the most common auxiliary families used in agrochemical formulations

Antifreeze	Dye
Wetting agent	Antifoam
Emulsifier	Sticker
Binder	Preservative
Buffer	Diluent/carrier
Solvent	Deodorant
Dispersing agent	Thickener
Emetic	Propellant
Stabilizer	Repellent

stability of heterogeneous systems, such as suspension and emulsion concentrates. They are, for example, necessary to achieve good wetting of the dispersed phase of a formulation, to prevent particle re-agglomeration, and to emulsify oil/water systems. Also, choice of the right surfactant can often optimize biological activity and thus help to reduce dose rates. Their surface activity improves wetting of the treated target, reduces run-off or spray drift, and may help solubilization and penetration of the active ingredient into the target. Surfactants are sometimes sold separately as adjuvants for direct addition to the spray tank in order to enhance the bioactivity of agrochemical formulations (2).

In 1992, the total amount of surfactants used in agrochemical formulations worldwide was around 150,000 tons (3). This represents around 2% of the overall world consumption of surfactants. Compared to other auxiliaries, the cost of surfactants is high (in the order of $2000/ ton), therefore the formulation chemist needs to balance this cost with the improvement obtained in quality and efficacy. The concentration in agrochemical formulations is generally in the order of 5–15%.

The most common surfactant types used in agrochemical compositions are the nonionic and anionic groups. Nonionic surfactants have no electrical charge on the molecule, while anionic surfactants have a negative charge on the hydrophilic part of the molecule (Table 2).

The structure and ionic nature can affect significantly the orientation and packing of a surfactant at the interface. This in turn has an influence on the effectiveness of the surfactant, and consequently on such characteristics as foaming and wetting. Generally, anionic surfactants are more efficient wetters than nonionic surfactants.

The most fundamental property of a surfactant is its ability to reduce surface tension between two phases. This is essential not only for physical stability of heterogeneous phase formulations, such as suspension concentrates, but also for the rapid wetting/dispersion of the formulation in the spray tank and the correct distribution/retention of the spray. In addition, low surface tensions may help to transport the active ingredient toward the site of action. Effective and rapid penetration is not important only for bioactivity but also for reducing the risk of being washed

Table 2 Examples of nonionic and anionic surfactants used in agrochemical products

Nonionic surfactants

$$R\text{---}\langle\text{phenyl}\rangle\text{---}O\text{---}(CH_2\text{---}CH_2\text{---}O)_x\text{---}H \qquad R\text{---}O\text{---}(CH_2\text{---}CH_2\text{---}O)_x\text{---}H$$

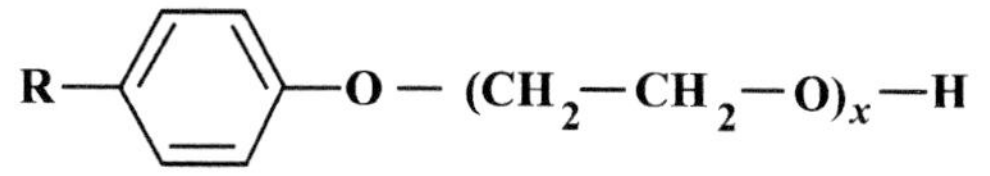

Alkyl phenol ethoxylate Alcohol ethoxylate

$$R\overset{\displaystyle O}{\overset{\|}{\text{---}C}}\text{---}O\text{---}(CH_2\text{---}CH_2\text{---}O)_x\text{---}H \qquad R\overset{\displaystyle O}{\overset{\|}{\text{---}C}}\text{---}O\text{---}CH_2\text{---}CH(OH)\text{---}CH_2\text{---}OH$$

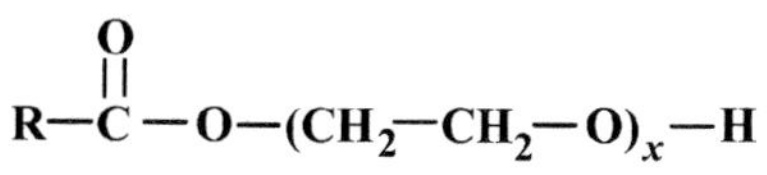

Fatty acid ethoxylate Glycerol ester

Anionic surfactants

$$R\text{---}\langle\text{phenyl}\rangle\text{---}SO_3^-\cdot Na^+$$

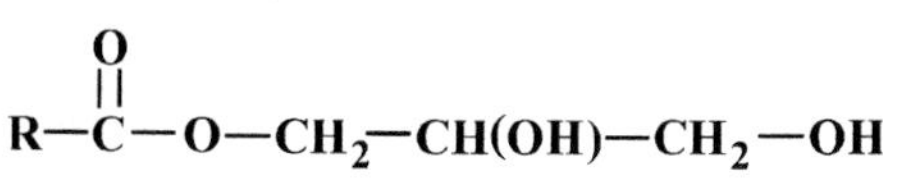

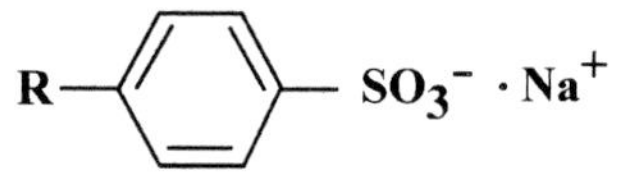

Alkyl aryl sulphonate

$$R\text{---}O\text{---}\underset{\overset{\|}{O}}{C}\text{---}CH_2\text{---}\underset{\overset{|}{SO_3^-\,Na^+}}{CH}\text{---}\underset{\overset{\|}{O}}{C}\text{---}O\text{---}R$$

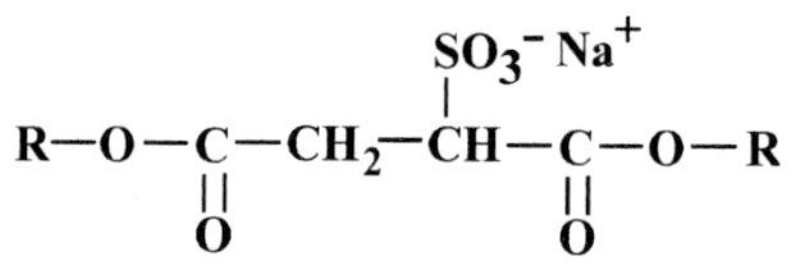

Di-alkyl sulphosuccinate

$$R\text{---}O\text{---}(CH_2\text{---}CH_2\text{---}O)_x\text{---}SO_3^-\cdot Na^+$$

Alkyl ether sulphate

$$R\text{---}O\overset{\displaystyle O}{\underset{\displaystyle OH}{\overset{|}{\underset{|}{\text{---}P\text{---}}}}}O^-\,Na^+$$

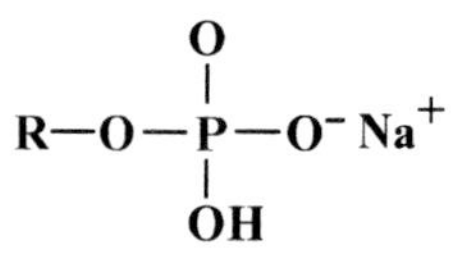

Phosphate ester

off by rain soon after application. In order to provide all the characteristics required of the formulation, it is often necessary to add two or more surfactants in the composition, mostly as nonionic/anionic combinations.

While certain groups of surfactants (for instance, alkyl phenol ethoxylates) are currently being reviewed for use in agrochemical formulations by environmental agencies in developed countries, other new surfactant families are appearing, such as sucroglycerides, organosilicone derivatives, and fluorocarbon surfactants.

Solvents

Health, safety, and environmental factors are gradually driving down the use of organic solvents in agrochemical compositions. Nevertheless, the percentage of solvent-containing formulations remains high. In 1994, more than 30% of the world market was formulated as emulsifiable concentrates (ASTM study, 1998). As the concentration of solvents in formulations is often high (as much as 70%), cost becomes an important issue in their selection as a formulation auxiliary. Solvents must have not only high dissolving power of the active ingredient, but also high flash point (for storage), low toxicity (for safety), low phytotoxicity (to avoid crop damage), low cost (to reduce formulation costs), and compatibility with the product pack (to avoid leakages). The choice of useable solvents that meet all these criteria is limited, and therefore mixtures of solvents may be necessary. The most commonly used solvents for emulsifiable concentrates are the aromatic hydrocarbons (relatively cheap and effective). However, for poorly soluble active ingredients, it is often necessary to add a more expensive polar co-solvent such as a ketone, an alcohol, or an ester.

The dissolution of the active ingredient in a surfactant/solvent phase often enhances its bioactivity. Different mechanisms take part in the transport of the active ingredient to the site of action, but the essential step is the diffusion of the active ingredient through the plant cuticle, a complex structure with waxes embedded in it. Use of lipophilic surfactant/solvent systems can facilitate the transport of the active ingredient through the cuticular layer. The downside is that toxicity and phytotoxicity can be increased at the same time, so the formulation chemist aims to select only those ingredients that preferentially favor the transport in the target species.

Diluents

Despite being described as inert ingredients, diluents and carriers play an important role in the quality and application of the formulation. They are used in the compositions of powder and granular products, and typically have the following characteristics:

- Free-flowing powder or granule
- High absorptive capacity for granule application carrier
- Rapid wetting and dispersion in the spray tank
- Chemical compatibility with the active ingredient
- Cheap and locally available

Typical classes of fillers used in agrochemical formulations include silicate clays (e.g., kaolin, talc, attapulgite), sand, and products of plant origin (e.g., corn cob grits). In the case of solid presentations such as dusts, wettable powders, and water-dispersible granules, the diluent generally has a low absorptive capacity and is used to maintain a consistent concentration of active ingredient. Critical properties are bulk density, particle size, and flow. In the case of impregnated granular formulations for local or total application, the key issues are the absorptive capacity of the granular carrier, which is highest in the case of silicate clays, and cost, as the application rate of granules is generally high.

Stickers

In addition to those additives that play a role in improving active ingredient transfer to the target site, other auxiliaries are used in agrochemical compositions to provide a mechanical resistance to natural factors that may reduce activity. These include UV light inhibitors, spray drift control agents, leaching-rate modifiers, and rain-fastness stickers. The latter group is comprised of chemicals that improve the adherence of spray droplets to the target surface, and their effectiveness is measured in terms of the deposit resistance to time, rain, and mechanical and chemical action. Their ability to improve rain-fastness is of great interest to the formulation chemist for use in product applications intended to remain on the target surface for a long period, such as protective fungicides. Typical groups of stickers include natural and synthetic polymers (latex, polyvinyl alcohol, polyvinylpyrrolidone), mineral or vegetable oils, and natural cellulose derivatives.

Other Auxiliaries

Apart from those auxiliaries included to optimize quality/bioavailability characteristics, agrochemical formulations may also contain others that simply make the product more user-friendly, more easily identified, and safer from accidental or deliberate ingestion:

- Deodorants, such as perfume components, are sometimes added to organophosphate insecticide

Table 3 Typical costs of active ingredient and auxiliaries used in agrochemical products

	Concentration (%)	Cost ($/kg)	Cost ($/kg formulation)
Active ingredient	5–90	10–100	0.50–90
Surfactants	5–15	3	0.15–0.45
Fillers/carriers	5–50	0.25	0.01–0.10
Solvents	10–50	0.25–1	0.02–0.50
Other additives	1–20	0.20–1	0.002–0.20

formulations in order to mask the unpleasant odor of mercaptan impurities.

- In order to identify seed treated with agrochemical products and to indicate clearly the uniformity of coverage, dyes are included in seed treatment formulations. At concentrations of up to 20% in the formulation to give the required color on the seed, the formulation chemist is concerned with cost, chemical compatibility, and legislation (only those for cosmetic use are permitted).
- Some agrochemical formulations contain repellents and emetics in order to deter ingestion by humans and animals of the formulation itself or of treated crops.

FUTURE TRENDS IN USE OF AUXILIARIES[a]

Globally, the chemical crop protection market is reaching maturity with very little growth in developed countries. Consequently, the total consumption of auxiliaries is not expected to change significantly in the near future. It is expected however that a gradual change in the spectrum of formulation types (and thus auxiliaries) will take place. With increasing pressure to provide safer and more environment-friendly products, and with higher research costs leading to fewer new active ingredients on the market-place, product differentiation by formulation innovation is becoming more important. The need to combine effectiveness, competitive advantage, and environment-friendly features in new agrochemical formulations means that the role of the formulation chemist takes on increased significance in the development of new agrochemical products. The inclusion of specialized auxiliaries in a formulation as a way of decreasing the dose/acre of active ingredient, or even the number of applications, is of special interest, particularly as the cost of auxiliaries is generally low compared to the cost of the active ingredient (Table 3). It is nevertheless an objective of both the industry and the regulatory authorities that the overall concentration of auxiliaries in crop protection products be kept to a minimum.

During the 1990s, regulation of auxiliaries has undergone significant change in Europe, and a more harmonized approach has taken place. All member states of the EU now conform to the revised Directive 91/414/EEC, which has become law since 1993. Country approval requirements vary, but all require statement of composition, information on the auxiliaries used, their role in the formulation, and perhaps the manufacturer's safety data sheet. Substances used in agrochemical compositions, other than the active ingredient, must be identified by CAS or EINICS numbers, or else new notification is necessary. Biodegradability of the inert auxiliary is a prime concern for the regulatory authorities in reviewing approval for use in agrochemical formulations. Close collaboration between the agrochemical industry, the suppliers of auxiliaries, and the regulatory authorities is essential to achieve the common goal of providing new, safe, efficient, and cost-effective products for the agrochemical pest control market.

REFERENCES

1. Tadros, T.F. *Surfactants in Agrochemicals*; Marcel Dekker, Inc.: New York, 1995; 248.
2. Foy, C.L.; Pritchard, D.W. *Pesticide Formulation and Adjuvant Technology*; CRC Press: Boca Raton, FL, 1996; 323–352.
3. *Surfactants and Other Additives in Agricultural Formulations*; Hewin International, Inc.: Amsterdam, 1993; 202.

[a]See also *Formulation*, pages 309–312; *Optimizing Pesticide Applications*, pages 547–549.

BACTERIAL ENDOPHYTES

B

Chris P. Chanway
University of British Columbia, Vancouver, British Columbia, Canada

INTRODUCTION

It has long been known that tissues of healthy plants can be colonized internally by nonpathogenic bacteria. Such microorganisms are often referred to as bacterial endophytes. Defined literally, an endophyte is an organism that lives inside a plant (i.e., the term ''endo'' is derived from the Greek word ''*endon*,'' meaning ''within,'' and ''phyte'' from the Greek ''*phyton*,'' meaning ''plant'') (1). Notwithstanding their discovery almost 50 years ago, comparatively little was known about bacterial endophyte diversity, population dynamics, and effects on host plant growth until recently. However, many common soil bacteria genera possess strains that have been isolated from internal plant tissues (Table 1) (1, 2). In contrast, there are fewer plant genera with species known to harbor bacterial endophytes (Table 2), but this is likely due to a lack of research rather than any special characteristic of these plants that precludes endophyte colonization. In addition to the plant tissues listed in Table 2, endophytic bacteria have also been isolated from within plant seeds of several plant species.

COLONIZATION SITES AND EFFECTS ON HOST PLANTS

Bacterial endophytes generally colonize intercellular spaces within plant tissues as well as the lumen of vascular tissues without interfering with the normal movement of metabolites within the plant (2). Colonization of vascular tissues likely contributes to the ability of some endophytes to move within the plant systemically, as certain bacterial endophytes may be found in root, stem, and leaf tissues of the same plant. There have been occasional reports of intracelluar colonization by bacterial endophytes, but it is not clear if plant cells colonized in this way are viable (2).

Experimentation within the past decade has indicated the potential of some bacterial endophytes to promote host plant growth. Though the mechanisms by which this happens are not fully understood, fixation of biologically significant amounts of nitrogen in situ is very likely involved in some plant x microbe interactions, for example, *Acetobacter diazotrophicus* and sugarcane in Brazil (1, 3). However, one of the most promising characteristics certain endophytic bacteria possess is their ability to act as biocontrol agents, either through direct antagonism of microbial pathogens (4) or by inducing systemic resistance to disease-causing organisms in the host plant (5, 6).

DIRECT ANTAGONISM OF PLANT PATHOGENS

Microorganisms capable of colonizing internal plant tissues systemically are ideally situated to suppress or halt the proliferation of fungal and bacterial diseases through direct antagonism of the disease-causing pathogen (1, 2). For such biocontrol agents to be effective, one would expect that a comparatively large endophytic bacterial population would need to develop for effective pathogen suppression. This certainly can be the case as bacterial endophyte populations exceeding 10^7 colony forming units (cfu) per gram plant matter have been reported within tissues of plant species such as red clover (*Trifolium pratense* L.) (7). However, population sizes between 10^2 and 10^6 cfu per gram (wet weight) of root, stem, or leaf tissue are more commonly observed (1, 2).

Examples of fungal pathogens that have been suppressed through direct interaction with bacterial endophytes in vivo include *Ceratocystis fagacearum, Fusarium oxysporum* F. sp. *vasinfectum, Verticillium albo-atrum, Rhizoctonia solani*, and *Sclerotium rolfsii* on plant hosts such as live oak (*Quercus fusiformis* Small), cotton (*Gossypium hirsutum* L.), potato (*Solanum tuberosum* L.), and bean (*Phaseolus vulgaris* L.) (2, 7, 8). Endophytic bacteria have also been shown to possess antagonistic activity against bacterial pathogens such as *Clavibacter michiganensis* ssp. *sepedonicum* on potato (2). The mechanisms by which bacterial endophytes directly antagonize microbial pathogens are likely similar to those hypothesized for plant growth promoting rhizobacteria (PGPR), the external root

Table 1 Examples of genera of nonpathogenic bacteria that have been isolated from internal plant tissues

	Bacterial genera	
Acetobacter	*Clavibacter*	*Pasteurella*
Achromobacter	*Comamonas*	*Photobacterium*
Acidovorax	*Corynebacterium*	*Phyllobacterium*
Acinetobacter	*Curtobacterium*	*Providencia*
Actinomyces	*Deleya*	*Pseudomonas*
Agrobacterium	*Enterobacter*	*Psychrobacter*
Alcaligenes	*Erwinia*	*Rahnella*
Arthrobacter	*Esherichia*	*Rhizobium*
Azoarcus	*Flavobacterium*	*Rhodococcus*
Azorhizobium	*Herbaspirillum*	*Rickettsia*
Azospirillum	*Kingella*	*Serratia*
Bacillus	*Klebsiella*	*Shewanella*
Bordetella	*Lactobacillus*	*Sphingomonas*
Burkholderia	*Leuconostoc*	*Staphylococcus*
Capnocytophaga	*Methylobacterium*	*Stenotrophomonas*
Cellulomonas	*Micrococcus*	*Variovorax*
Chryseobacterium	*Moraxella*	*Vibrio*
Citrobacter	*Pantoea*	*Xanthomonas*
		Yersinia

colonizing bacteria that can also act as biocontrol agents (3). These mechanisms include the production of antibiotics, competition for infection sites and nutrients (particularly Fe^{3+} through siderophore production), parasitism through lysis of fungal cell walls, and production of hydrocyanic acid. Further research is needed to determine which mechanism(s) are commonly employed by endophytic bacteria and which are most effective.

INDUCED SYSTEMIC RESISTANCE

While many endophytic bacteria are capable of inhibiting pathogen growth or activity directly, inoculation of plants with endophytic bacteria can also result in a different, and arguably, more effective control of pathogens. They do this by activating the plant's natural defense mechanism(s) before the pathogen contacts the plant, a phenomenon commonly referred to as induced systemic resistance (ISR) (2, 5).

It should also be noted that ISR is not the sole province of endophytic bacteria. Many external root tissue colonizing bacteria, that is, rhizobacteria and PGPR, have been shown to elicit the same response in host plants (4, 6), and indeed, much of what is known about ISR has resulted from studies of such external root tissue colonizing bacteria. However, based on results of studies of bacterial endophytes and disease resistance, the principles of PGPR-induced systemic resistance appear to be very similar to those governing systemic resistance induced by their endophytic counterparts.

NATURE OF THE RESPONSE

ISR is a particularly effective disease control mechanism for pathogens that are able to avoid triggering natural plant defense responses upon infection, but disease severity can also be reduced when caused by pathogens that are able to suppress or evade plant defense responses even after they have been triggered (2, 5). Bacteria that are capable of inducing systemic resistance usually do so without causing any visible symptoms of injury on the host plant, though certain anatomical changes associated with disease resistance may occur within plant tissues (5) (see below).

INDUCED SYSTEMIC RESISTANCE VERSUS SYSTEMIC ACQUIRED RESISTANCE

The morphological and physiological characteristics of systemic acquired resistance (SAR), which is typically

Table 2 Examples of plant species and internal tissues from which nonpathogenic bacteria have been isolated

Plant species	Tissue
Alfalfa (*Medicago sativa* L.)	Root
Coffee (*Coffea arabica* L.)	Root and stem
Cameroon grass (*Pennisetum purpureum* Schumach)	Stem and leaves
Corn (*Zea mays* L.)	Root and stem
Cotton (*Gossypium hirsutum* L.)	Root and stem
Cucumber (*Cucumis sativis* L.)	Root
Grapevine (*Vitis spp.*)	Stem
Hybrid spruce (*Picea glauca x engelmannii*)	Root
Kallar grass (*Leptochloa fusca* [L.] Kunth)	Root
Live oak (*Quercus fusiformis* Small)	Stem and branch
Lodgepole pine (*Pinus contorta* Dougl. Ex Loud)	Root, stem, and needle
Potato (*Solanum tuberosum* L.)	Tuber
Red clover (*Trifolium pratense* L.)	Stem, root, and leaf
Rice (*Oryza sativa* L.)	Root and stem
Rough lemon (*Citrus jambhiri* Lush.)	Root
Sorghum bicolor (L.) Moench	Shoot
Sugar beet (*Beta vulgaris* L.)	Root
Sugarcane (*Saccharum officinarum* L.)	Root and stem
Teosinte (*Zea luxurians* Itins and Doebley)	Stem
Western red cedar (*Thuja plicata* Donn)	Root, stem, and leaf

induced by pathogen infection and maximally expressed with the appearance of necrotic tissue, have been compared with those of ISR in several studies (5). Though both ISR and SAR are thought to produce a general resistance response, which is effective against a wide range of pathogens, there appear to be significant differences between ISR and SAR in addition to the SAR-tissue induced necrosis mentioned previously. The most notable of these relates to the production of pathogeneisis related (PR) proteins such as β-1,3 glucanases and chitinases, which are capable of hydrolyzing fungal cell walls (5). These are produced routinely during an SAR response, likely in response to salicylic acid (SA) production by the infecting microorganism. In contrast, ISR does not result in the production of PR proteins, even in the presence of exogenous SA that can act as an elicitor of ISR in certain cases (see below).

In addition, the production of other plant defense compounds, such as phytoalexins, has not been observed in the ISR response. However, structural changes involving epidermal and cortical cell wall fortification through the deposition of callose and phenolic materials has been documented and may constitute an important component of the ISR response (4, 5).

ELICITORS OF ISR

It is well known that the outer membrane lipopolysaccharide of pseudomonad PGPRs is an effective inducer of systemic resistance in plants, and the same has been shown to be true for endophytic pseudomonads capable of ISR (2, 5). Other compounds known to be important for induction of systemic resistance by rhizobacteria include siderophores, such as pyoverdin, and in some cases, SA, though the involvement of the latter appears to vary with the specific microoganism–plant host interaction. Bioactive oligosaccharides such as chitosan, a deacetylated chitin derivative that occurs in the cell wall of many fungi, have also been shown to be effective elicitors of ISR in plants (9). In addition, there is evidence that ethylene-dependent signaling is required at the bacterial colonization site for ISR to occur (5).

DEMONSTRATING ISR

Because endophytic bacteria capable of biocontrol are often able to suppress pathogens directly through antagonistic interactions such as production of chitinolytic compounds, as well as by inducing systemic resistance within the host plant, determining exactly which mechanism is at work is difficult. To unequivocally demonstrate ISR as the mechanism by which a bacterial endophyte controls a disease, it must be shown that no contact occurs between the inducing bacteria and the disease-causing pathogen in the plant and that no symptoms of disease develop. In addition, the possibility that biocontrol results from translocation of pathogen-inhibiting substances produced by endophytic bacteria in situ must also be eliminated. Use of heat-killed endophytic bacteria as controls as well as purified bacterial components known to induce systemic resistance, such as lipopolysaccharide, have helped demonstrate ISR by endophytic bacteria. It is also reassuring if the inducing microorganism shows no sign of pathogen inhibition or antagonism in vitro (2).

The ability of an endophytic bacterium to protect host plants through ISR was first demonstrated in 1991 using *Pseudomonas fluorescens* strain WCS417r inoculated onto root systems of carnation (*Dianthus caryophyllus* L.) where it colonized internal root tissues (10). One week later, plants were challenged with *Fusarium oxysporum* f. sp. *dianthi* on stems, and plants treated with the bacterial endophyte developed disease symptoms less frequently and with less intensity than controls. Because strain WCS417r could not be isolated from stem tissues which also displayed the protective effect, the biocontrol mechanism was concluded to be ISR.

Since that initial report, there have been several demonstrations of ISR by inoculation with endophytic bacteria (2). There is also some evidence that ISR resulting from inoculation with endophytic bacteria affords a degree of protection from plant-parasitic nematodes, but little work has been done in this area (2).

More recent research indicates that ISR efficacy can be significantly enhanced when plants are cotreated with endophytic bacteria and chemical elicitors of ISR (9). For example, tomato (*Lycopersicon esculentum* Mill.) resistance against *Fusarium oxysporum* f. sp. *radicis-lycopersici* was significantly enhanced when treated with the bacterial endophyte *Bacillus pumilus* SE34 in combination with chitosan. Such an approach, involving a combination of biotic and abiotic ISR elicitors, may prove to be an effective adjunct to purely biological means of pathogen control and warrants further study.

FUTURE PROSPECTS AND CONCERNS

Inoculation of plants with endophytic bacteria capable of reducing or eliminating disease in the absence of organic pesticides would seem to be an excellent strategy to attain sustainable and environmentally benign increases in economic yield of agricultural or forestry crops. Because it is unclear which mechanism would be best for this purpose, that is, direct antagonism of pathogens versus

ISR, development of single or mixed culture inoculants capable of both direct pathogen antagonism and inducing systemic resistance could be the best strategy. If naturally occurring endophytic bacteria are screened and selected for this purpose, there would appear to be little risk associated with such an approach, other than the possibility that pathogens will ultimately evolve resistance to these endophytic biocontrol agents and the systemic resistance they induce in host plants.

REFERENCES

1. Chanway, C.P. Bacterial endophytes: ecological and practical implications. Sydowia **1998**, *50* (2), 149–170.
2. Hallmann, J.; Quadt-Hallmann, A.; Mahaffee, W.F.; Kloepper, J.W. Bacterial endophytes in agricultural crops. Canadian J. Microbiol. **1997**, *43* (10), 895–914.
3. Baldani, J.I.; Caruso, L.; Baldani, V.L.D.; Goi, S.R.; Dobereiner, J. Recent advances in BNF with non-legume plants. Soil Biol. Bioch. **1997**, *29* (5/6), 911–922.
4. Kloepper, J.W. Plant growth-promoting rhizobacteria as biological control agents. In *Soil Microbial Ecology: Applications in Agricultural and Environmental Management*; Metting, F.B., Jr., Ed.; Marcel Dekker, Inc.: New York, 1993; 255–274.
5. van Loon, L.C.; Bakker, P.A.H.M.; Peiterse, C.M.J. Systemic resistance induced by rhizobacteria. Annu. Rev. Phytopathol. **1998**, *36*, 453–483.
6. Kloepper, J.W.; Tuzun, S.; Zehnder, G.W.; Wei, G. Multiple disease protection by rhizobacteria that induce systemic resistance—historical precedence. Phytopathology **1997**, *87* (2), 136–137.
7. Sturz, A.V.; Christie, B.R.; Matheson, B.G.; Nowak, J. Biodiversity of endophytic bacteria which colonize red clover nodules, roots, stems and foliage and their influence on host growth. Biol. Fertil. Soils **1997**, *25* (1), 13–19.
8. Brooks, D.S.; Gonzalez, C.F.; Appel, D.N.; Filer, T.H. Evaluation of endophytic bacteria as potential biological control agents for oak wilt. Biol. Control **1994**, *4* (4), 373–381.
9. Benhamou, B.; Kloepper, J.W.; Tuzun, S. Induction of resistance against fusarium wilt of tomato by combination of chitosan with an endophytic bacterial strain: ultrstructure and cytochemistry of the host response. Planta **1998**, *204* (2), 153–168.
10. Van Peer, R.; Niemann, G.J.; Schippers, B. Induced resistance and phytoalexin accumulation in biological control of *Fusarium* wilt of carnation by *Pseudomonas* sp. Strain WCS417r. Phytopathology **1991**, *81* (7), 728–734.

BANDING OF PESTICIDES

Sharon A. Clay
South Dakota State University, Brookings, South Dakota, U.S.A.

B

INTRODUCTION

Pesticides are applied to field crops to reduce insect, disease, and weed losses. In 1996, more than 97% of United States row crops, such as corn, *Zea mays*, and soybean *Glycine max*, were treated with one or more pesticides (1) that were often applied in a uniform, broadcast pattern. The extensive use of pesticides is costly and of environmental concern. Band placement of pesticides is an alternative application practice that places chemicals only over crop rows, leaving interrows untreated. Banding reduces pesticide usage in proportion to the band width; treated areas typically range from 25 to 75% of the total field acreage. Banding also reduces pesticide input costs and environmental impacts. Insecticides are band applied more commonly than herbicides. Control of pests outside the application band may still need to be accomplished. For example, cultivation and residue management are non-chemical techniques used in interrows for weed control.

BAND APPLICATION USAGE

Chemicals applied in a band are restricted to a linear strip on or along a crop row as opposed to broadcast applications, where chemicals are applied uniformly over the entire area. About 20% of the total 1996 U.S. crop acreage was treated with a band application of insecticide over the row or soil injected next to the row to control insects that attack the crop seed, roots, or foliage (2). The importance of banding varied by crop with 10% of soybean acreage, 60% of corn acreage, and 86% of tomato, *Lycopersicon esculentum*, acreage treated with a band application. Insecticides can be band applied by several methods (3, 4). Seed treatments protect crop seeds from foraging insects such as the fire ant, *Solenopsis invicta*. Pre-emergent band applications may be applied in the open furrow (T-band) during planting or prior to crop emergence as an in-furrow spray and are used to reduce losses from soil dwelling insects such as corn rootworm *Dibrotica* spp., and sugar beet (*Beta vulgaris*) root maggot, *Tetanops myopaeformis*. Soil applied insecticides also may be systemic to control foliage feeders. Postemergent foliar band applications control stem and leaf pests such as aphids and mites in potato, *Solanum tuberosum*, beetles and loopers in soybean, and bores and earworms in corn.

In contrast to insecticides that are banded to protect only the crop, herbicides generally are applied uniformly as a broadcast treatment to control all weeds. Herbicide band application has been proposed for row crops but is not a standard practice in United States crop production (Table 1). An average of 10% of all herbicide treated acres had a band application in 1996 (2), with 5% of soybean acres, 9% of the corn acres, and 38% of cotton *Gossypium hirsutum*, acres treated with a band. Herbicide banding was much greater in cotton grown in southern U.S. regions (40%) compared to cotton in the western region (9%). The overall low usage of herbicide band application may be due to the fact that weeds, even if limited to interrow areas, have the potential to reduce yield by competition or cause problems at harvest by slowing combine speed or plugging combines.

Advantages of Banding Pesticide Applications

Banding reduces chemical amounts applied to fields. For example, if a band covers 38 cm of a 76-cm row, 50% of the pesticide will be applied compared to a broadcast application reducing chemical costs. Pest control in the band is comparable to control in broadcast areas. Crop yields from band and broadcast herbicide treatments are similar if interrow weeds are controlled in a timely manner (5, 6).

Another major advantage of band applications is the reduction of environmental impact through leaching and runoff losses. The U.S. Geological Survey (7) estimated that more than 100,000 metric tons of soluble pesticides were applied to the U.S. Mississippi River Basin in 1996. Of this amount, most were herbicides, and more than 60% were soil applied. A problem with soil-applied pesticides

Table 1 Percent of herbicide treated acres using herbicide broadcast vs. band application methods and percent of all planted acres using mechanical cultivation for weed control in the United States

Crop	Region	Broadcast	Band	Cultivation
		(% of herbicide-treated acres)		(% of planted acres)
Corn	Northeast	83	2	
	North Central	85	9	
	South	82	9	
	All corn states	85	9	51
Soybean	North Central	89	3	
	South	83	12	
	All soybean states	88	5	29
Cotton	South	43	40	
	West	71	9	
	All cotton states	45	38	89

(From Ref. 1.)

is that only about 10% of the applied amount reaches the target. The rest of the pesticide is needed to counteract sorption to soil, microbial degradation, volatilization, photodecomposition, and soil dilution. However, a portion may cause environmental problems from off-site movement. The frequency of detection, maximum concentration, and mass loss of pesticide have been reported to be three to four times lower from band applications where 50% of the area was treated compared to losses from broadcast applications (5).

Disadvantages of Band Applications

Band applications may allow pests to take shelter in interrow areas causing crop damage or pest populations to increase. For example, corn rootworm larvae may not attack roots directly under insecticide bands. However, as the roots expand into untreated areas, larvae may feed on these roots resulting in stressed plants and an adult population, although yield loss may not occur. Untreated interrow areas also may become refuges for novel pests permitting new outbreaks. The threat of outbreaks must be balanced against control and potential crop loss costs.

Supplemental methods, such as mechanical cultivation done once or twice prior to crop canopy closure, generally are needed to control weeds outside the herbicide band application area. Interrow cultivation of row crops is more common than banded herbicide applications. In 1996, 51, 29, and 89% of all corn, soybean, and cotton acres, respectively, were cultivated at least once for weed control (Table 1). Cultivation and band applications are compatible and may help explain the high frequency of band applications in southern U.S. cotton. However, a large discrepancy between cultivated and banded acres exists and points to other disadvantages of band applications.

If herbicide bands are too narrow, or cultivation not wide enough, strips of weeds may remain (8). However, cultivation too close to rows may prune crop roots possibly resulting in yield loss. The timing of cultivation is important. Adverse weather may cause unsuitable conditions for cultivation when weeds are vulnerable to mechanical control. Interrow weeds growing unrestrained for whatever reason often result in reduced yields, harvest problems, and increased weed seeds in the soil that perpetuate weed problems.

High fuel prices increase cultivation costs so that band plus cultivation cost may be similar if not more expensive than a broadcast application. This may discourage the use of banding techniques. However, interrow seeding of cover crops, such as short season annual medic (*Medicago* spp.) varieties that suppress weeds (9), or residue management may be alternatives to cultivation.

Equipment Requirements

Broadcast applications rely on spray patterns that overlap for uniform application rates. Broadcast nozzles are flat fan types with feathered spray patterns resulting in more output directly below the nozzle than on either side. Even nozzles are used for band applications and provide uniform output across the spray pattern. Spray from adjacent nozzles should not overlap, as output in overlap areas would be double the amount desired.

Spray boom height must be appropriate and consistent, as boom height and nozzle output angle regulate band width. The band becomes wider as boom height or nozzle output angle increases. A 5-cm increase in boom height increases band width of either a 80° or 95° series nozzle by about 10 cm. The boom must be 2.5 cm higher for an 80° than a 95° nozzle to get the same band width.

Band applications of granular or dry material can be applied in several ways. Some pesticides can be applied as a seed treatment. Other banding methods include shanking the pesticide into the soil as a liquid, placing the pesticide in the seed furrow (T-band) that is subsequently covered with soil, or applying dry granules to the soil surface.

FUTURE APPLICABILITY OF BANDING PESTICIDES

Reliance on soil-applied pesticides has been reduced, but not eliminated, by the introduction of genetically modified crops for insect resistance and postemergence herbicide tolerance. However, soil-applied pesticides, which are often targeted for band application, will continue to be used for early control of crop pests or in situations where weather limits postemergent options. Band application use will increase despite concerns of uncontrolled pests and need for supplemental control methods. Band applications reduce offsite leaching into tile lines and groundwater, and runoff into surface water (6). Nonpoint source reductions in pesticide residues are needed to tackle major environmental problems such as aquifer and surface water quality, and pesticide residue and exposure regulations, based on U.S. Food Quality Protection Act.

See also *Crop Rotations (Plant Diseases),* pages 172–173; *Crop Rotations (Insects and Mites),* pages 169–171; *Principles of Pest Management with Emphasis on Plant Pathogens,* pages 666–669; *Site-Specific Farming/Management (Precision Farming),* pages 771–772; *Integration of Tactics,* pages 416–419; *Ground Sprayers,* pages 351–353.

REFERENCES

1. Fernandez-Cornejo, J.; Jans, S. *Pest Management in U.S. Agriculture*; ERS/USDA: Washington, DC, 1999. www.ers.usda.gov/epubs/pdf/ah717 (accessed Dec 2000).
2. Padgitt, M.; Newton, D.; Penn, R.; Sandretto, C. *Production Practices for Major Crops in U.S. Agriculture, 1990–97*; ERS/USDA, SB-969, 2000. www.ers.usda.gov/epubs/pdf/sb969/index.htm (accessed Dec 2000).
3. Drees, B.M.; Cavazos, R.; Berger, L.A.; Vinson, S.B. Impact of seed-protecting insecticides on sorghum and corn seed feeding by red imported fire ants (Hymenoptera: Formicidae). J. Econ. Entomol. **1992**, *85* (3), 993–997.
4. Woodford, J.A.; Gordon, S.C.; Foster, G.N. Side-band application of systemic granular pesticides for the control of aphids and potato leafroll virus. Crop Prot. **1988**, *7* (2), 96–105.
5. Clay, S.A.; Clay, D.E.; Koskinen, W.C.; Berg, R.K., Jr. Application method: impacts on atrazine and alachlor movement, weed control, and corn yield in three tillage systems. Soil and Tillage Res. **1998**, *48* (3), 215–224.
6. Anderson, J.L.; Allmaras, R.R. Tillage systems and agricultural management: water quality effects. Soil and Tillage Res. **1998**, *48* (3), 141–257.
7. Thurman, E.M.; Meyer, M.T. *Herbicide Metabolites in Surface Water and Groundwater*; ACS Symposium Series 630, American Chemical Society: Washington, DC, 1996; 320.
8. Paarlberg, K.R.; Hanna, H.M.; Erbach, D.C.; Hartzler, R.G. Cultivator design for interrow weed control in notill corn. Appl. Eng. Agric. **1998**, *14* (4), 353–361.
9. Vos, R.J. *Effect of Spring-seeded Annual Medics on Weed Management in Zea mays Production*; Ph.D. Dissertation, South Dakota State University: Brookings, SD, 1999; 173.

B

BIOACCUMULATION

Tomaz Langenbach
Universidade Federal do Rio de Janeirollha do Fundão, Rio de Janeiro, Brazil

INTRODUCTION

Bioaccumulation is the equilibrium process in which the uptake of substances in biota reaches much higher concentrations than those occurring in the environments. In literature, a clear definition of the characteristics that identify the substances that are able or unable to bioaccumulate is not found. The most common use of this term is related to hazardous substances such as heavy metals or predominantly nonpolar xenobiotics. Many of these substances are in low concentrations in the environment and the severity of the poisoning effect is mainly due to bioaccumulation up to high concentrations in the organism. The poisoning effects in animals are quite diverse but the most common ones observed are related to the nervous system, which can lead to death.

MOLECULAR PROPERTIES FOR BIOACCUMULATION

Bioaccumulation occurs only with molecules with low degradability and correlates with their grade of lipophilicity (1). Organic substances with main bonds of aliphatic and aromatic C—C, C—H, and C—Cl (or other halogens) are predominantly nonpolar molecules (Lipophilic) with low water solubility and high stability. They are less susceptible to chemical reactions of hydrolysis, oxidation, and enzymatic attack (2). On the other hand, bonds with different functional groups with O, P, N, and other elements turn molecules more polar, soluble, and degraded more easily. Bioaccumulation can occur with molecules between 100 and 600 units of molecular weight with the maximum of 350 (3). Probably this is related to membrane permeability capacity.

A common feature of bioaccumulation is the molecular stability of lipophilic organic substances and the nondegradability of heavy metals. The severity of heavy metals is due to many factors.

1. Metals with Hg, Cd, Zn, Cu, and Pb are the most toxic and most studied types followed by metals containing Ni, Al, As, Cr, and other elements (4). Bioaccumulation can also occur also with essential metals such as Fe, Zn, Cu, Mo, Na, and Ca.
2. Speciation is the anions or other components that constitute the heavy metal molecules. This is important in defining solubility that, for example, is high for sulfate and low for sulphide (5). Heavy metals bound to organic molecules such as methyl, ethyl or other aliphatic or aryl groups increase penetration capacity through membranes and consequently, have a poisoning effect.
3. The sensivity to the toxic effects of heavy metals and other xenobionts is dependent on the biological material being a microorganism, plant, animal, or type of tissue.

BIOACCUMULATION AND THE ENVIRONMENT

The pollution sources can be released by discharge of substances with uneven distribution in air, water, and soil. The movement of these substances up to bioaccumulation can occur by different routes mainly mediated by the food chain. This process can involve water, suspended particles, sediments, food, soil, and air particles (Fig. 1). An important part of these substances can be concentrated in nonliving components. From these sources, persistent organic pollutants (POP) or heavy metals can be released to biota (6). The final distribution presented by the mass balance in the environment with a group of organochlorines and polyaromatics, shows that most are found in soil or sediment, whereas for highly volatile substances, most

Fig. 1 Bioaccumulation in the environment. The red points represent the pollutant molecules.

remain in the atmosphere. Less than 0.7% of the total remains in vegetation and no more than $2 \times 10^{-3}\%$ can be found in the aquatic biota (7). The relationship between biota and environment shows that concentrations of bioaccumulated heavy metals in organisms are always higher than in water, but are usually lower than in sediments (4).

Water Environment

After pollution reaches water bodies, different processes can occur to incorporate it into nonliving components as sediments and biota represented by microorganisms, plants, crustaceans, fishes, etc. The route of pollutant uptake in the biota if from waterborne, adsorption, filtration, or by food chain is an important factor in bioaccumulation (8). Along the food chain the step-wise increase of concentration from lower to higher trophic levels, called biomagnification, can reach the bioconcentration factor (BCF) up to 100,000. In this process terrestrial animals as well as birds can be heavily contaminated by eating polluted fish.

In the global marine environment the apparent final fate of persistent organic pollutants (POP) is in the flora, fauna, and sediment of the abyss (9). The main transport of POP follows the downstream movement of the organic flow in the water and in the long term these chemicals are incorporated in the sediment that function as final sink. It was observed that the bioaccumulation in the deep water fishes are up to 10 times higher than in surface water fishes (9).

In the flora and fauna some heavy metals can bioaccumulate up to threshold values and others maintain a correlation with the concentration in the environment (4). In aquatic plants, fish, and other metazoarians the distribution of the substances are quite distinct between tissues (10). Lipophilic substances are preferentially found in adipose tissues with high lipid content.

In the Soil

Soil is polluted in large areas by pesticides application or by discharge as final disposal in landfills of industrial products. These lipophilic substances move in soil rather slowly by leaching, runoff, and volatilization. The main factor that influences bioaccumulation process in the soil is the biodisponibility. This property is conditioned by the adsorption/desorption capacity of the different soils and by the chemical nature of the pollutant. This process is driven by the stronger or weaker binding forces involved, which influence the amount that is bioavailable for plant uptake of these lipophilic pollutants. The main flow of POP generally occurs toward organic matter from soil particles and not to biota, a process called preferential partition. A negative correlation was observed between the adsorbtion coefficient related to soil organic carbon (*K*oc) and the bioaccumulation factor by plants. This means that in organic rich soils, bioaccumulation in plants is rather small (2). A similar situation occurs with microorganisms in which previous bioaccumulated organochorines can move out from the cell to the soil (11). The preferential partition toward soil can be the reason why a lack of toxicity on soil microorganisms by pesticide applications was frequently observed, even in high concentrations of pesticides. Little information about bioaccumulation in soil could be observed, but nevertheless cotransport of some organochlorine accumulated in microorganisms in sand aquifers with low organic content was reported (12).

Soil invertebrates such as earthworms, beetles, slugs, and others can bioaccumulate lipophilic pesticides. The bioaccumulation process could be seen as a soil to soil–water equilibrium followed by a soil–water to worm equilibrium (2). Consumers of this biota in animals of higher trophic levels such as birds can biomagnificate these chemicals. Plants can adsorb and bioaccumulate products from the soil with incorporation of residues mainly in the root. The translocation from root to foliage depends on plant species and on the chemical properties of the pollutant. Several evidences show that lipophilic compounds are sorbed onto the outer surface of roots of several plants, and in this case translocation is very low.

Bioaccumulation in plants can also occur with heavy metals. As safety rules, domestic waste and sludge from wastewater treatment stations with heavy metal contamination can be applied on soil for agriculture up to limited amounts to avoid pollution with hazardous toxicological effects. Contaminated grass, grain, and fruits can be accumulated by biomagnification when consumed by mammals, insects, and birds. Terrestrial animals have a plant mediated relationship with soil contamination.

In the Air

The main sources of atmospheric pollution are pesticide spraying with the reverse process of evaporation from soil to air, poliaromatics produced by burning of fuels, and plastic incineration. The rate of entry to the atmosphere and the distance of movement are principally dependent on the vapor pressure of the pollutants and metereological conditions. In some cases movement occurs on a global scale. The dynamic nature of the atmosphere can dilute pollutants in the air to exceptionally low concentrations and in these cases no significant bioaccumultion occurs (13). Nevertheless urban and industrial areas, as well as the margins of roads with intense traffic, can have high concentrations of pollutants. Plants exposed to xenobiotics in the form of vapor, particules, aerosol or larger droplets, can undergo a passive process of foliage adsorption with an uptake mainly in the wax cuticle (2). Bioaccumulation in plants can result in damage and can also affect higher trophic levels that consume these vegetables.

Direct contact of animals with these chemicals can enter by the respiratory organs, in mammals, or the outer body surface, mainly in insects.

BIOACCUMULATION MECHANISM IN BIOTA

In terrestrial animals, pollutants can enter by dermal contact, respiration, and food consumption. Atmospheric pollutants move to the lungs, where an equilibrium is difficult to be established, while generally atmosphere dilutes pollutant concentrations, unless there is an exposure to constant pollution sources. Chemicals move from lungs to circulatory fluid (plasma) and can be metabolized with further excretion. Another route is the storage mainly in rich lipid bodies such as brain and eggs in bird's (2). If the entrance is by food consumption, the gastrointestinal tract can eliminate (10) these substances or degrade than to more polar compounds with further excretion, or can promote adsorption by plasma following the same route described earlier.

The uptake of heavy metals in microorganisms can occur by bioadsorbtion in capsular polysaccharides and cell-wall polymers, or cross these layers and cell membranes by an active enzymatic process involving phosphatases, reaching to the cell interior (14). Some authors define bioaccumulation as only the process in which molecules reach cell interior. Many cells from animals, plants, fungi, yeast, and bacteria have metal-binding proteins with low molecular weight called metallothioneins. These proteins bind mainly to Cd and Zn and constitute a protection mechanisms to the toxic effects of these substances. Some other cell protective mechanisms exist such as enhancement of efflux from cell to the outside. Metals bind on different macromolecules and change enzymatic activities with inhibition or stimulation effects.

Lipophilic hydrocarbons in microorganisms cross polysaccharides from capsule and cell wall polymers with adsorption mainly by the lipids of the membranes (15). Compounds that are inserted in cell lipids are more difficult to be degraded by chemical or enzymatic processes getting higher persistence (16). The probable mechanism of action seems to be nonspecific, this means not related to a specific target.

APPLICATIONS AND FUTURE PERSPECTIVES

From the scientific point of view a better understanding of the integration between the different environment compartments and biota including modeling systems is an important approach that needs more development. Another possibility is to use bioaccumulation for environmental monitoring, based on the accumulation capacity of many pollutants in specific plants or animals, allowing chemical measurements that otherwise in water or air are below the analytic detection capacity (17). This method has the possibility to integrate all pollutant exposure of plants or animals in a specific environment and can be in the future an important parameter for ecotoxicological evaluations.

After the disaster of the mercury pollution in the Minamata Bay in Japan in which more than 630 people died and many became physically and mentally disabled, the magnitude of poisoning effects due to bioaccumulation was recognized for the first time (18). This was the beginning of a scientific research that produced a large amount of information. With this knowledge it became clear that bioaccumulation is a natural process that cannot be stopped by man but can be avoided with a more efficient control of pollutant release. To overcome the economic, social, and political difficulties for better pollution control together with the development of more ecological technologies are our challenge for today and for the future.

REFERENCES

1. Connell, D.W. General Characteristics of Organic Compounds which Exhibit Bioaccumulation. In *Bioaccumulation of Xenobiotic Compounds*; Connell, D.W., Ed.; CRC Press: Boca Raton, Florida, 1990; 47–57.
2. Connell, D.W. Bioamagnification of Lipophilic Compounds in Terrestrial and Aquatic Systems. In *Bioaccumulation of Xenobiotics Compounds*; Connell, D.W., Ed.; CRC Press: Boca Raton, Florida, 1990; 145–185.
3. Brooke, D.N.; Dobbs, A.J.; Williams, N. Octanol: Water partition coefficients (*P*): measurement estimation and interpretation, particularly for chemicals with $P > 10^5$. Ecotoxicol. Eviron. Saf. **1986**, *11*, 251.
4. Goodyear, K.L.; McNeill, S. Bioaccumulation of heavy metals by aquatic macro-invertebrates of different feeding guilds: a review. Sci. Total Environ. **1999**, *229*, 1–19.
5. Bourg, A.C.M. Speciation of Heavy Metals in Soils and Groundwater and Implications for Their Natural and Provoked Mobility. In *Heavy Metals: Problems and Solutions*; Salomons, N., Förstner, U., Mader, P., Eds.; Springer Verlag: Berlin, 1995; 17–31.
6. Tsezos, M.; Bell, J.P. Comparison of the biosorption and desorption of hazardous organic pollutants by life and dead bioamass. Wat. Res. **1989**, *23* (5), 561–568.
7. Connell, D.W.; Hawker, D.W. Predicting the distribution of persistent organic chemicals in the environment. Chem. Aust. **1986**, *53*, 428.
8. Carbonell, G.; Ramos, C.; Pablos, M.V.; Ortiz, J.A.; Tarazona, J.V. A system dynamic model for the assessment of different exposure routes in aquatic ecosystems. Sci. Total Environ. **2000**, *247*, 107–118.
9. Froescheis, O.; Looser, R.; Cailliet, G.M.; Jarman, W.M.; Ballschmiter, K. The deep-sea as a final global sink of semivolatile persistent organic pollutants part I: PCBs in surface and deep-sea dwelling fish of the north and south atlantic and the Monterey bay canion (California). Chemosphere **2000**, *40*, 651–660.
10. Lin, K.H.; Yen, J.H.; Wang, Y.S. Accumulation and elimination kinetics of herbicides butachlor, thiobencarb and Chlomethoxyfen by *Aristichthys nobilis*. Pestic. Sci. **1997**, *49*, 178–184.
11. Brunninger, B.M.; Mano, D.M.S.; Scheunert, I.; Langenbach, T. Mobility of the organochlorine compound dicofol in soil promoted by *Pseudomonas fluorescens*. Ecotoxic. Environ. Safety **1999**, *44*, 154–159.
12. Jenkins, M.B.; Lion, L.W. Mobile bacteria and transport of polynuclear aromatic hydrocarbons in porous media. Appl. Environ. Microbiol. **1993**, *59* (10), 3306–3313.
13. Connel, D.W. Environmental Routes Leading to the Bioaccumulation of Lipophilic Chemicals. In *Bioaccumulation of Xenobiotic Compounds*; Connell, D.W., Ed.; CRC Press: Boca Raton, Florida, 1990; 59–73.
14. Gomes, N.C.M.; Mendonca-Hagler, L.C.S.; Savvaidis, I. Metal bioremediation by microorganisms. Rev. Microbiol. **1998**, *29*, 85–92.
15. Mano, D.M.S.; Langenbach, T. [^{14}C]Dicofol association to cellular components of azospirillum. Pestic. Sci. **1998**, *53*, 91–95.
16. Mano, D.M.S.; Buff, K.; Clausen, E.; Langenbach, T. Bioaccumulation and enhanced persistence of the acaricide dicofol by *Azospirillum lipoferum*. Chemosphere **1996**, *33* (8), 1609–1619.
17. Maagd, P.G.J. Bioaccumulation test applied in whole effluent assessment: a review. Env. Toxic. Chem. **2000**, *19* (1), 25–35.
18. Takashi, H.; Tsubaki, T. Epidemiology: Mortality in Minamata Disease. In *Recent Advance in Minamata Disease Studies; Methylmercury Poisoning in Minamata and Niigata Japan*; Kodansha Ltd.: Tokyo, 1986; 1–23.

BIOFERTILIZERS

J. Renato de Freitas
University of Saskatchewan, Saskatoon, Saskatchewan, Canada

INTRODUCTION

Biofertilizers include microorganisms and their metabolites that are capable of enhancing soil fertility, crop growth, and/or yield. These include both indigenous microbes and microbial inoculants, that is, microorganisms that replace fertilizers or increase a crop's fertilizer use efficiency. Soil microorganisms such as bacteria, ectomycorhiza, arbuscular mycorrhizal fungi, and soil algae, especially the N_2-fixing cyanobacteria have potential as biofertilizers. Nitrogen-fixing inoculants based on *Rhizobium* species were among the first biofertilizers introduced into agroecosystems back in the nineteenth century. In the twenty-first century, biofertilizers will become an increasingly important area of research and development (1). The use of fertilizers and pesticides has increased steadily since the 1970s; consequently, concerns about the impacts of these chemicals on land, air, and water have become significant environmental issues. Biofertilizers provide an alternative to agricultural chemicals as more sustainable and ecologically sound practices to increase crop productivity. Biofertilizer sales forecasts in the United States for the years 2001 and 2006 represent $690 million and $1.6 billion, respectively. Examples of some biofertilizers currently in use worldwide are shown in Table 1 (2).

MARKET FOR BIOFERTILIZERS

The market potential for biofertilizers includes the high value vegetable industry. A comparison of the base value of various crops and the increased value that can be obtained as the crop yield rises is illustrated in Table 2. Due to high nutrient requirements and high susceptibility to diseases, vegetable growers spend substantial amounts to protect this valuable produce. For example, average broccoli and tomato crops grown in California require ca. $62 and $170 worth of fertilizer and/or fungicide per acre, respectively. When the U.S. government prohibits the use of methyl bromide as a soil fumigant, as anticipated in 2005, development of biological products will be stimulated as an alternative to the use of chemicals (3).

MECHANISMS OF GROWTH PROMOTION

The mechanisms covered in this paper are those that show commercial market potential; thus, it does not include all modes of action by which biofertilizers promote crop growth. Biofertilizers promote crop growth using several mechanisms with the primary one varying as a function of environmental conditions. Although the mechanisms of commercially available biofertilizers are not always entirely understood, growth promotion has been classified as the result of indirect or direct mechanisms. Indirect plant growth promotion may be associated with the repression of negative effects caused by phytopathogenic organisms, that is, biological control. Conversely, direct growth promotion mechanism may either provide some compound essential to crop development and/or stimulate nutrient uptake. Biofertilizers based on biological control agents *Mycorrhizae* and *Rhizobium* will be discussed in more detail elsewhere in this encyclopedia (e.g., Biological Pest Controls; Mycorrhiza; Rhizobia).

Phytohormones

Production of phytohormones is a commonly noted direct mechanism of plant growth promotion (4). The nature of growth response may be the result of phytohormone production in the rhizosphere. Phytohormones are produced by many biofertilizers and include a list of plant growth regulators that are important in the plant's metabolism. For example, auxins such as indole-3-acetic acid are known for their ability to stimulate root cell division, differentiation, and promote cell elongation. Other phytohormones such as cytokinin, gibberellin, and ethylene also play key roles in plant development and have been reported to increase the growth of various commer-

Table 1 Organisms, mode of action, crops, and producers of biofertilizers currently in use for agriculture

Type	Mode of action	Crop	Used in
Rhizobium spp.	N_2 fixation	Legumes	Russia; several countries
Cyanobacteria	N_2 fixation	Rice	Japan; several countries
Azospirillum spp.	N_2 fixation	Cereals	Several countries
Mycorrhizae	Nutrient acquisition	Conifers	Several countries
Penicillium bilaii	P solubilization	Cereals, legumes	Western Canada
Directed compost	Soil fertility	All plants	Several countries
Earthworm	Humus formation	Vegetables, flowers	Cottage industry

(Adapted from Ref. 2. Copyright 1998 Elsevier Science.)

cial crops. The horticulture market for biofertilizer products based on gibberellin and other auxins is currently estimated at $600 million per year.

Plant Nutrient Acquisition

Several direct mechanisms are responsible for increased nutrient acquisition.

Biological nitrogen fixation (BNF)

Nitrogen (N) is an essential macronutrient, that is, it is the key building block of proteins, thus an indispensable component of the protoplasm of microorganisms, animals, and plants. The supply of biologically available N to agriculture through BNF represents ca. 140×10^6 ton/year, globally (1). Therefore, BNF represents an economy of millions of dollars. N_2-fixation by free-living bacteria such as *Azospirillum, Azotobacter, Bacillus,* and *Derxia* species have been exploited in agricultural systems for many decades and constitute an important source of N input into agroecosystems (5). Other BNF associations include the water fern *Azolla* that forms a symbiosis with the heterocystous cyanobacterium *Anabaena azollae.* The *Azolla–Anabaena* system has been used as a biofertilizer in Vietnam and China for rice production and has the potential to supply the entire N requirement (30–50 kg N/ha) for a rice crop during the growing season (6). Another diazotroph, the N_2-fixing actinomycete *Frankia,* forms nodules (actinorrhizae) in ca. 17 genera of nonlegume wood species with *Alnus* (alder) and the genus *Casuarina* being the most important for forestry and agriculture. Estimates of total N_2 fixed range between 50–250 kg/ha/year, depending on the plant species and region. However, in some cases, inoculation with *Frankia* is necessary for nodulation to occur. Actinorrhizal plant species have been successfully inoculated with *Frankia* on a large scale. For example, millions of actinorrhizal trees, especially *Alnus* spp., inoculated with *Frankia* were used in land reclamation programs established in Canada (7).

Phosphorus solubilization

Certain microorganisms are very effective in solubilizing phosphorus (P) from insoluble phosphate compounds such as hydroxyapatite through the action of organic acids. Numerous claims have been made about biofertilizers that can enhance plant growth by solubilizing P. A classical example is the bacterium *Bacillus megaterium,* which was formulated into an inoculant under the name of Phosphobacterin in the former Soviet Union. A similar biofertilizer

Table 2 Market price and potential price increments with yield increases of 5% and 25% in selected vegetable crops

		Price increments ($ per acre)	
Vegetable crop	Market price ($ per acre)	5% yield increase	25% yield increase
Carrot	4,520	226	1130
Cauliflower	4,179	209	1045
Celery	10,132	507	2533
Cucumber	3,296	165	824
Lettuce	5,882	294	1471
Tomato	9,966	498	2492

(Adapted from Ref. 3.)

based on P-solubilizing fungi is currently marketed in Canada as JumpStart® for use on wheat, canola, mustard, and N_2-fixing legumes.

Microbial siderophore uptake

Iron (Fe) is an important plant micronutrient. Plants assimilate iron by acidifying the rhizosphere and/or secreting phyto-siderophores with subsequent reassimilation of the iron–siderophore complex. However, plants also may benefit from the direct uptake of microbial siderophore–iron complexes. For example, some biofertilizers synthesize siderophores that can solubilize and sequester Fe from soil and provide it to plant cells, thus contributing to the nutrition and development of crops. In fact, studies demonstrate that ferric pseudobactin 358 may stimulate chlorophyll synthesis in carnation and barley (8).

Other nutrients

Studies with *Azospirillum* spp. and plant growth-promoting rhizobacteria (PGPR) have demonstrated the ability of these biofertilizers to promote enhancement of nutrient and water uptake into the plant. For example, inoculation of winter wheat seeds with pseudomonad PGPR stimulated the uptake of soil-Fe and fertilizer-^{15}N by winter wheat cultivated in two Canadian soils (9). Similarly, inoculation of canola seeds with a *Pseudomonas putida* increased phosphate uptake from nutrient solution (10). In these cases, the authors speculated that plant growth regulators produced by the biofertilizers in the plant's rhizosphere stimulated root development which, in turn, enhanced nutrient acquisition.

FUTURE RESEARCH DIRECTIONS

It is clear that commercial crops can benefit directly from biofertilizers. Certainly, with the development of molecular biology and genetic manipulation of biofertilizers to improve N_2-fixation, rhizosphere competence and ability to be used together with specific chemicals, will contribute to an integrated strategy to reduce the total amount of chemicals used in agriculture. Although biofertilizer products are currently available on the market, consistency is still the major factor that limits their use. Elucidation of mechanisms, development of stable formulations, effective delivery systems, and field demonstration of effective biofertilizers will definitely improve reliability and enhance their use as commercial biofertilizers.

REFERENCES

1. Killham, K. *Soil Ecology*; Cambridge University Press: Cambridge, UK, 1994; 242.
2. Tengerdy, R.P.; Szakács, G. Perspectives in agrobiotechnology. J. Biotechnol. **1998**, *66*, 91–99.
3. USDA. USDA Economics and Statistics System. National Agricultural Statistics Service; Cornell University. http://mann77.mannlib.cornell.edu/reports/nassr/fruit/pvg-bban/vegetables_annual_summary-01.16.98 (accessed June 1999).
4. Glick, B.R. The enhancement of plant growth by free-living bacteria. Can. J. Microbiol. **1995**, *41*, 109–117.
5. Pankhurst, C.E.; Lynch, J.M. The role of soil microbiology in sustainable intensive agriculture. Adv. Plant Pathol. **1995**, *11*, 230–247.
6. Zuberer, D.A. Biological Dinitrogen Fixation: Introduction and Nonsymbiotic. In *Principles and Applications of Soil Microbiology*; Sylvia, D.M., Fuhrmann, J.J., Hartel, P.G., Zuberer, D.A., Eds.; Prentice Hall: Upper Saddle River, NJ, 1998; 295–321.
7. Périnet, P.; Brouillette, J.G.; Fortin, J.A.; Lalonde, M. Large scale inoculation of actinorrhizal plants with *frankia*. Plant Soil **1985**, *87*, 175–183.
8. Duiff, B.J.; de Kogel, W.J.; Bakker, P.A.H.M.; Schipper, B. Significance of Pseudobactin 358 for the Iron Nutrition of Plants. In *Improving Plant Productivity with Rhizosphere Bacteria*; Ryder, M.H., Stephens, P.M., Bowen, G.D., Eds.; Third International Workshop on Plant-Growth Promoting Rhizobacteria, Adelaide, Australia, March 7–11, CSIRO Division of Soils: South Australia, 1994; 142–144.
9. De Freitas, J.R.; Germida, J.J. Growth promotion of winter wheat by fluorescent pseudomonads under growth chamber conditions. Soil Biol. Bioch. **1992**, *24*, 1127–1135.
10. Lifshitz, R.; Kloepper, J.W.; Kozlowiski, M.; Simonson, C.; Carlson, J.; Tipping, E.M.; Zaleska, I. Growth promotion of canola (rapeseed) seedlings by a strain of *pseudomonas putida* under gnotobiotic conditions. Can. J. Microbiol. **1987**, *33*, 390–395.

BIOLOGICAL CONTROL OF INSECTS AND MITES

Ann E. Hajek
Cornell University, Ithaca, New York, U.S.A.

B

INTRODUCTION

Biological control is defined as the use of natural enemies to suppress a pest population, making the pests and their associated damage less abundant. Natural enemies were first used to control insect pests when farmers in ancient China and Yemen moved colonies of predaceous ants to control pests of tree crops. Today, the natural enemies used to control insect and mite pests include a diversity of predators, parasitoids, and pathogens. Specific strategies have been developed for release of natural enemies or enhancement of their persistence and activity. Biological control has been used very successfully for permanent suppression of introduced pests. Among natural enemies applied for shorter-term control, in 1990 even the most widely used biological control agent, *Bacillus thuringiensis*, accounted for <1% of the insecticide market. However, biological control agents are widely used for control in environmentally sensitive areas or controlled environments and constitute important components of integrated pest management programs.

STRATEGIES FOR USING BIOLOGICAL CONTROL

Natural enemies can be used in a variety of very different ways. The first major uses of natural enemies for pest control were directed at control of introduced insect pests. Natural enemies from the land of origin of introduced pests were released in areas of pest introduction. This strategy, called classical biological control, now also includes introduction of exotic natural enemies to control native pests. In all cases, a high degree of host specificity is required in the natural enemies to be introduced. After the exotic natural enemy is established in the new location, its effectiveness is based on population increases in response to increasing densities of pest populations. Classical biological control can be dramatically effective, with 34% of insect natural enemies that are released becoming established and 17% completely controlling devastating pests. Classical biological control is known to be extremely cost effective with cost benefit estimates of up to 200:1, if a program is successful at establishing an effective natural enemy.

A second strategy, augmentation, involves releasing natural enemies for pest control, usually in instances where natural enemies can be effective but are not sustained in the environment at high enough densities to provide control. Inundative augmentation is used when only the natural enemies that are released in high numbers are expected to exert control. Under inoculative augmentation, control effects are more delayed and are predominantly exerted by the progeny of the released organisms. Natural enemies used for augmentation are often mass reared, so understanding requirements for mass production of high quality natural enemies that are healthy and active after shipping and release is critical for the use of this strategy.

The third major strategy, conservation, involves manipulations to enhance the persistence and activity of natural enemies already occurring in the environment. This strategy takes on a diversity of forms based on requirements of the individual natural enemies. To cause less mortality of natural enemies, use of synthetic chemical pesticides that kill natural enemies can be altered in different ways ranging from eliminating their use to selecting pesticides with less impact on natural enemies to timing pesticide applications to minimize the effect on natural enemies. Alternatively, natural enemy populations can be increased by maintaining or improving the environment to provide ideal conditions. For example, irrigating, strip-harvesting, intercropping, retaining vegetation adjacent to crops, and planting cover crops all have been shown to provide favorable habitats and food to maintain or increase populations of natural enemies. In a program to control the brown planthopper on rice in south and southeast Asia, the activity of a suite of native natural enemies, aided by host plant resistance and application of insecticides only when absolutely necessary, provided better control than pesticides alone.

TYPES OF NATURAL ENEMIES

Predators

Predators are generally larger than their prey and each usually consumes several prey individuals either for growth of immatures or for subsistence and reproduction of adults. The predatory life style is very common among insects and mites but predators with the most importance for biological control belong to four insect orders (Hemiptera, Coleoptera, Diptera, and Hymenoptera) and eight mite families (Fig. 1). Predators feed on a diversity of prey life stages, from eggs to adults, and display a range of host specificity, from feeding only on one prey species to generalized feeding on many prey species. One of the most famous examples of classical biological control is the introduction of the highly host specific Vedalia beetle that was imported from Australia and released against outbreak populations of the introduced cottony cushion scale threatening the southern California citrus industry. After the 1888–1889 releases of this predator, cottony cushion scale populations decreased dramatically and, by 1890, scale populations had been decimated. The immense success of this early program was instrumental in building interest in use of natural enemies for biological control. As a second example, in more recent years, phytoseiid mites attacking tetranychid spider mites have been developed for mass release in greenhouses or on some outdoor crops. Pesticide-resistant mite strains have also been developed for use against spider mites attacking tree crops.

Fig. 1 An adult of the multicolored Asian lady beetle (*Harmonia axyridis*), which was introduced from Asia for control of aphids and scales (length ca. 1 cm). Both larvae and adults are predatory. (Photo by J. Ogrodnick.)

Fig. 2 An adult of the parasitoid *Muscidifurax raptor* (length ca. 2 mm) parasitizing house flies. Females lay eggs in fly puparia, larvae grow while consuming the fly pupae and winged adults then emerge to mate and find more hosts. These flies can be purchased and released to augment naturally occurring populations. (Photo by S. Long.)

Parasitoids

Parasitoids develop at the expense of a single host and usually kill their hosts. Parasitoids have been used extensively for biological control because, in contrast to predators, the impressive degree of host specificity often characteristic of parasitoids leads to sensitive responses to changes in host density. Parasitoids used for biological control are predominantly in the Order Hymenoptera (Fig. 2) with less common use of Diptera. The immature parasitoid is usually a featureless larva associated with the host while winged adults disperse to mate and find new hosts. To enable their close association with hosts, parasitoids have adopted amazing and diverse life cycles. Different species of parasitoids attack different life stages of hosts (egg through adult) and can develop either externally on hosts or internally within hosts. One to many parasitoid individuals of one or more species can develop within a host. Parasitoids have been more widely used for classical biological control than either predators or pathogens. In recent years, the tiny wasp *Epidinicarsis lopezi* was released by land and air in 34 countries in Africa to control the introduced cassava mealybug. Due to the activity of this wasp, cassava mealybug is no longer considered a problem, saving African farmers hundreds of millions of dollars in reduced crop losses. Some parasitoids widely used for augmentative biological control are tiny species of *Trichogramma* attacking eggs of Lepidoptera and *Encarsia formosa*, a member of the Aphelinidae that attacks whiteflies. Both of these tiny wasps are mass produced in insectaries and shipped to users for release against pest populations threatening crops.

Pathogens

Microorganisms that are parasitic, referred to as pathogens, are masters at exploiting insect and mite hosts (Fig. 3). Pathogens important for biological control include a diversity of viruses, bacteria, fungi, protozoa, and nematodes. This range of types of pathogens exhibits a comparable medley of diverse interactions with their hosts. Of primary importance, viruses, bacteria, and most protozoa must be ingested by hosts in order to infect, while fungi and some protozoa can penetrate directly through the host cuticle. The nematodes of greatest importance to biological control, *Steinernema* and *Heterorhabditis*, can enter hosts through body openings although some possess the ability to penetrate directly through the cuticle. While some pathogens have mechanisms for active dispersal to find new hosts, host finding is generally not directed and these pathogens rely principally on their production of huge numbers of progeny in order to be assured of locating healthy hosts. Associations between pathogens and hosts range from facultative to obligate but pathogens important for biological control are all specialized for infecting only insects and mites. Pathogens have been used for classical biological control relatively infrequently although some programs have provided complete control. Much of the development of pathogens has been directed toward inundative augmentation of mass-produced microbes. The bacterium *Bacillus thuringiensis* is applied more than any other biological control agent. Strains of this bacterium predominantly kill Lepidoptera, Diptera, and Coleoptera through the activity of a toxin destroying the integrity of the gut. For many years, this bacterium was applied principally as a spray but recently several crop plants have been engineered to express genes encoding the toxin.

BIOLOGICAL CONTROL IN PRACTICE

There is great demand for use of biological control programs to eliminate insect and mite pests, especially in environmentally sensitive areas and areas where humans live. Use of natural enemies to control pests can be highly effective due to the diversity of types of natural enemies and approaches. However, because biological control involves management of living organisms, it can be somewhat unpredictable. Therefore, biological control programs generally are tailored to specific pest systems and to optimize control, often require knowledge of the biology and ecology of the insect or mite host. Biological control

Fig. 3 Gypsy moth (*Lymantria dispar*) larvae killed by the entomopathogenic fungus *Paecilomyces farinosus* (each larva ca. 3 cm in length). When infecting, this fungus penetrates externally through the larval cuticle, then increases within the host and, after host death, grows out through the integument to produce spores that will infect healthy hosts. (Photo by T. Ebaugh.)

Table 1 Characteristics of systems and conditions more commonly associated with successful biological control

Highly efficient natural enemy
Less mobile pest living in an exposed location
Perennial crop, natural habitat, or controlled environment, for example, a greenhouse
Crop or environment where some pest damage is tolerated
Controls for other pests do not interfere with the activity of natural enemies

has proven to be most effective under certain conditions (Table 1), although these generalities should not prevent investigations of use of biological control for alternative situations.

Introduced pests are not insignificant, comprising 39% of the 600 major arthropods pests in the United States. By 1990, more than 4300 introductions of exotic parasitoids and predators had been made to control insect pests, many of which were introduced. Due to its low cost and permanent effectiveness when successful, classical biological control continues as one of the first control strategies to be investigated after a new pest has been introduced.

Augmentation and conservation are now often employed as important parts of integrated pest management programs. Although since the late 1940s (the start of the DDT era), for most pests synthetic chemical pesticides have been the first control strategy considered, use of natural enemies for control is increasing, especially for specific applications and systems.

See also *Biological Control Successes and Failures*, pages 81–84; *History of Biological Controls*, pages 373–375; *Genetic Improvement of Biocontrol Agents*, pages 329–332; *Conservation of Biological Controls*, pages 138–140; *Fungal Control of Pests*, pages 320–324; *Nematode Control of Pests*, pages 526–529; *Protozoan Control of Pests*, pages 673–676; *Virus Control of Pests (Insects and Mites)*, pages 888–891.

BIBLIOGRAPHY

Barbosa, P., Ed.; *Conservation Biological Control*; Academic Press: San Diego, 1998; 396.

Bellows, T.S., Fisher, T.W., Caltagirone, L.E., Dahlsten, D.L., Gordh, G., Huffaker, C.B., Eds.; *Handbook of Biological Control: Principles and Applications of Biological Control*; Academic Press: San Diego, 1999; 1046.

Evans, H.F., Ed.; *Microbial Insecticides: Novelty or Necessity?*; British Crop Protection Council: Surrey, UK, 1997; 301.

Flint, M.L.; Dreistadt, S.H. *Natural Enemies Handbook: An Illustrated Guide to Biological Pest Control*; University of California Press: Berkeley, CA, 1998; 154.

Greathead, D., Waage, J., Eds.; *Insect Parasitoids*; Academic Press: London, 1986; 389.

Lacey, L.A.; Goettel, M.S. Current developments in microbial control of insect pests and prospects for the early 21st century. Entomophaga **1995**, *40* (1), 3–27.

Lynch, J.M., Hokkanen, H.M.T., Eds.; *Biological Control: Benefits and Risks,* Cambridge University Press: Cambridge, UK, 1995; 304.

New, T.R. *Insects As Predators*; New South Wales University Press: Kensington, Australia, 1991; 178.

Van Driesche, R.G.; Bellows, T.S., Jr. *Biological Control*; Chapman & Hall: New York, 1996; 539.

BIOLOGICAL CONTROL OF NEMATODES

B

Simon Gowen
The University of Reading, Earley Gate, Reading, United Kingdom

INTRODUCTION

Nematodes are a difficult group of pests to manage because generally they are inhabitants of soil and roots and are not easily influenced by soil treatments or cultural practices. The interest in the exploitation of natural enemies of nematodes has increased in recent years because of the demise of soil fumigants and nematicides through restrictions and withdrawals of registration of some products.

Many pathogens and predators of nematodes are known but few have the necessary characteristics of specificity, mobility, or speed of colonization to have a significant influence on a pest population. Attempts at their commercial exploitation as field treatments have not been successful largely because of their inconsistency. Understanding the subtleties associated with the deployment of biocontrol agents will require considerable research effort. Additionally, the recommended rates of application and the formulation on suitable carriers and nutrient sources poses a problem in practicability and in the interpretation of the biological processes involved.

Contemporary research has shown that natural control does exist and that in certain crop/nematode/pathogen situations nematode populations will decline as they are attacked by components of the soil microflora. Soils where this occurs are known as suppressive, but well-documented examples of naturally occurring suppressiveness to particular nematode pests are uncommon.

During the life of a crop the population densities of many of the serious nematode pests can increase by 1000-fold. Economic damage may result from initial population densities of one nematode per gram of soil. To be effective therefore a biocontrol agent must have an impact on the numbers of nematodes that would invade a host and not simply eliminate the surplus individuals that may never locate or invade a root. This being so, those pathogens and predators that are relatively unspecific (trapping, ingesting, or parasitizing all types of free-living nematodes in soil) may be considered less promising than those that parasitize specific pests.

Significant progress has been made in the recognition and deployment of such microorganisms parasitic on some of the species of sedentary nematodes such as the root-knot nematodes, *Meloidogyne* spp., and some of the cystnematodes, *Heterodera* spp., and *Globodera* spp.

Root-knot and cyst nematodes produce eggs either in clusters on roots or contained within or attached to the cuticle of the female nematode. Biocontrol agents that prevent these nematodes from reproducing may have more impact from an epidemiological point of view than those that kill the free-living individuals in the soil.

BIOCONTROL AGENTS SPECIFIC TO CERTAIN NEMATODE PESTS

Verticillium chlamydosporium is a facultative, soil-dwelling fungus that parasitizes eggs in egg masses exposed on the root surface. Under the right conditions, such fungi will have a significant effect on nematode populations. The efficacy of *V. chlamydosporium* is partly dependent on its root colonizing ability; this can vary according to the plant host. Skill is required in selecting crops that support and/or increase the root colonization by the fungus but are also less favored hosts of root-knot nematodes. *V. chlamydosporium* may be less effective when it is deployed with plants that are highly susceptible and large galls are produced in response to the nematode infection. In such cases, many egg masses may not be exposed on the root surface and so escape infection.

Paecilomyces lilacinus is another fungus commonly found infecting the eggs of sedentary nematodes such as the root-knot and the cyst nematodes, and, like *V. chlamydosporium* being relatively easy to produce on defined growth media, has good potential for commercial development.

Pasteuria penetrans, an obligate bacterial parasite of root-knot nematodes begins its life cycle on free-living juveniles in the soil. Spores attach to the juveniles as they move in search of host roots. Parasitic development begins after the nematode enters a root and continues in synchrony with that of its host. The nematode eventually is overcome by its parasite; it fails to produce eggs; and its body, filled with the spores of the bacterium, eventually ruptures releasing spores into the soil.

The efficiency of *P. penetrans* as a biocontrol agent of root-knot nematodes depends on the concentrations of spores in the soil, the chances of contact with the juvenile stage, and the specificity of the particular *P. penetrans* population. Commercial success will depend therefore on finding techniques for mass-producing the bacterium and on developing populations with a broad spectrum of pathogenicity.

Other *Pasteuria* species parasitic on some sedentary *(Heterodera)* and migratory *(Pratylenchus)* species have been described.

Biological control agents such as *V. chlamydosporium*, *P. lilacinus*, and *P. penetrans* could provide an adequate replacement for nematicides in some cropping systems but the lack of immediate effects, such as are provided by nematicide or fumigant treatments, is a disadvantage. Protection is normally needed in the early stages of plant growth such as in nursery beds. In this situation, integration with other practices such as nematicides, rotation, solarization, and mulches is necessary.

There are several reports of the successful deployment of these biocontrol agents. Small field plots treated once with *P. penetrans* spores (produced by an in vivo system) caused a decline in numbers of root-knot nematodes and increases in yield over a series of crop cycles using root-knot nematode susceptible crops. In other locations, where treatments with *P. penetrans* were combined with *V. chlamydosporium*, organic manures and grass mulches showed similar declines in nematode populations. These two organisms acted against root-knot nematodes in a complementary fashion. As part of this strategy, root systems containing spore-filled cadavers were deliberately left to disintegrate in the soil after each crop. No field treatments were effective after only one crop indicating that some crop loss must be expected during the development of suppressiveness. *P. penetrans* was also effective when used in combination with a nematicide in permanent beds within a plastic polytunnel. Better control of root-knot nematodes was achieved if the biocontrol agent was combined with other control strategies. With such treatments, beneficial effects may develop over one crop cycle.

The chlamydospores of *V. chlamydosporium* do not have the persistence of the spores of *P. penetrans*, which can remain viable for many years.

NONSPECIFIC BIOCONTROL AGENTS

There is a long history of interest in the fungi that trap nematodes in soil such as species of *Arthrobotrys*. These are commonly found in all soils but despite much research effort the problems of the unreliability of soil applications have not been solved and none have become established as successful commercial products.

There are several rhizosphere colonists that have potential for alleviating nematode damage. The precise mechanisms are not clear. Some produce toxins but others may affect root exudation and thus indirectly the attractiveness of roots to nematodes. Experiments have shown that strains of *Pseudomonas fluorescens* can reduce root invasion by different plant parasites but as with the trapping fungi, poor consistency hinders successful development of these microorganisms as commercial products.

FUTURE PROSPECTS

Recently, the nematicidal (and insecticidal) effects of the toxins produced by the bacteria associated with entomopathogenic nematodes (*Photorhabdus* spp., *Xenorhabdus* spp., and *Pseudomonas oryzihabitans*) have been demonstrated.

Success in the commercial development of biocontrol agents does appear promising with those microorganisms that can be formulated as a standard product with proven reliability; others may have a future as single treatment introductions in the more intensively managed protected cropping systems but commercialization may be difficult.

Research is still needed to develop reliable methods of production, formulation, and application. The challenge is to provide a sufficient duration of protection. Such treatments will need to be part of a package of control measures.

BIBLIOGRAPHY

Aalten, P.M.; Vitour, D.; Blanvillain, D.; Gowen, S.R.; Sutra, L. Effect of rhizosphere fluorescent *Pseudomonas* strains on plant parasitic nematodes *Radopholus similis* and *meloidogyne* spp. Letters Appl. Microbiol. **1998**, *27*, 357–361.

Bourne, M.; Kerry, B.R.; De Leij, F.A.A.M. The importance of the host plant on the interaction between root-knot nematodes

(*Meloidogyne spp.*) and the nematophagous fungus, *Verticillium chlamydosporium* Goddard. Biocon. SciTech. **1996**, *6*, 539–548.

Crump, D.J. A method for assessing the natural control of cyst nematode populations. Nematologica **1987**, *33*, 232–243.

Gowen, S.R.; Bala, G.; Madulu, J.; Mwageni, W.; Trivino, C.T. In *Field Evaluation of Pasteuria penetrans for the Management of Root-Knot Nematodes*, The 1998 Brighton Conference on Pests and Diseases, 1998; 3, 755–760.

Kerry, B.R.; Jaffee, B.A. Fungi as Biocontrol Agents for Plant Parasitic Nematodes. In *The Mycota IV Environmental and Microbial Relationships*; Wicklow, D.T., Soderstrom, B.E., Eds.; Springer-Verlag: Berlin, Heidelberg, 1997; 204–218.

Samaliev, H.Y.; Andreoglou, F.I.; Elawad, S.A.; Hague, N.G.M.; Gowen, S.R. The Nematicidal Effects of the Bacteria *Pseudomonas oryzihabitans* and *Xenorhabdus Nematophilus* on the Root-Knot Nematode, *Meloidogyne javanica*. Nematology, *in press*.

Stirling, G.R. Biological Control of Plant Parasitic Nematodes. CAB International: Wallingford, U.K., 1991; 282.

Tzortzakakis, E.A.; Channer, A.G. de R.; Gowen, S.R.; Ahmed, R. Studies on the potential use of *Pasteuria penetrans* as a biocontrol agent of root-knot nematodes (*Meloidogyne spp.*). Plant Pathol. **1997**, *46*, 44–55.

Tzortzakakis, E.A.; Gowen, S.R. Evaluation of *Pasteuria penetrans* alone and in combination with oxamyl, plant resistance and solarization for control of *Meloidogyne* spp. on vegetables grown in greenhouses in Crete. Crop Prot. **1994**, *13*, 455–462.

Weibelzahl-Fulton, E.; Dickson, D.W.; Whitty, E.B. Suppression of *Meloidogyne incognita* and *M. javanica* by *Pasteuria penetrans* in field soil. J. Nematol. **1996**, *28*, 43–49.

BIOLOGICAL CONTROL OF PLANT PATHOGENS (FUNGI)

Timothy Paulitz
United States Department of Agriculture, Washington State University, Pullman, Washington, U.S.A.

INTRODUCTION

In the classical definition of biological control, certain fungi, termed biocontrol agents (BCAs), can reduce the amount of inoculum or disease-producing activity of a plant pathogen, usually another fungus (1). The net result is a reduction of plant disease and crop loss. This section will cover mechanisms of how these fungi antagonize the pathogen, what part of the pathogen life cycle can be targeted, how the BCAs can be applied, and examples of commercially available products. Within the past 20 years, there has been a tremendous increase in interest and research on the subject, spurred by a search for more environmentally benign methods of disease control. But fungal BCAs have limitations that have restricted the number of products that are currently on the market. Table 1 shows some products used against soilborne pathogens on the market as of January 1999.

MECHANISMS OF BIOLOGICAL CONTROL BY FUNGI

The strategy behind managing pathogens is to target or interrupt part of the pathogen life cycle (1, 2). Like any microbe, pathogens start from inoculum in the environment, which can be spores, mycelia, or other dormant survival structures. These germinate on the plant surface, penetrate and infect the plant, and reproduce and sporulate on the plant to produce new inoculum. Many pathogens can also grow saprophytically on dead organic matter and plant debris. In biological control, the pathogen can be targeted in three ways (1, 2). First, the inoculum of the pathogen can be reduced or destroyed. This is most effective for soilborne pathogens, where the inoculum is dormant in the soil and the monocyclic disease is determined by the initial inoculum present in the field. BCAs can also interfere with inoculum formation by pathogens growing saprophytically on organic matter and plant debris. However, this strategy is not very effective for foliar polycyclic diseases, where inoculum comes from outside the field and the initial inoculum has little effect on the final outcome of the disease. Another strategy is one of protection, where a population of the BCA is established on the infection site of the plant before the pathogen attacks, thus preventing the pathogen's entry. These infection sites can be on seeds, bulbs, roots, leaves, fruit, flowers, or wounds. Finally, nonpathogenic or avirulent fungi can stimulate the plant to a higher level of resistance to a later-attacking pathogen, a concept termed induced resistance.

The most direct way a fungus can attack a fungal pathogen is by mycoparasitism, where the BCA uses the pathogen as a source of food (3). The hyphae of the mycoparasite contact, penetrate, and colonize the hyphae, spores, or survival structures of the host fungus. Many of these mycoparasites produce enzymes that degrade the cell walls of the fungal host, including β-1-3 glucanase and chitinase. Most mycoparasites are necrotrophic and eventually kill their fungal host. Much of this research has focused on reducing the inoculum of soilborne pathogens. Classic examples include *Trichoderma* and *Gliocladium* spp. parasitizing *Rhizoctonia solani* and *Pythium* spp., which cause seed, seedling, and root rots (4–6). *Pythium* spp. such as *Pythium oligandrum* and *P. nunn* parasitize pathogenic species of *Pythium. Coniothyrium minitans* and *Sporodesmium sclerotivorum* parasitize sclerotia of *Sclerotinia* spp., such as the white mold pathogen *S. sclerotiorum*, which attacks hundreds of plant species (57). *Ampelomyces quisqualis* parasitizes cleistothecia of powdery mildews. Major limitations to this strategy are that mycoparasites are slow acting and large amounts of mycoparasite inoculum must be added to the soil to ensure it will encounter the propagules of the pathogen. However, a promising strategy demonstrated with *S. sclerotivorum* is to render a soil suppressive to the pathogen by an inoculative release at a lower inoculum density, and allowing the mycoparasite to build up over successive seasons, using

Table 1 Some commercial biocontrol products for use against soilborne crop diseases

Biocontrol fungus	Trade name	Target pathogen/disease	Crop	Manufacturer
Ampelomyces quisqualis M-10	AQ10 Biofungicide	Powdery mildew	Cucurbits, grapes, ornamentals, strawberries, tomatoes	Ecogen Inc., Langhorne, Pennsylvania
Candida oleophila I-182	Aspire	*Botrytis, Penicillium*	Citrus, pome fruit	Ecogen Inc., Langhorne, Pennsylvania
Fusarium oxysporum (nonpathogenic)	Biofox C	*Fusarium oxysporum*	Basil, carnation, cyclamen, tomato	S.I.A.P.A., Galliera, Bologna, Italy
Trichoderma harzianum and *T. polysporum*	Binab T	Wilt and root rot pathogens, wood decay pathogens	Fruit, flowers, ornamentals, turf, vegetables	Bio-innovation, Algaras, Sweden
Coniothyrium minitans	Contans	*Sclerotinia sclerotiorum* and *S. minor*	Canola, sunflower, peanut, soybean, lettuce, bean, tomato	Prophyta Biologischer Pflanzenschutz, Malchow/Poel, Germany
Fusarium oxysporum (nonpathogenic)	Fusaclean	*Fusarium oxysporum*	Basil, carnation, cyclamen, gerbera, tomato	Natural Plant Protection, Nogueres, France
Pythium oligandrum	Polygandron	*Pythium ultimum*	Sugar beet	Plant Protection Institute, Bratislavsk, Slovak Republic
Trichoderma harzianum and *T. viride*	Promote	*Pythium, Rhizoctonia, Fusarium*	Greenhouse, nursery transplants, seedlings	JH Biotech, Ventura, California
Trichoderma harzianum	RootShield, Bio-Trek T-22G, Planter Box	*Pythium, Rhizoctonia, Fusarium, Sclerotinia homeocarpa*	Trees, shrubs, transplants, ornamentals, cabbage, tomato, cucumber, bean, corn, cotton, potato, soybean, turf	Bioworks, Geneva, New York
Phlebia gigantea	Rotstop	*Heterobasidium annosum*	Trees	Kemira Agro Oy, Helsinki, Finland
Gliocladium virens GL-21	SoilGard (formerly GlioGard)	Damping-off and root pathogens, *Pythium, Rhizoctonia*	Ornamentals and food crops grown in greenhouses, nurseries, homes, interiorscapes	Thermo Triology, Columbia, Maryland
Trichoderma harzianum	Trichodex	*Botrytis cinerea, Colletotrichum, Monilinia laxa, Plasmopara viticola, Rhizopus stolonifer, Sclerotinia sclerotiorum*	Cucumber, grape, nectarine, soybean, strawberry, sunflower, tomato	Makhteshim Chemical Works, Beer Sheva, Israel
Trichoderma harzianum and *T. viride*	Trichopel, Trichoject	*Armillaria, Botryosphaeria, Fusarium, Nectria, Phytophthora, Pythium, Rhizoctonia*		Agrimm Technologies, Christchurch, New Zealand

This information was provided by the U.S. Department of Agriculture, Agriculture Research Service, Beltsville, Maryland, and was compiled by D. Fravel (http://www.barc.usda.gov/psi/bpdl/bodlpood/bioprod.htm).

the pathogen as a food source. This is similar to the classic predator–prey relationship found in insect biocontrol.

Some fungi can produce antibiotic compounds that are toxic to other microbes, including plant pathogens. *Trichoderma* spp. produce volatile and nonvolatile antifungal compounds, including peptiabols, pyrones, and terpenoid antibiotics (4). *Gliocladium virens* produces glioviren and gliotoxin that inhibit *R. solani* and *Pythium ultimum*. This mechanism is most effective when the BCA can grow to high populations and has an energy source to produce the antibiotic. An example would be *Trichoderma* or *Gliocladium* spp. applied to seeds or where a food base is added to the inoculum (4–5).

Plant pathogens require carbon, nitrogen, iron, and other nutrients to grow. Many spores have an exogenous requirement for these nutrients, supplied by the plant rhizosphere or phyllosphere, in order to germinate. BCAs can compete with the pathogen for these limiting nutrients. For example, nonpathogenic species of *Fusarium oxysporum* can compete with pathogenic forma speciales for these limiting nutrients, resulting in control of wilt diseases. Competition by yeasts or hyphal fungi may protect flowers and foliage against nectotrophic pathogens such as *Botrytis cinerea* by colonizing the senescent tissue or nutrient-rich flower petals (8–9). This mechanism, although difficult to prove experimentally, is probably one of the primary ways BCAs can protect a plant surface through preemptive exclusion of the pathogen.

Fungal biocontrol agents can also affect the pathogen by acting indirectly on the plant to make it more resistant to pathogen attack. Nonpathogenic microbes can induce a systemic resistance in plants (79). When the plant recognizes the inducing BCA, a signal is transduced systemically to the entire plant, bringing the defenses to a ''high state of alert,'' so that a subsequent challenge by a pathogen is reduced. Nonpathogenic isolates of *F. oxysporum* induce a defense reaction against pathogenic isolates of *F. oxysporum*. This mechanism has several advantages. Once induced, the resistance is systemic, the entire plant becomes more resistant, and high populations of the BCA do not need to be maintained. It can also protect parts of the plant that cannot be protected directly by the BCA, including new growth of shoots and roots. However, more research is needed to investigate the applicability of this technology under greenhouse and field conditions.

APPLICATION OF FUNGAL BIOCONTROL AGENTS

How are fungal BCAs applied? Most are applied in an inundative strategy in large amounts to build up the population of the BCA high enough to overwhelm and have an effect on the pathogen (1, 5). Most are also targeted toward soilborne pathogens. However, one limitation of this strategy is the large amount of inoculum that must be applied and the high cost of production of spores, conidia, biomass, or chlamydospores (5, 9). Another problem is the erratic performance of many biocontrol agents under field conditions, due to unfavorable environmental conditions for the BCA, and the problem of establishing the BCA in a niche already occupied by competing microflora. Therefore most of the commercially available products have targeted applications that avoid these problems. For example, the greenhouse and nursery markets are prime targets because of the controlled environmental conditions and the high economic value of the crops. Another method is to use a protective strategy and apply the BCA directly to the infection court when it is small. For example, high populations of *Trichoderma* or *Gliocladium* conidia can be coated onto seeds to protect against damping-off pathogens such as *R. solani* and *P. ultimum*. Transplant cuttings and bulbs can be treated with liquid suspensions of products before planting in the greenhouse or field. Products such as formulations of *G. virens* can be mixed directly into the soil or soilless mixes in the greenhouse. Granules of *Trichoderma* spp. can be added to seed furrows, mixed with seeds in a planter box, or added to a commercial seed slurry (4).

Since many pathogens gain access to plants through wounds, biocontrol agents can also be applied to transplant or pruning wounds. A classic example, and one of the first commercially used fungal biocontrol agents, is the application of *Phlebia* (= *Peniophora*) *gigantea* to cut pine stumps to prevent the stumps from being colonized by the pathogen *Heterobasidium annosum*. The pathogen can spread from these stumps to the entire plantation via the root system.

Postharvest pathogens are weak pathogens that require wounds on fruit to gain access. Yeast-like organisms such as *Candida* spp. can be applied to fruit during processing to exclude rot pathogens such as *Penicillium* spp. from colonizing wounds (8). However, these applications require more stringent testing for registration, since they are applied directly to a food product. Competition is preferable to antibiosis for this application, since antifungal compounds would also have to be tested for animal and human toxicity.

Roots are a difficult infection court to protect, since the susceptible tips are constantly growing and moving through space encountering new inoculum. One strategy is to treat the entire rooting medium in the greenhouse or nursery. Another approach is to use a fungus that can colonize the root system from a seed or furrow treatment and protect the expanding root surface. This characteristic, called rhizosphere competence, has been demonstrated in some strains of *Trichoderma* spp.

Foliar applications of fungi are the least common, although this is the most common method of fungicide application. One example is *Pseudozyma* (= *Sporothrix*) *flocculosa*, a yeast-like fungus which is being developed for control of powdery mildews on greenhouse roses and cucumbers in Canada. *Trichoderma* and *Gliocladium* spp. can be applied to foliage and flowers and can prevent infection by necrotrophic fungi such as *B. cinerea* and *S. sclerotiorum*.

FUTURE OF BIOLOGICAL CONTROL BY FUNGI

In conclusion, the full potential of controlling plant diseases with fungi still has not been realized. Only a small number of products are on the market, but this is a vast improvement compared to only five years ago. There are still many economic constraints in terms of the cost of development and registration of products and the low cost production of organisms in liquid fermentation or solid on substrates (5, 9). Like chemicals, the risks of fungal BCAs need to be addressed, including the displacement of nontarget microbes, allergenicity to humans and other animals, and toxigenicity and pathogenicity to nontarget organisms (10). However, there are no existing chemical controls for many diseases, because of deregistration of pesticides, pathogen resistance to pesticides, and environmental concerns. These diseases may be the niches for fungal biocontrol agents. It is unlikely that biological control will succeed alone, but it needs to be integrated with other disease management strategies, including cultural control and genetic disease resistance.

B

REFERENCES

1. Cook, R.J.; Baker, K.F. *The Nature and Practice of Biological Control of Plant Pathogens*; American Phytopathological Society Press: St. Paul, MN, 1983; 539.
2. *Principles and Practice of Managing Soilborne Plant Pathogens*; Hall, R., Ed.; American Phytopathological Society Press: St. Paul, MN, 1996; 442.
3. *Fungi in Biological Control Systems*; Burge, M.N., Ed.; Manchester University Press: Manchester, 1988; 269.
4. *Trichoderma and Gliocladium*; Harman, G.E., Kubicek, C. R., Eds.; Taylor and Francis: London, 1998; 1 and 2, 278–393.
5. *Pest Management: Biologically Based Technologies*; Vaughn, J.L., Lumsden, R.D., Eds.; American Chemical Society: Washington, DC, 1993; 435.
6. *Biological Control of Plant Diseases. Progress and Challenges for the Future*; Papavizas, G.C., Cook, R.J., Tjamos, E.C., Eds.; Plenum Press: New York, 1992; 462.
7. Whipps, J.M. Biological control of soil-borne plant pathogens. Advances in Botanical Res. **1997**, *26*, 1–134.
8. *Plant-Microbe Interactions and Biological Control*; Kuykendallm, L.D., Boland, G.L., Eds.; Marcel Dekker, Inc.: New York, 1998; 442.
9. *Integrated Pest and Disease Management in Greenhouse Crops*; Lodovica Gullino, M., van Lenteren, J.C., Elad, Y., Albajes, R., Eds.; Kluwer Academic Publishers: Dordrecht, The Netherlands, 1999; 545.
10. *Biological Control: Benefits and Risks*; Hokkanen, H.M.T., Lynch, J.M., Eds.; Cambridge University Press: Cambridge, 1995; 304.

BIOLOGICAL CONTROL OF PLANT PATHOGENS (VIRUSES)

Hei-Ti Hsu
Beltsville Agricultural Research Center, Beltsville, Maryland, U.S.A.

INTRODUCTION

Prevention of virus and viroid infections in plants is based on biological means rather than chemical measures. In principle, there are no chemicals available for controlling plant diseases caused by viruses and viroids. The most feasible approaches for combating viruses and viroids are the elimination of source inoculum, prevention of secondary spread, cross protection, and use of crops bearing resistance traits.

ECONOMIC LOSS

Damage to crop plants due to virus and viroid infections is difficult to assess. The actual figures for global crop loss are not available. Plant disease losses are estimated at $60 billion annually. Losses due to virus and viroids have been considered second to those caused by fungi. Unlike diseases caused by fungi, bacteria, and nematodes, where control measures using chemical, biological, and integrated pest management approaches have been effective, diseases caused by viruses or viroids are far more difficult to manage.

Economic crop loss resulting from virus and viroid disease is due to the reduced growth and vigor of infected plants which, in turn, causes a reduction in yield. In some instances, a virus infection may kill a plant. Apart from yield reduction, the quality and market value of commercial end products may be affected. There are also costs of attempting to maintain crop health such as vector control, production of pathogen-free propagation materials, and quarantine and eradication programs. In addition, resources are being diverted to research, extension, and education as well as toward breeding for resistance to virus or viroid infection.

WORLD IMPACT

No single country is exempt from crop losses. Production of food, fiber, and horticultural crops are seriously affected worldwide by virus or viroid infection of plants (1). This is even more so in developing countries that depend on one or a few major crops; for example, *Cassava mosaic virus* in cassava plants in Kenya, *Citrus tristeza virus* in citrus trees in Africa and South America, and *Cacao swollen shoot virus* in cacao trees in Ghana. Recently, *Papaya ringspot virus* (PRV) infection has affected every region where papaya plants are grown. The virus induces a lethal disease in papaya. The widespread aphid-transmitted PRV has changed the way papayas are grown in many parts of the world. Normally, papayas are produced annually for a number of years over the life of the papaya plant. For proper management of the disease due to PRV infection, papaya has now become an annual crop in which healthy seedlings are planted each year. Even so, productivity is still below the average yield obtained before PRV became a problem. Viroids infect a limited number of crops when compared with viruses. However, they can cause severe problems in specific crops, for instance, cadang-cadang disease of coconuts, potato spindle tuber disease, and chrysanthemum stunt disease.

CONTROL MEASURES

No direct chemical control means are available to combat virus infections in plants. Control of viral diseases is achieved primarily by sanitary practices that involve reducing sources of inoculum from outside, preventing spreading within the crop, and limiting the population of insects, mites, nematodes, and fungi that may serve as vectors for many plant viruses (1). Virus disease testing programs are now common in many parts of the world where the economic importance of growing virus-free plants is recognized. Although seeds and seedlings certified as virus-free are more expensive than those that have not been tested for certain viruses, testing provides assurance of virus-free production materials. Early detection of virus in a field and removal of the infected plants minimizes spread of the virus.

Plants may be protected from development of severe disease symptoms by first introducing a mild strain of virus into a healthy plant. A plant systemically infected with a mild strain of virus is protected from infection by a severe strain of the same virus. This phenomenon in called "cross protection" and has been observed for many plant viruses (2). It is also observed to occur between viroids or plant virus satellites. In practice, cross protection is of great interest since it has been utilized to protect plants against severe virus strains (*Citrus tristeza virus, Papaya ringspot virus, Zucchini yellow mosaic virus, Tomato mosaic virus*, etc.) in the field.

Another approach toward controlling plant virus diseases is to develop resistant or tolerant plants (3). Historically, long-term manipulation of crop plants through breeding has produced many valuable commercial varieties resistant to plant viruses. Breeding plants resistant to vectors may also offer control of the virus they transmit. Conventional breeding of crossing and back crossing commercial varieties with plants bearing virus resistance traits takes years to develop. In order for a new variety to be commercially acceptable, undesirable traits from the resistant parent breeding line must be selected out. The process is labor intensive and time consuming. Advances in science have allowed new technology to precisely manipulate resistance genes at the molecular level (4). Biotechnology represents the fastest growing area of biological research. The application of biotechnology in breeding for resistance to virus infection is a major area of research. Successful control of viral disease through resistance breeding will undoubtedly reduce the use of synthetic pesticides for vector control (5).

Table 1 Genes that contribute or may contribute toward control of virus diseases in plants

Virus-derived gene sequences
Coat proteins
Replicases
Movement proteins
Polyprotein proteases
Sense RNAs
Antisense RNAs
Plant host-derived transgenes
Pathogen-related proteins
Anti-viral proteins
Proteinase inhibitors
Natural resistance genes
Lectins
Other transgenes and sequences
Satellite RNAs
Virus-specific antibodies
Interferon-induced mammalian oligoadenylate synthetase
Insect toxins
Anti-viral ribozymes (catalytic RNA)

(From Ref. 1.)

Introducing virus resistance and vector resistance into a cultivar by gene transfer technology (genetic engineering) has been successful in combating plant viruses (6). The technology has several major advantages over conventional cross breeding. It is a relatively fast procedure. Desirable genes can be introduced without disturbing the balanced genome of target plants. Furthermore, there is no restriction on the source of the transgenes allowing the use of genes from other plant species or even from outside the plant kingdom (Table 1) (1, 7).

Several approaches for producing transgenic virus-resistant plants have been explored. Among these, plants expressing virus coat protein genes, parts of other viral genes, or virus satellite ribonucleic acids (RNAs) have been shown to offer the best control (2, 8, 9). Plants expressing antisense viral RNAs, ribozymes, pathogen-related proteins, or virus-specific antibody genes may also confer resistance to virus infection. Control of virus vectors by introducing insect toxins such as trypsin inhibitor, lectin, and *Bacillus thuringiensis* (Bt) toxin genes into plants would undoubtedly contribute toward achieving the goal of controlling plant virus diseases.

PROSPECTS

Use of resistant cultivars is considered the best approach to combat virus infection in plants. Biotechnology, no doubt, will play a significant role in the economic growth of many countries. Molecular breeding, however, will not replace but complement the efforts of conventional cross breeding. Much attention has been given to engineering resistance to plant viruses. Recently, genetic engineering of crop plants has been closely scrutinized and criticized due to increasing public concerns regarding human health and environmental impact. Careful assessment of the benefits and potential risks involving the release of genetically modified plants into the environment and their consumption is necessary before these crops become widely accepted by the public (10, 11).

REFERENCES

1. Khetarpal, R.K., Koganezawa, H., Hadidi, A., Eds. *Plant Virus Disease Control*; APS Press: St. Paul, MN, 1998, 1–684.
2. Beachy, R.N. Coat-protein-mediated resistance to tobacco mosaic virus: discovery mechanisms and exploitation. Phil. Trans. R. Soc. Lond. B. **1999**, *354*, 659–664.
3. Salomon, R. The evolutionary advantage of breeding for

tolerance over resistance against viral plant disease. Israel J. Plant Sci. **1999**, *47*, 135–139.
4. Kawchuk, L.M.; Prufer, D. Molecular strategies for engineering resistance to potato viruses. Can. J. Plant Pathol. **1999**, *21*, 231–247.
5. Barker, I.; Henry, C.M.; Thomas, M.R.; Stratford, R. Potential Benefits of the Transgenic Control of Plant Viruses in the United Kingdom. In *Plant Virology Protocols: From Virus Isolation to Transgenic Resistance*; Foster, G.D., Taylor, S.C., Eds.; Humana Press, Inc.: Totowa, NJ, 1998; 81, 557–566.
6. Dempsey, D.A.; Silva, H.; Klessig, D.F. Engineering disease and pest resistance in plants. Trends Microbiol. **1998**, *6*, 54–61.
7. Gutierrez-Campos, R.; Torres-Acosta, J.A.; Saucedo-Arias, L.J.; Gomez-Lim, M.A. The use of cysteine proteinase inhibitors to engineer resistance against potyviruses in transgenic tobacco plants. Nat. Biotechnol. **1999**, *17*, 1223–1226.
8. Maiti, I.B.; Von Lanken, C.; Hong, Y.; Dey, N.; Hunt, A.G. Expression of multiple virus-derived resistance determinants in transgenic plants does not lead to additive resistance properties. J. Plant Biochem. Biotech. **1999**, *8*, 67–73.
9. Prins, M.; Goldbach, R. RNA-mediated virus resistance in transgenic plants. Arch. Virol. **1996**, *141*, 2259–2276.
10. Hammond, J.; Lecoq, H.; Raccah, B. Epidemiological risks from mixed virus infections and transgenic plants expressing viral genes. Adv. Virus Res. **1999**, *54*, 189–314.
11. Kaniewski, W.K.; Thomas, P.E. Field testing for virus resistance and agronomic performance in transgenic plants. Mol. Biotechnol. **1999**, *12*, 101–115.

BIOLOGICAL CONTROL OF VERTEBRATES

Peter Kerr
Pest Animal Control Cooperative Research Center,
CSIRO Sustainable Ecosystems, Canberra, Australian Capital Territory, Australia

INTRODUCTION

Biological control can be defined as the use of one species to reduce the population of another species, in this case vertebrate pests. This definition could include the use of vertebrate predators to control other vertebrates, e.g., cats to control rodents. However, biological control is more commonly considered as the use of microparasites: viruses, protozoa or bacteria; or macroparasites: helminths and arthropods. Conceptually, biological control of vertebrate pests is a very attractive, cost-effective alternative to other methods of control such as poisoning, shooting, trapping, and exclusion. In support of this, theoretical models and practical experience show that pathogens and parasites can regulate host populations either by increasing mortality rates or by decreasing the reproduction rate. Unfortunately, most potential biological control agents identified in laboratory or small-scale field trials have been ineffective. This can be for several reasons depending on the epidemiology and distribution of the parasite and the population biology of the pest species. The organism may already be present and adapted to the pest species, it may become established at too low a prevalence to reduce the rate of increase of the host population, or it may fail to persist and transmit. Even following successful establishment of a biological control agent in a pest population, subsequent adaptation of host and parasite or specific aspects of the epidemiology of the pathogen may significantly reduce the impact of biological control over quite short time periods. The introduction of the viral disease myxomatosis to control European rabbits (*Oryctolagus cuniculus*), initially in Australia and later in Europe, is the best example of the success and limitations of biological control of vertebrates. In fact, the European rabbit is the only vertebrate pest for which biological control has had a major impact.

MYXOMATOSIS AND EUROPEAN RABBITS

The wild European rabbit was introduced into Australia in 1859 and over the next 60 years spread across the continent, establishing itself as the major vertebrate pest species and a highly significant cause of agricultural and ecological depredation. In 1995, it was estimated that effective rabbit control could be worth 3% of agricultural production in Australia. The ecological costs of rabbits to Australia have never been quantified. Myxoma virus (1) is a poxvirus of the South American forest rabbit (*Sylvilagus brasiliensis*). In this species, it causes a benign cutaneous fibroma at the site of inoculation. Mosquitoes or other biting arthropods have their mouthparts contaminated with virus when probing through this fibroma and then transmit the virus at subsequent feedings. In European rabbits, myxoma virus causes the lethal disease myxomatosis and its potential as a biological control agent for rabbits was recognized as early as 1919. Extensive testing demonstrated that myxoma virus could only infect lagomorphs. However, successful use of the virus in European rabbits initially proved elusive. In trials in Australia and Europe the virus spread locally but then died out. In retrospect, the reasons for this are obvious. The virus is rapidly lethal, there were no reservoir hosts and rabbits that survive infection are neither persistent carriers of the virus nor susceptible to reinfection. Thus the virus becomes extinct when there are too few susceptible animals to support the epidemic and no insect vectors to transmit virus into high density populations of susceptible rabbits.

In the summer of 1950–1951 a strain of myxoma virus succeeded in establishing in the Australian rabbit population. These rabbits had never been exposed to myxoma virus so all were susceptible and mosquito vectors were plentiful due to unusually wet seasons. During the next two years the virus spread to all the rabbit-infested areas of Australia. An estimated 99.8% of infected rabbits were killed and the rabbit population decreased by 95% across most of southern Australia. In the wool industry alone, it was estimated that myxomatosis allowed the production of 32,000,000 kg of additional wool in 1953. However, this high lethality was not maintained. Moderately attenuated strains of virus that killed 70–90% of infected rabbits were more efficiently transmitted than either highly virulent or highly attenuated strains and became predominate in the field. These attenuated strains of virus allowed survival of rabbits with a degree of innate resistance, which led to a rapid selection for resistance to myxomatosis.

Myxomatosis epidemics in the Australian rabbit population now provide an additional level of mortality above that due to other diseases and predation. It is difficult to quantify the current impact of myxomatosis. However, large-scale field immunization trials in Australia, and trials in Britain removing the flea vector of myxoma virus, indicated that the virus still has a substantial effect on rabbit populations. Some releases of virulent myxoma virus are still made by property holders and pest control agencies in Australia. These releases may have some local effect on reducing rabbit numbers, with the virus acting as a biocide, but are epidemiologically insignificant. Of more impact has been the deliberate establishment of two exotic arthropod vectors of myxomatosis in the Australian wild rabbit population. The European rabbit flea *Spilopsyllus cuniculi* acts as a vector in the temperate zones of Australia while the Spanish rabbit flea *Xenopsylla cunicularis* provides a vector in the arid and semi arid regions where mosquitoes may be absent for long periods. These fleas have significantly altered the epidemiology of myxomatosis by removing the reliance on mosquitoes for transmission of myxoma virus.

RABBIT HEMORRHAGIC DISEASE VIRUS

Rabbit hemorrhagic disease virus (RHDV), a calicivirus that appears to be specific for European rabbits, emerged in China in 1984. It has subsequently spread to most areas of the world where European rabbits are farmed or occur in the wild. The virus causes rapid death in rabbits more than 4–6 weeks old and is spread by contact and by insect vectors. RHDV was investigated as a potential biological control for rabbits in Australia. In October 1995, RHDV escaped from a quarantine site off the mainland of Australia and spread rapidly (more than 300 km in 2 weeks) through high density rabbit populations in South Australia. During the next 12 months it spread throughout the rabbit populations of southern Australia (2). RHDV has had a major impact on the rabbit population of Australia particularly in arid and semi-arid areas. In some areas population densities were reduced by 95% allowing regeneration of native trees and shrubs. So far, emergence of attenuated strains of RHDV or development of resistance in rabbits has not been documented in Australia and RHDV has not significantly altered the epidemiology of myxomatosis. However, some resistance to RHDV may be present in wild rabbits in Europe. There are also naturally attenuated strains of the virus present in Europe. These probably predate the RHDV epidemics and at least in Britain appear to have limited the spread of the virulent virus by immunizing the rabbit population. Recent serological evidence indicates that a related virus may be present in Australia and New Zealand. However, the epidemiological consequences of this are not yet known. The myxoma virus model suggests that some form of resistance to and attenuation of RHDV will eventually emerge in Australia but the time scale and selection pressures for these may be distinctly different from those operating for myxomatosis.

BIOLOGICAL CONTROL OF CATS USING FELINE PANLEUCOPENIA VIRUS

On the sub-Antarctic Marion Island, a population of feral cats (*Felis catus*), estimated at more than 3400 cats in 1977, was descended from five cats released in 1948–1949. These cats killed more than 450,000 seabirds in 1975. Feline panleucopenia virus is a parvovirus that is endemic in cat populations around the world but was not present in the cats on Marion Island. A virulent isolate of this virus was used to inoculate 93 trapped cats in 1977. Over the next five years the population decreased to an estimated 615 cats. But by 1982 the rate of decrease had slowed from 29% per year to 8%. Shooting and trapping to eradicate the cats followed the population knockdown (3). The release of panleucopenia virus to control cats on other islands has been ineffective. Surveys of feral cats in mainland locations have usually demonstrated that this virus is endemic and its release as a biological control is unlikely to have a significant impact.

BIOLOGICAL CONTROL OF RODENTS

Rodents are the major vertebrate cause of crop loss and spoilage worldwide. They are also significant vectors of human and animal diseases and when introduced onto islands can cause severe ecological damage. The difficulties of controlling rodents by conventional means, the development of resistance to rodenticides that has occurred, and the dangers posed to other species by many rodenticides makes biological control a very attractive option. This has been attempted with bacteria, viruses, protozoa, cestodes, and nematodes but without significant success (4).

GENETIC MODIFICATION OF BIOLOGICAL CONTROL AGENTS

Biological control agents that reduce reproduction rates have the potential to be just as effective at population reduction as lethal agents. A novel approach to population control is to genetically engineer viruses or other micro-organisms to deliver contraceptive vaccines to vertebrate pest species. In laboratory trials, mice inoculated with recombinant mousepox virus (5) or, recombinant murine cytomegalovirus expressing a murine oocyte antigen developed long-term infertility. Parallel studies are being undertaken with recombinant myxoma virus expressing a rabbit oocyte antigen. Whether it will be possible for such recombinant viruses to spread in the field remains to be determined. However, using molecular genetics, the potential for biological control of vertebrate pests may finally be realized.

REFERENCES

1. Fenner, F.; Ross, J. Myxomatosis. In *The European Rabbit. The History and Biology of a Successful Colonizer*; Thompson, H.V., King, C.M., Eds.; Oxford University Press: Oxford, 1994; 205–240.
2. Mutze, G.; Cooke, B.; Alexander, P. The initial impact of rabbit hemorrhagic disease on european rabbit populations in South Australia. J. Wildl. Dis. **1998**, *34*, 221–227.
3. Van Rensburg, P.J.J.; Skinner, J.D.; Van Aarde, R.J. Effects of feline panleucopaenia on the population characteristics of feral cats on Marion Island. J. Appl. Ecol. **1987**, *24*, 63–73.
4. Singleton, G.R. In *The Prospects and Associated Challenges for the Biological Control of Rodents*, Proceedings of the 16th Vertebrate Pest Conference, Halverson, W.S., Crabb, A.C., Eds.; University of California: Davis, CA, 1994.
5. Jackson, R.J.; Maguire, D.J.; Hinds, L.A.; Ramshaw, I.A. Infertility in mice induced by a recombinant ectromelia virus expressing mouse zona pellucida glycoprotein 3. Biol. Reprod. **1998**, *58*, 152–159.

BIOLOGICAL CONTROL OF WEEDS (INSECTS AND MITES)

Peter Harris
Agriculture and Agri-Food Canada, Lethbridge, Alberta, Canada

INTRODUCTION

Insects and other natural enemies are primarily used for controlling weeds by the classical or inoculative approach. This involves establishing natural enemies from other places for continuing suppression of the weed. It is done by government in the public interest, as it commonly takes 20 years to achieve control and once established the enemy finds the weed on its own. However, there are public concerns about safety, which change over time. No crops have suffered economic loss, few native plants have been attacked, and none have been exterminated but the record is not enough. The concerns must be addressed directly and this has driven the evolution of weed biocontrol.

The usual targets for classical biocontrol are introduced weeds of uncultivated land that lack enemies and have spread to dominate large areas. Such weeds cause huge habitat and forage losses, reduce diversity of the native flora and fauna, and displace rare species (1). The concern about establishing their natural enemies is that desirable plants might also be attacked. Testing to ensure that this does not happen costs about two scientist years per agent ($900,000). About a third of the agents fail to establish, a third remain scarce, and nearly half of those that become abundant do not achieve control. On the other hand, costs of biocontrol increase little with infestation size and the same agent may achieve control in climates as different as Australia and Canada. Thus, the cost per ha can be small and biocontrol is less damaging to nontarget plants than most alternatives. Release approval for the agents in North America rests with Animal and Plant Health Inspections Service-United States Drug Administration and the Canadian Food Inspection Agency who consult advisory groups.

HISTORY AND IMPACT OF CLASSICAL WEED BIOCONTROL

Classical weed biocontrol was the serendipitous result of introducing the mealybug (*Dactylopus ceylonicus* thinking it was *D. coccus*), to India in 1795 (2). The purpose was to start a dye industry on the impenetrable stands of the South American prickly pear cactus, *Opuntia vulgaris*. The result was rapid cactus kill and little dye production. After 68 years and several eradication attempts the government distributed the insect for *O. vulgaris* control.

Hawaii took the ball in 1902 by importing 23 insect species to control the introduced shrub *Lantana camara* (2). Their intention was to establish a complex to attack all parts of the plant. Only 12 species arrived alive in sufficient numbers for release and eight established. Success was claimed for a fly that destroyed 86% of the seed, but none of the criteria for success were demonstrated: that the weed is reduced to a low density in several locations, it remains at a low density, or it returns to a high density when protected from the agent (3).

The mealy bug *D. ceylonicus* practically eliminated *O. vulgaris* in Australia but did not affect two more abundant cacti, which by 1925 had infested 60 million acres. Concerns that introduced insects might attack cultivated plants were met with the "no-choice" test in which only species that starve on representative crop plants were released. They tested 49 insect species, released 24, of which 12 established. Control was attributed to just two: the mealy bug *D. opuntia* and the cladode-boring moth *Cactoblastis cactorum*, which, two years after increase, caused a crash of cactus from 5000 to 11 plants/acre (4).

Australia then tackled St. John's wort (*Hypericum perforatum*) (1). Eight insect species were released, four established and, up to seven years after release, the leaf-feeding beetle *Chrysolina quadrigemina* increased to control the weed. The four established insects were then released in California, where 40,500 ha of open range that were infested in northern California in 1929 had increased to almost 2 million ha in western North America by 1940. Increases of *C. quadrigemina* returned most of the infested area to a native bunch grass community with fair to good productivity and an increased diversity of plant species, and the weed stabilized at about 1% of its former density (5). The two beetle species were released in Canada in

1952. Up to 13 years later, *C. quadrigemina* increased to control the weed on summer-dry sites and *C. hyperici* on summer moist sites, but both failed above 900 m elevation (6). This program terminated chemical control of the weed in British Columbia and removed it from the Ontario noxious weed list.

Both *Chrysolina* spp. developed on Canadian *Hypericum* spp. in no-choice tests, but in the field they have not been attacked (6). The plant genus contains photosensitizing compounds lethal to sun exposed larvae. In nature the larvae feed at dawn and dusk on a winter mat of foliage, which is absent in the Canadian species. Two nontarget species are attacked, but are poor hosts: the introduced evergreen ornamental *H. calycinum*, which has hard mature foliage, and in California the native *H. concinnum*, which has diffuse foliage (presumably shading is reduced and the need for climbing by inactive larvae is increased).

It is more predictive to use the no-choice tests to show that the larval host range is restricted to a taxonomic group of plants rather than that individual crop plants are unacceptable. Some reviewers felt that tests of crop plants were still necessary for public confidence, but gradually these tests have been eliminated. The laboratory larval host range limit exaggerates the field host range, since congeners not attacked in nature usually support development in the test. The larva role is to stay on the plant on which it hatched and eat, which may involve distinguishing the host from intermingled vegetation, but not host finding. Host finding is done by the adult female using habitat and plant cues (7). The larval host range limit was a satisfactory measure of risk as long as native congeners were not an issue. This changed in 1997 when the attack of native *Cirsium* spp. by the seed-head weevil *Rhinocyllus conicus* was deplored (8). The 1967 petition for release of the weevil against the introduced thistle *Carduus nutans* stated that its host range included *Cirsium*, but this caused no concern. Today the view is mixed. Some deplore the weevil, but to many a reduction of native thistles is not a concern and the control of most stands of *C. nutans* with increased forage production, makes it an eminent success.

THREE CURRENT ISSUES

Host-Specificity Tests

Most introduced weeds in North America have native congeners. A total of 23 of 27 insect species established in Canada developed on congeners in laboratory no-choice tests, but only one species in Canada and three in the United States have attacked the North American relatives of the weed in the field. Exclusion of species with larvae that develop on native congeners in the laboratory will practically end weed biocontrol. A scientifically sound alternative is to base release approval on the adult host range. The larval no-choice test does not distinguish between species with adults that oviposit on the weed's relatives in the field from those that exclude them. However, adult tests in large field cages or open releases allow assessment of this choice. Thus, although the beetle *Altica carduorum* develops on all *Cirsium* spp., the adult only "sees" *C. arvense* in large arenas. This combined with a low suitability of other *Cirsium* spp. for the immature stages means there is selection pressure against host plant shifts and accounts for the field specificity of the beetle in Eurasia (9). However, the development of new host screening procedures are currently at an impasse because some reviewers still feel larval ability to complete development in the laboratory is an unacceptable risk even if the adult does not recognize the plant as a host.

Costs and Agent Success

Past cost increases of weed biocontrol have been partly compensated for by using fewer agents and by improving establishment success. The first Hawaiian and first two Australian projects tried to release 55 insect species of which 44% established and three reduced weed density. The current world establishment rate is 65% with the release of 4.1 species against each successfully controlled weed (10). The idea that all parts of the plant need to be attacked is not supported by the fact that 81% of the successes result from single species. Where several agents contributed, they tend to do so in different habitats such as *C. quadrigemina* and *C. hyperici*.

New host tests are likely to increase costs. An obvious solution is to eliminate the approximately 80% of agents released that contribute little. Present practice is to be more discriminating with expensive agents than those that

Table 1 Success of weed biocontrol agents released in Canada (1981–1990)

	Sponsor of pre-release studies	
	Canada	United States
No. released	18	6
% Established	73	33
Species contributing to control	8	1
Canadian costs/agent	2 SY[a]	0.04 SY
Canadian costs/success	6.6 SY	0.24 SY

[a]SY = scientist year, currently about $450,000.

are cheap simply because testing was done by another country (Table 1). However, establishing any agent may have ecological effects. Thus, it is prudent to reduce the number released to the necessary minimum. No agent trait associated with effectiveness has been found, although establishment is broadly correlated with a high rate of increase, long adult life, the number of generations a year, and small size. However, effectiveness may be plant related. For example, root feeders have been more effective on tansy ragwort, *Senecio jacobaea*, and leafy spurge, *Euphorbia esula*, than defoliators and seed feeders, which have been effective on other plants. If this were determined in preliminary studies, host specificity tests could be restricted to insects in the relevant feeding guild. This combined with studies of the insect's ecological requirements could halve the number of failures.

Legislation

The enabling legislation used for classical weed biocontrol is the Federal Plant Pest Act of 1957 in the United States and the Plant Protection Act of 1990 in Canada. The purpose of both is to prevent pest establishment and spread, so release of nonpests is allowed by reverse logic. However, with no explicit mention of biocontrol, other acts, with various purposes, come into play. For example, the U.S. Endangered Species Act of 1973 requires federal agencies to ensure their actions are not likely to jeopardize the continued existence of endangered or threatened species. The ability of larvae to develop on an endangered plant may indicate jeopardy, but the issue is complicated. The act can save habitat from human destruction, but it cannot stop invading weeds, such as leafy spurge, which are displacing the threatened northern prairie skink in Canada and the endangered western prairie fringed orchid in both countries. Biocontrol is the best hope for their survival; however, there are rare Florida spurges that will support larval development of leafy spurge agents. It is not possible under the act to weigh the benefits against the risks, although the risks are small if the agent has an obligatory winter diapause that prevents survival in Florida. Nevertheless, the approval of agents for the major North American leafy spurge problem is currently stalled. The Australian Biocontrol Act of 1984 solves the issue by instructing reviewers to approve the release of agents when the expected benefits outweigh the risks. A further need is to judge risk on a holistic assessment of adult and immature needs and habits. It would also be helpful to make weed biocontrol more open to public input and have procedures for resolving conflicts of interest.

REFERENCES

1. Pimentel, D.L.; Lach, L.; Zuniga, R.; Morrison, D. Environmental and Economic Costs Associated with Non-Indigenous Species in the United States. http://www.news.cornell.edu/releases/Jan99/species_costs.htm (accessed Feb 25, 1999).
2. Goeden, R.D. Biological Control of Weeds. In *Introduced Parasites and Predators of Arthropod Pests and Weeds: A World Review*; Clausen, C.P., Ed.; Part II, Agricultural Handbook 480, USDA: Washington, DC, 1978; 357–414.
3. Smith, H.S.; DeBach, P.; The measurement of the effect of entomophagous insects on population densities of their hosts. J. Econ. Entomol. **1942**, *3*, 845–849.
4. Crawley, M.J.; The success and failures of weed biocontrol using insects. Biocontrol. News Inf. **1989**, *10*, 2213–2223.
5. Huffaker, C.B.; Kennet, C.E.; A ten-year study of vegetation changes associated with biological control of klamath weed. J. Range Manage. **1959**, *12*, 69–82.
6. Harris, P. Status of Introduced and Main Indigenous Organisms on Weeds Targeted for Biocontrol in Canada. http://res.agr.ca/leth/weedbio.htm (accessed Dec 20, 1999).
7. Fox, C.W.; Lalonde, R.G.; Host confusion and the evolution of insects' diet breadths. Oikos **1993**, *67*, 577–581.
8. Louda, S.M.; Kindal, D.; Connor, J.; Simberloff, D. Ecological effects of an insect introduced for the biological control of weeds. Science **1997**, *277*, 1088–1090.
9. Harris, P. Evolution of classical weed biocontrol: meeting survival challenges. Bull. Ent. Soc. Can. **1998**, *30*, 134–143.
10. Julien, M.J. Biological control of weeds worldwide: trends, rates of success and the future. Biocontrol. News Inf. **1989**, *10*, 299–306.

BIOLOGICAL CONTROL OF WEEDS (MICROBES)

B

Karen L. Bailey
Susan M. Boyetchko
Agriculture and Agri-Food Canada, Saskatoon Research Centre, Saskatchewan, Canada

INTRODUCTION

Microscopic organisms such as bacteria, fungi, and viruses can cause diseases on many types of plants, including economically important crops and weeds. Under suitable environmental conditions, some plant pathogens can cause sufficient damage to susceptible weed hosts with the purpose of attaining weed control below an economic threshold level. This is biological control of weeds using microorganisms (1, 2). Several biological, environmental, technological, and commercial factors that currently challenge the availability and use of biological control agents will be discussed with regard to two different biological control strategies: classical and inundative.

WHY DEVELOP BIOCONTROL FOR WEEDS?

Worldwide, approximately $11.8 billion is spent on chemical herbicides each year. More than $12 million was spent annually on weed control in North America in the 1990s, with less than 1% being spent on biological weed control strategies. In North America, 70–90% of all agricultural land is treated with herbicides yet weeds in crops cause yield losses of about $1 billion per year. Weeds compete with crops for water, nutrients, and sunlight resulting in reductions in crop yield and quality. Yield losses in individual farmers' fields can vary considerably ranging from 1 to 60% or higher, leading to complete crop failure. Product quality grades are lowered when harvested crops are contaminated with weed seeds.

Despite efforts to control weeds with chemical herbicides, weeds remain a significant problem to farmers and are increasing in some agricultural areas due to changes in crop management practices. For example, improper use of chemical herbicides (i.e., not rotating with herbicides that have different mechanisms of action) has increased the incidence of herbicide resistance. Environmental concerns resulting from chemical residues have led countries such as Denmark, Sweden, the Netherlands, and some states and provinces in the United States and Canada, to impose legislated directives for 50% reduction in the use of chemical herbicides. Biological control may be applied in situations to complement chemical and cultural methods of weed control such as for management of herbicide-resistant weed populations, where chemical herbicide residues in soil or ground water cause problems for succeeding crops, in organic farming systems, on rangelands and roadsides, and in public/urban areas.

CLASSICAL AND INUNDATIVE BIOLOGICAL CONTROL

Biological control is the practice of using a living organism to manage a pest problem (3, 4). Biocontrol agents may either kill the weed, inhibit weed seed germination, or suppress weed growth and vigor making it less competitive with the crop. In some cases, microorganisms may promote weed germination and growth so that other control measures such as tillage or herbicide applications are used at the most appropriate time. Realistically, weed management with biological control is the result of the dynamic interchange of three biological systems (weed pest, biological control agent, and crop) compounded by other biological and environmental interactions (5).

Biological control strategies may be classified under two main categories: classical and inundative (Table 1). Classical biocontrol involves the mass release of a biocontrol agent, relies on natural dispersal and infection of the microorganism to spread and infect throughout the area of the weed infestation, and may take several years for weed control to be achieved. Classical biocontrol does not eradicate a weed population but maintains it at an ecological threshold level so both the host and pathogen coexist. Inundative biological control entails an artificially high level of microbial inoculum applied to a localized area for control of the weeds over a short period of time. Ideally, there is no natural spread or carryover of the microorganism from year to year. This type of bio-

Table 1 Features of classical and inundative biological control

Classical approach	Inundative approach
Importation of foreign pathogens from geographical area of coevolution with host	Indigenous and exotic microorganisms used
Release of agents into new geographical location where target weed is a problem and natural enemies are absent	Artificial increase and release of high inoculum levels of agent
Natural spread, dispersal, and disease development; one time release	Artificial application of agent to weeds at a specific location; one or more times throughout growing season
Weed control expected to be long term	Weed control expected to be short term
Ecological approach	Economic threshold level approach
Biocontrol agent coevolves with host	Host exposure to biocontrol agent is for short period of time
Suitable for control of perennial weeds that grow in dense stands and infest large areas of land with difficult terrain; eradication of weed is not desirable	Suitable for control of annual and perennial weeds in defined areas of land used with products of high cash or social value (e.g., agriculture, forestry, and turf)

control is often referred to as the bioherbicide approach and may involve one or more applications during a growing season. However, biological weed control should not be considered as a substitute for herbicides. It provides a new option with potentially novel mechanism(s) for weed management and should be integrated with other practices such as tillage, rotation, sanitation, and chemical herbicide use.

EXAMPLES OF BIOLOGICAL CONTROL AGENTS

Biological control of weeds using microorganisms is a relatively recent technology and currently holds less than 1% of the traditional herbicide marketplace. There are several examples of successful classical and inundative biological control agents (Tables 2 and 3) (6, 7). In the 1970s, the rust fungus, *Puccinia chondrillina* Bub. and Syd., was successfully introduced to manage skeleton weed in Australia and later the western coast of the United States. The success of this classical biological control application was attributed to the management and manipulation of initial dispersal of the rust fungus by aircraft (2 g/ha) and by applying only when environmental conditions were conducive to rust infection. Skeleton weed has been reduced by 16–70% and the rust is still providing control almost 30 years after its introduction. It was estimated that it cost $2.6 million to find, develop, and implement this program. Many other classical biocontrol agents, mostly rust fungi, have been successfully introduced against *Acacia* spp., musk and nodding thistle, and diffuse knapweed.

Inundative biological control of weeds was introduced in the 1980s with the commercial development of DeVine® (*Phytophothora palmivora* Butler) for control of stranglervine and Collego® [*Colletotrichum gloeosporioides* (Penz.) Penz. and Sacc. f.sp. *aeschynomene*] for control of northern jointvetch. The success of the first two commercially available biocontrol agents in the United States may be attributed to their consistent and reliable performance that led to a highly effective product, the availability of an economical method to manufacture, and consumer demand for a particular product. It was estimated that the costs associated with the research and deve-

Table 2 Examples of classical biological control

Target weed	Pathogen	Country of importation
Skeleton weed	*Puccinia chondrillina* (rust fungus)	Australia, United States
Musk thistle, nodding thistle	*Puccinia carduorum* (rust fungus)	Canada, United States
Diffuse knapweed	*Puccinia jaceae* (Centaurea rusts)	Canada (B.C.), United States (Washington, Oregon, Idaho, Montana, South Dakota)
Acacia spp.	*Uromycladium tepperianum* (Acacia gall rust)	South Africa
Blackberry (*Rubus* spp.)	*Phragmidium violaceum* (Blackberry rust)	Chile, Australia

Table 3 Examples of inundative biological control

Target weed	Pathogen	Trade name	Comments
Stranglervine	*Phytophthora palmivora*	DeVine®	First commercially available bioherbicide; small niche market, made to order
Northern joint vetch	*Colletotrichum gloeosporioides* f.sp. *aeschynomene*	Collego®	Commercially available from Encore Technologies (Minnesota)
Yellow nutsedge	*Puccinia canaliculata*	Dr. BioSedge®	Registered product but not commercially available
Round-leaved mallow	*Colletotrichum gloeosporioides* f.sp. *malvae*	BioMal®	First registered bioherbicide in Canada
Annual bluegrass	*Xanthomonas campestris* pv. *poae*	Camperico®	Registered by Japan Tobacco for use in golf courses and turf
Weedy hardwood tree species	*Chondrostereum purpureum*	BioChon®	Natural wood decay promoter; commercially available in Netherlands
Black and golden wattle	*Cylindrobasidium laeve*	StumpOut®	Registered in South Africa

lopment for Collego® was $2 million in comparison to $30 million for research and development of a chemical herbicide. Other biocontrol agents that are commercially available include the product BioChon® for use against weedy hardwood tree species in the Netherlands and StumpOut®, a registered product for control of black and golden wattle in South Africa. StumpOut® is produced in a small factory in the Plant Pathology Research Institute, Weed Biocontrol Unit, Stellenbosch, and distributed to clients on demand as an oil formulation product that is painted onto freshly cut surfaces of tree stumps. BioMal® was the first registered bioherbicide in Canada for use against round-leaved mallow. It was discovered by scientists at Agriculture and Agri-Food Canada and, although not commercially available to date, a company has licensed the microorganism to develop it for commercial use. A variety of bacterial agents that control weeds include the foliar-applied bacterium, *Xanthomonas campestris* pv. *poae* for control of annual bluegrass, *Pseudomonas syringae* pv. *tagetis* that causes apical chlorosis in Canada thistle and several deleterious rhizobacteria that inhibit root growth and development in several grass and broadleaved weeds.

CHALLENGES IN BIOLOGICAL WEED CONTROL

Biological weed control should be judged for its own unique merits and properties and not those necessarily designed for chemical models. Hurdles limiting the development and release of microbial products for weed control have been their high level of host specificity that limits the number of weed species that can be controlled by a single agent, sensitivity to the environment (e.g., moisture, temperature, and radiation), monetary and labor costs associated with initial research and development prior to commercialization, technological challenges associated with mass-production and delivery, and regulatory requirements. Economic feasibility for classical and inundative biocontrol strategies is governed by: 1) whether the weed is perceived to be an economic problem; 2) the market potential or social value placed on the weed problem; 3) the availability and/or cost of existing control methods of the target weed; 4) the effectiveness, reliability and speed of control of the agent; and 5) the direct or indirect costs or opportunities provided to the end-user for employing biological control.

FUTURE PROSPECTS

Research in biological control of weeds has made significant progress over the past several decades. There are more than 23 fungal pathogens that have been studied for introduction as classical biological control agents, and more than 100 fungal agents for use as bioherbicide products. Also, bacteria are seriously being considered for application to the foliage and/or to roots of various weed species.

Even though there are several challenges facing the introduction and development of microorganisms for biological weed control, there are committed teams of researchers in a variety of disciplines around the world who continue to find new microorganisms and develop new strategies that will make biological control easy to use (8). New technologies in screening, fermentation, and formulation are providing more effective ways to use these microorganisms and there is a great deal of potential for using biological controls for increasing crop production and applying them in an integrated pest management program. Currently, there are several patents granted or submitted for biological control of weeds such as purple nutsedge, pigweed, grass weeds, and forest weeds around the world. We are just beginning to realize the extent of the possibilities of controlling weeds with microorganisms that nature has already provided.

REFERENCES

1. Boland, G.J.; Kuykendall, L.D. *Plant-Microbe Interactions and Biological Control*; Marcel Dekker, Inc.: New York, 1998; 1–442.
2. TeBeest, D.O. *Microbial Control of Weeds*; Chapman and Hall: New York, 1991; 1–284.
3. http://ipmwww.ncsu.edu/biocontrol/biocontrol.html (accessed March 2000).
4. http://pest.cabweb.org (accessed March 2000).
5. Kremer, R.J. Microbial Interactions with Weed Seeds and Seedlings and Its Potential for Weed Management. In *Integrated Weed and Soil Management*; Hatfield, J.L., Buhler, D.D., Stewart, B.A., Eds.; Ann Arbor Press: Ann Arbor, MI, 1998; 161–179.
6. Boyetchko, S.M. Innovative Applications of Microbial Agents for Biological Weed Control. In *Biotechnological Approaches in Biocontrol of Plant Pathogens*; Mukerji, K. G., Chamola, B.P., Upadhyay, R.K., Eds.; Kluwer Academic/Plenum Publisher: New York, 1999; 73–97.
7. TeBeest, D.O. Biological Control of Weeds with Plant Pathogens and Microbial Pesticides. In *In Advances in Agronomy*; Sparks, D.L., Ed.; Academic Press: Toronto, 1996; 56, 115–137.
8. Kennedy, A.C. Molecular Biology of Bacteria and Fungi for Biological Control of Weeds. In *Molecular Biology of the Biological Control of Pests and Diseases of Plants*; Gunasekaran, M., Weber, D.J., Eds.; CRC Press: Boca Raton, FL, 1996; 155–172.

BIOLOGICAL CONTROL SUCCESSES AND FAILURES

B

Heikki Hokkanen
University of Helsinki, Helsinki, Finland

INTRODUCTION

Biological control of pests has been actively practiced for the control of pests, weeds, and plant diseases for more than 100 years, and it has had some 150 spectacular successes (1), which in economic terms have been just as impressive as in ecological terms: the calculated return for investment is 32:1, while for other control methods the ratio is around 2.5:1 (2, 3). However, the obtained successes are only the tip of the iceberg of all the work carried out in the field. To date, more than 6000 introductions of alien natural enemies have been carried out, worldwide (4). It is estimated that only about 35% of all introduced biocontrol agents have become ecologically established in the target ecosystem, and only 60% of these have provided any economic or biocontrol success (3, 5). Of all the individual biocontrol projects, only 16% have resulted in complete control of the target pest (6). A major ecological and economic challenge is to improve the ratio of successes in biological control, while retaining the excellent safety record of this approach to pest control.

GENERAL PRINCIPLES

While it has been shown that biological control can be effective in any climate, ecosystem, and crop, the factors determining success or failure remain largely unknown, and often are economic rather than ecological in nature (1). Very few general principles to improve the efficacy and predictability of biological control have emerged; these include better ecological background knowledge, genetic improvement (in particular, genetic engineering) of biocontrol agents, and the utilization of new ecological associations in selecting the biocontrol agents (7). The genetic engineering of biocontrol agents, especially insects, is still in its infancy and cannot be expected to improve the success ratios in the foreseeable future. In contrast, the new association principle has—usually unknowingly—been used for a long time, and is increasingly employed to find more effective natural enemies for current biological control programs.

NEW ASSOCIATIONS

The standard biological control principle is to reestablish the ecological balance between an exotic pest and its natural enemies occurring in their country of origin (the "old association approach") (2, 3). It has been argued, however, that this is an inefficient way of practicing biological control, because due to an evolved long-term equilibrium between the pest and the natural enemy, the control agent only seldom is very efficient (6). To find more effective enemies one should search among agents that do not share an evolutionary history with the target pest (the "new association" approach). Such natural enemies can be found, for example, for the target pest in areas where the pest has been introduced only recently, or among enemies attacking related species in other geographical areas (6).

EVIDENCE FOR IMPROVED EFFICACY

Analysis of past biocontrol successes and failures have indicated that when employing the new association principle it is possible to increase the success ratio by at least 75% (6). More detailed studies showed that some natural enemy groups may be particularly attractive as new association agents (Table 1). Such analyses are, however, often confounded by the fact that new association agents seldom have been considered as the primary choice in biological control, and consequently, usually five to seven old association agents are introduced before a new association agent is tried. In addition, on average much greater numbers (two- to fourfold) of old association agents are normally introduced (Table 1), which further increases the probability of biocontrol success, and biases the analyses against new association agents. Therefore, the estimate for improving the success ratio appears to be conservative.

Several spectacular, well-documented biocontrol successes that have employed new association control agents are known, and these include the complete control of serious pests such as the sugarcane borer *Diatraea saccharalis* in the Caribbean, coconut spike moth *Levuana*

Table 1 Comparisons of biological control introductions with old and new association control agents utilizing Tachinidae, Braconidae, and Eulopidae

	Proportion (%) of successes of all cases (introductions)		Total number of cases		Bias in the release numbers[a]
	Old	New	Old	New	
Tachinidae	10.9	17.1	92	41	3.8-fold
Braconidae	17.2	14.4	169	97	1.6-fold
Eulopidae	28.6	35.7	56	28	1.7-fold
Overall	17.4	18.7	317	166	

[a]Indicates how many more individuals on average of old association agents were released in the introduction projects, compared with new association agents. In the case of Tachinidae and Braconidae the mean number of released new association agents was below 5000 individuals, which is considered to be the necessary number to ensure a fair chance for the natural enemies to establish themselves. (From Hokkanen, H. M. T., unpublished data.)

iridescens in Fiji, southern green stink bug *Nezara viridula* in Hawaii, the moth *Oxydia trychiata* in Colombia, and several scale insect species in California, Greece, and Australia (7). Further, more detailed examples will be given below on new research with good prospects of success utilizing this approach.

RECENT CASES EMPLOYING NEW ASSOCIATIONS

Eurasian Watermilfoil

The Eurasian watermilfoil (*Myriophyllum spicatum*) was introduced into North America several decades, possibly 100 years, ago. It grows rapidly, forms a dense canopy on the water surface, and often interferes with recreation, inhibits water flow, and impedes navigation. Herbicides and mechanical harvesting have been used to control infestations, costing \$150–\$2000 per acre annually in Minnesota (8).

Sometimes naturally occurring declines of the watermilfoil have been observed. The main causal agent proved to be a native beetle *Euhrychiopsis lecontei*, the milfoil weevil, which subsequently has shown control potential in controlled field experiments. The weevil is a specialist herbivore of watermilfoils, but prefers the Eurasian to its native host, the northern watermilfoil (*M. sibiricum*). Research is in progress to use the milfoil weevil effectively as a biocontrol agent against the Eurasian watermilfoil in North America (8).

Lantana

Lantana camara is a serious weed of Mexican or Caribbean origin, affecting cropping lands and forest areas in 47 countries. Lantana was the focus of the first weed biocontrol effort in history (1902), and there is an enormous literature on Lantana biocontrol. Several complexes of herbivores have been credited for exerting some degree of biocontrol of the weed (e.g., in Hawaii), many employing new association agents jointly with old association agents. Latest research gives data on the good efficacy and release in Australia of the moth *Ectaga garcia* originating from South America, where it feeds on the related weed *Lantana montevidensis* (9).

Triffid (Siam) Weed

The triffid weed (*Chromolaena odorata*) is a perennial shrub native of tropical America. In recent decades it has become a serious pest of humid tropics around the world (10). It spreads rapidly in lands used for forestry, pasture, and plantation crops and can reach a height of three meters in open situations and up to eight meters in forests. For more than two decades the triffid weed has been the subject of intensive research as a target for biological control. However, so far all attempts at biocontrol of *C. odorata* have failed. Recently the new association biological control agent, arctiid moth *Pareuchaetes aurata aurata* collected from *C. jujuensis* in South America, was considered as more promising than the related moth *P. pseudoinsulata*, an old association control agent previously thought of as one of the best biocontrol candidates (10).

Southern Green Stink Bug

The biological control of the southern green stink bug (the green vegetable bug) (*Nezara viridula*) in Australia, New Zealand, and Hawaii has been heralded as a landmark example of classical biological control (11). An egg parasitoid—old association agent—*Trissolcus basalis* and a tachinid fly—new association agent—*Trichopoda pen-*

nipes have jointly provided these successes. Control by the fly has been considered as relatively more important, and indeed, in Australia where the fly has failed to establish, the control is poor and the bug remains a serious pest. Currently in Australia another new association tachinid fly, *Trichopoda giacomellii*, is being released after research showed it has excellent potential for control (12).

Citrus Leafminer

Citrus agroecosystems have numerous potentially damaging pests often maintained under substantial to complete biological control by both old and new association agents. The citrus leafminer *Phyllocnistis citrella*, native to Asia, has spread rapidly throughout the citrus growing areas of the world in recent years (13). It arrived in Florida in 1993 and in less than one year invaded and colonized the entire state. An old association parasitic wasp *Ageniaspis citricola* was introduced in 1994, and after establishment it has held the pest under control with significant help from native parasitoids such as *Pnigalio minio* (new association agent) (13). In some other areas native parasitoids similarly have shown significant control effect on the citrus leafminer (e.g., in Italy). This example illustrates well the fact that invading species often do not become pests, because effective local natural enemies keep them in check.

Tarnished Plant Bug

An ongoing study in the United States has identified as the most important parasitoid of the native pest *Lygus lineolaris*, the tarnished plant bug, the exotic species *Peristenus digoneutis*, originally introduced for the control of related introduced mirid plant bugs (14). This example serves well to point out the importance of native pests, which in most if not all areas form the majority of all pest species. As old association biological control agents seldom can be utilized for the control of native pests, their biocontrol by introduced natural enemies has attracted relatively little attention and, indeed, only three decades ago was considered an impossible task. Several recent examples, usually utilizing new association control agents, show that biological control can work against native pests just as well as against exotic ones.

FUTURE PROSPECTS

Compared with chemical control, the success rates of biological control are outstanding. While only about one out of 15,000 tested chemicals ends up as a chemical pesticide meeting the requirements of efficacy and safety, approximately one out of seven introductions of natural enemies has been successful using old associations (15). Using new association control agents this rate could still be increased to about one out of four, while the array of potential natural enemies is also substantially larger providing a wider choice. In addition, the potential uses for natural enemy introductions are broadened to include the control of native pests.

A major concern with respect to all biological control introductions is the question of nontarget safety. Biological control has an excellent record of safety (3, 16) and it covers the new association agents as well: there have been some 1500–2000 introductions already (out of 6000) that have involved new association agents (7). Those extremely few cases where a negative nontarget effect has been suspected as a result of biological control, all involve old association agents; therefore it is clear that new associations can safely be used to help obtain biological control successes at an increasing rate.

REFERENCES

1. Hokkanen, H.M.T. Success in classical biological control. CRC Crit. Rev. Plant Sci. **1985**, *3*, 35–72.
2. Cullen, J.M.; Whitten, M.J. Economics of Classical Biological Control: A Research Perspective. In *Biological Control: Benefits and Risks*; Hokkanen, H.M.T., Lynch, J.M., Eds.; Cambridge University Press: Cambridge, U.K., 1995; 270–276.
3. *Biological Control: Benefits and Risks*; Lynch, J.M., Hokkanen, H.M.T., Eds.; Cambridge University Press: Cambridge, U.K., 1995.
4. Waage, J. In *Agendas, Aliens and Agriculture*; Global Biocontrol in the Post UNCED Era, Cornell Community Conference on Biological Control, http://www.nysaes.cornell.edu/ent/bcconf/talks/waage.html (accessed Jan 5, 1999).
5. Hokkanen, H.M.T. Pest Management, Biological Control. In *Encyclopedia of Agricultural Science*; Academic Press, Inc.: San Diego, 1994; 3, 155–167.
6. Hokkanen, H.M.T.; Pimentel, D. New associations in biological control: theory and practice. Can. Entomol. **1989**, *121*, 829–840.
7. Hokkanen, H.M.T. New Approaches in Biological Control. In *CRC Handbook of Pest Management in Agriculture*, 2nd Ed., Pimentel, D., Ed.; CRC Press: Boca Raton, FL, 1991; II, 185–198.
8. Newman, R.M. Biological Control of Eurasian Watermilfoil.http://www.fw.umn.edu/research/milfoil/milfoilbc.html (accessed Feb 5, 1999).
9. Day, M.D.; Wilson, B.W.; Latimer, K.J. The life history

and host range of *Ectaga garcia*, a biological control agent for *Lantana camara* and *L. montevidensis* in Australia. BioControl **1998**, *43*, 325–338.

10. Kluge, R.L.; Caldwell, P.M. The biology and host specificity of *Pareuchaetes aurata aurata* (Lepidoptera: Arctiidae), a "New Association" biological control agent for *Chromolaena odorata* (Compositae). Bull. Entomol. Res. **1993**, *83*, 87–94.
11. Caltagirone, L.E. Landmark examples in classical biological control. Annu. Rev. Entomol. **1981**, *26*, 213–232.
12. Coombs, M. *Biological Control of Green Vegetable Bug in Australia and PNG*; Pest Management Current Programs and Projects. http://www.ento.csiro.au/research/pestmgmt/pmp16.htm (accessed Oct 5, 1999).
13. Timmer, L.W. *Citrus Leafminer Proves to be an IPM Success*; IPM Florida, 1996 *Winter*. http://www.ias.ufl.edu/~FAIRSWEB/IPM/IPMFL/v2n4/leafminer.htm (accessed Feb 5, 1999).
14. Day, W.H. Host preferences of introduced and native parasites (Hymenoptera: Braconidae) of phytophagous plant bugs (Hemiptera: Miridae) in Alfalfa-Grass Fields in the Northeastern USA. BioControl **1999**, *44*, 249–261.
15. Hokkanen, H.M.T. Role of Biological Control and Transgenic Crops in Reducing Use of Chemical Pesticides for Crop Protection. In *Techniques for Reducing Pesticide Use*; Pimentel, D., Ed.; John Wiley & Sons: New York, 1997; 103–127.
16. *Evaluating Indirect Ecological Effects of Biological Control*; Scott, J.K., Quimby, P.C., Wajnberg, E., Eds.; CABI Publishing: Wallingford, U.K., 2001; 261.

BIOPESTICIDES

B

Gavin Ash
Charles Sturt University, Wagga Wagga, New South Wales, Australia

INTRODUCTION

A biopesticide is a type of augmentative biological control agent in which an inundative application of a living organism is used to kill the target pest. In this type of strategy, massive amounts of inoculum of the organisms (usually fungi, nematodes, or bacteria) are applied in an effort to manage the target. This is a general term applied to a range of pests and control organisms. Furthermore, this term does not include the use of toxins or secondary metabolites alone applied as pesticides (1). These chemicals, although derived from micro-organisms, are simply analogous to synthetic pesticides and do not contain a living organism as an active ingredient. A parallel term to biopesticide, used to describe the suppression of the pest, is biopestistat (Fig. 1). These types of biological control agents, when applied to a pest, suppress the population to below an economic threshold or injury level. There is a growing interest in the use of these types of organisms in conjunction with competitive crops to control weeds (2). However, they do not kill the target organism per se, and so are excluded from the remaining discussion of the term biopesticide.

Biopesticides have been used to control a range of pests including insects, weeds, and diseases. In all of these cases, the biopesticide is packaged, handled, stored, and applied in a fashion similar to that of traditional pesticides (3). The success of this type of control revolves around the cost of production, the quality of the inoculum, and the field efficacy of the organism. In comparison to classical control, in which the cost of research and development is borne by the community, biopesticides are usually developed by commercial companies in an expectation that they will recoup their costs by sale of the product. This type of strategy can be used against both native and introduced pests.

Biopesticides may be further subdivided based on the type of target pest. For example, a bioherbicide is a biopesticide developed to kill weeds. Further subdivision based on the type of agent used is also common. The term mycoherbicide is used widely to describe the formulation of a fungal agent in a bioherbicide.

Bioinsecticides

More than 1500 species of pathogens have been shown to attack arthropods and include representatives from the bacteria, viruses, fungi, protozoa, and nematodes (4). Diseases caused by insects have been known since the early 1800s with the first attempts at inundative applications of fungi to control insects being developed in 1884, when the Russian entomologist, Elie Metchnikoff, mass-produced the spores of the fungus *Metarhizium anisopliae*. The majority of entomopathogenic fungi belong to the Deuteromycotina, a group of fungi without known sexual stages. These fungi have been developed as biopesticides, primarily in tropical regions due to their requirements for high humidity for infection.

Bacteria that attack insects can be divided into nonspore-forming and spore-forming bacteria. The nonspore-forming bacteria include species in the Pseudomonaeae and the Enterobacteriaceae. The spore-forming bacteria belong to the Bacilliaceae and include species such as *Bacillus popilliae* and *Bacillus thuringiensis*. *B. thuringiensis* (Bt) has primarily been developed as a biopesticide to control Leptodopteran larva. However, other serotypes of Bt produce toxins that kill insects in the Coeloptera and Diptera as well as nematodes. Commercial formulations of the bacteria contain living spores of the bacteria.

Entomopathogenic nematodes of the families Steinernematidae and Heterorhabditidae, in conjunction with bacteria of the genus *Xenorhabdus*, have been used successfully deployed as biopesticides, for example BioVECTOR (4). They are usually applied to control insects in cryptic and soil environments. The nematodes harbor

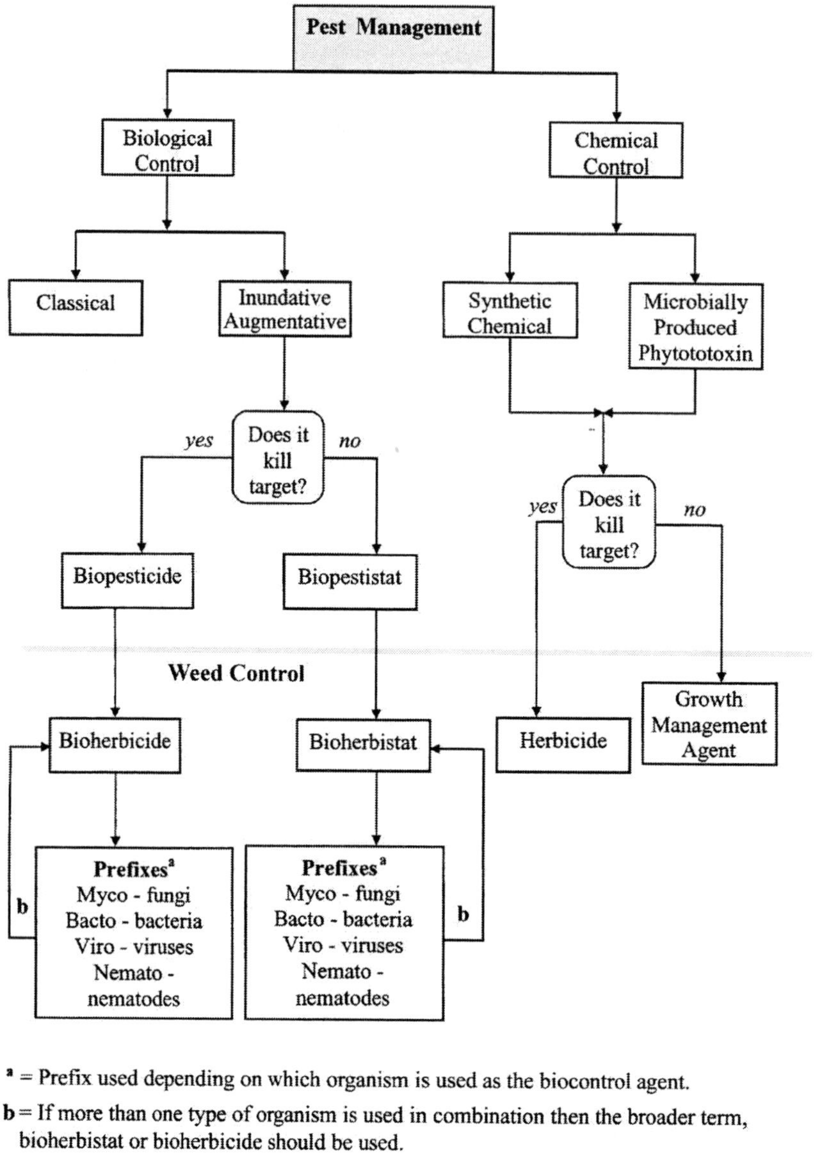

Fig. 1 A hierarachy of terminology in pest manangement with emphasis on weed control. (From Ref. 1.)

the bacteria in their intestines. The infective third-stage larvae enter the host through natural openings and penetrate into the haemocoel. The bacteria are voided in the insect and cause septicemia, killing the insect in approximately 48 h.

Biofungicides

Biological control of fungi that cause plant disease can be accomplished by a number of mechanisms including antibiosis, hyperparasitism or competition. Additionally, weak pathogens may induce systemic acquired resistance in the host, giving a form of cross-protection. Biofungicides have been used in both the phylloplane and rhizosphere to suppress disease. A biological control agent for the control of foliar pathogens in the phylloplane must have a high reproductive capacity, ability to survive unfavorable conditions, and the ability to be a strong antagonist or be very aggressive. A wide range of bacteria and fungi are known to produce antibiotics that affect other microorganisms in the infection court. Most often these organisms are sought from a soil environment, as this environment is seen as the richest source of antibiotic producing species. Species of *Bacillus* and *Pseudomonas* have been successfully used as seed dressings to control soil-borne plant diseases (5). Fluorescent Pseudomonads are also often seen as a component of suppressive soils. These bacteria may prevent the germination of fungi by the induction of iron competition through the production of siderophores (iron chelating compounds). These are effective only in those soils where the availability of iron is low. Control of foliar and fruit pathogens such as *Botrytis cinerea*, a pathogen of strawberries, has been accomplished by the foliar application of the soil inhabiting fungus *Trichoderma viride* (6). This fungus inhibits *Botrytis* using a combination of antibiosis and competition. On grapevines, *Trichoderma harzianum* competes with *B. cinerea* on senescent floral parts, thus preventing the infection of the ovary. It has also been shown to coil around the hyphae of the pathogen during hyperparasitism (7).

Bioherbicides

Fungi are the most important group of pathogens causing plant disease. Therefore, fungi are most commonly used as the active ingredient in bioherbicides and as such the formulated organism is referred to as a mycoherbicide (1). The aim of bioherbicide development is to overcome the natural constraints of a weed–pathogen interaction, thereby creating a disease epidemic on a target host(s) (8). For example, the application of fungal propagules to the entire weed population overcomes the constraint of poor dissemination. After removal of the host weed, the pathogen generally returns to background levels because of natural constraints on survival and spread.

The first commercially available biopesticide for the control of weeds was DeVine®, a mycoherbicide for the control of stranglervine in citrus groves in the United States. It was released in 1981 (8). In 1982, a formulation of *Colletotrichum gloeosporioides* f.sp. *aeschynomene*, was released to control northern jointvetch in soybean crops in the United States. Since these early examples there have been a range of organisms investigated with a view to the production of a commercial bioherbicide in Canada, United States, Europe, Japan, Australia, and South Africa. Necrotrophic or hemibiotrophic fungi are usually used as the basis of mycoherbicides, as they can be readily cultured on artificial media and so lend themselves to mass production. Other desirable characteristics of fungi under consideration as mycoherbicides include the ability to sporulate freely in artificial culture, limited ability to spread from the site of application, and genetic stability. In most cases these biopesticides are applied in a similar fashion to chemical herbicides using existing equipment, although the development of specialized application equipment and formulation may improve their efficacy and reliability.

The development of biopesticides relies on agent discovery and selection, development of methods to culture the pathogen (or nematode), creation of formulations that protect the organism in storage as well as aid in its delivery, studies of field efficacy, and methods of storage. Each biopesticide is unique, in that not only will the organism vary but so too will the host, the environment in which it is being applied, and economics of production and control.

There are a number of advantages of the use of biopesticides over the use of conventional pesticides, including the minimal residue levels, control of pests already showing resistance to conventional pesticides, host specificity, and the reduced chance of resistance to biopesticides. This indicates an emerging, strong role for biopesticides in any integrated pest management strategy and an important involvement in sustainable farming production systems in the future.

FUTURE DIRECTIONS

There have been some spectacular successes in the use of biopesticides, despite the perceived constraints to their deployment (9). In the past, biopesticides have been expected to behave in the same way as synthetic pesticides. For the ultimate success of biopesticides, microorganisms developed for biological control must be viewed by researchers, manufacturers, and end-users in a biological paradigm rather than a chemical one.

The efficacy and reliability of biopesticides is largely determined by environmental parameters in the period between application and infection. In this period, the availability of moisture is of the utmost importance. Furthermore, the narrow host range of many pathogens may restrict their commercial attractiveness. Both of these issues can be addressed by research into the use of genetic engineering and formulation. As research into the molecular basis of host specificity and pathogenesis continues, it will become possible to produce more aggressive pathogens with the desired host range for biological control. The survival and efficacy of these pathogens will be enhanced through the use of novel formulations.

REFERENCES

1. Crump, N.S.; Cother, E.J.; Ash, G.J. Clarifying the nomenclature in microbial weed control. Biocontrol Sci. Tech. **1999**, *9*, 89–97.
2. Mortensen, K.; Makowski, R.M.D. Tolerance of strawber-

ries to *Colletotrichum gloeosponriodes* F. sp *malvae*, a mycoherbicide for control of round-leaved mallow (*Malva pusilla*). Weed Sci. **1995**, *43* (3), 429–433.
3. Van, Drieshe R.G.; Bellows, T.S. *Biological Control*; Chapman & Hall: New York, NY, 1996.
4. Kaya, H.K.; Gaugler, R. Entomopathogenic nematodes. Ann. Rev. of Entomol. **1993**, *38*, 181–206.
5. Johnsson, L.; Hokeberg, M.; Gerhardson, B. Performance of the *Pseudomonas chlororaphis* biocontrol agent ma 342 against cereal seed-borne diseases in field experiments. Eur. J. Plant Pathol. **1998**, *104* (7), 701–711.
6. Sutton, J.C.; Peng, G. Biocontrol of *Botrytis-cinerea* in strawberry leaves. Phytopathology **1993**, *83* (6), 615–621.
7. Oneill, T.M.; Elad, Y.; Shtienberg, D.; Cohen, A. Control of grapevine grey mould with *Trichoderma harzianum* t39. Biocontrol Sci. & Tech. **1996**, *6* (2), 139–146.
8. TeBeest, D.O.; Yang, X.B.; Cisar, C.R. The status of biological control of weeds with fungal pathogens. Ann. Rev. of Phytopath. **1992**, *30*, 637–657.
9. Auld, B.A.; Morin, L. Constraints in the development of bioherbicides. Weed Tech. **1995**, *9* (3), 638–652.

FURTHER READING

Baker, K.F.; Cook, R.J. *Biological Control of Plant Pathogens*; WH Freeman and Company: San Francisco, CA, 1974.

Chen, J.; Abawi, G.S.; Zuckerman, B.M. Efficacy of *Bacillus thuringiensis*, *Paecilomyces marquandii*, and *Streptomyces costaricanus* with and without organic amendments against *Meloidogyne hapla* infecting lettuce. J. Nematol. **2000**, *32* (1), 70–77.

Lacey, L.A.; Goettel, M.S. Current developments in microbial control of insect pests and prospects for the Early 21st Century. Entomophaga **1995**, *40* (1), 3–27.

Sticher, L.; Mauchmani, B.; Metraux, J.P. Systemic acquired resistance. Ann. Rev. Phytopathol. **1997**, *35*, 235–270.

BIOREMEDIATION

B

Ragini Gothalwal
P.S. Bisen
Institute of Microbiology and Biotechnology,
Barkatullah University, Bhopal, Madhya Pradesh, India

INTRODUCTION

Rapid industrialization and urbanization during the past decades has revolutionized and affected society in many ways. However, the ever-increasing industrial affluence has led to the generation of effluents that are posing a threat to the environment. Thus, it can be said that the boon of modernization has come along with the curse of pollution. The conventional methods employed for wastewater treatment include the use of chemicals and resins. However, these methods are not ecofriendly and in addition they are expensive and cumbersome. In such a scenario, the microbial biomass for bioremediation is an attractive option due to its easy operation and economic viability.

PESTICIDE POLLUTION

A pesticide is a double-edged sword. If used properly it is a boon to humanity but can involve hazards if carelessly handled. There has been an enormous increase in the use of various synthetic pesticides that contribute to the spectacular increase in the crop yield. The soil, groundwater, and sediments are the ultimate sink for the pollutants, where they either are broken down to simpler forms or remain persistent polychlorinated biphenyls (e.g., DDT). Some of the pesticides are susceptible to bioaccumulation/biomagnification and cause more danger to the environment. The safe and economical disposal of excess pesticide waste is a problem of considerable magnitude.

Some of the pesticides include nitroaromatic compounds, polycyclic aromatic hydrocarbons (PAH), derivatives of benzene, and phenolic compounds. Toxicity and bioaccumulation potential of chlorophenol increase with the degree of chlorination of lipophilicity. The constituents of many pesticides are generally derivatives of benzene. The aromatic ring has large negative resonance energy and for this benzene and its derivatives are a stable group of compounds (1). Recent research has revealed a number of microbial systems capable of biodegradation of organic compounds.

MICROBIAL POTENTIAL OF PESTICIDE DEGRADATION

The natural capacity of microorganisms to degrade a huge variety of herbicides, pesticides, and some inorganic compounds (2) is the essence of the microbial method for degradation of soil contaminants (bioremediation)—the basis for "green" technologies (3). Another very interesting feature of microorganisms is that they degrade organic substances that are produced only synthetically. Although degradation does not always lead to detoxification, in many cases the products are less hazardous and/or become susceptible to further degradation.

A relatively enormous catalytic power, large surface volume action, and rapid rate of reproduction of microorganisms contribute to the major role they play in the chemical transformation. Generally there are three possibilities by which the microbial community acquires the ability to degrade pesticide: 1) some organisms require adaptation before attacking a pollutant having degradative enzymes of a constitutive type; 2) some through random mutation; and 3) some have adaptive enzymes that are induced in the presence of a particular pollutant. The most important environmental factors affecting biodegradation are temperature, pH, water, and oxygen content. The biochemical processes induced by microorganisms under aerobic and anaerobic conditions are mineralization, detoxification, cometabolism, and activation. The microbes engaged in degradation of pesticides are bacteria, actinomycetes, fungi, algae, and cyanobacteria.

Bacteria and Actinomycetes

Twenty-eight genera of bacteria that utilize aliphatic hydrocarbon have been isolated (Table 1).

Table 1 Examples of bacteria, actinomycetes, and fungi engaged in pesticide degradation

Organism	Organic compounds or pesticides
Bacteria and Actinomycetes	
Alcaligens denitrificans	Fluoranthene (PAH)
Alcaligens faecalis	Arylacetonitriles
Archrombacter	Carbofuran
Arthrobacter	EPTC, glyphosate, pentachlorophenol
Bacillus sphaericue	Urea herbicides
Brevibacterium oxydans IH 35A	Cyclohexylamine
Burkholderia sp. P514	1,2,4,5-TeCB
Clostridium	Quinoline, glyphosate
Comomonas testosteroni	Arylacetonitriles
Corynebacterium nitrophilus	Acetonitrile, carboxylic acid, ketones
Dehalococcoides ethenogenes 195	Trichloroethylene (TCE)
Desulfitobacterium dehalogenes	Hydroxylated PCBs
Desulfovibrio sp.	Nitroaromatic compound
Flavobacterium	Pentachlorophenol
Geobacter sp.	Aromatic compound
Klebsiella pneumoniae	3&4 hydrobenzoate
Methylococcus capsulatus (Bath)	Trichloroethylene
Methylosinus trichosporium OB 3b	1,1,1-trichloroethane (TCA)
Moraxella	Quinoline, glyphosate
Nitrosomonas europaea	1,1,1-trichloroethane
Nocardia	Quinoline, glyphosate
Pseudomonas aeruginosa	Nitriles, biphenyl, parathion
Pseudomonas sp.	Quinoline, glyphosate
Pseudomonas stutzeri	Parathion
Pseudomonas cepacia	2,4,5-T
Pseudomonas paucimobilis	PCP
Pseudomonas putida 6786	Propane
Pseudomonas striata	Propham, chloropham
Rhodococcus chlorophenolicus	Pentachlorophenol
Rhodococcus corallinus	S-triazines
Rhodococcus rhodochrous	Propane
Rhodococcus sp.	Propane, 1,1,1,-trichloroethane
Rhodococcus UM1	Pyrene
Fungi	
Aspergillus flavous	DDT
Aspergillus paraceticus	DDT
Aspergillus niger	2,4-D
Candida tropicalis	Phenol
Chrysosporium lignorum	3,4-dichloroaniline
Fusarium solani	Acylamilide
Fusarium oxysporum	DDT
Hendesonula toruleidea	2,4-D
Hydrogenomonas + *Fusarium sp.*	DDM, nitrile
Mucor alterans	DDT
Penicillium	Acylamilide
Penicillium megasporum	2,4-D
Phallinue weirii	DDT
Phanerochaete chrysosporium	PAH, 2,4,6-trinitrotoluene, pentachlorophenol, DDT, 2,4,5-T and lindane
Pleurotus ostreatus	DDT
Polyporue versicolor	DDT
Pullularia	Acylamilide

(Continued)

Table 1 Examples of bacteria, actinomycetes, and fungi engaged in pesticide degradation (*Continued*)

Organism	Organic compounds or pesticides
Fungi	
Rhodotorula	Benzaldehyde
Stereum hirsistum	Phenanthrene
Trametes versicolor	Dieldrin
Trichoderma sp.	Nitrile
Trichoderma viride	DDT
Trichospron cutaneum	Phenol
Yeast	Paraquat
Ectomycorrhizal fungi	
Tylospora fibrillosa	Mefluidide
Thetophora terrestris	Mefluidide
Suillus variegatus	Mefluidide
Suillus granulatus	Mefluidide
Suillus luteus	Mefluidide
Hymenoscphur ericae	Mefluidide
Paxillus involutus	Mefluidide

Fungi

There are two main ways by which fungi can be utilized to degrade environmental pollutants. The first approach is to modify sewage treatment systems, the second is "bio-augmentation." Fungi can utilize several cosubstrates for energy. Hence, fungi can effectively degrade several high-strength industrial wastes more efficiently than bacteria (4).

The ecto and ericoid mycorrhizal fungi have been shown to degrade a wide range of persistent organic pollutants such as PCBs, atrazine, 2,4-D, chlorophenol, 2, 4,4-trinitrotoluene, making the pollutants suitable target organisms for facilitating a bioremediation program. The sustainability of the organisms favor their use in bioremediation over white rot fungi that require oxidative enzymes (5) to facilitate remediation (Table 1).

Algae

Microalgae have received more attention in recent years as an alternative secondary biosystem for waste treatment. The success of an algal system relies on its ability to take up inorganic nutrients such as N and P from wastewater and assimilate them for growth, for example, *Chlorella vulgaris, Scendesmus* spp., *Chlamydomonas humicola,* and *Chlorella minesstissima* could rapidly utilize different organic compounds (acetate and glucose).

Cyanobacteria

Utilization of cyanobacteria in effluent treatment is a recent phenomenon. It has great potential to take up external nutrients, hence, it can be a good candidate for tertiary treatment of industrial effluents, in turn helping to solve the problem of eutrophication. The common cyanobacterial strain used in wastewater treatment includes *Anabaena doliolum, Anabaena* CH3, *Lyngbya gracilis, Phormidium faveolarum, Oscillatoria animalis, Oscillatoria pseudogeminata, Phormidium laminosum,* and *Spirulina maxima*. Cyanobacterial filters comprising of *Oscillatoria annae* and *Phormidium tenue* proved to be most suitable for sewage water treatment.

CONSTRUCTED STRAINS

In the natural environment a mixed culture plays an important role in the degradation of xenobiotic compounds. Some members of the culture might be able to provide important degradative enzymes whereas others supply surfactants or recombinant biocatalysts to ensure the bioavailability of pollutants (6, 7). A consortium of *Arthobacter ilicis* and *Agrobacterium radiobacter* mineralized ethylene glycol nitrate. The *Arthrobacter* strain was the actual degrading organism, although the second microbe facilitated the mineralization. The bacterial degradation of Benzene, Toluene, Ethylene, and Xylene (BTEX) has used microbial consortia since no pure culture is known to degrade all the components of BTEX efficiently.

The enzymes involved in the catalysis of many of these pesticides generally can be grouped into two classes namely hydrolases and oxidoreductases. Microorganisms have evolved the degradative capacity by altering the substrate specificity of an enzyme (6) already encoded in

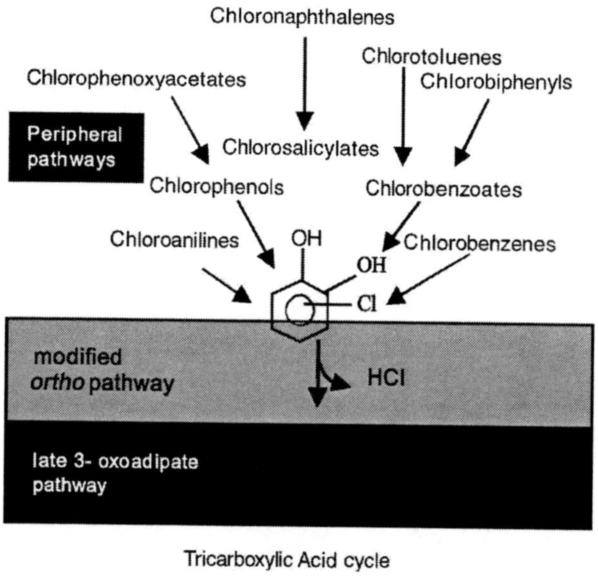

Fig. 1 Combination of pathway segments to develop a hybrid pathway for the mineralization of chloroaromatic compounds.

the genome, as seen in the evolution of the 3-chlorobenzoate pathway (Fig. 1). It is possible nonetheless, to use selective evolution (through the use of a chemostat) to develop a strain capable of degrading recalcitrant 2,4,5,-T (6). This type of evolution is facilitated by the presence of a gene pool, so an organism can readily take up foreign DNA. However, recent advances in genetic techniques have opened up new avenues to move toward the goal of genetically engineered microorganisms to function as "designer catalysts," in which certain desirable biodegradation pathway or enzymes from different organisms are brought together in a single host with the aim of performing specific detoxification (6, 7). These applications will ultimately create "super biocatalysts" capable of degrading several pesticides rapidly and cost effectively (8). The transfer of plasmid pJP4 from *Alcaligens eutrophus* JMP134 into *Pseudomonas cepacia* AC1100 was performed and constructed strain RHJ 1 was shown to efficiently degrade a mixture of 2,4-D and 2,4,5-T. Such strains in many instances may be better suited than

Table 2 Degradative plasmids with relevance to the construction of chloroaromatics-degrading hybrid strains

Plasmid	Size (kb)	Substrate	Host
Peripheral pathways			
TOL	117	Xylenes, toluene, toluate	*Pseudomonas putida*
NAH7	83	Napthalene via salicylate	*Pseudomonas putida*
pWW60-1	87	Napthalene via salicylate	*Pseudomonas sp.*
pDTG1	83	Napthalene via salicylate	*Pseudomonas putida*
SAL1	85	Salicylate	*Pseudomonas putida*
pKF1	82	Biphenyl via benzoate	*Acinetobacter sp.* (reclassified as *Rhodococcus globerulus*)
pWW100	~200	Biphenyl via benzoate, methylbiphenyls via toluates	*Pseudomonas sp.*
pWW110	>200	Biphenyl via benzoate, methylbiphenyls via toluates	*Pseudomonas sp.*
pCITI	100	Aniline	*Pseudomonas sp.*
pEB	253	Ethylbenzene	*Pseudomonas fluorescens*
pRE4	105	Isopropylbenzene	*Pseudomonas putida*
pWW174	200	Benzene	*Acinetobacter calcoaceticus*
pHMT112	112	Benzene	*Pseudomonas putida*
pEST1005	44	Phenol	*Pseudomonas putida*
pVI150	Mega	Phenol, cresol, 3,4-dimethylphenol	*Pseudomonas sp.*
Central pathways			
pAC25	117	3-Chlorobenzoate	*Pseudomonas putida*
pJP4	77	3-Chlorobenzoate, 2,4-D	*Ralstonia eutropha* (formerly *Alcaligenes eutrophus*)
pBR60	85	3-Chlorobenzoate	*Alcaligens sp.*
pRC10	45	2,4-D	*Flavobacterium sp.*
pP51	100	1,2,4-Trichlorobenzene	*Pseudomonas sp.*
pMAB1s	90	2,4-D	*Burkholderia*(formerly *Pseudomonas*) *cepacia*

(From Reineke, W. Development of hybrid strain for the mineralization of chloroaromatics by patchwork assembly. Annu. Rev. Microbiol. **1998**, *52*, 287–331.)

microbial consortia for use in the degradation of certain toxic chemical mixtures (Table 2).

BIOREMEDIATION TECHNOLOGY

To ensure the success of any implemented bioremediation process, bench scale experiments are usually required to determine the natural and augemented microbial degrading activities in the contaminated media. Biotreatability tests determine the amount of specific microbial metabolic activity, stimulated by the addition of complementary microorganisms and/or nutrients and the optimization of environmental factors (9).

The immobilization of cells offers various advantages and the process is cost effective. Though a majority of the work pertains to entrapment of cells in natural or synthetic polymers, a few are cross-linked with various compounds. The rotating biological contactor (RBC) is an effective reactor configuration and when operated sequentially, it can degrade even 500 mg/l phenol in 24 h (5). Enhanced removal efficiency can be obtained by repeating anaerobic and aerobic treatment processes in combination.

Biocapsules have been tested for various applications and can be produced for site-specific application (9). They are products that surround microorganisms and/or microbial components (protein or enzymes) with stabilizers, nutrients, and other materials to ensure their survival during storage and function on dispersal in various environments.

In situ groundwater biodegradation in the United States has been carried out through various processes by numerous companies namely Biogenesis Technologies, Biopim, ABB Environmental Service, Bioremediation System, Remediation Technologies Inc., IT Corporation, Geo-Microbial Technologies Inc., etc. (10). These companies use natural biological ingredients that are harmless. The contaminants that can be treated are BTEX, TPH, TCE, TCA, phenol, organic acid, monochlorobenzoate, and toxic radioactive materials (at contaminated military sites). The removal effectiveness can approach 100%.

FUTURE DEVELOPMENTS

The understanding of biotransformation pathways is necessary for designing bioremediation strategies and wastewater treatment processes. Enzymes with broad substrate specificity and microorganisms with degradative competence against a range of substrates offer the possibility that one or a few microbial cultures can degrade all the important wastes in a complex mixture. The simplest strategy is improving the biodegradation performance of a consortium through the addition of a "specialist" organism.

ACKNOWLEDGMENT

The authors gratefully acknowledge the financial assistance from University Grants Commission, New Delhi, under the SAP program.

REFERENCES

1. Paul, J.; Varma, A. Microbial Degradation of Xenobiotics—An Overview. In *Microbes: For Health, Wealth & Sustainable Environment*; Varma, A., Ed.; Malhotra Publishing House: New Delhi, 1998; 413–432.
2. Eccles, H. Treatment of metal contaminated wastes: why select a biological process. Trends Biotechnol. **1999**, *17*, 462–465.
3. Karamarev, D.G. Biodegradation of soil contaminants. J. Sci. and Ind. Res. **1999**, *58*, 764–772.
4. Spain, J.C. Biodegradation of nitroaromatic compounds. Annu. Rev. Microbiol. **1995**, *49*, 523–555.
5. Manimekatai, R.; Swaminathan, T. Biodegradation of phenolic compounds using *Phaenrochate chrysosporium*. J. Sci. and Indust. Res. **1998**, *57*, 833–837.
6. Timmis, K.N.; Pieper, D.H. Bacteria designed for bioremediation. Trends Biotechnol. **1999**, *17*, 201–204.
7. Timmis, K.N.; Steffan, R.J.; Unterman, R. Designing microorganisms for treatment of toxic wastes. Annu. Rev. Microbiol. **1994**, *48*, 525–557.
8. Chen, W.; Mulchandani, A. The use of live biocatalysts for pesticide detoxification. Trends Biotechnol. **1998**, *16*, 71–75.
9. Bioremedation Applied Biosciences. http://www.bioprocess.com (accessed Dec 1999).
10. Bioremedation, 1999. http://www.in weh.unu.edu/447/lectures/bioremedation.htm (accessed Dec 1999).

BIOTIC COMMUNITY BALANCE UPSETS

David P. Kreutzweiser
Canadian Forest Service, Sault Ste. Marie, Ontario, Canada

INTRODUCTION

A biotic community can be defined as an assemblage of plant or animal species utilizing common resources and cohabiting a specific area. Examples could include a fish community of a stream, an insect community of a forest floor, and a plankton community of a pond. Interactions among species provide ecological linkages that connect food webs and energy pathways, and these interconnections provide a degree of stability to the community. Toxic effects of pesticides can disrupt these linkages and cause community balance upsets. This can occur, for example, when a pesticide is directly toxic to certain species in a community, and other unaffected species respond to the reduced competition for food resources or habitat niches.

MEASURING BIOTIC COMMUNITY DISTURBANCE

Detecting effects of pesticides on biotic communities typically involves repeated, quantitative sampling and a comparison of community attributes among contaminated and uncontaminated sites over time. Traditional measures of community attributes have focused on structure, and have been expressed in terms of single-variable indices such as species richness, diversity, or abundance. These indices are useful descriptors of community structure but suffer from the fact that they reduce complex community data to a single summary measure, and may miss subtle changes in species composition across sites or times. Over the past decade, ecotoxicologists have increasingly turned to multivariate statistical techniques for analyzing community response data.

Multivariate statistical methods are usually considered superior for the analysis of community data because they retain and incorporate the spatial and temporal multidimensional nature of biological communities. Outputs from ordination analysis, for example, can be plotted to provide graphical representation of community structure in which points that lie close together in ordination space represent communities of similar composition (richness, abundance), while communities with dissimilar species composition are plotted far apart (1). Fig. 1 illustrates the use of an ordination analysis for detecting differences among aquatic insect communities of control and insecticide-treated streams. At both concentrations of the insecticide, the community structure of stream insects clearly shifted away from the natural community composition in control streams as depicted by the separation of treated streams (M and H) from controls (C) in the 3 D ordination plot. The moderate (M) and high (H) concentration streams also tended to separate along axes 1 and 2, indicating a differential response by the insect communities to the two test concentrations.

PESTICIDE EFFECTS ON BIOTIC COMMUNITIES

Aquatic Communities

Some of the best examples of biotic community upsets caused by pesticide applications are found in aquatic studies. Aquatic communities are usually contained within distinct boundaries or systems and this generates a high degree of connectivity among species, and a susceptibility to pesticide-induced disturbances at the community level. Applications of the insecticide diflubenzuron to small ponds reduced populations of several aquatic invertebrate species. This in turn resulted in indirect effects on algae (increased productivity because of release from grazing pressure by the invertebrates) and on juvenile fish populations (reduced production because of limited invertebrate prey availability) (2). Methoxychlor applied to forest streams altered the community structure of stream invertebrates to the extent that in-stream processes were affected. The invertebrate community shifted from one

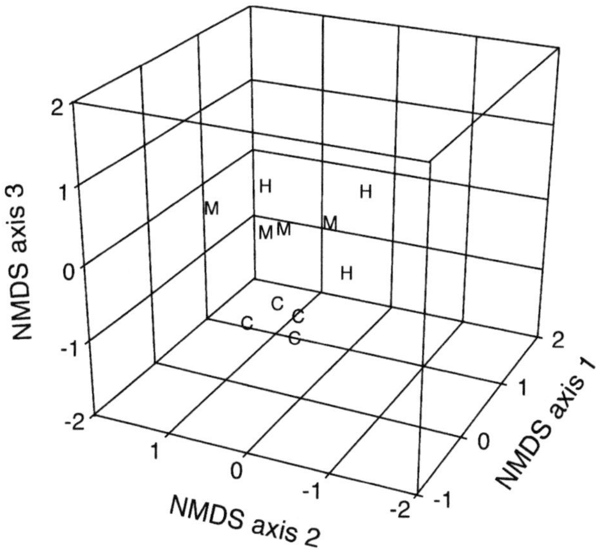

Fig. 1 Ordination by nonmetric multidimensional scaling of aquatic insect communities in stream channels. Each point represents the community structure of control channels (C) and channels treated with a neem-based insecticide at a moderate (M) and high (H) concentration. (From Kreutzweiser, D.P.; Capell, S.S.; Scarr, T.A. Community-level responses by stream insects to neem products containing azadirachtin. Environ. Toxicol. Chem. **2000**, *19*, 855–861.)

dominated by small numbers of large detritivorous insects to one dominated by large numbers of small collector and predator type insects and non-insects. This resulted in a reduction of detritus processing within the stream, thus disrupting the natural energy flow (3). Differential responses among species caused community balance upsets among aquatic insects of stream channels exposed to neem-based insecticides. Reductions in some key species of treated streams caused significant shifts in community structure as compared to the natural community composition among control streams (Fig.1). However, the effects of neem were measured under experimental conditions and only occurred at concentrations well above expected environmental concentrations.

Community disturbance as a result of pesticide exposure has also been demonstrated at lower trophic levels in aquatic systems. Herbicides including atrazine, hexazinone, and copper sulfate have been shown to be directly toxic to most species of phytoplankton (waterborne algae). After herbicide applications, reductions in phytoplankton of contaminated ponds or littoral areas have caused secondary reductions in herbivorous zooplankton, resulting from a depleted food source for the zooplankton. Investigators have also shown that direct adverse effects on phytoplankton can cause disruptions in the bacterial-based energy pathways as well by reducing carbon flow from phytoplankton to bacteria, and ultimately to grazing protozoans and zooplankton (4). A molt-inducing insecticide, tebufenozide, was applied to enclosures in a small lake and was found to be selectively toxic to small crustacean zooplankton (cladocerans) while larger zooplankton (copepods) and small noncrustacean zooplankton (rotifers) were not directly affected. Reductions in cladocerans, however, caused secondary increases in rotifer populations resulting from reduced competition with cladocerans for a common food source. This community balance upset was sufficient to cause measurable changes in system-level metabolism of the lake enclosures (5).

Terrestrial Communities

Pesticides are less likely to cause biotic community balance upsets in terrestrial ecosystems, especially at higher trophic levels. Individuals or populations of nontarget species may be affected by pesticide contamination, but because habitat boundaries are usually less confining or restrictive than in aquatic systems, community-level disturbances in terrestrial environments are often alleviated by immigration or recruitment from adjacent populations. Nevertheless, significant pesticide effects can occur in terrestrial environments. Organophosphate and carbamate insecticides have been shown to directly kill or seriously impair the behavior of wild birds after applications to control insect pests in agricultural or forest regions resulting in changes to native bird populations. Herbicides have reduced the quality of habitat for some birds and other wildlife that rely on specific plant species, and this has resulted in altered behavioral patterns and population dynamics. In some extreme cases pesticide poisoning has resulted in large bird kills, but it is not known what the effects of these kills, and those of more subtle, chronic exposures to sublethal concentrations, are on whole bird communities (6, 7).

Pesticides can affect community balance at lower trophic levels in terrestrial ecosystems, but effects at the community level are often short-lived because of immigration from nearby populations. Synthetic pyrethroids used for control of tsetse flies have had significant adverse effects on nontarget arthropod communities. Some arthropod species have been shown to be more sensitive to the pyrethroids than others, causing shifts in the community structure, but recovery to pretreatment conditions was usually apparent within 2 or 3 months after the applications. Similarly, field trials conducted to examine effects of neem-based insecticides on nontarget invertebrate communities showed that neem was selectively toxic among nontarget species, and that this caused some changes in community structure. Recovery was generally documented within a few months after the treatments (8). On the other hand, pesticides with longer residual action can sustain community balance upset for longer periods. Dimilin was applied to deciduous forests to control gypsy moth defoliation, and the diversity of species among canopy arthropods was reduced for the duration of a 27-month study period. During this period, other species of canopy arthropods were unaffected causing an imbalance in the forest arthropod community that was over and above the natural variation in community structure (9).

Earthworm species have been shown to be variously affected by pesticides, either through direct toxicity or sublethal effects on growth and fertility, and can be useful indicators of disturbance to soil macroinvertebrates. Certain insecticides in particular are highly toxic to some species, including organochlorine insecticides lindane and endrin, the organophosphate phorate, and the carbamate insecticide carbofuran. On the other hand, natural and synthetic pryrethroids appear to be much less toxic to earthworms. Many studies have demonstrated that pesticide applications can decrease the species diversity and abundance of spiders. Community balance is affected when various species of spiders respond differently to pesticide contamination. Indirect effects such as reductions in prey availability, repellent activity, and changes in habitat through herbicide applications can also cause disruptions in community structure. As a further example, soil macroinvertebrate communities have been disturbed through differential sensitivities among soil-dwelling Diptera to a number of pesticides (10).

Community balance upsets have been demonstrated among soil microarthropods as well. Various pesticides, for example, have selectively affected species among oribatid and astigmatid mites, with responses including population reductions, no changes, and population increases. Differences in population dynamics between these two groups have contributed to community balance upsets. Oribatids have longer life cycles (more than one year) and therefore have limited capacity to respond rapidly to perturbation. In contrast, astigmatid mites have shorter life cycles and are capable of exploiting disturbed microhabitats. A general resistance of astigmatids to pesticide disturbance has been documented. Life cycle differences also played a role in the susceptibility of other mites and aphids to pesticide exposure. Species with higher reproductive potential showed greater resiliency to pesticide effects at the population level (11). Studies such as these demonstrate that differential population dynamics or life cycle characteristics among species can induce community-level responses to pesticides. These are traits that must be considered when evaluating the risk of pesticide effects on biotic communities.

FUTURE DIRECTIONS IN DETECTING COMMUNITY BALANCE UPSETS

Environmental concerns about the use of widely effective, broad-spectrum pesticides have led to increased emphasis on the development and use of narrow-spectrum materials targeted at particular pest species. As pesticides become more pest-specific, their effects on nontarget biotic communities are likely to become more subtle and difficult to detect. This will require increased reliance on more sophisticated biological sampling and data analysis techniques to determine community balance upset in response to pesticide exposure. Advances in detecting community-level effects of pesticides will improve our capabilities in assessing and predicting the environmental risks of pesticide use (12, 13), and will contribute to ensuring the sustainability of biotic communities in areas subjected to pest management programs. Advances in the development of more pest-specific and environmentally benign control products are likely to decrease the potential for community balance upsets as a result of pesticide applications.

REFERENCES

1. van Wijngaarden, R.P.A.; van den Brink, P.J.; Oude Voshaar, J.H.; Leeuwangh, P. Ordination techniques for analyzing the response of biological communities to toxic stress in experimental ecosystems. Ecotoxicology **1995**, *4*, 61–77.
2. Boyle, T.P.; Fairchild, J.F.; Robinson-Wilson, E.F.; Haverland, P.S.; Lebo, J.A. Ecological restructuring in experimental aquatic mesocosms due to the application of diflubenzuron. Environ. Toxicol. Chem. **1996**, *15*, 1806–1814.
3. Cuffney, T.F.; Wallace, J.B.; Webster, J.R. Pesticide manipulation of a headwater stream: invertebrate responses and their significance for ecosystem processes. Freshwat. Invert. Biol. **1984**, *3*, 153–171.
4. deNoyelles, F., Jr.; Dewey, S.L.; Huggins, D.G.; Kettle, W. D. Aquatic Mesocosms in Ecological Effects Testing: Detecting Direct and Indirect Effects of Pesticides. In *Aquatic Mesocosm Studies in Ecological Risk Assessment*; Graney, R.L., Kennedy, J.H., Rodgers, J.H. Jr., Eds.; Lewis Publishers: Boca Raton, FL, 1994; 577–603.
5. Kreutzweiser, D.P.; Thomas, D.R. Effects of a new molt-inducing insecticide, tebufenozide, on zooplankton communities in lake enclosures. Ecotoxicology **1995**, *4*, 307–328.
6. Tome, M.W.; Grue, C.E.; Henry, M.G. Case Studies: Effects of Agricultural Pesticides on Waterfowl and Prairie

Pothole Wetlands. In *Handbook of Ecotoxicology*; Hoffman, D.J., Rattner, B.A., Burton, G.A, Jr., Cairns, J., Jr., Eds.; Lewis Publishers: Boca Raton, FL, 1995; 565–576.
7. Canadian Wildlife Service. Pesticides and Wild Birds. In *Hinterland Who's Who*; #CW69-4/98-1998E, Canadian Wildlife Service, Environment Canada: Ottawa, ON, 1998. http://www.ec.gc.ca/cws-scf/nwrc/pesticid.htm (accessed Nov 1999).
8. Stark, J.D. Comparison of the impact of a neem seed kernel extract formulation, ''margosan-o'' and chlorpyrifos on non-target invertebrates inhabiting turf grass. Pestic. Sci. **1992**, *36*, 293–299.
9. Butler, L.; Chrislip, G.A.; Kondo, V.A.; Townsend, E.C. Effect of diflubenzuron on nontarget canopy arthropods in closed, deciduous watersheds in a central Appalachian forest. J. Econ. Entomol. **1997**, *90*, 784–794.
10. Paoletti, M.G. Invertebrate biodiversity as bioindicators of sustainable landscapes. Agric. Ecosys. Environ. **1999**, *74* (special issue), 1–441.
11. Walthall, W.K.; Stark, J.D. A comparison of acute mortality and population growth rate as endpoints of toxicological effect. Ecotoxicol. Environ. Saf. **1997**, *37*, 45–52.
12. Matthews, R.A.; Matthews, G.B.; Landis, W.G. Application of Community Level Toxicity Testing to Environmental Risk Assessment. In *Risk Assesment: Logic and Measurement*; Newman, M.C., Strojan, C.L., Eds.; Ann Arbor Press, Inc.: Ann Arbor, 1998; 225–253.
13. van den Brink, P.J.; ter Braak, C.J.F. Principle response curves: analysis of time dependent multivariate responses of biological communities to stress. Environ. Toxicol. Chem. **1999**, *18*, 138–148.

BIRD DAMAGE AND CONTROL

James J. Matee
Tropical Pesticides Research Institute, Arusha, Tanzania

INTRODUCTION

Birds have coexisted with human beings since time immemorial: some have been beneficial and others harmful (1). Birds that are in conflict with human beings in a number of ways are termed as pests. Worldwide bird pests are endemic in localized geographical areas. They cause damage to crops in many parts of the world such as corn in northern America, wheat in New Zealand, orchards in Australia, and grain in Africa (2). The most serious bird pests are gregarious and migratory, moving from one area or country to another. The red-billed quelea (*Quelea quelea*), known to be the most numerous and destructive in the world (3), is confined to Africa in about 25 countries south of Sahara, on the Sahel and Savannah regions (1, 4). Its population is estimated to be at least 1.5 billion; it is highly nomadic, and has extensive breeding grounds, estimated to cover 2 million square kilometres. Quelea birds cause extensive damage to cereal grains, especially wheat, rice, barley, sorghums, and millets at milky (dough) and ripening stages.

Control of quelea birds has evolved from traditional methods of using scarecrows, noise-making devices, and slings to flame throwers (5), burning of roosts and colonies, cutting trees harboring roosts and colonies, and ultimately the use of chemical poisons like parathion and fenthion to kill the birds wherever they are located. Despite the annual destruction of millions of quelea birds every year by use of pesticides, damage has continued to increase annually.

DAMAGE BY QUELEA BIRDS

Bird pests of economic importance tend to be communal or flock feeders (gregarious). Quelea birds are waste feeders, they feed and destroy grains equivalent to their own body weight (average 18 grams) per day. Thus a flock of 2 million birds (a common phenomenon in Eastern Africa) can destroy up to 50 tons of grain in a day (or 1500 tons within 30 days between dough and ripening stage) worth an equivalent of $600,000.

Although crop losses due to quelea birds may not exceed 5% on a national scale, quelea can cause catastrophic damage to grain crops locally in some years. Systematic assessments for the five eastern African countries (Somalia, Kenya, Tanzania, Ethiopia, and Sudan) have indicated an annual total loss of grain worth $15 million. Such losses can be equated to the amount that could be used for importing food to replace the deficit (1). Combined damage by other bird species is insignificant compared to that of quelea birds.

DAMAGE BY NONQUELEA BIRDS

Apart from birds being pests on cereal grains, damage is also experienced in fruits, sown seeds, and young seedlings. Parrots cause damage to fruits; red-winged blackbirds (*Agelaius phoeniceus*), common grackles (*Quiscalus quiscula*), starlings (*Sturnus vulgaris*), and crows (*Corvus* spp.) cause damage to sown seeds and maize during the ripening stage.

Birds of the families Anatidae, Rallidae, Phasionidae, Gruidae, and Charadriidae cause damage to sown seeds and seedlings of rice crops.

BIRD CONTROL

Reduction of bird damage varies from one locality to another depending on the magnitude, nature of damage, and species involved. Quelea control has varied from traditional methods used by subsistence farmers to the use of pesticides and modern spray techniques. The common method of applying pesticides (fenthion) is by spraying roosts and colonies at night using aircraft (6).

Quelea Control in Africa

Bird control in cereal-growing countries in Africa was in the past carried out using the organochlorine parathion and has now been replaced with fenthion.

Spray quantities of this avicide have been reduced from more than 30 L/ha to 10–15 L/ha when applied by fixed-wing aircraft and 1–2 L/ha when applied by helicopter with ULV spraying equipment.

The current quelea control strategy is by aerial spraying of roosts and colonies by use of organophosphorus pesticides (fenthion 60% active ingredient), which is effective and less harmful to the environment (4, 6, 7).

ALTERNATIVES TO CHEMICAL CONTROL

Repellents as Bird-Control Alternatives

Chemical repellents may provide an effective and humane method of reducing bird damage to crops via modification of the feeding behavior of the target species (8–10).

Bird-control measures include the use of chemical repellents as a preplanting seed treatment against birds like red-winged blackbirds (*Agelaius phoeniceus*) and boat-tailed grackles (*Quiscalus major*).

Application of repellents such as methiocarb on the edges of the field and in spots that are being damaged have shown some success in reduction of bird damage to wheat in eastern Africa, although the presence of weeds in a field can completely negate any repellent effect of a chemical.

Sticky repellents with or without toxic chemicals can be painted on tree barks or branches. For example, lassa made out of jackfruit milk, gum arabic, commercial coal tar and a silica-based paint are common products used as stickers to repel birds (11).

Cultural Control

Growing crops less susceptible or less preferred by birds, i.e., breeding for resistant varieties, dark-seeded, tannin rich, and poorly digestible by birds; trapping quelea birds en mass and processing them for poultry feeds; and early harvesting of crops before damage threatens to be serious, reduce the use of chemicals (12).

Recorded distress calls amplified over speakers to drive away perching birds from fields is more effective in orchards and smaller crop areas. Reflective ribbons—polypropylene, metallic, shiny red, and silvery white strips are also useful. The reflection of bright sun rays and the humming sound produced by the wind over the strips scare birds from the fields.

ENVIRONMENTAL IMPACT OF BIRD CONTROL

Quelea birds roost communally in strategic sites for security. These sites may be in inaccessible thorny acacia, tall eucalyptus species, or reeds in swamps covering areas ranging up to 25 ha thus necessitating aerial spraying of pesticides. Although the use of pesticides is varied, direct mortality of birds in and around treated fields, roosts, or colonies is one of the most visible signs of pesticide impact.

Secondary poisoning occurs when predators consume prey contaminated by these pesticides thus reducing the local population of such species (nontarget), as in the case of quelea sprays (6, 13–16).

The gregarious behavior of quelea birds tends to attract other species in the process of feeding and roosting. During spray operations such species are also killed (7).

Massive killings of quelea birds attract local communities in the area to collect the dead birds for consumption, thus creating health hazards. Although organophosphate spraying for controlling bird pests has proved successful, nontarget fauna have also been affected indiscriminately (7).

Target-specific pesticides must be chosen in order to make pest-control programs compatible with conservation of wildlife and the environment.

Effective control operations may lead to resurgence of species that initially were not pests—especially the weaver birds (Ploceidae)—during quelea control operations. Where spraying of quelea birds is carried out in swampy areas, the environmental implication is the killing of other organisms and aquatic vertebrates like fish. Beneficial insects are also killed during these operations. In order to overcome the impact on the nontarget species, control operations have to be limited to only those concentrations that are adjacent to vulnerable crops and only those concentrations that are causing economic damage to crops, hence damage assessment has to be carried out before control is conducted.

FUTURE CONCERNS

Pesticides should be used only when pest populations are likely to cause economic damage and nonchemical methods will not be effective enough. Quelea control operations must be targeted only to roosts causing economic damage to crops.

Further research on safer products like pyrethroids and nonchemical alternatives must be given more emphasis so as to preserve biodiversity.

REFERENCES

1. Elias, D.J. Pests with backbones. CERES **1988**, *21* (2), 122.

2. Hone, J. Bird Damage to Crops. In *Analysis of Vertebrate Control*; Press Syndicate of the University of Cambridge: Cambridge, 1994; 40–45.
3. Anon. Lost crops of Africa. BOSTID **1996**, 1—Grains, Appendix A.
4. Allan, R. *The Grain-Eating Birds of Sub-Sahara Africa: Identification, Biology and Management*; Natural Resources Institute (NRI): Chatham, U.K., 1997; 199.
5. Mathur, L.M.L. Save Your Crop from Bird Damage. In *Environmentally Sound Technologies for Women in Agriculture (IIRR)*. http://www.oneworld.org/globalprojects/humcdrom (accessed Sept 4 1999).
6. Roux, D.; Jooste, S.; Truter, E.; Kempster, P. An aquatic toxicological evaluation of fenthion in the context of finch control in South Africa. Ecotox. and Environ. Safe. **1995**, *31* (2), 164–172.
7. Mullie, W.; Diallo, A.O.; Gadji, B.; Ndiaye, M.D. Environmental hazards of mobile ground spraying with cyanophos and fenthion for quelea control in Senegal. Ecotox. and Environ. Safe. **1999**, *43* (1), 1–10.
8. Gill, E.L.; Feare, C.J.; Cowan, D.P.; Fox, S.M.; Bishop, J.D.; Langton, S.D.; Watkins, R.W.; Gurney, J.W. Cinnamamide modifies foraging behaviour of free-living birds. J. Wildl. Manage. **1998**, *62* (3), 872–884.
9. Avery, M.L.; Humphrey, J.S.; Primus, T.M.; Decker, D.G.; McGrane, A.P. Anthraquinone protects rice seed from birds. Crop Prot. **1998**, *17* (3), 225–230.
10. Stevens, G.R.; Clark, L. Bird Repellents: Development of avian specific tear gases for resolution of human–wildlife conflicts. Int. Biodeterior. Biodegrad. **1998**, *42* (2–3), 153–160.
11. Cowan, D. Newslines—The End of Bird Scarers. In *SPORE, Bull. of the CTA*; U.K., 1995; 56.
12. FAO. FAO in action. CERES **1985**, *106*.
13. Keith, J.O.; Bruggers, R.L. Review of hazards to raptors from pest control in Sahelian Africa. J. of Raptor Res. **1998**, *32* (2), 151–158.
14. Hunt, K.A.; Bird, D.M.; Mineau, P.; Shutt, L. Secondary poisoning hazard of fenthion to American kestrels. Arch. of Environ. Contamination and Toxicol. **1991**, *21*, 84–90.
15. Hunt, K.A.; Bird, D.M.; Mineau, P.; Shutt, L. Selective predation of organophosphate exposed prey by American kestrels. Animal Behav. **1992**, *43* (6), 971–676.
16. Mineau, P.; Keith, J.A. *Pesticides and Wildlife: A Short Guide to Detecting and Reducing Impact of Pesticides Use in Developing Countries (IDRC)*, Proceedings of a Symposium, Ottawa, Canada, Sept 17–20, Forget, G., Goodman, T., de Villiers, A., Eds.; IDRC: Ottawa, 1990; 56.

BIRD IMPACTS

Pierre Mineau
Canadian Wildlife Service, Ottawa, Ontario, Canada

B

INTRODUCTION

Birds inhabiting our farmland are in decline. This is known to be the result of agricultural intensification in which pesticide use plays a large direct and indirect role. Most pesticides used in developed countries no longer accumulate in birds but they can poison birds or make them more susceptible to other causes of mortality; pesticides are known to alter birds' basic requirements of food and shelter and may be affecting their reproduction on a large scale. Several different strategies are employed to study bird impacts, ranging from close monitoring of pesticide applications to surveys of birds in farms subject to different pesticide regimes.

HOW SERIOUS IS THE IMPACT?

Most of our farmland bird species appear to be declining globally and even common species are experiencing long-term declines, both in North America and in Europe (1, 2). For example, 76% of common grassland species in Canada are declining. The proportion of species declining or showing range contractions in the United Kingdom is higher still. Less is known about common farmland bird species outside of Europe or North America but farming worldwide has been implicated in declines of specific groups of birds such as raptors. It is difficult to isolate the specific factors responsible for these declines: it is likely that a combination of factors is to blame and each species must be considered on a case-by-case basis. Agricultural landscapes have changed dramatically in the twentieth century. There has been a shift from mixed agriculture, including row crops, field crops, and livestock, to more specialized farming where monocultures are regionally dominant. Field size has increased to accommodate larger machinery and this increase often has been at the expense of marginal noncrop habitats such as fencelines, ditches, hedgerows, windbreaks, and remnant woodlots. In Europe especially, a shift to autumn sowing of grain crops has meant that much of the waste grain traditionally available to birds postharvest and throughout the lean winter months is no longer available.

Agricultural inputs in the form of synthetic fertilizers and pesticides have increased dramatically also and, increasingly, have been found to be contributing to bird declines. Two decades ago, a long-term study of declining grey partridge (*Perdix perdix*) populations in the United Kingdom identified insect prey reductions resulting from both insecticide and herbicide use as the main contributing factor. More recently, researchers in several regions of North America and Europe have shown that organic farms tend to support a higher diversity and abundance of birds even when matched for habitat characteristics. The reproductive success of some farmland species such as the Eurasian skylark (*Alauda arvensis*) is higher on organic or reduced input farms than it is on more "conventional" ones. Taken as a whole, these results implicate current agricultural practices, and pesticide use in particular, in the decline of several farmland species. This is doubly unfortunate because, with a few exceptions, birds can play a useful role in integrated pest management systems (3).

TYPES OF BIRD IMPACTS

There are several mechanisms through which pesticides can affect birds (4). The case of the grey partridge, Eurasian skylark, and other European species has shown that the effect can be an indirect one, mediated through "weed" removal and loss of insect biomass at critical times of the breeding season (5). Herbicide use has increased dramatically in the past decades and herbicide sales far surpass insecticide sales in North America and Europe at least. However, several direct mechanisms through which birds are impacted are also recognized.

Persistent Organochlorine Pesticides

Historically, several species of raptors such as the Eurasian sparrowhawk (*Accipiter nisus*) and peregrine falcon (*Falco peregrinus*) as well as fish-eating species such as the brown pelican (*Pelecanus occidentalis*) faced serious

Encyclopedia of Pest Management

difficulties and regional extinction as a result of persistent organochlorine pesticides such as DDT, aldrin/dieldrin, chlordane, and heptachlor. These were poorly metabolized and poorly excreted by birds and accumulated in fatty tissue. The impact of such substances were twofold. Some, such as aldrin and dieldrin caused frequent poisonings, especially during lean times when birds metabolized their fat reserves and the pesticides reached extreme concentrations in the brain. Others, such as DDE, a breakdown product of DDT, interfered with the bird's ability to lay eggs with normal shells. These substances were banned or severely restricted in most of the developed world in the early '70s. Yet, lower reproduction of birds breeding in areas with high historical usage is still being documented because of long persistence in soils. Several of these pesticides continue to be used massively in parts of the world such as the Indian subcontinent although current impacts on bird life are poorly documented. By and large, modern pesticides do not show such extreme persistence, at least in warm blooded organisms.

Lethal Effects

Two groups of pesticides, the organophosphorus and carbamate insecticides were introduced to replace persistent organochlorines. Unfortunately, they proved particularly toxic to birds (6). Their mode of action (inhibition of the enzyme acetylcholinesterase in the nervous system and at neuromuscular junctions) is not specific to the pests and affects a broad range of vertebrates and invertebrates alike. Birds are especially vulnerable because their ability to detoxify these pesticides is generally much lower than that of mammals. The more toxic products such as the carbamate insecticide carbofuran can kill thousands of individuals in a single application. Recent reports of large numbers of North American birds being poisoned on their wintering grounds in Latin America by the insecticide monocrotophos have emphasized the need to consider bird impacts in a hemispheric, if not global, context (7). The poisoning of birds is largely inevitable where acutely toxic pesticides are registered at high rates of application and used broadly. Of particular concern have been granular insecticides and seed treatments because birds are often attracted to them (8). Pesticide poisonings can be a significant source of mortality relative to other factors, especially in the case of long-lived species such as birds of prey.

Secondary Poisoning

Secondary poisoning occurs when predators, such as hawks or owls, consume prey contaminated by pesticides. Such predators are few because of their position at the top of the food chain. Therefore, the death of one predator may constitute a significant reduction in the local population of that species. Historically, researchers have associated secondary poisoning with persistent organochlorine insecticides and other substances that are not readily metabolized and that accumulate in tissues. However, other currently registered pesticides can cause secondary poisoning when the predator encounters the pesticide in a high concentration on the surface or in the gastrointestinal tract of its prey. Also, predators capture birds debilitated by insecticides much more easily.

Sublethal Effects and Delayed Mortality

Many pesticides can affect the normal functioning of exposed individuals at doses insufficient to kill them directly. At high doses, the organophosphorus and carbamate pesticides previously described cause respiratory failure and death. However, wild birds exposed to these agents in lesser amounts have experienced impaired coordination, weight loss, an inability to maintain body temperature, and loss of appetite. Also, exposed birds may spend less time at the nest, provide less food for their young, be less able to escape predation, and be more aggressive with their mates. Finally, exposure to some pesticides may reduce resistance to disease.

Effects on Reproduction

A high proportion of pesticides currently registered have the potential to affect reproduction by reducing egg production, hatching, or fledging success although the extent to which this actually happens in the wild is not known (9). A few products cause embryonic mortality when sprayed directly onto eggs.

MEASURING BIRD IMPACTS

Field Testing—Active Monitoring

Carrying out a field study to measure the impact (or lack of impact) of a specific pesticide usually consists of the surveillance of a group of birds prior to, during, and after the application of the pesticide according to label instructions. Researchers observe or count individuals of one or more species within and outside the treated area and record their behavior. Frequently, they search for carcasses in order to determine the extent of pesticide-induced mortality. They may capture birds to ascertain the health of individuals or to collect samples, for example, blood for biochemical assays (10) or feathers and foot rinses for residue determinations. Agricultural engineering studies (e.g., measurements of granular insecticides or treated seed remaining on the surface) or monitoring of pesticide residues remaining on avian food items over time provide valuable information on expected exposure levels. The most sophisticated field studies will involve monitoring nests, as well as banding, marking, or radio-tagging individuals in order to assess turnover rates and help locate sick and dead birds. Rare, vulnerable, or ecological keystone species can be used as indicator species where relevant and feasible.

It is not always feasible to investigate the effects of a single pesticide. In a number of cropping situations, several pesticides are used as a mix or in quick succession making the identification of compound-specific impacts difficult. In agricultural systems, the mosaic of treated fields can be so complex as to make it difficult to assess exposure to any one pesticide. Two approaches then suggest themselves: 1) treated sites or landscapes are compared to nontreated areas provided those can be found, and 2) the ''severity'' of treatment (the a priori expectation of toxicity) for any given site is used as a variable against which a number of different parameters (such as reproductive success) are regressed. Great care must be taken in comparing treated to nontreated areas because they are likely to differ in other ways as well.

Surveys

Data from regional or national surveys of bird population levels are rarely adequate to demonstrate specific pesticide impacts although surveys can point to a general situation of bird declines in farmland. In order to carry out wildlife monitoring in treated areas, it is necessary to have a good knowledge of the normal complement of species for the area of concern and to be able to assess the vulnerability of each of these species during and after pesticide treatments. The diversity or abundance of species may already have been affected by previous pesticide use so that only a complement of the more insensitive species remains available for testing.

Regardless of the strategy employed, more attention needs to be paid to the impact of pest control practices on bird species if we are to reverse the current trends of population declines.

See also *Wildlife Kills*, pages 899–903; *Bioaccumulation*, pages 50–53; *Environmental and Economic Costs of Pesticide Use*, pages 237–239.

REFERENCES

1. Askins, R.A. Population trends in grassland, shrubland, and forest birds in Eastern North America. Current Ornithology **1993**, *11*, 1–34.
2. Sirawardena, G.; Baillie, S.R.; Buckland, S.T.; Fewster, R.M.; Marchant, J.H.; Wilson, J.D. Trends in the abundance of farmland birds: a quantitative comparison of smoothed common birds census indices. J. Appl. Ecol. **1998**, *35*, 24–43.
3. Kirk, D.A.; Evenden, M.D.; Mineau, P. Past and current attempts to evaluate the role of birds as predators of insect pests in temperate agriculture. Current Ornithology **1997**, *13*, 175–269.
4. Grue, C.E.; DeWeese, L.R.; Mineau, P.; Swanson, G.A.; Foster, J.R.; Arnold, P.M.; Huckins, J.N.; Sheehan, P.J.; Marshall, W.K.; Ludden, A.P. In *Potential Impacts of Agricultural Chemicals on Waterfowl and Other Wildlife Inhabiting Prairie Wetlands: An Evaluation of Research Needs and Approaches*, Transactions of the 51st North American and Natural Resources Conference, Wildlife Mgmt. Inst.: Washington, DC, 1986; 357–383.
5. Campbell, L.H.; Avery, M.I.; Donald, P.; Evans, A.D.; Green, R.E.; Wilson, J.D. In *A Review of the Indirect Effect of Pesticides on Birds*, Joint Nature Conservation Committee Report No. 227, JNCC: Peterborough, U.K., 1997.
6. Mineau, P. Difficulties in the Regulatory Assessment of Cholinesterase-inhibiting Insecticides. In *Cholinesterase Inhibiting Insecticides—Their Impact on Wildlife and the Environment*; Mineau, P., Ed.; Elsevier: Amsterdam, 1991; 277–299.
7. Hooper, M.J.; Mineau, P.; Zaccagnini, M.E.; Winegrad, G.W.; Woodbridge, B. Monocrotophos and the Swainson's hawk. Pestic. Outlook **1999**, *10* (3), 97–102.
8. Stafford, T.R.; Best, L.B. Bird response to grit and pesticide granule characteristics: implications for risk assessment and risk reduction. Environ. Toxicol. Chem. **1999**, *18* (4), 722–733.
9. Mineau, P.; Boersma, D.C.; Collins, B. An analysis of avian reproduction studies submitted for pesticide registration. Ecotoxicol. Environ. Saf. **1994**, *29*, 304–329.
10. Mineau, P.; Peakall, D.B. An evaluation of avian impact assessment techniques following broadscale forest insecticide sprays. Environ. Toxicol. and Chem. **1987**, *6* (10), 781–791.

BIRDS IN PEST MANAGEMENT

M.L. Avery
National Wildlife Research Center, Gainesville, Florida, U.S.A.

INTRODUCTION

As human activity encroaches on wildlife habitats and natural food sources become increasingly scarce, it is no wonder that birds turn to habitats such as crop fields and aquaculture ponds. Where else can large flocks of birds find sufficient, nutritious, readily available food? Agricultural crops create ideal foraging sites for gregarious bird species, and virtually everything that humans grow or raise for food is subject to some level of bird damage. For most farmers, bird damage is a fact of life, but not a major concern. The unlucky few producers for whom depredations are severe, however, do incur substantial financial losses.

ESTIMATING BIRD DAMAGE

There is little research on the economic thresholds of bird damage. The point at which the grower can no longer tolerate such loss varies according to numerous factors, and might be as much psychological as it is economical. Reliable, practical methods of measuring bird damage are not well developed for many crops. Damage measurement techniques that are available are time-consuming and often produce very large confidence intervals (1, 2).

Rather than measure damage directly, an alternative is to calculate crop loss from food habits and dietary information, bird population size, and residency period. Such calculations generally yield estimated losses that are small compared to the overall size or value of the crop. Overlooked by indirect estimates of damage, however, is the fact that bird damage is distributed unevenly (Fig. 1). The percentage of the crop damaged by birds might be less that 5% overall, but this means little to an individual producer who loses 20–25% of the crop.

CONSTRAINTS TO CONTROLLING BIRD DAMAGE

There are a number of constraints to the successful management of bird damage problems.

1. Private companies have few financial incentives to invest in the development of bird control chemicals because available markets are small (4). The potential demand for a chemical bird repellent is insufficient to entice most companies to spend the money to meet the data requirements imposed by regulatory agencies.
2. Production cost are crucial in agricultural, so many effective bird control methods that are available are unsuitable because they are too costly.
3. Avian behavior and population dynamics make controlling bird damage particularly challenging. Almost uniformly, depredating species are group foragers. Feeding flocks can number in the hundreds of thousands, and birds can travel great distances to exploit food resources that become available. Birds' ability to exploit crops is further enhanced by their social habit of roosting in large numbers at night. Members of the roost that foraged successfully that day recruit other roost members to the site on subsequent days, thereby increasing pressure on the crop.
4. Public sentiment places constraints on the types of damage management methods that can be imposed. As a result, lethal control methods usually are more difficult to implement than nonlethal methods.

REDUCING THE IMPACT OF BIRDS IN CROPS

A vineyard, orchard, or field of corn represents the best possible return for a bird's foraging effort because food items are readily available and superabundant. To reduce bird damage to crops, it is necessary to increase the cost or reduce the benefit to the birds, thereby reducing profitability of feeding on the crop (5). Only when the costs of finding, handling, eating, and digesting the food items increase sufficiently to exceed the nutritional benefits, will birds shift to other food sources.

In bird damage control, the ultimate reduction in benefit is to exclude birds from the crop with a net or other

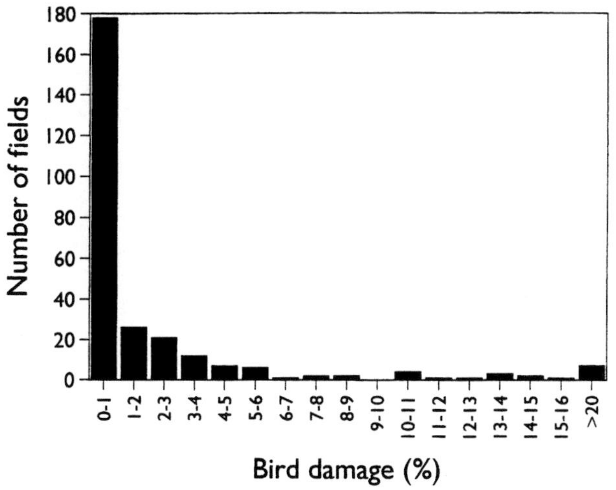

Fig. 1 Bird damage measured in sunflower fields in North and South Dakota, 1996–1998. (Redrawn from data in Ref. 3.)

physical barrier. This also is one of the costlier methods for a producer to employ. Nevertheless for high-value commodities, including tropical fish, early-ripening blueberries, and wine grapes, netting is a cost-effective approach. Netting is impractical for field crops such as corn and rice where hundreds or thousands of acres must be protected. In such cases, the cost per unit area of netting far exceeds the value of the crop.

When exclusion is not feasible, alternate means of increasing costs to birds must be sought. These methods include making the crop unpalatable or more difficult to eat, or making the food items harder to handle. Approaches that address this concept include genetic alteration to produce bird-resistant varieties (6) or the application of chemical feeding deterrents to the crop (7).

Increasing the birds' perceived risk of predation through the use of visual and auditory scare devices will also increase the cost of feeding on the crop. Such scare tactics are usually only effective in the short term, however, as birds readily perceive the emptiness of the implied danger. The effectiveness of nonlethal scare devices might be improved by augmenting these tactics with lethal control but no data are available to confirm this.

BIRDS AND INSECT PESTS

Whereas some species of birds are detrimental to agriculture, there are other situations in which birds' feeding activity is beneficial. Birds are opportunistic feeders, and they can respond facultatively to sudden increases in prey. Individual birds respond by taking greater numbers of the prey species as prey density increases (functional response). Furthermore, the number of avian predators can increase, through immigration or reproduction, with the expanded prey population (numerical response).

Outbreaks of some forest insect pests can be devastating. During the peak of a large-scale population surge of an insect such as the Eastern spruce budworm *Choristoneura fumiferana*, avian predators have relatively little impact. The impact of bird predation may be crucial, however, at the beginning or declining phase of an outbreak and would serve to limit the severity and extent of the damage (8). Predation by birds is also likely to be critical in maintaining insect populations at low levels and in extending the period between population irruptions. Managing forests to increase habitat favorable for insectivorous birds would be likely to increase further the positive effects of bird predation on forest pest species (9).

The abundance of insects in a crop can affect the level of bird damage. Red-winged blackbirds are attracted in greater numbers to plots of corn with high infestations of European corn borer (*Ostrinia nubialis*) than to similar plots with lower infestations (10). As a result, high infestation plots suffer greater bird damage than do the other plots. Similarly, when insect populations in cornfields are reduced with carbamate pesticides, blackbird numbers and bird damage to corn are also reduced. These results suggest that one effective means to lessen the impact of blackbirds, at least in corn, is to reduce the availability of insect prey that might attract birds to the field. Additional studies of this concept in other crops are warranted.

REFERENCES

1. Stone, C.P.; Mott, D.F.; Besser, J.F.; DeGrazio, J.W. Bird damage to corn in the United States in 1970. Wilson B **1972**, *84*, 101–105.
2. Hothem, R.L.; DeHaven, R.W.; Fairaizl, S.D. Bird damage to sunflower in North Dakota, South Dakota, and Minnesota, 1979–1981. *Fish and Wildlife Technical Report 15*, U.S. Department of the Interior Fish and Wildlife Service: Washington, DC, 1988; 1–11.
3. Linz, G.M.; Homan, H.J.; Bleier, W.J. In *Blackbird Densities and Sunflower Damage in North Dakota and South Dakota: 1996–1998*, Proceedings of the Sunflower Research Workshop, 1999; 21, 136–139.
4. Mason, J.R.; Clark, L. In *Nonlethal Repellents: The Development of Cost-effective, Practical Solutions to Agricultural and Industrial Problems*, Proceedings of the Vertebrate Pest Conference, 1992; 15, 115–129.
5. Pyke, G.H.; Pulliam, H.R.; Charnov, E.L. Optimal foraging: a selective review of theory and tests. Q. Rev. Biol. **1977**, *52*, 137–154.
6. Bullard, R.W.; Gebrekidan, B. Agronomic Techniques to

Reduce Quelea Damage to Cereals. In *Quelea Quelea–Africa's Bird Pest*; Bruggers, R.L., Elliott, C.C.H., Eds.; Oxford University Press: Oxford, England, 1989; 281–292.

7. Avery, M.L.; Decker, D.G.; Humphrey, J.S. Development of seed treatments to control blackbirds. Proceedings Vertebrate Pest Conference **1998**, *18*, 354–358.
8. Crawford, H.S.; Jennings, D.T. Predation by birds on spruce budworm *Choristoneura fumiferana*: functional, numerical, and total responses. Ecology **1989**, *70*, 152–163.
9. Holmes, R.T. Ecological and evolutionary impacts of bird predation on forest insects: an overview. Stud. Avian Biol. **1990**, *13*, 6–13.
10. Straub, R.W. Red-winged blackbird damage to sweet corn in relation to infestations of european corn borer (Lepidoptera: Pyralidae). J. Econ. Entomol. **1989**, *82*, 1406–1410.

BOTANICALS

B

Bhupinder P.S. Khambay
IACR-Rothamsted, Harpenden, Hertfordshire, United Kingdom

INTRODUCTION

The term "botanical pesticides" refers to those compounds or materials derived from plants (excluding synthetic analogues) that demonstrate useful toxicity to pests involved in agriculture, horticulture, animal husbandry, and human health problems. The plant material is used either directly (e.g., powdered bark) or as a source of extracts for use in formulations that are often combined with other pesticides, synergists, and formulants to increase efficacy and speed of action. Their main commercial application is in the control of insects, though some are active against fungi, weeds, or rodents (1–5). In comparison to synthetic pesticides, the market for botanical pesticides is very small (less than 2% of the total chemical crop protection market of nearly $30 billion). They are also generally less effective and persist only briefly under field conditions. While this is an advantage with respect to residue levels, most agricultural pest control requires greater persistence. However, public concern over the impact of synthetic pesticides on the environment and nontarget species (though perhaps unwarranted at times) is likely to generate increased interest in the use of natural pesticides, e.g., by organic farmers.

DISCUSSION

Botanical pesticides can be considered under two broad categories: 1) those produced as commercial products, and thus subject to registration procedures, and 2) those prepared locally on farms (mainly by small-scale farmers) whose efficacy is derived from traditional knowledge. In most developed countries, registration requirements are applied equally to botanical and synthetic pesticides, but botanicals can receive preferential consideration in some countries (e.g., the United States) so the registration process is quicker. Although several highly effective plant extracts have already been identified, cost of registration will remain as the main limitation to their commercialization. When the products are based on food-grade materials, e.g., essential oils from fruits, then a clear advantage applies and in many countries registration requirements are substantially reduced or even nil.

The production of commercially successful botanical pesticides is highly dependent on several factors, for example a regular and reliable supply of plant material and the ability to ensure batch to batch consistency (levels of active compounds in plants may change with season, age of plant, growing conditions, etc.).

The most important of these pesticides are listed in Table 1.

BOTANICAL TYPES AND THEIR EFFECTS

Pyrethrum extract contains a mixture of six active compounds that combine rapid knockdown action with lethal activity, a property particularly useful in control-ling flying insects. Its main limitations include high toxicity to aquatic organisms, and eye and skin irritation in some humans. Its lack of adequate photostability prevents wide-scale use in the field. The structure of the most active component, pyrethrin I, was used as a template to develop a new class of highly effective synthetic insecticides (the pyrethroids) for field use; these account for nearly one-third of the global insecticide market. Both mammalian toxicity and production cost of the natural extract are significantly higher than the most important synthetic analogs. Nonetheless, pyrethrum extract is currently by far the most important commercial botanical insecticide.

Although known for centuries for its pesticidal properties and used traditionally by farmers in the tropics, neem extract has only recently been commercialized in the United States (but not yet in the United Kingdom). The compounds it contains are reputed to exhibit several modes of action including antifeedant behavior and growth inhibition of the larval stage of insects. Both properties are insect-specific so, as long as the toxins also present initially are removed, very low mammalian toxicity is observed for the extract. In contrast to pyrethrum extract, it acts slowly (several days) and has a different spectrum of activity. The active ingredients are of much higher molecular weight and therefore do not exhibit vapor action. In addition, the photostability of the extract has been increased significantly, without the loss of efficacy, through improved formulations and simple chemical conversion of the active ingredient(s) (hydrogenation of one or more

Table 1 Important botanical pesticides

Botanical pesticide	Source	Examples of application
Pyrethrum extract	*Tanacetum cinerariaefolium* (Asteraceae)	Wide range of insects and mites in public and animal health, stored product pests, fruits, and vegetables
Neem extract	*Azadirachta indica* (Meliaceae)	Wide range of insects and some fungi
Rotenone	*Derris* and *Lonchocarpus* (Papilionaceae) spp.	Sucking pests on crops, animal health, indoor pests, and controlling fish populations
Nicotine	*Nicotiana* (Solanaceae) spp. e.g., *N. rustica*	Insects on a range of crops, especially in confined spaces
Strychnine	*Strychnos* and other Loganiaceae spp.	Rodents and moles
Fatty acids	Wide range of plant species	Insects and fungi on vegetables, fruits, and ornamental plants; weeds and mosses in lawns
Ryania	*Ryania* (Flacourtiaceae) spp. e.g., *R. speciosa*	Insects on maize and some fruit trees
Essential oils (e.g., containing monoterpenes)	*Citrus* and other species	Insects, fungi, and weeds

alkene bonds). Interestingly some countries are prepared to accept such semi-synthetic materials as "natural" for registration purposes. The extract of a related plant species, *Melia azadarach*, in which the active ingredient is structurally different than that of *A. indica*, has been commercialized in China. Demand for neem will increase as registration is obtained in additional countries.

Rotenone shares many of the market applications of pyrethrum extract but has much higher mammalian toxicity. It was a traditional fish poison, and is still used to control fish populations in some countries. Ryania (also registered in the United States but not the United Kingdom) can be applied under field conditions, especially on maize, but otherwise its use is restricted to selected fruit trees. Nicotine has broad-spectrum activity and is especially effective as a fumigant. However, its high mammalian toxicity severely restricts its use. Strychnine is highly poisonous to mammals and therefore its use is strictly controlled. In the United Kingdom it can be used only in underground worm baits to control moles. Fatty acids have also been used for a long time; recent interest has focused on the control of whiteflies, an increasingly important pest. A few essential oils (e.g., with limonene as the active component) exhibit pesticidal activity, but are usually less active than other classes. However, some have been commercialized for domestic use.

CONCLUSION

Pyrethrum extract, rotenone, nicotine, strychnine, and ryania are known to act primarily at individual target sites on the nervous system. However, neem and essential oils contain mixtures of several types of active compounds that often elicit more than one type of action (e.g., behavior modification and growth inhibition). Several reports have suggested that botanical pesticides are less susceptible than synthetics to the development of resistance. This arises primarily because they are less persistent, so selection pressure is low. In addition, the use of natural mixtures rather than pure compounds, e.g., neem extract vs. azadirachtin, may present a greater challenge to insects and has been shown to slow resistance development in the laboratory. A further consideration is whether cross-resistance has developed. For pyrethroids, insect populations highly resistant to synthetic pyrethroids are less resistant to the natural extract, possibly because the resistance mechanism is particularly sensitive to structural features only present in the synthetics.

REFERENCES

1. Hall, F.R.; Menn, J.J. *Biopesticides: Use and Delivery*; Humana Press: New Jersey, 1999; 626.
2. Copping, L.G. *The BioPesticide Manual*; British Crop Protection Council: Farnham, Surrey, U.K., 1998; 333.
3. Pachlatko, J.P. Natural products in crop protection. Chimia **1998**, *52*, 29–47.
4. Rong, T.; Coats, J.R. Starting from nature to make better insecticides. Chemtech **1995**, 23–28, (July).
5. *Farm Chemicals Handbook 2000*; Meister Publishing Company: Willoughby, Ohio, 2000; 850.

CANCERS FROM PESTICIDES

Jan Dich
Karolinska Institutet and Karolinska University Hospital, Stockholm, Sweden

INTRODUCTION

Chemical exposure from pesticides has been connected worldwide with increased risk not only of acute poisoning, but also in recent decades regarding long-term occupational exposure with increased risk of chronic diseases such as cancer. Pesticides are designed to destroy or control living organisms. They constitute a potential health hazard also for other organisms than those that are intended to be controlled. The health hazard of pesticides is due to the toxic properties both of the active and inert (nonactive) ingredients or contaminants (1, 2).

RISK OF CANCER

About 100 active ingredients in pesticides have been found to cause cancer in experimental animals or humans. Human data, however, are limited by the small number of studies that evaluate pesticides (1). Most public concern about pesticides is directed at the toxicity, health and environmental impacts of active ingredients, but in some cases, the toxic (e.g., carcinogenic) properties derive from the inerts and not from the active ingredients. These compounds are approximately two times the volume of active ingredients (1). Until recently there exists only scarce information concerning these compounds. After 1987, the U.S. EPA has begun to require that manufacturers using the most toxic chemicals (Table 1, List 1) either state it on the label or withdraw it from use (3). The number of inerts in the United States, however, has almost doubled since 1987 from 1200 to more than 2300. The number of active ingredients is in the magnitude of 700. As many as 20 inerts have been listed as known or suspected carcinogens (3, 4). Examples are included in Table 1. In addition, more than 1500 inerts are listed as of unknown toxicity.

The International Agency for Research on Cancer (IARC) and the U.S. National Cancer Institute have tested and evaluated more than 70 pesticides and also intermediate products and contaminants in pesticides. Arsenic compounds and ethylene oxide have been classified as human carcinogens by the IARC. Some other pesticides have been classified as carcinogens based on animal studies (Table 1). Human data, however, are limited by the small number of studies that evaluate individual pesticides (5).

PESTICIDE EXPOSURE

The main cancer sites associated either directly to pesticide exposure or to occupations involving exposure to pesticides are malignant lymphoma (non-Hodgkin's lymphoma and Hodgkin's disease), soft tissue sarcoma, testis cancer, prostate cancer, and brain cancer. The incidence of cancer among occupations in agriculture, forestry, and gardening has been shown to be low for lung and liver cancer, but higher or equal to the general population for other tumors such as lip, skin, malignant lymphoma, soft tissue sarcoma, and multiple myeloma (1, 6–8). These findings draw the attention to the increasing use of pesticides and other chemicals among these occupations as a plausible risk factor. Increased risk of lip cancer, however, depends on exposure to sunlight.

ANIMAL STUDIES

The reliance upon experimental animal data, in the continuing absence of human biomarkers, of known carcinogenic predictivity, is inevitable in the absence of sufficient human evidence. Carcinogenic pesticides may increase the risk of cancer through a variety of mechanisms, including genotoxicity, tumor promotion, hormonal action, and immunotoxicity (1, 5, 9). Complete carcinogens have both genotoxic and tumor-promotive properties. Of pesticides, arsenic compounds are examples of nongenotoxic carcinogens. In addition, chemicals with weak genotoxicity such as chlordane and 2,3,7,8-tetrachlorodibenzo-para-dioxin (TCDD) have been evaluated as carcinogenic based on their tumor properties (10). Phenoxy acid herbicides do not appear to be

Encyclopedia of Pest Management

Table 1 Examples of pesticide chemicals classified by the International Agency for Research on Cancer (IARC) or U.S. EPA as carcinogenic, probably carcinogenic, or possibly carcinogenic to humans

	IARC Category (1)	U.S. EPA Category (2)	U.S. EPA Inerts Category (3)
1. Active ingredients			
Ethylene oxide	1	—	
Captafol	2A	B2	
Creosote	2A	B1	
Nonarsenical insecticides, occupational use	2A		
Chlorophenols	2B	—	
DDT	2B	B2	
Dichlorvos	2B	B2	
Lindane (*γ*-Hexachlorocyclohexane)	2B[a]	B2	
Phenoxy acid herbicides	2B	—	
Sulfallate	2B	—	
2. Inert ingredients and contaminants			
Benzene	1	A	
2,3,7,8-Tetrachlorodibenzo-p-dioxin (TCDD)	1		—
Epichlorohydrin	2A	B2	2
Formaldehyde	2A	B1	1
N-nitrosodimethylamine (NDMA)	2A	B2	2
Trichloroethylene	2A		—
1,4-Dichlorobenzene	2B		2
Dichloromethane	2B	B2	2
Di(2-ethylhexyl)-phthalate	2B	B2	1
Ethylene thiourea (ETU)	2B	B	2
Hexachlorobenzene	2B	B2	2
Isophorone		C	1

Category (1): 1 = Sufficient evidence of carcinogenicity in humans; 2A = Probably carcinogenic to humans; 2B = Possibly carcinogenic to humans.
Category (2): A = Human carcinogen; K = Known carcinogen; B1 and B2 = Probable human carcinogen; C = Possible human carcinogen; — = No available evaluation.
Category (3): 1 = Of toxicological concern; 2 = High priority for testing. — = No available evaluation.
[a]Refers to the whole group hexachlorocyclohexanes.

genotoxic or carcinogenic in experimental animals, but based on epidemiological studies, they are classified by IARC as possibly carcinogenic to humans (5). Several organochlorine pesticides and contaminants in some pesticides with hormonal properties recently have been reviewed with respect to carcinogenicity. The dominating tumor type in rodents exposed to these organochlorins are liver tumors. Bisdithiocarbamates (e.g., maneb and mancozeb) including ethylene thiourea (ETU) as contaminant are examples of suspected carcinogens with extensive use in developing countries (11–13).

CANCER IN HUMANS; EPIDEMIOLOGICAL STUDIES

Until now, the majority of exposure data derives from animal studies, but epidemiological studies are in strong progress and supplements laboratory investigations, e.g., biomarkers as tools to predict carcinogenic properties. Epidemiological studies have in some studies (but not all) linked exposure to phenoxy acid herbicides as risk factors for malignant lymphoma and soft tissue sarcoma (1, 7). These studies had major implications since these chemicals were widely used as pesticides all over the world. In many countries (e.g., the United States, New Zealand, and Sweden) it was found that old formulations mainly of 2,4,5-T contained TCDD and other dioxins. Other studies could not find any increased risk for non-Hodgkin's lymphoma or soft tissue sarcoma and only a small, nonsignificant risk for Hodgkin's disease. Some studies have reported risk for non-Hodgkin's lymphoma among Vietnam war veterans, particularly those exposed to "Agent Orange" (a defoliator consisting of the phenoxy acid herbicides 2,4-D and 2,4,5-T). Most studies, however, did not find an increased risk. Some epidemio-

C

logical studies conducted on malignant lymphoma are inconclusive, which may depend on diagnostic difficulties to separate different entities.

The unclear etiology of the increasing incidence of prostate cancer has provoked much attention and several studies found an association between farming and prostate cancer. A study of Swedish pesticide applicators, who are more exposed to pesticides than farmers in general, found a small increase in risk of prostate cancer compared to farmers. Exposure to certain aromatic solvents (e.g., toluene and xylene with benzene as contaminant) has been associated with leukemia in humans (1, 5).

Epidemiological studies provide limited support of an association between exposure to pesticides, mostly wood preservatives and insecticides of organochlorine origin, and risk of cancer. Well-designed, long-term studies of some old pesticides (among them MCPA) show no carcinogenicity in animals. Compared to other chemicals the documention is extensive and the overall risk of cancer after exposure to pesticides seems to be low. The potential risk of cancer will be reduced if the user wears protective equipment in an appropriate manner (1).

REFERENCES

1. Dich, J.; Zahm, S.H.; Hanberg, A.; Adami, H-0. Pesticides and cancer. Cancer Causes Control **1997**, *8*, 420–443.
2. Maroni, M.; Faith, A. Health effects in man from long-term exposure to pesticides. A Review of the 1975–1991 Literature, International Centre for Pesticide Safety. Toxicology **1993**, *78*, 1–174.
3. US EPA. In *List of Pesticide Product Inert Ingredients*; Office of Prevention, Pesticides and Toxic Substances, U.S. Environmental Protection Agency: Washington, DC, 1995, 34, May 1995.
4. Marquardt, S.; Cox, C.; Knight, H. *Toxic Secrets. Inert Ingredients in Pesticides 1987–1997*, Northwest Coalition for Alternatives to Pesticides (NCAP) and Californians for Pesticide Reform, 1998.
5. International Agency for Research on Cancer (IARC). In *IARC Monographs on the Evaluation of Carcinogenic Risks to Humans*; IARC Press: Lyon, 1983–1997; Vols. 30, 41, 53, 58, 60, 62, 69.
6. Blair, A.; Zahm, S.H. Cancer among farmers. Occup. Med. **1991**, *6*, 335–354.
7. Hardell, L.; Eriksson, M.; Axelson, O.; Zahm, H. S. Cancer Epidemiology. In *Dioxins and Health*; Schecter, A., Ed.; Plenum Press: New York, 1994; 525–547.
8. Sanner, T. Laboratory for Environmental and Occupational Cancer, Institute for Cancer Research. In *Classification of Carcinogens, Mutagens and Substances Toxic to Reproduction by Different Countries and Organisations*; Ver 3.0, The Norwegian Radium Hospital, N-03 10 OSLO Norway, 1999; 214.
9. Ahlborg, U.G.; Lipworth, L.; Titus-Ernstoff, L.; Hsieh, C-C.; Hanberg, A.; Baron, J.; Trichopoulos, D.; Adami, H-0. Organochlorine compounds in relation to breast cancer, endometrial cancer, and endometriosis: an Assessment of the biological and epidemiological evidence. Crit. Rev. Toxicol. **1995**, *25*, 463–531.
10. Kogevinas, M.; Becher, H.; Benn, T.; Bertazzi, P.A.; Bofetta, P.; Bas Bueno-de-Mesquita, H.; et al. Cancer mortality in workers exposed to phenoxy herbicides, chlorophenols, and dioxins. Am. J. Epiderniol. **1997**, *145*, 1061–1075.
11. Vainio, H.; Matos, E.; Kogevinas, M. Chapter 4: Identification of Occupational Carcinogens. In *Occupational Cancer in Developing Countries*; Pearce, N., Matos, E., Vainio, H., Bofetta, P., Kogevinas, M., Eds.; International Agency for Research of Cancer: Lyon, 1994, IARC Sci. Pub. No. 129, 41–59.
12. Wesseling, C.; McConnell, R.; Partanen, T.; Hogstedt, C. Agricultural pesticide use in developing countries: health effects and research needs. Int. J. Health Services **1997**, *27*, 273–308.
13. WHO. In *International Programme on Chemical Safety*; WHO-Environmental Health Criteria Series, WHO: Geneva, Switzerland, 1976–1999.

CANOPY MANAGEMENT

Victor O. Sadras
CSIRO Land & Water, Glen Osmond, South Australia, Australia

INTRODUCTION

Plant canopies have been present in marine habitats since the Cambrian, in freshwater habitats since the Silurian, and in terrestrial habitats since the Lower Devonian when they produced fairly dense photosynthetic covers of relatively branched organisms (1). About 400 million years ago, therefore, the scene was set for the development of evolutionary and ecologically complex interactions among terrestrial plant canopies and arthropod herbivores. Phytopathogens and organisms at the third thropic level added further complexity to the system. For plants, canopies have two key roles: capture of light for assimilation of carbon and nitrogen, and storage of nutrients—chiefly nitrogen in the form of Rubisco[a] (2, 3). Canopies also work as evaporative surfaces and support critical reproductive processes such as floral display, pollen, and fruit production and dispersion. For herbivores, canopies are a potential source of food, refuge, and sites for mating and egg laying. For instance, mortality of *Helicoverpa* spp eggs caused by wind, rainfall, and predation is substantial under the poor shelter of young cotton crops but decreases markedly in large, well-established canopies (4).

NEGATIVE EFFECTS OF PESTS ON PLANT CANOPIES

Herbivores and pathogens have the potential to affect canopy size, uniformity, form, and function (Table 1). *Canopy size* is commonly quantified as leaf area index (LAI), the area of all leaves (one side only) per unit area of ground. For analytical purposes, LAI can be split into four factors:

$$\mathrm{LAI} = P \times B_p \times L_b \times L \quad (1)$$

where P is plant population density (plants per m^2), B_p is number of branches or tillers per plant, L_b is number of leaves per branch, and L is average leaf size (m^2). Pests can affect all four components of LAI (Table 1). Insects and seedling pathogens are important sources of mortality at early stages of crop establishment, leading to potentially important stand reductions. Pests can reduce branching by killing meristems or reducing growth, and the number of leaves per branch by reducing the rate of leaf unfolding, increasing the rate of senescence or both. Smaller leaves can result from direct loss of leaf tissue to defoliators, or indirectly by altered water relations, or shortage of resources for expanding leaves (6).

Owing to the nonuniform spatial distribution typical of most pests, reduction in plant population often goes together with reduced *crop uniformity* (11). The ability of plants to fill gaps left by missing neighbors is greater for systematic than for random plant loss (5). Damage that does not kill but slows down the growth of some plants in the crop also contributes to crop heterogeneity by changing competitive relationships among neighboring plants (5, 12).

Stem sawfly (*Cephus* spp., *Trachelus* spp.) or Hessian fly (*Mayetiola destructor*) causing lodging in severely infested wheat crops are examples of pests affecting *canopy form*. More subtle effects have been found when loss of reproductive organs in indeterminate species, as caused by *Helicoverpa* spp. and sucking mirid and pentatomid bugs (Hemiptera: Miridae and Pentatomidae) in cotton and soybean, allows for continued vegetative growth. As a consequence, damaged crops can be taller and with greater extinction coefficients[b] than their undamaged counterparts (13). In soybean, extensive feeding on reproductive organs may cause foliar retention and not only yield losses but also problems at harvest.

A large number of herbivores and pathogens affect *canopy function*. Impaired photosynthesis has been reported in many systems, e.g., cotton/spider mites (14), potato/*Verticillium dahliae* (9), wheat/aphids (5). Pests that accelerate leaf senescence, such as *Verticillium dahliae* or

[a]Ribulose-1,5-bisphosphate carboxylase/oxygenase

[b]The extinction coefficient of a canopy (k) relates the fraction of light intercepted (f) to the leaf area index (LAI) according to $f = 1 - \exp(-k\ \mathrm{LAI})$.

Table 1 Examples of canopy features affected by pests

Canopy feature	Crop	Pest	Ref.
Size[a]			
Plant population density	Wheat	*Delia coarctata, Eurygaster integriceps, Opomyza florum*	5
Branches per plant	Wheat[b]	*Mayetiola destructor*	5
Leaves per branch	Cotton	*Tetranychus urticae*	6
Leaf size	Sunflower	*Verticillium dahliae*	7
Uniformity[c]	Potato	*Phoma foveata*	8
Form[d]	Wheat	*Cephus* spp., *Trachelus* spp., *Mayetiola destructor*	5
Function[e]			
Capture of light	Potato	*V. dahliae*	9
Nitrogen store	Cotton	*T. urticae*	10

[a]Canopy size is often quantified as leaf area index, the area of all leaves (one side only) per unit area of ground. Eq. 1 (see text) shows the relationship between leaf area index and plant population density, branches or tillers per plant, leaves per branch, and average leaf size.
[b]Hessian fly (Biotype L) reduced tillering of wheat ''Monon'' irrespective of infestation intensity. In wheat ''Newton,'' light to moderate infestation stimulated tillering.
[c]Crop uniformity could be assessed, for instance, using the coefficient of variation of plant size.
[d]Crop form or architecture could be assessed with the canopy coefficient of light extinction.
[e]Key canopy functions are light harvesting for nitrogen and carbon assimilation, and storage of nitrogen—particularly important for grain filling in annual species.

spider mites, have strong effects on the distribution and amount of nitrogen stored in the canopy (10).

POSITIVE EFFECTS OF INJURY ON PLANT CANOPIES

In most cases, pests affect negatively one or more canopy features, leading to substantial reduction in crop yield, quality, or both. There are significant exceptions, however, whereby pests have neutral or positive effects on crops; these need explicit consideration in IPM strategies, as illustrated in the following examples.

Canopy size and yield can be enhanced by insects damaging vegetative meristems of indeterminate plant species, such as soybean, pidgeon pea, and cotton (12, 15, 16). This benefit arises from profuse branching following the release of apical dominance.[c]

Yield can increase by some degree of heterogeneity as caused be uneven loss of apical meristems in cotton (12) or gangrene (*Phoma foveata*) in potato (8). Population-level compensation, which occurs when damage to one individual allows a competing neighbor to grow faster, accounts for this effect (19).

[c]Apical dominance is the control exerted by apical shoot portions over the axilary buds. Variation in the degree of apical dominance has dramatic effects in the adaptation of plants to herbivory (17, 18).

Yield of rye and triticale can be enhanced by early defoliation (5). This benefit was most evident in years when late emerging ears in defoliated crops avoided frost injury that severely affected earlier, noninjured crops.

Importantly, the putative benefits of damage are strongly dependent on the interactions between factors related to the plant (e.g., growth habit, cultivar), the pest (type, timing, intensity, and duration of damage), and the environment, i.e., other biotic stresses, weather, supply of resources and time available for crop recovery (20–22). Removal of the growing tip increased yields in ''Hampton,'' an indeterminate soybean variety, but not in ''Bossier,'' a determinate one (15).

MULTIPLE EFFECTS OF PESTS AND MULTIPLE STRESSES

The previous sections outlined the main effect of single pests on plant canopies. It is worth noting, however, that a given pest can cause more than one effect on plant canopies and also that plants in the field are often exposed to multiple stresses whose interactions are hard to predict. The two-spotted spider mite (*Tetranichus urticae*), for instance, can reduce both canopy size and photosynthetic capacity in cotton with greater reductions measured in fully irrigated crops than in their water-stressed counterparts (23). Concurrent infestations of potato leafhopper (*Empoasca fabae*), early blight (*Alternaria solani*), and Verticillium wilt (*Verticillium dahliae*) resulted in foliage reduction that was less than the sum of losses caused by each organism acting in isolation (24). Additive (25) and nonadditive (26) effects on crop yield have been reported for combined stresses involving plant injury and weed interference.

CONCLUSION

Herbivores and pathogens have the potential to negatively affect key canopy features, including size, uniformity, form, and function. In many cases, these effects lead to yield loss, quality reduction, or both. Plants are not passive victims of pests, however, and crop yield could be unaffected or even increased by mild, timely damage. Understanding the role of defenses and compensatory mechanisms underlying these responses could greatly improve the design of IPM strategies by taking explicit advantage of the ability of plants to naturally cope with pests.

REFERENCES

1. Raven, J.A. Evolution of Plant Life Forms. In *On the Economy of Plant Form and Function*; Givnish, T.J., Ed.; Cambridge University Press: New York, 1986; 421–492.
2. Monteith, J.L. Climate and efficiency of crop production in Britain. Phil. Trans. Royal Soc. London **1977**, *281* (Series B), 277–294.
3. Thomas, H.; Sadras, V.O. The capture and gratuitous disposal of resources by plants. Funct. Ecol., *in press*.
4. Dillon, G.E.; Fitt, G.P.; Forrester, N.W. In *Natural Mortality of Helicoverpa Eggs in Cotton*, 7th Australian Cotton Conference, Broadbeach, August 10–12, 1994; 75–80.
5. Sadras, V.O.; Fereres, A.; Ratcliffe, R.H. Wheat Growth, Yield and Quality as Affected by Insect Herbivores. In *Wheat: Ecology and Physiology of Yield Determination*; Satorre, E.H., Slafer, G.A., Eds.; Food Product Press: New York, 1999; 183–277.
6. Redall, A.A. Ph. D. Thesis. In *Physiological and Morphological Responses of Cotton to Damage by the Two Spotted Spider Mite,* (Tetranychus urticae, *Koch*); University of New England, 2000.
7. Sadras, V.O.; Quiroz, F.; Echarte, L.; Escande, A.; Pereyra, V.R. Effect of *Verticillium dahliae* on photosynthesis, leaf expansion and senescence of field-grown sunflower. Ann. Bot. **2000**, *86*, 1007–1015.
8. Hide, G.A.; Welham, S.J.; Read, P.J.; Ainsley, A.E. Influence of planting seed tubers with gangrene (*Phoma foveata*) and of neighbouring healthy, diseased and missing plants on the yield and size of potatoes. J. Agric. Sci. **1995**, *125*, 51–60.
9. Bowden, R.L.; Rouse, D.I. Effects of *Verticillium dahliae* on gas exchange of potato. Phytopathology **1991**, *81*, 293–301.
10. Sadras, V.O.; Wilson, L.J. Nitrogen accumulation and partitioning in shoots of cotton plants infested with two-spotted spider mites. Aust. J. Agric. Res. **1997**, *48*, 525–533.
11. Hughes, G. Incorporating spatial pattern of harmful organisms into crop loss models. Crop Prot. **1996**, *15*, 407–421.
12. Sadras, V.O. Population-level compensation after loss of vegetative buds: interactions among damaged and undamaged cotton neighbours. Oecologia **1996**, *106*, 432–439.
13. Sadras, V.O. Cotton responses to simulated insect damage: radiation-use efficiency, canopy architecture and leaf nitrogen content as affected by loss of reproductive organs. Field Crops Res. **1996**, *48*, 199–208.
14. Sadras, V.O.; Wilson, L.J. Growth analysis of cotton crops infested with spider-mites. I. Light interception and radiation-use efficiency. Crop. Sci. **1997**, *37*, 481–491.
15. Tayo, T.O. The response of two soya-bean varieties to the loss of apical dominance at the vegetative stage of growth. J. Agric. Sci. **1980**, *95*, 409–416.
16. Tayo, T.O. Growth, development and yield of pigeon pea [*Cajanus cajan* (L.) Millsp.] in the lowland tropics. 3 effects of early loss of apical dominance. J. Agric. Sci. **1982**, *98*, 79–84.
17. Sadras, V.O.; Fitt, G.P. Apical dominance: variability among gossypium genotypes and its association with resistance to insect herbivory. Environ. & Exp. Bot. **1997**, *38*, 145–153.
18. Aarssen, L.W. Hypotheses for the evolution of apical dominance in plants: implications for the interpretation of overcompensation. Oikos **1995**, *74*, 149–156.
19. Crawley, M. *Herbivory. The Dynamics of Animal-Plant Interactions*; Blackwell Scientific Publications: London, 1983; 10.
20. Sadras, V.O. Compensatory growth in cotton after loss of reproductive organs. A Review. Field Crops Res. **1995**, *40*, 1–18.
21. Oesterheld, M.; McNaughton, S.J. Effects of stress and time for recovery on the amount of compensatory growth after grazing. Oecologia **1991**, *85*, 305–235.
22. Sadras, V.O. Cotton compensatory growth after loss of reproductive organs as affected by availability of resources and duration of recovery period. Oecologia **1996**, *106*, 432–439.
23. Sadras, V.O.; Wilson, L.J.; Lally, D.A. Water deficit enhanced cotton resistance to spider mite herbivory. Ann. Bot. **1998**, *81*, 273–286.
24. Johnson, K.B. Evaluation of a mechanistic model that describes potato crop losses caused by multiple pests. Phytopathology **1992**, *82*, 363–369.
25. Helm, C.G.; Kogan, M.; Onstadt, D.W.; Wax, L.M.; Jeffords, M.R. Effects of velvetleaf competition and defoliation by soybean looper (Lepidoptera: Noctuidae) on yield of indeterminate soybean. J. Econ. Entomol. **1992**, *85*, 2433–2439.
26. Sadras, V.O. Effects of simulated insect damage and weed interference on cotton growth and reproduction. Ann. Appl. Biol. **1997**, *130*, 271–281.

CATTLE PEST MANAGEMENT (ARTHROPODS)

C

Carl J. Jones
University of Tennessee, Knoxville, Tennessee, U.S.A.

INTRODUCTION

Cattle and dairy products annually account for more than one-third of the total agricultural sales in the United States (1). In this chapter the major arthropod pests of cattle, the economic losses caused by these pests, and the technology used to reduce these losses will be discussed. Feedlots, dairies, and pastures are all home to cattle. Many cattle are on relatively small facilities where the loss of even a single animal's productivity is unacceptable. For this reason and other reasons discussed in this chapter, development of threshold values for cattle systems is far behind that of crop systems.

INSECTS

Dipterans are the most important arthropods causing economic damage to cattle. Their high mobility combined with the advantages of complete metamorphosis give them the opportunity to exploit many habitats and, often, many hosts.

Myiasis-Causing Flies

Myiasis is the infestation by dipteran larvae of a living host. Obligate myiasis in cattle occurs primarily with *Hypoderma* species and with the screwworm. *Hypoderma* are large flies that as adults resemble honeybees. The nonfeeding adults deposit their eggs on a bovine host; the larvae burrow into the skin and spend most of the next year burrowing through the host. Emerging from the backs of animals in the Spring as larvae, they quickly burrow into the soil and pupate, emerging as adults in the summer to complete their life cycle. Most control technology for these flies revolves around treatment of the host with systemic insecticides, although an economically sound program of genetic and chemical control worked successfully on an area wide basis.

Eradication of the screwworm, *Cochliomyia hominovorax*, is the most famous success story in livestock pest management. Native to North and South America, female adult screwworms deposit their eggs on or near the broken skin of a host. After hatching, the larvae feed on the living flesh for 5–7 days. They fall to the ground, pupate in the soil, and complete their life cycle in about 24 days. Females may lay as many as 10 egg masses, of 200–450 eggs each, over a 30-day span of adult life. Because they increase the size of the wound as they feed, screwworms can cause devastating damage to individual hosts. Secondary infections frequently lead to death for untended cattle. The direct benefit of screwworm control to the American livestock industry was estimated at more than $715 million in 1992, with an annual cost of about $35 million for the joint United States–Mexican eradication program. The suppression of a 1988 outbreak in North Africa was valued at $230 million.

Screwworms are highly mobile insects, commonly moving dozens of kilometers searching for hosts. The use of the sterile insect technique, proposed originally by E.F. Knipling, successfully eradicated screwworms from Florida in the late 1950s. The technique capitalized on the fact that the fly populations did not survive freezing temperatures and that after eradication of the Florida population, the southern (and eventually northern) states in the United States were repopulated with screwworms annually from Mexico. Release of millions of sterile flies over large areas of the United States and northern Mexico eventually created an "autocidal" barrier zone through which fertile flies did not pass. Enforcement of legislation restricting movement of animals from south of the clean area, constant surveillance, and the use of baited toxic traps all combined to free virtually all of North America from the menace of the screwworm. Current efforts are directed at eradication in Central America.

Farmyard Flies

Farmyard flies include *Musca domestica*, the house fly, and *Stomoxys* species, the stable flies, represented by *S. calcitrans* in many temperate regions. House fly adults may vary in size from 5–12 mm in length, with a black and gray thorax. Stable flies, though darker and somewhat smaller (7–8 mm), with a distinctive bayonet-like proboscis that points forward, are commonly mistaken for house flies. The painful bite of stable flies often results in their being labeled "biting house flies."

Manure, urine-soiled straw or hay, and other decaying vegetable matter, including compost piles, are excellent larval development sites for both species. Recently stable flies have gained a reputation for feeding on pastured cattle under conditions that are currently undefined, perhaps related to changes in agricultural practices, such as the introduction of large round hay bales that have proven to provide excellent larval development sites for stable flies under some circumstances. The life cycle, from egg to adult, can be as short as 10 days for house flies and about 3 weeks for stable flies.

Damage

Damage caused by the house fly is difficult to define. Recognized as a vector of many zoonotic diseases under modern conditions, the house fly primarily has become the focus of negative interactions between cattle owners and spreading suburban populations. Here the nuisance and public health concerns may lead to legal action aimed at closing the agricultural facility when flies disperse into communities. Stable fly adults may feed 2–3 times a day, causing substantial irritation and pain as well as blood loss. They also serve as vectors for parasitic nematodes and surra. Animals under attack by stable flies may bunch together to avoid being bitten. The resulting heat stress and lack of feeding can lead to weight gain losses as high as 9 kilograms in the summer and milk production losses of 6%, with estimates of up to 40% milk loss reported. Economic thresholds, estimated as low as five flies per front leg, vary depending upon weather and animal nutrition.

Management

House fly resistance to most conventional insecticides has resulted in increased use of integrated pest management (IPM) procedures. Improved sanitation can result in nearly 70% reductions in house fly and stable fly populations. Housing procedures such as using wood chips instead of straw as bedding materials have also resulted in measurable declines in house fly and stable fly numbers. Parasitoid use is becoming more common, but appears to work best when used with previously mentioned technology or insect growth regulators (IGRs). Traps are often used as an adjunct to other technology. Several pasture traps are available. Electric and pheromone-assisted traps are used as adjuncts to other control technology in modern barns.

Face Fly

Musca autumnalis, the face fly, and *Haematobia* species (*H. irritans*, the horn fly, in much of the world) are the common pasture flies. Adult face flies, similar in appearance to house flies, feed primarily on bovine exudates such as nasal mucus, saliva, and lacrimal fluids. They will ingest blood when it is available, but have sponging mouthparts and cannot penetrate the host's skin. The horn fly, contrary to the name, is usually found along the backline of animals. When numbers are high, they may cover much of the body, looking somewhat like a blanket. Extreme environmental circumstances may drive adult flies onto the belly. Their mouthparts are similar to those of stable flies, although their palps are usually easily seen near the bayonet-like proboscis.

Larval development for both face flies and horn flies occurs only in fresh bovine dung, where egg-to-adult development is completed in about 2 weeks.

Damage

Damage from the face fly is most likely to occur because of its habit of feeding near the eyes of its host. Although it is not the only vector, bovine keratoconjuctivitis or pinkeye, can easily be transmitted among cattle by this species. As a result, the economic threshold, when pinkeye is present, is considered to be one fly. Other forms of damage include transmission of *Thelazia* eyeworms in the United States, and *Parafilaria bovicola* in Europe. Horn fly damage is primarily through blood loss, with reduced milk production or growth declining almost 15% when horn flies are not controlled. Additionally, horn flies may serve as a vector of *Stephanofilaria stilesi*, a nematode skin parasite of cattle. Economic thresholds have been estimated at 50–200 flies per animal. Nutrition, climate, and weather play important roles in the way cattle respond to horn flies.

Management

The face fly is most likely to be controlled by insecticides, either in ear tags or by spray. There are few known natural enemies. Its highly calcareous puparium protects it from the parasitoids and predators normally found in pastures. The horn fly has been the subject of much controversy since it was determined that resistance to pyrethroid ear tags had become a worldwide problem. "Walk-through traps" leading to water or grain have proven to be effective at keeping populations under economic thresholds. Experimental work on the use of predatory or dung-burying beetles shows promise for the future. Innate bovine resistance to horn flies has long been recognized and is being quantified for prediction between breeds and reproductive lines.

Tabanidae

Horse flies and deer flies are robust flies with biting and sponging mouthparts. Adults slash quickly (and painfully) through the host's integument and then insert a small amount of anticoagulant before feeding. These flies may range in color from light grey or yellow to black. Chrysops species, or deer flies, have characteristic color patterns on their wings and are often called pictured-wing flies.

Tabanidae oviposit between 100 and 1000 eggs on vegetation that hangs over or grows in water. Larvae develop for several months in moist to wet substrate but some species may take a year or more to complete development in temperate climates. As a result, many species are active as adults for only a few weeks each year. Horse fly larvae are predaceous, while deer fly larvae feed on detritus. As adults, both sexes feed on plant nectar and other liquids, but the females require a blood meal for egg development, except in a few species in which a single batch of eggs may be deposited before blood seeking begins.

Damage

Damage from Tabanids often results from blood loss associated with their feeding. Estimates of 100 ml of blood lost per day to Tabanid attack support estimates of 10 kg or more of cattle weight gain lost to these flies. In addition, facultative blood-feeding insects may feed at bite sites, and myiasis-causing flies may use these sites for oviposition. Disease transmission includes mechanical transmission of anthrax, anaplasmosis, bovine leukemia, and trypanosomiasis.

Management

Management technology is severely limited. Because economic thresholds have not been established, control efforts have recently revolved around repellant effects of pyrethroids. Area-wide control techniques would probably be superior to animal-based techniques, but larval control is unlikely because they often live in ecologically sensitive wetlands. It is not known just how important a role the larvae of these species play in the maintenance of wetland ecosystem diversity.

Simuliidae

Black fly or buffalo gnat adults are less than 5 mm long with a humped thorax and short, rather broad, wings. They may range in color from yellow to black, although most are dark. They are found worldwide. Both males and females drink nectar, but the females require blood meals for egg development. Large swarms may occur close to flowing bodies of water.

The female deposits her eggs on vegetation or rocks at the water's edge in groups of 150–600. After approximately 1–2 weeks, the larvae emerge and enter the water attached to silken threads spun from the salivary glands. The threads and single caudal suckers anchor the larvae in the streams over the next few weeks to a year while the larvae feed on detritus in this highly oxygenated flowing water. Strong fliers, black flies can move hundreds of kilometers from their larval development sites (2).

Damage

Damage to cattle most commonly occurs due to feeding by extremely high numbers of black flies. Blood loss and irritation have caused an estimated 45% decline in milk production and have been implicated in deaths of pastured cattle from anemia. A filarial nematode of cattle, *Onchocerca gutterosa*, is transmitted by black flies. The nematode produces skin lesions that cause hide damage, resulting in unspecified losses.

Management

Management of black flies is generally aimed at the relatively immobile larval stage. Confined to reasonably small areas, they have proven susceptible to a variety of insecticides, including biological toxins, pulsed strategically into water. Most black fly control, however, is aimed at public health concerns rather than bovine health. Short periods of protection for individual cattle are afforded by the use of pyrethroids.

Culcidae

Culicidae, or mosquitoes, are as common of pests to cattle as they are to most other mammals. A 3–9 mm length, wing scales, and a prominent elongated proboscis clearly differentiate them from similar forms of Diptera.

Mosquito eggs may be deposited on land, in treeholes, on the surface of water, in plants, and in a variety of other places. Some species require relatively pristine waters for larval development, while others may develop in sewage treatment lagoons. With few exceptions, populations are multivoltine. Males are strictly nectar feeders, while the females of most species require a single blood meal prior to oviposition. A notable exception occurs throughout the genus *Toxorhynchites*, the largest mosquitoes, in which nutrition sufficient for egg development is acquired during its predaceous larval stages, and both male and female adults feed on nectar.

Damage

Blood loss to mosquitoes has been estimated to cost the U.S. cattle industry more than $50 million a year in lost poundage. However, the difficulty in measuring losses to mosquitoes without the combination of other flying, blood-feeding pests complicates the issue. Because they do not serve as primary vectors to bovine disease in most of the world, little work has been done to develop economic thresholds.

Management

As major public health and nuisance problems, mosquitoes have received more attention than any other blood-sucking arthropod. Biological control technology alone has been the subject of numerous recent texts. The list of mosquito control technology that is currently in use is far too long for discussion here. It could be said, correctly, that the technology available is extremely dependent on the larval habitat and adult habits of each species of mosquito.

Ceratopogonidae

The biting midges are compact flies 0.6–5.0 mm in length and generally considered weak fliers. Nevertheless, they may travel several miles from their larval development site and have been carried much further by air currents. The two most common genera of interest are *Culicoides* and *Leptoconops*. Members of the first genus are generally crepuscular feeders, while *Leptoconops* are more likely to be diurnal feeders. Most adult females are opportunistic feeders within a class of animals. Mammalophilic species often feed on a variety of livestock.

Larval development of these flies take place in moist soil with vegetative matter, most often near bodies of water, although some species are found in desert areas, where eggs are deposited in soil cracks. Larval development may take between 2 weeks and 7 months, depending on species and substrate suitability. In severe circumstances, larvae may remain in diapause for up to 6 years.

Arboviruses, for which ceratopogonids serve as vectors, include bluetongue, akabane, bovine ephemeral fever and epizootic hemorrhagic disease. Losses to bluetongue include both direct damage and (because bluetongue can be transmitted vertically by cattle) loss of germplasm sales; an important consideration in today's global market. Additionally, the nematode *Onchocerca gibsoni* may be transmitted to cattle by *Culicoides* sp, and livestock can develop intense immune responses to their salivary secretions. Direct losses annually exceed $135 million in the United States.

Damage from disease transmission by ceratopogonidae is more important than blood loss, making it imperative to stop the adults from feeding on livestock (3). The application of insecticides, especially pyrethroids, to cattle has been effective in ceratopogonid control. Unfortunately, insect death often occurs after a feeding, which allows disease transmission. Because ceratopogonids generally fly only short distances to hosts, it is often possible to identify the larval development sites. A program of source reduction and chemical control can be effective in depressing insect numbers below disease transmission levels.

Anoplurans

Lice

There are two groups of lice that affect cattle: biting and sucking. Both cause irritation and itching in the host. The single species of biting lice found on cattle, *Bovicola bovis*, does not take blood, but feeds on skin fragments. The sucking lice are all blood feeders and cause direct damage from blood loss when populations are seasonally high in the winter. Blood loss to sucking lice may exacerbate other underlying conditions in cattle. High populations have been associated with death under some severe weather or disease conditions. Thresholds for this group have not been developed. Control technology at the present time is virtually all insecticidal. However, it should be noted that because infestations generally result from direct contact with an infested host or handling facilities through which an infested host recently passed, isolation of uninfested herds has proven to be highly efficacious in preventing louse infestations.

Acarina

Ticks of all life stages, except eggs, are blood feeders. However, most of the 800 species of ticks spend 90% or more of their life off their hosts, digesting blood meals, seeking a host, or ovipositing. Life cycles vary from a few months to several years. Their complex interactions with native hosts, climate, and vegetation make them a challenging group to control.

Damage

Damage from ticks can take many forms. Direct blood loss, leading to decreased meat or milk production, is the most common. However, ticks are vectors of many viruses, protozoa, rickettsia, and bacteria. After the tick has finished feeding, the bite site may become an oviposition site for myiasis-causing flies. In addition, some tick species can cause a rare paralysis of hosts through the delivery of a neurotoxin while feeding. In the United States, about one-quarter of the more than 80 species known have an effect on the livestock industry. In 1991, two of these species were estimated to cause more than $100 million damage on U.S. range cattle.

Management

In the past 15 years, resurgent interest in tick-borne diseases has led to intense scrutiny of tick population dynamics. As a result, there are computer models available for examining the effects of climate, weather, host density, wildlife, and control technology on tick populations. Although these have not been tied to economic thresholds developed in the 1970s, they contain most elements needed for development of an IPM program flexible enough to be used throughout the range of each species. Insecticides with variable application times and rates along with vegetative management are among the current technology available for use in the model developed by Mount and his colleagues at the USDA in Gainesville, Florida (4). Other programs include control of alternative hosts such as white-tailed deer, controlled burns, and natural seasonal population changes. Still in the experimental stages for tick control are vaccines, self-treatment devices for wildlife, and the development of genetic strains of cattle with natural resistance to tick attachment or feeding. Currently, neither parasitoids nor other forms of biological control have been successful outside of the laboratory. Experiments on aggregation pheromones and other baited trapping procedures have shown great promise in the control of a few species of ticks, but are currently unavailable for practical use.

Mites

The other Acarina often associated with cattle are mites. These microscopic pests can cause severe problems, particularly with young and weak animals. Mites cause intense itching of animals by feeding. The resultant scratch lesions can cause severe loss of plasma or lymph. Mites are invariably controlled through use of acaricides (5). In the United States federal legislation dictates that cattle infested with *Psoroptes ovis* be quarantined and prescribes treatment regimens. Although annual losses of $58 million were estimated in 1987, mites probably cause considerably more damage in parts of the world where quarantine is not enforced.

REFERENCES

1. *Research and Extension Needs for Integrated Pest Management for Arthropods of Veterinary Importance*; Hogsette, J.A., Geden, C.J., Eds.; Proceedings of a workshop, 1999, http://www.ars-grin.gov/ars/SoAtlantic/Gainesville/cm_fly/Lincoln.html.
2. Jones, C.J.; Isard, S.A.; Cortinas, M.R. Dispersal of synanthropic diptera. Lessons from the past and technology for the future. Ann. Ent. Soc. Am. **2000**, *92* (6), 829–839.
3. Mullens, B.A. In *Integrated Management of Culioides variipennis: A Problem of Applied Ecology, Bluetongue*, African Horse Sickness, and Related Orbiviruses, Proceedings of the 2nd International Symposium; Walton, T.E., Osburn, B.I., Eds.; CRC Press, Inc.: Boca Raton, FL, 1999; 896–904.
4. Mount, G.A.; Haile, D.G.; Barnard, D.R.; Daniels, E. New version of LSTSIM for computer simulation of *Amblyomma americanum* (Acari: Ixodidae) population dynamics. J. Med. Entomol. **1993**, *30* (5), 843–857.
5. Wall, R.; Shearer, D. *Veterinary Entomology*; Chapman and Hall: New York, 1997; 439.

CENTERS OF ORIGIN

Amor Yahyaoui
ICARDA, Aleppo, Syria

INTRODUCTION

The disruption of natural plant flora began with the evolution of effective agriculture. Although the onset of plant cultivation remains unknown, signs of plant manipulations emerging 8000–9000 years ago in the Near East are well documented. The discussion traces the origin of crop species and crop diseases.

ORIGIN OF CROP SPECIES

Plants were domesticated primarily in their centers of origin then spread to secondary centers. Harlan (1) described the establishment of an agriculture system. Sites in the Middle East represent the nuclear areas of origin for many species. It took centuries for these species to spread to Greece, to the Mediterranean shores, then to Ethiopia. Vavilov developed the concept of centers of origin, named them and listed a series of plant species for each of them. More than 700 crop species were distributed among Vavilov's defined centers. At least 30% of the annual crop species defined by Vavilov originated in western Asia, the Near East, and the Middle East. It was well documented that major food crops have come from isolated mountain areas that were characterized by a highly diversified habitat. Annual crops such as wheat, barley, oats, lentils, chickpea, pea, root crops, *Aegilops* species (goat grass) and spices originated in western Asia and the Near East. These species were among the first domesticated crop plants. Other crops such as lettuce, vetch, and vegetables originated in the Mediterranean basin. Barley, and tef (*Eragrostis tef*) originated in Ethiopia (the Abyssinian center). Some wheat species and also onion were primarily found in southwest Asia. Several species of wheat originated in the Near East (*Triticum dicoccoides, Triticum urartu, Triticum boeoticum*). Apparently, it is in this part of the world that agriculture evolved over several thousand years before its spread to the New World. It is conceivable that host–pathogen relationships were first established among these crops. Although studies on plant/pest coevolution are recent, there is little doubt that the same principles apply to the evolution in fungi as in the flowering plants. In this respect climate plays an important role in host–pathogen relationships and in the evolutionary trends of fungal species and consequently of plant diseases similar to animal and human diseases. The process by which pathogen populations evolved is not well understood, but it must be linked to human activity. Johnson (2) described this phenomenon as a "man-guided" evolution.

EVOLUTION OF CROP DISEASES

The earliest cells in disease organisms had no nuclei. They were known as prokaryotes, and include bacteria and blue green algae. More evolved organisms such as eukaryotic fungi, similar to plants and animals, have a membrane-bound nuclei that contains the chromosomes, and hence the necessary genetic information that allows for change. For plant pathologists, the evolution of eukaryotes from prokaryotes is the most significant evolutionary event. The lower fungi, the *Myxomycetes* group of fungi, are generally aquatic and somehow less evolved, even though they have spores that are similar to those of true fungi. The primitive aquatic fungi are long-lived pathogens, but their flagellate spores require free water for their mobil-ity, which therefore limits their fitness and their impact as a disease of field crops. There have been no major epidemics caused by this group of fungi since the late blight of potatoes that disrupted the well being of the Irish society in the mid 1800s. Following the epidemic of the blight disease of potatoes caused by *Phytophtora infestans*, the epidemiology of the disease has become better understood and pest management practices were put in place, based both on host plant resistance and on cultural practices.

The true fungi are evolved higher fungi that occur in terrestrial environments. They have a wide host range and their cosmopolitan attributes allow them to adapt and survive under harsh environmental conditions. Fungal diseases caused by these fungi are abundant in tropical regions. Many pathogen species have responded to ecological pressures and developed specialized features through genetic recombination that allowed them to adapt to new hosts and varying environmental conditions. Environmental variation led to the appearance of new fungal forms with morphological and genetic characteristics that enabled them to survive and reproduce.

Sexual recombination in fungal pathogens is an important evolutionary trend in the host–pathogen relationships, enabling an easy discrimination between lower and higher fungi. In evolutionary terms, sexual recombination is very important. It enhances the variability and fitness of the pathogen. A fungal organism that does not have a sexual stage is considered less evolved. It simply produces identical copies of itself and thus has a limited potential for genetic change. Pathogens that belong to the Fungi Imperfecti group are readily affected by environmental changes. They are generally restricted in their distribution and easily contained. No major epidemic was attributed to fungi in this group. Soil-borne diseases such as take-all disease of wheat (*Gaeumannomyces graminis* var. *tritici*); cereal root rot complex: dryland root rot (*Fusarium culmorum*), common root rot (*Bipolaris sorokiniana*), crown rot (*Fusarium graminearum* group I); and wilt disease of lentils (*Fusarium oxysporum*) are soil inhabitants and usually are restricted in their dissemination. A buildup of antagonists will reduce the concentration of fungal spores and infectious mycelium in the soil and hence reduce their impact on the host crop.

DISEASE EPIDEMICS AND CROP MANAGEMENT

Diseases have had a powerful effect on plant evolution even under conditions of endemic balance where diseases do little damage. Disease epidemics are spectacular results of a disturbed ecosystem balance. No disease epidemic has occurred yet on a crop species in its center of origin. Climatic changes that influenced evolutionary imbalance took place over a long period of time. Humans, however, initiated changes that took only few thousand years for some crops and only few hundred years for others. In ancient agriculture systems, large numbers of plant species were harvested in the wild and used as food sources, so that crop species provided only a portion of the diet. The fields were small and often scattered. The crops, generally made of heterogeneous landraces, were often grown in mixtures in the same field. Farming communities were small, separated, and generally mobile, and with population increase and mobility new land was often brought into cultivation. These features provide some sort of defense against disease epidemics. A dynamic equilibrium of wild ecosystems was maintained. This situation is still encountered with subsistence farming systems practiced in various parts of the world. In Eritrea, for example, small farmers still grow mixtures of annual crops and practice multiple cropping on small pieces of land. Hanfetse is an Eritrean cereal crop that consists of a physical mixture of *Triticum aestivum* and *Hordeum vulgare*. The barley in the mixture reduces the incidence of leaf rust disease (*Puccinia recondita*). Wheat grown alone will rarely make it to maturity to be harvested for grain. Wheat in the mixture plays the role of tutor plant to barley that is susceptible to lodging. Hanfetse out-yielded pure crop stands of both barley and wheat in farmers' fields as well as in experimental stations in Eritrea.

With the evolution of social structures, villages became more permanent and farmers had to cultivate larger fields, using few homogeneous crops. This resulted in a sharp decline of wild plants as a food source and a consequent reduction of genetic variability among cultivated crop species. Hence, the opportunity for host–pathogen interaction increased and the consequences were obvious: 1) few species are cultivated over a large area, and 2) species must adapt to new environments to survive. Wheat and barley originally cultivated in areas of winter rainfall and

Table 1 Differential responses of *Aegilops* species to artificial infection by yellow rust (*Puccinia striiformis* f.sp. *tritici*) and stem rust (*Puccinia graminis* f.sp. *tritici*) of wheat (*Triticum aestivum*)

		Reaction to rust	
Aegilops species	Collection site (country)	Yellow rust	Stem rust
Ae. ovata 400881	Jordan	R	R
Ae. ovata 401667	Morocco	R	R
Ae. ovata 400165	Portugal	R	R
Ae. ovata 400027	Syria	S	R
Ae. ovata	Tunisia	S	S
Ae. ovata	Morocco	R	S
Ae. ovata	Morocco	R	R
Ae. caudata 400901	Syria	S	S
Ae. caudata 400973	Syria	R	R
Ae. triuncialis	Tunisia	R	R
Ae. triuncialis 400221	Tunisia	R	S
Ae. triuncialis	Morocco	S	S
Ae. squarrosa	Lebanon	S	S
Ae. longissima 400541	Unkown	R	S
Ae. ventricosa 401430	Algeria	S	S
Ae. geniculata 401657	Morocco	R	R
Ae. cylindrica 400420	Turkey	S	S

R = Resistant; S = Susceptible.
(From Yahyaoui, A. *Sources of Resistance to Rust Among Wild Relatives*, unpublished.)

summer drought moved to regions with summer rainfall. Wheat growers in North America were faced by a series of stem rust epidemics in the mid 1900s. The Ethiopian plateau has a cool climate and conditions that would favor foliar disease pathogens. Yet, no epidemics of rusts or leaf spot diseases have been recorded there, as indigenous species were able to respond using deployment of genetic defenses. Resistance to viral and fungal diseases has been found in Ethiopian barley. The Abyssinian center is considered the secondary center of origin of such crop species as barley, wheat, local grains (tef), and spices. The endemic crop plants are characterized by complex traits, which themselves are connected with distinct areas. *Triticum aestivum* types with stiff spikes and minimal threshability were common in southwestern Asia. Easy-to-thresh *Triticum* types are particular to areas of less continental climates. The stiff spike trait offers a physical barrier to seed-borne pathogens that usually infect plants at flowering stages. The enclosed glumes obstruct infection.

Farmers continue to grow crops that both meet their changing needs and adapt to their environments. Genetic diversity plays a major role in helping meet the future challenges of sustainable food production and avoidance of severe disease epidemics. Tapping the gene pool among wild relatives of wheat has been very useful in identifying new sources of resistance to cereal diseases. Among tested accessions of *Aegilops* species for resistance to yellow and stem rusts (Table 1), differential reactions to rust were observed. Certain *Aegilops* species carry effective resistance to yellow and stem rust of wheat. These two rusts rarely occur together. Yellow rust is prevalent in cool winter environments and its development is hindered by high temperature regimes. Stem rust appears late in the season when temperatures are relatively high. Several strategies have been proposed that might control diseases and reduce the chances of epidemic occurrence. The exploitation of inherent resistance factors in wild relatives and landrace cultivars is highly recommended by plant pathologists. It allows maintaining a high level of variability for disease resistance among cultivated crops. Cereal breeders have developed the technology for gene transfer and they favor the introgression of new sources of resistance, which thus broaden the genetic pool. In annual crops such as cereals, the use of multilines, synthetic cultivars and mixtures would be a stabilizing strategy to influence the direction of balanced host–pathogen interactions. Breeding systems for disease resistance are faster and more convenient. Durable resistance has been identified for several foliar diseases. Wheat has been the basic element for human food for more than 8000 years, and is cultivated nearly all over the world. The resulting range of genetic variation in wheat is astonishing. The formation of new intraspecific varieties, the exploitation of genetic variability inherent in wild relatives, and the implementation of an integrated disease management are sets of dynamic strategies that will allow the sustainable improvement in productivity of food crops.

REFERENCES

1. Harlan, J.R. *Crops and Man*; Wis. Crop Sci. Soc.: Madison, WI, 1975; 295.
2. Johnson, T. Man-guided evolution in plant rusts. Science **1961**, *133*, 357–362.

FURTHER READING

Harlan, J.R.; Zohary, D. Distribution of wild wheats and barley. Science **1966**, *153*, 1074–1080.

Vavilov, N.I. In *The Process of Evolution in Cultivated Plants*, Proceedings of the VI International Congress of Genetics, Ithaca, NY, 1932; 1, 331–342.

Zeven, A.C.; de Wet, J.M.J. *Dictionary of Cultivated Plants and Their Regions of Diversity*; Pudoc: Wageningen, The Netherlands, 1982.

CHAGAS' DISEASE

C

Jorge Rabinovich
Universidad de Buenos Aires, Buenos Aires, Argentina

INTRODUCTION

Chagas' disease results from infection with the protozoan parasite *Trypanosoma cruzi*. In humans it is characterized by an initial acute phase with parasites circulating in blood. As parasitaemia declines the infection becomes latent and an acquired, humoral, and cell-mediated immune resistance modulates the host/parasite relationship, with no clinical manifestation. A chronic phase follows, with cardiac, brain, digestive, and other clinical signs. No safe and effective chemotherapy exists, so early case detection and treatment are of little value. Chagas' disease affects about 250 million people in Latin America, and approximately 24 million are seropositive for *T. cruzi*. Sixty-five million people have been estimated to be at risk in endemic areas (1). There is no reliable number for new cases each year, but crude estimates suggest rates of up to 850,000 new infections per year. Serological surveys indicate that between 5 and 10% of national populations are infected.

THE VECTORS

Triatomine bugs are people's main source of infection with *T. cruzi*, a relationship well established by Chagas' in Brazil in 1909 and by Mazza in Argentina. Triatominae are one of more than 20 subfamilies of the family Reduviidae, and are known in the English literature as "cone-nose" or "kissing" bugs, but carry a variety of names in Latin America: "chinchas" in Mexico, "chipo" in Venezuela, "barbeiro" in Brazil, "uluchi" in Bolivia, and "vinchuca" Argentina. There are more than 114 species of triatomine bugs, grouped in five tribes and 14 genera. However, 90% of the species belong to two tribes: Triatomini (six genera, 88 species) and Rhodniini (two genera, 15 species), that include the two most important vector genera, *Triatoma* (66 species) and *Rhodnius* (12 species). Four of the five tribes occur exclusively in the neotropical region. These hemimetabolous insects have five wingless nymphal stages and an adult stage with two pairs of functional wings. Adult size varies in length from 5 (*Alberprosenia goyovargasi*) to 45 mm (*Dipetalogaster maxima*), although most of the species are about 20–25 mm in length. There are complete keys to genera and species in English, Spanish, and Portuguese (2). All developmental stages of all species are haematophagous, feeding solely on vertebrate blood (mainly on mammals and birds and occasionally on reptiles), and they have a remarkable ability to withstand several months of starvation. Biting is a relatively painless process, and people—particularly children—are rarely wakened by triatomine bites when asleep (3).

THE TRANSMISSION OF TRYPANOSOMA CRUZI BY TRIATOMINES

The transmission of *T. cruzi* from the insect to the host is very inefficient, for it occurs through the insects' feces, thus requiring that bugs defecate while feeding on the human host. An average of more than 1000 bites of positive *T. infestans* are required to infect a person (4). Triatomines' diet has epidemiological importance. *T. infestans* and *T. dimidiata* feed frequently on cats and dogs (important domestic reservoirs of *T. cruzi*) while *R. prolixus* and *P. megistus* rarely do so. On the other hand domestic populations of *T. maculata* and *T. sordida* feed predominantly from chickens (uninfectable by *T. cruzi* as all birds are), and consequently are unimportant epidemiologically in domestic transmission. All species of triatomines can become infected with *T. cruzi* at any stage in their development, and almost always remain infected for life, without being adversely affected. *T. cruzi* prevalence in the vector is usually 20–40%, but occasionally may rise to about 90% levels by the time they become adults. The sylvatic cycle involves many insect and host species. Several marsupials, particularly *Didelphis* spp., play an important role in the wild cycle epidemiology for they are frequently infected with *T. cruzi*, show almost permanent parasitaemia, and often visit peridomestic and domestic habitats. On the other hand, sylvatic triatomines, sometimes infected with *T. cruzi*, may also fly into houses drawn toward lights or carried in with palm fronds used to repair roofs. Thus there is an effective *T. cruzi* circulation among houses, and between houses and the sylvatic environments. However, when a bug population becomes established in a house *T. cruzi* circulates dominantly from human to insect to human.

LIFE CYCLE, DISTRIBUTION, AND ECOLOGY

Depending on species, females lay 300–800 eggs (either singly or in small batches) about 10–20 days after copulation. Unmated females normally lay a fewer number of infertile eggs. Eggs hatch after 10–30 days. Nymphs molt only after at least one full blood meal. The egg to adult cycle usually takes between 100 and 800 days depending on species and temperature. Triatomines are typical of tropical and subtropical areas, and range from Central America to Argentina, although some species of the genus *Triatoma* range from the southern United States to Patagonia (3). The primary sylvatic habitats are palm fronds, epiphytic bromeliads, fallen logs, and hollow trees where they find burrows, shelters, and nests of a variety of wild host species (marsupials, edentates, rodents, carnivores, bats, birds, and lizards). Some species only colonize peridomestic areas (chicken houses, pig sites, stables, and cattle enclosures), but a few species of the genus *Triatoma, Rhodnius and Panstrongylus* colonize huts and shanty houses typical of Latin American peasants. Species of these three genera play a dominant role as vectors of Chagas' disease to man. There are plenty of food sources (people, dogs, and chickens) and the type of construction material of rural houses (thatched roofs, mud, and unplastered or adobe brick walls) provide an ideal microhabitat; as a result *T. infestans* and *R. prolixus* tend to build up large domestic populations. In an extreme case in Venezuela, in a house inhabited by 11 people and harboring about 8000 *R. prolixus* bugs, an average of 58 bites per person per day was estimated, producing a blood loss in the order of 100 cm^3 per person per month (5). Triatomines prefer dark places and hide during the day, although when densities are high they can be seen active at any time. Bedrooms are the preferred resting place in houses, bugs frequently being found close to the site where hosts sleep at night. Other species of *Triatoma* such as *T. dimidiata, T. pallescens, T. pallidipennis, T. phyllosoma*, and *T. brasiliensis*, as well as *Panstrongylus megistus*, are also of some medical importance (2). Adults, despite having functional wings, are not good fliers, and probably take off mainly when nutritional state and weight is low, seeking new habitats with new hosts on which to feed. However, passive transport by human activities is responsible for most of the dispersal that takes place between houses.

SAMPLING

Since a simple and cheap method for detecting the presence of domestic triatomines was proposed in 1965 (6), several adaptations have proven successful for standard and regular use for control including a recently provided robust calibration procedure for a rapid, cheap, and reliable sampling method for *R. prolixus* (7).

CONTROL MEASURES

Chemical Control

Triatomine population control has largely relied on house-spraying using organophosphates (Malathion, Fenitrothion). However, due to permanent recolonization by active and/or passive dispersal, the residual effects of pyrethroids (e.g., deltamethrin, cyfluthrin, and lambda-cyhalothrin) have proved more economical and effective options. Resistance to insecticides has been very limited, mainly due to the long generation time of most triatomine species.

Biological Control

Parasitic mites (*Pimeliaphilus zeledoni*) and predatory spiders (*Theridion rufipes, Scytodes longipes*, and *S. fusca*) have been found as natural enemies under sylvatic and domestic conditions. Additionally, three species of micro-hymenopteran parasitoids are known to parasitize the eggs of triatomines: *Telenomus costa-liami* (Scelionidae) and *Ooencyrtus trinidadenisis* var. *venatorius* (Encyrtidae) attack eggs of *R. prolixus* (8), while *Telenomus fariai* (Scelionidae) attacks the eggs of several species of the genus Triatoma (9). Natural parasitism by these parasitoids is relatively low (between 1 and 10% parasitization, and rarely reaching 30%) and the few tests for biological control carried out in domestic habitats failed due to the strong phototropism of female wasps: populations get diluted through the windows in the predominantly dark houses of rural peasants.

House Improvement

Houses with smooth plastered walls and tiled roofs are seldom colonized by triatomines. Thus, improving standards of both new and existing houses plays an important role in diminishing *T. cruzi* transmission levels. Individual or community-based self-help measures combined with simple and cheap methods to improve existing rural houses has shown to be a good alternative to chemical control.

Plastering of mud walls with a blend of earth, lime, and/or cattle dung, and replacing palm-thatched roofs with corrugated zinc or plastic sheets result in a striking reduction of crevices and cracks and a decrease of suitable habitat space (10), the outcome being very light infestation or no infestation at all. Peridomiciliary areas are one of the main sources of reinfestation after spraying, so improving chicken houses' structures is also important (11). However, house improvements—unless carried out on a extensive scale—will generally be insufficient to interrupt disease transmission. To be really effective it will have to be combined with: 1) public health awareness based on sanitary education, 2) some degree of insecticide use, especially at the start, and 3) a permanent vigilance to avoid possible reinfestations, with community participation (12).

Male Sterilization Technique

Although males usually live longer than females the male sterilization technique of control showed no promising results because: 1) laboratory reared insects do not survive well in the wild, 2) sterile males do not compete well with normal males for copulation after ionizing radiation and chemosterilants use, and 3) females generally have multiple matings during their life span.

PRESENT SITUATION

Argentina, Bolivia, Brazil, Chile, Paraguay, and Uruguay, where *T. infestans* is the main vector of *T. cruzi*, have established the "Southern Cone Initiative" (13), integrating local and central health institutions, community-based groups, and educational institutions with systematic residual action pyrethroid-based spraying activities. Between 1992 and 1995, in those six countries 382,334 houses were sprayed in the attack phase, and 282,246 houses in the vigilance phase, covering 80% of the scheduled houses, resulting in a strong reduction in infestation or complete negativization. Although spraying activities are relatively expensive, a cost-benefit analysis (considering as benefits the dollar values of avoidance of premature death and of supportive treatment and care) showed a high internal rate of return (14, 15). However, despite a limited gene flow between domestic and sylvatic populations of two species of *Rhodnius*, possibilities of extending this control scheme to northern South America and Central America are still remote, due to important sylvatic residual foci of *T. dimidiata* and of several species of *Rhodnius*, the main vectors in those regions.

REFERENCES

1. Schofield, C.J.; Dujardin, J.P. Chagas' disease vector control in central America. Parasitology Today **1997**, *13*, 141–144.
2. Lent, H.; Wygodzinsky, P. Revision of the triatominae (Hemiptera, Reduviidae), and their significance as vectors of Chagas' disease. Bull. Am. Mus. Nat. Hist. **1979**, *163*, 127–520.
3. Minter, D.M. Cimicid and Triatomine Bugs. In *Manson's Tropical Diseases*; Cook, G.C., Ed.; 20th Ed., W.B. Saunders Company, Ltd. London, 1996; 1721–1736.
4. Rabinovich, J.E.; Wisnivesky-Colli, C.; Solarz, N.D.; Gürtler, R.E. Probability of transmission of Chagas' disease by *Triatoma infestans* (Hemiptera, Reduviidae) in an endemic area of Santiago del Estero, Argentina. Bull. World Health Org. **1990**, *68*, 737–746.
5. Rabinovich, J.E.; Leal, J.A. Feliciangeli de Piñero, D. domiciliary biting frequency and blood ingestion of the Chagas' disease vector *Rhodnius prolixus* Stahl (Hemiptera: Reduviidae), in Venezuela. Trans. Roy. Soc. Trop. Med. Hyg. **1979**, *73*, 272–283.
6. Gómez, Núñez, J.C. Desarrollo de un nuevo método para evaluar la infestación intradomiciliaria por *Rhodnius prolixus*. Acta Cient. Venez. **1965**, *16*, 26–31.
7. Rabinovich, J.E.; Gürtler, R.; Leal, J.A. Feliciangeli de Piñero, D. Density estimates of the domestic vector of Chagas' disease, *Rhodnius prolixus* Stål (Hemiptera: Reduviidae), in rural houses in Venezuela. Bull. World Health Org. **1995**, *73*, 347–357.
8. Feliciangeli de Piñero, M.D. Parasitismo de huevos de *Rhodnius prolixus* por los microhimenópteros *Telenomus costa-liami* y *Ooencyrtus trinidadenisis var. venatorius*. Bol. Dir. Malar. San. Amb. (Venezuela) **1977**, *XVII*, 131–139.
9. Rabinovich, J.E. La dinámica poblacional de *Telenomus fariai*, parásito de los vectores de la enfermedad de Chagas. Acta Cient. Venez. **1972**, *23* (3), 79–83.
10. Cedillos, R.A. The effectiveness of design and construction materials in Chagas' disease vector control. Rev. Argent. Microbiol. **1988**, *20* (1), 53–57.
11. Cecere, M.C.; Gürtler, R.E.; Canale, E.; Chuit, R.; Cohen, J.L. The role of the peridomiciliary area in the elimination of *Triatoma infestans* from rural Argentine communities. Pan. Am. J. Public Health **1997**, *1*, 273–279.
12. Dias, J.C.; Borges Dias, R. Participation of the community in the control of Chagas' disease. Ann. Soc. Belg. Med. Trop. **1985**, *65* (1), 127–135.
13. Schofield, C.J.; Dias, J. The southern cone initiative against Chagas' disease. Adv. Parasitol. **1999**, *42*, 2–27.
14. Basombrio, M.A.; Schofield, C.J.; Rojas, C.L.; del Rey, E.C. A cost-benefit analysis of Chagas' disease control in North-Western Argentina. Trans. Roy. Soc. Trop. Med. Hyg. **1998**, *92*, 137–143.
15. Schofield, C.J.; Dias, J.C.P. A cost-benefit analysis of Chagas' disease control. Mem. Inst. Oswaldo Cruz **1991**, *86*, 285–295.

CHEMICAL SEX ATTRACTANTS

Wendell L. Roelofs
Cornell University, Geneva, New York, U.S.A.

INTRODUCTION

The chemical communication system used by insects for finding mates not only has been exploited by using synthetic sex attractants in monitoring, mass trapping, and mating disruption programs in pest management programs, but also this system has proved to be extremely valuable in the quest for knowledge on the sense of smell and how odorous signals are processed. Basic research on this communication system is generating novel strategies for disrupting mating in pest species, and the emerging technologies of molecular biology have been combined with those of neurophysiology, biochemistry, behavior, and ecology to address fundamental questions in the field of olfaction.

INSECT CHEMICAL COMMUNICATION

Animals use a variety of sensory modalities for mate attraction, but the most common form of long-distant sex attraction in insects (and many other animal species) involves invisible odors (Fig. 1). This phenomenon of using ''smell'' was noted in the literature more than a century ago, and in the nineteenth century it already was proposed that chemists should be able to aid the entomologists by imitating the powerful animal secretions for use in insect control.

Decoding the chemical signal involved in insect sex attraction was not accomplished until 1959 with the characterization of the sex pheromone for the domesticated silkworm, *Bombyx mori*, after extraction of a half million female silkworm pheromone glands and 30 years of classical chemical analyses by German scientists. This work showed that there was nothing magical about these sex signals and that perhaps this essential mate-finding process could be manipulated for insect control if the signals were decoded for pest species. Advances in instrumentation had made the characterization of these signals from other species a reality.

Research initially was focused on the overt chemical signals used in the mating process and resulted in the characterization of sex attractants for more than 1600 insect species from more than 90 Families in nine Orders, with the main emphasis on Lepidoptera (1). The variety and complexity of the 300-plus defined chemicals attest to the insect's amazing ability to sequester and synthesize unique structures and blends, mainly composed of acetogenins and mevalogenins (Fig. 2).

Studies on the chemistry of sex attractants then led to more in-depth research on the underlying biochemical, hormonal, neurobiological, behavioral, and genetic mechanisms of this chemical communication system, which, in turn, led to new hypotheses on the evolution of this communication system and its role in speciation (2).

Sex Attractant Structures

The identification of pheromone structures in several species of beetles and moths initially by classical chemical methods of isolation and structural analyses allowed chemists to synthesize these attractants as well as many structurally related chemicals. Field tests in many countries throughout the world with these libraries of chemicals revealed attractant structures for hundreds of additional species of beetles and moths, many of which were pest species. Pheromone identifications also were aided with the development of the electroantennogram (EAG) technique in which the male's antennae were used as detectors for bioassays in the isolation and purification of pheromone extracts. The EAG technique also was used with moth species to predict pheromone structures by generating EAG response profiles to series of geometric isomers of varying carbon-chain length and functionality and determining the most active compound within the most active series. The male antennal receptors proved to be extremely sensitive and specific to their own pheromone components, and so the EAG became a powerful tool in pheromone research.

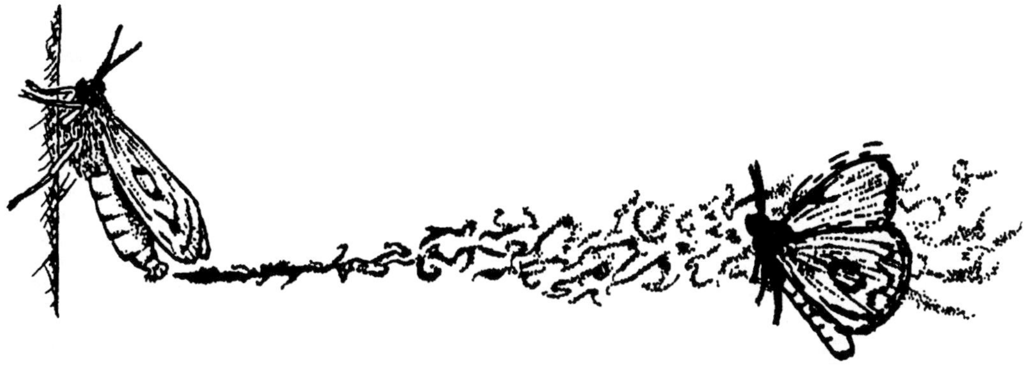

Fig. 1 The sex pheromone communication system in moths showing a female moth in a "calling" position as she releases her pheromone in an odorous plume for response by a con-specific male moth.

The development of capillary gas/liquid chromatography (GC) provided a sensitive technique to separate minute amounts of closely related isomers. Using the EAG as a detector for the GC provided the means to detect all components (usually two–seven) of a pheromone blend from individual moth pheromone glands. It soon became obvious that many species use similar pheromone components, but that specificity was determined by the exact ratio and composition of the blend.

Characterization of chemicals in the pheromone gland was only the first step in defining a pheromone blend. To complete the process it was necessary to determine what chemicals were released by the female and used by con-specific males to elicit the appropriate mate-finding responses. Airborne collection devices were used to define the chemical composition of effluvium from calling females, and male behavioral responses were analyzed with a variety of olfactometers, flight tunnels, and moth tracking techniques in the laboratory and the field.

Sex Attraction Responses

In-depth behavioral studies on how sex attractants (pheromones) actually function in "attracting" mates became critical to the development of strategies for manipulating male insects in pest management systems. These studies revealed that attraction induced by the pheromone blends is merely the end result of a series of behavioral reactions to the wisps of fluctuating pheromone filaments in the odorous plume. Thus, attraction is not a behavior, but rather a displacement through space as a result of behavioral responses (maneuvers) of the moths, measured on the order of milliseconds, consisting of changes in the course angle and airspeed of the responding insect. These maneuvers provide the basis for the two main mechanisms used for pheromone source location: optomotor anemotaxis (steering with respect to the wind) and self-steered counterturning (zigzags across the windline,

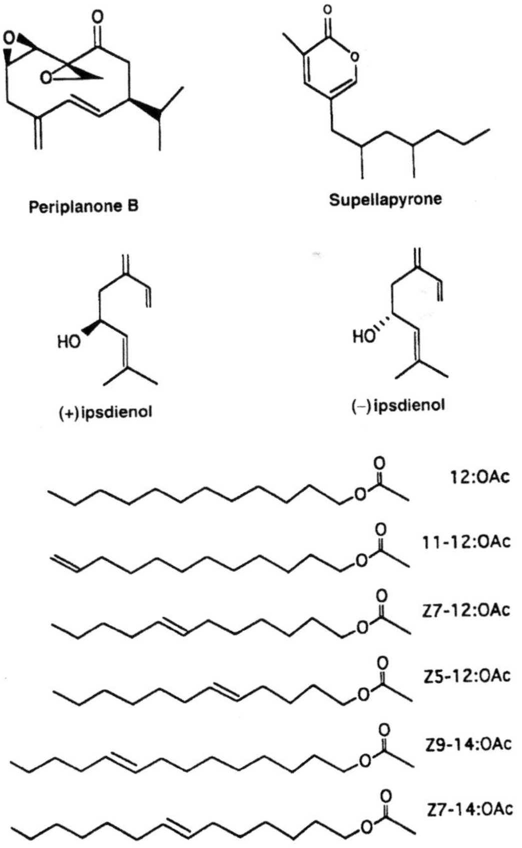

Fig. 2 Pheromone structures of the American cockroach (periplanone B), the brownbanded cockroach (supellapyrone), bark beetles (ispdienol-enantiomers), and the cabbage looper (six acetates).

especially obvious when moths start casting moments after losing contact with the pheromone plume) (3).

Further studies showed that male responses were elicited at the lowest release rates when exposed to a pheromone blend ratio of all components. These data have importance in using pheromone blends for monitoring and for mating disruption (4, 5). A monitoring trap that releases a blend at a rate and component ratio approximating that of a calling female will provide the most sensitive and specific monitoring device for that particular species. Behavioral sensitivity to the complete blend at the lowest thresholds also implies that the complete blend will have more input at the neuronal level of the central nervous system (CNS) than individual components and should be more effective in mating disruption programs at lower concentrations than partial blends.

PHEROMONE ANTAGONISTS

Moth pheromone systems appear to be very tightly controlled blends and not subject to significant changes in blend ratios (6). However, prezygotic reproductive isolation appears to be reinforced in some situations with the selection for what have become known as behavioral antagonists. Species that cohabit with other species utilizing common components in their multicomponent blends could experience a decrease in specificity and sensitivity for their conspecific mating signals. Thus, a change in neuronal processing could occur to switch a stimulatory input from the receptors of one of the pheromone components to an inhibitory input. Perception of a blend containing this component would result in arrested upwind flight, because now this compound would function as a behavioral antagonist (7).

An example of the above scenario is with the soybean looper and cabbage looper moths. They both use (*Z*)-7-dodecenyl acetate as their main pheromone component, and males of both species give low-level responses to that compound alone. However, soybean looper males are much more responsive to their own four-component blend, and do not respond to the cabbage looper blend, even though it contains the main component. Neurophysiological studies revealed that the soybean looper males possess an olfactory receptor neuron on their antennae that is very sensitive to (Z)-5-dodecenyl acetate, which is a minor component of the cabbage looper pheromone blend (Fig. 2). This compound is not part of the pheromone released by soybean looper females, but it is a strong antagonist to the soybean looper males. When added to the soybean looper blend, it causes a dramatic reduction in trap catch and a significant level of arrested upwind flight in flight-tunnel studies. Since this compound is part of the cabbage looper pheromone blend, it could be the salient factor responsible for reducing the response of soybean looper males to cabbage looper females. There are antennal receptors on soybean looper males for this antagonistic compound, which indicates that it could have been part of the soybean looper pheromone blend at one time, but competition for this chemical channel could have resulted in a reduction in production of this compound by female soybean loopers and a switch to an antagonistic response with the males.

Many examples exist among sympatric species wherein a pheromone component of one species is antagonistic to the male response of another species when added to the pheromone mixture of that species. Since these antagonistic compounds cause the males to cease upwind flight, it was proposed that they should be good agents to use for mating disruption in the field. However, it was determined that an antagonist must be emitted from the same point source as the pheromone blend to be optimally effective in reducing the amount of upwind flight and source location. In a study on the limits of the male moth's ability to resolve individual strands of odor within finely structured plumes, it was found that the corn earworm male can distinguish strands of pheromone from those of a behavioral antagonist that are separated by no more than 1 mm and temporally by at most 0.001 sec. Thus, in general, the behavioral antagonists have been found to be ineffective in mating disruption programs because male moths are able to locate their conspecific pheromone plumes in an atmosphere permeated an antagonist.

FUTURE CONCERNS

Chemical attractants are known for many pest species and are being used in pest management systems in monitoring and mating disruption programs. However, much more basic research needs to be carried out to build on the base of information already generated on this chemical communication system. Novel strategies of pest management could be generated, for example, from basic research wherein: 1) biosynthetic pathways involved in producing the pheromone compounds have been defined, and characterization of the gene structure of key enzymes involved in these pathways could allow for their use in disrupting the biosynthetic process (8); 2) the discovery of neuropeptides that regulate the stimulation of this biosynthesis resulted in the characterization of genes for these hormones, which could be inserted in baculoviruses to enhance activity (9); and 3) an understanding of how olfactory information is processed (10) by a combination of neurophysiological studies of moth olfaction and upwind flight behavior in moths could lead to better methods of manipulating olfactory pathways to prevent mate-finding, oviposition, and other odor-mediated behaviors.

REFERENCES

1. Mayer, M.S.; McLaughlin, JR. *Handbook of Insect Pheromones and Sex Attractants*; CRC: Boca Raton, FL, 1991; 1–1083.
2. Roelofs, W.L. In *Chemistry of Sex Attraction*, Proceedings of the National Academy of Science USA, 1995; 92, 44–49.
3. Baker, T.C. Sex pheromone communication in the lepidoptera: new research progress. Experientia **1989**, *45*, 248–262.
4. Cardé, R.T.; Minks, A.K. *Insect Pheromone Research: New Directions*; Chapman & Hall: New York, 1997; 1–684.
5. Ridgway, R.L.; Silverstein, R.M.; Inscoe, M.N. *Behavior-Modifying Chemicals for Insect Management*; Marcel Dekker, Inc.: New York, 1990; 1–761.
6. Roelofs, W.L.; Glover, T.J.; Tang, X.H.; Sreng, I.; Robbins, P.; Eckenrode, C.J.; Löfstedt, C.; Hansson, B.S.; Bengtsson, B.O. In *Sex Pheromone Production and Perception in European Corn Borer Moths Requires both Autosomal and Sex-linked Genes*, Proceedings of the National Academy of Science USA, 1987; 84, 7585–7589.
7. Linn, C.E., Jr.; Roelofs, W.L. Pheromone Communication in Moths and Its Role in the Speciation Process. In *Speciation and the Recognition Concept*; Lambert, D. M., Spencer, H.G., Eds.; Johns Hopkins University Press: Baltimore, MD, 1995; 263–300.
8. Knipple, D.C.; Miller, S.J.; Liu, W.; Tang, J.; Ma, P.W.K.; Roelofs, W.L. In *Cloning and Characterization of a cDNA Encoding a Pheromone Gland-specific acyl-CoA 11-desaturase of the Cabbage Looper Moth, Trichoplusia ni*, Proceedings of the National Academy of Science USA, 1998; 95, 15287–15292.
9. Ma, P.W.K.; Davis, T.R.; Wood, H.A.; Knipple, D.C.; Roelofs, W.L. Baculovirus expression of an insect gene that encodes multiple neuropeptides. Insect Bioch. Mo-lec. Biol. **1998**, *28*, 239–249.
10. Todd, J.L.; Baker, T.C. The cutting edge of insect olfaction. Amer. Entomol. **1997**, *43*, 174–182.

CHRONIC HUMAN PESTICIDE POISONINGS

Birgitta Kolmodin-Hedman
Karolinska Institute, Stockholm, Sweden

INTRODUCTION

According to the United Nations Food and Agriculture Organization (FAO), a pesticide is "any substance or mixture of substances intended for preventing, destroying, or controlling any pest, including vectors of human or animal disease, unwanted species of plants or animals causing harm during, or otherwise interfering with the production, processing, storage, transport or marketing of food, agricultural commodities, wood and wood products, or animal foodstuffs for the control of the insects, arachnids, or other pests in or on their bodies."

Pesticides comprise various groups of compounds according to their use (Table 1).

BACKGROUND

Pesticides have been used predominantly during the past 100 years with increased use after World War II. DDT was introduced in 1939 and extensively used in the mid-1940s to combat malaria and typhus and in the developing countries. Due to its persistence in the biological change—ecotoxicological aspects—a reduction has been undertaken.

Organophosphorus compounds (OPs) were developed for use as war gases. They all exert their main effect by inhibiting the cholinesterases (AcChE and BuChE) in the body. They were introduced in 1960 and are still used worldwide.

Of the herbicides, the phenoxy acids were introduced in the late 1940s. In various parts of the world, a dominance of either 2,4-D, or MCPA have occurred for forestry use; 2,4-D, and 2,4,5-T were the main products. During the past 10 years glyphosate has replaced use of phenoxy acids in many countries.

In certain parts of the world (such as South Africa and East Asia), paraquat has been used widely. Synthetic pyrethroids have been used since the end of the 1970s. Some parts of the world still can't afford these products. In China the agricultural use is widespread.

In general, pesticide use has more economic importance in tropical climates with high temperatures and high humidity compared to countries with a more temperate climate.

POSSIBLE LONG-TERM EFFECTS OF PESTICIDES

The possible long-term effects of pesticides are as follows:

- Metabolic degradation of persistent compounds, such as degradation of DDT to stable DDE.
- Slow toxicokinetic properties of compounds, i.e., phenoxy acid, MCPA, 2,4-D, and 2,4,5-T.
- A cumulative effect from repeated exposure; for instance, stepwise reduction of cholinesterase by organophosphates.
- Additive effects of organic solvents.
- Contamination of commercial products, for example, dioxin in 2,4-D, and 2,4,5-T.

STATISTICS CONCERNING CHRONIC INTOXICATION

According to statistics from the World Health Organization (WHO), 200,000 cases of acute poisoning occur each year. The main cause of acute poisoning is intake during suicide.

Estimated chronic effects, besides cancer, are as high as 735,000 a year. It is very difficult to estimate the chronically exposed population (exposure via environmental contamination or unintentional exposure through food intake). Also, in various parts of the world, different amounts of pesticides are used (1). More insecticides are used in the tropical agricultural areas than in temperate zones (Fig.1)(2).

Estimates of chronic poisonings connected to pesticides are hard to find in the research literature. Various literature retrieval systems have been searched from 1990 to 1999. Literature case reports with poisoning followed for months

Table 1 Pesticides grouped by use

Insecticide	Compounds that combat insects
Fungicide	Compounds that combat fungi
Herbicide	Compounds that kill weeds
Rodenticide	Compounds that combat rats and mice
Acaricide	Compounds that combat mites

afterwards are reported and could be characterized as rest conditions after an acute poisoning. Groups with small repeated exposure are mostly described less extensively.

GROUPS POISONED CHRONICALLY

Exposure to pesticides occurs during concentrate mixing and equipment loading. Pesticides can be applied by air spraying, tractor-driven equipment, and backpack spraying. Beside mixing and spraying, contaminated surfaces present a problem during equipment repair and cleaning. People can also be contaminated during re-entry into sprayed crops or greenhouses.

For the general population, contamination of food and water is the dominant poisoning method.

During occupational spraying, the most common exposure is dermal. Due to product volatility and product administration, inhalation sometimes can occur.

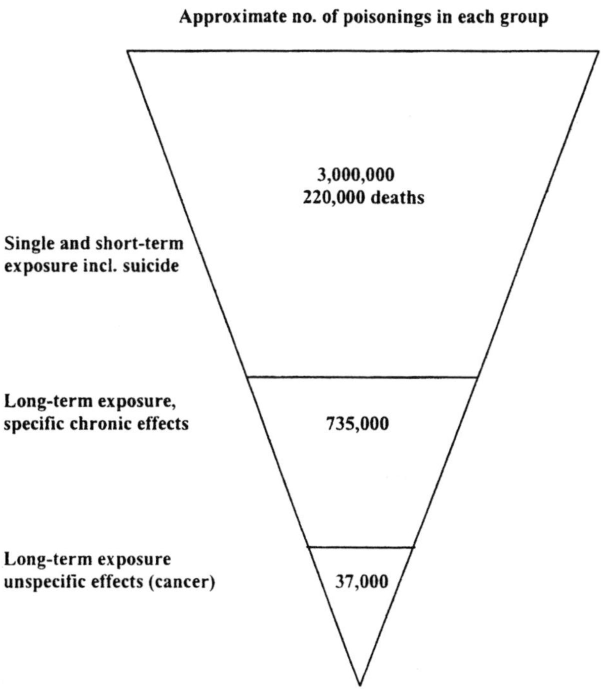

Fig. 1 Estimated overall annual public health impact of pesticide poisonings. (From Ref. 1.)

INSECTICIDES

Organophosphorus Compounds (OPs)

As Table 2 shows, inborn toxicity varies 100-fold, and volatility of compounds differ.

OPs are mainly excreted in urine as alkyl phosphates or alkyl thiophosphates.

Mechanisms of Action

Acetylcholine access gives symptoms in parasympatic nerve endings, cholinergic nerve synapses in CNS, and autonomic ganglia as well as symptoms from the motor end-plate. Various organophosphates do, to a certain extent, inhibit acetylcholinesterase or buturylcholinerase. Some OPs interact with an enzyme site in an irreversible way called aging. New synthesis, if the acetylcholine enzyme has been depressed 100%, occurs within one month. Symptoms in humans do not mostly develop until a 70% reduction of enzyme activity occurs. The enzyme is stepwise inhibited so a subclinical effect can be detected by repeated enzyme analyses compared to the person's pre-exposure value.

Symptoms of acute cholinesterase intoxication are miosis, tears, salivation, sweat, nausea, diarrhea, mucus increase in bronchi, drop in blood pressure, headache, and muscular fasciculation.

Metabolism of organophosphates is a degradation to various alkyl phosphates that are excreted in urine. For parathion a defined metabolic product is p-nitrofenol.

Organophosphorus pesticides may enter the organism through inhalation, ingestion, and percutaneous absorption.

The relative importance of these routes of entry varies according to the physiochemical properties of the OPs, the exposure conditions, and the nature of formulation of the pesticides. Inhalation is mainly important during the production of pesticides, their formulations, and application by aerial spray.

In agricultural situations, dermal absorption may occur during dilution of concentrates when pesticides are handled improperly—without protection. Oral intake might occur from occupational exposure and employees' poor

Table 2 Some data of commonly used organophosphorus insecticides

Toxicity	LD 50 mg/kg^{-1}	Sat. conc. at 20°C	Metabolism
Parathion	10	3,78	p-nitrofenol
Malathion	1000–2000	2,26	DMP, DMTP
Diazinon	220–600	1,39	DEP, DETP
Dichlorvos	55–80	145	DMP, DETP

personal hygiene or from contaminated areas where the general public is exposed. Intake from suicide is not described in this paper.

Intermediate Syndromes

After an acute intoxication, delayed symptoms in various muscles, such as the respiratory muscles, arms, and legs may result. A mechanism by calcium-mediated muscular necrosis has been discussed.

OPIDIN

Lotti et al. (3) have studied delayed organophosphate-induced neuropathy after exposure, for instance, to metamidophos. A preceding inhibition of lymphatic neuroesterase (NTE) could be found. However, most clinical intoxications reported in the literature occur after very high doses are taken to commit suicide.

In animal experiments, the hen is the most sensitive animal. OPIDIN occurs independently of the previously described acute intoxication with possible chronic effects (4, 5).

Effects of Repeated Exposure to OPs

Another effect after repeated exposure to OPs could be impairment of vibration thresholds, which has been found in workers in Nicaragua.

In 1991 Rosenstock et al. (6) investigated pesticide workers in Costa Rica and Nicaragua who had been using a mixture of OPs. Researchers gave them various psychomotor tests and found some evidence that they have increased vibration thresholds. The ethiological mechanisms, however, are unclear (7). Confounding factors from such studies are protein deficiency, presence of other illnesses, diabetes, other forms of polyneuropathy, alcoholism, and nerve conduction studies strongly related to age. A Canadian review committee report from 1999 had mostly critical comments on the presented studies.

SYNTHETIC PYRETHROIDS

Common generic names for these are permethrin, cypermethrin, and deltamethrin. They have an intermediate, acute toxicity and are metabolized, degraded, and excreted in urine. They have a high toxicity for fish. Fish have died after ponds were polluted.

Chronic intoxication in highly exposed agricultural workers in China has been reported. Some of the cases were hospitalized with symptoms such as skin irritation, paresthesia (tingling) in arms and legs, and reversible paresis of limbs. Very few persons presented persistent symptoms. Synthetic pyrethroids' effects are caused by peripheral irritation of nerve endings after skin absorption.

No allergic, teratogenic, or genotoxic effects have been reported for synthetic pyrethroids.

ORGANOCHLORINE INSECTICIDES (DDT, DDD)

DDT is metabolized by liver microsomal enzymes to DDT and to DDE. This can be analyzed in blood and fat. High exposure to DDT is metabolized and predominantly excreted in urine as DDA. With more chronic exposure, it is metabolized to DDE, which can be found in fat and blood long after DDT exposure. Breast milk might be contaminated and give up-take in the infant. Lately a hormonal effect has been discussed—that DDD has been shown to have estrogenic effects for the developing of breast cancer in men. In groups exposed to LH, testosterone and estradiaol can be checked. In one group of workers exposed to a mixture of carbamates, organophosphates, phenoxy acids, and pyrethroids, 24 pesticide workers had a higher level of testosterone.

HERBICIDES

Phenoxy Acids

Phenoxy acids are chlorinated compounds such as 2,4-D and 2,4,5-T. In MCPA, the chlorine has been changed to a methyl group. Phenoxy acids are bound to plasma proteins with a slow excretion in urine. Half-lives in men range from 12 to 24 h. Dermal absorption is slow. One dermal dose of 1 mg could be followed for one week in urinary levels (8).

The carcinogenicity of phenoxy acids is being debated. No clear evidence was shown during the 1980s and 1990s. A contamination with dioxin (2,4,7,8-tetrachlorodioxin) as the cause to this has been discussed. In Sweden, however, case-reference studies have shown a five-fold increase in sarcomas and also an increase in the frequency of morbus Hodgkin's and non-Hodgkin's lymphoma. In cohort studies mostly negative results were found. The frequency of brain tumours caused by the phenoxy acids has been discussed.

Glyphosate

No long-term effects were reported.

Paraquat

Paraquat exposure has been studied in a group of South African orchard sprayers. Lung function and spirometry were somewhat decreased (10–15%) by long-term exposure. One effect was less oxygen transport in exercise tests, indicating impaired gas exchange (9). Also, researchers measured diffusion capacity for CO; 1.2 million agricultural workers potentially have been exposed in South Africa. Cases of pulmonary obstructive disease appearing two to three weeks after acute ingestion is well known.

OTHER EFFECTS OF PESTICIDES

Other effects of pesticides are as follows:

Reproductive effects. Reproductive effects have been studied by evaluating the sperm quality of Chinese pesticide factory workers.

Genetic effects. Genotoxicity tests in patients should also be considered.

Enzyme variation. Studying either a liver microsomal enzyme, such as CYP2E, or the formation of hemoglobin adducts might check individual differences in metabolic rates (10).

Individual sensitivity.

EPIDEMIOLOGICAL STUDIES

In epidemiological studies in general, it is important to present confounding factors and to have large groups of exposed persons. Case reports, cohorts, and case referent studies could be used (1).

WHO has summarized the expected cases in agriculture. Koh and Jeyaratnam described pesticide hazards in East Asia (2).

PREVENTION

In the meantime, prevention programs for better handling practices and structured information of health hazards is very important. Here linguistic difficulties might be found. In some areas illiteracy is common, and in other areas many pesticide workers have a rather short school education. One way to avoid exposure is to change the formulations to low-volume preparations, wettable powders, and baits to bring both acute and chronic intoxication risks down. Another way is to improve the methods for crop protection that avoid or use less pesticide or fight insects in other ways entirely, for instance by physical means such as oxygen deprivation.

A network of research and administrative persons in the occupational and environmental medicine fields should be established to monitor the regulation of pesticides, and the information from producers. Information about acute, intermediate, and chronic intoxications should be gathered into a generally accessible database.

CONCLUSIONS

A survey of literature for chronic human exposure to pesticides is difficult to conduct. The populations potentially exposed are difficult to identify in countries where no good epidemiological data exists.

For some countries, the use of pesticides is presented in yearly statistics. As many as 30 compounds might be used in farming and horticulture. It is difficult to identify combined exposure from mixtures of pesticides.

REFERENCES

1. WHO and UNEP. *Public Health Impact of Pesticides Used in Agriculture*; WHO: Geneva, 1990.
2. Koh, D.; Jeyaratnam, J. Pesticides hazards in developing countries. Science Total Environ. **1996**, *188* (Suppl 1), 78–85.
3. Lotti, M.; Moretto, A.; Zoppellari, R.; Dainese, R.; Rizzuto, N.; Barusco, G. Inhibition of lymphocytic neuropathy target esterase predicts the development of organophosphate-induced delayed polyneuropathy. Arch. Toxicol. **1986**, *59*, 176–179.
4. Echobichon, D.J. Organophosphorus ester Insecticides. In *Pesticide and Neurological Diseases,* 2nd Ed.; Ecobichon, D.J., Roy, R.M., Eds., CRC Press: Boca Raton, FL, 1994; 381.
5. Eyer, P. Neuropsychopathological changes by organophosphorus compounds—a review. Human and Expt. Toxicol. **1995**, *14*, 857–864.
6. Rosenstock, R.; Keifer, M.; Daniell, W.; McConell, R.; Claypoole, K. Chronic central nervous system effects of acute organophosphate pesticide intoxication. Lancet **1991**, *338*, 223–227.
7. McConell, R.; Téllez-Delgado, E.; Cuadra, R.; et al. Organophosphate neuropathy due to methamidophos: biochemical and neurophysiological markers. Arch. Toxicol. **1999**, *73*, 296–300.
8. Kolmodin-Hedman, B.; Höglund, S.; Svensson, Å.; Åkerblom, M. Studies on phenoxy acid herbicides. II. Oral and dermal uptake and elimination in urine of MCPA in humans. Arch. Toxicol. **1983**, *54*, 267–273.
9. Arroyave, M.E. Pulmonary Obstructive Disease in a Population Using Paraquat in Colombia. In *Impact of Pesticide Use on Health in Developing Countries*; Forget, G., Goodman, T., de Villiers, A., Eds.; Proceedings of a Symposium in Ottawa, Canada, Sept 17–20, 1990. International Development Research Centre: Ottawa, 1990; 333.
10. Lewalter, J.; Leng, G. Consideration of individual susceptibility in adverse pesticide effects. Toxicol. Lett. **1999**, *107*, 131–144.

COMPATIBILITY OF CHEMICAL AND BIOLOGICAL PESTICIDES

Barbara Manachini
University of Milan, Milan, Italy

INTRODUCTION

In the majority of crop systems today, emphasis is still placed on single technologies such as the use of pesticides, host plant resistance, and biocontrol, consideration rarely being given to their interaction (1). However, an important approach that could be taken in integrated pest management (IPM) programs is the use of biological pesticides together with a rational use of chemical pesticides. In fact, when a range of pests is present, or when only one method is not efficient, there may often be economic and environmental advantages in combining two or more control methods. Such methods need to be compatible with each other, as incompatibility can lead to loss in effectiveness, increased toxicity to humans and other nontarget organisms, the development of pesticide resistance, major product loss, and crop injury. Some information on the selectivity of most pesticides to natural enemies of pests is already known, but data on the compatibility of chemical and specific biopesticides are often limited and are sometimes conflicting (2).

POSSIBILITY OF INTEGRATION: BENEFITS AND LIMITATIONS

Increasing problems with chemical pesticides have stimulated the search for alternative control measures, such as the use of biological pesticides including viruses, bacteria, fungi, protozoans, and nematodes. A common advantage of biological pesticides is that they target a narrow range of pests, and therefore minimize unintended adverse effects on beneficial organisms (3). Moreover some biological pesticides work better in controlled environments such as greenhouses where consistent results are more likely than in exposed field environments. As the target for biological pesticides is quite narrow, there are still many situations for which they are not available. The use of chemical control agents is still probably the most frequently used and widespread means of achieving effective and reliable pest reduction. However, nonselective chemical pesticides can eliminate natural enemies of pests, and induce other problems such as secondary pest outbreaks and pest resurgence. Intensive use of most pesticides will often also lead to pesticide resistance in target pests.

After considering the costs, benefits, timing, as well as the ecological and environmental impacts, the effects of economically damaging pest populations may be reduced to acceptable levels through biological and chemical control measures applied individually or in combination, especially in the presence of more than one pest (1, 4).

Although some data on the compatibility of certain pesticides and natural enemies is available (2, 5), there have been no detailed reviews on the compatibility of chemical and biological pesticides. Nonetheless, chemical pesticides have been reported to have negative, neutral or positive, and synergetic effects on biological pesticides (Table 1).

Negative Interactions

Biological and chemical pesticides can be incompatible as chemical compounds can severely reduce the activity of the live organisms used in biological pesticides, inducing mortality, low reproduction rates, reduced infectivity capacity, and changes in host searching behavior.

Some chemical insecticides (chlorpyrifos, propetamphos, and cyfluthrin) have been found not to affect the conidial germination of the fungus *Metarhizium anisopliae* used against the German cockroach (*Blatella germanica*), but they do have an adverse effect on growth and sporulation of the fungus. Because of the direct relationship found between *M. anisopliae* growth and insecticide concentration, it has been necessary to greatly limit the in-field use of these insecticides (6).

Whereas mortality levels from direct chemical exposure have been measured and tested for several species, using laboratory bioassays, possible sublethal effects resulting from indirect effects or secondary compounds are relatively unknown. For example in the presence of sunlight, xanthene-based insecticides have been found to have a negative effect on useful species of fungi such as *Beauveria bassiana, M. anisopliae*, and *Penicillium fumosoroseus* (7).

Table 1 Examples of interactions between chemical and biological pesticides

Chemical pesticides	Biological pesticides	Type of interaction	References
Insecticides (Tetrachlorvinphos, Etrimfos, Quinalphos)	*Bacillus thuringiensis*	Negative: Inhibition of replication but probably due more to the presence of emulsifiers.	16
Insecticide (Allosamidin)	*Bacillus* sp.	Negative: Inhibition of the activity of the chitinase.	17
Xanthene dyes	Entomopathogenic fungi	Negative: Photodynamic inhibition of conidial germination in fungi by xanthene dyes.	7
Insecticides (Diflubenzuron, Carbaryl, Cyhalothrin) Fungicides (Benlate - Benomyl) Herbicide (Bentazone)	Entomopathogenic fungi *Nomuraea rileyi*	Negative: Fungicide and herbicide were not compatible with the fungus. Insecticides, had lower inhibitory effect on the fungus.	18
Insecticides (Carbofuran, Fonofos, Terbufos)	*Heterorhabditis bacteriophora*	Negative: Lowered vitality and change in behaviour of *H. bacteriophora*.	19
Insecticides: (Carbofuran, Fenoxycarb)	*Steirnenema carpocapsae* (All strains) *S. feltiae* (Umea strain)	Neutral–Negative-: Both insecticides caused mortality of IJ. The infectivity of IJ that survived exposure to of the carbamates was not compromised by treatments.	20
Fungicides (Aromex, Captafol, Captan, Chlorothalonil, Dinocap, Metalaxyl, Sulphur, Triadimefon)	*Bacillus* sp.	Neutral: Activity of chitinase was not affected.	17
Insecticides (Acephate, Chloropyriphos, Cypermethrin, Diclovorus, Dimethoate, Methomyl, Malathion, Methylparathion, Monocrotophos, Allosamidin)		Negative: Inhibition of the activity of the chitinase for allosamidin.	
Insecticide (Acephate)	*B. thuringiensis*	Neutral: no inhibition effect on replication, spore germination, and crystal size of the bacterium.	21
Insecticides (Methyl Parathion, Permethrin)	*B. thuringiensis* var. *kurstaki*	Neutral: Compatible but without synergetic effects.	22
Insecticides (Triterpene Azidirachtin)	Nuclear polyhedrosis virus	Positive: Control of *Lymantria dispar* larvae was enhanced.	23
Insecticide (Tefluthrin)	*Steirnenema carpocapsae* (Mexican strain) *Heterorhabditis bacteriophora* (Lewiston strain)	Positive: Increase mortality of *Diabrotica virginifera*. Synergy of the effects.	9
Insecticide (Carbofuran)	*Beauveria bassiana*	Positive: Increase mortality of *Ostrinia nubilalis*.	24
Insecticides [Actellic 50EC(R), Aqua Resigen(R), Resigen(R), Fendona SC(R)]	*B. thuringiensis* var. *israelensis*	Positive: Recommended to add Bt to the chemical insecticides for control of both larvae and adults of *Aedes*.	25

Another problem is that incompatibility often occurs only for some species or strains used in biological pesticides, and compatibility is recorded for relatively close species. For example, there are many insecticides (e.g., chlorpyriphos, animazine, carbofuran, carbaryl) that are incompatible with the use of entomopathogenic nematodes (EPN), negative effects having been recorded especially for some species and strains of *Heterorhabditis* (8). However some synergy of EPN use, together with chemical pesticides, has been recorded (see Positive Interactions below). These examples show the need for detailed research on possible interactions before chemical and biological pesticides can be integrated effectively.

Positive Interactions

A good example of what can be accomplished in the area of IPM with biological and chemical pesticides is the use of tefluthrin with the EPN *Steirnenema carpocapsae*. Control of larvae of *Diabrotica virginifera* was enhanced by their combination, resulting in a synergistic response and an increase in expected mortality of 24%; the combined effect of the insecticides plus EPN was greater than either product applied on its own (9).

The activity of EPN is not limited by some common agrochemicals so the two can be easily integrated into standard chemical control practices. In fact, several authors have reported that rather than view EPNs as a replacement for chemical insecticides, there may be advantages in integrating their use with chemical control methods (10). In some cases the combination of EPN and chemical insecticides has resulted in a synergy of insect suppression. The mechanism for increased nematode efficacy is not fully known, although it has been suggested that it could be due to a reduction in the host insect's immunity and activity even at sublethal dosages (8–10).

There is some evidence that sublethal insecticide doses may have synergetic effects in combination with biological pesticides, probably because the pest is already debilitated by the chemical, thus allowing the biological agent to work better. Extensive experiments with mixtures of *B. thuringiensis* ssp. *kurstaki* (Dipel) and pyrethroids (cypermethrin, deltamethrin, fenvalerate, and permethrin) have shown the efficacy of these mixtures against the winter moth, *Operophtera brumata*, and their compatibility with integrated mite control in apple orchards, especially for cypermethrin. In fact the level of winter moth damage to harvested fruit was found to be just as low with Dipel and pyrethroid mixtures as with full-rate pyrethroid treatments (11).

Table 2 Practices to promote compatibility of chemical and biological pesticides

1. Use physiologically selective chemical pesticides to support biological pesticides
2. Evaluate the probable effect of secondary compounds
3. Reduce dosages of pesticides
4. Reduce the area of applying chemical pesticide (e.g., treatments in alternate rows)
5. Reduce the contact between chemical and biological pesticides (e.g., applications in strips or inside traps)
6. Time the application of pesticides considering the least amount of interference possible if they are not totally compatible
7. Avoid the periods of greatest susceptibility of organisms composing the biological pesticides
8. Use and eventually create through biotechnology pesticide-resistant organisms for biological pesticides

Furthermore, several authors highly recommend the judicious use *B. thuringiensis* (*Bt*) products combined with Insect Growth Regulators (IGRs) and other biotechnology techniques, to achieve the most effective and sustainable use of *Bt*. The IGR tebufenozide was found to be compatible with fenoxycarb and natural enemies in controlling *Cydia pomonella* in apples, as well as enhancing the biological control on phytophagous mites by predatory ones (12).

The above examples demonstrate that the integration of appropriate chemical compounds and biological pesticides can sometimes provide better control than using one method alone. Table 1 shows other examples of positive, negative, and neutral interactions.

PRACTICAL APPROACHES TO INTEGRATION

Although the replacement of chemical pesticides by biological pesticides would be auspicious it is not always possible, such replacement depends on the type of pests, the crop, the costs involved, and the social and cultural possibility of the countries. Furthermore what Goettel claimed must be kept in mind "if we are to succeed in long sustainable agriculture, we must move swiftly to put the 'I' back into IPM" (13).

In fact it is essential that an IPM strategy integrate all the techniques into a single coordinated program. Table 2 gives some indications for the integration of chemical and biological pesticides. IPM should not be "chemically dependent" but neither should it neglect the judicious use of chemical pesticides. Consideration must always be given to the effectiveness of combinations of chemical and biological pesticides in controlling a target pest and the toxicity of this "combined product" to nontarget pests. Therefore it is essential to verify the selectivity of chemical compounds and their compatibility with biological pesticides before using them (14). Moreover, by exposing the pest to multiple control tactics, pest adaptability should be reduced. Thus, the compatibility of chemical and biological pesticides is useful for insecticide resistance management programs (15), and their combined use can also result in the reduction of chemical doses, saving environment and biodiversity.

See also *Biopesticides,* pages 85–88; *Insect Growth Regulators,* pages 398–401; *Fungicides,* pages 325–328; *Crop Rotations (Insects and Mites),* pages 169–171; *Modeling Pest Management,* pages 500–503.

C

REFERENCES

1. Thomas, M.B. In *Ecological Approaches and the Development of "Truly Integrated" Pest Management*, Proceedings of the National Academy of Sciences USA, 1999; 96, 5944–5951.
2. Croft, B.A. *Arthropod Biological Control Agents and Pesticides*; John Wiley & Sons: New York, 1990.
3. Van Driesche, R.G.; Bellows, T.S. *Biological Control*; Chapman & Hall: New York, 1996.
4. Saucke, H. Selective bioinsecticides, selective chemical insecticides: important options for integrated pest management (IPM) in cabbage. Harvest **1994**, *16* (1/2), 16–19.
5. Sterk, G.; Hassan, S.A.; Baillod, M.; Bakker, F.; Bigler, F.; Blumel, S.; Bogenschutz, H.; Boller, E.; Bromand, B.; Brun, J.; Calis, J.N.M.; Coremans-Pelseneer, J.; Duso, C.; Garrido, A.; Grove, A.; Heimbach, U.; Hokkanen, H.; Jacas, J.; Lewis, G.; Moreth, L.; Polgar, L.; Roversti, L.; Samsoe-Peterson, L.; Sauphanor, B.; Schaub, L.; Vogt, H. Results of the seventh joint pesticide testing programme carried out by the IOBC/WPRS-working group pesticides and beneficial organisms. Biocontrol **1999**, *44* (1), 99–117.
6. Pachamuthu, P.; Kamble, S.T.; Yuen, G.Y. Virulence of metarhizium anisopliae (*Deuteromycotina: Hyphomycetes*) strain ESC-1 to the german cockroach (*Dictyoptera: Blattellidae*) and its compatibility with insecticides. J. Econ. Entomol. **1999**, *92* (2), 340–346.
7. Krasnoff, S.B.; Faloon, D.; Williams, J.E.; Gibson, D.M. Toxicity of xanthene dyes to entomopathogenic fungi. Biocontrol Sci. Technol. **1999**, *9* (2), 215–225.
8. Rovesti, L.; Heinzpeter, E.W.; Tagiente, F.; Deseo, K.V. Compatibility of pesticides with the entomopathogenic nematode *Heterorhabditis bacteriophora* poinar (*Nematoda: Heterorhabiditidae*). Nematologica **1988**, *34*, 462–476.
9. Nishimatsu, T.; Jackson, J.J. Interaction of insecticides, entomopathogenic nematodes, and larvae of the western corn rootworm (*Coleoptera, Chrysomelidae*). J. Econ. Entomol. **1998**, *91* (2), 410–418.
10. Ishibashi, N. Integrated Control of Insects Pests by *Steirnenema carpocapsae*. In *Nematodes and Biological Control of Insect Pests*; Bedding, R., Akhurst, R., Kaya, H., Eds.; CSIRO: Canberra, Australia, 1990; 105–113.
11. Hardman, J.M.; Gaul, S.O. Mixtures of *bacillus thuringiensis* and pyrethroids control winter moth (*Lepidoptera: Geometridae*) in orchards without causing outbreaks of mites. J. Eco. Entomol. **1990**, *83* (3), 920–936.
12. Valentine, B.J.; Gurr, G.M.; Thwaite, W.G. Efficacy of the insect growth regulators tebufenozide and fenoxycarb for lepidopteran pest control in apples, and their compatibility with biological control for integrated pest management. Aust. J. Exp. Agric. **1996**, *36* (4), 501–506.
13. Goettel, M. Whatever happened to the "I" in "IPM"? Soc. Invertebrate Pathol. Newslett. **1992**, *24*, 5–6.
14. Hokkanen, H.M.T. Role of Biological Control and Transgenic Crops in Reducing Use of Chemical Pesticides Protection for Crop. In *Techniques for Reducing Pesticide Use: Environmental and Economic Benefits*; Pimentel, D., Ed.; John Wiley & Sons: Chichester, UK, 1997; 103–127.
15. Lacey, L.A.; Gottel, M.S. Current developments in microbial control of insect pests and prospects for the early 21st century. Entomophaga **1995**, *40* (1), 3–27.
16. Kuzmanova, I. Study on the compatibility of *Bacillus thuringiensis* berliner with three organophosphours insecticides. Gradinarska i Lozarska Nauka **1981**, *18*, 23–27.
17. Bhushan, B.; Hoondal, G.S. Effect of fungicides, insecticides and allosamidin on a thermostable chitinase from *Bacillus* sp. BG-11 world. Journal of Microbiology & Biotechnology **1999**, *15* (3), 403–404.
18. Terribile, S.; Monteiro de Barros, N. Compatibility between pesticides and the fungus *Nomuraea rileyi* (Farlow) Samson. Revista de Microbiologia **1991**, *23* (1), 48–50.
19. Rovesti, L.; Heinzpeter, E.W.; Tagiente, F.; Deseo, K.V. Compatibility of pesticides with the entomopathogenic nematode *Heterorhabditis bacteriophora* poinar (*Nematoda: Heterorhabditidae*). Nematologica **1989**, *34* (4), 462–476.
20. Gordon, R.; Chippett, J.; Tilley, J. Effects of two carbamates on infective juveniles of *Steinernema carpocapsae* all strain and *Steinernema feltiae* umea strain. Journal of Nematology **1996**, *28* (3), 310–317.
21. Morris, O.N. Effect of some chemical insecticides on the germination and replication of commercial *Bacillus thuringiensis*. Journal of Invertebrate Pathology **1975**, *26* (2), 199–204.
22. Habib, M.E.M.; Garcia, M.A. Compatibility and synergism between *Bacillus thuringiensis* (Kurstaki) and two chemical insecticides. Zeitschrift fur Angewandte Entomologie **1981**, *91* (1), 7–14.
23. Cook, S.P.; Webb, R.E.; Thorpe, K.W.; Podgwaite, J.D.; White, G.B. Field examination of the influence of azadirachtin on gypsy moth (*Lepidoptera: Lymantriidae*) nuclear polyhedrosis virus. Journal of Economic Entomology **1997**, *90* (5), 1267–1272.
24. Lewis, L.C.; Berry, E.D.; Obrycki, J.J.; Bing, L.A. Aptness of insecticides (*Bacillus thuringiensis* and carbofuran) with endophytic *Beauveria Bassiana*, in suppressing larval populations of the european corn borer. Agriculture Ecosystems & Environment **1995**, *57* (1), 27–34.
25. Seleena, P.; Lee, H.L.; Chiang, Y.F. Compatibility of *Bacillus thuringiensis* serovar israelensis and chemical insecticides for the control of *Aedes* mosquitoes. Journal of Vector Ecology **1999**, *24* (2), 216–223.

CONSERVATION OF BIOLOGICAL CONTROLS

Douglas Landis
Michigan State University, East Lansing, Michigan, U.S.A.

Steve Wratten
Lincoln University, Canterbury, New Zealand

INTRODUCTION

Biological control is the use of living natural enemies to control pest species and is generally accomplished via importation, augmentation or conservation of natural enemies (1). Conservation biological control seeks to manipulate the environment to enhance the survival, fecundity, longevity, and behavior of natural enemies to increase their effectiveness in controlling pests (2, 3). This differentiates it from the numerical augmentation of natural enemies through inundative or inoculative releases, or the importation of exotic natural enemies sometimes referred to as classical biological control. All forms of biological control ultimately depend on natural enemies finding a favorable environment in which to impact the pest, thus, conservation of natural enemies should be a critical consideration in all biological control efforts (4–7).

REDUCING HARMFUL CONDITIONS

Conservation efforts may be focused on either reducing factors that are harmful to natural enemies, or on enhancing those environmental attributes that are favorable to natural enemies (2). Reducing harmful conditions is the first step in an attempt at conservation biological control. This must be considered at the time natural enemies first enter the environment as well as throughout the remainder of their life cycle.

Pesticides

Pesticides are perhaps the most well-known factor inhibiting natural enemy effectiveness. Insecticides, herbicides, fungicides, and antibiotic products may all interfere with the success of natural enemies. Insecticide applications can directly kill insect predators or parasitoids and may have prolonged toxicity due to their residual activity. For example, in greenhouse production systems where insect natural enemies are routinely augmented, producers can consult information on the relative toxicity of many pesticides to common natural enemies and estimates of the length of time before natural enemies can be successfully re-established. However, this type of information is less available for other types of agroecosystems, particularly those that rely on naturally occurring enemies (4).

Insecticides may also have indirect effects on the success of arthropod natural enemies. Even if natural enemies survive initial pesticide application(s), they may find unfavorable conditions due to a reduction in the number or availability of their hosts or prey. Some insecticides may have sub-lethal impacts that reduce fecundity or induce sterility in natural enemies. Others may alter the behavior of arthropod natural enemies so that they are less effective at searching for hosts or are repelled from treated surfaces.

Fungicides and other chemical agents may have direct negative effects on beneficial fungi or bacteria that help to control pests. Poorly timed fungicide applications can exacerbate certain pest populations via suppression of insect pathogenic fungi that would naturally suppress pest numbers. Herbicides can also have negative impacts on certain arthropod natural enemies. While only a few are directly toxic, others may act as repellents or irritants increasing natural enemy emigration from treated areas. Perhaps more important, by reducing weed density, herbicides alter the crop environment in other ways that may be detrimental to natural enemies (see ''Alternative Food Sources and Shelter and Microclimate'' below).

Cultural Practices

Various cultural practices such as tillage or burning of crop debris can kill natural enemies or make the crop habitat unfavorable (8). Ceasing to burn crop residues increased parasitism of the sugarcane leafhopper and reduced pest density in India. In annual crops, the repeated disturbances of plowing, planting, and cultivation may make the habitat unsuitable for ground-dwelling predators such as spiders and carabid beetles. Providing small habitats as refuges from these disturbances has been suggested as one effective conservation measure. In orchards, repeated tillage may create dust deposits on leaves, killing small predators and parasites and causing increases in certain insect and mite pests (9).

Host Plant Effects

Host plant effects, such as chemical defenses that are harmful to natural enemies but to which pests are adapted, can reduce the effectiveness of biological control. Some pests are able to sequester toxic components of their host plant and use them as defense against their own enemies. In other cases, physical characteristics of the host plant, such as leaf hairiness, may reduce the ability of the natural enemy to find and attack its prey or hosts. Selection of appropriate cultivars can enhance natural enemy survival and reduce pest damage (4).

Secondary Enemies

All natural enemies are themselves attacked by other predators, pathogens, and parasitoids. These organisms can in some cases suppress the natural enemy population sufficiently to inhibit effective biological control. Controlling secondary enemies such as hyperparasitoids is yet another conservation tool. Perhaps the most effective means of doing so is to prevent secondary enemies from becoming established in the first place. This is a regular part of importation biological control programs, where quarantine of imported material is required to assure it is free of unwanted secondary enemies.

ENHANCING FAVORABLE CONDITIONS FOR NATURAL ENEMIES

Ensuring that the ecological requirements of the natural enemy are met in the agricultural ecosystem is the other major means of conserving natural enemies. While this may seem self-evident, it is not a trivial matter. To be effective, natural enemies may need access to alternative prey or hosts, adult food resources, overwintering habitats, constant/alternative food supply, and appropriate microclimates. Identifying the ecological factor(s) necessary to favor even a few key natural enemies can be a time-consuming research endeavor. Ensuring that these resources are available in the agroecosystem at the correct time(s) and spatial scale(s) is then required to complete the task (6, 7).

Alternative Prey or Hosts

Generalist natural enemies may require a succession of different prey/hosts in order to remain in a given environment, and manipulating the presence of alternate prey/hosts within a crop before the arrival or seasonal increase of targeted pests may be necessary. This may be done by managing alternative prey in surrounding vegetation, on weeds within the crop, or in plant residue/organic matter maintained on the soil surface. Providing alternative hosts for more host- or habitat-specific natural enemies may be more challenging. The classic example is *Anagrus* (Hymenoptera: Mymaridae) parasitoids of the grape leafhopper that must overwinter on plant hosts outside the grape vineyard. Provision of wild or cultivated plants, which support overwintering eggs of alternative leafhopper hosts, increases *Anagrus* parasitism and contributes to control of grape leafhopper. Manipulating the dispersal of natural enemies from habitats containing alternative prey via carefully timed mowing or herbicide treatment of these habitats may also be necessary to encourage natural enemy movement into the desired crop (6).

Alternative Food Sources

Many natural enemies benefit from the availability of food sources other than their prey/hosts. Plant nectar is consumed by many parasitoid species and can enhance activity, longevity, and rates of parasitism. The presence of honeydew-producing insects is another means by which some parasitoids obtain access to plant sugars. Many natural enemies including parasitoids, certain lady beetles, syrphid fly adults, and predaceous mites consume pollen. For some species pollen may simply provide nutrition during periods of low prey availability, while in other species access to pollen may be necessary to complete egg development.

Provision of flowering plants that provide pollen and nectar to enhance natural enemies has received considerable attention. In Europe, the North American annual plant *Phacelia tanacetifolia* Bentham has been extensively used since it produces large quantities of pollen and nectar, and can enhance the activity and fecundity of syrphid flies that are important aphid predators. Lists of plants that are known to be attractive to a variety of predators and parasitoids are available. Often these are perennial species that flower over extended periods of time, produce large quantities of pollen, or have accessible nectaries. While the presence of flowering plants almost always augments natural enemy populations, there is less evidence that they increase biological control in adjacent crop areas, due in part to the difficulties in conducting appropriate field experiments. An alternative to using plants to attract and enhance natural enemies is that of artificial food sprays. While this approach has proven effective in some cases, it may be more costly and most appropriate for high-value crops (6).

Shelter and Microclimate

Most natural enemies require sheltered sites in which to overwinter, or to escape disruptive cultural practices and pesticide applications. Effective conservation may require that such habitats are present within the agroecosystem. An early example of habitat management was the provision of a refuge for natural enemies of alfalfa pests displaced by cutting. Strip-harvesting of alfalfa was used to provide temporary refuges from the disruption of harvesting and encourage recolonization of the regrowing crop. Overwinter survival of many natural enemies can be encouraged by providing undisturbed habitats in or adjacent to crop fields. The example of grassy ''beetle banks'' for the conservation of ground-dwelling arthropods has been adopted in several European countries. Overwintering predator populations exceeding 1500 individuals/m^2 have been reported after two years of beetle bank establishment (10). Intercrops during the growing season and cover crops during the noncrop period can also be used to enhance natural enemy habitat (7).

ECOLOGICAL INFRASTRUCTURE IN AGROECOSYSTEMS

Many of the difficulties that natural enemies experience in annual crops can be traced to the frequent and intense disturbances that characterize these agricultural ecosystems (9–11). In these simplified ecosystems, many ecological services associated with the maintenance or enhancement of biodiversity, such as soil and water erosion prevention, nutrient cycling, biological control, pollination etc., are also compromised. The concept of restoring these functions by managing the ecological infrastructure of landscapes shows promise in alleviating some of these problems. For example, landscape features such as windbreaks or riparian buffers that are primarily used to reduce soil erosion, prevent pesticide or nutrient runoff, and maintain water quality may also be managed to provide needed resources to support natural enemies for biological control. The cost of this ecological infrastructure is thus spread across several segments of society. Taking a broader approach and seeking to manage agricultural landscapes to provide multiple ecosystem benefits may be one key to the further implementation of conservation biological control.

REFERENCES

1. Van Driesche, R.G.; Bellows, T.S., Jr. *Biological Control*; Chapman Hall: New York, 1996; 539.
2. Rabb, R.L.; Stinner, R.E.; van den Bosch, R. Conservation and Augmentation of Natural Enemies. In *Theory and Practice of Biological Control*; Huffaker, C.B., Messenger, P.S., Eds.; Academic Press: New York, 1976; 233–254.
3. Van den Bosch, R.; Telford, A.D. Environmental Modification and Biological Control. In *Biological Control of Pests and Weeds*; DeBach, P., Ed.; Reinhold: New York, 1964; 459–488.
4. Barbosa, P. *Conservation Biological Control*; Academic Press: San Diego, 1998; 396.
5. Gurr, G.M.; Wratten, S.D. *Biological Control: Measures of Success*; Kluwer Academic Publishers: Dordrecht, The Netherlands, 2000; 429.
6. Landis, D.; Wratten, S.D.; Gurr, G. Habitat manipulation to conserve natural enemies of arthropod pests in agriculture. Annu. Rev. Entomol. **2000**, *45*, 173–199.
7. Pickett, C.H.; Bugg, R.L. *Enhancing Biological Control: Habitat Management to Promote Natural Enemies of Agricultural Pests*; University of California Press: Berkley, 1998; 422.
8. Altieri, M.A.; Letourneau, D.K. Vegetation management and biological control in agroecosystems. Crop Protection **1982**, *1*, 405–430.
9. Landis, D.A.; Marino, P.C. Landscape Structure and Extra-field Processes: Impact on Management of Pests and Beneficials. In *Handbook of Pest Management*; Ruberson, J., Ed.; Marcel Dekker, Inc.: New York, 1999; 74–104.
10. Thomas, M.B.; Wratten, S.D.; Sotherton, N.W. Creation of island habitats in farmland to manipulate populations of beneficial arthropods: predator densities and species composition. J. Appl. Ecol. **1992**, *29*, 524–531.
11. Menalled, F.D.; Marino, P.C.; Gage, S.H.; Landis, D.A. Does agricultural landscape structure affect parasitism and parasitoid diversity? Ecol. Applic. **1999**, *9*, 634–641.

CONSERVATION OF NATURAL ENEMIES

Cetin Sengonca
University of Bonn, Bonn, Germany

C

INTRODUCTION

After the successful utilization of two methods of biological control, classical biological control (importation and establishment of exotic natural enemies against either exotic or native pests) and augmentation of natural enemies (either inundative or inoculative releases of mass reared natural enemies), the third method—conservation and enhancement—has become more and more important during recent years (1, 2). Conservation of natural enemies is probably the most important concept in the practice of biological control and, fortunately, is one of the easiest to understand and readily available to growers. Most authors consider conservation as an environmental modification to protect and enhance natural enemies (3). This definition will be the main subject of this paper.

Natural enemies of arthropod pests, also known as biological control agents, include predators and parasitoids that occur in all production systems from commercial fields to backyards where they have adapted to the local environment and target pests. Their conservation is generally noncomplicated and cost-effective (4). With relatively little effort the activities of these natural enemies can be observed. Natural control agents are a major factor in controlling agricultural pests and need to be considered when making pest management decisions. Today, therefore, the conservation of natural enemies is considered inseparable from enhancement and together they represent a successful biological control method.

Natural enemies suffer severely from the use of broad-spectrum pesticides in agroecosystems and absence or destruction of food resources, shelter sites, egg-laying places, etc., during the vegetation period. Furthermore, unfavorable environmental conditions and lack of overwintering shelters during hibernation are causing high mortality among beneficial insects. The careful use of pesticides and the proper modification of environmental conditions may conserve natural enemies and increase their efficacy.

AVOID HARMFUL PRACTICES OF PESTICIDES

Pesticides used in agriculture are not only killing target pests but they also can have direct effects on natural enemies by killing them, or indirectly by eliminating their hosts or preys and causing them to starve. In contrast, the conservation concept of natural enemies attempts to avoid the application of particularly broad-spectrum, highly disruptive pesticides. Applying selective or specific and beneficially safe pesticides may contribute much toward preserving natural enemies. Pesticide selectivity to beneficial arthropods has been broadly classified into two forms. The first of these is physiological selectivity, that is, pesticides are less toxic to natural enemies than to their target pest when applied at the recommended rate. The second form is ecological selectivity that pertains to the means and domains in which pesticides are used. Systemic pesticides killing leaf-feeding herbivores, for example, may have little or no effect on the many natural enemies that have contact only with the leaf surface. In some cases, pesticides can be successfully integrated into pest management systems with little or no detrimental effect on natural enemies, and this trend is likely to increase substantially in the future. Nowadays, the pesticide industry places increasing emphasis on the development of beneficially safe and environmentally friendly pesticides that exhibit greater selectivity for natural enemies and have minimal environmental impacts. In the same way, governmental regulatory agencies increasingly consider the adverse effects of pesticides on natural enemies in their registration process for pesticides, reflecting the growing concern over negative effects on beneficial insects. Despite these important steps and the great progress that has been made, the latent effects of pesticides and the impact of ''cocktail applications'' on natural enemy populations are still not fully understood. Further research and implementation of research results are urgently needed.

Alternatively, when selective pesticides are unavailable, recommended conservation tactics usually involve exact timing of pesticide applications. Careful forecasting and observation of the occurrence and growth of pest populations can substantially reduce the number of pesticide applications. Forecasting systems should be based on a defined economic threshold for each pest, considering also the presence of natural enemies (5). Another approach is the selective placement of pesticides in agrcultural fields. Limiting pesticide application only to infested parts of the field will reduce costs and also conserve natural enemies. An ideal alternative approach is the use of microbial insecticides, such as commercially available *Bacillus thuringiensis* and fungal and viral products that have little adverse effects on natural enemies and the environment.

HABITAT AND ENVIRONMENTAL MANIPULATION

Another form of natural enemy conservation is habitat and environmental manipulation. The agricultural landscape is currently so intensively managed that the species diversity of many natural habitats has disappeared or become endangered. A similar reduction can be observed among natural enemies too. It has been strongly suggested by many experts that natural enemies also can be conserved by simply encouraging vegetational diversity of the agroecosystem. In this context, hedgerows, cover crops, strips inside and bordering fields, and even in-field balks provide important refuges for parasitoids and predators of many pest species. There they find and benefit from safe shelters, sources of pollen and nectar, and also alternative prey or hosts in case of food scarcity in cultivated fields. And thus, at such sites, a long-lasting and self-regulating biocoenosis will develop. A higher acceptance of weeds in agricultural crops may also increase the efficacy of natural enemies. Similarly, mixed plantings, for example, Umbelliferae, mustard, and *Phacelia tanacetifolia* (6), growing weed strips even within fields, and providing flowering field borders significantly increase habitat diversity. *P. tanacetifolia* has been cultivated widely between crops in the production system in Germany for more than a decade. At the same time this will provide shelters and alternative food sources for natural enemies. Another important concept in conservation by habitat manipulation is that of connectivity. Natural or less disturbed habitats are often scattered and isolated within the agricultural landscape. Connecting these habitats, for example, by hedgerows and woods, will establish a continuous network of corridors allowing movement of natural enemies between fields.

Experiments have shown that a constant population of natural enemies can be established and conserved by releasing their prey or hosts during periods of scarcity, for example, releasing the red mite, *Tetranychus urticae* Koch, to support the establishment of its predatory mite, *Phytoseiulus persimilis* Athias-Henriot, in cucumber and bean cultures widely in greenhouses in middle Europe (7). Similarly, distributing sterilized *Eupoecelia ambiguella* Hb. eggs between the two generations of this lepidopteran pest preserved its egg parasitoid *Trichogramma semblidis* (Auriv.) in the Ahr valley in Germany. A classical example is the black scale, *Saissetia oleae* (Oliv.), which interrupts its development for a short period during summer in the hot arid areas of central California. Planting irrigated oleander plants adjacent to citrus orchards allowed the black scale a continuous development. As a result, this enabled its specific parasitoid *Metaphycus helvolus* (Comp.) to maintain its population, particularly during the hot summer months (7). In another example, it has been found that preservation of nettles, *Urtica* spp., an important host plant of *Aglais urticae* L., can enhance the efficacy of *T. semblidis* on the second generation of *E. ambiguella.* The reason is that the parasitoid maintains its population on this alternative host during the two nonoverlapping generations of *E. ambiguella* (8).

The manipulation of some simple cultural measures can also conserve natural enemies. A famous example is strip harvesting hay alfalfa, allowing mobile natural enemies to disperse from cut strips to half-grown strips. Similarly, *Trissolcus vasilievi* (Mayr) and *T. semistriatus* Nees, two important egg-parasitoids of the sunn pest, *Eurygaster integriceps* Put., were successfully conserved in Turkey by growing shade trees in hot arid areas, providing shade and thus suitable climatic conditions for these parasitoids.

OVERWINTERING AND SHELTER SITES

Natural enemies build up considerably high population densities during summer periods, but then suffer from lack of overwintering or shelter sites and unfavorable climatic conditions during winter. As a result of these detrimental environmental factors, extreme low entomophagous arthropod densities are often present the following year. This permits pest populations to explode and in consequence requires more pesticide applications. In contrast, by preserving existing or providing artificial hibernation sites or shelters, natural enemies can be conserved during overwintering periods. For example, planting of trees and perennial bunch grasses near agricultural sites, the use of burlap or cloth trees, and wrap and stones provided as hiding places at overwintering time allow coccinellid lady beetles higher survival rates during hibernation. In the same way "trunk traps" and "trap bands" can be used as artificial overwintering sites for predatory bugs and lacewings (9), and felt belts can be wrapped around the trunks of fruit trees and vines for the predatory mite *Typhlodromus pyri* Scheuten. The green lacewing, *Chrysoperla carnea* (Stephens), overwinters as an adult in barns, roof trusses, houses, and under the bark of trees, where mortality rates during hibernation in middle European climatic conditions may still reach 60–90%. By using specially designed, simple wooden shelters (hibernation boxes) the overwintering mortality was reduced to only 4–8% (10). On this ground, these hibernation boxes are now being accepted and commonly used by farmers, gardeners, and also environmental protectionists in Germany and Switzerland.

FUTURE CONCERNS

The majority of pest problems in agriculture are due to the elimination of natural enemies by the indiscriminate and intensive use of pesticides. Improper habitat manipulation and mismanagement of ecosystems have further intensified this problem by reducing the available flora and fauna. Conservation and enhancement of natural enemies is the easiest and least costly method of biological control offering solutions to most pest problems without harming and disturbing the natural ecosystem. Unfortunately, however, this field has received little attention and very little investment has been made in research. There is a serious need for research into the areas of direct conservation and enhancement of natural enemies during the vegetation period and hibernation. In a self-regulating mechanism focusing on conservation and enhancement, natural enemies can keep agricultural pests below their economic threshold and help to reduce the number and frequency of pesticide applications. Furthermore, integration of conservation and enhancement of natural enemies into existing IPM programs (11) will lead to a more sustainable and cost-effective agriculture system.

See also the *Cosmetic Standards*, pages 152–154; *Conservation of Biological Controls*, pages 138–140; *Augmentative Controls*, pages 36–38; *Biological Controls...*, pages 57–60; pages 61–63; pages 64–67; pages 68–70; pages 71–73; pages 74–76; pages 77–80; pages 81–84.

REFERENCES

1. Ehler, L.E. Conservation Biological Control: Past, Present and Future. In *Conservation Biological Control*; Barbosa, P., Ed.; Academic Press: New York, 1998; 1–8.
2. Bugg, R.L.; Pickett, C.H. Introduction: Enhancing Biological Control–Habitat Management to Promote Natural Enemies of Agricultural Pests. In *Enhancing Biological Control*; Pickett, C.H., Bugg, R.L., Eds.; University of California Press: Berkeley, 1998; 1–23.
3. DeBach, P. *Biological Control of Insect Pests and Weeds*; Chapman & Hall: New York, 1964; 844.
4. Barbosa, P. Agroecosystems and Conservation Biological Control. In *Conservation Biological Control*; Barbosa, P., Ed.; Academic Press: New York, 1998; 39–59.
5. Sengonca, C. Conservation and enhancement of natural enemies in biological control. Phytoparasitica **1998**, *26* (3), 187–190.
6. Sengonca, C.; Frings, B. Einfluss von phacelia tanacetifolia auf schaedlings- und nuetzlingspopulation in Zuckerruebe. Pedobiologia **1988**, *32* (5/6), 311–316.
7. Krieg, A.; Franz, J.M. *Lehrbuch der biologischen Schaedlingsbekaempfung*; Verlag Paul Parey: Berlin, 1989; 302.
8. Schade, M.; Sengonca, C. Foerderung des traubenwicklereiparasitoiden *Trichogramma semblidis* (Auriv.) (Hym., Trichogrammatidae) durch bereitstellung von ersatzwirten an brennesseln im weingebiet ahrtal. Vitic. Enol. Sci. **1998**, *53* (4), 157–161.
9. Beane, K.A.; Bugg, R.L. Natural and Artificial Shelter to Enhance Arthropod Biological Control Agents. In *Enhancing Biological Control*; Pickett, C.H., Bugg, R.L., Eds.; University of California Press: Berkeley, 1998; 240–253.
10. Sengonca, C.; Frings, B. Enhancement of green lacewing *Chrysoperla carnea* (Stephens) by providing artificial facilities for hibernation. Turk. Entomol. Derg. **1989**, *13* (4), 245–250.
11. *CRC Handbook of Pest Management in Agriculture*; Pimentel, D., Ed.; CRC Press: Boca Raton, FL, 1991; 2, 757.

CONTROL OF HOUSE FLIES AND "FILTH" FLIES

Tove Steenberg
Jørgen B. Jespersen
Danish Pest Infestation Laboratory, Lyngby, Denmark

INTRODUCTION

The house fly (*Musca domestica*), and a number of related flies within the families *Muscidae, Calliphoridae* (blowflies) and *Sarcophagidae* (flesh flies), are generally referred to as filth flies and are of worldwide public health importance. These flies breed in human and animal filth, and feed on filthy substrates as well as on human food.

BIOLOGY

All filth flies have developmental cycles with four life stages [egg, larva (maggot), pupa, and adult] (Fig. 1). The adults are medium to large-sized flies with lapping-sponging or piercing mouthparts. House flies and stable flies are 5–8 mm long with dull grayish colors, flesh flies are also grayish but are larger (11–13 mm), and blowflies are mostly metallic blue, green, black, or coppery and the size of flesh flies or larger (1). The reproductive potential of filth flies is high, and the time of development depends strongly on temperature. For the house fly it may take only 10 days from egg to adult, and many generations of flies will be produced in a short time. Depending on the climate, the length of the fly season may be restricted to some months, and in areas with a very hot and dry summer the high temperatures may cause a reduction in fly development. Decaying organic material of various kinds provide breeding substrates for filth flies, and include animal and human feces, food wastes, silage, sewage, and soil contaminated with any of these substances [see (2) for an extensive list of breeding sources]. Blowflies and flesh flies almost exclusively breed in decaying matter of animal origin such as carcasses, animal feces, and wounds and sores. Fly breeding sites may be found in various places in urban and rural areas of developed and developing countries (e.g., garbage cans, refuse dumps, manure piles, septic tanks, and open latrines).

Adult filth flies, especially the blowflies and flesh flies, have developed a strong perception of odors emitted from suitable breeding and feeding material. Once suitable substrates are located, flies feed on the moist surface or regurgitate their stomach contents onto substrates to liquefy the food before sucking it up. Filth flies also frequently defecate on food substrates or when resting on surfaces, e.g., food utensils. In addition to the transmission of disease-causing organisms via the digestive tract, filth flies may harbor these organisms externally on bristles, hairs, feet, and mouthparts. House flies tend to rest on sunny surfaces in the daytime, although in warm climates they may seek cooler areas. In the night they often enter buildings or other structures to rest. Blowflies and flesh flies are active in sunny weather and rest on the vegetation at night. Adult house flies are generally found within 500 m of breeding sources, although they are able to fly much longer distances. They may also be transported passively, e.g., via vehicles. Blowflies and flesh flies are good fliers, and breeding sites may be located far away from where adults are found. However, in urban areas they frequently aggregate around breeding sites in refuse. While house flies frequently enter houses and food production facilities, blowflies and flesh flies do not generally enter into dwellings or structures, except for oviposition. However, some species may enter homes for overwintering, and other species may occur in large numbers in structures when they develop in dead rodents or birds hidden in the structure.

PUBLIC HEALTH IMPORTANCE

Filth flies may carry disease-forming organisms ranging from viruses and bacteria to microscopic roundworms. House flies have been demonstrated to carry the causative agents of diarrhea, typhoid, cholera, dysentery, and eye diseases, as well as hookworm, whipworm, tapeworm, and a range of other disease-causing organisms. Other filth

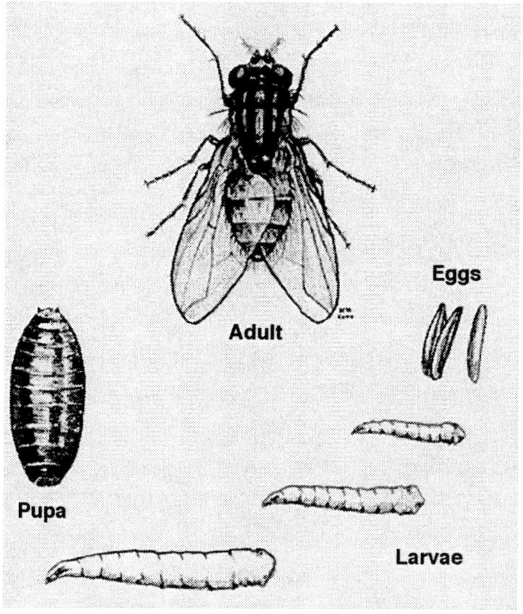

Fig. 1 Life cycle of the house fly (*Musca domestica*). (Drawing by K.-M. Vagn Jensen.)

flies also have the potential to mechanically vector a range of disease-causing organisms to humans. However, those fly species that are attracted the most to human and animal excreta as well as to human food are generally the most important as public health pests. The house fly is a very abundant and widespread species, and is considered the main species involved in transmission of enteric diseases, while the bazaar fly (*Musca sorbens*) is the primary species involved in the transmission of eye diseases (trachoma and others). As a consequence of the breeding and feeding habits of filth flies, flies come into contact with substrates that may harbor pathogens, and with humans, their food and eating utensils. Although flies are not the only factor in disease transmission, the prevalence of enteric and eye infections is frequently associated with the local or seasonal abundance of flies, and in some cases the prevalence of enteric disease and trachoma has been reduced by fly control (2–5). Further, the nuisance caused by large populations of filth flies can be considerable and may impede the quality of life. Fly control is therefore imperative in areas to be developed for recreational use, and livestock farms preferably should be located at distance from human dwellings and food manufacturers. In addition, some species of blowflies also have medical and veterinary importance because they may cause facultative myiasis (infestation by fly maggots) in humans and livestock, while other filth flies may cause accidental myiasis in humans (1).

PREVENTION AND CONTROL

Fly control is obtained primarily through sanitation of breeding sites and exclusion of flies from humans, food, and eating utensils; if necessary these methods may be supplemented with physical and chemical control methods. The starting point in fly control is to identify the fly species, and to locate its breeding, feeding, and resting sites before control measures are initiated. In well-known problem areas fly populations should be monitored continuously in order to be able to initiate control as soon as fly numbers increase. In general, the larval stage should be the main target for control in order to tackle the problem at its base. Methods for fly management in the urban environment include: 1) environmental sanitation and hygiene, 2) physical control, and 3) chemical control. For management of fly problems in livestock production, including the use of biological control agents, see the article *Livestock Pest Management (Insects)* in this encyclopedia.

Environmental Sanitation and Hygiene Measures

Garbage and other organic waste in urban areas should be eliminated by proper collection, storage, transportation, and disposal. Garbage should be placed in closed plastic bags or dustbins, and in warmer climates collections should be made twice a week, as some fly larvae leave the garbage for pupation after only 3–4 days. Any residue left in the bottom of the container should be cleaned out. In case garbage is not burned but taken to a refuse dump, it must be compacted and covered daily with a solid layer of soil. In rural areas, or in the absence of such systems, garbage can be burned or disposed in a dug pit that is covered with a fresh layer of soil as often as necessary. In general, rubbish heaps should be eliminated, and efficient disposal of household and industrial sewage and sanitary wastes organized. Many fly problems will be prevented by installation of fly-proof latrines, and in case defecation in the field cannot be eliminated, feces should be left uncovered in warmer climates for quick dessication and at least 500 m downwind from settlements (2). Hygiene measures include prevention of contact between flies and sick people and other sources of pathogenic organisms, protection of food and eating utensils by placement in fly-proof containers, and protection of infants by nets. Screening of buildings can effectively prevent flies from entering, but may reduce ventilation and light and may be expensive. Flies can also be prevented from entering buildings by closing cracks and ill-fitting doors and windows, or by the establishment of an air current in doorways.

Physical Control

Management by physical methods include the use of sticky traps, light traps with electrocutors, and fly swatters. Fly traps are easy to use in hospitals, offices, hotels and shops, but they are not very effective when fly densities are high. Outdoors, odor-baited traps can catch a large number of flies when placed in bright sunlight, away from shadows of trees.

Chemical Control

Chemical pesticides may be absolutely necessary for suppressing adult fly populations in some situations, e.g., during outbreaks of cholera, dysentery, or trachoma, but they should not be used as a substitute for prevention of fly problems through the elimination of breeding sites. In many parts of the world flies, especially house flies, have become resistant to several of the insecticides used for fly

Table 1 Main types of application for chemical control of flies (with focus on house flies and the urban environment), and their main advantages and disadvantages[a]

	General description	Advantages	Disadvantages
Larvicides	Larvicides are applied topically to fly breeding sites, or used as feed-throughs. To prevent resistance development, IGRs should be used, although also some OPs and CAs can be used.	Tackles the problem at its base None or low resistance to IGRs	Reapplication every 1–3 weeks OPs and CAs often kill natural enemies, and they may favor development of resistance IGRs relatively expensive
Toxic baits	Toxic baits are sugar (or other material) containing insecticide, often in combination with a fly attractant. Toxic baits are applied at fly feeding sites. Various OPs and CAs are used as insecticides.	Cheap and easy to use Low to moderate risk of resistance development	Not effective at high fly densities Some types of baits require frequent application
Fumigation	DDVP vaporizers release dichlorvos slowly over a period of up to three months when hung in rooms.	May be effective for 2–3 months	Only to be used in places with little ventilation Possesses health hazard to humans, and the use may be restricted
Space sprays	Mists or aerosols of insecticide solutions or emulsions are sprayed indoors, outdoors or directly on fly aggregations. Various OPs and PYs are used.	Immediate effect Quick control in emergency situations Low to moderate risk of resistance development	Short-lasting effect Not to be used where meals are being prepared or served Indoor ventilation has to be turned off for 1/2 h after spraying Outdoors effectiveness will depend on air currents
Impregnated materials	Insecticide impregnated materials are hung where flies are resting. Cords, strings, bands, strips, bednets and curtains are used for impregnation with various OPs, CAs, and PYs.	Cheap and with long-term residual effect Low risk of resistance development	Initially fly numbers are reduced slowly Cannot be used in places with an air draught High concentrations may be repellent to flies
Residual treatment	Insecticide solution is sprayed onto surfaces where flies rest to obtain a long-lasting effect. Various OPs and PYs are used.	Immediate and long-term residual effect	High risk of resistance development in flies

[a]Larvicides are used against larvae, while the other five types are directed against the adult flies.
IGR: Insect growth regulator; PY: Pyrethroid; OP: Organophosphorous compound; CA: Carbamate; DDVP: Dichlorvos

control (6), resulting in control failure or use of extensive amounts of insecticide to control the flies. In such regions the resistance levels to commonly used insecticides should be documented, and a resistance management strategy developed and implemented as an integrated part of a regional fly control strategy. There are six main types of application for chemical fly control, one of which is directed against the larvae (2, 7, 8). An overview of these methods describing their main advantages and disadvantages is given in Table 1.

Successful fly control always requires a management approach that integrates those control methods that are suitable for the setting in which the fly problem occurs. This selection of methods also will have to take into account the economic costs of using the different methods, as well as different local constraints. Further details may be found in (2, 7, 8).

REFERENCES

1. Busvine, J.R. *Insects and Hygiene: The Biology and Control of Insect Pests of Medical and Domestic Importance*; 3rd Ed., Chapman and Hall: New York, 1980; 568.
2. Keiding, J. *The Housefly—Biology and Control. Training and Information Guide*; World Health Organization: Geneva, 1986, 63Publication WHO/VBC/86.937.
3. Chavasse, D.C.; Shier, R.P.; Murphy, O.A.; Huttly, S.R.A.; Cousens, S.N.; Akhtar, T. Impact of fly control on childhood diarrhoea in Pakistan: community-randomised trial. Lancet **1999**, *353*, 22–25.
4. Emerson, P.M.; Lindsay, S.W.; Walraven, G.E.L.; Faal, H.; Bøgh, C.; Lowe, K.; Bailey, R.L. Effect of fly control on trachoma and diarrhoea. Lancet **1999**, *353*, 1401–1403.
5. Cohen, D.; Green, M.; Block, C.; Slepon, R.; Ambar, R.; Wasserman, S.S.; Levine, M.M. Reduction of transmission of shigellosis by control of houseflies (*Musca domestica*). Lancet **1991**, *337*, 993–997.
6. Rozendaal, J. *Vector Control: Methods for Use by Individuals and Communities*; World Health Organization: Geneva, 1999; 412.
7. Chavasse, D.C.; Yap, H.H. *Chemical Methods for the Control of Vectors and Pests of Public Health Importance*; World Health Organization: Geneva, 1997; 129.
8. Keiding, J. Review of the global status and recent development of insecticide resistance in field populations of the housefly, *Musca domestica* (*Diptera: Muscidae*). Bull. Ent. Res. **1999**, *89* (Suppl.), 7–67.

CONTROLLED DROPLET APPLICATION

G.A. Matthews
Imperial College of Science, Technology, and Medicine, Berkshire, United Kingdom

INTRODUCTION

Controlled droplet application (CDA) occurs when the droplets in a spray are all of a very similar size, the size being selected according to where the droplets should be deposited (1). This contrasts with hydraulic nozzles that produce a spray containing a much wider range of droplet sizes, which is generally acknowledged to be effective but inefficient, requiring high spray volumes (Fig. 1). The idea of controlling the spray droplet size range is that waste of pesticide in droplets that are too large or too small is minimized. This allows minimization of both the chemical dosages and environmental contamination. CDA evolved from application of ultra-low volume (ULV) sprays developed using an oil-based formulation of insecticide against locusts (2) (see the article on ULV application in this encyclopedia).

DISCUSSION

Initially CDA required a volume median diameter (VMD) to number median diameter ratio (NMD) of <2, but as NMD values are not measured directly with light diffraction laser droplet sizing equipment, it is usual to refer to the relative span, defined as: $D_{90} - D_{10}/D_{50}$, where D_{50} = VMD. Typically the narrow spectrum of a CDA spray is indicated by a relative span of less than 1 (3).

The CDA/ULV technique was developed to apply insecticides to protect crops, especially cotton on small farms in semi-arid regions, where water supplies were poor (4, 5). This was made possible by the availability of hand-carried, battery-operated sprayers with a rotary atomizer to produce a narrow droplet spectrum (6, 7) (Fig. 2). Relatively involatile formulations were applied in sprays with a VMD generally in the range of 70–100 μm. The CDA concept was extended later to the application of herbicides by using water-miscible formulations with larger 250 μm-diameter droplets (8).

Subsequently conventional insecticide formulations were also used diluted in water at very low volumes (VLV) to allow greater flexibility in the choice of chemical in integrated pest management programs (9, 10). VLV sprays are applied in droplets of 100–150 μm; the increase in droplet size is necessary to minimize evaporative loss from droplets between the atomizer and the crop.

Droplets of 50–120 μm are the optimum size for applying CDA sprays on plants, especially when it is possible to deposit an insecticide directly on insects resting on foliage. Smaller droplets will be transported further by the wind and are liable to be lost by convective air movement. Apart from extensive use on cotton, this droplet size range has been particularly relevant for controlling locusts both with chemical and bioinsecticides (11). When applying the myco-pesticide *Metarhizium anisopliae* var. *acridum*, too small a droplet will not contain sufficient spores, but once the droplets are larger than 100 μm diameter (i.e. a volume of 524 picolitres), each droplet contains too many spores, so waste occurs. Some downwind movement of the spray occurs with droplets of 50–100 μm diameter, and this can be exploited in locust control to provide wide swaths.

CDA herbicide sprays are applied with larger droplets >200 μm, in order to minimize downwind spray drift. These large droplets fall more vertically to treat soil surfaces or weed foliage close to the ground. In contrast, small flying insects, such as adult mosquitoes and whiteflies, are effectively controlled by applying extremely small droplets (c. 10–30 μm) that remain airborne. This technique is optimal for applying low dosages of nonresidual insecticides, as the aim is to kill only insects in flight at the time of application. These small droplets can be obtained by using a very high speed rotary nozzle, but are usually obtained by fogging. There is an inhalation hazard with droplets smaller than 20 μm diameter, especially droplets <10 μm.

TYPES OF APPLICATION

The principle method of producing CDA sprays is by using a rotary atomizer. The size of droplets thrown centrifugally from a rotating surface is inversely proportional to the angular velocity of the rotating surface (12). At low

C

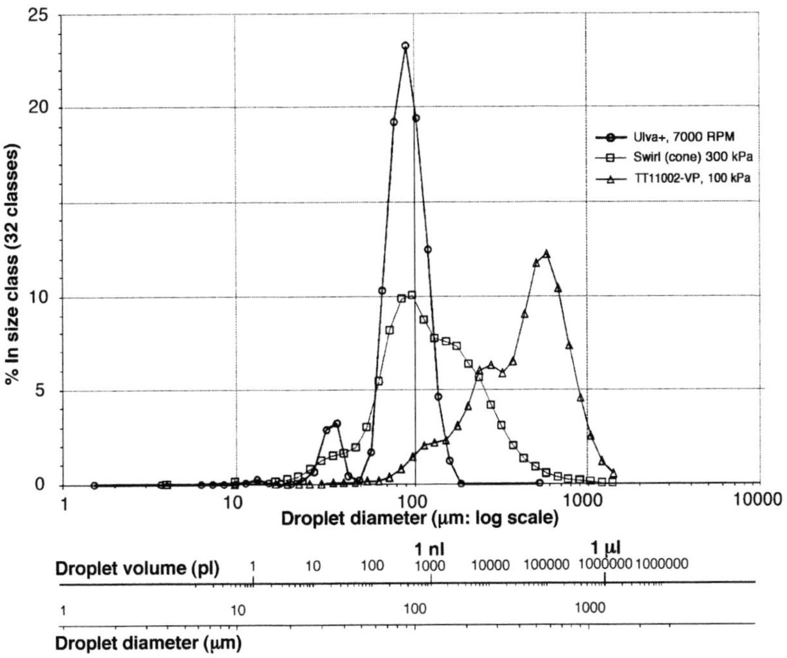

Fig. 1 Comparison of droplet spectra between rotary atomizer and hydraulic nozzle.

flow rates, individual droplets are produced. When the flow of liquid is increased, ligaments are formed (Fig. 3). These ligaments produce smaller droplets, without a change in rotational speed, as droplets are formed from the ligaments, which become thinner as they leave the disc. Some smaller satellite droplets are also formed from ligaments, so there is bimodal droplet distribution. An increase in flow rate with ligament formation will increase droplet size, but if too much liquid is fed on the rotating surface, the ligaments coalesce and control of droplet size is lost. Some rotary atomizers have a series of grooves cut on the inside of a cup- or saucer-shaped disc so that liquid is fed uniformly to teeth around the edge of the disc. The purpose of the teeth is to provide the minimum surface on which liquid can adhere, thus facilitating droplet formation. Uniform distribution of liquid on the rotating surface provides a narrower droplet spectrum.

CDA sprays can also be produced with an electrostatic nozzle known as the "Electrodyn" (13), with which droplet size is determined by the applied voltage and flow rate. An increase in voltage decreases the droplet size and an increase in flow rate produces larger droplets.

Development of CDA equipment has primarily been with manually carried sprayers due to the logistical energy saving advantages of ULV and VLV application (Fig. 1). Equipment to apply insecticides has a tubular handle in which a series of D-sized batteries are carried to provide power to a small DC electric motor that drives the rotary atomiser. Spray liquid is carried in a small container attached to the spray head and is fed to the spinning disc atomiser by gravity through a restrictor. Choice of restrictor will depend on flow rate required and the viscosity of the spray liquid. Insecticides can be applied at ultra-low volume (<5 L/ha) using oil-based pesticide formulations with droplets in the 50–100 μm range or at very low volume (5–15 L/ha) with water-based formulations.

The atomizer is held above the crop foliage to allow the spray cloud to form and disperse downwind utilizing air turbulence to deposit droplets within the crop canopy. The user walks along a series of passes through a field, moving progressively upwind. This enables the user to avoid walking through treated foliage to ensure operator safety (14). The treated swath will vary depending on the height of the atomizer, droplet size and fluctuations in wind speed and direction. Users normally adopt a specified track spacing between passes to allow for a low wind speed. While ULV sprays were highly successful, especially in treating cotton crops in the semi-arid areas of Africa, the change to VLV sprays, apart from allowing greater flexibility in the choice of insecticide in relation to pest incidence, reduced costs with use of less expensive formulations. Similar equipment is used to apply herbicides, but the atomizer is held close to the ground, preferably behind the user, and disc speed is slower to produce droplets >200 μm to ensure minimal downwind displacement of spray.

A

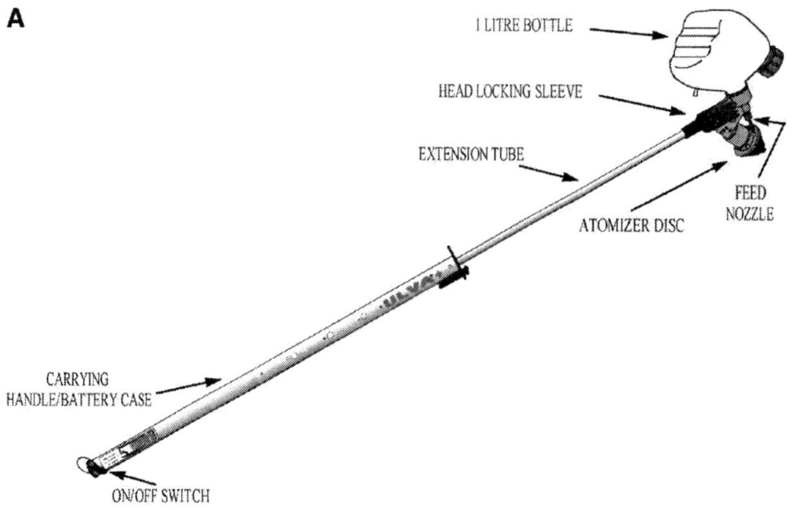

B

Fig. 2 ULVA + sprayer.

An adjuvant may be added to the spray to reduce evaporation and stabilize droplet size, while other adjuvants are used to improve spread of discrete droplets over surfaces. Some specially formulated pre-mixed pesticides, especially herbicides, are supplied in containers that connect directly to the sprayer to reduce user exposure during

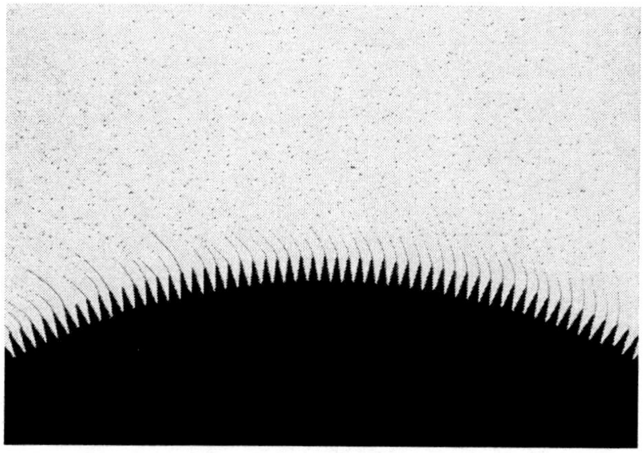

Fig. 3 Ligament formation around a spinning disc.

preparation of a spray. A larger container can be carried on the user's back to refill the container on the handle.

CDA equipment can also be used on tractor-mounted sprayers. Early studies (15) examined the use of a series of small shrouded spinning discs mounted on a common shaft so that, by adjusting their speed of rotation, droplets of 150, 250, and 350 μm diameter were produced. Studies on the effect of different herbicides on broad-leaved and grass weeds showed that when volumes as low as 15 litres per hectare were applied, weeds were not so effectively controlled with contact herbicides using 250 μm droplets. More consistent control required 30–60 litres per hectare to provide sufficient coverage of the weed foliage.

Vehicle-mounted equipment with a series of multiple discs on a shaft is used to treat locusts, but a larger cup-shaped disc was developed for tractor booms (16). This atomizer fitted with a shroud has been used to treat pavements and curbs with herbicides in urban areas and in tree crops. Other CDA atomizer attachments have also been used on standard (axial fan) airblast orchard sprayers, while a portable air–assisted sprayer has been developed to treat bush crops, such as coffee (17).

Rotary atomizers have been used on aircraft, such as in migrant pest control and forestry, but the droplet spectra are also affected by shearing effects of the high-speed air flow across the atomizer. The CDA quality of the spray is thus less than when operated in low air velocities.

REFERENCES

1. Matthews, G.A. CDA—controlled droplet application. PANS **1977**, *23*, 387–394.
2. Sayer, H.J. An ultra low volume spraying technique for the control of the desert locust schistocerca gregaria. Bull. Ent. Res. **1959**, *50*, 371–386.
3. Bateman, R. Simple, standardized methods for recording droplet measurements and estimation of deposits from controlled droplet application. Crop Protection **1993**, *12*, 201–206.
4. Matthews, G.A. Ultra low volume spraying of cotton in Malawi. Cotton Growing Rev. **1973**, *50*, 242–267.
5. Cauquil, J. Cotton-pest control: a review of the introduction of ultra-low volume (ULV) spraying in sub-Saharan French-speaking Africa. Crop Prot. **1987**, *6*, 38–42.
6. Bals, E.J. In *Design of Rotary Atomisers*, Proceedings of the 4th International Agricultural Aviation Congress, 1969; 156–165.
7. Bals, E.J. Some observations on the basic principles involved in ultra-low volume spraying applications. PANS **1973**, *19*, 193–200.
8. Bals, E.J. Development of CDA herbicide handsprayer. PANS **1975a**, *21*, 345–349.
9. Clayton, J. In *New Developments in Controlled Droplet Application (CDA) Techniques for Small Farmers in Developing Countries—Opportunities for Formulation and Packaging*, Brighton Crop Protection Conference, 1992; 333–342.
10. Cauquil, J.; Vaissayre, M. Protection phytosanitaire du cotonnier en Afrique tropicale. Constraintes et perspectives des nouveaux programmes. Agriculture et Developpement **1995**, *5*, 17–29.
11. Bateman, R.P.; Matthews, G.A.; Hall, F.R. Ground-Based Application Equipment. In *Field Manual of Techniques in Invertebrate Pathology*; Lacey, L., Kaya, H., Eds.; Kluwer Academic Publishers: Dordrecht, The Netherlands, 2000; 77–112.
12. Walton, W.H.; Prewett, W.C. Atomization by spinning discs. Proc. Phys. Soc. **1949**, *B62*, 341–350.
13. Coffee, R.A. In *Electrodynamic Energy—A New Approach to Pesticide Application*, Proceedings of the 1979 British Crop Protection Conference—Pests and Diseases, 1979; 777–789.
14. Thornhill, E.W.; Matthews, G.A.; Clayton, J.C. In *Potential Operator Exposure to Insecticides: A Comparison Between Knapsack and CDA Spinning Disc Sprayers*, Proceedings of the Brighton Crop Protection Conference—Pests and Diseases, 1996; 1175–1180.
15. Taylor, W.A.; Merritt, C.R.; Drinkwater, J.A. An experimental tractor-mounted machine for applying herbicides to field plots at very low volumes and varying drop sizes. Weed Res. **1976**, *16*, 203–208.
16. Heijne, C.G. In *A Study of the Effect of Disc Speed and Flow Rate on the Performance of the "Micron Battleship,"* Proceedings of the 1978 British Crop Protection Council Conference—Weeds, 1978; 673–679.
17. Povey, G.S.; Clayton, J.C.; Bals, T.E. In *A Portable Motorised Axial Fan Air-assisted CDA Sprayer: A New Approach to Insect and Disease Control in Coffee*, Proceedings of the Brighton Crop Protection Conference—Pests and Diseases, 1996; 367–372.

C

COSMETIC STANDARDS (BLEMISHED FOOD PRODUCTS AND INSECTS IN FOODS)

Kelsey A. Hart
David Pimentel
Cornell University, Ithaca, New York, U.S.A.

INTRODUCTION

The American marketplace features nearly perfect fruits and vegetables. Gone are apples with an occasional blemish and fresh spinach with a leafminer. This increase in the "cosmetic standards" of fruits and vegetables has resulted from the efforts of the Food and Drug Administration (FDA) and the U.S. Department of Agriculture (USDA) to limit the levels of insects and mites in produce, and new standards established by food wholesalers, processors, and retailers. Meeting more stringent standards has led to significant increases in the amounts and toxicity of pesticides used in crops. Increased pesticide use has negative environmental and public health consequences. In comparison, the health risks from consuming herbivorous insects/insect parts in food do not exist and certainly do not justify the increase in pesticide use and the associated problems. Recent research indicates that pesticide use can be reduced by 35% to 50% without any substantial increase in food prices or loss of crop yields (1). Surveys suggest that the public would support relaxation of cosmetic standards if it decreases pesticide residues in its food.

HISTORY OF COSMETIC STANDARDS

The FDA sets defect action levels (DALs) for insects and mites allowed in fruits and vegetables and in products made from them. These DALs were established to reduce insect and mite infestation in foods to a reasonable and safe level, because their presence in food products was thought to indicate that crops had insufficient insect and mite control, were improperly washed, were unsatisfactorily inspected, and contained insects and mites harmful to health. Besides visual prejudice against insects in food, there is the well-placed concern that insects such as non-herbivorous houseflies and cockroaches may transmit disease.

During the past 40 years, the FDA has steadily lowered DALs (2). For example, a five-fold decrease in the number of leafminers permitted in spinach occurred from 1930 to 1974 (1). As tolerance levels for insects in food have fallen, wholesalers, processors, and retailers have increased their "cosmetic standards" for produce and other food products so that most marketed U.S. produce is visually perfect. Produce distributors encourage high cosmetic standards because their contracts enable them to visually inspect produce before buying and reject it when the supply is excessive. Growers are motivated to produce cosmetically perfect produce to ensure its sale.

However, to meet these increasingly stringent regulations, farmers have had to use greater amounts of increasingly toxic pesticides and implement other pest control strategies. Synthetic pesticide use in the United States has increased about 33-fold since 1945, and the toxicity of pesticides used has increased 10- to 100-fold in the past 25 years (Fig. 1) (1) . More pesticides need to be used to produce the blemish-free produce distributors and consumers expect.

There is little evidence that eating herbivorous insects or insect parts is hazardous to human health (3). However, solid data suggest that the adverse health and environmental impacts of pesticide exposure are substantial. Given the direct correlation between increases in cosmetic standards and increases in pesticide use, why are cosmetic standards and DALs growing increasingly severe and perpetuating further increases in pesticide use?

The increase in cosmetic standards and more stringent DALs are based on the premise that consumers demand unblemished, insect-free food. Clearly, cosmetic appearance of produce is a primary factor consumers use in assessing the quality of produce. Unfortunately, this assessment is often made without more substantive quality information, such as nutritional values or pesticide residue levels. Recent evidence suggests that when consumers are aware of the trade-offs between blemish-free produce and

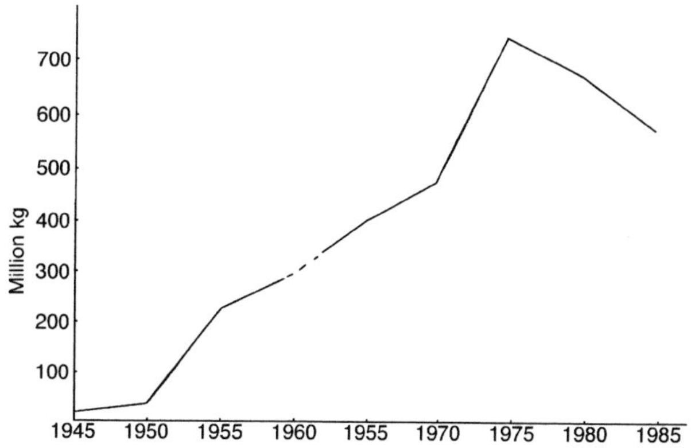

Fig. 1 The amount of synthetic pesticides produced in the United States. About 90% is sold in the United States. The decline in total amount produced since 1975 is in large part due to the 10- to 100-fold increased toxicity and effectiveness of the newer pesticides (1).

pesticide use, they will purchase produce that is not cosmetically perfect because it has less pesticide residue.

ENVIRONMENTAL AND HEALTH EFFECTS OF PESTICIDE EXPOSURE

An estimated 625,000 tons of more than 600 different kinds of pesticides are used annually in the United States, at a cost of approximately $9 billion (4). Still, pests such as insects, plant pathogens, and weeds destroy 37% of all potential food and fiber crops (5). Typically, each dollar invested in pesticides returns about $4 in crops saved.

However, this economic evaluation does not take into account the impacts of pesticide use on public health and the environment. Approximately 0.1% of applied pesticides reach target pests, leaving 99.9% of the pesticides to impact the environment (5). Environmental effects of pesticides can be significant: Domestic animals and wildlife can be poisoned or adversely affected by pesticide exposure; beneficial natural enemies of harmful pests can be destroyed by pesticide use; heavy pesticide use can result in pesticide resistance and subsequently even heavier or more toxic pesticide use; and already limited natural resources such as soil and ground and surface water can be contaminated by pesticide residues or drift (5).

The human health effects of pesticide exposure through food are also diverse and significant. About 35% of foods purchased by American consumers have detectable levels of pesticides, and about 1–3% of these foods have residue levels over the legal tolerance level (5). These estimates are conservative because detection methods currently detect only about one-third of the pesticides now in use in the United State. The contamination rate is undoubtedly higher for fruits and vegetables because they receive the highest levels of pesticides. One USDA study indicates that some pesticide residue remains in produce even after it is washed, peeled, and cored.

Both the acute and chronic health effects of pesticide exposure are significant. Worldwide, about 26.5 million acute pesticide poisonings occur each year, resulting in about 3 million hospitalizations, approximately 220,000 fatalities, and 750,000 cases of chronic pesticide-related illness (4). Chronic effects can adversely affect most systems of the human body. U.S. data indicate that 18% of all insecticides and about 90% of all fungicides are carcinogenic. Many pesticides are also estrogenic, linked to increased breast cancer among some women in the United States. Pesticide exposure can also damage the respiratory and reproductive systems, leading to conditions like asthma and infertility.

The negative health effects that pesticides can have are more significant in children. Children have higher metabolic rates than adults, and their ability to detoxify and excrete toxic compounds is different. Also, because of their smaller size, children are typically exposed to higher levels of pesticides than adults. Finally, certain types of pesticides, such as carbamates and organophosphates, are more dangerous for children than adults (4).

Given the significant environmental and public health impacts that pesticides can have, it appears desirable to limit pesticide exposure to minimize these adverse effects. However, the increasingly stringent DALs and cosmetic standards have resulted in considerable increases in pesticide use. Do the health effects of eating herbivorous insects, insect parts, or blemished produce warrant the risks and the substantial health consequences of increasing pesticide exposure to meet these standards?

HEALTH EFFECTS OF EATING INSECTS/INSECT PARTS IN FOOD

Even under the current stringent DAL regulatory guidelines, a few insects and mites do remain in or on produce. For instance, the DAL for apple butter is an "average of 5 whole insects or equivalents per 100 grams not counting mites, aphids, thrips, or scale insects." DALs for many other food products are similar. Many insects commonly found in foods and food products are so minute in size that they are practically impossible to eliminate. Although the numbers of insects are strictly limited by FDA regulations, some do remain and are eaten.

This, however, is not a cause for concern. In contrast with the well-documented acute and chronic negative health effects resulting from pesticide exposure, there is not one known case of human illness from ingesting insects and mites in or on foods. In addition, though some insects do carry disease or present health risks—houseflies, for example—all herbivorous insects/mites found on harvested produce are harmless to humans.

While ingesting insects or insect parts in our food may seem distasteful to many Americans, many cultures eat insects by choice. Insects are a substantial source of protein, with digestible protein content ranging from 40% to 65% (6). And insects, shrimp, lobster, and crawfish are all arthropods; the latter three are often considered food delicacies.

Given that herbivorous insects found on produce are not a health hazard, consumers must decide whether they are willing to tolerate the presence of a few insects rather than insisting on visually ''perfect'' produce that requires high levels of pesticides.

CONCLUSIONS AND FUTURE DIRECTIONS

If the health effects of herbivorous insects found on or in food products are not cause for concern, then the need for strict DALs might be relaxed. Relaxing cosmetic standards for some fruits and vegetables might be feasible. Approximately 10% to 20% of pesticides applied to fruits and vegetables are used only to comply with the current strict cosmetic standards established by the FDA, USDA, wholesalers, and retailers that result in blemish-free produce. Rigorous cosmetic standards are probably unnecessary, since surface blemishes on fruits and vegetables generally do not affect nutritional content, storage life, or flavor. However, will the American public purchase produce that appears less than perfect?

Research on public preferences shows that 97% of Americans prefer food without pesticide residues. In addition, 50–66% are willing to pay more for food with less pesticide residue (7). It is estimated that in the United States, pesticide use can be reduced by about 50% without reducing crop yields. The estimated increase in the consumer's food costs would be only 0.6% (4). This marketplace cost increase does not take into account the positive environmental and health benefits that would be realized if pesticide use were reduced. The small increase in consumer cost would be more than offset by these benefits.

Therefore, given the environmental and health tradeoffs related to high cosmetic standards for produce, it appears that human health and the environment would be best protected by less stringent DALs and relaxed cosmetic standards for produce, to minimize unnecessary pesticide use and related adverse effects.

REFERENCES

1. Pimentel, D.; Kirby, C.; Shroff, A. The Relationship Between ''Cosmetic Standards'' for Foods and Pesticide Use. In *The Pesticide Question: Environment, Economics, and Ethics*; Pimentel, D., Lehman, H., Eds.; Chapman & Hall: New York, 1993; 85–105.
2. FDA. In *The Food Defect Action Levels: Levels of Natural or Unavoidable Defects in Foods That Present No Health Hazards for Humans*; Department of Health and Human Services, Public Health Service, Food and Drug Administration, Center for Food Safety and Applied Nutrition: Washington, DC, 1995.
3. Defoliart, G.R. The human use of insects as food and as animal feed. Bull. Entomol. Soc. Am. **1986**, *35*, 22–35.
4. Pimentel, D.; Hart, K.A. Ethical, Environmental, and Public Health Implications of Pesticide Use. In *New Perspectives in Bioethics*; Galston, A., Ed.; Johns Hopkins University Press: Baltimore, MD, *in press*.
5. Pimentel, D.; Greiner, A. Environmental and Socioeconomic Costs of Pesticide Use. In *Techniques for Reducing Pesticide Use*; Pimentel, D., Ed.; John Wiley & Sons: Chichester, UK, 1997; 51–78.
6. Gorham, J.R. Foodborne filth and human disease. Journal of Food Protection **1989**, *52*, 674–677.
7. Anon. Appearance of produce versus pesticide use. Chemecology **1991**, *20* (4), 11.

ADDITIONAL READINGS

Anon. Amounts and kinds of filth in foods and the parallel methods for assessing filth and insanitation. Food Purity Perspect. **1974**, *3*, 19–20.

EPA. In *An Overview of Food Cosmetic Standards and Agricultural Pesticide Use*; Office of Policy, Planning, and Evaluation, U.S. Environmental Protection Agency: Washington, DC, 1990.

Pimentel, D.; Culliney, T.W.; Bashore, T. Public Health Risks Associated with Pesticides and Natural Toxins in Foods. University of Minnesota: Twin Cities, MN. http://www.ent.agri.umn.edu/academics/classes/ipm/chapters/pimentel.htm (accessed Oct. 18, 1996).

van den Bosch, R.; Brown, M.; McGowan, C.; Miller, A.; Moran, M.; Petzer, D.; Swartz, J. *Investigation of the Effects of Food Standards on Pesticide Use*; U.S. Environmental Protection Agency: Washington, DC, 1975.

COVER CROPS

C

Zeyaur R. Khan
International Centre of Insect Physiology and Ecology (ICIPE), Nairobi, Kenya

INTRODUCTION

Cover crops are low-growing plant species cultivated for the purpose of protecting a bare soil surface as a tactic for conserving soil and water and maintaining soil productivity. The basic benefits of cover crops include the prevention of soil detachment by wind, rainfall, and splash erosion and the return of cover crop roots and tops to replenish the soil's organic matter. Many crops can serve this purpose, but some give direct economic returns. Cover crops are also grown by farmers to suppress weeds. They are either sown between rows of the main crop species or are grown during the fallow periods. Highly competitive cover crops are generally grown as "smother crops" during the fallow periods or as a phase within a rotational cropping system. In addition, cover crops can be effective in decreasing crop losses due to insect pests and diseases. A number of impressive examples of the use of cover crops for both weed and insect control are available (Tables 1 and 2).

COVER CROPS FOR WEED MANAGEMENT

The recent focus on nonchemical means of pest control has revived interest in the use of cultural weed control. Cultural weed control methods include techniques that supplement crop competition in order to shift the balance in favor of the main crop or to provide growing conditions unfavorable for weeds. There are many examples of how cover crops have been used successfully for weed suppression and maximizing crop yields under conditions of strong competitive pressure from weeds (Table 1). However, it should be noted that intersown cover crops could also greatly reduce the yields of the main crop if competition for water and/or nutrients is strong. The ability of cover crops to compete with the main crop could be reduced by use of less aggressive cover crop cultivars, simultaneous planting dates for main and cover crop species, and pruning or mowing of the cover crop. An understanding of how cover crops and weeds respond to changeable environmental and cultural factors is needed to design an effective, weed-suppressive, and productive intercrop system (1).

The weed-suppressive cover crops compete for light, soil moisture, and nutrients with weeds, or may be allelopathic to weeds. Although allelopathy—the toxic effect of one plant species on another—has usually been considered a problem for agriculture, there is evidence that in some circumstances it might provide an excellent opportunity to help manage weed problems (2, 3). Approaches that have been explored include the use of allelopathic rotational or companion cover plants in annual or perennial cropping systems. Studies show that several cover crops may have significant suppressive effects on weeds that can not be attributed to either the physical presence of mulch, or to the lack of tillage. However, despite considerable evidence of allelopathy, the effect is difficult to separate from plant competition or the physical effect of a mulch cover on weeds. A major challenge for the exploitation of allelopathy for weed control is to minimize its negative impacts on the main crop (4).

Parasitic Weeds

Parasitic weeds are interesting because of their complex life cycles and intricate interactions with hosts. But they are also pernicious pests of a wide range of crops, as they deprive the host plants of water and nutrients. Among the recorded parasitic weeds of crop plants, the plants belonging to the genus *Striga* (Family Scrophulariaceae) are some of the most important as they affect the lives of more than 100 million people in Africa and infest 40% of arable land in the savanna region, causing an annual loss of $7–13 billion. Infestations by weeds of this genus have resulted in farmers' abandoning much of the arable land

Table 1 Examples of weed suppressing effects of cover crops reported from various tropical and temperate countries[a]

Main crop	Cover crop	Country	Weed suppression	Effect on yield of main crop
Maize	Soybean	India	+	
	Mung bean	Trinidad	+	
	Cowpea	Brazil	–	
		Nigeria	+	
Pigeon pea	Groundnut	Nigeria	+	
	Mung bean	Trinidad	+	
	Rye	Germany	+	Yield reduced
		Canada	+	
	Centrosema pubescens	Nigeria	+	
	Psophocarpus palustris	Nigeria	+	
	Arachis repens	Nigeria	+	
	Desmodium triflorum	Nigeria	–	
	Indigofera spicata	Nigeria	–	
	Desmodium uncinatum	Kenya	+	
	Axonopus spicata	Nigeria	–	
	Crotalaria mucronata	Brazil	–	
	Trifolium incarnatum	Canada	+	Yield reduced
	Trifolium alexandrinum	Canada	+	
Sweet maize	*Vicia villosa*	United States	–	
	Rye	United States	+	Yield reduced
Sorghum	Fodder cowpea	India	+	
	Grain cowpea	India	+	
	Mung bean	India	+	
	Peanut	India	+	
	Soybean	India	+	
Barley	Ryegrass	United Kingdom	+	
	Red clover	United Kingdom	+	
Fava bean	Ryegrass	United Kingdom	+	
	Red clover	United Kingdom	+	
Cassava	*Desmodium heterophyllum*	Colombia	+	
	Bean	Colombia	+	
	Groundnut	Nigeria	+	
	Cowpea	Nigeria	+	
Rice	*Azolla pinnata*	Philippines	+	
	Lolium multiflorum	Brazil	+	
	Vicia sativa	Brazil	+	
	Lotus corniculatus	Brazil	+	
Broccoli	Soybean	United States	+	Yield reduced
	Setaria italica	United States	+	Yield reduced
Cabbage	Rye	United States	+	Yield reduced
Tomato	Rye	United States	+	
Sugarcane	Oat	Hawaii	–	
Coffee	*Desmodium ovalifolium*	Nicaragua	?	
	Aracis pintoi	Nicaragua	+	
	Desmodium sp.	Kenya	+	
	Commelina sp.	Cuba	?	Yield reduced
	Zebrina pendula	Cuba	+	
Coconut	*Pueraria javania*	France	+	
	Centrosema pubescens	France	+	
	Calopogonium caeruleum	France	+	
	Psophocarpus palustris	France	–	
Vineyards	*Lolium perenne*	Italy	+	

+, Weeds reduced; –, no significant reduction; ?, effect not clear.
[a]Information compiled from published literature on use of cover crops to suppress weeds.

Table 2 Examples of insect pest suppression and enhancement of natural enemies by cover crops reported from various tropical and temperate countries[a]

			Insect pest		Natural enemy	
Main crop	Cover crop	Country	Species	Suppression	Species	Enhancement
Sorghum	Cowpea	Kenya	*Chilo partellus*	Yes	—	
			Chilo partellus	Yes	*Denticasmias busseolae*	Yes
		Burkina Faso	*Atherigona soccata*	?	*Neotrichoporoides* sp.	Yes
		India	*Empoasca kerri*	Yes		
Maize	*Vicia faba, Cucurbita moschata*	Mexico	*Spodoptera frugiperda*	No effect	—	—
			Empoasca kraemeri	Yes	—	—
	Cowpea	Kenya	*Chilo* sp., *Sesamia calamistis*	Yes	Egg parasitism	Yes
					Larval parasitism	No
	Melinis minutiflora	Kenya	Stemborers	Yes	*Cotesia sesamiae*	Yes
Carrot	*Medicago littoralis*	Sweden	*Psilia rosae*	Yes	—	—
Cabbage	*Trifolium repens*	Poland	*Plutella maculipennis*	No effect	—	—
			Pieris rapae	No effect	—	—
			Phyllotreta sp.	Yes	—	—
Cotton	*Trifolium pratense*	USA	*Lygus hesperus*	Yes	—	—
			Helicoverpa zea	Yes	—	—
			Heliothis virescens	Yes	—	—

[a]Information compiled from published literature on use of cover crops to suppress insect pests.

in Africa. *Striga* infestation continues to extend to new areas and a further 40% of arable land may become infested in the next 10 years.

Recommended control methods to reduce *Striga* infestation include heavy applications of nitrogen fertilizer, crop rotation, use of trap crops and chemical stimulants to abort seed germination, hoeing and hand pulling, herbicide application, and the use of resistant or tolerant crop varieties. All these methods, including the most widely practiced hoe weeding, have a serious limitation in the reluctance of farmers to accept them, for both biological and socioeconomic reasons.

To put *Striga hermonthica* control within the reach of African farmers, use of cover crops belonging to the genus *Desmodium* has been recommended. *Desmodium uncinatum* or *D. intortum,* when intercropped with maize, inhibits parasitization of maize by *S. hermonthica* and therefore increases maize yields significantly. As an additional benefit, these leguminous plants are of great economic importance to farmers in eastern Africa as livestock fodder. Intercropping of *Desmodium* plants with maize to manage *S. hermonthica* offers an excellent option for cereal farmers who practice mixed farming in eastern Africa and is showing great potential in farmer participatory on-farm trials in controlling *S. hermonthica.* Investigations are underway at the International Centre of Insect Physiology and Ecology (ICIPE) to find out if the suppression mechanism associated with *Desmodium* is physical or allelochemical, or both.

COVER CROPS FOR INSECT PEST MANAGEMENT

Several theories have been put forward to account for the effects of cropping systems on the population dynamics of insect pests. Deterrence of colonization through intrafield diversity is probably one of the most promising means of controlling pests. Host plant location by insect pests can be disturbed by certain nonhost cover crops (which provide camouflage, and/or diversionary or repellent volatile compounds). The release of repellent or odor-masking substances into the air by nonhost cover crops can confer some protection to the host plant by decreasing insect pest attack below that found in monocropping systems. There are numerous examples in the literature of a decrease in insect pest attack on host plants due to intercropping of various cover crops (Table 2).

Some cover crops act as a suitable alternative host for a particular pest, and may reduce the feeding damage to the main crop by diverting the pest (5). However, this kind of diversionary effect is likely to be extremely complex and there may be an equal likelihood of increasing loss due to the pest as of reducing it.

Other studies advance the natural enemy hypothesis, which states that beneficial insects find greater resources in a diversified cropping system, and thus act more effectively than in a monoculture system (Table 2). Such resources include food (especially nectar), physical cover and alternate prey (6, 7). Ground cover provided by intercrops can be valuable to predators, particularly those that feed by night and hide by day. Predators and parasites can usually be more effective in controlling pests if alternate prey or hosts are available to permit them to survive and reproduce during times when the pest population is low. Many cover crops can serve as the plant hosts for herbivorous insects, which in turn serve as alternate hosts/prey for natural enemies. However, sometimes crop diversification can have as great an inhibitory effect on natural enemies as it does on pests, because predators and parasites also respond to a variety of stimuli when orienting to their prey. This is particularly the case if a predator or parasite orients to its insect host by the odor of the plant on which the host feeds.

A recent study has reported the effectiveness of intercropping maize with the nonhost molasses grass (*Melinis minutiflora*) for reducing stemborer damage on maize (8). In field trials, molasses grass showed no colonization by stemborers, and when used as a cover crop with maize, significantly reduced stemborer infestation of the main crop. A significant increase in parasitism of stemborers by the larval parasitoid *Cotesia sesamiae* was also observed. Volatile agents produced by the molasses grass repelled the female stemborers but attracted foraging *C. sesamiae*. Female *C. sesamiae* were attracted to (E)-4,8-dimethyl-1,3,7-nonatriene, one of the volatile components released by intact molasses grass. Nonatriene has been implicated as an SOS signal for recruiting predators and parasitoids and is also produced by stemborer-damaged maize plants. The production of such a compound by molasses grass provides an explanation for the increased parasitism observed in the maize-molasses grass intercrop. Intact plants with an inherent ability to release such attractive stimuli could be used in new crop protection strategies. As well as serving as an effective cover crop, the molasses grass at the same time provides good fodder for livestock. The grass is now being tested in on-farm trials in Kenya to control stemborers on maize.

CONCLUSION

The economic gain from the use of cover crops depends on the balance between a lowered cost of control of weeds and insect pests and the increased cost of maintaining an intercropped field, along with any decrease in yield of the main crop from greater plant competition. Net profit can be increased if the cover crop favorably changes the balance between income and costs. Economic data assessing the financial returns as well as the biological effects is therefore most useful in making decisions on the use of cover crops for insect pest and weed control.

REFERENCES

1. Liebman, M. Ecological Suppression of Weeds in Intercropping Systems: A Review. In *Weed Management in Agroecosystems: Ecological Approaches*; Altieri, A.A., Liebman, M., Eds.; CRC Press: Boca Raton, FL, 1988; 197–212.
2. Putnam, A.R. Allelopathy: Problems and Opportunities in Weed Management. In *Weed Management in Agroecosystems: Ecological Approaches*; Altieri, A.A., Liebman, M., Eds.; CRC Press: Boca Raton, FL, 1988; 77–88.
3. Weston, L.A. Utilization of allelopathy for weed management in agroecosystems. Agron. J. **1996**, *88*, 860–866.
4. Gill, K.S.; Arshad, M.A.; Moyer, J.R. Cultural Control for Weeds. In *Techniques for Reducing Pesticide Use: Economic and Environmental Benefits*; Pimentel, D., Ed.; John Wiley & Sons: England, 1997; 237–275.
5. Bugg, R.L. Cover Crops and Control of Arthropod Pests of Agriculture. In *Cover Crops for Clean Water: The Proceedings of an International Conference*; Hargrove, T.L., Ed.; Soil and Water Conservation Society, West Tennessee Experimental Station: Ankeny, 1991; 157–163.
6. Cromartie, W.J., Jr. The Environmental Control of Insects Using Crop Diversity. In *Handbook of Pest Management in Agriculture*; Pimentel, D., Ed.; CRC Press: Boca Raton, FL, 1981; 1, 223–251.
7. Gurr, G.M.; van Emden, H.F.; Wratten, S.D. Habitat Manipulation and Natural Enemy Efficiency: Implications for the Control of Pests. In *Conservation Biological Control*; Barbosa, P., Ed.; Academic Press: San Diego, 1998; 155–183.
8. Khan, Z.R.; Ampong-Nyarko, K.; Chiliswa, P.; Hassanali, A.; Kimani, S.; Lwande, W.; Overholt, W.A.; Pickett, J.A.; Smart, L.E.; Wadhams, L.J.; Woodcock, C.M. Intercropping increases parasitism of pests. Nature (London) **1997**, *388*, 631–632.

CROP COVERS FOR WEED SUPPRESSION

C

Fred Thomas
CERUS Consulting, Chico, California, U.S.A.

INTRODUCTION

A cover crop is a sown but nonharvested plant or mixture of plants that is used to add fertility and tilth to the soil, fix atmospheric nitrogen, reduce erosion, suppress weeds, trap and recycle excess soil nitrates, and provide other specific management benefits for the commercial crop. An annual cover crop such as vetch can be planted every year during the winter to provide these benefits for the succeeding summer crop that will be more productive as a result. Other cover crops for orchards and vineyards such as New Zealand white clover are often long-lived perennial ground covers grown on the floor to provide the same benefits, but planted only every 10 or 15 years.

USAGE IN WORLDWIDE AGRICULTURE

As a fertility and erosion control tool, cover crops are used extensively in both temperate and tropical agriculture. The use of cover crops as a method of competitive weed control is a practice that has been used since antiquity with reports of the Romans using bell beans between the rows of their vineyards to fix nitrogen and improve soil fertility for the harvested grape crop. The use of commercial fertilizers and chemical herbicides during the twentieth century has displaced this once common practice and only recently has the use of cover crops returned as an important agronomic tool.

U.S. Usage of Cover Crops

Within the past decade the use of cover crops in the United States has increased substantially as the reliance of modern agriculture on synthetic fertilizers and chemical herbicides has resulted in soil erosion and loss of soil structure. Conservation tillage as it is practiced in the Midwest has encouraged farmers to experiment with cover crops to reduce soil erosion. In California many vineyard districts from Northcoast winegrapes to San Joaquin raisin production report up to 40 percent of the farmers using cover crops for agronomic benefit. Another reason cover crop use has increased is its role in preventing ground and surface water pollution by leaching contaminants and preventing off-site movement of soil and chemicals caused by intensive farming practices. The interest in sustainable agriculture has resulted in several USDA programs such as the Environmental Quality Incentives Program (EQIP) that offer cost-sharing to farmers for practices that improve environmental quality including the establishment of cover crops as well as planted buffers around field edges.

Cover Crop Systems for Annual Crops

The most common form of cover cropping is the planting of either a legume, such as vetch, or grains, such as oats, barley, and rye, or mixtures of both. These are referred to as green manures and are usually planted in the fall after the commercial crop is harvested, and then turned under in the spring prior to the planting of corn, sunflowers, tomatoes, or melons. When properly established these green manure cover crops usually suppress winter weeds competing for sunlight, nutrients, and moisture that would otherwise allow the weeds to flourish. While much of the United States uses the winter window for growing cover crops, other milder and more subtropical states such as California, Texas, and Florida have the option of growing summer cover crops (see Fig. 1) in a short rotation post-harvest to smother weeds and produce organic matter.

Led by Drs. Abdul-Baki and Teasdale, USDA Agricultural Research Service, Beltsville, Maryland, new advances in conservation tillage during the past 10 years have created innovative systems that allow the winter cover crop to be killed and left on the surface while the commercial crops of tomatoes or vegetables are mechanically transplanted into the mulch with specialized equipment. These ''mulch'' systems provide excellent summer-long weed control.

The concept of permanent vegetation on a grassed roadway or filter strip has also been recently applied at the edges of annual cropping systems. Many farmers are

Fig. 1 A sudangrass cover crop following melons that was planted to suppress field bindweed (*Convolvulus arvensis* L.) and build soil organic matter. Fresno County, CA.

establishing perennial vegetation at the field borders to intercept the off-site movement of soil, fertilizer, and chemicals that would otherwise end up in waterways and riparian areas. Often a 30-foot wide grassed roadway is all that is needed to result in an 80 percent reduction in the off-site movement of pollutants. The perennial border for the filter strips keeps noxious weeds from moving into the field and becoming established.

Cover Crop Systems for Perennial Crops

For perennial crops such as stone and pome fruit and vineyards there is a wide range of cover crop usage and methods. Many orchard and vineyard growers use the same plant species as annual crop farmers, planting green manures after harvest that are grown during the winter and incorporated in the spring. Many growers also use long-lived perennial or permanent covers, such as perennial ryegrass, dwarf orchard grass, trefoil, or New Zealand white clover as part of their orchard floor management system. With this system, the cover crop is established in the alleys and borders of the tree or vine crop and remains a weed-suppressive ''living mulch'' for many years with occasional mowing (see Fig. 2). From the farmer's perspective there are numerous reasons besides suppressing weeds to establish a perennial cover crop system compared to tillage between the trees. The main benefits to the orchardist or vineyard manager are improved water infiltration, preservation of surface feeder roots that are otherwise killed by disking, orchard and vineyard access for winter operations, and a reduction in labor and energy costs associated with mowing versus disking.

Costs, Benefits, and Problems of Using Cover Crops

The actual cost of cover crop seed can be as little as U.S. $15 per acre, when barley, wheat, or sudangrass are used as the cover crop, or it can be as high as U.S. $70 per acre when long-lived plants such as perennial ryegrass or sheep fescue are established in an orchard for a 15-year lifespan. The best way to budget the cost to the farm operation is to amortize the cost according to its usefulness to the commercial crop. Most people assign a two-year benefit to a green manure crop with the benefits of reduced erosion, weed suppression, and nitrate trapping the first year and the benefits of improved tilth, nitrogen release, soil microbial bloom for nutrient release, and disease reduction the second year. For a standard Mid-

Fig. 2 A perennial cover crop of New Zealand white clover and strawberry clover established in a new kiwi vineyard is a suppressive ''living mulch'' effectively smothering all weeds. Butte County, CA.

western cover crop like cereal rye and hairy vetch the cost of U.S. $25 per acre can be amortized over two years.

For perennial crops the seed and establishment cost of hard fescue, trefoil, or white clover can be amortized over the 15-year life of the cover. When comparing the cost of planting a green manure annually to a perennial crop the actual cost savings also includes the benefits of reduced compaction, less dust, fewer mites, and better winter access.

There are many benefits to growing a cover crop, but there are also problems that can develop from introducing an additional management system along with the commercial crop. Some of the more common problems associated with growing a cover crop for annual systems include missing the spring planting window because of wet weather and an overabundance of material from the cover crop. Problems with wireworms or seed-corn maggots can also occur when planting too soon after incorporation of the cover crop. The problems most often associated with perennial cropping systems include the increased chance of spring frost resulting in damage or loss of the grape or fruit crop, increased water use with up to 30 percent more irrigation required, extra nitrogen requirements for grass sods, and increased problems with rodents, particularly gophers and voles which can be destructive to new and established trees. Most problems associated with cover crops are either accepted as a trade-off for the benefits, or the cover crop is managed to minimize or avoid potential problems.

BIBLIOGRAPHY

Abdul-Baki, A.; Teasdale, J.R. *Sustainable Production of Fresh-Market Tomatoes and Other Summer Vegetables with Organic Mulches* Bulletin No. 2279, USDA/ARS: Beltsville, MD, 1997; 23.

Bowman, G. *Steel in the Field: A Farmer's Guide to Weed Management Tools* USDA-Sustainable Agriculture Network (SAN): Burlington, VT, 1997; 128.

Bugg, R.L.; Zomer, R.J.; Auburn, J.S. *The U.C. SAREP Cover Crops Database* University of California Sustainable Agriculture Research and Education Program: Davis, CA, 1996. http://www.sarep.ucdavis.edu/ccrop (accessed Feb. 26, 2001).

Clark, A. *Managing Cover Crops Profitably,* 2nd Ed.; USDA-Sustainable Agriculture Network (SAN): Burlington, VT, 1998; 212.

Cover Cropping in Vineyards: A Growers Handbook Ingels, C. A., Bugg, R.L., McGourty, G.T., Christensen, L.P., Eds.; Publication 3338, University of California Division of Agriculture and Natural Resources: Oakland, CA, 1998; 162.

Miller, P.R.; Graves, W.L.; Williams, W.A.; Madson, B.A. *Cover Crops for California Agriculture* Publication 21471, University of California Division of Agriculture and Natural Resources: Oakland, CA, 1989; 24.

CROP DIVERSITY FOR PEST MANAGEMENT

Marianne Karpenstein-Machan
Maria Finckh
University Kassel, Witzenhausen, Germany

INTRODUCTION

Improvement in crop management by the use of modern machinery equipment, fertilizers, and pesticides and progress in plant breeding within a few select crops has led in recent decades to highly specialized agricultural practices on farms. It is estimated that about 7000 crops are known worldwide; however, only seven crops are cultivated on more than 90% of the arable land. A general impoverishment of plant diversity and a high degree of genetic erosion in several important crops for human nutrition is documented worldwide.

Simplification of the ecosystem by using one-sided crop rotations, monoculture, and crop plants of uniform genotype and the elimination of weeds with herbicides result in a strong selection for adapted pests, pathogens, and weeds leading to frequent resistance breakdown and severe weed infestations (1–3).

The chances that pests and weeds will develop resistance to any one of the control agents are reduced by increasing the diversity of selection pressures acting on the species. The establishment of diversity in the crop ecosystem with adapted crops in sequential cropping, multiline cultivars, and variety mixtures results in less damage by pests, pathogens, and weeds than when grown in mono cultures. In a diverse crop rotation, weeds are less abundant but with a greater diversity in weed species. Outbreaks of weed, pest, and disease epidemics and the probability of losses are reduced in diverse rotations (2–4).

In this article, diversity by planting sequence (crop rotation), crop-border diversity, and crop-weed diversity are discussed.

THE INFLUENCE OF CROP ROTATION ON DISEASES AND PESTS

The positive effects of a diverse crop rotation on yield and agricultural stability have been well known since ancient times (Table 1). The risk of infestation by weeds, diseases, and pests can be decreased by a well-organized crop rotation. Crop rotation is most effective in the case of specialized pathogens and pests that are dependent on a host crop or have a narrow host range. For example, many soil-borne pathogens or insects survive on root residues and require the cultivation of a susceptible crop for continuous survival. The length of host crop interval for disease control will depend on how quickly the pest or pathogen can be destroyed by antagonistic effects (1). When developing a crop rotation, certain criteria should be considered: maintenance of high levels of organic matter in the soil, tillage to expose the residues to weathering, and inclusion of crops which do not stimulate subsequent growth of the pathogen or crops that have direct negative effects on pests and pathogens (e.g., producing toxins).

As an example, the pigeon pea wilt fungus, *F. oxysporum* f. sp. *udum*, is specific to the pigeon pea. Many other legumes, such as cowpea and soybean, which are non-hosts, are able to stimulate the germination of this fungus. If one of these legumes is cultivated as a subsequent crop, the chlamydospores will germinate. In soils with high microbiological activity due to rich organic matter, the fungus may be destroyed by antagonists, but in the case of low biological activity the fungus may survive by forming secondary chlamydospores (1, 2). Cereals stimulate chlamydospores less compared with legumes. In addition, root exudates contain toxic components that may kill germinated chlamydospores before they can form spores again (2, 3).

THE INFLUENCE OF DECOY AND TRAP CROPS ON PESTS

Decoy crops are nonhost crops that are sown to stimulate the activation of dormant propagules of the pathogen in the absence of the host. In this way the soil-borne pathogens waste their inoculum potential. For example, *Lolium* spec., *Papaver rhoeas*, and *Reseda odorata* can act as decoy crops for the pathogen *Plasmodiophora brassicae* in *Brassica* (3).

Table 1 Crop diseases and pests in which crop rotation has a major effect in disease and pest control

	Wheat	Barley	Sorghum	Maize	Rye	Oats	Crucifers	Beta-Beet	Soybean	Other legumes	Potatoes	Tobacco
Virus												
Beet necrotic yellow vein virus (BNYVV)								×				
Barley yellow mosaic virus (BYMY)		×										
Bacteria												
Scab (*Streptomyces scabies*)											×	
Fungi												
Take all and eyespot diseases of cereals (*Gaeumannomyces graminis, Pseudocercosporella herpotrichoides, Rhizoctonia cerealis*)	×	×			×							
Leaf and ear diseases of cereals (*Typhula incarnata, Rhyncho-sporium secalis, Septoria nodorum, Septoria tritici*)	×	×			×							
Fusarium diseases of cereals (*Fusarium avenaceum, F. culmorum, F. nivale*)	×	×		×	×							
Brown spot (*Septoria glycinea*)									×			
Fusarium root rot complex of legumes										×		
Stark rot (*Fusarium moniliforme*)			×									
Stem canker (*Leptosphaeria maculans*)							×			×	×	
Corn smut (*Ustilago maydis*)				×								
Cercospora leaf spot (*Cercospora beticola*)								×				
Black root rot (*Thielaviopsis basicola*)												×
Root rot (*Meloidogyne hapla*)			×									
Nematodes												
Cereal cyst nematode (*Heterodera avenae*)	×	×		(×)[a]	×							
Beet cyst nematode *Heterodera schachtii*)							×	×				
White and golden cyst nematode (*Globodera pallida, G. rostochiensis*)											×	
Stem and bulb nematode (*Ditylenchus dipsaci*)				×[b]	×[b]	×[b]	×[b]	×[b]				
Cyst nematode (*Heterodera glycinea*)									×			
Insects												
Springtails (*Collembola* spp.)								×				
Pigmy mangold beetle (*Atomaria linearis*)								×				
Haplodiplosis marginata	×	×			×							
European corn borer (*Ostrinia nubilalis*)									×			
Cabbage stem flea beetle (*Psylliodes chrysocephalus*)							×					
Swede seed midge (*Dasineura brassicae*)							×					

[a]Infection without strong effect on crop yield.
[b]Host specific races.

In the case of trap crops, the crop is host to the pathogen (nematode). The trap crop attracts nematodes to infect, but before the nematode can complete its life cycle the crop is harvested or destroyed. Therefore in Germany it is recommended to sow crucifers and plow before the beet cyst nematode can fully develop its life cycle (5).

THE INFLUENCE OF CROP ROTATION ON WEED ABUNDANCE

The kind of crop rotation influences weed density and abundance of weed species. Main effects are caused by the sowing time of crops (winter or spring crops), sequence and placement of crops in the crop rotation, competitive ability of different crops against weeds, and weed management methods (herbicides and/or mechanical control) (6).

In monoculture and pure grain rotations weed infestation reaches a high level and weeds are more difficult to control due to herbicide resistance. Long-term trials in Germany have shown that in pure grain rotations and cereal monocultures the weed density was two and three times higher compared to a rotation in alternation with dicotyledonous plants, for example field beans. Especially the degree of infestation of problem weeds, e.g. *Apera spica venti, Viola tricolor arvensis*, and *Matricaria* spp., increases considerably (7).

Allelopathic effects of certain crops (e.g., barley, rye, maize, sorghum, sudangrass, buckwheat, sunflowers, rape, soybeans, alfalfa, and hemp) may reduce germination of different weeds. The root exudates of barley, wheat, and rye have especially strong negative effects on the germination and development of *Stellaria media, Capsella bursa pastoris*. Wheat and rye hinder germination of *Anthemis arvensis* and *Matricaria perforata* (8).

THE INFLUENCE OF BORDER DIVERSITY ON PESTS

The abundance and diversity of entomophagous insects within a field are closely related to the character of the surrounding vegetation. There are many examples that indicate that crops cultivated near hedgerows or uncultivated fields with flowering weeds sustain less damage by pest and disease organisms than crops cultivated in the absence of these flowering weeds. Brassicas were less damaged by the diamondback moth, *Plutella xylostella* (L.) and the cabbage aphid, *Brevicoryne brassicae* (L.) due to increased predation and parasitism caused by flowering weeds in bordering uncultivated fields. Less intensive management of hedgerows tends to increase the proportion of predacious insects (9).

Potatoes grown near woods were less infected by the Colorado potato beetle *Leptinotarsa decemlineata* (Say), due to increased predation on the larvae.

The presence of nectar source plants in sugarcane fields in Hawaii allows for the development of higher levels of a sugarcane weevil parasitoid, *Lixophaga sphenophori*, thus increasing parasitoid efficiency. A complex vegetation in the crop borders was also the reason that the mean number of species of both herbivore and predator parasitoids per habit space in soybean fields was higher at the edge than in the center of the fields (9).

Sometimes the neighboring vegetation can also contain host plants for diseases and crop pests (4). An astute management of the surrounding crop fields is an option to reduce the amount of damage to crop plants.

THE INFLUENCE OF WEED DIVERSITY ON PESTS

Weeds not only compete with crops; they are also hosts and intermediate hosts for diseases and parasites and offer with their flowers and leaves the basic nutrition for many predators in an agricultural ecosystem. Within monocropped fields, weeds increase diversity and may be useful to improve the stability of the agricultural ecosystem. Certain weeds (mostly *Umbelliferae, Leguminosae*, and *Compositae*) play an important ecological role by supporting a complex of beneficial arthropods that aid in suppressing pest populations and improve the chances of the crop to escape the pest damage. Strip management with weeds or flowering crops influenced the rates of colonization of natural enemies within the fields, as an example from Switzerland shows. Weed strips sown between winter cereals increased ground beetle densities and the number of species considerably by providing these beneficial arthropods with better food supplies and more suitable overwintering sites, from which they can colonize cereals in spring. In spring their potential as pest control agents is greatest. A reduction of insect pests (aphids) due to the en hancing effect of beneficial arthropods was shown (10, 11).

In numerous insect studies with crops such as cotton, sugarcane, alfalfa, soybeans, and corn in addition to others, the reduction of population density of insect pests in weedy crops compared to weed-free fields has been verified (2). In most cases natural enemies regulated pest populations, acting as predators and parasitoids.

FUTURE CONCERNS

Diversity provides an essential key to reduce the risk of losing crop yield to pest damage. Many field trials worldwide have shown that with an astute management of diversity improvement of agricultural stability is achievable. The realization of greater crop diversity by crop rotation, trap crops, and surrounding vegetation and high production of corn, wheat, soybean, cotton, and other crops on a large scale is feasible. The implementation of concepts based on crop diversity will preserve long-term stability and productivity of agricultural land and minimize environmental problems caused by intensive agriculture, e.g. soil erosion, groundwater and air pollution with nutrients and pesticides, and genetic erosion of both flora and fauna (12).

See also *Intercropping for Pest Management*, pages 423–425, 100000391.

REFERENCES

1. Altieri, M.A. Biodiversity and Biocontrol: Lessons from Insect Pest Management. In *Advances in Plant Pathology*; Andrews, J.H., Tommerup, I., Eds.; Academic Press: London, 1995; 11, 191–209.
2. Baliddawa, C.W. Plant species diversity and crop pest control—an analytical review. Insect Sci. Appl. **1985**, *6* (4), 479–487.
3. Chaube, H.S.; Singh, U.S. Adjustment of Crop Culture to Minimize Disease. In *Plant Disease Management: Principles and Practices*; Chaube, H.S., Singh, U.S., Eds.; CRC Press: Boca Raton, FL, 1991; 199–214.
4. Finckh, M.R.; Wolfe, M.S. Diversification Strategies. In *The Epidemiology of Plant Diseases*; Jones, D.G., Ed.; Kluwer: Dordrecht, 1998; 231–259.
5. Mueller, J.; Steudel, W. The influence of cultivation period of different trap crops on the abundance of heterodera schachtii schmidt. Nachrichtenbl Deutscher Pflanzenschutzd **1983**, *35*, 103–108, (in German).
6. Liebmann, M.; Dyck, E. Crop rotation and intercropping strategies for weed management. Ecological Appl. **1993**, *3*, 92–122.
7. Kreuz, E. Late weed infestation in winter wheat stands in relation with intensification of the cultivation and with crop rotation. Archiv. Phytopath. Pflanz. **1993**, *28*, 379–388, (in German).
8. Narwal, S.S. Allelopathy: Future Role in Weed Control. In *Allelopathy in Agriculture and Forestry*; Narwal, S.S., Tauro, P., Eds.; Scientific Publishers: Jodhpur, 1994; 245–272.
9. Pollard, E.; Hedges, V.I. Habitat diversity and crop pests. A study of *brevicoryne brassicae* and its syrphid predators. J. Appl. Ecol. **1971**, *8*, 751–780.
10. Wyss, E.; Niggli, U.; Nentwig, W. The impact of spiders on aphid populations in a strip management apple orchard. J. Appl. Entomol. **1995**, *119*, 473–478.
11. Hausammann, A. The effects of weed strip-management on pests and beneficial arthropods on winter wheat fields. J. Plant Dis. Prot. **1996**, *103*, 70–81.
12. Karpenstein-Machan, M. *Chances of Pesticide Free Cultivation of Energy Crops for Thermal Uses*; DLG-Verlags-GmbH: Frankfurt, 1997; 183, (in German).

C

CROP PEST CONTROL CONSULTANTS

Marjorie A. Hoy
University of Florida, Gainesville, Florida, U.S.A.

INTRODUCTION

Crop pest control consultants are specialists in pest management, many with M.S. or Ph.D. degrees in agronomy, horticulture, entomology, plant pathology, weed science, or nematology. Some are employed by large agricultural companies to provide pest management advice for the company's agricultural production systems. Others may go into business for themselves and provide consulting services to a number of agricultural producers. In some cases, companies hire a number of consultants in order to provide services to multiple farm managers or farmers. Crop pest consultants provide a service to farm managers who must deal with diverse and complex pest and disease problems as well labor, economics, agronomy, and marketing problems. The benefits derived from crop pest consulting services are expected to include improved crop production and/or reduced pest management costs in order to justify the cost of the services.

SPECIALIZATION

Many crop pest control consultants specialize in insect and mite problems, although they also must be knowledgeable about all aspects of the relevant crop production practices, including fertilization, irrigation, weed, nematode, and disease control. Crop pest control consultants should be knowledgeable about the insect and mite pests in the crop so that they can accurately identify them and their natural enemies, monitor their population fluctuations, estimate when the economic injury level (if known) will be exceeded, and they should have the ability to evaluate all pest management options available to them. Consultants need to be familiar with the most recent research and improvements in pest management.

MONITORING

The crops must be monitored in a timely and efficient manner so that the consultant can inform the farm manager of the need for some type of intervention in order to reduce the potential for negative economic effects on crop production. Monitoring may involve sampling plant parts, soil, or traps for the pests at the appropriate season. Adequate sample sizes must be taken so that reliable conclusions can be made. Most consultants will describe the options available to remedy a pest problem, although many pest control consultants consider it unethical to provide the pest control service (such as pesticide applications) because this could be construed as a conflict of interest.

PEST CONTROL OPTIONS

Pest control consultants should know what management options are available and suitable for each pest management situation. This includes knowing which pesticide products are registered for use on the crop, the proper application methods, and the possible unintended side effects of the pesticidal product, such as toxicity to natural enemies, phytotoxic effects on specific crop varieties, compatibility with other pest control practices, and unintended effects on the environment. If augmentative biological control is an option, then the consultant should provide information on the appropriate natural enemy to release for the target pest, effective release rates and methods, and monitoring methods for evaluating the efficacy of the natural enemy. If mass trapping is an option, then the consultant should provide information on effective trap placement, number of traps necessary, and sampling schedules.

PROFESSIONAL QUALIFICATIONS

In some states within the United States, pest control consultants are licensed, but not all states require this. As a result, growers may find it difficult to evaluate the qualifications of consultants with whom they have no prior experience. In many states with a licensing requirement, the pest control consultant is required to obtain continuing training in order to maintain the license. Approved training may be available from seminars/courses offered by cooperative extension agents and state or federal departments of agriculture.

Pest control consultants have several professional organizations where a variety of professional issues can be discussed. For example, many pest control consultants would like to be able to obtain professional insurance that would provide assistance if legal actions are taken against them should crop losses occur. Ethical issues also are discussed and most have a code of ethics in which guidelines are given regarding the consultant's relationship to the public, to the employer/client, and to each other. For example, the National Alliance of Independent Crop Consultants' code of honor includes statements about confidentiality, lack of bias in promoting a particular product or vendor "except on the basis of advantage to the client."

Many pest control consultants have degrees in specific disciplines (such as entomology), but most must still obtain a considerable amount of 'on the job' training because agricultural pest problems and solutions can be quite site specific. Thus, many new consultants work for an established consultant to gain experience with local pest problems for a season or two before striking out on their own.

Outstanding pest control consultants will be highly knowledgeable about all pest control options, including cultural controls, host plant resistance, biological control, and least toxic methods of chemical control. The consultant, or their assistants, will monitor the fields sufficiently often that solutions can be provided in a timely manner. Ideally, pest problems will be anticipated and, in consultation with the farm manager, cultural controls or other controls will be introduced before there is a problem. For example, spider mites are often more serious pests in the margins of crops grown along dusty roads; reducing dust can reduce significantly the pest spider mite populations, so the pest control consultant may recommend paving a road or other practices that will reduce dust.

Pest control consultants need to have or develop excellent communication skills because most farm managers want the best and most complete information transmitted to them in a clear and timely manner. This can take the form of newsletters, personal reports, written reports, and web-based newsletters. Sometimes the consultant must simply indicate that the information is unavailable or inappropriate for the particular situation facing the farm manager. Such areas of uncertainty must be described in an open and honest manner so that the farm manager can trust the advice of the consultant.

Often farm managers will hire a pest control consultant because they wish to change their pest management practices. Perhaps the manager wishes to change from a pesticide-based management strategy to an integrated pest management (IPM) strategy, in which pesticides are used less often and only when other management tactics are unable to maintain low pest populations. Such a change in strategy must proceed carefully, with both the farm manager and consultant listening carefully and learning to trust each other. Unfortunately, IPM is sometimes as much art as science because pest problems are a little different in every field in every year. Cropping systems are dynamic with weather and other components, such as the economic, environmental, and legal issues also changing. Thus, most consultants find that a hands-on, one-on-one, field-by-field approach is necessary if the grower is to adopt an IPM-based pest management strategy. A change in strategy may require several years to achieve.

Several professional organizations exist in the United States, including the National Alliance of Independent Crop Consultants (NAICC), which was founded in 1978 and consists of more than 450 members in 40 states and several foreign countries. The Certified Professional Crop Consultant (CPCC) and the Independent Certified Professional Crop Consultant (CPCC-I) programs are administered by the NAICC.

CHOOSING A CROP PEST CONTROL CONSULTANT

How should farm managers choose their crop consultant? Several questions should be asked. References of individual consultants or of companies should be evaluated by the farm manager. If the consultant is part of a company, the farm manager can ask if staff members belong to professional societies such as NAICC; hold certifications such as Certified Professional Agronomist, Certified Professional Crop Scientist, Certified Professional Crop Consultant—Independent Certified Crop Advisor; and attend scientific meetings. How long has the company or individual consultant been in business? What is the company's training program? If information is not available, does the company carry out research to develop recommendations? Does the company replicate the research, using controls and randomized plots? Does the company receive compensation for recommending a product? Are there incentives to find a problem? What is the relationship between the recommended products and the pest problems?

Crop consulting is a challenging and demanding profession, requiring considerable knowledge, interpersonal skills, and site-specific experience. The goal of the consultant is to help the farm manager grow the best crop possible under the conditions in that specific field.

SELECTED RELEVANT WEB SITES

1. Association of Applied Insect Ecologists (AAIE). http://aaie.com/index.html (accessed Jan 2001).
2. Iowa Independent Crop Consultants Association. www.pme.iastate.edu/IICCA (accessed July 2000).
3. Minnesota Independent Crop Consultants Association. www.mnicca.org (accessed Jan 2001).
4. National Alliance of Independent Crop Consultants (NAICC). www.naicc.org/ (accessed Jan 2001).
5. NebGuide. How to Hire a Crop Consultant. lanr.uni.edu/pubs/FarmMgt (accessed July 2000).
6. University of Minnesota. Role of the Private Crop Consultant in Implementation of IPM. http://ipmworld.umn.edu/chapters/mjones.htm (accessed Jan 2001).
7. Mellinger, Charles. How to Choose a Crop Consultant. www.pmac.net/charlie.htm (accessed Jan 2001).

CROP ROTATIONS (INSECTS AND MITES)

C

Michael E. Gray
University of Illinois, Urbana, Illinois, U.S.A.

INTRODUCTION

The pest management benefits of crop rotation have long been established for certain insect pests; however, for some pests, the effects of rotation are neutral or even negative. Insect species most likely to be affected positively by crop rotation include those that are soil dwelling, host specific, nonmobile, and overwinter as eggs or partially developed larvae. In addition to some insects, certain plant pathogens and weeds often may be more effectively managed when several crops are rotated. For instance, the rotation of corn with soybean may reduce the likelihood of several diseases on corn including eyespot, anthracnose, and southern corn leaf blight. The severity of several soybean diseases (including soybean cyst nematodes, *Alternaria* leafspot, anthracnose, bacteria blight, *Cercospora* leafspot, pod and stem blight, mildews, *Septoria* brown spot, and stem canker) also is lessened when soybean is grown in rotation with another crop such as corn. With regard to weed management, the rotation of crops may allow growers to utilize different cultivation practices as well as several herbicides resulting in more efficient weed control for a given species in one crop as compared with another. The maintenance of soil structure, tilth, moisture, and fertility also are added benefits that accompany the rotation of crops, particularly when legumes are included. Many economic, environmental, and agronomic factors must be considered in selecting an optimal crop rotation system such as potential for soil erosion, tillage options, potential for herbicide carryover, commodity prices, governmental policies, production costs, and likelihood for increasing or lessening potential pest injury.

SUCCESSFULLY MANAGING PESTS WITH CROP ROTATION

The annual rotation of corn and soybean has been a tremendous success story with regard to the management of western and northern corn rootworms. Both western and northern corn rootworms have an annual life cycle. The winter is passed in the egg stage and larvae have a narrow host range. Soon after eggs hatch in late spring, larvae must secure a grass root, preferably corn, to complete their larval development. By rotating corn (the preferred oviposition site) with soybean, producers have been able to effectively break the life cycle (larvae cannot survive on soybean roots) of this key insect complex and save millions of dollars annually by not using soil insecticides. The economic and environmental benefits that have accrued for decades because of the effectiveness of crop rotation against corn rootworms are staggering.

The rotation of corn and soybean also has generally led to reduced densities of other soil insect pests such as wireworm and white grub species. Other successful crop rotation strategies include the rotation of small grains with crops such as alfalfa, flax, mustard, soybean, and sweet clover to reduce densities of false wireworms, white grubs, brown wheat mites, and wheat stem sawflies. Densities of some mite species such as the winter grain mite, a pest on several small grains, also may be reduced by rotation to a nongrass crop every third year. The use of a 3-year cotton/sorghum/cotton rotation has proven effective in enhancing the biological control of certain insect pests of cotton and also in suppressing some soil pathogens. Other 3-year rotations with pest management benefits include cotton/small grain/cotton and cotton/sorghum/small grain. Separating corn and cotton fields by as much distance as possible is a good pest management strategy. This reduces the movement of large numbers of bollworms from corn to cotton. Sorghum does not serve as a reservoir of cotton pests; however, it does support large densities of beneficial insects that can move to cotton fields and lessen the pest pressure caused by aphids, mites, and lepidopteran insects. The rotation of cotton and sorghum across the agricultural landscape is a superior pest management approach to that of creating a patchwork of corn and cotton fields.

The use of crop fallowing, a form of crop rotation, can offer pest management benefits for certain insect pests. A fallow field refers to a field taken out of production for a given period of time, often a single growing season. This agronomic practice is more common in arid regions such as western Kansas and Nebraska. The use of summer fallow practices has been attributed to reductions in wireworm densities in some arid crop-growing regions of Canada. Livestock that are removed from tick-infested pastures may benefit when these pastures are taken out of production for a given period of time (several months), a practice known as "pasture spelling." This pest management technique was particularly effective against the cattle tick. For other species of ticks, such as the lone star stick, the rotation of cattle to another pasture is less effective due to the wider host range of this tick species.

NEUTRAL OR NEGATIVE PEST MANAGEMENT EFFECTS OF CROP ROTATION

Crop rotation has little to no influence on the densities of many important insect pests, especially migratory insects such as fall armyworms, corn earworms, potato leafhoppers, green cloverworms, and corn leaf aphids. The development of economic infestations of these insect pests is more dependent upon planting date, suitability of the host plant, abiotic factors (weather), and the level of natural control provided by diseases, predators, and parasitoids. Other highly mobile insects such as European corn borers and bean leaf beetles are unaffected by the rotation of crops. European corn borer moths emerge in the spring from cornfield residue and can easily fly several miles within a single day, most often selecting the earliest-planted cornfields for egg-laying purposes. Bean leaf beetles leave soybean fields in late summer and fly to nearby wooded areas that serve as overwintering sites. In early spring, bean leaf beetles fly to alfalfa fields and subsequently infest the earliest planted soybean fields. Densities of other highly mobile insects with a very diverse host range (such as many grasshopper species) also generally are unaffected by crop rotation. For instance, some species of grasshoppers such as two-striped and differential grasshoppers lay their eggs at field edges, along roadsides, grass waterways, or ditch banks. Following egg hatch, the highly active grasshopper nymphs may begin to feed on vegetation in uncultivated areas and eventually move to adjacent crops. Other grasshopper species such as the migratory grasshopper lay their eggs directly into fields devoted to crop production. However, migratory grasshoppers also are relatively unaffected by crop rotation since following egg hatch, the nymphs can feed on a wide range of host crops.

Densities of some insect pests may be increased due to certain types of crop rotations. Economic infestations of black cutworms are more common in cornfields that have been rotated with soybean as compared with continuous cornfields. Black cutworm moths have a slight preference for soybean stubble versus corn debris for ovipositional (egg-laying) purposes. However, the presence of winter annual weeds in corn or soybean fields is far more important in attracting egg-laying black cutworm moths in the spring than the sequence of crops. Densities of some soil inhabiting insect pests such as wireworms and white grubs are increased by rotations that include grasses and perennial legumes. Adults of wireworms (click beetles) and white grubs lay their eggs most often in grasses, perennial legumes, or uncultivated areas. Therefore, corn planted after sod greatly increases the likelihood that economic infestations of wireworms, white grubs, and sod webworms will occur. Corn that follows red clover also is very likely to support large densities of grape colaspis. Other insects such as true armyworms are more apt to be favored when corn is planted after a small grain. For instance, some of the most severe infestations of true armyworms result when corn is planted no-till into a rye cover crop.

ADAPTATION TO CROP ROTATION

Although the entomological literature is replete with examples of insect pests that have developed resistance to a great variety of insecticides, there are fewer instances in which insects have adapted to crop rotation. The annual rotation of corn and soybean fields has and continues to afford most corn producers protection from infestations of northern and western corn rootworms. However, in the mid-1980s, very sporadic cases of "first-year" corn rootworm larval injury occurred in rotated cornfields of several states such as Illinois, Iowa, Minnesota, and South Dakota. The cause for this pest management failure of crop rotation was prolonged diapause of a very small percentage of northern corn rootworm eggs enabling them to survive for more than a single winter. Since 1995, farmers in east central Illinois and northern Indiana have witnessed the near complete loss of crop rotation as a viable pest management practice for the control of western corn rootworms. Although an exceedingly small percentage of western corn rootworm eggs has been documented to prolong diapause under laboratory conditions, the underlying mechanism for the recent collapse of crop rotation as a pest management tactic is a change in the ovipositional behavior of this rootworm species. Instead of laying eggs exclusively in cornfields, western corn rootworm adults in east central Illinois and northern Indiana are now utilizing soybean fields for ovipositional purposes. Preliminary evidence also suggests that other crops such as alfalfa may be equally suitable for oviposition. The net result of this unique adaptation to crop rotation is the escalation of soil insecticide use at planting by producers. The hypothesis used to explain this remarkable shift in egg-laying behavior is the intense selection pressure that was placed on the western corn rootworm population for decades with the limited two-crop rotational scheme in the eastern Corn Belt. For those western corn rootworm adults that continued to lay eggs in cornfields, the penalty was severe for their offspring since they could not survive on soybean roots. Elsewhere in the Corn Belt where continuous corn (nonrotated) is more common, the rotation of corn with another crop continues to provide rootworm control. Current models suggest that the new strain of western corn rootworm will continue to expand its range from east central Illinois and northern Indiana towards the north and east. Corn producers in southern Michigan and western Ohio are already beginning to report corn rootworm larval injury in their rotated cornfields.

BIBLIOGRAPHY

Food, Crop Pests, and the Environment; Fry, W.E., Zalom, F.G., Eds.; The American Phytopathological Society Press: St. Paul, MN, 1992; 179.

Gray, M.E.; Levine, E.; Oloumi-Sadeghi, H. Adaptation to crop rotation: western and northern corn rootworms respond uniquely to a cultural practice. Recent Res. Dev. Entomol. **1998**, *2*, 19–31.

Henn, T.; Weinzierl, R.; Gray, M.; Steffey, K. *Alternatives in Insect Management: Field and Forage Crops*; Circular 1307, University of Illinois: Urbana-Champaign, 1991; 26.

Horn, D.J. *Ecological Approach to Pest Management*; The Guilford Press: New York, 1988; 285.

Introduction to Insect Pest Management; Luckmann, W.H., Metcalf, R.L., Eds., 3rd Ed., John Wiley & Sons, Inc.: New York, 1994; 650.

Pedigo, L.P. *Entomology and Pest Management*; MacMillan Publishing Company: New York, 1981; 646.

CROP ROTATIONS (PLANT DISEASES)

Jeffrey P. Wilson
University of Georgia Coastal Plain Experiment Station, USDA-ARS, Tifton, Georgia, U.S.A.

INTRODUCTION

Crop rotation is the practice of growing different crops in sequence on a given piece of land to maintain its productivity. When a particular crop is consistently grown on a site, certain weeds and pathogenic microorganisms that can limit production or profitability will frequently increase. As an agronomic practice, crop rotation can have several beneficial effects, such as enhancing structure, fertility, and water retention capacity of soils, and reducing losses to soil-borne pests. As a disease control practice, crop rotation aims to minimize the populations of pathogenic nematodes, fungi, bacteria, and viruses in the soil (1). Most soil-borne pathogens have a limited capability for aerial dispersal, therefore, the extent of crop damage from disease is related to the quantity and distribution of pathogen propagules, or inoculum, in the soil. Rotations minimize the amount of inoculum in the soil by cultivating appropriate crops that do not serve as host to the target pathogen(s) for an appropriate length of time between susceptible crops. Crops are chosen based upon the inability of the pathogen to reproduce on them. Rotation intervals are determined by attrition of the pathogen in the soil environment. Rotations will effectively manage disease if the rate of inoculum attrition is greater than the rate of inoculum production. Several factors interact to affect the rates of inoculum production and attrition in the soil.

PRODUCTION OF PATHOGEN INOCULUM

Limiting reproduction of soil-borne pathogens is achieved by growing disease resistant varieties, or by growing crop species that are not hosts for the target pathogens. The choice of crops used in rotations should be based on a knowledge of the host range of the target pathogen. If plants used in rotations are nonhosts of the target pathogen, infection is prevented, reproduction of the pathogen is limited, and increase of inoculum in the soil is restricted.

Complete inhibition of reproduction of the pathogen is unlikely to occur unless it is an obligate parasite with very specific host requirements. The alternative crop used for rotation may be a poor host that permits a low level of reproduction of the pathogen (2). Pathogens can reproduce on volunteer plants, or sometimes on weeds. Alternatively, crop residue can often serve as a substrate for growth of a variety of facultative pathogens (3). Given the proper environmental conditions, facultative parasites can increase by saprophytic growth on residues in the absence of suitable, susceptible hosts. The importance of crop residues on pathogen reproduction is dependent on several factors, including the pathogen, the environment, and agronomic practices.

The length of growing season and cropping systems are important when considering pathogen dynamics. Seasonal differences become less apparent closer to the equator, and the crop rotation and double-cropping systems in temperate climates transition into multiple cropping systems in tropical climates. In the longer growing seasons of the subtropical and tropical climates, pathogens have a longer period for reproduction than in the shorter growing seasons of the temperate latitudes (2). Most pathogens will actively grow and reproduce if favorable environmental conditions and suitable hosts are present.

ATTRITION OF PATHOGEN INOCULUM

Pathogen population dynamics are influenced by the rate of death, or attrition of inoculum as well as by the reproduction rate. As a general rule, longer rotation intervals between cultivation of susceptible crops are more successful in managing soil-borne diseases. Longer intervals allow greater time for inoculum attrition due to hyperparasitism, ingestion by soil microfauna, dessication, or loss of viability from age (1, 3).

Crop residues can be an important reservoir of pathogen inoculum (2). Rotations are particularly beneficial for controlling disease in conservation tillage systems, in which crop residues cover at least 30% of the soil surface. Conventional tillage practices tend to bury most of the residue in the soil, where decomposition is hastened. In conservation tillage systems, crop rotations allow for more thorough decomposition of infested residue between susceptible crops.

Certain crops grown in rotations reduce the severity of soil-borne diseases beyond the rate attributable to normal attrition of pathogen inoculum. In these cases, chemical compounds produced by a crop or formed during decomposition of residues either inhibit pathogen germination or growth, or the compounds may kill certain pathogens. These allelopathic compounds can be an important factor in disease control.

FUTURE TRENDS

Crop rotation has been practiced for thousands of years (1, 2). Although many of its benefits are well documented, continuing research is still necessary. New crop species will continue to be integrated into nontraditional production areas. Diversified production systems enhance economic stability of agricultural communities, and shifting population demographics has resulted in an increased demand for a greater variety of produce and plant products in all regions. Research will be necessary to evaluate cultural and biological compatibility as new crops are integrated into existing agricultural systems.

Identifying new crops causing allelopathic effects toward soil-borne pathogens should have a high priority in rotation studies. Crops could be grown in rotations based upon known allelopathic reactions, or tailored through plant breeding to express greater levels of the effective allelopathic compounds. Crop rotation as a disease-control tactic would enter a new dimension from the relatively slow and passive approach of manipulating pathogen population dynamics to an active, crop-mediated method of biological control of some soil-borne diseases.

Several benefits in addition to disease control are associated with crop rotation, including improved structure, fertility, and water retention capacity of soils. As such, it is an environmentally beneficial method of disease control. Nevertheless, detrimental environmental effects may occur if cultivation practices required for growing the crops included in the rotation have such effects.

Rotations with perennial crops merit greater scrutiny. Most rotation studies have examined rotations of annual crops, and relatively few have assessed perennial crops as practical alternatives. Rotations using perennial forage and hay grasses have been quite effective for reducing diseases of annual crops. Incorporating grasses in rotations can be extremely beneficial to the environment, since soil erosion and nutrient runoff is minimal in properly managed pastures. The environmental benefits of rotations with perennial forage and hay grasses should be examined to further develop practical and economical systems.

Crop rotation is just one component in the development of sustainable agricultural production systems. Disease control based on several integrated practices provides an inherently stable buffer against losses due to soil-borne pathogens. Rotation is frequently practiced when producing crops of low cash value for which more expensive controls such as soil fumigation, nematicides, or fungicides are not economical. As use of chemical controls become limited due to legislation or impractical due to increasing cost, alternative methods will be required for disease control in all crops. Crop rotation is a fundamental, time-tested strategy that is indispensable for effective management of soil-borne diseases.

REFERENCES

1. Sumner, D.R. Crop Rotation and Plant Productivity. In *Handbook of Agricultural Productivity*; Rechcigl, M., Jr., Ed.; CRC Press: Boca Raton, FL, 1982; *I*, 273–313.
2. Rush, C.M.; Piccinni, G.; Harveson, R.M. Agronomic Measures. In *Environmentally Safe Approaches to Crop Disease Control*; Rechcigl, N.A., Rechcigl, J.E., Eds.; Agriculture and Environment Series, CRC Press LLC: Boca Raton, FL, 1997; 243–282.
3. Parker, C.A.; Rovira, A.D.; Moore, K.J.; Wong, P.T.W.; Kollmorgen, J.F. *Ecology and Management of Soil-borne Plant Pathogens*; APS Press: St. Paul, MN, 1985.

CULTURAL CONTROLS

B. Rajendran
Tamil Nadu Agricultural University, Cuddalore, India

INTRODUCTION

Among the several pest management techniques, cultural control envisages the adoption of methods and strategies to suppress the pest population taking into consideration ecological, economic, and social acceptance. The ''pesticides rush'' has come to an appropriate end in several countries after the excessive and indiscriminate use of hazardous pesticides with the sole aim of obtaining a quick kill of the pests. In cultural control most of the suitable and appropriate techniques are blended in a harmonious pattern with minimum disruption to the ecosystem. Cultural control has been implemented in crop production by the farming community themselves in one way or other as an age old indigenous farming practice to break the sustenance of the pest chain in their crops. There are many merits for this strategy as compared to other means of pest control (Table 1).

CULTURAL CONTROL FOR INSECT PESTS OF CROPS

Several control practices like adjusting the times of sowing/planting, seed selection, mulching, proper irrigation, drainage, intercropping, crop rotation, fallowing, etc., have been found to be useful in the suppression of many crop pests.

In Tamil Nadu, a major rice growing region of India, the optimum time of planting the rice crop is found to be the first week of August, first and third weeks of October in ''Kharif'' season to ward off green leaf hopper (*Nephotettix sp.*), brown plant hopper (*Nilaparvata lugens*) and leaf folder (*Cnaphalocrocis medinalis*) (1). Rice transplanted during the middle of May is observed to be highly infested even up to 30% by the rice mealy bug *Brevennia rehi*, which is attributed to an increase in temperature and wind velocity during that period. The incidence of the pest is minimal in July transplanted rice (2). Alternate flooding and draining of irrigation water reduces the populations of rice leaf hopper and plant hopper (3).

The investigations in Punjab, India, on the effects of sowing dates and spacing practices on the incidence of key pests of cotton under unsprayed conditions revealed that the population of jassids is slightly higher in late than earlier sowings. Wider spacing also slightly reduces the infestation of bollworms (4). Intercropping of cowpea in cotton crop reduced damage to 23% as against the highest damage of 65% by bollworms in sole cotton crop (5). In Tamil Nadu, India, intercropping of sunflower or green gram or black gram in cotton reduced the incidence of Cicadellid, *Amrasca biguttulla biguttulla*, significantly. Among the crop combinations, cotton intercropped with green gram realized the highest gross income (6). Similarly sowing of cotton in late August rather than September lowers the incidence of bollworms damage to cotton and subsequent boll rot incidence (7).

Early planting (December–January) reduces sugarcane shoot borer, *Chilo infuscatellus*, an important pest causing severe yield loss in late season crops. The cultural practice of trash mulching through the spreading of a thick blanket of trash (dried older leaves of sugarcane) soon after germination greatly helps in the suppression of shoot borer and weeds (Fig. 1). Similarly a light partial earthing up given in the field at 45 days after planting serves to reduce shoot borer (8). The trials conducted from 1991–1994 at the Sugarcane Research Station, Cuddalore, India (9), revealed that sugarcane intercropped with green gram had reduced shoot borer infestation (19.32%) as against sugarcane sole crop (39.76%) (10). Intercropping in sugarcane also realized economic returns over sole crop as per cost–benefit analysis (11, 12) (Table 2). Removal of lower and older leaves referred as detrashing during the fifth and seventh month stage of crop growth in sugarcane is effective in suppressing the activity of internode borer, stalk borer, scale insect, and mealy bugs. Studies undertaken during 1989–1995 in India also clearly indicated

Table 1 The relative merits of the different pest management strategies

Pest management method	Ease of application	Readiness of availability	Durability of effect	Safety to environment	Cost
Cultural	**	***	*	***	**
Mechanical	**	***	*	***	*
Varietal	***	*	**	***	***
Biological	***	*	**	***	**
Chemical	**	***	*	*	**
Pheromones	**	*	*	***	*

* Less favorable.
** Moderately favorable.
*** More favorable.
(Courtesy of the Sugarcane Breeding Institute, (ICAR), Coimbatore, India.)

that the sugarcane whitefly, *Aleurolobus barodensis* and leaf hopper *Pyrilla perpusilla* could be well managed by detrashing of older leaves of cane during the fifth and seventh months after planting (13).

Trash burning during the middle of April in sugarcane ratoon fields resulted in the reduction of blackbug (*Cavelarius sweetii*) population and as a result, higher yields were realized (14). This practice of burning ratoon field and subsequent sowing of wheat followed by a green manure and then planting sugarcane is advocated for suppressing sugarcane scale insect, *Melanaspis glomerata* (15).

Fig. 1 Mulching reduces weeds and shoot borer pests in sugarcane.

The effect of planting dates on the incidence of foliar pests and yield losses were studied on groundnut during 1988–1991 in India. More leaf miners were recorded on plants sown on July 15th, while sucking pests like *Empoasca kerri* and thrips *Frankliniella schultzei*, *Scirtothrips dorsalis*, and *Caliothrips indicus* were recorded on plants sown on June 15th (16). In Orissa state, groundnut sown from the third week of June to the first week of July recorded lower pests. Rainfall at sowing and occasional rains with sunshine during the growth stages are responsible for poor germination, higher pest incidence, and a low yield in late-sown (August) crops (17).

Intercropping pearl millet in groundnut crop at a ratio of 8:1 lowers the incidence of pests since this system favors more parasitoids and predators (18). Intercropping of sorghum, maize, cowpea, and sunflower in groundnut was reported to reduce the larval population of leaf miner, *Aproaerema modicella* as compared to the non-intercropped groundnut (19). Similarly intercropping of sesame at ratio of 1:4 with pearl millet reduced the damage by sesame shoot webber, *Antigastra catalaunalis* (20).

In a five-year study in Burkina Faso, termite (Isoptera) damage to groundnut was directly related to delayed harvest. Less than 4% was damaged when harvested 70 and 90 days after sowing; 15 and 46% was damaged when harvested at 110 and 125 days respectively. Aflatoxin concentration in seeds increased with increased age at harvest and was correlated with pod damage by Isoptera and hence enhanced invasion by *Aspergillus flavus* (21).

The need for sound ecofriendly pest management strategies in vegetable crops is of high priority. Cultural

Table 2 Effect of cultural practices like intercropping, trash mulching in the management of shoot borer of sugarcane during 1991–1994 seasons in Sugarcane Research Station, Cuddalore, India

Treatments	Cumulative mean of shoot borer incidence (%) 1991–1994	Decrease over control (%)	Economic shoot count in 1000's/ha	Cane yield (t/ha)	C:B ratio
Sugarcane alone	39.76e	—	100.9	108.2b	1.92
Sugarcane + Black gram	27.60c	30.6	104.8	116.5a	2.16
Sugarcane + Green gram	19.32b	51.4	107.7	115.5a	2.15
Sugarcane + Sunnhemp	25.92c	34.8	107.3	115.8a	2.01
Sugarcane + Soyabean	32.62d	17.9	103.9	115.4a	2.13
Endosulfan 0.07% twice	27.86c	29.9	112.8	116.7a	2.02
Trash mulching — a week after planting	9.97a	74.9	105.9	118.2a	1.89

In a column, mean followed with the same alphabets do not differ significantly ($p = 0.05$) by DMRT.

pest control strategies like healthy seeds, growing of marigold as a trap crop for 14 rows of tomato against *Helicoverpa armigera*, mustard grown as a repellant crop and other organic manuring practices resulted in the low incidence of pests and diseases in vegetables like potato, tomato, onion, cabbage, french bean, and cucumber (22).

CULTURAL CONTROL OF DISEASES OF CROPS

Cultural control techniques of diseases are not so widely practiced as compared to insect pest control in many crops. Rice blast disease can be managed by lesser application of nitrogenous fertilizers. Split application of fertilizers along with potash reduces brown spot disease. Stem rot (*Sclerotium oryzae*) a notorious problem in heavy soils of tropical Asia can be culturally managed by avoiding stagnation of water and burning off stubbles after harvest to kill scelerotia. Many bacterial diseases are culturally checked by the burning of crop remains to destroy the bacteria presented in infected straw and stubble (23).

Intercropping strategy is advised for groundnut to reduce the problems by diseases like *Phaeoisariopsis personata* (*Mycosphaerella berkeleyi*) and *Puccinia arachidis* (24). Cotton wilt, *Verticillium dahliae*, can be culturally managed by rotation with corn, wheat, sorghum, safflower, or rice (25). Crop rotation with nonsolanaceous host crops reduces the incidence of Fusarium wilt and Anthracnose diseases in tomato.

In sugarcane, selection of disease-free nursery, crop rotation with rice, fertilizer and water management strategies are advocated for the management of red rot disease, *Colletotrichum falcatum* (25).

In banana, leaf spot diseases could be well managed by cultural operations like improved drainage, weed control, correct spacing, removal and destruction of trash (badly spotted leaves) by burning. Fusarium wilt of banana spreads by high soil moisture, bad drainage, strong development of new roots after a nematode attack, and by high soil inoculum. Any measures taken to counteract these will help to prevent infection (26).

CULTURAL CONTROL OF WEEDS

Certain generally applied cultural practices for the control of weeds in many crops are crop rotation and deep ploughing after fallowing. Impounding of water in rice fields discourages weed growth (23). Water management is the most important cultural factor for the successful control of many rice weeds like *Echinochloa sp.* Continuous impounding of water to a depth of 4–6 inches effectively checks grass weeds. In sugarcane, weed control could be achieved by trash mulching, crop rotation with rice, and clean cultivation. Crop rotation with rice makes the soil puddle well and thereby monocot weeds are controlled. Intercropping of pulses and groundnut—short duration intercrops grown on the ridges adjoining sugarcane plants—reduces weed population in sugarcane (27). Trash mulching and press mud application on the ridges soon after germination of sugarcane plants affords substantial weed control efficiency.

CULTURAL CONTROL OF NEMATODES

Crop rotation is one of best cultural strategies for the effective management of plant parasitic nematodes in soil. Marigold (*Tagetes erecta*) intercropping with tomato at

1:4 and 1:6 ratios and mustard (*Brassica spp.*) at 1:2 ratio is reported to be effective in reducing root gall, egg masses, and nematode population and significantly better in terms of cost–benefit ratios varying from 1.064 to 1.836 as against 0.515 obtained in chemical control (28). Crop rotation with rice is recommended for the control of sugarcane and banana nematodes. Addition of organic manures, compost, and press mud in soil checks nematode multiplication.

FUTURE OUTLOOK

The use of pesticides leading to resistance buildup in target insects and biomagnification at different trophic levels in the food web is alarming and has global implications. Hence, high priority should be given for sound eco-friendly pest management techniques. Cultural control is one such for both sustainability and economic viability. Field application of this measure requires active participation by farmers and the ultimate success depends on the proper execution of the technique. Advance planning of cultural control is also essential as it deals with seasons, seed rate, time of sowing, rotation of crops, intercropping, etc. The strategy also widely varies with location and region, and is highly influenced by environment. Success depends on carefully monitoring the local region over a period of years. A cost–benefit analysis also might be taken into consideration for the implementation of this technique.

REFERENCES

1. Karuppuchamy, P.; Gopalan, M. Influence of time of planting on the incidence of rice pests. Madras Agric. J. **1986**, *73* (11), 606–609.
2. Gopalan, M.; Coumararadja, N.; Balasubramanian, G. Effect of different planting dates and irrigation regimes on the incidence of rice mealy bug, *Brevennia rehi* Lindinger. Madras Agric. J. **1987**, *74* (4-5), 226–227.
3. Karuppuchamy, P.; Uthamasamy, S. Influence of flooding, fertilizer and plant spacing in insect pest incidence. Int. Rice Res. Newsletter **1984**, *9* (6), 17.
4. Butter, N.S.; Brar, A.S.; Kular, J.S.; Singh, T.H. Effect of agronomic practices on the incidence of key pests of cotton under unsprayed conditions. Ind. J. Entomol. **1992**, *54* (2), 115–123.
5. Suresh, S.; Rajavel, D.S.; Narasimhan, C.R.L.; Muthuswamy, P. Against bollworm in cotton. Madras Agric. J. **1993**, *80* (3), 172.
6. Venkatesan, S.; Balasubramanian, G.; Sivaprakasam, N.; Narayanan, A.; Gopalan, M. Effect of intercropping of pulses and sunflower on the incidence of sucking pests of rainfed cotton. Madras Agric. J. **1987**, *74* (8-9), 364–368.
7. Ilango, L.; Uthamasamy, S. Influence of sowing time on the incidence of bollworms and its influence on boll rot complex of cotton. Madras Agric. J. **1989**, *76* (10), 571–573.
8. David, H.; Easwaramoorthy, S.; Jayanthi, R. Tactics in Sugarcane Pest Management—Cultural and Mechanical Methods. In *Integrated Pest Management in Sugarcane*; Sugarcane Breeding Institute (ICAR): Coimbatore, India, 1991; 8.
9. David, H.; Sithanandam, S. Cultural and Mechanical Practices. In *Sugarcane Entomology in India*; David, H., Easwaramoorthy, S., Jayanthi, R., Eds.; Sugarcane Breeding Institute (ICAR): Coimbatore, 1986; 464–467.
10. Rajendran, B. Management of insect pests of sugarcane through cultural practices. Coop. Sugar **1999**, *30* (10), 961–964.
11. Raja, J.; Rajendran, B. In *Intercropping in Sugarcane for the Management of Shoot Borer, Chilo infuscatellus Snell*, Lead paper presented at the 30th Meeting of Sugarcane Research and Development Workers held at Erode, Tamil Nadu, India, July 3–4, 1994; Sugarcane Breeding Institute: Coimbatore, India, 1997; 44.
12. Rajendran, B. Management of insect pests of sugarcane through cultural practices. Ind. Sugar **1999**, *XLIX* (4), 271.
13. Rajendran, B. Management of whitefly *Aleurolobus barodensis* mask in sugarcane. Insect Environ. **1997**, *3* (3), 68–69.
14. Mrig, K.K.; Chaudhary, J.P. Effect of date of trash burning on population of black bug, *Cavelarius sweeti* slater and miyamota in sugarcane ratoon. Coop. Sugar **1994**, *26* (2), 111–112.
15. Varun, C.L.; Singh, H.N. Cultural control of sugarcane scale insect (*Melanaspis glomerata* green) in bhat soil. Ind. J. of Entomol. **1993**, *55* (2), 219–220.
16. Shetgar, S.S.; Bilapate, G.G.; Londhe, G.M. Effect of sowing dates on pest incidence and yield losses due to foliage pests on groundnut. Ind. J. of Entomol. **1994**, *56* (4), 441–443.
17. Dash, P.C.; Santokke, B.K. Effect of sowing dates on insect pest incidence and yield of kharif groundnut under Western Orissa conditions. Orissa J. Agric. Res. **1994**, *7* (3–4), 63–66.
18. Kennedy, F.J.S.; Balagurunathan, R.; Christopher, A.; Rajamanickam, K. Insect pest management in peanut: a cropping system approach. Trop. Agric. **1994**, *71* (2), 116–118.
19. Rajagopal, D.; Hanumanthasamy, B.C. Effect of intercropping on the incidence of groundnut leaf miner, *Aproerema modicella* Deventer (*Gelechidae: Lepidoptera*). Madras Agric. J. **1999**, *86* (4–6), 461–464.
20. Baskaran, R.K.M.; Mahadevan, N.R.; Thangavelu, S. Influence of intercropping on infestation of shoot webber (*Antigastra catalaunalis*) in sesame. Ind. J. of Agric. Sci. **1991**, *61* (6), 440–442.

21. Lynch, R.E.; Dicko, I.O.; Some, S.A.; Ouedraogo, A.P. Effect of harvest date on termite damage, yield and aflatoxin contamination in groundnut in Burkina Faso. Int. Arachis Newsletter **1991**, *10*, 24–25.
22. Chinnakonda, Dilip; Ravi, K.; Lanting, H.H. In *Organic Vegetable Production—A Case Study with Special Reference to Pest Management*, Proceedings of I National Symposium on Pest Management in Horticultural Crops, Bangalore, India, Reddy, P. Paravatha, Krishnakumar, N., Verghese, Abraham, Eds.; 1998; 145–148.
23. Susan, D.F. Diseases. In *Pest Control in Rice*; Susan, D. F., Ed.; PANS Manual No. 3, Centre for Overseas Pest Research: London, 1974; 42–68.
24. Muthiah, C.; Senthivel, T.; Venkatakrishnan, J.; Sivaram, M.R. Effect of intercropping on incidence of pest and disease in groundnut (*Arachis hypogaea*). Ind. J. of Agric. Sci. **1991**, *61* (2), 152–153.
25. Singh, R.S. Wilt of Cotton. In *Plant Diseases*; Singh, R. S., Ed.; Oxford IBH Publishing Co. Pvt. Ltd.: New Delhi, India, 1990; 422.
26. Susan, D.F. Diseases. In *Pest Control in Bananas*; Susan, D.F., Ed.; PANS Manual No. 1, Centre for Overseas Pest Research: London, 1972; 15–35.
27. Sundara, B. Weed Management. In *Sugarcane Cultivation*; Sundara, B., Ed.; Vikas Publishing House Pvt. Ltd.: New Delhi, India, 1998; 286.
28. Rangaswamy, S.D.; Reddy, P.; Paravatha; Nanjegowda, D. Management of root knot nematode *Meloidogyne incognita* in tomato by intercropping with marigold and mustard. Pest Management in Hortic. Ecosys. **1999**, *5* (2), 118–121.

CURE (THERAPEUTICS)

C

Larry P. Pedigo
Iowa State University, Ames, Iowa, U.S.A.

INTRODUCTION

As opposed to preventive pest management, "cure," or therapeutics, involves developing and implementing an effective treatment plan for pests to avoid continued crop losses. This article discusses examples of pest outbreaks and treatments and their economic and ecological implications.

DISCUSSION

Although prevention should form the basis and first line of defense in the ideal Integrated Pest Management (IPM) program, prevention alone is usually not adequate. Cure or therapy also is necessary in most IPM programs. Cure seeks to cure an acute or chronic crop disorder. It differs from preventive pest management in that a pest population is present and injury is occurring when management decisions are made. In other words, cure is applied after pest sampling is conducted and an assessment of pest status indicates that economic damage is imminent. Although cure is considered distinct from prevention, there still is a preventive aspect to it. This is because future losses must be prevented by use of a cure; otherwise, the cure would not be economically feasible.

Cure often constitutes the first phase of IPM development because therapeutic programs can be formulated quickly to alleviate ongoing problems. From an IPM developmental standpoint, cure may be the final phase of program development for some occasional pests. In such instances, the general equilibrium position of the pest is suppressed naturally, and there may be no need for preventives because nature supplies them. Here, all that may be needed is an infrequent "correction" of population density to achieve the management goal. With these occasional pests, the judicious use of pesticides may be all that is required because such use is effective and usually would not result in ecological backlash. Examples of such situations include spider mite and grasshopper outbreaks associated with drought in the upper Midwest.

The ecological objective of therapy is to interrupt ongoing pest population growth and associated injury. The goal of cure can be achieved either by dampening pest population peaks or by truncating increasing injury by altering crop exposure to the pest. Dampening pest density peaks is usually accomplished by proper application of a pesticide to kill pests and/or early harvest to reduce crop exposure. An example of the latter approach is found with dense populations of alfalfa weevil, *Hypera postica* (Gyllenhal) (Coleoptera: Curculionidae), in alfalfa stands with advanced growth. In this instance, early cutting, which terminates further foliage loss, is often economically advantageous.

For key insect pests, cure or therapeutics is envisioned as being used in conjunction with a preventive pest management program. Here, if the preventives fail, cure serves as a "correction" in the system. In such IPM systems, population sampling and consultation of economic thresholds are the primary activities to determine need for the correction.

CONCLUSION

Several tactics are available for use as IPM cures. Some of these include selective traditional pesticides, fast-acting nonpersistent biological controls such as microbial insecticides, early harvest, and mechanical removal of pests (e.g., hand picking, pruning infested branches). Of these tactics, traditional pesticides currently are the most practical for IPM cure.

BIBLIOGRAPHY

All, J.N. In *Importance of Designating Prevention and Suppression Control Strategies for Insect Pest Management Programs in Conservation Tillage*, Proceedings of the 1989 Southern Conservation Tillage Conference, University of Florida: Tallahassee, FL, 1989; 1–3.

Cate, J.R.; Hinkle, M.K. *Integrated Pest Management: The Path of a Paradigm*; National Audubon Soc. Special Rept., National Audubon Society: Washington, DC, 1993; 43.

Pedigo, L.P. Closing the gap between IPM theory and practice. J. Agric. Entomol. **1995**, *12*, 171–181.

Pedigo, L.P. *Entomology and Pest Management*, 3rd Ed., Prentice-Hall: Englewood Cliffs, NJ, 1999; 691.

Rabb, R.L. A sharp focus on insect populations and pest management from a wide area. View. Bull. Entomol. Soc. Am. **1978**, *24*, 55–61.

Stern, V.M.; Smith, R.F.; van den Bosch, R.; Hagen, K.S. The integrated control concept. Hilgardia **1959**, *29*, 81–101.

DAIRY PEST MANAGEMENT (ARTHROPODS)

Phillip E. Kaufman
Cornell University, Ithaca, New York, U.S.A.

D

INTRODUCTION

There are 10 major arthropod pest species affecting dairy cattle including six fly and four louse species. Damage inflicted by arthropods falls into two categories: direct damage and indirect damage. Direct damage includes blood loss, introduction of salivary secretions, tissue damage, reduced value of saleable animal products, and annoyance. Indirect damage includes transferring pathogenic organisms and decreased vigor.

Pest management procedures follow integrated approaches discussed elsewhere in this publication. To be successful, the tactics utilized must disrupt the arthropod life cycle at several places.

At any given time, cattle can be infested with several species of arthropods and as parasite levels rise and fall with the seasons, animals often experience compensatory gains offsetting earlier damage. However, when considering losses in milk production, compensatory gains are not applicable. Determining which species is responsible for a given amount of damage is difficult. A consensus on the losses as well as the costs attributed to arthropod infestations on dairy cattle is not available. The monetary figures presented here have been estimated for all cattle (beef and dairy) in the United States (1). Most studies investigating the effects of individual parasite species have been limited to beefcattle.

FLIES AFFECTING CONFINED ANIMALS

The house fly, *Musca domestica* (L.), and stable fly, *Stomoxys calcitrans* (L.), are the primary pests of confined dairy cattle (2, 3). House flies are nonbiting insects that breed in animal droppings, manure piles, decaying silage, bedding, and other organic matter (3). Their reproductive potential and movement to off-farm locations makes fly management imperative. House flies have been documented to mechanically transmit more than 100 known disease organisms. Monetary damage estimates have not been calculated for the house fly, however, losses can be expected to increase as urban expansion continues to encroach on traditionally agricultural areas.

The stable fly is a blood-feeding insect that causes considerable distress to cattle (2). It is often observed feeding on the lower legs of cattle and its presence is indicated when animals stomp. Stable flies and house flies breed in similar material, however, stable flies also breed in grass clippings and other types of decaying vegetative matter (3). Reported losses (milk production, butterfat, etc.) attributed to stable fly infestation have varied from significant damage to no observed effect. Additionally, observed weight losses have been shown to be offset by supplementing animal diets with grain. It has been estimated that stable flies annually result in losses of more than $398 million.

House fly and stable fly populations can be monitored in the barn using spot cards (3). Stable flies can also be monitored by counting the number of flies on all four legs of at least 10 animals. Flies can be managed successfully by integrating intensive manure management with cultural, biological, and chemical controls (2, 3). Manure management is the primary method of confined fly control. Removing or drying breeding areas can considerably reduce fly abundance. In areas that cannot be cleaned regularly (such as calf pens and silage storage areas) or in outdoor hay storage areas, fly management can be aided with good water drainage. Utilizing augmentative biological control, including parasitoid releases, in areas traditionally difficult to clean can help to reduce fly populations further. The correct species of parasitoid to be used depends greatly on the region of the United States and facility type.

When needed, chemical control should be used with an overall IPM approach in mind. Fogs or space sprays containing pyrethrins should be the first choice (3). Pyrethrin fogs and fly baits are compatible with biological control. Residual premise sprays should be reserved for emergency and late season uses only.

FLIES AFFECTING PASTURED ANIMALS

Two flies, the horn fly, *Haematobia irritans* (L.) and the face fly, *M. autumnalis* De Geer, are the primary pests of pastured cattle in most areas of the United States (2). Both flies breed only in freshly deposited cattle dung and neither are pests of confined cattle. Horn flies are blood-feeding insects that are found in constant association with cattle. The economic losses in the United States associated with horn flies have been estimated at more than $730 million annually (1, 4).

The nonbiting female face fly visits cattle for short periods of time where she consumes animal nasal and ocular secretions (2). Fly feeding habits, irritancy, and disease aspects result in economic loss. The face fly has been shown to alter the time of grazing (2), which in turn can alter feed efficiency. While feeding, the fly can mechanically transmit the causative agent of pink-eye and *Thelazia* eyeworms (3). Most research has demonstrated that face fly infestations do not affect milk secretion (5). Annual losses to cattle have been estimated to surpass $50 million, predominantly due to pink-eye transmission (1).

Management of these two pests has relied heavily on insecticide applications (2). Currently, few successful, nonchemical options are available (3). The use of sticky fly traps, walk-through traps, and introduction of natural enemies has met with mixed success. Cattle dung supports a large number of arthropod species that either prey on, compete with, or alter the dung environment (3). These arthropods can affect pest fly populations, and current recommendations, designed to conserve natural enemies, are to avoid systemic and feed-through insecticide applications during the spring and summer seasons.

CATTLE GRUBS

Two species of cattle grubs occur in the United States: the northern cattle grub, *Hypoderma bovis* (L.), and the common cattle grub, *H. lineatum* (Villers) (2, 3, 6). Cattle grubs are usually observed during their larval stage in warbles on the backs of animals. These flies are most often pests of young stock, as older animals develop a degree of immunity. Economic losses to cattle grubs result from several forms of attack. First, gadding behavior in response to adult *H. bovis* oviposition activity alters grazing efficiency and increases the risk of self-inflicted injuries. The migration of the larvae through animal tissues results in losses in weight gain, delayed time to first lactation and long-term production losses. Finally, breathing holes cut by larvae into the animal's hide severely reduce the value of the leather. Meat surrounding the warbles is discolored and often must be trimmed further reducing the carcass value. Unlike damage inflicted by other cattle pests, a portion of the economic damage inflicted by grubs (hide damage and systemic insecticide-related paralysis and death) to dairy cattle can be directly compared with that observed in beef cattle. Annual losses to the cattle industry are estimated to surpass $66 million (excluding control costs), however, because many dairy cattle are no longer pastured and thus not infested with grubs, this figure is predominantly associated with beef cattle (7).

Very few natural enemies of cattle grubs have been reported (6). Management of cattle grubs is difficult without chemical control; however, flies will not enter darkened buildings (3). Therefore, animals confined in barns will not have cattle grubs. Systemic insecticides should not be used for control of cattle grubs on lactating animals. Because of larval migration through sensitive areas (esophagus or spinal column), systemic insecticide applications should not be made if larvae have entered these areas (date is dependent on latitude and elevation). Treatments made following the suggested treatment period may result in paralysis, bloat, and death.

CATTLE LICE

Lice are the primary, permanent ectoparasites of dairy cattle. These include three sucking lice species (*Haematopinus eurysternus* (Nitzsch); *Linognathus vituli* (L.); *Solenopotes capillatus* (Enderlein) and the cattle chewing louse, *Bovicola bovis* (L.) (2, 3). All four species cause extreme annoyance to cattle. Milk production declines in heavily infested cattle (3). Hair loss, reduced feed conversion, and general unthriftiness also result from louse infestations. Populations of lice on adult animals are generally highest in the winter months while young stock housed in barns show high levels of infestation throughout the year (3). Cows in stanchion barns and calves housed communal pens are much more likely to be infested than cows in free stalls and calves in hutches, respectively. Initial infestations of lice primarily occur from direct animal to animal contact. Because of the environmental conditions produced in a barn, dairy cattle often carry heavy louse populations longer than beef cattle. Losses associated with cattle lice have been estimated to surpass $125 million annually (1).

No louse predators or parasitoids have been reported and pathogens have not been shown effective (2). Infestations of cattle lice must be controlled with insecticide applications; however, good management practices using cultural control can reduce the chance of reinfestation and thus make subsequent insecticide applications unnecessary (3). Following treatment of a herd, all new animals should be treated and quarantined prior to introduction into the general population. Careful and regular monitoring will allow producers to detect louse infestations and take action before populations get out of control. Calves can be housed outdoors in hutches to reduce chance of infestation by 90%.

COMBINED INFESTATIONS

Because dairy cattle are seldom affected by a single pest, studies documenting production losses due to multiple species attack are valuable. However, mixed results have also been obtained with these studies (8). Protection from horn fly and stable fly infestations have shown either milk production losses attributed to fly attack or no benefit to pest control. Obtaining loss data from dairy cattle is difficult due to the fact that milk production can significantly be affected by such variables as pest infestation level, climate, breed and age of host, and animal husbandry practices.

REFERENCES

1. Drummond, R.O.; Lambert, G.; Smalley, H.E., Jr.; Terrill, C.E. Estimated Losses of Livestock to Pests. In *Handbook of Pest Management in Agriculture*; Pimentel, D., Ed.; CRC Press: Boca Raton, FL, 1981; 111–127.
2. Schmidtmann, E.T. Arthropod Pests of Dairy Cattle. In *Livestock Entomology*; Williams, R.E., Hall, R.D., Broce, A.B., Scholl, P.J., Eds.; John Wiley & Sons: New York, 1985; 223–238.
3. Rutz, D.A.; Geden, C.J.; Pitts, C.W. *Pest Management Recommendations for Dairy Cattle*; Cornell University and Penn State Coop Ext: Ithaca, NY, 1994; 11.
4. Palmer, W.A.; Bay, D.E. A review of the economic importance of the horn fly, *Haematobia irritans irritans* (L.). Prot. Ecol. **1981**, *3*, 237–244.
5. Schmidtmann, E.T.; Berkebile, D.; Miller, R.W.; Douglass, L.W. The face fly (Diptera: Muscidae): effect on holstein milk secretion. J. Econ. Entomol. **1984**, *77*, 1200–1205.
6. Scholl, P.J. Biology and control of cattle grubs. Ann. Rev. Entomol. **1993**, *39*, 53–70.
7. Drummond, R.O. Economic Aspects of Ectoparasites of Cattle in North America. In *The Economic Impact of Parasitism in Cattle*; Leaning, W.H.D., Guerrero, J., Eds.; Proceedings of the MSD AGVET Symposium, 23rd World Vet. Congr., Montreal, 1987, MSD AGVET: Rahway, NJ, 1987; 9–24.
8. Schwinghammer, K.A.; Knapp, F.W.; Boling, J.A. Physiological and nutritional response of beef steers to combined infestations of horn fly and stable fly (Diptera: Muscidae). J. Econ. Entomol. **1987**, *80*, 120–125.

DECISION MAKING

Leon G. Higley
University of Nebraska, Lincoln, Nebraska, U.S.A.

Robert K.D. Peterson
Dow AgroSciences, Indianapolis, Indiana, U.S.A.

INTRODUCTION

Decision making in pest management is about determining the need for using technologies against pests. Unlike simple pest control, which begins with an assumption that pests are intolerable, pest management depends on the recognition that not all potential pests require management. The ability to define tolerable and intolerable pests is at the heart of what distinguishes pest management from other approaches. Decision making, particularly through the use of economic injury levels and economic thresholds, addresses the question of when action against pests is justified. Most decisions reflect economic (cost/benefit) considerations, but, in some instances, decisions may also reflect aesthetics or environmental risk. Also, in addition to decisions for therapeutic action against pest attack, approaches for preventive action can be used, such as epidemiological models to predict the need for preventive actions against plant pathogens. Beyond decisions about whether or not to use a tactic, other types of decision making include issues such as choice of tactic, optimization in use of tactics, multidimensional decisions, and risk assessment.

HISTORICAL PERSPECTIVES

The development and application of technology to prevent pest attack has been the key principle guiding human relationships with pests throughout history, particularly since the invention of agriculture. Largely, this has been a process of identifying new technologies for use against pests, and the challenge for thousands of years has been inventing or recognizing tools or techniques for pest control. Moreover, it is only within the past 200 years that many pests, plant pathogens, and insect vectors of disease, for instance, have been recognized.

The rise of scientific agriculture and other aspects of the scientific revolution starting in the early 1800s began to change the nature of human efforts against pests. The germ theory of disease, the theory of natural selection, and the genetic basis of inheritance all provided the theoretical background for revolutionizing biology, including the biology of pest species. Concurrently, industrialization and advances in other areas of science, especially chemistry, led to the explosion in agricultural productivity of the 1900s. Similarly, the availability of resources to address pests led to exponential growth in the 1900s. In less than 150 years, the effort of thousands of years to find methods to deal with pests not only succeeded, but succeeded beyond all expectation.

It is against the backdrop of this history that pest management, and especially decision making in pest management, must be understood. When few tools were available, the interest was in identifying new tools. But as options for addressing problems became available, even abundant, then the nature of the question changed. Rather than only asking, "What tool can I use to deal with pests?" it was now possible to consider "How can I use this tool?" or even "Should I use this tool?" The development of pest resistance to insecticides was a further impetus to the issue of how and when to use pest management tactics.

ECONOMIC DAMAGE AND ECONOMIC INJURY LEVELS

Although others had discussed this question, it wasn't until the late 1950s that four entomologists, Stern, Smith, van den Bosch, and Hagen, formally addressed how and when to take action against pests. They argued that pests did not require management action unless their numbers were sufficient to cause economic losses equal to or in excess of the costs of management. This is a form of cost/benefit assessment in which the benefit of action (preventing economic loss from pests) is weighted against costs of action (costs of a parasitoid release or, most commonly, cost of a pesticide application). They defined economic damage as the value of potential loss equal to management economic damage, and further defined an economic injury level (EIL) as the pest population necessary to cause economic damage. The EIL provides a criterion for judging the need for action against pests.

In defining economic damage and the EIL, Stern and colleagues formally defined what a pest is and is not. Combined with other contributions on integration of tactics and perspectives on managing pest populations (from the Australian scientists Geier and Clark), these concepts of economic damage, integration, and management form the foundation of what is now called integrated pest management. Despite the theoretical importance of this work, however, practical application lagged. Informal procedures for assessing pest populations undoubtedly followed work by Stern and his colleagues (and in some instances likely preceded their efforts), with what are now called nominal thresholds (assessments based on opinion rather than calculated values). But it wasn't until the early 1970s that a mathematical description of economic damage and the EIL was provided by Stone and Pedigo, who described methods to calculate both economic damage and the EIL. The key limitation in developing a mathematical description has to do with relating biology (specifically yield loss from pests) to economic variables, and Stone and Pedigo presented a solution to this barrier.

After the mathematical description of economic damage and the EIL, development of EILs proceeded rapidly for insect pests. With the development of these decision tools, the process of pest management became one of assessment (through sampling), evaluation (through use of EILs and related decision guidelines), and appropriate action (or inaction). In conjunction with procedures such as sequential sampling, it is possible to greatly improve the efficiency in the assessment and evaluation process. Extensions to the cost/benefit approach to pest management decision making have been proposed and used to consider such issues as aesthetic injury, assessing environmental risk (in economic terms), and multiple pest species. Also, EIL criteria have been incorporated in many more sophisticated and complex management models. Despite these approaches, however, the value of the EIL approach has been its simplicity and demonstrated value in making rational, economically sound decisions.

OTHER APPROACHES TO PEST MANAGEMENT DECISION MAKING

In entomology, EIL-based decision making is commonplace and well developed, but for other pests, EILs are less developed and perhaps less applicable. If a pest will always exceed economic levels (as is often the case with weeds and with disease epidemics), then assessment of the need for action becomes trivial. Moreover, EIL decision making depends on making pest assessments and taking action before significant injury (yield loss) has occurred. For example, EILs for weeds were not developed until a viable management tactic, postemergence herbicides, was available. With plant pathogens, management often depends on preventive action, such as plant resistance or preventive fungicides. Consequently, decision making for many plant diseases came to focus on models for predicting the likelihood of disease occurrence. These epidemiological models consider host susceptibility, pathogen load, and abiotic factors in predicting the need for preventive action. Through such models, reductions in use of fungicides and more efficient use of fungicides were achieved.

Modeling in pest management has seen many applications beyond epidemiology. The focus of most of these efforts is to more comprehensively represent the array of factors faced by producers in making a management decision. Consequently, models have been developed to evaluate multiple tactics, multiple pests, natural enemies, and crop production practices. Where models have a management (rather than research) focus, the emphasis is often on identifying more cost effective, environmentally sustainable management options. Typically, regional or season-long considerations will be an integral feature of such models, as opposed to EILs that are essentially a single event decision tool. Also, through more sophisticated models it has been possible to consider such issues as optimizing pesticide treatments. Modeling approaches range from simple regression models to simulation models and expert systems, and many approaches have seen application in pest management systems.

Another arena in which modeling plays an important role concerns risks associated with pesticides. Regulatory decisions on uses for pesticides are based in part on a process of risk assessment. Risk assessment is a modeling effort in which the physical and biological properties of a compound are evaluated on the basis of expected environmental exposures in juxtaposition with toxicity to different types of organisms. Controversies about the suitability of risk assessment tend to focus on the validity of assumptions made as a part of the modeling process (questioning assumptions is key to evaluating all types of models). Nevertheless, risk assessments provide a valuable approach for predicting and comparing potential environmental impacts of various chemicals.

An emerging area in decision making is the development of resistance management plans or models. More properly, these might be termed "resistance avoidance plans," given that once resistance to a tactic has developed in a pest population, we cannot mitigate expression of that resistance (other than by not using the tactic). Resistance avoidance plans entail modeling the selection pressure posed by a given tactic and considering the genetics of resistance in an exposed pest population. The objective behind such efforts is to identify tactic use options that will prevent or delay the onset of resistance in the pest population.

ADOPTION, LIMITATIONS, AND THE FUTURE OF DECISION MAKING

In entomology, the EIL and associated economic threshold are by far the most widely used decision tools. The EIL works well with the most commonly used therapeutic tactic, insecticides, and it provides an easy means of evaluating the need for action. Unfortunately, the lack of alternatives to insecticides for therapeutic management and limitations in therapeutic agents for weeds and diseases have similarly limited the usefulness of EILs and related decision tools. In many situations, decision making does not play a prominent role either because management alternatives are not available or because pest pressure is far beyond tolerable levels. Another barrier to the adoption of more sophisticated management tools has been the complexity of these tools. In particular, data requirements for more sophisticated management models may seriously limit the adoption and use of such models by producers. Advances in automated data acquisition, as may be possible through some approaches in precision agriculture, may help address this problem of acquiring data for more sophisticated management models. A perennial limitation in the development and use of decision models for preventive pest management has been the dependence of many of these models on weather data. The application and improvement of preventive management tools depends on improved weather predictions. As long as our ability to make weather predictions is limited, so will many of our decision models be similarly limited.

Decision making in pest management has had a long history of adopting new technologies and novel approaches. Undoubtedly, as future new options arise, these also will find application for pest management. For genetically modified organisms risk assessment is likely to play an increasingly important role. The growth of geographic information systems and other tools in precision agriculture will lead to more site specific management decisions, especially with pests such as weeds. Also, as our understanding of how pests cause yield losses improves, new opportunities for assessment and decision making based on plant characteristics, such as leaf area development, are likely. However, the backbone of pest management decision making has been the simple cost/benefit approach codified in the EIL, and, at least with insects, it seems likely that this tool will continue to dominate practical decision making. Moreover, with the development of new management tactics and an increasing emphasis on sustaining tactics and avoiding unacceptable environmental risk, it seems likely that decision tools addressing these issues will also play an increasingly important role in the practice of pest management.

BIBLIOGRAPHY

Teng, P.S., Ed. *Crop Loss Assessment and Management.* APS Press: St. Paul, MN, 1989; 270.

Pedigo, L.P., Higley, L.G., Eds. *Economic Thresholds for Integrated Pest Management.* University of Nebraska Press: Lincoln, NE, 1996; 327.

Cothern, C.R. *Handbook for Environmental Risk Decision Making: Values, Perceptions, and Ethics*; Lewis Publishers: Boca Raton, FL, 1996; 408.

Geier, P.W.; Clark, L.R. In *An Ecological Approach to Pest Control*, Proceedings of the 8th Technical Meeting of the International Union for Conservation of Nature and Natural Resources, 1960; Warsaw, Poland, 1961; 10–18.

Higley, L.G.; Pedigo, L.P. Economic injury level concepts and their use in sustaining environmental quality. Agric. Ecosys. Env. **1993**, *46*, 233–243.

Norton, G.A.; Mumford, J.D. *Decision Tools for Pest Management*; CAB International: Wallingford, Oxon, UK, 1993; 279.

Pedigo, L.P.; Hutchins, S.H.; Higley, L.G. Economic injury levels in theory and practice. Annu. Rev. Entomol. **1986**, *31*, 341–368.

Stern, V.M.; Smith, R.F.; van den Bosch, R.; Hagen, K.S. *The Integrated Control Concept*; Hilgardia, 1959; 29, 81–101.

Stone, J.D.; Pedigo, L.P. Development of economic-injury level of the green cloverworm on soybean in Iowa. J. Econ. Entomol. **1972**, *65*, 197–201.

DEER DAMAGE AND CONTROL

Paul D. Curtis
Cornell University, Ithaca, New York, U.S.A.

D

INTRODUCTION

Dramatic changes in the relationship between people and white-tailed deer have occurred in the United States during the past three decades. Deer have made a remarkable recovery since the early 1900s, when there were perhaps no more than 500,000 white-tails nationwide. While virtually extirpated early this century, numbers of white-tailed deer now likely exceed 20 million across the country (1). Most eastern states have experienced dramatic increases in deer abundance during the past 15 years, particularly at the suburban–rural interface surrounding many communities (2).

Wildlife damage to agricultural crops has been estimated at $500 million annually, and deer were most often responsible for these losses (3, 4). In forested landscapes of the northeast and northwest, deer damage to timber productivity is conservatively estimated to exceed $750 million per year. In addition, losses of landscape and garden plants in suburban areas may exceed $250 million, with homeowners spending an additional $125 million per year on deer damage control. Annual estimates of deer damage are reported to exceed $2 billion nationwide (5).

CAUSES FOR INCREASES IN DEER ABUNDANCE

The recovery of deer populations is a wildlife management success story throughout much of the country. However, many landowners and homeowners increasingly view this situation with mixed feelings. Many factors are associated with increases in deer abundance (2). Farm abandonment and reversion of agricultural fields to shrub and forest have created ideal deer habitat across broad landscapes. In much of the northeast, mild weather has resulted in high fawn survivorship for many years. Snow depth and duration have been severe enough to limit herd growth during only a few winters in the past 20 years. Gardens and shrubs associated with residential development at the fringe of metropolitan areas have provided a rich and diverse forage base for deer. Landowner decisions to prevent deer hunting, and restrictions on the use of firearms in suburban areas have created refuges where deer numbers may grow rapidly. Deer populations can double in size every two to three years if sufficient food is available and there is no hunting mortality. Elimination of large carnivores, such as mountain lions and wolves, in the eastern United States by the early 1900s removed a potential check on deer abundance. These changes have been gradual, and even with foresight, it is unlikely that anyone could have altered the course of events that have resulted in the current level of deer conflicts. In reality, the wide range of plants and plant parts eaten, their nighttime foraging habits, and their ability to adapt to man-made ecosystems all serve to make the white-tailed deer one of the most problematic and economically important wildlife species in North America.

FORAGING BEHAVIOR

While deer are known to eat more than 500 different kinds of plants, they are often selective feeders that forage or browse on plants and plant parts with considerable discrimination (6). Whether or not a particular plant species or variety will be eaten depends upon the deer's nutritional needs, previous feeding experience, plant palatability, weather conditions, and the availability of alternative forage. Most damage to fruit trees and woody ornamentals occurs during winter when snow cover reduces the availability of other foods. Damage to soybeans, corn, and other row crops during summer is associated with deer density and the amount of brush land or woodlot area near the fields. Deer require escape and resting cover, and crop losses are often most severe near field edges.

ECONOMIC IMPACTS CAUSED BY DEER

Agricultural Crop Losses

In recent surveys, about two-thirds of farmers reported problems with deer, and about one-third indicated their crop losses exceeded $1000 per year (7). When combining damage to field crops, fruits, and vegetables, a conservative estimate of deer damage to agriculture would be $100 million per year in the United States. Limited data are available to estimate impacts to tree nurseries and the horticulture industry. However in one New York study, nursery producers reported total losses of $519,000 in a five-county area. Three producers reported more than $150,000 in deer damage, and the average loss for all growers exceeded $20,000 per nursery (8).

Damage to Garden and Ornamental Plants

Deer damage to ornamental plants is widespread, but is not evenly distributed across the landscape. Impacts are often most intense near the suburban–rural fringe of metropolitan areas, or near urban parks and preserves. A survey of homeowners in the United States indicated that 2.4 million households experienced deer damage, resulting in approximately $251 million per year in plant losses (5).

More detailed mail surveys of nursery producers and homeowners in suburban areas of New York State indicated higher levels of deer damage to landscape plants. The average replacement costs for trees and shrubs was nearly $500 per household with deer damage (8). A survey in Westchester County, New York, indicated that more than 40% of respondents reported plant damage caused by deer. Average annual cost of plant replacement for households with deer damage averaged $94 for vegetables, $102 for flowers, $156 for fruit trees, and $635 for shrubbery. Estimated replacement costs for plants in northern Westchester County ranged from $6.4 to $9.5 million in 1987 (9).

Forest Regeneration Failures

Deer foraging on tree seedlings either kills them, or reduces their growth, increasing the number of years it takes for trees to reach a marketable size. In eastern forests, many high-value tree species (e.g., sugar maple, oak, ash, etc.) used for saw-timber are also highly palatable to deer (10). In Pennsylvania alone, deer damage to commercial timber species may exceed $367 million. At the Quabbin Reservation in Massachusetts, deer have greatly reduced oak regeneration and the natural process of forest succession (11). Growing forests in many part of the northeast contain an inadequate diversity of seedlings and saplings (10). In the Pacific Northwest, deer reduced the survival of newly planted Douglas fir trees by 20% and the height of remaining trees by 24%. In addition, the survival and height of ponderosa pine trees was reduced by 31 and 22%, respectively (12). Estimates of the total cost for these tree losses ranged from $118 to $378 million. Given the limited data available in the United States, deer damage to forest regeneration may well exceed $750 million annually (5).

DAMAGE PREVENTION: AN INTEGRATED APPROACH

Damage Management Methods

There is no simple solution for controlling deer conflicts. For example, repellent applications or low-cost fence designs may work well for protecting a few shrubs or a backyard garden (6). However, usually these techniques are not practical for commercial agricultural fields. The larger an area a grower may want to protect, the more persistent deer may be in attempting to cross a physical or behavioral barrier (13). Also, techniques that work well in summer when natural forage is abundant may fail completely during winter when snow limits food availability. Consequently, an integrated pest management approach is needed and growers or homeowners must choose the best combination of management options to suit their specific situation.

Population Control

Hunting has been the traditional tool for regulating herd size, and is the only practical option for managing deer populations at the landscape scale (14). Where legally possible, landowners should require that hunters harvest female deer, as removing breeding-age females from the herd is essential for reducing deer numbers and damage. For high-value crops with heavy losses, state wildlife agencies may issue special permits to kill deer outside of the hunting season. Options for controlling herd growth are very limited in suburban landscapes closed to hunting. Hiring professional sharpshooters to cull female deer has reduced densities in some communities. Experimental fertility-control drugs can reduce fawn production by almost 90%. However, the contraceptive vaccines are not commercially available, and dart-gun delivery can be very expensive.

Physical Barriers

Fencing provides the most cost-effective, long-term control of deer damage in many circumstances, and woven-wire fences eight or more feet high are the most effective barriers (13). Electric fence designs have been developed that are less costly and almost as effective as woven-wire barriers. Wire cages and plastic netting offer effective control for small planting beds or individual shrubs or trees.

D

Chemical Repellents

A variety of commercial repellents are labeled for preventing deer feeding (6). Repellents are most effective when applied prior to the occurrence of damage, and in situations where damage is expected to be light to moderate. Research has shown that egg-based products provide the most reliable control, however even the best materials seldom provide more than 40–45 days of protection under high feeding pressure from deer (15). As a general rule of thumb, odor-based repellents provide more consistent plant protection than taste-based materials.

FUTURE TRENDS

Overabundant deer populations currently cause substantial economic losses in many parts of the United States. The problems are particularly severe in eastern states where the forage and cover available near suburbia, and protection from hunting in many residential areas or parks, have provided an ideal situation for deer populations to expand rapidly. Hunting will continue to be the most cost-effective management tool for controlling herds in areas where sportsmen can gain access to the deer.

The specialized management needed for regulating the growth of suburban deer populations is quite different from conventional hunting programs. Alternative approaches (such as sharpshooting deer over bait or fertility control) will continue to be tested and refined in parks and communities where hunting is prohibited. These high-technology approaches are very expensive, and it is unclear how many communities will be willing to pay the costs for development and long-term application of these techniques.

In summary, the biological and social dimensions of deer management will continue to pose a tremendous challenge for wildlife professionals. There is no panacea that will resolve deer–human conflicts. Rapid growth of deer herds will continue as long as people choose to limit mortality factors (i.e., hunting, predation, etc.) and quality forage is available to maintain high fawn production.

REFERENCES

1. McCabe, T.R.; McCabe, R.E. Recounting Whitetails Past. In *The Science of Overabundance: Deer Ecology and Population Management*; McShea, W.J., Underwood, H.B., Rappole, J.H., Eds.; Smithsonian Institution Press: Washington, DC, 1997; 11–26.
2. Curtis, P.D.; Richmond, M.E. Future challenges of suburban white-tailed deer management. Tran. North Am. Wildl. Nat. Resour. Conference **1992**, *57*, 104–114.
3. Conover, M.R. Perceptions of grass-roots leaders of the agricultural community about wildlife damage on their farms and ranches. Wildl. Soc. Bull. **1994**, *22*, 94–100.
4. Conover, M.R.; Decker, D.J. Wildlife damage to crops: perceptions of agricultural and wildlife professionals in 1957 and 1987. Wildl. Soc. Bull. **1991**, *19*, 46–52.
5. Conover, M.R. Monetary and intangible valuation of deer in the United States. Wildl. Soc. Bull. **1997**, *25*, 298–305.
6. Curtis, P.D.; Richmond, M.E. *Reducing Deer Damage to Home Gardens and Landscape Plantings*; Cornell Cooperative Extension, Department of Natural Resources, Cornell University: Ithaca, NY, 1994; 22.
7. Wywialowski, A.P. Agricultural producers' perceptions of wildlife-caused losses. Wildl. Soc. Bull. **1994**, *22*, 370–382.
8. Sayre, R.W.; Decker, D.J.; Good, G.L. Deer damage to landscape plants in New York State: perceptions of nursery producers, landscape firms, and homeowners. J. Environm. Hortic. **1992**, *10*, 46–51.
9. Connelly, N.A.; Decker, D.J.; Wear, S. *White-Tailed Deer in Westchester County, New York: Public Perceptions and Preferences*; Human Dimensions Research Unit Series Number 87-5, Department of Natural Resources, Cornell University: Ithaca, NY, 1987; 80.
10. Tilghman, N.G. Impacts of white-tailed deer on forest regeneration in northwestern Pennsylvania. J. Wildl. Manage. **1989**, *53*, 524–532.
11. Healy, W.M. Influence of Deer on the Structure and Composition of Oak Forests in Central Massachusetts. In *The Science of Overabundance: Deer Ecology and Population Management*; McShea, W.J., Underwood, H.B., Rappole, J.H., Eds.; Smithsonian Institution Press: Washington, DC, 1997; 249–266.
12. Black, H.C.; Dimock, E.J. II; Rochelle, J.A. *Animal Damage to Coniferous Plantations in Oregon and Washington: Part 1. A Survey 1963–1975*; Research Bulletin 25; Forestry Research Laboratory, Oregon State University: Corvallis, 1979; 44.
13. Curtis, P.D.; Fargione, M.J.; Richmond, M.E. In *Preventing Deer Damage with Barrier, Electrical, and Behavioral Fencing Systems*, Proceedings of the 16th Vertebrate Pest Conference, Santa Clara, California, March 1–3, 1994; Halverson, W.S., Crabb, A.C., Eds.; University of California: Davis, CA, 1994; 223–227.
14. Curtis, P.D.; Moen, A.N.; Enck, J.W.; Riley, S.J.; Decker, D.J.; Mattfeld, G.F. *Approaching the Limits of Traditional Hunter Harvest for Managing White-Tailed Deer Populations at the Landscape Scale*; Human Dimensions Research Unit Series Number 00-4, Department of Natural Resources, Cornell University: Ithaca, NY, 2000; 17.
15. Sayre, R.W.; Richmond, M.E. In *Evaluation of a New Deer Repellent on Japanese Yews at Suburban Home Sites*, Proceedings of the 5th Eastern Wildlife Damage Control Conference, Ithaca, NY, Oct 6–9, 1991; Curtis, P., Caslick, J., Eds.; Cornell University: Ithaca, 1992; 38–43.

DEPOSITION

G.A. Matthews
Imperial College of Science, Technology, and Medicine, Berkshire, United Kingdom

INTRODUCTION

Deposition of a pesticide on a surface will be affected by a large number of factors, including the spray droplet size and velocity, the properties of the spray liquid, and the position and characteristics of the target surface. Static or dynamic surface tension and viscosity of the liquid alter with the formulation of the pesticide and the concentration at which it is applied.

Most pesticides are formulated so that when mixed in water they can be applied through a hydraulic pressure nozzle. Such nozzles produce a wide range of droplet sizes (1). Droplets falling in still air accelerate downward due to the force of gravity until aerodynamic drag forces prevent further acceleration and then fall at their terminal velocity. Small droplets, generally below 100 μm have a low terminal velocity (< 0.3 m/s) and become airborne, moving within any natural or artificial air flow. Thus, the trajectory of individual droplets within a spray cloud is immediately influenced by air movements occurring naturally or created by the passage of the sprayer. Larger droplets, with a terminal velocity greater than 0.3 m/s, will fall more or less vertically to the ground, if not intercepted by an object. Horizontal leaves at the top of a crop canopy will intercept a high proportion of spray. Excess liquid will drip from the surface and be lost due to run-off.

Deposition of droplets influenced by gravity is by sedimentation. Even the smallest droplets will eventually sediment in still conditions on the upper surface of leaves or other horizontal surfaces. Droplets moving within air flows are deposited by impaction, unless carried around objects. Deposition is affected by the relative velocity of the droplet and object and by droplet size (2). Collection efficiency is defined as the proportion of droplets approaching an object that are deposited on it, and is increased on narrow surfaces.

The proportion of a pesticide spray that can be biologically effective will depend on how it is deposited on the intended target. When the soil surface has to be treated with an herbicide, a high proportion in large droplets (> 200 μm) will land on the soil. However, spraying foliage is generally less efficient because leaf surfaces have a water-repellent surface (3–5). The angle of contact of water droplets will vary significantly between different plant surfaces. Some weeds and crops, e.g., peas, have particularly waxy leaves and are very difficult to wet. The largest droplets can bounce from plant surfaces that are difficult to wet or may shatter on impact into smaller droplets. Retention is improved with smaller droplets. Deposition within a crop canopy is also limited by the extent to which spray droplets penetrate the canopy. Insect movement increases secondary pick-up from treated leaf surfaces and some deposits are ingested.

IMPROVING DEPOSITION

A pesticide is usually formulated with a surfactant to reduce the surface tension of the spray liquid and reduce the risk of droplet bounce (Fig. 1). It is also possible to mix an adjuvant into a pesticide spray to alter its physical characteristics and improve deposition. Some application equipment (6) is provided with a fan to create an airflow that projects droplets into a crop canopy and increases their velocity to improve deposition. Spray droplets are deposited more efficiently on small, thin leaves, such as those on pine trees, which filter the airborne spray, and when droplets are carried in a turbulent air flow that causes leaves to flutter.

Deposition on the undersides of leaves, especially on broad-leaved plants, can be poor with downwardly directed sprays and larger droplets. The addition of an electrostatic charge usually effective only with small droplets can increase underleaf coverage, provided there is sufficient space between plants for a charged cloud of droplets to reach different parts of the crop canopy. How-

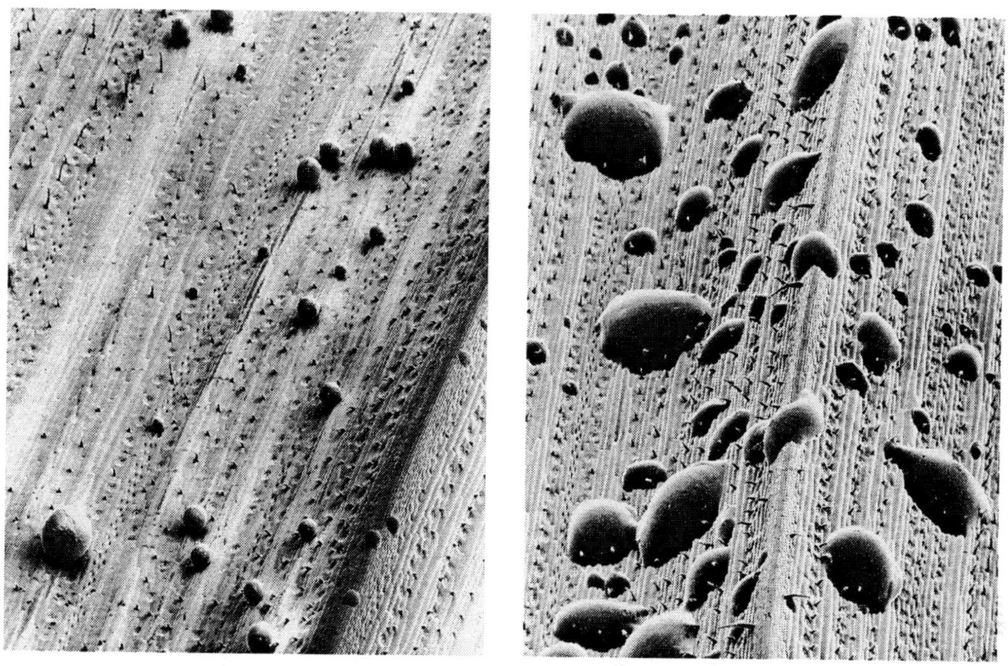

Fig. 1 Droplets on a leaf surface. (**A**) Droplets without surfactant not wetting leaf surface. (**B**) Droplets with surfactant.

ever, good control of pests and pathogens situated underneath leaves generally requires the upward projection of spray droplets with nozzles positioned in the interrow.

Concern about the adverse effects of pesticides deposited outside a treated area due to downwind spray drift has led to greater use of coarse sprays (Fig. 2) with the volume median diameter of the spray larger than 200 μm, for example, by using air induction nozzles (Fig. 3). However, when larger droplets are applied, fewer droplets are deposited per unit area, unless the spray volume is significantly increased. Doubling the droplet diameter requires an eight-fold increase in spray volume to produce the same number of droplets. Large droplets in a coarse spray can be used to apply systemic or translocated pes-

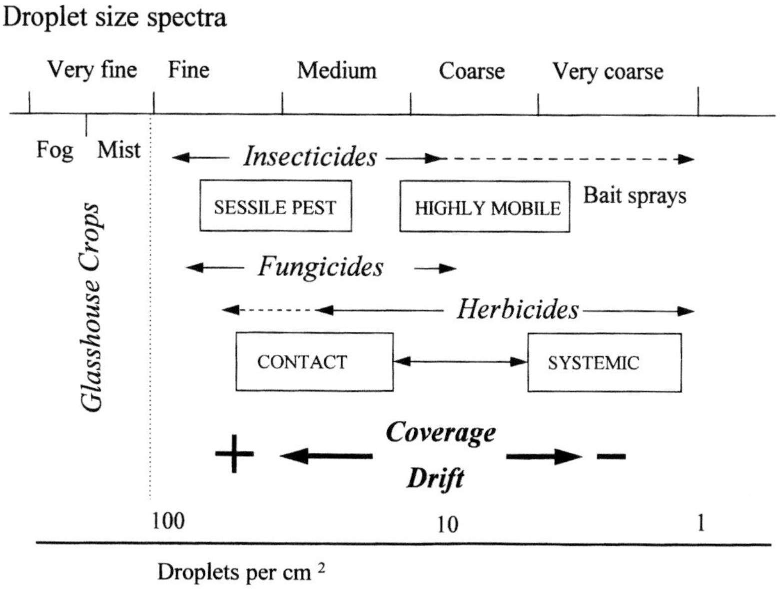

Fig. 2 Spray quality diagram. This diagram illustrates selection of droplet size in relation to different pesticides.

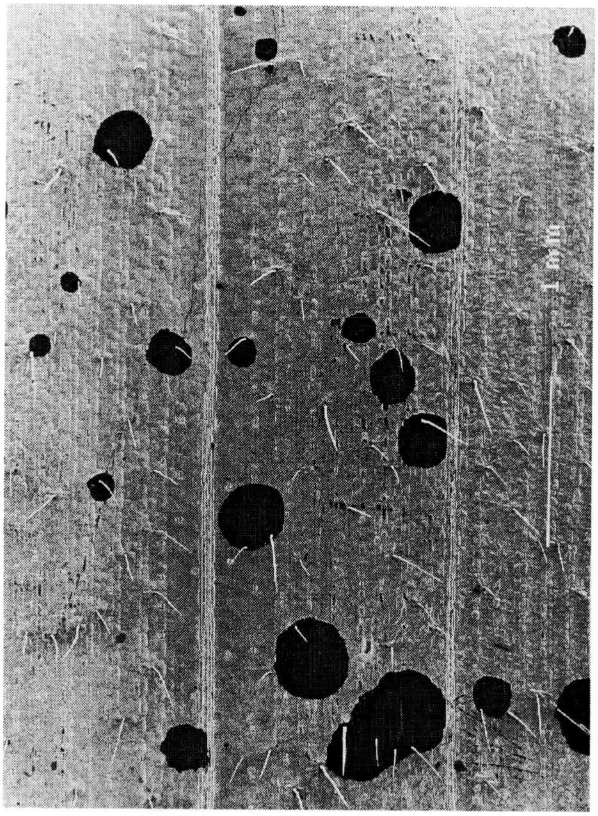

Fig. 3 Droplet deposition with an air induction nozzle. This photo shows large droplets and relatively poor coverage of leaf surface.

ticides, as the chemical can be redistributed within the treated plant.

Advice on spray quality and driftability of droplets from different hydraulic pressure nozzles is now available (7). Spray not retained on foliage either drifts downwind or contaminates the soil within a treated field, exposing nontarget organisms to the pesticide. Spray drift of an herbicide can kill or adversely affect susceptible plants downwind of a treated area and adversely affect hedgerow/field margin habitats of beneficial insects. Similarly bees and other beneficial insects can be killed by drift of an insecticide spray. Particular attention has been given to the need to avoid pollution of water surfaces; thus, for some pesticides, a no-spray buffer zone to protect aquatic life is required.

REFERENCES

1. Matthews, G.A. *Application of Pesticides to Crops*; IC Press: London, 1999; 340.
2. Bache, D.; Johnstone, D.R. *Microclimate and Spray Dispersion*; Ellis Horwood: Chichester, U.K., 1992.
3. Hartley, G.S.; Brunskill, R.T. Reflection of Water Drops from Surfaces. In *Surface Phenomena in Chemistry and Biology*; Danelli, J.F., Pankhurst, K.G.A., Riddiford, A.C., Eds.; Pergamon: London, 1958; 214–223.
4. Reichard, D.L.; Cooper, J.A.; Bukovac, M.J.; Fox, R.D. Using a videographic system to assess spray droplet impaction and reflection from leaf and artificial surfaces. Pestic. Sci. **1998**, *53*, 291–299.
5. Webb, D.A.; Western, N.M.; Holloway, P.J. Modelling the impaction behaviour of agricultural sprays using monosized droplets. Aspects of Applied Biology **2000**, *57*, 147–154.
6. Matthews, G.A. *Pesticide Application Methods*; 3rd Ed.; Longman: New York, 2000; 432.
7. Southcombe, E.S.E.; Miller, P.C.H.; Ganzelmeier, H.; Miralles, A.; Hewitt, A.J. In *The International (BCPC) Spray Classification System Including a Drift Potential Factor*, Proceedings of the Brighton Crop Protection Conference, 1997; 371–380.

DISPERSAL OF PLANT PATHOGENS

Paul D. Hildebrand
Agriculture and Agri-Food Canada, Kentville, Nova Scotia, Canada

D

INTRODUCTION

Dispersal of plant pathogens is critical to their survival as species. After they parasitize and consume their hosts they must disperse to new susceptible tissue in order to survive. If new tissue is not found they must use an alternative means of survival or they die. Dispersal is important not only for the maintenance of the population within its existing geographic region, but also for its extension into new regions and for its genetic development as it interacts with indigenous strains of the new region. To accomplish this, fungal, bacterial and viral pathogens display a wide range of mechanisms for short- and long-distance dispersal. Fungi often have active dispersal mechanisms mediated through weather factors, whereas bacteria, viruses, and nematodes are usually associated with hosts, host debris, soil particles, or vectors and are less often affected by weather factors in their dispersal.

WIND

Wind is perhaps the most important means by which fungal spores are dispersed. The dispersal process of airborne fungal spores is characterized by three phases: discharge, transport, and landing (1, 2). The discharge phase can be either active or passive. In either case, spores that are produced on the ground or on leaves need to escape the stagnant boundary layer of air that exists near these surfaces in order to be swept up and dispersed by air currents and eddies (3). An example of active spore discharge occurs in some ascomycete fungi. Their spores (ascospores) are contained in microscopic sacs or asci that line the inside of a cup-shaped apothecium. The asci burst open in response to a sharp decrease in relative humidity, ejecting the spore contents resulting in a visible smoke-like ''puff'' of spores (see Fig. 1). Spores may be discharged as much as 2.0-5.0 cm from an apothecium (1). Other fungi, such as those belonging to the Peronosporales that cause downy mildew diseases, have spores borne on sharply pointed microscopic tree-like sporophores that quickly twist and fling off spores due to rapid changes in temperature and relative humidity. Infrared radiation at sunrise may also contribute to the build-up of electrostatic charges on the sporophores causing spores to eject. Fungi with passive discharge mechanisms rely on wind to dislodge their spores from diseased plant parts. In many genera, the sporophore has an erect stalk that raises the spores slightly (0.1–2 mm) above the substratum, increasing their chances of take-off under the influence of wind eddies that may break into the boundary layer. Spores may also be jarred loose or abraded off by the action of leaf flutter or by the ''tapping'' action of rain drops (1, 4, 5).

Once airborne, most spores are redeposited within 100 m of their source often resulting in focal disease development (5). However, a large spore cloud emanating from a diseased crop may occasionally be carried aloft by convective air currents. Depending on surface temperatures and topography, spores may be carried as high as 3 km in cumulus-type cloud systems or much higher in cyclonic air systems and transported thousands of kilometers in continental and intercontinental air streams (1, 2, 6) (Fig. 2). The *Puccinia* rust (*Puccinia graminis* f.sp. *tritici*) pathway in North America is an example of this that occurs annually when spring-sown wheat in Canada receives spore showers from diseased autumn-sown wheat in Mexico (4). Rust spores advance northward by a succession of jumps on intervening wheat crops in the mid United States or they may be transported more than a thousand kilometers in a single flight. Similarly, winter wheat in the south becomes infected during the autumn by spore showers from the north. Spores of pathogens that successfully travel long distances typically are thick walled and darkly pigmented enabling them to withstand dry conditions, low temperatures, and UV radiation.

Spores can be brought back to earth after short- or long-distance transport by vertical air mixing and subsequent impaction onto plant surfaces. Large spores with greater momentum impact more efficiently onto leaves and stems than do small spores (2, 5). Deposition may also occur under the influence of gravity in relatively still

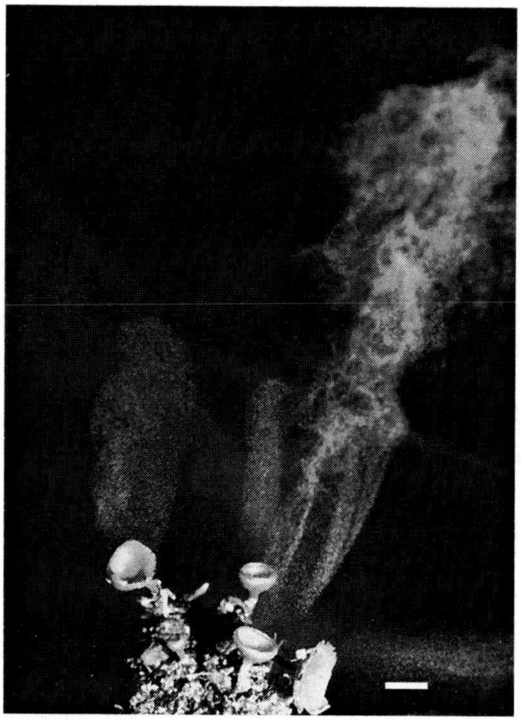

Fig. 1 Ascospore discharge from apothecia of the fungal pathogen *Monilinia vaccinii-corymbosi*, the causal agent of mummy berry disease of blueberry. Scale bar = 1 cm.

air, or spores may be deposited very quickly by the action of rain that has scrubbed spores from the atmosphere. Spores at high altitudes may also act as nuclei for the formation of raindrops (6).

RAIN

Some fungal and bacterial pathogens rely on rain for their dispersal. Rain consists of droplets 0.2 — 5 mm in diameter falling at terminal velocity capable of causing splash droplets when impacting a leaf or the ground (5, 7). Drops falling onto water films 0.1 – 1.0 mm deep can produce up to 5,000 splash droplets 5–2400 μm in diameter. Splash droplets follow a ballistic trajectory and travel only about 50–75 cm, but if high winds accompany rain, dispersal may occur over several kilometers. Small splash droplets tend not to carry fungal spores but they may carry bacterial cells. Droplets formed on leaves by rain or mist may roll off carrying substantial numbers of spores or bacteria allowing for redistribution and subsequent splash within the crop. Spores of the genera *Colletotrichum* and *Fusarium* and bacteria of the genera *Pseudomonas* and *Xanthomonas* tend to be dispersed by rain splash and are typically produced in mucilage or polysaccharides. These substances, when dissolved, leave a suspension of pathogen cells that is readily washed off or splash dispersed by rain (1). Some fungi may be actively triggered to release their spores in response to rain. For example, the ascospores of the fungus *Venturia inaequalis* that causes apple scab, are actively discharged from a flask-shaped perithecium embedded in overwintering leaves in response to as little as 0.2 mm rain. When wetted, its asci swell rapidly by osmotic pressure squirting out the spore contents (1, 2).

SEED

Dispersal of pathogens in seed is a common and highly efficient means of spread. Seeds may survive for long periods; be transported by humans, animals, and birds over long distances; and introduce pathogens into a crop from the beginning of its development. As few as 3–5 seeds in 10,000 infected with bacteria may cause an epidemic (4, 7). About one-fifth of the known viruses are transmitted through seed (8). Seedborne viral pathogens colonize their hosts and infect the seed as it forms whereas fungal and bacterial pathogens may exit their hosts for a time. For example, spores of fungal species within the genus *Tilletia*, that cause bunt diseases of cereal crops, are carried on infested kernels and infect young seedlings. The pathogens grow systemically in the plant and eventually colonize the internal tissue of the kernel producing masses of spores. At harvest the seed coat shatters releasing the spores that contaminate healthy seed or the spores may be dispersed long distances by wind. Similarly, the bacterium *Xanthomonas campestris* pv. *campestris* that causes black rot disease of crucifers is present in the seed coat and enters the seedling through stomata of the cotyledon to which the infected seed coat is attached. Visible lesions on mature leaves appear later and the bacteria are subsequently rain splashed to healthy leaves or to seed stalks where invasion of developing seed coats occurs.

INSECTS

Insect vectoring is the most common method of dispersal by viruses but it also may occur with fungal and bacterial pathogens. There are more than 350 species of insects in nine orders that transmit viruses with aphids, whiteflies, and leafhoppers being most important (8). Insect-borne viruses are classed as nonpersistent or persistent. Nonpersistent viruses are acquired by insects with piercing stylet mouthparts as they probe plant tissues for food. Aphid stylets can become contaminated by viruses from

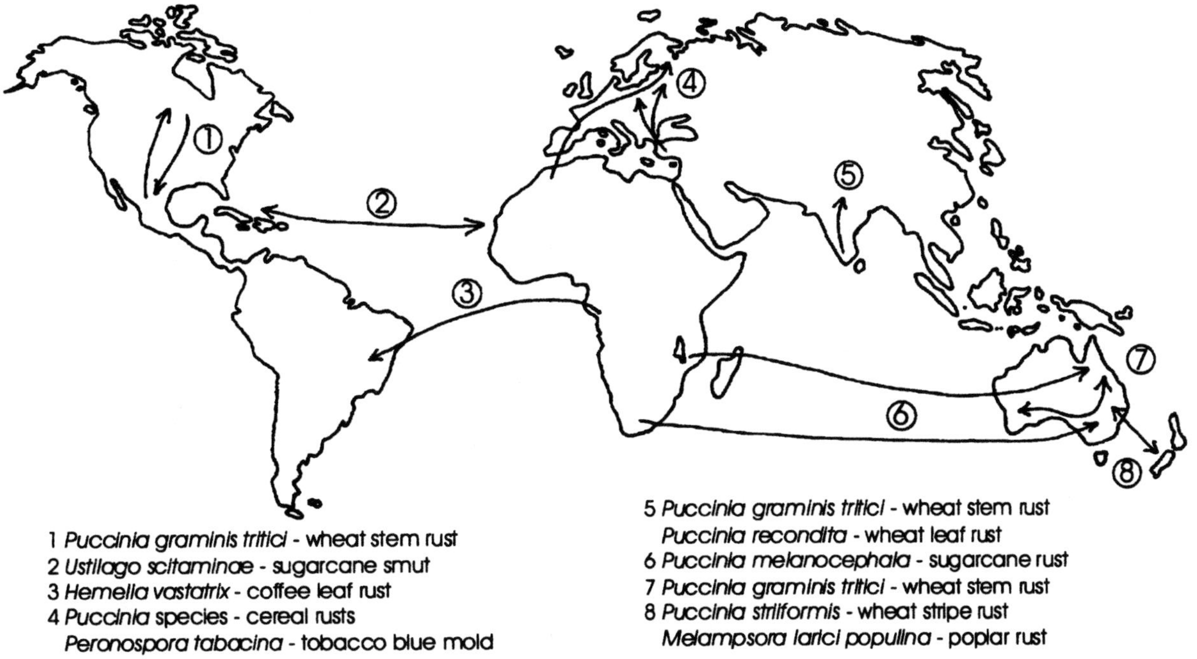

Fig. 2 Long-distance dispersal routes of some fungal pathogens. (Adapted from Ref. 6.)

an infected epidermal cell within seconds, but the aphid is viruliferous for only a few minutes to a few hours afterward. The virus is lost from the stylet tip through scouring as the aphid probes subsequent host and nonhost plants. Nevertheless, these nonpersistent viruses can spread quickly within a crop if the vectoring insect population is high. Persistent viruses are acquired by insects when they penetrate their stylets deeply to the phloem tissue and withdraw virus-laden plant sap into their gut. The viruses may persist in the insect for its life and may even multiply making the insect's subsequent feeding more inoculative. Aphids frequently probe many different plant species in an exploratory manner that creates extensive opportunity to acquire viruses and disperse them to other species. Winged aphids may carry viruses long distances over land by successive colonizations or more than 2000 km in a single flight when carried by large weather systems. Such is the case with barley yellow dwarf virus. Cereal aphids that harbor this virus may be blown each spring from an overwintering region in Texas through the U.S. midwest to the Canadian prairie provinces.

HUMANS

Human activity regularly aids pathogen dispersal locally and has also served to spread pathogens to many parts of the world where they would not otherwise have arrived. Pathogens frequently are mechanically spread from plant to plant by worker activity in fields and by movement of farm implements and soil within fields and from infested to healthy fields. Contaminated irrigation water may also serve as a vehicle for local spread of pathogens. On a global scale, seed, bulbs, rootstocks, and cuttings harboring all types of pathogens can be distributed easily often with disastrous consequences. A classic example is the global spread of potato late blight caused by *Phytophthora infestans*. The original hosts of this pathogen were wild *Solanum* species occurring in Mexico, while the cultivated forms originated in South America and were never exposed to the pathogen. The first epidemic outside of Mexico developed rapidly along the eastern seaboard of the United States in 1843. The fungus was likely imported on wild *Solanum* species from Mexico and spread as airborne spores. The pathogen was then observed in Belgium in 1845 and likely arrived on infected cultivated tubers. From there, the pathogen spread quickly throughout Europe by wind and destroyed the potato crop in Ireland reducing the human population by more than half due to starvation and emigration (9).

Dispersal of a plant pathogen is no less important than any other phase of its life cycle for without dispersal there would be no potential for the development of an epidemic. Within the context of disease management, a sound understanding of the dispersal mechanism of a given pathogen is essential. Growers, pest managers, and regulatory agencies need to know how pathogens are dispersed and what practices are important to minimize this process on a local and global scale.

REFERENCES

1. Ingold, C.T. *Fungal Spores—Their Liberation and Dispersal*; Clarendon Press: Oxford, 1971; 1–302.
2. Gregory, P.H. *The Microbiology of the Atmosphere*; John Wiley & Sons: New York, 1973; 1–377.
3. Aylor, D.E. The role of intermittent wind in the dispersal of fungal pathogens. Annu. Rev. Phytopathol. **1990**, *28*, 73–92.
4. Campbell, C.L.; Madden, L.V. *Introduction to Plant Disease Epidemiology*; John Wiley & Sons: New York, 1990; 1–532.
5. Lacey, J. Spore dispersal—its role in ecology and disease: the British contribution to fungal aerobiology. Mycol. Res. **1996**, *100* (6), 641–660.
6. Nagarajan, S.; Singh, D.V. Long-distance dispersion of rust pathogens. Annu. Rev. Phytopathol. **1990**, *28*, 139–153.
7. Hirano, S.S.; Upper, C.D. Ecology and epidemiology of foliar bacterial plant pathogens. Annu. Rev. Phytopathol. **1983**, *21*, 243–269.
8. Matthews, R.E.F. *Plant Virology,* 3rd Ed.; Academic Press: Toronto, 1991; 1–835.
9. Goodwin, S.B. Molecular tracking new migrations of an old pathogen, the re-emergence of potato late blight. Phytoprotection **1999**, *80* (2), 85–95.

DISPOSAL OF PESTICIDES AND PESTICIDE CONTAINERS[a]

D

Nancy Fitz
Jude Andreasen
U.S. Environmental Protection Agency, Washington, D.C., U.S.A.

INTRODUCTION

Unwanted or obsolete pesticide stocks and empty pesticide containers must be disposed of in an environmentally protective manner. U.S. regulations vary since some, but not all, pesticides are classified as hazardous wastes when disposed. Research on innovative disposal methods is principally focused on hazardous waste disposal, and pesticide-specific research tends to focus on remediation of soil and water contaminated by pesticides. In developing countries, the problem is particularly acute, and efforts are underway to resolve it and prevent future accumulation of obsolete stocks.

U.S. LAWS AND REGULATIONS

Federal Pesticide Law

Under the Federal Insecticide, Fungicide, and Rodenticide Act (FIFRA), which is implemented by the U.S. Environmental Protection Agency (EPA), all pesticide products must have instructions for the disposal of pesticides and pesticide containers on their labels, and users are required to follow all instructions. For container disposal, labels generally instruct the user to rinse the container or empty it completely and then offer alternative management options, such as recycling, reconditioning, landfilling, incineration or open burning, depending on the type of container. For agricultural pesticide products, it is recommended that the rinse water be added to the pesticide mixture or applied to the same crop. For residential products, rinsing is not always recommended due to the rinse water disposal problem. The labels (except for household pesticide labels) state "Do not contaminate water, food, or feed by storage and disposal" and also include another statement if the pesticide is classified as a hazardous waste.

Federal Waste Law

Once the decision is made to dispose of or discard a pesticide, it becomes a solid waste regulated under the federal solid and hazardous waste law, the Resource Conservation and Recovery Act (RCRA) (1). The term "solid waste" includes solid, semi-solid, liquid and contained gaseous materials. The RCRA regulations require the person disposing of the waste (the "generator") to determine if the solid waste is a hazardous waste. Hazardous wastes are defined in several different ways, including being specifically listed in EPA regulations as hazardous wastes, or exhibiting one of the four hazardous waste characteristics. EPA has published a handbook to help small businesses understand the hazardous waste rules (2).

If a pesticide is classified as a solid waste (but not a hazardous waste) when disposed, it can be managed as "regular trash" and can be placed in a solid waste landfill or sent to a nonhazardous waste incinerator. If a pesticide is classified as a hazardous waste when disposed, it must be treated or disposed of at a RCRA-permitted facility and must meet all relevant standards. State hazardous waste regulations can be more stringent than federal regulations, and state regulatory agencies can provide specific disposal guidance. EPA maintains an information hotline on solid and hazardous waste requirements (3).

U.S. DISPOSAL TECHNOLOGY

Disposal techniques include biological, chemical, and physical methods (4). The EPA web site states that 214 million tons of hazardous waste (of which 97% was wastewater) were generated in the United States during 1995–1996. Most of the waste (68%) was managed in wastewater treatment units. Eleven percent was stored in deep well injection, less than 0.5% was landfilled, and 2% (4 million tons) was incinerated by high temperature

[a]The material in these articles has been subject to Agency technical and policy review. The views expressed in this article are those of the authors and do not necessarily represent policies of the U.S. Environmental Protection Agency.

combustion. The remaining 36 million tons were subjected to other types of disposal or treatment. EPA's Technology Innovation Office (5) evaluates and publishes information on current disposal research.

NONREGULATORY PESTICIDE COLLECTION AND DISPOSAL PROGRAMS

Many states and counties have established voluntary waste pesticide collection and disposal programs, commonly called Clean Sweeps. These programs provide a simple way for people who have unwanted agricultural pesticides to properly dispose of them at little or no cost to the participants. Each state or county designs its program to fit its own needs and funding sources, so there is no single format for a Clean Sweep program. However, a common format is for a state department of agriculture to organize single-day collection events, where farmers in the area bring their unwanted pesticides to a central site. The state contracts with a waste management company to collect the pesticides and properly dispose of them. In the past decade, more than 22 million pounds of unwanted pesticides have been collected and disposed through Clean Sweep programs nationwide, and 70% of the states have conducted Clean Sweep programs in the past three years. Most pesticides collected by Clean Sweep programs are disposed of in high temperature incinerators.

Clean Sweep programs generally focus on agricultural pesticides, although some programs allow users such as golf courses and road maintenance departments to participate. Household pesticide users in many municipalities can take their unwanted pesticides to household hazardous waste (HHW) collection programs. In 1997, there were more than 3330 HHW collection programs in the United States.

PESTICIDE CONTAINER PROGRAMS AND ALTERNATIVES TO DISPOSAL

Empty pesticide containers can often be recycled, reconditioned, or returned to the manufacturer. Iron and steel scrap metal recyclers may accept metal pesticide containers that have been properly rinsed. Recycling of properly rinsed plastic containers is also possible in many areas. Most states have empty pesticide container collection and recycling programs, which are run in conjunction with the Ag Container Recycling Council (ACRC) (6). The ACRC is a nonprofit cooperative effort and is funded by pesticide manufacturers, packagers, dealers, and distributors and others in the agricultural industry. The organization supports pesticide container collection programs, coordinates with state regulatory agencies and works directly with recycling contractors.

Many pesticide manufacturers and formulators distribute their products in packaging other than the traditional metal drums or plastic jugs to reduce the amount of waste produced each year. For example, pesticides that are used in relatively large volumes are often distributed in refillable and reusable "minibulk" containers, which range in size from 15 gallons to about 300 gallons. Because minibulk containers can be refilled, they can displace hundreds or even thousands of single-use disposable containers. As another example, some pesticides that are formulated as water-dispersible granules are distributed in water-soluble packaging, a plastic film packet surrounding the pesticide. The whole packet is placed in the application equipment and the plastic film dissolves in the water, releasing the pesticide. This type of packaging also has a moisture-protective outer layer, such as a foil pouch, which can be handled as solid waste because it has not been in contact with the pesticide.

ACCUMULATION OF OBSOLETE PESTICIDES WORLDWIDE

Scope of the Problem in Developing Countries

Many countries have accumulated large quantities of pesticides that are obsolete, unwanted or unusable for their intended purpose (to control pests). The Food and Agriculture Organization of the United Nations (FAO) (7) estimates that there are several hundred thousand tons of obsolete pesticide stocks worldwide, with more than 100,000 tons in developing countries. FAO estimates there are 20,000 tons in African countries, at least 60,000 tons in Poland, and more than 20,000 tons in the Ukraine. In Africa in particular, enormous quantities of pesticides have been donated to control outbreaks of migratory pests such as locusts. Some stocks are highly toxic, persistent pesticides that are banned or have deteriorated due to prolonged storage.

Causes of the Problem in Developing Countries

Common causes of pesticide obsolescence and specific examples include:

- Poor local management (inadequate storage facilities, damage during transport, ineffective distribution systems);
- Regulatory policy (pesticide banned while in storage, lack of enforcement infrastructure);

Poor stewardship (manufacturer provides incomplete labels, inappropriate packaging, short shelf-life products in tropical conditions); and

External factors (lower than expected pest incidence, dumping as a pretext of donations, well-meaning donations in inappropriate quantities, aggressive profit motive by vendors).

DISPOSAL OPTIONS IN DEVELOPING COUNTRIES

There are limited options for disposing of pesticide wastes in developing countries. Burial and long-term storage are not recommended due to the high value given to containers, the potential for leakage, and subsequent soil and ground water contamination. Destruction in a dedicated high-temperature incinerator is generally considered to be the safest way to dispose of obsolete pesticides, usually requiring that the pesticides be repackaged and transported in accordance with United Nations recommendations (8) to a country with a hazardous waste destruction facility. However, under certain circumstances, incineration may be accomplished in a cement kiln, which reaches temperatures of up to 4000°F (2200°C). The waste pesticide may be used as fuel in the cement production process if it is combustible, has a significant energy content, and has a low chlorine and metals content.

Although many developing countries have cement kilns, these must be modified and staff trained if pesticide waste is to be used as fuel. Emission and monitoring systems must be installed and waste input must be carefully regulated. Although cement kilns in developing countries (Pakistan, Tanzania) have been retrofitted to successfully burn waste pesticide in pilot projects, community opposition has interfered with such operations in other countries (Mozambique).

FAO has made addressing the problem of obsolete pesticide stocks a priority. They have conducted or supported inventories of stocks in African and Near Eastern countries, developed and distributed videotapes of disposal operations and guidance documents on inventory, prevention and disposal, and conducted training sessions. FAO supports and monitors disposal operations (e.g., Yemen, Zambia, the Seychelles, Uganda), and estimates that it currently costs between $3000–$4500 per ton to remove obsolete stocks from Africa and dispose of them. The first disposal operations in Africa were much more expensive, but experience has increased efficiency and lowered cost. According to FAO, since 1994 approximately 3000 tons of obsolete pesticides have been disposed of in 14 African and 2 Near East countries at a cost of $24.4 million. The cleanup has been mainly funded by the Netherlands, Denmark, Germany, South Africa, the United States, and Finland. At the current rate, FAO calculates that it will take more than 30 years to finish the clean-up of obsolete stocks in Africa and the Near East, not including the more difficult disposal of contaminated soil.

The U.S. EPA developed a training module entitled "Pesticide Disposal in Developing Countries," using FAO guidelines as resource materials. The course has been given to Central American participants in Honduras and national participants in Indonesia, and is included in the EPA Catalogue of International Training Modules (9) with descriptive brochures in English, French, and Spanish available at the EPA web site (10).

The Global Crop Protection Federation (GCPF) (11), an industry association of pesticide manufacturers, has provided technical assistance and funding to pesticide disposal operations in the Gambia, Tanzania, and Madagascar.

PREVENTIVE MEASURES

In order to prevent accumulation of obsolete stocks, governments, manufacturers, industry, aid agencies, development banks, nongovernment organizations, and farmers must work together to follow the FAO guidelines for the safe management and proper storage and disposal of pesticides. Training is needed at all levels, from the highest government officials who approve pesticide purchases and negotiate donations to the individuals supervising the storage area and the workers handling the containers.

In September 2000, FAO, the Organization for Economic Cooperation and Development (OECD), and the United Nations Environment Program (UNEP) organized a workshop on obsolete pesticides, hosted by the EPA in Virginia. Representatives from 30 countries and numerous nongovernmental and industry organizations participated, exploring the causes and solutions to the problem and ways to prevent its recurrence in the future.

To prevent the accumulation of obsolete stocks, actions that governments could take include refusing excessive donations, anticipating the effects of product bans, and maintaining appropriate handling, storage, and transport infrastructures. Actions that aid agencies could take include coordination to ship realistic quantities, promotion of integrated pest management (IPM), and promoting lower toxicity pesticides. Actions industry could take include practicing product stewardship, providing high quality products in durable containers with long-life labels, and practicing FAO's recommendation to take back unwanted or unused products. Actions farmers could take include purchasing appropriate products, package size, and quantity and seeking training on sustainable agricultural methods.

REFERENCES

1. EPA. In *Identifying Your Waste: The Starting Point*; 1997; EPA530-F-97-029. http://www.epa.gov/epaoswer/hazwaste/id/wasteid.pdf (accessed Jan 2001).
2. EPA. In *Understanding the Hazardous Waste Rules—A Handbook for Small Businesses*; 1996; EPA530-K-95-001. http://www.epa.gov/epaoswer/hazwaste/sqg/sqghand.htm (accessed Jan 2001).
3. EPA, Resource Conservation and Recovery Act/Underground Storage Tanks (RCRA/UST), Superfund, and Emergency Planning and Community Right-to-Know Act (EPCRA) Hotline: (703)-412-9810 or (800)-424-9346.
4. Bourke, J.; Felsot, A.; Gilding, T.; Jensen, J.; Seiber, J. *Pesticide Waste Management Technology and Regulation*; American Chemical Society Symposium Series 510, American Chemical Society: Washington, DC, 1992.
5. EPA Technology Innovation Office web site. http://www.epa.gov/swertio1/index.htm (accessed Jan 2001).
6. Ag Container Recyling Council web site. http://www.acrecycle.org (accessed Jan 2001).
7. FAO web site. http://www.fao.org/WAICENT/ FAOINFO/AGRICULT/AGP/AGPP/Pesticid/Disposal/default.htm (accessed Jan 2001).
8. U.N. In *United Nations Recommendations on the Transport of Dangerous Goods*, 11th Ed., United Nations, N.Y., 1999.
9. EPA training catalogue.http://www.epa.gov/oia/modules.htm (accessed Jan 2001).
10. EPA Office of Pesticide Programs, Pesticide Disposal in Developing Countries web site. http://www.epa.gov/oppfead1/international/disposal.htm
11. Global Crop Protection Federation web site. http://www.gcpf.org (accessed Jan 2001).

DOMESTIC ANIMAL IMPACTS

D. Swarup
Indian Veterinary Research Institute, Izatnagar, UP, India

D

INTRODUCTION

Up to 90% of the total consumption of pesticides is used in modern agriculture to protect crops, pastured land, vegetables and fruits, grain, etc., from pests. Pesticides are also used in veterinary practice for treatment and control of tick, mange, lice, and fly infestations. More than 600 different active ingredients were registered in the United States to produce nearly 40,000 different pesticide brands. Of these, some pesticides are selective and effective against a narrowly defined group of organisms (target organisms), whereas some are broad spectrum and toxic to a wide variety of species including nontarget organisms. It is stated that only $<1\%$ of the pesticide applied to crops reaches to the target pest and $>99\%$ enters different components of the environment and contaminates them (1). Thus, despite obvious beneficial effects of boosting food production, indiscriminate and extensive use of pesticides has also caused serious environmental impacts, including health hazards in humans, domestic animals, and wildlife. For example, human deaths due to malaria were reduced significantly from 6 million in 1939 to 2.5 million in the 1970s following introduction of DDT to control malaria vector (2). But, the overuse of DDT was associated with the well-known ecological disaster, the "thin egg-shell syndrome", causing heavy mortality of fish-eating birds in the United States in 1960.

ANIMAL HEALTH IMPACTS

Domestic animals are often raised concurrently with crop production on the same premises. This makes them highly vulnerable to being exposed to farm chemicals. Among all farm chemicals, pesticides, owing to their toxic potential, pose the greatest hazards and are incriminated as the most common cause of poisoning in animals. Pesticides accounted for 85 (17.65%) of the 487 reported cases of poisoning in animals globally during 1986–96 (Table 1). Besides specific acute or chronic clinical toxicities, pesticides are ubiquitous contaminants of animal food products. Further, the animal studies suggest that several pesticides are capable of inducing immunotoxic, carcinogenic, and teratogenic effects.

Insecticide Poisoning

Among all categories of pesticides, synthetic insecticides, represented by organophosphate (OP), carbamate and chlorinated hydrocarbon (organochlorines) compounds, are the major toxicologic concern to animals accounting for nearly 70% of total pesticide poisoning reported in literature between 1986 and 1996. Even in the developed countries death of animals due to insecticide poisoning are quite significant. An economic loss of $160,000 ($20,000 annually) was estimated due to death of 258 farm animals by insecticides during 1982–1989 in Southern Ontario, Canada (3). Grazing recently sprayed areas, intake of insecticide treated herbage or grains, accidental incorporation of granular insecticides in animal feed, and use of concentrations higher than the optimum levels of insecticides in spray, dipping, and pour-on applications meant for direct use in animals to control parasitic infestations are the main sources of insecticide exposure. Frequent repetition of insecticide applications at short intervals and use of oily-based OP products directly on animals is often associated with poisoning. Rooting animals such as pigs often contract toxicity of insecticides used to control soil pests before sowing of maize and potato crops. Fields in which heptachlor has been used to control soil pests remain hazardous to grazing cattle for a considerable period (4). Accidental release of insecticides from manufacturing sources has caused mass poisoning in animals. It is reported that 1047 animals died due to accidental leakage of methyl-iso-cyanide gas from an insecticide manufacturing unit in 1984 in Bhopal (India). The well-publicized nerve gas poisoning (Utah sheep kill

Table 1 Reported incidence by cause of poisoning in animals globally between 1986 and 1996

	No. of incidences		
Cause	Farm animals	Pet animals	Total
Pesticides	44	41	85 (17.45%)
Insecticides	36	24	60
Herbicides	3	–	3
Rodenticides	2	16	18
Other	3	1	4
Organic chemicals and drugs	46	28	74 (15.19%)
Inorganic chemicals	102	14	116 (23.81%)
Plants, mycotoxins, and zootoxins	190	22	212 (43.53%)
Overall	382	105	487

(From Index Veterinarius 1987, 55; 1990–92, 58–60; 1995–96, 62–64; and Veterinary Bulletin 1988–89, 58–59; and 1994, 64.)

syndrome) in the United States in 1969 was also an outcome of accidental release of OP from an army unit. OP insecticides are highly toxic and all species of animals are susceptible to them. Carbamates are potentially less dangerous than OPs. Acute poisoning by DDT and benzene hexachloride is rare due to environmental pollution, but aldrin and dieldrin have caused fatal poisoning in animals.

Herbicide Toxicity

Due to increasing popularity of chemical control of noxious weeds in the past three decades, the quantity of herbicides used in agriculture exceeds that of insecticides and frequency of herbicide poisoning in animals appears to be greater than insecticide poisoning, mainly in developed countries. For example, herbicides accounted for 26.4% and insecticides for 20.1% of the cases of pesticide poisoning in animals in France (5). Accidental ingestion of concentrated chemicals in undiluted form due to improper storage or careless disposal of herbicide containers exposes animals to herbicides, generally. Water contaminated by runoff from agricultural application and purposely or inadvertently contaminated forage may be hazardous, but poisoning by these sources is rare. On an average proper application rates of generally 0.45–4.55 kg herbicide per hectare provides the maximum dosage of 7–70 mg per kg body weight to animals being fed on treated forage; whereas the toxic dosage of the most herbicides range between 10 and 250 mg per kg body weight. However, highly toxic herbicides such as CDAA (2-chloro *N,N*-diallyl-di-2-propenylacetamide), bensulide, pebulate, EPTC (*S*-ethyl dipropyl thiocarbamate), atrazin, and simazin have been reported toxic to animals even at normal rates of application (6).

Rodenticide Toxicity

Some rodenticides are highly toxic to animals and instances of poisoning due to strychnine, fluoroacetate, warfarin, and zinc phosphide are encountered due to accidental exposure in farm animals. Instances of poisoning in dogs from cholecalciferol rodent bait are also reported. If handled properly, most rodenticide applications are not hazardous to animals. Even the ingestion of baits by cattle normally does not cause poisoning because small packages of baits do not contain the quantity of rodenticides sufficient to induce poisoning. However, in some countries such as France, rodenticides accounted for the highest percentage (42.5%) of all pesticide poisoning cases in animals (5).

PESTICIDE RESIDUES IN ANIMAL FOOD PRODUCTS

Extensive use of pesticides has precipitated the problem of pesticide residues in food commodities, particularly those of animal origin. Many pesticides, especially organochlorine insecticides, are chemically more stable and persist in the environment. They have affinity to fat and tend to bioconcentrate in the fat depots in the animal body. Milk and milk products from the exposed animals are more frequently contaminated, because lactating animals excrete an appreciable concentration of ingested/stored pesticide in the milk. A report of the U.S. Department of Agriculture described polyhalogenated hy-drocarbons including pesticides such as heptachlor and dieldrin as the second most contentious chemical residues in foods of animal origin (7). Interestingly, EPA banned dieldrin from use in the United States in 1974.

Despite low consumption of pesticides (nearly 20% of the total world production), the magnitude of pesticide residues in food items in general and animal products in particular is very high in developing countries. In India, 37% of milk samples were reported to contain DDT and HCH residues in concentration higher than maximum residue levels of 0.05 and 0.01 mg/kg, respectively (8) and in Egypt, lindane (62.1%) and beta-HCH (51.7%) were the most frequently distributed pesticide contaminants of milk during 1988–1990 (9). The presence of pesticide residues in animal products is a potential source of the entry of pesticides or their metabolite in human diet. However, recent reports indicate global decline in the level of pesticide contamination probably due to change in pesticide usage practices.

FUTURE NEEDS

The concept of Integrated Pest Management (IPM) has provided approaches to control ''pests'' in an eco-friendly manner. Advances in plant breeding to evolve pest-resistant crop varieties are aimed at reducing the demand of chemical pesticides. Biopesticides, which occur naturally or are derived from natural products including products of genetic engineering, are rapidly biodegradable and possess better specificity. They are likely to offer the best alternative to conventional synthetic chemical pesticides with low risk of exposure of nontarget groups and long-term contamination of the environment. It is also extremely important to avoid use of ''banned'' insecticides directly for animal treatment. Raising insect/disease resistant breeds of animals has been successful in reducing use of insecticides in veterinary practice. For example, introduction of cattle breeds with 50% *Bos indicus* blood in most tick-infested areas of Australia was reported beneficial in reducing frequency of chemical insecticide use to control ticks (4).

REFERENCES

1. Dhaliwal, G.S.; Arora, R. *Principles of Insect Pest Management*, 1st Ed.; National Agricultural Technology Information Centre: Ludhiana, India, 1996; 374.
2. Harte, J.; Holdren, C.; Richard, S.; Shirley, C. *Toxics A–Z: A Guide to Everyday Pollution Hazards*; University of California Press: Berkeley, CA, Bishen Singh Mahendra Pal Singh (Indian Edition): Dehradun, India, 1993; 112–140.
3. Frank, R.; Braun, H.E.; Wilkie, I.; Ewing, R.A. A review of insecticide poisoning among domestic livestock in Southern Ontario, Canada 1982–89. Can. Vet. J. **1991**, *32* (4), 219–223, 226.
4. Radostits, O.M.; Blood, D.C.; Gay, G.C. *Veterinary Medicine*; Baillière Tindall: London, 1994; 1507–1524.
5. Lorgue, G.; Lechenet, J.; Riviere, A. *Clinical Veterinary Toxicology*; Blackwell Science: Oxford, 1996; 210.
6. Palmer, J.S. Organic Herbicides. In *Current Veterinary Therapy: Food Animal Practice*; Howard, J.L., Ed., 1st Ed., W.B. Saunders Co.: Philadelphia, 1984; 481–487.
7. United States Department of Agriculture. In *Domestic Residue Data Book. National Residue Programme, 1998–89*; Food Safety and Inspection Service, USDA: Washington, DC, 1990.
8. Swarup, D.; Dwivedi, S.K. Research on effects of pollution in livestock. Indian J. Anim. Sci. **1998**, *68 (8)*, 814–824.
9. Fayed, A.E.; Zidan, Z.A.; Abou-Arab, A.A.K.; El-Nockrashy, S.A. Incidence of some environmental pollutants in milk and its products at great Cairo markets. Ann. Agric. Sci. Cairo. **1993**, *1* (special issue), 85–95.

DORMANT TREATMENTS

Frank G. Zalom
University of California, Davis, California, U.S.A.

INTRODUCTION

Deciduous perennial plants are hosts for many pests that persist from year to year. If permitted to survive unchecked, some of these pest species will increase to damaging levels through time. Plant diseases (1) and arthropods (2) including scale insects and certain species of mites are examples of pests that commonly persist on perennial crop hosts throughout the year. Pests that overwinter on their deciduous host plant are typically synchronized to one life stage, and are concentrated on the bark or buds. If the overwintering life stage is susceptible to pesticides, treatment applied during the dormant season can provide effective control. Dormant sprays are used both in agricultural crops and for pests of ornamental plants in the landscape.

ADVANTAGES OF DORMANT SPRAYS

Spraying during dormancy has several advantages (3):

- Greater synchrony of life stages makes it possible to control a pest with a single treatment.
- Coverage of the tree bark is more complete than is possible during the growing season when leaves on the tree prevent materials sprayed at the tree from reaching the interior leaves and bark, therefore better control is achieved.
- Beneficial arthropods are less affected by sprays applied during the dormant period as many species leave the plant to seek refuge for the winter, or overwinter as a form less susceptible to the pesticides.
- Several pest species may be controlled simultaneously with a single application.
- Fewer conflicts exist with other cultural practices, making precise timing of applications less important than during the growing season.
- Other than pruning and orchard sanitation, fewer workers enter fields during the dormant season resulting in reduced exposure to treatments.
- Because there is no crop on the plants, pesticide residues are not an issue.

The advantages of dormant sprays for specific diseases and arthropods lead to their widespread use in many parts of the United States and in several areas of the world. Dormant sprays were considered a good integrated pest management practice because the single preventative spray applied to control diseases or arthropods in the dormant season reduced the number of in-season applications that were often applied by farmers, while preserving natural enemies and avoiding certain worker exposures.

DISEASES CONTROLLED BY DORMANT SPRAYS

Liquid lime sulfur is applied as a dormant spray for the control of many plant diseases. Perhaps its most widespread use is for control of peach leaf curl. Peach leaf curl is ubiquitous on peaches, causing blistering and deformation of leaves, premature defoliation, and weakening of trees. One dormant season application is sufficient in many growing areas, but a second treatment is needed in warmer regions with high rainfall or where leaf curl has become an increasing problem (1). Fixed copper fungicides or copper applied as a Bordeaux mixture during dormancy will also control peach leaf curl, as will several conventional fungicides including chlorothalonil and ziram. A related disease on plums referred to as plum pockets, which causes blisters on fruit that enlarge to form a hollow cavity within the fruit, is also controlled by a dormant lime sulfur spray. Anthracnose of cane berries and grapes; spur blight of cane berries, blueberries, and Phomopsis cane; and leaf spot and twig blight of grapes and blueberries are also controlled by lime sulfur sprays applied just before buds begin to break in the spring. Development of powdery mildew epidemics on grapes in New York State has been delayed by over-the-trellis sprays of lime sulfur during dormancy. In the ornamental landscape, lime sulfur is recommended as an eradication

Encyclopedia of Pest Management

treatment for control of overwintering spores of rose canker, black spot, rust, and powdery mildew on roses, and is applied before buds begin to swell.

Copper fungicides (Bordeaux mixture, copper hydroxide, and other forms) are also used during dormancy for control of certain fungal and bacterial diseases. Coryneum blight of peaches and cherries and bacterial canker, in addition to peach leaf curl, is such an example.

Sometimes dormant sprays must be used in combination with other control approaches to be effective. Black knot on cherry and plum is characterized by production of elongated swellings or knots on the limbs of susceptible tissue. It is controlled by employing a combination of canker removal by pruning and a dormant application of lime sulfur. Fireblight is a serious bacterial disease of apples and pears that kills blossoms, branches and limbs, and entire trees of susceptible varieties in seasons when the disease is especially severe. Pruning infested tissue or removing severely infected trees is critical for control of the disease, but a dormant spray application of a copper fungicide will help to control the disease.

ARTHROPODS CONTROLLED BY DORMANT SPRAYS

Horticultural oils applied as dormant sprays can control many important species of scale insects, certain species of phytophagous mite eggs, and eggs of some aphids. Several types of oil are available, with ''superior'' oils having a viscosity rating of 60–70 being considered safer to use on plants than the heavier dormant oils with a rating of 90 or above that were used many years ago. When sufficient oil is utilized and excellent coverage is achieved, a single annual dormant spray can control moderate populations of these arthropods without further treatment. Scale insects controlled by oil include San Jose scale, white peach scale, and Italian pear scale in addition to soft scales. When populations of these scales are high or when conditions dictate that lower rates of oil be used, an organophosphate insecticide is added to the dormant spray to improve efficacy. Mite and aphid species that overwinter on tree bark or buds as eggs are suppressed by dormant oil applications. Economically damaging species controlled by a single dormant oil spray include European red mite and brown mite. Pear rust mite females overwinter beneath bud scales, and are controlled by dormant lime sulfur sprays.

Peach twig borer is a serious pest of stone fruit and almonds. Its larva overwinters in a frass tube called a hibernacula. Where this insect occurs at damaging levels, a majority of orchards are treated in the dormant season with oil in combination with an insecticide. Organophosphates have been used for this purpose for many years, but pyrethroids and insecticides of other chemical classes may also be used. Pear psylla and certain mealybug species such as grape mealybug may also be suppressed with dormant sprays of oil applied in combination with organophosphates or other insecticides.

PROBLEMS ASSOCIATED WITH DORMANT SPRAYS

Problems reported with dormant sprays include lack of efficacy, phytotoxicity, and environmental issues. Problems with efficacy are usually associated with improper coverage. Coverage is better with ground sprays than aerial sprays. Ground applications of oil at higher volumes of water (3500 L/ha) often provide better control than do lower-volume applications. When aerial sprays are applied, the sprays must target deposition on branches in the plant canopy. Studies have shown that improper nozzling, speed, and elevation can result in as much as 90% of the pesticide applied being deposited on the ground rather than on the plant. Mineral oils, if improperly applied during the dormant season, can cause terminal and/or branch dieback that will be apparent after leaves appear in the spring. Such phytotoxicity occurs when plants are water stressed due to lack of rainfall or irrigation during the winter prior to application, or when oils are applied when temperatures are below freezing. Oils applied in combination with lime sulfur can also be quite phytotoxic when used as a single spray. Organophosphates applied as dormant sprays have been associated with air and water quality concerns because they occur coincidentally. For example, organophosphates have been found in the thick fog that occurs in the winter in California's San Joaquin Valley (4). Diazinon and chlorpyriphos, the primary organophosphates used as dormant sprays, have been found following winter rains in the San Joaquin and Sacramento River drainages (5). Because dormant sprays are applied at a time of the year when rainfall may occur, there is always a concern for runoff. Practices that keep pesticides on site and from moving into surface waters should be used.

ALTERNATIVES TO DORMANT SPRAYS

Dormant sprays are a preventative pest management approach. As such they are not traditionally applied with existing knowledge of pest densities. For certain arthropod pests, history of pest incidence and monitoring during the dormant season may indicate that a dormant spray is not necessary (3). For example, if scales or insects such as peach twig borer have not been a problem, monitoring prunings for scales or peach twig borer hibernaculae in the winter may indicate that damage is unlikely to occur from resident populations. Applying disease or arthropod control treatments in-season based on risk assessment and/or monitoring of pest densities is also possible, but multiple applications may be necessary and other advantages associated with the use of dormant sprays would be lost.

For specific pests, management options that do not require conventional dormant sprays are possible. Perhaps the best example is peach twig borer, that can be controlled with *Bacillus thuringiensis* (Bt) sprays applied during bloom (6). When high label rates of oil are applied during the dormant season together with bloom-time Bt sprays, dormant and in-season sprays may not be needed to prevent economic damage. Mating disruption using sex pheromones is another approach that is possible for control of peach twig borer.

It has been recognized for many years that the best time to treat peach twig borer and San Jose scale with an organophosphate (OP) insecticide (diazinon, chlorpyriphos, methidathion, phosmet, and others) and oil mixture is during dormancy.

REFERENCES

1. Ogawa, J.M.; Zehr, E.I.; Bird, G.W.; Ritchie, D.F.; Uriu, K.; Uyemoto, J.K. *Compendium of Stone Fruit Diseases*; APS Press: St. Paul, MN, 1995; 98.
2. Rice, R.E.; Jones, R.A.; Black, J.H. Dormant sprays with experimental insecticides for control of peach twig borer. California Agric. **1972**, *26* (1), 14.
3. University of California. *Integrated Pest Management for Stone Fruits*; Publication 3389, University of California Division of Agriculture and Natural Resources: Oakland, CA, 1999; 264.
4. Glofelty, D.E.; Seiber, J.N.; Liljedahl, L.A. Pesticides in fog. Nature **1987**, *325*, 602–605.
5. Domagalski, J.L.; Dubrovsky, N.M.; Kratzer, C.R.J. Environ. Qual. **1997**, *26*, 454–465.
6. Barnett, W.W.; Edstrom, J.P.; Coviello, R.L.; Zalom, F.G. Insect pathogen bt controls peach twig borer on fruits and almonds. California Agric. **1993**, *47* (5), 4–6.

DURABLE RESISTANCE IN CROPS TO PESTS

Angharad M.R. Gatehouse
University of Newcastle, Newcastle upon Tyne, United Kingdom

INTRODUCTION

Worldwide an estimated 30% of crops grown are lost to pests and diseases. A fundamental objective of plant breeders is therefore to produce crops that are inherently resistant to attack by pests and diseases, and which will form a major component of any integrated pest management (IPM) program (1, 2). In order to increase the effective lifespan of resistant crops it is desirable that resistance should be durable, that is, sustainable. Durable resistance can be achieved in a number of different ways, and although it is normally associated with polygenic resistance, major gene resistance can also be used to achieve this end. Possible strategies involve, for example, particular spatial and temporal patterns of resistance deployment. Alternatively, and perhaps of greater importance, is the use of gene pyramiding to produce durable resistance in crops.

CROP LOSSES

One of the major constraints on agricultural production worldwide is that of loss due to insect pests and plant pathogens, responsible for estimated crop losses of up to 30%. Thus, crop protection must play a vital and integral role in modern-day agricultural production. Recent figures show that the supply of food is rapidly being outstripped by demand for it, with a projected increase in world population of 9–10 billion (3) over the next four decades; thus an immediate priority is to achieve maximum production of food and other products. However, the ever-increasing demands on yield and intensification of farming practice, particularly in the developed world, have increased the problem of pest damage and hence control. It has long been recognized that extensive cultivation of certain crops, to the exclusion of other plants, that is, monoculture, may favor drastic increases in pests that feed upon these crops. Not only does this situation occur in the field, but also during storage where large quantities of seed or other products ensure that any pest that can utilize them as a food source is almost certain to undergo a population explosion, thus resulting in significant damage or loss. Although the buildup of pests to a specific host plant is generally reduced in diverse plant communities, agricultural practices such as combining crops where each acts as host for the same pest, can have the opposite effect, thereby exacerbating the situation.

CROP PROTECTION

At present crop protection in intensive agricultural systems relies predominantly on the use of agrochemicals, although cases exist where host plant resistance and biological control are successfully employed. Insecticide usage by pest order shows that Lepidoptera (37%) are the major target for insecticides, followed by Homoptera (26%), Coleoptera (11%), and mites (11%) (4). Although the agrochemical industry is investing in the production of safer and more environmentally benign pesticides, much higher levels of protection are still required. Both conventional breeding and biotechnology are very important to this end. It is interesting to note that there has been a 600-fold return for research investment in varietal resistance compared to only a fivefold return for chemical pesticides. Thus, the enlightened grower is looking toward integrated pest management that combines practices such as the judicious use of pesticides, crop rotation, field sanitation, but above all, the use of pest resistant cultivars.

TYPES OF RESISTANCE

Genetic host plant resistance may be defined as the collective heritable characteristics by which that plant reduces the possibility of successful utilization as a host by a pest species. Thus, in agronomic terms, it represents the inherent ability of crop plants to limit or overcome pest infestations, thereby improving the yield and or quality of the harvestable crop product. Resistance may be governed either by polygenes or major genes (5). Polygenic (horizontal) resistance is a quantitative trait governed by a large number of genes, with the individual genes each making a small contribution to the overall levels of resistance; resistance levels do not usually approach the very high levels often found with major gene resistance, but can nevertheless reduce or eliminate the need for pesticides (2). Moreover, as no single gene exerts a high selection pressure on the pest, resistance is likely to be more stable than with major gene resistance. Major gene (vertical) resistance, on the other hand, is exhibited by a gene-for-gene relationship between genes for resistance in the host and genes for virulence in the pest and involves simple inheritance of qualitative traits; such forms of resistance exhibit high levels of resistance and tend to exert high levels of selection pressure on pest populations.

PYRAMIDING GENES FOR DURABLE RESISTANCE

A major objective of plant breeding for varietal resistance in crops is to produce high levels of sustainable resistance that is compatible with optimizing crop yield and quality. Given the time and expense required in breeding resistance genes into crop plants without losing other desirable agronomic traits by conventional breeding techniques, it is necessary, wherever possible, that resistance should be durable. In this context, the emergence of technologies that have allowed plants to be transformed stably with foreign genes are beginning to make a major contribution to plant breeding (3, 4).

Several strategies have been proposed for achieving durable resistance, and while both polygenic and major gene resistance can be used to achieve this end, durable resistance is more often associated with the former. Possible strategies involve particular spatial or temporal patterns of resistance deployment. Spatial patterns include variety mixtures or multilines using either different types of resistant varieties or resistant and susceptible varieties (6). While the use of variety mixtures, each having different resistance genes, will enhance genetic heterogeneity, and thus reduce the potential for development of resistance in pest populations, a disadvantage is that all other characteristics need to be uniform such as maturation time, phenology, yield and quality, et cetera. Such otherwise isogenic lines are characteristic of transgenic host plant resistance. Spatial patterns of resistance deployment have been successfully adopted for the growing of insect resistant transgenic crops expressing genes encoding δ-endotoxins from Bacillus thuringiensis (Bt) (7). In this instance the transgenic crop is bordered by a refuge of a susceptible, nontreated variety of the crop. This strategy, together with a high "dose," that is, high expression levels of the endotoxin, should help delay the buildup of resistance in the pest population. Temporal patterns of resistance deployment include variety rotations from one season to the next, or the substitution of one variety for another, but only when virulent genotypes start to increase within the pest population.

Another strategy advanced for producing crops with durable resistance to pests is that of pyramiding genes for resistance (5). This involves the combination of two or more major genes into the same variety. Such varieties are likely to have a longer useful lifespan (i.e., be more durable) as the development of new pest biotypes will be slower. For example, it has been estimated that simultaneous release of two genes for resistance to the Hessian fly of wheat would result in better durability than sequential release. This strategy has been employed in order to produce rice with durable resistance to rice brown planthopper. However, a major limitation of this approach, particularly when using conventional plant breeding techniques, is the ability to distinguish plants containing both genes as opposed to those containing one or the other in a breeding program; this can be achieved by tagging the genes of interest with molecular markers.

The pyramiding of genes expressing agriculturally desirable traits, including that of durable resistance, is perhaps more readily achieved via recombinant DNA technology. In order to increase the effective lifespan of Bt genes in crop protection, it has been proposed that two or more such genes should be incorporated into the target crop in question, particularly for resistance to lepidopteran pests; however recent work has demonstrated that this approach may not necessarily be more advantageous in that cross-resistance can occur. For example, one gene in diamondback moth confers resistance to four Bt toxins (8). However, this technology can readily be used to increase the protective efficacy, spectrum of activity, and durability of resistance by introducing cassettes of different genes into crops. It is important that the individual gene products introduced should each act against different targets within the insect pest, thus mimicking the multimechanistic resistance that occurs in nature. Among those gene products being developed for control of insect pests, protease inhibitors should be particularly valuable in this respect since, apart from their inherent insecticidal effects, they would protect other gene products from premature digestion in the insect gut. For example, trypsin inhibitors have been shown to have a marked potentiating effect on Bt toxins.

The first demonstration of gene pyramiding for insect resistance using molecular techniques was the introduction into tobacco of both a trypsin inhibitor gene from cowpea (CpTI) and a lectin gene from pea that resulted in significantly enhanced levels of resistance compared to the performance of either individual gene product (4). In this example the two genes were engineered independently into different lines, the progeny of which were subsequently crossed. However, later examples of pyramiding genes to confer insect resistance have been introduced using constructs containing the different coding sequences. Such examples include expression of two or more different plant derived genes such as those encoding snowdrop lectin (GNA) and chitinases (BCH) for enhanced aphid resistance (9). Other examples include expression of up to two different Bt genes (Cry1Ac, Cry2A) together with GNA in rice, resulting in 100% mortality of two lepidopteran pests of rice. The use of fusion proteins is another strategy being used to introduce multimechanistic resistance into crops. It must be emphasized that proof of concept for pyramiding transgenes for durable resistance has yet to be demonstrated in the field. However, given that this approach works for conventional plant breeding, there is no reason why it should not be equally successful for transgenic crops.

REFERENCES

1. Smith, C.M. Plant Resistance to Insects. In *Biological and Biotechnological Control of Insect Pests*; Rechcigl, J.E., Rechcigl, N.A., Eds.; CRC Press: Boca Raton, FL, 1999; 171–208.
2. Van Emden, H.F. In *The Interaction of Host Plant Resistance with Other Control Measures*, Proceedings of Brighton Crop Protection Conference, Farnham, U.K., 1990; 3, 939–949.
3. Gatehouse, J.A.; Gatehouse, A.M.R. Genetic Engineering of Plants for Insect Resistance. In *Biological and Biotechnological Control of Insect Pests*; Rechcigl, J.E., Rechcigl, N.A., Eds.; CRC Press: Boca Raton, FL, 1999; 211–241.
4. Gatehouse, A.M.R. Biotechnological Applications of Plant Genes. In *Global Plant Genetic Resources for Insect Resistant Crops*; Clement, S.L., Quisenberry, S.S., Eds.; CRC Press: Boca Raton, FL, 1998; 263–280.
5. Panda, N.; Khush, G.S. *Host Plant Resistance to Insects*; CAB International: Wallingford, U.K., 1995; 431.
6. Wilhoit, L.R. Evolution of Herbivore Virulence to Plant Resistance: Influence of Variety Mixtures. In *Plant Resistance to Herbivores and Pathogens; Ecology, Evolution and Genetics*; Fritz, S.R., Simms, E.L., Eds.; Chicago University Press: Chicago, 1992; 91–120.
7. Gould, F. Sustainability of transgenic insecticidal cultivars: integrated pest genetics and ecology. Annu. Rev. Entomol **1998**, *43*, 701–726.
8. Tabashnik, B.E.; Liu, Y.-B.; Finson, N.; Masson, L.; Heckel, D.G. In *One Gene in Diamondback Moth Confers Resistance to Four Bacillus thuringiensis* Toxins, Proceedings of the National Academy of Sciences U.S.A., 1997; 94, 1640–1644.
9. Gatehouse, A.M.R.; Davison, G.M.; Newell, C.A.; Merryweather, A.; Hamilton, W.D.O.; Burgess, E.P.J.; Gilbert, R.J.C.; Gatehouse, J.A. Transgenic potato plants with enhanced resistance to the tomato moth *Lacanobia oleracea* growth room trials, Molec. Breeding **1997**, *3*, 49–63.
10. Dent, D. *Insect Pest Management*; CAB International: Wallingford, U.K., 1991; 604.
11. Smith, C.M. *Plant Resistance to Insects: A Fundamental Approach*; John Wiley & Sons: New York, 1989.

ECOLOGICAL ASPECTS OF PEST MANAGEMENT

E

David J. Horn
Ohio State University, Columbus, Ohio, U.S.A.

INTRODUCTION

Every animal and plant population is surrounded by an interactive biotic and physical environment, and these enormously complex interactions often stabilize population densities to reduce the probability of a pest outbreak. However, ecological interactions are often disrupted, simplified, or overridden in managed ecosystems, reducing the impact of naturally-occurring pest population regulation and leading to outbreaks. Understanding how ecological interactions may either cause or prevent pest outbreaks can lead to crop and landscape management activities that achieve relative stability of pest populations below damaging levels without resorting to widespread and environmentally disruptive intervention. Increasing environmental complexity can enhance pest management. Intelligent environmental management with due regard for the place of a pest within a complex and interconnected ecosystem reduces pest outbreaks when attention is given to 1) increasing species diversity within cropping or landscaped ecosystems, 2) crop rotations and use of short-term, rapidly-maturing cultivars, 3) reducing field size and encouraging intervening areas of uncultivated and undisrupted vegetation, 4) reduced tillage and tolerance of weedy backgrounds, and 5) increased genetic diversity within the crop.

PEST POPULATION DYNAMICS AND SPECIES DIVERSITY

Apparently simple ecosystems such as annual agricultural crop monocultures are complex, and the relative impact of alternate crops, weeds, natural enemies, competitors, and associated organisms on pests may be highly variable. Pest populations and their effective environments are constantly changing in space and time, and events impacting one population at one location and time interval usually do not duplicate events in the same population at another time and place. Despite ecosystemcomplexity, seasonal agricultural crops that are periodically disrupted due to harvesting and tilling may never achieve a steady, sustained state typical of later stages in ecological succession. Annual or more frequent disruption selects for pests that can locate and exploit resources quickly and efficiently. Colonizing species of plants and insects display rapid dispersal and an ability to increase numbers quickly when suitable habitat is located. The ancestors of many crop plants are typical of early stages of ecological succession, as are their associated insect pests. Conventional agriculture invites early-successional species that are likely to undergo uncontrolled population outbreaks. Populations of such pests are rarely in equilibrium at any given place and time, but are maintained by a loose balance of colonization and extinction over a large geographical area. Equilibrium is more likely to be reached in populations occupying longer-lasting ecosystems such as orchards and forests. Population fluctuations in these more complex ecosystems are partly buffered by the complex interactions within food webs, so there is less likelihood of outbreak of any particular pest species.

Species diversity is formally measured as an index combining numbers and proportion of each species present in an ecosystem. A species diversity index reflects the number of links in a food web, and the overall stability of any ecosystem is partly a function of the number of interactions among plant, pests, natural enemies, and pathogens. Ecologists and pest managers debate over whether there exists a direct relationship between species diversity and stability of individual populations within an ecosystem. It is often presumed that pest outbreaks are suppressed in more complex (and therefore more diverse) ecosystems. The so-called "diversity–stability hypothesis" holds that ecological communities with a higher species diversity are more stable because outbreaks of pest species are ameliorated by the checks and balances and alternative pathways that exist within a large and integrated food web. Evidence supporting this view comes from experiments that mix several plant species with the primary host of a specialist herbivore, resulting in reduced populations of the specialist herbivore. This observation is termed the *resource concentration hypothesis*, which holds that insect herbivores are more likely to locate and to remain on hosts growing in dense or pure stands, and the most specialized species frequently attain highest densities in ecosystems with low plant species diversity. Biomass becomes concentrated in a few species, with a concomitant decrease in species diversity of herbivores in monocultures. Increases in herbivore populations in crop monocultures generally result from higher rates of colonization and reproduction along with reductions in dispersal, predation, and parasitism. As species diversity increases in agroecosystems, more internal links result within food webs and these links promote greater stability, resulting in fewer pest outbreaks. Structural diversity is an important physical component of overall diversity; for instance, cropping systems with taller plants (such as corn among beans and squash) present more physical space to arthropods and this increases the variety of prey and provides greater shelter for predators.

MONOCULTURE AND POLYCULTURE

Monoculture is the planting of a single species of crop plant, which often results in increased populations of specialist herbivores, a result consistent with the resource concentration hypothesis. Polyculture, the planting of more than one crop species in the same local area (often the same field), may reduce impact of herbivorous pests because the presence of a variety of plants disrupts orientation of specialist herbivores to their hosts. For example, cabbage flea beetles and cabbage aphids that locate their hosts via specific chemical cues (such as the alkaloid sinigrin) are less effective in locating their host plants when these are intermingled with a variety of other plant species. The result is lower populations of the flea beetles and aphids. Local movement of cucumber beetles and lady beetles is enhanced when cucumbers are interplanted with corn and beans as compared with these insects' movement in monocultures, where they tend to remain on individual plants. Numbers of specialist herbivores (cabbage aphids, diamondback moth, and imported cabbageworm) on collards planted in weedy backgrounds are lower than populations of these same herbivores on collards planted against bare soil or plastic mulch. This influence of weeds intensifies once the weeds become as tall as the collards, effectively allowing the collards to escape herbivory by concealment among the weeds. The frequent replacement of one crop by another in crop rotation maintains populations of specialist herbivores below damaging levels. Field crop producers in the midwestern U.S. can prevent the increase of corn rootworm populations by rotating from corn to soybeans every 2 or 3 years.

OPEN AND CLOSED ECOSYSTEMS

Ecosystems may be considered to be *open* (subsidized) or *closed* depending on the amount of nutrient and energy exchange with ecosystems outside themselves. Open ecosystems are characterized by regular input of energy and nutrients, followed by removal of a large proportion of nutrients. In maize fields in the midwestern U.S. heavy importation of mineral fertilizer occurs at planting and subsequent energy is input when the crop is tilled and pesticides are applied. Most of the nutrients in a maize field are removed at harvest, as yield or crop residue. In landscaping towns and suburbs we fertilize (providing input) and rake leaves and remove mowed grass (exporting nutrients) to maintain a pleasing appearance.

The assemblage of species within a maize field is artificial, with novel interspecific associations. Many of the major insect pests are of exotic origin, and the association with maize is relatively recent. For instance, the European corn borer arrived in North America around 1910, before which maize (native to Mesoamerica) had no ecological association with the corn borer. It may take many years for native natural enemies to expand their host or prey range to include exotic organisms. Most anthropogenic agroecosystems and landscaped ecosystems are artificial assemblages and open ecosystems. The species assemblage in planned landscapes often includes a preponderance of exotic species.

In a closed ecosystem, such as a deciduous forest, most nutrient movement remains localized with little import and export. For example, the nutrients and energy in the forest canopy fall to the ground as leaves and frass, or leaves are converted into caterpillars, which in turn are eaten by insectivorous birds, predatory insects, and parasitic wasps. There is little ''leakage'' of nutrients from closed ecosystems into surrounding environments. The species assemblages of many closed ecosystems have been associated for millennia, resulting in multiple trophic links and close ecological associations among mostly native species. Such ecosystems are relatively ''immune'' to invasion by exotic species (although there are exceptions, such as the successful invasion of the gypsy moth into the forests of eastern North America).

EXAMPLES

The Mexican bean beetle overwinters in hedgerows and along field edges, so that soybean fields nearest overwintering sites are likely to become infested earlier and bean beetle populations subsequently will be higher. Soybeans located near bush and pole beans (which are more suitable hosts for the bean beetle) are also likely to develop economically damaging infestations earlier. Natural enemies often move from unmanaged field edges and nearby hedgerows and forests into adjacent farm fields and the nature of this movement may be very important to local suppression of pests. Many studies have shown that there are increased numbers and activity of natural enemies near field borders when there is sufficient natural habitat to provide cover and alternate prey and hosts, as well as food in the form of nectar and pollen. This function of wild border areas significantly enhances biological control.

Weedy vegetation in or near crop fields support a diverse fauna, including natural enemies of pests on the crop plants. This depends on the species of weeds present; for instance, if the weeds were particularly attractive to aphids and their natural enemies, this enhances aphid control on a commercial crop. Weeds such as pigweed (*Amaranthus*), lambs quarters (*Chenopodium*), and shepherd's purse (*Capsella*) when heavily infested with aphids serve as "nurseries" for production of aphid predators and parasitoids, which move onto neighboring crop plants when these are located near these weeds.

Growing several crops in the same space reduces pest problems relative to monocultures of the same species. When blackberries are planted among grapevines in central California, the parasitoid *Anagrus epos* attacks the eggs of both grape leafhopper and the leafhopper *Dikrella cruentata* on blackberry. By encouraging blackberries between alternate grape arbors, a constant supply of eggs of both leafhoppers are available to the parasitoid, which persists in populations high enough to bring the grape leafhopper under biological control.

Floral undergrowth in orchards provides resources to adult parasitic wasps and flies and increases parasitism of phytophagous insects (particularly Lepidoptera) on the trees. The presence of nectar and pollen along with alternate prey is particularly favorable to populations of generalist predators such as the lady beetle *Coleomegilla maculata* resulting in lower aphid populations. In relay cropping, two (or more) different crops are grown on the same area in successive seasons. The seasonal change from one crop to another prevents the increase of specialist pests especially if the crops are evolutionarily distantly related (such as legumes and grasses). When soybeans are relay cropped after winter wheat, pests of soybeans are less abundant than when soybeans are cropped alone.

Planting and harvesting corn and beans in alternating plots rather than solid monocultures reduces pest numbers on both crops, and this management approach is widely practiced in traditional agriculture. Where alfalfa can be grown throughout the year it is possible to harvest on a 3–4 week rotation when half the field is cut in strips. Natural enemies of the alfalfa weevil, alfalfa caterpillar, and aphids are conserved in the regrowth, and there are alternative food sources and hiding places for these predators and parasitoids all year, so they are always present to suppress pest populations below damaging levels. By planting alfalfa adjacent to cotton, control of *Lygus* bugs is achieved by allowing increase of natural enemies of *Lygus* in the alfalfa. These natural enemies move into the cotton and control *Lygus* bugs there.

BIBLIOGRAPHY

1. Altieri, M.A. *Biodiversity and Pest Management in Agroecosystems*; Food Products Press: New York, NY, 1994; 185.
2. Collins, W.W.; Qualset, C.O. *Biodiversity in Agroecosystems*; CRC Press: Boca Raton, FL, 1999; 334.
3. Horn, D.J. *Ecological Approach to Pest Management*; Guilford Press: New York, NY, 1988; 285.
4. National Research Council. *Ecologically Based Pest Management: New Solutions for a New Century*; National Research Council: Washington, DC, 1996; 160.

ECOLOGICAL CAUSES OF PEST OUTBREAKS

Mark D. Hunter
University of Georgia, Athens, Georgia, U.S.A.

INTRODUCTION

The population size of any organism can change only by four processes; birth, death, immigration, or emigration. It follows that the ecological causes of pest outbreaks must operate through increases in rates of birth, decreases in rates of death, increases in immigration, or decreases in emigration. These processes are not mutually exclusive and the growth of a pest population can arise, for example, because both rates of mortality and rates of emigration decline. Although changes in these four critical rates are central to pest outbreaks, it remains a challenge to untangle the various ecological factors that cause changes in the numbers of pests in space and time. The challenge arises, in part, because so many different ecological forces can impinge upon the population dynamics of organisms. Variation in soil quality, predation pressure, agents of disease, the density of competitors, levels of precipitation, fluctuations in temperature, and the quality of food are just some of the factors that can influence rates of birth, death, and movement.

SOME FUNDAMENTALS OF POPULATION CHANGE

The dual processes of positive and negative feedback are central to the dynamics of pest organisms (1). Positive feedback occurs when, as a population increases, its growth rate also increases (Fig. 1, line A). Although all populations have the potential for exponential increase, positive feedback adds to this runaway process and can result in the dramatic population growth that typifies some pest outbreaks. For example, many species of bark beetle are better able to overcome the defenses of their host trees if they attack en masse. As a result, their rates of population growth actually rise as population density increases. Negative feedback, in contrast, helps to explain why bark beetle populations do not continue to grow to overwhelm the planet. Negative feedback, also generally referred to as density dependence, occurs when population growth declines as the population increases (Fig. 1, line B). Negative feedback generally operates through resource limitation (declines in the quality or quantity of resources) or the action of natural enemies, and acts to bring pest populations toward some equilibrium density.

CYCLIC AND ERUPTIVE DYNAMICS

Cyclic Dynamics

It is necessary to distinguish between cyclic and eruptive population dynamics to discern among ecological causes of pest outbreaks. Cyclic dynamics are recognized as predictable oscillations in pest density (Fig. 2, line A) that may bring the pest population above some economic threshold with regular frequency. Cyclic dynamics generally occur when negative feedback processes operate on some time delay (3). For example, it may take several generations for a predator to respond to increases in the density of a pest population that represents a major source of food. The delay in the response of the predator can generate predator-prey cycles (4). One ecological cause of some pest outbreaks, therefore, is a time-lag in the action of a density-dependent process such as predation or resource limitation. Although it is sometimes possible for an abiotic factor such as rainfall to occur in cycles and to generate predictable patterns in pest dynamics (5), most cyclic outbreaks are probably associated with biotic variables such as resources and enemies.

Eruptive Dynamics

In contrast, if the frequency of a pest outbreak does not remain constant over time, the pest is said to exhibit eruptive dynamics (Fig. 2, line B). By their nature, eruptive dynamics are difficult to predict and it may be harder to identify the ecological causes of their occurrence. Nonetheless, there are two main processes that can

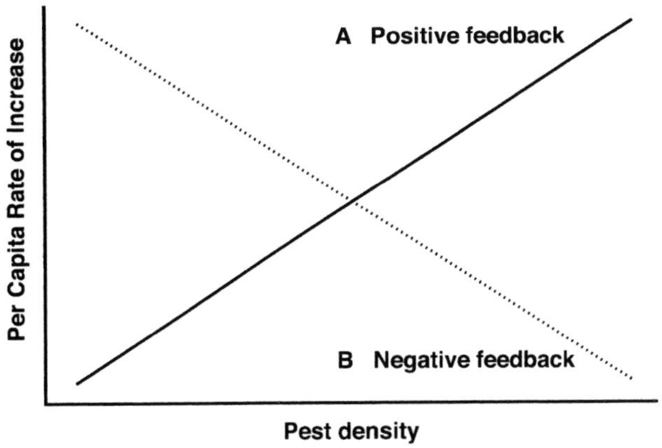

Fig. 1 Positive feedback (A) occurs when the growth rate of the pest population rises as the population density increases. Negative feedback (B) reflects a decline in population growth as pest density increases.

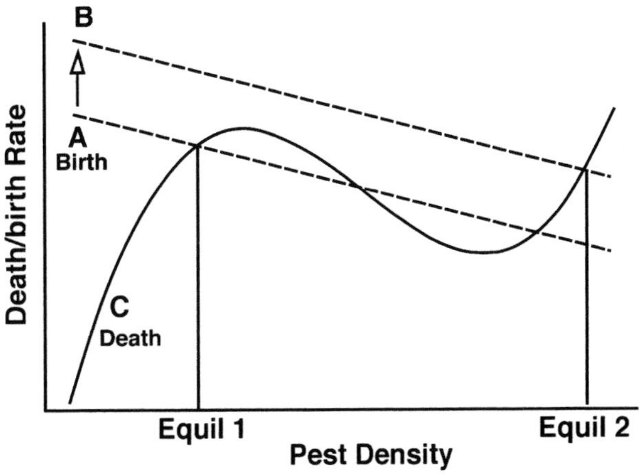

Fig. 3 Eruptive dynamics can occur when a pest population escapes low-density regulation (Equil 1) and reaches a higher equilibrium density (Equil 2). In this hypothetical example, an increase in birth rate from A to B permits the pest population to escape low-density regulation. Further details on such effects can be found in Ref. 7.

generate eruptive dynamics. First, the pest organism may track changes in some aperiodic ecological variable such as temperature, rainfall, or disturbance. Such exogenous forces, if they are themselves unpredictable, may generate eruptive dynamics. For example, hurricane damage to forests can result in dramatic increases in the density of defoliating insects and subsequent leaf damage to trees (6). Because hurricane damage in a particular forest is an unpredictable event, the dynamics of the pest insects that respond to such disturbances are clearly eruptive.

The second major process that can lead to eruptive dynamics is escape from low-density regulation. The populations of some pest species can be regulated at low density by one ecological force such as predation. If the pest population escapes that regulatory process, it will increase until some other force such as competition or disease acts to stabilize or reduce pest numbers (Fig. 3). If the ecological process that permits escape from low-density regulation is an unpredictable event (e.g., drought or fire) then the subsequent dynamics may be eruptive. Table 1 in Mattson and Haack (8) provides a list of insect pests whose eruptive outbreaks are associated with drought. They include species of beetles, aphids, sawflies, Lepidoptera, and grasshoppers. One of the most convincing examples of eruptive dynamics based on release from regulation at low density comes from the work of Belovsky and Joern (9). Their studies of rangeland grasshoppers indicate that increases in plant quality associated with nutrient availability and rainfall can release grasshoppers from regulation by their natural enemies. Feeding on high quality plants both decreases death rates and increases birth rates such that natural enemies are overwhelmed and grasshopper populations erupt. Plant quality varies both in space and time, resulting in a spatiotemporal mosaic of population dynamics that includes endemic and epidemic populations. Of course, human-induced disturbance can also contribute to escape by pests from regulation. The application of pesticides, for example, can generate eruptions of pest species by reducing the densities of natural enemies. Outbreaks of bollworm and bud worm on cotton can result

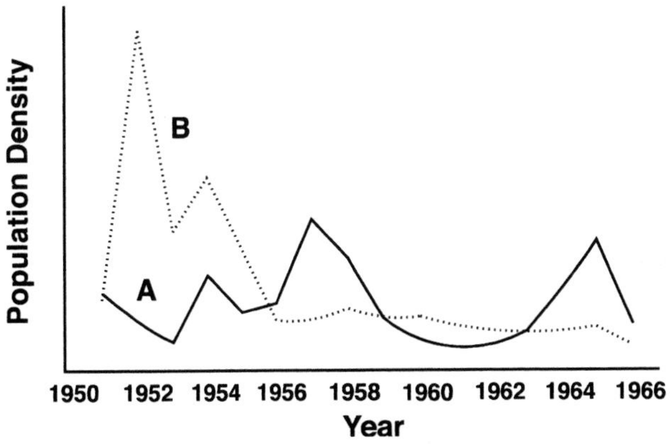

Fig. 2 Changes in the densities of cyclic pests such as *Operophtera brumata* (Lepidoptera: Geometridae) (A) are more predictable than are changes in the densities of eruptive pests such as *Tortrix viridana* (Lepidoptera: Tortricidae) (B). (Adapted from Ref. 2.)

from pesticide applications targeted at cotton boll weevil that unintentionally reduce the natural enemy pool.

FOOD QUALITY AND PEST OUTBREAKS

Outbreaks of pests associated with changes in food quality are not limited to rangeland grasshoppers. Indeed, an entire field of research on pest ecology has been generated by the "stress hypothesis of insect outbreak" (10, 11). White has suggested that many plants respond to drought stress by mobilizing nitrogen in their tissues and, given the strong association between nitrogen availability and herbivore performance, outbreaks of pests can result. Certainly, some plants do change their internal nitrogen dynamics in response to stress (Table 5 in 8). However some insect pests, including those that form galls on plant tissue, appear to do better on vigorous than stressed plants, and their outbreaks are associated with increases in plant quality (5, 9).

THE ECOLOGY OF PRODUCTION SYSTEMS

An ecological approach to understanding pest outbreaks of food and forage crops has much to offer when we consider the ecology of the production systems themselves. Fields, orchards, chicken houses, and hog farms are all ecosystems, though certainly simpler than most natural ones (4). Our food production systems generally are designed to maximize production and minimize the costs of management and harvesting. As a consequence, production ecosystems usually are characterized by high densities of particular plants or animals in single-species populations. Such monocultures, be they wheat or chickens, have clear consequences for pest outbreaks.

Monocultures

Returning to the four parameters of population change (birth, death, immigration, and emigration), pests in monocultures are likely to exhibit higher rates of birth and immigration and lower rates of death and emigration. For many pests, a significant proportion of their mortality occurs during dispersal and host location (12). Monocultures both reduce the need for dispersal (emigration) and decrease the likelihood of failing to find a suitable host. Moreover, the energy saved in host location may be allocated to reproduction. With the added effect that the probability of failing to find a suitable mate is low, rates of birth are likely to increase. Finally, monocultures present large target areas for colonizing pests (it's easier to spot the mall than the corner store) and result in increased rates of immigration. Many production systems also minimize the age structure of the commodity (e.g., single-age stands of trees) and reduce the genetic diversity of the crop or livestock. Both have a similar effect on the critical rates of pest population growth as do monocultures, and add to the simplification of production ecosystems.

Genetic Diversity

Indeed, reduced genetic diversity is a fundamental feature of most farm and forest production with a number of consequences for pest outbreaks. First, genotypic variation in resistance is much reduced and a strain of pests that is adapted to the chosen genotype(s) has an immense pool of resources to exploit. Second, we often choose or breed our crops and livestock for features such as high yield (e.g., rice) or low fat (e.g., chicken) and, in the worst cases, the traits that we select in our production systems are also attractive to pests. Finally, crops that are grown outside of their native range present novel genotypes to local pests. Rather than conferring resistance, history tells us that native pests are often well equipped to attack novel crops (4). More than 98% of crops are introduced species (or genotypes) whereas 60% of pests are of local origin. If the crops are grown outside of their optimal climates (e.g., apples in the southern United States or *Pinus caribbeae* in the Philippines) this can further increase their susceptibility to pests (10, 11).

At their heart, all pest outbreaks are ecological phenomena and can be best understood through the analysis of ecological principles. These include the fundamental principles of population growth that can be applied to the pests themselves, and the principles of community ecology in which production systems are seen as simplified ecosystems.

REFERENCES

1. Berryman, A.A. *Principles of Population Dynamics and Their Application*; Samuel Thornes: London, 1999.
2. Hunter, M.D.; Varley, G.C.; Gradwell, G.R. In *Estimating the Relative Roles of Top-down and Bottom-up Forces on Insect Herbivore Populations: A Classic Study Revisited*, Proceedings of the National Academy of Science, 1997; 9176–9181.
3. Turchin, P. Rarity of density dependence or population regulation with lags. Nature **1990**, *344*, 660–663.

4. Speight, M.R.; Hunter, M.D.; Watt, A.D. *The Ecology of Insects: Concepts and Applications*; Blackwell's Scientific: Oxford, 1999.
5. Hunter, M.D.; Price, P.W. Cycles in insect populations: delayed density dependence or exogenous driving variables. Ecol. Entomol. **1998**, *23*, 216–222.
6. Hunter, M.D.; Forkner, R.E. Hurricane damage influences foliar polyphenolics and subsequent herbivory on surviving trees. Ecology **1999**, *80*, 2676–2682.
7. Hunter, M.D. Incorporating Variation in Plant Chemistry into a Spatially Explicit Ecology of Phytophagous Insects. In *Forests and Insects*; Watt, A.D., Stork, N.E., Hunter, M.D., Eds.; Chapman Hall: London, 1997; 81–96.
8. Mattson, W.J.; Haack, R.A. The Role of Drought Sress in Provoking Outbreaks of Phytophagous Insects. In *Insect Outbreaks*; Barbosa, P., Schultz, J.C., Eds.; Academic Press: New York, 1987; 365–410.
9. Belovsky, G. E.; Joern, A. Regulation of Rangeland Grasshoppers: Differing Dominant Mechanisms in Space and Time. In *Population Dynamics: New Approaches and Synthesis*; Cappuccino, N., Price, P.W., Eds.; Academic Press: San Diego, 1995; 359–386.
10. White, T.C.R. A hypothesis to explain outbreaks of looper caterpillars, with special reference to populations of *Selidosema suavis* in a plantation of *Pinus radiata* in New Zealand. Oecologia **1974**, *16*, 279–301.
11. White, T.C.R. The abundance of invertebrate herbivores in relation to the availability of nitrogen in stressed food plants. Oecologia **1984**, *63*, 90–105.
12. Hunter, M.D. Differential susceptibility to variable plant phenology and its role in competition between two insect herbivores on oak. Ecol. Entomol. **1990**, *15*, 401–408.

ECOLOGICAL ROLE OF PESTS

Douglas A. Landis
Michigan State University, East Lansing, Michigan, U.S.A.
Deborah A. Neher
University of Toledo, Toledo, Ohio, U.S.A.

INTRODUCTION

Humans apply the term ''pest'' to those organisms that compete with people for resources, or occur in a place or in numbers that they find undesirable. In agroecosystems, pests are generally considered to be those arthropods (insects, mites, spiders), plants (i.e., weeds), plant-pathogenic bacteria, fungi and nematodes (i.e., pathogens), and vertebrates that interfere with the production of food and fiber. However, the definition of a pest is subjective and depends on both the observer and context. For example, a wasp may be considered a pest in a house, but may be considered beneficial outdoors, where it attacks caterpillars. Similarly, volunteer seedlings of last season's valued corn crop, are often considered weeds in the following year's soybean planting. Therefore, labeling an organism as a pest is based largely on human values and not always on the organism's ecological or agricultural role. Indeed, organisms that are at times considered pests may play vital roles in agroecosystems.

ECOLOGICAL ROLE OF ARTHROPOD PESTS

Arthropods are among the most abundant and diverse forms of life on Earth. They inhabit nearly every ecosystem and play a variety of ecological roles. In agroecosystems, pest and beneficial arthropods occur in soil, in water, in and on plants, and in the air. Arthropods are vital in decomposition processes and nutrient cycling. Herbivores may initiate the process of plant decomposition by removing or otherwise damaging plant tissue or creating routes of entry for plant pathogens. Senescing or dead plant material in or on soil is further fragmented by the feeding of detritovores such as springtails and mites, some of which are also sometimes classified as pests. These microarthropods reduce particle size and increase the surface area of decaying plant material making it accessible to soil microorganisms to further decompose it and return nutrients to soil. Herbivore activity may increase nutrient turnover rate, by making plant nutrients available directly to soil microbes via excretion of readily decomposable frass (insect feces) (1).

Pest arthropods provide food for many other organisms. Arthropod predators and parasitoids consume herbivores and are, in turn, consumed by other predators and hyperparasitoids. Many vertebrates such as rodents, amphibians, and birds readily accept arthropod pests as a food source. Finally, many pathogenic organisms such as nematodes, fungi, protozoa, bacteria, and viruses rely on arthropod pests as a host and/or food source.

Pest arthropods may also act as selective forces by removing less fit individuals from the gene pool. Slow growing or poorly defended plants may be more vulnerable to pest damage and, therefore, not contribute to the next generation. The role of arthropods as vectors of diseases in plants and animals may also contribute to such selection. Plants infected with a virus or bacterium may senesce prematurely, contributing organic matter and nutrients to the soil earlier than do healthy individuals. This may provide a unique resource for exploitation by other organisms.

ECOLOGICAL ROLE OF WEEDS

A weed has been defined as ''any plant that is objectionable or interferes with the activities or welfare of humans.'' If the human value judgement is removed, the ecological role of these plants can also be seen (2). All plants including weeds contribute to soil health in several ways. Root systems penetrate and help to loosen compacted soil and transfer nutrients from lower depths into upper soil layers. By constant sloughing of dead cells and exudation of fluids, roots contribute to the food base for microorganisms that live in the rhizosphere and are vital to nutrient cycling. Plant roots serve as a host for symbiotic mycorrhizal fungi. These fungi obtain carbon from the plant and, in turn, provide fungal-derived nutrients to the plant that can improve plant growth. When weeds die, both above and below ground plant parts are decomposed by soil flora and fauna contributing to soil organic matter.

Weeds may modify the microclimate of a crop altering its suitability for use by other organisms. Increased weed cover can favor certain soil-dwelling insects that prey on arthropods and weed seeds. Selective feeding by these and other seed predators may change weed abundance in the seed bank and may lead to subsequent shifts in plant communities. Weeds may also provide cover, encouraging birds, rodents, and beneficial arthropods to utilize the crop habitat.

By producing pollen, nectar, and plant sap, weeds may also provide needed resources for arthropod predators and parasitoids. Plant pollen is required by some syrphid flies to increase fecundity. Many parasitoids require access to plant nectar, sap, or homopteran honeydew (sugary secretions of aphids and scale insects) for energy to fuel flight and host search.

ECOLOGICAL ROLE OF PATHOGENS

Pathogens differ from other pests by causing disease, which occurs when a susceptible host, virulent pathogen, and favorable environment coincide (3, 4). Pathogens may affect outcomes of inter- and intraspecific competition, distribution of plant species, genetic structure of populations, and diversity of plant communities. In an ecosystem where individuals of different species compete for limiting resources, a disease that affects one species more than another may be expected to change the competitive interaction between them. Disease may reduce the number, growth, longevity, reproductive output, and competitive vigor of a host, thereby influencing intraspecific competition. Single or multiple diseases may build up over successive seasons. Successive or chronic attacks may result in progressive changes in crop distribution and diversity. Pathogens can be used as pest management tools themselves, such as in biological control of weeds and arthropods.

Most plant populations have the potential to increase in size until restrained by physical and chemical limitations of their environment. Where pathogens are present, high-density, single-species stands that result from such unrestricted growth are particularly vulnerable to infection and disease. If a pathogen causes a reduction in size and density of host population, any partly unoccupied resource niche may be colonized by other (more resistant) plant species or by novel resistant genotypes within the same host population.

Disease and disease resistance develop through a process of coevolution between host and pathogen. Gene-for-gene models imply that for every gene (cluster) that accounts for resistance in the host, there is a corresponding gene (cluster) that accounts for virulence in the pathogen. Under strong natural selection pressure, a resistance gene(s) may be overcome by a gene(s) conferring virulence. If a new resistance gene(s) arises, the cycle can perpetuate itself. There may be inherent fitness costs to the host to develop and maintain resistance genes and costs to the pathogen to develop and maintain virulence genes. Costs may include energy diverted from growth and reproduction to defense, production of secondary metabolites, new cell division or differentiation. Sexual reproduction is a source of genetic variation and new (resistance or virulence) genes. However, many pathogens reproduce asexually, produce more progeny, and have shorter generation times than do their hosts. Therefore, pathogens may have an evolutionary advantage under stable environmental conditions where asexual reproduction is favorable.

Increasing complexity of agroecosystems is one strategy for managing disease. Complexity may be increased by spatial and temporal heterogeneity that creates patchiness and often a complex age structure. Other ways to increase complexity are to increase intra- and interspecific diversity. Such diversity encompasses variation within landraces, varietal mixtures of single crops, intercrops, and polycultures that create multiple host–pathogen interactions. Complexity often reduces selection pressure and disease severity relative to a simple, pure stand. Replacement of susceptible plants by resistant ones reduces the amount of susceptible tissue, and may also reduce the amount of inoculum available for dispersal. Increased distance between susceptible individuals increases the average distance that inoculum has to travel before initiating successful infections. Resistant plants interfere with movement of inoculum between susceptible individuals. The usefulness of disease control by cultivar or plant mixtures is influenced by the pathogen and its means of dispersal. Factors thought to be involved include physical barriers to the spread of aerial pathogens and their vectors, altered microclimates, and specific host–pathogen interactions such as induced resistance. An understanding of alternative hosts among a complex vegetation system is critical to avoid accentuating disease progress. For example, weeds have been demonstrated to serve as alternative hosts.

ECOLOGICAL ROLE OF VERTEBRATE PESTS

Vertebrate pests of agriculture include birds, rodents, and deer. Because of their larger size and mobility, many vertebrates are only transient visitors to agricultural crops. When present, birds and rodents may feed heavily on weed seeds, selectively removing those of a certain size or palatability and, therefore, changing the weed community. Some of these seeds pass through the digestive systems intact, or are otherwise redistributed, and may change the plant community where they are deposited. Some vertebrates prey on insect pests as well. Crows and certain woodpeckers are known to glean fields in winter months and selectively consume European corn borer larvae from their overwintering sites in corn stalks. Rodents contribute to soil aeration and inversion via their tunneling activity and provide habitat for other organisms in abandoned nests and burrows.

CONCLUSION

Organisms that are sometimes considered pests play a variety of vital roles in agricultural ecosystems. They increase the biodiversity of these systems with consequences for the structure of ecosystems and interactions that occur (5). Some provide shelter, and all provide food for other organisms and contribute to nutrient cycling. While the full breadth of these interactions is beyond the scope of this entry, it is safe to say that without such "pests," the ecological function of agroecosystems would be simplified greatly and, perhaps, be less efficient.

REFERENCES

1. Price, P. *Insect Ecology*, 3rd Ed.; Wiley: New York, 1997.
2. Radosevich, S.; Holt, J.; Ghersa, C. *Weed Ecology*; John Wiley & Sons, Inc.: New York, 1997.
3. Burdon, J.J. *Diseases and Plant Population Biology*; Cambridge University Press: Cambridge, 1987.
4. Lenné, J.M.; Teverson, D.M.; Jeger, M.J. Evaluation of Plant Pathogens in Complex Ecosystems. In *Ecology of Plant Pathogens*; Blakeman, J.P., Williamson, B., Eds.; CAB International: Wallingford, 1994; 63–77.
5. Tilman, D. Biodiversity population versus ecosystem stability. Ecology *77*, 350–363.

ECONOMIC AND SOCIAL ASPECTS OF PEST MANAGEMENT

E

Frederick H. Buttel
University of Wisconsin, Madison, Wisconsin, U.S.A.

INTRODUCTION

Economic and social assessment of any agricultural or nonagricultural technology is a challenging task. Technologies typically are so closely intertwined with social institutions and social practices—not to mention social interests, ideologies, and concerns—so that disaggregating the costs, benefits, and other implications of a particular technology or class of technologies is difficult. And, as will be stressed below, there are few technologies in which economic and social assessment is more complex and contested than is the case with pest management technologies and practices.

FROM GREEN REVOLUTION TO ENVIRONMENTAL REVOLUTION

The twentieth century has witnessed an epoch-making agricultural change—the "Green Revolution," understood in the broadest sense of the term (see below). And very likely a second major agricultural–environmental transition—the environmentalization of agriculture—has begun to emerge during the late twentieth century.

By the Green Revolution, I mean the transition that began in the industrial world early in the twentieth century and spread to the developing world from the 1950s through the 1980s (1). The Green Revolution was a plant-breeding-driven, output-enhancing technological transition typified by the development of hybrid corn in the United States and by the widespread cultivation of irrigated wheat in the Punjab. In addition to the role played by improved varieties (as well as by parallel advances in animal agriculture), the essence of the Green Revolution was the displacement of traditional peasant production systems within which plant protection and minimization of risk were provided (albeit imperfectly) by biological diversity (e.g., genetic diversity of cultivars, polyculture, and crop rotation). The very expression "plant protection," which was a product of the Green Revolution age, conjures up the notion that we need to protect new and improved, but vulnerable plant varieties that have improved yield-related characters from the ravages of pests and pathogens that threaten to reduce the scope of production gains.

The use of chemical plant protection substances increased very rapidly in Western agricultures after World War II for a variety of reasons. In addition to the fact that modern/improved varieties generally were more vulnerable to pests than traditional cultivars (especially when cultivated under monoculture and continuous-cropping), chemical pesticides also tended to be highly consistent with increased scale and specialization, to have favorable returns to investment, to reduce the risks of agricultural protection, and to be labor saving (2). In addition, use of these plant protection chemicals was occasionally mandated by processors through production contracts. The commercial success of industrially produced pesticides at the farm level complemented the overall scientific consensus of the time that agriculture would necessarily continue to be based on chemical pesticides for the foreseeable future (3). Improved varieties, chemical pesticides, and mechanization were increasingly seen as constituting a natural "package" of "modern technologies" (4, 5).

This initial agricultural–environmental transition of the twentieth century led to some sizable social benefits. Chief among these benefits was the fact that at a global level agricultural output increased at the historically unprecedented rate of about 2.4% per year from 1950 to 1985. Accordingly, food prices declined and the share of world income devoted to food expenditures declined significantly (especially in the advanced industrial countries). These benefits were captured disproportionately by the world's poor and hungry, though many of the direct costs of the new technology were also borne by the poor and hungry (6). Green Revolution technology proved to be relatively compatible with mechanization, and as a result these two agricultural–technical clusters led to an enormous decline in the share of labor forces in agriculture (currently $< 2\%$ in the United States for example).

Just as pesticides were central to the first agricultural–environmental transition, the ecological implications of these substances and other Green Revolution practices have been central to the environmentalization of agriculture. The environmentalization of agriculture involves three separate but related processes. The first is that environmental problems of agriculture (e.g., health risks of certain plant protection substances, pesticide resistance, soil erosion, waterlogging and siltation caused by irrigation) constrain agricultural production in one way or another. The second meaning of the notion of the environmentalization of agriculture is that social movements emerge to challenge prevailing agricultural research, education, policy making, and production institutions in order to reduce the environmental problems of agriculture. The third sense of the environmentalization of agriculture is that environmental scrutiny of agriculture has been reflected in and driven by laws and other public policies, particularly important examples of which are the Federal Insecticide, Fungicide, and Rodenticide Act (FIFRA) and the Food Quality Protection Act (FQPA).

One should not exaggerate the degree to which the second agricultural–environmental transition currently is a decisive break with the first one of this century. The overall use of pesticides in the United States has never been higher (7). In addition, the notion of integrated pest management (IPM) began during the 1960s, the very heyday of the Green Revolution.

Nonetheless, the key issues of this most recent agricultural–environmental transition are largely fourfold, and revolve around whether and how much the use of pesticides ought to be restricted or otherwise reduced. First, what are the social costs and social benefits of employing alternatives to chemical plant protection substances (IPM, biologically based pest management, biological control, organic farming, sustainable agriculture)? Put somewhat differently, do the costs of pesticide use reductions tend to be in excess of the benefits? Second, how can and should pesticide regulation (8) and "regulatory science" (9) best be undertaken to enhance human welfare and environmental quality? Third, are the most recent generations of plant protection chemicals sufficiently superior to previous ones so that scientists, policy makers, and citizens ought to cease to referring to them collectively as "pesticides"? Fourth, what are the implications for the use of chemicals of some of the newest innovations in crop production practices (e.g., GMOs and other biotechnology products, precision farming)? In other words, will these new innovations serve to substitute for chemicals, or reinforce the use of chemicals and the problems that result from synthetically compounded biocides (10, 11)?

SOCIAL ASSESSMENT OF PEST MANAGEMENT SYSTEMS

Earlier I noted that agricultural technology assessment in general, and the assessment of agricultural chemicals technologies in particular, is highly complex. This is nowhere better illustrated than in the case of studies of the social implications of pesticide reduction and of alternatives to chemical pesticides. Studies of this sort are of more than passing academic interest. FIFRA requires EPA to conduct impact studies that weigh the estimated risks to human health against the predicted economic benefits of pesticide use when making decisions about the registration of a particular agricultural chemical. Thus, results of these types of studies have very considerable public policy, environmental, and human welfare implications.

There is a tendency for studies that draw alarmist conclusions to be published in response to proposed public policies that involve a likelihood of significant constraints on the manufacture and use of plant protection chemicals. In the late 1980s and early 1990s, for example, a number of European countries as well as Ontario province in Canada passed laws involving one or another provisions that would lead to reduced use of plant protection chemicals, especially herbicides and insecticides. The late 1980s momentum behind restricting the use of plant protection chemicals prompted several studies (12, 13) of the economic and social impacts of reducing chemical usage. In the United States the FQPA (passed by Congress in 1996 and which has been undergoing implementation since that time) has prompted further studies (14, 15) that have involved forecasts of major adverse impacts and social dislocations that could result from far tighter restrictions on the use of agrochemicals. The results of studies such as these also seem to be consistent with the lay instinct to assume that pesticide reduction policies involve environmental and economic trade-offs, and that economic losses are likely to be in excess of environmental benefits in most cases.

In recent years, however, there has been an accumulation of evidence that studies such as these have a number of policy-relevance and scientific shortcomings (2). In terms of policy relevance, for example, the studies that report large social costs to pesticide reduction usually assume that pesticide reduction will take the form of *total bans* on all forms of pesticides (16). In practice, even the most far-reaching pesticide reduction reforms in the world (especially those in Northern Europe) rely on a wide range of policy instruments (pesticide taxes, expansion of public funding of IPM and biological control research, stricter food residue tolerances, and stricter pesticide registration and cancellation procedures) that are *incremental and non-draconian*, and that lend themselves to farmer, scientific, and broader food system innovation in pest management. Relatedly, the studies forecasting major dislocations from reduced use of pesticides also tend to assume that there is little or nothing farmers can do to adapt to or cope with restrictions on the use of pesticides. But there is sound evidence that farmers (and pest-control suppliers) individually and collectively are enormously innovative in responding to changes in the policy and biosphysical environments of farming (16, 17). Third, these studies tend to drastically *undervalue* the environmental and other benefits of pesticide reduction (16). Finally, there is a *tendency to exaggerate the costs* of pesticide reduction. The typical finding that there are widespread farm income losses as a result of pesticide use restrictions is largely inconsistent with the long-accepted generalization that because most agricultural commodities have low price and income elasticities of demand, reduced aggregate output will lead to disproportionate increases in production prices and to an overall increase in net farm income (4). In addition, a comprehensive study of 49 studies of the economics of IPM (18) has shown that in seven out of eight studies chemical use was reduced *and* net returns were increased.

As noted earlier, newly emerging trends in agricultural technology suggest strongly that pestmanagement chemical use will continue to be a central issue in agricultural research, technology, and regulatory policy. Biotechnology and precision farming, both of which are growing very rapidly in terms of the share of U.S. crop production, have been heavily promoted, in part, because of their possible benefits in terms of reducing pesticide use. By contrast, these technologies have generated considerable social movement as well as scientific scrutiny as to their implications for overall chemical usage, and the environmental and other social impacts of these chemicals (see, for example, Refs. 10 and 19 from the university agricultural science community, and Ref. 20 from a social movement perspective that draws on scientific studies). Socioeconomic assessment studies of biotechnology, precision farming, and pest management practices thus will very likely become increasingly central in shaping public agricultural research and environmental policy toward agriculture.

REFERENCES

1. Buttel, F.H. Twentieth century agricultural–environmental transitions, a preliminary analysis. Res. Rural Sociology and Dev. **1995**, *6*, 1–21.
2. Buttel, F.H. Socioeconomic Impacts and Social Implications of Reducing Pesticide and Agricultural Chemical Usage in the United States. In *The Pesticide Question*; Pimentel, D., Lehman, H., Eds.; Chapman and Hall: New York, 1993; 153–181.
3. Busch, L.; Lacy, W.B.; Burkhardt, J.; Lacy, L.R. *Plants, Power, and Profit*; Basil Blackwell: Oxford, 1991; 275.
4. Cochrane, W.W. *The Development of American Agriculture*; University of Minnesota Press: Minneapolis, 1978; 464.
5. Fliegel, F.C. *Adoption Research in Rural Sociology*; Greenwood Press: Wesport, CT, 1993; 132.
6. Lipton, M.; Longhurst, R.F. *New Seeds and Poor People*; Johns Hopkins University Press: Baltimore, 1989; 473.
7. Lin, B.H.; Padgitt, M.; Bull, L.; Delvo, H.; Shank, D.; Taylor, H. *Pesticide and Fertilizer Use and Trends in U.S. Agriculture*; Agricultural Economic Report No. 7171, Economic Research Service, U.S. Department of Agriculture: Washington, DC, 1997; 47.
8. Zilberman, D.; Schmitz, A.; Casterline, G.; Lichtenberg, E.; Siebert, J.B. The economics of pesticide use and regulation. Science **1991**, *253*, 518–522.
9. Rothstein, H.; Irwin, A.; Yearley, S.; McCarthy, E. Regulatory science, europeanization, and the control of agrochemicals. Social Studies of Science **1999**, *24*, 241–264.
10. Pierce, F.J.; Nowak, P.J. Aspects of precision agriculture. Advances in Agronomy **1999**, *67*, 1–84.
11. Krimsky, S.; Wrubel, R. *Agricultural Biotechnology and the Environment*; University of Illinois Press: Urbana, IL, 1996; 294.
12. Council on Agricultural Science and Technology (CAST). *Pesticides and Safety of Fruits and Vegetables*; Report No. 1990-1, CAST: Ames, IA, 1990; 15.
13. Knutson, R.D.; Taylor, C.R.; Penson, J.A.; Smith, E.G. *Economic Impacts of Reduced Chemical Use*; Knutson & Associates: College Station, TX, 1990.
14. California Department of Food and Agriculture. In *Methyl Bromide: An Impact Assessment*; Office of Pesticide Consultation and Analysis, California Department of Food and Agriculture, State of California: Sacramento, CA, 1996.
15. Council on Agricultural Science and Technology (CAST). *Agricultural Impact of the Sudden Elimination of Key Pesticides Under the Food Quality Protection Act*; CAST: Ames, IA, 1998; 14.
16. Jaenicke, E. *The Myths and Realities of Pesticide Reduction*; Henry, A., Ed.; Wallace Institute for Alternative Agriculture: Greenbelt, MD, 1997; 35.
17. Benbrook, C.M.; Groth, E., III; Halloran, J.M.; Hansen, M.K.; Marquardt, S. *Pest Management at the Crossroads*; Consumers Union: Yonkers, NY, 1997; 272.
18. Norton, G.; Mullen, J. *Economic Benefits of Integrated Pest Management: A Brief Review*; Department of Agricultural and Applied Economics, Virginia Polytechnic Institute and State University: Blacksburg, VA, 1993.
19. Gould, F. Sustainability of transgenic insecticidal cultivars integrating plant genetics and ecology. Annu. Rev. Entomol. **1998**, *43*, 701–726.
20. Rissler, J.; Mellon, M. *The Ecological Risks of Engineered Crops*; MIT Press: Cambridge, MA, 1996; 168.

ECONOMIC BENEFITS OF PEST MANAGEMENT

Rajinder Peshin
Sher-e-Kashmir University of Agricultural Sciences & Technology, Jammu, India

INTRODUCTION

The pest outbreaks in agriculture have caused severe economic and human losses from time to time. Potato blight epidemic in Ireland during 1845 and 1846 caused the death of more than a million people (1). Downey and powdery mildew in 1872 devastated the French wine industry and leaf spot disease of rice in 1942 caused Bengal famine in India. The greatest crop loss of about $1 billion was caused by leaf blight in the United States in 1970 (2). Despite the advancement in agricultural sciences, insect pests, diseases, weeds, and birds cause specific crop losses varying between 10 and 90% (3) and on an average 35–40% of all potential food and fibre crops are lost to pests. The crop losses in 1991 were valued around $300 billion per year, which was equal to 30% of potential global food, fibre, and feed production (4) and presently losses due to pest are valued over $500 billion per year. This is despite the yearly application of 2.5 million metric tons of active ingredient pesticides (5), valued at $31.25 billion during 1996 (Table 1).

IMPACT OF PESTICIDE USE

The discovery of Bordeaux mixture in 1882 and Paris green in 1870, proved successful against grape diseases and potato beetle, respectively (2). Sulphuric acid was used as a selective herbicide in cereals in France during 1931 and in the United Kingdom the same practice was adopted in 1932 (6). The discovery of insecticide properties of DDT in 1940s started the era of synthetic pesticides. The global pesticide market has increased 54 times since 1960 (Table 1).

Though pesticides are not yield-enhancing chemicals like fertilizers, but they reduce the damage caused by insect pests, diseases, and weeds. It is estimated that each dollar invested in pesticide control gives a benefit of approximately $4 in crops saved, and overall losses to pests would increase by 10% if no pesticides were used at all and specific crop losses would range from zero to nearly 100% (7). In Germany the benefits of pesticide control were approximately one billion dollars (at 1997 rates) and after deducting the negative side effects, the social benefit cost ratio was 1:1.47. The total ban of pesticides in Germany would cause a loss of $581 million per year (8). In the United States, an investment of approximately $4 billion in pesticide control saves approximately $16 billion (9). In Indian state of Punjab, the cost benefit of insecticides in specific crops range from 1: 4–26 (10). The overall losses in India from insect pests, weeds, and diseases are 35–40%, and estimated to value $2 billion in 1995 (11). With a tenfold increase in insecticide use in the United States from 1945 to 1989, the losses from insect pests have gone up from 7 to 13% (9) but without the use of pesticides the losses from weeds are supposed to increase from 8 to 9%, diseases from 12 to 15%, and insect pests from 13 to 18% (12). With the increase of 234% in insecticides, 548% increase in fungicides, and 5414% increase in herbicide over a period of 15 years (1964–79) in Brazil, there was only 16.8% increase in production of 15 major crops (13). In India, the losses from insect pests in rice, wheat, sorghum, maize, cotton and sugarcane have increased by 15, 2, 20, 30–33, 32, and 10%, respectively, during the last three decades, even though the consumption of technical grade pesticide increased by 45 times. Losses to pests in different agricultural systems are 20–50%, as were before the advent of the synthetic pesticide era (14).

BENEFITS OF OTHER CONTROL MEASURES

Nonchemical pest control measures are highly effective. Host plant resistance is one of the most important components of pest management. Insect resistant cultivars during the last two decades have prevented the application of more than 6 million tons of insecticide onto cropland in the United States (15). Rice cultivars with multiple resistance to insect pests are grown in more than 20 million

Encyclopedia of Pest Management

Table 1 Worldwide pesticide market

Class	Year				
	1960[a]	1970[a]	1980[a]	1993[a]	1996[b]
Insecticides (%)	36.5	37.1	34.7	30.0	29.0
Fungicides (%)	40.0	22.2	18.8	18.7	21.0
Herbicides (%)	20.0	34.8	41.0	45.9	44.0
Others (%)	3.5	5.9	5.5	5.4	6.0
Value ($ billion)	0.58	2.7	11.6	25.3	31.25

[a]From Madhusoodanan, S. Post GATT Scenario—Prospects of Developing New Molecules. The Pestic. World 1996, March, 71–73.
[b]From Pestic. Inf. **1998**, *24* (1), 46.

hectares in Asia and Central and South America (16). More than 500 cultivars resistant to key pests in important crops are grown in many countries (17).

For each dollar invested in research and development of Sorghum midge resistant variety, benefits were $24.2 when no insecticide was used. The average value of benefit was $9.9 for each dollar spent with zero, three, and five applications of insecticides (18). In China with a 10% increase in host plant resistance in rice cultivars could save $0.305–$7.55 million per year on insecticide (19). The development of green bug resistant sorghum hybrid resulted in a $389 million net benefit to the United States (20).

Crop rotation was reported to control maize root worm on about 60% of the United States maize area (21). Citrus and olive crop (0.8 million ha) have a number of pests, which were controlled by natural enemies (22). In Peru, cotton yields decreased by 55% from 728 kg to 332 kg ha^{-1}, even with 16 pesticide applications. With the number of cultural control measures: Sowing of early maturing varieties, prohibition of ratoon cotton, crop rotation, change in dates of planting, sanitation, and planting of maize and wheat as host crops for building natural enemy populations resulted in the increase of yield to 800 kg ha^{-1}. The average number of pesticide applications dropped to 2.3 (4).

Insecticides used in response to scouting reports cost an average of $3.11 ha^{-1} less than no insecticide application and both these were $20.91 and $17.80 ha^{-1} less costly, respectively, than routine prophylactic use of insecticides in managing potato leaf hopper and alfalfa weevil in alfalfa crop in Maryland, United States (23). In the United States, pheromones and sticky trap cards were used to monitor infestation levels (1966–1982/83) of California red scale at a cost of $730,000 (1967 rates), showed a benefit of $2.5–$3.6 on each dollar spent on research and a savings in resources (24). A phytosanitory eradication program in the Netherlands for glasshouse crops showed a benefit of $3.66–$12.84 million at the cost of $0.22 million (25).

Developing and cultivating resistant varieties have mainly controlled plant diseases. In India, one of the severe diseases of rice bacterial leaf blight is controlled by cultivating resistant cultivars to the bacterial leaf blight and manipulation of cultural practices. In India, manual mechanical wheat control losses in cereals are 0.4–15.4% compared to 1.1–31.5% in chemical weed control (26). Kariba weed in northwest Zimbabwe was controlled using biocontrol agents and the cost ratio of biological control to chemical control was 1:10.6 (27). Tansy ragwort weed control in Oregon (US) was achieved by biological control, which provides an estimated benefit of more than $5 million with minimum cost benefit ratio of 1:13, and herbicide use was reduced by $0.85 million per year and the cost of control was $5 ha^{-1} (28).

ECONOMIC IMPACT OF INTEGRATED PEST MANAGEMENT

IPM began in the 1970s in an attempt to counter the damage of the synthetic pesticide era. Presently IPM is practiced in more than 50,000 communities throughout the world. Using IPM has shown a seasonal profit increase of 30%, rise in yields from 1% to over 10%, and reduction in amount and cost of synthetic pesticides used between 30% and 95%, and most of these communities grow rice and cotton (29). In Latin American countries, the adoption of IPM has decreased the crop protection costs and pest-induced losses with an increase in yield. In Nicaragua, the yields dropped by an average of 15–30% despite 28 insecticide applications. By manipulating dates of transplanting to avoid boll weevil and boll worm attack and trap cropping (planting small cotton plots at the beginning and at the end of growing seasons to attract and concentrate weevils and then killing them with insecticides) resulted in yield increase and drop in pesticide applications (30). In Costa Rica by stopping aerial application of dieldrin granular pesticides over 12,000 ha the outbreak of six major pests was suppressed as a natural enemy population started colonizing in the area (31). Use of *Bacillus thuringiensis* and release of natural enemies like *Trichogramma* spp. in the tomato crops of Colombia grown on an area of 2000 hectares reduced the pesticide applications from 20–30 to 2–3, thereby saving $650 ha^{-1} (32).

In Sudan and Egypt, employing a number of cultural, mechanical, and biological control measures checked the declining cotton yield due to bollworm and white fly. The cost of plant protection in Sudan reduced from 33.3%

Table 2 Economic impact of integrated pest management practices

Crop	Country	Pest management practice	Pest	CB ratio/Economic benefit	Reference
Sorghum hybrid	USA	Resistant variety	Green bug	Profit increase by $33.94/ha	33
Sorghum	USA	Resistant variety	Green bug	Net profit $389 million	20
Sorghum	USA	Resistant variety	Sorghum midge	1:9.9	18
Citrus	USA	Sex pheromones	California red scale	1:2.5–3.6	24
Apple & peach	USA	Biological control	Comstock mealy bug	1:24–135	34
Lemon	USA	Biological control	Comstock mealy bug	1:14–79	34
Different crops	Africa	Biological control	Cassava mealy bug	1:149	35
Mainly cotton	USA	IPM programs	Multiple	4.6:7.3	36
	India	IPM programs	Multiple	1:2.7	37
Cotton	India	IPM programs	Multiple	1:2.4	
				Yield increase (%)	
	Bangladesh,			17	
	Sri Lanka,			9–25	
Rice	Philippines,	IPM programs	Multiple	17.44	29
	Korea,			2.46	
	China,			4.0	
	Vietnam			4.0	

(1985–86) to 19.3% (1988–89) associated with yield increase per unit area (4).

In some Asian countries, after adopting IPM practices in the rice crop, the yield per unit area rose by 2.46–25% (see Table 2) and pesticide cost reduced by 27–86%. IPM programs in India showed that the cost benefit of pest control was raised to 1:2.4 compared to 1:1.8 in prophylactic pesticide control in cotton, and in rice the cost benefit was 1:2.7 compared to 1:2.2 in pesticide control only (37) and saving $30 million as annual subsidy for insecticides (29). The economic implications of several IPM programs in the United States indicate reduced pesticide use, energy savings, and increased farm profit. A short season cotton production in Texas showed a 30% yield increase, 33% reduction in the cost of cultivation and a profit increase from $30 to $257 per hectare—a total of $11 million increase in production profit. Other IPM programs with emphasis on cotton production show a benefit cost ratio of 7.3 and 4.6 (36).

Pesticides have played and will continue to play an important role in reducing the losses caused by pests in combination with improved mechanical, cultural, biological practices, and the use of resistant cultivars, and it is estimated that pesticide applications can be reduced by 50% at a cost of $ one billion (38).

REFERENCES

1. Large, E.C. *The Advances of the Fungi*, 3rd Ed.; Jonathan Cape: London, 1950; 488.
2. Upadhyay, R.K.; Mukerji, K.G.; Rajak, R.L. Integrated Pest Management. In *IPM Systems in Agriculture, Principles and Perspective*; Upadhayay, R.K., Mukerji, K.G., Rajak, R.L., Eds.; Aditya Books Private Limited: New Delhi, India, 1996; *1*, 1–23.
3. Youdeowei, A. Major Arthropod Pests of Food and Industrial Crops of Africa and Their Economic Importance. In *Biological Control: A Sustainable Solution to Crop Pest Problems in Africa*; Yaninek, J.S., Harren, H.R., Eds.; IITA: Ibadan, Nigeria, 1989; 210.
4. Oudejans, J.H. *Agro-Pesticides Properties and Functions in Integrated Crop Protection*; United Nations Economic and Social Commission for Asia and Pacific: Bangkok, 1991; 329.
5. Pimentel, D. Pest Management in Agriculture. In *Techniques for Reducing Pesticide Use: Environmental and Economic Benefits*; Pimentel, D., Ed.; John Wiley & Sons: Chichester, U.K., 1997; 1–12.
6. Anonymous. *A Practical Guide to Sulphuric Acid Spraying in Agriculture*; National Sulphuric Acid Association Ltd. U.K., *undated*.
7. Pimentel, D.; Acquay, H.; Biltonen, M.; Rice, P.; Silva, M.; Nelson, J.; Lipner, V.; Giordano, S.; Horowitz, A.; D'Amore, M. Environmental and economic eosts of pesticide use. BioScience **1992**, *42* (10), 750–760.
8. Waibel, H.; Fleischer, G.; Becker, H.; Runge, M.A. Social costs and benefits of chemical pesticide use in German agriculture. Agrarokonomische Monographien und Sammelwerke **1998**, *254*.
9. Pimentel, D.; McLaughlin, L.; Zepp, A.; Lakitan, B.; Krans, T.; Kleinman, P.; Vancini, F.; Roach, W.J.; Graap, E.; Keeton, W.S.; Selig, G. Environment and Economic Impact of Reducing U.S. Agricultural Pesticide Use. In *Handbook of Pest Management in Agriculture*; Pimentel, D., Ed.; CRC Press: Boca Raton, FL, 1991; 679–718.

E

10. Bakhetia, R.C.; Udean, A.S. Safe and judicious use of pesticides. Pestic. World **1998**, *3* (1), 20–26.
11. Gautam, K.C.; Mishra, J.S. Problems, prospects, and new approaches in weed management. Pestic. Inf. **1995**, *21* (1), 7–19.
12. Pimentel, D.; Krummel, J.; Gallahan, D.; Hough, J.; Merill, A.; Schreiner, I.; Vittum, P.; Koziol, F.; Back, E.; Yen, D.; Fiance, S. Benefits of Pesticide Use in U.S. Food Production. In *Pesticides: Role in Agriculture, Health and Environment*; Sheets, T.J., Pimentel, D., Eds.; Humana Press: Clifton, NJ, 1978.
13. FAO. *Pests in Agricultural Environmental Protection and Productivity: Conflicting Goals*, Series: Plant Production and Protection; FAO Regional Office for Latin America and the Caribbean, Chile, 1986.
14. Hobbelink, H. *Biotechnology and Future of World Agriculture*; Zed Books: London, 1991.
15. Smith, C.M. Global Aspects of Insect Resistant Crop Plants in Integrated Pest Management: An Ecological Perspective of Plant–Insect Interactions. In *Ecological Agriculture and Sustainable Development*; Dhaliwal, G.S., Randhawa, N.S., Arora, R., Dhawan, A.K., Eds.; CRRID: Chandigarh, India, 1998; 37–52.
16. Pathak, M.D.; Khan, Z. *Insect Pests of Rice*; International Rice Research Institute: Los Baños, Philippines, 1994.
17. Uthamasamy, S. Insect Resistant Crop Varieties and Integrated Pest Management. In *Ecological Agriculture and Sustainable Development*; Dhaliwal, G.S., Randhawa, N.S., Arora, Rajesh, Dhawan, A.K., Eds.; CRRID: Chandigarh, India, 1998; 53–64.
18. Ervin, R.T.; Khalema, T.M.; Peterson, G.C.; Teetes, G.L. Cost–benefit analysis of sorghum hybrid resistant to sorghum midge. Southwestern Entomologist **1996**, *21* (2), 105–115.
19. Widawsky, D.; Rozelle, S.; Jin, S.; Huang, J. Pesticide productivity, host-plant resistance and productivity in China. Agricultural Econ. **1998**, *19*, 203–217.
20. Eddleman, B.R.; Chang, C.C.; McCarl, B.A. *Economic Benefits from INTSORMIL Grain Sorghum Variety Improvement in the United States*; Tex. Agric. Exp. Stn., Report TAMRF9045, 1991.
21. Pimentel, D.; Shoemaker, C.; LaDue, E.L.; Rovinsky, R.B.; Russel, N.P. *Alternatives for Reducing Insecticides on Cotton and Corn: Economic and Environmental Impact*, Report on Grant No. R82518-02; EPA147, Washington DC, 1977.
22. van den Bosh, R.; Messenger, P.S. *Biological Control*; Intext Educational Publishers: New York, 1973.
23. Lamp, W.O.; Nielsen, G.R.; Dively, G.P. Insect pest induced losses in alfalfa: patterns in Maryland and implications for management. J. Econ. Entomol **1991**, *84* (2), 610–618.
24. Gardner, P.D.; Ervin, R.T.; Moreno, D.S.; Baritelle, J.L. California red scale (Homoptera: Diaspididae): cost analysis of a pheromone monitoring program. J. Econ. Entomol **1983**, *76* (3), 601–604.
25. Roosjen, M.G.; Bouwman, V.C.; Barwegen, J. Development of a cost/benefit model for phytosanitary eradication campaigns. OEPP/EPPO Bul. **1996**, *26*, 605–667.
26. Sahoo, K.M.; Saraswat, V.N. Magnitude of losses in yields of major crops due to weed competition in India. Pestic. Inf. **1988**, *14* (1), 2–9.
27. Chikwenhere, G.P.; Keswani, C.L. Economics of biological control of kariba weed (*Salvinia molesta*) at Tengwe in North-Western Zimbabwe—a case study. Int. J. of Pest Manage. **1997**, *43* (2), 109–112.
28. Coombs, E.M.; Radtke, H.; Isaacson, D.L.; Snyder, S.P.; Hoffman, J.H. In *Economic and Regional Benefits from Biological Control of Tansyregwort, Senrio jacobaea, in Oregon*, Proceedings of the 9th International Symposium on Biological Control of Weeds, Stellenbosch, January 19–26, 1996; South Africa, 1996; 489–494.
29. Kenmore, P.A. A perspective on IPM. ILEIA Newsletter **1997**, *13* (4), 8–9.
30. Swezey, S.L.; Murray, D.L.; Daxol, R.G. Nicaragua's revolution in pesticide policy. Environment **1986**, *28*, 6–9.
31. Stephens, C.S. Ecological upset and recuperation of natural control of insect pests in some Costa Rican banana plantations. Turrialba **1984**, *34*, 101–105.
32. Belloti, A.C.; Cardona, C.; Lapointe, S.L. Trends in pesticide use in Colombia and Brazil. J. Agril. Entomol. **1990**, *7*, 191–201.
33. Dharmarante, G.; Stool, J.R.; Lacewell, R.D.; Teetes, G. *Economic Impact of Greenbug Resistant Grain Sorghum Varieties*; Texas Blackland Tex. Agric. Exp. Stn., MP-1585, 1986.
34. Ervin, R.T.; Moffit, L.J.; Meyerdirk, D.E. Comstock mealybug (Homoptera: Pseudococcidae), cost analysis of a biological control program in California. J. Econ. Entomol **1983**, *76* (3), 605–609.
35. Norgaard, R.B. The biological control of cassava mealybug in Africa. Amer. J. Agric. Econ. **1988**, *70*, 366–371.
36. Lacewell, R.D.; Taylor, C.R. In *Benefit-cost Analysis of Integrated Pest Management Programs, Pest and Pesticide Management in Caribbean*, Proceedings of Seminar and Workshop CICP-USAID, November 3–7, 1980; 2284–302.
37. FAO (Food and Agriculture Organization). *Country Briefs*, Programme Advisory Committee (PAC) Meeting; FAO Intercountry Programme for IPM in Asia, February 06–09, 1996; Hyderabad, India, 1996.
38. Pimentel, D.; McLaughlin; Zepp, A.; Lakitan, B.; Kraus, T.; Kleinman, P.; Vancini, F.; Roach, W.J.; Graap, E.; Keeton, W.; Selig, G. Environmental and economic effects of reducing pesticide use. Bioscience **1991**, *41* (6), 403–409.

ECONOMIC DECISION LEVELS

Robert K.D. Peterson
Dow AgroSciences, Indianapolis, Indiana, U.S.A.

Leon G. Higley
University of Nebraska, Lincoln, Nebraska, U.S.A.

INTRODUCTION

The Economic Injury Level (EIL) and the Economic Threshold (ET) are arguably the most important concepts in Integrated Pest Management (IPM). IPM is posited on the premise that certain levels of pests are tolerable. The EIL and ET represent a practical underpinning for the theory of IPM because they provide information on how much pest injury and how many pests are tolerable. And, indeed, without EILs and ETs, it is difficult, if not impossible, to distinguish between the mere presence of a pest species and the need to manage that species based on the injury that it causes. Therefore, economic decision levels are responsible for the evolution of pest management technology from identify-and-spray approaches to IPM approaches.

THE EIL AND ET CONCEPTS

The concept of tolerating pest injury was discussed by scientists as early as 1934. However, the original concepts of the EIL, the ET, economic damage, and pest status were first proposed in a landmark paper by Stern et al. in 1959. The EIL can be thought of as the lowest population density of pests that will cause economic damage. Economic damage is defined as the amount of injury that will justify the cost of control. The ET can be defined as the density of pests at which control measures should be taken to prevent the pest population from reaching the EIL. Although they are not the only decision-making criteria in pest management, the EIL and ET have become the most commonly used.

The EIL

The EIL concept is quite simple. It is merely a cost-benefit equation in which the costs (losses associated with managing the pest) are balanced with the benefits (losses prevented by managing the pest). Even though the original concept of the EIL was published in 1959, it would be another 13 years before a formula for calculating the EIL was promulgated. The EIL actually represents a level of injury, not a density of pests. However, numbers of pests per unit area is often used because injury is difficult to sample and measure. Consequently, pest numbers commonly are used an index for injury. Using pest numbers as an index, EILs may be expressed as "larvae/plant," "beetles/sweep," "milkweed plants/square meter," or "moths/trap."

The most frequently used equation to determine the EIL is $EIL = C/VIDK$, where C = management costs per production unit (e.g., \$/ha), V = market value per production unit (\$/kg), I = injury per pest equivalent, D = damage per unit injury (kg reduction/ha/injury unit), and K = proportional reduction in injury with management.

Despite the simplicity of the EIL concept, there is considerable complexity in the economic and biologic components of the equation. This is especially true for the biological components because injury per pest (I) and yield loss per unit injury (D) are difficult and costly to determine. It is crucial that these components are determined so that pests can be managed effectively within an IPM program. However, because of the difficulty of determining I and D separately, both components often are determined together as simply loss per pest.

The ET

The EIL determines the injury level or pest density at which the pest management cost equals the cost from yield loss if no management occurs. However, the decision to initiate management activities must be made before the EIL is reached so that economic damage can be prevented. In many IPM programs, the decision to initiate management action is based on the ET. Indeed, the ET is the most widely used decision tool in IPM. Although defined in terms of pest density at which management action should be taken, the ET, sometimes called the action threshold, is actually an index for when to implement management activities. For example, if an EIL is 10 insect larvae per plant, then an ET may be 8 larvae per plant. Action would be taken when eight larvae per plant are found not because that density represents an economic loss, but rather because it provides a window of time to take action before the pest density or injury increases to produce economic damage.

DEVELOPMENT OF EILs AND ETs

Development of calculated EILs has increased dramatically since 1972, the year the first calculated EIL was presented. Today, there are more than 200 published articles on EILs and ETs. Economic decision levels have been determined for more than 40 commodities, including most of the world's major crops. Greater than 80% of those articles are on insect pests, with about 10, 6, and 4% on mites, weeds, and plant pathogens, respectively.

Why is there a preponderance of decision levels for arthropods, as evidenced by the scientific literature? The EIL and ET were first defined and used by entomologists primarily because of the pressing need for insecticide resistance management and conservation of natural enemies. Also, arthropods are largely amenable to curative management techniques, making economic decision levels relatively easy to incorporate into IPM programs. Plant diseases and weeds traditionally have been managed preventively, hindering the use of economic decision levels. However, in recent years, new pesticides and technologies, such as herbicide-tolerant crops, have hastened the development of thresholds for diseases and weeds.

LIMITATIONS

Economic decision levels traditionally have been best suited for IPM approaches that involve curative action, such as pest suppression. Consequently, the use of thresholds has been closely tied to the use of pesticides that can be used therapeutically. The EIL and ET concept can be used for other pest management technologies, but, to date, few efficacious and economic alternatives to pesticides have been developed.

The use of EILs and ETs is much more limited when pests almost always cause economic damage, when reliable sampling of injury or pests is difficult, or when curative action is difficult or not available. Most pests that almost always cause economic damage, often called perennial pests, directly infest the marketable product, such as fruits and vegetables. The damage they produce, in addition to the high market value for the commodities, results in EILs that are so low they are practically unusable. Sampling is difficult for many pests, such as plant pathogens. Difficulties in quantifying plant pathogens, in addition to few therapeutic management options, have hindered EIL development.

Difficulties in assigning economic values to the market value of the host have resulted in additional limitations to EIL and ET development. For example, the economic value of a reduction in a host's aesthetic appeal is subjective and difficult to quantify. Other limitations include the relatively high cost of conducting the research necessary to determine EILs, and a lack of knowledge about the interaction between biotic and abiotic stresses on the host.

Development of economic decision levels for veterinary pests has proven difficult because quantifying injury is difficult. Further, ETs for livestock pests, such as house or stable flies, may be considerably above levels that would result in intolerably annoying effects of those pests on humans. Economic decision levels for pests that impact human health, such as disease-carrying mosquitoes and flies, are virtually impossible to use because they would require a market value for human life. Therefore, ETs would be based on the acceptance of a level of morbidity or mortality before management is taken. This is clearly unacceptable from ethical and moral perspectives.

CURRENT AND FUTURE APPROACHES

Advances in the EIL concept have occurred primarily through extensions of the model that Pedigo et al. discussed in 1986. Aesthetic injury levels have been determined based on attributes not readily definable in economic terms, such as form, color, texture, and beauty. To ascertain value of these attributes, researchers have obtained input from the owner or general public using techniques such as the contingent-valuation method. Examples of resources in which aesthetic injury levels could be used are lawns, ornamental plants, homes, and public buildings.

In recent years, environmentally based EILs have been developed and discussed. Each variable in the EIL equation reflects management activities that could be manipulated to potentially enhance environmental sustainability. In particular, researchers have suggested incorporating environmental costs into the management cost variable, C. The resulting EILs have been termed Environmental Economic Injury Levels (EEILs).

Important conceptual advances in the ET have occurred in the past two decades. Perhaps the foremost advance has been the conversion of insect-population estimates into insect-injury equivalents. An insect-injury equivalent is the potential total injury potential of an individual pest if it were to survive through all injurious life stages. Injury equivalents are determined from estimates of pest-population structure, pest density, and injury potential. Incorporation of insect larval survivorship has led to a further refinement of the injury equivalency concept.

Researchers have continued to refine the ET concept, primarily through delineating and naming categories of ETs. Subjective ETs are not based on EILs, whereas objective ETs are based on calculated EILs; therefore, the ETs change with the changing parameters in the EIL equation. Three types of objective ETs have been defined: fixed, descriptive, and dichotomous. Fixed ETs are set at some percentage of the EIL. Descriptive ETs are based on estimates of pest population growth. Consequently, sampling techniques are designed to evaluate the growth potential of the pest population. Dichotomous ETs use statistical techniques to classify a pest population as economic or noneconomic.

Probably the most elusive goal for IPM is the establishment of multiple-species EILs. These EILs represent a significant advance because they provide decision makers with the ability to manage a complex of biotic stressors instead of merely managing single pest species. The primary advances in this area have involved integrating pest injury from different species by characterizing the homogeneities of physiological impact on hosts. If different pest species produce injuries resulting in similar physiological responses, then the pest species can be grouped into injury guilds. The injury guilds can then be used to characterize damage functions. One approach to determine multiple-species EILs has been to combine the injury guild concept with the injury equivalency concept. To develop multiple-species EILs using this concept, pest species must 1) produce a similar type of injury; 2) produce injury within the same physiological time-frame of the host; 3) produce injury of a similar intensity; and 4) affect the same plant part.

Although the EIL concept is biologically based and can accommodate variations in control cost, market value, and damage functions, research to develop EILs and ETs has not kept pace with other pest management areas, such as pesticide resistance management or biological control. Progress in EIL development will occur only with concomitant progress in a conceptual approach to IPM that includes a recognition of the importance of tolerating levels of pest injury. Economic decision levels are central to IPM if a primary goal of IPM is environmental and economic sustainability.

BIBLIOGRAPHY

Higley, L.G.; Pedigo, L.P. Economic injury level concepts and their use in sustaining environmental quality. Agric. Ecosys. Environ. **1993**, *46*, 233–243.

Hutchins, S.H.; Higley, L.G.; Pedigo, L.P. Injury equivalency as a basis for developing multiple-species economic injury levels. J. Econ. Entomol. **1988**, *81*, 1–8.

Ostlie, K.R.; Pedigo, L.P. Incorporating pest survivorship into economic thresholds. Bull. Entomol. Soc. Am. **1987**, *33*, 98–102.

Pedigo, L.P. *Entomology and Pest Management*, 3rd Ed.; Prentice-Hall: Englewood Cliffs, NJ, 1999; 691.

Pedigo, L.P., Higley, L.G., Eds.; *Economic Thresholds for Integrated Pest Management;* University of Nebraska Press: Lincoln, NE, 1996; 327.

Pedigo, L.P.; Hutchins, S.H.; Higley, L.G. Economic injury levels in theory and practice. Annu. Rev. Entomol. **1986**, *31*, 341–368.

Peterson, R.K.D. The Status of Economic-Decision-Level Development. In *Economic Thresholds for Integrated Pest Management*; Higley, L.G., Pedigo, L.P., Eds.; University of Nebraska Press: Lincoln, NE, 1996; 151–178.

Peterson, R.K.D.; Danielson, S.D.; Higley, L.G. Yield responses of alfalfa to simulated alfalfa weevil injury and development of economic injury levels. Agron. J. **1993**, *85*, 595–601.

Stern, V.M.; Smith, R.F.; van den Bosch, R.; Hagen, K.S. The integrated control concept. Hilgardia **1959**, *29*, 81–101.

Stone, J.D.; Pedigo, L.P. Development of economic-injury level of the green cloverworm on soybean in Iowa. J. Econ. Entomol. **1972**, *65*, 197–201.

EFFECTS OF PESTICIDES

Ana María Evangelista de Duffard
Universidad Nacional de Rosario, Rosario, Argentina

E

INTRODUCTION

When an organism's homeostasis is altered, the animal's life is in danger. Neurobehavioral toxicology is specially concerned with the behavioral plasticities involved in normal adjustments—homeostatic processes—to the constantly changing physical and psychosocial environments, through habituation, tolerance, learning, and memory development (1). Behavior in the wild is adaptative and context dependent. Parent behavior, for example, is a complex group of disparate activities (nest building, retrieval of young, defense from predators, feeding, etc.). Failure to perform optimally in any of these activities will result in decreased survivability of the young.

The chemicals that constitute the class of pesticides with potential neurotoxicity vary widely in structure and include chlorinated hydrocarbons, organophosphates, carbamates, and pyrethroids. Although most of them are insecticides, some are also used as rodenticides, acaricides, fungicides, or herbicides.

Studies of captive and free-living birds provide support about a change in behavior can be expected when brain acetylcholinesterase (AChE) activity falls below about 50% normal. Organophosphates and carbamates (2) have been related to avian mortality (50% chronic and 80% acute of AChE inhibition). Bobwhite quail (*Colinus virginianus*) treated with methyl parathion were more likely to be caught and killed by a domestic cat introduced into an observational field. In addition, a decrease of falcons' nest-defense behaviors has been shown to be an important factor in the decline of several species of birds of prey, and was correlated with both the degree of eggshell thinning and egg residue organochlorine levels in the eggs (2).

Because bees often come in close proximity to pesticides and they are divided into different age-based castes and labors, they can also be adversely affected. Bee dancing is disrupted by exposure to methylparathion or permethrin, a change that can reduce a bee colony's chances of survival and produce a substantial economic impact for those that depend on bees as crop pollinators (3). In addition, exposure to atrazine or diuron can affect various behaviors of fish by altering the chemical perception of natural substances of eco-ethological importance (4).

Biocides pass through the placenta and have been found in the milk of both animals and human beings. In many developing countries, nursing infants are potentially ingesting organohalogens at a ratio many times that of the acceptable daily intakes as estimated by the Food and Agricultural Organization (5). Nutritional animal state (specifically malnutrition), is another variable that may affect an organism's susceptibility to pesticides. Supporting that, offspring from mother rats fed with dieldrin at levels often found in the environment and with a low protein diet showed altered behavioral effects (6).

The actions of xenobiotics on an immature brain is very different from their action on adult animals. In the development of a mammal there is a period of rapid brain growth that may be critical for its normal maturation of axonal and dendritic outgrowth and the period when the synaptogenesis takes place. These periods are strongly modified by the timing and duration of chemical exposure as well as by the dose. Neonatal exposure to a single low oral dose of DDT (1.4 μmol/kg b.wt.) can lead to a permanent hyperactive condition in adult mice (7). This amount of DDT is of physiological significance, since it is of the same order of magnitude as that to which animals and man can be exposed during the lactation period. The consequence of the early exposure is quite different from that reported for animals exposed to DDT as adults. The DDT dose required in adult animals to provoke symptoms and effects such as ataxia, tremor, increased activity in open-field test and avoidance responding, is more than 50–200 times the dose for neonatal exposure to produce toxic effects. In addition, behavioral alterations induced in rats by a pre- and postnatal exposure to the herbicide 2, 4-Dichlorophenoxyacetic acid (2, 4-D) was demonstrated in our laboratory (8).

Behavioral assessment is important because it is often the case that one of the earliest indications of exposure to neurotoxicants is subtle behavioral impairment such as paraesthesia or short-term memory dysfunction; frequently, such behavioral effects precede more obvious and frank neurological signs. In humans, volunteer studies and case reports have shown that exposure to dicofol results in disturbance of equilibrium, dizziness, confusion, headaches, tremors, fatigue, vomiting, twitching, seizures, and loss of consciousness (9).

To assess the effects of toxicants on the nervous system two levels of sensitivity/complexity were described by Tilson et al. (10). At the first level of behavioral analysis, those procedures that require little or no training of experimental animals have the capacity to include large numbers of subjects. They provide an indication that a neurobehavioral deficit is present or absent, and permit scientists to quantify any deficits as precisely as possible, and are referred to as screening techniques. Examples of such tests include simple measures of locomotor activity, sensorimotor reflexes, and neurological signs. Moser's laboratory has been evaluating neurotoxicants as well as several nonneurotoxicants (negative controls) in the Functional Observational Battery (FOB) to establish selectivity, reliability, redundancy, specificity, and sensitivity of the individual tests as well as the battery as a whole (11). FOB is a series of tests to assess sensory, neuromuscular, and autonomic functions in animals and is similar to clinical neurological examinations in humans in that it rates the presence and, in some cases, the severity of behavioral and neurological signs (12). In addition, a continuous recording of a rat's activity using a residential maze in which infrared optical gates are connected to a digital computer was used by Elsner et al. (13) to register the effects of lindane, dichlorvos, etc., on the spontaneous rat's behavior.

Sex and the physiological state of an animal are other important factors. We demonstrated that oral administration of 2, 4-D butyl ester (2, 4-Dbe) to nulliparous females had no effects on either open field (OF) and rotarod performance. By contrast, dams treated with 2,4-Dbe during pregnancy and intact male rats exhibited impairments of OF activity and rotarod endurance (14).

At the second level of evaluation, tests requiring extended or special training, frequent evaluation (i.e., daily sessions) and/or manipulation of motivational factors such as food deprivation or electric footshock are used. These procedures may be useful in estimating environmentally acceptable limits such as nonobservable-adverse-effect level (NOAEL) or lowest-observed-adversed-effect level (LOAEL). Examples of these tests include discriminated conditioned response methodologies to assess specific sensory or motor dysfunction and procedures to measure chemical-induced alterations in cognitive function (15). The data available suggest that organochlorine pesticides may disrupt performance differentially. Burt (16) and Desi (17) observed that dieldrin decreased overall rates of fixed-interval responding and disrupted the within-interval pattern of responding in rats and Japanese quail maintained under fixed-interval schedules of food reinforcement. Dietz and McMillan (18) compared the effects of daily administration of mirex and chlordecone on the performance of rats under several schedules of reinforcement. Both pesticides produced delayed disruption in performance inversely proportional to the dosage administered daily. Social interaction, plus-maze behavior, and one-way passive avoidance were studied in rats orally treated with fenvalerate to determine the anxiolitic effect of this pesticide (19).

The concept that a challenge to a system may overcome compensatory mechanisms and thereby reveal otherwise hidden neurotoxicant induced damage is used as a method of assessment in neurobehavioral toxicology. We demonstrated hidden 2,4-D latent psychotic effects, the Serotonergic Syndrome, through amphetamine to an organism previously exposed to the 2,4-D. These rats also showed some dopaminergic behavior such as rearing and catalepsy if they were challenged with haloperidol in a noncataleptic dose of this drug (20).

Since behavior is the net result of integrated sensory, motor, and cognitive function occurring in the nervous system, pesticide-induced changes in behavior may be a relatively sensitive indicator of nervous system dysfunction, which may hamper survivability possibilities in wildlife.

REFERENCES

1. Russel, R.W.; Singer, G. Neurobehavioral toxicology: a view from "down under," Neurobehav. Toxicol. Teratol. **1982**, *4*, 5–7.
2. Peakall, D. Disrupted patterns of behavior in natural populations as an index of ecotoxicity. Neurobehav. Toxicol. **1996**, *104* (Environmental Health Perspectives Suppl.), 331–335.
3. Cohn, J.; MacPhail, R. Ethological and experimental approaches to behavior analysis: implications for ecotoxicology. Neurobehav. Toxicol. **1996**, *104* (Environmental Health Perspectives Suppl.), 299–305.
4. Saglio, P.; Trijasse, S. Behavioral responses to atrazine and diuron in goldfish. Arch. Environ. Contam. Toxicol. **1998**, *35*, 484–491.
5. FAO/WHO. Guidelines for predicting the dietary intake of pesticide residues. Bull. WHO **1988**, *66*, 429–434.
6. Olson, K.L.; Boush, G.M.; Matsumura, S. Pre- and postnatal exposure to dieldrin: persistent stimulatory and behavioral effects. Pestic. Biochem. Physiol. **1980**, *13*, 20–33.

7. Eriksson, P.; Archer, T.; Fredriksson, A. Altered behaviour in adult mice exposed to a single low dose of DDT and its fatty acid conjugate as neonates. Brain Res. **1990**, *514*, 141–142.
8. Bortolozzi, A.; Duffard, R.; Evangelista de Duffard, A.M. Behavioral alterations induced in rats by a pre- and post exposure to 2,4-dichlorophenoxyacetic acid. Neurotol.Toxicol. **1999**, *21*, 451–465.
9. Hayes, W.J., Jr. *Pesticides Studies in Man*; Williams and Wilkins: Baltimore, 1982; 180–208.
10. Tilson, H.A.; Cabe, P.A.; Burne, T.A. *Experimental and Clinical Neurotoxicology*; Ch. 51, Williams and Wilkins: Baltimore, 1980; 758–766.
11. Moser, V.C. Screening approaches to neurotoxicity: a functional observational battery. J. Am. Coll. Toxicol. **1989**, *8*, 85–93.
12. Tilson, H.A.; Moser, V. Comparison of screening approaches. Neurotoxicology **1992**, *13*, 1–14.
13. Elsner, J.; Loosed, R.; Zbinden, G. Quantitative analysis of rat behavior patterns in a residential maze. Neurobehav. Toxicol. **1979,** (Suppl. I), 163–174.
14. Evangelista de Duffard, A.M.; Orta, C.; Duffard, R. Behavioral changes in rats fed a diet containing 2,4-dichlorophenoxyacetic butyl ester. Neurotoxicology **1990**, *11*, 563–572.
15. Tilson, H.A. Neurobehavioral methods used in neurotoxicological research. Toxicol. Lett. **1993**, *68*, 231–240.
16. Burt, G.A. Use of Behavioral Techniques in the Assessment of Environmental Contaminants. In *Behavioral Toxicology*; Weiss, B., Laties, V.G., Eds.; Plenum Press: New York, 1975; 241–263.
17. Desi, I. Neurotoxicological effects of small quantities of lindane. Animal Studies. Int. Arch. Arbeitsmed. **1974**, *33*, 153–162.
18. Dietz, D.D.; McMillan, D.E. Comparative effects of mirex and kepone on schedule-controlled behavior in the rat. II. spaced-responding, fixed-ratio, and unsignalled avoidance schedules. Neurotoxicology **1979**, *1*, 387–402.
19. De Spouzza Spinoza, H.; Silva, Y.M.; Nicolau, A.A.; Bernardi, M.M.; Luciano, A. Possible anxiogenic effects of fenvalerate, a type II pyretroid pesticide in rats. Physiol. Behav. **1999**, *67*, 611–615.
20. Evangelista de Duffard, A.M.; Bortolozzi, A.; Duffard, R. Altered behavioral response in 2,4-dichlorophenoxyacetic acid treated and amphetamine challenged rats. Neurotoxicology **1995**, *16*, 479–488.

EFFECTS OF TRANSGENIC (Bt) CROPS ON NATURAL ENEMIES

Salvatore Arpaia
Metapontum Agrobios, Metaponto, MT, Italy

INTRODUCTION

Transgenic plants resistant to insects (TPRI) via expression of Cry toxins derived from *Bacillus thuringiensis* Berl. (Bt) genes, are being cultivated every year on a larger scale. The rate at which these new varieties are being adopted is quite different among world geographic areas; in the United States and in China for instance, TPRIs seem to benefit from a wider acceptance by policymakers and consumers, while in Europe the number of transgenic crops still waiting for approval is increasing. Apart from the worry of potential dangers of transgenic food to the human health, there is great concern about the impact of TPRIs on the environment.

One of the perceived threats of use of insect-resistant transgenic crops is the possible effect on nontarget insect biodiversity. To benefit optimally from biological control, natural enemies ideally should not be adversely affected in the long run when feeding on herbivorous prey that have ingested Bt toxins.

It must be admitted that, in spite of seeing TPRIs quite widespread in agricultural landscapes, a sound scientific base for resolving their impact in agroecosystems is still lacking.

While insect pests' natural enemies belong to different taxonomic groups that could be affected by Cry toxins (1), we will focus our attention on predator and parasitoid insects, which comprise the largest number of species active in agroecosystems.

CRY PROTEIN TOXICITY FOR HERBIVORE NATURAL ENEMIES

There is some rationale that lies behind the worry of negative side effects due to the presence of Cry toxins in transgenic plants. While Bt-based insecticides are correctly considered environmentally safe biopesticides, because of their specific mode of action and their fast degradation in the environment, this scenario will likely be different in transgenic fields where the plants will produce, under constitutive promoters, Cry toxins in high concentrations in every plant tissue all season long. Therefore, the probability that the toxin will be ingested by pests' natural enemies will become more frequent.

The main adverse impact of TPRIs on the third trophic level may be due to the depletion of nutritional sources as the prey become intoxicated by the foreign protein expressed in host plants. As it concerns predators, unintended toxin ingestion may occur both upon eating herbivore larvae that had been feeding on transgenic plants, and when fed high concentrations of Cry toxins directly (e.g., via pollen ingestion). For instance, laboratory experiments showed that Cry1Ab can cause increased mortality in lacewing larvae directly, although it does so only at high concentrations; when the toxin was provided via prey to the lacewing larvae, much lower toxin concentrations resulted in similar mortality (2). What is the explanation for these findings is not completely known, as accurate information about processing and possible concentrations of toxins in nontarget insects is lacking.

Parasitoids might also be affected in their action by the presence of transgenic Bt plants. This action may be exerted through an unbalanced development of parasitoid larvae in nonoptimal prey, and, consequently, negative effects on adult fecundity. More subtle negative effects may be due to the change of attractiveness of the prey for the parasitoid, and the possible changes of plant volatiles that are important in prey searching behavior of parasitoids and predators as well.

Also, other herbivores represent potential toxin carriers to the next trophic level. Nontarget herbivores will remain in a Bt-crop field, where they constitute prey for natural enemies, but are also candidates for becoming a secondary pest. The latter will depend, among other factors, on the capacity of the natural enemies to keep these herbivores in check in a Bt-crop field. Therefore, the degree to which natural enemies may be affected by toxic host plants will depend primarily on two factors: if, and to what extent, are the remaining nontarget herbivores affected by the toxin and if, and to what extent, the Cry toxin expressed in TPRI-fed herbivore prey affects natural enemies directly.

A further important point is the prey range of natural enemies, since these negative effects are likely to be "diluted" in the case of more generalist predators. Predators like lacewings and ladybirds may feed on eggs and young larvae of the target insect (e.g., Colorado potato beetle, European corn borer, etc.), but they can find alternative prey (e.g., aphids) for their development. The presence of Cry toxin-bearing pollen in the agroecosystem, which can be ingested by adult predators, makes the final outcome even more complicated and unpredictable.

PREDICTING INSECT POPULATION DYNAMICS IN TRANSGENIC AGROECOSYSTEMS

Due to the complex interactions involved at several trophic levels, the prevision of the effect due to the introduction of Bt-expressing TPRIs in an agroecosystem is not an easy task.

As we have learned from "traditional" host plant resistance studies, interactions between natural enemies and plants resistant to insects may lead to considering the two as compatible, synergistic, or alternative strategies.

Laboratory studies have sometimes identified possible negative effects on predators. When these negative effects were found, the toxins led to dramatic reduction in adult emergence [78–90% in *Myiopharus doriphorae* on Bt-treated Colorado potato beetle larvae (3)], delayed development [*Chrysoperla carnea* on Bt-fed European corn borer, but not for Bt-fed *Spodoptera littoralis* (4)], lower parasitism rate [*Campoletis sonorensis* on *Heliothis virescens* young larvae in short trials (5)], and increased larval mortality [lacewing larvae reared on Bt-fed *Ostrinia nubilalis* and *Spodoptera littoralis* (4)]. It is to be noted however, that some authors working with the same insects did not achieve similar results in other feeding trials (6).

Conversely, field studies so far have showed comparable dynamics only between natural enemy populations in Bt and non-Bt plots of experimental fields (Table 1).

It must be recognized, though, that the number and the length of these field surveys is still too low to consider those examples as really compatible interactions between TPRIs and natural enemies. Also, at this early stage, it may be wise to start considering some possible tactics in order to avoid or to contain negative effects on the third trophic level. Basically, we could resort to the same shrewdness adopted in order to delay pest adaptation to Cry toxins expressed in transgenic plants. The rationale of these approaches in fact, is to limit unwanted exposure to Cry toxin with the aim of preserving some individuals

Table 1 Field studies on population dynamics of target insect's natural enemies in transgenic agroecosystems

Host plant	Target insect	Beneficial	Effect	Location
Tobacco (low expression)	*Heliothis virescens*	*Campoletis sonorensis*	Increased pest mortality	United States (5)
		Cardiochiles nigriceps	No different mortality	
Corn	*Ostrinia nubilalis*	*Coleomegilla maculata*	No different predation	United States (8)
		Orius insidiosus	No different predation	
		Various parasitoids	No different parasitism	
Corn	*Ostrinia nubilalis*	*Coleomegilla maculata*	Comparable predator abundance and survival	United States (6)
		Orius insidiosus	Comparable predator abundance and survival	
		Chrysoperla carnea	Comparable predator abundance and survival	
Corn	*Ostrinia nubilalis*	Coccinellids and parasitoids	Similar abundance of beneficials	Italy (9)
Cotton	*Pectinophora gossypiella*	Various predators	Similar abundance	United States (10)
Eggplant	*Leptinotarsa decemlineata*	*Chrysoperla carnea*	Similar abundance of predators	United States (11)

from its toxic effects. We could, for instance, evaluate the efficacy of refuges as possible in-field areas where more healthy prey available for predation will be preserved.

Conversely, the use of transgenic plants expressing sublethal doses of Cry toxins or the obtainment of transgenic plants with inducible expression of Cry toxins seems less practical because of the technical constraints that discourage their use as resistant varieties (7).

FUTURE RESEARCH NEEDS

As mentioned before, a possible negative effect of Cry toxins on some predator larvae is real. Therefore, the first objective to be achieved is a better understanding of the mode of action of Cry toxins in predators. This involves testing for binding and pore formation capacity of these toxins to epithelial gut membrane vesicles from predator larvae since the occurrence of both phenomena is correlated with the activity of a Cry toxin in a particular insect.

Another major gap that needs to be filled is the realization of pluriannual surveys of the entomophauna in transgenic agroecosystems where comparative predator/prey dynamics and ratios should be provided. Further studies, such as the chemical evaluation of the "green odor" of transgenic plants, may lead to less unpredictable foresight about the effects of TPRIs on higher trophic levels.

Generally speaking, it is wise to suggest a deep analysis of every single agroecosystem of interest in order to gain a more comprehensive picture of the sustainability of each TPRI as means of pest control based on sound scientific information.

REFERENCES

1. Paoletti, M.G.; Pimentel, D. The environmental and economic costs of herbicide resistance and host-plant resistance and host-plant resistance to plant pathogens and insects. Technological Forecasting and Social Change **1995**, *50*, 9–23.
2. Hilbeck, A.; Moar, W.J.; Pusztai-Carey, M.; Filippini, A.; Bigler, F. Toxicity of *Bacillus thuringiensis* Cry1Ab toxin to the predator *Chrysoperla carnea* (Neuroptera: Chrysopidae). Environ. Entomol. **1998**, *27* (5), 1255–1263.
3. Lopez, R.; Ferro, D.N. Larviposition of myiopharus doriphorae (Diptera: Tachinidae) to Colorado potato beetle (Coleoptera: Chrysomelidae) larvae treated with lethal and sublethal doses of Bacillus thuringiensis Berl. subsp. *Tenebrionis*. J. Econ. Entomol. **1995**, *88*, 870–874.
4. Hilbeck, A.; Baumgartner, M.; Fried, P.M.; Bigler, F. Effects of transgenic Bt-corn fed prey on immature development of *Chrysoperla carnea*. Environ. Entomol. **1998**, *27* (2), 480–487.
5. Johnson, M.T. Interactions of resistant plants and wasp parasitoids of tobacco budworm (Lepidoptera: Noctuidae). Environ. Entomol. **1997**, *26* (2), 207–214.
6. Pilcher, C.D.; Obrycki, J.J.; Rice, M.E.; Lewis, L.C. Preimaginal development, survival and field abundance of insect predators on transgenic *Bacillus thuringiensis* corn. Environ. Entomol. **1997**, *26* (2), 446–454.
7. Gould, F. Sustainability of transgenic insecticidal cultivars: integrating pest genetics and ecology. Annu. Rev. Entomol. **1998**, *43*, 701–726.
8. Orr, D.B.; Landis, D.A. Oviposition of european corn borer (Lepidoptera: Piralidae) and impact of natural enemy populations in transgenic versus isogenic corn. J. Econ. Entomol. **1997**, *90* (4), 905–909.
9. Lozzia, G.C.; Rigamonti, I.E. Prime osservazioni sull'artropodofauna presente in campi di mais transgenico. Atti Giornate Fitopatologiche **1998**, 223–228.
10. Wilson, F.D.; Flint, H.M.; Deaton, W.R.; Fischhoff, D.A.; Perlak, F.J.; Armstrong, T.A.; Fuchs, R.L.; Berberich, S.A.; Parks, N.J.; Stapp, B.R. Resistance of cotton lines containing *Bacillus thuringiensis* toxin to pink bollwon (Lepdoptera: Gelechidae) and other insects. J. Econ. Entomol. **1992**, *85* (4), 1516–1521.
11. Arpaia, S.; Acciarri, N.; Di Leo, G.M.; Mennella, G.; Sabino, G.; Sunseri, F.; Rotino, G.L. In *Field Performance of Bt-expressing Transgenic Eggplant Lines Resistant to Colorado Potato Beetle*, Proceedings of the Xth EUCARPIA Meeting on Capsicum and Eggplant, Avignon, Sept. 7–11, 1998; INRA: Paris, 1998; 191–194

ENVIRONMENTAL AND ECONOMIC COSTS OF PESTICIDE USE

E

Kelsey A. Hart
David Pimentel
Cornell University, Ithaca, New York, U.S.A.

INTRODUCTION

The use of pesticides makes a significant contribution to maintaining world food production. It is estimated that if no pesticides were used at all, losses to pests would increase by 10%; specific crop losses could range from zero to nearly 100% (1). In general, each dollar invested in pesticide control returns about $4 in crops saved—each year in the United States, pesticides generally provide about $16 billion per year in saved crops (2). However, most of the commonly cited benefits of pesticides are based only on these direct crop returns; such assessments do not consider the indirect, but substantial, environmental and economic costs associated with pesticide use. It has been estimated that only 0.1% of applied pesticides reach the target pests, while the bulk of each pesticide application (99.9%) is left to impact the surrounding environment (1). These environmental impacts—and their related economic costs, which have been estimated to total about $8 billion per year (2)—must be closely examined to facilitate the development and implementation of an environmentally sound policy of pesticide use.

ENVIRONMENTAL IMPACTS AND RELATED COSTS

Destruction of Beneficial Natural Predators and Parasites

In both natural and agricultural ecosystems, many species of predators and parasites assist in controlling herbivorous populations. These naturally beneficial species can help ecosystems remain foliated and "green," which is vital to maintaining environmental health and clean air. Beneficial parasites and natural predators help to keep herbivore populations at low levels and limit plant losses. Like pest species, though, these beneficial natural enemies are adversely affected by pesticides. For example, the following pests have reached outbreak levels in cotton crops following the destruction of their natural enemies by pesticides: bollworm, tobacco budworm, cotton aphid, spider mite, and cotton looper. Significant pest outbreaks also have occurred in other crops.

When outbreaks of secondary pests occur because their natural enemies have been destroyed by pesticides, additional—and sometimes more expensive and toxic—pesticide treatments are often required to sustain crop yields, raising overall costs and exacerbating pesticide-related problems. In fact, it is estimated that the destruction of natural enemies by pesticides, the subsequent crop losses, and additional pesticide applications cost the United States $520 million every year (Table 1) (1).

Pesticide Resistance in Pests

The extensive use of pesticides can also result in the development of pesticide resistance in insect pests, plant pathogens, and weeds. In a report by the United Nations' Environment Program, pesticide resistance was ranked as one of the top four environmental problems in the world. About 504 insect and mite species, a total of nearly 150 plant pathogen species, and about 273 weed species are now resistant to pesticides (2).

Increased pesticide resistance in pest populations frequently results in the need for several additional applications of commonly used pesticides to maintain expected crop yields. These additional applications can cause increased environmental contamination, as well as significant economic losses due to increased expenditures and crop losses from pesticide contamination of products. One study reported a yearly loss of $45 to $120/ha to pesticide resistance in California cotton alone (3). Thus, approximately $348 million of the California cotton crop was lost to resistance. Since $3.6 billion worth of U.S. cotton was harvested in 1984, the loss due to resistance for that year was approximately 10%. Assuming a 10% loss in other major crops that receive heavy pesticide treatments in the United States, crop losses due to pesticide resistance are estimated to be about $1.4 billion/yr (2).

Crop and Crop Product Losses

Pesticides are applied to protect crops from pests in order to increase harvests, but sometimes the crops themselves are damaged by pesticide treatment. Damage occurs when 1) the recommended dosages suppress crop growth, development, and yield; 2) pesticides drift from the targeted crop to damage adjacent crops; 3) residual herbicides either prevent chemical-sensitive crops from being planted in rotation or inhibit the growth of crops that are planted; and/or 4) excessive pesticide residues accumulate on crops, necessitating the destruction of the harvest. Crop losses translate into financial losses for growers, distributors, wholesalers, transporters, retailers, food processors, and others, as potential profits as well as investments are lost.

The costs of crop losses increase even further when the related costs of investigations, regulation, insurance, and litigation are added. When crop seizures, insurance, and investigation costs are added to the costs of direct crop losses due to the use of pesticides in commercial crop production, the total monetary loss is estimated to be about $959 million annually in the United States alone (2). These huge financial losses hurt everyone—farmers, consumers, and the overall economy.

Groundwater and Surface Water Contamination

Most soluble pesticides applied to crops eventually drift and/or erode into ground and surface waters, and some pesticides even have been detected in rain and fog. It has been estimated that the total runoff of all pesticides from nonirrigated farmland is 1%, runoff from irrigated farmland is 4%, and the runoff from airplane-applied pesticides is 33% (2).

Pesticide contamination of surface water—lakes, rivers, and streams—is a serious concern because surface water is used extensively for drinking and recreation. It is particularly alarming that conventional drinking water treatment is not designed for, *and therefore does not remove*, pesticides. One study showed that after conventional treatment, 90% of drinking water samples contained at least one pesticide, while 58% of the samples contained at least four different pesticides (4). In addition, estimates suggest that nearly one-half of the groundwater and well water in the United States is or has the potential to be contaminated. Specifically, groundwater contaminated with pesticides is a huge problem for two major reasons: first, about one-half of the population obtains its water from wells, and second, once groundwater is contaminated, the pesticide residues remain for long periods of time. It would cost an estimated $1.3 billion annually, though, if well and groundwater were monitored for pesticide residues in the United States. Including an additional $500 million for clean-up costs, the total cost regarding pesticide-polluted groundwater is estimated to be about $1.8 billion annually (2).

Wild Birds and Mammals

Like fish, wild birds and mammals are also damaged by pesticides; these animals make excellent "indicator species"—species whose numbers, if falling, can indicate an environmental problem. Deleterious effects on wildlife include death from direct exposure to pesticides or secondary poisonings from consuming contaminated prey; reduced survival, growth, and reproductive rates from exposure to sublethal dosages; and habitat reduction through elimination of food sources and refuges. In the United States, approximately 3 kg of pesticide per hectare are applied on about 160 million ha/yr of land (5). With such a large portion of the land area treated with heavy dosages of pesticide, the impact on wildlife is expected to be significant. The full extent of bird and mammal destruction is difficult to determine exactly, though, because animals are often secretive, camouflaged, highly mobile, and live in dense grass, shrubs, and trees.

Nevertheless, many wildlife casualties caused by pesticides have been reported. Some highly toxic or heavily applied pesticides kill birds—and other animals—directly. In addition to directly killing birds, the indirect effects of pesticide exposure—such as the elimination of food supply or habitats, and/or reproductive effects—can also adversely affect bird and wildlife species.

Although gross economic values for wildlife are not available, the money that humans spend on wildlife-related activities is one measure of the monetary value of

Table 1 Total estimated environmental and social costs from pesticides in the United States[a]

Costs	Millions of $/year
Public health impacts	933
Domestic animal deaths and contamination	31
Loss of natural enemies	520
Cost of pesticide resistance	1400
Honey bee and pollination losses	320
Crop losses	959
Surface water monitoring	27
Groundwater contamination	1800
Fishery losses	56
Bird losses	2100
Government regulations to prevent damage	200
Total	8,346

[a]See Refs. 2 and 5 for a detailed description of how the values in Table 1 were determined.

wild birds and mammals. If we assume that the damages pesticides inflict on birds occur primarily on the 160 million hectares of cropland that receive most of the pesticides applied, and if the bird population is estimated to be 4.2 birds per hectare of cropland (6), then 672 million birds are directly exposed to pesticides. If it is conservatively estimated that only 10% of the bird population is killed, then the total number killed each year is 67 million birds. This is considered a conservative estimate because secondary losses to pesticide-related reductions in invertebrate prey poisonings are not included in the assessment. Estimates for the value of all types of birds range from \$0.40 to more than \$800/bird (5). Assuming the average value of a bird is \$30, then an estimated \$2 billion in birds are destroyed by pesticides each year.

Domestic Animal Poisonings and Contaminated Animal Products

Several thousand domestic animals are poisoned by pesticides each year, mainly dogs and cats. Estimates indicate that about 20% of the total monetary value of animal production, or about \$4.2 billion, is lost to all animal illnesses, including pesticide poisonings (7). Specifically, about 0.5% of animal illnesses and 0.04% of all animal deaths reported to a veterinary diagnostic laboratory were due to pesticide toxicosis, so about \$22 and \$9.5 million are lost to pesticide poisonings and pesticide-related deaths, respectively (8).

Contamination of animal products—milk, meat, and eggs—with pesticides can lead to even more significant monetary losses. Pesticides applied to feed crops, farm buildings, or directly to the animal for pest control can build up in animal products and warrant them unsafe for human consumption. Animal products sold for human use are inspected for contamination by the National Residue Program (NRP). However, of the more than 600 pesticides now in use, the NRP tests only for the 41 pesticides that the FDA and EPA have determined to be a public health concern. Animal products determined to contain pesticide residues must be disposed of, sometimes at a tremendous financial loss; the total costs attributed to domestic animal poisonings and contaminated animal products is estimated at about \$32 million annually (2).

Honey Bee and Wild Bee Poisonings and Reduced Pollination

Honey and wild bees are vital for pollination of fruits, vegetables, and other crops. It has been estimated that production of approximately one-third of all human food is dependent on bee pollination. The direct and indirect benefits of bees to agricultural production range from \$10 to \$33 billion each year in the United States (2). Because most agricultural insecticides are toxic to bees, pesticides have a major impact on both honey bee and wild bee populations.

D. Mayer at Washington State University estimates that approximately 20% of all honey bee colonies are adversely affected by pesticides. Based on analyses of honey bee and pollinator losses caused by pesticides, pollination losses attributed to pesticides are about 10% of pollinated crops, at a total yearly cost of about \$200 million. The combined annual costs of reduced pollination and direct loss of honey bees due to pesticides can be estimated at about \$320 million (2).

CONCLUSIONS AND FUTURE DIRECTIONS

In total, the environmental and public health impacts of pesticides cost more than \$8 billion each year. While the economic return from crops saved by pesticides is significant, the environmental and economic costs of pesticide use can not be ignored. The judicious, responsible use of pesticides, through competent pesticide regulations and integrated pest management programs, can maximize the benefits of pesticides in crops saved while limiting the costs and protecting public health and the environment.

REFERENCES

1. Pimentel, D.; Greiner, A. Environmental and Socio-Economic Costs of Pesticide Use. In *Techniques for Reducing Pesticide Use*; Pimentel, D., Ed.; John Wiley & Sons: Chichester, UK, 1997; 51–78.
2. Pimentel, D.; Hart, K.A. Ethical, Environmental and Public Health Implications of Pesticide Use. In *New Perspectives in Bioethics*; Galston, A., Ed.; Johns Hopkins University Press: Washington, DC, 1997, *in press.*
3. Carrasco-Tauber, C. *Pesticide Productivity Revisited*; M.S. Thesis, University of Massachusetts: Amherst, MA, 1989.
4. Kelley, R.D. In *Pesticides in Iowa's Drinking Water,* Proceedings of a Conference: Pesticides and Groundwater: A Health Concern for the Midwest, Navare, MN, October 16–17, 1989; 121–122.
5. Pimentel, D.; Acquay, H.; Biltonen, M.; Rice, P.; Silva, M.; Nelson, J.; Lipner, V.; Giordano, S.; Horowitz, A.; D'Amore, M. Assessment of Environmental and Economic Impacts of Pesticide Use. In *The Pesticide Question: Environment, Economics and Ethics*; Pimentel, D., Lehman, H., Eds.; Chapman & Hall: New York, 1993; 47–84.
6. Blew, J.H. Breeding bird census; 92 conventional cash crop farm. J. Field Ornithology **1990**, *61* (Suppl.), 80–81.
7. Gaafar, S.M.; Howard, W.E.; Marsh, R. *World Animal Science B: Parasites, Pests, and Predators*; Elsevier: Amsterdam, 1985.
8. Colvin, B.M. Pesticide uses and animal toxicoses. Vet. Human Toxicol. **1987**, *29* (Suppl. 2), 15.

ENVIRONMENTAL CONTROLS

Jeffrey A. Lockwood
University of Wyoming, Laramie, Wyoming, U.S.A.

INTRODUCTION

Environmental control of pests is a strategy that consists of a diverse set of tactics intended to supplement, augment, or otherwise work with natural processes to create a state of ecological stability and resilience that represents a balance between the interests of agricultural production and the needs of a dynamic and healthy ecosystem. These methods emphasize prevention or suppression of pests rather than curative intervention. Some practitioners contend that the notion of "environmental control" is oxymoronic, that to work with—rather than against—natural processes is to replace the notion of control with that of management, collaboration, or partnership. For such methods to be environmentally based they must be integrated with the other ecological and sociological elements. However, fully integrated pest management approaches rarely have been achieved in practice, and perhaps no two systems will ultimately employ precisely the same methods in the same spatiotemporal context. Thus, the tactics can be presented as a set of tools that must be selected, integrated, and employed to meet the complex ecological and sociological (historical, legal, political, cultural, and economic) conditions of particular agricultural systems.

PREVENTION TACTICS

The purpose of these tactics is to prevent the potential pest from arriving, becoming established, or persisting in the agroecosystem. The concept of an economic injury level (or threshold) is inapplicable, because in this approach the agriculturalist attempts to avoid having a potentially damaging species present. Purely preventative tactics exclude the pest via regulatory procedures, eradication, or ecological evasion. To the extent that any of these tactics fail to prevent a pest species from occurring in a system, the methods become avoidance tactics (see later).

Regulatory and Quarantine Methods

Perhaps the most common and effective means of preventing pest damage is to establish and enforce regulations that preclude the introduction of new, potential pests into agricultural regions. The U.S. Department of Agriculture's Animal and Plant Health Inspection Service (Plant Protection and Quarantine and Veterinary Services) is primarily responsible for assuring that exotic pests are not introduced into the United States; other countries have similar quarantine services.

Eradication Methods

Absolute eradication

If a new pest is found in an area, it may be possible to eradicate the incipient population before it becomes established and spreads (e.g., isolated infestations of the Mediterranean fruit fly in the continental United States). In other cases, areawide eradication is used to eliminate a pest that is already established (e.g., screwworm fly and the ongoing cotton boll weevil project), but success in these instances depends on a complex set of ecological and operational conditions.

Relative eradication

Local eradication of a pest can be pursued through eliminating the organism, alternate hosts, or suitable environmental conditions. A variety of plant diseases can be extirpated via removal of infected crop refuse (often costly), burning (timing is critical), flooding (often costly, short-lived, and damaging to the soil), pruning (in urban settings and orchards), and roguing (purging infected plants or alternate hosts, as in the case of eradicating barberry to eliminate stem rust of cereals).

Evasive Methods

Host phenology and temporal evasion

Some herbivores and plant pathogens can be extirpated (true eradication is rare) by altering agricultural practices so that the host is absent during a critical period of the pest's life history. Both within-year (planting or harvesting dates) and between-year (rotation) practices evade a pest. Fly-free planting dates for wheat virtually eliminated Hessian fly infestations and several plant pathogens can be essentially eliminated by planting nonhost crops for a period of time.

Host selection and spatial evasion

Many pests can be locally eliminated by moving hosts to other areas or extirpating alternate hosts. For example, tick populations can be eliminated in particular areas by keeping livestock out of infested areas, and the tsetse fly was eliminated from regions of Africa by destruction of wild hosts. Although exotic introductions are risky, pine trees are essentially free of pests in Australia and other regions where these plants are novel and have no native herbivores or pathogens.

AVOIDANCE TACTICS

The purpose of these tactics is to prevent an established species from increasing in population to the point of becoming a pest. In this approach the agriculturalist accepts the presence of the potential pest in the agroecosystem but attempts to avoid conditions that would allow the organism to reach the economic threshold, which would require curative tactics. Avoidance tactics involve management of host ecology (including evasive methods discussed earlier), alterations in pest movement, and conservation of biological control agents.

Host Ecology

Host health

The damage caused by many pest insects, weeds, and plant pathogens can be substantially reduced or kept below an economic threshold by maintaining healthy plants and animals. For example, it is a common practice to fertilize with potassium to reduce the severity of many tomato diseases. However, simply adding nutrients is not sufficient as nutrient excesses (as well as deficiencies) and imbalances can increase the vulnerability of plants to particular pests.

Host resistance

Classical breeding and genetic engineering can enhance resistance of hosts to arthropods and pathogens. Resistance is manifest as nonpreference or antixenosis (e.g., cucurbits with low levels of allelochemicals that attract spotted cucumber beetles), antibiosis (e.g., maize with high levels of cyclic hydroxamic acids interfere with pupation of European corn borers), and tolerance (e.g., maize genotypes that repair root damage by western corn rootworms).

Host diversification

Spatial polycultures can reduce pest damage by increasing populations of natural enemies (e.g., diverse agroecosystems provide higher levels of alternate prey and hosts) and by reducing colonization of pests (e.g., obscuring of host-finding cues by other plants), establishment (e.g., trapping of pathogen spores by nonsusceptible plants), and damage (e.g., diversion of feeding to tolerant or less valuable plants). Crop rotations or "temporal polycultures" diversify hosts in time.

Movement and Dispersal

Traps

Populations of insects and nematodes can be attracted and retained in trap crops, thereby decreasing colonization of the "protected" crop and concentrating pest populations for application of curative measures (e.g., alfalfa has been used as a trap crop for lygus bugs in cotton production systems). Pheromones are most often used to bait traps for monitoring, but these chemicals can be used to trap colonists, bait toxins, disrupt mate finding, and otherwise alter pest movement.

Barriers

Contact between the pest and the commodity can be limited with physical or chemical barriers. Because of their size, vertebrate pests are amenable to management by physical barriers (e.g., rat proofing), and environmental barriers (e.g., cleared land) can reduce contact between arthropod pests and livestock. Chemical barriers may protect crops from invading insects (e.g., insecticide barriers are used to reduce the movement of rangeland grasshoppers into adjacent crops).

Conservation Biological Control

Pest management practices can be timed, located, or reduced to protect naturally occurring or introduced biological control agents. For example, in orchards and vineyards, the substitution of ground cover for clean cultivation can increase the ratio of natural enemies to pests, and strip-cutting alfalfa can preserve the spatial synchrony between hymenopteran parasitoids and their more mobile lepidopteran hosts.

Classical Biological Control

Introduction of an exotic organism (predator, parasite, or pathogen) to control an exotic pest (insect or weed) has proven to be a viable strategy (neoclassical biological control, the use of exotic agents to control native pests, may have greater environmental risks). For example, the introduction of insects to control Klamath weed in California and prickly pear cactus in Australia reestablished ecological and evolutionary associations and virtually eliminated weed infestations.

SUPPRESSION TACTICS

The purpose of these tactics is to reduce the density of a pest population that has exceeded the economic threshold. This approach is utilized when prevention and avoidance methods have failed, although some of these previous tactics may be consistent with curative approaches (e.g., particular chemical formulations may reduce pest populations and conserve natural enemies). Suppression tactics involve interventions that directly reduce the pest populations while minimizing disruption to other elements of the agroecosystem and avoiding the creation of new problems.

Augmentative Biological Control

The addition of predators, parasites, or pathogens may reduce a pest infestation. Inundation with pathogens produced ex situ is often the most successful approach, as these organisms can be applied and/or can reproduce sufficiently to control the host (e.g., fungal bioinsecticides for control of locusts in Australia and Africa). Habitat modification may also augment populations of natural enemies (e.g., bands on peach trees provide overwintering refuges for predaceous mites).

Chemical Agents and Formulations

Judicious selection of insecticides and formulations can reduce environmental disruption (e.g., harm to beneficial and nontarget species). Both physiological selectivity (mode of action) and ecological selectivity (dosage and carrier) can minimize adverse effects. For example, particular insecticides and fungicides can be used in orchards with little adverse effect to predaceous mites, and insecticidal baits can control ants and fruit flies with little harm to other species.

Chemical Delivery Systems

Insecticides can be applied in time and space so as to provide acceptable control of pest populations while minimizing harm to other components of the ecosystems. This approach can be integrated with other tactics, as with the application of low rates of insect growth regulators to widely spaced swaths for management of rangeland grasshopper outbreaks, a method that relies on both movement of the pest into treated swaths and conservation of natural enemies.

"NONACTION" TACTICS

A final approach to environmental pest management is the decision to take no action. This may allow natural regulatory processes to develop (e.g., native organisms may suppress an introduced pest without competition from classical biological control agents) or may avoid a problem (e.g., not applying an insecticide may avert secondary pest outbreaks). Tactical inaction requires research (e.g., assessing long-term costs of industrial food production) and education (e.g., eliminating cosmetic criteria for pest damage). Finally, reperceiving the place of humans, the importance of biodiversity, the value of the land, and the obligations to future generations is essential to environmental pest management.

See also *Biotic Community Balance Upsets*, pages 94–97; *Conservation of Natural Enemies*, pages 141–143; *Ecological Aspects of Pest Management*, pages 211–213; *Integration of Tactics*, pages 416–419; and *Pest Status*, pages 590–592.

BIBLIOGRAPHY

Biotechnology and Integrated Pest Management; Persley, G.J., Ed.; CAB International: Wallingford, 1996.

National Research Council. *Ecologically Based Pest Management*; National Academy Press: Washington, DC, 1996.

Pimentel, D. *Handbook of Pest Management in Agriculture*; CRC Press: Boca Raton, FL, 1981.

ENVIRONMENTAL IMPACTS OF THE RELEASE OF GENETICALLY MODIFIED ORGANISMS

Sheldon Krimsky
Tufts University, Medford, Massachusetts, U.S.A.

INTRODUCTION

Within a decade after recombinant DNA technology was discovered in 1973, newly formed biotechnology companies began genetically modifying crops (GMC), microorganisms (GEM), and animals for agricultural and industrial applications. Microorganisms were genetically engineered for extracting ores in deep mining operations, for degrading persistent pollutants in soil and water, and for reducing frost damage to fruits and vegetables. Crops were genetically modified for disease, pest, and herbicide resistance, as well as altered nutritional qualities. Domesticated animals and wildlife have been genetically altered for the production of food and pharmaceuticals. The release of these transgenic organisms into the environment introduces potential ecological hazards.

MICROBES

In *Diamond v. Chakrabarty* (1980), the U.S. Supreme Court approved the first patent application exclusively for a living organism. The patent described a strain of *Pseudomonas* with at least two stable extrachromosomal DNA segments that gave the organism the capacity to degrade crude oil. The patented microbe, however, was never developed beyond laboratory use in part because of the ecological uncertainties associated with releasing large quantities of the GEMs into the environment, the deficiency of a regulatory regime to evaluate and manage large scale releases, and the limited effectiveness of the organism's oil-degrading properties. In 1987, after five years of regulatory review and ecological assessment, the U.S. Environmental Protection Agency (EPA) approved the first field release of a GEM. The microorganism, known as ice minus, was constructed from the bacterium *Pseudomonas syringae* by removing genes that synthesize ice-nucleating proteins. Both in the laboratory and in preliminary field tests, ice minus was shown to inhibit frost damage to plants when the temperature drops a few degrees below freezing. Despite early signs that the genetically engineered *P. syringae* might prove successful in reducing frost damage to certain crops in the Frost Belt, no commercial product was developed from this strain (1).

GM PLANTS

From 1986 to 1994 there were more than 1500 field tests involving genetically modified organisms (GMOs) carried out throughout the world, most occurring in North America and Europe and more than 95% involving plants. In 1998 there were more than 1000 approved field tests in the United States. During the 1990s, major research and development investments were made in genetically modified seeds that conferred a pesticidal property to staple crops such as corn, soybeans, potatoes, alfalfa, peanuts, canola, barley, and cotton. The primary traits conferred to GMCs in the United States between 1987 and 1998 as a percentage of all GMCs were herbicide tolerance (29%); insect resistance (24%); product quality, such as high lysine soybeans or modified-oil soybeans (20%); and disease resistance (15%). The herbicides most frequently targeted for crop resistance are glyphosate, sulfonylurea, bromoxynil, glufosinate, and bialaphos. Glyphosate-resistant soybeans were the first of the family of herbicide resistant (or tolerant) crops (HRCs) commercialized. Genetically modified tomatoes reached consumers in 1994 while herbicide-tolerant soybeans, cotton, and canola entered commercial agriculture in 1996. By 2000 approximately one fifth of the U.S. corn acreage, over half of the soybean acreage, and three quarters of the cotton acreage, comprising nearly 30 million hectares, were planted with crops genetically engineered to be resistant to insects and herbicides.

ECOLOGICAL IMPACTS

Ecological concerns over the release of HRCs include the following: 1) Transgenic crops might invade natural habitats if their germination, root growth, resistance to abiotic stresses, or dispersal has been enhanced; 2) Genes transplanted to the crop for herbicide tolerance might transfer to other plants, thereby spreading herbicide tolerance in ways that are ecologically undesirable. The probability of gene flow is greatest when transgenic crops are grown in close proximity to sexually compatible wild relatives; 3) Successes of HRCs can result in the increased use of herbicides and/or compromise efforts toward incorporating integrated pest management; 4) By building herbicide resistance into a few widely used low toxicity herbicides, the rate of weed resistance is likely to increase requiring the use of more toxic herbicides.

Postulated benefits of genetically modified crops include the transition to low toxicity herbicides such as glyphosate, lower poundage of herbicide use, reduction in the use of insecticides, improvement of crops against viral diseases, and the nutritional enhancement of crops.

In a large-scale, plant-population field test to evaluate the invasiveness of a transgenic crop, Crawley et al. (2) compared conventional and genetically engineered oilseed rape (canola). No evidence was found that the transgenic rape was invasive in undisturbed natural habitats or more persistent in disturbed habitats than the nontransgenic rape. Parker and Kareiva (3) identified limitations in the study because the engineered traits were not expected to enhance plant performance or alter biotic interactions in the experimental conditions examined.

The transfer of herbicide tolerant traits by pollen has been established. Oilseed rape that was genetically engineered to be tolerant to glyphosate was planted in proximity to a closely related *Brassica camprestris.* Within second-generation plants, there was a rapid spread of genes from oilseed rape to its weedy relatives. The transfer of herbicide resistance traits to feral plant populations depends on airborne dispersal rates of pollen and the distance between the GMOs and related species. Deshayes (4) reported that at 50 m from a small plot with genetically modified plants of oilseed rape containing an herbicide resistance gene, about one out of ten thousand seeds produced by the surrounding nongenetically modified oilseed rape plants showed resistance to the herbicide. While it is uncommon for pollen to be transported more than a few kilometers, it does occur during unusual weather conditions when the pollen is swept high enough in the atmosphere (5).

In field studies, Bergelson et al. (6) compared transgenic varieties of *Arabidopsis thaliana* with wild varieties on their outcrossing characteristics. The investigators found that the transgenic variety of *A. thaliana* was 20 times more likely to outcross compared to ordinary mutant strains. While no explanation could be given for the difference in this characteristic, it raises questions about the unanticipated spread of transgenic plant varieties. From growth-chamber studies Snow et al. (7) found that a transgenic variety of *Brassica nupus* with glufosinate resistance introgressed into populations of *Brassica rapa* even when herbicides were not applied. The studies also found a group of transgenic progeny that were less fit (e.g., reduced transmission and lower fecundity) than those without the transgene.

One of the world's largest selling herbicides and widely acclaimed also to be among the safest is glyphosate, marketed as Roundup. It is used in conjunction with Roundup Ready seeds. Lappé and Bailey (8) reported that glyphosate fed to animals at high levels was shown to cause liver toxicity. In a case control study in Sweden published in the journal *Cancer*, Hardel and Eriksson (9) found that exposure to glyphosate revealed increased risks for non-Hodgkin's lymphoma. The introduction of glyphosate tolerance to crops has expanded the use of this herbicide. These preliminary studies on glyphosate could be a forewarning that unsuspected hazards may accompany glyphosate resistant crops.

The primary ecological concerns of introducing insect toxins into the genomes of plants is that it would accelerate insect resistance to those toxins, that the toxins would be spread to other plants for which insect resistance is not desired, and/or that the toxins would exhibit secondary effects on beneficial insects that fed on an insect pest that fed on a pesticide-expressing plant. The first generation of crops genetically modified for insect resistance involved the transfer of the genes that synthesize delta-endotoxins of *Bacillus thuringiensis* (Bt). A major advantage of Bt over synthetic insecticides is that it is selectively toxic to certain herbivorous insects. In May 1995, the EPA approved as the first commercial release of an insecticidal transgenic crop, Bt potatoes created by Monsanto for control of the Colorado potato beetle. In August of that year commercial approval was given to Bt corn developed for control of the European Corn Borer.

Losey et al. (10) discovered that pollen from Bt corn can be hazardous to monarch butterflies. In laboratory studies, investigators dusted pollen from Bt corn on milkweeds, a food source of monarch butterflies. The monarch larvae reared on the milkweed ate less, grew more slowly, and suffered higher mortality than the controls, according to the study. Other studies have shown that Bt toxin can harm carnivorous predators of insect pests such as lacewing insects a beneficial insect that feeds on the European corn borer, and that aphids are capable of extracting toxins from agricultural crops and are using them to attack their predators—a species of beetles.

The use of transgenes to confer disease resistance to crops represents another possible ecological risk. If genes that code for viral coat proteins of pest plant viruses are transferred to crops, there is a potential that the genes will recombine in wild plants creating new plant viruses of increased disease severity and/or host range. Risk assessment studies have lagged in this area, in part, because U.S. regulatory agencies have exempted plants genetically engineered for disease resistance from government oversight (11).

TRANSGENIC FISH

Application of genetic engineering techniques to animals has expanded beyond domesticated species to marine organisms. New strains of transgenic fish are being developed for restocking natural waterways and for use in aquaculture. The traits of current interest include accelerated growth, size enhancement, disease resistance, and phenotypes suitable for survival in expanded habitats, such as tolerance for low temperatures. Whether for sport or commerce, once introduced into the environment, transgenic fish may become permanently and irreversibly established in natural aquatic communities. Mass gene transfer methods such as lipofection, particle bombardment, and electroporation of embryos and sperm cells are being developed to genetically modify large numbers of eggs (12). Potential ecological risks of releasing transgenic marine organisms into the natural environment have been widely discussed in the scientific literature (13–15). For transgenic fish, studies have found greater vulnerability to predation (16), altered foraging behavior (17), increased ability to compete for food (18), and swimming deficits (19).

To date, research on transgenic releases has concluded: case-by-case studies are important in predicting ecological outcomes (20, 21); Bt crops will increase insect resistance unless mitigating actions such as refugia (zones where non-transgenic crops are planted near the transgenic crops) are taken (22); pollen is an important source of transmission for transgenes; field tests are not likely to reveal the magnitude of risks of weed or insect resistance that may result when large areas are cultivated with transgenic crops (23), particularly when the transgenes affect biotic interactions (protection against natural enemies); release of transgenic marine organisms could destabilize aquatic ecosystems; the eggs of transgenic fish used in aquaculture could escape into local waters if confinement of the facility is not failsafe. Finally, the experience with nonindigenous species introductions can be useful in studying the release of transgenic plants, microbes, insects, and animals into the environment.

E

REFERENCES

1. Krimsky, S.; Plough, A. The Release of Genetically Engineered Organisms into the Environment: The Case of Ice Minus. In *Environmental Hazards*; Auburn House Pub. Co.: Dover, MA, 1988; 75–129.
2. Crawley, M.J.; Hails, R.S.; Rees, M.; et al. Ecology of transgenic oilseed rape in natural habitats. Nature **1993**, *363*, 620–623.
3. Parker, I.M.; Kareiva, P. Assessing the risks of invasion for genetically engineered plants: acceptable evidence and reasonable doubt. Biological Conservation **1996**, *78* (Oct–Nov), 193–203.
4. Deshayes, A.F. Environmental and Social Impacts of GMOs: What Have We Learned from the Past Few Years. In *The Biosafety Results of Field Tests of Genetically Modified Plants and Microorganisms*; Jones, D.D., Ed.; Proceedings of the 3rd International Symposium, Monterey, CA, Nov 13–16, 1994, Division of Agriculture and Natural Resources, The University of California: Oakland, CA, 1994; 5–19.
5. Campbell, I.D.; McDonald, K.; Flannigan, M.D.; et al. Long distance transport of pollen into the arctic. Nature **1999**, *399*, 29–30.
6. Bergelson, J.C.; Purrington, B.; Wichmann, G. Promiscuity in transgenic plants. Nature **1998**, *393* (Sept), 25.
7. Snow, A.A.; Anderson, B.; Jørgensen, R.B. Costs of transgenic herbicide resistance introgressed from *Brassica nupus* into weedy *Brassica rapa*. Mol. Ecol. **1999**, *8* (April), 605–615.
8. Lappé, M.; Bailey, B. *Against the Grain*; Common Courage Press: Monroe, MN, 1998.
9. Hardel, L.; Eriksson, M.A. Case-control study of non-Hodgkins lymphoma and exposure to pesticides. Cancer **1999**, *85* (March), 1353–1369.
10. Losey, J.E.; Rayor, L.S.; Carter, M.E. Transgenic pollen harms monarch larvae. Nature **1999**, *399* (May), 214.
11. Kesser, D.A.; Taylor, M.R.; Maryanski, J.H.; et al. The safety of foods developed by biotechnology. Science **1992**, *256* (June), 1747–1749.
12. Sin, F.Y.T. Transgenic fish. Fish Biology and Fisheries **1997**, *7* (Dec), 417–441.
13. Hallerman, E.H.; Kapuscinski, A. Transgenic fish and

public policy: regulatory concerns. Fisheries **1990**, *15* (Jan–Feb), 12–19.

14. Jonsson, E.; Johnsson, J.I.; Bjornsson, B.T. Growth hormone increases predation exposure of rainbow trout. Proc. R. Soc. Lond. B Biol. Sci. **1996**, *263* (May), 647–651.
15. Muir, W.M.; Howard, R.D. Possible ecological risks of transgenic organism release when transgenes affect mating success: sexual selection and the Trojan gene hypothesis. Proc. Nat. Acad. Sci. U.S.A. **1999**, *96* (Nov), 13853–13856.
16. Jonsson, E.; Johnsson, J.I.; Bjornsson, B.T. Growth hormone increases predation exposure of rainbow trout. Proc. R. Soc. Lond. B Biol. Sci. **1996**, *263* (May), 647–651.
17. Abrahams, M.V.; Sutterlin, A. The foraging and antipredator behavior of growth-enhanced transgenic Atlantic salmon. Animal Behavior **1999**, *58* (Part 5, Nov), 993–942.
18. Devlin, R.H.; Johnsson, J.I.; Smailus, D.E.; et al. Increased ability to compete for food by growth hormone-transgenic coho salmon *Oncorhynchus kisutch*. Aquatic Research **1999**, *30* (July), 479–482.
19. Farrell, A.P.; Bennett, W.; Devlin, R.H. Growth-enhanced transgenic salmon can be inferior swimmers. Can. J. Zool. **1977**, *75* (Feb), 335–337.
20. Dommelen, Ad van. *Hazard Identification of Agricultural Biotechnology*; International Books: Utrecht, The Netherlands, 1999.
21. *Assessing Ecological Risks of Biotechnology*; Ginburg, Lev R., Ed.; Butterworth-Heinemann: Boston, 1991; 15.
22. Krimsky, S.; Wrubel, R. *Agricultural Biotechnology and the Environment*; University of Illinois Press: Champagne, IL, 1996; 66.
23. Rissler, J.; Mellon, M. *The Ecological Risks of Engineered Crops*; The MIT Press: Cambridge, MA, 1996; 24.

ERADICATION OF PESTS

Alexei Sharov
Virginia Polytechnic Institute and State University, Blacksburg, Virginia, U.S.A.

INTRODUCTION

Eradication is the elimination of reproducing populations of a pest species from a geographic area that is sufficiently isolated to prevent rapid recolonization. Most eradication programs target recently introduced exotic pest species, although, some are targeted at well-established species. Various pest management techniques can be used to achieve eradication including pesticides, sterile insect release (SIR), and application of pheromones. The chances of successful eradication increase if species have a limited range and can be easily managed at low-density levels. If eradication appears infeasible or unprofitable, the project should assume a containment and/or suppression strategy.

EXAMPLES OF ERADICATION PROGRAMS

Klassen documented 41 eradication projects, 21 of which were considered successful, 10 partially successful, and 10 unsuccessful. An eradication program is more likely to be successful if initiated promptly after the arrival of a pest species. A species can be eradicated more easily if it occupies a limited area and is poorly adapted to its new environment.

Examples of successful eradication include the oriental fruit fly, *Bactrocera dorsalis*, which was eradicated on the Rota Island using toxic baits and on Guam using SIR in the 1960s. Citrus canker, a bacterial disease of oranges, grapefruits, and lemons was successfully eradicated in Florida from 1914 to 1943. Eradication was achieved via destruction of more than 250,000 fruit-bearing trees. The malaria mosquito, *Anopheles gambiae*, was eradicated from Brazil in 1939–1940 using pesticides (including arsenic). The black striped mussel was eradicated in Australia by applying of 285 tons of chlorine to the Cullen Bay in 1999. Although this project is considered a success, it may have negative ecological consequences in the long term because it completely destroyed the entire ecosystem.

The elimination of the screwworm, *Cochliomyia homonovorax*, is a classical example of eradication success in a large area. This cattle parasite caused economic losses as large as $400 million annually in southern states from Florida to Texas. In 1958–1960, it was eradicated from Florida using SIR. Millions of male flies were reared, sterilized by irradiation, and released from airplanes. Females that mated with sterilized males produced nonviable progeny, and populations declined until extinction. By 1966 the screwworm was eliminated from the entire United States. Later, the program was expanded to Mexico and Central American countries. The project of screwworm eradication from the United States and Mexico was estimated to have cost $750 million.

Examples of unsuccessful eradication projects in the United States are attempts to eliminate gypsy moths, Japanese beetles, fire ants, and Dutch elm disease. Eradication was not possible because these pest species have spread over a large area and there was no effective method for population reduction.

ONGOING ERADICATION PROJECTS

The boll weevil, *Anthonomus grandis*, entered the United States from Mexico in 1892. It became established in all cotton-growing areas, causing an estimated 8% loss of cotton yield. An eradication project was launched in 1978 in the northeastern part of the geographical range of the boll weevil. Eradication strategy was a two-step process. In the first year, population numbers were suppressed by insecticides and cultural control measures. Then, in the second year, populations were eliminated using pheromone traps and SIR. In subsequent years the program has expanded into all cotton-growing states. Eradication was completed in Virginia, North Carolina, South Carolina, California, Arizona, and portions of Florida, Georgia, and Alabama. However, further progression of eradication was more difficult than expected. Heavy use of insecticides created environmental problems, increased insect resistance, and decreased the abundance of natural enemies that control other pest species. As a result, the damage from secondary pest species (beet armyworm and tobacco budworm) has increased. Also the cost of the program became too high. Cotton growers in several areas decided to quit the program. Apparently, this project has evolved from eradication to containment and suppression with a lower use of pesticides.

The Asian longhorn beetle, *Anoplophora glabripennis*, is a polyphagous woodboring pest native to China, Japan, and Korea. In the United States it was first discovered in 1996 in Brooklyn, New York. This infestation was eradicated by destruction of a large number of trees. Later, another infestation was found in Chicago, but despite an intensive suppression program, new infested trees are found there every year. Existing sampling methods are not sensitive enough for detecting low-density populations of the Asian longhorn beetle. Thus, the actual spatial distribution of this species may appear wider than now considered. The beetle is easily transported with wood packaging materials and has been detected at 26 warehouse sites in 14 states. The ban on the import of hardwoods from China, including packaging materials, which has been enforced by the USDA Animal and Plant Health Inspection Service (APHIS) since 1998, may reduce the risk of future introductions.

ERADICATION METHODS

Any pest-management method (alone or in combination with other methods) can be used for eradicating a pest species. For insect pests, eradication methods include chemical and biological insecticides, cultural control measures, traps, baits, mating disruption with pheromones, and SIR. However, the success of eradication projects depends largely on the ability to manage low-density populations. It is hardly possible to kill every target organism; but if pest populations can be suppressed below the density threshold that supports mating and population growth, then they become extinct naturally.

SIR and mating disruption (confusion) are preferred eradication methods because their effect increases as population density decreases. The effect of SIR may be reduced if sterile males are less competitive than feral males (e.g., in screwworm, fruit flies, and gypsy moth). If pest population density is high, then it should be reduced by pesticides before using SIR or mating disruption. However, SIR and mating disruption are effective only against particular groups of pest species. SIR is effective mostly in Diptera, and mating disruption is possible only in species that use largely pheromone communication for finding a mate.

EVALUATING THE SUCCESS OF ERADICATION

Highly sensitive monitoring methods are needed to confirm the success of eradication. The method should be capable of detecting individuals at all densities that support mating success and population growth. The use of ineffective sampling techniques often resulted in erroneous reports of successful eradication. For example, several projects on eradication of the gypsy moth in Michigan in the 1960s and 1970s were considered a success based on moth counts in traps baited with racemic disparlure. But in 1980, when new traps were deployed with (+)-enantiomer of disparlure that is several times more attractive to moths than racemic disparlure, gypsy moth was discovered in 37 counties of Michigan indicating the failure of eradication.

Eradication can be confirmed historically if the pest species has not been detected for many years since the eradication effort. Historical evaluation does not require highly sensitive sampling methods, but it can not be used for an instant evaluation of an eradication program.

DECISION MAKING IN ERADICATION PROJECTS

Eradication programs in the United States are usually initiated by APHIS after detecting a reproducing population of a potentially dangerous pest species. The cost/benefit analysis of the project is seldom possible at the initial stage, because little information is available on the biology of the species, damage costs, and effectiveness of control. But as information accumulates, it becomes sufficient to make a reasonable evaluation of costs and benefits of the project. Eradication costs are not limited to manufacturing and application of pesticides or other control agents but also include pest monitoring, public relations, potential lawsuits, risk to human health, and other indirect costs. It may be difficult to predict ecological consequences of eradication (e.g., decline of biodiversity and outbreaks of secondary pest species). An eradication project is profitable if its economic benefits are higher than costs. Benefits from eradication extend to large (sometimes infinite) time intervals; thus, future benefits should be adjusted for inflation.

If an eradication project is found to be unprofitable, it should be terminated. Because this decision is often politically charged, the benefits of some projects may have been overestimated and costs underestimated in order to keep these projects going. But it is wrong to view the termination of an eradication project as a total failure. Even if eradication was not achieved, the project might have postponed the spread of the pest species by several years. Because the cost of damage is adjusted for inflation, postponed costs are valued less than present costs. Thus, postponing or slowing the spread of a pest species generates economic benefits. Transition from an eradication objective to containment and/or suppression is an adjustment in a pest control strategy that may be justified economically in a given situation.

Eradication is not a panacea in pest management of exotic species. Successful biological control may be a viable alternative to eradication. For example, attempts to eradicate the citrus blackfly, *Aleurocanthus woglumi*, failed in the United States. However, after introduction of several species of parasitoids, population densities of the citrus blackfly were reduced considerably, and eradication was no longer needed. Eradication of the gypsy moth in the United States is no longer feasible, but containment and slowing the spread of this pest are still economically justified.

See also *Areawide Pest Management*, pages 28–32 and *Sterile Insect Technique*, pages 788–791.

BIBLIOGRAPHY

Ecology of Biological Invasions of North America and Hawaii; Drake, J.A., Mooney, H.A., Eds.; Springer-Verlag: New York, 1986; 321.

Enserink, M. Biological invaders sweep. Science **1999**, *285*, 1834–1836.

Eradication of Exotic Pests: Analysis with Case Histories; Garcia, R., Dahlsten, D.L., Eds.; Yale University Press: New Haven, CT, 1989; 296.

Klassen, W. Eradication of introduced arthropod pests: theory and historical practice. Misc. Publ., Entomol. Soc. Am. **1989**, *73*, 1–29.

Knipling, E.F. Eradication of plant pests. Pro. Bull. Entomol. Soc. Am. **1978**, *24*, 44–52.

Krafsur, E.S. Sterile insect technique for suppressing and eradicating insect population: 55 Years and Counting. J. Agric. Entomol. **1998**, *15*, 303–317.

Myers, J.H.; Savoie, A.; van Randen, E. Eradication and pest management. Annu. Rev. Entomol. **1998**, *43*, 471–491.

Sharov, A.A.; Liebhold, A.M. Bioeconomics of managing the spread of exotic pest species with barrier zones. Ecol. Appl. **1998**, *8*, 833–845.

Smith, J.W. Boll weevil eradication: area-wide pest management. Ann. Entomol. Soc. Am. **1998**, *91*, 239–247.

E

ETHICAL ASPECTS OF BIOTECHNOLOGY

Kathrine Hauge Madsen
Peter Sandøe
The Royal Veterinary and Agricultural University, Frederiksberg, Denmark

INTRODUCTION

Genetic engineering of crops and other food products has resulted in widespread opposition, particularly in Europe. Partly in response to the concerns of the general public, a system has been set up to make scientifically based risk assessments prior to the release of genetically modified organisms (GMOs) and the marketing of GM products. However, these risk assessments do not seem fully to address worries commonly felt by members of the general public. The aim of this entry is to define the issues that are at stake in a broad ethical discussion concerning GM food.

PUBLIC CONCERNS

The public's concerns about the genetic modification of crops and other applications of gene technology to food production have been voiced in some parts of the world. In Europe, a large-scale survey, the so-called ''Eurobarometer,'' conducted in 1999 showed that on average only 31% of those asked ''mostly or totally agreed'' that using modern biotechnology in the production of foods should be encouraged. Compared to the previous survey, made in 1996, this represents a fall of 13% (1). Public concern in Europe has led to a temporary ban on new releases of transgenic crops until new and stricter legislation is in place. In the United States, by contrast, the technology has been positively received and widely adopted with less of public debate.

The extensive Eurobarometer surveys strongly suggest that lack of knowledge about gene technology does not explain this opposition to GM food. Indeed increased knowledge serves to merely polarize those who have not yet made up their minds into positive or, more frequently, negative attitudes. Furthermore, the surveys show that resistance is not directed towards the technology as such. Most of those interviewed welcomed the medical progress brought about by developments in genetic engineering. Thus, it is not the process of genetic engineering per se that engenders opposition to it, but rather specific applications of it.

The authors of the Eurobarometer surveys give the following interpretation of what will be required if the people of Europe are to accept applications of gene technology outside the medical sphere:

> First, usefulness is a precondition of support; second, people seem prepared to accept some risk as long as there is a perception of usefulness and no moral concern; but third, and crucially, moral doubts act as a veto irrespective of people's views on use and risk.

Consumer choice is a further requirement that would seem to play a role in encouraging acceptance of the technology.

The remainder of this entry looks in more detail at these four requirements for acceptability: usefulness, low risk, consumer choice, and no ''moral'' objections. However, a few words should first be said about the involvement of ethics in this area.

THE ROLE OF ETHICS

To understand ethical concerns is to open an agenda and encourage discussion. To set out the ethical issues concerning the development and use of gene technology, a distinction needs to be made between two types of ethical question. The first and most fundamental sort of question raises the issue: *What sorts of concern are morally relevant when deciding whether or not it is morally acceptable for us to use gene technology in agriculture*? Suppose, for instance, that we are trying to decide whether it is ethically acceptable to create herbicide-resistant crops. Some people think that as long as these crops do not give rise to environmental hazards and are safe to eat there is no ethical problem. Others, however, think that, besides environmental and food safety issues, other concerns are morally relevant. For example, some people object to the creation of genetically modified crop plants because they think that modifying living organisms in this way violates the order of nature or involves an attempt to modify the work of the Creator, thus amounting to an objectionable exercise in "playing God."

When discussing ethical issues in biotechnology, it is crucial that an effort should be made to distinguish these sorts of concerns. In too many discussions, those involved simply fail to see that their disagreement arises from the fact that they take different concerns to be morally relevant. Where this happens, the moral debate about biotechnology is apt to make less progress than it otherwise might.

Of course, recognition that the biotechnology debate involves a range of distinct concerns is no guarantee that further discussion will result in consensus. But parties to the argument will at least know what it is that they disagree about at the fundamental level, and thus gain some sort of understanding of why other people reject views that they take to be settled truths. It may, furthermore, be possible to argue rationally that some concerns genuinely are more important than others, or that some alleged moral concerns are not morally relevant at all.

The second sort of ethical question addresses the problem of *how to weigh the different ethical concerns against each other*. This presupposes that the first question has already been answered and it is known what the relevant concerns are. Because these issues are separable, two people may agree entirely on what the relevant concerns are and nevertheless continue to disagree over the use of gene technology in food production, because they disagree over what weight to assign to their agreed concerns.

USEFULNESS

The usefulness of gene technology is obvious to the agrochemical industry and plant breeding companies, as a way of extending the market of existing products by developing herbicide-resistant crops rather than searching for new products. For crop seed companies, the technology has reduced the time frame for development and enabled the production of varieties with an entirely new set of traits, something, which could not as easily have been achieved using traditional breeding methods.

As far as farmers are concerned, the genetically engineered crops currently in production have already displayed a number of advantages, such as increased yield levels, reduced pesticide use, reduced input costs, and more flexibility. However, many farmers are concerned that it may become increasingly difficult to handle and sell genetically engineered products as a result of increasing opposition towards these crops.

To date, the usefulness of gene technology in food production has not been obvious to the general public in Europe. Cheaper food is not really an issue in this part of the world. Many people think that food prices are already low: for example, only 8% of Danes are willing to accept the case for genetically modified food products on the basis of cost savings alone (2). Rather, people tend to be concerned about the undesirable effects of modern intensive farming. However, some adherents of gene technology argue that herbicide-resistant crops may be useful from the point of view of environmental protection because herbicides may be applied at a later stage of growth leaving food for insects and small mammals in the field, thus enhancing wildlife.

One reason why the environmental argument has not gained public support in Europe and among influential nongovernment organizations is that in Europe there is widespread scepticism about the usefulness of pesticides in general because of possible unwanted side effects on human health and the environment. Many see the development towards nonchemical farm management practices as the very same thing as sustainable farming. The introduction of herbicide-resistant crops is based on the use of chemicals, and it is therefore automatically viewed as a nonsustainable path for the future.

Scepticism may change if benefits of applying gene technology to food production become more obvious. If so, the next issue to be discussed is whether there are significant risks involved in the application of gene technology to food production, and whether these risks are worth taking.

RISK

Many biotechnologists see gene technology as a precise method by which one or more identifiable genes are inserted in plants or animals. This contrasts with conventional breeding methods, where thousands of random genes are transferred. It is therefore easier to predict the toxicological and ecological effects of inserting the genes. Opponents of the technology, however, argue, that inserting genes from distant organisms, which would never interbreed under natural circumstances, might have completely unknown effects that may only be identified through long-term monitoring. Furthermore, ecological risk assessment has not yet been able to quantify the risk of growing genetically engineered crops because biological systems are complex.

In order to address public concerns about environmental risks, the so-called "Precautionary Principle" has gradually been adopted into international environmental legislation. Within the European Union, the principle creates actionable rights in connection with products or processes carrying unacceptable risks, provided that potential hazards have been identified, and that it is not possible to make a scientific risk assessment and thereby define risk with adequate precision.

Despite the general doubts already mentioned, it is fair to say that a number of GM crops appear to be safe in terms of ecological risk assessment, judged by ordinary scientific standards. Furthermore, from the point of view of health, no serious hazards have been identified as regards the use of GM crops for food production. However, it is important to be aware that the general population and its spokespeople may have a broader notion of what counts as a hazard than the one taken for granted in ordinary risk assessment.

There may also be hazards of a socio-economic nature. One argument often heard in the debate draws attention to the monopoly enjoyed by the multinational agrochemical industry. These companies have bought several plant-breeding companies to link seed production to agrochemicals. Many people resent this development towards monopolization. Instead they wish to secure influence on development at the community level—a development that is based on "need, not greed," and that allows the consumer free choice.

CONSUMER CHOICE

Consumers object to food produced by genetic engineering if they have no freedom of choice not to buy the product. Such an opportunity can be achieved only by labeling genetically modified food products—something already regulated by the novel food EU-directive. From the consumer's point of view, it may be desirable to extend the labeling requirements to include food, which, while it no longer contains genetically engineered compounds, was produced using this technology. Stringent labeling requirements could, however, discourage retailers from stocking genetically engineered food products. This is because, following the imposition of such labeling requirements, there would probably be increased costs, because now two selections of each product may have to be available: one produced with, and one produced without, the use of genetic engineering. At this point it is difficult to predict which strategy will be more profitable in the long run to the producer.

MORAL OBJECTIONS

Many people see genetically modified food as a "threat the to the natural order of things," because the processes, or results of gene technology are somehow "unnatural," and for that reason objectionable. Some people have also felt that "scientists should not play God." According to this allegation, scientists involved in biotechnology are playing God in the sense that they are trying to modify the work of the Creator. In so doing, they are, perhaps, assuming the role of divine creators themselves. If this allegation were taken quite literally, it would force mankind to stop doing many things that have been essential to human survival. For several thousand years agriculture has been based on the principle of selection, and thus on deliberately changing species. Few think this is a bad thing, even if it involves modifying the work of the creator, so it is not really clear why more advanced techniques that aim at similar ends should be considered objectionable exercises in "playing God."

The first allegation, that the processes or results of gene technology are "unnatural" faces problems as well. First of all, it will have to be explained what is meant by "unnatural." And this explanation needs to rule out the inappropriate kinds of "advanced" biotechnology but not the more traditional farming methods—methods which almost everyone accepts and which cannot plausibly be claimed to be morally dubious, such as the use of ordinary selective breeding.

Secondly, even if such an explanation could be provided, it would still have to be shown that unnatural in the specified sense is also immoral. This point is particularly important since, as is well known, throughout history the characterization of certain acts or practices as unnatural has often revealed no more than stubborn prejudice. Discrimination against homosexuals is a case in point.

Even though these allegations are difficult to sustain, the fact that they are widely treated with sympathy is significant. It shows, among other things, that there is a need for continuing public discussion of nature and humankind's place in it.

REFERENCES

1. Anonymous. Eurobarometer 52.1, *The Europeans and Biotechnology,* Report by INRA (Europe)—ECOSA; Directorate-General for Research, Directorate B—Quality of Life and Management of Living Resources Programme and Directorate-General for Education and Culture "Citizens Centre" (Public Opinion Analysis Unit); Brussels, Belgium, 2000; 85.
2. Anonymous. *Negative Holdning (Negative Attitude),* A Survey by the Danish Technology Council; AC Nielsen and AIM A/S; Politiken 15 November 1998; 2 (in Danish).

ETHICAL ASPECTS OF PESTICIDE USE

E

Hugh Lehman
University of Guelph, Guelph, Ontario, Canada

INTRODUCTION

While use of pesticides has yielded significant benefits for human beings both in control of diseases and in efficient food production, in recent years many critics have alleged that use of pesticides is objectionable on moral grounds (1). A large number of arguments to this effect have been advanced. Production and use of pesticides has been criticized as harmful to farm workers, as well as to people involved in pesticide production or who live near to production facilities (2). Use of pesticides has been criticized as leading to contamination of air, water, and soil and thereby as harmful to humans as well as to other creatures (3). Further, use of pesticides has been criticized as being part of our industrialized agricultural system which, it is alleged, is not sustainable (4). Assessment of these and related criticisms requires reflection both on basic ethical principles and on factual matters relating to pesticide production and use. In this article we call attention to basic elements of ethical reflection and indicate how such elements bear on ethical controversies regarding pesticide use. Naturally, in this brief report, we can do no more than sketchily indicate basic patterns of ethical thought. For more complete discussion and further references see the articles by Lehman (5–7).

PESTS

Pests are living organisms that inhibit the fulfillment of human objectives. Examples are weeds, insects, and fungi. Pesticides are chemical substances used to eliminate or to reduce the numbers of pests, that is, to control pests. In some contexts pests can be controlled without the use of pesticides, for example, through use of mechanical methods of pest control or through the use of organisms that prey on the pests or destroy the reproductive capacity of the pests. In other contexts organisms that would be pests can be tolerated through changing the manner in which humans live or produce crops, for example, by using alternative crops or modifying the timing of the harvest. Where pests cannot be controlled or eliminated without use of pesticides, it may be that satisfactory control of pests may be achieved through minimizing the amount of pesticide that is used or by using pesticides that are reduced to harmless chemicals quickly after application to the pest (8–10).

ETHICAL QUESTIONS

One can raise ethical questions with respect both to matters of character and to human actions. Since we are concerned here with use of pesticides, we are concerned with human actions. With respect to actions, ethical questions concern whether particular acts are permitted or required on moral grounds. One can ask whether use of any pesticide is permitted, or whether use of some pesticide is permitted subject to particular conditions. For example, in a time of famine it might be permissible ethically to use DDT, whereas in other times use of this pesticide might be morally unacceptable. Ethical grounds for determining whether a particular action is permissible should be distinguished from legal grounds. Use of a pesticide might well be prohibited by the laws of some country even though moral considerations would imply that use of the pesticide should be permitted. Conversely, the laws of a country might permit uses of a pesticide that would be intolerable ethically.

ETHICAL GROUNDS

Ethical grounds for determining the permissibility or acceptability of an action may be divided into two categories, consequentialist and nonconsequentialist grounds. According to the former, an action in a context is permissible if there is no alternative action that could be performed in the context that would produce a greater amount of good consequences (benefits) or a lesser amount of bad consequences (harms). While there are egoistic, nationalistic, or other restricted versions of consequentialism, where the harms or benefits are totaled only for some individual or some limited group such as a nation or an ethnic group, such restrictions are arbitrary and therefore unacceptable as the basis of ethical judgement. Plausible versions of consequentialism are universalistic. That is, benefits or harms of all beings capable of being benefited or harmed must be included in the totals. Forms of universalistic consequentialism may be distinguished by reference to the way(s) that benefit or harm are determined. For example, according to many consequentialist views pain is regarded as a harm while satisfaction of desires is often regarded as being good or as a benefit. Many ethical discussions concerning the acceptability of various pesticide uses presuppose some form of consequentialism.

Application of consequentialist views to determining the permissibility of the use of a pesticide can be enormously complex. Complexity may arise both from the difficulty in determining all of the consequences that result from the use of the pesticide and from the difficulty in measuring the amount of benefit or harm of those consequences. It is common to assess amounts of benefits and harms by reference to profits and costs. For example, a benefit deriving from the use of a pesticide would be a reduction in the cost of producing a desired crop or an increase in the profits of such production. However, use of costs and profits resulting from an action does not, on plausible views of the nature of benefit or harm, yield a complete assessment of benefits or harms. Benefits or harms that have no monetary value will be omitted from the assessment. (In many publicly proclaimed cost-benefit assessments, even costs or benefits that have monetary value are omitted from consideration. Such costs or benefits are called externalities. To improve ethical assessments of pesticide use, externalities should be taken into account.) Further, the measurement of benefit or harm that is required for consequentialist analyses may be impossible in principle. For the present at least, references to amounts of pleasure or satisfaction of desires are often, at best, highly speculative.

Nonconsequentialist ethical grounds are often expressed by reference to general moral principles. For example, a crop producer may claim to have a moral right to use a particular pesticide on the general ground that he (or she) has a right to determine what is done on his own property. Nonconsequentialist ethical grounds include assumptions of basic moral rights such as are specified in the Bill of Rights of the United States. Such grounds may be traced to moral thought such as that of the philosophers John Locke and Immanual Kant. One way of expressing the basic moral principle of Kant's thought is to maintain that human beings are never to be treated merely as means to the fulfillment of another person's purposes. This principle implies that humans must be allowed to act voluntarily rather than being induced to act through use of deceit or force. This does not mean that human activities cannot be regulated by law. It does, however, imply that laws that are implemented to regulate human behavior must be accepted by the humans who are subject to those laws through a fair democratic process; that is, a process which, to a sufficient extent, allows those humans to exercise their capacity to determine the conditions of their life.

Determination of the acceptability of pesticide use through nonconsequentialist considerations is no less complex than making such determination on consequentialist grounds. Where such grounds are expressed through appeal to claims of moral rights, complexity arises as a result of difficulty in determining the limits of the alleged rights. Such determination is required in order to resolve controversies that arise from appeal to more than one alleged right. Again, use of laws to regulate pesticide use is a form of coercion. According to the Kantian principle, any such appeal to law must be assessed to determine whether it is compatible with the principle that humans are never to be used merely as means to serve the purposes of other people. Many aspects of the production and use of pesticides are open to serious plausible ethical challenge on nonconsequentialist grounds. For example, serious question should be raised concerning the degree of exposure of farm workers to pesticides. Similarly, serious question should be raised concerning the exposure of poor people to toxic pesticides that occurs as a consequence of the location of pesticide production facilities. In modern societies, farm workers, and the poor generally, often have so little control over the conditions under which they have to live or to earn a living that it is reasonable to conclude that they are being used merely as means to serve to purposes of the other members of our societies.

Consequentialist patterns of thought about ethical issues conflict with nonconsequentialist patterns in some fundamental patterns. In consequence we may expect that some controversies concerning our moral obligations may not yield solutions. This is a matter of disagreement among ethical philosophers. However, even if there are some fundamental disagreements in ethical theory, it does not follow that use of ethical reasoning cannot yield solutions to many ethical controversies at less fundamental levels, for example, in regard to whether pesticide use is morally acceptable under particular conditions. Very often, in regard to such practical matters, thorough analyses based on consequentialist ethical assumptions yield the same result as do careful nonconsequentialist analyses. For example, protecting either the poor or farm workers from risk of serious illness or death through reducing exposure to toxic pesticides would appear to be ethically required regardless of whether the ethical analysis rested on consequentialist or Kantian assumptions.

REFERENCES

1. Aiken, W. Ethical Issues in Agriculture. In *Earthbound: New Introductory Essays in Environmental Ethics*; Regan, T., Ed.; Random House: New York, 1984; 247–288.
2. Pimentel, D.; Greiner, A. Environmental and Socio-Economic Costs of Pesticide Use. In *Techniques for Reducing Pesticide Use: Economic and Environmental Benefits*; Pimentel, D., Ed.; John Wiley & Sons: New York, 1997; 51–78.
3. Edwards, C.A. The Impact of Pesticides on the Environment. In *The Pesticide Question: Environment, Economics and Ethics*; Pimentel, D., Lehman, H., Eds.; Chapman & Hall: New York, 1993; 13–46.
4. Brown, L.R. Sustaining World Agriculture. In *State of the World 1987*; Brown, L.R., Chandler, W.U., Flavin, C., Jacobson, J., Pollock, C., Postel, S., Starke, L., Wolf, E.C., Eds.; Norton: New York, 1987; 122–138.
5. Lehman, H. Values, Ethics and the Use of Synthetic Pesticides in Agriculture. In *The Pesticide Question: Environment, Economics and Ethics*; Pimentel, D., Lehman, H., Eds.; Chapman & Hall: New York, 1993; 347–379.
6. Lehman, H. *Rationality and Ethics in Agriculture*; University of Idaho Press: Moscow, ID, 1995.
7. Lehman, H. Environmental Ethics and Pesticide Use. In *Techniques for Reducing Pesticide Use: Economic and Environmental Benefits*; Pimentel, D., Ed.; John Wiley & Sons: New York, 1997; 35–50.
8. Pettersson, O. Swedish Pesticide Policy in a Changing Environment. In *The Pesticide Question: Environment, Economics and Ethics*; Pimentel, D., Lehman, H., Eds.; Chapman & Hall: New York, 1993; 182–205.
9. Surgeoner, G.A.; Roberts, W. Reducing Pesticide Use by 50% in the Province of Ontario: Challenges and Progress. In *The Pesticide Question: Environment, Economics and Ethics*; Pimentel, D., Lehman, H., Eds.; Chapman & Hall: New York, 1993; 206–222.
10. *Environmental and Economic Impacts of Reducing U.S. Agricultural Pesticide Use*; Pimentel, D., McLaughlin, L., Zepp, A., Lakitan, B., Kraus, T., Kleinman, P., Vancini, F., Roach, W.J., Graap, E., Keeton, W.S., Selig, G., Eds.; Chapman & Hall: New York, 1993; 223–278.

E

FALLOWS

Les Robertson
Sugar Research and Development Corporation, Innisfail, Queensland, Australia

F

INTRODUCTION

Fallows, or temporary breaks in cropping, have long been used to control weeds, pests, and diseases of crops, and to improve soil fertility. Fallowing is commonly used to suppress the diseases and arthropod pests that increase under crop monocultures. Traditionally, fallows were maintained by periodic tillage to control weeds. Cover crops that are not harvested, or temporary pastures (ley fallows), are also used as a break from continual cropping to control pests and diseases, and to improve soil fertility. Improved fallows in forestry plantations utilize plant species such as *Sesbania*. Natural fallows (including ''bush'' fallows in Africa) involve little or no management. Herbicides, or combinations of herbicide and tillage, have been used more recently to maintain fallows. Alternatives to fallowing include rotation of botanically diverse crops, planting of varieties resistant to pests and diseases, and pesticides.

Bare fallowing can result in serious degradation of soil due to erosion by wind and water. Conservation tillage is defined as fallow management that leaves at least 30% soil cover from residues of the previous crop (also known as stubble or trash). This may be achieved using no-tillage (or zero-tillage) and use of herbicides to control weeds or by using cultivating implements (e.g., chisel plows) that do not invert the soil or bury the crop residues.

EFFECT OF FALLOWING PRACTICES ON PEST INCIDENCE

Weeds

Agricultural weeds tend to be annuals that produce abundant seed and are able to colonize disturbed habitats including cultivated cropping land. Annual grasses and broadleaf weeds thrive despite tillage during fallows. Weeds that are able to propagate vegetatively, from stolons for example, are ineffectively controlled by tillage. Perennial weeds are more readily controlled by tillage.

Retention of crop residues, combined with zero-tillage and herbicide application, has resulted in an overall reduction in weeds in Australian grain and sugarcane crops. The development of resistance to herbicide in ryegrass (*Lolium* sp.), however, is challenging the sustainability of fallows maintained by herbicides, and some grain producers have reverted to tillage for grass weed control.

Diseases and Nematode Pests

Diseases reduce crop yields particularly under monoculture, even with fallows between growing seasons. In Australian wheat crops, diseases such as bare patch (*Rhizoctonia solani*) and take-all (*Gaeumannomyces graminis*), and nematodes including root-lesion (*Pratylenchus thornei*) and cereal cyst (*Heterodera avenae*) persist under conventional tillage to affect the next crop (1). High levels of the disease can be maintained in crop residues during the fallow. Burning of stubble, or incorporation and rapid decomposition of stubble, may reduce the incidence of disease. Persevering with no-tillage fallows, retention of crop residues and no-till planting increases the biodiversity and abundance of beneficial soil flora and fauna, and results in biological suppression of plant diseases and nematodes through an increase in antagonists of the pests (2).

Soil-Dwelling Arthropod Pests

Tillage and residue management during a fallow can directly affect pests that have at least one life stage living on or in the soil. Soil-dwelling insects include many serious pests such as wireworms, whitegrubs, and rootworms that feed on roots, germinating seeds, or other underground parts of the plant. Surface-active pests including crickets, beetles, cutworms, millipedes, and mites attack cotyledons, stems, and leaves, and can be particularly damaging to newly emergent crops. The effectiveness of fallows in reducing incidence of arthropod pests relies on several factors, including: 1) length of time without a suitable plant host, including weed hosts; 2) type and intensity of tillage; 3) retention or removal of the previous

Table 1 Response of two pests of seedling crops to different fallow management strategies

Pest (no./m^2)	Zero tillage	Reduced tillage	Intensive tillage
Wingless cockroach, *Calolampra elegans*	5.3	1.7	0.7
Black field earwig, *Nala lividipes*	Zero	2.7	5.3

(From Ref. 6.)

crop residues; 4) susceptibility of the pest species, and its natural enemies, to tillage; and 5) mobility of the pest and its rate of reproduction.

The complexity of interactions between tillage, crop residues, pest incidence, and crop damage is evident from the often-conflicting reports on the effects of reduced versus conventional tillage on pest incidence (3). False wireworms (*Gonocephalum* and *Pterohelaeus* spp.) serve to illustrate the complexity of interactions in Australian crops. When burning of crop residue was discontinued and all residues retained to reduce soil erosion, the incidence of false wireworm attack on crop seedlings increased (4). During the fallow, false wireworm beetles feed on crop residues on the soil surface. When crop residues are incorporated prior to planting, false wireworms switch from feeding on crop residues to feeding on emerging seedlings. Direct drilling of crops (zero-tillage planting) results in reduced damage to seedlings from false wireworms as they continue to feed on the surface-retained crop residues (4).

Soil-inhabiting pests respond in different ways to changes in fallowing practices. Some species of wireworms (Elateridae) selectively lay eggs in tilled soil. The black field earwig (*Nala lividipes*) thrives under intensive tillage with or without crop residues. Conversely, wingless cockroaches (*Calolampra* spp.) are favored by no-tillage with retention of crop residues on the soil surface (Table 1) and can cause severe damage to seedling crops (5). The population density of *Calolampra elegans* declined after several years of continual no-tillage cropping and retention of all residues, possibly due to the concomitant increase in predators, particularly centipedes, in this system (6).

Above-Ground Arthropod Pests

Many above-ground crop pests have a soil-living stage, and tillage during critical periods in the life cycle of the pest has been used to reduce damage in subsequent crops. *Heliothis* and *Helicoverpa* spp. pupae overwinter in soil and can be destroyed by shallow cultivation. This strategy is recommended where large numbers of insecticide-resistant individuals have survived in intensively managed crops such as cotton.

Residues of the previous crop apparently repel aphids when crops are direct drilled. Crop residues retained on the soil surface increase the survival of sorghum midge (*Contarinia sorghicola*) and European corn borer (*Ostrinia nubilalis*) between crops. Armyworms (*Leucania* and *Mythimna* spp.) are attracted to decaying crop residues and the incidence of damage to sugarcane increased after widespread adoption of residue retention. Minimum tillage and mulching create cooler soil conditions compared to conventionally cultivated fallows. This may slow the rate of reestablishment of pests including Colorado potato beetle (*Leptinotarsa decemlineata*), or delay the emergence of soil-living pests such as rootworms (*Diabrotica* spp.).

EFFECT OF FALLOWING PRACTICES ON BENEFICIAL SOIL BIOTA

Tillage and the removal of crop residue both reduce soil biodiversity including beneficial soil flora (7). Vesicular-arbuscular mycorrhizal fungi (VAM) form symbiotic relationships with many plants to aid nutrient uptake. Bare

Table 2 Effect of fallow management strategies on numbers of soldier fly larvae and their natural enemies in the next sugarcane crop

Fallow treatment	Soldier fly larvae (no./m^2)	Predatory beetles (no./m^2)	*Metarhizium* spores $\times 10^4$/g soil
Herbicide/one tillage pass before replanting	Zero	100	9.3
Soybean cover crop/5 tillage passes	Zero	25	3.1
Short fallow/6 tillage passes	17	38	3.0
Long fallow/10 tillage passes	one	38	4.5

(From Ref. 8.)

Table 3 Benefits and costs of fallow management options

Fallow management practice	Advantage (benefit)	Disadvantage (cost)
Bare fallow — tilled	Weed, disease control Fewer surface-active pests	Machinery and fuel costs; soil degradation; loss of soil biodiversity; possible resurgence of pests without biological control
Bare fallow — herbicide	Weed, disease control Fewer surface-active pests	Herbicide costs; soil degradation; loss of soil biodiversity; possible resurgence of pests
Retained crop residues — reduced tillage	Reduced soil degradation	Crop residue may maintain diseases and pests
Retained crop residues — zero tillage	Soil conservation; soil biodiversity; fewer annual weeds and soil-dwelling pests	Herbicide costs; perennial weeds and above-ground pests may increase
Cover crop	Soil conservation; soil biodiversity; weed, disease, and insect control	Cost of sowing and removing cover crop
Ley pasture	Soil conservation; soil biodiversity; weed, disease, and insect control	Cost of grass control before replanting to crop

fallowing reduces VAM levels. In Australia, a condition known a ''long fallow disorder'' is common where VAM-dependent crops including sunflowers suffer severely depressed yields when sown after one or more years of fallow.

Pest incidence can be affected indirectly by the effects of tillage and residue management practices on natural enemies. Higher population densities of predators and pathogens of pests are generally maintained in reduced and zero-tillage fallows compared to conventionally tilled fallows, resulting in fewer pests attacking subsequent crops. Greater levels of insect pathogenic fungi (8) and other entomopathogens persist from one crop cycle to the next with zero-tillage whereas frequent tillage suppresses incidence of these pathogens (Table 2).

Soil-living Diptera appear to be particularly susceptible to general predators, including carabid and staphylinid beetles, in relatively undisturbed fallows. Crops following conventionally tilled fallows tend to have higher densities of seed-corn maggots (*Delia* spp.) compared to crops following reduced tillage (9). Fallows managed with herbicides have been shown to control larvae of sugarcane soldier fly (*Inopus rubriceps*), while maintaining high numbers of the predators and a pathogen of the pest (Table 2) (9). Fumigation of sugarcane soils leads to a resurgence of soldier fly, and also root-lesion nematode (*Pratylenchus zeae*), due to suppression of natural enemies.

Weedy fallows support high population densities of predatory insects including carabids, and this may aid natural control of some pests (3). Weedy fallows may also maintain crop pests and diseases at high population levels between cropping cycles. Grass weeds present during the fallow host diseases and nematode pests of cereals and sugarcane. Improved fallow trees including *Sesbania*, *Gliricidia*, or *Tephrosia* may also host insect pests and diseases of agroforestry crops.

RATIONAL FALLOW MANAGEMENT

The ecology and life cycle of pests should be considered when managing fallows for pest control (Table 3). The impact of tillage and residue management on soil degradation also needs to be considered. Management of fallows using aggressive tillage regimes or fumigation may provide short-term benefits of pest control, but resurgence of pests to high levels is common following reestablishment of crops. Replacing tillage operations with herbicides, and retaining all crop residues, can result in increased soil biodiversity and biological control of pests with a subsequent reduction in insecticide use. Herbicide use however needs to be managed carefully to prevent the development of resistance in weed species.

REFERENCES

1. Neate, S.M. Soil and Crop Management Practices that Affect Root Diseases of Crop Plants. In *Soil Biota: Management in Sustainable Farming Systems*; Pankhurst, C.E., Doube, B.M., Gupta, V.V.S.R., Grace, P.R., Eds.; CSIRO: Melbourne, 1994; 96–106.
2. Roper, M.M.; Gupta, V.V.S.R. Management practices and soil biota. Australian J. Soil Res. **1995**, *33*, 321–339.

3. Stinner, B.R.; House, G.J. Arthropods and other invertebrates in conservation-tillage agriculture. Annu. Rev. Entomol. **1990**, *35*, 299–318.
4. Wildermuth, G.B.; Thompson, J.P.; Robertson, L.N. Biological Change: Diseases, Insects and Beneficial Organisms. In *Sustainable Crop Production in the Subtropics: An Australian Perspective*; Clarke, A.L., Wylie, P.B., Eds.; Department of Primary Industries: Brisbane, 1997; 112–130.
5. Robertson, L.N.; Kettle, B.A.; Simpson, G.B. The influence of tillage practices on soil macrofauna in a semi-arid agroecosystem in Northeastern Australia. Agric. Ecosys. Environ. **1994**, *48*, 149–156.
6. Robertson, L.N. Influence of Tillage Intensity (Including No-Till) on Density of Soil-Dwelling Pests and Predatory Animals in Queensland Crops. In *Pest Control and Sustainable Agriculture*; Corey, S.A., Dall, D.J., Milne, W.M., Eds.; CSIRO: Melbourne, 1993; 349–352.
7. Rovira, A.D. The Effect of Farming Practices on the Soil Biota. In *Soil Biota: Management in Sustainable Farming Systems*; Pankhurst, C.E., Doube, B.M., Gupta, V.V.S.R., Grace, P.R., Eds.; CSIRO: Melbourne, 1994; 81–87.
8. Samson, P.R.; Phillips, L.M. In *Farming Practices to Manage Populations of Sugarcane Soldier Fly*, Inopus rubriceps *(Macquart), in Sugarcane, Soil Invertebrates in 1997*, Proceedings of the 3rd Brisbane Workshop on Soil Invertebrates, Brisbane, July 15–17, 1997, Allsopp, P.G., Rogers, D.J., Robertson, L.N., Eds.; Bureau of Sugar Experiment Stations: Brisbane, 1997; 96–101.
9. Hammond, R.B. Long-term conservation tillage studies: impact of no-till on seedcorn maggot (*Diptera: Anthomyidae*). Crop Prot. **1997**, *16* (3), 221–225.

FEDERAL INSECTICIDE, FUNGICIDE, AND RODENTICIDE ACT

F

Michael E. Gray
University of Illinois, Urbana, Illinois, U.S.A.

INTRODUCTION

The Federal Insecticide, Fungicide, and Rodenticide Act (FIFRA) was passed by the U.S. Congress in 1947. Prior to this sweeping legislation, the Federal government's role in pesticide regulation was directed through the Insecticide Act (1910) that sought primarily to protect the interests of farmers against those making fraudulent claims regarding product efficacy. Because of the passage of the FIFRA, the Federal government's role in pesticide regulation was forever expanded and continues to evolve.

HISTORY

Initially, both the U.S. Department of Agriculture (USDA) and the Food and Drug Administration were required to implement all aspects of the FIFRA and the Federal Food, Drug, and Cosmetic Act (FFDCA) (1906, 1938). This included requiring the registration of pesticides prior to their entry into interstate commerce, as well as assuring consumer protection and pesticide applicator safety. In 1970, Congress mandated that the administration and enforcement of the FIFRA be transferred from the USDA to a new Federal body, the Environmental Protection Agency (EPA). Two years later, the passage of the Federal Environmental Pesticide Control Act (1972) amended the FIFRA; thus began a continuing trend toward the regulation of pesticides aimed principally at reducing unreasonable risks to humans and the environment resulting from pesticide applications. This was a significant shift in Federal policy that had sought previously to control pesticides from the vantage of promoting reasonably safe use by producers. Since the early 1970s, the FIFRA has been amended many times outlining new pesticide registration, regulation, and enforcement responsibilities of the EPA. The passage of the FIFRA amendments during the past three decades has resulted in an overarching policy moving toward reduction of environmental and human health and safety risks associated with pesticide use and away from issues of efficacy.

KEY PROVISIONS OF FIFRA

According to FIFRA, a nonregistered pesticide may not be sold, distributed, or used. Pesticides must be registered by the EPA and the product label approved. Pesticides are defined as ''any substance or mixture of substances intended for preventing, destroying, repelling, or mitigating any pest'' and these may include ''any substance or mixture of substances intended for use as a plant regulator, defoliant, or desiccant.'' The use of pesticide labels serves as a key element in the enforcement of the FIFRA by the EPA. As defined by the FIFRA, the term ''label'' is defined as ''the written, printed, or graphic matter on, or attached to, the pesticide or device or any of its containers or wrappers.'' The term ''labeling'' refers to ''all labels and all other written, printed, or graphic matter accompanying the pesticide or device at any time; or to which reference is made on the label or in literature accompanying the pesticide or device.'' Information that must be included on a pesticide's label includes: 1) specific and detailed instructions regarding safe application procedures, 2) explicit information on product ingredients, 3) toxicological details of the product, 4) first-aid and applicator protective equipment instructions, 5) statements of potential environmental hazards, 6) reentry instructions and harvest restrictions, 7) transportation requirements, 8) instructions on the proper storage of pesticides, and 9) instructions on the safe and proper disposal of pesticides, containers, and wastes. Those who attend pesticide applicator training programs sponsored by Federal and State authorities are vigilantly reminded that the pesticide label is the law. According to the FIFRA, the EPA is required to categorize each pesticide as either ''general use'' or ''restricted use'' or both. Pesticides that are classified as ''general use'' may be purchased and applied by anyone; however, ''restricted use'' products may only be used by certified applicators or by individuals under the direct supervision of a certified applicator. States may conduct certification programs once approved by the EPA. After attending a certification program and successfully passing an examination an individual may become certified as a private or commercial applicator. Private or commercial applicators who use a pesticide in violation of

Encyclopedia of Pest Management

the FIFRA and "in a manner inconsistent with its labeling" are subject to civil penalties. The EPA may immediately suspend a pesticide's registration if use of a product is deemed hazardous to the environment that "includes water, air, land, and all plants and man and other animals living therein, and the interrelationships which exist among these." The EPA also can remove a pesticide from the marketplace if a manufacturer chooses not to seek reregistration. Special uses of pesticides, as outlined by the FIFRA, may be granted by the EPA and include "experimental use" permits, "special local needs" [section 24(c)], and "emergency exemptions" (section 18). In addition to registration of all pesticides, the FIFRA requires the EPA to register all establishments that manufacture pesticides or active ingredients used within products. Many unlawful acts are outlined in the FIFRA and most apply to pesticide registrants, manufacturers, and distributors. However, some of the unlawful provisions apply to producers and commercial applicators. In an effort to promote more judicious use of pesticides, the FIFRA requires the EPA in conjunction with the USDA to promote the accessibility of integrated pest management (IPM) information through the Cooperative Extension Service. Integrated pest management is defined as a "sustainable approach to managing pests by combining biological, cultural, physical, and chemical tools in a way that minimizes economic, health, and environmental risks."

REGULATING PESTICIDE RESIDUES ON AGRICULTURAL COMMODITIES

The regulation of dietary exposure to pesticide residues falls under legal requirements of the FIFRA and the FFDCA. The Food Quality Protection Act (FQPA) signed into law by President Clinton, August 3, 1996, greatly reformed U.S. food safety laws and significantly amended the FFDCA so that the Delaney Clause (1958) no longer affects pesticides. The Delaney Clause mandated a zero cancer risk standard for residues of pesticides on processed foods as compared with a negligible risk standard on raw commodities. Prior to the passage of the FQPA, strict adherence to the Delaney Clause resulted in registration denials for experimental products due to increasingly sophisticated residue detection techniques. This perpetuated the use of many older products rather than substituting them with newer and supposedly more benign compounds. This situation was referred to as the Delaney Paradox. By amending the FIFRA and the FFDCA, the FQPA established a single safe standard of a reasonable certainty of no harm to consumers through their dietary consumption of raw and processed foods as well as their overall aggregate exposure to pesticide residues. Key provisions of the FQPA include: 1) strict limits on exemptions from established standards of safe pesticide residues, 2) supplemental safety factors (10-fold) for infants and children, 3) recognition of states' rights to require labels on food products that have been treated with pesticides, 4) mandates that current tolerances of pesticide residues be reviewed within 10 years, 5) a mechanism for quicker reviews and registrations of pesticides deemed to be of lower risk, 6) the creation of minor-use pesticide programs in the EPA and the USDA to coordinate effective registration policies, and 7) the consideration of common toxicity of pesticide groups and aggregate exposure in the establishment of safe pesticide tolerances.

TRANSGENIC INSECTICIDAL CULTIVARS AND THE FIFRA

New and emerging pest management approaches, such as the use of transgenic insecticidal cultivars will require that the FIFRA continue to be an evolving piece of legislation. Currently, transgenic crops that express insecticidal proteins such as delta-endotoxins produced by various strains of *Bacillus thuringiensis*, fall under the regulations of the FIFRA because these proteins are viewed by the EPA as plant pesticides. Continuing discussions among scientists within industry and academia have focused on the development of resistance management plans to prolong the potential usefulness of these transgenic crops. The EPA has issued provisional registrations (through the FIFRA) for certain transgenic events that require industry to implement effective resistance management plans.

BIBLIOGRAPHY

Becker, W.J. *The Federal Insecticide, Fungicide, and Rodenticide Act as Amended*; Section 1–31, Circular 1076, Florida Cooperative Extension Service, University of Florida: Gainesville, FL, 1992.

The Food Quality Protection Act of 1996. *The FQPA*, 1998. http://www.ecologic-ipm.com/fqpa.html (accessed June 1, 1999).

Olexa, M.T. *Laws Governing Use and Impact of Agricultural Chemicals: Registration, Labeling, and the Use of Pesticides*; Fact Sheet FRE–71, Florida Cooperative Extension Service, University of Florida: Gainesville, FL, 1995.

Pedigo, L.P. *Entomology and Pest Management*; Macmillan Publishing Company: New York, 1989; 646.

U.S. Code, Title 7, Ch. 6. *Insecticides and Environmental Pesticide Control*, 1998. http://www4.law.cornell.edu/uscode/7/ch6.text.html#PC6 (accessed June 1, 1999).

FERTILIZER NUTRIENT MANAGEMENT

F

Joseph M. DiTomaso
University of California, Davis, California, U.S.A.

Carl A. Bruice
John Taylor Fertilizers, Wilbur-Ellis Company, Sacramento, California, U.S.A.

INTRODUCTION

Weeds often accumulate more nutrients than crops, thus gaining a competitive advantage for other resources, particularly light (1). Furthermore, the nutrient levels in the soil will likely be depleted more rapidly in the presence of weeds, causing additional losses in crop yield (2). Thus, it is important to develop fertilization or crop production strategies with three primary objectives in mind: maximize crop competitiveness, minimize weed competitiveness, and lessen the opportunity for nitrate contamination of groundwater. This can include such practices as banding fertilizers in the crop row as opposed to broadcast applications; carefully controlling the timing of nutrient application; and using nitrification inhibitors, slow release fertilizers, composts, or manures. In addition to manipulating fertilization strategies, nutrient use efficiency can be enhanced through the choice of appropriate cultivars and the employment of cultural practices that maximize nutrient uptake by crops.

COMPETITION FOR NUTRIENTS AMONG WEEDS AND CROPS

Numerous studies (3) have shown that crop yields improve following the application of nutrients to the soil, particularly nitrogen (N), potassium (K), and phosphorus (P). However, while nutrients clearly promote crop growth, the application of fertilizers often benefits weeds more than crops (2). This is usually due to an increase in the innate ability of weeds to accumulate minerals (1). Increased uptake of minerals in weeds can result in a significant competitive advantage over crop species (shading, moisture depletion). For example, the addition of N fertilizer to wheat infested with wild oat (*Avena fatua*) or green foxtail (*Setaria viridis*) increased the weed density and decreased the crop grain yield. Conversely, banding P in the seed row increased the crop's competitiveness, thus increasing grain yields while depressing wild oat.

Competition among weeds and crops for nutrients is not independent of competition for other resources. The ability of a species to better utilize available nutrients can also provide an advantage in competition for water and light. Increasing N in rice benefited purple nutsedge (*Cyperus rotundus*) more than the crop (4). The subsequent increase in nutsedge growth reduced light transmission to the crop, thus reducing rice grain yield.

FERTILIZATION STRATEGIES FOR IMPROVED WEED CONTROL AND ENVIRONMENTAL QUALITY

By adopting fertilizer best management practices it is possible to simultaneously reduce the harmful effect of weeds on crop growth and yield, maximize the competitive ability of crops, and reduce the potential for groundwater degradation. This can be accomplished through a number of strategies, including fertilizer banding, use of nitrification inhibitors and alternative nitrogen sources, and fertilizer application timing (5).

Banding

Despite the tremendous advantages associated with the use of inorganic nutrients, particularly N, widespread water quality problems have been attributed to their use. In a national survey, the Environmental Protection Agency estimated that 52% of the 94,600 community water system wells and 57% of the 10.5 million rural domestic wells in the United States contained levels of nitrate above the

National Pesticides Survey minimum reporting limit of 150 ppb (6).

One approach to enhancing fertilizer applications is to band nutrients in the crop row to maximize crop uptake and minimize accessibility to weeds. This strategy serves to improve nutrient uptake efficiency, reduce nitrate leaching potential, and suppress weed growth both in and between the banded row. Banding fertilizers within the row of many legume and grass crops not only lowered weed populations as compared to broadcast applications, but also increased crop yield (7, 8). When the nutrients were deep-banded in the crop row below the weed seed level the crop gained a significant competitive advantage and the rate of fertilizer was reduced.

Nitrification Inhibitors and Alternative Nitrogen Sources

All forms of organic or inorganic nitrogen when applied to soil may eventually be converted to the nitrate ion. Nitrate is a negatively charged compound that moves freely in the soil profile with the direction of water movement and is therefore subject to leaching out of the crop root zone and possibly into groundwater. In some cases, enhanced N use efficiency is possible through the utilization of nitrification inhibitors or alternative sources of N, such as composts or manures, that slow the rate of conversion to nitrate (5).

In certain cases weed species can vary in their response and sensitivity to various forms of N. For example, witchweeds (*Striga* spp.) are parasitic weeds that can markedly reduce crop yields. Although mechanical and chemical control methods have not proved successful in controlling these weeds, urea and to some extent ammonium reduced witchweed infestations by inhibiting germination and radicle elongation (9). In another example, the growth response to nitrate and ammonium differed between corn andredroot pigweed (*Amaranthus retroflexus*). While corn shoot dry weight was unaffected by the two forms of N plus a nitrification inhibitor, ammonium plus a nitrification inhibitor not only caused chlorosis and crinkling in redroot pigweed leaves, but also dramatically reduced pigweed shoot dry weight and total N accumulation compared to nitrate treatment (10). Thus, in some situations, enhancing the proportion of N as ammonium may not only provide more effective control of ammonium-sensitive weeds, but also reduce N loss from the crop root zone, and reduce leaching of N into groundwater.

Timing of Fertilization Applications

The critical timing for utilization of nitrogen may differ markedly between crops and weeds. For example, demand for N in wheat is critical early in the season when tillering begins. By comparison, annual ryegrass produces tillers continuously and is stimulated by late N applications (11). Therefore, early N application reduces the competitive effect of annual ryegrass on wheat yield. In another example (12), N application at winter wheat planting did not increase yield in the presence of downy brome (*Bromus tectorum*). However, N application during a fallow period in a fallow winter wheat cycle reduced downy brome interference with wheat. This response occurred because the N penetrated deeper into the soil profile by the time wheat was planted. The shallow root system of downy brome was unable to access the bulk of the N, whereas wheat absorbed N at greater depths in the soil.

Plant growth stage can also have a dramatic effect on the utilization of available nutrients. Phosphorus uptake in a variety of weeds and crops varied with age and species. In the perennial weed purple nutsedge, uptake of P was rapid until plants were 24 days old (13). In contrast, the demand for P in other weeds and a number of crops was higher after 60 days, during the flowering and fruiting stages. However, this practice can lead to a conflicting situation. Due to the very low mobility of P in soil, the vast majority of P applications are made before or during planting. This augments early season P that is reduced in availability due to cold soil temperatures.

METHODS OF ENHANCING NUTRIENT USE EFFICIENCY

Nutrient-Efficient Crop Cultivars

Crop cultivars vary in a number of developmental characteristics, including stature, canopy development, and leaf orientation. These qualities can have a dramatic effect on weed competition. Although few studies have focused on this aspect of crop cultivar selection, some studies have demonstrated improved weed control by increasing N use efficiency by selecting appropriate crop cultivars. The development of better N efficient crop cultivars would be of great value in tropical environments where poor nutrient soils are common and the economics of the region prohibit extensive use of fertilizers and pesticides.

Green Manures

Legume species (e.g., clovers) grown as green manures are generally used to provide N to the subsequent crop. In this capacity, they can substitute for synthetic fertilizers and promote crop growth and yield. As an additional benefit, green legume manures can also compete with weeds and reduce reliance on herbicides. For example, in a doublecropping system of crimson clover followed by sweet corn, corn yields were similar to a conventional system, but lambsquarters populations were dramatically reduced using the doublecropping approach (14).

Effective Weed Control

In most instances, excessive fertilizer application rates can provide sufficient nutrients for season-long growth of both the crop and weeds. However, in other cases, the availability of nutrients for crop growth is dramatically reduced in the presence of weeds. To maintain adequate mineral uptake in crops under these conditions, one or more weed control measures must be employed (5). Thus, a situation often exists in which fertilizer application increases the dependence on other weed control measures by enhancing the competitive effect of weeds. Thus, maintaining adequate weed control can greatly enhance the uptake and efficiency of fertilizers in crops. In a California rangeland, combining a selective broadleaf herbicide with a liquid fertilizer carrier provided excellent control of yellow starthistle and dramatically stimulated grass forage production. In contrast, the fertilizer applied alone increased starthistle production without a concomitant increase in grass forage.

Row Spacing and Seeding Rate

By manipulating crop row spacing, seeding rate, or seeding method it is possible to improve crop nutrient efficiency (5). For example, reduced row spacing, increased seeding rate, or cross-sowed wheat seed significantly increased uptake of N, P, and K in the crop, and reduced nutrient uptake in grass and broadleaf weeds.

FERTILIZERS AS HERBICIDES

In certain cropping systems, nitrogen applications may actually serve the dual role of fertilizer and contact herbicide. Liquid ammonium nitrate has been known for years to be an effective broadleaf herbicide when applied as an over the top spray to crops with thick, waxy cuticles such as onion, cabbage, or broccoli.

REFERENCES

1. Vengris, J.; Drake, M.; Colby, W.G.; Bart, J. Chemical composition of weeds and accompanying crop plants. Agronomy Journal **1953**, *45*, 213–218.
2. Carlson, H.L.; Hill, J.E. Wild oat (*Avena fatua*) competition with spring wheat: effects of nitrogen fertilization. Weed Science **1986**, *34*, 29–33.
3. Loomis, R.J.; Connor, D.J. Soil Management. In *Crop Ecology: Productivity and Management in Agricultural Systems*; Cambridge University Press: New York, 1992; 319–348.
4. Okafor, L.I.; DeDatta, S.K. Competition between upland rice and purple nutsedge for nitrogen, moisture, and light. Weed Science **1976**, *24*, 43–46.
5. DiTomaso, J.M. Approaches for improving crop competitiveness through the manipulation of fertilization strategies. Weed Science **1995**, *43*, 491–497.
6. United States Environmental Protection Agency. *National Survey of Pesticides in Drinking Water Wells*, Phase I Report; U.S. Department of Commerce: Springfield, VA, 1990; 15.
7. Cochran, V.L.; Morrow, L.A.; Schirman, R.D. The effect of N placement on grass weeds and winter wheat responses in three tillage systems. Soil Tillage Research **1990**, *18*, 347–355.
8. Kirkland, K.J.; Beckie, H.J. Contribution of nitrogen fertilizer placement to weed management in spring wheat (*Triticum aestivum*). Weed Technology **1998**, *12*, 507–514.
9. Mumera, L.M.; Below, F.E. Role of nitrogen in resistance to *Striga* parasitism in maize. Crop Science **1993**, *33*, 758–763.
10. Teyker, R.H.; Hoelzer, H.D.; Liebl, R.A. Maize and pigweed response to nitrogen supply and form. Plant Soil **1991**, *135*, 287–292.
11. Davidson, S. Wheat and ryegrass competition for nitrogen. Rural Research **1984**, *122*, 4–6.
12. Wicks, G.A. Integrated systems for control and management of downy brome (*Bromus tectorum*) in cropland. Weed Science **1984**, *32* (Suppl. 1), 26–31.
13. Pandey, H.N.; Misra, K.C.; Mukherjee, K.L. Phosphate uptake and its incorporation in some crop plants and their associated weeds. Annals of Botany **1971**, *35*, 367–372.
14. Dyck, E.; Liebman, M.; Erich, M.S. Crop-weed interference as influenced by a leguminous or synthetic fertilizer nitrogen source. I. Doublecropping Experiments with Crimson Clover, Sweet Corn, and Lambsquarters. Agriculture Ecosystems and Environment **1995**, *56*, 93–108.

F

FIELD CROP PEST MANAGEMENT (INSECTS AND MITES)

Steve Wratten
Lincoln University, Canterbury, New Zealand

Geoff Gurr
University of Sydney, Orange, New South Wales, Australia

INTRODUCTION

Despite current pest management attempts, each of the world's major field crops is subject to significant losses from insect and mite pests. These pests may chew or bore into their host plants (e.g., moth larvae) or suck phloem sap or individual cell contents (aphids and mites, respectively). According to 1988–1990 data, pests cause losses ranging between 8.8% in barley to 20.7% in rice (Fig. 1). Since the introduction of organochlorine pesticides (such as DDT) in the 1940s and 50s, reliance on chemical control has tended to be high, but increasing numbers of growers now employ integrated pest management (IPM). This approach uses a combination of techniques to maintain pest densities below levels that cause unacceptable crop loss. Pesticides remain important, though those compounds now in use tend to be more target-specific, less environmentally hazardous, and are applied at lower rates than those previously used. Pesticide applications are now often made in response to monitoring ("scouting") having shown that pest numbers are escalating, rather than a calendar-date-driven, prophylactic approach. Nonchemical methods for insect and mite control in field crops include cultural techniques such as rotations and tillage; biological control using predators, parasites, or pathogens of the target pest and the use of pest resistant varieties. The latter approach has recently involved genetic engineering of crops to produce compounds toxic to major pests such as heliothis moth (*Helicoverpa* spp.), European corn borer [*Ostrinia nubilalis* (Hubner)], and Colorado potato beetle [*Leptinotarsa decemlineata* (Say)]. Genetically modified insect-resistant maize and cotton are now grown over large areas, particularly in the U.S. Making crops resistant, however, can lead to the pests' developing resistance to the crops' defences.

ECONOMIC IMPACT OF PESTS

Good historical data exist for insect pest losses in U.S. cotton, maize, potato, and cabbage crops and allow an analysis of trends (1). Averaged over these crops, annual losses have fluctuated between 9 and 18% with no reduction over the 1900–1960 period. Increased knowledge of pests and technological advances—including massive increases in pesticide inputs—have been offset by planting more susceptible varieties, reduced use of cultural pest control practices (especially tillage, sanitation, and rotations), pest resistance to pesticides, increased cosmetic standards, and accidental introductions of exotic pests (1). More recent worldwide data indicate that losses due to pests have increased over the period 1965 to 1988–1990 in most crops (2) (Fig. 1). These data include losses due to viruses, which is appropriate because many viral diseases are vectored by insects, especially those—such as aphids—with sucking mouthparts.

COTTON PEST CASE STUDY

Worldwide, a total of 1326 species of insect have been recorded from cotton (3) but only the minority are significant pests, many being beneficial predators or parasites of pest species. Despite the numerical dominance of beneficial species, most cotton-growing systems worldwide have several key pests that require intervention every year, as well as some well established secondary pests that have appeared as a result of their natural enemies being killed by long term insecticide use.

Few cotton pests are stem and root feeders but foliar pests and especially fruit/boll feeding species are important. In cotton, high levels of damage are common because the plant has a prolonged period of fruiting, is a

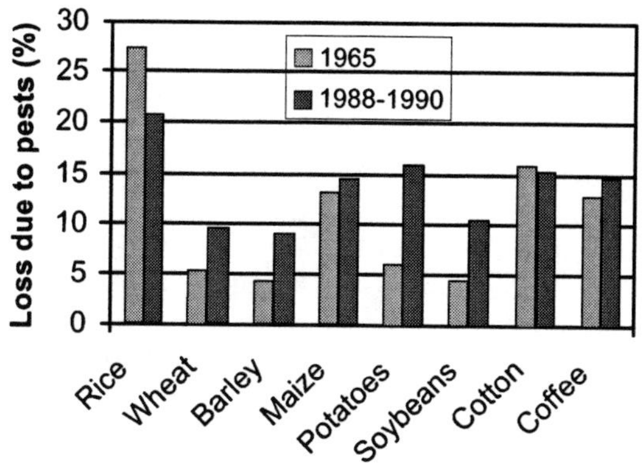

Fig. 1 Field crop losses caused by animal pests (and viruses). (Adapted from Ref. 2.)

preferred host of many pest species and because many key pests feed preferentially upon the fruiting structures and younger fruits ("squares"). The plant readily sheds these, even when lightly damaged (4). Cotton is also a high value crop and this has justified, in economic terms, high levels of pesticide inputs.

Most cotton pests cause quantitative damage (yield loss) which ranges from 9.3% in European crops to 19.7% in African crops (2). Some other pests cause qualitative damage (decreased quality). The honeydew produced by aphids and whiteflies (*Bemisia* spp.) can contaminate the lint and reduce its value and the bug *Dysdercus* (the "cotton stainer") stains the fibre by secondary infection of the boll by the fungus *Netospora*.

Important ecological characteristics of many cotton pests are: (1) an *r*-selected bionomic strategy, including an ability to migrate to the crop from alternative hosts and (2) facultative diapause to survive between periods of host plant availability. *Helicoverpa* spp. possess both attributes and are key worldwide pests in cotton.

Damage by *Helicoverpa* and other lepidopteran pests is often caused by early instars and this led to the formerly common practice of routine, prophylactic pesticide applications. Current best practice is to spray only when monitoring indicates unacceptable densities of eggs or young larvae present within the crop. Because such monitoring is time consuming and requires training, species-specific sex pheromone traps may be used to capture adults and indicate when the crop is likely to be attacked by resulting larvae. *Helicoverpa* populations, like those of many other pest species, have developed resistance to commonly used insecticide groups. This phenomenon has led to widespread adoption of insecticide resistance management (IRM) programs. In these, certain insecticide groups are used only during defined periods ("windows") of the growing season and other active ingredients are used on a rotational basis. IRM is best used within an IPM approach because nonchemical methods are important in destroying the resistant survivors of insecticide applications. Examples from the very wide range of nonchemical control methods for use against cotton pests are given in Table 1.

F

CONCLUSION

Many of the nonchemical methods in Table 1 remain experimental and further research and development is required if they are to be adopted in a widespread fashion for use in cotton and other field crops. Potentially, the use of nonchemical pest management techniques can allow great reductions in the need for pesticides. A good example is the IPM system used for rice in Indonesia. Following the introduction of high yielding varieties in association with synthetic fertilizers and insecticides during the "Green Revolution," increases in rice production were achieved, but only with dramatic increases in insecticide use (Fig. 2). The brown plant hopper (*Nilaparvata lugens*) became the major pest over this period, primarily because of pesticide-induced mortality of its natural enemies. When, in the mid-1980s the majority of insecticides were banned and funds previously used for pesticide subsidies were diverted into training, IPM became widely used. The reestablishment of effective biological control of the brown plant hopper in this IPM program has allowed rice production to increase further whilst pesticide inputs have declined by 60%, saving $120 million per annum (12).

The potential contribution of biological control to IPM is particularly high in tropical field crops because of high arthropod diversity and year-round activity. In temperate systems with greater levels of temporal disturbance, cultural practices may be used in "habitat management" to enhance the impact of natural enemies (13) but there will also be a need for integration of other methods to achieve adequate levels of pest management. In the future, part of this need will be met by genetically engineered pest resistant varieties. These are likely to be grown on a progressively larger area despite objections by some on ethical or technical grounds. Certainly such varieties are no panacea to pest problems. Protecting their efficacy from "erosion" by resistance in target pest populations will depend upon their being used in combination with other methods. This will make IPM the dominant paradigm for field crop pest management for the foreseeable future.

Table 1 Examples of nonchemical pest management methods used in cotton

Management method	Target pest (reference)
Cultural methods	
Control of cruciferous weeds in the vicinity of planned cotton crops	False cinch bugs, *Nysius raphanus* Howard (5)
Preplanting soil cultivation to destroy soil-dwelling pupae	Bollworm, *Helicoverpa* spp. (6)
Avoiding use of safflower and other preferred winter/spring crops	Mirids, e.g., *Creontiades dilutus* Stal (6)
Dry fallowing	False wireworms, *Blapstinus rufipes* Casey (5)
Plowing of a steep-sided trench around crops to trap migrating larvae	Saltmarsh caterpillar, *Estigmene acrea* Drury (5)
Removal of volunteer cotton plants from ditches and roadsides	Cabbage looper, *Trichoplusia ni* Hubner (5)
Delaying planting until weather favorable for rapid plant growth	Seedcorn maggot, *Delia platura* Meigen (5)
Flood irrigation	Wireworm, *Limonius* spp. (5)
Control of dust from roads by sprinklers or sealing	Twospotted mite, *Tetranychus urticae* Koch (5)
Sprinkler irrigation to seal pupating insects in the soil	Bean thrips, *Caliothrips fasciatus* Pergande (5)
Rotation with nonhost crop (alfalfa)	Wireworm, *Limonius* spp. (5)
Use of alfalfa strips in cotton as a preferred "decoy" crop	Lygus bugs, *Lygus hesperus* Knight (5)
Early and thorough shredding and plowing-in of crop residue	Boll weevil, *Anthonomus grandis grandis* Boheman (5)
Destruction of field margin strips of trap crop cotton planted two weeks earlier than the main crop	Boll weevil, *Anthonomus grandis grandis* Boheman (5)
Maintenance of an open crop canopy (leading to prompt desiccation of infested shed squares)	Boll weevil, *Anthonomus grandis grandis* Boheman (5)
Retention of Russian thistle during early- mid-summer to avoid relocation of pests to nearby cotton	Stinkbugs, e.g., *Euschistus* spp. (5)
Application of oviposition inhibitors to the crop	Bollworm, *Helicoverpa* spp. (7)
Mating disruption using sex pheromones	Bollworm, *Helicoverpa* spp. (8)
Use of plant growth regulators and judicious application of nitrogenous fertilizer and irrigation to avoid late maturation of crops	Bollworm *Helicoverpa armigera* Hubner (6)
Early planting and harvest combined with regional synchronization of crops	Pink bollworm, *Pectinophora gossypiella* Saunders (5)
Mass release of sterilised adult males to reduce production of fertile eggs (sterile insect technique)	Pink bollworm, *Pectinophora gossypiella* Saunders (5)
Biological control methods	
Naturally occurring predators such as spiders	Lygus bugs, *Lygus hesperus* Knight (5)
Naturally occurring parasitoids such as *Hyposter exiguae* (Vier.)	Beet armyworm, *Spodoptera exigua* Hubner (5)
Naturally occurring entomopathogenic nematodes	Field crickets, *Gryllus assimilis* Fabr. (5)
Naturally occurring (and induced through inoculation with infected host cadavers) epizootics of polyhedrosis virus disease	Cabbage looper, *Trichoplusia ni* Hubner (5)
Enhancement of arthropod natural enemies (conservation biological control) by intercropping with alfalfa	Bollworm, *Helicoverpa* spp. (5)
Enhancement of natural enemies (conservation biological control) by avoiding pesticidal disruption of natural enemies	Saltmarsh caterpillar, *Estigmene acrea* Drury (5)
Attraction of natural enemies by food spray application to the crop	Bollworm, *Helicoverpa* spp. (9)
Inundative release of *Trichogramma* spp. parasitoid wasps	Bollworm, *Helicoverpa* spp. (10)
Application of *Bacillus thuringiensis* ("Bt") formulations	Loopers, e.g., *Trichoplusia ni* Hubner (5)
Soil cultivation to expose egg masses of orthopterans to insectiverous birds	Grasshoppers, *Melanoplus* spp. (5)
Host-plant resistance	
Conventionally bred trait for less preferred red stem and leaf color	Boll weevil, *Anthonomus grandis grandis* Boheman (5)
Conventionally bred trait for "okra leaf condition" accelerating desiccation of shed, infested squares	Boll weevil, *Anthonomus grandis grandis* Boheman (5)
Conventionally bred trait for "frego bract condition" inhibiting feeding	Boll weevil, *Anthonomus grandis grandis* Boheman (5)
Conventionally bred trait for absence of extra-floral nectaries	*Helicoverpa* spp. (11)
Conventionally bred trait for reduced plant pubescence	*Helicoverpa* spp. (11)
Transgenic resistant varieties expressing "Bt" toxin	*Helicoverpa* spp. (11)

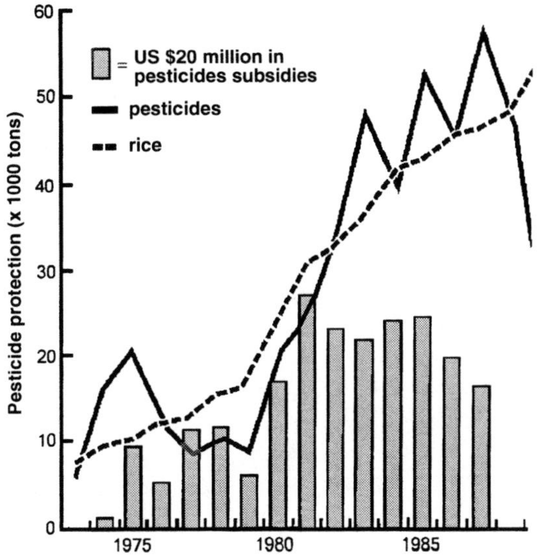

Fig. 2 Adoption of IPM in Indonesian rice allowing increased production and greatly reduced use of pesticides.

See also *Sampling*, pages 745–746; *Biological Control of Insects and Mites*, pages 57–60; *Transgenic Crops (Annuals)*, pages 846–849; and *Cultural Controls*, pages 174–178.

REFERENCES

1. Metcalf, R.L. Arthropods as Pests of Plants: An Overview. In *Handbook of Pest Management*; Ruberson, J.R., Ed.; Marcel Dekker, Inc.: New York, 1999; 377–394.
2. Oerke, E.C.; Dehne, H.W.; Schonbeck, F.; Weber, A. *Crop Production and Crop Protection: Estimated Losses in Major Food and Cash Crops*; Elsevier: Amsterdam, 1994.
3. Hargreaves, H. *List of Recorded Cotton Insects of the World*; Commonwealth Institute of Entomology: London, 1948.
4. Hearn, A.B.; Fitt, G.P. Cotton Cropping Systems. In *Field Crop Ecosystems*; Elsevier: Amsterdam, 1992; 85–142.
5. Leigh, T.F.; Goodell, P.B. Insect Management. In *Cotton Production Manual*; Johnson Hake, S., Kerby, T.A., Hake, K.D., Eds.; University of California: Oakland, CA, 1996; 260–293.
6. Fitt, G.P. Cotton pest management: part 3. an Australian perspective. Ann. Rev. Entomol. **1994**, *39*, 543–562.
7. Mensah, R.K. Suppression of *Helicoverpa* spp. (Lepidoptera: Noctuidae) oviposition by use of the natural enemy food supplement envirofeast (R). Aust. J. Entomol. **1996**, *35* (4), 323–329.
8. Betts, M.D.; Gregg, P.C.; Fitt, G.P.; MacQuillan, M.J. A Field Trial of Mating Disruption for *Helicoverpa* spp. in Cotton. In *Pest Control and Sustainable Agriculture*; Corey, S.A., Dall, D.J., Milne, W.M., Eds.; CSIRO: East Melbourne, Australia, 1993; 298–300.
9. Mensah, R.K. Local density responses of predatory insects of *Helicoverpa* spp. to a newly developed food supplement ''envirofeast'' in commercial cotton. Aust. Int. J. Pest Manage. **1997**, *43* (3), 221–225.
10. Knutson, A. *The Trichogramma Manual; Texas Agricultural Extension Service*; Texas A&M University, USA, *undated*.
11. Smith, C.W. History and status of host plant resistance in cotton to insects in the United States. Adv. Agron. **1992**, *48*, 251–296.
12. Gullan, P.J.; Cranston, P.S. *The Insects: An Outline of Entomology*, 2nd Ed.; Blackwell Science: Malden, USA, 2000.
13. Landis, D.; Wratten, S.D.; Gurr, G.M. Habitat management for natural enemies. Ann. Rev. Entomol. **2000**, *45*, 175–201.

FIELD CROP PEST MANAGEMENT (PLANT PATHOGENS)

Paul Vincelli
University of Kentucky, Lexington, Kentucky, U.S.A.

INTRODUCTION

Disease management in field crops relies heavily on cultural practices such as crop rotation, and on host plant resistance. Except for seed treatment, pesticides play a minor role in managing field crop diseases. Biological control through intentional introduction of antagonists is rarely a significant disease management strategy in field crops. Most disease control decisions in field crops are preplant decisions. Once a crop is sown, most or all of a disease management program is essentially in place, for better or worse. Some management practices are highly effective against certain diseases; an example is selecting a wheat variety with complete resistance to prevailing races of stem rust. However, in most cases, individual disease management practices for field crops are valuable but may only be partially effective by themselves. Thus, the most effective disease management program requires integration of several appropriate management practices.

CULTURAL PRACTICES

Crop Rotation and Other Methods of Field Sanitation

Crop rotation is a fundamental management practice for many diseases of field crops. Because many plant pathogens have narrow host ranges, rotating among crops deprives pathogens of a food source and exposes them to starvation. Furthermore, as infested host residues decompose, pathogen propagules are exposed to antagonism by native soil microbiota. These mechanisms have the effect of naturally eradicating pathogen propagules from the soil. Less aggressive pathogen strains may also predominate under rotation than under monoculture (1, 2).

Other cultural practices can be important in eradicating pathogen propagules from a field. For example, weed control can contribute to disease management, when a weed species is also host to a pathogen and therefore can be a source of *inoculum* (pathogen propagules that initiate disease). For example, in winter-wheat production areas in the Great Plains, "volunteer" wheat that grows during the summer fallow period can provide a source of inoculum for wheat streak mosaic epidemics in the autumn-planted crop. Eradication of millions of barberry plants from the Great Plains has helped in management of stem rust of wheat by depriving the pathogen of the host on which it completes sexual recombination, thus reducing genetic diversity of the pathogen population. Crop residue can be burned, thereby reducing levels of residue-borne inoculum, although environmental concerns limit the use of this practice.

Tillage

Conservation tillage practices have important agronomic and soil conservation benefits but they can enhance the severity of certain diseases. For pathogens that survive in crop residue, especially those that attack above-ground plant parts, leaving host residue on the soil surface can greatly increase primary inoculum. There are exceptions, of course. Some diseases of field crops are not significantly influenced by tillage practices, and certain diseases can actually be lessened under conservation tillage, such as stalk rot diseases of sorghum in the Great Plains. However, adoption of conservation tillage often increases the need for other disease management practices, especially crop rotation and resistant varieties.

High-Quality, Pathogen-Free Seed

Many important pathogens of field crops are seedborne. Thus, use of seed that has been grown or treated in such a way as to minimize or eradicate these pathogens can be important. For example, seed of dry beans is typically grown in arid regions under furrow irrigation to minimize seed contamination by bacterial pathogens. The growing crop is usually inspected and the seed may be laboratory-tested so as to receive certification. Such seed certification does not guarantee freedom from pathogens but it provides some assurance that pathogens are below threshold levels. Choosing seed of high vigor may also contribute to disease management; in corn, seed that has been aged or stressed in some way is more susceptible to *Pythium* infection after sowing.

Other Agronomic Practices

Many agronomic practices can influence diseases positively or negatively. For example, excessive or inadequate fertility, especially nitrogen, can enhance susceptibility of the host to infection. Powdery mildews and downy mildews are classic examples of diseases favored by excessive nitrogen. Achieving a plant population appropriate for the cultivar and site can be important for disease management. For example, stalk rots of corn are commonly more severe under excessive populations, which stress the plants and enhance their susceptibility. Changes in row spacing may also influence diseases. For example, white mold of soybean is more severe in fields with narrow row spacing, which accelerates canopy closure and increases humidity under the canopy. Selection of planting date can also be important; for example, newly sown corn is highly susceptible to Pythium seedling diseases when planted into cool, wet soils, so sowing is delayed until soil temperatures exceed 10°C. Proper field drainage, subsoiling, and similar practices can reduce diseases favored by excessive soil moisture, such as Phytophthora root rot of soybeans. Diseases favored by inadequate soil moisture, such as charcoal rot of soybean and common root rot of wheat, are more difficult to manage agronomically, since irrigation is often too costly or impractical for field crops. Timely harvest of forage crops can help reduce losses in yield and quality caused by foliage diseases, by capturing leaves in the forage before they are rotted or abscised due to disease.

HOST PLANT RESISTANCE

The use of crop varieties with resistance to diseases is, along with crop rotation, a foundation for managing many diseases of field crops. In fact, only a small number of diseases may actually pose an economic threat in most fields in any given year, since most cultivars of field crops have significant levels of resistance to most of the pathogens known to attack our crop species. Cultivars in any given field crop pathosystem typically exhibit a spectrum of responses ranging from highly susceptible to immune. Resistance can be described as either partial or complete, denoting the level of disease control expected with that cultivar against a particular disease (Fig. 1).

Complete resistance has been used very successfully for a variety of diseases, such as several rust diseases and powdery mildews of small grain crops, Phytophthora root rots of soybeans and tobacco, and several virus diseases of bean. Complete resistance is often controlled by one or very few tightly linked genes that are inherited in simple Mendelian fashion, and it provides essentially a complete control program for the target disease. Unfortunately, for some field crop diseases, complete resistance is often short-lived, and new pathogen *races* (strains capable of infecting a cultivar previously considered highly resistant) often emerge following deployment of this type of resistance, sometimes within only a few years. Although genes

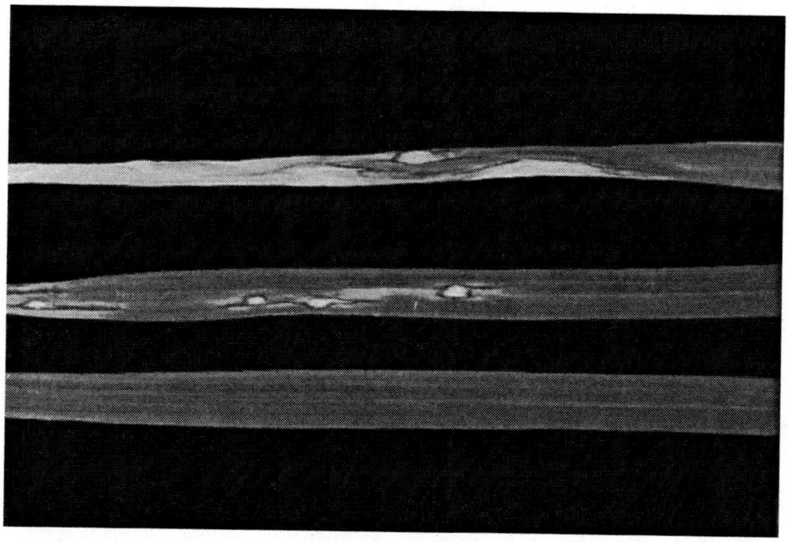

Fig. 1 Leaves of rice inoculated with *Pyricularia grisea*, the cause of the rice blast. Top leaf: completely susceptible reaction; middle leaf: partial resistance reaction; bottom leaf: complete resistance reaction. (Photo courtesy of Mark L. Farman.)

conferring complete resistance sometimes are long-lasting, some field crop breeders have abandoned the use of such "major genes" in favor of those that confer partial resistance.

Partial resistance is the most common type of resistance in field crops, since no cultivar is highly susceptible to all the diseases known to attack that crop. Cultivars with partial resistance are those that, under uniform conditions, exhibit less disease than some standard cultivar or host line. Disease will still develop on a partially resistant cultivar (a point that producers should understand when sowing the cultivar), but typically it is slower to develop and less severe. This usually reduces the impact of the disease on crop development and yield. Partial resistance is often conferred by several to many genes and, commonly, new races that attack the resistant cultivar do not develop quickly. However, partial resistance does not confer complete disease control. Therefore, environmental conditions highly conducive to disease or high inoculum levels limit the value of partial resistance, and other disease control practices are often needed.

Sometimes cultivars of field crops may exhibit *tolerance*, which is the ability to yield well in spite of the presence of disease. For example, certain wheat varieties can grow nearly as well when infected by barley yellow dwarf virus as when virus-free, whereas other cultivars are much more sensitive to virus infection.

CHEMICAL CONTROL

Fungicidal treatment of seed is a common practice with field crops, and it represents one of the most economical and environmentally acceptable uses of disease-control chemicals. Seed treatments can provide protection of most crop seeds and seedlings against infection by *Pythium* species and other plant pathogens found in agricultural soils. Seed treatment can also eradicate certain seedborne fungal pathogens, although fungal infections beneath the seed coat must be treated with systemic fungicides (those that penetrate and translocate within plant tissues). Seed treatments may also provide short-term protection of the young plant against diseases caused by airborne pathogens, such as powdery mildews of small grains.

Treatment of foliage with fungicides can be effective in controlling many fungal diseases of foliage. However, the economic return on such practices typically is not sufficient in field crops to offset the costs of purchasing and applying such materials. Fields used for seed production or high-yield, intensive farming systems sometimes are treated with foliar fungicide applications, although these represent a small proportion of the total acreage of most field crops. For the same reason, preplant fumigation of soil, or postplanting treatment of soil with fungicides or nematicides for root diseases, are uncommon practices in field crops.

SCOUTING

The value of scouting for management of field crop diseases is primarily in developing site-specific knowledge of diseases that may pose a threat to future crops. This can be important information in deciding future crop rotations, cultivar selection, and other agronomic practices. Scouting can also be valuable when valid action thresholds and a postplanting management option exist. However, those conditions are not met for most diseases of field crops in most locations.

POSTHARVEST DISEASE CONTROL

Proper handling and storage of the harvested commodity is necessary to prevent the development of molds that can caused loss of quality and, in some cases, the accumulation of mycotoxins. Proper ventilation and control of temperature and moisture are key factors. Proper handling can also be important to reduce breakage that can create sites where infection may occur. Disease-control products are rarely applied postharvest on field crop commodities.

DIVERSITY

Natural ecosystems usually comprise diverse communities of plant species, which reduce activity of many plant diseases. Monoculture of field crops provides many significant agronomic advantages, but these accrue at the expense of host plant diversity. Increased host plant diversity does not always lead to reduced disease, and in some cases the increased spatial diversity of crop species with strip intercropping can enhance certain diseases. However, major pandemics in the United States, such as the Victoria blight of oat in 1946 and southern corn leaf blight in 1970, are striking reminders of the risks of widespread genetic uniformity in modern cropping systems.

Mixing seed of different cultivars of small grains has been practiced widely in recent years in certain production areas of Europe and North America (3). Cultivars must, of course, have complementary growth traits, including time of maturity, and mixing cultivars does not necessarily result in less disease. However, a number of important foliar diseases on several crops have been reduced through host mixtures. Several mechanisms are thought to contribute to disease reduction, including interception of inoculum by resistant plants and induced resistance (3). Multilines (a mixture of lines bred to be near-isogenic except for resistance) have been used to reduce rust diseases in several small grain crops. Multilines provide diversity for resistance genes but uniformity for all other traits.

Sowing several cultivars in different fields instead of a single cultivar across an entire farm can reduce the risk of yield loss from biotic or abiotic stresses, since different cultivars tend to respond differently to the vagaries of weather. Such a practice can also introduce a measure of disease avoidance. For example, sowing several wheat cultivars having different flowering dates reduces the risk of catastrophic loss from head scab more than sowing only one cultivar, which could be severely affected should conditions be highly favorable for infection while it is flowering. Different resistance genes can be deployed in different geographic regions, as was done for crown rust of oat, in order to limit the virulence of wind-dispersed inoculum among regions.

More information on diseases of field crops can be found in the Compendium of Plant Disease Series, published by APS Press, American Phytopathological Society, St. Paul, and in Martens, J. W.; Seaman, W. L.; Atkinson, T. G. Diseases of Field Crops in Canada. The Canadian Phytopathological Society: Harrow, Ontario, 1988. 160 pp.

REFERENCES

1. Cook, R.J.; Veseth, R.J. *Wheat Health Management*; APS Press: St. Paul, 1991; 152.
2. Martens, J.W.; Seaman, W.L.; Atkinson, T.G. *Diseases of Field Crops in Canada*; The Canadian Phytopathological Society, Harrow: Ontario, 1988; 160.
3. Garrett, K.A.; Mundt, C.C. Epidemiology in mixed host populations. Phytopathology **1999**, *89*, 984–990.

FIELD CROP PEST MANAGEMENT (WEEDS)

Doug Derksen
Agriculture and Agri-Food Canada, Brandon, Manitoba, Canada

INTRODUCTION

Managing weeds in field crops minimizes economic losses caused by weeds from weed–crop competition, losses while harvesting, and contamination in harvested grain. Typically weeds are removed using tillage, herbicides, or biological control agents. However, these tools need to be part of a systems approach to weed management with three goals: 1) to minimize weed densities as crops emerge, 2) to give crop plants a competitive advantage over weeds, and 3) to alter selection pressures to keep weed communities off balance so that problem species do not become dominant. Integrating weed management strategies in this way reduces the likelihood of failure from overreliance on any one tool. This avoids the problems of soil erosion and degradation from tillage, weed community shifts toward difficult-to-control species resulting from monocultural cropping, and resistance to chemical or biological control agents from their overuse.

MINIMIZING WEED DENSITIES AS CROPS ARE ESTABLISHING

Weed competition causes yield losses in field crops when crops are seedlings, therefore, minimizing weed densities at this time is critical. Weeds germinating after this ''critical period'' do not affect yield unless they impede harvest. Keeping crops weed free during critical periods is a recent thrust in field crop production. The length of critical periods varies by crop species, but competition is always reduced if crops emerge before weeds. Seeding crops shallowly, using high quality seed, providing the crop but not the weeds with a good seedbed, planting crops shortly after tillage or herbicide application, and planting well before or after the peak emergence time of weeds will reduce competiton. When weeds are present herbicides, biocontrol agents, postemergence harrowing, and interrow cultivation can be used to keep crops weed free during the critical period.

Soil residual herbicides, such as atrazine, applied before crops emerge, remove weeds during and after critical periods, but their persistent nature has led to weed resistance and, in some cases, problems of ground water contamination. Recently, the use of herbicides applied after crop emergence has become common. With postemergence herbicides producers can determine if weed densities are above threshold values for yield loss, and decide if herbicide usage is economical for a field or for areas within a field. Unfortunately, the overuse of postemergence herbicides has also led to similar resistance and environmental problems reenforcing the need for integrated approaches that do not rely solely on one means of weed control.

Fresh or decomposing crop residues, especially rye, can be used as a mulch to reduce weed densities by suppressing weed growth and emergence chemically, physically, and/or biologically. In zero tillage, residues are left on the surface perpetually thereby impeding weed growth between rows and providing a habitat for biological control organisms, such as carabid beetles.

Since most weeds come from seeds produced during the previous year, minimizing weed seed production can reduce the number of subsequent weed seedlings. Chaff carts that collect and remove weed seeds with the chaff can be pulled behind harvesters. Removing crops as silage for livestock feed prevents weeds from going to seed since harvesting occurs before weed seeds mature. Killing weed seeds by composting manure prior to spreading it on fields prevents the addition of viable weed seeds through manure application. However, focusing on reducing seedbank densities may not be always be useful, since the proportion of weeds emerging from seedbanks can be very low for some species and will vary greatly from year to year.

GIVING GROPS A COMPETITIVE ADVANTAGE OVER WEEDS

Improved crop competition against weeds increases crop yield, reduces weed growth within corps, reduces harvest problems, and reduces weed seed production. Increasing crop competition involves good agronomy. For example, zero-tillage systems provide a good seedbed for crops by planting in a notch free of last year's crop residues and a poor seedbed for weeds because of crop residues left on the surface between new crop rows. Selectively fertilizing crops rather than weeds further promotes crop competitiveness. For instance, broadcasting nitrogen fertilizer on the soil surface allows weeds to access nutrients before crops, while banding nitrogen gives the crop the advantage. Banding nitrogen beside corn rows has been common and the practice has recently been adopted in small grain production, particularly in conservation-tillage systems. Research and farm experience has shown reduced weed growth in these situations.

Selecting competitive crops, such as fall rye, for seeding into weedy fields helps suppress weeds and minimize yield losses. Furthermore, selecting the most competitive crop variety is important for any crop. Plant breeders have focused on breeding for disease and insect resistance, but have generally ignored breeding for weed "resistance." Losses from weeds are usually greater than losses from other pests for many field crops. Consequently, herbicide usage is greater than the use of other pesticides. Breeding for competition against weeds is beginning because of a renewed interest in organic farming, integrated weed management, and reducing herbicide usage.

Higher seeding rates and narrower row spacings reduces the time it takes for crops to close their canopies, which inhibits weed emergence and growth. Research has shown that increasing seeding rate has a greater impact than narrowing rows. The potential for increased weediness in wider row spacings used for easier seeding in zero tillage may be counteracted by the suppressant effect of crop residues and a reduction in weed germination caused by tillage.

KEEPING WEED COMMUNITIES OFF BALANCE

Weeds communities are the end result of selection pressures operating within a field that favor some species over others. Therefore, it is important to identify the selection pressures that have led to weedy situations and then to vary selection pressures enough to keep weed communities off balance so that problem species do not become dominant. This involves using a multiyear cropping system. The type and timing of herbicide/biocontrol agent application can be varied among preseeding, in-crop, preharvest, and postharvest application windows. Crop seeding date can be varied within a field and between fields over time. Varying early and late season crops, full and short season crops, competitive and noncompetitive crops, and crops with summer annual, winter annual, and perennial life cycles can further keep weeds off balance.

The key to managing problem weeds, therefore, is to diversify cropping systems; use interventive measures such as tillage, herbicides, and biological control agents as part of the varied selection pressure rather than as the primary means of weed control; and give crops a competitive edge by employing good agronomic practices. This is particularly important for the sustainability of production systems where the use of tillage or herbicides is restricted. For example, the absence of tillage in conservation-tillage systems to conserve soil moisture and increase soil quality has led to concerns about increased herbicide usage. Similarly, the absence of herbicides in organic farming has led to concerns about excessive tillage causing soil erosion and degradation. In both cases the integration of weed management principles with weed management tools is important to avoid selecting for difficult-to-control weeds and to reduce the overall density of weeds so that returns can be maximized in a sustainable way. Furthermore, the use of an integrated approach to weed management in conservation production systems will increase their sustainability by reducing their reliance on both herbicides and tillage.

BIBLIOGRAPHY

Altieri, M.A.; Liebman, M. *Weed Management in Agroecosystems: Ecological Approaches*; CRC Press: Boca Raton, FL, 1988; 354.

Derksen, D.A.; Watson, P.R.; Loeppky, H.A. Weed community composition in seedbanks, seedling, and mature plant communities in a multiyear trial in Western Canada. Aspects App. Biol. **1998**, *41*, 43–50.

Radosevich, S.; Holt, J.; Ghersa, C. *Weed Ecology Implications for Management*, 2nd Ed.; John Wiley & Sons: New York, 1997; 265.

Smith, A.E. *Handbook of Weed Management Systems*; Marcel Dekker, Inc.: New York, 1995; 741.

FLAME WEEDING IN CORN

Gilles D. Leroux
Université Laval, Québec, Ontario, Canada

INTRODUCTION

Flaming is an old method of weed control. It was widely used in cotton in the early 1940s but was abandoned in the mid-1960s with the advent of selective herbicides (1). There is currently a renewed interest in weed flaming in various heat-tolerant crops. Flaming combined with mechanical cultivation is an alternative to chemical weed control in corn. Flaming kills weeds by rupturing the plasmalemma, due to the expansion of intracellular contents (2). As a result, plants lose water by excessive transpiration and they dry out in a few days.

Flaming can be a selective weed control method in corn, provided that there is a period when both sensitivity of weeds and tolerance of corn to heat are at maximum. Laboratory experiments were conducted to determine the threshold temperature at various growth stages of corn and weeds. From these results, nonchemical weed control strategies were developed in corn using combinations of flaming, rotary hoeing, and mechanical cultivation.

LABORATORY EXPERIMENTS

Experimental Setup

As in the field, two round burners were used for post-emergence band flaming in the laboratory (Fig. 1). They were set 8 cm above the soil surface and they were spaced 20 cm apart, with a 30° angle from the vertical axis. These parameters were found optimal for the field cultivator that we used. Accordingly, the temperature rise was obtained by varying the propane pressure and the exposure time, therefore the ground speed of the burners.

Effects of Flaming on Corn

Four growth stages of corn were tested: coleoptile, 5–8 cm, 20–25 cm, and 50 cm high (Fig. 2). Corn was exposed to ten temperatures varying between 110°C and 390°C. The tolerance to flaming is greatest from pre-emergence up to the coleoptile stage, when the growing point lies below the soil surface, thus being fully protected from heat. Corn is very sensitive to heat between the 5 cm (1 leaf) and 50 cm (6 leaves) stage. Flaming can be directed toward the base of corn plants when they start to elongate (50 cm), with no injury to the leaf canopy (Fig. 1). Our results agree with those reported earlier (3).

Effects of Flaming on Weeds

Six annual broadleaved weed species and three annual grass weed species commonly found in corn fields were studied: red root pigweed (*Amaranthus retroflexus* L.), wild mustard (*Sinapis arvensis* L.), common lambsquarters (*Chenopodium album* L.), wild buckwheat (*Polygonum convolvulus*), common ragweed (*Ambrosia artemisiifolia*), velvetleaf (*Abutilon theophrasti*), scentless chamomile (*Matricaria maritima*), barnyard grass (*Echinochloa crusgalli*), green foxtail (*Setaria viridis* L. Beauv.), and wild proso millet (*Panicum milliaceum*). Each weed species was treated at three growth stages: 0–2 leaf, 4–6 leaf and greater than 8 leaf stage. The temperatures tested were the same as those used for corn.

Regardless of the weed species, flaming at 0–2 leaf stage provided almost complete killing when used at temperatures equal or greater than 110°C (Fig. 2). At the 4–6 leaf stage, flaming at 175°C was required to reduce the weed dry biomass by 80% compared to the untreated control. At the 8 leaf stage, flaming was only effective (85%) at temperatures as high as 350°C (Fig. 1). Ascard (4) reported similar results.

Flaming was also tested on three perennial weed species, quackgrass (*Elytrigia repens*), ox-eye daisy (*Chrysanthemum leucanthemum*), and tufted vetch (*Vicia cracca*), at two growth stages (1–3 and 4–6 leaves). At least three or four consecutive flamings were needed to sup-

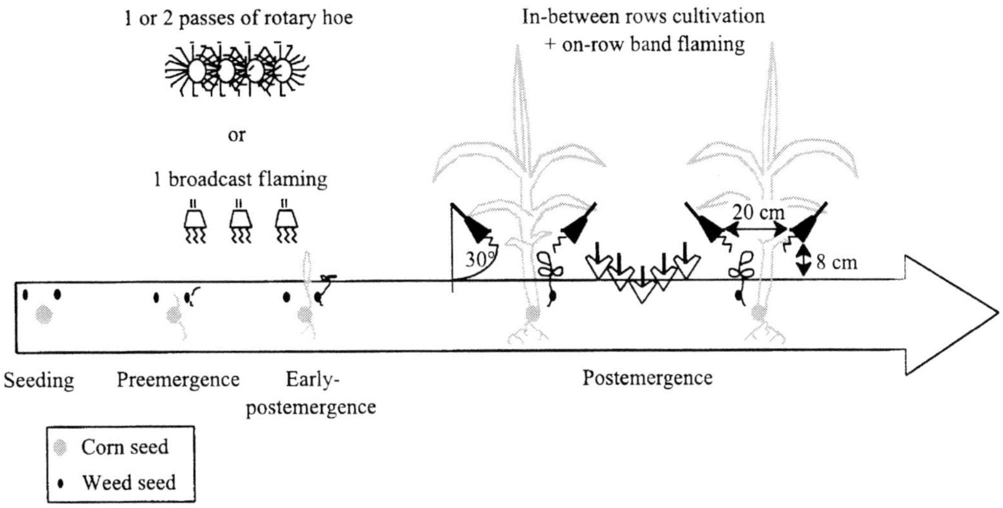

Fig. 1 Diagram of the two nonchemical weed control strategies compared in corn.

press the regrowth of perennial weeds, but no treatment succeeded to completely deplete the viability of the rhizomes and roots.

In summary:

Corn tolerates temperature of 175°C at both coleoptile and 50 cm stages of growth;

Most annual weeds tested are sensitive to 175°C temperature up to the 6 leaf stage.

FIELD EXPERIMENTS

Two nonchemical weed control strategies combining flaming, rotary hoeing, and mechanical cultivations were designed based on the principles that preemergence or early-postemergence weed control must create a growth differential between on-row weeds and corn. This differential is required to flame small weed seedlings at the base of 50 cm high corn plants (6 leaf stage).

We thus compared three weed control treatments: 1) rotary hoeing in preemergence and early postemergence of corn, followed by postemergence band flaming and mechanical cultivation between corn rows; 2) broadcast flaming from preemergence to coleoptile emergence followed with band flaming as in number 1; and 3) standard preemergence application of metolachlor and atrazine. Timing of weed control strategies is illustrated in Fig. 1 and treatments are detailed in Table 1.

No phytotoxicity to corn was recorded for any of the treatments. Early in the season (before the postemergence weed control operation), annual broadleaf weed control, as measured by reduction in dry biomass, did not differ among the three treatments tested. But, annual grass control was lower with broadcast flaming (treatment

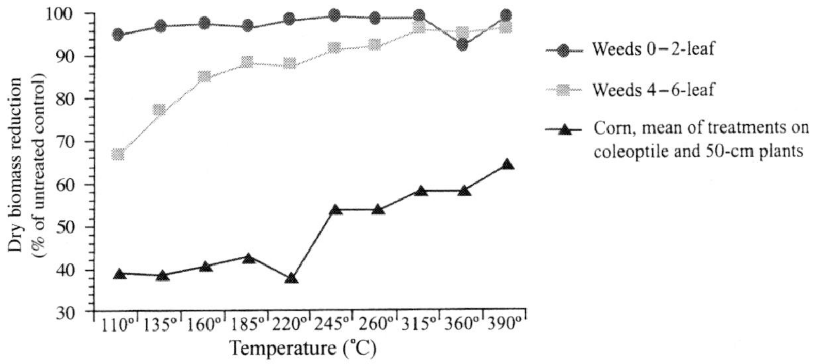

Fig. 2 Mean heat temperature effects on weed dry biomass for flaming at 0–2 leaf and 4–6 leaf stage compared to that on corn treated at coleoptile and 50 cm stage.

Table 1 Weed control treatments tested in the field

Treatment	Application
1. Weedy control	
2. Broadcast herbicides (metolachlor @ 2 kg/ha + atrazine @ 1125 kg/ha)	Preemergence
3. Rotary hoe[a] and band flaming/ mechanical cultivation[b]	Preemergence @ coleoptile Postemergence @ 50 cm of corn
4. Broadcast flaming[c] and band flaming/ mechanical cultivation[b]	Preemergence Postemergence @ 50 of cm corn

[a]Rotary hoe: 15 km/h.
[b]Band flaming/mechanical cultivation: two round burners at 175°C, 6 km/h, propane pressure of 367 kPa. Burners toward the corn rows and in-between rows mechanical cultivation.
[c]Broadcast flaming: flat burners at 175°C, 6 km/h, propane pressure of 367 kPa.

number 2) compared to rotary hoeing and chemical weed control. Weeds must be emerged at time of flaming. While most broadleaf weed seedlings were emerged, annual grasses were not and, thus, they could escape flaming. At mid-growing season, weed control did not differ between the nonchemical strategies. Weed dry biomass was reduced by 85% compared to the weedy untreated control. There was no difference in weed control and grain yield between the nonchemical strategies and the standard herbicide treatment (data not shown).

Preemergence or early-postemergence broadcast flaming has some advantage over hoeing because it can be used when soil conditions are wet. Rotary hoeing is effective against weed seedlings by pulling their roots out of the soil, thus causing their desiccation. Inadequate efficacy is obtained when the soil is wet. The climatic conditions prior to corn emergence will dictate whether flaming or rotary hoeing must be used. If soil conditions are dry, the rotary hoe provides excellent results and is less expensive than broadcast flaming.

Under dry conditions, rotary hoeing is advantageous because it is used at a greater speed than the flamer (15 km h^{-1} vs. 6 km h^{-1}). But if the soil is wet or the weeds are emerged, flaming will provide better results. In addition, flaming does not stimulate weed seed germination compared to hoeing. Soil disturbance exposes weed seeds to light, thus triggering germination. Because most weed species are photosensitive (i.e., they need a short flash of light to germinate) soil cultivation, while killing emerged weed seedlings, stimulates the emergence of new weeds.

Rotary hoeing or flaming does not have any residual activity against weeds. Postemergence weed control is thus needed to prevent competition during the critical period of corn growth.

ADVANTAGES, DISADVANTAGES, AND COSTS OF FLAMING

The main advantage of flaming is that there is no residual pollution of crop, soil, or ground water compared to herbicide use. Crop rotation remains flexible, as there is no herbicide carry-over effect on subsequent crops. Flaming can control a wide array of annual species including some that are resistant to herbicides. The activity of flaming on weeds is less affected by the climatic conditions than is that of herbicides, mostly those that are soil applied. As mentioned above, flaming does not stimulate weed seed germination and can be implemented when soil conditions are too wet for rotary hoeing. Flaming is the sole nonchemical method for selective control of on-row weed seedlings and it fits into an integrated weed control program. Overall flaming preserves the species biodiversity because it has no effect on beneficial organisms.

The disadvantages of flaming include the absence of soil residual activity against weeds, thus calling for repeated applications. The equipment and the labor costs are higher than those of chemical weed control mainly because of high machinery costs and low field capacity due to narrower boom width and generally slower ground speed across the fields. The slower ground speed is a deterrent in large-scale enterprises. As such, the cost of propane compares to that of herbicides and thus is not the most expensive part. But flaming must be custom-applied as many growers feel uncomfortable with this technology. Some feel that compared to mechanical cultivation, flaming is dependent on propane, an external resource not found on the farm.

SUMMARY

Flaming can be used selectively in corn. The selectivity is based on the differential tolerance to heat existing between weeds and corn. Corn tolerates flaming at temperatures up to 175°C until the coleoptile is fully developed. Small annual weed seedlings are killed at that temperature with a 90% efficacy. But flaming does not provide adequate control of perennial weeds at any stage of growth, except seedling.

The tolerance of corn plants to flaming resumes again when they reach a height of 50 cm. At this stage, annual weeds must be small (4 leaf or 10 cm) for effective flaming. In order to meet these conditions, weeds must be controlled using preemergence or early-postemergence rotary hoeing, or preemergence broadcast flaming. Both nonchemical weed control strategies can be completed by selective postemergence on-row flaming combined with in-between-rows mechanical cultivation. Selectivity is greatest when corn plants are 50-cm tall (6 leaves). Corn is very sensitive to flaming between the 1-leaf and the 5-leaf stage. As a nonchemical weed control method, flaming is unique in providing selective postemergence control of weeds present in corn rows. Both strategies that were compared had weed control levels and grain yields equivalent to that of standard preemergence herbicide treatment.

REFERENCES

1. Vester, J. Flame Cultivation for Weed Control, Two Year's Results. In *Weed Control in Vegetable Production*; Cavalloro, R., El Titi, A., Eds.; Proceedings of a Meeting of the EC Experts' Group, Stuttgart, Oct. 28–31, 1986; A.A. Balkema Brookfield: Rotterdam, 1986; 153–167.
2. Ellwanger, T.C.; Bingham, S.W.; Chappel, W.E. Physiological and cytological effects of ultra high temperatures on corn. Weed Sci. **1973**, *21*, 296–303.
3. Lalor, F.W.; Buchele, F.W. Effects of thermal exposure on the foliage of young corn and soybean plants. Trans. of the ASAE **1970**, *13* (4), 534–537.
4. Ascard, J. Effects of flame weeding on weed species at different developmental stages. Weed Res. **1995**, *35*, 397–411.

FURTHER READING

Ascard, J. Comparison of flaming and infrared radiation techniques for thermal weed control. Weed Res. **1998**, *38*, 69–76.

Ascard, J. Flame weeding: effects of burner angle on weed control and temperature patterns. Acta Agric. Scand., Sect. B, Soil and Plant Sci. **1998**, *48*, 248–254.

Ascard, J. Flame weeding: effects of fuel pressure and tandem burners. Weed Res. **1997**, *37*, 77–86.

Laguë, C.; Gill, J.; Lehoux, N.; Péloquin, G. Engineering performances of propane flamers used for weed, Insect Pest, and Plant Disease Control. Appl. Eng. in Agric. **1997**, *13* (1), 7–16.

Leroux, G.D.; Douhéret, J.; Lanouette, M. Pyrodésherbage dans les cultures de maïs. La Lutte Physique en Phytoprotection INRA **2000**, 41–55, Chapter 3.

Parish, S. In *Investigations into Thermal Techniques for Weed Control*, Proceedings of the 11th International Congress on Agricultural Engineering (CIGR), Dodd, V.A., Grace, Eds.; Dublin, Ireland, 1989; 2151–2156.

Rahkonen, J.; Vanhala, P. In *Response of Mixed Weed Stands to Flaming and Use of Temperature Measurements in Predicting Weed Control Efficiency*, Communications de la Quatrième Conférence Internationale I.F.O.A.M, Thomas, J.-M., Ed.; ENITA: Dijon, France, 1993; 167–171.

F

FLEAS

Rory Karhu
Stanley Anderson
University of Wyoming, Laramie, Wyoming, U.S.A.

INTRODUCTION

More than 2,400 species of fleas exist worldwide. Of these, less than 20 species have been documented to readily bite humans. The primary host of these species can be livestock, dogs, cats, wild rodents, or humans themselves as in the case of *Pulex irritans*. Because *P. irritans* is now uncommon, the most important flea-related problems encountered by humans and domestic animals today are more centered around cat and dog fleas (*Ctenocephalides felis* and *C. canis*), Northern rat fleas (*Nosopsyllus fasciatus*), Oriental rat fleas (*Xenopsylla cheopis*), chigoe fleas (*Tunga penetrans*), European chicken fleas (*Ceratophyllus gallinacea*), sticktight fleas (*Echidnophaga gallinacea*), Western chicken fleas (*Ceratophyllus niger*), and several species of wild rodent fleas.

Cat and dog flea infestations of human dwellings and domestic cats and dogs account for the most widespread flea problem in the United States. Northern rat fleas and Oriental rat fleas are important vectors of plague. Female chigoe fleas, unlike any other species, will burrow under the skin of their host (mainly pigs or humans) to lay their eggs. European chicken fleas, sticktight fleas, and Western chicken fleas, will all feed on humans and can be serious pests of poultry operations. The fleas *Oropsylla montanus* and *Hoplopsyllus anomalus*, both important plague vectors of ground squirrels, and *Oropsylla hirsuta* and *O. tuberculata cynomuris*, fleas of prairie dogs, are all capable plague vectors in North America.

IMPORTANT FLEA-RELATED PROBLEMS

Nuisance

Because all fleas must feed on blood to survive, they present a biting nuisance to any animal or human that might serve as a suitable host. Generally fleas will bite the ankles or lower legs of humans, but they will feed on any part of the body especially when the unsuspecting host is sleeping. The reaction of human antibodies to flea bites can sometimes lead to considerable irritation and swelling as well as secondary infection due to scratching of the bites. Female chigoe fleas can also cause serious irritation and infection as a result of burrowing under the skin of the host.

Disease

Of equal or greater importance than the biting nuisance of fleas is the possibility of disease transmission. Fleas can transmit flea-borne (murine) typhus (*Rickettsia typhi*), tapeworms (*Dipylidium caninum*), myxomatosis, and plague (*Yersinia pestis*) to their hosts. Flea-borne typhus in humans occurs when infected flea feces contaminate a cut or abrasion. Symptoms of flea-borne typhus are relatively mild and rarely fatal in healthy individuals. Tapeworm infection is quite rare since the infected adult flea must be swallowed to infect a human. Myxomatosis is a viral infection that affects only rabbits, and is often fatal to them. Plague, responsible for the "Black Death" and various other pandemics, is spread by fleas. A host is infected when a flea regurgitates *Y. pestis* into the bite wound. Northern and Oriental rat fleas are the main vector of plague transmission from rat to humans in the Old World. As urban rat populations declined due to waste clean-up efforts and better hygiene, the incidence of urban plague in human populations has declined. While small plague outbreaks still occur in underdeveloped countries, the majority of plague cases in the United States are isolated and tied to wild rodent populations, especially ground squirrels (*Spermophilus* spp.) and prairie dogs (*Cynomys* spp.) in the Southwestern states of New Mexico, Arizona, and California. The incidence of sylvatic plague (wild-borne plague) has actually increased in the United States in the past 40 years. This is probably due to increased human and domestic pet encroachment into ground squirrel and prairie dog habitats.

Effects on Poultry Production

In the south and southwest United States the sticktight flea will attack not only chickens, but horses, rabbits, cats, dogs, and humans. Chickens can become so heavily infested that young birds die of blood loss. In addition, heavy infestations can reduce egg production and suppress physical growth of immature birds.

FLEA CONTROL

Chemical Control

For centuries humans have experimented with different ways to reduce flea populations, employing everything from natural plant derived repellents to flea traps that were worn around the neck. In the 1950s through the 1970s the answer to flea control problems were often sought in the form of chlorinated hydrocarbon insecticides like Aldrin, Dieldrin, Heptachlor, and Dichlorodiphenyl-trichloroethane (DDT). While chlorinated hydrocarbon insecticides were found to be quite effective at controlling flea populations, they were eventually banned when it was discovered that they were detrimental to many nontarget organisms. With the ban of chlorinated hydrocarbons, it became apparent that alternative methods of flea control needed to be developed and tested.

Most flea control products available today contain pyrazole, pyridylmethylamine, pyrethroid, organophosphate, carbamate, borate, or citrus fruit-derivative insecticides available as emulsifiable concentrates, topical (pour on) solutions, wetable powders, or dusts that kill fleas primarily by disrupting their nervous system. While most of the insecticides have demonstrated relatively poor residual control and require frequent application, recently developed flea control products containing pyrazole or pyridylmethylamine-class insecticides only require treatment once monthly. These products attempt to control flea infestations by killing adult fleas on pets (dogs and cats) before they have a chance to lay eggs. Other flea control products have combined the adulticide action of the insecticides with an insect growth regulator (IGR) such as fenoxycarb, metoprene, lufenuron, or pyriproxyfen. IGRs reduce flea populations by interfering with the physiological development of egg, larval, and pupal stages or chitin formation (lufenuron). The combination of both an insecticide and an IGR provides a more complete flea control system because it not only kills adult fleas directly, but continues to control flea populations by reducing successful maturation and reproduction of immature fleas. IGRs are desirable because they generally exhibit greater residual effectiveness compared to insecticides and are considered quite safe for humans and their pets. Pyriproxyfen and fenoxycarb have the most desirable properties of the IGRs because they are more stable in UV light and thus, are useful for both outdoor and indoor application. The combination of an insecticide and an IGR has proven most feasible for control of cat and dog fleas in domestic situations where the primary habitat (house and\or yard) is known, but can be useful for other applications as well. Use of an IGR alone should be limited to situations where immediate control of the adult flea population is not of concern. Using pyriproxyfen for treatment of rodents and rodent burrows to control plague outbreaks has met with very limited success and is not yet a feasible flea control alternative. Diazinon and pyrethroid based dusts are still the best products for controlling flea numbers in burrowing rodent populations.

Preventative Control

Preventative measures are important to any flea control management program, especially in human dwellings. Sleeping areas of domestic cats and dogs, as well as carpeted floors of human dwellings, provide an ideal habitat for cat and dog fleas. The most effective preventative measures include physical removal of fleas through direct combing of pets, vacuuming of carpets, and mopping of floors. Keeping animal facilities as clean as possible will also help reduce flea infestations of livestock operations.

FUTURE CONCERNS

As new flea control products become available it will become harder for the consumer to determine the best flea control program for their situation. Vigilance and education will be required for consumers to keep up with the overall value, safety, and effectiveness of available flea control products.

As humans and their pets continue to expand into areas with dense ground squirrel and prairie dog populations, the risk of contracting sylvatic plague will also increase. Control of rodent populations themselves is undesirable not only from an ecological perspective, but because fleas will leave dead or removed animal burrows and search out other hosts. Thus chemical control of fleas in these areas will become increasingly more important. Ongoing research will need to continue to focus on developing safe and effective means to control flea populations in ground squirrel and prairie dog communities. In addition, medical doctors in these regions need to be able to diagnose plague cases when they occur.

BIBLIOGRAPHY

Beard, M.L.; Rose, S.T.; Barnes, A.M.; Montenieri, J.A. Control of *Oropsylla hirsuta*, a plague vector, by treatment of prairie dog burrows with 0.5% permethrin dust. J. Med. Entomol. **1992**, *29* (1), 25–29.

Burgess, N.R.H.; Cowan, G.O. *A Colour Atlas of Medical Entomology*; Chapman & Hall: London, 1993; 144.

Hinkle, N.C.; Koehler, P.G.; Patterson, R.S. Residual effectiveness of insect growth regulators applied to carpet for control of cat flea (Siphonaptera: Pulicidae) larvae, J. Ecol. Entomol. **1995**, *88* (4), 903–906.

Hinkle, N.C.; Rust, M.K.; Reierson, D.A. Biorational approaches to flea (Siphonaptera: Pulicidae) suppression: present and future. J. Agric. Entomol. **1997**, *14* (3), 309–321.

Karhu, R.K.; Anderson, S.H. The effects of pyriproxyfen spray, powder, and bait treatments on the relative abundance of fleas in black tailed prairie dog (Rodentia: Sciuridae) towns. J. Med. Entomol. **2000**, *37* (6), *in press.*

Koehler, P.G. *Fleas: ENY–087*; University of Florida Institute of Food and Agricultural Services. http://edis.ifas.ufl.edu BODY_IG087 (accessed April 3, 2000).

Kuepper, T. *The Complete Book of Flea Control for You, Your Pet, and Your House*, 3rd Ed.; TK Enterprises: Oxnard, California, 1995; 78.

Miller, B.E.; Forcum, D.L.; Weeks, K.W.; Wheeler, J.R.; Rail, C.D. An evaluation of insecticides for flea control on wild mammals. J. Med. Entomol. **1970**, *7* (6), 697–702.

Palma, K.G.; Meola, R.W. Field evaluation of nylar for control of cat fleas (Siphonaptera: Pulicidae) in home yards. J. Med. Entomol. **1990**, *27* (6), 1045–1049.

Pedigo, L.P. *Entomology and Pest Management*; Macmillan Publishing: New York, 1989; 646.

Potter, M. *Ridding Your Home of Fleas, ENTFACT 602*; University of Kentucky College of Agriculture. http://www.uky.edu.Agriculture/Entomology/entfacts/struct/ef602.htm (accessed April 3, 2000).

FLOODING (WEED CONTROL IN RICE)

F

Alan K. Watson
International Rice Research Institute,
Makati City, Philippines

INTRODUCTION

Rice is a semiaquatic plant and one of only a few food crops that can be grown in standing water as well as in dry land conditions. Most varieties of rice grow better and produce higher yields when grown in flooded soils than when grown in nonflooded soils. Because rice tolerates flooding better than most weeds, flooding is an effective method of cultural control for many weed species. Maintenance of standing water in the rice field is an age-old traditional weed control practice and water management continues to be a very important weed control tool in modern rice production.

RICE PRODUCTION SYSTEMS (1)

Rice is the staple food for nearly one half of the world's population, most living in Asia and many of them among the poorest in the world. Rice is grown primarily in the humid and subhumid tropics and subtropics. More than half of the harvested area of rice worldwide is grown under irrigated conditions and the irrigated rice ecosystems provide 75% of global production. Although not truly an aquatic plant, rice thrives in waterlogged soils and produces higher grain yields when grown in flooded soil than when grown in dry soil. Flood water provides the plant's evapotranspiration needs, increases the availability of many nutrients, and suppresses many weed species.

ECOLOGY OF WEEDS IN RICE (2, 3)

Weeds are the major biological constraint in most rice-growing areas in the world. Weed species that cause problems in rice vary with soil, temperature, latitude, altitude, rice culture, seeding method, water management, fertility, nutrient management, and weed control technology. Cultural practices such as methods of direct seeding, nutrient management, and water management all influence weed competition and weed species composition. For example in much of Southeast Asia, the replacement of transplanting by direct wet seeding resulting in the lack of a continuous flood has allowed the annual grassy weeds, especially *Echinochloa crus-galli* and *Leptochloa chinensis*, to become major weed problems. However, in California, water seeding suppresses *Echinochloa* spp., *Leptochloa chinensis* and some other grass species, but large-seeded *Echinochloa* spp. and a complex of less competitive obligate aquatic weeds subsequently invade the fields.

More than 350 species in more than 150 genera and 60 families have been reported as weeds of rice. The majority of the weeds are in the Poaceae and Cyperaceae families. *Echinochloa crus-galli* (barnyard grass) is the world's worst weed in rice and its close relative, *Echinochloa colona* is the second worst weed in rice. Other weeds of world importance in rice culture include *Cyperus difformis, Cyperus iria, Cyperus rotundus, Eleusine indica, Fimbristylis miliacea, Leptochloa chinensis, Ischaemum rugosum, Monochoria vaginalis, Sphenoclea zeylanica*, and weedy forms of *Oryza sativa* including red rice, wild rice, and off types. In addition to *Echinochloa crus-galli, Cyperus difformis, Oryza sativa* (red rice), and *Monochoria vaginalis, Echinochloa oryzoides, Echinochloa phyllopogon, Eleocharis* spp., *Rotala indica, Sagittaria* spp., *Scirpus* spp., *Leersia oryzoides, Paspalum distichum, Heteranthera limosa*, and *Potomogeton distinctus* are common weeds of rice in temperate regions. *Aeschynomene virginica* and *Sesbania exaltata* are common in the southern United States.

PROBLEMS OF INFESTATIONS (4)

Weeds cause greater economic losses than all other pest problems in rice culture. Losses due to weeds in rice are not only their adverse impact on yield, but weeds also restrict flow of irrigation water, harbor insect pests and diseases of rice, slow harvest operations, increase grain-drying costs, and reduce grain quality. In the United States, weeds are estimated to cause average yield losses in rice of 17% (ranging from 12 to 34%) compared to 8% and 7% for insects and diseases, respectively. Yield losses in India are estimated to be 10%, and in the Philippines, 11% in the dry season and 13% during the wet season. In direct dry seeded rice, grain losses of 10–70% can occur depending on the weed species present and weed density.

Individual weed species differ substantially in their competitive ability with rice. Yield losses in dry-seeded irrigated rice from season-long single species infestations were 82% for red rice (*Oryza sativa*), 70% for barnyard grass (*Echinochloa crus-galli*), 36% for bearded sprangletop (*Leptochloa fascicularis*), 32% for broadleaved signal grass (*Brachiaria platyphylla*), 21% for duck salad (*Heteranthera limosa*), and 17% for northern jointvetch (*Aeschynomene virginica*).

USE OF FLOODING TO CONTROL WEEDS (5, 6)

The use of flooding for weed control is almost as old as the culture of rice itself. The first step in traditional weed control was the selection of a submerged rice culture; the second step was transplanting of rice seedlings in flooded fields. Prior to the introduction of herbicides, weed control was achieved by land preparation, land levelling, seed selection, levee construction, water management, and mechanical and hand weeding. Rice grown under submergence competes better with weeds than dryland rice in which weed problems are more severe. Because rice tolerates low oxygen (hypoxia) conditions better than most weeds, flooding is an effective method of cultural control for many weed species. If fields are kept flooded, weed populations are generally kept at a minimum.

For most weeds, there is an optimal time, duration, and flooding depth for which weed growth is suppressed. Problem weeds generally decrease with increasing water depth. Although continuous flooding kills some weeds, aquatic weed species such as *Monochoria, Pistia, Salvinia*, and *Eichhornia* can thrive. The growth of *Monochoria vaginalis* can be suppressed by water levels below 2.5 cm, and draining in mid season may control some of these aquatic weeds. Mid-season draining may harm the rice, and depending on the length of drain, may encourage germination of many grass weeds. Intermittent flooding kills some weeds, but alternate drying and flooding will encourage yet the emergence of other weed species. Rapid draining and reflooding done carefully and precisely (pinpoint flooding) maintains soil capillaries filled with water, excluding atmospheric oxygen and maintains enforced dormancy of germinable red rice and other weed seeds below the soil surface. Alternate flooding and drying does not adversely hamper productivity if the drainage interval between infiltration and irrigation does not exceed a few days.

Water management can effectively control some of the major weeds in rice. Water seeding and continuous flooding have achieved red rice (*Oryza sativa*) control. The worst weed in rice, *Echinochloa crus-galli* and some other grass weeds can be controlled by continuous flooding at 10–20 cm depth. Continuous flooding at the deeper depths, however, often stresses the rice with stems becoming etiolated and weak; and algae, *Ammannia, Heteranthera, Cyperus, Scirpus*, and *Eleocharis* spp. problems increase. In Japan deep flooding is not an option for *Echinochloa crus-galli* and other grass weed control because levees are only high enough to flood at 3–5 cm depths. These lower water levels are required for effective application and dispersion of granular herbicides.

WEED CONTROL WITH LESS LABOR AND LESS WATER (7)

Water is a scarce and declining resource. Rapid industrialization and urbanization compete with agriculture for limited water resources. In addition, many irrigation infrastructures are also poorly maintained, causing water shortages in rice production. The use of water in traditional rice culture is highly inefficient. Water is required for land preparation to saturate the soil (land soaking) and to maintain a water layer for plowing, harrowing, puddling, and leveling before transplanting. During crop development, water levels are often maintained at relatively high water depths to control weeds and to reduce the frequency of irrigation so as to reduce labor costs. Percolation losses increase with greater water depths.

Increased irrigation efficiency in rice can be attained during the land preparation phase by eliminating or reducing soil crack formation and water flow into soil cracks and by reducing the land preparation time. During the crop growth phase, the rice field can be maintained at near saturation level and weeds controlled by herbicides resulting in a 45% reduction in water use and no loss in yield compared to the standard continuous shallow (2–5 cm depth) submergence. Precision field leveling greatly improves the farmer's ability to manage water.

Water conservation measures in rice production, such as intermittent flooding and shallow water depths, generally make weed control more difficult. With labor and water shortages and the shift to direct seeding, farmers have few weed control alternatives other than to increase herbicide use. Increased herbicide is being faced with social and environmental problems including herbicide resistance, herbicide drift damage to nontarget crops, and contamination of soil and water. Populations of *Echinochloa crus-galli* in various parts of the world are resistant to propanil and several weed species are resistant to bensulfuron. In China *Echinochloa crus-galli* is coresistant to butachlor and thiobencarb. Herbicide use in California rice fields has resulted in fish kills and an off-taste in potable water supplies. The approach to mitigate herbicide contamination in California has been to lengthen the water-holding time in rice fields to allow more time for physiochemical and biological breakdown of the chemicals. The resulting recirculating, pollution abatement, and water conservation system is not possible in the tropics with little storage area and high rainfall.

Of all cultural practices to control weeds in rice, water management is the most important. Even so, water management is but one aspect of an integrated approach that will involve preventative, cultural, and chemical practices to solve weed problems in rice.

REFERENCES

1. De Datta, S.K. *Principles and Practices of Rice Production*; John Wiley & Sons: New York, 1981.
2. Smith, R.J., Jr. Weeds of Major Economic Importance in Rice and Yield Losses Due to Weed Competition. In *Weed Control in Rice*; International Rice Research Institute: Los Baños, Philippines, 1983; 19–36.
3. Moody, K. Weed Management in Rice. In *Handbook of Pest Management in Agriculture*; Pimentel, D., Ed.; 3rd Ed., CRC Press Inc.: Boca Raton, FL, 1991; 301–328.
4. Moody, K. Yield Losses Due to Weeds in Rice. In *Crop Loss Assessment in Rice*; International Rice Research Institute: Los Baños, Philippines, 1990; 193–202.
5. Hill, J.E.; Smith, R.J., Jr.; Bayer, D.E. In *Rice Weed Control: Current Technology and Emerging Issues in the United States, Temperate Rice—Achievements and Potential*, Temperate Rice Conference, Griffith, NSW, Australia, February 21–21, 1984; Humphreys, E., Murray, E.A., Clampett, W.S., Lewin, L.G., Eds.; Temperate Rice Conference Organizing Committee: Griffith, 1994; 377–391.
6. Williams, J.F.; Roberts, S.R.; Hill, J.E.; Scardaci, S.C.; Tibits, G. Managing water for weed control in rice. Californian Agric. **1990**, *44* (5), 7–10.
7. Bhuiyan, S.I.; Tuong, T.P.; Wade, L.J. Management of Water as a Scarce Resource: Issues and Options in Rice Culture. In *Sustainability of Rice in the Global Food System*; Dowling, N.G., Greenfield, S.M., Fischer, K.S., Eds.; Pacific Basin Study Center: Davis, CA, 1998; 175–192, International Rice Research Institute, Los Baños, Philippines.

FOOD CONTAMINATION WITH PESTICIDE RESIDUES

Denis Hamilton
Animal and Plant Health Service, Brisbane, Queensland, Australia

INTRODUCTION

> "How much pesticide residue did I eat today?"
>
> "No more than necessary, and less than would be detrimental to your health."

Government authorities must be able to support the answer to the consumer's question with scientific data and valid scientific studies (1). The "no more residues than necessary" concept comes from the principle of good agricultural practice, which implies that the desired effect (pest control) will be achieved without leaving more residues than necessary in the food.

BEFORE REGISTRATION (2, 3)

Pesticide residue evaluation and risk assessment prior to registration are summarized in Fig. 1. Risks to the environment and to the user are also evaluated, but are not considered further under the present topic—*food contamination with pesticide residues.*

Metabolism studies on a pesticide in crops and farm animals identify the nature of the residue. The residue may consist of parent compound or metabolites or a mixture. In some cases different pesticides produce the same metabolites; in other cases the metabolite of one pesticide is another pesticide. Some crops genetically modified for herbicide resistance achieve their resistance by metabolizing the herbicide to a derivative with no herbicidal activity.

The acceptable daily intake (ADI) of a pesticide is an estimate of the amount that can be ingested daily over a lifetime without appreciable health risk to the consumer on the basis of all the known facts at the time. The ADI is based on animal feeding studies that find the daily dose over a lifetime resulting in no observable effect on the most sensitive animal species tested. Then a margin of safety (safety factor, commonly 100) is applied to allow for extrapolation from animals to humans and the possible wide variation in sensitivity among humans.

The acute reference dose (acute RfD) of a pesticide is an estimate of the amount that can be ingested over a short period of time, usually during one meal or one day, without appreciable health risk to the consumer on the basis of all the known facts at the time of the evaluation. The acute RfD is also based on the results of animal dosing studies with a suitable safety factor.

The maximum residue limit (MRL), synonymous with "tolerance," is the maximum concentration of pesticide residue legally permitted in or on food commodities. The MRL usually applies to the commodity of trade, which may or may not be the same as the edible portion. For a fruit such as apples it is the same, while for bananas the MRL applies to the whole banana, but only the pulp is eaten. An MRL provides a division between food that is legally acceptable or unacceptable. Foods derived from commodities complying with the relevant MRLs are intended to be toxicologically acceptable, but the MRL is not a dividing line between safe and unsafe.

Risk assessment tells us whether or not the amounts of residue are likely to be safe for consumers (4, 5). We estimate dietary intake (also referred to as "dietary exposure") of pesticide residues by multiplying the level of residue in the food ready for consumption by the amount of the food consumed. For chronic risk assessment we compare the sum for all foods of expected long-term average intake with the ADI for the pesticide. For acute risk assessment we compare possible intake from high consumption of a food in one meal or in one day with the acute RfD.

AFTER REGISTRATION

The design of monitoring studies for residues in food commodities depends on the purpose: random survey of food consignments (surveillance), targeted enforcement sampling where a residue problem is suspected (compliance), export monitoring to meet trade requirements, and total diet studies.

Encyclopedia of Pest Management

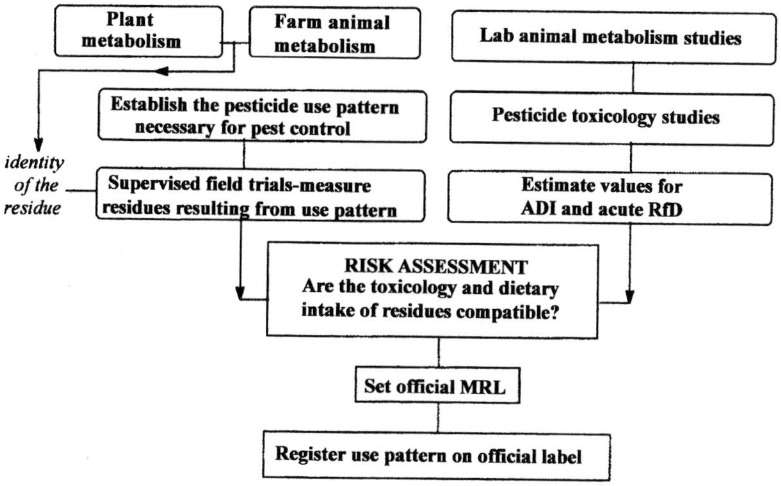

Fig. 1 Risk assessment process before registration for pesticide residues in food. ADI: acceptable daily intake; Acute RfD: acute reference dose; MRL: maximum residue limit or tolerance.

Government authorities regularly survey agricultural and animal produce for levels of pesticide residues. If the label directions were based on reliable and representative field trials and if users are faithfully following label directions then residues will be within the legal MRLs. Most surveys have demonstrated a high level of compliance.

Total diet studies measure which pesticides and what quantities people are actually consuming. Food purchased in the marketplace is prepared by peeling and cooking as in the normal household and is then subjected to residue analysis. Amounts of foods consumed are known from specially designed food surveys for subpopulations such as adult males and females, children, toddlers and infants, as well as for ethnic groups and regions or localities. Dietary intakes for populations and subpopulations are calculated from the diets and the residue levels found by analysis. Commonly, total diet studies demonstrate intakes much less than the ADI.

FOOD PROCESSING (6)

Food processing usually reduces pesticide residue levels because of the washing or cleaning, peeling, milling, juicing, cooking, or baking. Residue levels may increase in some processed commodities because the residue tends more to one fraction than another. For example, residues on the surface of a wheat grain will find their way into the bran fraction with little in the flour. Residues of an oil-soluble pesticide will find their way mainly into the vegetable oil fraction from an oil seed such as soybean.

In particular cases a food process can change the nature of the residue. For example, ethylenebisdithiocarbamate fungicides are converted, on cooking, to ethylenethiourea, which is more toxic than the parent pesticide. Fortunately, ethylenebisdithiocarbamates are essentially surface residues and levels can be substantially reduced by thorough washing before a cooking or blanching step.

TRADE ISSUES

MRL values derived from good agricultural practice are, by their nature, local. A pesticide is used in the best way within local cultural practices to control a specific pest and the rate of pesticide disappearance depends on local environmental conditions. Comparisons among countries of national MRLs and tolerances will frequently reveal substantial differences. Table 1 shows the range of MRLs for ethephon in 17 countries for each of four crops.

The differences pose problems for international trade in food commodities. The importing country may reject shipments of food that do not comply with its national MRLs. It is attractive for some lobby groups and some governments to use national differences in MRLs as a barrier to trade.

Where no MRL or tolerance has been set for a pesticide on a food, some governments apply a ''zero tolerance,'' that is, the MRL is assumed to be zero unless otherwise stated. The reason no MRL is set could simply be that the pest problem does not occur or that the crop is not produced locally; for example, cold temperate countries cannot produce pineapples, so there will be no local uses or local MRLs. Again, zero tolerances lend themselves to use as trade barriers.

Table 1 National ethephon MRLs and tolerances (mg/kg) in 1999

	Peppers	Tomatoes	Pineapples	Grapes
Argentina	2	2		
Australia		2	2	10
Brazil		1.5	0.5	
Canada		2		
France				0.05
India		2	2	
Ireland	3	3		
Italy		3		0.05 wine grapes 3 table grapes
Korea		3	1	2
Netherlands	3	3		
New Zealand		1		
Poland		3		
Portugal	3	3		
South Africa			1	5
Taiwan		2	2	2
United Kingdom	2	3		
United States of America	30	2	2	2

The Codex Alimentarius Commission was established in 1961 to implement the FAO/WHO Food Standards Program. A purpose of the program is to protect the health of consumers and to ensure fair practices in the food trade. The Codex Committee on Pesticide Residues (CCPR) has the responsibility to establish Codex MRLs for food commodities in international trade.

CCPR relies on data supplied by member governments and has established more than 2500 MRLs. Methods of data evaluation in Codex are very similar to methods in Europe, the United States, Japan, Australia, and Canada; Codex draws on the expertise of scientists from these and other countries around the world. Member government acceptance of Codex MRLs for food commodities in international trade is reducing the incidence of trade barriers based on national MRLs.

Developing countries repeatedly suffer pesticide residue trade difficulties because a lack of resources makes it difficult to monitor their exports effectively to ensure compliance with the importing country MRL requirements and precludes data generation and supply to CCPR to obtain Codex MRLs related to their pesticide uses on their crops.

ANALYTICAL METHODS FOR PESTICIDE RESIDUES (7)

Analytical methods for pesticide residues in food typically rely on gas–liquid chromatography (GLC) or high performance liquid chromatography (HPLC) in the final measurement step following extraction from the sample and a sequence of clean-up steps. Multiresidue methods include many residues in the one procedure for the sake of economy. Monitoring usually requires the detection and quantitative measurement of residue levels down to concentrations of around 0.01–0.05 mg/kg. Laboratories must validate their procedures down to the required level, that is, prove they can identify and measure with a specified precision residues down to a required "limit of quantitation" (LOQ).

The LOQ is important in the interpretation of monitoring data. An analytical result reported as "less than LOQ" or sometimes as "no detectable residue" could possibly mean no residue or a residue at a level too low for the method.

Not all pesticide residues are amenable to inclusion in multiresidue methods; they may need separate analysis, which becomes expensive. Reports of monitoring data should state explicitly which residues would have been detected if present above stated LOQs.

FUTURE

The science of acute risk assessment will be further developed in the near future. As new transgenic crops are developed and introduced, their influence on the nature and levels of residues will be fully investigated and MRLs will be set accordingly. Trade issues will continue to be problematic while national governments and lobby groups see residues in food as a mechanism for trade protection. Exporters will need to monitor residues in a high percentage of their exports to meet the requirements of their customers.

REFERENCES

1. In *Pesticide Chemistry, Advances in International Research, Development, and Legislation*, Proceedings of the Seventh International Congress of Pesticide Chemistry (IUPAC), Hamburg, 1990; Frehse, H., Ed.; VCH Verlagsgesellschaft mbH: Weinheim, Germany, 1991; 361–601.
2. FAO. *Submission and Evaluation of Pesticide Residues Data for the Estimation of Maximum Residues Levels in Food and Feed*; FAO Plant Production and Protection Paper, 2002; *170*, 1–192.
3. *Pesticide Residues and Food Safety: A Harvest of Viewpoints*; Dishburger, H.J., Ballantine, L.G., McCarthy, J., Murphy, J., Tweedy, B. G., Eds.; ACS Symposium Series 446, American Chemical Society: Washington, DC, 1991; 1–348.
4. Hamilton, D.J.; Holland, P.T.; Ohlin, B.; Murray, W.J.; Ambrus, A.; De Baptista, G.C.; Kovacicová, J. Optimum use of available residue data in the estimation of dietary intake of pesticides. Pure Appl. Chem. **1997**, *69*, 1373–1410.
5. WHO. *Guidelines for Predicting Dietary Intake of Pesticide Residues (Revised)*, WHO/FSF/FOS/97.7; Program of Food Safety and Food Aid, World Health Organization, Switzerland, 1997; 1–33.
6. Holland, P.T.; Hamilton, D.; Ohlin, B.; Skidmore, M.W. Effects of storage and processing on pesticide residues in plant products. Pure Appl. Chem. **1994**, *66*, 335–356.
7. In *Pesticide Chemistry and Bioscience: The Food-Environment Challenge*, Proceedings of the 9th IUPAC International Congress of Pesticide Chemistry, Brooks, G.T., Roberts, T.R., Eds.; Royal Society of Chemistry: Cambridge, UK, 1999, 339–360.

FOOD LAWS AND REGULATIONS

Ike Jeon
Kansas State University, Manhattan, Kansas, U.S.A.

INTRODUCTION

The food supply in the United States is considered the safest in the world, although chemical residues such as pesticide residues on food have been of great concern to the consumer. Food labels on consumer packages do not contain any statements relative to the pesticide residues or other matters such as insect fragments. This is because foods the consumer buys at supermarkets or grocery stores should be free from these contaminants. In a practical sense, however, producing foods absolutely free from chemical residues or insect fragments is not possible with the practices of modern agricultural production. The U.S. basic food law (Food, Drug, and Cosmetic Act) allows the regulatory agencies to establish tolerance limits for various food products. A tolerance limit is defined as the maximum quantity of a substance allowable on food. There are two major categories for tolerance limits. One is for poisonous or deleterious substances (e.g., pesticide residues) in human food and animal feed, and the other is for natural or unavoidable defects (e.g., insect fragments) in foods that present no health hazards for humans. These tolerance limits are enforced by the federal government during food processing, packaging, and distribution.

TOLERANCE LIMITS FOR PESTICIDE RESIDUES

The responsibility for ensuring that pesticide residues in foods are not present above the limits is shared by three major government agencies (1). The Environment Protection Agency (EPA) determines the safety of pesticide products and sets tolerance levels for pesticides. The Food and Drug Administration (FDA) enforces the tolerances in all foods except meat and poultry products. The U.S. Department of Agriculture's Food Safety and Inspection Service (FSIS) regulates commercially processed egg, meat, and poultry products including combination products (e.g., stew, pizza). In addition, any products containing 2% or more poultry or poultry products, or 3% or more red meat or red meat products are also under jurisdiction of the FSIS. The pesticides of concern usually include insecticides, fungicides, herbicides, and other agricultural chemicals. Table 1 illustrates examples of tolerance levels for pesticide residues in several food categories (2, 3). These tolerance levels are extremely low, usually below parts per million, but do not represent permissible levels of contamination where it is avoidable. In addition, blending of a food (or feed) containing a substance in excess of an action level or tolerance with another food (or feed) is not permitted, and the final product from blending is unlawful, regardless of the level of the contaminant.

Regulatory Inspection and Enforcement

The FDA monitors the levels of pesticide residues in processed foods. For imported products, the FDA checks a sample of the food at entry into the United States and can stop shipments at the entry. If illegal residues are found in domestic samples, FDA can take regulatory actions, such as seizure or injunction.

The U.S. Department of Agriculture also monitors pesticide residues in food (4). The Department was charged in 1991 with implementing a program to collect data on pesticide residues on various food commodities. The program has become a critical component of the Food Quality Protection Act of 1996 and currently is known as the Pesticide Data Program. The data on pesticides in selected commodities are used by the EPA to support its dietary risk assessment process and pesticide registration and by the FDA to refine sampling for enforcement of tolerances.

If a product is in violation of the tolerance limits, it is *adulterated* under the food law. The product may be destroyed or recalled from the market by the manufacturer or shipper. The recall may be initiated voluntarily by the manufacturer (or shipper) or at the request of the regulatory agency. The responsible agency also may seize the product on orders obtained from the Federal courts and may prosecute persons or firms responsible for the violation.

F

Table 1 Examples of tolerance limits for pesticide residues in human food

		Action level	
Substance	Commodity	(Parts per million)	Remark
Aldrin & dieldrin	Asparagus	0.03	
	Fish	0.3	Edible portion
	Peanuts	0.05	
Chlordane	Carrots	0.1	
	Fish	0.3	Edible portion
	Lettuce	0.1	
	Poultry	0.3	Fat basis
DDT[a]	Carrots	3.0	
	Citrus fruits	0.1	
	Tomatoes	0.05	
Lindane	Beans	0.5	
	Corn	0.1	
	Milk	0.3	Fat basis
	Beef	7.0	Fat basis

[a]Dichlorodiphenyltrichloroethane.
(From Refs. 2 and 3.)

TOLERANCE LIMITS FOR INSECT FRAGMENTS

Many food materials may contain natural but unwanted debris that cause no health hazards for humans. These debris may include insects, insect fragments, and rodent hairs and are considered unavoidable defects in foods with the current agricultural practices. In fact, the use of chemical substances to control insects, rodent, and other contaminants has little, if any, impact on natural and unavoidable defects in foods. The FDA contends that the use of pesticides does not effectively reduce the presence of these food defects. This has led the regulatory agencies to establish maximum levels of natural or unavoidable defects allowable in foods for human use. The FDA currently lists over 100 products from fruits to fish (5), and Table 2 shows only several examples. If no defect action level exists for a product, the FDA evaluates and decides on a case-by-case basis using criteria of reported findings such as length of hairs and size of insect fragments.

The FDA sets these action levels under the premise that it is economically impractical to grow, harvest, or process raw products that are totally free of nonhazardous, naturally occurring, unavoidable defects. It is incorrect, however, to assume that because the FDA has an established defect action level for a food, the manufacturer needs only keep defects just below that level. The defect levels do not represent averages of the defects that occur in any of the products. The levels represent limits at which FDA will regard the food product as *adulterated* and, therefore, subject to enforcement action. Like pesticide residues, blending of food with a defect at or above the current defect action level with another lot of the same or another food is not permitted. That practice renders the final food unlawful regardless of the defect level of the finished food.

RESPONSIBILITY OF FOOD MANUFACTURERS

Food manufacturers are required to follow the standard manufacturing procedures under a federal regulation,

Table 2 Examples of tolerance limits for natural or unavoidable defects in foods

Product	Defect	Action level
Sweet corn, canned	Insect larvae	2 or more 3 mm or longer larvae
Macaroni	Insect filth	225 insect fragments or more per 225 g
	Rodent filth	4.5 rodent hairs or more per 225 g
Peaches, canned and frozen	Mold/insect damage	Wormy or moldy on 3% or more fruits
	Insects	1 or more larvae and/or larval fragments whose aggregate length exceeds 5 mm in 12 one-pound cans
Peanut butter	Insect filth	30 or more insect fragments per 100 g
	Rodent filth	1 or more rodent hairs per 100 g
Popcorn	Rodent filth	1 or more rodent excreta pellets or rodent hairs in 1 or more subsamples
Tomato juice	Drosophila fly	10 or more fly eggs per 100 g
	Mold	24% of mold counts in 6 subsamples
Wheat flour	Insect filth	75 or more insect fragments per 50 g
	Rodent filth	1 or more rodent hairs per 50 g

(From Ref. 5.)

known as Good Manufacturing Practice (GMP), during food production (6). The GMP guidelines imply that all food materials used must not exceed the tolerance limits set for pesticide residues or any other poisonous or deleterious substances. The GMP also calls for the same regulatory requirement for natural or unavoidable defects in all food materials. The food materials susceptible to contamination may be tested for compliance or relied on a supplier's guarantee or certification that they are in compliance. In addition, the GMP regulation stipulates that food manufacturers and distributors must utilize at all times quality control operations that reduce natural or unavoidable defects to the lowest level feasible with the current technology.

POTENTIAL CONSUMER BENEFITS

Through conducting a monitoring program, the federal government agencies work together to improve consumer protection. The EPA will continue to review scientific data on all pesticide products, while the FDA and U.S. Department of Agriculture will closely monitor levels of pesticide residues in all foods including both domestic and imported products. The U.S. Department of Agriculture's data for 1998 suggest that violation of the pesticide tolerance limits was very low in all raw products including fruit and vegetable, wheat, and milk samples. In 1993, the FDA reported that no pesticide residues were found in infant formulas, and no residues over EPA tolerances or FDA action levels were found in any of the foods that were prepared as consumers normally would prepare them at home (7).

ACKNOWLEDGMENT

Contribution No. 00-231-B, Kansas Agricultural Experiment Station, Manhattan, Kansas 66506, U.S.A.

REFERENCES

1. FDA. *FDA's Food and Cosmetic Regulatory Responsibilities*; U.S. Food and Drug Administration: Washington, DC, 1998; 1–5, http://vm.cfsan.fda.gov/~dms/regresp.html (Accessed June 2000).
2. FDA. *Action Levels for Poisonous or Deleterious Substances in Human Food and Animal Feed*; U.S. Food and Drug Administration: Washington, DC, 1998; 1–17, http://vm.cfsan.fda.gov/~lrd/fdaact.html (accessed June 2000).
3. USDA. *Domestic Residue Book (Appendix I)*; U.S. Department of Agriculture, Food Safety and Inspection Service: Washington, DC, 1998; 1–30, http://www.fsis.usda.gov:80/OPHS/redbook1/appndx1.htm (accessed June 2000).
4. USDA. *Pesticide Data Program Annual Summary—Calendar Year of 1998*; U.S. Department of Agriculture, Agricultural Marketing Service: Washington, DC, Jan. 2000; 1–19.
5. FDA. The Food Defect Action Levels—Levels of Natural or Unavoidable Defects in Foods that Present No Health Hazards for Humans. In *FDA/CFSAN Food Defect Action Level Handbook*; U.S. Food and Drug Administration: Washington, DC, 1998; 1–36, http://vm.cfsan.fda.gov/~dms/dalbook.html (accessed June 2000).
6. CFR. Current Good Manufacturing Practice in Manufacturing, Packing, or Holding Human Food. In *Code of Federal Regulations, Title 21, Part 110*; U.S. Government Printing Office: Washington, DC, 1999; 206–215.
7. FDA. *FDA Reports on Pesticides in Foods*; U.S. Food and Drug Administration: Washington, DC, 1993; 1–5, http://vm.cfsan.fda.gov/~lrd/pesticid.html (accessed June 2000).

FOOD QUALITY PROTECTION ACT

F

Christina DiFonzo
Michigan State University, East Lansing, Michigan, U.S.A.

INTRODUCTION

The Food Quality Protection Act (FQPA) was signed into law on August 3, 1996, with broad-based support from industry, agricultural commodity, environmental, and consumer groups. FQPA amends the two most important laws regulating pesticides in the United States: the Federal Insecticide, Fungicide, and Rodenticide Act (FIFRA), which sets guidelines for pesticide use, registration, classification (general versus restricted), and applicator certification; and the Federal Food, Drug, and Cosmetic Act (FFDCA), which regulates the setting of tolerances for pesticides used on food crops. Some of the major issues addressed by FQPA are residue tolerances, children's health, endocrine disruption, and consumer right-to-know with regard to pesticides. The Environmental Protection Agency (EPA) is responsible for interpreting and implementing FQPA.

TOLERANCES

FQPA fundamentally changes the way EPA sets tolerances for pesticide residues in food. EPA must review all (nearly 10,000) pesticide tolerances under new FQPA guidelines. The tolerance assessment schedule developed by EPA called for examining 33% within 3 years after August 1996, 66% within 6 years, and 100% within 10 years. EPA initially took a ''worst-first'' approach, to review the pesticides it considered to be of greatest risk, particularly to children, by August 1999. Three major pesticide groups, organophosphates (OPs), carbamates, and probable human carcinogens (B2s), were targeted under the worst-first approach. OPs and carbamates, the majority of which are insecticides, are neurotoxins structurally related to nerve gas. They affect the enzyme acetylcholinesterase in animals, including humans. B2 carcinogens are pesticides classified by EPA as having sufficient evidence for causing cancer in lab animals (usually at very high dose levels), but human evidence is lacking. Several important fungicides, plus a few herbicides and insecticides, are classified in this category.

Before FQPA, a single tolerance was established for each pesticide/crop combination, based only on dietary exposure to pesticide residue. Under FQPA, EPA must consider the combined (aggregate) exposure to a pesticide through dietary, drinking water, and nondietary sources (for example, structural, turf, garden, and pet uses) as well as the cumulative exposure to related pesticides with a common mechanism of toxicity. Furthermore, FQPA directs EPA to consider sensitive subpopulations, especially children, when setting tolerances. To insure that sensitive groups are adequately protected, EPA can require a safety factor of up to 10-fold on existing tolerances.

THE RISK CUP

An analogy of a ''risk cup'' is used by EPA to explain changes in the establishment of tolerances under FQPA. Before FQPA, there was a separate risk cup for each pesticide/crop combination, containing only dietary exposure to residue. FQPA creates a separate risk cup for *each group* of related pesticides with common toxicity. Multiple pesticides, as well as multiple residues from all sources—food, water, and nonfood—of each pesticide, go into the same cup. Under this scenario, the cup gets crowded, and individual tolerances for each pesticide/crop combination in the group must get smaller. Furthermore, safety factors for children may reduce the overall size of each cup, potentially by a factor of 10.

ENDOCRINE DISRUPTION

Under FQPA, all pesticides and pesticide additives must be tested for effects on the endocrine system. This may require in vitro and in vivo screening for three different types of endocrine effects: estrogenic (mimics or blocks estrogen), androgenic (mimics or blocks androgens), and thyroid. Of the potential targets of a screening program, these three hormone groups are important in human development, are fairly well studied, and some laboratory methodology is already available to detect changes in level and function. Estimates are that up to 70,000 pesticides and other chemicals will be screened under FQPA and a second law, the Safe Drinking Water Act.

CONSUMER RIGHT-TO-KNOW

Another issue addressed in FQPA is consumer right-to-know about pesticide residues in food. FQPA mandated that EPA create a brochure to inform consumers about pesticide risks and benefits, and ways to remove residues from food they purchase. The brochure was completed and distributed to supermarkets in early 1999. However, FQPA did not mandate that stores actually display the publication.

POTENTIAL IMPACTS OF FQPA

Pesticides that do not meet FQPA standards must either be mitigated (use patterns changed) or eliminated (some or all uses dropped). Thus, as FQPA is implemented, it potentially will have a tremendous impact on American agriculture.

- Changes in labeling or use patterns (number, frequency, and timing of applications) of pesticides to mitigate residue.
- Loss of critical pesticide uses, particularly for so-called minor (specialty) crops. These commodities represent smaller markets for pesticide manufacturers and thus are often "expendable."
- Increases in production costs. Traditional broad-spectrum products might be replaced by more expensive, reduced-risk alternatives that control a narrower range of pests.
- Increased complexity of production and pest management systems. Broad-spectrum pesticides will be replaced by narrower spectrum tactics that require better knowledge and more intense management of the production system on the part of the producer.
- Potential for pesticide resistance. Loss of certain classes of pesticides could lead to resistance to remaining products, which are being relied on too heavily.

BIBLIOGRAPHY

Colborn, T.; Dumanoski, D.; Myers, J.P. *Our Stolen Future*; Penguin Books U.S.A.: New York, 1996; 316.

Public Law 104-170 to Amend the Federal Insecticide, Fungicide, and Rodenticide Act and the Federal Food, Drug, and Cosmetic Act, and for Other Purposes, H.R. 1627; Federal Register: Washington, DC, Aug. 3, 1996.

Proceedings National Pesticide Impact Assessment Program Workshop, USDA CSREES NAPIAP Program, Sacramento, CA, May 5–7, 1998; Melnicoe, R., Ed.: Washington, DC, 1998; 76.

National Research Council. *Pesticides in the Diets of Infants and Children*; National Academy Press: Washington, DC, 1993; 386.

U.S. Environmental Protection Agency, Office of Pesticide Programs. *The Food Quality Protection Act (FQPA) of 1996.* http://www.epa.gov/oppfead1/fqpa/

FORECASTING PEST OUTBREAKS

Roland Sigvald
Swedish University of Agricultural Sciences, Uppsala, Sweden

F

INTRODUCTION

The economic importance of pests and diseases of agricultural crops; availability of new, highly effective pesticides; and the negative effects of insecticides, fungicides, and herbicides have focused attention on forecasting pest outbreaks. Forecasting systems will require better knowledge of the dynamics of insect populations, viruses, and fungal pathogens as well as economic threshold values for the damage they cause. Forecasting pest outbreaks during the growing season is very important not only in conventional farming but also in alternative agriculture (ecological farming) to increase profitability and at the same time minimize the negative effects on flora, fauna, and groundwater.

The occurrence of a pest or a disease is often correlated to damage or yield loss at harvest. To prevent the insect or pathogen from reaching the economic damage threshold it is often necessary to act at an earlier developmental stage, the action threshold. Long-term forecasts are also useful. An example of this is planning of pesticide imports for the coming growing season.

FORECASTING APHIDS ON CEREALS

Aphids are very important on a great number of crops, not only because of the direct damage they cause but also indirectly by transmitting virus. In Sweden *Rhopalosiphum padi* is the most important aphid species on cereals, especially on oats and spring barley. Aphicide treatments have been profitable on about 70% of national barley and oat crops during years with heavy outbreaks of *R. padi*, but less than 10% in years with low infestations. Because of the great variation in infestation intensities between years accurate forecasting methods are needed.

Aphid migration has been studied by using suction traps for more than 20 years in Sweden, partly to develop the capability for making long-term forecasts to estimate the general need of chemicals long before the actual spraying (several months), but also for short-term forecasts. The number of autumn migrators give rather good information on the risk for attack the following year, but the relationship between number of overwintering eggs and proportion of fields above the damage threshold is stronger (r^2 = 0.4), but even better between spring migration and proportion of fields above the damage threshold (r^2 = 0.75). In some years unfavorable weather (i.e., low temperatures and rainfall) makes it difficult for the aphids to migrate to spring cereals and during such years the severity of attack is much lower than predicted based on the winter eggs counts. Under such conditions suction traps give good information about the risk for attack by *R. padi* on spring barley and oats. The forecasts together with a good warning system and observations in the specific field gives the farmer good bases for decision making.

FRIT FLY ON OATS

The frit fly, *Oscinella frit*, is a stem-boring fly that causes damage to cereals and grasses. In Sweden it is an important pest on oats, and may cause losses of up to 50%. Chemical control is based on the prevention of egg-laying by application of a pyrethroid before the two leaf stage. Therefore, the farmer has to decide whether to spray early in the cropping season.

A method for forecasting the infestation of frit flies has been developed in Sweden. Important factors include timing between insect and plant development, population level, and the weather during egg-laying period (1). The method is based partly on meteorological data, biological observations, and sampling. Sunshine hours the previous year and suction trap catches give a good estimate of the present population level (2). During spring and early summer the temperature sum is calculated using a base temperature of +8°C, which gives an accurate prediction of when migration of the frit fly will take place. Weather data such as temperature is also considered, because it has a great influence on frit fly activity and egg laying.

A practical method based on this research has been validated in more than 800 oat fields. There is very good agreement between risk points in the specific field and the proportion of main stems infested by the larva. The method is now widely used by the farmers. They get information via leaflets, but also interactively via the internet from the Swedish University of Agricultural Sciences.

POTATO VIRUS Y

In Sweden and many other countries in northern Europe, potato virus Y (PVY) is one of the most important virus diseases on potatoes. There are several strains of PVY of which PVY° is the most important one in Sweden. A simulation model has been developed that predicts the extent to which the proportion of progeny tubers infected with PVY° will increase during late summer (3). Proportion of virus sources, latent period, mature plant resistance, vector efficiency, and cultivar susceptibility are some factors in the model. There is a close correlation between forecasts and tested samples from specific potato fields (more than 800 fields).

The PVY° simulation model can be used in forecasting the risk for virus spread. Seed potato growers would be able to take prophylactic measures at an early stage and would also benefit by being able to predict the proportion of progeny tubers infected in late summer. If there is a great risk that the level of infection of the tuber yield will exceed the threshold for seed potatoes, it may be more profitable to delay haulm destruction and market the potatoes for consumption or industrial use (4). The method could be used both by conventional and ecological farming. The method could also be used to decide need of postharvest testing of seed potatoes.

SCLEROTINIA STEM ROT

Sclerotinia stem rot, *Sclerotinia sclerotiorum* is one of the most important diseases on spring sown oilseed rape. During years with high humidity, yield reduction can reach 60% in heavily infested fields and in such years it is profitable with chemical treatment in 40–60% of the fields. Sclerotinia stem rot can be effectively controlled by fungicide treatment during full flowering. Routine spraying is not profitable since the cost for chemical treatment is high and disease incidence varies greatly between years and regions and also between fields within a region, thus motivating a forecasting system.

A method for forecasting the incidence of Sclerotinia stem rot has been developed in Sweden (5). Besides field experiments and laboratory studies, data from more than 800 fields have been collected to improve the method as well as for validation of the risk assessment. The method is mainly based upon a number of risk factors such as crop density, crop rotation, level of previous Sclerotinia infes-

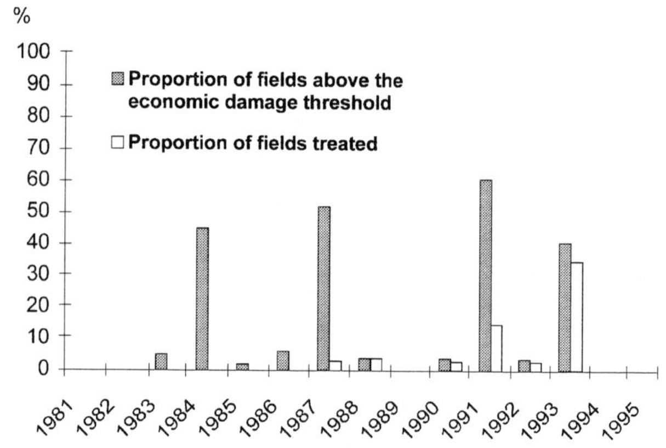

Fig. 1 Proportion of fields of spring-sown oilseed rape in the province of Uppland in central Sweden that were profitable to spray and the proportion of fields that actually have been sprayed.

tation (estimation of inoculum in soil), time for apothecia formation from sclerotia, rainfall during early summer and during flowering, and weather forecast. An initial risk assessment showed very good agreement between risk points and Sclerotinia stem rot incidence. To further improve the model, specific field data were analyzed by logistic regression (6). The results showed that the model could be simplified and still give very good or even better predictions.

The method has been available in paper form, but during the past years also interactively via the world wide web. Since the method was initially introduced about 10 years ago the proportion of fields requiring treatment that actually received treatment has increased, and today there is a very good agreement between the need for sprays and the actual treatments (Fig. 1). A contributing factor to the success of this forecasting method is participation of the advisory service and farmers during data collection, validation, and implementation.

FUTURE ASPECTS ON FORECASTING PESTS AND DISEASES

Further improvements of forecasting systems for pests and diseases on field crops depend on a close collaboration between researchers, the advisory service, and farmers, especially for validation and implementation of methods. The research and development of current forecasting methods have included validation using field specific data from more than 3000 fields of different crops. In addition weather data has been included in the models to improve them. In the near future the availability of forecasting methods via the world wide web will increase and thereby improve the possibility for farmers to have direct access to new models for specific fields.

REFERENCES

1. Lindblad, M.; Sigvald, R. A degree-day model for regional predictions of first occurrence of frit flies in oats in Sweden. Crop Prot. **1996**, *15* (6), 559–565.
2. Lindblad, M.; Solbreck, C. Prediction oscinella frit population densities from suction trap catches and weather data. J. Appl. Ecol. **1998**, *53*, 871–881.
3. Sigvald, R. Forecasting the Incidence of Potato Virus Y^o. In *Plant Virus Epidemics—Monitoring, Modelling and Predicting Outbreaks*; McLean, G.D., Garett, R.G., Ruesink, W.G., Eds.; Academic Press: Australia, 1986; 419–441.
4. Sigvald, R. Forecasting Aphid-borne Virus Diseases. In *Plant Virus Disease Control*; Hadidi, A., Khetarpal, R.K., Koganezawa, H., Eds.; APS Press, The American Phytopathological Society: St. Paul, MN, 1998; 172–187.
5. Twengström, E.; Sigvald, R.; Svensson, C.; Yuen, J. Forecasting sclerotinia stem rot in spring sown oilseed rape. Crop Prot. **1998**, *17* (5), 405–411.
6. Yuen, J.; Twengström, E.; Sigvald, R. Calibration and verification of risk algorithms using logistic regression. European J. of plant pathol. **1996**, *102*, 847–854.

F

FOREST LOSSES TO PEST INSECTS/MITES AND PLANT PATHOGENS

Richard S. Hunt
Natural Resources Canada, Canadian Forest Service, Victoria, British Columbia, Canada

INTRODUCTION

Quantification of worldwide losses to forest pests has never been attempted. The total number of tree species is roughly estimated at 20,000 distributed among 1500 genera. Host indices and entomology reference texts suggest that each genus or family has 2–5 insect pests or pathogens causing serious economic loss. In general, insect-caused losses occur in cycles, so in a particular area losses may be negligible for several years then severe for a few years. Most important plant pathogens are insidious, and constantly cause minor mortality and severe growth loss. On average, annual estimated losses to plant pathogens and insects are similar. Losses to mites are considered negligible.

Worldwide losses caused by insect pests and pathogens in commercial forests is about 650×10^6 m^3 per year, or about 18×10^9 dollars. Pests and diseases contribute to the biodiversity of forests, so total control of losses is undesirable.

WORLDWIDE IMPACT

Volume losses in many parts of the world mean fuel woods have to be gathered from further afield, or alternative fuels sought. Increasing human populations exacerbate the problem, resulting in deforestation. Flooding, soil loss, species loss, infringement on national parks, and desertification may also result from deforestation (1).

Many native diseases and insects impact commercial forests in native habitats. Often moving desirable crop species to different continents results in increased volume production because many of the debilitating diseases and insects have been left behind.

It is usually too expensive to transport logs long distances; thus, an assured local log supply is usually necessary to keep mills operating, and small communities viable. Salvage logging on a large scale is sometime necessary to harvest insect-killed timber before it decays. Such activity disrupts the long-term log supply, and thus contributes to temporary or permanent mill closures. Forest losses caused by insects and pathogens are more important if they impact local log supplies than if they attack less accessible timber. Similarly, losses are more acute if there has been an investment into forest management, such as planting, pruning, spacing, and fertilization, compared to native forests where little, or no investments have been made.

Counting dead trees in plots to estimate mortality over a wider area frequently depends on choosing plots representative of particular ecological zones, and it may depend on additional sampling by aerial surveys. Ecological zones often need better delineation, and aerial surveys are sometimes omitted because they are expensive.

Growth loss estimates frequently depend on sampling representative trees at various heights for increment and volume loss, then applying the values to wider geographic areas. This has been done in detail only for a few insect pests such as the western spruce budworm; thus, growth loss estimates are extremely conservative.

The importance of the disease or insect is dependent on the product expected from the crop. Forests destined to produce saw logs usually need to be spaced to encourage good incremental growth on crop trees. Some diseases, such as stem rusts, can act as natural spacing agents. However, losses by natural thinning agents in small diameter materials destined for crops such as firewood or fence posts could be important. Some diseases, therefore, can be beneficial in certain situations, but detrimental if they affect crop trees. Although annual volume loss estimates are sometime made, in many instances the volume could not have been utilized because the timber was inaccessible. As timber becomes more accessible, and the demand for different types of wood products from the same forest increases, the impacts of disease and insect losses are expected to rise.

Quarantines designed to prevent the introduction of nonindigenous diseases and insects require knowledgeable inspectors and diligence. There is increasing concern about introductions as world trade increases. Some quarantines impair trade, which at times may be advantageous to the importing country and detrimental to the exporting country (2).

North American Losses

As an example, annual Canadian forest losses due to disease and insects have been estimated at 128×10^6 m^3, which, if the lost volume was available and marketable, could employ about 200,000 persons (3). Diseases and insects often interact: for example, bark beetle populations are usually maintained on root-diseased trees. Often such losses are unrealistically attributed to just disease or insect. Estimating losses depends on good forest inventories. Because trees are harvested after long rotations and ground measurements are infrequent, inventories themselves must be regarded as estimates.

In urban forests, property values decline when large specimen trees are killed by insects or disease. Some pathogens particularly those causing root, butt, or stem decay may make a tree hazardous. Trees affected by these pathogens may topple, resulting in property losses, death, insurance claims, and court costs. Recently, these losses have been reduced substantially by proper tree assessment and hazard tree abatement. On the other hand, such trees are desirable for wildlife habitat. Some are retained on harvest sites according to strict safety protocols. In some instances, healthy trees are purposely damaged or infected to produce desirable wildlife habitat.

Landscapes can be dramatically altered when insects, such as bark beetles, kill trees over vast areas. Such areas may be more prone to forest fires. Forests may be changed in age or species structure, which in turn changes the abundance and species structure of animal populations (2–4).

Some native species, such as chestnuts, Fraser fir, and white pines have suffered extreme losses and have been expiated from segments of their historical range by the introduction of exotic diseases or insects. Besides direct mortality losses, there are ecological losses when this happens. Bears with cubs still prefer sleeping at the bases of old-growth eastern white pine because the ferruled bark makes them easy to climb when danger threatens, and raptors find the high canopy branches ideal perches to await the movement of prey below. The copious nut crops of the chestnut formerly sustained many wildlife species, including passenger pigeons. These trees have been replaced by less economically valuable species, with reduced ecological significance (5, 6).

Control

Pesticides are rarely used in the forest. The North American Forestry Commission (United States [National Forests Lands only], Canada, and Mexico) estimates that in 1997–98 chemical insecticides were used on only 18,150 ha and fungicides on 80 ha, accounting for $< 0.1\%$ of the chemical pesticides used in North America. Additionally, stump tops were treated with borax on 15,600 ha to control *Heterobasidion annosum*, the causal agent of Annosus root disease, and 156 ha were treated to control mistletoes. Also fungicides, insecticides, and fumigants are used to control insects and pathogens in forest seed orchards and nurseries (7).

Genetic controls are slowly gaining in use, particularly for pines and spruce. Biological control through the introduction of natural predators, parasites, and disease has been used for important exotic insect pests. The North American Forestry Commission estimates that *Bacillus thuringensis* formulations were used against moths on 75,320 ha in 1997–98. Occasionally, virus preparations, such as TM BioControl® for the Douglas-fir tussock moth, are used against moths. Pheromones are used to attract insects; for example, baited trees lure bark beetles, which are then destroyed before the next generation emerges. Particularly in Europe, commercial preparations of the fungus *Phlebia gigantea*, are applied to pine stumps to inhibit colonization by the Annosus root disease pathogen.

Silvicultural controls are widely used to control forest diseases and insect pests. This includes: 1) scheduled harvesting (particularly of hazard rated stands); 2) species and tree selections during spacing, commercial thinning, harvesting, and planting; 3) prunings; and to reduce root disease inoculum; 4) stump removal, pop-up spacing, bridge-tree removal, and push-over logging (8).

Ambrosia beetles can be controlled in logs by sprinklers, ponding, or prompt milling. Sometimes pesticides are used to control blue-stain fungi in lumber. Minimizing wane and kiln-drying controls insects and pathogens, particularly in lumber bound for export.

FUTURE CONCERNS

The introduction of exotic insect pests and diseases is the greatest threat to the world's forests (2, 4, 9). There is particular concern for importation of known major pests and pathogens, and often quarantines or certification programs are in place to reduce their importation. However, minor pests and pathogens associated with new hosts can be devastating. Quarantines to restrict the movement of such minor insects and pathogens are rare because the damage they cause in new hosts is unanticipated. This has been particularly noted in recent years with the introduction of pine wilt nematode to the Orient and dogwood leaf blight to the southern United States. Now, goods and services move regularly within countries, and quarantines within a country are less common than external quarantines. There are differences in insect and pathogen populations in different parts of the same continent and the mixing of their genes may produce more devastating insects or pathogens. Some insects and diseases endemic to restricted areas may produce epidemics when moved to an area where new hosts are encountered. For instance, the pitch canker disease, which is native to pines in the eastern United States, recently has become a major exotic disease in Monterey pine in California (10). Forest tree species are often keystone to particular ecosystems and their loss greatly effects biodiversity. The forest industry is concerned because direct losses may ensue, and because the choice of species for planting at particular sites is reduced.

REFERENCES

1. Kersten, I.; Baumbach, G.; Oluwole, A.F.; Obioh, I.B.; Ogunsola, O.J. Urban and rural fuelwood situation in the tropical rain-forest area of South-West Nigeria. Energy-Oxford **1998**, *23*, 887–898.
2. Niemelä, P.; Mattson, W.J. Invasion of North American forests by European phytophagus insects: legacy of the European crucible. Bioscience **1996**, *46*, 741–753.
3. Whitney, R.D.; Hunt, R.S.; Munro, J.A. Impact and control of forest diseases in Canada. Forestry Chronicle **1983**, *59*, 223–228.
4. Wilcove, D.S.; Rothstien, D.; Dubow, J.; Phillips, A.; Losos, E. Quantifying threats to imperiled species in the United States: assessing the relative importance of habitat destruction, alien species, pollution, overexploitation, and disease. Bioscience **1998**, *48*, 607–615.
5. Monnig, E.; Byler, J. *Forest Health and Ecological Integrity in the Northern Rockies*; FPM Report 92-7, USDA Forest Service, 1992.
6. White Pine Symposium Proceedings. Baughman, M.J., Stine, R.A., Eds.; University of Minnesota: St. Paul, MN, 1992; 202, NR-BU-6044-S.
7. Wanner, K.W.; Kostyk, B.C.; Helson, B.V. Recommendations for control of cone and seed insect pests of black spruce, picea mariana (Mill.) B.S.P., with insecticides. Forestry Chronicle **1999**, *75*, 685–691.
8. Thies, W.G.; Sturrock, R.N. *Laminated Root Rot in Western North America*; PNW-GTR-349, USDA Forest Service, 1995.
9. Filip, G.M.; Morrell, J.J. Importing Pacific rim wood: pest risks to domestic resources. J. For. **1996**, *94* (Oct), 22–26.
10. Storer, A.J.; Gorden, T.R.; Wood, D.L.; Bonello, P. Pitch canker disease of pines: current and future impacts. J. For. **1997**, *95*, 21–26.

FOREST PEST MANAGEMENT (INSECTS)

F

James A. Allen
Paul Smiths College, Paul Smiths, New York, U.S.A.

INTRODUCTION

Insects are a vital component of forest ecosystems, where they provide services ranging from pollination to breaking down the wood of dead or dying trees. They can also be very serious pests, however, and are capable of causing catastrophic tree mortality and a wide range of less visible impacts. In addition to the enormous economic damage insects cause every year in terms of the direct loss of forest products (through means such as tree mortality, damage to wood, and reduced growth), they also can reduce the recreational or aesthetic value of forests, alter hydrological properties of watersheds, impact the value of forests for wildlife or grazing, and increase the susceptibility of forests to other damaging agents, such as diseases or wildfire. Insects can also affect forest management through their damage to related operations, such as tree seedling nurseries and seed orchards.

FOREST INSECT PEST IMPACTS

Insect pests cause enormous damage each year in the world's forests. According to one recent study, just the exotic forest arthropod pests (primarily insects) cause $2.1 billion losses and damage to forests each year in the United States. In addition to losses and damages, the cost of control can often be significant. At least $11 million is spent every year in the United States just to control one species—the gypsy moth (*Lymantria dispar*). Table 1 provides an estimate of the costs associated with damage and control of forest insect pests for one medium-sized U.S. state, during a year with only moderate insect infestations.

In addition to more typical year-to-year infestations, large-scale outbreaks of forest insect pests are a well-known phenomenon, and can cause enormous disruptions of forest management efforts. The eastern spruce budworm (*Choristoneura fumiferana*), a species native to eastern Canada and the northeastern U.S., is especially well known for its periodic, highly destructive outbreaks. Large outbreaks occurred beginning in 1807, in 1878, from 1910 to 1920, and from 1945 to 1965. The most recent outbreak occurred from about 1970 to 1985, and impacted millions of hectares in Maine and eastern Canada. In the state of Maine, an estimated 21% of all balsam fir trees were killed by 1982, more than 25 million m^3 of spruce and fir was killed directly by the budworm, and an additional 29 million m^3 was lost to blowdown in stands opened up by budworm mortality. In addition to the losses of wood, the outbreak caused major disruptions of forest management programs, forcing many landowners to increase their harvesting dramatically to salvage dead trees or cut trees before they were killed.

EMERGING FOREST INSECT PEST PROBLEMS

While many of the most serious forest insect pests have been causing problems for many years, new problems are continually emerging, mostly due to the introduction of insects into new geographic areas. The Asian long-horned beetle (*Anoplophora glabipennis*) was reportedly introduced to the United States as larvae inside wood shipping pallets from China. This insect, which feeds on approximately 100 species of hardwoods, was found infesting trees in New York City in 1996 and in Chicago in 1998. It appears poised to become a very serious new pest in North America, initially in urban areas but perhaps eventually in wildland and managed forests. Its presence has triggered vigorous eradication efforts, which has included the cutting, removal, and incineration of many valuable hardwoods along city streets and in private yards. Other species, such as the hemlock woolly adelgid (*Adelges tsugae*) in eastern North America, the pink hibiscus mealybug (*Maconellicoccus hirsutus*) on Caribbean islands, the gypsy moth in eastern Europe and Turkey, the leucaena psyllid (*Heteropsylla cubana*) in eastern and southern Africa, and the giant wood moth (*Endoxyla cinereus*) in Queensland, Australia, have also been reported as serious new pests in natural forests or plantations. Recent developments in the international trade of forest products threaten to introduce new insect pests. Increasing export of

Table 1 Estimates of forest insect damage and control costs for the state of Georgia, USA, in 1997

Insect	Cost of damage	Cost of control	Total cost
Pine tip moths	$4,650,000	$1,780,000	$6,430,000
Southern pine beetle	$2,342,000	$1,023,000	$3,365,000
Ips spp. and black turpentine beetle	$2,880,000	$465,000	$3,345,000
Defect and degrade-causing insects	$2,870,000	$100,000	$2,970,000
Seed and cone insects	$2,700,000	$86,000	$2,786,000
Reproduction weevils	$1,340,000	$965,000	$2,305,000
Other insects	$1,250,000	$86,000	$1,336,000
Gypsy moth[a]	nd	$180,000	$180,000
Total	$18,032,000	$4,685,000	$22,717,000

[a]The gypsy moth is not generally established in Georgia, but isolated infestations have been detected and controlled. No damage estimates are available for this species.
(Adapted from the University of Georgia Department of Entomology (www.bugwood.org/sl97/forest97.htm).)

wood products from Siberia, for example, may provide a means of introduction into North America for several serious new insect pests of firs, larches, pines, and other conifers.

While most emerging forest insect pest problems are due to exotic species, native insects occasionally become pests in new situations. An example from the southern United States is the fruit-tree leafroller (*Archips argyrospila*), a species long known as a pest in orchards and of some native hardwoods, which has recently emerged as a new and serious pest of the conifer species bald cypress.

FOREST INSECT PEST MANAGEMENT STRATEGIES

Forest insect pests may be controlled by: 1) mechanical and physical methods, 2) chemical control, 3) biological control, 4) integrated pest management (IPM), and 5) prevention/quarantine methods.

Mechanical and physical methods can be as simple as hand-picking of insects, such as during the 1990 outbreak of *Dendrolimus punctatus* in Vietnam, when 1800 workers collected an estimated 50 tons of larvae from young pine plantations. Likewise, cones infested with *Sinorsillus piliferus* are selectively removed from seed orchards in China. In regions with high labor costs, such approaches are generally not economically feasible, but other methods, such as piling and burning of infested trees or slash to prevent bark beetle outbreaks, spraying piles of stored logs with water to prevent wood borer and ambrosia beetle attacks, and heat treating of logs during shipping are often employed.

Chemical control of forest insect pests is a widely used, but controversial, approach. In contrast to agricultural crops, which are essentially two-dimensional, forests have four dimensions, the length, breadth, height, and multiyear duration of the tree crop. This, plus complicating factors such as use of forests for recreation and their rich biodiversity, makes chemical control problematic or even unacceptable in many situations.

The use of chemicals has evolved dramatically over the last century. As an example, in the late 1800s and early 1900s, heavy-metal-based compounds such as Paris green and lead arsenite were used to control the gypsy moth in the northeastern U.S. Beginning in the 1940s, DDT became the chemical of choice, and was eventually sprayed over 3.7 million ha of forest land in the northeast. By the early 1960s the nontarget effects of DDT and other chlorinated hydrocarbons were better understood and alternatives such as organophosphates (e.g., Orthene) and carbamates (e.g., Sevin) came into favor. These in turn are gradually being replaced by newer chemicals that have shorter half-lives or fewer nontarget impacts. Included among these newer chemicals are several, known as semiochemicals, which are synthetic versions of the natural pheromones used in insect-to-insect communication. Semiochemicals can be used to lure insects into traps, disrupt mating, disrupt or redirect insect aggregration, or inhibit feeding.

Classical approaches to biological control involve the use of predators, parasatoids, and disease-producing micro-organisms to reduce pest populations. One advantage of this approach is that it can be self-perpetuating; once a control agent is established it may persist in the environment and result in effective permanent control of the pest. Over the last century, there have been several outstanding examples in North America of predators or parasatoids being used to control introduced forest insect pests, such as the winter moth (*Operophtera brumata*),

European spruce sawfly (*Diprion hercynia*), and the larch casebearer (*Coleophora laricella*).

Other biological control techniques include the development of pest-resistant plant varieties and the use of biological insecticides such as the bacteria *Bacillus thurigiensis* (*Bt*), the fungus *Beauvaria bassiana*, nuclear polyhedral viruses (NPVs), and entomopoxviruses (EPVs). As with chemical control methods, the use of biological control may have unintended consequences, such as when *Bt* kills nontarget insects, or introduced predators or parasatoids reduce the numbers of beneficial native insects. Biological control, therefore, must always be used with care and only after serious efforts have been made to identify possible nontarget effects.

Integrated Pest Management (IPM) is an approach to lower pest populations to a level where their fluctuations will not result in serious economic impacts. Although the use of chemical pesticides or biological control can be an important part of this overall approach, IPM attempts to integrate *all* suitable methods. In forestry applications, the use of silvicultural techniques is an important part of most IPM programs. In the Pacific Northwest, for example, thinning of Douglas fir stands is done as part of a larger strategy to reduce bark beetle outbreaks. Thinning is effective because it reduces the resource base for the bark beetles, reduces competition among the remaining trees (thereby increasing their vigor and ability to resist bark beetle attacks), and increases stand temperatures to levels that can reduce beetle survival. Most well-designed IPM programs also invest a significant amount of effort into monitoring insect pest populations and sometimes, populations of their natural enemies also.

In dealing with exotic forest insect pests, prevention of their arrival in a new region is by far the preferred approach to control. One key to this approach is an effective means of identifying potential pests, a process that can be difficult because insects that are not major pests in one region may become very serious pests when introduced into a new area. The development of efficient pest risk assesment techniques, which attempt to predict whether insect species will become pests if introduced to a new region, is currently an active area of research. In addition to effective means of identifying potential pests, prevention of their arrival relies on the implementation of effective inspection and quarantine techniques.

BIBLIOGRAPHY

Edmonds, R.L.; Agec, J.K.; Gara, R.I. *Forest Health and Protection*; McGraw-Hill: New York, 2000.

Food and Agriculture Organization of the United Nations. *The State of the World's Forests 1999*; FAO: Rome, Italy, 1999.

Goyer, R.A.; Wagner, M.R.; Schowalter, T.D. Current and proposed technologies for bark beetle management. J. of For. **1998**, *96*, 26–33.

Liebhold, A.M.; MacDonald, W.L.; Bergdahl, D.; Mastro, V.C. Invasion of exotic forest pests: a threat to forest ecosystems. For. Sci. Monogr. **1995**, *30*, 1–49.

McFadden, M.W.; Dahlsten, D.L.; Berisford, C.W.; Knight, F.B.; Metterhouse, W.M. Integrated pest management in China's forests. J. For. **1981**, *79*, 722–725.

Pedigo, L.P. *Entomology and Pest Management*; Prentice Hall: Upper Saddle River, NJ, 1999.

Pimentel, D.; Lach, L.; Zuniga, R.; Morrison, D. Environmental and economic costs of nonindigenous species in the United States. Bioscience **2000**, *50*, 53–65.

Speight, M.R.; Wainhouse, D. *Ecology and Management of Forest Insects*; Clarendon Press: Oxford, UK, 1989.

Speight, M.R.; Wylie, F.R. *Insect Pests in Tropical Forestry*; Oxford University Press: Oxford, UK, *in press*.

Turgeon, J.J.; Roques, A.; de Groot, P. Insect fauna of coniferous seed cones: diversity, host plant interactions, and management. Annu. Rev. of Phytopathol. **1994**, *39*, 179–212.

FOREST PEST MANAGEMENT (PLANT PATHOGENS)

Ken I. Mallett
Natural Resources Canada, Canadian Forest Service,
Edmonton, Alberta, Canada

INTRODUCTION

Forests have covered a significant portion of the world landmass for much of the earth's recent history. Today they cover about one-third of the earth's surface, with the largest forested areas occurring in Brazil, Canada, and Siberia. Forests are plant communities predominated by trees and woody plants. They can be generally classed as the rural forests (those outside urban areas) and urban forest (trees found in the urban landscape including parks). Rural forests are under increasing pressure from expanding agriculture, urbanization, and the demand for fiber. Many physical factors such as geographic location, climate, and fire have and do contribute to the forest structure and development; however, biological forces such as insects and diseases have an immense impact on forest species' composition, structure, development, and productivity. Urban forests face similar physical and biological forces that rural forests do, except that in many instances they are subjected to stresses caused by close proximity to humans such as soil and root compaction, as well as air pollution.

It is well known that disease losses of forest trees, especially growth losses, are equal to if not greater than losses caused by insects and are certainly greater than losses due to fire over the long term. Despite this, herbivory of forest trees (especially losses caused by insect pests) has received much greater attention from forest managers than damage caused by diseases. This is in part due to the macroscopic nature of insects and the damage that they cause. Insect outbreaks develop rapidly and are easily seen. Diseases of biotic origin tend to be incited by microscopic organisms and are often insidious in nature.

There are several unique attributes of trees and forests that affect the management of diseases. Trees are long-lived, in many cases exceeding one hundred years. They therefore experience a full range of climatic and environmental conditions that many herbaceous plants would not. The silvics of a tree species have to be considered before deciding on a disease control strategy. Rural forests are often managed for multiple uses such as fiber, wildlife, and recreation. Disease management may be contraindicated among uses. In some cases, diseases in rural forests may be viewed as being important for the natural development or succession of the forest. Rural forests often are large in size and have extensive management practices, making it difficult and costly to apply certain disease control measures. Whereas trees growing in the urban setting or in intensively managed forest plantations may be more highly valued, easier to treat, and therefore the cost of a control measure may be economically justified. Physical, chemical, biological (including genetic resistance), and regulatory control measures have been or are being used to a certain extent in rural and urban forests.

The tree part affected by a disease, the age of the tree, the location of the tree, the season of the year, and the disease itself will have a major influence on the type and extent of the control measure applied. For example, a foliar disease on a boulevard tree may be easily treated by a chemical application during the growing season. A stem rust found on a pine branch should be physically removed before it sporulates. The best disease-control strategy to follow in both rural and urban forests is prevention. Provide optimal conditions for tree growth including matching species to site and follow good planting and silvicultural practice. Insects and pathogens that attack trees are often inextricably linked and a sound integrated pest management plan should be developed for both rural and urban forests.

REGULATORY CONTROL

Exotic (nonnative) tree diseases are perhaps the most important and potentially devastating diseases to both the rural and urban forests. These diseases are brought into a forest through importation of live trees or tree products. Examples of exotic diseases that have caused severe economic damage and near extirpation of species are Dutch elm disease (*Ophiostoma ulmi* [sensus lato]) on American elm (*Ulmus americana* L.), smooth-leaved elm (*U. Carpinifolia* Gleditsch), and Wych elm (*U. glabra* Huds.); white pine blister rust (*Cronartium ribicola* J.C. Fisch.) on eastern white pine (*Pinus strobus* L.), and western white pine (*P. monticola* Dougl. ex D. Don); and chestnut blight [*Cryphonectria parasitica* (Murrill) Barr] on American chestnut [*Castenea dentata* (Marsh.) Borkh.]. The most effective control measure for such diseases is regulatory control (prevention). National, state, or provincial governments can demand a phytosanitary certificate with a shipment of live plant material from countries, states, or provinces that are known to have or suspected of having certain diseases. Regulations regarding the treatment of wood products such as dunnage or logs before importation may be necessary. Enforcement of regulations, inspection, and testing are critical for this control strategy to succeed.

PHYSICAL CONTROL

Physical control methods such as pruning, thinning, and tree removal for disease control have been widely practiced in forests for centuries. Pruning of tree branches can be done to eliminate diseased branches in the case of canker diseases. It may also be done to improve the vigor of a tree and eliminate potential infection courts (wounds, etc.). Pruning is done primarily to high-value trees such as those in the urban forest, seed orchards, and high-value plantations. Thinning, the removal of a prescribed number of immature trees within a stand, is often practiced in rural forests and high-value plantations to increase the diameter growth or form of the remaining trees. Trees that have noticeable disease on the stems such as stem cankers or rusts can be removed in the thinning operation. Whole tree removal is often necessary for certain diseases such as Dutch elm disease, especially in the urban forest. Generally the stem and branches are taken down removed and destroyed through burning or burial. Occasionally roots are removed to prevent transmission of the disease through root grafts. Stumping, pushover logging, and pop-up harvesting are techniques that are being used in western North American forests to control root diseases. Roots and stumps serve as food bases for many root disease-causing fungi such as *Armillaria* species. Removal of the roots and stumps from the soil can help to reduce the inoculum available for infection of new trees. Prescribed fire treatments have been used to control certain foliar disease and to reduce inoculum of certain disease-causing fungi.

CHEMICAL CONTROL

Fungicides are rarely if ever used in rural forests. This is in part due to the extensive nature of these forests, but also to public concern of using pesticides in so-called wild lands. Fungicides are used in urban forests on high-value trees to control foliar and some vascular diseases. They are often applied as a foliar spray, but some are injected into the base of the tree.

BIOLOGICAL CONTROL

There are examples in the forest pathology literature of biological control of forest diseases using other fungi, bacteria, or insects. Few have been put into wide-scale practice. This is not to say that this may not be an important natural control mechanism in forests. Hyperparasites of rust fungi and certain foliar diseases have been identified, but their importance in controlling epidemics of forest diseases is not well understood. Genetic resistance has been identified for certain diseases and may be used practically in urban forests and high-value plantations.

BIBLIOGRAPHY

Butin, H. *Tree Diseases and Disorders*; Oxford University Press: New York, 1995.

Castello, J.D.; Leopold, D.J.; Smallridge, P.J. Pathogens, and processes in forest ecosystems. BioScience **1995**, *45*, 16–24.

Manion, P.D. *Tree Disease Concepts*, 2nd Ed.; Prentice-Hall: Engelwood Cliffs, NJ, 1991.

Schmidt, R.A. Diseases in Forest Ecosystems: The Importance of Functional Diversity. In *Plant Disease: An Advanced Treatise, Vol II. How Disease Develops in Populations*; Horsfall, J.G., Cowling, E.B., Eds.; Academic Press: New York, 1978; 287–315.

Tattar, T.A. *Diseases of Shade Trees*, Rev. Ed.; Academic Press: San Diego, CA, 1989.

FOREST PEST MANAGEMENT (WEEDS)

Michael A. Valenti
Delaware Forest Service, Dover, Delaware, U.S.A.

INTRODUCTION

A weed can be defined as a plant inhabiting an area where it is considered undesirable and thus hinders a manager from attaining a desired goal (1, 2). Weeds are becoming increasingly problematic in forest resource management regardless of whether the goals are timber or fiber production, aesthetics, recreation, wildlife habitat, preservation, or any combination thereof. The escalation of weed problems in forest management can, to a large degree, be ascribed to the loss of historical control options (especially for native competing vegetation) and an accelerating increase in the number and distribution of invasive exotic species. Estimates of annual economic impacts of all weeds on the United States economy are $20 billion with losses in forestry conservatively estimated at $60 million (3).

NATIVE COMPETING VEGETATION

Many native plant species interfere with the stated objectives of forest management plans. After a natural or human-caused disturbance in a forest, such as a fire or prescribed silvicultural treatment, native competing species often become established and quickly dominate a site. These plants compete with more desirable tree species for sunlight, growing space, water, and nutrients, preventing or prolonging the natural forest successional sequence or inhibiting growth of tree seedlings (4). This type of domination, in which competing vegetation prevents the establishment of desirable tree species, inhibits tree growth, and reduces tree survival (5–7), has been termed ''inhibition succession'' (8) and greatly increases the costs of forest management activities.

Historical control methods applied to native competing vegetation, e.g., mechanical removal, prescribed fire, and herbicide treatments (9, 10), have associated drawbacks such as high costs, poor efficacy (11), damage to tree seedlings (12), and environmental risks including adverse effects on nontarget organisms. Restrictions on controlling undesirable vegetation have significantly increased the total area of competing vegetation on federal lands in the United States (13). The escalating controversies over current direct control techniques (especially herbicides and fire) and the increase in demand for more environmentally sound control methods has increased the need for alternatives such as using native herbivores to mitigate the effects of native competing vegetation (14).

Native plants that are considered to be weeds present a great challenge in forest management. With a limited number of high-cost control options available, managers often fall back to a ''do nothing'' strategy and allow the plants to grow as they may. For timber production this often translates to failed regeneration efforts or a prolonged harvest rotation, thus a significant economic loss will be incurred. For other forest management options, such as recreation and aesthetics, the ecological effects would be similar but the economic impacts (losses) are more difficult to determine. In either case, however, the costs associated with losses caused by native competing vegetation are significant.

EXOTIC INVASIVE WEEDS

Plant species that are not native to a particular geographic area and that have a significant impact on the natural environment are termed exotic invasive species. Like native competing vegetation, some exotic weeds also outcompete desirable vegetation for sunlight, growing space, water, and nutrients. However, exotic weeds have the potential to spread over vast areas and cause extensive damage in their new environment because they have, in effect, escaped from their natural regulating agents in their native ranges. Invasive plants profoundly affect forest environments because they create fire hazards, cause serious problems in forest nurseries, adversely affect young pine plantations, and contribute to a decline in forest health (15). Exotic species threaten the integrity of various forest ecosystems and represent an enormous economic burden to the people living in the areas affected (16–18).

Forest land managers commonly use herbicides and mechanical removal to control exotic invasive species. However, unless the target weed is eradicated, control measures (and thus the costs associated therein) must continue on into perpetuity. The most promising control method for invasive exotic weeds in forest management is, arguably, biological control. This form of control where one living organism is used to control another living organism [e.g., an insect against a weedy plant (19) is cost-effective because, once initiated, control is often self-sustaining. There have been many spectacular successes in pitting one living organism against another (20) and the biological control of weeds is an area of science that is gaining more and more attention (21).

Exotic invasive weeds represent a much more significant problem to the forest manager than their native counterparts because of the potential inherent aggressive nature of exotic species. Invasive weeds that threaten forest ecosystems, such as mile-a-minute (*Polygonum perfoliatum*) and kudzu (*Pueraria montana*), may cover extensive geographic areas, leaving the forest manager with no cost-effective method for control. As world trade increases we can expect the establishment of more exotic species outside their natural ranges. A portion of these species undoubtedly will become problematic in forested ecosystems, representing an arduous challenge to forest managers worldwide.

WHAT THE FUTURE HOLDS

Weeds inevitably will play an increasingly important role in forest management for years to come. Cooperative efforts are currently underway to reduce the negative impacts of invasive plants. A national strategy for invasive plant management includes proactive steps necessary to prevent an establishment in the first place and reactive steps for effective control and environmental restoration, once an invasive plant has become established (22). Forest land managers will continue to direct the succession of forest ecosystems (23) in the future and will rely on the continued development of a technologically advanced and cost-effective arsenal of biological, chemical, and cultural control methods.

REFERENCES

1. Zimdahl, R.L. *Fundamentals of Weed Science*; Academic Press: New York, 1999; 556.
2. McNabb, K.; South, D.B.; Mitchell, R.J. Weed Management Systems for Forest Nurseries and Woodlands. In *Handbook of Weed Management Systems*; Smith, A.E., Ed.; Marcel Dekker, Inc.: New York, 1995; 667–711.
3. Bridges, D.C. Impact of weeds on human endeavors. Weed Tech. **1994**, *8*, 392–395.
4. McDonald, P.M.; Fiddler, G.O. *Ponderosa Pine Seedlings and Competing Vegetation: Ecology, Growth, and Cost*; Research Paper PSW-199, USDA Forest Service, Pacific Southwest Forest and Range Experiment Station: Berkeley, CA, 1990; 1–10.
5. Conard, S.G.; Radosevich, S.R. Growth responses of white fir to decreased shading and root competition by Montana chaparral shrubs. Forest Sci. **1982**, *28*, 309–320.
6. Lanini, W.T.; Radosevich, S.R. Response of three conifer species to site preparation and shrub control. Forest Sci. **1986**, *32*, 61–77.
7. Radosevich, S.R. Interference Between Greenleaf Manzanita (*Arctostaphylos patula*) and Ponderosa Pine (*Pinus ponderosa*). In *Seedling Physiology and Reforestation Success*; Duryea, M.L., Brown, N., Eds.; Dr. W. Junk Publisher: Boston, MA, 1984; 259–270.
8. Connell, J.H.; Slatyer, R.O. Mechanisms of succession in natural communities and their role in community stability and organization. Am. Naturalist **1977**, *111*, 1119–1144.
9. Bentley, J.R.; Graham, C.A. *Applying Herbicides to Desiccate Manzanita Brushfields Before Burning*; Research Note PSW-312, USDA Forest Service, Pacific Southwest Forest and Range Experiment Station: Berkeley, CA, 1976; 1–8.
10. Paley, S.M.; Radosevich, S.R. Effect of physiological status and growth of ponderosa pines (*Pinus ponderosa*) and greenleaf manzanita (*Arctostaphylos patula*) on herbicide selectivity. Weed Sci. **1984**, *32*, 395–402.
11. Lanini, W.T.; Radosevich, S.R. Herbicide effectiveness in response to season of application and shrub physiology. Weed Sci. **1982**, *30*, 467–475.
12. Newton, M. Some herbicide effects on potted Douglas fir and ponderosa pine seedlings. J. For. **1963**, *61*, 674–676.
13. Walstad, J.D.; Newton, M.; Boyd, R.J., Jr. Forest Vegetation Problems in the Northwest. In *Forest Vegetation Management for Conifer Production*; Walstad, J.D., Kuch, P.J., Eds.; Wiley: New York, 1987; 15–53.

14. Valenti, M.A.; Berryman, A.A.; Ferrell, G.T. Potential for biological control of native competing vegetation using native herbivores. Agricu. Forest Entomol. **1999**, *1*, 89–95.
15. Westbrooks, R.G. *Invasive Plants: Changing the Landscape of America: Fact Book*; Federal Interagency Committee for the Management of Noxious and Exotic Weeds (FICMNEW): Washington, DC, 1998; 1–109.
16. Liebhold, A.M.; MacDonald, W.L.; Bergdahl, D.; Mastro, V.C. Invasion by exotic forest pests: a threat to forest ecosystems. Forest Sci. Monogr. **1995**, *30*, 1–49.
17. http://www.news.cornell.edu/releases/Jan99/species_costs.html.
18. Wallner, W.E. Invasive pests (''biological pollutants'') and US forests: Whose problem, who pays? OEPP/EPPO Bull. **1996**, *26*, 167–180.
19. Room, P.M. Ecology of a simple plant-herbivore system: biological control of *Salvinia*. Trends Eco. Evol. **1990**, *5*, 74–79.
20. Caltagirone, L.E. Landmark examples in classical biological control. Annu. Rev. Entomol. **1981**, *26*, 213–232.
21. McFadyen, R.E.C. Biological control of weeds. Annu. Rev. Entomol. **1998**, *43*, 369–393.
22. Federal Interagency Committee for Management of Noxious and Exotic Weeds. *Pulling Together: A National Strategy for Management of Invasive Plants*; U.S. Government Printing Office, 1998; 1–22.
23. Luken, J.O. *Directing Ecological Succession*; Chapman & Hall: New York, 1990; 251.

FORMULATION

Peter Messenger
Aventis CropScience France S.A., Villefranche-sur-Saône, France

F

INTRODUCTION

Crop protection products are available as many different types of liquid or solid formulations. Each formulation type is composed of a blend of one or more active ingredients and several ''inert'' auxiliaries. The composition is developed by the formulation chemist in order to obtain the necessary quality, stability, biological activity, and application characteristics. In the early days of formulation technology, products were formulated mainly as aqueous solutions, emulsifiable concentrates, and powders. These were often bulky presentations that were difficult to manipulate and presented risks to user and environment. Increasing economic, social, and environmental constraints have since created a need to develop new formulation types and packaging systems in order to improve safety and ease-of-handling and to optimize biological activity. Nowadays, as the global agrochemical market reaches maturity, companies are looking toward formulation technology innovation as a way of extending product life and providing product differentiation. The development of new formulation technologies however leads inevitably to more complex compositions and manufacturing processes, and the benefits have to be balanced against increased R&D and production costs.

FORMULATION TECHNOLOGY

In the 1960s and 1970s, the formulation was seen essentially as a means of transferring the active ingredient onto the target. Crop protection products were generally formulated as aqueous solutions, emulsifiable concentrates, or wettable powders, and applied at high dose rates. For selective agrochemical formulations, the priority was often to have as wide a pest control spectrum as possible while maintaining a tolerable level of crop damage. This approach naturally enhanced the risk not only to the user but also to the environment, particularly with respect to pack handling and disposal.

By the 1980s, with research costs and competition between the major crop protection companies rising, formulation technology started to play an important part in product differentiation. The ease-of-use of the formulation, and the optimization of the biological activity, became important factors. Aqueous suspension concentrates, with their advantage of handling and particle size, were natural replacements for wettable powders. With a greater range of ''inert'' ingredients available, the formulation chemists began to examine their impact on the activity/selectivity of the active ingredient. Studies on the effect of formulation type and composition on leaching, rainfastness, spray drift, run-off, leaf penetration, and bio activity became common-place in the development of new crop protection products.

In the 1990s, formulation changes were predominantly driven by market forces and, to a lesser degree, by product performance. There was increasing pressure for the crop protection industry to develop new formulation technologies with improved user safety, and reduced environmental risk. The concentration, type, and environmental impact of inert auxiliaries started to come under closer scrutiny by regulatory authorities in developed countries. Container rinsing requirements are now being imposed as low as 0.0001% retained after washing, which means that package disposal/recycling has become a major issue.

Environmental pressure to reduce the quantity of organic solvents used in crop protection products is increasing. In developed countries, this means a gradual introduction of new formulation technologies such as water-dispersible granules to replace suspension concentrates and wettable powders, water-soluble bags to reduce risk of user contact, and emulsion concentrates to reduce organic solvent concentrations. The advantages and disadvantages of different formulation types are shown below (Table 1).

Although the advances made in formulation technology over the past 30 years have therefore significantly improved user and environmental safety, volumes of the original types of formulation, such as ECs and WPs, remain high for technical and economic reasons. For exam-

Table 1 Formulation types: Points for and against

	Ease of use	Worker exposure	Container disposal	Solvent
Emulsifiable concentrate (EC)	0	–	–	–
Wettable powder (WP)	–	–	0	+
Suspension concentrate (SC)	0	0/+	–	+
Emulsion in water (EW)	0	0/+	–	0/+
Suspoemulsion (SE)	0	0/+	–	0/+
Water-dispersible granule (WG)	+	+	0/+	+
Water-soluble bag (SB)	+	+	+	+

+ = favorable impact; – = unfavorable impact; 0 = neutral impact.

ple, the physico-chemical properties of the active ingredient (chemical stability in water, melting point ...) may mean that a switch from an emulsifiable concentrate to a suspension concentrate or a water-dispersible granule is not technically possible. Equally, as the improvement in formulation safety has led to more complex technology and thus an increase in process, development, and ingredient costs, the choice of a water-dispersible granule may be less attractive economically than a wettable powder. (See Fig. 1)

Emulsifiable Concentrates (EC)

In 1994, more than 30% of the world market was formulated as emulsifiable concentrates, which have the advantage that they are easy to handle, can be measured volumetrically, are simple and cheap to produce, and may improve bioactivity. However, ECs contain high concentrations of organic solvents (up to 70%), which not only present a fire hazard, but may be toxic and contribute to atmospheric volatile organic emissions (VOC). They may also be phytotoxic to the crop (1). The most commonly used solvents are the aromatic hydrocarbons (cheap and effective), however, for poorly soluble active ingredients, more expensive polar cosolvents such as ketones, alcohols, or esters are often added. Emulsifiable concentrates contain from 3–10% surfactants in order to achieve a stable emulsion upon dilution in the spray tank.

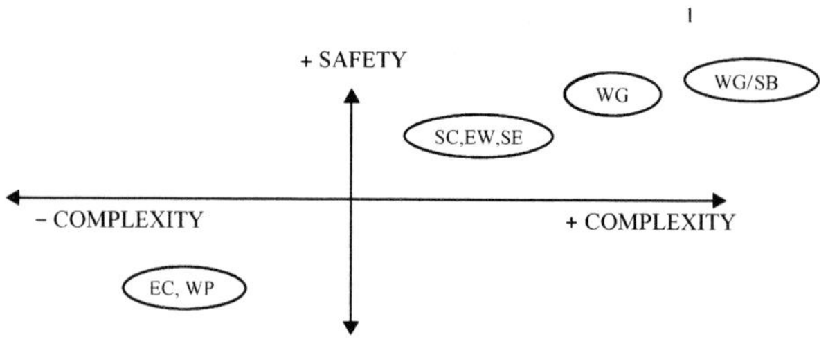

Fig. 1 Safety vs. complexity.

Wettable Powders (WP)

Wettable powders generally contain an active ingredient concentration of more than 50%. The remainder of the composition consists of wetting/dispersing agents to ensure a rapid and stable suspension upon dilution in water and an inert diluent, such as a silicate clay (2). They are prepared by dry-milling down to a particle size around 5–10 micron. Despite their poor rating for ease of use and worker exposure, more than 15% of the world market in 1994 was commercialized in this form as the WP remains a cheap formulation to produce and gives very low phytotoxicity to the crop. Nevertheless, because of their dusty nature and difficulty to mix into the spray tank, WPs are gradually being replaced by suspension concentrates and water-dispersible granules.

Suspension Concentrates (SC)

Suspension concentrates (commonly known as flowables) consist of insoluble active ingredients, milled to particle sizes averaging around 3 micron and dispersed in a liquid phase. Generally, these are aqueous-based and contain little or no organic solvents. The benefits of suspension concentrates are that they are pourable, are easy to measure, generate no dust, and give rise to no storage flammability problems. For these reasons, suspension concentrates have shown a rapid development since their introduction, particularly in Europe, where in 1994, they represented 14% of the market. Apart from their high concentrations of active ingredients (usually 50–80%), aqueous SCs typically contain 5–10% surfactants for dispersing and suspending the solid particles, 5–10% antifreeze (for storage at low temperature), up to 1% antifoam (to prevent formation of excessive foam during production and application), 0.1–5% thickeners (to prevent sedimentation), and water.

F

The main disadvantage of SCs is that the solid phase, being generally denser than water, will tend to sediment on long storage. In order to prevent this, additives to increase the viscosity are necessary, but this makes pack rinsing more difficult and thus increases pack disposal problems.

Emulsions in Water (EW)

EW's are stabilized, concentrated emulsions in water. The interest in this type of formulation is on the increase as, compared to emulsifiable concentrates, a significant part of the organic solvent is replaced by water, and the concern over the toxicological and environmental impact of the solvent diminishes. In addition, this provides a higher flashpoint of the finished product (therefore safer for transport and storage) and also reduces formulation/packaging costs. The disadvantage of EWs is that the conception of a stable emulsion is difficult and depends on the suitability of the active ingredient to this type of technology. Mechanisms involved in their destabilization include flocculation, Ostwald crystal, ripening and emulsion droplet coalescence (3). EW formulations typically contain 5–50% active ingredient, 0–40% solvent and 3–7% emulsifier. The other ingredients are water, antifreeze (for low temperature storage), antifoam (to prevent foam build-up) and thickener (to prevent sedimentation). Suspoemulsions (SE) consist of a blend of a suspension concentrate (SC) and emulsion in water (EW) in a single liquid preparation.

Water-Dispersible Granules (WG)

Water-dispersible granules combine all the advantages of solid formulations (long-term physical stability, low residue in pack, no splashes when emptying pack) with many of the advantages of liquid formulations (no dust, good flowability, ease of measurement). In order to achieve these characteristics, the conception of the WG composition and process tends to be more complex than for other presentations, and this increases the development and production costs. Water-dispersible granules are agglomerates of milled ingredients that are rapidly redispersible after dilution in the spray tank. The manufacturing process consists essentially of a milling stage, an agglomeration stage, and a drying stage. The bulk density, redispersibility, and resistance to attrition depend on the type of process technology chosen. The size of the granules is usually around 0.2–1 mm, although for extruded granules, this can be as high as 5 mm. A WG composition typically contains 50–90% active ingredient, 5–20% surfactants for redispersing rapidly the WG in the spray tank, antifoam (to prevent excessive foam formation) and binding/disintegrating agents (for agglomeration and redispersion).

Water-Soluble Bags (SB)

Water-soluble films (based on polyvinyl alcohol) that have sufficient tensile strength and water solubility to be used as primary packaging for powders and WG's are now commercially available. All risk of contact to the user is thus eliminated as the bag is added directly to the spray tank at which point the film dissolves, thus liberating the product inside. The advantages are obvious in terms of user safety, however, the chemical nature of the film means that compatibility with the product is a major issue, and full development programs have to be carried out to ensure the chemical and physical stability, while maintaining adequate application characteristics. Other disadvantages are the higher formulation and packing costs, and the need to treat the crop with unit doses.

EVOLUTION OF TECHNOLOGY AND MARKET TRENDS

Globally the agrochemical market is reaching maturity, in particular in developed countries. Apart from other contributing factors such as patent expiry and generic competition, this reduction in growth has naturally led to more inter-company competition where the formulation technology plays an important role in product differentiation and market segmentation. In addition, over the past decade, there has been an impetus given by the customer and by regulatory authorities to introduce more user- and environment-friendly presentations. Development programs by commercial crop protection companies have provided technologies to reduce the use of organic solvents, eliminate user contact with the product, reduce the dose of active ingredient required per acre, and meet regulatory pack disposal requirements. Despite this, studies show that there is only a very slow overall change in the spread of formulation types used (Table 2). The reason for this is generally either technical (the physico-chemical

Table 2 Global crop protection formulation volumes by type

	% EC	% WP	% SC	% WG	% Total
1990	33.5	18	11	2	64.5
1994	36	19	9	2	66
1999	25	13	16	5	59

properties of the active ingredient limit the choice of formulation type) or economic (presentations that are more user and environment friendly are more complex in conception and production, and therefore cost more).

For the future, it is expected that there will be only a small shift from emulsifiable concentrates to emulsions-in-water/suspoemulsions, but that the gradual replacement of wettable powders (and to some extent suspension concentrates) will continue. Water-soluble packing is viewed positively and is an option for extending the commercial life of wettable powder formulations. As the main impetus for formulation and packaging technology changes will probably come from regulatory authorities, particularly in developed countries, a harmonized approach involving the crop protection industry will be necessary to continue the improvement in standards and technology.

See also *Auxiliaries: Deodorants, Diluents, Solvents, Stickers, Surfactants, and Synergists,* pages 39–42.

REFERENCES

1. Foy, CL.; Pritchard, D.W. *Pesticide Formulation and Adjuvant Technology*; CRC Press: Boca Raton, FL, 1996; 33–68.
2. *Surfactants and Other Additives in Agricultural Formulations*; Hewin International, Inc.: Amsterdam, 1993; 202.
3. Tadros, T.F. *Surfactants in Agrochemicals*; Marcel Dekker, Inc.: New York, 1995; 107–119.

FRUIT CROP PEST MANAGEMENT (INSECTS AND MITES)

F

Serge Quilici
CIRAD-FLHOR Réunion, Saint-Pierre Cedex, France

INTRODUCTION

Fruit crops are particularly important worldwide in terms of both quantity and value of production. Furthermore, they play a major role in the nutrition of populations in both developed and developing countries by providing essential nutrients such as vitamins and mineral salts.

However, an important proportion of fruit production is damaged by insect and mite pests. This proportion is highly variable depending on the fruit crop, pest complex, socio-economic conditions of producers, and objective of production (fresh fruit or processing).

FRUIT PRODUCTION, CONSUMPTION, AND MARKETING

The total world fruit production reached 434.7 million M tons in 1998, for a production area of 47.9 million hectares. Table 1 gives recent production, export, and acreage data for the main temperate fruits, tropical fruits, and citrus. Among temperate fruits, grape and apple rank far ahead of pear, peach-nectarine, and plum.

Mostly grown in subtropical or mediterreanean climates, citrus alone represent 102.8 million M tons, with a majority of oranges followed by tangerines and clementines, lemons and limes, then pummelos and grapefruits. In terms of production, orange ranks first among all types of fruit.

Banana is the main tropical fruit (58.6 million M tons in 1998). Among other tropical fruits, plantain ranges second, but is generally considered as a food crop. Other important tropical fruits are mango, papaya, and avocado.

IMPACT OF INSECT AND MITE PESTS

Most fruit crops are usually attacked by a few (1–3) key pests, several secondary pests, and a large number of occasional or minor pests whose identities may vary depending on the geographical region.

Apple will be considered here as an example of temperate fruits. The codling moth, *Cydia pomonella* is a key pest of apple worldwide except in parts of Asia. Depending on the region, additional key pests include other Lepidoptera (tortrix moths), weevils, Diptera and aphids. Various Lepidoptera are also secondary pests, while occasional or secondary induced pests include scale insects, aphids, leafhoppers, and phytophagous mites.

In many producing areas, the key pests of citrus consist of mites (for example the citrus rust mite, *Phyllocoptruta oleivora*, or the broad mite, *Polyphagotarsonemus latus*), thrips, fruit flies, and various Lepidoptera species. In certain areas, vectors of diseases include key pests, like psyllid vectors of citrus greening disease in many African or Asiatic countries. Depending on the local situation, scale insects may become major or occasional pests. Minor pests include aphids, whiteflies, weevils, some Lepidoptera, and tetranychid mites.

On banana, the weevil *Cosmopolites sordidus* is generally considered the main insect pest, together with nematodes. Though present in all producing areas, its damage is particularly important for small holding production of plantain. Some thrips species may also be key pests.

Specific complexes of pests are associated with the main tropical fruit crops. Mango is subject to attacks by flower or leaf midges (Diptera Cecidomyidae), leafhoppers, scale insects, bugs, thrips or fruit flies. The mango weevil, *Sternochetus mangiferae* may be important as a quarantine pest. Moreover, in particular areas, fruit-piercing moths (Lepidoptera: Noctuidae) can totally destroy the production of mango or other tropical fruits. High levels of damage can be caused to mango production by pests attacking either flowers (bugs, thrips, midges) or fruit (fruit flies). Papaya may be damaged by various mites (the broad mite, various tetranychids) while avocado is attacked by scolytid beetles or other Coleoptera, thrips, fruit flies, and various Lepidoptera, scale insects and whiteflies. Main litchi pests include mites, Lepidoptera, and fruit flies.

Table 1 World production, export, and acreage for the main fruit crops

Fruit crop	Production[a] (10^6 Mt)	Export[b] (10^6 Mt)	Area[a] (10^6 ha)
Temperate fruits			
Apple	56.1	5.4	7.1
Grape	57.4	2.2	7.3
Pear	14.4	1.7	1.8
Peach-nectarine	11.1	—[c]	1.7
Plum	8.0	0.4	1.7
Citrus			
Lemon, lime	9.3	1.4	0.7
Orange	66.2	4.5	3.8
Pummelo, grapefruit	5.1	1.2	0.2
Tangerine, clementine, satsumas	18.0	2.5	1.6
Total citrus	102.8	—[c]	—[c]
Tropical fruits			
Banana	58.6	14.1	3.9
Mango	23.5	0.5	2.7
Papaya	4.8	0.1	0.3
Avocado	2.3	0.2	0.3
Total world (all fruits)	434.7	—[c]	48.0

[a]Data for 1998.
[b]Data for 1997.
[c]Not available.
From FAO-STAT: http://apps.fao.org (accessed June 1999).

INTEGRATED PEST MANAGEMENT (IPM) TACTICS

Temperate Fruit: Apple

Considerable progress has been achieved during the past 20 years in the use of more selective insecticides and reduction of the number of treatments (1–3). In Europe, guidelines for integrated fruit production on pome fruits issued by IOBC/WPRS prohibit the use of pyrethroid insecticides or miticides (4). Many nonchemical alternative methods have also been developed to control specific apple pests.

Regarding codling moth, sexual confusion has provided promising experimental results, while Sterile Insect Technique (SIT), though effective, is often considered not economical. Good results have also been obtained with baculoviruses. When selective chemical or nonchemical methods may be used against this key pest, many secondary pests are regulated by their natural enemies, though this approach is less effective where the moth is multivoltine. Successful integrated mite control programs, based on sampling of phytophagous mites and phytoseiid predators, have been developed in different countries.

However, because of low tolerance to insect damage on fruit highly rated for its consumer appeal, and the inability of natural enemies in certain cases to keep some key pests below damaging levels, pesticide-dominated programs continue until now to be the mainstay of deciduous fruit pest control in many producing areas.

Citrus

Biological control traditionally has been an important component of IPM on citrus in many countries. The famous early success of biocontrol, the importation of *Rodolia cardinalis* from Australia to California at the end of the last century to control *Icerya purchasi*, is considered a case study of classical biocontrol. It paved the way to various other attempts during this century. Among other successes are the good results in the biocontrol of various whiteflies like the woolly whitefly, *Aleurothrixus floccosus*, in many countries. In island situations, outstanding results sometimes have been obtained: one of the psyllid vectors (*Trioza erytreae*) of greening disease has been eradicated from Reunion Island (in the Indian Ocean) in the seventies following the introduction of an exotic eulophid parasitoid, *Tamarixia dryi*. Many good results have also been obtained worldwide with biocontrol against scale insects (5, 6).

IPM programs for citrus, more or less based on biocontrol, have been developed in many countries (United States of America, Europe, Australia, Brazil, South Africa etc.) (5, 6). In Mediterranean conditions, IPM programs frequently are based on a small number of selective mineral oil sprays controlling scale insects and mites (7). Oils are also an important component of most programs worldwide. Monitoring techniques, based on visual control or trapping systems, and intervention thresholds have been developed for the major pests, allowing a reduction in pesticide use (7). Moreover, considerable progress has been achieved in the knowledge of selectivity of pesticides. While side-effects of organophosphorous compounds and pyrethroids have been known for some time, recent studies showed that an excessive use of IGR (Insect Growth Regulators) could also lead to outbreaks of secondary pests.

The development of IPM programs frequently is hindered by some key pests difficult to control with selective pesticides. Such cosmetic pests as citrus thrips or citrus rust mite fall in this category and their control strategy will differ greatly if citrus are produced for fresh fruit market or processing. Thrip control usually is difficult as a limited number of nonselective compounds have a good efficacy on them. In certain cases some botanicals, though less effective, allow a good compatibility with IPM.

Fruit flies are among the pests whose control is critical within an IPM program. Ground-bait sprays of a mixture of food attractant and insecticide limit the side-effects of sprays on beneficials. When applicable, other control methods compatible with IPM programs may be used singly or in combination: SIT, MAT (Male Annihilation Technique), or inundative releases of parasitoids. Efforts are also underway to develop biotechnical control methods based on selective female traps (Fig. 1). Orchard sanitation still remains an important fruit fly control component on citrus and tropical fruits.

Fig. 1 A sexual trap used for monitoring fruit fly (*Ceratitis* spp.) populations in orchards in Reunion Island (Photo: D. Vincenot, SUAD/CIRAD).

Fig. 2 Beating of flowers for monitoring thrip populations on mango in Reunion Island (Photo: D. Vincenot, SUAD/CIRAD).

Tropical Fruit

Like nematodes, the banana weevil has been for a long time mainly controlled by soil chemical sprays in large-scale plantations. Sampling methods may allow for a decrease in the number of sprays. Various cultural practices, such as the use of uninfested material, can also be used by small-holding producers. In recent years, efforts have been focused on more environmentally friendly methods of production. Promising results on pheromones and kairomones may be used for monitoring or to develop new control methods (use of microbiological control in bait stations) compatible with IPM. Efforts toward classical biocontrol of this key pest gave poor results. On the other hand, microbiological control, which gave very interesting experimental results particularly with nematodes but also entomopathogenic fungi, still has to be extended on a larger scale.

Efforts have also been devoted to develop IPM for the main tropical fruit crops. On mango, classical biocontrol gave good results, particularly for scale insects (Fig. 2). Fruit fly control is also critical to develop a global IPM (see Citrus). However IPM research on mango is relatively young, and further research is needed (monitoring methods, population dynamics, natural enemies, varietal susceptibility etc.) in order to improve existing programs.

Comparatively, avocado and litchi have generally smaller pest complexes and IPM programs should be easier to develop on these crops.

IPM methods for fruit crops have been the subject of considerable interest and progress during the past 20 years. However, their development faces various constraints, and much research still needs to be done. Some key pests difficult to control may constitute the limiting factor. Continuous improvements still have to be provided by entomologists for the control of each major fruit pest. Also, a systemic approach requires specialists able to embrace the problem from its early definition to real field implementation (8, 9).

REFERENCES

1. Metcalf, R.L.; Luckmann, W.H. *Introduction to Insect Pest Management,* 2nd Ed.; John Wiley & Sons: New York, 1982; 577.
2. Croft, B.A. *Arthropod Biological Control Agents and Pesticides*; John Wiley & Sons: New York, 1990; 723.
3. International Conference on Integrated Fruit Production Proceedings Meeting, Cedzyna, Poland, August 28–September 21995; Polesny, F., Müller, W., Olszak, R.W., Eds.; *IOBC/WPRS Bull.*, 1996; 19 (4), 442.
4. *Guidelines for Integrated Production of Pome Fruits in Europe. Technical Guideline III*; Dickler, E., Cross, J.V., Eds., 2nd Ed.; *IOBC/WPRS Bull.*, 1994; 17 (9), 40.
5. *Integrated Pest Management for Citrus*; Kobbe, B., Ed.; Publ. 3303, University of California, Statewide IPM Project, Division of Agriculture and Natural Resources, 1984; 144.
6. Knapp, J.L.; Noling, J.W.; Timmer, L.W.; Tucker, D.P.H. Florida Citrus IPM. In *Pest Management in the Subtropics. Integrated Management—A Florida Perspective*; Rosen, D., Bennett, F.D., Capinera, J.L., Eds.; Intercept, Ltd.: Andover, 1996; 578.
7. In *Integrated Pest Control in Citrus-Groves*, Proceedings of the Experts' Meeting, Acireale, March 26–29, 1985; Cavalloro, R., Di Martino, E., Eds.; A.A. Balkema: Rotterdam, 1986; 600.
8. Dent, D. *Insect Pest Management*; CAB International: Wallingford, 1991; 604.
9. Norton, G.A.; Mumford, J.D. *Decision Tools for Pest Management*; CAB International: Wallingford, 1993; 279.

FUMIGANTS

Nick Price
Central Science Laboratory, Ministry of Agriculture, Fisheries, and Food, Sand Hutton, York, United Kingdom

F

INTRODUCTION

Fumigants are chemicals in gas form kept at specific temperatures and pressures. *Fumigation* is the use of these chemicals for the disinfestation of soil, commodities, or building structures. Target pests include insects, mites, and soilborne fungi; bacteria, viruses, and nematodes; as well as weeds (1). The term fumigation is sometimes used to describe treatments with pesticides in the form of smokes, fogs, or mists. These are not true fumigants since they may have residual action which true gases do not have, and do not have the penetrative power of true gases. Some contact insecticides may have sufficient vapor pressure to exert some of their action in the vapor phase but again, these are not included in the definition of true fumigants. True fumigants may be formulated as solids (which emit gas under the influence of environmental conditions), as liquids under pressure, or as gases in cylinders. Fumigation has been practiced over many centuries, and wide range of gases have been used as fumigants, but in modern times the laudable stringent regulation of pesticides introduced in much of the world has limited gaseous pesticides to a handful of compounds. The two most common fumigants, phosphine (hydrogen phosphide) and methyl bromide (bromomethane), are both at risk for technical and political reasons and alternatives are being sought (2).

FUMIGATION CONTEXT AND METHODS

Pesticides are often applied as a protective or preventative measure, giving residual action and ensuring that potential infestations do not occur. Fumigants are not used in this way but as disinfestants. They may be used in situations that make other treatments difficult or impossible, such as the treatment of large bulks of commodities in ship holds, or when other measures have failed and ''fire brigade'' pest control is required. The gaseous nature of fumigants allows rapid and deep penetration of the active material and after the treatment period airing off of the commodity, structure, or soil removes the active material. In general residues of fumigants in the treated material will be low and offer no protection against reinfestation, though residues can arise due to sorption of the gas to the treated material or reaction with it to form new compounds. The extent and nature of these residues will depend on the nature of the gas used. The major situations in which fumigation is used include horticulture, transport, food storage, buildings and other structures. Many countries have quarantine regulations that require the fumigation of infestable material prior to importation.

In horticulture, soil is sterilized prior to planting of a wide range of crops, both under glass and in open fields. This is done for a variety of pest or disease management reasons including insects, nematodes, fungi, bacteria, viruses, and weeds. The method usually involves sheeting the soil with one of a range of polymer sheeting materials and then ''injecting'' the fumigant of choice (which may be liquid, solid, or gaseous formulation) into the soil beneath the sheets. After the required exposure period the sheets are removed and the gas ''aired off'' prior to planting. Common applications include: preplant treatment for cut flowers, strawberries, tobacco, cucurbits, tomatoes, and peppers; replant treatment for fruit trees; and seedbed treatments for a wide range of seeds.

For durable commodities fumigation is used primarily to control insect pests. Common uses include cereal grain and related commodities, dried fruit and nuts, and timber and wooden artifacts such as furniture and museum specimens. In the case of durable food commodities treatments may be required under quarantine or phytosanitary regulations. Fumigation of durable commodities may be carried out in containers, bag stacks in warehouses, stacks (either bagged or loose) outdoors, vehicles, or in indoor floor stores (loose). Commodities are made as gas-tight as possible in their particular situation prior to fumigation. This usually involves the use of polymer sheets and the sealing of all possible cracks in walls, bulkheads, sheeted stacks, etc. The gas is then introduced beneath the sheeting via tubing if the gas is in a liquid or gaseous formulation, or by placing solid formulations prior to final sealing of the commodity.

Fumigation is also used to eradicate insect pests in some perishable commodities, such as fresh fruit, vegetables, and ornamental plants. Such treatments are usually required to satisfy official quarantine or local contractual regulations, and are carried out in a similar way to that of durable commodities.

Fumigation is also used to disinfest buildings or vehicles. Flour mills and similar premises are often fumigated against a range of stored product beetles and moths that may have become resident in the fabric of the building living off the cereal residue. In other applications, whole buildings may be fumigated to control insects that damage the structure, such as wood-boring beetles, ants, or termites. Ship holds, freight containers, or railway trucks may be fumigated either empty or containing a durable cargo to control insect or rodent pests. In some countries aircraft are routinely fumigated, usually for rodent control but occasionally for insect control. A key to the success of structure fumigation is the gas tightness of the structure or part of the structure being treated. Great emphasis is placed on good sealing techniques prior to the introduction of gas.

Environmental applications of fumigation include the treatment of the underground burrows of vertebrate pests such as rabbits.

COMPOUNDS USED

A wide range of chemical types have been used or experimented with as fumigants. These include halogenated hydrocarbons, esters, sulphides, fluorides, phosphides, aldehydes, ketones, cyanides, and organometallics. Some plant-derived volatiles, primarily terpenes, are widely reported to have fumigant properties but little good scientific data is available to make practical assessments of this claim. In general the properties required of a good fumigant occur in a narrow band. The compound must be a gas at normal temperatures and pressures; and, since volatility is related to molecular mass, fumigants are generally simple molecules. Simple molecules that are also biocides tend to have nonspecific toxic properties and thus many compounds have undesirable chronic or acute toxicological properties that render them unsuitable. The use of many compounds has been discontinued on these grounds, resulting in a wide range of minor fumigants that have been tried, discarded, or banned. Because of the toxic nature of fumigants their production, labeling, and use is strictly controlled, with their application in many countries being restricted to trained professional operatives.

The efficacy of fumigants is not determined solely by concentration but also by the exposure time. In general any combination of concentration and time giving the same arithmetic product ($C \times T = k$) will produce the same mortality in a given population of insects. Thus high concentrations over short time periods will give the same kill as low concentrations for long time periods. This "*CT* rule" holds true within certain limits for most gases except phosphine for which time of exposure seems to be a critical factor.

The main fumigants currently in use or under investigation for possible use are shown in Table 1.

In the past, halogenated hydrocarbons such as carbon tetrachloride, 1,2-dichloroethane, and 1,2-dibromoethane were also used but have been discontinued due to adverse

Table 1 Main fumigants

Soil	Commodities	Structures
	Existing Fumigants	
Methyl bromide	Methyl bromide	Methyl bromide
Methyl isothiocyanate	Phosphine	Phosphine
1,3-Dichloropropene	Ethylene oxide	Sulfuryl fluoride
	Ethyl formate	Hydrogen cyanide
	Hydrogen cyanide	
	Carbon disulphide	
	Potential Fumigants	
Methyl iodide	Methyl iodide	Controlled atmospheres
Propargyl bromide	Methyl isothiocyanate	
Ozone	Methyl phosphine	
Cyanogen	Sulfuryl fluoride	
	Controlled atmospheres	
	Carbonyl sulphide	

toxicological properties. The major current fumigants worldwide are methyl bromide and phosphine.

The pesticidal action of methyl bromide, (bromomethane) was first reported in 1932 and it has been in wide commercial use for more than 50 years. About 69,000 tons of methyl bromide was produced for fumigation in 1996 of which 76% was for soil treatment, 15% for durable commodities, and 9% for perishable commodities and structures. The gas is normally supplied as a liquid under pressure in cylinders (it boils at 4°C). As a potent methylating agent, methyl bromide has been shown to methylate many biochemical components and thus its mode of action is believed to be complex, though it may involve SH groups in vital enzymes (3). In 1992 methyl bromide was listed as an ozone depleting agent under the Montreal Protocol, and after a number of years of political and scientific debate, was scheduled for phase out in developed countries by January 2005 (4).

Phosphine (hydrogen phosphide) has been used as a fumigant for more than 40 years and is usually generated by the action of atmospheric moisture on aluminium or magnesium phosphide. This can be formulated as tablets, pellets, sheets or ''blankets.'' More recently cylinder-based formulations have been developed that may provide a more even gas distribution. Phosphine is used primarily for the disinfestation of durable commodities. It is believed to act as a respiratory inhibitor specifically inhibiting cytochrome-C oxidase, though its reaction with a number of other metalloenzymes, especially catalase, may also be a factor in its toxicity. In a global survey of resistance carried out more than 25 years ago, resistance to phosphine was identified (5). Since then there have been many reports of resistance to phosphine in a wide range of insect species from all over the world. This widespread resistance, combined with disputed evidence that it can induce transient chromosomal rearrangements, is causing some regulatory authorities to look closely at the status of phosphine.

ALTERNATIVES

It is clear for the reasons given above that alternatives to existing fumigants are desperately needed. There are several schools of thought on the way forward. One view is that there will be no chemicals whose properties fit the narrow band of requirements for a fumigant. However, chemical diversity is vast and candidates may well exist that have not yet even been considered. Against this is the likelihood that such chemicals, to be effective against the range of pests and diseases that current fumigants are, will be broad-spectrum poisons less likely to gain acceptance than in the past. Simple gaseous replacements currently under investigation include methyl iodide, cyanogen, carbonyl sulphide, ozone, and methyl phosphine. Aromatic plants produce volatile oils (usually terpenoids) that have been shown to have some fumigant properties. The breakdown of glucosinolates from plants such as brassicas, by the action of the enzyme myrosinase, gives rise to isothiocyanates that have been proposed as ''biofumigants.'' The development of controlled or modified atmospheres has received much attention. In this technique atmospheric oxygen is replaced, rendering the fumigated space incapable of supporting aerobic life. Replacement may be achieved by the burning of oxygen resulting in a nitrogen/carbon dioxide atmosphere, or in small volumes by the addition of oxygen scavengers. In some techniques, direct displacement of oxygen by an inert gas is used in place of removal of oxygen by burning.

In yet another view, the future will not involve gases at all but will rely on Integrated Pest Management using a combination of cultural practices, biocontrol, and physical methods such as heating or freezing, all combined with careful monitoring. In the short to medium term it is likely that gases themselves can play a significant part in integrated strategies (6).

REFERENCES

1. Bond, E.J. *Manual of Fumigation for Insect Control*; FAO: Rome, 1984.
2. Bell, C.H.; Price, N.R.; Chakrabarti, B. *The Methyl Bromide Issue*; John Wiley: Chichester, UK, 1996.
3. Price, N.R. The mode of action of fumigants. J. Stored Prod. Res. **1985**, *21* (4), 157–164.
4. UNEP. *Assessment of Alternatives to Methyl Bromide*; Report of the Methyl Bromide Technical Options Committee, UNEP: Nairobi, Kenya, 1998.
5. Champ, B.; Dyte, C.E. *Pesticide Susceptibility of Stored Grain Pests*; FAO: Rome, 1976.
6. Bell, C.H. In *Current and Future Prospects for Stored Product Protection Using Fumigants and Gases*, Proceedings of the International Forum on Stored Product Protection and Post-Harvest Treatment of Plant Products, Council of Europe: Strasbourg, 1995; 61–88.

FUNGAL CONTROL OF PESTS

David Moore
CABI Bioscience UK Centre (Ascot), Berkshire, United Kingdom

INTRODUCTION

The earliest recorded observations on entomopathogenic fungi appear to date from around 4700 BP (1). Oriental accounts describe silkworms infected with *Beauveria bassiana* and also examples of the genus *Cordyceps*. During the 18th century the Italian scientist A. Bassi pioneered the germ theory of disease, using *B. bassiana* to infect silkworms and E. Mechnikoff worked on *Metarhizium anisopliae*, stimulating an extensive study of fungi to control agricultural pests that waned by 1925. A resurgence of interest from the 1960s resulted in major advances in understanding, especially during the 1990s.

BIOLOGICAL FEATURES

More than 700 species of fungi, mostly Hyphomycetes and Entomophthorales from about 90 genera, are pathogenic to insects. The major genera are noted in Table 1 (1, 2). Groups such as the Laboulbeniales and the genus *Septobasidium* may not be directly pathogenic but may be significant in pest population control by reducing reproductive capacity: nonlethal effects of insect diseases are increasingly being acknowledged as important in control.

Some species have a very wide host range, although individual isolates are selective; for example, *Metarhizium anisopliae* has been isolated from more than 300 species of Lepidoptera, Coleoptera, Orthoptera, and Hemiptera but isolates of *M. anisopliae* var *acridum* act largely on orthopteran targets. *Aschersonia aleyrodis* infects only soft scale insects and whitefly.

The nature of the entomopathogen flora varies according to ecosystem 2. Aquatic ecosystems are predominately associated with the lower fungi (Mastigomycotina, Zygomycotina), probably because these produce motile spores. Entomopathogenic fungi may be exerting a high level of control in long-lasting ecosystems such as forests that have stable microclimates with many humid niches, well protected from sunlight. Disturbed ecosystems such as agricultural lands, may demonstrate erratic, but devastating epizootics. Consequently, humid, tropical forests may carry many species of *Cordyceps*, a genus that is relatively rare in degraded forests or agricultural land; pastures may represent an intermediate degree of stability where *Cordyceps* species are more common than in arable systems. Members of the Entomophthorales cause epizootics in both forest and arable ecosystems; in the latter their effect appears to depend on circumstances of microclimate, especially humidity and insect populations.

Entomopathogenic fungi infect insects directly through the cuticle, using a process that begins with adhesion of the spore to the host. This is followed by germination, penetration through combined chemical and physical mechanisms, and then development inside the host, causing death. External or internal sporulation may then follow in days or months respectively, releasing new agents capable of infecting new hosts, potentially increasing the duration and degree of control.

STRATEGIES FOR USE

The devastating effects of epizootics, in a range of habitats from aquatic through to agricultural, demonstrate the potential of entomopathogenic fungi to control insect (and mite) populations. Large-scale use of *Beauveria bassiana* occurred in the centrally planned economies of the Soviet Union and the People's Republic of China in the 1960s. Brazil makes extensive use of *Metarhizium anisopliae*, marketed as the product *Metaquino*, for control of *Mahanarva posticata* in sugarcane and applied annually to around 1.6×10^5 hectares (3). There are examples in many countries of specialized use of fungi in niche markets [*Beauveria brongniartii* gives long-term control of *Melolontha melolontha* in Switzerland (4)] but few countries, with the notable exception of Cuba, base their crop protection on biological pesticides. Control comparable to that of chemical insecticides, but without the as-

Table 1 Major taxonomic groups containing entomopathogenic species and their reported hosts

Subdivision	Class	Order	Genera	Notes
Mastigomycotina	Chytridiomycetes	Chytridiales	*Coelomycidium*	Diptera
			Myiophagus	Scales, mealybugs
	Chytridiomycetes	Blastocladiales	*Coelomomyces*	Mosquitoes, Hemiptera
	Oomycetes	Lagenidiales	*Lagenidium*	Mosquitoes
			Aphanomycopsis	
	Oomycetes	Saprolegniales	***Leptolegnia***	Eggs of midges
			Couchia	Mosquitoes
Zygomycotina	Zygomycetes	Mucorales	*Sporodiniella*	Membracidae
	Zygomycetes	Entomophthorales	*Conidiobolus*	Entomophthorales infect families in Hemiptera, Homoptera, Diptera, Lepidoptera, Coleoptera, Orthoptera, and Hymenoptera
			Entomophaga	
			Entomophthora	
			Erynia	
			Massospora	
			Meristacrum	
			Neozygites	
Ascomycotina	Hemiascomycetes	Endomycetales	*Blastodendrion*	Entomopathogenic yeasts in Endomycetales, midges, mealybugs, and cockroaches reported as hosts
			Metschnikowia	
			Mycoderma	
			Saccharomyces	
	Plectomycetes	Ascosphaerales	*Ascosphaera*	Bees
	Pyrenomycetes	Hypocreales	*Cordyceps*	*Cordyceps* known from many insect orders as well as arthropods in general (Araneida)
			Torrubiella	
			Nectria	Other Hypocreales important pathogens of scales and whiteflies
			Hypocrella	
			Calonectria	
	Laboulbeniomycetes	Laboulbeniales	*Filariomyces*	Most common hosts of Laboulbeniales are Coleoptera
			Hesperomyces	
			Trenomyces	Not implicated in host mortality but may reduce reproductive capacity
	Loculoascomycetes	Myriangiales	*Myriangium*	Scale insects
	Loculoascomycetes	Pleosporales	*Podonectria*	Scale insects

(*Continued*)

Table 1 Major taxonomic groups containing entomopathogenic species and their reported hosts (*Continued*)

Subdivision	Class	Order	Genera	Notes
Mitosporic fungi	Hyphomycetes		Acremonium	
			Akanthomyces	
			Aschersonia[a]	Scale insects, whiteflies
			Aspergillus	
			Beauveria	Wide host range
			Culicinomyces	Mosquitoes
			Engyodontium	
			Fusarium	
			Gibellula[b]	Spiders
			Hirsutella[c]	Mites
			Hymenostilbe[c]	
			Metarhizium	Wide host range
			Nomuraea[c]	Lepidoptera
			Paecilomyces[c]	Wide host range
			Paraisaria[c]	
			Pleurodesmospora	
			Polycephalomyces	
			Pseudogibellula	
			Sorosporella	
			Sporothrix	
			Stilbella	
			Tetranacrium	
			Tilachlidium	
			Tolypocladium	
			Verticillium[c]	Mosquitoes, aphids, scales
Mycelia sterilia			Aegerita	
Basidiomycotina	Phragmobasi-diomycetes	Septobasidiales	Filobasidiella	
			Septobasidium	Symbiotic relationships; trade-off between loss of individual scales and benefit to colony
			Uredinella	Mortality factor, attacks individual scales. Limits reproduction in diaspid hosts

Genera in bold have been researched extensively for mycopesticide potential.
[a]*Hypocrella* anamorphs.
[b]*Torrubiella* anamorphs.
[c]*Cordyceps* anamorphs.
(Adapted from Refs. 1 and 2.)

sociated problems, is possible; present thinking is moving away from mycoinsecticides as biological analogues of chemicals and toward an appreciation of the potential superiority of biologicals.

Fungi can be used in many ways, including conservation and stimulation of natural levels of fungal activity, classical introductions by exploiting endophytic activity as mycopesticides, and by utilizing nonlethal effects.

Conservation involves minimizing adverse and maximizing beneficial effects of the natural or managed environment, on natural levels of control agents. Within the context of agriculture, a significantly interventionist force on the environment, many practices influence natural levels of entomopathogenic fungi. The consequences of practices such as burning, plowing, liming, monoculturing, and pesticide use will be significant but difficult to generalize.

The value of pathogens for classical biological control in agricultural systems has been doubted, because they are considered most effective only at high host densities and when suitable climatic conditions prevail. Many diseases have been distributed worldwide over the ages and exert a continued, yet often unrecorded degree of control. *Erynia radicans* was introduced into Australia in 1979 to control the aphid *Therioaplis trifolii* form *maculata*: although its efficacy is debated vis à vis the actions of a parasitoid, there is no doubt that establishment was successful and a continual level of control occurs.

Endophytes grow systemically in certain plant groups and can provide protection against invertebrate predators, by repellence and causing direct mortality or reduced insect performance (5). Most work has been on grasses, especially those containing *Acremonium* species, and conifers. Plant breeding and selection for cultivars with effective insect control without unwanted side effects (for example poisoning of grazing animals) has shown marked success, but major advances could come with extensive research on cereals and greater understanding of fungal associations in other plant groups. These may include mycorrhizal associations, which can increase plant hardiness and can result in enhanced plant defenses, directly effective against insects or in greater tolerance of insect attack.

Mycoinsecticides represent the technological end of the pathogen-use spectrum, mirroring formulation, application, and control requirements of chemical insecticides (6–8). Although low-input systems of production and use are possible, increasingly quality and technological standards are being raised to optimize efficacy.

A number of commercial products have been developed (9) to control pests such as whitefly, thrips, vine weevil, colorado beetle, cockchafer, and locusts and grasshoppers (10). These have not yet made a major impact into any market despite being, in some instances, superior to the competing chemical pesticide (11).

With all strategies, non- and pre-lethal effects of pathogens are important. Minimizing feeding or movement of a pest and reducing its subsequent populations by decreased fecundity demonstrate control without mortality and these features are slowly being understood.

CONSTRAINTS

Despite significant research into entomopathogenic fungi there has been limited success in their practical use. Mycopesticides have been unreliable. Field use has sometimes been very successful, but often control has been erratic, heavily dependent on prevailing microclimate, with high humidity, and low ultra-violet irradiation being required. Until the late 1980s emphasis was on the active ingredient, the fungal spore, but during the 1990s there were major advances based on recognition of the need to optimize the agent, formulation, and application (the delivery system). Even more recently, increased understanding of the vital importance of the pest/pathogen/environment interactions has been shown to be critical (12).

Some constraints are psychological (for example, beliefs that pathogen use is restricted by environment and that mycoinsecticides should be mimics of chemicals with similar qualities). It is now possible to kill insect pests efficiently in hot dry climates, such as African deserts, by formulating the fungal spores in oil and maximizing dose transfer. With many other insects relatively minor research inputs would provide practical solutions (13). Perceptions of effective pest control, such as high levels of quick kill, need addressing. Control by macrobial and microbial agents (plus abiotic factors) kill vast numbers of insects and prevent many pest attacks, even if each component only causes a small and unnoticed mortality or other chronic effect. Mycopesticides act in days and weeks rather than minutes and hours (significantly many chemical insecticides also work slowly) and consequently nuisance pests such as houseflies that require immediate control are not likely targets. However, population control of the same insects could be feasible if breeding sites could be treated.

A major constraint is commercial. Agrochemical companies are reluctant to research and market mycopesticides, possibly because of the perception that markets will be niche and small. Allied to this, mycopesticides based on hyphomycetes are expensive, due to high production costs. Research into mass production would greatly reduce the price.

FUTURE DEVELOPMENTS

New paradigms are evolving: mycoinsecticides can be used in ways that provide control superior to that of chemical insecticides, but they must be regarded as separate, with their own specific requirements. Secondary cycling of fungal spores from cadavers can give long-term persistency. Pathogens can be applied for chronic rather than acute effect, that is with the intent of debilitation rather than kill (to reduce population growth for example), or vice versa. The importance of low-level intervention by natural pathogen levels or by endophytes is being recognized. Pathogens can reduce pest effect by reducing feeding fecundity or movement, and can weaken insects allowing later abiotic effects to effect kill.

Pathogens provide a complex of activities that can be managed; some almost completely with, for example mycoinsecticides; less predictably with endophytes or introductions or in almost unmeasurable ways by conservation of natural pathogen activity.

Future research must concentrate on improving the economics of mass production and exploring specific host/pathogen interactions. Generic principles must be established so, for example, many soil pests could be controlled if the basic features of pathogen persistence and dose transfer in the soil were established. Delivery systems, formulation, and application, have to be optimized.

REFERENCES

1. Tanada, Y.; Kaya, H.K. *Insect Pathology*; Academic Press: San Diego, 1993; 666.
2. Samson, R.A.; Evans, H.C.; Latgé, J.-P. *Atlas of Entomopathogenic Fungi*; Springer-Verlag: Berlin, 1988; 187.
3. Mendonça, A.F. Mass Production, Application and Formulation of *Metarhizium Anisopliae* for Control of Sugarcane Froghopper, *Mahanarva Posticata* in Brazil. In *Biological Control of Locusts and Grasshoppers*; Lomer, C.J., Prior, C., Eds.; CAB International: Wallingford, 1992; 239–244.
4. Keller, S.; Schweizer, C.; Keller, E.; Brenner, H. Control of white grubs (*Melolontha melalontha*) by treating adults with the fungus *Beauveria brongniartii*. Biocontrol Sci. and Technol. **1997**, *7*, 105–116.
5. Siegel, M.R.; Latch, G.C.M.; Johnson, M.C. Fungal endophytes of grasses. Annu. Rev. of Phytopathol. **1987**, *25*, 293–315.
6. Bateman, R. Delivery Systems and Protocols for Biopesticides. In *Methods in Biotechnology: Biopesticides: Use and Delivery*; Hall, F.R., Menn, J.J., Eds.; Humana Press: Totowa, NJ, 1998; 5, 509–528.
7. *Formulation of Microbial Biopesticides*; Burges, H.D., Ed.; Kluwer Academic Publishers: Dordrecht, The Netherlands, 1998; 412.
8. Feng, M.G.; Poprawski, T.J.; Khachatourians, G.G. Production, formulation and application of the entomopathogenic fungus *Beauveria bassiana* for insect control: current status. Biocontrol Sci. and Technol. **1994**, *4*, 3–34.
9. Butt, T.M.; Harris, J.G.; Powell, K.A. Microbial Biopesticides. The European Scene. In *Methods in Biotechnology: Biopesticides: Use and Delivery*; Hall, F.R., Menn, J.J., Eds.; Humana Press: Totowa, NJ, 1999; 5, 23–44.
10. *Microbial Control of Grasshoppers and Locusts*; Johnson, D.L., Goettel, M.S., Eds.; Memoir of the Entomological Society of Canada, 1977; 171–400.
11. Langewald, J.; Ouambama, Z.; Mamadou, A.; Peveling, R.; Stolz, I.; Bateman, R.; Attignon, S.; Blanford, S.; Arthurs, S.; Lomer, C. Comparison of an organophosphorus insecticide with a mycoinsecticide for the control of *Oedaleus senegalensis* (Orthoptera: Acrididae) and other Sahelian grasshoppers at an operational scale. Biocontrol Sci. and Technol. **1999**, *9*, 199–214.
12. Blanford, S.; Thomas, M.B. Host thermal biology: the key to understanding host–pathogen interactions and microbial pest control. Agric. and Forest Entomol. **1999**, *1*, 195–202.
13. Maniania, N.K. Potential of some fungal pathogens for the control of pests in the tropics. Insect Science and Its Application **1991**, *12*, 63–70.

FUNGICIDES

W.S. Washington
Institute for Horticultural Development, DNRE, Melbourne, Victoria, Australia

F

INTRODUCTION

Fungicides are a diverse range of compounds that can be used to manage fungal diseases that could otherwise damage food crops and other plant products worldwide. Before the 1930s, only sulphur and copper fungicides had any impact for plant disease control. Subsequently, many synthetic fungicides have been developed and now benefit crop production by ensuring reliable yields from crops that could otherwise be destroyed or seriously affected by plant disease. Most fungicides developed since the late 1960s are systemic (mobile within the plant) and have a selective mode of action on the fungus, greater efficacy at low rates, and can be used more flexibly. Unfortunately, this selectivity, which also confers greater safety to users and the environment, has led to problems of buildup of resistant populations of the target fungi. Of the 170 different fungicidal active ingredients now in use worldwide, most show low mammalian toxicity. Fungicides are used mainly as foliar sprays, and are less used as seed and soil treatments. Although fungicides are used throughout the world, the newer, highly active (and more expensive) products tend to be used mainly in high value crops that are grown in technologically sophisticated production systems usually in developed countries. Cheaper, less selective protectant products are used more commonly in developing countries.

TYPES OF FUNGICIDES

Fungicides may be classed as protectant, curative, eradicant, and systemic. Protectant fungicides do not enter plant tissue but remain on the surface where they can prevent the fungus spore from germinating and penetrating the plant tissue. They must be applied before infection has occurred. Curative fungicides can prevent a fungal infection after the fungus has begun to infect the plant but before the symptoms of disease are present. Eradicant fungicides can control disease after symptoms have appeared. Systemic fungicides (usually curative as well as protectant in activity) are absorbed into a plant and translocated away from the site of application. The majority of new fungicides have systemic activity because they are most profitable to produce and have many favorable properties including safety to users, consumers, and the environment; safety to the crop; and high activity at low use rates. Despite this, protectant fungicides including mancozeb, copper, chlorothalonil, and sulphur make up almost 20% of the global fungicide sales.

Plant pathogenic fungi can be allocated to four major groups. These are the Oomycetes (which include the water molds and downy mildews, e.g., *Phytophthora* and *Plasmopora* spp.), the Ascomycetes (which include powdery mildews and scab, e.g., *Erysiphe* and *Venturia* spp.), the Basidiomycetes (which includes the rusts and smuts, e.g., *Puccinia* and *Ustilago* spp.), and the anamorphic fungi (which include gray mold and root rot, e.g., *Botrytis* and *Fusarium* spp.). Some fungicides are active against fungi in most or all of these groups and can be called broad spectrum, for example, mancozeb, while others are narrower in their spectrum of activity, for example, iprodione, which is active mainly against fungi in the Basidiomycetes and anamorphic groups.

Until the late nineteenth century sulphur was probably the only widely recognized fungicide, active mainly against powdery mildew diseases. In 1885, Bordeaux mixture, a combination of copper sulphate and lime, was introduced by Millardet in France for the control of grape downy mildew and other diseases. Sulphur, copper, and related inorganic fungicides (comprising chemicals containing molecules without carbon) still remain a significant part of the total world fungicide usage. In 1913 the organomercury-based fungicides were introduced as seed treatment for control of bunt of cereals. From the 1930s a number of synthetic, organic (made up of complex molecules including carbon), protectant fungicides were developed. They include thiram (1934) and later, maneb, mancozeb, and other members of the important group

Table 1 Representative types of fungicides and their characteristics

Common chemical name	Chemical group	Date introduced	Type	Mode of action	Resistance risk	Spectrum of activity[a]
Bordeaux mixture (copper & lime)	Inorganic	1885	Protectant	Multi-site	Low	O A B AF[b]
Captan	Phthalimide	1952	Protectant	Multi-site	Low	O A AF
Mancozeb	Dithiocarbamate	1961	Protectant	Multi-site	Low	O A B AF
Benomyl	Benzimidazole	1968	Systemic	Microtubulin polymerization	High	A AF
Iprodione	Dicarboximide	1970	Protectant	NADPH-cytochrome C reductase	High	B AF
Propiconazole	Sterol biosynthesis inhibitor	1979	Systemic	C14 demethylation	Moderate	A B AF
Azoxystrobin	Strobilurin	1990	Systemic	Complex III inhibitor	High-moderate	O A B AF
Acibenzolar	Benzothiadiazole	1996	Activator of systemic acquired resistance	Plant defense activator	Low?	No direct effect

[a]Active against the major groups of plant pathogenic fungi: O = Oomycetes, A = Ascomycetes, B = Basidiomycetes, AF = anamorphic fungi.
[b]Active also against a wide range of plant pathogenic bacteria.
(Adapted from data in Ref. 3 and other sources.)

known as the dithiocarbamates, which are active against a wide range of fungal pathogens. The 1940s mark the period when the real potential of fungicide use in crop protection was realized, and significant compounds introduced since then include captan, organo-tins, streptomycin (antibiotic), dodine, chlorothalonil, and aromatic hydrocarbons including dicloran.

In the late 1960s, the first of many groups of successful systemic fungicides, the benzimidazoles (benomyl, carbendazim, and others active against a wide range of ascomycete fungi), the hydroxypyrimidines (e.g., dimethirimol, active against powdery mildews) and the oxathiins or carboxamides (including oxycarboxin active against rusts and smuts) were introduced. Other significant groups introduced from the 1970s include the dicarboximides (e.g., iprodione), the sterol biosynthesis inhibitors (a large group containing fenarimol, bitertanol, myclobutanil, propiconazole, triticonazole, and others, active against all fungal groups except the Oomycetes), the phenylamides (active against water molds and downy mildews) and the anilinopyrimidines (active against grey mold and apple scab), most of which are systemic in plants. Most systemic fungicides have only limited mobility in the plant, especially in woody hosts, and are mainly moved upward in the xylem. An exception are the phosphonates (fosetyl aluminum and phosphorous acid), which can move both upward (in the xylem) and downward (in the phloem) in the vascular system of plants. These compounds are active against *Phytophthora* and related soil pathogens, and provide the potential for improved control of previously intractable root rots of many crops. By the end of the twentieth century, new developments include the introduction of fungicide groups related to naturally occurring compounds (the strobilurins e.g. azoxystrobin and kresoxim methyl, which are active against fungi in all four groups) and compounds such as acibenzolar which induce systemic acquired resistance (SAR) in host plants (Table 1).

DISCOVERY

Early fungicides were found largely by chance. A classic example is the observation by Millardet in the Bordeaux region of France in 1882 that grapevines treated by farmers with a copper sulphate and lime-based repellent to discourage pilfering were also protected against the ravages of downy mildew. This led to his development and subsequent announcement of the now famous Bordeaux mixture in 1885. New fungicides generally are developed by large international companies that invest resources in screening candidate fungicides for control of significant diseases that have the potential to give an acceptable return on the investment. Of the thousands of species of plant pathogens and many crop types, only a small number are important enough globally to be targeted and therefore these candidate fungicides are screened against key pathogens using laboratory and glasshouse tests. Most fungicides are used on high value crops such as vegetables, temperate cereals, rice, grapevines, and pome fruit.

SAFETY AND ENVIRONMENTAL IMPACT

Health and environmental issues have become more important to consumers in recent decades, especially to those in developed countries who have ready access to cheap and reliable food supplies. As a result, regulatory requirements mean that toxicology, crop residue, and environmental fate studies now form a major component of the development costs of a new fungicide, which average about $200 million spread over an 8-year period. A range of animal tests are used to determine the toxicity of a fungicide, both acute (short-term) and long-term, and studies in soil and water determine its environmental fate. Crop residues are also determined in a range of environments and the potential hazard to sprayer operators is estimated. Environmental hazards from fungicides generally appear to be low, although copper accumulation in the soil of sprayed vineyards can interfere with soil fauna and other microorganisms. Similarly, benzimidazole fungicides can reduce earthworm populations in soil under treated crops and many fungicides can alter the balance of nontarget leaf or soil microflora.

RESISTANCE MANAGEMENT

Despite the extensive use of fungicides over the past 90 years, resistance has only emerged as a significant problem since 1970, shortly after the introduction of the first systemic fungicides. Resistance is a consequence of the more specific mode of action of systemic fungicides compared with that of the protectant fungicides. The latter are generally multisite inhibitors that act against a number of biochemical sites in the target fungus. By contrast the systemic fungicides tend to act at one or at most a few sites, and resistance to such fungicides is common because the mutations required to overcome the toxicant are fewer. High resistance-risk compounds include members of the benzimidazole, dicarboximide, and acylalanine groups where field resistance can be selected after relatively few applications. Moderate risk groups include the sterol biosynthesis inhibitors. Low risk groups include the multisite general cell toxicants that comprise most of the protectant fungicides, many of which have been used without any evidence of resistance for more than 50 years. In order to minimize the risks of crop loss and loss of use of a particular fungicide group, resistance management strategies have been developed. These include limiting the use of high- or moderate-risk fungicides, combining or alternating them with low-risk fungicides, and reducing disease pressure by using nonfungicidal methods such as sanitation. As cross-resistance occurs between fungicides with a common mode of action, the chemical groupings to which fungicides belong are important in applying resistance management strategies.

BACTERICIDES

The term fungicide is often used to refer to compounds that control both fungal and bacterial plant diseases. This is technically incorrect as the term bactericide is preferred when referring to a compound that controls bacteria. Copper fungicides, with their broad spectrum of activity against both fungi and bacteria, are one of the few groups of bactericides used worldwide, although phytotoxicity can be a problem on some crops. Antibiotics (e.g., streptomycin) are also used in some countries for control of diseases such as fireblight of pome fruit, although this use is often severely restricted or banned in many countries due to concerns of antibiotic resistance spreading to human pathogens.

FUTURE TRENDS

Current estimates indicate that up to 20% of crop yield would be lost without the use of fungicides and in some cases total crop loss could occur. Fungicides are used throughout the world to minimize these losses and to help stabilize food production that in the past has been devastated by plant diseases, for example the Irish potato famine in the mid 1800s. They also benefit food quality by controlling many fungi that produce harmful mycotoxins. Although it is likely that fungicides will remain a significant component of crop disease management, their future is expected to be influenced by the following:

- increased adoption of integrated crop management systems including improved disease diagnosis and use of decision support systems;
- development of novel compounds with useful characteristics including such products as the plant defense activator acibenzolar;
- expiry of existing fungicide patents that will make the development of new compounds less commercially attractive;
- increasingly strict legislation concerning the safety and environmental effects of fungicides (e.g., the Food Quality Protection Act in the US);

- development and acceptance of genetically modified crops with disease resistance; and
- development of effective biological control agents.

BIBLIOGRAPHY

Fungicides Resistance Action Committee (FRAC), A Specialist Technical Group of the Global Crop Protection Federation, Belgium, http://www.gcpf.org/frac/frac.html (accessed Feb 2001).

Hewitt, H.G. *Fungicides in Crop Protection*; CAB International: Wallingford Oxon, U.K., 1998; 221.

Fungicidal Activity. Chemical and Biological Approaches to Plant Protection; Hutson, D., Miyamoto, J., Eds.; John Wiley & Sons: Chichester, U.K., 1998; 254.

Knight, S.C.; Anthony, V.M.; Brady, A.M.; Greenland, A.J.; Heaney, S.P.; Murray, D.C.; Powell, K.A.; Schulz, M.A.; Spinks, C.A.; Worthington, P.A.; Youle, D. Rationale and perspectives on the development of fungicides. Ann. Rev. Phytopathol. **1997**, *35*, 349–372.

Kraska, T. *The Plant Pathology Internet Guidebook*; University of Bonn, Germany. http://www.ifgp.uni-hannover.de/ppigb/ppigb.html (accessed Feb. 2001).

Modern Fungicides and Antifungal Compounds II, 12th International Reinhardsbrunn Symposium, May 24th–29th, 1998; Lyr, H., Russel, P.E., Dehne, H.-W., Sisler, H.D., Eds.; Intercept Limited: Andover, U.K., 1998; 505.

The Pesticide Manual, A World Compendium; Tomlin, C.D.S., Ed., 12th Ed.; The British Crop Protection Council Surrey, U.K., 2000; 1250.

GENETIC IMPROVEMENT OF BIOCONTROL AGENTS

Marjorie A. Hoy
University of Florida, Gainesville, Florida, U.S.A.

INTRODUCTION

Biological control of pest arthropods or weeds involves the use of natural enemies to suppress pest populations. Many arthropod populations are effectively suppressed by a complex of naturally occurring parasitoids, predators, pathogens (viruses, bacteria, fungi, protozoa), or entomopathogenic nematodes. In other cases, natural enemies must be introduced (especially if an alien pest arthropod has invaded a new geographic area without its natural enemy complex) in a method called classical biological control.

Natural enemies also can be mass reared and released to suppress pest arthropods, which is especially useful in ephemeral cropping systems in which the natural enemies are unable to suppress pests quickly enough to avoid economic injury; this tactic is called augmentation. It also is possible to modify cropping practices, such as crop rotation, maintenance of ground covers, or modification of pesticide application methods (concentrations applied or application sites), to preserve natural enemies so that they are more effective (conservation biological control).

Despite this diversity of tactics, the efficacy of natural enemies sometimes is limited by their intrinsic genetic characteristics (1). Under these circumstances it may be appropriate to enhance their effectiveness by genetic manipulation. A primary example is the development of pesticide-resistant predators and parasitoids so that they can survive to suppress pests while other pests are controlled with a pesticide (2). However, it will be ethically undesirable to develop resistant natural enemies if this increases the use of toxic pesticides with their undesirable environmental and human health effects. Thus, genetic manipulation requires that we understand why a natural enemy is ineffective and have a mechanism with which to alter it (Table 1).

HISTORY OF GENETIC IMPROVEMENT OF PREDATORS AND PARASITOIDS

Genetic manipulation of natural enemies (parasitoids or predators) was first discussed by Mally in 1918. He argued that, like the breeding of crops and domestic animals for artificial agricultural ecosystems, genetic selection could make natural enemies more effective. The first projects focused on determining whether it was, in fact, possible to select natural enemies for specific traits such as increased temperature tolerance, developmental rate, altered diapause and pesticide resistance, but the "improved" strains were not evaluated in the field or employed in practical pest management programs. Thus, the value of genetically modified natural enemies was doubted by most pest management specialists for many years. After 1973 several projects began to focus more on learning how to deploy the genetically modified natural enemies in pest management programs. Most involved selecting a predator or parasitoid of a secondary pest for resistance to pesticides and the resistant natural enemy strain was released so that it could establish and survive the pesticides that were applied to control a primary pest that was not controlled by any other method (3).

TWO MAIN IMPLEMENTATION STRATEGIES

Implementation of genetically improved natural enemies typically has used one of two strategies: the improved natural enemy is deployed by inoculation, in which the new strain is released one or more times into the environment where it is expected to establish and persist. This strategy has the advantages of reducing rearing costs by requiring fewer individuals for release and the likelihood that fitness of the individuals reared in the field will be high. The inoculation strategy has at least three possible population genetic mechanisms by which it can be achieved: 1) the new strain is released into a new environment where no native populations occur (open niche), and it establishes and provides long-term control; 2) the new strain is released into the environment after the native population is greatly reduced (e.g., after pesticide applications or winter) and the new strain replaces the old, providing long-term control, especially if the wild population is unable to recolonize the release sites; or 3) the new strain is released into the environment and through interbreeding and selection in the field (perhaps

Table 1 Important questions to answer when developing a genetic manipulation project if it is to be deployed successfully

Phase I. Defining the problem and planning the project
- What genetic trait(s) limit effectiveness of beneficial species or might reduce damages caused by the pest?
- Can alternative control tactics be made to work effectively and inexpensively, and are they environmentally friendly?
- Can agencies be found to support the high costs and long duration of genetic manipulation projects?
- How will the genetically manipulated strain be deployed?
- What risk issues, especially of transgenic strains, should be considered in planning?
- What advice do the relevant regulatory authorities give regarding your plans to develop a transgenic strain?

Phase II. Developing the genetically manipulated strain and evaluating it in the laboratory
- Where will you get your gene(s)?
- Is it important to obtain a high level of expression in particular tissues or life stages?
- How can you maintain or restore genetic variability in your selection or transgenesis program after obtaining the pure lines?
- What methods can you use to evaluate "fitness" of the modified strains in artificial laboratory conditions that will best predict effectiveness in the field?
- Do you have adequate containment methods to prevent premature release of the transgenic strains into the environment?
- Do you have adequate rearing methods developed for carrying out field tests?
- What release rate will be required to obtain the goals you have set?
- Have you tested for mating biases, partial reproductive incompatibilities, or other population genetic problems?
- If the strain is transgenic, have you obtained approval from the appropriate regulatory authorities to release the strain in the environment or greenhouse?
- How will you measure effectiveness of the modified strain in the field trials?

Phase III. Field evaluation and eventual deployment in practical pest management project
- If the small-scale field trials were promising, what questions remain to be asked prior to deploying the manipulated strain?
- If permanent releases are planned, have all the risk issues been evaluated?
- How will the program be evaluated for effectiveness and potential environmental or other risks?
- Will the program be implemented by the public or private sector?
- What did the program cost and what are the benefits?
- What inputs will be required to maintain the effectiveness of the program over time?

with a pesticide), a new hybrid strain is produced that can persist.

The second strategy, augmentation, involves mass rearing the new strain and releasing it periodically. Augmentation is particularly appropriate for greenhouse systems or other crops of high value and short persistence so that the high costs of periodic releases can be justified and achieved effectively. However, it is difficult to imagine that natural enemy augmentation would be cost-effective or technically feasible over vast acreages of relatively low-value crops. Present-day rearing and release technologies would make such releases prohibitively expensive.

Several genetically modified parasitoid and predator species have been used in practical pest management programs, including the use of pesticide-resistant predatory mites to control spider mites in almonds, apples, citrus, and pears, and the use of pesticide-resistant parasitoids to control aphids in walnuts and California red scale insects in citrus (1, 2). The predatory mite (*Metaseiulus occidentalis*) implemented in a California almond IPM program resulted in a significant cost-benefit analysis because fewer pesticides were applied to control spider mites. Unfortunately, relatively few cost-benefit analyses have been conducted on genetically modified natural enemy programs. Genetic manipulation thus remains a minor component of biological control research tactics.

TRANSGENIC TECHNOLOGIES

A benefit of transgenic technology is the ability to insert cloned genes from any prokaryotic or eukaryotic species so that we are no longer limited by the intrinsic genetic variability within that species (4, 5). There also may be disadvantages to using transgenic methods because the steps involved in developing and implementing a pest management program must be modified to deal with the potential risks associated with the release of transgenic arthropods into the environment (6–8).

Elegant molecular genetic methods have been and are being developed for inserting exogenous DNA into arthropods other than *Drosophila melanogaster* (4). Most methods use viral or transposable element vectors to carry the foreign DNA into the chromosomes. Interesting genes and regulatory elements that have the potential for conferring useful traits upon specific transgenic arthropods in a tissue-specific manner have been developed for a few pest arthropods, but are lacking for arthropod natural enemies. Ideally, the completion of the *Drosophila* Genome Project will make identifying useful genes for natural enemy improvement much easier.

The breakthroughs in transforming arthropods by recombinant DNA technology have not been matched with significant breakthroughs in our ability to deploy them (8). Significant efforts will have to be made to develop the data and resources to deploy transgenic strains. Relevant issues include: What is the probability that the transgenic natural enemies (released permanently into the environment) will create future environmental problems? Will transgenes inserted into natural enemies somehow be transferred horizontally through currently unknown mechanisms to pest species? Can we develop mitigation methods or methods for retrieving transgenic natural enemies from the environment should they perform in unexpected ways? The concerns about potential risks will require both researchers and regulatory agencies to accept new responsibilities and conduct creative research (8, 9).

The first experimental field release of a transgenic natural enemy, a strain of the predatory mite *Metaseiulus occidentalis* carrying a molecular marker (lacZ construct with a *Drosophila heatshock 70* promoter), was authorized in March 1996 by the U.S. Department of Agriculture Animal and Plant Health Inspection Service (USDA–APHIS). The temporary releases took place on the campus of the University of Florida after officials of the USDA–APHIS, Florida Department of Agriculture and Consumer Services, the University of Florida Biosafety Committee, and the U.S. Department of Fish and Wildlife also evaluated potential risks to threatened and endangered species and the environment. The releases took place without incident in March 1996 and no environmental harm was detected. Subsequent releases of transgenic arthropods may receive equivalent or greater scrutiny, especially if the goal is to establish the new strain permanently in the environment. Specific concerns include the potential for horizontal gene transfer (movement of the foreign gene from the transformed species to another species by a variety of little known mechanisms). Other potential risks include changes in host range or host specificity (10, 11).

If we assume that every field release of a transgenic arthropod should be conducted only after thorough peer review by scientists and regulatory agencies, appropriate efforts to contain transgenic insects and mites in the laboratory prior to their purposeful release into the environment should be made (9). At present there are no guidelines regarding methods to contain transgenic arthropods in the laboratory, but we know that greenhouses or other general laboratory facilities usually are inadequate to prevent accidental releases. Suggestions have been made that the containment facilities for classical biological control projects, which are certified by the USDA–APHIS and state departments of agriculture, would be suitable for transgenic arthropods. Such facilities and procedures were developed to prevent the escape of undesired organisms, including pest arthropods, plant pathogens, and hyperparasitoids. Personnel working in these facilities adhere to specific handling and disposal procedures designed to prevent accidental releases.

CONCLUSIONS

The genetic manipulation of arthropod natural enemies remains a minor component of IPM tactics. Under certain circumstances, such programs have been effective and cost-effective. Recombinant DNA techniques offer exciting new opportunities for improving arthropod natural enemies, but also make genetic manipulation more complex and expensive, primarily because potential risks associated with releases of transgenic arthropods into the environment need to be considered.

The most readily implemented pest management projects employing genetically manipulated natural enemies are those where augmentative releases can be conducted and the organism used in relatively small areas such as temporary cropping systems, or where the natural enemy has a low dispersal rate and can be established in individual orchards, or where the natural enemy is released into a geographic region where the wild strain does not occur. The most difficult projects to implement are those in which the new biotype is expected to replace the endemic population. This is due to the difficulty in developing mass rearing technology, quality control methods, and lack of information on population structure and hidden partial reproductive isolation mechanisms.

One field release of a transgenic natural enemy has occurred but there are no guidelines yet as to how to evaluate the risks of permanent releases of transgenic arthropods and it could take five to ten years of evaluating short-term releases of transgenic arthropods before permanent releases are permitted if we follow the paths taken by transgenic plants and microorganisms.

It is difficult to anticipate the opportunities that might arise over the next few years as improved methods are developed for genetic manipulation of arthropods by recombinant DNA methods. However, deploying a transgenic arthropod natural enemy in a pest management program will remain a challenge, requiring risk assessments, detailed knowledge of the population genetics, biology, and behavior of the species in the field, as well as coordinated efforts between molecular and population geneticists, ecologists, regulatory agencies, and pest management specialists.

REFERENCES

1. Hoy, M.A. Pesticide Resistance in Arthropod Natural Enemies: Variability and Selection Responses. In *Pesticide Resistance in Arthropods*; Roush, R.T., Tabashnik, B.E., Eds.; Chapman and Hall: New York, 1990; 203–236.
2. Hoy, M.A. Almonds: Integrated Mite Management for California Almond Orchards. In *Spider Mites, Their Biology, Natural Enemies, and Control*; Helle, W., Sabelis, M.W., Eds.; Elsevier: Amsterdam, 1985; 1B, 299–310.
3. Hoy, M.A. In *Novel Arthropod Biological Control Agents, Biotechnology and Integrated Pest Management*, Proceedings of a Bellagio Conference on Biotechnology for Integrated Pest Management, Lake Como, Italy, October 1993; Persley, G.J., Ed.; CAB International: Wallingford, U.K., 1996; 164–185.
4. Ashburner, M.; Hoy, M.A.; Peloquin, J. Transformation of arthropods—research needs and long term prospects. Insect Molec. Biol. **1998**, *7* (3), 201–213.
5. Hoy, M.A. Insect Molecular Genetics. In *An Introduction to Principles and Applications*; Academic Press: San Diego, 1994; 1–540.
6. Hoy, M.A. Biological control of arthropods: genetic engineering and environmental risks. Biol. Control **1992**, *2*, 166–170.
7. Hoy, M.A. Criteria for release of genetically improved phytoseiids: an examination of the risks associated with release of biological control agents. Exp. Appl. Acarol. **1992**, *14*, 393–416.
8. Hoy, M.A. Impact of risk analyses on pest management programs employing transgenic arthropods. Parasitol. Today **1995**, *11* (6), 229–232.
9. Hoy, M.A.; Gaskalla, R.D.; Capinera, J.L.; Keierleber, C. Laboratory containment of transgenic arthropods. Amer. Entomol. **1997**, *43* (4), 206–209, 255–256.
10. Tiedje, J.M.; Colwell, R.K.; Grossman, Y.L.; Hodson, R.E.; Lenski, R.E.; Mack, R.N.; Regal, P.J. The planned introduction of genetically engineered organisms: ecological considerations and recommendations. Ecology **1989**, *70*, 298–315.
11. Purchase, H.G.; MacKenzie, D.R. *Agricultural Biotechnology Introduction to Field Testing*; Office of Agricultural Biotechnology, U.S. Department of Agriculture: Washington, DC, 1990; 1–58.

GENETICALLY MODIFIED FOODS

Jim M. Dunwell
The University of Reading, Reading, United Kingdom

G

INTRODUCTION

Food derived from genetically modified (GM) or ''transgenic'' crop plants (1, 2) has been on the market since 1995 in the United States and for a shorter period in other parts of the world. To date, most of these GM crops, particularly corn and soybeans, include genes that provide resistance to herbicides or insects, and they are grown widely in North America. Despite general acceptance of GM food by consumers in this region, there has been much more concern (3) in Europe where the safety of, and necessity for, such foods is being questioned by many. In the future it is likely that GM foods with specific consumer benefits, such as increased nutrient composition, will become available.

BACKGROUND

The first GM product that was approved for sale was the fresh ''Flavr Savr'' tomato marketed by Calgene in the United States. Although it was not a commercial success, it was soon followed by approval of GM field crops such as corn (maize), soybeans, and cotton that had been modified to provide tolerance to a nonselective herbicide (glyphosate) and/or resistance to the European corn borer, the major insect pest of corn. These products were designed to reduce the costs of agricultural production by reducing the need either for preemergent herbicide treatment or for expensive, and often ineffective, insecticide application. Such economic benefits led to a rapid increase in acreage of GM crops from a few percentage of the total in 1995–1996 to more than 50% of the total areas in some regions; occasionally it reached 100% in smaller areas. At the same time, food manufactured from these crops was approved for sale by the European Union and was therefore included in a wide range of processed food products such as soups, confectionery, and ready prepared meals; no attempt was made by the producers or importers to separate the GM material from its non-GM equivalent. Initially, the retailers accepted the consequence of nonsegregation, but they soon came under pressure from consumers and environmental campaigners to reverse this stance. Much of this campaign was driven by the precedent of the bovine spongiform encephalitis (BSE) crisis and was fueled by perceived fears of ''Frankenfood.'' Its result has been a withdrawal of all GM products from the shelves of the major supermarkets in the United Kingdom and a search for ''identity preserved'' non-GM sources of supply. Similarly, there has been an expansion in the demand for technical methods to detect the presence of GM protein and/or DNA in food at the various stages of processing, cooking, and sale in shops and restaurants.

Despite the uneven acceptance of GM food at the present time, large numbers of field trials of GM crops are still being conducted in many parts of the world (Table 1) (4).

THE CONCEPT OF SUBSTANTIAL EQUIVALENCE

Before being approved for sale, all GM products are subject to a range of regulatory inspections, which vary from country to country. In the United Kingdom, the relevant committee is the Advisory Committee for Novel Foods and Processes, and it is likely that this group will become part of the proposed Food Standards Agency. A key feature of most of these regulatory reviews is the concept of ''substantial equivalence.'' In other words, if the new product can be considered as substantially the same as the existing equivalent, then is likely to be approved for sale. Clearly, this definition is subjective, and the concept of using ''equivalence'' in this context has been severely criticized, particularly by those who consider that GM foods should be regulated in the same way as new pharmaceutical products. There is also potential for conflict between the concept of substantial equivalence and agreements such as the Montreal Protocol that endorse the precautionary principle.

Encyclopedia of Pest Management

Table 1 Summary of ten most frequent GM crops for which field trials have been conducted in the United States and Europe

Ranking	United States	Country/region Europe (EU)
1	Corn (2388)	Corn (409)
2	Potato (649)	Oilseed rape (310)
3	Soybean (504)	Sugar beet (220)
4	Tomato (494)	Potato (162)
5	Cotton (350)	Tomato (70)
6	Oilseed rape (158)	Tobacco (54)
7	Tobacco (158)	Chicory (26)
8	Melon/squash (122)	Cotton (25)
9	Rice (92)	Fodder beet (25)
10	Wheat (92)	Beet (16)
Total	5007	1317

BENEFITS

Although food derived from the present generation of GM crops cannot be considered to have any direct benefit for the consumer in terms of its nutritional composition, it is possible that there will be lower amounts of residues from some of the agrochemical products used on their non-GM equivalent. The more direct benefits are likely to come in the next generation of products that are being designed and tested at the present time (Table 2) (5, 6). Such novel products include crops with modified protein, starch, or oil content, as well as those with increased amounts of vitamins or other micronutrients. For example, a gene encoding an enzyme involved in the synthesis of vitamin A has been transferred from *Narcissus* to rice and has been shown to improve the vitamin composition of the rice grain (7). This may be of value in reducing the level of juvenile blindness (caused by vitamin A deficiency) in those populations that rely on rice as a staple crop. An additional type of GM food product already under development is that with a reduced content of specific allergenic protein. This concept has already been demonstrated with the production of hypoallergenic rice, and discovery of the allergenic epitopes on the surface of the most important peanut allergenic has facilitated research to remove this potentially lethal compound from peanuts.

Other GM food products under development (Table 2) include those with an added content of specific "nutraceuticals," that is compounds with claimed health benefits. Examples of these include antioxidants to reduce heart disease and/or cancer, iron-containing compounds to reduce anemia, and compounds to reduce levels of cholesterol.

Much attention has been given recently to the production of antibodies and antigenic material in GM food (2, 5, 8). Among the former projects is one in which an antibody to a cell wall component of the bacterium causing dental caries has been isolated and expressed in plants. When consumed, this GM plant material inhibits growth of the bacteria over a prolonged period. Other successful tests have been conducted on plants that produce vaccines and potentially could give immunity to a range of food poisoning and other important viral diseases. In the longer term this approach of "edible vaccines" in fruit such as bananas may help in those countries where provision and transport of conventional vaccines is expensive and unreliable.

RISKS

Despite many attempts to suggest otherwise, to date there is no proven risk associated with any GM food product (9). However, there are several theoretical risks that have been raised by critics of the technology. Probably the first aspect of GM food to raise alarm was the use of antibiotic resistance marker genes used to identify the transgenic cell(s) during the laboratory phase of research. The marker of choice used in many of the early experiments, including that used to produce the Flavr Savr tomato, was the neomycin phosphotransferase gene that provides resistance to the antibiotic kanamycin. It was argued that this marker gene, which is expressed by the GM product, could be transferred from the plant to a human pathogenic bacterium and could thus prejudice the use of therapeutic antibiotics. However, expert advice from the World Health

Table 2 Examples of experimental GM food either with altered nutritional profile or expressing additional compound(s)

Crop	Nutritional profile	Additional compound
Brassica		
Coffee	Reduced caffeine	
Lettuce		Protein sweetener
Oilseed rape	Low saturated fat/oil	
Peanut	Reduced allergen	
Potato	Increased solids	Fructans, sweeteners
Rice	Reduced allergen	Vitamin A (7), ferritin
Soybean	Amino acid, fatty acid	
Spinach	Reduced oxalate	
Strawberry		Protein sweeteners
Sugar beet		Fructans
Sunflower	Amino acid	
Tomato		Protein sweeteners

G

Organization and elsewhere, confirmed that this gene was already widespread in the environment, that there was no known mechanism of transfer from plant to microbe, and that kanamycin was an outdated product and only used occasionally for topical application. Recently, more serious objections were raised to use of the ampicillin resistance marker (as used in direct gene transfer techniques), and it is generally accepted that inclusion of such markers in GM products should be avoided, even though there is no significant risk associated with their use.

Among the other specific dangers suggested are those resulting from the increased residues of herbicides used on herbicide tolerant crops. Previously, these specific agrochemical products could not be used directly on crops, as they were nonselective. As a consequence of the use of the new GM crops and the associated "over the top" herbicide treatment, there has been a relaxation in the allowable residues of certain herbicides found in some fresh or processed food products.

Another oft-claimed area of danger relates to the accidental introduction of allergens into GM food. Transfer of a methionine-rich Brazil nut thionin into soybean is the usual example quoted, though this particular type of GM crop was never considered as a potential commercial product.

Much recent attention has been focussed on the apparent toxicity of GM potatoes modified to include the snow drop lectin, a compound known to inhibit feeding by aphids. It has been claimed that rats fed on GM tubers showed symptoms of toxicity not founds in animals fed on control tubers spiked with an equal amount of the same lectin. However, interpretation of these experiments is confounded by the small sample size, the unfortunate lack of many relevant controls, and the omission of any assays of known toxic compounds such as the alkaloids found in potato. The most recent study has not shown any ill-effect from feeding GM potatoes to rats (10).

Nonspecific and nonpredictable alteration in the amount(s) of other potentially toxic compounds is another general area of concern to GM critics. Although there is no evidence for unexpected alteration in composition of any GM product approved for sale to date, it is likely that new products, particularly those with deliberate changes in composition, will be subject to additional scrutiny. Any additional analysis in this context will probably involve use of improved analytical techniques such as global studies of the proteome and metabolome (i.e., a quantitative analysis of very large numbers of proteins and small molecular weight metabolites) (11).

The potential for environmental risk is not within the scope of this brief review and has been discussed elsewhere (12).

PROSPECTS

At present the status of GM food is uncertain, at least in those parts of Europe where the shadow cast by mad cow disease is still in place and the consumer has become sensitized to the issue of food safety. Despite reassurances from many experts, GM food is being avoided and its commercial success is in some doubt. Whether this position will hold, or whether the generally positive experiences from the United States will prevail, can only be determined over time. It is likely that self interest will predominate but that will be an option only when the consumer (3) is offered a GM food that provides some tangible personal benefit. Many of the large number of field trials of GM crops (Table 1) (4) have that specific objective.

REFERENCES

1. Dunwell, J.M. How to engineer a crop plant. Pestic. Outlook **1998**, *9*, 29–33.
2. Dunwell, J.M. Novel food products from genetically modified crop plants: methods and future prospects. Int. J. Food Sci. Tech. **1998**, *33*, 205–213.
3. Dale, P.J. Public reactions and scientific responses to transgenic crops. Curr. Opin. Biotechnol. **1999**, *10*, 203–208.
4. International Field Test Sources. http://www.nbiap.vt.edu/cfdocs/globalfieldtests.cfm (accessed May 2000).
5. Dunwell, J.M. Transgenic crop plants: the next generation or an example of 2020 vision. Ann. Bot. **1999**, *84*, 269–277.
6. Dunwell, J.M. Transgenic approaches to crop improvement. J. Exp. Bot. **2000**, *51*, 487–496.
7. Ye, X.; Al-Babili, S.; Kloti, A.; Zhang, J.; Lucca, P.; Beyer, P.; Potrykus, I. Engineering the provitamin A (beta-carotene) biosynthetic pathway into (carotenoid-free) rice endosperm. Science **2000**, *287*, 303–305.
8. Dunwell, J.M. Immunization: Vaccines in Food. In *The Hutchinson Encyclopedia of Science*; Helicon: England, 1998; 391.
9. Henney, J.E. Are bioengineered foods safe. FDA Consum. **2000**, *34* (1), 18–23.
10. Hashimoto, W.; Momma, K.; Yoon, H.J.; Ozawa, S.; Ohkawa, Y.; Ishige, T.; Kito, M.; Utsumi, S.; Murata, K. Safety assessment of transgenic potatoes with soybean glycinin by feeding studies in rats. Biosci. Biotechnol. Biochem. **1999**, *63* (11), 1942–1946.
11. Noteborn, H.P.; Lommen, A.; van der Jagt, R.C.; Weseman, J.M. Chemical fingerprinting for the evaluation of unintended secondary metabolic changes in transgenic food crops. J. Biotechnol. **2000**, *77* (1), 103–114.
12. Bright, S.W.J.; Greenland, A.; Halpin, C.; Schuch, W.; Dunwell, J.M. Environmental impact from plant biotechnology. Ann. N.Y. Acad. Sci. **1996**, *792*, 99–105.

GEOGRAPHIC LOCATION OF CROP

Bruce D. Gossen
Karen L. Bailey
Agriculture and Agri-Food Canada, Saskatoon Research Centre, Saskatoon, Saskatchewan, Canada

INTRODUCTION

Geographic location can viewed on a large scale (continents and agroecological regions) or on units as small as sites within fields. The small-scale aspects of location and topography are covered in other sections of this encyclopedia (see Site-specific Farming, Landscape Patterns, and Pest Problems), so this section focuses on location on a regional scale.

The optimum combination of management techniques for a particular pest or pest complex can differ from site to site, or can be similar across vast regions, depending on a complex interaction among the pest, the crop, and the ecology of the region. Location influences this interaction through factors such as climate and soils. Also, geographic features such as mountains, oceans, and deserts contribute to the isolation of populations of both plants and microorganisms.

CLIMATE

Geographic location and climate are used to define agroecological regions. Regions with frequent precipitation have a higher probability of foliar disease outbreaks on most crops than low-rainfall areas. In higher rainfall areas, growers must use a combination of various management techniques such as crop rotation, tillage, sanitation, resistant cultivars, disease-free seed, and optimal fertility to provide adequate disease control, whereas in regions where the risk is lower, simpler management approaches may be adequate (1, 2). In contrast, warmer and drier regions have a higher risk for outbreaks of many insect pests, and so a more integrated approach is required than in low-risk (cooler, wetter) areas.

Pest species may have worldwide, regional, or local distribution and importance. Some are adapted across broad geographical areas, while others are restricted to a specific climatic or ecological niche. For example, the Russian wheat aphid (*Diuraphis noxia*) causes significant damage on wheat in the American midwest, but causes much less damage in Canada because it generally cannot overwinter in the region. In Canada, the fungal pathogens that cause leaf spot diseases on barley have slightly different temperature optima: scald (*Rhyncosporium secalis*) does best at 15–20°C, net blotch (*Pyrenophora teres*) at 20–25°C, and spot blotch (*Helminthosporium sativum*) at 25–30°C. As a result, the relative prevalence of these species differs among regions and seasons, with scald more common in northern regions and in years where mean temperatures are low, and spot blotch more common in southern regions and in years of above-normal temperature.

SOILS

Soils are influenced by factors such as climate (arid, semiarid, tropical), vegetation (grasslands, forest, tundra, desert), soil texture (sand, silt, clay, or loam), and fertility (3). It is difficult to evaluate the relationship between soil type and pest problems because weather has a more direct impact on an outbreak. However, pest problems and yield losses are exacerbated by low soil fertility and poor soil texture, structure, and moisture holding capacity (4). On poor soils, crops are more susceptible to weed competition and diseases, and less tolerant of insect damage.

QUARANTINE

Historically, each crop species developed in a specific geographical region, which is known as its center of origin. A complex ecology of insect pests, diseases, and weeds, together with beneficial insects and microorganisms, developed along with each crop. Many or most of the species that made up these ecological systems were specific to the region, limited either by their association with the crop or by geographical barriers such as mountains, deserts, and oceans. Over the past 200–300 years, many of these barriers have been bypassed. Crop species have been moved to new regions and their pest species have moved with them. A classic example of this movement is the introduction of the potato to Ireland in the 1800s. The population in Ireland rapidly came to depend on this productive and nutritious crop. When the fungal pathogen *Phytophthora infestans* was inadvertently introduced, it caused widespread destruction of the potato crop, resulting in the Great Potato Famine.

Unfortunately, the beneficial organisms that are also a part of the ecological complex are often left behind when the crop (and its pest species) are moved to new regions. As a result, much of the research into biological control has focused on identifying these beneficial organisms in the centers of origin and establishing them in new regions to reduce problems with introduced pests.

Many pest species have not yet followed their crop host into all of the regions where the crop is grown and so most countries have pests they wish to exclude. Quarantine policies and regulations are used to reduce pest immigration (5, 6). Quarantine practices include field surveys and inspection of goods, as well as fumigation, seizure, or destruction of suspect samples. Breaches of quarantine are increased with export of goods harboring pests and poor mechanisms for disinfestation of goods in transit and storage. Australia and New Zealand, which are protected from natural immigration of many pests by their geographic isolation, have stringent quarantine policies that have been effective in limiting the introduction of many pest species. Where borders are political rather than geographical, quarantine can delay, but probably not completely prevent, pest introductions.

REFERENCES

1. Agrios, G.N. *Plant Pathology*, 4th Ed.; Academic Press: San Diego, CA, 1997; 1–635.
2. Howard, R.J.; Garland, J.A.; Seaman, W.L. *Diseases and Pests of Vegetable Crops in Canada*; Canadian Phytopathological Society and Entomological Society of Canada: Ottawa, Canada, 1994; 1–554.
3. Brady, N.C. *The Nature and Properties of Soils*, 8th Ed.; Macmillan Publishing: New York, 1974; 1–639.
4. Bruehl, G.W. *Soilborne Plant Pathogens*; Macmillan Publishing: New York, 1987; 1–368.
5. http://www.aphis.usda.gov/ (accessed March 2000).
6. http://www.cfia-acia.agr.ca (accessed March 2000).

GIS AND GPS SYSTEMS IN PEST CONTROL

Shelby J. Fleischer
Pennsylvania State University, University Park, Pennsylvania, U.S.A.

INTRODUCTION

Integrated pest management uses field information to make management decisions. Information about pest density or occurrence of specific life stages can be used to optimize timing of management inputs. Similarly, information about spatial patterns can optimize placement of management tactics. Global positioning systems (GPS) and geographic information systems (GIS) are powerful tools for acquiring and synthesizing spatial information.

GLOBAL POSITIONING SYSTEMS

GPS is an information technology for determining one's position anywhere on earth at any time (1, 2). It is a massive investment—more than \$12 billion—created by the U.S. Department of Defense using 24 satellites orbiting 20,000 km above the earth monitored continuously from five stations on earth. Ephemeris information is uploaded to the satellites at $\leq$ eight-hr intervals. Russia maintains GLONASS, a similar global navigation system.

A GPS receiver on earth determines its location based upon its distance from four or more satellites. The GPS receiver generates a psuedo-random code, and receives the same code from a satellite. Both codes are generated at the same time based upon atomic clocks in the satellites and internal clocks in the receivers. Code-matching is used to identify the satellite and when the signal was generated. The transmission moves at the speed of light (the velocity), and the difference between the time of signal generation and reception (transmission time) is used to estimate distance (velocity $\times$ time = distance). The technology operates with extremely low power: the satellite signals, carried on high-frequency radio waves, do not register above the inherent radio noise of the earth.

With information from one satellite, the GPS receiver is located on a sphere in the universe centered on the satellite, with a radius equal to the distance to that satellite. Information from two satellites narrows location to a circle defined by the intersection of two spheres. Information from the third satellite defines two points that intersect three spheres—and often only one point is a realistic solution (the other may not be on earth). The fourth satellite is needed to correct for imprecision from the GPS receiver's clock.

Errors also come from the troposphere and ionosphere (influencing transmission time); type of hardware, software, and antenna; length of observations; number of satellites used for calculating position; and "selective availability." Selective availability refers to purposeful introduction of errors into the satellite's psuedo-code by the U.S. Department of Defense to help ensure security. The Department of Defense in 1999 or 2000 turned off selective availability, but retains the ability to reintroduce this form of error. A GPS receiver to estimate location without any correction is sufficient (e.g., $\pm \sim$100 m) for some pest control operations.

It is possible to calculate and correct the error using a second GPS receiver set at very precisely known locations. Receivers and software can support this correction capability by downloading error correction files from a site within $\sim$100–150 km. This is called "postprocessing." Correction data is also broadcast via radio, cellular phone, or totally separate satellite transmissions to the GPS receiver, and used simultaneously with the GPS signals to determine position. This is termed "real-time differential correction," and is commonly used in precision agriculture with broadcasts from the Coast Guard or commercial subscriptions. Subscriptions via satellite cover most of the earth, and are termed "wide area differential correction."

GPS receivers are used in precision agriculture as handheld units, backpack models, and mounted on farm machinery. Receivers that communicate with up to 12 satellites, support real-time differential correction, and achieve <1 m accuracy at <1 s time intervals are common on yield monitors, variable-rate applicators, and guided soil samplers.

GEOGRAPHIC INFORMATION SYSTEMS

GIS is an integrated system of hardware, software, and organizational infrastructure to acquire, store, manage, transform, analyze, and display spatially referenced data (3). The technology evolved from computer-aided design (CAD), computer cartography, and techniques for handling remote sensed satellite imagery. A complete GIS can acquire data from point sources (e.g., GPS field measures) or as digital surfaces from sensors, aerial photographs, satellite imagery, spatial databases, or other maps. Data are stored as points, lines, polygons, rasters (grid cells), nodes, arcs, polyhedra, or voxels (row, column and layer) within a database that preserves topology information about where each feature is relative to every other feature and to a coordinate system on earth. Attribute data—information about a feature not related to position—is preserved along with topology. Modern database management software that can manipulate spatial functions is integral to any GIS. The spatial functions support transformation, such as interpolation from points to a surface model, or algebraic functions to determine where any given combination of attributes exists. GIS are increasingly being integrated with statistical, geostatistical, and fuzzy logic tools for analyses, and with simulation models. GIS supports a wide range of display capabilities, so that the map becomes a particular view of the spatial database at a point in time influenced by the GIS programmer. Availability of GIS advanced with the increase in computer power, decreased cost, and advances in information technologies. In 1995, more than 93,000 sites were using GIS. One trend relevant to precision agriculture and pest control is toward diversification of a fully functional GIS into specialty subsystem components that handle specific needs, such as interpolation, or interpretation of remote-sensed imagery.

USE OF SPATIAL INFORMATION IN PEST CONTROL

Pest populations tend to be spatially variable at multiple scales (4). Field studies have shown insect density in large portions of potato fields were below economic threshold even when average densities exceeded threshold (5), weed species occupied only 20% of a grain field (6), and plant pathogenic nematode density was spatially variable within potato fields (6). This spatial variation implies that targeted pest control could reduce pesticide inputs.

In precision agriculture, GIS and GPS are combined to target pesticides (and nutrients or other agricultural inputs), which is termed "variable-rate application." GPS and GIS (or system components) are used to map where inputs are needed. Real-time differential GPS mounted on the farm machinery informs an on-board GIS, which then controls the application. Studies in Nebraska fields showed potential reductions of 71% for broadleaf and 94% for grass herbicides. In potato trials, insecticide input declined by $\sim 40\%$ (5), but the amount and type of reduction differed among insect species. Variable rate application also preserved beneficial species, reduced the selection pressure, and slowed the development of pesticide resistance because temporally dynamic unsprayed refuges within fields were preserved (7). Targeted placement may make expensive pest control options, including inundative release of biocontrols, more feasible, because the area treated would be reduced and the survivorship of the biocontrols enhanced.

Spatial characterization of pests is improving our understanding of population dynamics and ecology. Maps of overwintered corn rootworms correlated with topography (6). Potato leafhopper densities were higher along field edges in alfalfa, with the degree of spatial autocorrelation increasing with time after each harvest (8). This new information creates new ideas for management. Spatial variation in the crop cultivar and plant population density has been proposed for managing certain plant pathogens.

Maps expressing spatial variation of pests change with time. Maps that are more stable or less expensive to obtain, but that correlate with more dynamic maps, may be useful for pest control. For example, maps expressing spatial variation of soil texture, organic matter, or weed density and species will typically be more stable than maps of flying insect stages. Therefore, variable rate application of herbicides may advance more quickly than insecticides. An example of spatially variable application showing success in cotton involves crop growth maps acquired with remote sensing that correlate with insect density and are being used to create targeted sampling and insecticide application (9, 10).

Differential GPS is also used to guide aerial application of pesticides in forests, and in vector control programs along rivers. Differential GPS is beginning to be applied to aerial application of pesticides in agroecosystems, which increases farmworker safety.

GIS and GPS use in pest control are currently more common at larger scales. GIS is used for data organization, program management, and analyses for areawide management and eradication efforts of insects, pathogens, and weeds; and is used in forest and rangeland pest management programs. GPS is used for data acquisition in these programs.

FUTURE CONSIDERATIONS

Increasing access to GIS and GPS, along with remote sensing, is leading to futuristic ideas about geospatial technologies in pest management (11). One can envision crop consultants downloading maps about weather, regional pest pressure, and field and crop conditions from WEB-GIS (which is another current trend in GIS), and adjusting scouting and pest control efforts accordingly (12). Scouting could be targeted and navigated with GPS. Pest control inputs would vary within fields where spatial variation was significant. Maps would be used to plan next year's crop, including plans related to pest control (e.g., cultivar and plant density). Crop consultants would organize information from multiple clients within a GIS and be able to visualize regional patterns. However, increased management adds cost, and variable rate pesticide applications have not yet become economically feasible. Cost and return will vary dramatically among systems and scales. Research is optimizing pesticide placement for urban and structural pest control using GIS/GPS, and areawide and eradication programs are routinely using these geotechnologies. Variable rate application is moving fastest with weed control. The current major cost constraint is data acquisition—making the pest map. Sampling must be reconsidered when the objective is to make a map (13). In precision agriculture, soil sampling typically occurs at a spatial resolution of about a 1-hectare grid, but sampling for weeds, diseases, and insects require a finer resolution. Maps have been developed that express the probability of exceeding economic thresholds based on categorical (presence/absence) sampling. Spectral signatures are being defined to create weed maps in corn and to create maps of crop conditions that correlate to insect density maps in cotton. Data acquisition about pests in the future may come with sensors mounted on platforms ranging from tractor booms to blimps to small aircraft to satellites.

REFERENCES

1. Hurn, J. *GPS: A Guide to the Next Utility*; Trimble Navigation, Ltd.: Sunnyvale, CA, 1989.
2. Tyler, D.A.; Roberts, D.W.; Nielsen, G.A. Location and Guidance for Site-Specific Management. In *The State of Site Specific Management for Agriculture*; Pierce, F.J., Sadler, E.J., Eds.; American Society of Agronomy: Madison, WI, 1997; 101–130.
3. Burrough, P.A.; McDonnell, R.A. *Principles of Geographic Information Systems*; Oxford University Press: New York, 1998; 333.
4. Fleischer, S.J.; Weisz, R.; Smilowitz, Z.; Midgarden, D. Spatial Variation in Insect Populations and Site-Specific Integrated Pest Management. In *The State of Site Specific Management for Agriculture*; Pierce, F.J., Sadler, E.J., Eds.; American Society of Agronomy: Madison, WI, 1997; 101–130.
5. Weisz, R.; Fleischer, S.J.; Smilowitz, Z. Site-specific integrated pest management for high value crops: impact on potato pest management. J. Econ. Entomol. **1996**, *89*, 501–509.
6. Ellsbury, M.M.; Clay, S.A.; Fleischer, S.J.; Chandler, L.D.; Schneider, S.M. Use of GIS/GPS Systems in IPM: Progress and Reality. In *Emerging Technologies for Integrated Pest Management: Concepts, Research and Implementation*; Sutton, T.B., Kennedy, G.C., Eds.; APS Press: St. Paul, MN, 2000.
7. Midgarden, D.M.; Fleischer, S.J.; Weisz, R.; Smilowitz, Z. Impact of site-specific IPM on the development of esfenvalerate resistance in Colorado potato beetle (Coleoptera: Chrysomelidae) and on population densities of natural enemies. J. Econ. Entomol. **1997**, *90*, 855–867.
8. Emmen, D. *Colonization Patterns of the Potato Leafhopper, Empoasca fabae (Harris) (Homoptera: Cicadellidae) in Alfalfa*; Ph.D. Thesis, Pennsylvania State University: University Park, PA, 1999.
9. Willers, J.L.; Seal, M.R.; Luttrell, R. Remote sensing, line-intercept sampling for tarnished plant bugs (Heteroptera: Miridae) in mid-south cotton. J. of Cotton Sci. **1999**, *3*, 160–70.
10. Dupont, J.K.; Willers, J.L.; Campanella, R.; Seal, M.R.; Hood, K.B. In *Spatially Variable Insecticide Applications Through Remote Sensing*, Proceedings of the Beltwide Cotton Conference, San Antonio, TX, 2000.
11. National Research Council. *Precision Agriculture in the 21st Century: Geospatial and Information Technologies in Crop Management*; National Academy Press: Washington, DC, 1997; 149.
12. Allen, J.C.; Kopp, D.D.; Brewster, C.C.; Fleischer, S.J. 2011: An agricultural odyssey. Am. Entomologist **1999**, *45*, 96–104.
13. Fleischer, S.J.; Blom, P.E.; Weisz, R. Sampling in precision IPM: when the objective is a map. Phytopathology **1999**, *89*, 1112–1118.

GLASSHOUSE CROP PEST MANAGEMENT (INSECTS AND MITES)

G

Gillian Ferguson
Ontario Ministry of Agriculture, Food, and Rural Affairs, Harrow, Ontario, Canada

Graeme Murphy
Ontario Ministry of Agriculture, Food, and Rural Affairs, Vineland Station, Ontario, Canada

INTRODUCTION

It is estimated that, worldwide, the area of protected or greenhouse (glasshouses, plastic houses, and tunnels) crops is 307,000 ha, with vegetables occupying 65% (200,000 ha) of this area, and ornamentals, 35% (107,000 ha) (1). Growing conditions within the protected environment of greenhouses are highly favorable to arthropod pests, the most important of which include thrips (*Frankliniella occidentalis, Thrips tabaci*), whiteflies (*Trialeurodes vaporariorum, Bemisia* spp.), spider mites (*Tetranychus urticae*), aphids (e.g., *Aphis gossypii, Myzus persicae*), leafminers (*Liriomyza* spp.), and several species of caterpillar. Greenhouse conditions consist of year-round warm temperatures conducive to insect/mite development, relatively high humidities, and an abundant food supply. The damage inflicted by arthropod pests on greenhouse crops varies with the pest, geographic region, and season. The level of damage that can be tolerated is greatly dependent on the type of crop. Producers of vegetable crops generally can accept a higher level of damage than those of ornamental crops that are produced for their aesthetic value. Although there are many similarities between the crops in terms of pest incidence there are some differences in the kind of losses that are incurred and in how the pests are managed.

In both vegetables and ornamentals, the current trend is to use integrated pest management (IPM) for pest control, incorporating monitoring for pests with a range of control strategies that are generally categorized into cultural, physical, biological, and chemical controls. Good sanitation practices are a key cultural control measure that is strongly advocated and commonly practiced to reduce carryover of pest populations between crops. Examples of physical controls include colored sticky traps, light traps, and insect barriers. For biological control a range of predators and parasitoids is available for many of the major pests, and to minimize the impact of pesticide applications use of compatible pesticides is increasingly being sought.

GREENHOUSE VEGETABLES

Crop Losses

The predominantly grown greenhouse vegetable crops are tomatoes, cucumbers, and peppers. Crop losses due to arthropod pests vary considerably among the major greenhouse areas of the world. Generally, losses tend to be more in greenhouses located in warmer climates than in those located in temperate regions. For example, crop loss in the Mediterranean area can be up to 100% due to the whitefly-transmitted tomato yellow leaf curl virus (2). In temperate areas such as Canada, many growers report yield losses of 1% or less (Ferguson, unpublished data), however, much larger losses have also been recorded. For example, in 1985, western flower thrips (*F. occidentalis*) caused an estimated 20% yield loss in greenhouse cucumbers in British Columbia (3). More recently in 1997, losses in a pepper crop due to aphids were estimated at CD $50,000 (U.S. $32,500) per hectare, and losses due to *Echinothrips americanus* in several pepper crops during the same year were estimated at CD $60,000 (U.S. $39,000) per hectare (4). One of the highest losses was reported in 1996 and was caused by a new pest, the potato psyllid. Losses inflicted by this pest were valued at CD $115,900 (U.S. $75,000) per hectare (4). In U.K. glasshouses, Lewis (5) reports that the Western flower thrips can cause up to 90% loss of summer replanted cucumbers worth up to U.S. $75,000 per hectare each year.

Cost of Pest Control

The cost of controlling pests in greenhouse vegetable crops is variable. Many growers in Canada currently spend on average, CD $10,000–20,000 (U.S. $6500–13,000) per hectare per year for pest control (biological control agents and selective pesticides) (Ferguson, unpublished data). In The Netherlands, the average cost for controlling pests in a program that integrates biological control agents with selective pesticides was U.S. $2200–5000 per hectare per year in 1997 (6). In pest control programs that focussed on biological control agents and avoided pesticides, the cost was U.S. $5500–9000 per hectare per year. The cost of pest control programs in the United Kingdom is somewhat higher. The total cost for pest control (biological control agents, pesticides, monitoring, and labor) was estimated at 7200–15,000 British pounds (U.S. $10,800–22,500) per hectare per season (7). Generally, the cost is lower for tomato crops and higher for cucumber and pepper crops. This cost differential results from the larger number of pests that have to be controlled in the latter two crops.

Control Strategies

Generally, there is strong reliance on use of biological control agents for managing pests particularly in greenhouses in temperate regions. In 1997, the total market for natural enemies for use in greenhouses was estimated at more than U.S. $30 million, and currently, greenhouse vegetables account for more than 90% of this market (6). The protected environment within greenhouses, combined with lack of infestation from outdoor crops during the cooler times of the year, favors the use of biological control. In addition, widespread use of bumble bees for pollination in tomatoes and peppers has provided a particularly strong incentive for use of biological controls. Other factors that have promoted adoption of biological control in greenhouse vegetable crops include pesticide resistance, a relatively high tolerance for plant damage, consumer concerns for pesticide residues, relatively reliable supplies of a wide range of natural enemies accompanied by technical support, and an increasing availability of more selective pesticides.

There are numerous benefits associated with use of biological control as the major control strategy. One important benefit is reduced dependence on pesticides that assists in delaying development of resistance to pesticides. Other benefits include reduced phytotoxic effects, a more pleasant working environment, and lack of restrictions on reentry intervals. Indeed, minimal pesticide input is increasingly used as a selling point for such products. However, biological control as a major strategy is often inadequate in situations where the pest population becomes overwhelming, such as often occurs because of influxes from outside during warmer seasons or in warmer climates.

GREENHOUSE ORNAMENTALS

Crop Losses

The current worldwide value of the greenhouse ornamental industry is estimated to be at least U.S. $10 billion, given that the Dutch industry alone was valued at U.S. $2.9 billion in 1991 (8), and the United States industry at $3.5 billion in 1992 (9). Given the intensive nature of production and the year-round threat of insect and mite pests, a conservative estimate of 1% losses would still mean overall losses in excess of $100 million per annum. Accounts of individual losses (due to specific pests) demonstrate the high risks associated with ornamental crop production. For example, in Pennsylvania between 1989–90, plants infected with the thrips-borne tomato spotted wilt virus, and with a retail value of U.S. $675,000 were destroyed (5) and Parrella (10) reported complete losses of gloxinia and chrysanthemum crops. Between 1987 and 1990, the Finnish government spent U.S. $390,000 to eradicate the western flower thrips from greenhouses, and estimated the cost of not doing so at U.S. $7.5 million over four years (5). In the early 1980s, the leafminer, *L. trifolii*, resulted in losses to California growers of U.S. $18–21 million per year (23% of the total crop) (11).

Cost of Pest Control

The cost of controlling pests in ornamental crops is variable, depending on crop and the pest itself. In The Netherlands in 1997, the average cost of an IPM program (including pesticides and biological control) per hectare per year was U.S. $10,000 in roses, U.S. $4000 for gerbera, and U.S. $2500 for poinsettia (6). Jacobson (7) gave the cost of pest control in bedding plants in the United Kingdom (including labor and monitoring), at approximately U.S. $4000–6000 per hectare per season (converted from UK currency). Murphy and Broadbent (12) reported that the pesticide costs in cut chrysanthemum were as high as CD $11,300 (approximately U.S. $7500) per hectare per year. Even higher costs of an estimated U.S. $14,825 per hectare per year were reported in California during the 1980s for control of leafminer in chrysanthemum (11). Despite the apparent high cost of pest control, pesticides in general are not considered a major cost, estimated on average at between 1–2% of total crop production costs (13, 14).

Control Strategies

Although greenhouse ornamentals are one of the most heavily targeted crops for pesticide use, generalizations can be misleading. Hundreds of different crop species are broadly grouped under the ''ornamentals'' heading. Some of these are excellent hosts to many different insect/mite pests and may receive many pesticide applications during a crop cycle. Others may have relatively few pests and require very few applications. For most growers, there is still strong reliance on the use of pesticides as the major component of their pest management programs (15). The benefits of using pesticides relate to their cost and effectiveness. In many situations, the amount spent on pesticides could be considered to return almost the entire value of the crop with a benefit/cost ratio of 50:1 (16), (given a pesticide cost of 2% of cost of production). This is not an unreasonable comment because there are many insect/mite pests that could result in complete loss of the crop if not controlled.

Since the mid–late 1980s, there has been an increasing trend in the industry toward the use of IPM strategies in producing greenhouse ornamental crops. Integrated pest management can be implemented in greenhouse ornamentals in three phases: 1) establishing a monitoring program, 2) modifying pesticide usage, and 3) biological control (12). Indeed, the use of monitoring alone can lead to significant reductions in pesticide usage (8, 12, 17–19). Murphy (20) has estimated that almost 90% of the Ontario, Canada greenhouse ornamentals industry uses a structured pest monitoring program with considerable reductions in pesticide use as a consequence.

Biological control is still a relatively minor component of IPM in greenhouse ornamentals, although its use is increasing. There are several reasons for the low level of adoption of biological control in ornamentals compared to vegetables:

- Low tolerance for damage (zero tolerance in export situations);
- Greater number of registered products and less incentive to reduce pesticide usage on a nonfood crop;
- Continuity of crops that does not allow the greenhouse to be emptied and cleaned of pests before introducing a new crop.

Despite these concerns, the interest in the use of biological control is high. In The Netherlands, it was estimated that biological control was used on 1% of the production area in 1992, and increased to 5% in 1995 (21). In Denmark, it is estimated that approximately 30% of flower growers currently use biological control (22). Similarly in Ontario, Canada, a survey of growers in 1999 found that 26% were using some form of biological control and an additional 27% had used the strategy previously (Murphy, unpublished data). The major stumbling block for further increases in biological control is the western flower thrips. Natural enemies are commercially available for this pest, however, their performance is inconsistent and there are few pesticides for thrips control that are also compatible with biological control agents.

FUTURE TRENDS

For both vegetables and ornamentals, efforts will continue in the development and refinement of key IPM strategies. For example, increasing interest in the use of biological control will provide the stimulus for increased research in this area, and an incentive to commercial insectaries to increase the range of biological control agents. At the same time, there is a need for compatible pesticides that have minimal deleterious effects on beneficial arthropods. Already in development are novel compounds that have lower residual activity or exhibit greater physiological selectivity (23). Also, increased use of physical controls, such as insect barriers, will continue, particularly where other control strategies either do not work, or have very limited effectiveness. In addition, development of cultivars that are resistant to diseases and arthropod pests will make a major contribution to successful implementation of IPM. Overall, it is anticipated that adoption of an IPM approach in the management of glasshouse pests will increase, if the glasshouse or protected crops industry is to remain a viable one. The historical consequences of chemical overuse dictate that we cannot do otherwise.

REFERENCES

1. Gullino, M.L.; Albajes, R.; van Lenteren, J.C. Setting the Stage: Characteristics of Protected Cultivation and Tools for Sustainable Crop Protection. In *Integrated Pest and Disease Management in Greenhouse Crops*; Albajes, R., Gullino, M.L., van Lenteren, J.C., Elad, Y., Eds.; Kluwer Academic Publishers: Norwell, 1999; 1–15.
2. Lapidot, M.; Friedmann, M.; Lachman, O.; Yehezkel, A.; Nahon, S.; Cohen, S.; Pilowsky, M. Comparison of resistance level to tomato yellow leaf curl virus among commercial cultivars and breeding lines. Plant Dis. **1997**, *81*, 1425–1428.
3. Anon. *Frankliniella occidentalis* (Perg.). OEPP/EPPO Bull. **1989**, *19*, 725–731.

4. *BC Hot House Growers Association*; Report on Disease and Insect Losses: Surrey, BC, 1998; V3S 8E7.
5. Lewis, T. Pest Thrips in Perspective. In *Thrips as Crop Pests*; Lewis, T., Ed.; University Press: Cambridge, 1997; 1–13.
6. Bolckmans, K.J.F. Commercial Aspect of Biological Control in Greenhouses. In *Integrated Pest and Disease Management in Greenhouse Crops*; Albajes, R., Gullino, M.L., van Lenteren, J.C., Elad, Y., Eds.; Kluwer Academic Publishers: Norwell, 1999; 310–318.
7. Jacobson, R.J. Integrated Pest Management (IPM) in Glasshouses. In *Thrips as Crop Pests*; Lewis, T., Ed.; University Press: Cambridge, 1997; 639–666.
8. Fransen, J.J. Development of integrated crop protection in glasshouse ornamentals. Pestic. Sci. **1992**, *36*, 329–333.
9. Miller, M.N. Looking at the big picture. Grower Talks **1995**, *59*, 8–16.
10. Parrella, M.P. IPM—Approaches and Prospects. In *Thrips Biology and Management*; Parker, B.L., Skinner, M., Lewis, T., Eds.; Plenum Press: New York, 1995; 357–363.
11. Newman, J.P.; Parrella, M.P. A license to kill. Greenhouse Manager **1986**, *5* (3), 86–92.
12. Murphy, G.D.; Broadbent, A.B. Development and implementation of IPM in greenhouse floriculture in Ontario, Canada. Bull. Int. Org. Biol. Control/Western Palaearctic Regional Section **1993**, *16* (2), 113–116.
13. Parrella, M.P. Biological control in ornamentals: status and objectives. Bull. Int. Org. Biol. Control/Western Palaearctic Regional Section **1990**, *13* (5), 161–168.
14. Mumford, J.D. Economics of integrated pest control in protected crops. Pestic. Sci. **1992**, *36*, 379–383.
15. van Lenteren, J.C. Biological control in protected crops: where do we go. Pestic. Sci. **1992**, *36*, 321–327.
16. Gullino, M.L.; Wardlow, L.R. Ornamentals. In *Integrated Pest and Disease Management in Greenhouse Crops*; Albajes, R., Gullino, M.L., van Lenteren, J.C., Elad, Y., Eds.; Kluwer Academic Publishers: Norwell, 1999; 486–506.
17. Newman, J.; Tjosvold, S.; Robb, K.; King, A. Implementation of IPM scouting programs for ornamental crop production. Bull. Int. Org. Biol. Control/Western Palaearctic Regional Section **1996**, *19* (1), 111–113.
18. Guldemond, J.A. Preliminary results on density and incidence counts of aphids in cut chrysanthemums in the greenhouse. Bull. Int. Org. Biol. Control/Western Palaearctic Regional Section **1993**, *16* (8), 92–97.
19. Price, J.F.; Overman, A.J.; Englehard, A.W.; Iverson, M.K.; Yingst, V.W. In *Integrated Pest Management Demonstrations in Commercial Chrysanthemums*, Proceedings of the Florida State Horticultural Society, 1980; *93,* 190–194.
20. Murphy, G. Pest management survey of ontario growers. Greenhouse Canada **1998**, *18* (7), 8.
21. Fransen, J.J. Recent trends in integrated and biological pest and disease control in The Netherlands. Bull. Int. Org. Biol. Control/Western Palaearctic Regional Section **1996**, *22* (1), 65–68.
22. Enkegaard, A.; Funck Jensen, D.; Folker-Hansen, P.; Eilenberg, J. Present use and future potential for biological control of pests and diseases in Danish glasshouses. Bull. Int. Org. Biol. Control / Western Palaearctic Regional Section **1999**.
23. Ishaaya, I.; Horowitz, A.R. Insecticides with Novel Modes of Action: An Overview. In *Insecticides with Novel Modes of Action*; Ishaaya, I., Degheele, D., Eds.; Springer-Verlag: Heidelberg, Berlin, 1998; 1–24.

GLASSHOUSE CROP PEST MANAGEMENT (PLANT PATHOGENS)

G

Piara S. Bains
Mohyuddin Mirza
Alberta Agriculture Food and Rural Development, Edmonton, Alberta, Canada

INTRODUCTION

Glasshouses or greenhouses are structures covered with a transparent material for the purpose of admitting natural light for plant growth. These structures usually have automatic control over heating, cooling, ventilation, shading, air movement, and humidity. Glasshouses are used to grow a variety of vegetables, ornamentals, bedding plants, cut flowers, tree seedlings, herbs, and other plant material from tissue culture sources. In colder climates, glasshouses provide a significant amount of fresh vegetables during the time when outdoor production is not possible. The total world area under glasshouses is estimated to be 306,500 ha (1). There has been a revolution in the glasshouse production technology during the past couple of decades. Precise control of environmental conditions and nutritional status of plants has provided increased yields and powerful disease management tools (2).

For optimal plant growth in a glasshouse, plants require an optimal level and quality of light, conducive day and night temperatures, relative humidity adjusted for adequate transpiration, carbon dioxide level higher than ambient air, all the essential nutrients, and adequate supply of high quality water (3, 4). Unfortunately, however, the environmental conditions provided for optimal plant growth in a glasshouse also generally are very favorable for development of many plant diseases. Any one of many stresses, including low or high temperature, over- or under-watering, low or high light intensity, too little nutrients, or too high salinity could predispose a plant to a disease (5). Furthermore, dense plant communities and generally inadequate systems of air circulation and removal of moist warm air from a glasshouse create a microclimate conducive for development of foliar diseases for example gray mold (*Botrytis cinerea*) and white mold (*Sclerotinia sclerotiorum*) of cucumber. Intensive plant handling activities by workers also spread the pathogens and promote the development of epidemics. Hydroponic cultures provide an excellent system for excluding soilborne diseases. The system, however, also provides a conducive medium for large-scale multiplication and a rapid plant to plant spread of many devastating pathogens like *Pythium* spp. causing damping off of many greenhouse-grown plant species. *Greenhouses: Advanced Technology for Protected Agriculture* (6) and *Managing Diseases in Greenhouse Crops* (7) are two excellent books on this subject.

DISEASE MANAGEMENT

Cultural Control Strategies

For effective management of plant diseases in a glasshouse, an integrated approach for disease control is required. The approach includes the use of resistant cultivars, eradication of primary sources of inoculum, creation of an environment unsuitable for pathogen growth, and use of chemical and biological control measures (8).

Resistant plant material is the first line of defense against pathogens and is the most economic and environmentally friendly method of plant disease control. It is important to select a cultivar that is comparatively resistant to the diseases in the area.

A disease in a glasshouse could be initiated by a pathogen present in any one of many primary inoculum sources including growing medium, water, seed or seedlings, plant residue, weeds surrounding the glasshouse, insect vectors, and airborne inocula (7). For effective management of a disease it is important to exclude inoculum from these sources. Soil could be pasteurized. Soilless medium containing peat may be infested with *Fusarium* and *Pythium* species and might require the use of a fungicide or a biological control agent for controlling these pathogens. Rockwool, Perlite, and polystyrene are heat sterilized during manufacturing. Vermiculite and sawdust are not inert, but none of these substrates need sterilization, if not reused. In hydroponic systems, several strategies are used to avoid contamination. Nutrient solution for a hydroponic system can be passed through ultraviolet irradiation system for sterilization (9). The process, however, causes precipitation of iron and manganese.The solution, therefore, should be very carefully monitored for status of these nutrients.

The use of ozone to sterilize irrigation water or recirculating nutrient solution to get rid off pathogens like *Pythium* and *Fusarium* spp. is gaining popularity. Ozone is a strong oxidizing agent and can precipitate iron and manganese out of the nutrient solution. It, therefore, is very important to monitor these elements in the nutrient solution (10).

All seeds and seedlings used should be checked for infection using routine culture-indexing methods for fungi and bacteria, and by enzyme-linked immunosorbent assay (ELISA) for important viruses. Although use of clean seed is the best choice, heat therapy and fungicidal treatment of seeds could be used to inhibit the development of seedling infection by some fungal and viral pathogens. Angular leaf spot of cucumber caused by the bacterium *Pseudomonas syringae* and alternaria infection of various bedding plant species by *Alternaria* spp. are examples of seedborne diseases (11).

Cleanliness of the glasshouse is very important. All residues should be discarded at least 10 m away and the glasshouse structure should be disinfected between crops. Plant residues are known to become an important source of inoculum for the development of a disease. Late blight of tomato caused by *Phytophthora infestans* and gummy stem blight of cucumber caused by *Didymella bryoniae* are examples of diseases in which infected plant residue play an important role in the development of the disease (11).

Weeds provide overwintering sites for many pathogens including *B. cinerea* and powdery mildew pathogens (11). An area of about 10 m around the glasshouse should be kept weed free. Use insect-proof meshes on vents to keep vectors of various disease pathogens, especially viruses, out of the glasshouse.

Environment Management Strategies

Management of the glasshouse environment to make it nonconducive for growth of a pathogen is another effective method of inhibiting the development of a disease. There are three important environmental factors that can be manipulated to manage diseases. These are humidity, temperature, and light. With recent advances in technology these factors can be very accurately monitored and adjusted.

Humidity strongly influences some of the economically important diseases of glasshouse crops. Botrytis gray mold (*B. cinerea*), a serious disease of many glasshouse-grown plant species causes significant economic losses under cool and damp conditions that are very favorable for growth, sporulation, spore release, and spore germination of the pathogen and for establishment of infection. Glasshouses that have poor air circulation and a poor mechanism to exhaust moist, warm air frequently have *B. cinerea* infection. *Cladosporium fulvum* causes leaf mold of glasshouse-grown tomatoes under similar conditions. Any improvement in the glasshouse air circulation and exhaust system will reduce the incidence of botrytis gray mold and leaf mold of tomatoes. Powdery mildew disease of cucumber, tomato, pepper, begonia, and poinsettia, on the other hand, is more prevalent under drier conditions. Overventilation in glasshouses should, therefore be avoided to control this disease (4).

Plants grown at suboptimal environmental conditions undergo stress and become susceptible to various pathogens. It is, therefore, very important to grow various plant species at temperatures that are most conducive for their growth and nonconducive for growth of their potential pathogens. Many plant pathogens have very specific temperature requirements. Both *Pythium ultimum* and *Pythium aphanidermatum* cause damping off and root rot of many plant species. *P. ultimum* is the main cause of the disease when temperature of the growing medium is below 12°C, while *P. aphanidermatum* mainly causes this disease above 25°C. Adjusting the temperature of the growing medium to the range given for a plant in Table 1 could be an effective method to avoid the development of this disease in these plants. Some other pathogens including *Fusarium, Verticillium, Rhizoctonia*, and *Sclerotinia* attack basal stems of various plants and cause diseases like wilts and rots. These pathogens can grow at much broader range of temperature than that required for optimal growth of majority of the plants. It becomes rather difficult to control these pathogens with simple temperature adjustments. Avoiding the development of condensation on stems by improving air circulation and temperature ramping will inhibit the development of these diseases.

Light intensity has largest effect on photosynthetic activity and in turn on the carbohydrate status of plants.

Table 1 Growing medium temperatures for avoidance of damping off and root rot caused by *Pythium ultimum* or *Pythium aphanidermatum*

Crop	Temperature of the growing medium (°C)
Cucumber	20–24
Lettuce	20–21
Pepper	18–20
Poinsettia	14–18
Tomato	18–20
Bedding plants	18–20
Tree seedlings	18–20

G

Table 2 Number of seedlings per square meter

Crop	Seedlings/(m^2)
Tomatoes	20–22
Cucumbers	16
Peppers	20
Lettuce	100–120

Low light intensities and unlimited water and nutrients, especially under hydroponic systems, produce plants that are spindly, possess inadequate carbohydrate reservoir, and are more susceptible to many pathogens. *Fusarium* and *Verticillium* species cause wilts under low carbohydrate status, whereas *B. cinerea* attacks carbohydrate-rich plants. Furthermore, carbohydrate status of plant parts changes according to various crop growing practices. Removal of fruit shifts the movement of photosynthates to stems and roots making them more susceptible to certain pathogens. In order to produce healthy plants in winter light conditions, maximize natural light by removing shades from the roof of the glasshouse and by installing high pressure sodium lights.

Overcrowding of plants should be avoided. It creates a microclimate of low light intensity, high humidity, reduced air circulation, and increased physical contact between plants. These conditions are ideal for development and spread of many plant diseases in glasshouses. Table 2 includes recommended densities of seedlings for some common glasshouse grown crops.

Plant health is also greatly influenced by the fertilizer regime of the growing medium. High nitrogen causes the plants to be succulent and more susceptible to infection by some pathogens, for example *B. cinerea* causes of grey mold on many glasshouse-grown plants. *Alternaria solani* causing early blight of potatoes, on the other hand, is more aggressive on plants with low nitrogen status.

Fungicidal and Biological Control Strategies

In additions to the cultural methods, fungicides and biological control agents are commonly used to manage the diseases of glasshouse-grown crops. Fungicides inhibit growth of fungal pathogens and protect healthy plant tissues from infection. Fungicides could be incorporated into the growing medium for protection from medium-borne pathogens, applied to soil and seeds for control of seedling infection, and sprayed on above-ground parts to inhibit the development of the disease. Development of resistance in pathogen populations to fungicides is a well-known phenomenon. To manage the development of resistance, it is advisable to include a number of fungicides that select for different resistance mechanisms.

The development of resistance and concerns of fungicides in food chain have generated considerable interest in the development of biological methods for controlling plant diseases. Precise control of environmental conditions in a glasshouse renders them much more suitable than a field for use of biological methods for control of plant diseases (2). A list of biocontrol products (containing living organisms) against soilborne crop diseases ''Commercial Biocontrol Products for Use Against Soilborne Crop Diseases'' maintained by United States Department of Agriculture has 41 entries of fungal and bacterial biocontrol agents (12). Target pathogens of these products cause many economically important diseases. Fourteen of these products include *Pythium* as one of the target pathogens. Sales of biological control agents have increased significantly in recent years.

REFERENCES

1. Jensen, M. In *Hydroponics Worldwide*, Proceedings of the International Symposium on Growing Media and Hydroponics, Windsor, Canada, May 19–20, 1997; Papadopoulos, A., Ed.; Acta. Hort.: **(1999)**, 481, 719–729.
2. Jarvis, W. Managing diseases in greenhouse crops. Plant Disease **1989**, *73*, 190–194.
3. Anonymous. *Floriculture Production Guide for Commercial Growers*; Ministry of Agriculture Fisheries and Food: Victoria, Canada, 1996.
4. Portree, J. *Greenhouse Vegetable Production Guide for Commercial Growers*; Ministry of Agriculture, Fisheries and Food: Victoria, Canada, 1997.
5. Chase, A.; Poole, R. Investigations into the roles of fertilizer level and irrigation frequency on growth, quality and severity of *Pythium* root rot of *Peperomia obtusifolia*. J. Amer. Soc. Hort. Sci **1984**, *109*, 619–622.
6. Hanan, J. *Greenhouses Advanced Technology for Protected Agriculture*; CRC Press: Boca Raton, FL, 1998.
7. Jarvis, W. *Managing Diseases in Greenhouse Crops*; APS Press: St. Paul, MN, 1992.
8. Cherif, M.; Belanger, R. Use of potassium silicate amendments in recirculating nutrient solutions to suppress *Pythium ultimum* on long English cucumber. Plant Dis. **1992**, *76*, 1008–1011.
9. Runia, W. Elimination of root-infecting pathogens in recirculation water from closed cultivation systems by ultra-violet radiation. Acta. Hort. **1994**, *361*, 361–371.
10. Runia, W.; van Os, E.; Bollen, G. Disinfection of drain water from soilless cultures by heat treatment. Neth. J. Agric. Sci. **1988**, *36*, 231–238.
11. Howard, R.; Garland, J.; Seaman, W. *Diseases and Pests of Vegetable Crops in Canada*; The Canadian Phytopathological Society and Entomological Society of Canada; Ottawa, 1994; 554.
12. Commercial Biocontrol Products for Use Against Soilborne Crop Diseases. www.barc.usda.gov/bpdl/bpdlprod/bioprod.html (accessed July 2000).

GREENHOUSE CROP LOSSES (DISEASES)

Ruguo Huang
AMCO Produce, Inc., Leamington, Ontario, Canada

William R. Jarvis
Agri-Food and Agriculture Canada, Harrow, Ontario, Canada

INTRODUCTION

A very wide range of crops is now grown under protected cultivation, with the term "greenhouse" encompassing simple plastic walk-in tunnels to ranges of several hectares of glass or plastic covered structures with sophisticated, computer-assisted controlled environments. The greenhouse area worldwide is estimated to be 400,000 ha in 1998 with a total production value of more than $200 billion. Traditional glasshouses are widely used in northern Europe (41,000 ha), but plastic greenhouses have been readily adopted on all five continents and have expanded rapidly in recent years, especially in the Mediterranean region, China, Japan, Korea, and Ontario. Florida alone produces more than 500 species of ornamental foliage plants, worth more than $350 million. The variety of crops increases, and with it, the number and severity of diseases. Greenhouse crops contribute substantially to the agricultural economies of The Netherlands, the United Kingdom, the United States, Canada, Japan, eastern China, Korea, and Colombia (1–3). In The Netherlands, for example, 5% of cultivated land area is used for glasshouse culture, representing a production value of approximate NLG 13 billion, or 19% of the total production of Dutch agriculture. Because of its ability to conserve water, hydroponic greenhouse production is also important in arid areas such as the Middle East countries.

In greenhouses, crops are grown in soil, which may be sterilized or not, or in hydroponic systems with various substrates such as rock wool, vermiculite, sawdust, and coconut fiber or without substrate like the deep flow system and the nutrient film technique (NFT). The nutrient solution in hydroponics is recycled (closed hydroponics) or not (open hydroponics). In some countries, such as The Netherlands, release of water and fertilizers (especially nitrates) into the water table is not allowed.

In its simplest form, the greenhouse is a cover, generally conserving incoming radiant energy and protecting crops from rain and snow, and from low temperatures, enabling crops to be grown out of season and facilitating year-round continuity of produce to the consumers. In advanced forms, the greenhouse is a complex structure with computer-controlled climate management. Whatever its complexity, however, the greenhouse does not exclude pests and diseases, and some may even be exacerbated in the warm, humid, and generally wind-free environment (4).

The main crops grown commercially in greenhouses are vegetables (tomato, cucumber, sweet pepper, lettuce, spinach, and eggplant), fruit (dessert grapes, strawberries, cantaloupes, and musk melon), and florist crops (carnation, rose, chrysanthemum, gerbera, poinsettia, exacum, gladiolus, and tulip, to name but a few).

DISEASE IMPACT ON THE GREENHOUSE INDUSTRY

Reliable estimates of greenhouse crop losses caused by diseases are not available and greenhouse growers are notably reticent about their losses. Even with the extensive application of pesticides, diseases cause substantial losses in the yield and quality of greenhouse crops. It is generally accepted that direct losses caused by diseases range from 10 to 15%, costing $35–55 billion per year. If the cost of pesticide application ($2,000–5,000 per ha per year) is factored in, this figure would be even higher. Sometimes disease damage may necessitate replanting the whole crop.

A major problem in growing in soil is soilborne diseases. Many greenhouse crops are grown continuously year-round without interruption in production, and crop rotation is not a common practice as in open field production. Soil temperature and moisture in greenhouses are more constant than in the open field. All of these factors can lead to excessive buildup of soilborne pathogens. In China, for example, root diseases usually kill 10–30% of the newly transplanted plants of greenhouse crops. In some cases, the whole greenhouse has to be replanted. Cucurbit crops (cucumbers, cantaloupes, muskmelons, and watermelons) cannot be grown there without grafting onto resistant rootstocks, usually squash or gourd. Grafting is also a very common practice of disease control in Japan and Korea, where skilled labor is relatively cheap. Although hydroponic production has resulted in a decrease in diversity of root-infecting pathogens compared with conventional culture in soil, root diseases are still a threatening problem, and disease losses are occasionally greater than in soil. Because hydroponic systems generally lack the indigenous biological control microflora that nonsterile soil has, root diseases generally develop faster in hydroponics than in soil. Root rots caused by *Pythium* spp. or *Phytophthora* spp. are common diseases of hydroponically grown crops (such as lettuce, cucumber, pepper, and tomato), and damage caused by those diseases ranges from very severe (100% loss) to moderate. In contrast with soil culture, where older plants of cucumber and tomato are not as susceptible to damage by *Pythium*, damage can be sudden and quite severe on older plants in hydroponics. Severe root rot and subsequent plant death can occur in just a few days. Root pathogens, once introduced into the system, can spread quickly throughout the culture system, particularly in closed systems, where the nutrient solution is recycled. This is the most important reason why some growers are reluctant to change their open systems into closed ones, notwithstanding that they can save up to 40–70% of fertilizer and 20–50% of water, and moreover can reduce fertilizer pollution to the environment.

Most leaf diseases, with the exception of powdery mildew, are no longer a severe problem in modern greenhouses with good environment control facilities, especially those with computer-controlled environments. Water condensation on the leaf surface is rare in such greenhouses. In simple greenhouses lacking environment control facilities, however, humidity is usually very high and water condensation is common at night. Such an environment is very conducive to leaf and fruit diseases, such as those caused by *Botrytis cinerea* Pers.: Fr., *Alternaria* spp., *Colletotricum* spp. and *Phytophthora* spp. Powdery mildews, which thrive in dry conditions and on crops in full production, are important in many greenhouse crops, such as sweet pepper, cucumber, rose, and poinsettia. Intensive pesticide spray programs are required to control leaf diseases, but they add considerably to the cost of production and stress the plants severely, often resulting in overall crop loss. Moreover, most pathogens soon become genetically tolerant to pesticides, necessitating continual redesign of pesticide programs.

Because greenhouses are self-contained units, with an average area of less than one ha under one manager, disease epidemics tend to be restricted, particularly where greenhouses are remote from each other. However, in some areas, such as The Netherlands, southern Ontario, and Guernsey in the UK Channel Islands, there are dense concentrations of large, multispan greenhouses (southwestern Ontario has about 500 ha within a radius of 10 km, for example), and diseases tend to spread rapidly. In 1974, for example, Fusarium crown and root rot of tomato (*Fusarium oxysporum* f. sp. *radicis-lycopersici* W.R. Javis & Shoemaker) hitherto known only sporadically in Japan and California, appeared in almost all southwestern Ontario greenhouses, and for a time threatened the livelihood of tomato growers. It did not respond to fungicides, and attempts to reduce soilborne populations by steam sterilization or fumigation failed because its natural antagonists in soil were killed, exacerbating the disease. In many greenhouses, 70% or more of individual crops were destroyed. Subsequent research, however, developed control methods, using the allelopathic properties of members of the Lactuceae. Thus, interplanting tomatoes with lettuce or dandelion, and raising soil temperatures slightly, gave excellent control, and the disease is no longer a problem (4). Subsequently, genetic resistance became available, and grafting onto resistant rootstock was sometimes practiced.

In southwestern Ontario, there are about 2500 ha of processing tomatoes grown in the field. In most years, bacterial canker [*Clavibacter michganesis* ssp. *michiganesis* (Smith) Davis et al.] occurs, and the bacteria are carried in wind-blown dust into nearby greenhouses. There, the disease begins beneath the roof ventilators and near doorways, and the bacterium is spread on workers' fingers and tools along the plant rows, killing 10–50% of plants, and by infecting fruits, reduces marketability. In some cases, part or the whole crop has to be replanted. Specialist growers raise tomato transplants in the greenhouse for the field processing industry and since the bacterium perennates in greenhouses, there is a constant threat for field crops.

By the same token that fungi and bacteria cannot be excluded from greenhouses, neither can arthropods that vector virus diseases. Insect screens can be fitted to ventilator openings and doorways but none is fully effective without impeding air exchange between the interior and the exterior, which has to be maintained at 0.75–1.0 changes per hour in summer (5). Thus, the western flower thrip (*Frankliniella occidentalis* Pergande) transmits tomato spotted wilt virus to a range of vegetable and flower crops; the whitefly (*Trialeurodes vaporariorum* Westwood) transmits beet pseudo yellows virus to cucumbers; while various aphids transmit several viruses to several crops, including cucumber mosaic virus to cucumber and lettuce mosaic virus to lettuce. Once in the plant, there is no cure for viruses, and yields and quality are severely affected.

Modern technology has pushed productivity of vegetable crops to three or four times the yields obtained 20 years ago. This has been achieved in cultivars not greatly changed over that time. Consequently, those cultivars have been stressed by productivity, and new diseases have appeared that are demonstrably exacerbated by altered source-sink relationships of assimilates. One example of this is penicillium stem and fruit rot of cucumber (*Penicillium oxalicum* Currie & Thom), which has caused losses of 20–30%. Another example is Fusarium stem and fruit rot of sweet pepper caused by *Nectria haematococca* Berk. & Broom [anamorph *Fusarium solani* (Mart.) Sacc.]. Both of these diseases have been remarkably difficult to control.

While most vegetable cultivars are screened by breeders for disease resistance, new ornamental crop cultivars are rarely screened. They are introduced rapidly for novelty in a highly competitive market and left to find their own niche. Diseases are always prevalent, demanding constant fungicide spray programs, which, however, detract from marketability because of visible residues. With areas increasing to monocrop situations, new diseases also appear. Thus, powdery mildew is a recent threat to poinsettia production, and *Phytophthora* sp. to hibiscus.

FUTURE OUTLOOK

Our natural resources such as cultivable land, water, and fossil fuels are limited. This is accompanied by rising populations with increasing numbers of affluent and demanding consumers who insist on improved diets, and home and community beautification. The greenhouse area will continue to increase. Diseases remain the most important limitation for greenhouse crop production. With the increasing public concern of the pollution caused by fertilizer leaching and pesticide escape, reducing pesticide application and recycling wastewater and fertilizer will be required by local or national authorities. This will exacerbate diseases, especially those caused by root-infecting pathogens, in greenhouse crops unless the recycled water is properly disinfested.

See also *Crop Rotations (Plant Diseases)*, pages 172–173; and *Cosmetic Standards (Blemished Food Products and Insects in Foods)*, pages 152–154; *Crop Rotations (Insects and Mites)*, pages 169–171.

REFERENCES

1. Jensen, M.J. Hydroponics worldwide. Acta Horticulture **1999**, *481*, 719–729.
2. Gullino, M.L.; Albajes, R.; van Lenteren, J.C. Setting the Stage: Characteristics of Protected Cultivation and Tools for Sustainable Crop Protection. In *Integrated Pest and Disease Management in Greenhouse Crops*; Albajes, R., Gullino, M.L., van Lenteren, J.C., Elad, Y., Eds.; Kluwer Academic Publishers: Dordrecht, The Netherlands, 1999; 1–15.
3. Wittwer, S.H.; Castilla, N. Protected cultivation of horticultural crops worldwide. Hort. Technol. **1995**, *5* (1), 6–23.
4. Jarvis, W.R. *Managing Diseases in Greenhouse Crops*; APS Press: St. Paul, MN, 1992; 288.
5. Berlinger, M.J.; Jarvis, W.R.; Jewett, T.J.; Lebuish-Mordechi, S. Managing Greenhouse, Crop and Crop Environment. In *Integrated Pest and Disease Management in Greenhouse Crops*; Albajes, R., Gullino, M.L., van Lenteren, J.C., Elad, Y., Eds.; Kluwer Academic Publishers: Dordrecht, The Netherlands, 1999; 97–123.

GROUND SPRAYERS

Mohamed Khelifi
Université Laval, Québec, Canada

INTRODUCTION

Worldwide, chemical pesticides remain the most popular and powerful tool for pest management despite the implementation of other efficient alternative methods. In modern agriculture, most of the pesticides used for crop protection are mixed with water, which acts as the carrier for the chemical material. To apply these chemicals as a spray solution, different types of equipment ranging from small hand-carried compressed-air sprayers to large aircraft sprayers are used. Ground sprayers are the most widely used type of liquid applicators in agriculture. These sprayers are primarily used to control weeds, insect pests, and fungi in field crops. In some cases, they can also be used to apply liquid fertilizers.

TYPES OF GROUND SPRAYERS

A wide variety of ground sprayers are available for applying liquid sprays in agriculture: field sprayers, backpack sprayers, air-assisted sprayers, and air-blast sprayers (1, 2). The most popular ground sprayer is the boom sprayer, which is also known as the field sprayer. There are three types of field sprayers: trailed, mounted, and self-propelled. A field sprayer mainly consists of a tank, a power-take-off-driven pump, a pressure control valve, a spray boom that can be divided into a number of independent sections, flow control valves to supply the boom sections, and spray nozzles and filters (see Fig. 1).

Trailed Sprayers

These sprayers represent the most common type used on medium and large farms. They are mounted on a frame supported by two or four wheels and trailed by a tractor. Tank capacity can reach up to 4000 L, which allows large areas to be sprayed between refilling. Boom length can be in excess of 35 m. These sprayers are usually used to apply most of the pesticides in field crops. They are more stable than mounted sprayers on rough terrain and tractor hitching and unhitching is rapid and easy. On the other hand, trailed sprayers may not be easy to handle under difficult conditions and may cause more damage to the soil and crops when used in rough conditions. They are also more expensive than mounted sprayers.

Mounted Sprayers

Mounted sprayers are entirely supported by the tractor. The tanks may be mounted either on the rear three-point hitch system, at the front of the tractor by means of a frame specially designed for this purpose, or on each side of the tractor (saddle tanks) between the front and rear axle. Sprayers mounted on the back are less expensive, more maneuverable in the field, and easier to hitch to the tractor. However, the capacity of their tank is limited both by the loading capacity of the rear axle of the tractor and by weight transfer considerations. Sprayers mounted at the front allow for the simultaneous operations of other implements: tillage tools for single-pass incorporation of pesticides, seeding equipment coupled with spraying of pesticides or liquid fertilizers, and cultivation tools in row crops. However, the front tires of the tractor have to be larger to avoid excessive soil compaction. Saddle tank sprayers are also very maneuverable in the fields and present similar advantages to those of front mounted sprayers. Their main disadvantage is related to the mounting of the tanks, which can be very time consuming.

Tank capacity for mounted sprayers varies between 500 and 2000 L, which may be divided between one or more tanks. Mounted sprayers are used particularly to spray chemicals over row crops. They require more energy than trailed sprayers having the same characteristics. The lift capacity and the longitudinal equilibrium of the tractor are the major limitations of mounted sprayers. Hitching and unhitching of mounted sprayers is also more complicated than for trailed machines. For this reason, a tractor is often solely dedicated to the spraying operations during a complete spraying campaign.

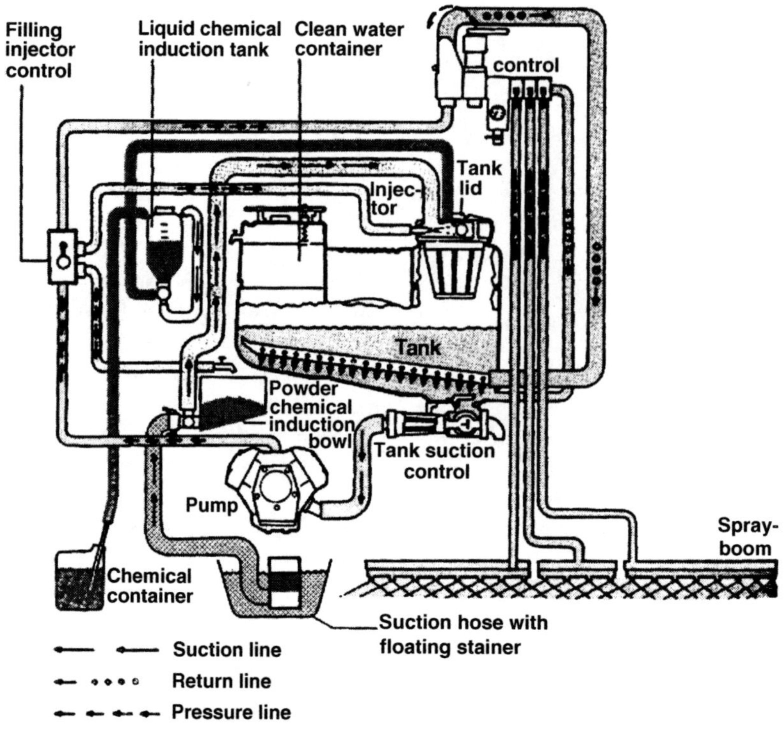

Fig. 1 Schematic diagram of the layout of a hydraulic sprayer. (From Ref. 2.)

Self-Propelled Sprayers

These high capacity machines are appropriate for large farms, custom operators, and tall crops. They are mounted on a motorized frame supported by tall and narrow tires that minimize soil compaction and reduce the damage inflicted to the crops. Such tires can also allow spraying under soil conditions that would not suit conventional tires. Usually, a wide range of height adjustments is provided for the boom. Self-propelled sprayers are both very autonomous and expensive. They are more adapted to difficult field conditions. These sprayers become non-operational following any mechanical problem to the engine, the gear box, the clutch, etc.

USE OF GROUND SPRAYERS IN AGRICULTURE

The potential for drift of chemical pesticides with aircraft is undoubtedly much greater than with most ground sprayers because of the lower boom height of the latter (3). In addition, the hazards of misuse of pesticides can be particularly acute in aerial applications. Also, aircraft operations are highly dependent on optimum weather conditions and are less efficient and more costly on relatively small areas. For these reasons and because of the great increase in public awareness and concern about environmental pollution, more than 80% of agricultural pesticides are applied by ground sprayers.

ON-TARGET SPRAY DEPOSITION FROM GROUND APPLICATIONS

Ideally, all the sprayed pesticide should land on the intended target rather than elsewhere. This is not the case in the field under real conditions (4). Indeed, some droplets released from the nozzles miss the target and end their trajectory on the ground. Other droplets hit the target and rebound as fine droplets under the impact of the high release pressure. These droplets can be easily carried away from the intended target or evaporated by the prevailing weather conditions. Also, small droplets carried by air may be deflected from their original course when the air splits near the target to flow around the obstacle. Furthermore, once the target surface is completely wetted and a certain amount of spray, corresponding to the maximum initial retention, is retained, the surplus quantities run off toward the ground.

The amount of pesticide lost to the ground of the treated area at the time of application has been estimated as a third of the spray applied to the crop. In some cases, even the pesticide that reaches the target could be washed off later by rain or following an overhead irrigation. Some estimates have suggested that up to 80% of the total pesticide applied to plants may eventually reach the ground (5). This can be easily conceived in the case of broadcast chemicals uniformly distributed in a field where only about 20% of the area is infested by weeds. In the case of fruit trees, for example, it was reported that as much as 80% of the spray was found on the ground beneath the trees when they were dormant against only 10% when they were in full leaf. This loss is caused not only by runoff, but also by spray passing between the branches. In some experiments, both the fine and the coarse spray were found to be wasteful, but in different ways. Before coming in contact with leaves, 70% of fine spray was lost as drift. On the other hand, almost no drift was observed with coarse spray at the same stage of development; however, 70% of the spray was lost to the ground underneath the trees.

FUTURE CONCERNS

Despite the recent improvements in spraying equipment and techniques, a high proportion of sprayed chemicals still remains out of control as it can easily reach the ground surface, evaporate, or move away from the intended target following spraying operations. Sometimes, it is necessary to completely cover the target area whereas in other cases, the target is only a portion of the plant, soil, or insect surface. In the field, it is very difficult if not impossible to get complete pesticide deposition on the target. The major factors enhancing the loss of pesticides include the density of the foliage, which varies considerably with the crop; the populations of weeds and insects, which are highly variable within the field; and the environmental conditions that alter the droplet movement to the target. Further research is therefore needed to increase the effectiveness of pesticide applications and to reduce to some extent spray losses.

REFERENCES

1. Kepner, R.A.; Bainer, R.; Barger, E.L. *Principles of Farm Machinery*, 3rd Ed.; Avi Publishing Company, Inc.: Westport, CT, 1978; 527.
2. Matthews, G.A. *Pesticide Application Methods*, 2nd Ed.; Longman Scientific & Technical: England, 1992; 405.
3. Gebhardt, M.R. Engineering Aspects of Pest Control. In *Introduction to Crop Protection*; Ennis, W.B., Jr., Ed.; The American Society of Agronomy, Inc., and The Crop Science Society of America, Inc.: Madison, WI, 1979; 267–292.
4. Khelifi, M. *Quantification and Prediction of On-Target Spray Deposition From Ground Applications*; M.Sc. Thesis, University of Guelph, School of Engineering, Guelph: Ontario, Canada, 1991; 226.
5. Courshee, R.J. Some aspects of the application of insecticides. Ann. Rev. Entomol. **1960**, *5*, 327–352.

HERBICIDE ALTERATION OF CROPS

Kelsey Hart
David Pimentel
Cornell University, Ithaca, New York, U.S.A.

H

INTRODUCTION

Herbicides are toxic chemicals designed to kill target weeds without killing the crop. Although herbicides do not kill the crop and may have no noticeable external impacts, still the herbicides can affect the physiology of crop plants. Altering the physiology of the crop plant itself can in turn influence the insects and plant pathogens feeding and attacking the crop plant. These physiological changes can result in altering the levels of amino acids, carbohydrate, protein, water, and nitrogen in a wide array of crop plants treated with recommended dosages of herbicides (1). Changes in the physiology of the crop can alter the plants' susceptibility to insect and plant pathogen attack.

EFFECTS OF HERBICIDES ON INSECT PEST POPULATIONS

Insects and other arthropods make up about 90% of all plant and animal species in nature. Insect interactions are vital to the functioning of both natural and agricultural ecosystems. For instance, insects play a vital role in the following: pollinating both crop plants and natural vegetation; controlling numerous pest arthropods and weeds; degrading and recycling wastes produced by society and natural ecosystems; and providing food for birds, fish, and various other organisms (1). However, even though most insects and related arthropods are beneficial, about 1% of arthropod species is considered pests of agriculture and humans. Despite the application of about 2.6 billion kg of pesticides per year worldwide, these estimated 10,000 species of insect and other arthropod pests destroy about 15% of food and fiber crops worldwide every year (2).

Some herbicides not only affect weeds and crop plants, but can also affect insect species, directly or indirectly. Direct effects of herbicides on insects are less common than indirect effects, but are observed. For instance, Baker et al. treated cotton with the herbicide monosodium methanearsonate and found that the numbers of thrips (*Frankliniella* spp.) and several other insect species were reduced (3). The direct reduction of insect numbers by herbicide exposure can be a problem when the species affected are beneficial insects, such as natural enemy or pollinator species. In no instance were insect numbers *directly increased* by herbicide exposure, but the indirect effects of herbicides on insect species can lead to increases in insect population numbers. As described above, crop plants can be influenced by the recommended application of herbicides, leading to various physiological changes in the plants. These changes in plant physiology caused by herbicide treatment may result in increased insect attack on crop plants (Table 1). Sometimes there can be as much as a threefold increase in abundance of insects on herbicide-treated crop plants. Note that in most of these cases, the herbicides that were applied at recommended dosages on the crops significantly increased insect attacks (1).

However, some reports suggest that insect pest numbers can also be reduced on herbicide-treated crop plants, again depending on the type of plants, herbicides, and insect species involved. The reduced insect numbers in these cases can be attributed to an increase in natural plant toxicants, such as cyanide and potassium nitrate, that were stimulated by crop exposure to herbicides (1).

In general, the response of insect populations to direct herbicide exposure seems to be a decline in numbers for some species exposed to certain herbicides. Insect population responses to indirect herbicide effects, including the physiological changes induced in herbicide-stressed plants, vary with the species of insect and particular herbicide applied. Sometimes insects seem to prefer to feed on the herbicide-treated vegetation—in some cases the exposed plants appear more nutritious and as a consequence, insect populations can increase to outbreak levels. However, in other cases, the physiological changes provoked in the herbicide-stressed plants make the plants less attractive to insect pests, and insect numbers on these plants typically decrease. The responses of insect populations to indirect herbicide exposure are most often

Table 1 Susceptibility to insect attack in plants exposed to various herbicides

Herbicide	Insect species	Plant species	Attack
Aciflurofen	*Geocoris punctipes*	Soybean	Increased eggs/hatch
Banvel D	English grain aphid	Barley	Increase
Banvel D	Oat bird-cherry aphid	Barley	Increase
Barban	English grain aphid	Barley	Increase
Barban	Oat bird-cherry aphid	Barley	Increase
Barban	Greenbug	Barley	Increase
Bentazon	*Geocoris punctipes*	Soybean	Increased eggs/hatch
Caragard Combi	Spider mite	Grape	Outbreaks
2,4-D	Sugarcane borer	Sugarcane	Increase
2,4-D	Wireworm	Wheat	Increase
2,4-D	Grasshopper	Weeds	Nymphs abundant
2,4-D (sublethal)	Aphid	Fava bean	Increase
2,4-D (sublethal)	Aphid	Broad bean	Increase
2,4-D	Rice stem borer	Rice	Increase
2,4-D	Corn leaf aphid	Corn	Threefold increase
2,4-D	Corn borer	Corn	28% increase
2,4-D	Yellow stem borer	Rice	Increase
Fluazfopbutyl	*Ceratoma trifurcata*	Soybean	Larvae abundant
Fluazfopbutyl	Corn earworm	Soybean	Larvae abundant
Fluazfopbutyl	Mexican bean beatle	Soybean, bean	Treatments preferred
Gramoxone	Spider mite	Grape	Outbreaks
MCPA	English grain aphid	Barley	Increase
MCPA	Oat bird-cherry aphid	Barley	Increase
Melfluidide	*Ceratoma trifurcata*	Soybean	Larvae abundant
Melfluidide	Corn earworm	Soybean	Larvae abundant
MSMA	Whitefly	Cotton	Increase
MSMA	Lygus bug	Cotton	Increase
Reglone	Spider mite	Grape	Outbreaks
Semparol	Spider mite	Grape	Outbreaks
Sethoxydim	*Ceratoma trifurcata*	Soybean, bean	Treated leaf preferred
Sethoxydim	Corn earworm	Soybean	Larvae abundant

(From Ref. 1.)

determined by the particular herbicide, plant species, and insect species involved.

EFFECTS OF HERBICIDES ON PLANT PATHOGENS IN CROPS

Similar to insects and other arthropods, treating crops with recommended herbicides can increase plant pathogen attacks in crops by altering the physiology of crop plants (4, 5). For example, treating field corn with recommended dosages of 2,4-D resulted in increasing the susceptibility of the corn to black-smut disease. In several of the cases, the corn smut disease organisms were five times larger than normal. The protein nitrogen level in the 2,4-D treated corn was found to be significantly higher. It has been reported that higher protein-nitrogen levels in crops has been found to increase the susceptibility of corn to smut and other diseases.

Another interesting finding with corn treated with recommended dosages of the herbicide, 2,4-D, was that the corn variety that was normally resistant to the Southern corn-leaf blight disease was now susceptible to the leaf-blight pathogen. The 2,4-D altered the physiology of the corn plant to make it more susceptible to the Southern corn-leaf blight. Whether this susceptibility was due to the high level of protein-nitrogen in the herbicide treatment or some other factor in the corn is unknown.

CONCLUSIONS AND FUTURE DIRECTIONS

Herbicides that have altered the physiologically of crop plants, even though the herbicides have no noticeable effects on crop plants, have in some cases made crops more susceptible to insect and other arthropod pest attacks as well as made crops more susceptible to plant pathogens. The physiological changes that make crop plants more susceptible to insect pests and plant pathogens are not understood because little biochemical research has been directed at investigating the causes of the insect and pathogen outbreaks in crops.

Another major concern related to insect pest and plant pathogen pest attacks due to the recommended use of herbicides in crops is the lack of investigations of herbicide impacts on crop pest problems. The reason for the lack of investigation is that weed control specialists seldom, if ever, examine the impacts of herbicide treatments on either insect pests or plant pathogen pests. At the same time, entomologists and plant pathologists seldom, if ever, examine the impacts of herbicides on insect pests and plant pathogens. Weed specialists, entomologists, and plant pathologists appear to remain quite narrow in their investigations and do not take a broad view concerning the impacts of chemicals, including pesticides, on the crop ecosystem.

It is hoped that in the future that weed specialists, entomologists, and plant pathologists will begin to take a broader perspective and investigate crop ecosystem changes, including those caused by the recommended use of herbicides in crops.

REFERENCES

1. Pimentel, D. Insect population responses to environmental stress and pollutants. Environ. Rev. **1994**, *2*, 1–15.
2. Pimentel, D.; Hart, K.A. Ethical, Environmental, and Public Health Implications of Pesticide Use. In *Perspectives in Bioethics*; Galston, A., Ed.; Johns Hopkins University Press: Baltimore, MD, *in press*.
3. Baker, R.S.; Laster, M.L.; Kitten, W.F. Effects of herbicide monosodium methanearsonate on insect and spider populations in cotton fields. J. Econ. Entomol. **1985**, *78*, 1481–1484.
4. Oka, I.N.; Pimentel, D. Herbicide (2,4-D) increases insect and pathogen pests on corn. Science **1976**, *193*, 239–240.
5. Oka, I.N.; Pimentel, D. Ecological effects of 2,4-D herbicide: increased corn pest problems. Contribution Center Research Institute of Agriculture. Bogor **1979**, *49*.

HERBICIDE-RESISTANT CROPS

Stephen O. Duke
United States Department of Agriculture, University, Mississippi, U.S.A.

INTRODUCTION

The term herbicide-resistant crops (HRC) (sometimes termed herbicide-tolerant crops) has come to mean crops that have been genetically altered by biotechnology to be resistant to herbicides to which they are normally susceptible. Until the advent of plant biotechnology, selective herbicides were designed to kill important weed species while causing limited injury to major crops. Nonselective herbicides that kill almost all plant species were used only at times and places where crop injury was not a concern or with complicated application methods that avoided contact with the crop. There was little success in breeding crops for herbicide resistance especially to nonselective herbicides. Crops with genetics altered by biotechnology to impart herbicide resistance now offer the farmer valuable new tools for weed management. During the past few years, herbicide-resistant canola, cotton, maize, and soybeans have been widely adopted in North America and a few countries outside of North America. This technology has had many critics who have pointed out an array of environmental, toxicological, and societal risks.

CURRENT IMPACT ON WEED MANAGEMENT

The largest segment of the transgenic crop market has been HRCs. Several HRCs are currently available in North America (Table 1). At this time, the most widely utilized HRCs are those that are resistant to two nonselective herbicides, glyphosate (e.g., Roundup®) or glufosinate (e.g., Basta®). Glyphosate-resistant crops in particular have been widely adopted in cotton, soybean, maize, and canola in North America. In 1999, 55 and 37% of the soybean and cotton acreage, respectively, in the United State was planted with glyphosate-resistant varieties. An even larger proportion of the soybean crop in Argentina was glyphosate resistant. The use of glyphosate-resistant maize grew from 950,000 acres in 1998 when it was introduced to 2.3 million acres in 1999. The rapid adoption of glyphosate-resistant crops in the United State (Fig. 1) indicates that farmers find this trait to be very valuable. Other HRCs, such as bromoxynil-resistant cotton (Fig. 1), have been useful in situations with special weed problems. At this time, HRCs are not available to European farmers because of public resistance to their use.

RISKS AND BENEFITS

Generalities regarding risks and benefits of HRCs are difficult to make, as what is true for one HRC can be quite different for another, and even different for the same crop at another place or time. Furthermore, risks and benefits must be considered within the context of current and predicted future farming practices. These products are relatively new, and there are relatively few data to support predicted risks and benefits. Nevertheless, an attempt will be made to point out likely potential benefits and risks of particular HRCs. Many of these risks are being considered by regulatory agencies in their regulation of HRCs.

Benefits and Risks for the Farmer

A major benefit of the HRCs that are resistant to nonselective herbicides (glyphosate and glufosinate) is that the herbicide kills all or almost all weeds. Thus, in these crops, one herbicide can substitute for several selective herbicides that were needed to manage an array of weed species. Furthermore, glufosinate and glyphosate are used as foliar sprays after the weeds have appeared. Theoretically, the farmer can avoid prophylactic herbicide treatments and only rely on the nonselective herbicide after the weed problem appears. Some weed species, however, require relatively high rates of glyphosate for adequate control. In these cases, farmers are finding that the most efficacious weed management with glyphosate-resistant crops sometimes requires use of a selective herbicide with glyphosate.

Perhaps one of the most attractive features of being able to apply nonselective herbicides directly on the crop is that it greatly simplifies weed management, eliminating or reducing the need for tilling, for applying preemergence herbicides, and for decisions as to which selective

Table 1 Herbicide-resistant crops available in North America

Herbicides	Crop	Year	Resistance mechanism
Bromoxynil	Cotton	1995	Enhanced degradation
Sethoxydim[a]	Maize	1996	Altered target site
Glufosinate	Maize	1997	Altered target site
	Canola	1997	Altered target site
Glyphosate	Soybean	1996	Altered target site
	Canola	1997	Altered target site and enhanced degradation
	Cotton	1997	Altered target site
	Maize	1998	Altered target site
Imidazolinones[a]	Maize	1993	Altered target site
	Canola	1997	Altered target site
Sulfonylureas	Soybean	1994	Altered target site
Triazines[a]	Canola	1984	Altered target site

[a]Not transgenic.

postemergence herbicides should be used. Management simplicity favors the small farmer who cannot afford crop protection consultants.

Despite the fact that U.S. farmers have had to pay for both the herbicide and a technology premium for the HRC seeds, glyphosate-resistant crops have significantly lowered the cost of weed management. In soybeans, these costs have been lower than conventional weed management costs, resulting in the prices of herbicides for use in non-HRC soybeans being substantially lowered. Thus, HRCs have lowered the cost of weed management for all soybean farmers, whether or not they plant a glyphosate-resistant crop.

Many selective herbicides are not entirely selective, causing some phytotoxicity to the crop at certain doses under some conditions. Farmers have learned to accept this because the crop usually outgrows the effect, and there is rarely any significant crop loss. Nevertheless, farmers prefer to have no crop injury from herbicides. HRCs eliminate or greatly reduce crop injury by herbicides at early stages of development. Whether occasional developmental abnormalities in later stages of glyphosate-resistant cotton are due to glyphosate or not has been a contentious issue.

Both glyphosate and glufosinate have activity against some fungi and microbes. There have been reports that, in addition to killing weeds, glufosinate can reduce certain plant pathogen damage to some HRCs. This type of unpredicted benefit has been understudied. There are some potential problems for farmers with HRCs. If a farmer rotates HRC crops (e.g., maize after soybeans) that are resistant to the same herbicide, the unharvested seed of the previous crop can result in a serious weed problem. Evolution of resistance to herbicides is a growing problem, although not to the extent of insecticide or fungicide resistance. Evolution of resistance to glyphosate has not been a significant problem, despite the heavy use of this herbicide over a long period. A bigger problem has been weed species shifts in glyphosate-resistant crops to those species that require higher doses for adequate management (e.g., *Amaranthus rudis* in soybeans). In some crops, the transgene may be introduced into a sexually compatible weedy relative (introgression), creating the need to use additional herbicides. This has not been reported yet, but it will occur eventually if reproductive barriers are not incorporated into certain HRCs.

Glyphosate and glufosinate are commonly sprayed over the tops of the HRCs as a foliar spray. Spray drift to nontarget plants, including other crops, has been a problem since selective herbicides such as 2,4-D were introduced. The potential adverse impact of spray drift is increased when nonselective herbicides are used, in that only the transgenic cultivars of the crop are resistant. The potential of a severe herbicide application error is compounded when the HRC and non-HRC varieties are grown in close proximity.

Adoption of HRCs largely has been driven by short-term economic advantage for the farmer. As mentioned above, the replacement of other herbicides by glyphosate has reduced the value and price of competing herbicides. Furthermore, the price of glyphosate has steadily declined due to the expiration of its patent. Herbicides are the largest segment of the pesticide market. Thus, a major portion of

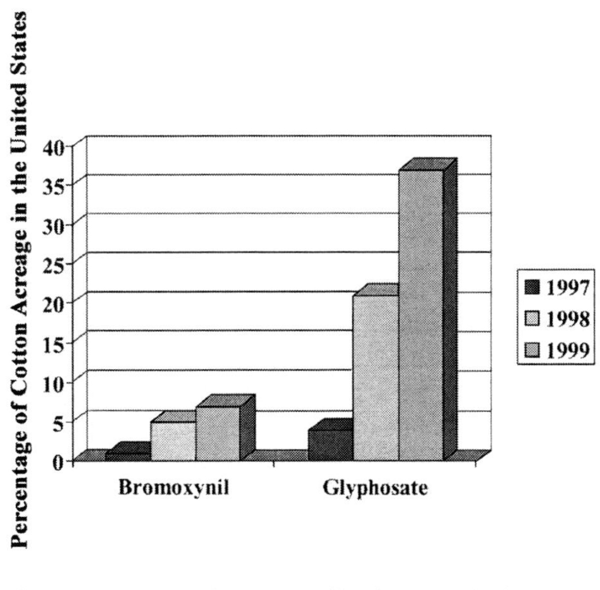

Fig. 1 Adoption of bromoxynil-resistant and glyphosate-resistant cotton in the United States during the last three years of the twentieth century.

the pesticide market has been significantly devalued, resulting in an escalation of the horizontal integration of the pesticide industry. Fewer companies and the devalued herbicide market will ultimately result in fewer herbicides from which to choose. The impact of this situation on farmers' abilities to cope with new weed problems and on the development of nonchemical weed management alternatives is difficult to predict.

Environmental Benefits and Risks

Other than removing land from its natural state, the primary long-term environmental damage of most agriculture has been soil erosion due to tillage. The soil that moves from plowed fields during rainfall events is often contaminated by pesticides, contributing to surface water contamination. The biggest hindrance to adoption of reduced and no-tillage agriculture has been inadequate weed management. The postemergence herbicides to which HRCs have been engineered allow farmers to reduce or, in some cases, eliminate tillage, thereby reducing soil erosion.

The leading HRCs are those resistant to glyphosate and glufosinate. These herbicides are among the most environmentally benign herbicides available. Both are amino acid analogues that degrade rapidly in the environment. Glyphosate is virtually inactivated upon contact with soil, due to its ability to bind soil components strongly. Both are toxicologically safer than most of the products that they replace, despite being relatively high dose rate herbicides.

A transgene that confers herbicide resistance represents a new potential threat to the environment. In some crops such as canola (*Brassica napus* L.), the transgene can introgress into weedy relatives. The herbicide resistance transgene confers no advantage to the weedy relative in a natural ecosystem. It can, however, favor the introgression of other transgenes of clear value in the wild (e.g., Bt toxin) that are packaged with the herbicide resistance gene. In the agricultural setting, all offspring of the weedy relative that are not a result of a cross are killed by the herbicide. Such a process could eventually lead to movement of trangenes of great survival value into natural populations, leading to significant natural ecosystem disruption. Reproductive barriers can be engineered into crops with introgression potential.

The vast majority of cropland in North America is devoted to agronomic crops and weed management in these crops is almost completely dependent on herbicides. Herbicides are expensive, but there are no economical alternatives to herbicides for weed management in these crops on the horizon. Thus, HRCs have simply substituted one herbicide for another. Studies done so far show little or no overall reduction in herbicide use rate (mass per unit area) with HRCs. However, the herbicides used with the most accepted HRCs are generally less environmentally suspect than the herbicides that they replace. For example, in maize and soybeans, glyphosate and glufosinate replace herbicides such as triazines and cloroacetamides that have generated environmental and toxicological concern.

THE FUTURE

Without public opposition, availability of currently available and future HRCs would eventually result in almost universal use of these products in all major crops. Growth of the adoption of HRCs will, however, depend on more than their value to the farmer. In a world economy, the rejection of transgenic crops by the European public could have a profound influence on their utilization in exporting countries such as the United States, even if there is relatively little opposition to their use where they are grown. Public opinion where they are now accepted could change. Whether HRC use increases or decreases is unlikely to significantly influence reliance on herbicides for weed management in major crops. New technologies such as precision agriculture and decision aid programs for weed management will, however, reduce both the volume of herbicides used and their environmental impact.

BIBLIOGRAPHY

Dekker, J.; Duke, S.O. Herbicide-resistant field crops. Adv. Agron. **1995**, *54*, 69–116.

Herbicide-Resistant Crops—Agricultural, Environmental, Economic, Regulatory, and Technical Aspects; Duke, S.O., Ed.; CRC Press: Boca Raton, FL, 1996.

Dyer, W.E.; Hess, F.D.; Holt, J.S.; Duke, S.O. Potential benefits and risks of herbicide-resistant crops produced by biotechnology. Hort. Rev. **1993**, *15*, 367–408.

Hess, F.D.; Duke, S.O. Genetic Engineering in IPM: A Case Study. Herbicide Tolerance. In *Emerging Technologies for Integrated Pest Management. Concept, Research, and Implementation*; Kennedy, G.G., Sutton, T.B., Eds.; American Phytopathology Society Press: St. Paul, 2000; 126–140.

HERBICIDES

Malcolm D. Devine
Aventis CropScience Canada Co., Saskatoon, Canada

H

INTRODUCTION

Chemical herbicides have had a tremendous impact on cropping practices, particularly in the developed world, by reducing the need for hand labor in weed control. The first chemicals used on a large scale for weed control in agriculture were inorganic compounds such as salt (sodium chloride), sulfuric acid, and arsenicals. In the 1920s, crude organic compounds (e.g., diesel oils) were first used. A major breakthrough came in the 1940s, with the development of selective herbicides (2,4-D and MCPA) for weed control in cereal crops. This ushered in the era of modern crop protection, using chemicals to provide effective weed control in a diverse range of crops.

HERBICIDE DISCOVERY

Traditionally, herbicides have been discovered through large chemical synthesis and screening programs. Given the high cost of discovery and development of new compounds, efforts have focused on herbicides that target major weeds in major world crops (e.g., corn, rice, wheat, soybean, cotton). However, these compounds often find uses in minor crops, also. The process normally involves a step-wise progression from greenhouse screening on a few crop and weed species, to more extensive indoor testing, and eventually to field trials in many locations examining the interactions between different weed species, soil types, weather conditions, etc. Toxicological testing and formulation improvement proceed concurrently, to ensure that regulatory and efficacy requirements are met.

A recent innovation is the use of combinatorial chemistry to identify lead compounds. Rather than being synthesized and evaluated in isolation, compounds are produced and screened as mixtures. When combined with in vitro screens (activity testing at the biochemical or cellular level) rather than whole-plant assays, this can allow the testing of 20- to 50-fold more compounds per year than a traditional herbicide discovery program.

"Rational" discovery of herbicides involves identification of a candidate herbicide target site in the plant, followed by design of inhibitors that specifically block that target. While this has led to the discovery of some novel enzyme inhibitors, to date no commercial herbicides have been discovered through this approach. One difficulty is that compounds predicted to have high activity might not penetrate the tissue satisfactorily or may be rapidly degraded inside the tissue.

Finally, herbicides can be developed from bacterial, fungal, or plant toxins. One commercial herbicide, glufosinate (=phosphinothricin), was developed from the bacterial toxin bialaphos from *Streptomyces hygroscopis.*

HERBICIDE CLASSIFICATION

Method or Timing of Application

Preplant incorporated herbicides are applied to the soil surface and mechanically incorporated into the upper 5–10 cm of the soil, in order to minimize photodecomposition and volatilization losses. Preemergence herbicides are applied to the soil surface and often rely on precipitation or soil moisture to transport them to the plant root or shoot for uptake. Postemergence herbicides are applied to exposed foliage after the plants have emerged.

Chemical Structure or Mode of Action

Herbicides within the same structural family usually have the same mode of action, with varying degrees of activity or selectivity depending on the structural variations between compounds. However, herbicides from different chemical families can have the same mode of action (see later).

COMPONENTS OF HERBICIDE ACTION

To be effective, herbicides must penetrate the tissue and reach the target site in sufficiently high concentrations to block its activity. The overall process of herbicide action can be separated into the following components (1):

Absorption: Herbicides can be absorbed by the roots or directly into the leaves. Root uptake occurs through mass flow of herbicide in soil moisture and is a function of root distribution in the soil, soil moisture status, the physical properties of the soil, and the behavior of the herbicide in the soil. Foliar absorption occurs following application to the leaves. To facilitate entry through the cuticle (the waxy layer on most leaf surfaces), herbicides are usually formulated with an array of inert ingredients including surfactants, emulsifiers, etc. Once inside the tissue, further penetration through cell walls and membranes usually occurs by simple diffusion.

Translocation: Long-distance transport of herbicides in the plant can occur in the xylem and/or phloem. In some instances, distribution through the plant is a critical component of overall activity. The amount of translocation depends on the plant stage of development, the physicochemical properties of the herbicide, and the rate at which herbicide injury slows down the movement of endogenous compounds.

Metabolic degradation: Plants have evolved various enzyme systems to detoxify potentially harmful compounds. Fortuitously, some of these enzymes (e.g., cytochrome P450 monooxygenases, glutathione S-transferases) can degrade herbicides to inactive compounds.

Interaction at the target site: Finally, all herbicides must interfere with some critical process in the plant. In most cases this involves binding to a protein (usually a functional enzyme, or a transport or structural protein) so that it cannot carry out its normal function. Over time, through the combined effect of this direct action and other indirect actions, the plant dies.

HERBICIDE SELECTIVITY

Most herbicides are selective, that is, they kill some plant species but not others. A few herbicides, on the other hand, are nonselective, and kill essentially all species. Selectivity is normally based on one of the following (1):

Failure to absorb the herbicide, due to either selective placement (e.g., directed spray on weeds growing between the crop rows) or failure of the herbicide to reach the roots of deep-rooted crops, while killing shallow-rooted weeds.

Enhanced rate of metabolic degradation of the herbicide in the crop. This is the most common basis of selectivity of agricultural herbicides. After entry into the plants, the crop metabolizes the herbicide to inactive compounds, whereas degradation does not occur or is slower in the susceptible weeds.

Differential sensitivity of the target site. This is particularly common in the case of herbicide-resistant weeds (see later).

HERBICIDE MODE OF ACTION

Herbicides kill plants by interfering with an essential process in the plant. The major modes of action of herbicides, the biochemical target sites, and some examples of chemical groups that interfere with those targets, are shown in Table 1.

In addition, many herbicides have been identified that interact with other, unique target sites in plants. However, most of the herbicides that have been developed over the past 50 years target about 15 distinct molecular targets.

HERBICIDE RESISTANCE IN WEEDS

The repeated use of herbicides can lead to the development of herbicide-resistant weed populations. The use of herbicides per se does not create herbicide resistance. Rather, the continuous selection pressure through herbicide use creates a niche, allowing resistant individuals to survive and increase in the population as susceptible weeds are killed. Thus, starting with an initial population of perhaps one resistant weed in a population of 10^6–10^9, resistance builds up over time until the resistant weeds become predominant in the field.

Herbicide resistance in weeds was first observed in the late 1950s, but was a minor problem until the mid-1970s, when resistance to triazine herbicides became a widespread concern. Since then the occurrence of resistance has increased dramatically, with >200 cases now reported (2).

In most cases, resistance is due to a point mutation in the gene coding for the herbicide target site (3). This alters the structure of the target protein in such a way that it can still perform its natural function, but herbicide binding is reduced. This type of resistance often confers cross-resistance to herbicides in the same chemical family or mode-of-action group, although exceptions exist and each case must be analyzed separately. However, target-site mutations do not alter the sensitivity of the weed to herbicides with other mechanisms of action.

Table 1 Modes of action of major herbicide groups[a]

Mode of action	Target site	Representative chemical groups
Inhibition of amino acid biosynthesis		
Branched-chain amino acids	Acetolactate synthase	Sulfonylureas, imidazolinones, triazolopyrimidines
Glutamine synthesis	Glutamine synthetase	Glufosinate
Aromatic amino acid biosynthesis	Enolpyruvylshikimate-phosphate synthase	Glyphosate
Photosynthesis		
Photosynthetic electron transport (PS II)	Q_B or D1 protein	*s*-Triazines, phenylureas, benzonitriles
Photosynthetic electron transport (PS I)	PS I electron acceptor	Bipyridiliums
Pigment biosynthesis		
Chlorophyll synthesis	Protoporphyrinogen oxidase	(Nitro)-diphenylethers
Carotenoid synthesis	Phytoene desaturase and others	Aminotriazole, clomazone
Lipid biosynthesis		
Fatty acid synthesis	Acetyl-coA carboxylase	Aryloxyphenoxypropionates, cyclohexanediones
Fatty acid elongation	''Elongase'' complex	Thiocarbamates
Cell division		
Spindle formation	β-Tubulin	Dinitroanilines, carbamates
Other		
Auxin disruption	Auxin-binding proteins (?)	Phenoxyacetic acids, benzoic acids
Homogentisate biosynthesis	4-hydroxyphenylpyruvate dioxygenase	Isoxazoles

[a]From Ref. 1.

Resistance can also be conferred by elevated activity of the enzyme(s) responsible for herbicide degradation. These weeds can be cross-resistant to other herbicides with different mechanisms of action. Again, the possibilities for cross-resistance have to be analyzed on a case-by-case basis.

Although herbicide resistance has become widespread, in almost all cases alternative control methods are available, through the use of other herbicides, changes in cropping or tillage practices, or some combination of these. Avoidance and management of herbicide resistance has become an integral part of good farming practices in modern agriculture.

HERBICIDE-RESISTANT CROPS

A recent development in selective weed control has been to create resistance to certain herbicides in crops where it did not exist previously. While this has been done primarily to extend the market share of certain products, it offers farmers the advantage of broad-spectrum weed control in crops with a single herbicide application (4). In some cases this has substantially reduced the total amount of herbicide required in a single season.

Herbicide-resistant crops can be produced by three methods (4):

1. Making crosses between the crop (sensitive to the herbicide) and a related, resistant species. This method was used to develop triazine-tolerant canola (*Brassica napus*).
2. Selecting resistant cells in tissue culture, through random mutation or by selecting somaclonal variants, and regenerating resistant plants from these cells. Corn lines resistant to the herbicide sethoxydim were generated in this way.
3. Transfer of an herbicide-resistance gene through genetic engineering. The gene is identified in an unrelated species (often a bacterium), cloned, and transferred into the crop species of interest. These procedures were used to develop canola varieties resistant to the nonselective herbicides glufosinate and glyphosate.

Herbicide-resistant crops have greatly facilitated weed control, but present some new research questions that have had to be addressed. These include the likelihood and long-term ecological consequences of gene flow to related species, and the need to control volunteer plants in the following season(s). These issues do not present insurmountable obstacles but, again, need to be dealt with on a case-by-case basis.

SAFETY AND ENVIRONMENTAL FATE OF HERBICIDES

Environmental safety is of prime concern in the development of new herbicides. This includes an understanding of herbicide toxicology, safe handling and application procedures, and environmental behavior and fate. In most countries, approval of herbicides by the relevant decision-making bodies is dependent on the registrant satisfying the regulatory requirements imposed by those countries.

Toxicological requirements vary from country to country, but usually include data on oral and dermal toxicity in a range of species, in tests of varying duration. Based on the data collected, maximum residue levels are established in food or food products. Data from these tests and field experiments are used to establish maximum application doses and safe intervals between product application and harvest or grazing. Data may also be required on effects on nontarget organisms and ecosystems that may be exposed to low herbicide doses.

Appropriate handling procedures are an important aspect of herbicide safety. The use of appropriate safety clothing (gloves, coveralls, masks, etc.), more benign formulations (e.g., dispersible granules rather than wettable powders), etc., contribute to reduced applicator exposure. Recently, novel formulations have been developed that further reduce applicator exposure when adding products to the spray tank.

Herbicide drift immediately after spraying can be a source of off-site contamination, resulting in injury to adjacent sensitive crops and other species. Various measures, including spraying only under calm conditions, use of wind deflectors, and avoiding very small droplets, can substantially reduce the risk of spray drift.

Herbicides in soil are lost by a combination of microbial and chemical degradation, plant uptake, and, in some instances, leaching or surface run-off. Stringent environmental regulations have been introduced in many countries to minimize the possibility of groundwater contamination. Herbicide residues in soil can provide extended weed control, but also may limit crop rotation options in future seasons. Field research is conducted to establish the risk of such carryover and the effects on future crops.

REFERENCES

1. Devine, M.D.; Duke, S.O.; Fedtke, C. *Physiology of Herbicide Action*; Prentice-Hall: Englewood Cliffs, NJ, 1993; 441.
2. Heap, I.M. International survey of herbicide-resistant weeds. http://www.weedscience.com (accessed April 2000).
3. Devine, M.D.; Eberlein, C.V. Physiological, Biochemical and Molecular Aspects of Herbicide Resistance Based on Altered Target Sites. In *Herbicide Activity: Toxicology, Biochemistry and Molecular Biology*; Roe, R.M., Burton, J.D., Kuhr, R.J., Eds.; IOS Press: Amsterdam, 1997; 159–185.
4. *Herbicide-Resistant Crops. Agricultural, Environmental, Economic, Regulatory and Technical Aspects*; Duke, S.O., Ed.; CRC Press: Boca Raton, FL, 1996; 420

HISTORICAL EPIDEMICS (e.g., IRISH POTATO FAMINE)

Phillip Nolte
University of Idaho, Idaho Falls, Idaho, U.S.A.

H

INTRODUCTION

There have been nearly 50 major famines resulting in significant loss of human life during the time period for which there are written records. The majority of crop failures that result in famines are due to unfavorable environmental conditions with either drought or flooding being the most common causes. Less frequent, but also occasionally significant, are famines due to insect or disease infestations. Insect pests, like locusts, that can destroy crops were readily recognized by ancient civilizations and reliable historical information on insects as causal agents of famine is fairly common. Obtaining historical information on plant diseases, however, is not nearly as straightforward. The Bible and other historical texts refer to blights and blasts that periodically decimated crops, but information on specific diseases is lacking. This is understandable since the vital connection between plant diseases and the microscopic agents, like fungi and bacteria, that cause disease remained unknown until the science of plant pathology was born in the mid-nineteenth century.

PLANT DISEASE AND FAMINE

In spite of these limitations, there is extensive information on several devastating famines that occurred within the past 150 years or so in which crop losses caused by plant diseases were major contributing factors. It is an oversimplification to assume that these famines occurred because the outbreak of severe crop disease meant that there was not enough food. In each of the three major famines discussed herein, there were social and economic circumstances that magnified simple "shortages of food" into full-blown disasters that resulted in wholesale loss of life. In addition, these famines directly influenced much more than large numbers of human deaths. Because of one famine, a war was lost. Because of another, governments were radically changed or even overthrown. This same famine was also responsible for one of the most massive human migration events in history.

The Great Irish Famine

Easily the most famous of all plant-disease-related famines was the Great Irish Famine. A major contributing factor was food shortages caused by failure of the potato crop over several growing seasons due to the late blight disease of potato (*Phytophthora infestans*), a newly introduced fungal disease. The real cause of the famine, however, was that over a 200-year period, from around 1600 when the potato was introduced into Ireland until 1845 when the famine began, the Irish people had gradually been maneuvered into a predicament where potatoes were virtually their sole source of food (1).

The structure of the Irish land management system had evolved into a series of layers with landowners, many of them "absentees" who lived in England, making up the very top layer. These landowners would subdivide their holdings and rent various parcels to middlemen who, in turn, subdivided and rented their parcels to yet another layer of land managers and so on. This system of dividing and subletting had grown to involve as many as five or more layers between the landowner and the final subdivision, small parcels of about one fourth of an acre in size, called "conacre," which were rented to the Irish peasants. Almost all of the agricultural products, including food like grain and cattle, represented income to the various middlemen and wealth to the absentee landlords and these products were exported to the mainland. Two exceptions to this rule were the pig that each tenant family raised and sold to pay the rent and the potatoes that each family grew on their quarter acre of land. Survival under these crushing economic and social conditions meant that the peasants had to eat prodigious amounts of potatoes, typically 8–12 pounds per person per day. The populace, and indeed the entire social system, was ripe for disaster when late blight came (1, 2).

The new potato disease first appeared in Ireland near the end of the 1845 growing season. Damage to the potatoes was substantial and about half of the crop was lost. It was a grim winter with widespread hunger and some starvation but the situation would get much worse.

In 1846 the disease appeared early, the weather was extremely favorable for blight and the potato crop failed totally. The winter of 1846–47 was horrifying with thousands starving and thousands more choosing immigration over almost certain death. The crop totally failed again in 1847 and, if anything, the winter of 1847–1848 was even worse. In a cruel policy difficult to imagine today, grain and cattle exports—wealth for the absentee landlords and middlemen—continued unabated for almost the entire duration of the famine. While hundreds of thousands starved, Ireland continued to export food!

The final number of deaths due to outright starvation or complications due to malnutrition over a five-year period beginning in 1845 are estimated at 1 million or more. Perhaps just as importantly, the potato shortages triggered a wave of mass migration out of Ireland to the United States and Canada. The final tally over the duration of the famine grew to an astounding 1.5–2 million immigrants! In just a few years, the population of Ireland was reduced from 8.5 million to about 4.5 million. The Irish system of land management received a badly needed overhaul and significant changes in English government policy also ensued.

Potato Shortages in Mainland Europe

Over the same five-year period, late blight also repeatedly destroyed potato crops all over Europe causing acute potato shortages on the continent as well. The peasants on the mainland enjoyed a more diversified food base than the Irish but potatoes made up a substantial portion of most diets and were eaten almost daily by vast numbers of poor Europeans. When the potato shortages occurred, starvation was fairly rare but hunger was not. Hunger fueled widespread discontent and the masses of common people became ever more openly critical of their respective governments. The turmoil came to a head in 1848 after several years of potato and food shortages triggered a severe economic depression. The ensuing political backlash led to abdication of the throne by Louis Philippe in France and either armed revolution or drastic changes in government in Hungary, Germany, Austria, and Italy (3).

Potato Famine in World War I Germany

A less publicized but also devastating famine was caused by the very same late blight disease of potato in 1916 Germany during WW I. Here again, the causes of the famine went much deeper than mere shortages of food due to the occurrence of a plant disease in the presence of favorable weather. Germany's wartime footing demanded that the vast majority of the country's products, including most of the food, went to the front lines for the war effort. In a situation very reminiscent of 1845 Ireland, the German people who remained behind at home were maneuvered into a predicament where their main source of food was the potato.

The good fortune of an incredibly large potato harvest in 1915 was marred by the lack of adequate storage facilities which, in turn, meant that a huge portion of the crop was wasted when storage decay set in. Cull potatoes, many of which had late blight, were dumped wherever disposal was convenient, providing sources of the late blight fungus virtually everywhere in the country when the 1916 growing season began. Weather extremely favorable for disease development in 1916 was followed by an epidemic of late blight and total failure of the potato crop. An effective preventative fungicide called "Bordeaux mixture," (copper sulfate and lime) had been in widespread use for late blight management since the 1890s but, tragically, all of the country's copper was earmarked for production of brass shell casings and other war-related uses so the German farmers were denied this effective remedy. Within a few months, more than 700,000 German peasants died from starvation or the complications of malnutrition. Germany's foes had found a most unlikely ally! This event had a profound effect on the morale of the German fighting men and almost certainly hastened the end of the war (3).

The Great Bengal Famine

The events leading to the Great Bengal Famine of 1943 started initially with a savage and very destructive typhoon. The storm and the ensuing flooding resulted in a considerable amount of crop damage. Following these events abnormally wet weather continued, providing environmental conditions perfect for an epidemic of brown spot of rice, a foliar disease caused by *Helminthosporium oryzae* (*Cochliobolus miyabeanus*). To make matters worse, all these problems occurred against the backdrop of a strained, wartime economy. From December 1942 through August 1943, the price of rice more than quintupled. Wages for the vast numbers of poor, meanwhile, did not increase to compensate and most simply could not afford to pay the inflated prices. Hordes of refugees from the countryside descended on the cities looking for food and work. The sudden onslaught of poor and hungry completely overwhelmed any possible effective or timely response by the government. Over the next few months, deaths from starvation and the complications thereof, including multiple outbreaks of cholera, reached truly staggering proportions. The actual number of deaths remains unknown but estimates place the toll at well over 2 million and range as high as 3 million (4, 5).

FUTURE CONCERNS

The causes of famine involve an interconnected combination of diverse elements like weather and biology as well as social and political factors. In the face of inevitable periods of weather unfavorable for crops or outright destructive to them, increasing instances of insect and plant pathogen populations that carry resistance to commonly used pesticides and political strife that limits the free movement of food, devastating famines could occur with frightening speed and regularity, even today. Understanding the complex nature of the great famines of the past can do much to help people avoid such disasters in the future.

REFERENCES

1. Large, E.C. *The Advance of the Fungi*; Dover Publications, Inc.: New York, NY, 1940; 488.
2. Redcliffe, N.S. *The History and Social Influence of the Potato*; Cambridge University Press: Cambridge, England, 1949; 685.
3. Carefoot, G.L.; Sprott, E.R. *Famine an the Wind*; Rand McNally and Company: Chicago, IL, 1967; 229.
4. Padmanabhan, S.Y. The great Bengal famine. Ann. Rev. Phytopathol. **1973**, *11*, 11–26.
5. Sen, A. *Poverty and Famines, An Essay on Entitlement and Deprivation*; Oxford University Press: Oxford, England, 1981; 257.

HISTORY

John H. Perkins
The Evergreen State College, Olympia, Washington, U.S.A.

INTRODUCTION

Pests are simply organisms that appear in places and times that are inconvenient or damaging to people. The invention of agriculture about 10,000 years ago intensified the competition among people and pests. For most of the time since the invention of agriculture, people coped as best they could. In about the mid-nineteenth century, professional scientists began to offer science-based technologies for managing pests. Today pest management science is the foundation for dealing with pests. Pesticides, chemicals that kill or repel pests, have been the major foundation of pest management since about the 1950s, but, despite the benefits, their uses have induced severe ecological consequences and sparked many controversies. Integrated control and integrated pest management were inventions designed to reduce reliance on pesticides in order to avoid their problems.

THE INVENTION OF AGRICULTURE AND PEST PROBLEMS

Uncertainty surrounds our understanding of why humans invented agriculture approximately 10,000 years ago. Encouragement of preferred plants and plant communities, through such means as deliberate fires, undoubtedly preceded agriculture. Farming, however, was a more intense operation in which existing plant communities were destroyed by clearing and new species were planted and protected. Animal agriculture brought the complete life cycle of an animal under control. A great deal of human labor was required for agriculture, especially if a community wanted or needed substantial yields.

Why would people go to the bother of farming? Was it that they saw an easier or more secure or more abundant life with domesticated plants and animals? Or did they conclude that supplies from subsistence hunting and gathering were inadequate to support their population? We may never fully understand the motivations involved in the transition (1), but we can have a clear understanding of the ecological consequences.

First, over long periods of time, the art of agriculture became sufficiently reliable to enable an expansion of the human population and the appearance of "civilization." Today, very few areas of the world support subsistence cultures of hunters, gatherers, and pastoralists. Most people in the world now are dependent on the yields of agriculture for their continued survival.

Second, agriculture increased the yields of nutrients not only for people but also for other organisms, particularly arthropod and microbial species. In addition, the clearing of endemic plant communities for sowing opened up a vast new territory for plants that thrived in disturbed environments. Thus were created the major categories of pests: insects, plant pathogens, and weeds. The ecology of the Earth's surface was substantially transformed by agriculture, and pests were now the constant companions of humans in the never ending efforts to capture the yields of agriculture.

Third, for many centuries the land available for agriculture was more than sufficient to supply both people and pests with their requirements for survival. Thus dealing with pests was predominantly a matter of farming enough land to supply people with their needs, despite the toll taken by pests. Today, however, human populations are sufficiently high that it is difficult to see how yields adequate for survival of all people can be procured without systematic efforts of plant protection. In short, support of the human population in the future will ecologically be partially dependent on pest management.

PEST CONTROL AND THE TRIPLE REVOLUTION: SCIENCE, INDUSTRY, AND CAPITALISM

For most of the existence of agriculture (about 10,000 Before Present to about 400 Before Present), the economy was agrarian, that is, most people farmed and farming provided most of the material wealth of the culture. Some forestry, fishing, mining, trade, and other arts supplemented, but the bulk of a community's riches came from agriculture.

About 400 years ago in Western Europe, a set of transformations completely changed economic life and, with it, pest control (2). New machines and new ways of making metals enabled industrialization. The new industrial processes were themselves linked to a new philosophy of nature, in which humans learned to manipulate natural processes more powerfully, particularly energy resources. Science became a way of looking at the world as matter in motion, not an organic being (3). Complementary to the emergence of industrialization and the new scientific philosophy was the emergence of capitalism, a new way of organizing and controlling the production and distribution of goods.

The interlocked industrial, scientific, and capitalist revolutions had profound implications for agriculture and pest control. Where precapitalist economies demanded sufficiency, capitalist enterprise also demanded efficiency. Where human labor (supplemented by animals) was normal in agrarian economies, supplanting human labor by machines became the norm. Where most human labor was invested in agriculture before, now increasing levels and eventually most human labor would go into non-agricultural pursuits. Where art and folk wisdom sufficed to control pests before, now science-based technologies would be judged necessary. Where agricultural yields sufficient for local subsistence were adequate before, now entrepreneurs sought to increase their wealth by raising large surpluses.

Modern economies thus depended upon and demanded a new level of efficiency in pest control. Farmers came under pressure not only to curtail pests but to control them to the maximum extent possible, based on the costs of control: if the farmer *could* control better and still make money, the farmer *must* control better or risk being driven out of business. Farming was no longer just a way of life; instead it became a highly competitive entrepreneurial business in which pests and pest control were components of the processes involved.

In the mid-nineteenth century, emerging applied life sciences offered the possibility of transforming the art of pest control into a science-based technology. At the same time, the emerging science of chemistry opened the door to making materials never before known. The applied life scientists were quick to perceive that some of these new materials were toxic to pest organisms. If the materials could be made cheaply enough and applied safely enough, then toxic chemicals offered the promise of pest control in ways that met the stringent efficiency demands of the new, industrialized, capitalist economies (4, 5).

PEST CONTROL SCIENCE AND SYNTHETIC PESTICIDES

Insects were the first of the major categories of pest organisms to come under the eye of applied biologists interested in control. By the early part of the twentieth century, American and European entomologists had classified, worked out the life cycles of, mapped the ranges of, and identified salient portions of the natural history and ecology of a substantial number of insects, including major pests of crops and livestock. These pioneer scientists had also organized themselves into regional and national professional societies, begun to staff universities and governmental laboratories, and initiated journals for the dissemination of research (5). Plant pathologists and weed scientists followed this same track in the following few decades. As a result, by the 1950s all of the applied pest control scientists were systematically working on the scientific aspects of pest control for insects, plant pathogens, and weeds.

Use of chemicals to kill pests began as an art form before pest control science. For example, the use of plant extracts such as pyrethrum and rotenone long predated the emergence of organized pest control science.In the 1860s, someone unknown tried Paris Green, an arsenical, against Colorado potato beetles in the American midwest and found that it worked.During the following few decades biologists and chemists produced workable insect-control practices with lead arsenate, calcium arsenate, paradichlorobenzene, and a few other compounds. Despite these successes, however, insect control remained only partially based on the use of synthetic pesticides. Of equal or more importance to entomologists were cultural and biological controls based on a knowledge of the natural history and ecology of pest insects.

Similarly, plant pathology's first successes were a mixture of practices, some based on chemicals and others on an understanding of the biologies of the pests and crop plants. Bordeaux mixture, a combination of copper sulfate and hydrated lime, was first used successfully to control downy mildew on grapes in the 1880s. In the following decades, fungicides began to include organic mercury compounds and the dithiocarbamate fungicides. In many crops, however, especially the grain crops, control of plant pathogens remained a matter of cultural practices and, starting in 1905, the use of genetic resistance to pathogen invasion.

Weed science was the last to emerge as a fully organized discipline. The traditional way of controlling weeds was plowing and hoeing. Weeds appeared like clockwork every year, and every year the farmer resorted to their physical destruction. Chemistry and biology had little to offer this never-ending struggle until the compound 2,4-dichlorophenoxy acetic acid was identified as herbicidal by the early 1940s.

A Swiss invention in 1939 began a profound transformation of pest control practices. Paul Herman Mueller, a staff scientist of the Geigy Company, identified dichlorodiphenyltrichloroethane (DDT) as insecticidal at the very time the entire world was descending into the tragedy of the Second World War. For the following six years, DDT's production and use were guided almost entirely by the war-time programming of the United States. Experimental quantities of the compound were released for civilian experimentation in 1943 and 1944. By the end of 1944, American entomologists concluded that ". . . never in the history of entomology had a chemical been discovered that offers such promise . .." (5).

It is no exaggeration to say that the invention of DDT was a turning point for the use of synthetic insecticides. Where fruits and vegetables had been almost the only arena for use of pre-DDT insecticides, DDT promised cost-effective control in virtually all crops, on livestock, in forestry, in the household, and for public health purposes. DDT's successes were the signal to the chemical industry that molecules with a high probability for commercial success could probably be found. Major companies invested, discovered, and marketed a host of new insecticides by the 1950s. Similarly new fungicides and herbicides of a wide range quickly followed.

Pest control science by the 1950s was virtually synonymous with applied toxicology and the use of synthetic pesticides. Chemical control was the foundation of pest management strategy. The major questions for science were which chemical, what dose, what formulation, how applied, and when applied. Entomologists more oriented to the biological dimensions of insects complained that anyone not working on the use of insecticides risked professional irrelevance. Research patterns shifted, and development of insecticidal methods for insect control rapidly rose to dominate the science of entomology.

Weed science had a similar story, driven largely by the phenoxyacetic acid herbicides and others. Herbicides reduced reliance on plowing and other means of the physical destruction of weeds. Weed science was largely herbicide science. Plant pathology, too, found a plethora of new fungicides and developed many uses for them. In many cases, however, plant pathologists continued to rely on genetic resistance in the crop plant to pathogens, especially on the major grain crops. Either no chemicals were technically suitable, or cost factors continued to demand the use of genetic resistance in place of more expensive fungicides.

THE CRISIS OF SYNTHETIC PESTICIDES AND A NEW SCIENCE OF PEST MANAGEMENT

Economic and technical successes of pesticides notwithstanding, chemicals as the basis of pest management began to collapse in the 1950s. The crisis emerged first among the entomologists, but by the 1970s the plant pathologists and weed scientists were beginning to feel the first winds of a need to reconsider and change. Two major factors drove the entomologists to seek new strategies for insect control (5).

First, numerous species of insects began to show genetic resistance to the chemicals. Entomologists observed the first case of resistance to insecticides in 1908 and articulated a theoretical framework for its importance by 1940. In 1954, resistance in boll weevils on cotton in the American southeast to chlorinated hydrocarbon insecticides created major concerns because no cost-effective alternatives to insecticides existed.

Second, entomologists observed that an insecticide aimed at a target insect also inadvertently killed many predatory and parasitic insects. As a result, the insecticide induced either enhanced outbreaks of the target insect (resurgence) or outbreaks of other insects previously innocuous (secondary pest outbreaks). Counterintuitively, the chemical thought to be the source of a solution to insect pests was itself the cause of a pest problem. Outbreaks of long-tailed mealybug, cottony cushion scale, and citrus red mite in California citrus in the 1940s established that destruction of predatory and parasitic insects was a serious economic problem.

These two problems, resistance and destruction of predators and parasites, were alarming to the entomologists who studied examples of their occurrences. Entomologists realized that pest control strategies based entirely on chemicals had a strong propensity to collapse, which left farmers with no viable alternatives. Collapse also left entomologists with no claim to credible knowledge and thus vulnerability to loss of funding and legitimacy.

In the 1950s, major new initiatives to create new strategies for pest management emerged. One, based in the U.S.Department of Agriculture, aimed to control insect pests by attacking them over their entire range. Sterile male releases, first demonstrated to be effective against the screwworm fly in the 1950s, were frequently at the heart of these wide-area programs. Use of insecticides to depress a target population to low levels before releasing sterile males was also common. Screwworm fly on cattle in the American southwest and boll weevil on cotton in the American southeast were the major venues for developing the first wide-area control programs.

Entomologists aimed for eradication of the insect pest in some wide-area initiatives (6, 7). The experimental evidence indicated substantial depression of the pest population, but claims for eradication remained controversial from the 1970s through the 1990s. High costs, regulatory complexity, and technical disputes about effectiveness plagued wide-area control programs, especially those aimed at eradication.

The second initiative for a new strategy developed from the concept of "integrated control" developed by entomologists at the University of California on alfalfa (5). From the 1940s through the 1950s, the concept was to integrate the use of biological suppression with chemical suppression of the pest. In the absence of crisis (resistance or destruction of predators and parasites), integrated control promised more effective and cost-efficient insect control.

In 1959, entomologists at the University of California were faced with a crisis from the spotted alfalfa aphid. Resistance and destruction of predators and parasites had combined to render both alfalfa growers and entomologists helpless against spotted alfalfa aphid. Entomologists used the concept of integrated control to find a successful combination of an insecticide, demeton, and biological control. Not only was this a technical success against one insect pest, the entomologists also expanded their notions into a fully articulated concept of integrated control. This work is generally cited as the beginning of integrated pest management or IPM.

From the 1960s on, IPM was the strongest theoretical framework that guided entomological research. Many definitions have appeared for IPM, but the heart of the original concept included the need to understand a) the ecologies of pest insects and their natural enemies, and b) the levels of pest populations that were economically worth treating. Biological and chemical controls were the first suppression practices to be "integrated," and in the 1960s genetic resistance to the pest and other methods were added.

Practices in the field by farmers, however, generally lagged behind the research work of the entomologists. Chemicals remained the technology of choice for most growers, because they were cost effective, generally effective biologically, easy to use, and reliable. IPM schemes are theoretically sound, but frequently the farmer was faced with the stark fact that no reliable, easy-to-use scheme existed for a specific insect control problem. A technology that was not reliable or cost effective had little chance of being adopted within the stringent constraints of modern agriculture.

POLITICS, VALUES, AND PEST MANAGEMENT

Resistance and destruction of predators and parasites were the main stimuli for development of both wide-area control and IPM. In the 1960s, however, a new factor joined in to stimulate further pressure for development of alternatives to chemicals. Rachel Carson's *Silent Spring* (1962) transformed the politics of pesticides. Previous to her work, only minor grumblings about the toxicity problems of pesticides were to be heard, and they enjoyed no widespread credibility. Carson changed that situation dramatically.Pesticides also lost social acceptability because of the campaign by the United Farm Workers against the occupational hazards created by the chemicals.

In the decade following Carson's book, the U.S. Congress completely overhauled its regulatory scheme for pesticides.Where previously pesticides were registered easily without extensive safety tests, after 1972 no pesticide could be sold or used unless it had undergone extensive tests for its environmental damages. Since 1954, pesticides leaving residues on food had been required to undergo safety tests for health hazards like cancer, but the new law in 1972 vastly expanded safety testing (8). Protection of occupational safety, however, tended to lag testing for environmental safety (9).

Since 1972, regulatory pressures against pesticides have continued to build. The general public now has a much more mixed view of the chemicals. Where before Carson most people considered pesticides benign and useful, now many harbor deep doubts. In 1993 the National Research Council concluded that pesticide safety testing was inadequate for protecting children (10). This finding led in 1996 to the near unanimous vote in the Congress to pass the Food Quality Protection Act. This new part of pesticide regulation makes uniform the consideration of all health problems and demands special cautions be directed toward issues of children's health.

THE FUTURE

At the end of the 1990s, synthetic pesticides remain entrenched as the technology most often selected by farmers and others to handle pest problems. This generalization holds in virtually all parts of the world. Despite the widespread use of pesticides, resistance is a problem now not only with insect pests but also with weeds and plant pathogens. In addition, political unhappiness with the dependency on pesticides is common throughout the world, and many environmental groups target pesticides as a major problem. Despite a widespread desire to have pest control less dependent upon chemicals, legislative bodies have been slow or unresponsive to the need for research on alternatives. In the meantime, farmers and other pest controllers remain primarily dependent upon synthetic chemicals.

Denmark, Sweden, Canada, Indonesia, and various American states, among others, have tried or considered mechanisms to reduce reliance on pesticides. These efforts have been successful to various degrees, suggesting that substantial reductions in pesticide use are possible. In addition, the Food Quality Protection Act may indirectly promote reduction in pesticide use in the United States.Economic dependency upon pesticides, however, is a strong force acting against rapid change. The fact that the human population in general is also dependent on high yields from agriculture also is a strong force demanding caution in how societies change their dependencies on chemicals. Nevertheless, all evidence indicates that less dependency on synthetic pesticides is both desirable biologically and technically feasible.

REFERENCES

1. Perkins, J.H. *Geopolitics and the Green Revolution. Wheat, Genes, and the Cold War*; Oxford University Press: New York, 1997; 337.
2. Perkins, J.H.; Patterson, B.R. Pests, Pesticides and the Environment: A Historical Perspective on the Prospects for Pesticide Reduction. In *Techniques for Reducing Pesticide Use: Economic and Environmental Benefits*; Pimentel, D., Ed.; John Wiley & Sons: Chichester, UK, 1997; 13–33.
3. Merchant, C. *The Death of Nature. Women, Ecology and the Scientific Revolution*; Harper, San Francisco: New York, 348.
4. Dunlap, T.R. *DDT, Scientists, Citizens, and Public Policy*; Princeton University Press: Princeton, NJ, 1981; 318.
5. Perkins, J.H. *Insects, Experts, and the Insecticide Crisis. The Quest for New Pest Management Strategies*; Plenum Press: New York, 1982; 304.
6. Knipling, E.F. *The Basic Principles of Insect Population Suppression and Management*; U.S. Government Printing Office: Washington, DC, 1979; 659.
7. Dahlsten, D.L.; Garcia, R.; Lorraine, H. *Eradication of Exotic Pests. Analysis with Case Histories*; Yale University Press: New Haven, CT, 1989; 296.
8. Bosso, C.J. *Pesticides & Politics. The Life Cycle of a Public Issue*; University of Pittsburgh Press: Pittsburgh, PA, 1987; 294.
9. Pulido, L. *Environmentalism and Economic Justice. Two Chicano Struggles in the Southwest*; The University of Arizona Press: Tucson, AZ, 1996; 282.
10. Wargo, J. *Our Children's Toxic Legacy. How Science and Law Fail to Protect Us from Pesticides*, 2nd Ed.; Yale University Press: New Haven, CT, 1998; 390.

HISTORY OF BIOLOGICAL CONTROLS

H

William L. Bruckart III
USDA-ARS, Foreign Disease–Weed Science Research Unit, Ft. Detrick, Maryland, U.S.A.

The words friend and foe, auxiliaries and ravagers, are here the mere convention of a language not always adapted to render the exact truth. He is our foe who eats or attacks our crops; our friend is he who feeds upon our foe. Everything is reduced to a frenzied contest of appetites. —J. Henri Fabre.

INTRODUCTION

Pests are organisms "out of place" in society as we know it. They have achieved "pest" status in a number of ways, including introduction into new habitats from international travel (e.g., colonization) and commerce, and through artificial agricultural practices that alter plants and ecosystems. Most major pest species are adventive, having been relocated, accidentally or intentionally, without the full complement of "natural enemies" that help regulate population levels in the pests' native habitats.

The biological control strategy pits one organism against another, ultimately to reduce a pest population to a manageable, i.e., economically or socially acceptable, level. The most important reason for biological control is the need to manage pests that reduce the value or quality of human products or activities. The history of biological control relates directly to this need for pest management (1).

Biological control is a knowledge-based approach to pest management, developed also out of curiosity about the biological processes it attempts to harness. Effective use of this strategy requires: 1) extensive knowledge about pest and biological control agent(s), 2) the interactions that occur between them, and 3) the potential benefit of these interactions in pest management. There also is the challenge of making it work within the economic, environmental, and social constraints of pest control.

MODERN EVENTS SUSTAINING DEVELOPMENT OF BIOLOGICAL CONTROLS

Concern about the environment, particularly the effects of chemical pesticides, has been a major factor sustaining recent development of biological controls (1, 2). Biological control organisms are part of the environment (i.e., natural) and thus do not present the image of "poison" associated with some chemical pesticides. Historically, biological controls were used where chemical and cultural strategies for pest control were inadequate. Issues associated with the introduction of pests into new habitats, such as rapid proliferation, extremely large infestations, and remote or inaccessible habitats of many pests, remain as major factors in sustaining research in biological control. These factors and developments, such as changes in agricultural practices that elevate the status of minor pests and the development of resistance to chemical pesticides, provide challenges anew for pest management and biological control (1).

HISTORY AND SCOPE OF BIOLOGICAL CONTROLS

The biological control strategy has developed through a seamless progression of events. Basic to biological control is the knowledge that some organisms feed on or otherwise debilitate others. This led to the concept that some interactions are potentially beneficial and even practical in pest management (1). Linnaeus observed in 1752 that, "every insect has its predator which follows and destroys it. Such predatory insects should be caught and used for disinfesting crop-plants" (1, 3). Although the concept is not new, it has been called "biological control" only during the last 100 years. It refers to the deliberate deployment of one (or more) organism(s) for pest control, as opposed to the use of chemicals, natural products, or other pest management practices.

Use of predaceous arthropods for pest control in China and Yemen predates development of natural history in Western Europe (1). Formal application of living organisms in modern pest management was not attempted until the middle of the nineteenth century, and the first successful control of Vedelia beetle occurred in the 1880s (see section "Introduced Insect Predators," given later). At this point, "knowledge, need, expertise, and providence came together" (1) giving economical biological control of a major pest. Since then, support of the concept has grown and now biological control is pursued at many research institutions around the world.

Pest organisms of every type, i.e., insects, plants, vertebrates, mollusks, and disease-causing microorganisms, have been targeted for biological control. And pests in every habitat, including aquatic, terrestrial, domestic, and urban environments, have been considered as targets of biological control. To combat this multitude of pests, representatives from all classes of living organisms have been evaluated as candidates for biological control (1, 3).

APPROACHES AND EARLY SUCCESSES

There are a number of approaches to biological control and early successes within each approach are described below. These successes have established biological control among the choices for pest management (1, 3). Success also has inspired pursuit of biological control in the areas of aquatic and forest weeds, plant diseases (root, foliar, flower, and postharvest), urban environments, and plant parasitic nematodes (1, 3–6).

Introduced Insect Predators

The first successful formal application of biological control was against the cottony cushion scale (*Icerya purchasi*), a major pest of citrus in California (United States) by 1886. The introduction of the Vedelia beetle (*Rodolia cardinalis*) from Australia brought the scale under control within two years of release (1).

Introduced Insect Parasites

The sugarcane leafhopper (*Perkinsiella saccharicida*) was causing significant damage to sugarcane production at the turn of the twentieth century in Hawaii. Between 1904 and 1916, six species of egg parasites (including *Anagrus optabilis*) from Australia were released, and these greatly reduced the problem. Full control was realized by subsequent introduction of an egg predator, *Tytthus mundulus* (1).

Introduced Insect Herbivores for Weed Control

Prickly pear cactus (*Opuntia inermis*) was one of several cacti introduced into Australia as an ornamental. By 1925, 24 million hectares of forest and grazing lands were rendered useless by the infestation. A moth, *Cactoblastis cactorum*, was introduced from South America in 1926, and within six years, there was a general collapse of cactus stands and a return of the land to grazing animals. The rapid decline in cactus populations resulted from interactions between the moth and plant pathogenic microorganisms associated with the cactus (1).

Introduced Insect Pathogens

The rhinoceros beetle (*Oryctes rhinoceros*), a major pest of coconut and oil palms, spread from southeast Asia to many of the Pacific islands. A baculovirus of larvae and adults was discovered in Malaysia in 1963. The virus was introduced into Western Samoa in 1967, where it subsequently controlled this pest. The virus has since spread to other islands with similar results (1).

Augmented Insect Parasitoids and Predators: Greenhouse Insect Control

A parasitic wasp (*Encarsia formosa*) of the greenhouse whitefly (*Trialeurodes vaporariorum*), discovered in 1926, was developed and used commercially until 1949. Then control shifted to new chemical insecticides. Improvements in production and shipment of the wasp and a predatory mite (*Phytoseiulus persimilis*), along with insecticide resistance problems, led to reemphasis of these arthropods in greenhouse pest management programs (1).

Augmented Insect Pathogens

Bacillus thuringiensis (Bt), causal agent of a bacterial disease of the flour moth, *Anagasta kuehniella*, was marketed in France in 1938, and thus is the first formulated micrrobial pesticide developed for commercial use. Several strains of Bt have been developed for use against other Lepioptera, and since 1980, against Diptera and Coleoptera (1).

Conservation Biological Control of Insects

In the 1960s and 1970s, changes in Asian rice production, fertilization, and pesticide use, among other practices, altered the status of the rice brown planthopper (*Nilaparvata lugens*) from occasional pest to one of major importance. Modifications in modern rice culture, particularly relating to the use of pesticides, resulted in the return to natural control of this insect (1).

Introduced and Augmented Weed Pathogens

Three fungi were discovered in the 1960s that established the use of plant pathogens in weed management. Evaluations of each in the decade to follow led to their use in biological control of weeds. The European rust fungus, *Puccinia chondrillina*, was developed for biological control of rush skeletonweed (*Chondrilla juncea*) in Australia and, later, in the United States. Significant reductions in stand density were noted within 10 years of release without any adverse effects. Resistance in certain forms of the weed to rust isolates used has been overcome in part by discovery of another strain of the fungus (6).

The other fungi, *Phytophthora palmivora* and *Colletotrichum gloeosporioides* f. sp. *aeschynomene*, are active ingredients in the registered pesticides DeVine (for control of stranglervine, *Morrenia odorata*, in citrus) and Collego (for control of northern jointvetch, *Aeschynomene virginica*, in rice and soybeans), respectively. These were the first microbial herbicides developed for commercial use against weeds, providing effective control of each target weed for the past 20 years (6).

ISSUES AND LESSONS

The use of biological controls, as with other pest management practices, is not risk free; each modern agent is subjected to an extensive risk assessment and regulatory review before use is permitted. Despite this, concerns about biological control persist. With foreign agents introduced for biological control, there is the added fact that release and establishment is an uncontrolled, irreversible process. Risk assessment and safety are the most intensely investigated aspects in the evaluation of candidate agents, and data about safety receive the greatest scrutiny in the regulatory process. This scrutiny, along with standards set by researchers developing the agents, has resulted in an excellent safety record with biological controls.

Biological control is not a panacea. In many situations, other pest management approaches are superior to biological control, e.g., the effective chemical control of multiple weed species in most cropping situations. Biological controls also must be integrated with other practices for successful pest control, particularly in situations where a crop pest and a biological control agent targeted for another pest are closely related. For example, control of a fungal plant disease must be integrated with use of a microbial herbicide for pest control in the same crop.

The inability to predict actual benefit and risk prior to release has presented a great challenge. For this reason, risk assessments have been very conservative, involving an extensive array of nontarget organisms to cover all conceivable scenarios. Foreign agents pose a particular problem because they cannot be field tested at the intended site of use without a permit for introduction. Lacking capability to predict efficacy before release means that foreign organisms must be approved for introduction based on data about safety; efficacy is determined after implementaion.

Advancement in the use of biological controls has been dependent upon a favorable social, economic, political, and philosophical environment. Adequate public support for collection, development, and utilization of biological controls is fundamental to continued success. Widespread acceptance and utilization also remains an issue within agriculture and industry. Actual control of a pest may take longer than with chemicals, and it will likely require modifications both in farming practices (e.g., use of scouts, specialized equipment, and flexibility in scheduling applications) and expectations in pest management. Agents developed for commercial sale also may require modification of industrial practices (i.e., mass production, formulation, delivery, etc.) to bring the product to market.

The chemical paradigm for pest control has been and is likely to remain a standard for efficacy and, to a certain extent, for regulatory decisions. Although the narrow spectrum of activity of many biological control agents provides assurances of safety, it also limits market size, economic returns, and therefore, interest in commercial development.

The ultimate lesson is that biological control works. Equally important is that biological control is safe when properly developed and delivered. Success and safety will remain as the key benefits of biological control of the future, a living component of pest management worldwide.

REFERENCES

1. Van Driesche, R.G.; Bellows, T.S. *Biological Control*; Chapman and Hall: New York, 1996; 1–539.
2. Carson, R. *Silent Spring*; Houghton Mifflin Co.: Boston, 1962; 1–368.
3. *Handbook of Biological Control*; Fisher, T.W., Bellows, T.S., Eds.; Academic Press: San Diego, 1999; 1–1046.
4. Campbell, R. *Biological Control of Microbial Plant Pathogens*; Cambridge University Press: Cambridge, UK, 1989; 1–218.
5. Stirling, G.R. *Biological Control of Plant Parasitic Nematodes: Progress, Problems, and Prospects*; C.A.B. International: Wallingford, Oxon, UK, 1991; 1–282.
6. *Microbial Control of Weeds*; TeBeest, D.O., Ed.; Chapman and Hall: New York, 1991; 1–284.

HISTORY OF PESTICIDES

Edward H. Smith
Cornell University, Ithaca, New York
George G. Kennedy
North Carolina State University, Raleigh, North Carolina, U.S.A.

And he gave it for his opinion, that whoever could make two ears of corn or two blades of grass to grow upon a spot of ground where only one grew before, would deserve better of mankind, and do more essential service to his country, than the whole race of politicians put together.
—Jonathan Swift, 1726

INTRODUCTION

Pests—insects, nematodes, plant pathogens, and weeds—destroy more than 40% of the world's food, forage, and fiber production. The struggle, pests versus people, grows ever more intense as population increases, arable land decreases, and human intervention disturbs biotic relationships on a global scale. Pesticides play a vital role in the struggle, but their use has not been without adverse ecological impacts and risk to the safety of those who apply them and those who consume treated products. This brief essay recounts the human experience with pesticides from the dawn of history to the dawn of the twenty-first century. It is the story of trial and error, old problems, and new lessons on nature's response to insult by ingenious synthetic molecules. It chronicles the intellectual probing and public debate of problems associated with the overreliance on chemical control that developed following World War II and led to adoption of integrated pest management (IPM) as an ecologically viable paradigm for crop protection. The success or failure of pest control programs may well hold the key to world order as the six billion peoples of the world compete for their place in the sun.

THE INEVITABLE CONFLICT

Agriculture, which dates back a mere 15,000 years, requires the modification of natural systems. Agricultural practice imposes ecological simplicity on biota driven by natural selection toward diversity. Agriculture swims against the ecological tide and crops must be protected from the Darwinian struggle for existence. Intervention is required in many forms, including the use of pesticides. The conflict is inevitable.

THE ASCENDANCY OF PESTICIDES

The biblical records provide insight into the philosophy surrounding humankind's encounters with pests. According to Judeo-Christian beliefs, man was accorded dominion over the plants and animals. Departures from the laws of God were punished by plague. ''I have smitten you with blasting and mildew your trees the palmer worm devoured them. Yet have ye not returned unto me,'' (Amos 4:9).

Development of the agricultural sciences progressed slowly. The emergence during the Renaissance of Natural Theology, which reconciled science and religion, followed by invention of the printing press, the microscope, and the Linnean system of biological nomenclature set in place the elements for rational thought and communication on natural history and on pest control.

The status of insect pest control in early nineteenth century Europe and North America is revealed by T.W. Harris's publication, *Report on the Insects of Massachusetts Injurious to Vegetation* (1841), prepared at the request of the state legislature. Harris, a Harvard librarian and meticulous scholar, drew upon European literature as well as American agricultural journals, which published pest control recommendations offered by their readers. His control measures included: hand picking; burning stubble and field refuse; smoke screens in orchards to drive moths away; running pigs in the orchards; poison baits; resistant varieties (wheat); favorable planting dates; dusting with ashes, quick lime, red pepper, sulfur, and tobacco; spraying with whitewash and glue; and encouraging woodpeckers in orchards. (How to do the latter was not specified.) While these measures were crude, they were based on reason and fragmentary knowledge of pest biology. They were free of the ridiculous nostrums proposed earlier out of ignorance, superstition, and fraud, and they represented early steps in cultural, biological, mechanical, and chemical control. A foundation had been laid drawing on knowledge from Europe and North America aided by state subsidy, a renowned educational institution, and an able scholar.

Recognition that abundant and reliable agricultural production was prerequisite to urbanization and industrial development helped to trigger the agricultural revolution in the United States, which began in the 1840s as canals and railroads linked the eastern population centers with expanding agricultural lands west of the Mississippi. The next great impetus to pest control in the United States came through congressional passage of the Morrell Act in 1862, establishing the Land Grant University System. This paved the way for professionals in the applied science of pest control, and was followed in 1887 by passage of the Hatch Act, establishing a coordinated system of State Experiment Stations devoted to the advancement of research. The Smith-Lever act of 1914 officially recognized the extension arm of the Land Grant University System and completed the American model that has proven to be one of the most innovative educational concepts of all time.

Annual reports on beneficial and injurious insects issued between 1856–1876 by early leaders such as Asa Fitch, (New York), B. D. Walsh (Illinois), and C. V. Riley (Missouri) became the backbone of applied entomology. These writers urged natural controls as the first line of defense and expressed their misgivings about the crude chemical controls of the time.

Pest control practices were strongly influenced by expanding commerce, which resulted in the introduction of exotic pests, and by the rapid, westward expansion of agriculture, which disrupted ecosystems and exposed crops to new pests. The Colorado potato beetle, *Leptinotarsa decemlineata* (Say), provides a prime example. It appeared as a devastating pest of potato in Iowa and Nebraska in 1861, having transferred from a native weed to an introduced relative, the potato. The beetle spread rapidly eastward, reaching the Atlantic coast in 1874, despite the use of traditional nonchemical means of control. In 1867, farmers in the west discovered that the Colorado potato beetle could be controlled with Paris Green, an arsenical. Paris Green was in general use by 1880 and became the first widely used pesticide in North America. Similar experiences followed with other major pests, such as the plum curculio *Conotrachelus nenuphar* (Herbst), boll weevil *Anthonomous grandis grandis* Boheman, gypsy moth *Lymantria dispar* (Linnaeus) and others.

During the first half of the nineteenth century, lime-sulfur and wettable sulfur gradually came into use for control of fungal pathogens, primarily of fruit trees and grapevines in Europe and the United States. In 1885, Pierre Milardet, professor of Botany at Bordeaux, France, demonstrated control of downy mildew, *Plasmopora viticola,* on grapevines using a mixture of copper sulfate and lime, subsequently known as Bordeaux mixture. The success of Bordeaux mixture led to efforts to improve upon its effectiveness and to the expanded use worldwide of it and its variants.

Other components—petroleum oil, nicotine, pyrethrum, and organomercury fungicides for seed treatment—were soon added to the pesticide arsenal. By 1910, the arsenicals Paris Green, lead arsenate, and calcium arsenate were the most widely used pesticides. Herbicides were notably absent; they did not appear until the discovery of plant growth hormones paved the way for the synthesis of stable synthetic hormone analogues (2,4-D and 2,4,5-T) in the 1940s.

Farmers, their advisors in the fledgling Land Grant Universities, and an emerging chemical industry rallied behind pesticides, especially insecticides, for one pragmatic reason; they provided a degree of reliability in control programs that was absent with other available methods. World War I stimulated pesticide use for food production. It also stimulated the production of insecticides, such as dinitrophenols (DNOC) and paradichlorobenzene (PDB), as by-products of the manufacture of explosives from coal tar.

On the eve of World War II, insecticides were the backbone of insect control but their use was fraught with unease. A host of problems surfaced. Control was marginal. British markets rejected U.S. apples because of the high arsenical residues. There were concerns for the health of workers and consumers. The codling moth *Cydia pomonella* (Linneaus) acquired resistance to arsenicals; excessive pesticide treatments were phytotoxic to foliage causing reduced yields, and there were concerns about the build-up of residues in the soil. To many entomologists, it appeared that they were losing the fight. They clung to the early, idyllic hope for control by natural means but in the crunch of practical experience, they turned to pesticides because they worked, not well, but better than the alternatives.

DDT: DISCOVERY, DEVELOPMENT, AND IMPACT

The discovery and introduction of DDT, while purely a commercial enterprise, became immediately enmeshed in the intrigue and urgency of World War II. The Swiss chemist Paul Mueller, an employee of J.R. Geigy Co., discovered the insecticidal property of DDT in September 1939; this event coincided with the Nazi invasion of Poland. DDT found a vital military role in the control of insect-borne diseases. When the war ended in 1945, DDT, the shining chemical sword of World War II, found extensive peacetime use. It was distributed quickly for testing through the well-organized network of Agricultural Experiment Stations. Data poured in confirming the effectiveness of DDT against a wide spectrum of insect pests of agricultural and medical importance. In striking contrast to the prewar pessimism, DDT produced hope that at last the age-old insect scourges could be controlled and perhaps eradicated.

Such optimism had a profound effect on the crop protection sciences. In entomology and weed science especially, research shifted focus away from pest biology and on to pesticide technology. At this point, the birthright of pest control scientists as biologists became endangered. Insecticide use soared, based on the promise of DDT and the related chlorinated hydrocarbon insecticides that followed. New classes of insecticides, the organophosphates and the methylcarbamates, were discovered and exploited. The success of 2,4-D for control of broadleaf weeds stimulated the development and use of chemical weed control. Similarly, the discovery of the dithiocarbamate fungicides during the 1930s led to the development of an array of very effective fungicides and increased fungicide use. All this was catalyzed by a powerful coalition: the chemical industry with its high capitalization and integrated skills in synthesis, testing, and marketing; the agricultural community with considerable political clout; and the Land Grant Universities with their triple mission of teaching, research, and extension. Pesticide use and reliance on pesticides for crop production increased steadily.

REBUFF AND REASSESSMENT

The euphoria that accompanied the dominance of chemical control in the 1950s was short-lived. By the end of the decade, warnings about the adverse effects of pesticides were being expressed by environmentalists and some pest control specialists, but these were largely ignored. There was fear within the crop protection disciplines, especially entomology, that reliance on pesticides was placing agriculture on a "pesticide treadmill." There were problems with resurgence of targeted pest populations and outbreaks of secondary pest populations following destruction of their natural enemies, and with the development of pesticide resistance, all of which necessitated additional applications of pesticides. Similar concerns surfaced regarding control of medical and veterinary pests.

In 1962, Rachel Carson's book *Silent Spring* galvanized public attention on the problems spawned by pesticide use. She made her case with poetic beauty sounding the alarm that "we have put poisonous and biologically potent chemicals indiscriminately in the hands of persons largely or wholly ignorant of their potentials for harm." What had been a debate among scientists became a public debate. Drawing on lessons of the civil rights movement, the antipesticide forces headed by the Environmental Defense Fund turned to litigation in defense of the right of citizens to a clean environment. After long, contentious hearings, the Environmental Protection Agency banned DDT in 1972. This landmark decision placed the issue of pesticides in the forefront of the greatly energized environmental movement.

Increased public activism over environmental and food safety issues, which began during the 1960s and continues today, led to dramatic changes in pesticide regulation and to restrictions on pesticide use in both the United States and Europe. These actions dramatically strengthened the environmental and toxicological standards that pesticides must meet before they can be approved for use. In doing so, these changes provided strong impetus for the development of safer and more environmentally friendly pesticides.

The regulatory framework for pesticides continues to broaden as new knowledge is acquired and perceptions change. For instance, in 1996 the U.S. Congress passed the Food Quality Protection Act, which established more stringent safety standards aimed at protecting infants, children, and other sensitive subpopulations from risks associated with pesticide residues on food. Subsequently, several major food processors imposed their own more stringent tolerances for pesticide residues on the produce that they purchase. The process is expected to continue in response to the ebb and flow of new findings and public concern.

In the late 1950s and early 1960s, growing awareness of the problems associated with pesticide use and the specter of faltering pest control, viewed in the context of decreasing availability of arable land and dwindling supplies of fossil fuel to drive the technology of agribusiness, stimulated a reassessment of pest control. Earlier work in biological control in several countries provided points of departure but it was the intellectual probing in entomology at the University of California (Berkeley and Riverside) that ignited a great debate, which in time involved pest control specialists the world over. The topic of debate was the concept of integrated pest management (IPM). (For more comprehensive treatment of the subject see entry by J. Perkins in this volume.) IPM emphasized that pest problems were under the influence of the total agroecosystem and that not all levels of pest abundance required treatment with pesticides. It also emphasized that pest management should be a multidisciplinary effort based on ecological principles and economic, social, and environmental considerations.

While the concept soon gained widespread acceptance, many factors impeded its implementation. The knowledge base was in most cases inadequate; the research and extension infrastructure required redirection; a corps of private consultants to supplement decision making by farmers had to be recruited and trained; and replacement of broad-spectrum pesticides was slow and costly. Federal and state governments were sold on the soundness of the concept and appropriated funds to overcome these constraints.

Four decades after the initiation of IPM, the steering mechanism for sound employment of pesticides, what is the score? The glass is half full. Great strides have been made on a worldwide scale. IPM has provided the framework to accommodate transition from singular reliance on broad-spectrum, long-residual pesticides to the use of highly selective, short-residual compounds as components of multifaceted crop protection programs, without an increase in losses to pests. Every phase of the University support network—teaching, research and extension—has been altered to reinforce the ecological foundations of IPM. The disappointing aspects are that adoption of programs has been slow, pesticides still predominate in many programs, overall use of pesticides has not declined, and successful interdisciplinary programs are few. Despite ongoing improvements in the characteristics of pesticides, the specter of pest resistance hangs like the sword of Damocles over the utility of pesticides.

Great challenges lie ahead as concepts of sustainable agriculture and the technology of genetic engineering meld with the ever-expanding scope of IPM. IPM has become a unifying catalyst, an intellectual quest that unites producers, plant protection disciplines, agribusiness, regulatory agencies, and the worldwide plant protection community concerned with the production of food and fiber for the six billion peoples of the world.

Genetic engineering technology will have broad application in control of pests of plants and animals. Using this technology, genes from one organism can be inserted into and expressed in totally different organisms. Genetic engineering is producing new kinds of insecticidal peptides and proteins, and is enabling plants, bacteria, and viruses to be used in novel ways to deliver toxins to targeted pests. The same technology has produced plants that are tolerant to broad spectrum, postemergence herbicides. The possibilities seem limitless. (For more comprehensive treatment of this subject see the entry by M.G. Paoletti in this encyclopedia.)

In a remarkably short span of three decades, the science of biotechnology became an applied technology and a new industry, involving new kinds of partnerships between university scientists and entrepreneurs. Overnight, genetically engineered crops were being planted on millions of acres in the United States.

The speed of scientific and technological advance in genetic engineering and the new partnership between universities and industry have given rise to a host of challenging issues: academic freedom in the context of university/industry partnerships; patenting of biological processes; monopolies; economic impact, particularly on developing countries; response of organisms to selective pressure (resistance). The most daunting questions focus upon risk assessment and regulatory procedures addressing the impact of organisms created outside the normal evolutionary pathways on the global biota.

Political debate on these issues has grown in intensity and rancor, first in Europe and then in the United States. While the time frame of debate and acceptance of genetic engineering is in doubt, it is clear that the tremendous pressures to meet food requirements for a world population of nine billion by 2050 are likely to force the incorporation of genetically engineered components into the arsenal of pest control.

The question germane to the present essay is what part will pesticides play in future IPM programs. They will be a vital component but in a modified role. Advances in toxicology, chemistry, biochemistry, physiology, molecular biology, and computer modeling are making possible the tailoring of pesticides to meet IPM requirements, which dictate low mammalian toxicity, high specificity conferring low environmental impact, and low residues on treated products. In the future, pesticides will be used with greater precision, made possible by improvements in pesticide application technology, pest and crop monitoring, weather prediction, and information processing, as well as by better understanding of population dynamics, microbial and weed ecology, and epidemiology.

It is important to note the influence of economics and elevated standards for pesticide potency and safety. These factors have dramatically increased the costs of pesticide discovery and development and have contributed to an internationalization and consolidation of the pesticide industry. They have also resulted in fewer new pesticides being introduced. While the agrichemical industry has not enjoyed a favorable public image in an era of environmental awareness, it should be remembered that it plays a vital role in the multifaceted IPM enterprise.

THE FUTURE

History should illuminate the future. We see pest control as a challenge woven into the economic, political, and social fabric of society. Major factors are shaping the new era of pest control.

Public Attitude

Growing environmental ills will further sensitize the public to problems arising from technology. This will find expression in stricter pesticide regulation and safer pesticides.

Global Commerce

Increased and increasingly rapid international movement of goods and people will intensify the introduction of exotic species and the spread of pesticide-resistant organisms.

Economics

The ever rising cost of developing new pharmaceuticals and pesticides will constrain research and product development, and the use of pesticides in developing countries.

Population Pressure

The environmental stress imposed by rapid growth of the human population will continue to exacerbate problems of agricultural production, including pest control. This is perhaps the most serious problem facing humankind, with no relief in sight.

Throughout the latter half of the twentieth century, pesticides contributed enormously to improvements in the quality and stability of the world's food supply and to the control of devastating insect-transmitted diseases of humans and livestock. Pesticides have also played a central role in fostering environmental awareness and public concern over food safety. The inevitable conflict between humans and pests will grow in intensity as the human population grows and arable land decreases. Pesticides, because of their ease and rapidity of use and the reliability with which they can rein in pest outbreaks, will continue to play an important role in IPM. Lessons having been learned, pesticides of the future will be safer and more environmentally friendly, and will be used more judiciously than in the past. Our crystal ball discerns no "silver bullet" of pest control, rather painstaking refinement of IPM, with further advances in established methods, including pesticides and biological control, and a melding of new technologies such as genetic engineering.

BIBLIOGRAPHY

Adler, E.F.; Wright, W.L.; Klingman, G.C. Development of the American Herbicide Industry. In *Pesticide Chemistry in the 20th Century*; Plimmer, J.R., Ed.; American Chemical Society: Washington, DC, 1977; 39–55.

Brent, K.J. In *One Hundred Years of Fungicide Use, Fungicides for Crop Protection 100 Years of Progress*, Proceedings of The Bordeaus Mixture Centenary Meeting, Smith, I.M., Ed.; British Crop Protection Council Publications: Croyden, UK, 1985 11–22, 1985; Monograph No. 31; 1.

Carson, R. *Silent Spring*; Houghton Mifflin: Boston, 1962.

Cassida, J.E.; Quistad, G.B. Golden age of insecticide research: past, present, or future. Annu. Rev. of Entomol. **1998**, *41*, 1–16.

Howard, L.O. *A History of Applied Entomology*; Smithsonian Institution: Washington, DC, 1930.

Knight, S.C.; Anthony, V.M.; Brady, A.M.; Greenland, A.J.; Heany, S.P.; Murray, D.C.; Powell, K.A.; Shulz, M.A.; Spinks, C.A.; Worthington, P.A.; Youle, D. Rationale and perspectives on the development of fungicides. Annu. Rev. of Phytopathol. **1997**, *35*, 349–372.

Lever, B.G. *Crop Protection Chemicals*; Ellis Horwood: New York, 1990.

Marco, G.J.; Hollingworth, R.M.; Plimmer, J.R. *Regulation of Agrochemicals: A Driving Force in Their Evolution*; American Chemical Society: Washington, DC, 1991.

Perkins, J.H. *Insects, Experts, and the Insecticide Crisis: The Quest for New Pest Management Strategies*; Plenum: New York, 1982.

Zimdahl, R. *Fundamentals of Weed Science*, 2nd Ed.; Academic Press: New York, 1999.

HOUSE DUST MITES

Matthew Colloff
CSIRO Entomology, Canberra, Australia

H

INTRODUCTION

Dust mites are more intimately associated with humans than virtually any other free-living animals. These minute arachnids, less than one-third of a millimeter long, are present in almost every home. They are in our beds, our clothing, and our carpets. The enzymes they produce in their guts in order to digest their food—fungi, bacteria, and shed human skin scales in house dust—are also potent allergens, responsible for triggering attacks of asthma and other allergic disorders. Dust mites are cosmopolitan, unseen, and ubiquitous in human dwellings. They are a major cause of morbidity for people who are atopic, that is, genetically predisposed to allergic reactions (1).

GLOBAL DISTRIBUTION, ABUNDANCE, AND DIVERSITY

The name ''house dust mite'' has been used for those members of the family Pyroglyphidae (order astigmata) that live permanently in house dust. Most of the 47 species of pyroglyphid mites are nest-dwellers or feather-associates of birds. Only *Dermatophagoides pteronyssinus, D. farinae, Hirstia domicola, Malayoglyphus intermedius*, and *Euroglyphus maynei* have been recorded from house dust consistently and throughout the world. *Dermatophagoides siboney* is known only from Cuba to date, and *D. microceras*, misidentified as *D. farinae*, is found predominantly in Europe. Specific allergens have been isolated and purified from all the *Dermatophagoides* species listed above and from *E. maynei*. ''Dust mite'' can also mean any mite species that lives permanently in house dust, regardless of whether it is a pyroglyphid or not. Important nonpyroglyphid dust mites include species traditionally regarded as ''stored products mites'' such as *Chortoglyphus arcuatus, Blomia tropicalis, Glycyphagus domesticus*, and *Lepidoglyphus destructor*.

Global distribution and abundance of dust mites is governed by the constraints imposed by their need to control body water loss. The point at which rates of loss and gain are equal—the critical equilibrium humidity (CEH)—is temperature-dependent. For *D. farinae* it varies from 55% RH at 15°C to 75% at 35°C. The CEH is approximately 70–75% RH at 25°C for all dust mite species investigated so far. Dust mites have evolved water regulation mechanisms that allow them to tolerate quite low humidities, such as uptake of water from unsaturated air via hygroscopic secretions (2). However, data on global distribution and abundance strongly suggest that highest population densities are found in places where macroclimate and, ultimately, indoor temperature and humidity are above the critical equilibrium humidity for most of the year (Fig. 1). Mite populations tend to be less dense in drier continental interiors, at high altitudes, and in subarctic regions and frequency of occurrence may fall from 95–100% of homes to less than 50%. Warm, moist conditions favor population increase and these macroclimatic conditions are typical where the majority of the world population lives: coastal areas, at altitudes of less than 100 m, the tropics, and the subtropics. The favorable microclimate within many homes in temperate regions ensures the range of *D. pteronyssinus* extends well into high latitudes. There is no doubt that in many such parts of the world houses are warmer, moister, and less well-ventilated than they were 30 years ago, partly due to double-glazing, central heating, and insulation.

The distribution and abundance of dust mites between homes is not uniform—houses next door to each other, and of the same design but with different indoor microclimates, can have vastly different mite population densities and species composition. Thus patterns of human exposure to mite allergens will vary accordingly. These different patterns, when extended regionally and globally, translate into epidemiological variables such as the proportion of people who develop mite-mediated allergies, the age at which symptoms are manifest, the severity of symptoms and their morbidity, and the risk of development of allergic diseases in newborn children.

Generally there is a close correlation between mite population density and the concentration of allergens in house dust. However, this is not the whole picture. The

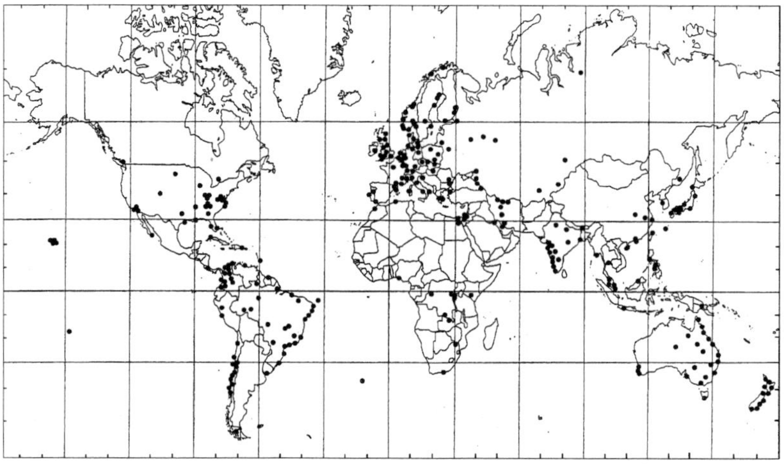

Fig. 1 Global distribution of *Dermatophagoides pteronyssinus*, one of the most allergenically important species of dust mite.

situation is complicated by massive seasonal fluctuations in numbers of dust mites. Concentrations of allergens, (being abiotic) although showing some seasonal fluctuation do so at different rates than mites.

Knowing how water balance and digestion are achieved is, without doubt, the key that is needed to attempt a more complete picture of global distribution of dust mites and the epidemiology of mite-mediated allergy. The kinds of houses the mites live in, their age, construction, design, number of occupants, age of beds and furniture, height of the dwelling above the ground and above sea level, climatic régimes, domestic activities, and patterns of usage of homes, even presence of companion animals and indoor plants have all at one time or another been used to try and explain the distribution and abundance of dust mites. In fact, many of these factors are secondary or tertiary ones that influence not the population of dust mites per se but the domestic microclimate. Humidity and temperature are the primary factors that determine where dust mites live, where they could live, how large their populations become, and how they fluctuate seasonally (3).

ALLERGIC REACTIONS

Symptomatic allergy to dust mites is likely to be experienced only by those people who are atopic, that is, genetically predisposed to become immunologically sensitized to common airborne allergens (such as those derived from molds, mites, pollen, cats and dogs, and cockroaches). If the exposure dosage to the allergen(s) is high enough, people with atopy respond with an allergic reaction, mediated by immunoglobulin E antibody, which may be manifest as inflammation of the skin giving rise to eczema, or of the airways resulting in asthma or rhinoconjunctivitis. There is relatively little global comparative data on the prevalence of atopy, though various population-based estimates quote figures in the region of 20–40%. It is important to note that many people with atopy will be asymptomatic, and that not all asthma is atopic asthma. However, the majority of cases of asthma, perhaps in as many as 80% of children, and 50% of adults, has be attributed to allergens of dust mites. Globally, the prevalence of asthma has been increasing exponentially

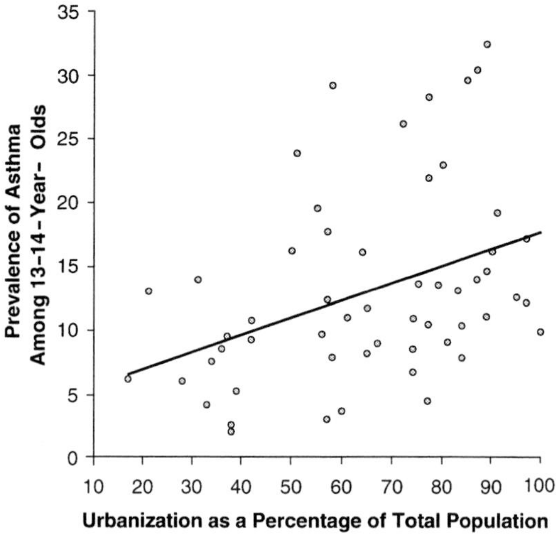

Fig. 2 The relationship between prevalence of asthma amongst 13- to 14-year-old adolescents and urbanization in 56 countries. ANOVA: $F = 9.03$, $P = 0.004$. Asthma prevalence data from Ref. 4; urbanization data from World Bank Group (cf. Table 1).

since the 1960s. The factors responsible for this increase are not known, but are under active investigation by members of the European Community Respiratory Health Survey (ECRHS) and the International Study of Asthma and Allergies in Childhood (ISAAC).

Asthma appears to be of greater prevalence in developed, industrialized societies than in developing countries (4) (Fig. 2), where other factors such as infections with intestinal parasites may possibly override the human immune response to dust mites. There is evidence too that bacterial infections in early childhood may have a protective role against allergic diseases, and that this early immune priming is less likely to happen in modern, urbanized societies with high standards of health care (5). Nevertheless, even though the prevalence of asthma may be lower in developing countries, because of their higher share of world population, the burden of morbidity is as high, if not higher, than in developing countries (Table 1).

About 12 different allergens from dust mites have been isolated and purified. The allergens are mostly associated with the fecal pellets of the mites and consist of serine and cysteine proteinases, amylase, trypsin, chymotrypsin, and other common digestive enzymes (6). During digestion, cells bud off from the wall of the mid-gut, engulf food particles, and travel along the gut lumen breaking down the food as they go (7). The products of digestion are absorbed through the gut epithelium into the haemolymph and the cells, by the time they reach the hind gut, start to dehydrate and die, packaging themselves into fecal pellets

Table 1 The top ten and bottom ten countries for prevalence of asthma amongst adolescent children

Country	Population size (millions)	Size of population aged 10–14 years (millions)	Prevalence of asthma among 13- to 14-year-olds (%)[a]	Estimate of numbers of 10- to 14-year-olds with mite-induced asthma (millions)[b]
United Kingdom	59.1	3.899	32.2	0.879
New Zealand	3.8	0.281	30.2	0.059
Australia	18.8	1.359	29.4	0.280
Ireland	3.7	0.297	29.1	0.057
Canada	30.6	2.075	28.1	0.408
Peru	24.8	2.973	26.0	0.541
Costa Rica	3.3	0.406	23.7	0.067
Brazil	165.9	17.427	22.7	2.769
United States	270.0	19.908	21.7	3.024
Paraguay	5.2	0.644	19.4	0.088
Total	**584.3**	**49.269**		**8.172**
Ethiopia	61.3	8.317	6.2	0.361
India	979.7	110.976	6.0	4.661
Taiwan	22.2	1.595	5.2	0.058
Russia	146.9	11.840	4.4	0.365
China	1238.6	119.880	4.2	3.525
Greece	10.5	0.577	3.7	0.015
Georgia	5.4	0.409	3.6	0.010
Romania	22.5	1.721	3.0	0.036
Albania	3.4	0.376	2.6	0.007
Indonesia	203.7	21.935	2.1	0.323
Total	**2694.2**	**277.479**		**9.361**

[a]From Ref. 4.
[b]Calculated as 70% of the numbers of 10- to 14-year-olds with asthma.
(Adapted from Country population data: World Bank Group. http://www.worldbank.org/data /countrydata/countrydata.html
Population age structure data: U.S. Census Bureau International Database. http://www.census.gov/ipc/www/idbsum.html)

surrounded by a peritrophic membrane. This slightly unusual mode of digestion results in relatively large quantities of enzymes accumulating in the fecal pellets. The pellets, some 20–50 μm in diameter, are egested and accumulate in the textiles that the mites inhabit. As a result of dusting, vacuuming, bed making, or any other activity that causes settled dust to become airborne, the fecal pellets become temporarily suspended in the air and inhaled or deposited on the skin.

CONTROL

Few, if any, methods of dust mite control can achieve total removal of mites and allergens without prohibitive expense. There are sound ecological reasons, primarily relating to the ease of recolonization, why total eradication strategies are ineffective. If one accepts this view, then it follows that the maintenance of dust mite populations at levels low enough to ameliorate clinical symptoms is a practical alternative. Good control has been achieved using physical methods—heating, drying, and freezing—followed by vacuum cleaning to remove dead mites, as well as methods that reduce habitat availability such as removal of carpets. Barrier methods, such as mattress and pillow covers reduce human exposure to mite allergens, and some success has been reported using the chemical acaricide benzyl benzoate (8).

REFERENCES

1. Platts-Mills, T.A.E.; Vervloet, D.; Thomas, W.R.; Aalberse, R.C.; Chapman, M.D. Indoor allergens and asthma: report of the third international workshop. J. Allergy and Clin. Immunol. **1997**, *100*, S2–S24.
2. Arlian, L.G. Biology and ecology of house dust mites *Dermatophagoides spp.* and *Euroglyphus spp.* Immunol. Allergy Clin. N. Am. **1989**, *9*, 339–356.
3. Colloff, M.J. Practical and theoretical aspects of the ecology of house dust mites in relation to the study of mite-mediated allergy. Rev. of Medical and Vet. Entomol. **1991**, *79*, 611–630.
4. The International Study of Asthma and Allergies in Childhood (ISAAC) Steering Committee. Worldwide variations in prevalence of asthma symptoms: the international study of asthma and allergies in childhood (ISAAC). European Resp. J. **1998**, *12*, 315–335.
5. Hamilton, G. Let them eat dirt. New Scientist **1998**, *July 18*, 26–31.
6. Stewart, G.A.; Thompson, P.J. The biochemistry of common aseroallergens. Clinical and Expt. Allergy **1996**, *26*, 1020–1044.
7. Colloff, M.J.; Stewart, G.A. House Dust Mites. In *Asthma*; Barnes, P., Grunstein, M., Leff, A.D., Woolcock, A.J., Eds.; Lippincott-Raven: Philadelphia, 1997; 1089–1103.
8. Colloff, M.J.; Ayres, J.; Carswell, F.; Howarth, P.; Merrett, T.G.; Mitchell, E.B.; Walshaw, M.J.; Warner, J.A.; Warner, J.O.; Woodcock, A.A. The control of allergens of dust mites and domestic pets: a position paper. Clinical and Expt. Allergy **1992**, *22* (Suppl. 2), 1–28.

HUMAN LICE

Davy Jones
University of Kentucky, Lexington, Kentucky

Richard J. Pollack
Harvard School of Public Health, Boston, Massachusets, U.S.A.

H

INTRODUCTION

People serve as the sole host for lice of three species, only one of which is of public health significance. The body louse (*Pediculus humanus*) sequesters mainly on clothes worn by those in particularly impoverished conditions, and serves as a vector of louse-borne or epidemic typhus (caused by *Rickettsia prowazekii*), epidemic or louse-borne relapsing fever (*Borrelia recurrentis*), and trench fever [*Bartonella (Rochalimaea) quintana*]. Epidemics of typhus and relapsing fever have decimated populations through the ages; millions more perished from these infections during major conflicts and famines of the past century (1). Body lice acquire these pathogens by feeding upon an infected person. Human infection occurs via inhalation or contact with crushed infected lice or their feces. Massive applications of DDT and lindane to control these lice have abated transmission and saved countless lives. Head lice (*Pediculus capitis*) and pubic lice (*Pthirus pubis*), in contrast, rarely cause disease and they do not serve as vectors. The medical community and the lay public in industrialized counties increasingly err by placing equal significance on each type of louse. An estimated $100 million is spent annually in the United States on combatting head louse infestations, and this sum may be dwarfed by losses from forced school absences and missed work time. The need for such expenditures, policies, or treatments is not well justified and remains controversial (2).

PREVALENCE OF LOUSE INFESTATIONS AND LOUSE-BORNE DISEASE

Each of the three types of human lice infest people worldwide, but the prevalence of these infestations is rarely well documented. Standardized methods are lacking for determining prevalence; thus it is difficult to compare such data between countries or cities, or even at the same location at different times. Although the incidence of typhus, relapsing fever, and trench fever are much reduced from the experiences earlier in the twentieth century, they continue to persist in eastern Africa and South America. Episodes of social strife and economic hardship have repeatedly led to conditions of overcrowding that then enhanced transmission of body lice and louse-borne disease. Modern tools, such as PCR, promise to aid in the detection of the etiological agents of these diseases (3).

SYMPTOMS OF LICE INFESTATION AND INJECTED SALIVARY TOXINS

The salivary components of biting flies, mosquitoes, bugs, fleas, and other blood-sucking insects have been much investigated in recent years (4). These components include anticoagulents, inhibitors of platelet aggregation, and vasodilators (5). The pharmacologically and immunologically active salivary components injected during feeding by human lice remain poorly understood (6). Human lice possess two salivary glands (''reniform'' and ''u-shaped'') from which components may be injected into the skin. The most obvious physiological sign that such active components are injected during feeding is a red spot, interpreted as vasodilation that typically appears at the bite site within a minute. Unlike the bites of other hematophagous insects, the bites of human lice often do not even induce pain, and in most cases, this erythema is not accompanied by swelling or itching, and then disappears within 1–2 h. Anticoagulant activity has been reported from combined extracts of the reniform and ushaped glands, but further characterization of the responsible molecules from either specific gland type has not been reported. An interesting activity in the saliva of the crab lice, localized to the reniform glands, leaves blue-colored marks at the bite site (''caerulae maculae''), but no further progress on the responsible agent has been forthcoming.

Less information is available on the identity of salivary components responsible for immunological reactions in lice hosts. Complicating older investigations was the failure to distinguish reactions to the salivary components versus reactions to louse feces or body parts that may contaminate the bite site. Isolated reports suggest IgE involvement in swollen cervical lymph nodes of louse-infested children. Laboratory rabbit hosts of human body lice appear to develop an antibody response to repeated feedings (7).

Because the reactions to bites by human lice are relatively mild (compared to other blood-feeding insects), the salivary components of these lice are poorly known. Human lice are obligate parasites of people, and seem to have coevolved to this relatively benign host-parasite relationship. Study of the louse's salivary components may lead to discovery of useful pharmacological agents that may have therapeutic value.

THERAPIES FOR LICE INFESTATION

Currently, pediculicidal agents are mainly limited to permethrin, pyrethrins, phenothrin, malathion, and lindane. Diverse antiparasitic agents, including ivermectin, insect growth regulators (IGRs), chitin synthesis inhibitors and certain antibiotics, have been proposed for treating human lice. Effective application strategies and safety and efficacy data for these products remain elusive.

Treatment should be considered solely after knowledgeable personnel have accurately identified lice and/or viable louse eggs (nits), as indicated in Table 1. Medical and public health concerns often justify interventions directed against body lice. Head lice and pubic lice are relatively innocuous; neither serving as vectors nor generally causing more than mild discomfort. Thus, interventions against head or pubic lice should be at the discretion of the patient (8).

Body Lice

Where louse-borne disease is endemic or potential for epidemic transmission exists (particularly in refugee camps and prisons), mass intervention campaigns directed at all individuals (infested or not) may save many lives. Improvements in personal hygiene (particularly the frequent and appropriate laundering of clothes) and mass treatment of residents and their clothing with certain insecticides may diminish risk.

Because body lice and their eggs are mainly restricted to the clothing, treatment of all clothing by steam or dry heat (>65°C) for 20 min is often efficacious. Quarantine of patients and contacts during epidemics of louse-borne infections is of questionable value and may be unjustified.

Head Lice

If live head lice are discovered, frequent use of a louse comb may effectively eliminate adult and nymphal lice, and some proportion of the eggs. The requirement to eliminate all eggs from the hair seems unnecessary but nonetheless is often fervently enforced by school personnel in North America and elsewhere. If desired, head louse eggs may be removed by protracted use of fine-toothed nit combs. When mechanical removal of lice is impractical, infestations may be effectively treated with over-the-counter or prescription-based pediculicides. Alternative ''natural'' remedies are becoming increasingly popular, but nearly all lack clinical testing. Insecticidal treatments to the environment (clothing, furniture, floor, classroom, and car) for head lice are unwarranted and counterproductive. Quarantine of infested people seems unjustified. Commonly held beliefs of ''fomite''-facilitated head louse transmission seem unsupported by objective findings.

Table 1 Treatment considerations

Type of louse	Age	Prevalence by socioeconomic class	Transmitted mainly by	Preferred intervention methods (s)
Head	School-aged children	All	Direct head-to-head contact	Mechanical removal, pediculicide applications
Body	Adults	Impoverished, homeless, inmates, refugees, war victims	Direct contact and sharing clothing or bedding	Improved hygiene, frequent clothing changes, pediculicide applications to person and clothing
Pubic	Adults	Sexually active	Sexual contact	Mechanical removal

Pubic Lice

These lice and their eggs are often best manually groomed from pubic, axillary, and facial hair (including eyelashes). As with head lice, pediculicides may be effective, but should be kept well away from the eyes.

RESISTANCE TO INSECTICIDES

Some degree of resistance by head and body lice to diverse classes of insecticides has been reported in parts of Europe, the Middle East, Africa, and the Americas, particularly where these products have been used intensively and fairly indiscriminately. Nonetheless, over-the-counter formulations of pyrethroids often continue to serve as a reasonable first-line treatment in newly diagnosed active infestations. When live lice persist following two treatments spaced about 10 days apart, applications of pediculicides based on malathion or lindane may be warranted. Increasing resistance to each of these insecticides has already significantly limited the effectiveness of these compounds in certain areas (9).

RESEARCH AND MANAGEMENT STRATEGIES FOR THE FUTURE

Efforts to elucidate the population biology and transmission dynamics of each type of louse will provide useful clues on which novel interventions may be practical and efficacious. Renewed attention to, and use of, modern immunological and molecular methods, may hold promise for the design of appropriate vaccines targeted at the louse-borne agents, and perhaps even to the lice themselves. Physiological inquiry aimed at developmental mechanisms offers hope of developing interventions based on IGRs and chitin synthesis inhibitors.

H

REFERENCES

1. Cloudsley-Thompson, J.L. *Insects and History*; Weidenfeld and Nicolson: London, 1976; 242.
2. Pollack, R.J.; Kiszewski, A.; Spielman, A. Overdiagnosis and consequent mismanagement of head louse infestations in North America. Pediat. Infect. Dis. J. **2000**, *19*, 689–693.
3. Gratz, N. In *Human Lice, Their Prevalence, Control and Resistance to Insecticides*; WHO 1997, WHO/CTD/WHOPES/97.
4. Ribeiro, J.M. Blood-feeding arthropods: live syringes or invertebrate pharmacologists. Infect. Agents Dis. **1995**, *4*, 143–152.
5. Mumcuoglu, K.Y.; Galum, R.; Kaminchik, Y.; Panet, A.; Levanon, A. Antihemostatic activity in salivary glands of the human body louse, *Pediculus humanus humanus* (Anoplura: Pediculidae). J. Insect Physiol. **1996**, *42*, 1083–1087.
6. Jones, D. The neglected saliva: medically important toxins in the saliva of human lice. Parasitology **1998**, *116* (Suppl.), S73–S81.
7. Jones, D.; Wache, S.; Chhokar, V. Toxins produced by arthropod parasites: salivary gland proteins of human body lice and venom proteins of chelonine wasps. Toxicon **1996**, *34*, 1421–1429.
8. Pollack, R.J.; Kiszewski, A.E.; Spielman, A. Head Lice Information. http://www.hsph.harvard.edu/headlice.html (accessed Sept 2000).
9. Pollack, R.J.; Kiszewski, A.; Armstrong, P.; Hahn, C.; Wolfe, N.; Rahman, H.A.; Laserson, K.; Telford, S.R., III; Spielman, A. Differential permethrin susceptibility of head lice sampled in the United States and Borneo. Arch. Pediat. Adolescent Med. **1999**, *153*, 969–973.

HUMAN PESTICIDE POISONINGS

Margareta Palmborg
Swedish Poisons Information Centre, Stockholm, Sweden

INTRODUCTION

The use of pesticides involves a risk of human exposure to substances with high toxicity. Acute pesticide poisoning is a global health concern. Morbidity and mortality are especially evident in developing countries. It has been estimated by the World Health Organization (WHO) that there are approximately three million cases hospitalized annually due to acute pesticide poisoning and approximately 220,000 deaths (1). Ninety-nine percent of fatal pesticide poisonings occur in developing countries (2). The problem has been well known for about 30 years, but although several projects of prevention have been undertaken to control the situation, there is still an urgent need of efforts to minimize the health hazards with pesticides.

WORLDWIDE KNOWLEDGE

There are no comprehensive figures on the total extent of pesticide poisoning worldwide and most published reports are reflecting local rather than nationwide conditions. The knowledge is based on estimates, which means that given numbers of poisonings and deaths are uncertain and there are many possibilities for miscalculation. The global estimates of acute pesticide poisoning are based mainly on hospital data and to a lesser extent on surveys among agriculture workers. Insufficient reporting of poisoning incidents among agriculture workers and misdiagnosis may lead to underestimates: hospital data mainly reflect the severe cases. However, studies from different parts of the world support the estimates that pesticide poisoning is a more evident problem in the developing part of the world.

Regional Reports

Sri Lanka, a country in the Asian–Pacific region with a population of more than 17 million inhabitants, is one of the few developing countries where basic epidemiological data on acute pesticide poisoning are available. In 1992, there were 17,636 patients with pesticide poisoning admitted to state hospitals in Sri Lanka and 1698 fatalities. Pesticide poisoning was reported as the fourth leading cause of death in state hospitals in Sri Lanka (3). A similar picture is given from Taiwan where pesticide poisoning was the leading cause of death in a study reported from the Taiwan National Poison Center during the period 1985–1993 (4). The incidence of pesticide poisoning in Costa Rica, a Central American developing country with a population of 2.4 million inhabitants has been studied with special attention to agriculture workers. One thousand eight hundred occupational pesticide poisonings were reported during 1986, and between 1980 and 1986 there were 3330 hospitalizations and 429 deaths (5). There are less data from African countries but there is one estimate that 11 million cases of pesticide poisonings occur annually in Africa. This estimate is based on reports from some African states from the 1980s and also includes minor cases that did not require hospital treatment (2). The information about acute pesticide poisoning in Africa during the 1990s is even more scarce—there are only a few regional reports in the literature. This reflects the situation in many African states with regard to health care without the possibility of epidemiological studies as well as insufficient systems for notification and surveillance of pesticide poisonings.

The health problem with pesticide poisoning in the developing world is heavily linked to the agriculture profile in the respective countries. From studies in Asia it is estimated that 3% of agriculture workers may suffer from pesticide poisoning each year, which would correspond to 25 million cases of agriculture pesticide poisoning in the developing countries alone (1). In another estimate based on a study from Costa Rica, the yearly incidence of occupational pesticide poisoning among agriculture workers was 4.5% (5). This figure from Costa Rica could be an overestimate as it is likely that medical care is more available there than in the Asian countries. This demonstrates the uncertainty of the estimates. In cases of occupational exposure the highest incidence of

Encyclopedia of Pest Management

poisoning is found among agricultural field workers while using the pesticides. In some cases the workers are not aware of the potential risk of the actual pesticide; in other situations there is adequate information but poor understanding of safety precautions. There may be lack of suitable protective clothing or the clothing is not used because it is uncomfortable. The pesticides are sometimes used in too high concentrations or with inadequate spray equipment. The most common substances involved in the cases of acute pesticide poisonings are organophosphorus insecticides and paraquat (3–5).

The situation in the developing countries is quite different from what is observed in the developed world. Despite the fact that more than 80% of the total amount of pesticides is used in the industrialized part of the world, acute pesticide poisoning is uncommon and only 1% of all fatal pesticide poisonings occur in these countries (2). In many developed countries there is more strict control and legislation of pesticides. However, also among developed countries, there are certain regions with a special agriculture profile and hence increased need for pest control, and in such regions there may be a higher incidence of pesticide poisoning. This is illustrated by the situation in California—a developed region with a warmer climate where an average of 665 incidents of acute pesticide poisoning among about 600,000 agriculture workers were reported each year between 1991 and 1996 (6). The very low incidence of acute pesticide poisoning reported from Finland, just 0.11% of all hospitalized poisoning cases 1987–1988, (7) may be due not only to strict legislation but also to the fact that there is less need of pesticides because of the climate. In Sweden there were three reported deaths due to acute pesticide poisoning during the years 1984–1994 (8). This low incidence is most likely true for many countries in the temperate zones.

Comparison

A few similarities can be seen between developing and developed countries. The fatal cases are mainly suicidal both in the developed and the developing world. One single pesticide, paraquat, is responsible for a high percentage of all suicides with pesticides and is a great problem worldwide. In Taiwan, paraquat (with 485 cases) was the leading cause of death among 1325 poisoning fatalities during 1985–1993, (4), and among deaths from pesticide poisoning in England and Wales in 1945–1989, paraquat was responsible for 56.3% (9). In the developing countries pesticides are extensively used as suicidal agents and in those countries about two thirds of all deaths due to pesticide poisoning are of suicidal origin. Although social and cultural factors contribute to the pattern of suicides in different countries, the availability of highly toxic pesticides is an important reason for the high rate of suicide in developing countries.

INTERNATIONAL SUPPORT

It is urgent to get control of the problem of pesticide poisoning. The national governments have the main responsibility for safe handling of pesticides. The international organizations have contributed in different ways to support individual countries to control the health problem with pesticide poisoning. Guidelines for classification of pesticides have been published by the World Health Organization within the International Programme on Chemical Safety (IPCS) with the purpose of achieving a more strict control and safer handling. Furthermore there are special training programs for agriculture fieldworkers elaborated by the International Labour Organisation (ILO) and by the United Nations Institute for Training and Research (UNITAR) (6, 10).

National authorities have an important role in legislation and control of pesticides. Highly toxic substances should be withdrawn from the market—if not absolutely indispensable—or when possible substituted with less toxic alternatives. The agrochemical industry can be involved in programs for research and development of new substances, as well as in safety work with protective measures. Education and training of workers must be maintained on a local basis but preferably with support from national authorities or international organizations. The Global Crop Protection Federation (GCPF), previously named the International Group of National Associations of Agrochemical Manufacturers (GIFAP), has initiated ''Pilot Safe Use Projects'' that have been working in Guatemala, Kenya, and Thailand (representating developing countries in Latin America, Africa and Asia). These projects have demonstrated that it is possible through integration among governments, industries, and local communities to achieve improvements in training of farmers, education of health personnel, and support for the development of the limited resources for acute medical treatment in the remote, rural areas of many developing countries with few hospitals and primary health care centers (10).

The lack of treatment facilities, drugs, and antidotes are factors contributing to the high morbidity and mortality from pesticide poisoning in the developing countries. Poison control centers can play an important role in the management of pesticide poisoning in providing adequate information on toxicity, acute risks, and treatment after exposure to pesticides. It has been possible to establish poison control centers in many developing countries with grants from international organizations and technical assistance from poison control centers in western countries. Special guidelines for poison control, particularly those encouraging developing countries to establish poison information centers, have been elaborated in a collaborative project between the International Programme on Chemical Safety (IPCS), the Commission of the European Communities (CEC), and the World Federation of Associations of Clinical Toxicology Centers and Poison Control Centers. The general public can receive advice on first-aid measures from the poison center in cases of poisoning accidents with pesticides and special treatment protocols can be elaborated for professional use in the hospital settings. Statistics from poison centers are also of importance in surveillance systems. Furthermore the poison centers have an important role in the prevention of poisoning by providing information on circumstances and trends in the pattern of poisoning (6).

REFERENCES

1. Jeyaratnam, J. Acute pesticide poisoning: A major global health problem. World Health Stat. Q. **1990**, *43* (3), 139–144.
2. Koh, D.; Jeyaratnam, J. Pesticides hazards in developing countries. Sci. Total Environ. **1996**, *188* (Suppl. 1), 78–85.
3. Fernando, R. Pesticide poisoning in the Asia–Pacific region and the role of a regional information network. J. Toxicol. Clin. Toxicol. **1995**, *33* (6), 677–682.
4. Yang, C.C.; Wu, J.F.; Ong, H.C.; Hung, S.C.; Kuo, Y.P.; Sa, C.H.; Chen, S.S.; Deng, J.F. Taiwan national poison center: epidemiologic data 1985–1993. J. Toxicol. Clin. Toxicol. **1996**, *34* (6), 651–663.
5. Wesseling, C.; Castillo, L.; Elinder, C.G. Pesticide poisonings in Costa Rica. Scand. J. Work Environ. Health **1993**, *19* (4), 227–235.
6. Internet links: European Chemical Industry Council (CEFIC) http://www.cefic.be/
   European Commission
   http://europa.eu.int/comm/index.htm
   Global Crop Protection Federation (GCPF)
   http://www.gcpf.org/
   Global Information Network on Chemicals (GINC)
   http://www.nihs.go.jp/
   International Labour Organisation (ILO)
   http://www.ilo.org/
   International Programme on Chemical Safety (IPCS)
   http://www.who.ch/pcs/
   International Union of Pure and Applied Chemistry (IUPAC) http://iupac.chemsoc.org/
   Organisation for Economic Cooperation and Development (OECD) http://www.oecd.org/
   Pesticide Action Network (PAN) Asia
   http://www.poptel.org.uk/panap/
   Pesticide Action Network (PAN) Europe
   http://www.gn.apc.org/pesticidetrust/
   Pesticide Action Network (PAN) North America
   http://www.panna.org/panna/
   United Nations Industrial Development Organisation (UNIDO) http://www.unido.org/
   United Nations Institute for Training and Research (UNITAR) http://www.unitar.org/
   World Health Organisation (WHO)
   http://www.who.int/ (accessed Oct. 1999).
7. Lamminpäa, A.; Riihimäki, V. Pesticide-related incidents treated in Finnish hospitals–a review of cases registered over a 5-year period. Hum. Exp. Toxicol. **1992**, *11* (6), 473–479.
8. Ekström, G.; Hemming, H.; Palmborg, M. Swedish pesticide risk reduction 1981–1995: food residues, health hazard, and reported poisonings. Rev. Environ. Contam. Toxicol. **1996**, *147*, 119–147.
9. Casey, P.; Vale, J.A. Deaths from pesticide poisoning in England and Wales: 1945–1989. Hum. Exp. Toxicol. **1994**, *13* (2), 95–101.
10. Ekström, G. *World Directory of Pesticide Control Organisations*, 3rd Ed.; The Royal Society of Chemistry: Cambridge, England, 1996; 12–18, 41–69.

HYGIENE (INCLUDING DESTRUCTION OF CROP REMAINS)

William W. Bockus
Kansas State University, Manhattan, Kansas, U.S.A.

H

INTRODUCTION

As a general rule, the discussions in this section concerning the effects of hygiene on pests refer to plant insect, weed, and pathogen pests. Many plant pests have stages that survive on crop residues or parts of the environment (such as storage bins) that can come into contact with plants at a later date. Residues and infested building areas play key roles in pest outbreaks by serving as reservoirs for the pests. The pests have colonized the plant tissue during the previous growing season and, when the crop matures, the residue serves as a refuge and food source for them during the period between hosts. They then reproduce and disseminate from the residue habitat and damage the succeeding crop. In many cases, therefore, significant control of pests can be accomplished by cultural practices associated with good hygiene practices. One practice to improve hygiene (= sanitation) is to eliminate or bury infected crop debris before the next crop is planted. Another is to disinfest areas of the environment that are contaminated by pests before healthy plants come into contact with them.

There are three components, referred to as the "Pest Triangle," that when brought together result in damage from pests. These are 1) a susceptible host, 2) a conducive environment for the pest, and 3) a virulent pest population. Removal of one part of the triangle results in control of pest problems. The goal behind hygiene practices is to eliminate the pest population, or reduce it below some threshold level so that little, if any, loss occurs. Numerous types of hygienic practices exist that can negatively impact pest populations (Table 1).

WHEN TO USE HYGIENE

Whether to adopt a certain hygiene practice should involve making an assessment of how it fits into a crop grower's production program. In addition to whether a producer has the proper equipment to perform a hygiene program, consideration should be given as to how it might affect other parts of the production scheme. Additionally, the cost/benefit ratio should be calculated because most hygiene practices have associated costs. In other words, one should determine the likely percentage damage to a crop by pests that might be controlled by the practice, the value of the increased production due to pest control, and whether the cost of the hygiene program justifies the benefit. Keeping these in mind, one can go through a hypothetical growing season and see the numerous places where hygiene can reduce the development of pest problems (Table 1).

EXAMPLES OF HYGIENE

Prior to beginning production of a crop, growers should consider whether there are nearby alternate hosts that can serve as sources of pest problems for the desired crop. These can be weed species, different crops, or other useful plants such as those in windbreaks. Good hygiene would involve the destruction of these alternate hosts in a large enough area surrounding where the desired crop will be grown so that pests are not spreading from other hosts into the crop. It should be noted, however, that such plant eradication programs are often difficult and expensive.

Immediately after harvest, decisions should be made as to what to do with the crop residue that is left. Options can include burial of the residue by such means as the use of a moldboard plow that inverts the soil profile. Tillage methods that use less fuel, such as using a disc, can also significantly incorporate the residue into the soil. When the residue is partially or completely buried, it is exposed to rapid microbial degradation in the moist soil environment. After the debris is degraded, the residue-borne pathogens die. Another residue management practice can involve the physical removal of crop debris from the field by gathering and baling it. More drastic measures might include burning the crop refuse, thus destroying any pests associated with it. Because residues can harbor many pests, using management practices that remove or destroy

Table 1 Hygienic practices associated with plant production and their effects on pests

Practice	Result
Removal of alternate hosts	Disruption of life cycle of pest
Tillage	Residue burial and exposure to soil microflora
Removal of crop residues	Physical removal of pests from field
Burning crop debris	Destruction of pest refuges
Crop rotation	Decomposition of crop debris harboring pests
Flooding the soil	Anaerobic conditions kill pests in soil
Solarization	Thermal inactivation of soilborne pests
Steam treatment of soil	Thermal inactivation of soilborne pests
Soil fumigation	Chemical inactivation of soilborne pests
Cleaning of bins and buildings	Removal of pests from farms and factories
Use of clean seeds	Inactivation of seed-borne pests
Pruning out infected plant parts	Removal of sources of pest infestations
Rogueing infected plants	Removal of sources of pest infestations
Chemical pesticides	Death or inactivation of pests in or on hosts
Biological pesticides	Death or inactivation of pests in or on hosts
Washing hands, tools, machinery	Reduction in human dissemination of pests

the residue will hinder pest proliferation, resulting in a healthier crop.

Whether to practice crop rotation is another decision related to hygiene. Rotation has long been known to affect pest populations. During the time that other crops are being produced, the infested residue from the previous crop will degrade, resulting in the death of the pests. However, to have a good hygienic effect, the other crops used in the rotation should not be hosts of the pests that are being controlled. Similarly, the pests cannot be ''soil inhabitants'' or they will not die out during the rotation period.

For certain cropping situations, more specialized treatments of the soil can be performed during the time between crops. These can include flooding fields so that the soil becomes anaerobic and kills any pests that are sensitive to lack of oxygen. Similarly the use of plastic mulches (solarization) can heat the soil sufficiently to kill many soil-borne pests. In greenhouse situations, the use of steam heat to sanitize soil is a common hygienic practice. In field conditions, there are chemical fumigants that can be injected into the soil that effectively kill many kinds of pests. These specialized types of hygienic soil treatments are usually expensive and only used on more high-value crops.

Sanitation in storage areas and plant-handling facilities can be practiced effectively between processing of plant products. While general cleanliness is important to remove waste debris that can harbor pests, it may also be important to use more intensive measures to thoroughly kill microorganisms. As an example, washing walls with dilute bleach solutions is inexpensive and effective for this purpose. Other methods may involve sealing areas and releasing a chemical fumigant. In either case, the important principle is to avoid a situation where healthy plant products come into contact with pests while in storage or being processed.

Another decision that occurs prior to planting that involves hygiene measures is the selection of seed or propagating material. Many pests are carried on or with the seed and it is important to start with pest-free seed. For most producers, this involves purchasing ''certified'' seed that meets certain requirements for seed-borne weeds, insects, and diseases. Pest-free seed is produced in two main ways; either by growing plants in an environment where seeds do not become infested with pests, or eradicating seed-borne pests prior to planting. The latter method usually involves exposure of seed to temperatures that are lethal to pests (but not the seed) or some sort of chemical seed treatment.

For perennial plants, a pruning program can be an effective way to remove pests from the area. For this practice, infected/infested parts of a plant are cut off and disposed of. While pruning can take place any time of the year, it is usually most effective if performed during that part of the year when plants are dormant and before spread of the pests has occurred. Similar to pruning is the sanitation practice of rogueing and disposing of entire plants that become infected. This prevents the spread of pests from the infected plants to neighboring, healthy ones. One must be careful to thoroughly dispose of the diseased plant material. For example, cutting up diseased trees and stacking the logs for firewood may not stop the insects or pathogens from multiplying in the logs and spreading to other trees. Instead, the logs should be immediately burned, buried, or removed a suitable distance from other host trees.

While the plants are actively growing, application of pesticides is the most common management practice that involves hygiene. Pesticides can be chemical or biological in nature. While pesticides may protect plants from attack by killing pests before they cause damage, many of them also have eradicative (''kick-back'') properties. That means that they are systemic so that they can enter into the plant tissue, kill pests, and actually ''heal'' plants.

Finally, good personal hygiene can be an important way to limit pest damage. For people who handle plants while they are growing, it is important to wash hands and frequently disinfest tools (such as pruning shears) while working. Many pests can be carried from an infected plant to healthy ones by hands or tools. A similar situation occurs in the field with regard to farm machinery that is moved from field to field. Such machinery can potentially transfer pests from one location to another. While this is usually not important for airborne pests, it can be important for those that are soilborne and of limited distribution. In that case, equipment should be cleaned after working in an infested field and prior to moving to a noninfested one.

SUMMARY

As seen by the examples above, hygiene practices should be considered year round. Employing good hygiene methods can greatly help to produce healthy plants. Hygiene is effective because it destroys pest populations or reduces them to manageable levels. Hygiene involves either the removal and destruction of infested crop debris or disinfesting building areas where there is contact with plants. Finally, by knowing the life cycles of the pests, their weaknesses can be observed and effective sanitation strategies devised and adopted.

BIBLIOGRAPHY

Bockus, W.W.; Shroyer, J.P. The impact of reduced tillage on soilborne plant pathogens. Annu. Rev. Phytopathol. **1998**, *36*, 485–500.

Cook, R.J.; Baker, K.F. *The Nature and Practice of Biological Control of Plant Pathogens*; APS Press: St. Paul, MN, 1983; 539.

Copping, L.G.; Hewitt, H.G. *Chemistry and Mode of Action of Crop Protection Agents*; Springer-Verlag: New York, 1998; 160.

Hardison, J.R. Fire and flame for plant disease control. Annu. Rev. Phytopathol. **1976**, *14*, 355–379.

Katan, J. Solar heating (solarization of soil for control of soilborne pests). Annu. Rev. Phytopathol. **1981**, *29*, 219–246.

Maude, R.B. *Seedborne Diseases and Their Control*; Oxford University Press: New York, 1996; 288.

Rechcigl, N.A.; Rechcigl, J.E. *Environmentally Safe Approaches to Crop Disease Control*; CRC Press: Boca Raton, FL, 1997; 400.

INORGANIC PESTICIDES

David B. South
Auburn University, Auburn, Alabama, U.S.A.

INTRODUCTION

Inorganic pesticides do not contain carbon as part of their chemical composition. These compounds usually have relatively low molecular weights and often contain less than 10 atoms. Several inorganic salts (often white and crystalline) are classified in this category of pesticides. A few inorganic pesticides have been used for more than 1000 years but their use increased dramatically from 1850 to 1950. The popularity of many inorganic pesticides declined after the development of more effective and less persistent organic pesticides.

Several natural inorganic compounds have toxic properties and, therefore, humans have been using them as pesticides for centuries (1). Romans used salt (the active ingredient in seawater) to keep weeds from growing on their roads and the Greeks fumigated their homes with sulfur (when burned it produces sulfur dioxide). Arsenic was used as an insecticide by the Chinese in 900 A.D. and was mined by the Egyptians and Greeks as well.

In the past, people were quick to adopt any treatment that would control pests and increase food supply. For example, in 1882 it was accidentally discovered that a mixture of hydrated lime and copper sulfate (Bordeaux mixture) would control downy mildew. This mixture had been applied on grapes to reduce pilfering by travelers alongside a highway in France. An astute traveler noted that all the vines in the orchard were infested with mildew except those treated alongside the road. News of this discovery spread quickly, its use increased rapidly and is still in use today.

Several inorganic pesticides are relatively stable and have longer half-lives than many organic pesticides produced today. Although this can be an advantage in regards to pest control, it also means the substance can persist on both the treated fruit and in the soil. Because of concerns about the persistence of residues of lead arsenate on apples and other food, a tolerance of 1.4 ppm of arsenic was established by the English government in the 1920s. The lead content of soil in some apple orchards is high today even though lead arsenate sprays ceased in the late 1940s after the introduction of the organic pesticide DDT. The popular belief that pesticides will accumulate in the soil over time is based in part from the frequent use of inorganic pesticides like lead arsenic and Bordeaux mixture. Today, however, organic pesticides usually degrade quickly, which partly explains why they have become more popular than older inorganic pesticides.

It is interesting to note that some "organic" farmers can use Bordaux mixture to control diseases and still certify their crop as "organically" grown. However, they are not allowed to use any synthetic pesticide (even those with a half-life of less than one month) without losing their certification. Zinc oxide (a fungicide) is sometimes applied to the skin as a sunblock and is considered safer than a short-term overexposure to ultraviolet radiation from the sun. From these observations, one might conclude that inorganic fungicides are safer than synthetic organic fungicides. But this would be an incorrect generalization. For example, the inorganic form of arsenic (As_2O_5 — LD_{50} 48 mg/kg) is more toxic when taken orally than organic forms of arsenic such as MSMA (700 mg/kg). It is therefore unwise to generalize about the relative safety of inorganic and organic pesticides. Although both are inorganic pesticides, sulfur is not very toxic while aluminum phosphide is very toxic when taken orally (Table 1). Likewise, the LD_{50} of the organic pesticide scilliroside is 0.4 mg/kg while the LD_{50} for the organic herbicide simazine exceeds 5000 mg/kg. Therefore, instead of generalizing about the relative safety of inorganic and organic pesticides, risks and benefits should be examined on a case by case basis.

INORGANIC HERBICIDES

Arsenic acid, copper sulfate, sodium chloride, sodium chlorate, sulfuric acid, iron sulfate, copper nitrate, and ammonium and potassium salts have all been used as herbicides (2). When applied in sufficient quantity, several of

Table 1 The chemical formula, oral toxicity (rat), and use of several inorganic compounds

Pesticide	Chemical formula	LD_{50} (oral) mg/kg (rat)	Major use
	Highly toxic		
Mercury bichloride	$HgCl_2$	1	Fungicide/bactericide
Sodium hypochlorite (12.5%)	NaOCl	5	Fungicide/bactericide
Arsenic oxide	As_2O_3	15	Rodenticide
Thallium sulfate	$Ti(SO_4)_x$	16	Rodenticide
Sodium azide	NaN_3	27	Fumigant
Zinc phosphide	Zn_3P_2	46	Rodenticide
Arsenic acid	AsH_3O_4	48	Rodenticide
	Moderately toxic		
Chromated copper arsenate	$As_2O_5+CrO_3+CuO$	48 : 80 : 470	Fungicide
Sodium fluoride	NaF	52	Insecticide
Sulfuryl fluoride	SO_2F_2	100	Fumigant
Lead arsenate	$PbAsO_4$	100	Insecticide
Mercurous chloride	Hg_2Cl_2	210	Fungicide/bactericide
Zinc oxide	ZnO	240	Fungicide
Bordeaux mixture	$CuSO_4+Ca(OH)_2$	300	Fungicide
Copper sulfate	$CuSO4 \bullet 5H_2O$	472	Fungicide
	Slightly toxic		
Copper hydroxidse	CuH_2O_2	1000	Fungicide
Sodium chlorate	$NaClO_3$	1200	Herbicide
Hydrogen peroxide (35%)	H_2O_2	1193	Fungicide
Ammonium sulfamate	$NH_4SO_3NH_2$	2000	Herbicide
Sulfuric acid	H_2SO_4	2140	Herbicide
Boric acid	H_3BO_3	2660	Insecticide
Sodium tetraborate	$Na_2B_4O_7 \bullet 5H_2O$	2660	Herbicide
Sodium chloride	NaCl	3320	Herbicide
	Low toxicity		
Sulfur	S	> 5000	Insecticide
Sodium fluoaluminate	AlF_6Na_3	> 10,000	Insecticide

(Adopted from various material safety data sheets.)

the salts will cause desiccation and cell plasmolysis. However, when applied at subtoxic levels, several of these salts are also fertilizers. As with most pesticides, the dose makes the poison. Sodium chlorate can defoliate cotton when applied at 6 kg/ha and can act as a soil sterilant at a rate of 225 kg/ha. Ammonium sulfamate was introduced in 1942 as a herbicide for use in controlling woody plants. It can either be used as a foliar spray or as crystals applied to cut surfaces in bark or freshly cut stumps. Foresters used ammonium sulfamate to kill or weaken undesirable tree species. The phytotoxicity of borate herbicides (sodium tetraborate, sodium metaborate, and amorphous sodium borate) is related to the amount of boron taken up by plants. Although boron is an essential element, for some species boron toxicity can result at rates of only a few kilograms per ha. Both sulfuric acid and nitric acid have sometimes been used as selective herbicides. Concentrated sulfuric acid is a broad-spectrum herbicide while solutions containing 3% or 4% sulfuric acid can be applied selectively to some crops like onions and cabbage. In general, the use of inorganic herbicides has declined but some compounds, like copper sulfate as an algicide in fish ponds, are still in use today.

INORGANIC INSECTICIDES

Several inorganic insecticides act as stomach poisons. Sulfur has been used for decades to control mites and insects and boric acid is still used against roaches. Arsenical insecticides (lead arsenate and calcium arsenate) were commonly used in agriculture before the adoption of organic insecticides. Wood that is placed in the ground is often treated with chromated copper arsenate (CCA) to protect against termites and fungi. Roaches and ants are sometimes controlled by fluorine insecticides such as barium fluosilicate, sodium silicofluride, and cryolite (sodium fluoaluminate). Diatomaceous earth is a fine dust that is comprised of the fossilized remains of diatoms (mostly silicon dioxide). The spines of the diatom skeletons can scratch insects, which then die due to dehydration (3). Diatomaceous earth can affect ants, cockroaches, earwigs, fleas, and other crawling insects and is sometimes fed to cattle, chickens, dogs, and cats to control internal pests.

INORGANIC FUNGICIDES

Sulfur, probably the oldest fungicide, is still used by gardeners today. However, during the late 1800s, it was discovered that metal salts were generally more effective in suppressing fungal growth. Copper sulfate (a main ingredient in Bordeaux mixture), copper hydroxide, and zinc compounds are still used in agriculture (4). During the 1970s about 20% of the fungicides used in the United States were inorganic copper compounds. Other metal salts with fungicidal activity include calcium, iron, cobalt, lead, nickel, cadmium, chromium, arsenic, and mercury. The use of inorganic fungicides containing mercury has declined because they are highly toxic to warm-blooded organisms and because organic fungicides are more effective.

Treating wood to prevent decay can increase the life expectancy by 5–10 times. In the past, organic compounds like creosote, ethyl mercury chloride, and pentachlorophenol were used as wood preservatives. But due to environmental concerns, a majority of treated wood now contains CCA (a combination of three metal salts: chromium, copper, and arsenic). The use of wood and wood substitutes would likely increase dramatically were it not for preservatives like CCA. Boric acid and borates are also sometimes used as wood preservatives.

Some common household products can be used as inorganic fungicides. Borax (used in washing clothes) is sometimes used by foresters to treat stumps to keep a root-rot fungus from spreading to uninfected trees. Borax can also be sprinkled in moist areas to prevent the growth of mold. Some nursery managers use diluted sodium hypochlorite (bleach) or hydrogen peroxide to reduce the amount of fungi on cuttings and tree seed (5).

INORGANIC FUMIGANTS

Fumigants in this group include chlorine gas, hydrogen sulfide, sulfur dioxide, phosphine gas, sodium azide, and ozone. Chlorine gas is produced when water is added to calcium hypochlorite. This compound is commonly used to treat water in hot tubs, swimming pools, and municipal drinking water. Ozone (O_3) is also used to treat water and can kill some bacteria faster than chlorine. Most bottled water is treated with ozone, and nearly 200 municipal water treatment plants in the U.S. employ ozone to help cleanse their drinking water. Ozone recently gained approval for use in the U.S. food processing industry as a disinfectant wash or spray to help rid food of dangerous bacteria, parasites, and fungi.

Aluminum phosphide is used in controlling insects and rodents. When added to water, it reacts to produce hydrogen phosphide gas (phosphine). This gas can be used to treat food in storage bins and also can be used outdoors to kill burrowing rodents. Sodium azide and potassium azide have potential as soil fumigants for suppression of fungi, weeds, and insects. After they are applied to the soil, hydrolysis results in the production of hydrozoic acid. Both compounds are dangerous to handle since they are potentially explosive and neither is currently registered in the United States.

REFERENCES

1. Ware, G.W. *The Pesticide Book*, 4th Ed.; Thomson Publications: Fresno, California, 1994; 386.
2. Matolcsy, G.; Nádasy, M.; Andriska, V. *Pesticide Chemistry*; Elsevier Science Publishers: New York, 1988; 808.
3. Marer, P.J.; Flint, M.L. *The Safe and Effective Use of Pesticides*, 2nd Ed.; Pesticide Application Compendium, University of California: Oakland, California, 2000; 352, Publication #3324.
4. Waxman, M.F. *Agrochemical and Pesticide Safety Handbook*; Lewis Publishers: Boca Raton, Florida, 1998; 616.
5. Olkowski, W.; Daar, S.; Olkowski, H. Some Useful Inorganics, Organics and Botanicals. In *Common-Sense Pest Control*; The Taunton Press: Newtown, Connecticut, 1991; 105–127.

INSECT GROWTH REGULATORS

Meir Paul Pener
The Hebrew University of Jerusalem, Jerusalem, Israel

INTRODUCTION

Insect growth regulators (IGR or IGRs) are compounds that interfere with insect specific physiological systems that do not exist in vertebrates (1, 2). A hard cuticle, serving as an exoskeleton, covers the body of insects and related arthropods. Once produced, the cuticle cannot grow; to allow growing of the insect, the old cuticle is shed and a new, larger, and often different, cuticle is formed. This change of the cuticle is termed molt or molting and a stage between two consecutive molts is named an "instar." To reach the adult stage, insects molt several times, i.e., they have several instars, and they undergo "metamorphosis" from larva or nymph to adult, or from larva to pupa and then from pupa to adult. Both molting and metamorphosis are regulated by hormones secreted by endocrine organs. These physiological systems are not shared by vertebrates, or differ from vertebrate systems, and they constitute targets to IGRs. Most IGRs used today in insect control either interfere with the formation of the new cuticle (3, 4) or disturb metamorphosis (5–7). IGRs deregulate rather than regulate insect development; therefore, the recently (though yet infrequently) used term "Insect Development Inhibitors" is more suitable.

IGR CATEGORIES

IGRs are usually classified according to their mode of action (1). Sometimes, however, terminology related to their chemical structure is also practiced; for example, most chitin synthesis inhibitors (see below) are benzoylphenyl urea derivatives and the term "benzoylphenyl ureas" is often used in the literature (1, 4).

Chitin Synthesis Inhibitors (CSIs) (1–4)

Chitin, a polysaccharide, is a very stable major component of the insect cuticle. It is a long-chain polymer of the amino-sugar, *N*-acetyl-d-glucosamine. The enzyme, chitin synthetase, is responsible for linking the individual amino-sugar molecules to produce the polymer. CSIs inhibit the action of chitin synthetase, resulting in abnormal, malformed, new cuticle. Consequently, the insect dies in the molt, or the abnormal cuticle causes mortality after the molt. Diflubenzuron (Fig. 1) was the first CSI-insecticide introduced into the market (under the trade name "dimilin"). Today there are many CSIs such as teflubenzuron, triflumuron, flufenoxuron, chlorfluazuron, hexaflumuron, and lufenuron (Fig. 1). A CSI that is not a benzoylphenyl urea derivative is buprofezin (Fig. 1).

CSIs are used mainly against insect larvae (larvicides). Adult insects do not molt and are not susceptible to CSIs, but egg development in CSI-treated adult females is disturbed (an ovicidal effect) in some instances. The usual mode of administration is by feeding, but some insects are susceptible to topical or contact treatments. Susceptibility much depends on the compound, on the kind of the insect pest (at the level of insect order, family, or even species), and sometimes also on the larval instar treated.

Juvenile Hormone Analogs (JHAs) (1, 2, 6, 7)

Insect metamorphosis is under hormonal regulation. Presence of the juvenile hormone (JH) in the nymphal or larval instars prevents metamorphosis to the adult. JH is secreted by a pair of endocrine glands, the corpora allata. In the last larval instar, and/or in the pupa, cessation of corpora allata activity leads to a temporary absence of JH, allowing completion of the metamorphosis. In the adult, the corpora allata resume activity and the JH is involved in the regulation of reproduction. There are several natural JHs in insect. The most common is JH-3 (Fig. 2), but all have the same sesquiterpenoid basic structure. JHAs (also termed "JH mimics" or "juvenoids") are compounds functionally similar to JH. Treatment of late nymphal or larval instars with JHAs prevents metamorphosis to adult. The chemical structure of JHAs may or may not resemble natural JH. There are hundreds of chemically different

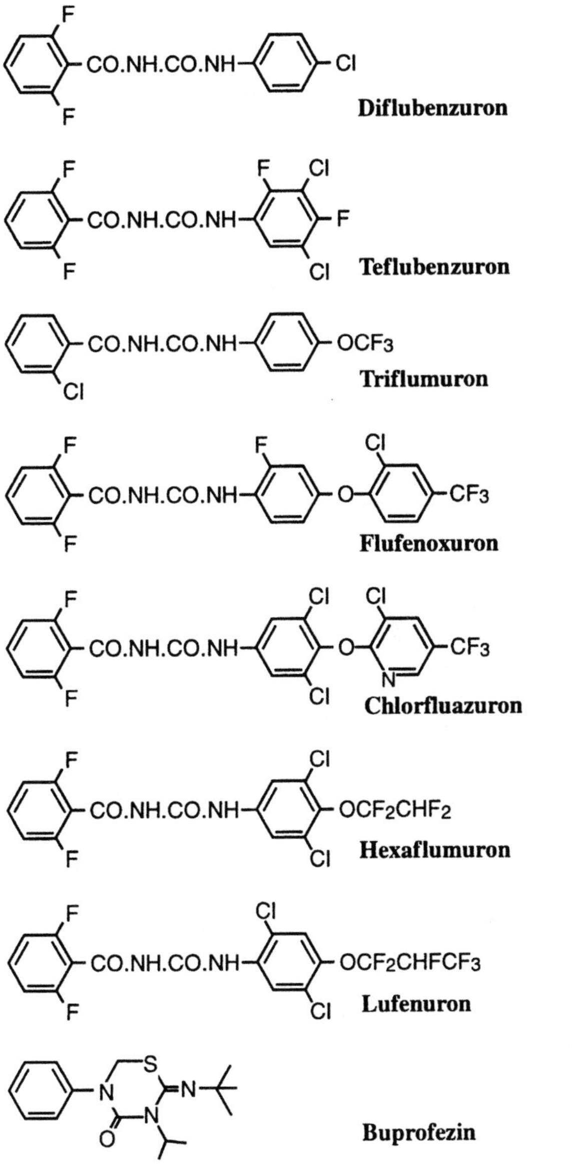

Fig. 1 The chemical structure of some chitin synthesis inhibitors.

compounds that have JHA activity. However, only few were developed as commercial insecticides. These include methoprene, hydroprene, fenoxycarb, pyriproxyfen, and diofenolan (Fig. 2), as well as the superseded triprene, kinoprene, and epofenonane.

In the course of normal development, JH disappears only in the last larval instar and this instar constitutes the target of JHAs. Nevertheless, treatment of earlier instars may also be effective because environmental degradation of JHAs takes some time and meanwhile the insect may reach the last larval instar. Also, metabolic inactivation of JHAs is usually slower than that of natural JHs and the JHA absorbed in an earlier instar is not necessarily inactivated before the insect reaches the last larval instar. JHAs are usually effective by contact; they penetrate the cuticle. JHAs prevent development of normal adults; larval-adult or pupal-adult intermediate creatures are obtained that die in the molt, or in a next (supernumerary) molt, or survive but do not reproduce. JHAs do not prevent damage caused by larvae; however, by preventing reproduction, they prevent development of the next generation. JHAs are extremely convenient if the larvae are harmless and only the adults are harmful (for example, mosquitoes). JHA treatment of adults often disturbs egg production and/or egg development and treatment of eggs disturbs embryonic development. Different orders, fam-

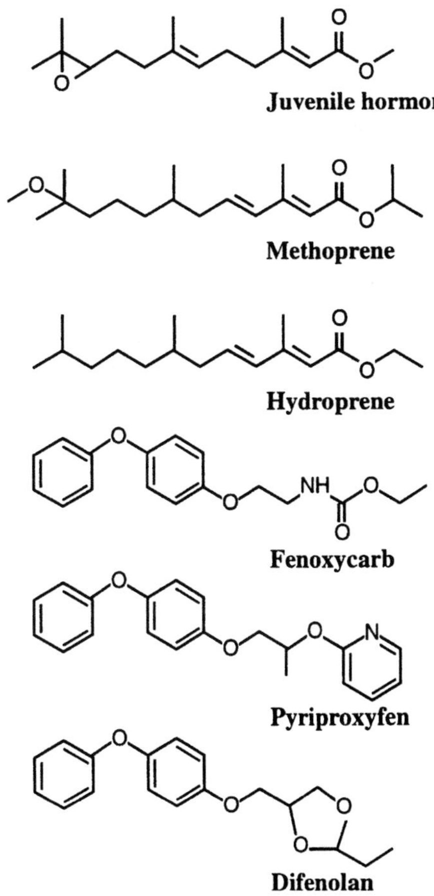

Fig. 2 The chemical structure of juvenile hormone 3 and some juvenile hormone analogs.

ilies, and even species of insects may show widely different susceptibility to the same JHA and to different JHAs.

Miscellaneous IGRs

Cyromazine (Fig. 3), a commercially produced IGR, disturbs cuticle formation of insects, but presumably not through inhibition of chitin synthetase. Its exact mode of action is unknown, but it causes mortality similar to that caused by CSIs. Di-*tert*-butyl phenols are compounds that also affect cuticle formation, resulting in mortality (3).

Precocenes (Fig. 3) are "anti-juvenile hormones" (5). They selectively destroy the JH-producing corpora allata, leading to JH deficiency. Treatment of early instars with precocenes induces precocious metamorphosis, resulting in miniature larval-adult intermediate creatures that do not reproduce (1, 5). Precocenes are not utilized for practical insect control because effective doses are high, the range of susceptible pests is narrow, and they are suspected to be carcinogenous. Some other anti-juvenile hormones are also known (5), but none of them has been developed for a commercial insecticide.

Ecdysteroids, especially ecdysone and 20-hydroxyecdysone, are the natural molting hormones of insects. Several bisacylhydrazine ecdysteroid agonists are known (7); tebufenozide and halofenozide (Fig. 3) are marketed commercially as effective IGRs. They induce a lethal molt in susceptible insect larvae.

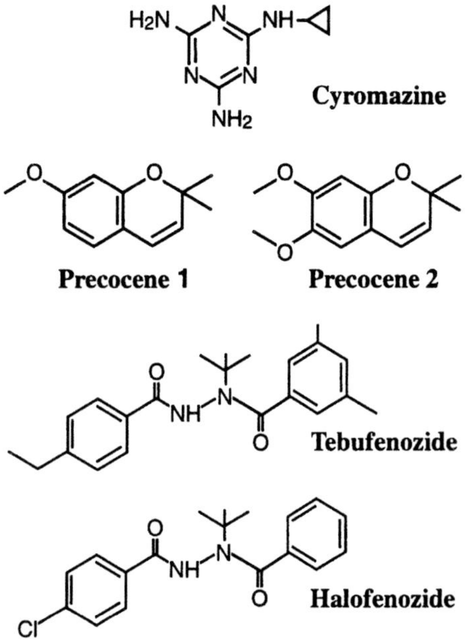

Fig. 3 The chemical structure of some miscellaneous insect growth regulators.

Azadirachtin, a natural product from the neem tree (*Azadirachta indica*) is an "antifeedant" (feeding deterrent) to many insects, but it is also a potent "anti-molting" agent, though the exact mode of its molt inhibitory action remains to be clarified. Azadirachtin-treated larvae often die in the molt or become "permanent" larvae that do not molt again and do not reproduce. Treatment of adults may inhibit reproduction. Azadirachtin has a complex chemical structure; it is not synthesized, but extracted from neem seed kernels. Its formulations are marketed by several commercial firms.

NAMING IGRs

IGR names may confuse nonprofessionals. In early stages of development, a commercial firm may use a code, usually constructed from the abbreviated company name and a product number. Later, a name (sometimes called "common name," and after commercial production also termed "active ingredient") is given to the compound and its "chemical name," reflecting its chemical structure, is revealed. The commercial product is a formulation which has a "trade name." Different formulations, often directed against different pests, but based on the same active ingredient, may have different trade names. For example, the JHA pyriproxyfen, was developed by Sumitomo Chemical Company, Japan, under the code "S-31183." It received the common name, "pyriproxyfen" and its chemical name is "2-[1-methyl-2-(4 phenoxy phenoxy)ethoxy] pyridine," or "4-phenoxyphenyl (*RS*)-2-(2-pyridyloxy)propyl ether," according to different naming systems of chemicals. Its formulations are registered under the trade names: "Sumilarv," "Knack," and "Admiral." The name of the active ingredient of a formulation can be found in *The Pesticide Index* and further details in *The Pesticide Manual* under the name of this active ingredient. The label on a formulation should include the name of the active ingredient.

ADVANTAGES AND DISADVANTAGES OF IGRs

IGRs are usually nontoxic or slightly toxic (often as low as at the g/kg level) to vertebrates. Therefore, their impact on the environment is minimal, though they may affect beneficial insects (bees, pollinators, insect predators, and parasites of pest) and also crustaceans and other arthropods. The selectivity of IGRs to a certain group of insects is an environmental advantage, but a disadvantage for marketing. A major disadvantage is that IGRs are effective only at special stages or instars (CSIs and ecdysteroid agonists before molt, JHAs at the last larval instar) and do not prevent some or all damage by the larvae. The belated mortality is an important psychological disadvantage; the farmer does not see dead insects immediately and consequently doubts the efficiency of the IGR. Outdoor stability of IGRs is usually satisfactory; if too short, encapsulation, allowing slow release, is practiced. Environmental degradation is also satisfactory. In contrast to some earlier beliefs, insects are able to develop resistance to IGRs. Development of new IGRs and their approval by the authorities are costly, especially in relation to marketing limitations. Nevertheless, new IGRs appear on the market and more effective IGRs, based on "anti-JH" or "anti-ecdysteroid" action may be developed in the future.

REFERENCES

1. Retnakaran, A.; Granett, J.; Ennis, T. Insect Growth Regulators. In *Comprehensive Insect Physiology Biochemistry and Pharmacology, Vol. 12. Insect Control*; Kerkut, G.A., Gilbert, L.I., Eds.; Pergamon Press: Oxford, 1985; 529–601.
2. Miyamoto, J.; Hirano, M.; Takimoto, Y.; Hatakoshi, M. Insect Growth Regulators for Pest Control, with Emphasis on Juvenile Hormone Analogs. In *Pest Control with Enhanced Environmental Safety*; Duke, S.O., Menn, J.J., Plimmer, J.R., Eds.; ACS Symposium Series No. 524, American Chemical Society: Washington, DC, 1993; 144–168.
3. Chen, A.C.; Mayer, R.T. Insecticides: Effects on the Cuticle. In *Comprehensive Insect Physiology Biochemistry and Pharmacology, Vol. 12: Insect Control*; Kerkut, GA., Gilbert, L.I., Eds.; Pergamon Press: Oxford, 1985; 57–77.
4. *Chitin and Benzoylphenyl Ureas*; Retnakaran, A., Wright, J.E., Eds.; Dr. W. Junk: Dordrecht, The Netherlands, 1987; X + 1–309.
5. Staal, G.B. Anti juvenile hormone agents. Annu. Rev. Entomol. **1986**, *31*, 391–429.
6. Grenier, S.; Grenier, A.-M. Fenoxycarb, a fairly new insect growth regulator: a review of its effects on insects. Ann. Appl. Biol. **1993**, *122*, 369–403.
7. Dhadiallah, T.S.; Carlson, G.R.; Le, D.P. New insecticides with ecdysteroidal and juvenile hormone activity. Annu. Rev. Entomol. **1998**, *43*, 545–569.

INSECTISTASIS AS A MEANS OF CONTROLLING PEST POPULATIONS IN THE STORAGE ENVIRONMENT

Hermann Levinson
Anna Levinson
Max-Planck-Institute of Behaviour Physiology, Seewiesen, Germany

INTRODUCTION

Storage insect pests include nearly four dozen abundant coleopteran and lepidopterous species capable of growing and reproducing on partly desiccated foodstuffs, which are maintained in spatially restricted and relatively dim warehouses. Predominant coleopteran storage pests comprise species of the families *Anobiidae, Bostrychidae, Bruchidae, Cucujidae, Curculionidae, Dermestidae, Silvanidae,* and *Tenebrionidae,* while the most common lepidopterous storage pests contain species pertaining to the families *Galleriinae, Gelechiidae,* and *Phycitidae* (cf. Table 1). The larvae of storage pests belonging to the above families are often polyphagous and consume a variety of dried seeds and plant and animal tissues, on which they can build up dense populations and thus become serious pests.

CURATIVE AND PREVENTIVE MEASURES OF INSECT CONTROL

Established methods of insect control in the storage environment may be broadly divided into *curative* and *preventive* measures. Prominent curative procedures involve the use of gaseous and contact insecticides, e.g., emission of hydrogen phosphide (Phosphine) from precursors, application of methyl bromide (monobromomethane)[a], Dichlorvos (DDVP)[b], Malathion[c], and certain pyrethroids[d] as well as the accumulation of carbon dioxide or nitrogen under hermetic conditions (1). Induction of insect sterility by gamma-irradiation or chemical sterilization, destructive concussion of insects by entoleters, as well as application of pathogens to insects are curative measures that have failed to become customary procedures in warehouses, possibly due to their high cost (2). The curative advantage of inorganic dusts (e.g., various sorts of calcium phosphate, chalk, kaolin, and silica aerogels) as grain additives (3) is lessened by their probable adverse influence on human health and abrasive effects on machinery (2).

Storage at lowered (4) or elevated (5) temperature as well as protection of desiccated foodstuffs by means of insect-resistant packing materials (6) are preventive procedures that are usefully employed. Potentially preventive measures could also be developed by utilizing hormone-mimetic agents, nutrient antagonists, and phagorepellents as well as semiochemicals.

[a]To be abandoned in the near future, because of its hazardous effects on the atmospheric ozone layer.
[b]*O, O*-Dimethyl *O*-(2,2-dichlorovinyl) phosphate.
[c]*O, O*-Dimethyl *S*-(1,2-dicarbethoxyethyl) phosphorodithioate.
[d]Synthetic compounds structurally related to natural pyrethrins.

PHEROMONES OF STORAGE INSECT PESTS

Utilization of insect pheromones in pest control is certainly a valuable contribution of insect physiology to applied entomology. Moreover, these semiochemicals provide an efficient tool for the restraint of insect populations infesting stored foodstuffs. The female-produced sex pheromones of ca. 15 coleopteran and ca. 7 lepidopterous species as well as the male-produced aggregation pheromones of ca. 16 coleopteran species injuring stored foodstuffs have been studied from chemical and physiological viewpoints, while the molecular and chiral structures of their main components were identified and synthesized (cf. 7, Table 1). Nanogram amounts of these pheromones transmit molecular messages triggering the reproductive instinct in responsive insects, being manifested by sexual attraction, mating, insemination, and fecundation.

Encyclopedia of Pest Management

Table 1 Important storage pests releasing pheromones of identified structure[a]

Order/family	Genus	Species	Species	Species	Emission	Pheromone type
Coleoptera:						
Anobiidae	*Lasioderma*	*serricorne*			f	s
	Stegobium	*paniceum*			f	s
Bostrychidae	*Prostephanus*	*truncatus*			m	agg
	Rhyzopertha	*dominica*			m	agg
Bruchidae	*Acanthoscelides*	*obtectus*			m	s
	Callosobruchus	*chinensis*	*maculatus*		f	s
Cucujidae	*Cryptolestes*	*ferrugineus*	*pusillus*	*turcicus*	m	agg
Curculionidae	*Sitophilus*	*granarius*	*oryzae*	*zeamais*	m	agg
Dermestidae	*Anthrenus*	*flavipes*	*verbasci*		f	s
	Attagenus	*elongatulus*	*megatoma*		f	s
	Dermestes	*maculatus*			m	agg
	Trogoderma	*granarium*	*glabrum*	*inclusum*	f	s, agg
Silvanidae	*Oryzaephilus*	*mercator*	*surinamensis*		m	agg
Tenebrionidae	*Tribolium*	*castaneum*	*confusum*		m	agg
Lepidoptera:						
Galleriinae	*Corcyra*	*cephalonica*			m	s
Gelechiidae	*Sitotroga*	*cerealella*			f	s
Phycitidae	*Anagasta*	*kuehniella*			f	s
	Cadra	*cautella*	*figulilella*		f	s
	Ephestia	*elutella*			f	s
	Plodia	*interpunctella*			f	s

[a] Pheromones produced by females or by males are designated by [f] or [m]; they act as sex pheromones [s] for conspecific mates or as aggregation pheromones [agg] for both sexes.

Bearing in mind that sensory responses of harmful insects to pheromones and food attractants, optical preferences for certain geometrical forms and thigmotactic responses, as well as a circadian periodicity are essential for mate finding and copulation in storage pests, highly efficient traps to detect and decimate the latter could have been designed (8–10). Various two- or three-dimensional adhesive traps and conical and multitubular pitfall devices releasing sex and/or aggregation pheromones were eventually employed to lure and retain one or both sexes of storage pest insects (11–14). The usefulness of sex and aggregation pheromones for the manipulation of storage pests was comprehensively reviewed (15–17); these semiochemicals are currently employed, in order to fulfil the following tasks:

- detection and surveillance of pest populations in food stores,
- mass trapping of pest insects and causation of insectistasis,
- proper timing of curative measures,
- checking the efficacy of curative measures,
- mating disruption in pest insects due to overdosed sex pheromones, and
- administration of pheromones together with insecticides as *attracticides* (18), in order to lure and kill harmful storage insects.

Male storage moths (*Phycitidae*) respond to the sex pheromone of their conspecific females along with figural stimuli, while both sexes of certain storage beetles (e.g., *Cryptolestes, Dermestes, Oryzaephilus, Prostephanus, Rhyzopertha, Sitophilus* and *Tribolium* spp.) usually respond to the aggregation pheromone of their conspecific males in presence of feeding stimuli (7, 16). Male phycitid moths have a remarkable visual acuity at a weak light intensity and steer upwind toward calling females guided by optical markers. Sexual calling (i.e. erection of the posterior abdomen and rhythmic exposure of the pheromone gland) by female *Anagasta kuehniella* usually occurs at dawn (ca. 0500–0700 h), by female *Cadra cautella* takes place after dusk (ca. 2200–2400 h), by female *Cadra figulilella* occurs around midnight (ca. 2300–0100 h), while sexual calling by female *Ephestia elutella* and female *Plodia interpunctella* is scattered throughout the night (ca. 2200–0500 h). Periods of maximal flight activity of male phycitids are in approximate agreement with the calling periods of conspecific females (19, 20).

The female-produced sex pheromones of the above phycitids comprise (*Z,E*)−9,12-tetradecadien-1-yl acetate (TDA) as a common and predominant component, acting as the main sex attractant for males of these sympatric species. The sex pheromone of female *Cadra cautella* also includes (*Z*)−9-tetradecen-1-yl acetate (TA), which synergizes the attractiveness of TDA for conspecific males and subdues the attractiveness of TDA for male *Anagasta kuehniella* and *Plodia interpunctella.* (Z,E)−9,12-tetradecadien-1-ol (TDO), an additional sex pheromone component of female *Anagasta kuehniella, Cadra cautella, Ephestia elutella*, and *Plodia interpunctella*, was found to enhance the attractiveness of TDA for male *Ephestia elutella*, and *Plodia interpunctella* as well as to suppress the responsiveness of male *Cadra cautella* to TDA and TA. Females of the gelechiid species *Sitotroga cerealella* emit (*Z,E*)−7,11-hexadecadien-1-yl acetate as the major sex pheromone component, which attracts conspecific males during the scotophase (cf. 7, 16).

Receptor potentials and nerve impulses recorded from the antennal olfactory sensilla of male *Anagasta kuehniella, Ephestia elutella*, and *Plodia interpunctella* following stimulation by the above pheromone components permit the conclusion that two receptor cells of differing specificity are available, one being selectively responsive to TDA and TA and the other being responsive to TDO alone (7, 21).

Receptor potentials recorded from the antennal olfactory sensilla of certain coleopterans (e.g., *Dermestes lardarius, D. maculatus, Sitophilus granarius, S. oryzae*, and *S. zeamais* as well as *Tribolium castaneum* and *T. confusum*) in response to their male-produced aggregation pheromones were considerably high in both sexes, while the receptor potentials following stimulation by female-produced sex pheromones were recorded from male beetles only (e.g., *Lasioderma serricorne*) (cf. 7; Table 1).

INSECTISTASIS

The dynamics of an insect population in the storage environment is manifested by periodic fluctuations between levels of low density (*insectistasis*) and high density (*causing economic damage*) which may be ascribed to the reproductive potential of the available pest species, *in*adequate or excessive food supply, *un*favorable or optimal temperature or to some other causes.

A relatively new strategy of pest control combines curative and preventive measures to cause *insectistasis* (stasis = stagnation), which denotes a state wherein the population density of harmful insect species is diminished to the extent of allowing food storage without significant impairment (22). Insectistasis can be achieved by progressively reducing the number of insects in a pest population by continual insect captures with attractant traps provided with sex or aggregation pheromones and food attractants together with an immobilizing device.

Fig. 1 reveals increasing and decreasing densities of an insect population infesting a spatially restricted food store, being recorded as successive catches in pheromone traps at regular time intervals. The depicted curve permits an arbitrary division of the pest population into three density levels: *insectistasis* (low population density), *intense growth* (intermediate population density) and *economic damage* (excessive population density). Administration of an insecticide (black arrow) is usually performed when the insect population has grown to the level of economic damage. The treated insect population will then be reduced to the level of insectistasis, which may eventually *re*increase to the level of economic damage.

In contrast to postponed curative control measures applied to an excessively dense pest population, the use of an insecticide preferably should be timed in accordance with the extent of trap catches (Fig. 2). As long as the pest population remains within the limits of insectistasis, monitoring is performed at long time intervals (white circles). When the pest population has grown beyond the level of insectistasis, as traced by larger and more frequent insect trappings (black circles), application of an insecticide is required in order to prevent a further increase of the pest population and to reduce its density to the level of insectistasis. Morever, when the prevailing conditions discourage rapid growth and reproduction of storage insects, continual mass trapping of the latter may be sufficient to maintain the pest population within the limits of insec-

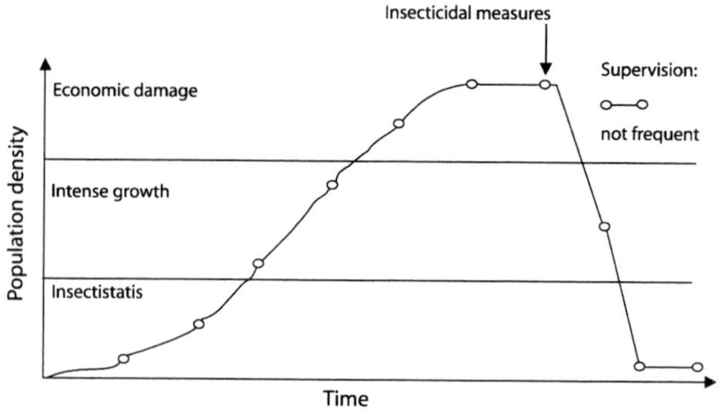

Fig. 1 Conventional use of insecticides in food stores. The main density levels of a pest population, viz. insectistasis, intense growth, and economic damage, are supervised by means of quantitative insect catches in pheromone traps (white circles). Insecticidal application is usually performed after the occurrence of economic damage. The time of insecticidal application is indicated by a black arrow above the curve plateau.

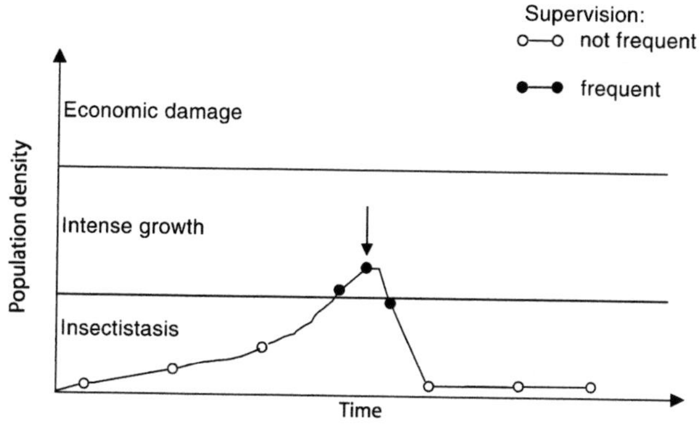

Fig. 2 Storage protection based on insectistasis. As long as the pest population remains within the limits of insectistasis, supervision of the population density (by quantitative insect catches in pheromone traps) is infrequently performed (white circles). If the population density surpasses the level of insectistasis, frequency of supervision is elevated (black circles). Insecticidal application should be timed in accordance with increasing insect catches in traps and be performed not later than during the stage of intense growth of the pest population (black arrow above the curve peak).

tistasis without application of an insecticide (e.g. 13, 23).

Another possibility of producing insectistasis in storage insects is prevention of mating by permeation of the atmosphere with overdosed sex pheromones. Disrupted mating may either result from confused recognition of females or neural habituation of conspecific males, due to excessive availability of pheromone sources. A severe infestation by *Cadra cautella* could be subdued to the level of insectistasis by a microencapsulated sex pheromone being amply dispersed in a food store (24), while a dense population of *Anagasta kuehniella* was markedly suppressed in a confectionery store due to continuous emission of TDA from numerous dispensers (25).

The attracticide method (18) involves the luring effect of a pheromone of some other semiochemical in combination with the lethal action of an insecticide. In a flour mill, long-term application of numerous laminar dispensers (20 × 20 mm) impregnated with TDA (2 mg) and Cypermethrin (5 mg), was found to suppress the male population of *Anagasta kuehniella* to the level of insectistasis (18, 26).

The main advantage of insectistasis is the maintenance of a minimized pest population combined with considerable restriction of insecticidal applications in food stores, whereby the risk of ecological contamination and development of insecticide-resistant strains would be diminished and the cost-benefit ratio of storage protection be improved. It may be adequate to close this essay by an exhortative statement of Smith and Reynolds (27): "The philosophy of pest control based on eradication of the pest species is the antithesis of integrated pest control ..."

REFERENCES

1. Adler, C. Vorratsschutz mit stickstoff und kohlenstoffdioxid. Mitt. Biol. Bundesanst. Land-Forstwirtsch **1998**, *342*, 277–293.
2. Levinson, H.Z.; Levinson, A.R. Integrated manipulation of storage insects by pheromones and food attractants—a proposal. Z. Angew. Entomol. **1977**, *84*, 337–343.
3. Fields, P.G.; Muir, W.E. Physical Control. In *Integrated Management of Insects in Stored Products*; Subramanyam, B., Hagstrum, D.W., Eds.; Marcel Dekker, Inc.: New York, 1995; 195–221.
4. Calderon, M. Aeration of grain–benefits and limitations. EPPO Bulletin **1972**, *6*, 83–94.
5. Plarre, R.; Halverson, S.; Burkholder, W.; Bigelow, T.; Misenheimer, M. Einsatz von energiereichen hochfrequenten mikrowellen zur selektiven bekämpfung von schadinsekten im getreide (Vorratsschutz). Mitt. Dtsch. Ges. Allg. Angew. Ent. **1999**, *12*, 155–160.
6. Wohlgemuth, R.; Reichmuth, Ch. Verpackung zum schutz von vorräten gegen insekten. Mitt. Biol. Bundesanst. Land-Forstwirtsch. **1998**, 325–341.
7. Levinson, A.; Levinson, H. Reflections of structure and function of pheromone glands in storage insect species. Anz. Schädlingskde, Pflanzenschutz, Umweltschutz **1995**, *68*, 99–118.
8. Burkholder, W.E. In *Application of Pheromones for Manipulating Insect Pests of Stored Products*, Proceedings of a Symposium on Insect Pheromones and Their Applications, Kono, T., Ishii, S., Eds.; Ministry of Agriculture and Forestry: Tokyo, 1976; 111–122.
9. Fleurat Lessard, F.; Pimaud, M.F.; Cangardel, H. Effets de Doses elevées de ZETA sur *Plodia interpunctella* Huebner (Lepidopterea, Pyralidae) dans les Stocks de Pruneaux d'Agen. In *Compte-Rendu de la Reunion sur Les Phéromones Sexuelles des Lépidoptères*; INRA Centre de Recherches de Bordeaux: Aquitaine, 1976; 163–169.
10. Levinson, H.Z. Possibilities of using insectistatics and pheromones in the control of stored product pests. EPPO Bull. **1974**, *4*, 391–416.
11. Barak, A.V.; Burkholder, W.E.; Faustini, D.L. Factors affecting the design of traps for stored product insects. J. Kansas Entomol. Soc. **1990**, *63*, 466–485.
12. Buchelos, C.Th.; Levinson, A.R. Efficacy of multisurface traps and lasiotraps with and without pheromone addition, for monitoring and mass-trapping of *Lasioderma serricone* F. (Col., Anobiidae) in insecticide-free tobacco stores. J. Appl. Entomol. **1993**, *116*, 440–448.
13. Trematerra, P. In *The Use of Sex Pheromones to Control Ephestia kuehniella Zeller (Mediterranean Flour Moth) in Flour Mills by Mass Trapping and Attracticide (Lure and Kill) Methods*, Proceedings of the 6th International Work-

ing Conference on Stored Products Protection, CSIRD: Canberra, 1994; 375–384.
14. White, N.D.G.; Arbogast, R.T.; Fields, P.G.; Hillmann, R.C.; Loschiavo, S.R.; Subramanyam, B.; Thorne, J.E.; Wright, V.F. The development and use of pitfall and probe traps for capturing insects in stored grain. J. Kansas Entomol. Soc. **1990**, *63*, 506–525.
15. Phillips, T.W. Semiochemicals of stored-product insects: research and applications. J. Stored Prod. Res. **1997**, *63*, 17–30.
16. Plarre, R. Pheromones and other semiochemicals of stored product insects—a historical review, current application and perspective needs. Mitt. Biol. Bundesanst, Land-Forstwirtsch **1998**, *342,* 13–83.
17. Trematerra, P. Integrated pest management of stored-product insects: practical utilization of pheromones. Anz. Schädlingskde, Pflanzenschutz, Umweltschutz **1997**, *70,* 41–44.
18. Trematerra, P.; Capizzi, A. Attracticide method in the control of *Ephestia kuehniella* zeller: studies on effectiveness. J. Appl. Entomol. **1991**, *111*, 451–456.
19. Traynier, R.M.M. Sexual behavior of the Mediterranean flour moth *Anagasta kuehniella*; some influences of age, photoperiod and light intensity. Can. Entomol. **1970**, *102*, 534–540.
20. Takahashi, F. Sex pheromones, are they really species-specific? Memoirs College Agric. Kyoto Univ. **1973**, *104*, 13–21.
21. Levinson, H.Z.; Levinson, A.R. Atractifs, repulsifs et phéromones en tant qu'insectistatiques dans le milieu de stockage. Le Cahiers de la Recherche Agronomie (Maroc) **1982**, *39*, 189–216.
22. Levinson, H.Z.; Levinson, A.R. Use of pheromone traps for the proper timing of fumigation in the storage environment. EPPO Bull. **1985**, *15*, 43–50.
23. Levinson, A.R.; Buchelos, C. Th. Population dynamics of *Lasioderma serricorne* F. (Col., Anobiidae) in tobacco stores with and without insecticidal treatments: a three-year survey by pheromone and unbaited traps. J. Appl. Entomol. **1988**, *106*, 201–211.
24. Prevett, P.F.; Benton, F.P.; Hall, D.R.; Hodges, R.J. Santos Serodio Dos, R. Suppression of mating in *Ephestia cautella* (Walker) (Lepidoptera: Phycitidae) using micro-encapsulated formulations of synthetic sex pheromone. J. Stored Prod. Res. **1989**, *25*, 147–154.
25. Süss, L.; Locatelli, D.P.; Marrone, R.V. Ulteriori conoscenze sulle possibilità di attuare la ''tecnica confusionale'' nei riguardi di *Ephestia kuehniella* (Zeller), Atti del 6. Simposio Difesa Antiparassitaria nelle Industrie Alimentari (Piacenza) **1997**, 135–142.
26. Trematerra, P. The use of attracticide method to control ephestia kuehniella zeller in flour mills. Anz. Schädlingskde, Pflanzenschutz, Umweltschutz **1995**; *68,* 69–73.
27. Smith, R.F.; Reynolds, H.T. In *Principles, Definitions and Scope of Integrated Pest Control*, Proceedings of the FAO Symposium on Integrated Control; FAO: Rome, 1966; 11–17.

INSECT-VECTORED CROP DISEASES

Peter A. Burnett
Lawrence M. Kawchuk
Agriculture and Agri-Food Canada, Lethbridge, Alberta, Canada

INTRODUCTION

Insect transmission of crop diseases is of economic importance because it is the mechanism by which many diseases are commonly spread in nature. Pesticides are often applied in an attempt to prevent disease transmission. In 1998, world trade in pesticides exceeded $23 billion. Plant viruses are the most common insect-transmitted pathogens and insect–virus interactions have been relatively well characterized. Examples of bacterial, fungal, and phytoplasma diseases vectored by insects are also discussed.

VIRUS DISEASES

At present there are approximately 1000 plant viruses that have been recognized but the proportion that are insect transmitted is not accurately documented. Virus diseases have been known for more than a century to reduce the yield and quality of crops (1). For example, barley yellow dwarf virus (BYDV), a common luteovirus of cereals and grasses causes worldwide yield losses of 1–3% but may cause yield losses of 20–35% when epidemics occur in specific areas.

Virus spread can be local or long distance and again can be illustrated by discussing BYDV. This virus is persistently transmitted by aphids. These aphids can either move locally or be transported long distances in jet streams. This means that an infective aphid leaving a crop in the southern United States could be deposited into emerging crops in Canada if ideal weather conditions existed.

Transmission of viruses from one plant to another by arthropods is the main means of spread in the crop for many viruses (2). There are 10 orders of phytophagous insects including Collembola, Orthoptera, Dermaptera, Coleoptera, Lepidoptera, Diptera, Hymenoptera, Thysanoptera, Hemiptera, and Homoptera. The first seven orders listed are all chewing insects that are capable of mechanical transmission. The Homoptera are the most important vectors perhaps because the mouthparts cause minimal damage while feeding, thereby minimizing damage of cells that are required for virus multiplication.

Transmission of plant viruses by insects can be classified as nonpersistent, semipersistent, or persistent to indicate the length of time that the vectors remain infective following acquisition of the virus. Nonpersistent transmission occurs within minutes, semipersistent transmission within hours, and persistent transmission over days, weeks, or months. Persistently transmitted viruses may be circulative or propagative within the vector. Circulative viruses do not multiply in their vectors and are not transmitted to progeny. Propagative viruses are capable of multiplication within the vector and some are acquired by progeny.

Insect Vectors

Aphididae

Aphids are the most important insect vectors of plant viruses, with more than 200 species transmitting almost 300 viruses. Aphid transmission of viruses is most often nonpersistent but circulative and propagative transmission also occurs. Examples of viruses that are transmitted in a nonpersistent manner are found in Potyvirus, Carlavirus, Caulimovirus, Cucumovirus, Alfamovirus, and Fabavirus groups. These viruses have helical and isometric particles, DNA and RNA nucleic acid, and mono-, bi-, and tripartite genomes. Aphids usually probe a leaf and can quickly contaminate the stylet tip, food canal, and foregut. There is no detectable latent period or evidence of virus in the hemocoel or salivary components. The virus is subsequently injected into healthy tissue in another exploratory feeding since aphids are capable of ingestion and regurgitation. The specificity sometimes observed between viruses and aphid vectors during nonpersistent transmission appears to be dependent on the coat protein and specific adsorption to aphid cells. Aphids often lose the ability to transmit within minutes following acquisition and this impacts spread of the virus. Some potyviruses and caulimoviruses require the presence of another virus to provide a helper component to allow transmission by aphids in a nonpersistent manner. A few aphid-transmitted viruses are referred to as semipersistent in that they have characteristics intermediate between nonpersistent and circulative transmission with acquisition periods of several minutes and retention periods of several hours.

Transmission of persistent viruses is usually by one or a few aphid species, a latent period follows acquisition, and virus is ejected in saliva from the maxillary saliva canal. The virus may replicate in the vector. Several rhabdoviruses are propagative within their aphid vector and have a long latent period. These viruses are found in many aphid tissues, can be serially injected from one aphid to another by injecting hemolymph, and are transmitted through eggs.

The luteoviruses represent the best characterized group of circulative viruses that do not replicate in the aphid. Luteoviruses are confined to the phloem tissues and penetration by the aphid of a sieve element is required for acquisition. The virus moves from the gut lumen to the hemocoel and a high level of specificity occurs at the accessory salivary glands. Luteoviruses are capable of phenotypic mixing during replication that results in the genome being encapsulated by another coat protein that confers transmission by another aphid species. Genetic engineering of luteovirus full-length infectious clones has identified specific viral genes such as the coat protein read through that are critical for transmission. Certain luteoviruses are also required as helper viruses for transmission of specific unrelated viruses via phenotypic mixing. Recently, endosymbiotic microbes within the aphid gut have been identified that are necessary for the transmission of luteoviruses.

Auchenorrhyncha

In contrast to aphids, planthoppers and leafhoppers transmit less than 50 viruses. There is no evidence for nonpersistent transmission but a few are transmitted in a semipersistent fashion. However, the majority of viruses that are transmitted by hoppers are circulative and transmission involves movement of the ingested virus to the salivary glands. A few geminiviruses are transmitted by leafhoppers in a circulative nonpropagative fashion. Transmission of geminiviruses is determined by the coat protein and the gut wall appears to regulate vector specificity. Several reoviruses, rhabdoviruses, tenuiviruses, and marafiviruses are transmitted by hopper vectors in a propagative manner. Transovarial transmission has been observed with many of these viruses and they are often found in many different tissues. Although acquisition occurs within minutes, the latent period may be several hours or months and nymphs are often more efficient vectors than adults.

Coleoptera

Beetles have biting mouthparts and vector several plant viruses (tymoviruses, comoviruses, bromoviruses, and sobemoviruses) that share many characteristics. All have icosohedral particles, possess a single-stranded positive-sense RNA genome, produce high titers, have relatively narrow host ranges, and are relatively stable and easy to transmit mechanically. Viruses are acquired quickly, do not have a latent period and may be retained for several days or longer periods at lower temperatures. Larvae and adults can acquire and transmit viruses as they regurgitate during feeding. Viruses are found in the hemolymph of some beetles but do not replicate. Vector specificity appears to result from interactions of the virus with the host plant and the enzymes contained within the regurgitant.

Aleyrodidae

Whitefly transmission of viruses is by nonpersistent, semipersistent, or persistent mechanisms. They have been reported to transmit geminiviruses, closteroviruses, carlaviruses, potyviruses, nepoviruses, and luteoviruses. There is no evidence for transmission through eggs or of multiplication in the vector. Specificity has been reported to involve the coat protein and sites within the salivary gland.

Arachnida

The eriophyid mites are capable of transmitting more than one virus simultaneously. Viruses are acquired by nymphs but not adults and remain infective for several days following virus acquisition. Lower temperatures can increase retention. There is evidence of a circulative relationship between virus and vector.

Thysanoptera

Thrips tabaci is an important vector of tomato spotted wilt tospovirus (TSWV). Several generations of winged adults are produced each year that feed on subepidermal cells but only larvae can acquire TSWV. Tospoviruses are persistently transmitted by thrips and multiply in the vector. Transmission of tobacco ringspot nepovirus and tobacco streak ilarvirus by thrips has been reported.

Coccoidea and Pseudococcoidea

Mealybugs are much less mobile than other vectors and have been reported to transmit only 10 plant viruses. They feed on the phloem and exhibit some similarities to nonpersistent aphid transmission.

Miridae

Mirids feed using a stylet and vector several sobemoviruses in a manner similar to that of beetles.

Piesmatidae

A piesmatid has been shown to vector a rhabdovirus in a propagative fashion but there is no evidence of transovarial transmission.

BACTERIAL DISEASES

Bacteria also cause serious diseases of many crops (3, 4). There are approximately 200 species of plant pathogenic bacteria. Bacteria are spread by splashing rain, plant to plant contact, plant debris, seed, animals including humans, and insects.

Bacteria can multiply rapidly inside plants where they may cause death of cells (necrosis), abnormal growth (galls), plugging of water-conducting tissue (wilting), or breakdown of tissue structure (e.g., soft rot). An example of an insect-transmitted bacterial disease is Stewarts bacterial wilt of corn. Infected corn wilts rapidly resembling plants suffering from drought. Leaves show linear pale green or yellow lines that may extend the length of the leaf. These lines soon become pale and brown. Infected plants may fill grain earlier. Sweet corn is generally more susceptible than most field corn.

A range of insect vectors is associated with the spread of this disease but the most important vector is probably the corn flea beetle. There are many other insect-transmitted bacterial diseases including common pathogens of potatoes and beans.

FUNGAL DISEASES

Fungi cause the most common and the most destructive diseases of cultivated crops. Spores are the principal inoculum of pathogenic fungi and these are extremely varied and adapted to various forms of dissemination. Wind, water, humans, and animals spread pathogenic fungi and relatively few depend on insects alone for transmission.

A good example of an insect-transmitted fungal disease is ergot *Claviceps purpurea* (5, 6). This is a disease of cereals and grasses which has been known since the ninth century and ergot poisoning has caused many health problems in both farm animals and humans. Historically, severe chronic ergot poisoning eventually would turn individuals into screaming, gibbering individuals whose fingers, toes, arms, and legs blackened and corroded away from dry gangrene. The name pertains to the conspicuous dark purple-black ergot body or sclerotium that is visible in infected grain head. However, it is the honeydew or asexual stage of this fungus that attracts insects and these transport the spores to other flowering plants. Fortunately the elucidation of the disease cycle of ergot has lead to successful control measures and, with the exception of the pandemic that is occurring in sorghum, ergot is essentially a rare disease.

PHYTOPLASMAL DISEASES

Phytoplasmas are intermediate in some respects between viruses and bacteria. Mycoplasma-like organisms (MLOs) of indefinite form that are surrounded by a membrane represent the main group of phytoplasmas. MLOs have been associated with approximately 200 yellow diseases (e.g., Aster Yellows) and these are typically transmitted by leafhoppers and restricted to sieve tubes (3, 4).

CONCLUSIONS

Additional pathogens and vectors continue to be discovered. Advances in understanding the vector–pathogen relationships will assist in determining epidemiology and developing disease control strategies. Considerable amounts of insecticides are used each year in an attempt to control viral, bacterial, and fungal diseases by reducing vector populations but these measures are only partially effective.

REFERENCES

1. Hadidi, A.; Khetarpal, R.K.; Koganezawa, H. *Plant Virus Disease Control*; APS Press: St. Paul, Minnesota, 1998; 684.
2. Nault, L.R. Arthropod transmission of plant viruses: a new synthesis. Ann. Entomolo. Soc. Am. **1997**, *90*, 521–541.
3. Carter, W. *Insects in Relation to Plant Disease*; John Wiley & Sons: New York, 1973; 759.
4. Harris, K.F.; Maramorosch, K. *Vectors of Plant Pathogens*; Academic Press: New York, 1980; 467.
5. Bandyopadhyay, R.; Frederickson, D.E.; McLaren, M.W.; Odvody, G.M.; Ryley, M.J. Ergot: A new disease threat to sorghum in the Americas and Australia. Plant Dis. **1998**, *82*, 356–367.
6. *Insect–Fungus Interactions*; Wilding, M., Collins, N.M., Hammond, P.M., Webber, J.F., Eds.; Academic Press: New York, 1989; 344.

INTEGRATED FARMING SYSTEMS

John M. Holland
The Game Conservancy Trust, Fordingbridge, Hampshire, United Kingdom

INTRODUCTION

The system of agricultural production has undergone immense changes in the past three decades (1). Crop yields have increased through a combination of plant breeding, intensive use of pesticides and inorganic fertilizer, and greater mechanization. Farming systems based on high inputs of agrochemicals and intensive tillage are now recognized as being unsustainable because they cause soil erosion, soil nutrient depletion, environmental pollution, increased pest problems, and public health hazards. In addition the profitability of cereal and oil crop production also declined at the end of the 20th century forcing farmers to examine ways of reducing their production costs (1). As a consequence farming is facing economic, political, environmental, and social pressures that are forcing changes in the methods of crop production. In response, a range of different production systems has evolved that try to address these issues. The most extreme are the organic or biological systems in which agrochemical use is almost eliminated. Various systems have been defined that are between the extremes, such as integrated farming systems, and these aim to use lower inputs and as such are regarded as being more sustainable (2). Many definitions exist for integrated farming but the most appropriate is all encompassed in the description by El Titi (3) ''*an holistic pattern of land use, which integrates natural regulation processes into farming activities to achieve a maximum replacement of off-farm inputs and to sustain farm income*.'' Integrated farming therefore incorporates the principles of integrated crop management (ICM) and integrated pest management (IPM) but takes a long-term, whole-farm approach that considers all aspects of crop production and land management (2). The emphasis is on preserving farm profitability by optimizing inputs, although consequently there may be ecological benefits and overall greater sustainability (4).

WHAT IS FORCING THE CHANGES?

The driving forces where integrated farming has been adopted vary between countries and farmers. The main influences are outlined below.

Legislation

The withdrawal of many pesticides (e.g., in Denmark) has led to some farmers having no option but to adopt integrated or organic systems. Some countries (e.g., The Netherlands) have set targets for national conversion to sustainable agriculture based on integrated farming (5).

Economic Pressure

Recent declines in cereal and oil seed values and price support has made European farmers look closely at their variable and fixed production costs and therefore a more integrated approach. Agrienvironment payments encourage better habitat management. Cross compliance, modulation, etc., if adopted may force more widespread changes.

Green Marketing Incentive

Crop assurance schemes may include integrated practices. These may provide a premium over conventionally produced food or are a requirement for the larger supermarket retailers (5).

Environmental

The widespread decline in farmland wildlife is now recognized by governments, farmers, environmental organizations, and the general public (1, 6). To mitigate some of these changes farmers are adopting some of the principles of integrated farming, especially with respect to pesticide use and management of their ecological infrastructure.

Agronomic

When pesticide resistance reaches such levels that chemical control is no longer possible or economical then farmers have to look for alternative solutions (7).

DEVELOPMENT OF INTEGRATED FARMING

The integrated approach incorporates all aspects of farm management and therefore requires knowledge of the interactions between components in the system, before the concept can be promoted to farmers. In many countries, long-term farm studies have been conducted to develop an integrated system and in some cases economic and environmental comparisons were made with the conventional approach (4, 8). Overall integrated farming proved equally profitable (4, 8). Lower yields could be expected, but this loss was compensated for by using lower inputs of agrochemicals and energy. In the United Kingdom results from nine farm-scale studies in which integrated and conventional farming was compared were pooled for analysis. In the integrated system substantial reductions in inputs were achieved without losing crop quality (Table 1).

The reduction in agrochemical inputs is likely, but as yet not satisfactorily proven to reduce the risk of pollution and toxic or sublethal effects to nontarget organisms. Nevertheless, some environmental benefits were attained (1, 6). Some studies reported greater numbers and diversity of soil fauna where an integrated system was used, although this was not always apparent. Beneficial insects vary naturally in their abundance between years, fields, and different crops and these fluctuations were usually greater than those between the two farming systems. In small-scale (often < 5ha) trials conventional inputs appeared to have few long-term, direct effects on nontarget arthropods, however, farm-scale monitoring in the United Kingdom for the past 20 years has indicated that many arthropods have declined since the advent of intensive farming and are showing little sign of recovering (1). In all farming system experiments weed, disease, and insect pest infestation levels were routinely monitored to aid crop protection decisions. Most studies reported lower disease and insect pest incidence, but higher weed burdens in the integrated system, although rigorous experimental appraisals were seldom given.

Table 1 Integrated farming inputs and their costs as a percentage of conventional systems

Input	% Change in quantity of use	% Change in cost
Fungicides	−50	−41
Herbicides	−42	−32
Insecticides	−40	−42
Fertilizers	−18	−15
Seeds	+10	+12
Operational	–	−9

(Adapted from *Integrated Farming: Agricultural Research into Practice*; MAFF Publications: London, 1998.)

PRINCIPLES OF INTEGRATED FARMING

Arable crops are complex ecological systems (1) and consequently a blueprint for their management using an integrated approach cannot be applied universally. Each farm, crop, and even each field must be considered separately based upon the previous history, current crop monitoring, and the farmers' individual objectives. There are, however, some broad principles that apply (2, 3, 5, 7, 9).

Multifunctional Crop Rotation

Rotation is the core of any integrated system with a minimum of four different crops, although a longer rotation may add more opportunities to build soil fertility and prevent disease and pest carryover. Although the choice of crops will depend on locality, markets, and prices, there are some broad principles. Crops that build soil fertility (leguminous crops and those with high inputs of organic matter, e.g., grasses) must be alternated with those that reduce soil physical and chemical fertility (root crops with low input of organic matter, e.g., potatoes and sugar beet). The sequence of each crop can also be used to ensure there are opportunities to control grass weeds in broadleaved crops and that pest organisms are kept below economic thresholds by ensuring that no single type or group of crops are grown in succession. There may have to be compromises and the rotation chosen will reflect the economic, agronomic, or environmental priorities of individual farmers.

Integrated Nutrient Management

Inputs of inorganic fertilizers are essential to maintain yield and quality but must be balanced to ensure supply does not exceed demand by more than 10%. Excessive nutrient supply encourages weeds, pests, and diseases and causes pollution. Application should maximize uptake by the crop and minimize loss to the environment. Where possible inorganic fertilizer should be replaced with animal and green manures. This has many benefits:

- Improved soil nutrient supply and physical structure
- Encouragement of beneficial organisms such as predatory invertebrates and microbials

Lower cost
Savings in nonrenewable resources

Optimizing nutrient supply through judicious use of off-farm inputs, organic manures wherever possible, and choice of a nutrient supplying crop rotation will generate economic benefits for the farmer and the environment, especially water quality.

Minimum Soil Cultivation

Soil cultivations require high inputs of energy and machinery, are damaging to beneficial soil fauna, and encourage erosion and nutrient leaching. Adopting a soil cultivation system to minimize these effects is an essential component of any integrated system. The type and intensity of cultivation must be chosen according to soil type and agronomic concerns (e.g., weed species composition). Conservation tillage/noninversion tillage in combination with mulching should be maximized thereby concentrating soil organic matter in the upper soil layers. This improves soil structure and workability, decreases erosion and leaching, improves drought tolerance, and creates more favorable conditions for beneficial predatory invertebrates, soil microorganisms, and earthworms. Weeds may be encouraged but these may be overcome by utilizing the benefits of rotation and targeted weed control. Considerable cost savings may be made and this is the driving force behind the recent interest in the noninversion tillage in western Europe.

Integrated Crop Management

Integrated crop management encompasses IPM but extends the principles to the control of weeds and diseases. The approach maximizes natural regulatory mechanisms through adoption of the above components and uses plant resistance, time of planting, economic thresholds, and the most environmentally safe pesticides. This requires more intensive crop monitoring and in some cases greater tolerance of pests. Mechanical weed control combined with low doses of herbicide may be appropriate. Encouragement of pest predators through provision of suitable habitat and trap cropping is also recommended.

Ecological Infrastructure Management

This is the provision of a suitable proportion of noncrop or ecological areas that if correctly managed can

Provide a source of beneficial species for pest control;
Act as buffer zones along ecologically sensitive areas such as watercourses preventing nutrient loss and soil erosion;
Provide a refuge from which reinvasion may occur after any adverse agricultural operations;
Prevent weed ingress into fields; and
Act as a source of biodiversity.

These areas may be field margins, hedgerows, fence lines, grass, or herbaceous strips arranged to ensure the maximum connectivity while keeping individual field size to a minimum. A minimum area of 5% of cropped land is recommended (3) although the ecological gains will be dependent on the quality of these areas. Research has identified the potential of different ecological areas to provide pest control and act as buffer zones.

Implementing an Integrated System

The approach requires considerable knowledge and commitment by the farmer. A whole farm evaluation helps the transition from high to lower inputs. On-farm trials may be needed to ascertain the most effective system. Extension services are needed for the transition period and afterward if it is to remain successful (2). New technology, such as decision support and precision farming techniques, are advocated.

REFERENCES

1. *The Ecology of Temperate Cereal Fields*; Carter, N., Darbyshire, J.F., Potts, G.R., Firbank, L.G., Eds.; Blackwell Scientific Publications: Oxford, UK, 1991; 469.
2. Vereijken, P. A methodic way to more sustainable farming systems. Neth. J. Agric. Sci. **1992**, *40*, 209–223.
3. El Titi, A.; Boller, E.F.; Gendrier, J.P. *Integrated Production—Principles and Technical Guidelines*; IOBC/WPRS Bulletin 16 (1), IOBC/WPRS: France, 1993; 96.
4. Jordan, V.W.L. The development of integrated arable production systems to meet potential economic and environmental requirements. Outlook on Agric. **1998**, *27* (3), 145–151.
5. Proost, J.; Matteson, P. Integrated farming in the Netherlands: flirtation or solid change. Outlook on Agric. **1997**, *26* (2), 87–94.
6. *Ecology and Integrated Farming Systems*; Greaves, M.P., Glen, D.M., Eds.; John Wiley & Sons: Chichester, UK, 1995; 329.
7. Tiwari, P.N. Integrated farming research for sustaining food production. J. Nucl. Agric. Biol. **1993**, *22* (1), 1–13.
8. Holland, J.M.; Frampton, G.K.; Çilgi, T.; Wratten, S.D. Arable acronyms analysed—a review of integrated arable farming systems in Western Europe. Ann. Appl. Biol. **1994**, *125* (2), 399–448.
9. Integrated Farming Systems. www.gct.org.uk/index.html (accessed July 2000).

INTEGRATED PEST MANAGEMENT

Helmut van Emden
The University of Reading, Reading, Berkshire, United Kingdom

INTRODUCTION

There have been many definitions of integrated pest management. The most practical operational definition is probably that by President Jimmy Carter of the United States in his 1979 environmental message: Integrated Pest Management "uses a system approach to reduce pest damage to tolerable levels through a variety of techniques... and, when necessary and appropriate, chemical pesticides." An excellent, more detailed definition is given by Flint and van den Bosch: "Integrated pest management (IPM) is an ecologically based pest control strategy that relies heavily on natural mortality factors such as natural enemies and weather and seeks out control tactics that disrupt these factors as little as possible. IPM uses pesticides, but only after systematic monitoring of pest populations and natural control factors indicates a need. Ideally, an integrated pest management program considers all available pest control actions, including no action, and evaluates the potential interaction among various control tactics, cultural practices, weather, other pests, and the crop to be protected" (1).

IPM as defined above represents, as a result of usage, a rather different concept from Apple and Smith's (2) original 1976 definition (see below) and is so well used that it is often written or spoken as IPM without further explanation. IPM is now the preferred option to replace both prophylactic (= insurance) reliance on insecticides and the concept of replacing insecticides with single-component biological control. Both these latter methods are now universally seen as components of IPM, though biological control alone still has its niche in low value subsistence agriculture and other situations where insecticide cost is prohibitive or where the pest has multiple tolerance to insecticides.

THE DERIVATION OF INTEGRATED PEST MANAGEMENT

Following on from problems created by the extensive use of pesticides in the 1940s and 1950s, particularly the selection of pest strains resistant to insecticides and damage to biological control, workers at the University of California at Berkeley evolved the concept of "integrated control." This was based on their work combining biological control with reduced and more selective pesticide doses against pesticide-resistant spotted alfalfa aphid (*Therioaphis maculata*), published in a landmark paper in the scientific journal *Hilgardia* in 1959 (3). The more selective pesticide doses improved the survival of natural enemies while still killing a proportion of the aphids. This principle was reflected in the definition of "integrated control" by the Californians as "pest control which combines and integrates biological and chemical control. Chemical control is used as necessary and in a manner which is least disruptive to biological control." This episode is the origin of the "I" in IPM.

Following on from an Australian concept of "managing" pest populations, the Canadian Beirne proposed "Pest Management" at a conference in Raleigh, North Carolina, in 1970 (4). It was an all-embracing expression for the selection of that combination of control measures or that single measure that was "in the best long term interests of mankind." Beirne envisaged that sometimes a single measure, for example, biological control alone, might be a better solution in environmental terms than the use of more than one measure implicit in "integrated control." Whether this represented any substantive change in thinking is debatable, especially as Beirne himself conceded that most often "integrated control" would be the preferred "pest management" solution. Be this as it may, the replacement of the word "control" with "management" was appealing and topical, and it caught on.

In 1976, there appeared a book (2) entitled *Integrated Pest Management* edited by Lawrence Apple of North Carolina State University and Ray Smith of The University of California at Berkeley. Their description of IPM ran as follows: "The concept of pest management has now been broadened to include all classes of pests (pathogens, insects, nematodes, weeds) and in this context is commonly referred to as IPM with the implication of both methodological and disciplinary integration." As pointed out earlier, the ''disciplinary integration'' implications have never been generally accepted. Today, the next step in the evolution of the terminology is to take IPM in Apple and Smith's sense and to see this as just one aspect of crop production. The term ''Integrated Crop Management'' now seems to be gathering momentum.

THE NEED FOR INTEGRATED PEST MANAGEMENT

The intensification of agriculture with the overuse of and overreliance on synthetic pesticides in the 1940s and 1950s caused serious side effects including the resurgence of the target pests, the appearance of new pests, and the development of tolerance to individual pesticides in many pest species. It was therefore clear that pests had been ''mismanaged,'' and the following failures to manage them correctly could be identified:

1. Overdosing with pesticides, leading to the development of tolerance;
2. Loss of biological control as result of pesticide use and a reduction of biodiversity in agroecosytems;
3. The introduction of genetically uniform, high-yielding but pest susceptible new cultivars over large areas of monoculture;
4. Agronomic changes such as abandonment of cultural controls and of mixed cropping.

These translate respectively into the four main building blocks of IPM (5):

(a) *Chemical control:* Decisions to spray are guided by economic thresholds, and selective materials are chosen where available;
(b) *Biological control:* Use of pesticides in a selective manner. Natural enemies can also be introduced or recolonized and/or promoted by habitat modification, including the planning of biodiversity into farm management;
(c) *Host plant resistance:* The introduction of crop varieties with at least partial plant resistance, based as far as possible on more than a single mechanism;
(d) *Cultural control:* Can break the life cycle of a pest or greatly improve conditions for natural enemies, but must be compatible with farm management systems.

Other components of IPM, such as insect growth regulators, the confusion pheromone technique, and sterile male release, are also available in relation to particular targets.

As IPM for a single target implies managing the pest population by more than one method, the following two ''golden rules'' of IPM are particularly important:

(a) If a single method gives adequate control on its own, then there is danger of a tolerant pest strain increasing in gene frequency and no opportunity to use a second method in addition. The method therefore needs to be made less efficient (reduced dose of pesticide, partial host plant resistance rather than immunity) for there to be value in introducing another control method to supplement it;
(b) Methods are increasingly worth combining to the extent that the control then achieved exceeds the additive effects of the two methods in isolation.

THE DEVELOPMENT OF PEST MANAGEMENT PACKAGES

There are three distinct philosophies to the synthesis of pest management packages (5).

Menu Systems

These provide considerable choice for a farmer to select control methods for combination in relation to the pest spectrum in the area. The results of many scientist-years of research into different methods enable menus to be provided but, as the choice of the farmer may vary, the methods will be highly target-specific and not dependent on interaction with each other. Often several methods may be available against one pest species. Cotton pest management is a good example of a menu system. Menu systems may be transferable to the same crop in another country or continent with little modification.

Computer-Derived Packages

If sufficient data are available, a computer should be able to select the components of a menu that are in the best interests of whatever selection criteria the computer is given. Some sophisticated models driven by crop growth via the weather are already available for orchard systems. Again, however, interactions between control methods are unlikely to be exploited, as they are rarely adequately understood.

Protocols

These are ''dish-of-the-day'' menus—that is, a complete package on a ''take it or leave it'' basis. These are characteristic of crop scenarios where there is a paucity of the detailed long-term research needed for a menu system. Protocols may be very simple. For example, quite effective IPM has been introduced in developing agriculture by little more than combining the introduction of economic thresholds with more selective pesticidal materials (e.g., bacterial toxins). Later stages may be the release of mass-produced beneficials (particularly the egg parasitoid *Trichogramma*) and the introduction of partial plant resistance against a key pest following cultivar testing and plant breeding at national research stations.

Protocols may also be derived directly by experimentation, without the need for the extensive background biological and ecological understanding needed for menu systems.

CONCLUSIONS

Apart from IPM being seen as one component of a holistic integration of crop production (integrated crop management), the concept is the widely accepted view of what pest control in modern crop production should be. In many countries around the world it has been declared as the national policy; it is also the policy of international agencies such as the Food and Agriculture Organization. It is the approach being used by many crop consultants and the method favored in developed countries by the large supermarkets, which have such an influence on farm management today.

Until relatively recently, the integrated approach to pest control was accepted temporarily by growers at times of a crisis of pest tolerance to insecticides, to be abandoned just as soon as a new and effective pesticide was marketed. However, IPM is now being introduced even where there is not a current crisis; at long last the message seems to have sunk in, that it is a way of preventing pesticide crises in the first place.

REFERENCES

1. Flint, M.L.; van den Bosch, R. *Introduction to Pest Management*; Plenum: New York, 1981; 1–240.
2. Apple, J.L.; Smith, R.F. *Integrated Pest Management*; Plenum: New York, 1976; 1–200.
3. Stern, V.M.; Smith, R.F.; van den Bosch, R.; Hagen, K.S. The integrated control concept. Hilgardia **1959**, *29* (2), 81–101.
4. Beirne, B.P. The Practical Feasibility of Pest Management Systems. In *Concepts of Pest Management*; Rabb, R.L., Guthrie, F.E., Eds.; North Carolina State University: Raleigh, NC, 1970; 158–169.
5. van Emden, H.F.; Peakall, D.B. *Beyond Silent Spring: Integrated Pest Management and Chemical Safety*; Chapman & Hall: London, 1996; 1–322.

INTEGRATION OF TACTICS

John All
University of Georgia, Athens, Georgia, U.S.A.

INTRODUCTION

For many people, using two or more insect management tactics in an integrated manner is common sense. Also, it is reasonable to believe that additive or even synergistic benefits can be achieved by blending multiple tactics in a coordinated program of Integrated Pest Management (IPM). Unfortunately, logic is not always the driving force of actions taken in pest management programs, and dollar losses and other damages occur by inefficient use of only one of an array of possible pest management strategies. The potential values of using IPM have been discussed by several authorities that basically agree on general concept, but place varying importance on particular methodologies such as pesticides, biological control, environmental management, etc.

PREVENTIVE AND SUPPRESSIVE PEST MANAGEMENT PHILOSOPHIES

IPM typically is a blend of preventive and suppressive (curative) pest management methods linked by the use of pest monitoring to determine when thresholds requiring remediation have been reached. Several preventive management tactics can be utilized in IPM, whereas application of insecticides is the most common method used for suppression of pest outbreaks. The manner in which preventive and suppressive control methods are deployed and the pest monitoring/threshold systems that are utilized in IPM programs vary according to many inherent and fluctuating factors unique to a commodity. These factors involve: 1) value relationships of the commodity, 2) hazard for individuals and complexes of resident and transient pests, 3) cost and effectiveness of available control and monitoring methods for pests, and 4) regulatory requirements for pest management.

Value relationships include costs for production of the commodity, value of individual units protected (for example, the unit value of an ear of sweet corn is higher than an ear of field corn), the economic tolerance of the commodity for the presence of pests or damage, and economic injury levels associated with pest damage to the commodity. Commodity value relationships are used in the development of economic thresholds for determining when remedial action is needed against pests.

The known risks for pest problems of a commodity are factors that vary widely in different environments. Pest risk involves a commodity's susceptibility to both resident and transient pests and the aggressiveness of the pest in producing injury. Resident pests build predictable infestations and may be more or less permanent inhabitants within commodity environments. These pests often build infestations under production conditions that favor proliferation of the pest population, such as monocropping of annual or perennial crops. Resident pests often have restricted preferences for certain crops and usually require one or more years per generation. Northern and western corn rootworms, wireworm, and grub species, etc., are examples of resident pests.

In contrast, transient pests are not as predictable as resident pests and produce periodic infestations that may be associated with poor weather or other environmental conditions that stress the commodity and/or favor the pest. Transient pests can have either restricted or polyphagous feeding habits and often have multiple generations annually, allowing populations to multiply rapidly within fields or enabling them to move from one crop to another. Corn earworm, tobacco budworm, armyworm species, aphids, bark beetles, etc., are examples of transient pests.

PEST MANAGEMENT TACTICS

There are many pest management methods available for IPM programs. For simplicity and practical consideration, the various pest management tactics can be classified in the following manner (emphasis is on plant pest management).

Cultural Control

Cultural control is any farm practice that is used with some consideration for IPM. Cultural control is applied ecology and involves the creation of artificial environments that either avoid producing high-risk situations for pest infestations or encourage conditions that are detrimental to pests. Cultural control practices are the foundation of preventive pest management and are often the most cost-efficient management tactics because the operations are usually essential for producing the commodity. Cultural control practices are usually not intended to produce dramatic suppression of pest outbreaks; they are designed to prevent problems from developing. A higher level of professional competence and patience is often required to implement and integrate cultural control tactics because detailed knowledge of biological interaction between pest and commodity is necessary. The many cultural control practices available for pests of different commodities include the following.

Crop and cultivar selection

Crop and cultivar selection involves considering the susceptibility of a crop or cultivar to pests, along with its yield and other desirable agronomic characteristics. It includes decisions for avoiding high-risk crops or cultivars or using pest-resistant crops or cultivars (including transgenic varieties) if available.

Crop rotation

Crop rotation is alternating crop species within an environment (crop field, greenhouse, etc.) to disrupt resident pests that are unable to survive or maintain damaging populations within both crops. Classically, crop rotation has been practiced with annual field crops, but the tactic is useable in multiple cropping situations in vegetable and greenhouse crops.

Sanitation

Sanitation is the destruction of the pest's habitat to deprive it of food, shelter, overwintering sites, etc., or direct removal or destruction of pest-infested debris that is a reservoir for infestation. Sanitation may involve postharvest crop destruction or destruction of weeds within or adjacent to fields that may be refuges for pests or that may be alternate hosts of disease pathogens that are vectored by insects.

Disruption of Phenological Synchrony

Disruption of phenological synchrony between the pest and crop is the manipulation of planting date or time of harvest or using cultivars of desirable maturity to avoid having susceptible crop stages available at the time when peak pest populations are present.

Manipulation of Tillage and Planting Operations

Manipulation of tillage and planting operations to avoid developing hazardous pest environments or to disrupt pest populations are cultural control operations directed at specific pests, and include surface tillage, conservation tillage, and the use of burning or other operations to destroy plant residues before planting crops, etc.

Deception and Concealment Practices

Deception and concealment practices are designed to lure pest populations away from a commodity or hide it from infestations and include the use of trap crops, intercropping, strip-cropping, companion planting, etc.

Pest Endurance Operations

Pest endurance operations are designed to maintain optimum commodity health or status so that pest infestations can be tolerated without economic damage. Use of vigorous seed, precision planting methods, remote monitoring of crop health, optimum fertilization, maximal irrigation, good weed management, pruning, thinning, etc., are practices that can ensure optimum commodity endurance of pest infestations.

Pest Preventive Maintenance Operations

Pest preventive maintenance operations are designed to avoid buildups of pests during static periods of commodity production, manufacture, and storage. Rapid commodity turnover operations after harvest, during processing, packaging, and transport and using a ''first to come in is the first to go out'' type inventory control philosophy are examples of cultural operations that avoid developing susceptible commodity reservoirs for pest infestations.

Biological Control

Biological control is the manipulation of natural enemies (parasitoids, predators, and pathogens, including genetically enhanced varieties) to control pests. Governmental organizations conduct searches, operate rearing activities, and coordinate distribution of exotic natural enemies for management of introduced or native pests (referred to as introduction type biological control) in large-scale regional programs. For pest managers, biological control is concerned with manipulating commodity environments to avoid conditions that are detrimental (conservation) or with using cultural practices that strengthen populations (enhancement) of natural enemies. Pest managers also use introductions of natural enemies in restricted environments to repopulate or rejuvenate existing populations (augmentation) or apply overwhelming numbers of natural enemies temporarily to suppress pests in a similar manner as conventional insecticides (inundation).

Mechanical Control

Mechanical control is the direct mechanical application of lethal physical forces that crush, chop, cook, irradiate, freeze, suffocate, etc., pests. Tactics also include devices and packaging designed to prohibit access of pests to commodities.

Pest Surveillance and Data Analysis for Economic Thresholds and Other IPM Decisions

Actions include direct monitoring (scouting) of pests and natural enemies or of commodity damage and use of indirect surveillance methods, including the various types of pest trapping. Economic thresholds are pest populations or injury levels that threaten commodity losses that are equal to or greater than the cost of pest suppression. In practice, economic thresholds are usually calculated directly from printed information and tables or through the use of computer aids, including artificial intelligence programs and IPM-dedicated websites.

Chemical Control with Enticements or Deterrents

Chemical control with enticements (attractive odors or feeding stimulants) or deterrents (repellent odors, masking agents, or feeding deterrents) are tactics that manipulate pest behavior using attractive odors such as sex pheromones to confuse mating between adults or utilization of baits containing insecticides or sterilants. Repellents and

Table 1 Pest and commodity interactions that must be considered in blending integrated tactics in various locations (example is for the Southeastern USA) and the general trends for utilization of control methods in different types of commodities

Commodities in descending order of value per unit of production	Pest hazards[a]	Economic threshold	Chemical control	Biological control	Cultural control	Mechanical control	Regulatory control
Ornamentals	1	↓	↑	↓	↓	↑	↑
Floriculture	1						
Turf	2						
Fruit	3						
Vegetables	3						
Tobacco	3						
Stored products	2						
Cotton	3						
Corn	2						
Sorghum	2						
Soybean	2						
Small grains	1						
Forage crops	1						
Forest crops	1						

[a]Relative insect problems with respect to actual economic impact.

feeding deterrents are chemicals used to mask commodities or to cause pest repulsion from treated surfaces. Specific examples include mating disruption of pink bollworm, *Pectinophora gossypiella* (Saunders), using sex pheromones in cotton, areawide management of the Mediterranean fruit fly, *Ceratitis capitata* (Wiedemann), with insecticide baits, and the necessity of insect repellents for hikers, hunters, and campers in recreation areas infested with biting flies.

Chemical Control with Insecticides

Chemical control with insecticides is the use of formulated substances that kill pests by intoxication of physiological processes or by inducing lethal disease infections. Insecticides consist of synthetic chemicals, natural products, or microbial pathogens formulated and applied in a variety of ways for prevention of pest infestations or suppression of outbreaks.

Regulatory Control

Regulatory control is participation by pest managers in governmental or institutional programs or adherence to policies governing pest management activities, including pest quarantine and areawide eradication of introduced and native pests. It also is abiding by laws and regulations, designed to protect or indemnify individuals or the public, that vary from obtaining a pesticide applicator's license to complying with refugia requirements for pest resistance management contracted by seed companies selling transgenic crops.

ADJUSTING INTEGRATED TACTICS TO MANAGE FIXED AND VARIABLE COMMODITY CONSTRAINTS FOR PEST MANAGEMENT ACTIONS

Decision making for integrating tactics is influenced by many factors besides the availability of effective pest management methods and hazards for pest problems inherent in a commodity (Table 1). In general, commodities of higher unit value have lower economic thresholds with greater use of insecticides or mechanical controls as protectants or suppressants. Use of these control practices is mediated on the one hand by consumer demands for zero-pest damage and on the other hand by higher regulatory and societal scrutiny for safety of the commodity for consumption. Cultural and biological tactics are usually integrated into preventive control programs for commodities of moderate to lower unit value, especially those where pest damage is somewhat tolerable. Narrow cost margins for pest management tactics occur with commodities of lower unit value because use of insecticides has a major impact on potential profits. Use of pest-resistant crop varieties is one of the first considerations by pest managers of moderate to lower unit value commodities and has served as a foundation for integrating other cultural and biological control methods. In the past, the cost of seeds of pest-resistant crops was not appreciably higher than for nonresistant cultivars, but this changed with the introduction in 1995 of the first transgenic crop varieties (cotton) possessing transgenes derived from the bacterium *Bacillus thuringiensis* (Berliner) (Bt). Pricing of transgenic Bt crop seed is competitive with insecticide costs.

Pest monitoring (scouting) for determination of action thresholds usually occurs with commodities that can tolerate a degree of pest injury, but are of sufficient value to justify one or more insecticide applications. One or more preventive management tactics are often integrated with monitoring and pesticide applications in these commodities. To a large extent, integration of pest management tactics is an art that involves utilization of published information on pests and management methods and reliance on experience in localized environments with pest management for specified commodities. As in all enterprises, the level of excellence at which an IPM program is conducted is dependent on the degree of professionalism exhibited by its creator.

BIBLIOGRAPHY

Kogan, M. Integrated pest management: historical perspectives and contemporary developments. Annu. Rev. Entomol. **1998**, *43*, 243–270.

National Academy of Sciences. *Insect Pest Management and Control*; Publ. 1695, National Academy of Sciences: Washington, DC, 1969; 508.

Pedigo, L.P. *Entomology and Pest Management*, 3rd Ed.; Prentice-Hall: Englewood Cliffs, NJ, 1999; 691.

Stern, V.M.; Smith, R.F.; van den Bosch, R.; Hagen, K.S. The integrated control concept. Hilgardia **1959**, *29*, 81–101.

INTERACTION OF HOST-PLANT RESISTANCE AND BIOLOGICAL CONTROLS

Helmut van Emden
The University of Reading, Reading, Berkshire, United Kingdom

INTRODUCTION

Although textbooks often cite host plant resistance to insects as ''compatible'' with other control measures, there will rarely be complete lack of interaction with biological control. The impact of biological control is inevitably an interaction with numerical changes in the density of the prey, the duration of the latter's life stages, and suitability as food nutritionally, toxicologically and of capture. All these may be affected by host plant resistance. The variations in host plant condition that can cause such interactions with biological control have recently been extended by the development of plant resistance based on transgenes from nonrelated sources such as quite unrelated plants, animals, and microorganisms. Interactions between two control methods are the foundation of the concept of IPM; it has been argued that plant resistance and biological control are often a foundation on which IPM packages can be developed (1). All effects of host plant resistance with biological control will result either from numerical (the density of natural enemies) or functional (prey killed per individual natural enemy) relationships of natural enemies to their prey. If these relationships are no more than proportional, that is there is the same percentage reduction in biological control mortality of the pest as the reduction the plant resistance causes in the number of prey, then there is no interaction. A negative or positive interaction (2) occurs when the proportional impact of biological control is respectively lower or higher than would be predicted from the percentage reduction in pest density (1).

NUMERICAL RELATIONSHIPS

These affect the numbers of natural enemies. They involve numbers arriving and leaving, surviving, and breeding. For those remaining to breed, generation time and fecundity are the key parameters.

Mortality of Natural Enemies

Where the plant resistance is due to a chemical toxic to the pest, the toxin may kill natural enemies when they consume the prey (3). Many secondary compounds in plants are considered to have evolved as defenses against insect herbivory and it has been necessary to reduce (often by several hundred times) their concentration to levels palatable and nontoxic to humans in the process of crop improvement by plant breeding over the centuries. Soybeans, corn, lupin, tomato, and tobacco all have examples of cultivars toxic to natural enemies (4). However, maize incorporating the bacterial Bt gene for pest resistance causes less mortality to lacewings than would insecticide protection of the crop, and the gene in brassicas seems not to cause mortality of parasitic Hymenoptera. There is also an example with aphid-resistant maize where the cultivars most toxic to aphids cause little mortality to ladybeetles because the aphids show reduced feeding and therefore little uptake of the toxin to pass on to the predator.

Mortality of small natural enemies is also common on plant varieties with glandular trichomes producing immobilizing fluids; the same mechanism that makes such varieties resistant to a range of pest species.

Shortage of Prey

Where plant resistance approaches immunity (as common with the transgenic approach), numbers of natural enemies may decrease from starvation or by dispersal from the area. The former has been observed in a different context in Africa, when the dramatic reduction of cassava mealybug by a parasitoid wasp caused a drop on other crops in ladybeetle populations, which had used the mealybug as a plentiful food source. Since there is no pest problem on such immune plants, the loss of natural enemies might appear not to matter. However, immunity brings the immediate danger of the selection of a strain of the pest adapted to the plant resistance, which would then multiply without a biological control restraint.

Breeding Rate

Even if not acutely toxic to natural enemies, pest-resistant crop varieties can reduce their fecundity. This may take the form of sterility, as with lacewings. Recent work has shown that ladybeetles fed on aphids from transgenic potatoes expressing snowdrop lectin had their longevity halved and their fecundity reduced by 20–40%.

More usually, fecundity reductions relate to the normally smaller size of pests on resistant plants. Although predators can compensate by eating more individuals, parasitoids develop on or within just a single individual. Thus, like their hosts, parasitoids on pest-resistant plants are smaller and less fecund (fecundity reductions in excess of 50% are known), as well as taking longer to develop. However, return to a pest-susceptible plant usually results immediately in normal fecundity.

Semiochemicals

Where natural enemies are attracted by semiochemicals, the potential exists for numbers arriving and staying on plants to be partially independent of the density of the prey. Many semiochemicals that attract natural enemies are produced by the prey or by the plant in response to pest attack, and such volatiles and arrestants (e.g., from aphid honeydew or caterpillar feces) will be reduced with the reduction in pest numbers on resistant varieties. However, where natural enemies respond to the odor of the undamaged plant, they may arrive equally at resistant and susceptible varieties if the volatiles from these are not too dissimilar. Equally, they may not recognize new pest-resistant cultivars as potential hosts of their prey. Parasitoids of aphids and diamondback moth are two examples where the emerging adult is attracted by the odor of the plant on which it developed. This can cause the adult parasitoids to discriminate between cultivars of the same crop, so that adults emerging from a pest-resistant cultivar will have some constancy for that cultivar.

FUNCTIONAL RELATIONSHIPS

These affect the pest mortality per individual natural enemy. The key components are the time spent searching rather than feeding, and the time the natural enemy spends with each pest individual before recommencing searching.

Pest Increase Rate

A major reason for positive interactions between host plant resistance and biological control is that the pest population increases more slowly on resistant varieties. The same voracity per natural enemy then enables even a smaller cohort to cause proportionally increased mortality on the pest. Part of the reduced increase rate results from decreased pest fecundity, but extended development time of the pest is also to be expected; each pest individual is then available for longer for stage-specific predation or parasitization.

Indirect Mortality from Natural Enemy Activity

It has now been established both for aphids and caterpillars (diamondback moth) that the presence of searching natural enemies causes an additional indirect mortality. The greater restlessness of insect pests on resistant plants causes a greatly increased rate of pests falling from the plants when disturbed by parasitoids and predators. Many fail to regain plants; some are taken by predators on the soil surface. In some aphid examples, this alone can result in just as many pests being lost from the total effects of biological control on pest-resistant cultivars as on susceptible ones. Since the former have smaller pest populations, the loss represents an increased percentage.

Pest Size

Pests are normally smaller on resistant plants. This leads to a higher proportional mortality from predators, simply because each predator will have to consume more individual pests to take in the same biomass of food before satiation is reached.

Additionally, there is evidence that large prey are better able than small prey to defend themselves by kicking at enemies or running away.

Searching Behavior of Natural Enemies

Hairs on leaves, often a characteristic used in resistance breeding as deterring landing and oviposition by small pests, impede the progress of searching natural enemies. A very nice example is the parasitoid *Encarsia formosa* parasitizing whitefly on cucumbers. On very hairy cultivars, whitefly populations are not controlled because searching by the natural enemy is greatly slowed; conversely on smooth-leaved cultivars, the parasitoid tends to move so rapidly it fails to detect its prey. Only intermediate cultivars are associated with successful biological control. Another example with peas has demonstrated that predation of aphids by ladybeetles is much lower on traditional pea varieties than on the ''leafless'' pea varieties, bred for resistance to mildew but also showing a little resistance to aphids. This is because the beetles tend to fall from the smooth leaves of traditional varieties, but have a much better foothold on the tendrils and stipules that replace the leaves on the mildew-resistant cultivars.

CONCLUSIONS

The generalization can be made that interactions with biological control will move from positive to negative progressively as the level of the plant resistance increases from very partial (which may be effective together with biological control) toward immunity. This will be especially true where the resistance is based on a toxin in the plant. Unfortunately, such toxin-based immunity is characteristic of the transgenic plant resistances currently under development.

Adaptation of pests to the resistance will also be slowed by a combination of partial plant resistance with biological control in comparison with the same level of pest control achieved by plant resistance alone (5).

REFERENCES

1. van Emden, H.F. Host Plant Resistance to Insect Pests. In *Techniques for Reducing Pesticide Use: Economic and Environmental Benefits*; Pimentel, D., Ed.; John Wiley & Sons UK, 1997; 129–152.
2. Thomas, M.; Waage, J. *Integration of Biological Control and Host-Plant Resistance Breeding*; ICTA: Wageningen, The Netherlands, 1996; 1–99.
3. Herzog, D.C.; Funderburk, J.E. Plant Resistance and Cultural Practice Interactions with Biological Control. In *Biological Control in Agricultural IPM Systems*; Hoy, M.A., Herzog, D.C., Eds.; Academic Press: Orlando, Florida, 1985; 67–68.
4. Hare, D.J. Effect of Plant Variation on Herbivore-Natural Enemy Interactions. In *Plant Resistance to Herbivores and Pathogens: Ecology, Evolution, and Genetics*; Fritz, R.S., Sims, E.L., Eds.; University of Chicago Press: Chicago, 1992; 278–565.
5. Gould, F.; Kennedy, G.G.; Johnson, M.T. Effects of natural enemies on the rate of herbivore adaptation to resistant host plants. Entomologia Experimentalis et Applicata **1991**, *58* (1), 1–14.

INTERCROPPING FOR PEST MANAGEMENT

Maria R. Finckh
Marianne Karpenstein-Machan
University of Kassel, Witzenhausen, Germany

INTRODUCTION

Until the past few hundred years, agricultural systems were based on large numbers of different crops, crop varieties, and landraces that were heterogeneous in genetic make-up. In addition, farming systems included both animals and plants further increasing diversity. As a result of increasing specialization, mechanization, and modern plant breeding, diversity on the farming system level and crop level has been drastically reduced worldwide at an ever accelerating speed especially over the past 100 years. Fewer and fewer varieties that are genetically homogeneous are being grown in ever-larger fields (1).

Monoculture refers usually to the continuous use of a single crop species over a large area. However, with respect to plant pathogens and pests it is important to differentiate between monoculture at the level of *species, variety*, or *resistance genes* (2). For example, within a species there may be many different genotypes with different resistances to a specific pest or pathogen and great variation with respect to competitiveness with weeds and other crops. Within a variety, there is usually no diversity (but see Table 1) for resistance or morphological traits. Resistance gene monocultures are more difficult to conceptualize. Many different varieties may exist; however, they may all possess the same resistance (or susceptibility) gene(s). For example, in the late 1960s, virtually all hybrid maize cultivars in the southeastern United States possessed the cytoplasmatically inherited Texas male sterility (*Tms*). Unfortunately, *Tms* is closely linked to susceptibility to certain strains of the pathogen *Cochliobolus carbonum* (syn. *Helminthosporium maydis*). The monoculture for susceptibility (while different varieties had been planted) led to selection for these strains and in 1970 the pathogen caused more than \$1 billion (= 10^9) losses (3).

Intercropping (4) can be practiced at the species, variety, and gene level (Table 1) with effects on pathogens (1, 2, 5), insect pests (6–8) and weeds (9, 10) (Table 2). One of the most important considerations for the successful design of intercropping systems for pest control is the achievement of *functional diversity*, i.e., diversity that limits pathogen and pest expansion and that is designed to make use of knowledge about host–pest/pathogen interactions to direct pathogen evolution (1, 2).

PROTECTION MECHANISMS ACTING IN INTERCROPPED SYSTEMS

Pathogens, insect pests, and weeds differ fundamentally in their biology and their effects on crops, and different protection mechanisms act with respect to these organisms (Table 2).

Pathogens are mostly dispersed through wind, water splash, soil, and animals (vectors). In intercropped systems, the most important mechanisms for disease control are mechanical distance and barrier effects. In addition, resistance reactions induced by a virulent pathogen strains may prevent or delay infection by virulent strains. A large percentage of the reduction of airborne diseases such as the powdery mildews and rusts in cereal cultivar mixtures has been shown to be due to induced resistance. The protection mechanisms are universal with respect to airborne, splashborne, and some soilborne, diseases. Mixtures of plants varying in reaction to a range of diseases will lead to a multitude of additional interactions and the overall response in such populations will tend to correlate with the disease levels of the components that are most resistant to these diseases. In addition, less affected plants may compensate for yield losses due to reduced competition from diseased neighbors (5).

In contrast to pathogens for which passive or vectored dispersal is the norm, insects often search actively for their hosts and behavioral, visual, and olfactory cues play an important role. While environmental factors and landing on a nonhost is likely the most important mortality factor for pathogens, natural enemies are at least as important for insect population dynamics (7, 8). Host dilution may affect an insect's ability to see and/or smell its hosts. Predators and parasitoids are dependent on the constant presence of prey and alternative food sources

Table 1 Possibilities for intercropping at three levels of uniformity on which monocultures are commonly practiced[a]

Level of uniformity	Intercropping possibilities
Species: Different individuals may differ in genetic make-up (resistance, morphology, etc.)	Arrangements among and within species, varieties, and resistances using intercropping
Variety: Usually genetically uniform, the same gene(s) in the same genetic background	Arrangements among and within varieties and resistances—includes variety mixtures, multilines, and populations
Resistance gene: The same gene may exist in different genetic backgrounds	Arrangements among resistances—multilines and populations

[a]From Ref. 2.

such as pollen and nectar in the absence of the hosts and it is critical that natural enemy populations are present in sufficient numbers to enable them to effectively control insect pests. The importance of natural enemies was often only recognized after insecticide applications induced pest resurgence due to the destruction of natural enemy populations. Intercrops and weeds therefore can play an important role in regulating insect pests.

Weeds usually are early successional plants adapted to colonize open, nutrient-rich spaces. Intercrops, especially cover and mulch crops directly compete with weeds for these spaces and also for light. As many weeds are adapted to certain crops and cropping patterns, changing these patterns (e.g., rotations) and management operations connected with different kinds of crops within the same field may make it difficult for weeds to cope (9, 10). An important consideration is that plants may be weeds only during certain phases of crop development. At other stages, the presence of the same "weeds" may be beneficial because they may provide food and habitat for beneficial insects and erosion control.

Besides the many positive effects of intercrops it is important to keep in mind that weeds may serve as alternative hosts for insect pests and pathogens and that insects often are disease vectors, especially for viruses that may reside symptomless in certain weeds.

Table 2 Mechanisms affecting pathogens, insect pests, and weeds in intercropped systems and selected additional interactions of importance

Mechanisms reducing disease
Increased distance between susceptible plants
Barrier effects of intercrop
Induced resistance
Selection for most resistant and/or competitive genotypes
Interactions among pathogen strains on host plants
Mechanisms reducing insect pests
Enhancement of natural enemies
Reduction of host density (reduced resource concentration)
Reduction of plant apparency (visual or olfactory cues reduced)
Alteration of host quality (with respect to the insect pest) through plant–plant interactions
Mechanisms reducing weeds
Reduction of bare soil and layering of crops (increased competition for light, water, and nutrients)
Variation in tillage needs and operations of intercrops may disturb weeds
Other beneficial interactions
Yield enhancement through niche differentiation of hosts
Compensation for yield losses by less affected hosts
Better soil cover with intercrop (soil and water conservation, microclimatic effects)
Possible unwanted interactions
Weeds may serve as alternate hosts for pathogens and insects
Interactions among virus vectors and weeds
Greater difficulty to specifically reduce weeds with herbicides or mechanically
Microclimatic effects may enhance certain problems

(From Refs. 5, 7, and 9.)

INTERCROPPING IN PRACTICE

Variety mixtures and multilines are used mainly to control diseases. For example, they are used in cereals on a commercial scale in the United States, Denmark, Finland, Poland, and Switzerland to control rusts, mildews, and certain soilborne diseases (e.g., *Cephalosporium* stripe). When barley cultivar mixtures were used on more than 300,000 ha in the former German Democratic Republic, powdery mildew of barley and consequently fungicide input was reduced by 80% within five years. Wheat cultivar mixtures and multilines are grown on several hundred thousand hectares in the US in the Pacific Northwest and in Kansas to protect against diseases and abiotic stresses. In Colombia, coffee multilines are grown on more than 400,000 ha to control coffee rust (2).

Attention has also been called to possible beneficial effects of greater intravarietal diversity in the oat-frit fly (*Oscinella frit* L.) system. The flies can attack the host plants only at a particular growth stage and a higher degree of variability within an oat crop could allow for escape from attack and subsequent compensation.

Cereal species mixtures for feed production are currently grown on more than 1.4 Mil ha in Poland and have been shown consistently to restrict diseases. In Switzerland, the ''maize-ley'' system (i.e., maize planted without tillage into established leys), which is being promoted to reduce soil losses and nutrient leaching, has been shown to reduce smut disease and attacks by European stem borer and aphids.

The deliberate planting or maintenance of flowering weeds and grass in established vineyards in Switzerland and Germany greatly increases natural enemies while reducing soil erosion. This practice is becoming increasingly popular in Californian wine growing areas and in apple production. In the United Kingdom, a newly developed low-input system for growing wheat with a permanent understorey of white clover (*Trifolium repens*) greatly reduces the major pest aphid species and slugs and there should also be reductions in splash-dispersed diseases, such as those caused by *Septoria* spp.

The required reduction in insect populations for effective reduction of insect-transmitted diseases may be beyond that which can be achieved by diversity alone. However, simultaneous reduction of insect vectors and disease inoculum can be effective (2).

FUTURE CONCERNS

There are many reasons why intercropping is not practiced more widely. First, modern crops are bred to be grown in monoculture and may not necessarily be well adapted to intercropping. Efforts of breeders to produce breeding lines adapted to intercropping need to be strengthened. Second, while intercropping clearly provides a means for reducing pesticide needs there is a lack of adapted machinery allowing for efficent management of intercropped cultures. Third, successful intercropping strategies have to be carefully designed as preventive measures while application of pesticides often can be done once a problem occurs; it is simple and usually very cheap. Fourth, a concern often raised is the quality of products raised as intercrops such as varietal mixtures of cereals. In some countries there is resistance in the food processing industry to such products. However, such problems could be overcome if breeders, producers, and processors work together. For example, in the 1980s, in the German Democratic Republic, first-quality malting barley was produced in large-scale mixtures in collaboration with breeders, growers, and processors.

REFERENCES

1. Finckh, M.R.; Wolfe, M.S. Diversification Strategies. In *The Epidemiology of Plant Diseases*; Jones, D.G., Ed.; Chapman and Hall: London, 1998; 231–259.
2. Finckh, M.R.; Wolfe, M.S. The Use of Biodiversity to Restrict Plant Diseases and Some Consequences for Farmers and Society. In *Ecology in Agriculture*; Jackson, L.E., Ed.; Academic Press: San Diego, 1997; 199–233.
3. Ullstrup, A.J. The impacts of the southern corn leaf blight epidemics of 1970-1971. Annu. Rev. Phytopathol. **1972**, *10*, 37–50.
4. Vandermeer, J.H. *The Ecology of Intercropping*; Cambridge University Press: Cambridge, New York, Melbourne, 1989; 235.
5. Wolfe, M.S.; Finckh, M.R. Diversity of Host Resistance within the Crop: Effects on Host, Pathogen and Disease. In *Plant Resistance to Fungal Diseases*; Hartleb, H., Heitefuss, R., Hoppe, H.H., Eds.; G. Fischer Verlag: Jena, 1997; 378–400.
6. Altieri, M.A.; Liebman, M. Insect, Weed and Plant Disease Management in Multiple Cropping Systems. In *Multiple Cropping Systems*; Francis, C.A., Ed.; Macmillan: New York, 1986; 183–218.
7. Andow, D.A. Vegetational diversity and arthropod population responses. Annu. Rev. Entomol. **1991**, *36*, 561–586.
8. Letourneau, D.K. Plant-Arthropod Interactions in Agroecosystems. In *Ecology in Agriculture*; Jackson, L.E., Ed.; Academic Press: London, New York, 1997; 239–290.
9. Liebman, M.L.; Dyck, E. Crop rotation and intercropping strategies for weed management. Ecol. Appl. **1993**, *3*, 92–122.
10. Liebman, M.L.; Gallandt, E.R. Many Little Hammers: Ecological Management of Crop-weed Interactions. In *Ecology in Agriculture*; Jackson, L.E., Ed.; Academic Press: San Diego, New York, London, 1997; 291–343.

INVASION BIOLOGY

Jennifer Ruesink
University of Washington, Seattle, Washington, U.S.A.

INTRODUCTION

Biological invasions occur when species or distinct populations breach biogeographical barriers and extend their ranges to areas where they were not historically present (1–3). Invasion biology concerns the causes and consequences of these new species, which are also referred to as invasive, introduced, alien, exotic, nonnative, or nonindigenous species.

Biological invasions occur in five steps: arrival, establishment, population growth, population spread, and impact (Table 1). Only a small proportion of species that arrive actually establish, and so forth; thus each step acts as a filter for the invasion process. Successful invasion depends on characteristics of the invading species, of the recipient environment, and of the process by which the two are brought together (4). However, no widely accepted method currently exists for identifying the characteristics that promote invasion a priori (5).

WHY DO NEW SPECIES INVADE?

Species have always expanded their ranges, but the pace of invasion has accelerated recently due to increased human travel and trade (5, 7). Humans transport species in three ways: 1) on purpose, with the intention that they will grow in outdoor environments (fish and game, plantation trees, and biological control agents), 2) on purpose but with no intention that they will establish (pets, horticultural and agricultural plants, aquaculture, and sterile releases), and 3) accidentally (hitchhiking on packages, live imports, and people). The contribution of each of these main pathways varies among taxa. Of South Africa's weeds, 89% were intentionally introduced (8), but only 11% of insect invaders in North America were intentional (9). Ducks, pheasants, pigeons, finches, and parrots have more introduced species worldwide than would be expected by chance because these bird families include many pet or game species (10).

WHICH SPECIES INVADE?

Propagule Pressure

Species can invade a new area if abiotic (especially climatic) conditions are suitable and an exploitable resource exists. Those species that can invade, will invade if given sufficient opportunity (11–13). High rates of arrival have been termed 'propagule pressure' and can occur either through numerous releases or releases of many individuals. Some of the best evidence that propagule pressure affects invasion comes from compilations of biocontrol introductions: the successful establishment of insect predators rose seven times when the number of introductions doubled, and releases $>31{,}200$ individuals were eight times more successful than those of <5000 individuals (14).

Species Traits

For a given propagule pressure, some species may be more likely to invade than others. Traits promoting invasion could include the ability to increase rapidly from low density, a generalist diet, and broad climatic tolerance. Although statistical relationships between species traits and invasibility are often weak (15), analyses of certain taxa introduced to particular environments have been successful (16, 17). For instance, invasive species of pines in South Africa tend to have small seeds, short intervals between reproductive bouts, and short times to maturity, whereas noninvasive species show the opposite traits (18). For woody plants, those that have become invasive in North America often reproduce vegetatively, germinate easily, and have a long fruiting period (19).

WHERE DO SPECIES INVADE?

In addition to species traits, characteristics of the recipient community may also influence the ease with which new species invade. Indeed, particular conditions in native environments (e.g., disturbance and species interactions) may select for species that are effective invaders of other

Table 1 Reasons for and responses to five steps of an invasion

Step	Why?	Control
Arrival	Intentional releases, intentional imports, accidental hitchhiking (by-product)	Risk assessment—choice of species, treatment or quarantine of vectors
Establishment	Suitable abiotic conditions, available resources, propagule pressure	Reduce pathways, containment
Population growth	Intrinsic rate of reproduction, apomixis/vegetative reproduction	Early eradication
Population spread	Mobility—dispersal, home range size, transport by humans	Eradicate new populations, reduce human transport
Impact	Density (abundant resources), few enemies, per capita effect (new role), alteration of resource base	Effective screening prior to introduction; mechanical, chemical, or biological control; make environment less suitable

areas, thus resulting in asymmetric patterns of invasion. For instance, European insects have invaded forests in North America but not vice versa (20), and there has been a unidirectional appearance of tillering grasses in bunchgrass habitats worldwide (21). Assemblages may be more resistant to invasion if they are undisturbed (22, 23) or contain many natural enemies (24, 25). For many years it was a rule of thumb that species-poor islands were invasible and species-rich tropics were not (1). However, surveys of plant assemblages indicate that areas of naturally high species richness tend also to have numerous invaders, perhaps because soil and climate conditions are generally conducive to plant growth (26). Although species richness per se may not influence invasion, loss of species could make systems easier to invade.

Habitat alteration (flooding, drought, fire, wind, eutrophication, and channelling) creates new conditions that are suitable for a new suite of species. These species are often introduced. For instance, western North America hosts many conspicuous invasive fishes in part because once fast-flowing rivers are now lakes separated by dams (27). Roadsides in North America contain a high proportion of European plant species, probably because European plants have had millenia to evolve to take advantage of disturbance, whereas disturbance in North American habitats has risen recently (28). Based on an Australian study of two invaders, both physical disturbance and nutrient addition improve the performance of invasive plants (29).

WHICH SPECIES HAVE EFFECTS AND WHERE?

Impacts of invaders relevant to pest management include the ecological and/or economic damage from pests and the effectiveness of biocontrol agents (30). Impacts are expected to be particularly pronounced when species reach high abundances or have high per capita effects (31). High abundances can occur when species escape limits to population growth (abundant resources or few enemies). High per capita effects may arise when a species plays a new role, especially by altering the resource base. For instance, many of the worst plant invaders of natural areas are nitrogen-fixers (which alter nutrients) or climbing vines (which alter light) (32).

A disproportionate number of invasive species have harmful effects, more so if introduced accidentally rather than intentionally. In Japan, for instance, 8% of native insects are considered pests, but 72% of introduced insects are pests (33). Of agricultural weeds in North America, 50–75% are nonindigenous (6). On the other hand, only 1–6% of nonindigenous plant species in Great Britain have become weedy or widespread (8). Some introduced species may be problematic due to an absence of natural enemies, but this escape from control does not appear to be entirely general. Based on a compilation of life tables for 124 holometabolous insects, mortality rates due to parasitoids, predators, and diseases do not differ for native and introduced species (34).

Species introduced accidentally often have harmful effects, and, conversely, species introduced intentionally have beneficial effects less often than desired. Of 463 grasses and legumes introduced to Australia to improve pasture, only 5% have raised productivity (35). About a third of established biocontrol insects actually reduce the target organism (36). Although impact in the native environment is not necessarily a useful indicator of what a species will do once introduced, one indicator of potential impact is the fate of prior introductions (19).

Pest status tends to be based on economic considerations, but introduced species also cause ecological damage. Invasive species contribute to endangerment of nearly 50% of species listed under the United States Endangered Species Act (37, 38), and they have dramatically altered the structure and function of ecosystems (39, 40). Ecological and economic effects have the same root causes—abundance and high per capita effects of invaders—but the affected habitats can be quite distinct. Only 25% of plants that cause problems in natural areas are also agricultural weeds (32). Thus, screening procedures that keep out economic pests would fail to restrict many species that cause ecological harm.

Time Lags and Surprises

Introduced species have occasionally surprised researchers by expanding to previously intolerable places or by irrupting after remaining localized and rare for many years. ''Boom and bust'' patterns have also been observed, in which an invader initially reaches high abundance and then declines, sometimes even going extinct. Tolerance of new conditions (e.g., temperature or host plants) may require genetic adaptation, which could result in time lags before invasion (41). Of 184 woody species currently considered invasive near Brandenburg, Germany, 51% did not appear to be invasive until > 200 years after their initial introduction (42). However, only 7% of 627 cases involving biocontrol introductions showed time lags before population increase, whereas 28% increased and 27% went extinct immediately (43). Regardless of frequency, cases in which invaders have unexpected impacts have become well-known, especially for biocontrol agents with nontarget effects such as feeding on endangered species (e.g., the weevil *Rhinocyllus conicus* on thistles and the moth *Cactoblastis* on *Opuntia* cacti) (44–46) or competing with natives (e.g., the ladybird beetle *Coccinella septumpunctata*) (5).

CAN INVASIONS BE CONTROLLED?

The first steps of an invasion can be controlled by limiting entry or by vigilantly eliminating newly established populations of invaders. For instance, an assessment of species associated with raw logs indicated a potential loss of billions of dollars due to forest pests if Siberian larch was not treated prior to import into the United States (47). For many years, medflies (*Ceratitus capitata*) have epitomized the notion that ''an ounce of prevention is worth a pound of cure.'' California spent $100 million to eradicate an incipient invasion in 1981, thereby preventing nine times that amount of crop damage. By 1996, however, medflies were apparently established and spreading. The extent of the invasion makes eradication unlikely (48). Humans simply have to learn to live with these naturalized species (49).

Control during the last steps of an invasion usually involves chemical, biological, or mechanical reduction of unwanted species in areas where effects are most serious. Control efforts at this stage would benefit from considerations of demography and behavior of invaders. For instance, seedling competition among annual plants is often fierce, so efforts to reduce seed production will not reduce plant numbers. Instead, control efforts should be directed at reducing seedling growth and survival (50). Knowing how insects move among microhabitats could aid in trap placement or in crafting habitats that promote desirable species and discourage undesirable ones (51). Knowing encounter rates and feeding rates could aid in calculating the number of consumers necessary for effective biological control.

Species invasions are a form of ecological gambling in which the consequences of any particular introduction are uncertain, despite an emerging framework of factors contributing to high-risk invasions. The influx of new species can be slowed by reducing pathways for introduction and by intentionally introducing species only when beneficial effects will be large and native alternatives do not exist (52). Distinct biotas are valuable and intriguing but increasingly difficult to maintain under pressures of globalization.

REFERENCES

1. Elton, C.S. *The Ecology of Invasions by Animals and Plants*; Chapman & Hall: London, 1985.
2. Williamson, M. *Biological Invasions*; Chapman & Hall: London, 1996.
3. Mack, R.N.; Simberloff, D.; Lonsdale, W.M.; Evans, H.; Clout, M.; Bazzaz, F.A. Biotic invasions: causes, epidemiology, global consequences, and control. Ecol. Appl. **2000**, *10* (3), 689–710.
4. Lodge, D.M. Biological invasions: lessons for ecology. Tr. Ecol. Evol. **1993**, *8*, 133–137.
5. Ruesink, J.L.; Parker, I.M.; Groom, M.J.; Kareiva, P.M. Reducing the risks of nonindigenous species introductions: guilty until proven innocent. Bioscience **1995**, *45* (7), 465–477.
6. Office of Technology Assessment. *Harmful Non-Indigenous Species in the United States*; U.S. Government Printing Office: Washington, DC, 1993.
7. Cohen, A.N.; Carlton, J.T. Accelerating invasion rate in a highly invaded estuary. Science **1998**, *279*, 555–558.
8. Crawley, M.J.; Harvey, P.H.; Purvis, A. Comparative ecology of the native and alien floras of the British isles. Phil. Trans. R. Soc. Lond. B. **1996**, *351*, 1251–1259.
9. Sailer, R.I. History of Insect Introductions. In *Exotic Plant*

Pests and North American Agriculture; Wilson, C.L., Graham, C.L., Eds.; Academic Press: New York, 1983; 15–38.
10. Lockwood, J.L. Using taxonomy to predict success among introduced avifauna: relative importance of transport and establishment. Conserv. Biol. **1999**, *13* (3), 560–567.
11. Veltman, C.J.; Nee, S.; Crawley, M.J. Correlates of introduction success in exotic New Zealand birds. Amer. Nat. **1996**, *147* (3), 542–557.
12. Green, R.E. The influence of numbers released on the outcome of attempts to introduce exotic bird species to New Zealand. J. Anim. Ecol. **1997**, *66*, 25–35.
13. Duncan, D.P. The role of competition and introduction effort in the success of passeriform birds introduced to New Zealand. Amer. Nat. **1997**, *149* (5), 903–915.
14. Beirne, B. Biological control attempts by introductions against pest insects in the field in Canada. Can. Entomol. **1975**, *107*, 225–236.
15. Goodwin, B.J.; McAllister, A.J.; Fahrig, L. Predicting invasiveness of plant species based on biological information. Conserv. Biol. **1999**, *13* (2), 422–426.
16. Panetta, F.D. A system of assessing proposed plant introductions for weed potential. Plant Prot. Quarterly **1993**, *8*, 10–14.
17. White, P.S.; Schwarz, A.E. Where do we go from here? The challenges of risk assessment for invasive plants. Weed Technol. **1998**, *12* (4), 744–751.
18. Rejmanek, M.; Richardson, D.M. What attributes make some plant species more invasive. Ecology **1996**, *77* (6), 1655–1661.
19. Reichard, S.H.; Hamilton, C.W. Predicting invasions of woody plants introduced to North America. Conserv. Biol. **1997**, *11* (1), 193–203.
20. Niemela, P.; Mattson, W.J. Invasions of North American forests by European phytophagous insects. Bioscience **1996**, *46* (10), 741–753.
21. Mack, R.N. Alien Plant Invasion into the Intermountain West. A Case History. In *Ecology of Biological Invasions of North America and Hawaii*; Mooney, H.A., Drake, J.A., Eds.; Springer-Verlag: New York, 1986; 191–213.
22. Orians, G.H. Site Characteristics Favoring Invasions. In *Ecology of Biological Invasions of North America and Hawaii*; Mooney, H.A., Drake, J.A., Eds.; Springer-Verlag: New York, 1986; 133–148.
23. Hobbs, R.J.; Huenneke, L.F. Disturbance, diversity, and invasion: implications for conservation. Conserv. Biol. **1992**, *6*, 324–337.
24. Goeden, R.D.; Louda, S.M. Biotic interference with insects imported for weed control. Annu. Rev. Entomol. **1976**, *21*, 325–342.
25. Mack, R.N. Predicting the identity and fate of plant invaders: emergent and emerging approaches. Biol. Conserv. **1996**, *78*, 107–121.
26. Stohlgren, T.J.; Binkley, D.; Chong, G.W.; Kalkhan, M.A.; Schell, L.D.; Bull, K.A.; Otsuki, Y.; Newman, G.; Bashkin, M.; Son, Y. Exotic plant species invade hot spots of native plant diversity. Ecol. Monogr. **1999**, *69* (1), 25–46.
27. Moyle, P.B.; Light, T. Fish invasions in California: do abiotic factors determine success? Ecology **1996**, *77* (6), 1666–1670.
28. Pysek, P. Is there a taxonomic pattern to plant invasions? Oikos **1998**, *82*, 282–294.
29. Hobbs, R.J. The Nature and Effects of Disturbance Relative to Invasions. In *Biological Invasions: A Global Perspective*; Drake, J.A., Mooney, H.A., di Castri, R., Kruger, F., Groves, R., Rejmanek, M., Williamson, M., Eds.; John Wiley & Sons: Chichester, U.K., 1986; 389–405.
30. Pimentel, D.; Lach, L.; Zuniga, R.; Morrison, D. Environmental and economic costs of nonindigenous species in the United States. Bioscience **2000**, *50*, 53–65.
31. Parker, I.M.; Simberloff, D.; Lonsdale, W.M.; Goodell, K.; Wonham, M.; Kareiva, P.M.; Williamson, M.H.; Von Holle, B.; Moyle, P.B.; Byers, J.E.; Goldwasser, L. Impact: toward a framework for understanding the ecological effects of invaders. Biological Invasions **1999**, *1* (1), 3–19.
32. Daehler, C.C. The taxonomic distribution of invasive angiosperm plants: ecological insights and comparisons to agricultural weeds. Biol. Conserv. **1998**, *84*, 167–180.
33. Morimoto, N.; Kiritani, K. Fauna of exotic insects in Japan. Bull. National Institute of AgroEnvironmental Sci. **1995,** (12), 87–120.
34. Hawkins, B.A.; Cornell, H.V.; Hochberg, M.E. Predators, parasitoids, and pathogens as mortality agents in phytophagous insect populations. Ecology **1997**, *78* (7), 2145–2152.
35. Lonsdale, W.M. Inviting trouble: introduced pasture species in Northern Australia. Australian J. Ecol. **1994**, *19* (3), 345–354.
36. Williamson, M.; Fitter, A. The varying success of invaders. Ecology **1996**, *77* (6), 1661–1666.
37. Foin, T.C.; Riley, S.P.D.; Pawley, A.L.; Ayres, D.R.; Carlsen, T.M.; Hodum, P.J.; Switzer, P.V. Improving recovery planning for threatened and endangered species. Bioscience **1998**, *48*, 177–184.
38. Wilcove, D.S.; Rothstein, D.; Dubow, J.; Phillips, A.; Losos, E. Quantifying threats to imperiled species in the United States. Bioscience **1998**, *48*, 607–615.
39. Vitousek, P.M.; D'Antonio, C.M.; Loope, L.L.; Westbrooks, R. Biological invasions as global environmental change. Am. Scientist **1996**, *84* (5), 468–478.
40. Mack, M.C.; D'Antonio, C.M. Impacts of biological invasions on disturbance regimes. Tr. Ecol. Evol. **1998**, *13* (5), 195–198.
41. Secord, D.; Kareiva, P. Perils and pitfalls in the host specificity paradigm. Bioscience **1996**, *46* (5), 448–453.
42. Kowarik, I. Time Lags in Biological Invasions with Regard to the Success and Failure of Alien Species. In *Plant Invasions—General Aspects and Special Problems*; Pysek, P., Prach, K., Rejmanek, M., Wade, M., Eds.; SPB Academic Publishing: Amsterdam, 1995; 15–38.

43. Crawley, M.J. The population biology of invaders. Phil. Trans. R. Soc. Lond. B. **1986**, *314*, 711–731.
44. Louda, S.M.; Kendall, D.; Connor, J.; Simberloff, D. Ecological effects of an insect introduced for the biological control of weeds. Science **1997**, *277* (5329), 1088–1090.
45. Johnson, D.M.; Stiling, P.D. Distribution and dispersal of *Cactoblastis cactorum* (Lepidoptera: Pyralidae), an exotic opuntia-feeding moth, in Florida. Florida Entomol. **1998**, *81*, 12–22.
46. Cory, J.S.; Myers, J.H. Direct and indirect ecological effects of biological control. Tr. Ecol. Evol. **2000**, *15* (4), 137–139.
47. U.S. Department of Agriculture. *Pest Risk Assessment of the Importation of Larch from Siberia and the Soviet Far East*; USDA Forest Service, Misc. Publ. No. 1495. U.S. Government Printing Office: Washington, DC, 1991.
48. Carey, J.R. The future of the Mediterranean fruit fly *Ceratitus capitata* invasion of California: a predictive framework. Biol. Conserv. **1996**, *78* (1–2), 35–50.
49. Myers, J.H.; Savoie, A.; Van Randen, E. The irradication and pest management. Annu. Rev. Entomol. **1998**, *43*, 471–491.
50. McEvoy, P.B.; Rudd, N.T. Effects of vegetation disturbance on insect biological control of tansy ragwort, *Senecio jacobea*. Ecol. Appl. **1993**, *3* (4), 682–698.
51. Holway, D.A.; Suarez, A.V. Animal behavior: an essential component of invasion biology. Tr. Ecol. Evol. **1999**, *14* (8), 328–330.
52. Ewel, J.J.; O'Dowd, D.J.; Bergelson, J.; Daehler, C.C.; D'Antonio, C.M.; Gomez, L.D.; Gordon, D.R.; Hobbs, R.J.; Holt, A.; Hopper, K.R.; Hughes, C.E.; LaHart, M.; Leakey, R.B.; Lee, W.G.; Loope, L.L.; Lorence, D.H.; Louda, S.M.; Lugo, A.E.; McEvoy, P.B.; Richardson, D.M.; Vitousek, P.M. Deliberate introductions of species: research needs. Bioscience **1999**, *49* (8), 619–630.

INVERTEBRATE PESTS

Thomas W. Culliney
Hawaii Department of Agriculture, Honolulu, Hawaii, U.S.A.

INTRODUCTION

Livestock (domesticated mammals of the orders Artiodactyla and Perissodactyla, and the gallinaceous birds) suffer attack from a wide variety of invertebrate animals. Relatively few of these, however, are of any economic significance. Of principal importance are the parasitic arthropods, especially ticks and flies, and nematode worms (Table 1), and economic losses due to these pests often are high. Loss is most commonly measured in terms of negative effects on production (i.e., feed conversion efficiency, reproductive efficiency) (1). Pests may affect production by causing annoyance (''worry,'' ''gadding'') or irritation, which interferes with feeding or resting, or, more seriously, by taking blood or damaging internal organs; by damaging hides, which devalues finished leather; or by causing reductions in wool, mohair, milk, or egg production, or weight gain in meat animals. Pest attack may leave livestock animals vulnerable to secondary infections, cause paralysis or allergic reactions, induce abortion, create unthriftiness, or cause death (1, 2).

WORLD LIVESTOCK LOSSES TO INVERTEBRATE PESTS

Worldwide, invertebrate pests are responsible for an estimated 4% annual loss to the livestock industry. This loss is valued at more than $57 billion. Arthropods contribute approximately 70% to total losses, with the remainder attributed to internal parasites. Arthropods take the highest percentage toll on swine and sheep (Table 2), whereas internal parasites have the most severe proportional impact on poultry (Table 3). Livestock losses to invertebrate pests in the United States amount to about $4 billion per year, almost half the estimated losses to the industry from all pests (3).

Losses to Arthropods

The greatest monetary loss to invertebrate pests occurs in cattle husbandry, almost all of which is attributable to arthropods. In the United States, cattle losses to arthropods amount to at least $2.4 billion per year. Losses to the Brazilian cattle industry from these pests exceed $1.2 billion (4). In Australia, arthropods are responsible for cattle losses exceeding $126 million (5, 6). The horn fly, *Haematobia irritans,* is considered to be the most important pest of cattle in the United States and elsewhere, (1) causing estimated annual losses in the United States of 2%, or $1.2 billion. High densities of *H. irritans*, both sexes of which are blood feeders, are associated with an average 14% weight loss in range cattle and 1% loss in milk yield in dairy cows (1, 7). Other flies, such as the face fly, *Musca autumnalis*, and the stable fly, *Stomoxys calcitrans*, cause production losses primarily by worrying stock; *S. calcitrans* may reduce milk yields by as much as 40–60% (8), and cause total losses to the U.S. cattle industry of $479 million per year. Mange, caused by various species of mite, costs another $300 million per year. Production losses from ticks may exceed $52 per infested calf in the United States (9). Ticks also cause high losses to the cattle industry in Nigeria (10).

Two insects, the sheep ked, *Melophagus ovinus,* and sheep nasal bot fly, *Oestrus ovis,* are major contributors to losses in U.S. sheep production, together costing the industry at least $47 million per year. In Australia, losses to the sheep industry caused by arthropods, including lice, mites, and the sheep blow fly, *Lucilia cuprina,* amount to at least $261 million (5, 6). Swine production in the United States suffers annual losses, particularly from lice and mites, amounting to $293 million.

Losses to Internal Parasites

A large proportion of the 5% of cattle, 15% of pigs, and 10% of sheep that die from diseases each year is attributable to infection by internal parasites (11). The bulk of livestock losses from internal parasites is caused by the gastrointestinal nematodes; liver fluke disease, caused by the helminth, *Fasciola hepatica,* also is a major problem in cattle and sheep (1). Annual losses of cattle, swine, sheep, equines, and poultry to parasites in the United States are valued at approximately $240 million, $439 million, $39 million, $84 million, and $504 million, respectively. In

Table 1 Major invertebrate pests of livestock

Livestock group	Pest species/parasitic condition	Order: Family	Control measures
Cattle	*Fasciola hepatica* L./liver fluke disease, fascioliasis	Echinostomida: Fasciolidae	anthelmintics (e.g., oxyclozanide, nitroxynil, albendazole, rafoxanide); destruction of intermediate hosts (via well-drained pasture, copper sulfate, sodium pentachlorphenate, *N*-tritylmorpholine)
	Dictyocaulus viviparus (Bloch)/verminous bronchitis, "husk," "hoose"	Strongylida: Dictyocaulidae	nematicides (e.g., avermectins); vaccine; rotational grazing
	Nematodes (e.g., *Ostertagia ostertagi* [Stiles], *Cooperia punctata* [v. Linstow], *Trichostrongylus vitrinus* Looss, *T. colubriformis* [Giles], *Oesophagostomum radiatum* [Rudolphil])/parasitic gastroenteritis	Strongylida: Trichostrongylidae, Strongylidae	nematicides (e.g., avermectins, benzimiadoles, levamisole); relieving overcrowding of young stock; good pasture maintenance
	Mites (e.g., *Sarcoptes scabiei* [De Geer], *Chorioptes bovis* [Hering], *Psoroptes ovis* [Hering], *Demodex bovis* Stiles)/mange	Acariformes: Sarcoptidae, Psoroptidae, Demodicidae	acaricides (e.g., amitraz, ivermectin)
	Ticks (*Boophilus* spp., *Amblyomma* spp., *Dermacentor* spp., *Rhipicephalus* spp., *Ixodes* spp., *Otobius megnini* [Duges])	Parasitiformes: Ixodidae, Argasidae	acaricides; rotational grazing
	Sucking lice (e.g., *Haematopinus eurysternus* [Nitzsch], *Linognathus vituli* [L.], *Solenopotes capillatus* Enderlein);	Anoplura: Haematopinidae, Linognathidae	insecticides (e.g., coumaphos, trichlorfon, amitraz, phosmet, chlorpyrifos); proper hygiene and husbandry
	Biting lice (e.g., *Bovicola bovis* [L.])	Mallophaga: Trichodectidae	
	Chrysomya bezziana Villeneuve, *Cochliomyia hominivorax* (Coquerel)/myiasis	Diptera: Calliphoridae	insecticide spray (e.g., coumaphos) or bait; avoidance or protection of wounds (e.g., tick bites); sterile insect technique
	Dermatobia hominis (L.)/myiasis	Diptera: Cuterebridae	insecticides (e.g., fenthion, trichlorfon)
	Hypoderma lineatum (de Villers), *H. bovis* (L.)/myiasis	Diptera: Hypodermatidae	systemic organophosphate insecticides
	Haematobia irritans (L.)	Diptera: Muscidae	systemic insecticides (e.g., coumaphos, crufomate, tetrachlorvinphos)
	Musca autumnalis De Geer	Diptera: Muscidae	systemic insectides (e.g., methoprene, tetrachlorvinphos, ivermectin)
	Stomoxys calcitrans (L.), *S. nigra* Macquart	Diptera: Muscidae	chemical control (e.g., residual sprays, traps); biological control; sanitation (elimination of larval breeding media)
Swine	*Ascaris lumbricoides* (L.)/ascariasis	Ascaridida: Ascaridae	nematicides (e.g., piperazine, thiabendazole, hygromycin); disinfection of pens

(Continued)

Table 1 Major invertebrate pests of livestock (*Continued)*

Livestock group	Pest species/parasitic condition	Order: Family	Control measures
	Nematodes (*Ostertagia* sp., *Hyostrongylus rubidus* [Hassall & Stiles], and *Oesophagostomum dentatum* [Rudolphi] [all stomach worms], *Metastrongylus* sp. [lung worm])	Strongylida: Trichostongylidae, Strongylidae, Metastrongylidae	nematicides (e.g., fenbendazole, tetramisole, thiabendazole, piperazine, dichlorvos in feed)
	Sarcoptes scabiei	Acariformes: Sarcoptidae	acaricides (e.g., phosmet, diazinon, bromocyclen & sulfur, amitraz)
	Haematopinus suis (L.)	Anoplura: Haematopinidae	insecticides (e.g., deltamethrin, permethrin, phosmet)
Sheep	*Fasciola hepatica*	Echinostomida: Fasciolidae	anthelmintics (e.g., diamphenethide); destruction of intermediate hosts
	Dictyocaulus viviparus	Strongylida: Dictyocaulidae	as for cattle
	Nematodes (*Haemonchus contortus* [Rudolphi], *Ostertagia circumcincta* [Stadelmann], *Trichostrongylus axei* [Cobbold], *Nematodirus spathiger* [Railliet], *Trichuris ovis* [Abildgaard], *Oesophagostomum columbiamum* [Curtice])/parasitic gastroenteritis	Strongylida: Trichostongylidae, Strongylidae; Enoplida: Trichuridae	nematicides (e.g., benzimiadoles, imidazothiazoles, tetrahydropyrimidines, organophosphates); rotational grazing
	Psoroptes ovis (Hering)/man.	Acariformes: Psoroptidae	acaricidal dips (e.g., flumethrin, diazinon, cypermethrin, amitraz)
	Ticks (*Ixodes* spp.)	Parasitiformes: Ixodi.	as for cattle
	Bovicola ovis (Schrank)	Mallophaga: Trichodectidae	insecticides (e.g., coumaphos, diazinon, dioxathion)
	Lucilia cuprina (Wiedemann), *Phaenicia sericata* (Meigen), *Calliphora stygia* (F.)/myiasis	Diptera: Calliphoridae	proper hygiene (e.g., preventing staining of wool by urine and feces); insecticides (e.g., diazinon, chlorfenvinphos, fenthionethyl); growth regulators
	Melophagus ovinus (L.)	Diptera: Hippoboscidae	insecticides (e.g., trichlorfon, carbaryl, dichlorvos, dimethoate)
	Oestrus ovis L./myiasis	Diptera: Oestridae	insecticides (e.g., trichlorfon, *Bacillus thuringiensis*)
	Hydrotaea irritans (Fallen)	Diptera: Muscidae	canvas head caps, housing injured animals
	Musca autumnalis	Diptera: Muscidae	as for cattle
	Wohlfahrtia magnifica (Schiner)/myiasis	Diptera: Sarcophagidae	insecticides (e.g., trichlorfon, crotoxyphos, dichlorvos, fenthion, propoxur, temephos, diazinon)
Equines	*Parascaris equorum* (Goeze)	Ascaridida: Ascaridae	nematicides (e.g., piperazine, thiabendazole); proper hygiene and husbandry
	Strongylus vulgaris (Looss)	Strongylida: Strongylidae	nematicides (e.g., thiabendazole); rotational grazing
	Gasterophilus intestinalis (De Geer), *G. nasalis* (L.), *G. haemorrhoidalis* (L.), *G. pecorum* (F.), *G. inermis* Brauer), *G. nigricornis* (Lowes)/myiasis	Diptera: Gasterophilidae	insecticides (e.g., carbon disulfide, dichlorvos, trichlorfon, ivermectin [internally, against larvae]; coumaphos, malathion [externally, against eggs])
	Rhinoestrus purpureus (Brauer)/ myiasis	Diptera: Oestridae	insecticides (e.g., trichlorfon)

(Continued)

Table 1 Major invertebrate pests of livestock (*Continued*)

Livestock group	Pest species/parasitic condition	Order: Family	Control measures
Poultry	*Davainea proglottina* (Davaine)	Cyclophyllidea: Davaineidae	flubendazole, piperazine
	Nematodes (*Ascaridia galli* [Schrank], *Capillaria* sp.)	Ascaridida: Heterakidae; Enoplida: Trichuridae	flubendazole, piperazine
	Mites (*Ornithonyssus sylviarum* [Canestrini & Fanzago], *O. bursa* [Berlese], *Dermanyssus gallinae* [De Geer])	Parasitiformes: Macronyssidae, Dermanyssidae	acaricides (e.g., carbaryl, applied to host or housing)
	Argas persicus (Oken)	Parasitiformes: Argasidae	acaricides (e.g., carbaryl on hosts; malathion, fenthion, tetrachlorvinphos, permethrin on housing
	Lice (*Menacanthus* spp., *Menopon gallinae* [L.], *Goniocotes* spp., *Lipeurus caponis* [L.], *Cuclotogaster heterographus* [Nitzsch])	Mallophaga: Menoponidae, Philopteridae	insecticides (e.g., carbaryl, coumaphos, malathion)
	Fleas (*Ceratophyllus gallinae* [Schrank], *Echidnophaga gallinacea* [Westwood])	Siphonaptera: Ceratophyllidae, Pulicidae	insecticides (e.g., permethrin)

(Adapted from Refs. 8, 12–14.)

Australia, sheep nematodes inflict the greatest costs on the grazing industries, with losses as high as $131 million (6).

CONCLUSIONS

Because of the often multiple and insidious effects pest invertebrates, particularly the internal parasites, have on the health of livestock, economic losses are difficult to estimate with any precision. Information is most forthcoming and most reliable from developed countries with large, well-managed livestock industries. Data from the less developed areas of the world largely are unavailable. For this reason, estimates of the magnitude of livestock losses to invertebrate pests undoubtedly are conservative. Underdeveloped countries may lack modern pest management technologies and the means for proper hygienic management of livestock. Government agencies and academic institutions within these countries may lack the resources necessary to undertake pest surveys and report their findings, and to extend the results of pest management research to farmers and ranchers. The challenge for the future is to improve the economic lot of the poorer countries so that livestock losses to pests can better be assessed and mitigation of losses significantly improved.

Table 2 Livestock losses to arthropod pests

Livestock group	% Loss	Value of loss ($ billions)
Cattle	4	31.8
Swine	6	2.5
Sheep	6	5.7
Total		40.0

(Adapted from Refs. 1, 15–19.)

Table 3 Livestock losses to internal parasites

Livestock group	% Loss	Value of loss ($ billions)
Cattle	0.4	3.2
Swine	9	3.8
Sheep	5	4.7
Equines[a]	0.5	1.9
Poultry[b]	12	3.7
Total		17.3

[a]Data based on expenditures for anthelmintics.
[b]Data for coccidiosis (a protozoan infection).
(Adapted from Refs. 1, 15–19.)

REFERENCES

1. Kunz, S.E.; Murrell, K.D.; Lambert, G.; James, L.F.; Terril, C.E. Estimated Losses of Livestock to Pests. In *CRC Handbook of Pest Management in Agriculture*, 2nd Ed.; Pimentel, D., Ed.; CRC Press, Inc.: Boca Raton, FL, 1991; 1, 69–98.
2. Drummond, R.O.; Bram, R.A.; Konnerup, N. Animal Pests and World Food Production. In *World Food, Pest Losses, and the Environment*; Pimentel, D., Ed.; AAAS Selected Symposium 13, Westview Press, Inc.: Boulder, CO, 1978; 63–93.
3. Pimentel, D.; Lach, L.; Zuniga, R.; Morrison, D. Environmental and economic costs of nonindigenous species in the United States. BioScience **2000**, *50* (1), 53–65.
4. Evans, D.E. Tick infestation of livestock and tick control methods in Brazil: a situation report. Insect Sci. Applic. **1992**, *13* (4), 629–643.
5. Ridsdill-Smith, T.J. A contribution to assessing the economic impact of redlegged earth mite on agricultural production in Australia. Plant Prot. Q. **1991**, *6* (4), 168–169.
6. McLeod, R.S. Costs of major parasites to the Australian livestock industries. Int. J. Parasitol. **1995**, *25* (11), 1363–1367.
7. Drummond, R.O.; Lambert, G.; Smalley, H.E., Jr.; Terrill, C.E. Estimated Losses of Livestock to Pests. In *CRC Handbook of Pest Management in Agriculture*; Pimentel, D., Ed.; CRC Press, Inc.: Boca Raton, FL, 1981; 1, 111–127.
8. Kettle, D.S. *Medical and Veterinary Entomology*; CAB International: Wallingford, UK, 1990; 658.
9. Barnard, D.R.; Ervin, R.T.; Epplin, F.M. Production system-based model for defining economic thresholds in preweaner beef cattle, *Bos taurus*, infested with the lone star tick, *Amblyomma americanum* (Acari: Ixodidae). J. Econ. Entomol. **1986**, *79* (1), 141–143.
10. Fasanmi, F.; Onyima, V.C. Current concepts in the control of ticks and tick-borne diseases in Nigeria: a review. Insect Sci. Applic. **1992**, *13* (4), 615–619.
11. *The Sheep Market for Animal Health Products*; Bird, J., Ed.; PJB Publications: Surrey, UK, 1989; 157.
12. Soulsby, E.J.L. *Helminths, Arthropods & Protozoa of Domesticated Animals*, 6th Ed.; Williams and Wilkins: Baltimore, MD, 1968; 824.
13. Lancaster, J.L., Jr.; Meisch, M.V. Ellis Horwood Series in Entomology. In *Arthropods in Livestock and Poultry Production*; Ellis Horwood, Ltd.: Chichester, UK, 1986; 402.
14. Sainsbury, D. *Animal Health: Health, Disease and Welfare of Farm Livestock*; Blackwell Science Ltd.: Oxford, UK, 1998; 253.
15. FAO **1998**, FAOSTAT. http://apps1.fao.org/servlet/XteServlet.j...n Livestock (accessed March 2000).
16. U.S. Dept. of Agriculture. *Agricultural Statistics 1998;* National Agricultural Statistics Service, USDA, U.S. Government Printing Office: Washington, DC, 1998.
17. USDA. Equine. U.S. Dept. of Agriculture. *National Agricultural Statistics Service*, Eq. 1 (3–99). http://usda.mannlib.cornell.edu/reports/nassr/livestock/equine/eqinan99.txt (accessed March 2000).
18. USDA. Poultry—Production and Value, 1998. Summary. *U.S. Dept. of Agriculture. National Agricultural Statistics Service. Pou 3–1 (99).* http://usda.mannlib.cornell.edu/reports/nassr/poultry/pbh-bbp/ (accessed March 2000).
19. USDA. Meat Animals: Production, Disposition, and Income, 1998 Summary. *U.S. Dept. of Agriculture. National Agricultural Statistics Service*, Mt An 1–1 (99)a. http://usda.mannlib.cornell.edu/reports/nassr/livestock/zma-bb/ (accessed March 2000).

LANDSCAPE PATTERNS AND PEST PROBLEMS

F. Craig Stevenson
Saskatoon, Saskatchewan, Canada

INTRODUCTION

Crop productivity most often varies across a field, with areas of relatively high and low crop yields. These landscape patterns for crop productivity may be associated with landscape patterns for differing levels of weed, disease, or insect pest pressure. An understanding of the landscape patterns for pests will be an integral component of effective and efficient pest management strategies designed to maintain high levels of crop production across all areas of a field. A path to such an understanding will require basic knowledge of where pests occur and processes controlling these landscape patterns.

PEST AND LANDSCAPE PATTERNS

Small Patches

Certain areas within a field may be more conducive to pest survival, establishment, and development relative to other areas. The size and proximity of localized infestations with a given field differs among the three major pest groups. Past research has shown that weed and disease patches generally have a radius of about 25 m (if you assume that they have a circular shape), but may be as large as 100 m (1–3). In the case of wild oat (*Avena fatua* L.), the patch may be associated areas that have higher soil water and nutrient availability (see Fig. 1). Weeds such as perennial sowthistle (*Sonchus arvensis* L.), with seeds adapted to wind dissemination, may occur in areas of higher elevation where seeds are trapped as they are blown across a field. Foliar disease patches generally are associated with low wet areas of fields or sheltered field margins that have higher canopy humidity and lower wind speeds. Landscape patterns for insects, however, are less persistent and harder to predict because of rapid reproductive rates and pest movement (4). To a lesser extent, this same level of complexity also occurs for polycyclic diseases that easily become airborne. However, the overwintering phases in northern agricultural regions and nonflying phases of insects may be associated with areas of a field with a more friable soil structure (e.g., grasshoppers) or other soil and microclimatic factors. In addition, for insects and to a lesser extent diseases, landscape patterns are complicated pest–predator interactions that can affect infestation of that pest in the given space in the next growing season. For example, the landscape pattern of nematodes has been shown to be partly a function of nematophagous fungi (5). Therefore, the proximity of areas more or less conducive for the pest is ultimately dependent on the topography of the field and spatial variation of microclimatic conditions, especially for those pests with restricted mobility.

Larger Scale Landscape Patterns

These patterns are seen where larger areas of a field, or numerous fields, are being infested by a pest. These types of landscape patterns generally are dependent on external factors that move the pest from a smaller patch to adjacent areas. Climatic processes, such as wind, that blow weed seeds, disease spores, and insects can rapidly increase the area of pest infestation. Farming practices such as tillage and harvest operations also cause larger scale landscape patterns. For example, a study showed that *Polygonum* spp. of weeds tended to vary with the direction of tramlines up to distances of 635 m, a distance about 600 m more than that observed in the direction perpendicular to tramlines (1). The spatial distribution of pests can be influenced by tillage practices and harvest operations that move crop residues and soil across a field. For example, nematodes are spread by cultivation from initial infestation foci (5). Also, grain combine harvesters and shank-type tillage implements can move weed seeds and other reproductive structures across a field quickly.

Temporal Stability of Landscape Patterns

Time tends to complicate and/or obscure distinct landscape patterns for pests within a field. Varying climatic patterns (rainfall, temperature, etc.) within and among growing seasons can have a profound effect on the ability of pests to survive, establish, and develop. This complexity, interacting with polycyclic and multigeneration reproductive strategies within a growing season and factors

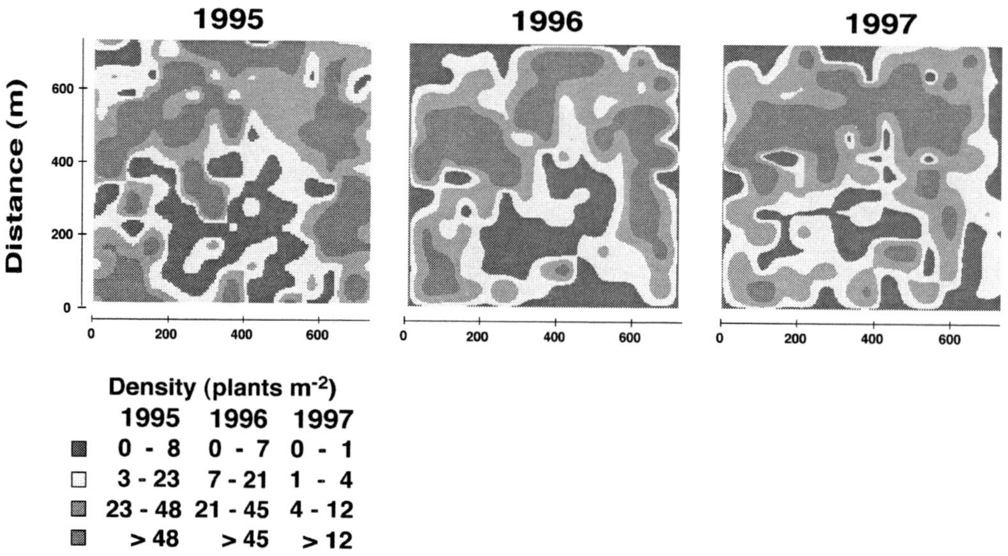

Fig. 1 Landscape pattern of wild oat in a 65 ha field. (From Ref. 6.)

moving pests across or between fields, poses a major hurdle to a holistic understanding of pests and landscape patterns. Extensive field research combined with predictive modeling may be a fruitful avenue for such a complex phenomenon.

FUTURE DEVELOPMENTS

The advent of global positioning systems (GPS), satellite imagery, and geographic information systems (GIS) software have heightened awareness and provided insight into landscape patterns and pests. Ultimately, these technologies may allow for the site-specific management of pests in accordance with their landscape patterns (7). For example, spatially referenced maps could be linked to GPS on pesticide applicators to reduce the total amount of pesticide applied, thus improving economic returns and resulting in farming systems less dependent on pesticides. Our current understanding clearly shows that pest infestations often occur in patterns across a field, however, a great deal more effort will be necessary to provide information to accurately predict where pests will occur and affect crop production most extensively. These challenges will be especially difficult considering the highly variable and dynamic nature of current climatic conditions and farm management systems.

REFERENCES

1. Nordbo, E.; Christensen, S.; Kristensen, K.; Walter, M. Patch spraying of weed in cereal crops. Asp. Appl. Biol. **1994**, *40*, 325–334.
2. Zadoks, J.C.; van den Bosch, F. On the spread of plant disease: a theory on foci. Ann. Rev. Phytopathol. **1994**, *32*, 503–521.
3. Zanin, G.; Berti, A.; Riello, L. Incorporation of weed spatial variability into the weed control decision-making process. Weed Res. **1998**, *38*, 107–118.
4. Hassell, M.P.; Comins, H.N.; May, R.M. Spatial structure and chaos in insect population dynamics. Nature **1991**, *353*, 255–258.
5. Webster, R.; Boag, B. Geostatistical analysis of cyst nematodes in soil. J. Soil Sci. **1992**, *43*, 583–595.
6. Thomas, A.G. *Agriculture and Agri-Food Canada*; Saskatoon, SK, Canada, 1999, unpublished data.
7. *The State of Site Specific Management for Agriculture*; Sadler, E.J., Pierce, F.J., Eds.; American Society of Agronomy, Inc., Crop Science Society of America, Inc., Soil Science Society of America, Inc.: Madison, WI, 1997; 423.

LAWN-CARE TREATMENTS (PLANT PATHOGENS)

W.J. Florkowski
University of Georgia, Griffin, Georgia, U.S.A.

INTRODUCTION

Pathogens are opportunistic organisms that persist in the environment and are more difficult to fight than weeds or insects. Pathogens include more than 100,000 fungal organisms and only a minute portion of them cause potential severe damage to lawns and turf (1). Many infect grass as a result of cultural practices, insect feeding, lawn use, and weather. Estimates of damage caused by pathogens are few and fragmented. Because of inadequate data collection, damages from diseases often go unreported and are difficult to isolate from the total observed pest damage. Furthermore, weather conditions determining the disease development are highly variable from year to year. Pathogen population growth can be disrupted by the application of fungicides and nematicides or by insecticides when insects serve as a vector in transmitting diseases. Diseases can damage a lawn overnight and lawn recovery is difficult. Typically, lawns and turf are treated with fungicides to prevent or to reverse damage. Fungicide applications aim at controlling rather than eradicating the diseases so the healthy grass can compete with pests.

Grasses are divided into cool and warm season grasses. Cultivated lawns and turf are concentrated in the temperate zone. Lawns enhance the value of real estate or use of recreational activities, for example, golf courses and soccer fields. In a recent survey of homeowners, 66% of participants answered that a lawn free of insect or disease damage was important or very important to them (2). Growing tourism and rising incomes contributed in recent years to the expansion of demand for lawns. Demand for lawn fungicides is derived from the demand for lawns. Demand for fungicide applications is highly income elastic, weather-dependent, and geographically concentrated in areas of upscale communities or on commercially operated recreational facilities. Demand for lawns rises as the global awareness of environmental damages increases. About 63% of homeowners want lawn care products to be environmentally friendly (2).

The use of fungicides on lawns and turf represents an increasing share of total pesticide applications as a result of pesticide reduction use programs in food crop production (3, 4). While in some countries the application of fungicides to lawns is prohibited (e.g., many European countries), other countries allow fungicide sales for commercial use or self-application by homeowners (e.g., the United States).

DISEASE PRESENCE AND DAMAGE

Improper maintenance by lawn and turf owners has been named by industry sources as the primary reason behind the disease damage (5). The substantial lag between the visible disease damage and the actual infection complicates the treatment and understanding of treatment needs by chemical applicators. The risk averse lawn and turf owners apply fungicides according to a preventative schedule. Companies serving commercial customers were found to purchase more fungicides than companies deriving their revenue primarily from serving homeowners (6).

In the southern United States, diseases that are the most difficult to control are brown patch (*Rhizoctonia solani*), dollar spot (*Sclerotinia homoeocarpa*), leaf spot (*Helminthosporium*) and melting-out (*Drechslera* spp.), *Fusarium* spp., *Anthracnose, Pythium* spp., and spring dead spot (*Leptosphaeria* spp.).

Pacific islands experience little damage from pathogens. Bermuda grass grown on golf courses (e.g., in Guam) does not suffer from the most prevalent diseases found in the southeastern United States, brown patch and dollar spot. Nematode damage is rare and fungicides are used sporadically. In northern Australia, Bermuda varieties suffer from Pythium and Rizoctonia aggravated by weather and cultural practices. In southern Australia, cool season grasses are common and suffer from Fusarium, spring dead spot, dollar spot, and white/black helmo.

Dollar spot disease occurs primarily in southern Europe (France, Italy, and Spain), while pink snow mold is a damaging disease in the United Kingdom, France, and Germany. Pythium and brown patch are not important in Europe.

In central and northern Europe the demand for home lawns has been increasing. Grass varieties are attacked by snow mold and leaf spot. Powdery mildew causes damage in partially shaded lawns, while *Puccinia striformis*, causing damage in the fall, is a relatively new disease observed on newly introduced varieties in Central Europe. Brown patch occurs during hot and wet summer weather, while fairy rings are common and observed on lawns rich in organic matter. About 50% of all grass seed produced in Europe is used for lawn and turf establishment. The annual production of grass seed in the European Union amounts to 130,000–135,000 tons and Denmark, Germany, and The Netherlands are the leading suppliers (7). Main varieties are *Lolium perenne*, Lolium multiflorum, and *Festuca rubra* spp. Among Central European countries, Poland produced an estimated 7800 tons of seed, of similar species, one-half of which was used on lawns (8).

U.S. LAWN LOSSES

A 1995 survey showed that diseases are second most important problem (following weeds) on lawns and turf according to lawn care service companies (9). Couch (10) provides an in-depth review of turn pathogens and fungicide treatments. Seven percent of American households purchased lawn fungicides in 1998 (11). The largest buyers were households with annual incomes above $100,000. Households in this income category represented a major portion of 14.0% of lawn care industry customers. In 1998, an average household spent about $190 on lawn care (11), but only a portion was spent on fungicides. The estimated expenditures on all fungicides applied solely by households amounted to $26 million in 1997 (12) and only a portion of 8 million pounds of active ingredients was applied to lawns. Commercial applicators applied additional amounts of fungicides.

The cost of fungicide application consists of the cost of the chemical and labor. The average lawn care firm in the southeastern United States reported an annual value of purchased fungicides at $452 or about 10% of the total value of purchased pesticides in 1989 (6). About 20% of lawn care and landscape maintenance firms applied fungicides on a preventive schedule (13), that is regardless of the diagnosed presence of a visible disease damage to a lawn. This risk averse approach is justified by the industry's view that diseases can rapidly overtake a substantial area once the proper combination of factors creates optimal growth conditions for pathogens. The cost of fungicide application ranged from $.10 to $.94 per 1000 square feet for a single application and was dependent on the type of fungicide applied (14). Because some fungicides control only a single disease, a mix of one or two may be applied, raising the costs. A single treatment may remain effective for five to 22 days depending on the fungicide type. The cost of labor, which ranged for a chemical applicator from $9.60 to $ 13.20 per hour in 1998 (5), discourages the use of commercial lawn services. However, where the turf is of essential value to a business, for example, a golf course, the use of fungicides during a growing season can be frequent. Turf diseases caused a loss of $78 million and cost an additional $48 million to control in Georgia alone (15). Damage inflicted by nematodes is generally small, sporadic, and treatable; for example, nematodes affected less than 3% of lawns and turf in Georgia in 1998.

Lawns and turf are not a tradeable commodity and, therefore, market forces do not price their value explicitly. Rather, lawn is a quality attribute of a composite good (e.g., a single family house) and its value is included in the total value of the good. Because local lawn and turf supply conditions determine the perceived value, estimates of losses from diseases are highly variable across local and disconnected markets. In contrast, application records and market discovered prices of row crops, fruits, or vegetables allow reliable disease loss estimation (16).

In 1997, about 26.5 million people played golf for recreation (17) and generated $5.7 billion of revenues at public golf courses alone. Public, private, and resort golf courses number in the thousands. Information about the disease damage and fungicide use is generally unavailable. In Georgia, golf courses reported the purchase of fungicides valued at $1.7 million in 1998 (18). The actual volume applied may be less because golf superintendents are trained to recognize conditions favorable for fungi growth. However, the use of fungicides on golf courses is substantial. Fungicides leave little residue and do not pose threat to surface water. According to a study in Germany, leaching was not a problem even from greens built on sand. Fungicides in Europe are sprayed two to three times during the season and only on greens and tees.

Aspelin and Grube (12) reported that the total fungicide expenditures worldwide were about $5.5 billion in 1997. The vast majority was applied to agricultural crops rather than lawns and turf. Assuming that two-thirds of all applied fungicides were applied in agriculture, approximately $1.8 billion were applied to nonagricultural uses including turf and lawns. Homeowners' fungicide use accounts for a fraction of this figure. Lawn care industry and recreational turf users and sod producers apply the sizable share of the balance. The value of purchased and applied fungicides represents the minimal value of losses caused by diseases, because it does not account for actual measurable damage to turf and labor and equipment cost.

FUTURE CONCERNS

Many countries developed national pesticide use reduction programs, which include taxation of chemicals. The programs have been less effective in reducing fungicide use than the use of herbicides or insecticides. The long-term solution is the selection and variety improvement through breeding grass lines resistant to particular diseases (19). Breeding programs for turf grasses are few and located in industrialized countries. The distinction between cool and warm season grasses dictates the scope of breeding programs. Furthermore, local growing conditions are highly variable and include grass tolerance for salinity, heat, moisture, and soil conditions. Major strides in the development of biological controls for diseases are dependent on basic research into characterization of the existing pathogens and their adversarial relationships. Efforts have been made to develop biological control methods for some fungi (20), but the cost effectiveness of these methods discourages their wide adoption.

Quality lawn as interpreted by the public is an artificial standard achieved through a combination of cultural practices including fungicide applications. A change in lawn and turf users' attitudes will lead to decline in fungicide use. Currently, consumers are unaware of organic or biological lawn maintenance options according to 85% of surveyed firms (9), although 69% of homeowners in the Atlanta metropolitan area, a major market for lawn care services, expressed interest in learning more about alternatives to chemical pesticides (2). Changing public attitudes through consumer education may increase awareness of the pros and cons of fungicide use, for example, consequences of pathogen resistance to fungicides. The search for innovative uses of cultural techniques such as changes in pH level, fertilization, proper mowing, and watering of soil also may provide alternatives to turf fungicide application and limit disease incidence.

REFERENCES

1. Koeller, W. Fungicide Resistance in Plant Pathogens. In *Handbook of Pest Management in Agriculture*; Pimentel, D., Ed.; CRC Press: Boca Raton, FL, 1991; 2, 679–720.
2. Jordan, J.L.; Florkowski, W.J.; Latimer, J.; Braman, S.K.; Varlamoff, S. *Georgia Homeowner Survey*; Georgia Agricultural Experiment Station, College of Agricultural and Environmental Sciences; University of Georgia, GA, 1999, *unpublished data*.
3. Sæthre, M.G.; Ørpen, H.M.; Hofsvang, T. Action. Action programs for pesticide risk reduction and pesticide use in different crops in Norway. Crop Prot. **1999**, *18*, 207–215.
4. Webster, J.P.G.; Bowles, R.G.; Williams, N.T. Estimating the economic benefits of alternative pesticide usage scenarios: wheat production in the United Kingdom. Crop Prot. **1999**, *18*, 83–89.
5. Florkowski, W.J.; Landry, G. *Survey of Lawn Care and Landscape Industry*; Georgia Agricultural Experiment Station, College of Agricultural and Environmental Sciences; University of Georgia: GA, 1999, *unpublished data*.
6. Florkowski, W.J.; Hubbard, E.E.; Landry, G.W.; Murphy, T.R. Impact of lawn-care firm characteristics on pesticide expenditures. HortScience **1994**, *29* (9), 1084–1086.
7. Kley, G. In *Seed Production in Grass and Clover Species in Europe, In Yield and Quality in Herbage Seed Production*, Proceedings of the 3rd International Herbage Seed Conference, Halle (Salle), Germany, June 18–23, 1995; 12–22.
8. Prończuk, S.; Prończuk, M. Nasiennictwo traw dla rekultywacji terenów trudnych. Grassland Sci. in Poland **2000**, *3*, 129–139.
9. Florkowski, W.J.; Robacker, C.; Braman, S.K.; Latimer, J.G.; Walker, J. The importance of pest management services among the Atlanta metro area lawn care and landscape maintenance firms. J. Georgia Green Ind. Ass. **1996**, *8* (3), 20–21.
10. Couch, H.B. *Diseases of Turfgrasses*, 3rd Ed.; Krieger Publishing Company: Malabar, FL, 1995; 421.
11. *National Gardening Survey, 1998–1999*. National Gardening Association, Inc.: Burlington, VT, 1999; 320.
12. Aspelin, A.L.; Grube, A.H. *Pesticides Industry Sales and Usage*; U.S. EPA: Washington, DC, 1999; 39.
13. Hubbell, B.J.; Florkowski, W.J.; Getting, R.; Braman, S.K. Pest management in the landscape/lawn maintenance industry: a factor analysis. J. Production Agric. **1997**, *10* (2), 331–336.
14. Brown, E.A. Turf. In *1998 Georgia Plant Disease Loss Estimates*; Williams-Woodward, J.L., Ed.; The University of Georgia Cooperative Extension Service, College of Agricultural and Environmental Sciences: Athens, GA, 1999a; 17.
15. Brown, E.A. *The Dollars and Sense of Disease Control*; The University of Georgia Cooperative Extension Service, College of Agricultural and Environmental Sciences, Dept. of Plant Pathology: Athens, GA, 1999b; 2.
16. Palm, E.W. Estimated Crop Losses without the Use of Fungicides and Nematicides and without Nonchemical Controls. In *Handbook of Pest Management in Agriculture*; Pimentel, D., Ed.; CRC Press: Boca Raton, FL, 1991; 2, 109–28.
17. *Statistical Abstract of the United States 1999*, 119th Ed.; U.S. Department of Commerce: Washington, DC, 1999; 1005.
18. Florkowski, W.J.; Landry, G. *Golf Course Superintendent Survey;* Georgia Experiment Station, College of Agricultural and Environmental Sciences. University of Georgia, 2000, *unpublished data*.
19. Landry, G.; Karnok, K.; Raikes, C.; Mangum, K. Bent (Agrostis spp.) cultivar performance on a golf course putting green. International Turfgrass Society Res. J. **1997**, *8*, 1230–1239.
20. Wilson, M.; Backman, P.A. Biological Control of Plant Pathogens. In *Handbook of Pest Management*; Ruberson, J.R., Ed.; Marcel Dekker, Inc.: New York, 1999; 309–335.

L

LEACHING

Lars Bergström
Swedish University of Agricultural Sciences, Uppsala, Sweden

INTRODUCTION

The contamination of groundwater by pesticides is of concern mainly because it may limit its use as drinking water. The extent of groundwater contamination to a large extent depends on the degree to which pesticides leach through the unsaturated zone of soils on which they have been applied (1). Pesticide movement through the unsaturated zone in tile-drained fields may also be a source of pesticides in surface waters, which support aquatic ecosystems and are used for drinking water in many areas of the world. Therefore, knowledge of pesticide movement in soil above the groundwater table is very important (2). This has also been the focus of a large number of studies performed during the past couple of decades (3, 4). The most important factors influencing pesticide leaching are soil properties, inherent properties of the pesticide molecules, climatic conditions, and management practices (3).

FACTORS INFLUENCING LEACHING

Soil and Hydrological Conditions

The rate and direction of water flow in the unsaturated zone are determined by the hydraulic gradient and the hydraulic conductivity. The presence of air-filled pores restricts the pathways through which water percolates downward, which means that the hydraulic conductivity in the unsaturated zone also depends on the level of water saturation. As the soil dries out, water becomes more strongly bound within the matrix of the soil, and the volume of water and the rate with which water percolates through soil decrease. This relationship between water retention and hydraulic conductivity varies considerably with soil type (soil texture and structure, and organic matter content) (5).

A complication, which has a major impact on pesticide leaching, is the fact that water, and pesticides dissolved in the water phase, often move through large pores in soil (e.g., earthworm and root channels, cracks etc.), a process commonly referred to as preferential flow (6, 7). Under such conditions, an equilibrium pesticide concentration throughout the soil profile cannot be obtained. This phenomenon primarily occurs in fine-textured soils with high clay contents, especially those that have the potential to swell and shrink. Through preferential flow, pesticides can be transported rapidly through large portions of the unsaturated zone and bypass biologically active layers in which they otherwise would be degraded or sorbed. Exposure to preferential flow is most pronounced soon after application of the pesticide, when high concentrations occur in the soil solution in upper soil layers, in combination with intensive rainfall (8). Once the pesticide is mixed in with the soil matrix, water moving through preferential flow paths does not interact with the soil, and leaching is therefore reduced. In other words, preferential flow can both increase and decrease pesticide leaching depending on the time when it occurs in relation to pesticide application. The final result is that (although transient flow peaks shortly after application causing elevated concentrations in water leaching through soils in preferential flow paths) the leaching loads over extended periods are typically quite small in such soils. Indeed, leaching loads are often larger in sandy soils, in which water and pesticide movement mainly occurs between individual soil particles within the main soil matrix (8). Pesticide concentrations are typically lower in sandy soils than in clay soils, but the water volumes displacing the pesticide are often much larger in sandy soils. The principal difference in pesticide leaching patterns in sand and clay soils is illustrated in Fig. 1. Irrespective of which leaching mechanism prevails, the total amount of the majority of pesticides that reach groundwater after normal agricultural use rarely exceeds 1% of the applied amount and is commonly well below 0.1%.

Pesticide Properties

The physicochemical properties of pesticides have a major impact on their leachability. In this context, the rate with which they are degraded and how strongly they are sorbed to soil are the most important factors. As a general rule of thumb, leaching decreases with increasing sorption affinity and faster degradation, and increases when the oppo-

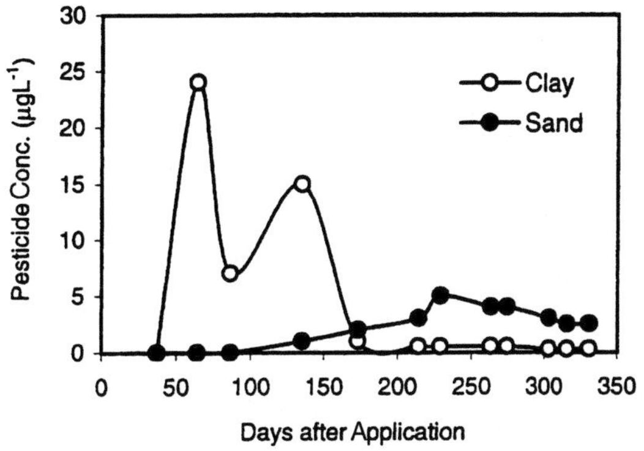

Fig. 1 Concentrations of a pesticide in water leaching from 1-m undisturbed soil columns of a clay and a sandy soil. (Modified from Ref. 8.)

site conditions prevail. However, it is important to note that degradation rates often become slower with residence time in the soil as a result of decreased availability due to sorption in the soil. This means that strongly sorbed pesticides, which are less mobile in soil than weakly sorbed compounds, are typically quite persistent. An example of a leaching classification scheme, based on the sorption strength of some pesticides, is shown in Table 1.

A factor that complicates the picture of pesticide movement in soil in relation to sorption affinity is the possibility that pesticide mobility is enhanced by adsorption to various mobile colloids, a process often referred to as ''facilitated transport.'' It is known that organic solutes, such as nonpolar pesticides with very low water solubilities, can form complexes with dissolved organic carbon and clay colloids that move through soil. Even though the role of colloids in facilitating pesticide transport is still relatively poorly understood, there is little doubt that failure to account for this mode of transport can lead to underestimates of both amounts and distances that strongly sorbed pesticides may migrate through the unsaturated zone.

Climatic Conditions

The amount and intensity of precipitation are the most important climatic factors influencing pesticide leaching. Water, in excess of evapotranspiration and what is required to maintain field capacity (i.e., the water content in soil when it is freely drained), leaches through the unsaturated zone and can thereby potentially move pesticides to groundwater. As mentioned above, in clay soils, it is primarily high intensity rainfall soon after application that may displace pesticides to depth in soil. In other words, the timing of precipitation is critical.

Soil temperature also has impact on leaching of pesticides, mainly by influencing the persistence of pesticides in soil and by affecting flow processes. Up to a certain level, degradation rates increase with increasing temperatures, which means that less of the compound will be available for leaching. Increasing temperatures will also increase evapotranspiration rates, which will reduce the amounts of water that can potentially move pesticides downward in soil. In climates with subzero temperatures in the winter season, soil will be frozen during extended periods. Under such conditions, pesticide movement in soil is very restricted, if it occurs at all. Leaching then will occur mainly during autumn and spring, when the soil is unfrozen and the evapotranspiration demand is low. Temperatures will also indirectly affect pesticide leaching by affecting the sorption/desorption process, although this influence is not yet thoroughly investigated and, therefore, less well recognized.

Management Practices and Strategies to Prevent Pollution

Management practices that have a major impact on the amount of pesticides that can move through soils can be grouped into the following categories: cropping/tillage, irrigation, and pesticide application practices.

Due to increased concern over soil erosion and input of pesticides to rivers and lakes, agricultural practices with reduced tillage or no-till management have been introduced. Such practices will also affect water infiltration rates, and therefore pesticide leaching through the unsaturated zone. In the short term, reduced tillage may decrease soil permeability compared with a conventionally tilled soil. However, over a whole growing season, infiltration rates tend to be higher under reduced tillage, especially in clay soils. This is largely due to the fact that

Table 1 Classification of pesticide mobility in soil based on their sorption strength, which in this case is expressed by their K_{oc} values (soil sorption coefficient, normalized to the soil organic carbon content)

K_{oc} value	Expected mobility	Type of pesticide
0–50	Very high	Bentazone, Dicamba
50–150	High	Atrazine, 2,4-D
150–500	Medium	Simazine, Metolachlor
500–2000	Low	Lindane, Linuron
2000–5000	Very low	Phenmedipham, Fenpropimorph
>5000	Immobile	DDT, Paraquat

(Modified from Ref. 9.)

reductions in tillage lead to less disruption of macropores in which pesticides can be rapidly transported through soil (see above). Reduced tillage also leaves more crop residues on the soil surface and contributes to reducing compaction of the subsoil caused by heavy equipment; both tend to increase permeability, and thus pesticide leaching. From the standpoint of reducing pesticide leaching, reduced tillage is therefore, in most cases, not a good management option.

As expected, irrigation increases leaching of pesticides by increasing the amount of water that potentially can move through soil. The amount of water, the rate at which it is applied, and the timing of irrigation are important for the same reasons as discussed previously for precipitation. Different irrigation methods (e.g., sprinkler, and drip and furrow irrigation) have also been shown to affect leaching.

Pesticide application strategies that influence pesticide residue levels in soil and thereby potential leaching include pre- vs. postapplication, split applications, placement methods, and use of different pesticide formulations. However, their influence on pesticide leaching is quite unclear and often overshadowed by other factors. Nevertheless, available data indicate that dividing the dose into two applications instead of one tends to reduce pesticide concentrations and the depth of migration in the subsoil. For similar reasons, pesticide leaching can be restricted by use of ''slow-release'' formulations in which the active ingredient is mixed with a solid matrix from which it gradually diffuses into the soil over an extended period. Placement of the pesticide instead of broadcasting, which reduces the soil surface area to which the pesticide is applied, also tends to reduce pesticide leaching.

FUTURE CONCERNS

Leaching of pesticides will undoubtedly continue to be of concern in the foreseeable future, and something that will be considered in various regulatory assessment schemes. In this context, it is important not only to look at the leachability, but also to evaluate the risks associated with leaching and the occurrence of pesticides in groundwater both from a human health and an ecotoxicological point of view.

In the future, there is reason to believe that fewer toxic compounds will be allowed, especially those that show high leachability in soil. There is also reason to believe that, in line with the increasing awareness of problems associated with leaching of pesticides, improved management strategies will be developed that reduce pesticide leaching further.

REFERENCES

1. Enfield, C.G.; Yates, S.R. Organic Chemical Transport to Groundwater. In *Pesticides in the Soil Environment: Processes, Impacts, and Modeling*; Cheng, H.H., Ed.; Soil Science Society of America Book Series: Madison, WI, 1990; 2, 271–302.
2. *The Lysimeter Concept—Environmental Fate of Pesticides*; Führ, F., Hance, R.J., Plimmer, J.R., Nelson, J.O., Eds.; ACS Symposium Series: Washington, DC, 1998; 699, 284.
3. Barbash, J.E.; Resek, E.A. *Pesticides in Ground Water—Distribution, Trends, and Governing Factors*; Ann Arbor Press: Chelsea, MI, 1996; 588.
4. Flury, M. Experimental evidence of transport of pesticides through field soils—a review. J. Environ. Qual. **1996**, *25* (1), 25–45.
5. Carter, A.D. Leaching Mechanisms. In *Pesticide Chemistry and Bioscience—The Food–Environment Challenge*; Brooks, G.T., Roberts, T.R., Eds.; Royal Society of Chemistry: Milton Road, U.K., 1999; 8, 291–301.
6. Bergström, L.F.; Jarvis, N.J. Leaching of dichlorprop, bentazon, and 36Cl in undisturbed field lysimeters of different agricultural soils. Weed Sci. **1993**, *41* (2), 251–261.
7. Brown, C.D.; Carter, A.D.; Hollis, J.M. Soils and Pesticide Mobility. In *Environmental Behaviour of Agrochemicals*; Roberts, T.R., Kearney, P.C., Eds.; John Wiley & Sons: Chichester, England, 1995; 9, 131–184.
8. Bergström, L.; Stenström, J. Environmental fate of pesticides in soil. Ambio. **1998**, *27* (1), 16–23.
9. Torstensson, L. Kemiska Bekämpningsmedel—Transport, Bindning och Nedbrytning i Marken. In *Aktuellt från Sveriges Lantbruksuniversitet*; Swedish University of Agricultural Sciences: Uppsala, Sweden, 1987; 357, 36, (in Swedish).

LEGAL ASPECTS OF PEST MANAGEMENT AND PESTICIDES

L

Michael T. Olexa
University of Florida, Gainesville, Florida, U.S.A.

INTRODUCTION

To promote public health, personal safety, and environmental protection, pesticides and their use are extensively regulated. Pesticide-related law is framed by both federal and state government. State laws typically supplement or duplicate federal laws; however, in some instances, state laws are more strict than federal laws. Consequently, compliance with state laws often assures compliance with federal laws. Both federal and state laws provide for criminal prosecution and can impose penalties such as fines or imprisonment. In addition, common law actions (lawsuits) also influence pesticide use. Common law actions are for civil wrongs. Such actions are initiated by a person who has suffered injury, or whose property has been damaged, as the result of the acts or omissions of a pesticide user (1).

STATUTORY LAW AND REGULATION

Statutory laws are the formal acts of federal and state legislatures. Although statutory laws often provide specific directions, in-depth details are often too technical or minute for the legislation to indicate. For this reason, a legislature typically empowers a regulatory agency to develop rules specifying the details of the statute's implementation. These rules have the force of law. Regulatory agencies implement their rules by requiring permits or licenses, and enforce those rules through civil and criminal penalties (2).

Federal Law

Pesticides were first subject to federal laws by the Insecticide Act of 1910 (3). This law protected farmers from sellers of adulterated or misbranded pesticide products. Following a surge in the development of new pesticides during the Second World War, Congress repealed the Insecticide Act of 1910 and enacted the Federal Insecticide, Fungicide and Rodenticide Act (FIFRA) of 1947 (4). FIFRA broadened federal control of pesticides by requiring the United States Department of Agriculture (USDA) to register any pesticide prior to its introduction into interstate commerce. In 1970, Congress transferred the administration of FIFRA to the newly created United States Environmental Protection Agency (EPA). Thereafter, Congress enacted the 1972 Federal Environmental Pesticide Control Act (FEPCA), to address public concerns about pesticides' environmental impact, and to ensure greater pesticide use regulation (5). As a result, federal policy shifted from regulation of pesticides for reasonably safe use in agriculture to regulation of pesticides to prevent unreasonable risks to people and the environment. Subsequent amendments to FIFRA (1988, etc.) have clarified the EPA's duties and responsibilities (6).

The Federal Food, Drug and Cosmetic Act (FFDCA), and 1954 Miller Amendments, required the government to establish tolerances (maximum allowable pesticide residue limits) for all pesticides used on food and feed crops (7). The 1996 Food Quality Protection Act (FQPA) amended the FFDCA and FIFRA, especially the process of establishing tolerances for pesticide residues in food and feed (8). For example, the EPA now uses a single standard to evaluate pesticide residues on raw and processed foods. FQPA represents the single largest shift in federal pesticide policy and process ever undertaken.

FQPA

- Establishes a *new safety standard* to be met when establishing tolerances for a reasonable certainty of no harm from *aggregate exposure.*
- Requires *all existing tolerances to be reassessed* under the new standard.
- Requires the EPA to make an explicit determination that tolerances for *residues in food are safe for infants and children* (8).

The heart of the regulatory scheme for pesticide application, storage, and disposal is the FIFRA provision making it unlawful to, ''. . .use any registered pesticide in a manner inconsistent with its labeling.'' (9). Thus, courts consider pesticide label instructions legislative regulations having the force of law. For pesticide users, *labeling* is the basis for enforcement of FIFRA. Hence, every pesticide applicator has a *legal obligation* to read and follow all label instructions attached to the product, and all product usage directions contained in any printed materials mentioned on the label.

Pesticide use enforcement

The EPA delegates enforcement of pesticide use violations to the states. Each state's agencies are responsible for administering pesticide regulations under state and federal statutes. States occasionally refer pesticide use violation cases to the EPA for enforcement (10).

Pesticide product registration

The EPA retains primary oversight for enforcing pesticide registration requirements. State agencies generally identify product registration violations during their inspections and refer them to EPA for enforcement (11).

Enforcement of tolerances for pesticides in food and feed

The Food and Drug Administration (FDA) monitors raw and processed commodities for compliance with residue tolerances. The USDA monitors meat, milk, and eggs for residue tolerance compliance. Both domestically produced and imported commodities are covered by the FDA and USDA programs. A few states (e.g., California and Florida) have additional residue monitoring and enforcement programs (12).

Common Law

Common law actions do not depend upon statutes for their authority. Instead, common law arises from the generalized legal duty individuals in a law-abiding society owe to one another. Every adult person is obligated to a certain duty of care for the personal and property rights of others. A violation of this obligation can become a basis for a common law action (lawsuit). Common law theories generally encountered in actions resulting from pesticide use include *negligence, trespass, nuisance,* and *strict liability* (13).

Negligence

This is a legal standard applied to an individual who fails to act in a reasonably prudent manner (14). A person found negligent is responsible for the damages caused by his or her act or omission. Pesticide users are liable for their negligence when their acts or omissions cause an injury that otherwise would not have happened.

Trespass

Trespass is an unauthorized entry onto the property of another by a person or thing that causes damage (15). Pesticide application can result in liability for trespass if the pesticide, its residue, or container becomes deposited on another's land (through dumping, drift, runoff, incineration, or other means) and causes substantial damage to the property.

Nuisance

A nuisance is substantial interference with another's use and enjoyment of land (16). A nuisance lawsuit requires no physical invasion, only a substantial interference with the possessor's enjoyment of land. Pesticide use resulting in offensive odors can be grounds for a lawsuit alleging nuisance. In almost all states, limited protection from nuisance actions is provided by state ''right-to-farm'' statutes. The farmer's ''right-to-farm'' defense is limited to nuisance actions (17).

Strict liability

Some states hold pesticide applicators absolutely responsible for their pesticide application activities, *regardless of fault,* without a showing of negligence. This is known as strict liability (18). Strict liability is normally associated with inherently dangerous or ultrahazardous activities. Persons who use explosives or keep wild animals are usually held strictly liable for their acts.

REGULATORY TRENDS

For the foreseeable future, current and emerging societal issues will continue to stimulate regulatory action. Principal EPA foci include (19) the following:

1. *Curtailing urban pesticide misuse.* The EPA's focus on pesticides in urban areas began with the misuse of an agricultural pesticide (methyl parathion) for urban pest control. Under the 1996 FQPA, EPA's risk assessment process centers on exposures that may occur from pesticide use in residences, lawns, and other areas where children may be in contact with pesticides.
2. *Minimizing occupational risks from pesticides.* Protection of agricultural workers and persons who mix, load, and apply pesticides is done through pesticide registration/reregistration, and regulatory programs, such as the Worker Protection Standard.

3. *Reducing environmental risks*. Protecting aquatic and nontarget organisms from pesticide use and misuse is emphasized.
4. *Protecting higher-risk humans* (*infants and children*). A primary purpose of the FQPA is to safeguard infants and children from exposure to pesticides in their diet. Members of this group are considered to have an increased risk of adverse health effects attributable to pesticide exposure, due to their metabolic rate, developing organs, and susceptibility to DNA mutations. FQPA's tolerance review process requires special consideration to be given to infants and children.

REFERENCES

1. Whitford, F.; Olexa, M.T.; Thornburg, M.; Gunter, D.; Ward, J.; Lejurne, L.; Harrison, G.; Becovitz, J. *Pesticides and the Law: A Guide to the Legal System*; Purdue University Cooperative Extension System, 1996; 12.
2. Olexa, M.T.; Leviten, A. *Circular 1139: Farm and Ranch Handbook of Florida Solid and Hazardous Waste Regulation*; University of Florida Cooperative Extension Service, 1999; 10.
3. *The Insecticide Act of 1910*, Pub. L. No. 6–152, 36 Stat. 331 (1910).
4. The Federal Insecticide. *Fungicide and Rodenticide Act of 1947*, Pub. L. No. 80–104, 61 Stat. 163 (1947).
5. *The Federal Environmental Pesticide Control Act of 1972*, Pub. L. No. 92–516, 86 Stat. 973, (codified at 7 U.S.C. Section 136 (1988); see also 7 U.S.C. Sections 135–135 y, as amended 7 U.S.C. Sections 136 et seq. (1994).
6. Olexa, M.T.; Kubar, S.; Cunningham, T.; Meriwether, P. *Bulletin 311: Laws Governing Use and Impact of Agricultural Chemicals: An Overview*; University of Florida Cooperative Extension Service, 1995; 7.
7. *21 U.S.C. Section 301 et. seq. (1994).*
8. *Food Quality Protection Act of 1996*, Pub. L. No. 104–170, 110 Stat. 1489 (1996).
9. *7 U.S.C. Section 136 j(a)(2)(G)(1994).*
10. Olexa, M.T.; Kubar, S.; Cunningham, T.; Meriwether, P. *Bulletin 311: Laws Governing Use and Impact of Agricultural Chemicals: An Overview*; University of Florida Cooperative Extension Service, 1995; 11.
11. Olexa, M.T.; Kubar, S.; Cunningham, T.; Meriwether, P. *Bulletin 311: Laws Governing Use and Impact of Agricultural Chemicals: An Overview*; University of Florida Cooperative Extension Service, 1995; 7.
12. Olexa, M.T.; Kubar, S.; Cunningham, T.; Meriwether, P. *Bulletin 311: Laws Governing Use and Impact of Agricultural Chemicals: An Overview*; University of Florida Cooperative Extension Service, 1995; 14–16.
13. Whitford, F.; Olexa, M.T.; Thornburg, M.; Gunter, D.; Ward, J.; Lejurne, L.; Harrison, G.; Becovitz, J. *Pesticides and the Law: A Guide to the Legal System*; Purdue University Cooperative Extension System, 1996; 12.
14. Keeton, et al. *Prosser and Keeton on the Law of Torts*, 5th Ed.; Section 32, 1984; 174.
15. Keeton, et al. *Prosser and Keeton on the Law of Torts*, 5th Ed.; Section 16, 1984; 67.
16. Keeton, et al. *Prosser and Keeton on the Law of Torts*, 5th Ed.; Section 87, 1984; 622.
17. Fischer, J.; Olexa, M. T.; Leviten, A.; Saju, A. *Circular 1224: Handbook of Florida Agricultural Laws*; University of Florida Cooperative Extension Service, 1999; 4.
18. Keeton, et al. *Prosser and Keeton on the Law of Torts*, 5th Ed.; Section 75, 1984; 536–537.
19. *1996 Food Quality Protection Act Implementation Plan*, U.S. EPA, Mar. 1997, 48.

LETHAL GENES FOR USE IN INSECT CONTROL

Thomas A. Miller
University of California, Riverside, California, U.S.A.

INTRODUCTION

The sterile insect technique (SIT), a genetic control strategy has been used for many years to control a variety of major pest insects (1). In SIT, target insects are mass-reared, then irradiated, usually with gamma rays from a Cobalt 60 radioactivity source. The doses selected cause multiple, random chromosome breakages that the nuclei of cells then attempt to repair. This results in a certain amount of debilitation due to incomplete chromosomal repair. But it also results in production of conditional lethal genes in the treated insects that when mated with wild insects produce no offspring. Because of the random nature of irradiation damage, the exact conditional lethal genes produced that cause sterility are never identified.

The sterilized SIT adults are transported to infestation areas and released in sufficient numbers compared to native populations to ensure a very high probability that reproduction will be unsuccessful. What the sterile insects lose in competitiveness due to radiation damage, they are thought to gain back by excessive numbers with 60:1 being a typical ratio of sterile to target insects. Because of the large numbers of sterile insects needed to ensure success in an SIT strategy, only the most economically compelling pest infestations can justify the high cost of this approach.

Improvements in SIT strategy have been slow in coming partly because of its already inherent simplicity and ease of use. But also it has not been possible to generate lines of insects that carry single conditional lethal traits without the drawbacks of nonspecific irradiation damage to the chromosomes. In addition, the application of genetic strategies beyond the SIT approach has remained largely theoretical until very recently.

Genetic transformation of vinegar fly, *Drosophila melanogaster*, by P elements showed that transposable elements (so-called jumping genes or mobile elements) were capable of providing the needed delivery vehicles for specific genes. And although P elements themselves proved to be useful for inserting genes in insects only very closely related to *D. melanogaster*, a growing number of transposable elements have now been discovered that have proven useful in genetic transformation of a growing number of insects aside from the drosophiloids (2), including lepidopterans (3).

Since most work has concentrated on the transformation method itself (2), the search for transgenes in insects that might confer population control has occupied comparatively less time. By contrast a large number of candidate genes that confer plants resistance to insects have now been tested (4). Still, a few genes have been proposed, and some exciting new transgenic insect control strategies have been discovered and are being developed.

Modern strategies for genetic control of insects (and diseases they vector) now fall into three discrete categories: transgenic insects, paratransgenic insects, and insect-resistant genes inserted into crop plants.

TRANSGENIC INSECTS

The *Notch* gene family has been thoroughly studied by developmental biologists. A vinegar fly strain, termed N^{60g11}, which carries two extra copies of a *Notch* gene mutation, was recently demonstrated to collapse a wild population of vinegar fly in three generations when the mutants were added in equal numbers to the wild types at each generation (5). This result is considerably better than the 60:1 ratio called for in the SIT strategy.

SEXUAL LETHAL GENES

Since 1983, attempts have been made at chemical selection of Mediterranean fruit flies (Med fly) for temperature-sensitive lethal genes (6–8). The purpose for these selections was to develop the ability to separate males and females. Great cost savings were expected if only one sex needed to be reared. Tests showed that releasing only sterile male Med flies into cages with wild populations caused more reproductive disruption compared to releasing both sterile males and females.

Separate efforts have been aimed at isolating color mutations that allowed the separation of females from males in culture. A strain of white female and brown male puparia were eventually developed, distinguished, and remained stable through several generations in mass culture before being lost due to both reduced viability of white pupae and a low level of recombination.

Chemical selection mechanisms have also worked with Anopheline mosquitoes, stable fly, and house fly when insecticide treatments killed female eggs allowing male eggs to survive. Due to a lack of transformation protocols until recently, none of these sexing strategies have been accomplished using genetic engineering. However, because of recent success at introducing genes into mosquito and Med fly, molecular approaches to sexual selection are certain to be advanced in the near future.

PARATRANSGENIC INSECTS

Paratransgenesis was the brainchild of the brilliant molecular parasitologist, Frank Richards, of Yale Medical School. Professor Richards and his colleagues conceived of the notion of disrupting insect-borne diseases by genetically altering the endosymbiotic bacteria present in the vector insect to produce a gene product toxic only to the disease agent (9). They chose Chagas' disease as a target that transmits the protozoan *Trypanosoma cuzi* delivered to victims by triatomine bugs. In a series of outstanding experiments, Ravi Durvasula of the Richards' team isolated and identified the endosymbiont bacterium, *Rhodococcus rhodnii,* as essential to growth of the vector insect, *Rhodnius prolixus*. He then worked out a genetic transformation system for *R. rhodnii* and the team proceeded to insert first cecropin A gene, and later a single-strand antibody gene into the endosymbiont.

By delivering genes via endosymbiotic bacteria in insects (10), paratransgenesis opens an entirely new route to control plant and animal diseases vectored by insects. Aside from the obvious hurdles of working with certain endosymbionts, which are often difficult to rear outside the host insect, the main advantage of this paratransgenic approach is the extremely short time needed to transform the endosymbiont itself. Compare, for example, the amount of time and effort needed to transform both plants and animals. In pink bollworm, one generation takes a month, and breeding to homozygosity easily could take a year to breed a pure transgenic strain. Endosymbionts, on the other hand, could be transformed in hours and delivered to target insects in a few days at most where further breeding is unnecessary, depending on the endosymbiont.

TRANSGENES IN PLANTS CONFERRING RESISTANCE TO INSECTS

Plant scientists have already tested a host of transgenes that show promising resistance to plant pest insects (11). Paramount among these are the endotoxin genes isolated from the bacterium *Bacillus thuringiensis* or simply *Bt* whose gene products are activated in certain insects that chew the plants and the cleavage products are thought to disrupt midgut membrane osmoregulation causing death.

A number of other gene products, derived from plants or microorganisms, have also shown promise in controlling insect feeding when their corresponding genes are inserted into plants. These include: cholesterol oxidase, vegetative insect protein, lectins, amylase inhibitors, proteinase inhibitors, peroxidases, cytokinins, and certain secondary plant metabolites. Lectins are carbohydrate-binding proteins that have lethal effects on certain plant pests.

REPRESSORS AND ACTIVATION FACTORS

Implicit in genetic strategies to control insects is not just inserting genes into insects or their endosymbionts, but designing various mechanisms to control the genes. One such repressor or activator factor is the well-known tet-repressor modifier derived from bacteria. This modifier is either held on or held off in the presence of the antibiotic tetracycline. Once the antibiotic is removed or added, the repressor or activation takes place and the gene is either turned off or its product is expressed.

DRIVING GENES INTO INSECT POPULATIONS

Scientists are just now beginning to uncover the complexities of interactions between plants, plant-sucking insects, plant virus transmissions, and the role played by endosymbionts in insects. In the few cases that are well studied, it appears that some endosymbionts have taken on the role of protecting plant viruses during their circulative route through the vector insect, which is essential for retransmission to uninfected plants. This research area is relatively new and challenging since it involves cooperative work between virology, entomology, plant pathology, immunology, and vector ecology.

One such group of endosymbionts, bacteria of the genus *Wolbachia*, confers unusual properties to infected insects. Some infected parasitic wasps are converted to purely parthenogenic reproductive behavior in which only viable infected female offspring are produced. Antibiotic treatment of these infected wasps restores sexual reproduction. In other insects, *Wolbachia* infections confer a condition known as ''Cytoplasmic Incompatibility'' that favors the production of infected offspring. Transgenic *Wolbachia* are widely held as an ideal mechanism with which to carry genes into insect populations. The main drawbacks preventing exploitation of *Wolbachia* as a vehicle to deliver insect or disease control strategies are lack of culture methods and lack of a transformation protocol.

The strategies described above for the genetic control of insects are not strictly the same as classical biological control that entails establishing pest-predator-parasite insect complexes designed, hopefully, to keep pest populations from exploding. Because it is so sophisticated, classical biological control is often not compatible with the monoculture methods in mainstream agriculture where they must compete with the predictable certainties of insecticide application.

Introducing paratransgenic insects into field populations might not require nearly the large numbers and logistics necessary for the sterile insect technique to function. Naturally, each pest–crop situation would be viewed as a unique challenge, and require its own particular approaches. The first applications of genetic control strategies beyond the sterile insect technique are still several years away; however, it is now clear that the means to deliver genes are in place and merely need to be exploited by the first pioneers.

REFERENCES

1. Bartlett, A.C. Insect Sterility, Insect Genetics, and Insect Control. In *Handbook of Pest Management in Agriculture*; Pimentel, D., Ed.; CRC Press: Boca Raton, FL, 1990; II, 279–287.
2. Atkinson, P.W.; O'Brochta, D.A. Genetic transformation of non-drosophilid insects by transposable elements. Ann. Entomol. Soc. Am. **1999**, *92* (6), 930–936.
3. Peloquin, J.J.; Thibault, S.T.; Staten, R.T.; Miller, T.A. Germ-line transformation of pink bollworm (Lepidoptera: Gelechiidae) mediated by the *piggyBac* transposable element. Insect Molecular Biology **2000**, *9* (3), 323–333.
4. Schuler, T.H.; Poppy, G.M.; Kerry, B.R.; Denholm, I. Insect-resistant transgenic plants. Trends in Biotechnol. **1998**, *16*, 168–176.
5. Fryxell, K.J.; Miller, T.A. Autocidal biological control: a general strategy for insect control based on genetic transformation with a highly conserved gene. J. Econ. Entomol. **1995**, *88*, 1221–1232.
6. In *Genetic Sexing of the Mediterranean Fruit Fly*, Proceedings of the Final Research Coordination Meeting, Organized by the Joint FAO/IAEA Division of Nuclear Techniques in Food and Agriculture, Colymbari, Crete, 1988; International Atomic Energy Agency (IAEA): Vienna, 1990; 221.
7. Crampton, J.M. In *Genetic Engineering of Insects and Applications in Basic and Applied Entomology*, Proceedings Series: Management of Insect Pests: Nuclear and Related Molecular and Genetic Techniques, (IAEA): Vienna, Austria, 1993; 33–47.
8. Crampton, J.M.; Warren, A.; Lycett, G.J.; Hughes, M.A.; Comley, I.P.; Eggleston, P. Genetic manipulation of insect vectors as a strategy for the control of vector-borne disease. Ann. Trop. Med. Parasitol. **1994**, *88*, 3–12.
9. Durvasula, R.V.; Kroger, A.; Goodwin, M.; Panackal, A.; Kruglov, O.; Taneja, J.; Gumbs, A.; Richards, F.F.; Beard, C.B.; Cordom-Rosales, C. Strategy for introduction of foreign genes into field populations of Chagal's disease vectors. Ann. Entomol. Soc. Amer. **1999**, *92* (6), 937–943.
10. O'Neill, S.L.; Hoffmann, A.A.; Werren, J.H. *Influential Passengers, Inherited Microorganisms and Arthropod Reproduction*; Oxford University Press: New York, 1997.
11. *Advances in Insect Control: The Role of Transgenic Plants*; Carozzi, N., Kozeil, M., Eds.; Taylor & Francis: Bristol, PA, 1997.

LICENSING AND CERTIFICATION OF PESTICIDE APPLICATORS

L

Patrick J. O'Connor-Marer
University of California, Davis, California, U.S.A.

INTRODUCTION

Pesticide applicators include people who mix, load, and apply pesticides as well as those who clean or repair pesticide-contaminated equipment, handle opened pesticide containers, or work as ground flaggers for aerial applications. In 1972 an amendment to the Federal Insecticide, Fungicide, and Rodenticide Act (FIFRA) mandated the certification of all pesticide applicators in the United States who handle or supervise the handling of especially hazardous pesticides classified by the U.S. Environmental Protection Agency (EPA) as *restricted use*. Specific criteria are established in federal regulations to certify both *private* and *commercial* applicators. EPA has the responsibility for assuring that certification plans submitted by states, territories, and tribes meet the FIFRA requirements.

NATIONAL REQUIREMENTS FOR CERTIFYING PESTICIDE APPLICATORS

The Federal Insecticide, Fungicide, and Rodenticide Act, signed into law in 1947 and amended in 1972, 1975, 1978, 1984, 1992, and 1996, is the basis for regulating the manufacture, sale, transportation, and use of pesticides in the United States. The enforcement of this Act is the responsibility of the U.S. Environmental Protection Agency (40 C.F.R. Parts 152 through 186). Many states have adopted supplementary laws that further restrict or regulate how pesticides are used in areas under their jurisdiction. Other federal agencies, including the Department of Agriculture (USDA), the National Institute for Occupational Safety and Health (NIOSH), and the Fish and Wildlife Service (FWS), set forth additional restrictions or advise EPA on certain types of pesticide hazards.

Incorporated into state, territorial, and tribal regulations are federal standards for certifying pesticide applicators who handle or supervise the handling of restricted-use pesticides (40 C.F.R. Part 171—*Certification of Pesticide Applicators*). Federal regulations establish requirements for both private and commercial pesticide applicator certification. A private applicator is a person who handles or supervises the handling and application of restricted-use pesticides for purposes of producing any agricultural commodity on property owned or rented by the applicator or the applicator's employer. A commercial applicator handles or supervises the handling and application of restricted-use pesticides for any purpose, usually for hire, on any property other than agricultural property owned or rented by the applicator or the applicator's employer.

Restricted-Use Pesticides

Federal regulations restrict the use of especially hazardous pesticides so that only certified commercial or private pesticide applicators can purchase, store, handle, or apply them. Some state, territorial, and tribal pesticide regulatory agencies restrict the use of additional pesticides that have not been classified as restricted-use by EPA. Criteria for restricting the handling of pesticides to certified applicators include:

- Danger to public health
- Hazard to applicators and farmworkers
- Hazard to domestic animals, honey bees, and crops from direct application or drift
- Hazard to the environment from drift onto streams, lakes, or wildlife estuaries
- Hazard related to persistent residues in the soil resulting in contamination of the air, waterways, estuaries, or lakes, causing damage to fish, wild birds, and other wildlife
- Hazard to subsequent crops due to persistent soil residues

The Certification Process

The handling and application of restricted-use pesticides requires that applicators learn certain technical skills through on-the-job training, attending educational programs, and studying pesticide manuals, pesticide labels, regulations, and other materials. The certification process provides a means for regulatory agencies to assess the competency of applicators to select, use, and handle these hazardous pesticides. People who are unable to pass the certification requirements are not allowed to handle restricted-use pesticides unless they are supervised by certified applicators.

Federal regulations require that written examinations be used to assess competency of individuals who desire to become certified commercial applicators. Private applicators must demonstrate their competency to an appropriate regulatory agency through written or oral examinations or other means. However, most state, territorial, and tribal

Table 1 Areas of competency for certified applicators

Area of competency	Private applicator	Commercial applicator
Label and labeling comprehension		
• Label format and terminology	•	•
• Instructions, warnings, terms, symbols, and other information	•	•
• Classification of product (general use or restricted use)	•	•
• Necessity for use consistent with the label	•	•
Pesticide safety factors		
• Toxicity, hazard to people, exposure routes	•	•
• Types and causes of common pesticide accidents		•
• Precautions needed to protect applicators and others	•	•
• Requirements for personal protective equipment	•	•
• Symptoms of pesticide poisoning	•	•
• First aid and decontamination procedures	•	•
• Identification, storage, transport, handling, mixing, and disposal methods and precautions		•
Protecting the environment		
• Understanding weather and other climatic conditions	•	•
• Understanding terrain, soil, or other substrate	•	•
• Presence of fish, wildlife, and other nontarget organisms	•	•
• Drainage patterns	•	•
Pest factors		
• Recognizing pest characteristics and damage	•	•
• Recognizing relevant pests		•
• Understanding pest development and biology for proper control		•
Pesticide factors		
• Types of pesticides		•
• Types of formulations		•
• Compatibility, synergism, persistence, and animal and plant toxicity		•
• Hazards and residues associated with use	•	•
• Factors that influence effectiveness or lead to problems such as resistance		•
• Dilution procedures	•	•
Equipment factors		
• Advantages and limitations of different types of equipment		•
• Equipment uses, maintenance, and calibration	•	•
Application techniques		
• Procedures used to apply various formulations and which method to use		•
• Proper placement of pesticide	•	•
• Preventing drift and loss of pesticide into the environment	•	•
Laws and regulations applicable to pesticide use	•	•

(From Code of Federal Regulations Section 40, Part 171.)

pesticide regulatory agencies use written examinations for certifying private applicators. Table 1 shows the areas of competency laid out in the federal regulations that a person must demonstrate to qualify to be a certified private or commercial applicator.

Pesticide applicator certification categories

Commercial pesticide applicators must be certified in one or more specialized categories as defined in the federal regulations. These 11 categories, identified in Table 2, correspond to the general types of work situations where restricted-use pesticides may be applied. Applicators must be qualified in a specific category before handling or supervising the handling of restricted-use pesticides in that work situation.

Many states have adopted other categories or subcategories for certified commercial applicators. Some states have as many as 35 categories. The federal regulations do not mandate separate categories for certified private applicators, although several states have adopted specific categories and require private applicators to become certified in one or more of these.

Recertification of pesticide applicators

All certified pesticide applicators must be recertified periodically. There are major differences among the time requirements and methods of recertification throughout the United States, although on average the recertification time ranges between two and five years for both private and commercial applicators. Recertification is accom-

Table 2 Certification categories for commercial applicators

Category	*Work situations*/Special knowledge requirements
Agricultural pest control (plant)	*Production agriculture, including commercial greenhouses and nurseries.* Knowledge of soil and water problems, harvest intervals, restricted-entry intervals, plantback restrictions, potential for environmental contamination.
Agricultural pest control (animal)	*Meat and dairy animals, horses, poultry, and others.* Knowledge of specific animals, associated pests, hazards related to formulation, impact of pesticides relating to application technique, age of animals, stress, and extent of treatment.
Forest pest control	*Forests, forest nurseries, Christmas tree farms, forest seed production.* Knowledge of pests, pest cycles, impact of pesticides on the environment and wildlife, specialized application equipment, and influences of weather on applications.
Ornamental and turf pest control	*Parks, school grounds, golf courses, cemeteries, residences, other.* Knowledge of production and maintenance of ornamental trees, shrubs, plantings, and turf; potential for phytotoxicity due to plant varieties, drift, and persistence; ways to minimize potential hazards to humans, pets, and other domestic animals.
Seed treatment	*Production of seeds for production agriculture and commercial nurseries.* Knowledge of types of seed requiring protection, factors that influence binding and germination, hazards with handling or misusing treated seed, proper disposal of unused treated seed.
Aquatic pest control	*Weed and other aquatic pest control in ponds, lakes, streams, rivers, and other bodies of water.* Knowledge of water use situations and potential downstream effects; impact of pesticides on plants, fish, birds, and other organisms; how to limit application areas.
Right-of-way pest control	*Highways, utility ways, railways.* Knowledge of impact of pesticide runoff, drift, and excessive foliage destruction; how to recognize target organisms; containing pesticides to target areas; impact on people and surrounding areas.
Industrial, institutional, structural, and health-related pest control	*Residences, commercial establishments, warehouses, other structures.* Knowledge of a wide variety of pests; ways to avoid exposing people or pets, contaminating food, or contaminating habitats.
Public health pest control	*Vector control agencies including mosquito abatement districts.* Knowledge of vector disease transmission; pests, life cycles, and habitats; mechanical and other non chemical control methods.
Regulatory pest control	*Public agencies responsible for exclusion, quarantine, and abatement of important pests.* Knowledge of factors influencing introduction, spread, and population dynamics of relevant pests.
Demonstration and research pest control	*Individuals demonstrating the safe and effective use of pesticides to others and those who conduct field research using pesticides.* Knowledge of all aspects of safe handling of pesticides; establishing research plots; proper ways to dispose of crops and crop residues treated with unregistered materials.

(From Code of Federal Regulations Section 40, Part 171.)

plished through testing, by attending a training meeting, by accumulating a specified number of continuing education hours, or through a combination of these methods, depending on where the applicator is certified. Among the states that require continuing education courses, some expect these courses to be specific to the category of certification.

FUTURE TRENDS FOR PESTICIDE APPLICATORS

Several states have pesticide regulations and applicator certification and training requirements that exceed the standards set forth by FIFRA and federal regulations. These stricter requirements provide glimpses of how applicators of all types of pesticides might be regulated in the future. For instance, California's pesticide regulations require that *all* pesticide applicators, working in any type of occupation and handling any type or category of pesticide, be given training each year by their employers. The training must be specific to the chemical classes of pesticides being handled; it must be documented and the training record signed by the employee, trainer, and employer; and training records must be retained by employers for two years.

The 1992 amendment to FIFRA, known as the federal Worker Protection Standard (WPS), was a major step at the national level in extending training beyond applicators of restricted-use pesticides. WPS regulations now require additional training at least every five years for people who handle any pesticides in production agriculture, commercial greenhouses and nurseries, and forests. (This is still considerably less restrictive than California's yearly training requirements for all pesticide applicators.) The WPS also requires training of fieldworkers and other employees at least every five years if these workers enter pesticide-treated areas any time within 30 days after the restricted-entry interval expires. This marks the first time that federal regulations require pesticide safety training for people who are not involved in handling pesticides.

In January of 1999 the U.S. EPA circulated a draft report for stakeholder review that presented the findings and proposals of the Certification and Training Assessment Group (CTAG). This report, titled *Pesticide Safety for the 21st Century*, proposes significant changes to the national pesticide applicator certification and training program by way of new amendments to FIFRA. State, territorial, and tribal pesticide laws and regulations throughout the United States would change as a result of these amendments.

CTAG recommendations include: (1) expand the applicator certification requirements that currently apply to restricted-use pesticides to all occupational uses of all pesticides; (2) extend the WPS pesticide safety education and training requirement into all occupational areas where pesticides might be applied rather than just production agriculture, commercial greenhouses and nurseries, and forests; (3) begin a national consumer education initiative designed to provide consumers and homeowners with more information and training on using pesticides; and (4) establish a four-tiered classification of pesticides rather than the current two-tiered (restricted- and general-use) classification. One of the proposed classifications is:

- consumer- and homeowner-use products general or unclassified pesticide products for occupational use
- restricted-use pesticides for occupational use
- *prescription* restricted-use pesticides for occupational use

Coupled with these proposed changes are plans to improve pesticide applicator education programs and materials throughout the United States, provide mechanisms to improve the skills of trainers, and use more sophisticated methods of developing, validating, and standardizing examinations.

REFERENCES

1. American Association of Pesticide Safety Educators (AAPSE), http://aapse.ext.vt.edu/.
2. CTAG. *Pesticide Safety for the 21st Century—The Findings and Proposals of the Certification and Training Assessment Group*; Draft Report, United States Environmental Protection Agency Office of Pesticide Programs: Washington, DC, 1999, *in review*.
3. O'Connor-Marer, P.J. *Pesticide Safety—A Reference Manual for Private Applicators*; University of California Division of Agriculture and Natural Resources: Oakland, CA, 1998; 120.
4. O'Connor-Marer, P.J. *The Safe and Effective Use of Pesticides*, 2nd Ed.; University of California Division of Agriculture and Natural Resources: Oakland, CA, 1999; 342.
5. US EPA. *The Federal Insecticide, Fungicide, and Rodenticide Act (FIFRA) and Federal Food, Drug, and Cosmetic Act (FFDCA) As Amended by the Food Quality Protection Act (FQPA) of August 3, 1996*; United States Environmental Protection Agency Office of Pesticide Programs: Washington, DC, 1997; 189.

LIGHT TRAPS

S. Mohan
Tamil Nadu Agricultural University, Coimbatore, India

INTRODUCTION

Insect vision is basic to orientation, movement, and consequently insect environment. General receptors such as simple dermal cells that respond to light occur in insects. Among the two types of eyes insects possess (simple eye and compound eye), compound eyes have been regarded as the most important sense organs that directly mediate attraction to light. Integrative action of the insect central nervous system also governs insect response to light. Vision together with smell, hearing, and other senses serve for the recognition of objects and orientation. Light allows this discrimination and recognition of objects since it transmits information about them. This phenomenon has been used for many years to collect insects. The Robinson brothers (1) thought that a source of bright light dazzles the insects and makes them lose coordination so that they involuntarily get closer to the light source. Lower luminosity of the environment on the sides of the light source also is a possible reason for the attracting efficiency of a light. The ultimate cause of the orientation of night flying insects towards artificial light is the need to find open flight routes and the proximate cause the unnatural radiance distribution caused by the attracting light source.

This attraction of adult insects to artificial light was observed and recorded by humans early in their history. The first such known record is that written by Greek poet Aeschylus (525–456 BC) in his Fragments 288 (450 BC) "The Fate of the Moth in the Flame" mentioned by Du Chanois (2). An open fire was probably the first artificial light source to attract and destroy flying insects, followed by the candle, kerosene lamp and lantern, acetylene lamp, gasoline lantern, carbon filament electric lamp, and finally by the electric lamps utilized today. Among the various sources of light, electric lights are considered as the best source for insect attraction.

TYPES OF LAMPS USED IN TRAPS

Currently numerous lamps are being used in traps (3) to determine the photo responses of various economically important insects. These lamps are selected to provide a wide range of radiation output in terms of quality (wavelength) and quantity (power). These lamps (Fig. 1) can be categorized broadly into three major groups as follows:

1. Incandescent lamp
2. Mercury vapor lamp
3. Black light

Incandescent Lamps

They produce radiation by heating a tungsten filament. The spectrum of lamp includes a small amount of ultraviolet, considerably visible, especially rich in yellow and red, and a peak of radiation in the infrared region that includes about three-fourths of the total lamp output.

Mercury Vapor Lamps

They produce primarily ultraviolet, blue, and green radiation with a little red. The proportion of infrared radiation is much smaller than that of incandescent lamps. A modification of this type of lamp is the fluorescent lamp. This is also called a mercury vapor lamp with the inner surface of the lamp coated with various phosphorus compounds that absorb short wave lengths and reradiate the energy at longer wave lengths. Thus the term fluorescent is used. The spectral output of this modified mercury vapor lamp includes the continuous spectra from the fluorescing phosphorus plus the bright lines as in the mercury vapor lamps.

Black Light

Black light is a popular name for ultraviolet radiant energy, with wave length ranging from 320 to 380 nm. This is the most common light source used in many insect control and survey programs in developed countries.

All available models of light trap use one of the above lamps as a source of attractant. The various models of light

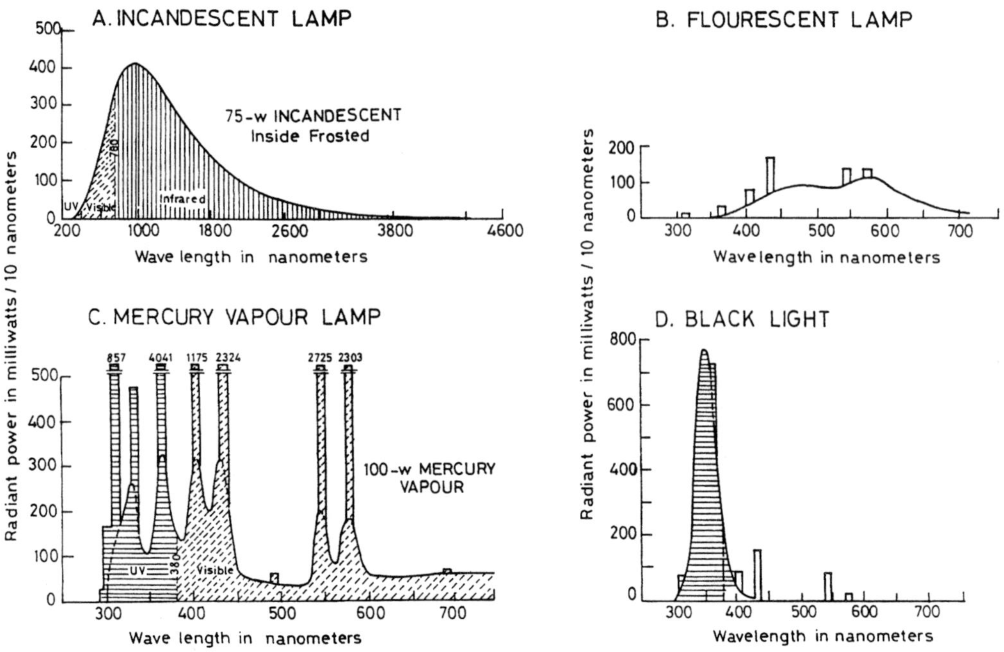

Fig. 1 Radiant power emitted from different light sources used in light traps. (From Ref. 3.)

trap (Figs. 2 and3) available are grouped according to the sources of light used and are described below.

TRAP MODELS

Incandescent Light Trap Models

Primitive light trap model: Montana light trap

The trap is made up of utensils commonly found on farms that served other purposes when not in use as light trap. The trap consisted of a galvanized iron wash tub and lantern. A galvanized iron arch fitted across the tub serves to deflect the moths and to hold the lantern. The lantern emits light rays consisting of visible spectrum and infrared rays as in the case of incandescent lamp.

Simple incandescent light trap

These traps have pans filled with water and a film of oil fastened below an electric lamp. This type of trap is very simple consisting of a reflector, lamp, and water pan. Portable models of this trap are also available. A low intensity incandescent light trap model for selective capture of delicate insects (4) has been widely used for monitoring insects in rice ecosystems and *Culicoides* sp.

Suction type model: Akins gnat trap

It has an inverted funnel-shaped reflector, a 100-W incandescent lamp, a thin sheet iron sleeve suspended about 10 in. from reflector by rods allowing clearance for the insects to come to the light, a small electric fan with motor, and a black muslin bag drawn over the lower end of the sleeve. The popular Communicable Disease Center (CDC) (5) miniature light trap is also of this type used for attracting mosquitoes. The CDC light trap was later modified to act as automatic chemosterlizing insect light trap. The insects entering a light trap are deposited on a chemosterilant. Tests with *Culex pipiens quinquefasciatus* Say showed 90% of the males and females entering the trap were sterilized by this technique. A very recent modification to CDC trap is the use of colored emitting diodes (6) as a light source in place of inside frosted incandescent lamps.

Mercury Light Trap Models

Robinson trap

The trap is the basic model designed by Robinson (7) in 1952. The trap proper is basically the upper structure consisting of an inverted truncated cone containing a number of radial vanes at the center, with a small light source lying just above the vanes through the top of the cone. The collection device is a parcel box in which cotton dipped in insecticides is generally placed inside the container. In order to prevent the escape of insects, the cone is constructed mainly of transparent material so that the point of exit may be invisible from the interior. This trap is currently used toward a wide range of Noctuides and

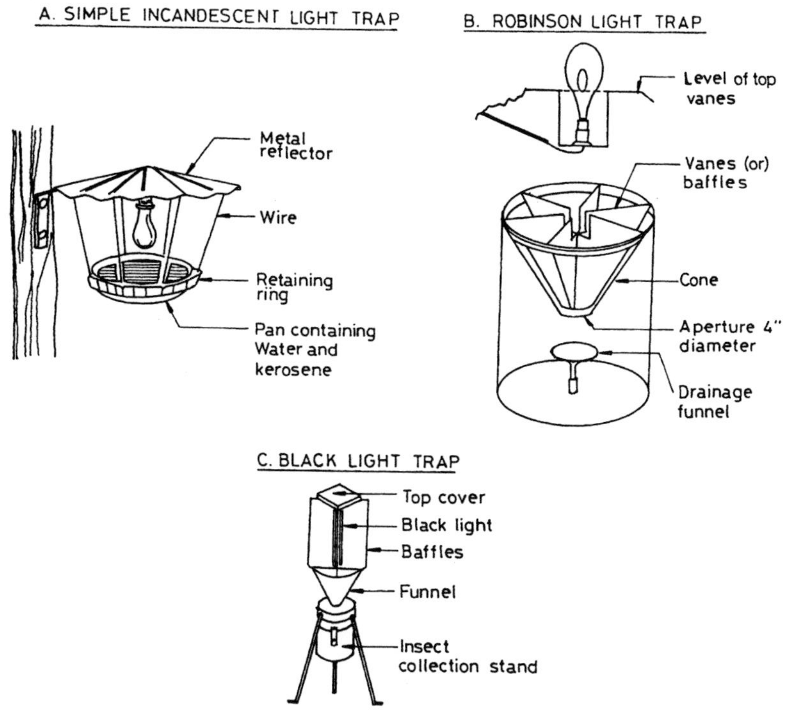

Fig. 2 Light trap models—simple to sophisticated. (From Ref. 3.)

other nocturnal flying insects. An automated modular Robinson trap design—described recently—widens the scope of use of this model in a pest monitoring program (8).

Other models of the mercury light trap include the Rothamsted light trap, the Muguga light trap, the Megaphone shaped trap, and a simple fluorescent unidirectional trap for trapping smaller insects attacking crops both in field and storage (9).

Black Light Trap Models

Black light insect survey trap (standard model)

The basic components of a black light trap model consist of a baffle assembly with four baffles (two crossed), a black light lamp (generally 15 W), a funnel (60° slope), a collection container, a tripod, and and electrical components. It can be used for obtaining indices of population levels for a wide variety of nocturnal insects.

Electrocuter grid insect trap

The electrocuter grid insect trap, known commonly as a "grid trap" (10), is used in applications intended to suppress nuisance insects around residences, business establishments, and food processing plants. Generally the black light is located relatively close to the charged grids so that attracted insects come into contact with the grid causing death.

Though numerous black light trap models are available, all have the same basic components as in the standard type. Some special models include a black light trap combined with a cage to hold virgin female moths and a black light trap with provisions for timely automatic separation of trapped insects.

USES OF LIGHT TRAPS

Light traps serve as an important and valuable method in collecting crepuscular and nocturnal insects for taxonomic purposes, for detection of the presence of insect pests, to determine population changes or trends, and to study the migratory strategies of several important agricultural pest species. Monitoring (11) insect biodiversity in various habitats is yet another important use of the light trap.

Diurnal species of insects are seldom taken in light traps. However the fact that a few species of butterflies belonging to the families Pieridae, Lycaenidae, and Nymphalidae are occasionally taken, particularly during migration periods, broadens the scope of using a light trap for taxonomic studies of butterflies, too.

Recording insect population data using a light trap is an established technique used by entomologists and ecol-

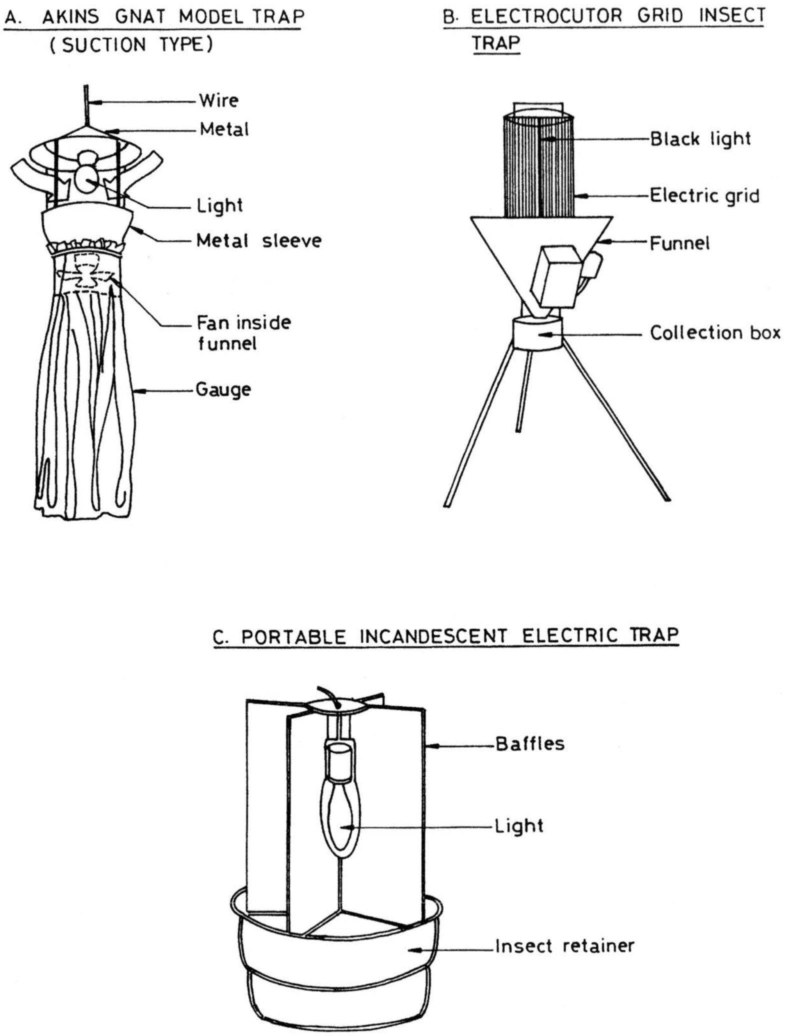

Fig. 3 Light traps—special models.

ogist. Specialized survey traps currently are available for individual insect species include the New Jersey mosquito trap, the CDC miniature light trap, the cigarette beetle trap, the Asiatic garden beetle trap, the pink bollworm trap, and the European chaffer beetle trap. Light traps are potential tools for estimating abundance of vector populations especially *Culicoides* and *Culicids* that transmit dreadful diseases in mammals. Light traps are also used to trap newly emerged cat fleas, *Ctenocephalides felis* (Bouche), in homes (12).

Light traps form an important component in the integrated pest management programs developed and adopted in many parts of the world for several important agricultural pest species (13). Forecasting and surveillance programs for many agricultural and medical insect pests include light traps as a potential tool. Instances have been recorded of major reduction in outdoor infestation of European corn borer and tobacco hornworm in small areas by use of light traps (3). Further illumination with yellow fluorescent lamps is known to discourage the entry of fruit-piercing moths into the orchard, thereby reducing their piercing activity (14).

Despite these advantages, use of light traps has some limitations. These include beneficial insect attraction to traps (15), and increase in crop damage around trap area especially when mercury lamps are used. However these limitations can be overcome by selective use of light sources and limiting the operations of traps during insect peak activity periods. It has been suggested that while using conical-type traps with mercury lamps the area surrounding the trap should be sprayed with insecticides (16).

Overall, light traps offer the promise of suppressing insect populations when the traps are used alone orin an integrated control program planned to reduce the required number of pesticide applications or to provide more effective control measures. Modern pest management programs can be operated successfully only by proper understanding of the behavior and activities of the pest species. In this connection light traps prove to be potential tools as they provide valuable information on insect population fluctuations.

REFERENCES

1. Robinson, H.S.; Robinson, J.J. The observed behaviour of lepidoptera in flight in the vicinity of light source. Entomol. Gaz. **1950**, *1*, 3–20.
2. Du Chanois, F.R. Relative periodic attractivity of light traps to adult mosquitoes as an index to mosquito annoyance intensity. N.J. Mosq. Exterm. Assoc. Proc **1959**, *46*, 170–180.
3. Hienton, T.E. Summary and investigation of electric insect traps. USDA ARS Tech. Bull **1974**, *1498*, 1–33.
4. Mohan, S.; Venugopal, M.S.; Janagarajan, A.; Rangarajan, M. A simple trap for *Culicoides* sp. TNAU Newsletter **1993**, *23* (4), 4.
5. Sudia, W.D.; Chamberlain, R.W. Battery operated light trap-an improved model. Mosq. News **1962**, *22*, 126–129.
6. Burkett, D.A.; Butler, J.F.; Klinel, Daniel. Field evaluation of coloured light-emitting diodes as attractants for woodland mosquitoes and other diptera in North Central Florida. J. Am. Mosq. Control Ass. **1998**, *14* (2), 186–195.
7. Robinson, H.S. In *On the Behaviour of Night Flying Insects in the Neighbourhood of a Bright Source of Light*, Proceedings of the Royal Entomological Society London (A), 1952; 27; 13–21.
8. White, E.G. An automated modular light trap design. New Zealand Entomol. **1996**, *19*, 81–85.
9. Rees, D.P. Review of the response of stored product insects to light of various wave lengths with particular reference to design and use of light traps for population monitoring. Trop. Sci **1985**, *25*, 197–213.
10. Stanley, J.M.; Webb, J.C.; Wolf, W.W.; Mitchel, E.R. Electrocutor grid insect traps for research purposes. Transaction of the ASAE **1976**, *20* (1), 175–178.
11. Intachat, J.; Woiwod, I.P. Trap design for monitoring moth biodiversity in tropical forests. Bull. Ent. Res. **1999**, *82* (2), 153–163.
12. Dryden, M.W.; Broce, A.B. Development of a trap for collecting newly emerged *Ctenocephalides felis* (Siphonaptera: Pulicidae) in homes. J. Med. Entomol. **1993**, *30* (5), 901–906.
13. Mohan, S.; Gopalan, M.; Babu, S.P.C.; Sreenarayan, V.V. Practical studies on the use of light traps and bait traps in the management of *Rhyzopertha dominica* (F.) in rice warehouse. Int. J. Pest Manage. **1994**, *40* (2), 148–152.
14. Nomura, K. Some considerations on the effect of orchard illumination against fruit-piercing moths. Tech. Bull. Fac. Hort. Chiba **1966**, *14*, 27–34.
15. Nabli-Henda; Bailey-Wayne, W.C.; Necibi-Semi, S. Beneficial insect attraction to light traps with different wave lengths. Biological Control **1999**, *16* (2), 185–188.
16. Mohan, S. *Studies on Light Traps with Reference to Rice Pests*; M.Sc.(Ag.) Thesis, TNAU: Coimbatore, India, 1982; 1–58.

LIVESTOCK PEST MANAGEMENT (INSECTS)

Richard Wall
University of Bristol, Bristol, United Kingdom

INTRODUCTION

Insect pests, particularly flies and lice, exert a major impact on the welfare of livestock and the economics of their husbandry, resulting in the loss of billions of dollars annually (1). These pests (2) may have a variety of direct and indirect effects, resulting in poor growth, loss of productivity, debilitation, and death. In the last 50 years or so, approaches to the control of these pests have centered on attempted eradication or the reduction of their numbers to some arbitrary very small number. This approach has been promoted by the availability of a range of very powerful synthetic chemical insecticides. However, the focus on killing insects and reliance on chemical insecticides to do this has been associated with a range of unwanted side-effects such as poisoning, environmental contamination, and the development of resistance. These problems have focused increasing attention on the use of a range of techniques to effect carefully-considered management, rather than attempted elimination of insect pest populations.The underlying principle of management is that pest populations should be dealt with in a way that minimizes their economic impact, usually by reducing abundance to below some economically defined threshold. Management does not necessarily imply killing insects directly.

THE EXTENT AND NATURE OF THE PROBLEM

Direct problems for livestock may be caused by blood-feeding adult insects. Although each individual removes only a small volume of blood, if pest numbers are high the amount of blood lost may be debilitating and anaemia is common in heavily attacked hosts. Larval flies are also important pests, infesting the living tissues of stock and causing considerable damage to the carcass or hide (3).

The feeding activity or presence of insects on the body or burrowing into the skin can stimulate keratinocytes to release cytokines, which leads to epidermal hyperplasia and cutaneous inflammation. The resulting irritation and scratching is often accompanied by hair and wool loss and occasionally by skin thickening. Lesions resulting from rubbing may become secondarily infected. The salivary and fecal antigens produced by ectoparasites can stimulate an immune response leading to hypersensitivity in some individuals.

The irritation caused as flies attempt to feed or oviposit commonly results in a variety of behaviors such as head shaking, stamping, skin twitching, tail switching, or scratching. These activities may result in reduced growth and loss of condition, because the time spent in avoidance behavior is lost from grazing or resting. The activity of particular pests, such as ovipositing adult warble flies (cattle grubs), may cause such dramatic avoidance responses in the intended host that serious self-injury may sometimes occur following collision with fences and other objects.

In addition, one of the most damaging roles of insects as pests is in their action as biological or mechanical vectors of pathogens. These pathogens include protozoa, bacteria, viruses, cestodes (tapeworms), and nematodes (round worms). The direct damage caused by most insect pests is directly proportional to their abundance. This is not necessarily the case for disease vectors, where even very low numbers of infected vectors may cause considerable economic and welfare problems.

CHEMICAL CONTROL

Early substances used against insect pests of livestock were found largely by trial and error or were derived from those developed for horticulture. They were often highly toxic, being based on arsenic and mercury, as well as tar, petroleum, and nicotine. Carbon disulphide and rotenone (extract of derris) were widely used. Sulfur, either as sublimed sulfur (lime sulfur, flowers of sulfur) or precipitated sulfur, has been incorporated in ointments, powders, and shampoos and is still widely used.

In the second half of the twentieth century, the development and use of neurotoxic insecticidal compounds, which affect the insect central nervous system, synapses, axons, or neuromuscular junctions, became widespread.

The first of these were the chlorinated hydrocarbons, such as DDT, γ-BHC, and dieldrin. They were gradually replaced by the carbamates and organophosphates. More recent advances have seen the development of the more insect–specific insecticides such as the pyrethroids, synthetic compounds based on the natural compound pyrethrum, the insect growth regulators (IGRs), which interfere with various aspects of arthropod growth and development, and the phenylpyrrazoles, which act by inhibiting the neurotransmitter γ-aminobutyric acid in the insect central nervous system.

The reliance on chemicals to attempt to eliminate insect pests has been associated with a wide range of side-effects or poisonings of livestock resulting from overdose, species sensitivity, breed sensitivity, or an interaction between administered medicines. In addition, the deliberate or inadvertent treatment of the wider environment with insecticides almost inevitably leads to undesirable effects on nontarget organisms (4). Environmental contamination and effects on nontarget animals have been well documented, particularly in the case of the organochlorine insecticides. Growing concern is associated with the organophosphates and pyrethroids.

At the recommended doses modern insecticides are highly effective at removing susceptible individuals, but they can impose strong selection pressure for the development of resistance (5). Often selection by one type of chemical hastens the development of resistance against other previously effective compounds; cross- and multiple-resistance in insect pests is now a growing problem, necessitating drug switching and the use of alternative chemicals (5). Resistance to existing chemicals is unlikely to be reversed and, indeed, will become increasingly widespread.There can be little doubt that new compounds developed in the future will also rapidly select for resistance.The same is likely to be true for any genetic modification or manipulation of hosts for parasite resistance.

Hence, the reliance on pesticides to eliminate pests is associated with a variety of major problems. These have focused increasing attention on the use of a range of techniques to manage pest populations.

THE MANAGEMENT OF LIVESTOCK PESTS

The management of insect pest populations usually aims to reduce or suppress populations to below an economically-defined threshold. One or several techniques may be used, but importantly these should be integrated with each other to form components of a general livestock pest management program. Livestock pest management is usually based on the use of control technologies that modify some aspect of the pest's environment, on or off the host, either to increase pest mortality, to reduce fecundity or to reduce contact between the pest and host (4, 6).

Physical Control

Simple modification of the environment may reduce pest abundance significantly. For example, many of the fly pests of cattle and horses have larval stages that develop in animal dung or decaying organic refuse. Management of dung, therefore, is of prime importance in their control and considerable success can be achieved simply by removing dung regularly from pastures or feed lots and dispersing it in such a way that it no longer attracts ovipositing flies. Biting and nonbiting flies also can be effectively controlled through simple procedures such as the removal of moist bedding and straw, food wastes, heaps of grass cuttings, and vegetable refuse in which they breed (6).

Similarly, changing the suitability of the on-host environment may help reduce the susceptibility of the host to pest attack. With sheep, for example, minimizing pasture worm burdens to reduce diarrhea, tail-docking (amputation of the tails), crutching, or dagging (the regular shearing of soiled wool from around the breech), all help to minimize the incidence of myiases by blowflies. These procedures reduce wool soilage and lower the humidity of the on-host environment, thereby reducing the availability of oviposition sites and suitability of the fleece for maggot survival (3, 7).

High pest abundance, in space or time, can be avoided by appropriate grazing practices that reduce contact with the pest. For example, avoiding low-lying pasture at particular times of the year or simply stabling horses during morning and evening may prevent or reduce the effects of biting midges during periods of high midge abundance and activity.

Barriers

Various types of physical barrier can be employed to protect stock from insect pests. These may be fine mesh screens on windows, plastic strips on milking parlour doors, or brow tassels for protection and to help dislodge insects that have alighted on the host (6). Such techniques may often be used in conjunction with an insecticide.

Biological Control

Organisms that are predators, parasites, competitors, and pathogens of some insect pests can be used as biological controls (6). For example, the nematode *Steinernema carpocapsae* has been shown experimentally to parasitize and kill the pupal stage of the flea *Ctenocephalides felis* and may be used as part of a flea control program where outdoor environmental treatment is important. House flies have been controlled in poultry facilities by the release of pupal parasites (8). The use of imported dung-burying beetles to remove pastureland dung and to increase the rate of removal of cattle pats has been attempted in Australia, Canada, and the U.S. to prevent the emergence of dung-breeding flies. Considerable interest is currently being given to the use of entomopathogenic fungi and the bacterium *Bacillus thuringiensis* as biopestides. Nevertheless, the use of these techniques can be a complex and costly operation and, as yet, cannot be attempted routinely in most circumstances.

Trapping

Insects use a complex interaction of olfactory, visual, and tactile cues to locate their hosts. If these cues can be identified and isolated, they can be selectively incorporated into trapping device at levels that produce exaggerated responses from the pest species.

Walk-through traps have been developed for the control of *Haematobia irritans*, stable flies, and face flies (6). A screwworm adult suppression system (SWASS) has been used to attract the new world screwworm fly *C. hominivorax* in North America.This combined an insecticide (2% dichlorvos) with a synthetic odor cocktail known as ''Swormlure'' to attract and kill adult flies (9). Field trials with the SWASS gave a 65–85% reduction in an isolated wild *C. hominivorax* population within three months. However, environmental concerns about the release of large quantities of dichlorvos resulted in the SWASS being largely abandoned as a control technique. The development of traps for the tsetse fly vectors of trypanosomyiasis in Africa has been highly successful, identifying and exploiting appropriate visual shapes and colors in combination with host-mimicking chemical odours to attract and catch flies (10). Traps baited with synthetic chemicals are also commercially available for the control of sheep blowfly in Europe and Australia (11). Trapping techniques hold considerable promise for future development.

Modeling and Forecasting

Allied to the use of the range of management techniques described, computer models and forecasting systems may be of particular value in helping to predict the seasonal and temporal patterns of abundance of particular insect pests and their economic consequences (12). The development of such predictive models will allow veterinarians, farmers, and entomologists to build effective pest management programs, incorporating the discriminating use of pesticides integrated with nonchemical control techniques, where appropriate.

FUTURE PROSPECTS

The design and application of effective management programs for livestock pests is a discipline still in its developmental stages. One key problem is that farmers can be unwilling to use management programs because their application is often initially difficult to plan and implement and, at least superficially, may appear less effective than chemical alternatives since pests are not eliminated. The wide availability of relatively cheap and highly effective chemical insecticides and the economic need to provide low-cost food and animal products do not promote long-term thinking about the pest control techniques employed. However, the many problems associated with reliance on pesticides suggest that practitioners of animal husbandry must begin to think long-term about sustainable insect pest management. This will require a reappraisal of the role of farming in the environment and the price consumers pay for their animal products, in the short-term in cash and the long-term in environmental impact.

REFERENCES

1. Fallis, A.M. Arthropods as pests and vectors of disease. Vet. Par. **1980**, *6*, 47–73.
2. Wall, R.; Shearer, D. *Veterinary Ectoparasites*, 2nd Ed.; Blackwell's Science: Oxford, 2001.
3. Hall, M.J.R.; Wall, R. Myiasis in humans and domestic animals. Adv. Parasitol. **1995**, *35*, 258–334.
4. Strong, L.; Wall, R. The chemical control of livestock parasites: problems and alternatives. Parasitol. Today **1990**, *6*, 291–296.
5. Denholm, I.; Rowland, M.W. Tactics for managing pesticide resistance in arthropods: theory and practice. Annu. Rev. Entomol. **1992**, *37*, 91–112.
6. Drummond, R.O.; George, J.E.; Kunz, S.E. *Control of Arthropod Pests of Livestock*; CRC Press, Inc.: Florida, 1988; 245.
7. Fenton, A.; Wall, R.; French, N.P. The effect of farm management strategies on the incidence of sheep strike in Britain: a simulation analysis. Vet. Par. **1998**, *79*, 341–357.
8. Axtell, R.C.; Arends, J.J. Ecology and management of arthropod pests of poultry. Annu. Rev. Entomol. **1990**, *35*, 101–126.
9. Krafsur, E.S.; Whitten, C.J.; Novy, J.E. Screwworm eradication in North and Central America. Parasitol. Today **1987**, *3* (5), 131–137.
10. Wall, R.; Langley, P.A. From behaviour to control: the development of trap and target techniques for tsetse fly population management. Agric. Zool. Rev. **1991**, *4*, 137–159.
11. Wall, R.; French, N.P.; Fenton, A. Sheep blowfly strike: a model approach. Res. Vet. Sci. **2000**, *68*, 1–9.
12. Thomson, M.C.; Connor, S.J. Environmental information systems for the control of arthropod vectors of disease. Med. Vet. Entomol. **2000**, *14*, 227–244.

LIVING MULCHES

John R. Teasdale
USDA-ARS, Beltsville, Maryland, U.S.A.

L

INTRODUCTION

Living mulches are plants grown in mixture with a cash crop. They are not grown for harvest or direct profit but, instead, to provide ecological benefits including protecting soils from erosion, improving soil fertility, providing traffic lanes, suppressing weeds, and reducing pest populations. Low-growing legumes and grasses typically are used for this purpose. Established living mulches protect crop plants by forming a barrier to pest and weed organisms originating in soils. Living mulches also create a more diverse community that can reduce insect pest levels by attracting natural enemies of pests or by creating an environment that makes it more difficult for pests to find and grow on crop plants. Competition between living mulch and crop plants for water and nutrients that often leads to lower crop yield is the major constraint on using living mulches. Creative management approaches are required to alleviate the detrimental effect of living mulches on crops while enhancing the benefits to pest and soil management.

Monoculture is generally considered the most efficient and economical form of crop production. However, plant mixtures either in the form of intercropping two or more crops simultaneously or one crop with a noncrop, or living mulch, can offer many benefits to agroecosystems. These include improving efficiency of resource utilization by crops, preventing erosion and building the fertility of soils, and reducing pest and weed pressure. Intercropping is practiced primarily in nonindustrialized areas of the world and is particularly suited to crop combinations where higher yields can be obtained from crop mixtures than from corresponding monocultures. Living mulches have been used most extensively in orchard production on a commercial scale but application to row crop production is still primarily experimental. Much of the early research on living mulches was associated with the development of reduced-tillage and reduced-input cropping systems (1, 2).

Many plant species have been used or tested as living mulches (3). Forage and turf species often are used as living mulches because their growth habit is lower than most crops and they are relatively easy to establish and manage. Most frequently tested species include the legumes white clover, various other clovers, alfalfa, annual medics, crownvetch, and hairy vetch, whereas the most commonly tested grasses include various fescue, ryegrass, and bentgrass species. The legumes are often included in cropping systems where improvements in soil fertility and quality are a primary goal, whereas the grasses are often included where durability and traffic ability are important.

INFLUENCES ON PEST AND WEED POPULATIONS

Because weed and living mulch plants compete for the same resources, weeds often are suppressed by the introduction of living mulches into cropping systems (4). The most effective living mulches for weed suppression are perennials or winter annuals that establish a complete ground-cover before weed emergence. However, living mulches planted either at the same time as the crop or after the last cultivation also can suppress weeds if they can establish faster than weeds. The presence of green vegetation covering the soil creates a radiation environment that is unfavorable for weed germination and emergence. Research conducted in Maryland (5) showed that several requirements for breaking dormancy and promoting germination of weed seeds in soils (light, high red-to-far red ratio, and high temperature amplitude) were reduced by living mulches more than by desiccated residue or no mulch (Table 1). In addition, an established, growing living mulch preempts the resources otherwise available to weeds. Reasons for weeds escaping suppression by living mulches include presence of gaps in the mulch canopy, presence of perennial or large-seeded weeds that have the resources to establish in the presence of a competitive mulch, and weed species with a similar phenology to the mulch that avoid suppression during establishment.

Living mulches can reduce the spread of plant diseases that originate from soil-borne inoculum. Research (6) has

Table 1 Influence of live and desiccated hairy vetch on environmental conditions in the vicinity of weed seeds at or near the soil surface at Beltsville, Maryland, in 1991

Hairy vetch treatment	Frequency of sites with <1% of sunlight transmitted(%)	Red to far-red ratio of transmitted radiation	Average daily soil temperature amplitude (°C)
None	0	1.15	15.3
Desiccated	4	0.83	8.5
Live	80	0.15	5.9

shown that dispersal of conidia from soils is reduced because of dissipation of rain droplet kinetic energy by living mulch vegetation (Table 2). On the other hand, diseases caused by viruses that are vectored by insects are influenced variably by living mulches depending on the variable responses of insects to the living mulch environment.

The response of insect pest populations to living mulches is a subtopic of the general subject of herbivore insect responses to increasing ecosystem diversity. A survey of this literature (7) suggests that, in the majority of cases, herbivore insect species are less abundant in diversified systems than in monocultures whereas natural enemies of herbivores are more abundant. In agroecosystems, there are numerous reports of reduced pest incidence in crop mixtures but very few reports of higher pest incidence. Diversification of the agroecosystem with living mulches has been shown to provide similar reductions in pest populations. A complex of interrelated processes account for pest reductions in diversified systems (7, 8). Polycultures can interfere with the capacity of pests to colonize hosts by imposing physical barriers, disrupting olfactory and visual cues, and creating diversions to noncrop hosts. Once established in a field, polycultures can interfere with pest populations by limiting dispersal, disrupting feeding, inhibiting reproduction, and enhancing mortality from predators and parasitoids. More research is needed to identify living mulch species and management systems that can be manipulated to limit pest populations. Specific approaches may include determining living mulch species and flowering patterns that attract natural enemies, identifying living mulches that interfere with pest olfactory and/or visual requirements, and understanding the optimum arrangement of crop and living mulch species in space and time to enhance pest management.

Table 2 Colonies of the pathogen *Colletotrichum Acutatum* dispersed by two rainfall intensities on bare soil and a 5-cm high sudangrass living mulch

Treatment	Colonies after an 11 mm/h rain (no. × 10^5)	Colonies after a 30 mm/h rain (no. × 10^5)
Bare soil	5.55	11.3
Living mulch	0.88	2.7

CONSTRAINTS

The major constraint that limits the use of living mulches in cropping systems is crop yield reduction because of competition for growth resources between the crop and living mulch (4). Generally, a living mulch that is competitive enough to suppress weeds will also suppress crop growth and yield. In a typical example, various forage legumes were tested as living mulches for small grains in New York State (9). These living mulches successfully reduced weed biomass by 33 to 95% but also reduced wheat yield by 34 to 85% (Table 3). Several approaches have been used to reduce competition by living mulches: 1) broadcast suppression of the living mulch with a nonlethal herbicide application, 2) band application of a lethal herbicide over the crop row, 3) strip tillage over the crop row, 4) mowing, 5) reducing crop row spacing and/or increasing crop population, 6) relay planting to avoid competition during the peak growth phase of the crop, and 7) supplementing

Table 3 Wheat and weed response to legume living mulches in New York state in 1984

Living mulch	Mulch biomass (t/ha)	Wheat grain yield (t/ha)	Weed biomass (t/ha)
None	—	2.08	1.04
Crownvetch	3.75	0.44	0.70
Birdsfoot trefoil	5.90	0.35	0.30
White clover	1.19	0.85	0.50
Ladino clover	3.78	0.32	0.41
Alfalfa	4.52	0.48	0.05
Red clover	3.46	1.38	0.37

L

Table 4 Cabbage head marketability, weight, and gross returns when grown as a monocrop or with a white clover or subterranean clover living mulch in the Netherlands in 1991

Living mulch treatment	Marketable heads (%)	Head weight (kg/head)	Gross return (Dfl./ha)
Monocrop	29	1.97	4,643
White clover	72	1.68	9,569
Subterranean clover	70	1.71	9,306

water and nitrogen to compensate for resources used by living mulch plants.

Living mulches are probably best suited to perennial orchard production and have been used in these systems for many years. Many grass and/or legume species have been used between crop rows to provide numerous services including soil improvement and reliable traffic lanes as well as pest management. Orchard production systems that include living mulches more closely resemble later stages of succession than do annual cropping systems and, therefore, may permit more stable management of the dynamic relationships between pest and predator populations (7). Orchard systems also allow more space to separate crop and living mulch plants thereby reducing crop losses from competition.

Horticultural cropping systems where quality of produce is more important than total biological yield also can benefit from living mulches. If marketable crop yield is improved by eliminating damage from pests, then living mulches may increase economic returns even if total yield is reduced by competitive interactions. For example, an experiment in the Netherlands (10) demonstrated that clover living mulches reduced populations of several pest species which resulted in less damage to cabbage heads by these pests. The higher percentage of marketable heads in the living mulch treatments more than compensated for the reduction in head weight caused by competitive interactions resulting in higher gross returns in the living mulch treatments (Table 4).

A profile of an ideal living mulch can be developed. Characteristics that would reduce competitiveness with the cash crop include prostrate growth habit, shallow root system, and a period of maximum growth that is dissimilar from that of the crop. Characteristics that would enhance weed suppression would be uniform and complete ground cover, dense intertwined vegetation, and capacity to establish more rapidly than weeds. Characteristics that would enhance pest suppression would be attraction of natural enemies, radiation reflection patterns that would interfere with pest visual requirements, release of volatiles that would interfere with pest olfactory requirements, and growth and flowering phenology that synchronizes these effects with periods of maximum pest pressure.

REFERENCES

1. *Crop Production Using Cover Crops and Sods as Living Mulches*; Bell, S.M., Miller, J.C., Eds.; Workshop Proceedings, International Plant Protection Center: Corvallis, OR, 1982.
2. Paine, L.K.; Harrison, H. The historical roots of living mulch and related practices. HortTechnology **1993**, *3*, 137–143.
3. *Anonymous. Managing Cover Crops Profitably*, 2nd Ed.; Sustainable Agriculture Network: Beltsville, MD, 1998. Handbook Series Book 3.
4. Teasdale, J.R. Cover Crops, Smother Plants, and Weed Management. In *Integrated Weed and Soil Management*; Hatfield, J.L., Buhler, D.D., Stewart, B.A., Eds.; Ann Arbor Press: Chelsea, MI, 1998; 247–270.
5. Teasdale, J.R.; Daughtry, C.S.T. Weed suppression by live and desiccated hairy vetch. Weed Sci. **1993**, *41*, 207–212.
6. Ntahimpera, N.; Ellis, M.A.; Wilson, L.L.; Madden, L.V. Effects of a cover crop on splash dispersal of *Colletotrichum acutatum conidia*. Phytopathology **1998**, *88*, 536–543.
7. Altieri, M.A. *Biodiversity and Pest Management in Agroecosystems*; The Haworth Press, Inc.: Binghamton, NY, 1994.
8. Trenbath, B.R. Intercropping for the management of pests and diseases. Field Crops Res. **1993**, *34*, 381–405.
9. White, J.G.; Scott, T.W. Effects of perennial forage-legume living mulches on no-till winter wheat and rye. Field Crop Res. **1991**, *28*, 135–148.
10. Theunissen, J.; Booij, C.J.H.; Lotz, L.A.P. Effects of intercropping white cabbage with clovers on pest infestation and yield. Entomologia Experimentalis et Applicata **1995**, *74*, 7–16.

LOSSES FROM AQUATIC WEEDS

Lori Lach
Cornell University, Ithaca, New York

INTRODUCTION

Worldwide only about 20 species of the more than 700 species of aquatic plants are considered serious weeds (1, 2). Several dozen other freshwater macrophytes may be considered weeds locally. An aquatic plant can be considered a weed when it interferes with the desired uses of a water body. Since a water body often supports multiple users with varied interests including food production, transportation, recreation, and energy production, a plant may be a weed to some stakeholders and not to others (3). Plants may achieve weed status when they are introduced outside of their native range without their natural enemies, or when the aquatic ecosystem changes, through eutrophication, for example, to allow their rapid proliferation. In general, aquatic weeds can be categorized as floating, submerged, and emergent (2–4). The types vary in the parts of the aquatic ecosystem they exploit, their productivity, the problems they cause, and the best management practices utilized for their control. Table 1 lists some of the world's worst aquatic weeds, their distribution and impacts.

IMPACTS OF AQUATIC WEEDS

Food Production

Though some level of aquatic vegetation is helpful for maintenance of healthy fish populations, when plants achieve weed status, they may negatively affect fisheries through changes in water chemistry and hydrology, and by limiting access to anglers. When floating or emergent weeds form dense mats, light penetration and water flow are affected, with consequent impacts on phytoplankton and zooplankton, leading to lower recruitment and loss of condition for fish (2, 3). Control methods or environmental conditions that cause rapid decay of weeds in a water body may result in severely depleted levels of dissolved oxygen leading to massive fishkills (3, 4). Aquatic weeds may also affect water level. Evapotranspiration rates over water hyacinth are three to six times that measured over open water. Some aquatic weeds are early successional species and unchecked infestation of them may lead to swamp conditions and eventually to a terrestrial habitat (2).

Rice is one of the world's most important crops and is often grown in flooded paddies. Aquatic weeds in rice fields interfere with seed germination and seedling growth, and compete with the crop for nutrients. *Monochoria vaginalis* is an important weed of paddy rice in 13 countries in Asia (1, 2). In the Philippines it has been shown to decrease rice yields by 35%. *S. molesta* is a major weed of rice in much of south and southeast Asia. In Africa and Australasia, weeds of rice are predominantly emergent grasses and sedges of the genera *Paspalum, Echinochloa*, and *Cyperus*. The latter two are responsible for losses of 35–70% in South American rice fields (1, 2).

Aquatic weeds may also effect food production indirectly, through impacts on irrigation. Submerged plants such as *Potamogeton* spp. and *Ceratophyllum demersum* are problematic in irrigation systems in both industrialized and non-industrialized countries throughout much of the world. Emergent plants such as *Typha* spp., *Phragmites* spp., and *Echinochloa stagnina* can invade from canal banks and can distort design features through sedimentation and by channeling flow (1, 2). Weeds also may occupy valuable canal volume; hydrilla occupied 50–75% of canal volume in an irrigation district in California before it was brought under control. Algae in the *Chara* genus are known to form dense monotypic stands in irrigation districts in South America (2). Filamentous algae and the submerged plants *Elodea canadensis* and *E. nuttallii* are problematic in irrigation and drainage channels in Europe. With some notable exceptions, free floating weeds are less problematic in irrigation systems (2).

Aquatic weeds are often a significant burden on channel maintenance budgets: in Egypt as much as 78% of the maintenance budget is spent on the removal and control of aquatic weeds. The economic impacts of aquatic weeds may be particularly severe in nonindustrialized

Table 1 Some of the world's worst aquatic weeds, their distribution, and examples of their impacts

Scientific name (common name)	Family and type	Presumed origin	Geographic distribution	Examples of impacts
Eichhornia crassipes (water hyacinth)	Pontedariaceae floating	Northern South America	Worldwide between lat. 40°N and 45°S	Widely rated as the world's worst aquatic weed because of its tendency to form large mats thereby altering hydrology and water chemistry
Salvinia molesta (water fern)	Salvinaceae floating	South America	Warmer regions of Asia, Africa, Australia, North America, South America	Coverage of 1000 km^2 of Lake Kariba, Zimbabwe; obstruction of 250 km^2 of Sepik River, Papua New Guinea
Pistia stratiotes (water lettuce)	Araceae floating	Uncertain	Pest status in Asia, Africa, Europe, North America, South America; also occurs in Australia	Completely covered irrigation channels totaling 45 km in southern Chad
Hydrilla verticullata (hydrilla)	Hydrocharitaceae submersed	Parts of Asia, Africa, and Australia	North America, Asia, Africa, Australia	The primary aquatic weed problem in Florida: established in 41% of waterways; \$55 million spent to control it from 1980–1991
Ceratophyllum demersum (common coontail)	Ceratophyllaceae submersed	Uncertain, possibly northern Europe	Worldwide in tropics and temperate climates	Shutdown of New Zealand hydroelectric plants
Myriophyllum spicatum (Eurasian watermilfoil)	Haloragaceae submersed	Europe, Asia, N. Africa	North America, Europe, Asia, Africa; problematic in 34 U.S. states and 3 Canadian provinces	Along with *M. exalbescens* accounts for 90–95% of all economic losses and control costs associated with aquatic vegetation in Canada
Alternanthera philoxeroides (alligator weed)	Amaranthaceae emergent	South America	Tropical and warm temperate regions of Asia, Australia, North America	Able to exist in a terrestrial form and resistant to 2,4-D; may invade where water hyacinth has been controlled

(From Refs. 1, 2, 4, 7 and 8.)

countries, which contain 73% of the world's gross irrigated area and often have inadequate resources to devote to irrigation system maintenance (2).

Public Health

Aquatic weeds threaten public health by altering water flow and by providing shelter, food, and oxygen to disease vectors and hosts (2–4). The most serious problems are in tropical and subtropical non-industrialized countries. *Anopheles, Aedes*, and *Culex* spp. mosquitoes, which together are vectors for malaria, dengue, yellow fever, filiariasis, and encephalitis, thrive in slow-moving, vegetation-clogged waterways. The larvae and pupae of two species of *Mansonia* mosquitoes, carriers of the pathogens that cause rural filiariasis and encephalitis, depend entirely on the tissues of *Pistia stratiotes* for oxygen (1, 2).

Aquatic snails, which are secondary hosts for several serious diseases including schistosomiasis, paragonimiasis, clonorchiasis, and fasciolopsiasis, also thrive in conditions created by aquatic macrophytes, such as *Eichhor-*

nia crassipes, Salvinia molesta, and *P. stratiotes*. Where aquatic vegetation has been controlled, reproduction of schistosome-host snails has declined (2).

Navigation, Recreation, and Energy

The world's waterways also are essential for transportation, recreation, and energy production, and unchecked nuisance aquatic macrophytes may physically interfere with all of these functions. Floating weeds in particular tend to severely impair water flow and boat traffic by formation of impenetrable dense mats of floating vegetation called sudds (2, 3). In large systems, sudds may reach several square kilometers in area and provide a substrate for other weeds. *E. crassipes, S. molesta*, and *P. stratiotes* are among the worst offenders particularly in Africa, Asia, and South America. The problem is especially severe in rural areas of non-industrialized countries where waterways are the main avenues for transport (2).

Impacts of aquatic vegetation on recreational opportunities mostly have been documented in industrialized countries. *Elodea canadensis* and *Lagarosiphon major* are problematic around boat ramps in Australia; in Canada, Eurasian watermilfoil threatens several lakes; in Florida, in addition to interfering with flood control, hydrilla adversely affects sportfishing, swimming, boating, and water-skiing (3–6). Economic impacts associated with lost recreational opportunities can provide political impetus for weed control. For example, failure to control Eurasian watermilfoil in Lake Okanagan in British Columbia was expected to result in a loss of $84 million in tourism revenues annually and a 10% devaluation in lakefront property values. Based on these costs, the benefit: cost ratio of the control program was calculated as 11.3:1 (5).

Dams may create ideal conditions for the proliferation of aquatic vegetation. *Egeria densa, Elodea canadensis, C. demersum*, and *Lagarosiphon major* infestations have resulted in turbine shutdown and lost energy production at several hydroelectric dams in New Zealand (2). The submerged weed *Ranunculus fluitans* blocks intake screens at Swiss hydroelectric power stations. Similar problems with weeds such as water hyacinth and water lettuce have been experienced in African and South American facilities. The buildup of mats of vegetation may stress dams to the point of bursting, resulting in massive flooding (2).

CONTROL

Control of aquatic weeds is particularly challenging because large scale removal or death over a short time period can be devastating to the aquatic system. Removal of one weed may result in proliferation of another (3, 4). Moreover, many aquatic plants reproduce vegetatively and/or produce hardy seeds. *S. molesta* is able to survive in fragments smaller than 1 cm; *E. crassipes* seeds can survive for 15 years (2, 6). To meet the challenge, globally, billions of dollars are spent annually on aquatic weed control. In North America, Washington state spends about $1 million per year to control Eurasian watermilfoil, as does Minnesota, Wisconsin, Vermont, New York, and British Columbia. Florida, South Carolina, and North Carolina spend $10, $2.5, and $0.5 million, respectively, on hydrilla control (4–7). Australia spends $1 million annually on chemical control to protect irrigation systems from aquatic weeds (2). Lack of economic analyses from non-industrialized parts of the world should not be taken as an indication that control efforts are less of a burden on available resources in those countries.

Generally, there are three options for control of nuisance aquatic macrophytes: chemical, manual or mechanical, and biological (3, 8). The appropriate strategy will depend on the desired uses of the waterway and the technological, labor, and economic resources available. In the United States mechanical control that was popular in the 1960s and 1970s shifted to chemical control. Australia also relies heavily on chemical control. However, with new concerns about environmental effects of herbicides, there has been a reduction in available chemicals approved for use. Canada and countries in Asia, Africa, South America, and parts of Europe rely largely on manual or mechanical removal of macrophytes (2). Given the ease with which many of the most serious weeds can resprout and proliferate, follow-up maintenance is required regardless of the initial removal strategy.

Biological control consists of introducing some of the plant's natural enemies, such as phytophagous fish or insects. The Chinese grass carp, *Ctenopharyngodon idella*, has been introduced widely as a biological control agent with mixed success (3). Though a generalist feeder, the fish does show preferences for certain vegetation, and may not prefer the target vegetation over other macrophytes. It is effective in controlling hydrilla in small aquatic systems. It does not feed on water hyacinth, *Salvinia*, or emergent weeds. In Asia and Africa grass carp may be fished out of the system before they can have an effect and they may adversely affect local fish populations (2). Insects also have been introduced. Release of *Cyrtobagous salviniae*, a weevil from Brazil, into the Sepik River system in Papua New Guinea resulted in a decrease of *S. molesta* infestation from 250 km^2 to 2 km^2 in 6 months (2). Although they may take longer to control a weed infestation than chemical or mechanical means, biocontrol agents will probably continue to receive research attention due to their relatively low cost and long-lasting effects.

Utilization of plant biomass is an area for future research and is in practice on a small scale in Asia and Africa. Water hyacinth is used for animal feed, cigar wrappers, mulch, compost, and biogas production. The economics of harvesting, drying, transporting, processing, and marketing the plants currently preclude large-scale use, but remain areas for future research (2).

REFERENCES

1. Holm, L.G.; Plucknett, D.L.; Pancho, J.V.; Herbeger, J.P. *The World's Worst Weeds*; University Press of Hawaii: Honolulu, 1997.
2. *Aquatic Weeds: The Ecology and Management of Nuisance Aquatic Vegetation*; Pieterse, A.H., Murphy, K.J., Eds.; Oxford University Press: Oxford, 1990.
3. *Aquatic Plant Management in Lakes and Reservoirs*; Hoyer, M.V., Canfield, D.E., Jr., Eds.; U.S. Environmental Protection Agency: Washington, DC, 1997.
4. University of Florida, IFAS, Center for Aquatic and Invasive Plants. http://aquat1.ifas.ufl.edu (accessed Oct. 3, 1999).
5. *International Symposium on the Biology and Management of Aquatic Plants*, July 12–16, 1992; Haller, W.T., Riemer, D.N., Bowes, G.E., Fox, A.M., Joyce, J.C., Madsen, T.V., Rattray, M.R., Eds.; J. Aquat. Plant Mgmt.: Daytona Beach, FL, 1993; 311–226, (special edition).
6. Schmitz, D.C.; Schardt, J.D.; Leslie, A.J.; Dray, F.A.; Osborne, J.A.; Nelson, B.V. The Ecological Impact and Management History of Three Invasive Alien Aquatic Plant Species in Florida. In *Biological Pollution: The Control and Impact of Invasive Exotic Species*; McKnight, B.N., Ed.; Indiana Academy of Science: Indianapolis, 1993; 173–194.
7. Aquatic Plants and Lakes, Washington State Water Quality Program. http://www.state.wa.us/ecology/wq/plants/index.html (accessed Oct. 3, 1999).
8. In *Aquatic Weeds Workshop,* Proceedings of the Weed Science Society of Victoria, Keith Turnbull Research Institute, February 12, 1999; Shepherd, R.C.H., Richardson, R.G., Eds.; Plant Protection Quarterly 1999; 14 (2), 77–80.

L

LOSSES TO NEMATODES

Robert McSorley
University of Florida, Gainesville, Florida, U.S.A.

INTRODUCTION

Plant-parasitic nematodes cause serious losses to a variety of agricultural crops worldwide (1–3). These unsegmented worms are often overlooked due to their microscopic size (most <2 mm long) and the fact that plant-parasitic species spend all or much of their life cycle in soil or plant tissues. Therefore development of the science of phytonematology, or plant nematology, is mainly a 20th century phenomenon. Today about 15,000 species of nematodes have been described, of which probably 1000 or less are plant parasites. Plant-parasitic nematodes share the soil habitat with many other nematode species, including bacterial feeders, fungal feeders, predators, and parasites of invertebrate or vertebrate animals.

NEMATODE DAMAGE AND SYMPTOMS

Nematodes damage plants by puncturing and feeding on individual plant cells. Ectoparasitic nematodes living and feeding at the root surface accomplish this without ever entering the root. Endoparasitic nematodes tunnel into and migrate within the root while feeding. Sedentary endoparasitic nematodes hatch from eggs in the soil, migrate to and within roots, and eventually establish permanent fee ding sites within the root. Plant cells at the feeding site may be modified and enlarged. Some species produce characteristic galls, knots, or swellings at infection sites. In general, endoparasites cause more damage than ectoparasites because of their increased disturbance of the root system. The plant-parasitic nematodes include several species of foliar nematodes that move from the soil into leaves, stems, or buds, resulting in distorted foliage. A few species of fungal-feeding nematodes, which are of little consequence on most crops, can cause serious problems in mushroom production.

The most obvious symptoms of nematode damage are root galls caused by root-knot nematodes or distorted foliage or buds from foliar nematodes. Peanuts, potatoes, carrots, and other below-ground plant parts may be misshapen or distorted. Severe root damage by the burrowing nematode can cause toppling of mature banana plants. Some nematode species cause visible lesions apparent on close examination of roots. Damage from many nematode species is subtle and may simply include reduced yield or other symptoms associated with a damaged root system such as wilting, yellowing, or nutrient deficiency symptoms. Accurate diagnosis usually requires a soil or plant sample. In addition, nematode infections may provide sites for entry of fungal and bacterial pathogens into root systems, and so the severity of many root diseases is increased when nematodes are present. Several species of nematodes are vectors for the transmission of virus diseases to plants.

LOSSES TO NEMATODES

Nematode damage to crops depends on both the kinds and numbers of nematodes present. Some of the most serious nematode pests are widely distributed (Table 1). In general, root-knot nematodes tend to be more problematic in warm temperate and tropical regions, whereas cyst nematodes are often the key nematode pests in cool temperate regions, although numerous exceptions occur. Many species of plant-parasitic nematodes have not been proven to cause economic damage to any plant species.

Nematode Numbers and Yield

Because of its minute size, an individual nematode is unlikely to cause economic damage to a crop plant. A minimum population level, called the tolerance limit or crop damage threshold, is needed before crop loss can be measured (4). Above this limit, crop yield decreases as nematode population size increases. The severity of damage also depends on environmental conditions, which affect both crop performance and nematode population growth. The reproductive potential of plant-parasitic nematodes is great; some species produce several hundred eggs per female, and typical generation times range from about three to six weeks. Generation time depends on

Table 1 Examples of key nematode pests

Type	Nematode	Main crops damaged	Main regions
Burrowing	*Radopholus similis*	Banana, plantain	Worldwide tropical
Citrus	*Tylenchulus semipenetrans*	Citrus	Worldwide subtropical, tropical
False root-knot	*Nacobbus* spp.	Sugar beet, some vegetables	Central and South America
Potato cyst	*Globodera* spp.	Potato	Europe, Andean South America
Reniform	*Rotylenchulus reniformis*	Wide host range	Worldwide subtropical, tropical
Root-knot	*Meloidogyne* spp.	Many vegetables, other crops	Worldwide
Soybean cyst	*Heterodera glycines*	Soybean	United States, Asia, South America
Stem and bulb	*Ditylenchus dipsaci*	Bulb crops, ornamentals	Europe
Sting	*Belonolaimus longicaudatus*	Turf, various crops	Southeastern United States
Sugar-beet cyst	*Heterodera schachtii*	Sugar beet, crucifers	Worldwide temperate

temperature, which is one reason why nematode problems in the tropics are especially severe.

Loss Estimates

Accurate data on nematode crop losses are difficult to obtain. Nematodes tend to be a site-specific problem. Even within a small geographic region, varying numbers of nematodes in different fields mean that losses could be severe or total in some fields and negligible in others. Relatively few comprehensive surveys documenting losses to nematodes are available, and so crop loss data are based on estimates. One survey of nematologists revealed an average loss estimate of 10.7% on a world basis, for 20 important food crops (5). Estimates ranged from 3.3% for rye to 19.7% for banana. Loss estimates within the United States tend to be lower than the world average, with the majority of estimates below 10%, depending on the crop (6, 7). United States loss estimates to soybean cyst nematode typically average 5–10% in some states, and losses to some vegetable crops from root-knot nematodes may be higher. Estimates may improve in the future as nematode damage becomes more widely recognized, and as more surveys of nematode incidence are conducted.

NEMATODE MANAGEMENT

Some of the more commonly encountered methods for managing nematodes are summarized (Table 2), but efficacy varies with method and situation (8). Unlike insects, nematodes have little opportunity to migrate to new sites on their own. Therefore quarantines and sanitation practices are particularly important for preventing introduction of nematodes into a new site. Once present in a site, nematodes are nearly impossible to eradicate, and future damage depends on the rise and fall of nematode numbers in the site. Many methods for nematode management are aimed at temporarily reducing nematode numbers for a cropping season. Occurrence of multiple species of nematode pests in a site can be problematic since different species may respond to management in different ways. For instance, a crop cultivar resistant to one nematode species might be a good host for another.

Table 2 Methods for managing nematodes

Biological control	Predators and parasites occur naturally, limited commercial use
Cover crops	Some crops suppressive to nematodes
Crop management	Reduce plant stress and treat symptoms through increased water, nutrition
Crop rotation	Nonhost crop to interrupt nematode cycle
Nematicides	Environmental issues with widely used fumigant methyl bromide; nonfumigants possible in some situations but may pose hazard to wildlife
Organic amendments	Variable performance against nematodes, may help with crop management
Quarantines	Prevent movement of nematode-infested material at national, state, or regional levels
Resistant cultivars	Important provided resistant germplasm is available
Sanitation	Use nematode-free soil and planting material, clean equipment, remove infected plants and weed hosts
Solarization	Soil under clear plastic heated to lethal temperature

FUTURE CONCERNS

Many losses to nematodes are currently prevented by nematicide usage. This approach may not be sustainable over the long term due to negative environmental implications and especially to the reliance on fossil fuel energy in product manufacture and application technology. Compared to nematicides, the effective use of alternative methods requires more detailed knowledge of the nematodes present and their ecology. Development of the information bases needed to manage specific nematodes on specific crops in specific regions will be critical for limiting nematode losses in the future.

See also *Antagonistic Plants*, pages 21–23; *Biological Control of Nematodes*, pages 61–63; *Cover Crops*, pages 155–158; *Cultural Controls*, pages 174–178; *Fumigants*, pages 317–319; *Nematode Management*, pages 530–532; and *Solarization*, pages 781–783.

REFERENCES

1. *Plant and Nematode Interactions*; Barker, K.R., Pederson, G.A., Windham, G.L., Eds.; American Society of Agronomy: Madison, WI, 1998; 1–771.
2. *Plant Parasitic Nematodes in Temperate Agriculture*; Evans, K., Trudgill, D.L., Webster, J.M., Eds.; CAB International: Wallingford, U.K., 1993; 1–648.
3. *Plant Parasitic Nematodes in Subtropical and Tropical Agriculture*; Luc, M., Sikora, R.A., Bridge, J., Eds.; CAB International: Wallingford, U.K., 1990; 1–629.
4. McSorley, R.; Duncan, L.W. Economic thresholds and nematode management. Adv. in Plant Pathol. **1995**, *11*, 147–170.
5. Sasser, J.N.; Freckman, D.W. A World Perspective on Nematology: The Role of the Society. In *Vistas on Nematology*; Veech, J.A., Dickson, D.W., Eds.; Society of Nematologists: Hyattsville, MD, 1987; 7–14.
6. Society of nematologists crop loss assessment committee. Bibliography of estimated crop losses in the United States due to plant-parasitic nematodes. Suppl. to the J. of Nematology **1987**, *1*, 6–12.
7. Koenning, S.R.; Overstreet, C.; Noling, J.W.; Donald, P.A.; Becker, J.O.; Fortnum, B.A. Survey of crop losses in response to phytoparasitic nematodes in the United States for 1994. Suppl. to the J. of Nematology **1999**, *31*, 587–618.
8. McSorley, R. Alternative practices for managing plant-parasitic nematodes. Am. J. of Alternative Agric. **1998**, *13*, 98–104.

LOSSES TO VERTEBRATE PESTS

Grant R. Singleton
CSIRO Sustainable Ecosystems, Australian Capital Territory, Canberra, Australia

L

INTRODUCTION

There are a wide variety of vertebrate pests that cause significant damage to agricultural produce. These range from elephants through their trampling of staple crops such as maize in India and East Africa; wild pigs through their rooting around at the base of banana plants, feeding on young plants and tropical fruits, and their trampling of crops such as sugar, rice, and maize in Australia (1), Indonesia (e.g. Sumatra), Pakistan, and United States (e.g. Hawaii, California) (2); rabbits through their damage to cereal crops in Australia (3); geese through their damage of autumn-sown cereals in Europe (4); birds such as quelea and their patchy devastation of staple crops in East Africa (5); and blackbirds in sunflower and maize crops in the United States (6). Although pests such as blackbirds in United States and rabbits in Australia may cause damage to agricultural crops around $10 million in some years, the vertebrates that cause the major damage to agricultural crops are those belonging to the order Rodentia—these include rats, mice, gophers, squirrels, porcupines, capybara, and beavers. It is the rodents that are the focus of this article.

RODENT PESTS—WHO ARE THEY AND WHAT IS THEIR IMPACT?

More than 42% of all mammal species are classified as rodents. They live in almost every habitat around the globe, often in close association with humans. Rodents have two major impacts. The first is as carriers of debilitating human diseases (rodent-borne zoonoses) such as the plague, rat-borne typhus, hantaviruses, and arenaviruses (7, 8). The second is the substantial impact they have on pre- and postharvest losses to agriculture.

Rodents have an enormous economic impact on stored grain in developing countries. Few studies have quantified the impact; however, reports of up to 20% postharvest losses of rice are not unusual. One estimate of impact is from the central Punjab in Pakistan, where for every person living in a village there were 1.1 house rats. Extrapolating the results from this regional study to the national level, it was estimated that 0.33 billion metric tonnes (rice, maize, and wheat) worth approximately $30 million were consumed by house rats in the villages of Pakistan every year (9). However, this is a conservative figure because rodents contaminate and damage more food than they consume and also cause major damage to the structure and electrical systems of grain stores (leading to increased weather and insect damage).

Globally, the main rodent species that cause losses to stored grain are the Norway rat, *Rattus norvegicus*, the roof rat or house rat, *Rattus rattus*, the house mouse, *Mus domesticus*, and the bandicoot rats, *Bandicota* species. The potential impact of rodents to stored grain is summarized in the following statistics: Rodents eat about 10–15% of their body weight each day, damage an amount equal to what they eat, and produce about 40 droppings per day. Therefore in one year, a population of 25 adult rats would eat and damage about half a ton of grain and produce about 375,000 droppings.

Often different rodent species reside in agricultural fields than in and around buildings. Moreover, there are marked differences between rodent species in their population biology, physiology, behavior, and genetic susceptibility to some rodenticides. Therefore any plan to manage pre- and post-harvest losses by rodents must consider the population biology of the rodent species that one is trying to control. This often leads to different control strategies for rodent pests around grain storage facilities and for those species that live in the surrounding areas.

Preharvest rodent damage is the most significant impact that vertebrate pests have on a global level. In some countries, such as Indonesia, rodent pests, primarily the ricefield rat, *Rattus argentiventer*, are economically the most important preharvest pest causing annual losses to rice production of 17%. It is estimated that the ricefield rat consumes enough rice to feed 25 million Indonesians

Encyclopedia of Pest Management

for a year—with rice providing 70% of their energy requirements. In Vietnam, preharvest damage to rice by rodents is one of three most important problems faced by the agricultural sector. There are two main pest species: the ricefield rat and the lesser ricefield rat, *Rattus losea*. In 1999, rodents severely damaged >700,000 ha of rice (10).

For Indonesia, Vietnam, and India (11), the rodent problems are chronic. In other countries such as Australia, China, Kenya, Laos, and Tanzania, rodent problems are linked to episodic eruptions of rodent populations. In Australia, house mice are the major rodent pests. In most years there are less than 10 mice per ha, in outbreak years there are >1000 mice per ha over areas of >10,000 km^2. During years of mouse plagues, losses to winter cereals and summer crops such as rice, sorghum, and maize are around $50 million (12). In China, there are many species of rodent pests across a range of agricultural ecosystems; some are chronic pests causing damage every year, while many cause sporadic, acute losses (10). In Laos, in the upland crops of rice, maize, and sorghum, the major rodent pests are *Rattus* spp., *Mus* spp., and the great bandicoot rat, *Bandicota indica*. In Kenya and Tanzania, the staple crop is maize and the principal rodent pest is the multimammate rat, *Mastomys natalensis* (13). For farmers in Laos, Kenya, and Tanzania, rodent damage is episodic and often patchily distributed. Given the average farm size is 0.5–5 ha, it is not unusual for a family to lose >70% of their standing crop to rodents. Such an impact for these largely subsistence farmers is a major concern for the financial security of a family, if it occurs two crops in a row then the situation becomes catastrophic.

In United States, rodent pests had little impact following the widespread use in the 1960s of herbicides and clean-farming practices. However, the advent of conservation tillage systems in the 1980s generally resulted in >60% of crop residues being retained in fields. This has led to a resurgence of rodents as major crop pests, especially in maize crops (14). The main pests are voles, *Microtus* spp., field mice, *Peromyscus* spp., kangaroo rats, *Dipodomys* spp., and ground squirrels, *Spermophilus* spp.

METHODS USED TO CONTROL RODENTS

A good knowledge of the ecology of a particular rodent pest often leads to more effective, cheaper, and more environmentally benign management. This knowledge also opens the way for management through manipulation of habitats where rodents survive the winter and/or habitats that are key breeding sites, the reduction of food supply at strategic times of the year, and the provision of alternative food at key times to deflect damage from high value crops (15).

Chemicals are commonly used for controlling pre-and postharvest damage by rodents. The chemicals are divided basically into two groups, the acute poisons from which a lethal dose will kill rodents within hours and the chronic poisons (anticoagulants) that generally take 3–7 days to kill rodents (16). A common concern with using rodenticides is that they are not specific for the species to be controlled. Primary poisoning (from directly feeding on the chemical) or secondary poisoning (from feeding on rodents that have eaten a rodenticide) can lead to nontarget deaths. It is therefore important to use an appropriate poison for the situation (crop, refuge habitat, grain store, etc.) and species that needs to be controlled. Also, how to present the poison, where to place it, and when to apply it, are important to minimize nontarget effects and to increase the benefit-cost of the control. The humaneness of the respective poisons also is becoming an important factor in the public acceptance of particular rodenticides.

Other methods of rodent control include snare and live-traps; digging or flooding of burrows to kill rats, often with the help of trained dogs; fumigation of burrows (e.g., sulfur gas in Indonesia); rat drives or battues organized at a community level (common in developing countries); physical barriers to deflect rats from a high value crop; physical barriers plus traps (multiple live-capture traps are placed intermittently adjacent to holes along a fence that is protecting a crop or stored grain); metal rat guards around the trunk of plantation trees to prevent rats climbing them; scaring devices (ranging from white cloth on poles to simulate the flapping of wings of birds of prey, to sophisticated electronic and magnetic devices); nest boxes and perches placed in and around crops to promote the predatory effects of birds of prey.

Bounty systems also are commonly used in developing countries. Communities use a combination of the above methods to catch rats and there is no denying that a bounty system can produce impressive returns of rodents—in 1998 in Vietnam, an estimated 82 million rats were killed. In one province, over 5 million rat tails were collected in nine months and it was estimated that there were well over 10 million rats; 10 rats for every person living in that province. These impressive figures need to be assessed against whether losses to agriculture have been significantly reduced; the opportunity-cost foregone through investing in bounty systems; the ability for populations of rodents to compensate for this ''harvesting'' through longer survival and higher reproductive rates of the remaining rats; whether a bounty system can be sustained (10).

CONCLUSION

The challenges confronting those farmers who are battling rodent pests are many. Probably the greatest of these is having a good knowledge of the ecology of the problem species in the agro-ecosystem. This applies equally to other vertebrate pests of agricultural systems. The principles of ecologically-based rodent management incorporate environmental and economic issues and requires a strong understanding of the population dynamics, habitat use, behavior, physiology, etc., of the pest that needs to be controlled.

REFERENCES

1. Choquenot, D.; McIlroy, J.; Korn, T. *Managing Vertebrate Pests: Feral Pigs*; Australian Government Publishing Service: Canberra, 1996; 163.
2. Sweitzer, R.A. In *In Conservation Implications of Feral Pigs in Island and Mainland Ecosystems, A Case Study of Feral Pig Expansion in California*, Proceedings of the 18th Vertebrate Pest Conference, Baker, R.O., Crabb, A.C., Eds.; UC Davis: California, 1998; 26–34.
3. Williams, K.; Parer, I.; Coman, B.; Burley, J.; Braysher, M. *Managing Vertebrate Pests: Rabbits*; Australian Government Publishing Service: Canberra, 1995; 284.
4. Patterson, I.J. Management of Pink-footed Goose Populations by the Use of Refuge Populations. In *Advances in Vertebrate Pest Management*; Cowan, D.P., Feare, C.J., Eds.; Filander-Verlag: Fürth, 1998; 297–307.
5. Bruggers, R.L.; Elliott, C.C.H. *Quelea quelea, Africa's Bird Pest*; Oxford University Press: Oxford, 1989; 402.
6. Linz, G.M.; Homan, H.J. In *Tracing the History of Blackbird Research Through an Industry's Looking Glass: The Sunflower Magazine*, Proceedings of the 18th Vertebrate Pest Conference; Baker, R.O., Crabb, A.C., Eds.; UC Davis: California, 1998; 35–42.
7. Gratz, N. Rodents as Carriers of Disease. In *Rodent Pests and Their Control*; Buckle, A.P., Smith, R.H., Eds.; CAB International: Wallingford, UK, 1994; 85–108.
8. Mills, J.N. The Role of Rodents in Emerging Human Diseases: Examples from the Hantaviruses and Arenaviruses. In *Ecologically-based Management of Rodent Pests*; Singleton, G.R., Hinds, L.A., Leirs, H., Zhang, Z., Eds.; ACIAR: Canberra, 1999; 134–160.
9. Mustaq-Ul-Hassan, M. *Population Dynamics, Food Habits, and Economic Importance of House Rat (Rattus rattus) in Villages and Farm Houses of Central Punjab (Pakistan)*; Doctoral Thesis, University of Agriculture: Faisalabad, Pakistan, 1992; 174.
10. Singleton, G.R.; Hinds, L.A.; Leirs, H.; Zhang, Z. Reevaluating Our Approach to an Old Problem. In *Ecologically Based Management of Rodent Pests*; Singleton, G. R., Hinds, L.A., Leirs, H., Zhang, Z., Eds.; ACIAR: Canberra, 1999; 17–29.
11. Parshad, V.R. Rodent control in India. Integrated Pest Management Reviews **1999**, *4*, 97–126.
12. Singleton, G.R.; Redhead, T.D. House Mouse Plagues. In *Mediterranean Landscapes in Australia: Mallee Ecosystems and Their Management*; Noble, J.C., Bradstock, R.A., Eds.; CSIRO: Melbourne, 1989; 418–433.
13. Makundi, R.H.; Oguge, N.O.; Mwanjabe, P.S. Rodent Pest Management in East Africa—An Ecological Approach. In *Ecologically Based Management of Rodent Pests*; Singleton, G.R., Hinds, L.A., Leirs, H., Zhang, Z., Eds.; ACIAR: Canberra, 1999; 460–476.
14. Hygnstrom, S.E.; VerCauteren, K.C.; Hines, R.A.; Mansfield, C.W. Efficacy of in-furrow zinc phosphide pellets for controlling rodent damage in no-till corn. Int. Biodeterioration and Biodegradation **2000**, *45*, 215–222.
15. Santini, L. Knowledge of Feeding Strategies as the Basis for Integrated Control of Wild Rodents in Agriculture and Forestry. In *Behavioral Aspects of Feeding*; Galef, B.G., Jr., Mainardi, M., Valsecchi, P., Eds.; Harwood Academic Publishers: Chur, Switzerland, 1994; 357–368.
16. Buckle, A.P. Rodent Control Methods: Chemical. In *Rodent Pests and Their Control*; Buckle, A.P., Smith, R.H., Eds.; CAB International: Wallingford, UK, 1994; 127–160.

LYME DISEASE

A. E. Kiszewski
Andrew Spielman
Harvard School of Public Health, Boston, Massachusetts, U.S.A.

INTRODUCTION

Lyme disease is a nonfatal but potentially debilitating tick-borne bacterial infection caused by the spirochete *Borrelia burgdorferi* that now accounts for more than 95% of disease episodes due to vector-borne pathogens reported in the United States. Transmission occurs throughout the Northern Hemisphere but most intensely in the northeastern and northcentral United States. Nearly 15,000 cases of Lyme disease were reported in the United States in 1999, maintaining a steady pattern of increase. Since the early 1980s, when the etiology of Lyme disease first became apparent, nearly 130,000 United State cases have been reported. Where transmission is particularly intense, as many as 3% of residents may be infected annually. People of any age can be affected, and reinfection occurs readily. The agent of Lyme disease mainly infects people who reside near or visit in enzootic foci maintained by certain woodland mice and *Ixodes* ticks.

HISTORY OF EMERGENCE

Lyme disease became notorious only recently, although human infections have been noted sporadically for at least a century. The spreading pathognomonic rash, now known as erythema migrans (EM), was described in Europe as early as 1909. The etiology of this condition had long proved mysterious beyond a likely bacterial association due to its antibiotic sensitivity. The condition went unreported in the United States until 1969, when an episode of human disease was described in Wisconsin. In 1975, a suspicious cluster of diagnoses of juvenile rheumatoid arthritis emerged in children living in south-central Connecticut, an affliction that normally affects just one child in 100,000. This summertime outbreak of arthritis centered around the eastern Connecticut community of Old Lyme and included a migratory rash, often appearing in a "bulls-eye" pattern. Infections did not appear to spread between familial contacts. Patients submitted ticks that had bitten them that were identified as *Ixodes dammini*, a member of the *I. ricinus* complex of ticks that includes the closely related *I. scapularis* (synonymous with *I. dammini* for some authors). Spirochetes detected in these ticks came to be designated as *Bo. burgdorferi*. Antigens associated with spirochetes isolated from a tick captured on Long Island reacted with sera prepared from the blood of juvenile arthritis patients, and the pathogen was cycled experimentally between deer ticks and rodents. The chain of causation was closed.

DNA analysis of museum specimens of rodents confirmed that these spirochetes had long been present in the United States, although their range had until recently been restricted. These spirochetes persisted in certain isolated island and woodland refugia among an intensely farmed landscape devoid of deer and trees.

Early explorers often commented on the open, park-like nature of the New England shore, due to Native American activities. Reforestation of farmland and a population explosion of white-tailed deer (*Odocoileus virginianus*) set the stage for a resurgence of deer ticks and their associated pathogens (1). As the full and expansive range of disease syndromes became recognized, the more general common name of Lyme disease was established for both the American infections as well as for the European EM.

MICROBIAL AGENT

The agent of Lyme disease is one of a diverse array of corkscrew-shaped bacteria known as spirochetes; many are free-living, symbiotic, or pathogenic. Pathogenic relatives include *Treponema pallidum*, the infectious agent of syphilis, *T. denticola*, an agent of periodontal disease, *Leptospira spp.*, which causes leptospirosis and *Bo. recurrentis*, the African agent of relapsing fever, which is transmitted by soft ticks. Although *Bo. burgdorferi* can infect many types of vertebrates; its perpetuation is generally maintained by a few species of rodents that are particularly competent as reservoirs. These include *Peromyscus spp.* in the Eastern and Midwestern United

States, *Neotoma* spp. on the U.S. Pacific coast, and *Glis* spp. and *Apodemus* spp. in Europe and Asia.

Vertical transmission of Lyme disease between ticks occurs rarely. Ticks generally acquire infections as larvae or nymphs by feeding on infected rodents. Upon infection of ticks and their subsequent molting to the next life stage, *Bo. burgdorferi* spirochetes enter a quiescent phase in which they remain inactive while appressed against the microvilli of the tick's midgut. They are released from dormancy by factors triggered by feeding on a host, after which they undergo amplification and migration through the haemocoel into the salivary gland acini. This process requires several days, during which an infected tick remains uninfectious. Thus, ticks removed within 48–72 hours of attachment generally will not have transmitted any spirochetes unless they had previously attached to a host and had their feeding interrupted.

Lyme borreliosis is distributed circumglobally but discontinuously through parts of North America, Europe, and Asia. Many strains of *Bo. burgdorferi* have been identified throughout the world through both serologic and genomic analysis. In Europe, three genomic groups with distinct clinical and ecological differences have been identified. These comprise *Bo. burgdorferi* sensu stricto, *Bo. garinii*, and *Bo. afzelii*. Although *Bo. burgdorferi* sensu stricto is the sole representative of this group in the United States; it is relatively rare in Europe, where *Bo. garinii* predominates.

PATHOLOGY AND TREATMENT

The syndrome of Lyme disease includes diverse symptoms that mimic various conditions, thus complicating prompt and accurate diagnosis and treatment. Upon inoculation by tick bite, spirochetes remain localized in the dermis for 2–12 days before disseminating from the site of inoculation. Although acute symptoms generally appear as the pathogen disseminates, they sometimes lag or go unnoticed. Early symptoms include a general malaise; fever; and joint, head, and muscle aches. Upon dissemination, spirochetes may reside in synovial fluid, neural or cardiac tissue where they may provoke a progression of symptoms over weeks and months that may include swelling and pain in the large joints, facial palsy, chorea, and other chronic neurological conditions. Cardiac complications can include atrioventricular block and, in rare cases, acute myopericarditis and cardiomegaly. In long-standing cases, severe neurological deficits from polyneuropathy and cognitive disorders from encephalopathy can arise. The latter condition is often associated with sleep disturbance, confusion, and personality changes. Skin disorders also arise occasionally as a consequence of long-term infections. Although rarely fatal, failure to promptly diagnose and treat Lyme disease can have serious consequences.

Coinfection of patients by the agents Lyme disease and babesiosis (due to infection by *Babesia microti*) can result in more severe disease than would occur as a result of either agent alone. Such coinfections are more readily acquired in regions where deer ticks are abundant and transmission is particularly intense. Particular genospecies of Lyme disease spirochetes have been associated with distinctly different clinical manifestations. *Bo. burgdorferi* sensu stricto tends to be associated with chronic arthritis, while *Bo. garinii* more commonly induces neurologic disorders and *Bo. afzelii* is linked to cutaneous manifestations.

When treated early, spirochetal pathogens generally respond readily to treatment with antibiotics. Difficulties ensue when spirochetes become systemic and penetrate regions of the body where standard drugs penetrate poorly. In such cases, intravenous therapy with specialized antimicrobial agents is warranted. Although not approved for humans or pets, topical antibiotic ointments applied to mice at the site of tick attachment have been shown to interdict progression of early infections. An FDA-approved vaccine based on the spirochetal OspA surface protein is currently available (Lymerix, SKB, Inc.). This vaccine attacks spirochetes within the midgut of the tick vectors, clearing infections before they can be transmitted.

ECOLOGY AND EPIDEMIOLOGY

The transmission of Lyme disease spirochetes to human hosts generally is associated with landscapes that juxtapose human habitations with transitional woodlands. This combination of habitats creates an interface between people and the woodland hosts that reservoir *B. burgdorferi* and supports the perpetuation of tick populations. Ecotonal boundaries between lawns or meadows and growing forests provide maximally productive forage for the mammalian hosts that drive enzootic cycles of Lyme disease.

Lyme disease spirochetes are transmitted by *Ixodes* ticks, a group that generally feeds on separate hosts between life stages. In North America the deer tick, *Ixodes dammini* (*I. scapularis*) serves as the vector in the eastern and north-central states while the western black-legged tick, *I. pacificus* transmits the infection in the northwestern part of the country (2, 3). In the northeast, deer ticks transmit an array of human pathogens including *Ba. microti* and *Ehrlichia microti*. Occasionally, all three pathogens may be carried by the same tick, especially in sites such as Nantucket where ticks are especially abundant and enzootic cycles of all three pathogens are well entrenched. Under such conditions, infection rates in nymphal ticks can sometimes reach 40%. The dominant vector in Western Europe is *I. ricinus* while *I. persulcatus* carries the infection in Eastern Europe and certain parts of Asia.

Although vertebrates can become infected by Lyme disease spirochetes, only a few provide reservoirs of infection sufficiently robust to maintain enzootic cycles. Lyme disease is maintained mainly by sylvatic rodents such as the white-footed mouse (*Peromyscus leucopus*) in the United States (4) and *Apodemus* spp., and edible dormice (*Glis glis*) in Europe. Once infected, rodent reservoir hosts tend to remain infected for life, with little if any pathology.

A close correlation between Lyme disease transmission and deer population density has been demonstrated. Although deer ticks may feed on other wild and domestic mammals, populations are greatest where white-tailed deer (*Odocoileus virginianus*) are abundant. Tick populations may linger at low levels if deer are removed, maintained by feeding on secondary hosts such as raccoons, opossums, cattle, etc. Nonrodents do not appear to be efficient reservoirs of the Lyme disease spirochete.

In North America, transmission of the Lyme disease spirochete generally is restricted to the coastal plains, river valleys, and lake shores of the northeastern and north-central states. The two-year life cycle of deer ticks in these regions enhances infection prevalence by allowing nymphs to infect reservoir hosts in early summer just prior the larval feeding season (5). Zooprophylactic absorption of infections by reptiles may further dampen the amplification of spirochetes in the southern United States. In Europe, transmission is widespread; but its intensity is highly focal and patchy. The progression of vector life stages in Europe generally follows the pattern of the American South.

Risk of acquiring Lyme disease is greatest at the intersection between suburban residential property and woodlands where the wildlife that maintain these enzootic cycles dwell. Thus, outdoor activities in pristine wild areas often pose less risk than work or recreation around homes (6). Ornamental borders or lawns bordering woodlands appear to offer the greatest threat of exposure. The peak season for risk in the United States coincides with the season of nymphal activity, which ranges from late May to early July, depending on location. Nymphs, which are no larger than a sesame seed and thus difficult to detect, account for most human cases. Adults, which are active in the fall and early spring, pose less of a threat because their larger size facilitates prompt detection and removal. Larvae, which hatch during late July to early August, generally transmit no infections.

PREVENTION AND INTERVENTION

Limiting outdoor activities during nymphal tick season, if not always practical, provides the simplest method of avoidance. Time spent scanning one's body for ticks after a possible exposure can be rewarded by interdicting potentially infectious bites, especially if they are found within the 48 hours or more of attachment that is required for a tick to transmit infection. The threat of exposure can be reduced by wearing light-colored long-sleeved clothing and tucking trousers into socks. For further protection, clothing can be treated with permethrin or DEET-based repellents can be applied to the skin.

Areas can be protected by distributing carboard tubes containing permethrin-treated cotton (Damminix) that white-footed mice carry back to their nests, thus clearing themselves and their nest-mates of deer tick infestations. Area treatment with granular formulations of insecticides such as Dursban is sometimes recommended. Clearing shrubs and other low-lying vegetation from the edges of lawns near woodlots removes questing substrates and can reduce exposure risk

Tick collars prevent pets from carrying deer ticks into residences. Outdoor/indoor cats have been particularly noted to increase infection risk among their owners due to the tendency for deer ticks to become disengaged from cats and reattach within homes. Avoiding Lyme disease requires recognition of its zoonotic nature and the intimate relationship between particular combinations of landscapes, wildlife, and tick fauna that foster transmission.

REFERENCES

1. Spielman, A. The emergence of lyme disease and human babesiosis in a changing environment. Ann. New York Acad. Sci. **1994**, *740*, 146–156.
2. Dennis, D.T.; Nekomoto, T.S.; Victor, J.C.; Paul, W.S.; Piesman, J. Reported distribution of *ixodes scapularis* and *Ixodes pacificus* (Acari: Ixodidae) in the United States. J. Med. Entomol. **1998**, *35* (5), 629–638.
3. Piesman, J.; Clark, K.L.; Dolan, M.C.; Happ, C.M.; Burkot, T.R. Geographic survey of vector ticks (*Ixodes scapularis* and *Ixodes pacificus*) for infection with the lyme disease spirochete, *Borrelia burgdorferi*. J. Vector Ecol. **1999**, *24* (1), 91–98.
4. Levine, J.F.; Wilson, M.L.; Spielman, A. Mice as reservoirs of the lyme disease spirochete. Am. J. Trop. Med. Hyg. **1985**, *34* (2), 355–360.
5. Piesman, J.; Mather, T.N.; Dammin, G.J.; Telford, S.R., 3rd; Lastavica, C.C.; Spielman, A. Seasonal variation of transmission risk of lyme disease and human babesiosis. Am. J. Epidemiol. **1987**, *126* (6), 1187–1189.
6. Smith-Fiola, D.C.; Hallman, W.K. Tick bite victims and their environment: the risk of lyme disease. New Jersey Med. **1995**, *92* (9), 601–603.

MALARIA

Mike W. Service
Liverpool School of Tropical Medicine, Liverpool, England

M

INTRODUCTION

Despite almost a century of control efforts, malaria remains the most important cause of fever and morbidity in much of the tropical world. It occurs in about 100 countries, ranging from Central and South America, subsaharan Africa, throughout Asia to Papua New Guinea, and associated islands. About 40% of the world's population remains at risk from the disease, with about 270 million of these actually infected. Some 1 million people, mainly children, die each year from malaria. About 90% of all malaria and 80% of deaths occur in subsaharan Africa (1).

MALARIA PARASITES

Four parasite species cause malaria in humans. The main two are *Plasmodium falciparum*, which is the most lethal and commonest species in Africa, although it occurs in other regions, and *P. vivax*, which is prevalent in most countries outside Africa, and historically was the most important species in Europe. *P. malariae* and *P. ovale* have more restricted distributions and are not common. Human malaria is transmitted by anopheline mosquitoes. When the female takes a blood meal malaria parasites are destroyed in the mosquito's gut except for the gametocyte stages. In the mosquito stomach male microgametocytes exflagellate and the extruded flagella (microgametes) fertilize female macrogametes to form zygotes, which elongate to become mobile oökinetes. These penetrate the stomach wall and form oocysts on its outer surface, which after 5–8 days release numerous slender spindle-shaped sporozoites, most of which migrate to the salivary glands. There are usually many oocysts on the stomach of an infected mosquito producing thousands of sporozoites. The duration of this cycle in the mosquito is termed the external incubation period and is 8–10 days at 28°C, longer at lower temperatures and cannot be completed below 15°C.

When the mosquito feeds, saliva containing sporozoites is injected into a person's blood but these disappear from the blood within 30 min. Some enter liver cells where they develop and divide to become schizonts (pre-erythrocytic schizogony). After 6–16 days merozoites are released and enter red blood corpuscles. They grow into young trophozoites (ring forms) that eventually develop into schizonts, each producing 8–20 small merozoites which, on release, invade more corpuscles. After some time merozoites in the blood produce microgametocytes (males) and macrogametocytes (females). Only when the mosquitoes ingest these stages can the malaria cycle be completed in the vector. Additionally to this cycle of development, with *P. vivax* and *P. ovale* infections, some exo-erythrocytic schizonts in the liver produce hypnozoites, which can remain dormant for many months before they mature and release merozoites into the blood. The interval between a person becoming infected with sporozoites and malarial parasites appearing in the blood (prepatent period) varies according to parasite species, being 5–7 days with *P. falciparum* and 6–8 days with *P. vivax*.

CLINICAL SYMPTOMS

The classical symptoms are intermittent and repeated cold and hot fevers followed by profuse sweating, alternating with periods free of symptoms. Fevers last 8–12 h. There can be short-term relapses (recrudescences) due to survival of erythrocytic parasites in the body that become active many months later, such as with some *P. falciparum* infections. Long-term relapses are caused by the release of merozoites from the liver after 3–8 years with infections of *P. vivax*, and up to 50 years with *P. malariae*.

Malaria often causes abortions and anemia, therefore pregnant mothers are at special risk. Immunity to malaria is most pronounced in *P. falciparum* infections. Acquired immunity is achieved after repeated infections, so that children surviving attacks until they are about 5 years old will have developed a high degree of immunity. After an absence of infections for about 5 years much of the immunity is lost, but is usually rapidly regained on re-exposure to infection. In addition there are several human genetic traits that afford protection against malaria, such as

Encyclopedia of Pest Management

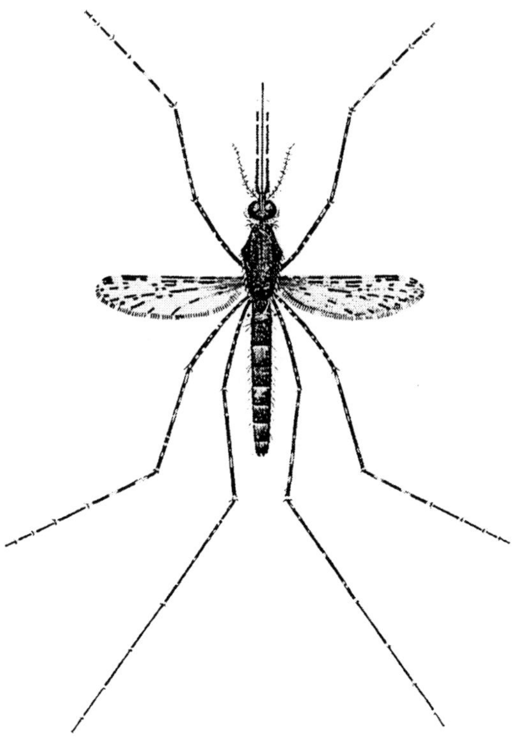

Fig. 1 The African mosquito, *Anopheles gambiae*. Probably the world's most dangerous malaria vector. (From Gillett, J.D. Common African Mosquitos and their Medical Importance; William Heinemann: London, 1972.)

red blood cells lacking the Duffy group antigens, people having abnormal hemoglobin (sickle cell trait), and those with G6PD deficiency.

PROPHYLAXIS AND TREATMENT

Individuals can obtain protection against developing malaria by taking prophylactic drugs regularly, the oldest of which is quinine. Later ones include proguanil (Paludrine), chloroquine, and pyrimethamine. Patients having malaria are also treated with drugs, again quinine being the oldest. Others include chloroquine, primaquine, or a combination of drugs such as sulfadoxine and pyrimethamine (Fansidar). But widespread drug resistance, mainly by *P. falciparum*, has made many drugs ineffective, which poses serious problems. Some of the more recent drugs, such as mefloquine (Lariam), can have serious side effects. Antimalarials are not widely used in malaria control programs.

MOSQUITO VECTORS

There are some 430 species of *Anopheles* (see Fig. 1) but only about 70 are malaria vectors, and of these probably only 40 are important. Larval habitats range from puddles, pools, ponds, rice fields, marshes, wells, and water tanks to water-filled tree holes. In the Neotropics larvae of a few vectors (e.g., *Anopheles bellator*) colonize leaf axils of bromeliads. Irrigated crops, especially rice (see Fig. 2), provide ideal habitats for some species and increased production of anophelines can result in increased malaria transmission. A few species breed in saline waters. Only adult females take blood and can be malaria vectors. Anophelines are crepuscular and nocturnal feeders. Some species, such as the African *A. gambiae*, feed predominantly on humans (anthropophagic) and are relatively long-lived,

Fig. 2 Flooded rice-fields in Sierra Leone that constitute a major larval habitat of the malaria vector *Anopheles gambiae*.

whereas other vectors such as the Asian species, *A. culicifacies*, feed mainly on cattle (zoophagic), but also to a lesser extent on people, and have a lower life expectancy. These attributes make *A. culicifacies* a less efficient vector than *A. gambiae*. Many species bite people predominantly in their houses (endophagic) but some feed on those outside their homes (exophagic). Some species usually rest indoors (endophilic) before or after blood-feeding, whereas others rest mainly out of doors (exophilic).

CONTROL

Most malaria control campaigns focus on the vectors. Prior to the DDT era, control was aimed at killing larvae by repeated applications of petroleum oils and Paris Green (copper aceto-arsenite) to larval habitats. Where feasible this was accompanied by environmental measures, such as draining or filling in larval breeding sites, or physically modifying them, such as increasing water flow through irrigation channels so making them inimical as breeding places (2). This strategy is impractical where vectors breed in temporary habitats like pools and puddles. When anopheline habitats are limited and easily found, larviciding with temephos or *Bacillus thuringiensis* ssp. *israelensis*, can give effective, but local, malaria control. Biological control, mainly with predatory fish such as *Gambusia affinis* and *Poecillia reticulata*, has long been practiced and received renewed interest in the 1960s when the limitations of insecticides were realized. Unfortunately, however, fish rarely have resulted in significantly reducing malaria transmission.

When DDT and other synthetic insecticides became available in the late 1940s control switched to spraying the interior walls and ceilings of houses with residual insecticides, which can kill indoor resting mosquitoes for 6 months. Consequently houses need to be sprayed only twice a year, or just prior to the monsoon period if transmission is restricted to a single wet season, to give protection against malaria. Although DDT is still used in some countries, organophosphate insecticides (e.g., malathion, fenitrothion) or the carbamates (e.g., carbaryl, bendiocarb) or the pyrethroids (e.g., permethrin, lambdacyhalothrin) are now mainly used, but must be sprayed every 3–4 months. If malaria vectors do not rest in houses this approach is of little use. Furthermore, spraying all houses in an area is expensive and requires considerable and sustained organization, and relatively few governments can afford this control strategy.

The most encouraging recent development in controlling transmission is the use of mosquito bednets (see Fig. 3) impregnated with permethrin, deltamethrin, or lambdacyhalothrin, all of which repel and kill mosquitoes and give protection against bites even when nets are torn (3). Their use can reduce substantially malaria morbidity and mortality. Communities, with little supervision, can impregnate their nets by dipping them in a pyrethroid insecticide, and these should remain effective for 4–6 months, or even a year, before needing to be reimpregnated. If, however, the pyrethroid is incorporated into polyethylene before it is formed into fibers for net-making, insecticide effectiveness can last 2, or possibly 4–5, years. Nets usually protect only the users and do not have a mass-killing effect on vector populations. Nets will afford little protection against vectors such as *A. albimanus* that in the tropical Americas bites people early in the evening before their bedtime.

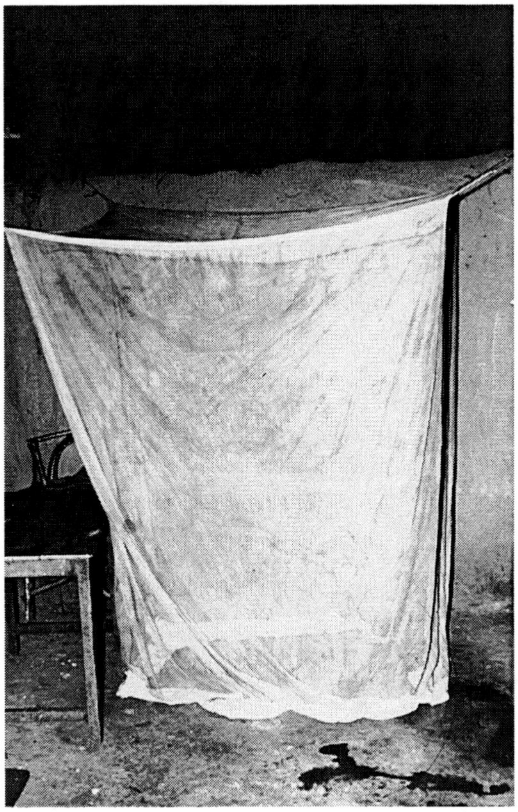

Fig. 3 A pyrethroid-impregnated bed net over a mat that is used as a bed in a Kenyan village house to protect against mosquitoes biting at night.

Because of insecticide resistance of the vectors and drug resistance of the parasites, malaria control is becoming increasingly difficult and consequently there is interest in alternative strategies. For example, one strategy is the production of transgenic mosquitoes that are incapable of transmitting malaria and when released into field populations of vectors will replace them (4). These techniques remain largely at the theoretical or laboratory stage. The only malaria vaccine that has been widely evaluated is the SPf66 vaccine, which has given substantial protection against *P. falciparum* in South America, although trials in Africa and Thailand have given conflicting results concerning its efficacy. Other candidate vaccines, including DNA ones, are under investigation or development.

The concept of malaria eradication promoted by the World Health Organization in the 1950s was abandoned in 1969. Now WHO supports a Global Roll Back Malaria partnership that aims to coordinate global action to fight malaria and help governments reach their own targets for control. The primary aim is to reduce child mortality and morbidity through using all appropriate control strategies.

COSTS

Days of disability per year per person caused by malaria range from 5 to 20, but it is difficult to get meaningful estimates. The malaria burden costs per capita about \$3–4 a year. The cost of spraying houses once a year with DDT is about \$4.50 per year, with other insecticides about \$8, compared to the annual cost of \$2 per person for protecting them with insecticide-treated bednets. Nets cost about \$5 and last for about 5 years, and the cost of each insecticidal treatment is about \$0.50.

REFERENCES

1. *Bruce-Chwatt's Essential Malariology,* 4th Ed; Warrell, D.A., Gilles, H.M., Eds.; Edward Arnold: London, *in press.*
2. World Health Organization. Vector control for malaria and other mosquito-borne diseases. WHO. Technol. Ser. **1995**, *857*, 1–91.
3. Greenwood, B.M. What's new in malaria control. Ann. Trop. Med. Parasitol. **1997**, *91*, 523–531.
4. Curtis, C.F.; Townson, H. Malaria: existing methods of vector control and molecular entomology. Brit. Med. Bull. **1998**, *54*, 311–325.

MECHANISMS OF RESISTANCE: ANTIBIOSIS, ANTIXENOSIS, TOLERANCE, NUTRITION

M

Helmut van Emden
The University of Reading, Reading, Berkshire, United Kingdom

INTRODUCTION

There are many plant anatomical, physiological and biochemical mechanisms (1) of host plant resistance to insects (Fig. 1). The first comprehensive treatment of the subject was by R.H. Painter of Kansas University, in his classic 1951 book *Insect Resistance in Crop Plants* (2). He grouped the various mechanisms into three types of host plant resistance: nonpreference, antibiosis, and tolerance. In simple terms, nonpreference is shown when the pest arrives, antibiosis is a resistance to the survival and multiplication of the pest once the plant has been colonized, and tolerance is the ability of the plant to cope in spite of the pest attack. This classification has stood the test of time and is still the basis of the entomologist's view of plant resistance, in spite of the later emergence of a genetically based classification for plant resistance to plant pathogens, a classification which is probably equally applicable in entomology. The only change to Painter's concept that has gained acceptance is the use of antixenosis (3) rather than nonpreference for Painter's first category (see below). Here we also separate out nutritional resistance as a separate category, though many authors include it as one mechanism of antibiosis.

Fig. 1 links the mechanisms of resistance with the classification described above, as far as possible in the order in which the pest encounters them (4).

ANTIXENOSIS

This is Greek for "resistance to outsiders," and refers to properties of plants that prevent or reduce colonization by a pest seeking food or oviposition sites. The term has two advantages over Painter's "nonpreference." First, like "antibiosis" and "tolerance" it is a property of the plant rather than of the insect and, second, it does not imply a choice situation. It is therefore not strictly synonymous with "nonpreference" for it also embraces the concept of "nonacceptance" in a no-choice situation.

Antixenosis is the first stage in the encounter between pest and plant. At this point the insect is most sensitive to variations in plant appearance and condition and, even on standard susceptible host plants, rejection rates of 50–70% have been recorded. Mechanisms of antixenosis include subtle variations in color (often based on different reflectivities for ultraviolet), waxiness or hairiness of leaves, and flavor as well as some morphological characters and defensive exudations of gums or resins.

A good example for the color mechanism, for it clearly distinguishes antixenosis and antibiosis, is that of red cabbages. Here the different color seems to deter colonization or oviposition by both cabbage aphids and cabbage white butterflies, though rearing them on red cabbages shows the latter are more nutritionally suitable. A glossy-leaved mutant of broccoli developed in the 1950s showed antixenosis to a wide range of brassica pests compared with standard waxy-leaved cultivars. As with red cabbage, the antixenotic broccoli appeared to possess no antibiosis. Later work on cereals with similar characteristics suggested that the glossy cultivars possess higher levels of pest-deterrent hydroxydiketones.

An increased density of trichomes on leaf surfaces deters many small insects such as aphids and whitefly. Some crops (e.g., potato) can have glandular hairs from wild relatives transferred by breeding. These hairs, when the liquid-filled sac at their apex is broken by contact with an insect, exude a fast-setting polyphenol that glues up the mouthparts and tarsi of small pests, including some as large as leafhoppers.

Secondary plant metabolites such as alkaloids and tannins are fundamental to insect host-plant choice, and therefore not surprisingly are involved in many examples of antixenosis (e.g., the phenolic DIMBOA in corn is antixenotic to the European corn borer *Ostrinia nubilalis*). An example of a morphologically based antixenosis is Frego bract cotton to bollworm; the moth is deterred from landing on these small bracts for oviposition. Another example is that cowpea cultivars that hold their pods apart and erect on long peduncles do not provide the shelter of

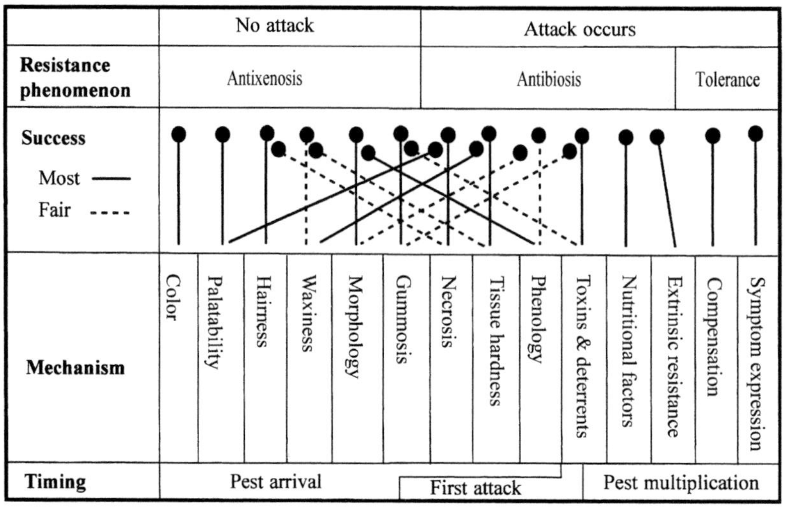

Fig. 1 Classification of mechanisms of plant resistance to pests. (From Ref. 1.)

pods touching each other or leaves, contact that encourages larvae of *Maruca* pod-borer to penetrate the pod.

Some antixenosis takes the form of ''escape,'' and is merely due to the crop not being in a suitable stage when the pest appears. An example is early flowering peas, which no longer have white flowers when the pea moth appears to oviposit on the soft receptacle below the attractive petals.

ANTIBIOSIS

This is Greek for ''resistance to life functions,'' and the word is well known in relation to the treatment of pathogenic bacteria in human medicine. In relation to plant resistance to insects, antibiosis is the ability of the plant to reduce survival, extend development time, or reduce growth or fertility. These are the elements that combine to determine the pest's potential rate of natural increase. This follows an exponential function, and can be translated into how long it takes the population to double under the given environmental conditions and in the absence of other mortality factors.

The term ''antibiosis'' suggests antibiotic chemicals, and certainly plant allelochemicals (often toxic) are important mechanisms of antibiosis. Common compounds involved in plant antibiosis to insects are inorganics (e.g., selenium), primary or intermediate metabolites such as cysteine and aromatic amino acids, and secondary compounds (particularly isoprenoids, alkaloids, protease inhibitors, glycosides, flavonoids, condensed tannins, and stilbenes). The many examples include DIMBOA in corn antibiotic to European corn borer and the alkaloid demissine, antibiotic in potatoes to Colorado beetle. Modern transgenic techniques now allow the direct transfer of single genes from totally unrelated organisms (including microorganisms and animals) for the expression of toxic proteins. Many crops (e.g., cotton, tobacco, corn, and canola) have now been genetically modified to express the toxin of the bacterium *Bacillus thuringiensis.* Clearly, unless the toxin is coupled with promoters so that expression is pest-induced, the production of such chemicals consitutively at levels giving virtual immunity to a pest must come at some cost to the plant, and the lower levels of antibiosis obtainable by traditional plant breeding, and by many other mechanisms, are likely to involve a lower cost.

Some crop varieties are antibiotic through what is actually the opposite phenomenon, hypersensitivity. Local necrosis, where damage to cells by insects brings together otherwise separated substrates and enzymes to produce toxic or distasteful polyphenols, is a powerful resistance mechanism against insects (especially sucking insects), nematodes and fungi. Even the collapse of whole plants when invaded by stem-borers can be a mechanism of antibiosis in crops such as cereals that are often sown above optimum density in terms of light capture by the canopy.

Plants are intrinsically defended from pest attack by the hardness of their surface tissues, and quite minor differences in stem solidity can confer antibiosis of cereals against borers such as the wheat stem sawfly. With antibiosis of rice to the pyralid stem-borer *Chilo*, the ''hardness'' is due to uptake of silica grains into the surface layers of the stem by the plant. These silica granules rapidly wear out the mandibles of the small larva trying to invade. The distribution of silica is as important as the amount; in rice cultivars where it is distributed in bands leaving unprotected strips, the antibiosis is lost.

Antibiosis, like antixenosis, can arise from failure of the pest to synchronize with the phenology of the crop in the field, though altering the synchronization (e.g., by sowing at a different time) shows the apparent antibiosis to be only "pseudoresistance." For example, some late flowering cultivars of faba bean receive very low oviposition by *Bruchus dentipes*, because the beetle's oviposition is drastically reduced in the absence of the nectar source provided by the flowers.

Apparent antibiosis may also result from the interaction of a particular crop variety with some extrinsic factor, usually mortality from natural enemies. For example in Australia, the larvae of a phytophagous ladybeetle switch to predation on young bollworm larvae on certain cotton cultivars that lack the leaf nectaries on which the ladybeetle larvae usually feed. More examples are given in the article "Interaction of Host-plant Resistance and Biological Control" in this encyclopedia.

TOLERANCE

Tolerance is the ability of a variety to show a reduced damage response (usually measured as yield) compared with a reference variety of the same crop suffering the same insect pest burden. Tolerance may therefore be envisaged as resistance to damage rather than to the damaging insect.

Examples of high levels of tolerance are, perhaps not surprisingly, relatively rare. Tolerance often represents some form of compensation. Thus, where leaf area ratios (i.e., crop leaf area/ground area) are supraoptimal for maximum light interception because the variety is very leafy, considerable tolerance to defoliating insects can be expected (e.g., some soybean and brassica cultivars). Indeterminate cowpea cultivars are relatively more tolerant than determinate ones to pod damage since, given nutrients and water, they will continue flowering and podding till a yield has been set.

Other examples of tolerance are much more specific. The shot-hole borer weakens the junction of branches of tea bushes with its gallery, so that large parts of the bush detach when the pickers move through the plantation. There are, however, tolerant tea varieties that strengthen the damaged junction with a support bracket of new callus growth, so that the branches can bend and spring back rather than fracturing.

Other tolerances can result from suppression of symptom expression. Certain cowpea varieties do not react with "hopperburn" (reddening of the foliage) to the toxic saliva of leafhoppers, though the biochemistry of this is not known. Certain wheat varieties owe their tolerance to aphids to the lack of curling of the flag leaf, which reduces the area of the flag leaf exposed to radiation. These varieties have already slowed their leaf growth by the time aphids form colonies on the lower surface of the flag leaf; the aphids interrupt the flow of materials on that side, but in these varieties there is no flow of growth substances to cause curling by continued expansion of the upper leaf surface.

NUTRITION

Insects seek nutrition from crop plants just as much as do humans and their livestock. Thus any poorer nutrition in terms of leaf protein, although conferring resistance to pests, would be counterproductive. However, there is no positive correlation between leaf protein and soluble nitrogen in the transport (phloem) vessels. It is thus possible to obtain resistance to sucking insects feeding in these vessels without impairing food quality for vertebrates. Cultivars do differ in the soluble nitrogen content of the phloem; this has been shown clearly in a study of the relative resistance of pea cultivars to pea aphid in Canada.

Resistance based on nutritional mechanisms is, however, particularly unstable in relation to changes in plant condition resulting from increased age and environmental conditions such as temperature, soil fertilization of major nutrients, and application of agrochemicals such as plant growth regulators and some insecticides.

CONCLUSIONS

Of the various categories of plant resistance, antibiosis is preferred in spite of the fact that it carries the greatest danger of selecting for adapted strains of the pest. Tolerance as a sole type of resistance has the danger that farmers will not control their pest populations adjacent to other cultivars of the crop on other farms, and antixenosis is rarely adequate on its own in a no-choice situation.

Fortunately, traditional plant breeding normally produces cultivars that combine more than one resistance mechanism as well as more than on category of resistance. Certainly the appearance of strains adapted to a plant resistance is very rare compared with similar events in the literature of plant pathology.

Insects can often identify the presence of antibiosis at the arrival stage and then show nonpreference and move elsewhere where survival selection will be for different attributes. It is also likely that some tolerance may be combined with other categories, but this is hard to determine experimentally since the effects of anitbiosis on pests make it impossible to preserve the same insect burden on antibiotic and nonantibiotic varieties.

REFERENCES

1. van Emden, H.F.; Peakall, D.B. *Beyond Silent Spring: Integrated Pest Management and Chemical Safety*; Chapman & Hall: London, 1996; 1–322.
2. Painter, R. H. *Insect Resistance in Crop Plants*; Macmillan: New York, 1951; 1–520.
3. Kogan, M.; Ortman, E.F. Antixenosis—a new term proposed to define painter's "non-preference" modality of resistance. Bull. Entomol. Soc. Am. **1978**, *24* (2), 175–176.
4. van Emden, H.F. Host Plant Resistance to Insect Pests. In *Techniques for Reducing Pesticide Use: Economic and Environmental Benefits*; Pimentel, D., Ed.; John Wiley & Sons: Chichester, UK, 1997; 129–152.

MINILIVESTOCK

Maurizio G. Paoletti
Padova University, Padova, Italy

Darna L. Dufour
University of Colorado, Boulder, Colorado, U.S.A.

INTRODUCTION

Small invertebrates such as insects, earthworms, molluscs, and spiders are the unconventional food resources considered minilivestock—these small animals are grown and/or gathered in most countries, especially in the tropics (1). They can provide proteins, fat, minerals, and vitamins complementing the diet and sometimes representing a considerable portion of the animal food in the diet. In some cases, insects provide 8–65% of the animal proteins, especially in difficult times of the year—for instance, during the rainy season in the Amazon areas (Table 1).

Worldwide, about 2000 insects (6) and possibly more than 400 different invertebrates including arachnids, molluscs, miriapods, earthworms, and crustaceans have been adopted, not only as nonconventional foods, but also for medicine (7–11). These numbers are probably underestimated if we consider that in many areas—for example, the Amazon—there is little current knowledge about the use of these creatures as protein resources for many human groups living there (12). In countries such as Mexico, Japan, China, Vietnam, Indonesia, Thailand, Ecuador, Colombia, and Papua New Guinea and in African countries (11, 13, 14), many different insects are traditionally eaten and marketed. For instance, the giant waterbug (*Lethocerus indicus*, family Belostomatidae) is exported from Thailand and is sold in California, USA (15). In Europe, few insects are traditionally exploited, except for the bee (*Apis mellifera*) and a cheese maggot (of the fly *Piophila casei* L.) that is appreciated in some seasoned local cheeses in Italy. In spite of the general Western attitude to refuse insects as food, some insects are becoming attractive to Western consumers, some canned or dry-smoked insect preparations are available in specialized food stores, a few restaurants are serving insects, and entomophagy clubs are being founded (16, 17).

DIVERSITY OF ORGANISMS

Worldwide, caterpillars, grasshoppers, beetles, termites, ants, and wasp-bees are the key groups utilized as minilivestock (10, 16, 18). Some aquatic insects have to be added to the list, such as Homoptera Belostomatidae, Trichoptera, Odonata, Ephemeroptera, Megaloptera, and Diptera and a few Dytiscidae and Elmidae among Coleoptera. Bees and wasps deserve a particular role as producers of honey, wax, and propolis but also, in many cases, of larvae and pupae.

Normally, gathering is done in the wild. In some cases, strategies to domesticate insects are present, particularly for bees. The domestication of insects has been obtained for honey bees and silkworms (7). Semidomestication has been accomplished for stingless bees (Meliponinae) (18), and many other more frequently gathered invertebrates are extensively managed. For instance, in China, the scorpion *Buhtus martensi* is grown on a large scale for medicinal and edible purposes. Edible earthworms (*An-*

Table 1 Proportion of insects as animal protein in Amazonian diets

Yanomamo (Amazonas, Venezuela) (2)	1–3% of animal proteins of mean diet, annually
Tukanoan Indians (Colombia) (3)	12% of animal proteins of men's diet, seasonally 26% of animal proteins of women's diet, seasonally
Piaroa, Rio Cuao, Alto Orinoco (4)	8% of animal proteins are invertebrates, annually
Guajibo (Alcabala Guajibo, Amazonas, Venezuela), rainy season (5)	60–70% of animal proteins, seasonally

Encyclopedia of Pest Management

diorrhinus sp.) [called *motto* by the Indians Ye'kuana (Alto Orinoco, Venezuela)] are actively disseminated to increase availability. Rain forest leaves and litter host the key invertebrates (such as caterpillars, *Atta* ants, *Syntermes* termites, and, in two cases, earthworms) traditionally eaten by Amerindians in Amazonia (see Fig. 1) (19).

In most cases, the larvae and pupae are the stages used as food, sometimes raw and consumed on the spot,

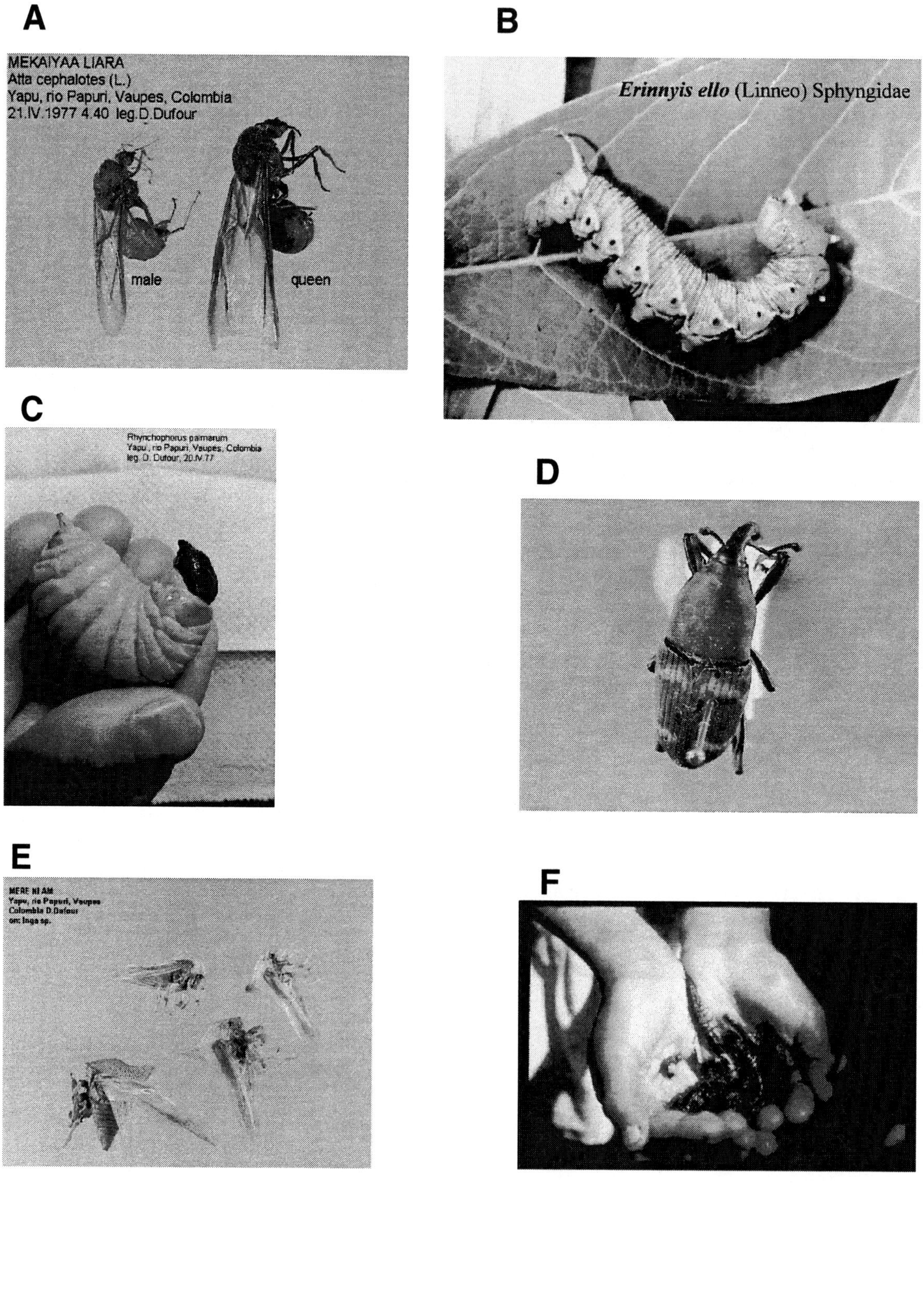

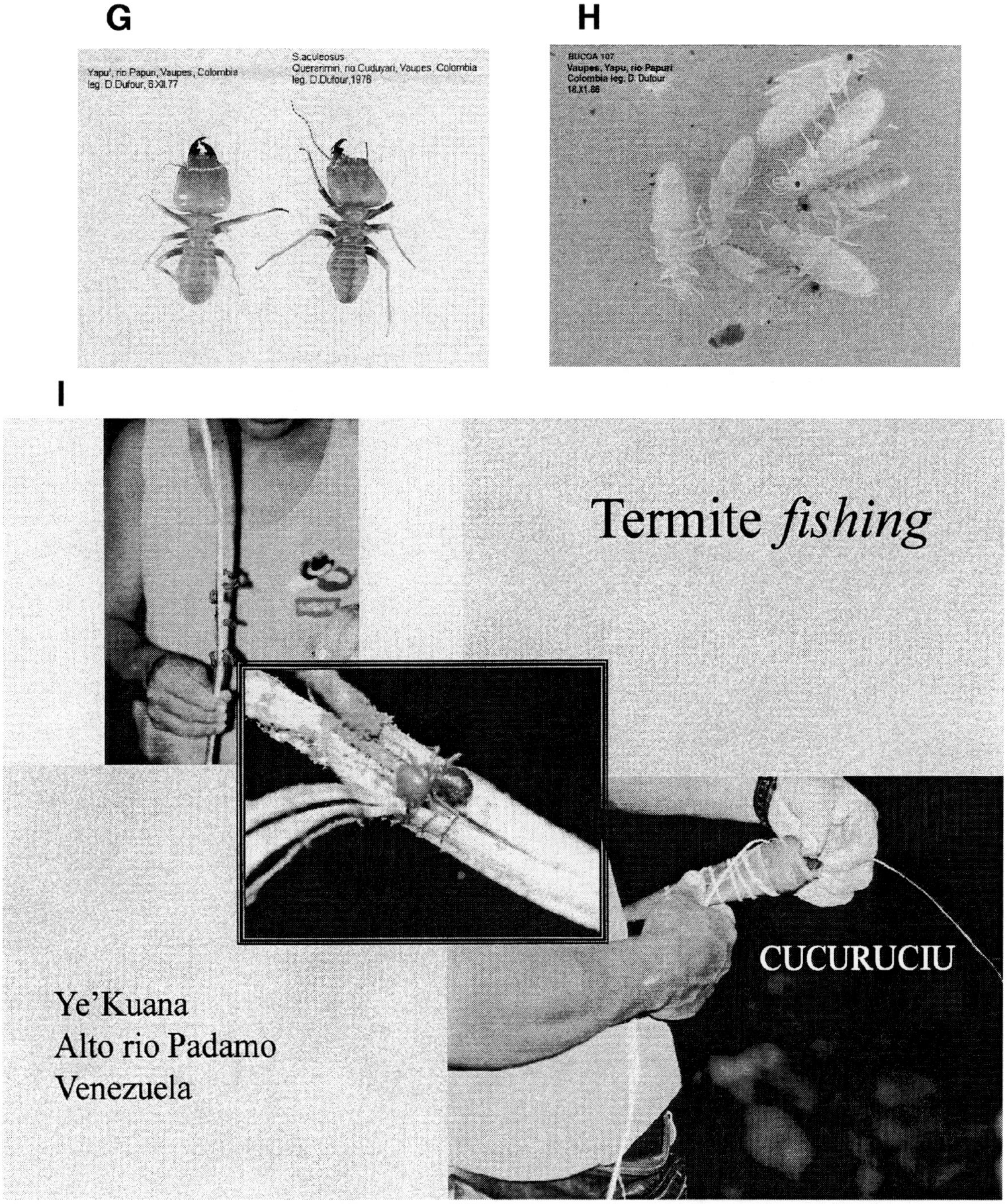

Fig. 1 Edible invertebrates. (**A**) The winged ants, *Atta cephalotes*, especially appreciated by native Amerindians. (**B**) Caterpillars of many species are adopted as food by Amerindians. *Erinnyis ello* is a key pest of yuca (*Manihot esculenta*). (**C**) Palm weevils are very appreciated in most tropical areas of the world. *Rhynchophorus palmarum* is a very good component of the palm association in South America. (**D**) *Metamasius* is another palm weevil component appreciated by Amerindians. (**E**) The small Membracidae *Umbonia spinosa* living on *Inga* sp. are very appreciated snacks in Ecuadorian and Colombian Amazon. (**F**) Aquatic insects are also important resources. Here they are represented by *Corydalus* sp. from Peruvian Andes. (**G—I**) Several termites are used as food, especially (**G**) soldiers and pupae in Amazon (Rio Padamo—Alto Orinoco). (**H**) Pupae of *Labiotermes labralis*, smaller than the previous group, are also foraged by Yanomamo and Tukanoans. (**I**) Soldiers of *Syntermes* sp. are ''fished'' inside the terrestrial nests using a probe. An efficient package is important to transport the living specimens to the village. (**J**) Grasshoppers such as *Tropidacris cristata* are collected and eaten by Guajibo in the Alto Orinoco, Venezuela. (**K**) Several wood-boring larvae are collected such as Cerambycidae. (L–N) Some earthworms are collected as well by different Amerindians (Piaroa and Ye'Kuana in the Alto Orinoco. The Ye'Kuana used to smoke the long Motto.

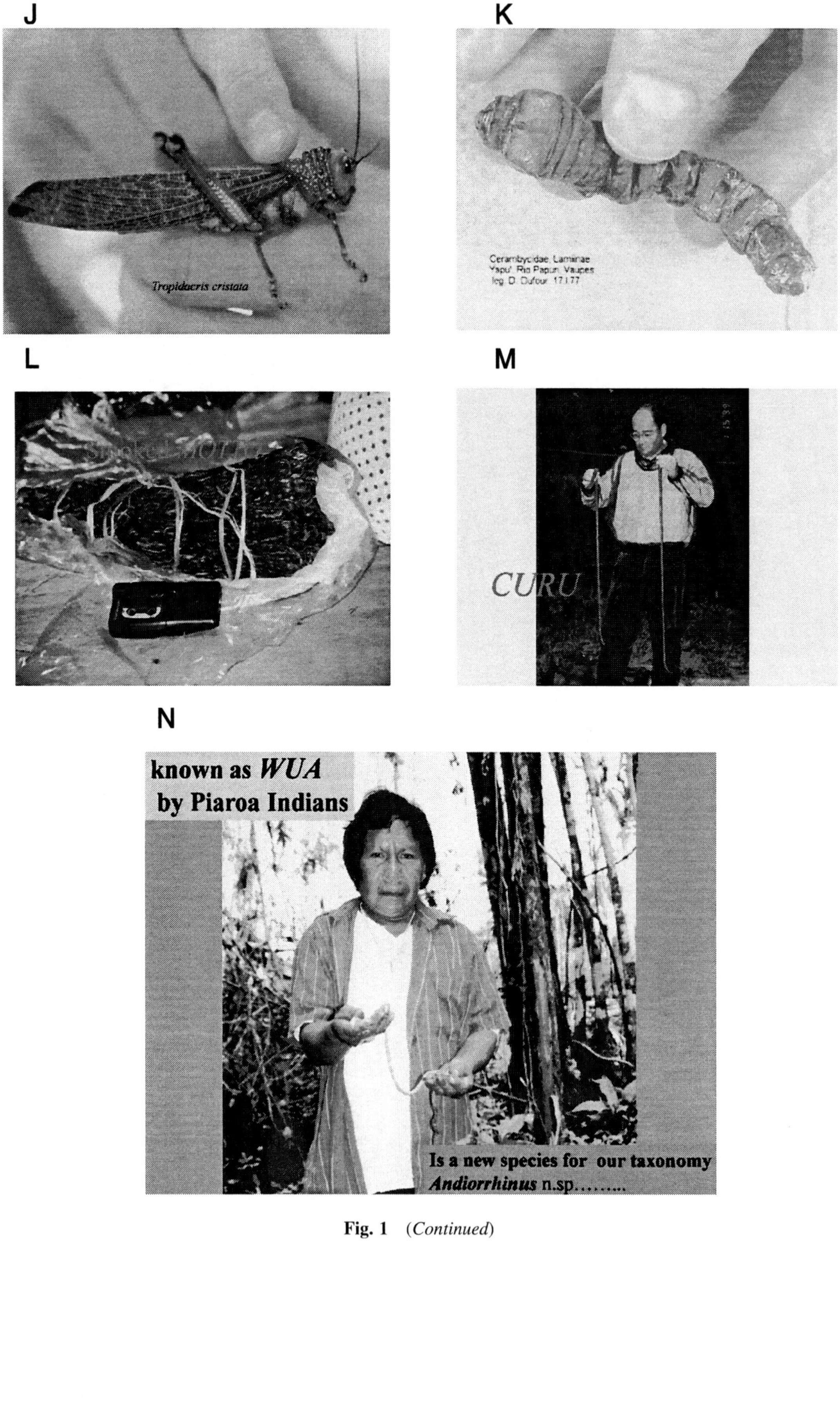
J
Tropidacris cristata
K
Cerambycidae, Laminae
Yapu', Rio Papuri, Vaupes
leg. D. Dufour 17.1.77
L
M
CURU
N
known as WUA
by Piaroa Indians
Is a new species for our taxonomy
Andiorrhinus n.sp.........

Fig. 1 (*Continued*)

Table 2 Mean foraging efficiencies for some invertebrates compared with fish and game in Amazonian countries

Animal	Foraging efficiency (g/h)	Foraging success rate[a] (%)	Source of the data
Earthworm *motto*	388	99	Ye'Kuana: Rio Padamo, Venezuela
Earthworm *curu*	446	99	Ye'Kuana: Buena Vista Rio Padamo, Venezuela
Syntermes spp.	300	100	Tukanoans (Dufour unpublished)
Syntermes aculeosus	200	99	Ye'Kuana: Rio Padamo, Venezuela
Atta spp.	200	99	Tukanoans: Vaupes, Colombia
Caterpillars	300	99	Tukanoans: Vaupes, Colombia
Fish	494	73	Tukanoans: Vaupes, Colombia
Fish	927	?	Amazon
Game	1,003	?	Amazon

[a]Success rate is the percentage of foraging trips with successful capture.

but in many cases cooked in different ways. Smoking and/or drying is often used for grasshoppers, termites, and caterpillars.

Elementary but effective techniques have been developed to obtain large amounts of the soldiers of the large termites (*Syntermes* in South America and *Bellicositermes* in Africa) and some ants of the genus *Atta* (in South America) that react by biting and remaining attached to the intruder enemy. A probe obtained from a palm-leaf rib is introduced in the subterranean galleries of the colony, and the soldiers biting it are easily ''fished'' out.

The time requirements to obtain insects is, in most cases, comparable to obtaining other larger animals (game, fish), with the advantage that in most cases the localization is less problematic than for most game and fish (Table 2).

The crude protein content varies from 22 to 91 g per 100 g of dry insect, and is similar to meats such as pork and beef that range from 45 to 75 g per 100 g dry weight (Table 3). However, the chitin-based exoskeletons of insects are not digestible in the human gastrointestinal tract because of the absence of relevant kitinase enzymes, and because the nitrogen in chitin is unavailable to humans (8) PAPAINA. However, fruits rich in protcolitic enzymes such as papaine can promote digestion of chitin. The digestibility of most insect protein varies, but in general it is more than 77% of the insect dry weight. Amino acids such as lysine and tryptophan are in most cases available in insects, making them an important complement to root crops or starchy foods (8). For instance, termites (*Bellicositermes bellicosus*) in Angola are rich in these two amino acids, and therefore complement maize, which is a poor source of lysine and tryptophan.

Termites and palmworms (*Rhynchophorus* sp.) are especially rich in fat, an important resource in tropical areas, and also contain essential fatty acids (linoleic and linolenic). Nutrients such as calcium and iron are sometimes more abundant in insects than in conventional meats. Vitamins, especially B vitamins, in most cases, appear higher in insects than in meat and bread.

LIMITING FACTORS

Westerners have a taboo against eating the small terrestrial invertebrates, and this attitude does not facilitate the research needed to develop this multifaceted resource (9). As soon as local communities using insects

Table 3 Nutritional value compared with other meats

Food	Moisture s(%)	Energy (kcal)	Protein (g/100 g)	Fat (g/100 g)
Female ant *Atta sexdens*	6.1	628	39.7	34.7
Female ant *Atta cephalotess*	6.9	580	48.1	25.8
Soldiers termite *Syntermes* sp.	10.3	467	58.9	4.9
Palmworm *Rhynchophorus palmarum*	13.7	661	24.3	55.0
Caterpillars various (smoke dried)	11.6	425	52.6	15.4
River fish (smoke dried)	10.5	312	43.4	7.0
Tapir (smoke dried)	10.3	516	75.4	11.9

Nutritional value of some insects consumed in tropical South America compared with other animal foods. Composition per 100 g edible portion. (From Dufour and Sander, 1999.)

as food become "civilized," they drop most of their traditions concerning insects and their consumption. This attitude, sometimes fostered by missionaries and educators, has to change.

A few antinutritional factors have been found in insects [for instance, in the African silkworm (*Anaphe venata*)]. It has also been argued that locusts contaminated with pesticide residues could contaminate people consuming them (8). The digestive tract is removed from some caterpillars in Amazonia [such as Cassava hornworm (*Erinnyis ello*)] and discarded before the caterpillars are roasted, seemingly to prevent consumption of residual toxicity of the host plant (*Manihot esculenta*), which contains variable amounts of cyanidric compounds).

Improving the existing knowledge of insects and other invertebrates as food and promoting their sustainable use could reveal an important hidden resource.

REFERENCES

1. Hardouin, J.; Stievenart, C. *Invertebrates (Minilivestock) Farming*; CTA (Center of Technical Cooperation): Wageningen, 1993; 223.
2. Lizot, J. Los Aborigenes de Venezuela. In *Ethnologia contemporanea*; Lizot, J., Ed.; Fundacion La Saille: Caracas, 1988; II, 479–513.
3. Dufour, D.L. Insects as food: a case study from northern Amazon. American Anthropologist **1987**, *89* (2), 383–397.
4. Zent, S. *Piaroa on the Cuao River, Venezuela*; Ph.D. Dissertation, Columbia University: New York, 1992.
5. Cerda, H.; Martinez, R.; Briceno, N.; Pizzoferrato, L.; Manzi, P.; Tommaseo, Ponzetta M.; Marin, M.; Paoletti, M.G. Palm worm (Insecta, Coleoptera, Curculionidae: *Rhynchophorus palmarum*) traditional food in Amazonas, Venezuela. Ecol. of Food and Nutr. **2001**, *40* (1), 13–32.
6. Ramos-Elorduy, J. Insects: a sustainable source of food. Ecol. of Food and Nutr. **1997**, *36*, 247–276.
7. Bodenheimer, F.S. *Insects as Human Food*; W. Junk Publ: The Hague, 1951; 239.
8. Bukkens, S.G.F. The nutritional value of edible insects. Ecol. of Food and Nutr. **1997**, *36*, 287–319.
9. De Foliart, G.R. Insects as food: why the western attitude is important. Annu. Rev. Entomol. **1999**, *44*, 21–50.
10. Dufour, D.; Sander, J. Insects as Food. In *The Cambridge History and Culture of Nutrition*; Kiple, K.F., Ed.; Cambridge University Press: Cambridge, 2000.
11. Malaisse, F. Se Nourrir en Foret Claire Africane. In *Approche Ecologique e Nutritionelle*; Les presses Agronomiques de Genbloux, A.S.B.L. Belgium, 1997; 384.
12. Paoletti, M.G.; Dufour, D. Edible invertebrates among Amazonian indians: A disappearing knowledge. Food Insect Newsletter **2000**.
13. Onore, G. A brief note on edible insects in Ecuador. Ecol. of Food and Nutr. **1997**, *36*, 277–285.
14. Paoletti, M.G.; Bukkens, S.G.F. Minilivestock. Ecol. of Food and Nutr. **1997**, *36* (2–4), 95–346.
15. Pemberton, R.W. The use of the Thai giant waterbug, *Lethocerus indicus* (Hemiptera: Belostomidae), as human food in California. Pan-Pacific Entomol. **1988**, *64* (1), 81–82.
16. Menzel, P.; D'Aluisio, F. *Man Eating Bugs*; Ten Speed Press: Berkeley, CA, 1998.
17. Comby, B. *Insetti che bontà*; Edizioni Piemme: Casale Monferrato, 1991; 158.
18. Posey, D.A. Ethnoentomological survey of Amerind Groups in lowland Latin America. Fla. Entomol. **1978**, *61*, 225–229.
19. Paoletti, M.G.; Dufour, D.L.; Cerda, H.; Torres, F.; Pizzoferrato, L.; Pimentel, D. Leaf- and litter-feeding invertebrates are important Amerindian sources of animal protein in Amazonia. Proceedings Royal Society-B **2000**, *267*, 2247–2252.

MINOR USE

Bent Bromand
Danish Institute of Agricultural Sciences, Slagelse, Denmark

M

INTRODUCTION

Information on disadvantages and possible consequences of the use of pesticides generated during the years have caused authorities in most countries to impose stricter rules for registration of pesticides concerning human health and environment. Reregistration programs of all previously registered pesticides have resulted in a great deal of pesticides being removed from the market leaving many cultures in small areas and attacks by sporadic pests without any means of control (minor uses). After reregistration in Denmark, only 78 out of 213 active ingredients were approved (Table 1). Development of a new pesticide is very costly and can easily run up to $150 million. Consequently, chemical companies are reluctant or not interested in development of new pesticides unless they can be used in major crops like wheat, corn, cotton, soybean, rice, and vine.

DEFINITION OF MINOR USES

Minor uses include crops with small acreage; areas with diseases, pests, and weeds for which no trials are being made due to economic reasons; and crops where pesticides are not approved against all pests or where control is not effective. Important minor crops are most vegetables, some fruits, flax, hops, nuts, spices, medicinal plants, tobacco, ornamentals, forestry, seed crops, and protected crops. The decision on what is or is not a minor use is a national decision, that is, a decision made by the designated national authority. However, generally speaking a minor use is any use of a pesticide for which the volume is insufficient to justify the cost by a commercial company to obtain a registration.

IR-4 PROGRAM, UNITED STATES

Interregional Research Project No. 4 (IR-4) was organized in 1963 specifically to address the problem of minor use registrations on food and feed crops. In 1977, the project was expanded to include pesticide clearances for nursery and ornamental crops, forest seedlings, and turfgrass. A further expansion of objectives in 1982 included the registration of microbials and biochemicals to the IR-4's mission. The problem has been greatly intensified by the requirement to reregister all pesticides first registered prior to November 1984, which has increased the need for financial support for the IR-4 program. The IR-4 National Headquarters, located at the New Jersey Agricultural Experimental Station, Rutgers University, provides the overall coordination for the diverse components of this national program. The objectives of the IR-4 minor use program are

- to obtain minor use and specialty use pesticide clearances and assist in the maintenance of current registrations, and
- to further the development and registration of microbial and specific biochemical materials for use in pest management systems.

The IR-4 is willing to sponsor research to establish efficacy, crop safety, and magnitude of the residue data provided the company agrees to make a properly labeled commercial product available. The pesticide under consideration must have at least one tolerance or exemption established by the Environmental Protection Agency (EPA) to assure that the necessary core data requirements (chemistry, toxicology, and environmental fate) are complete and have been accepted by EPA. All trials must be performed according to good laboratory practice (GLP). The total federal funding of the IR-4 program has increased substantially over the years and in 1996 it was $8.3 million. However, a fully implemented strategic plan will require a budget of $33 million by 2002. The program has resulted in 4400 food use clearances, 3600 ornamental registrations, and 70 biopesticide registrations during its first 33 years. Regulatory barriers arise because adding of a large number of minor uses for a widely used pesticide may cause the acceptable daily intake to be exceeded; and

Table 1 Reregistration of pesticides in Denmark

Decisions on reassessed active ingredients	213
Missing responses/reassessments not desired	60
Rejected due to insufficiencies of documentation	29
Approval withdrawn or application withdrawn	16
Prohibited or strictly regulated	30
Approved	78

(Adapted from Status of the Minister for the Environment's Action Plan for Reducing the Consumption of Pesticides. Danish Environmental Protection Agency, Pesticide Division, November 1997.)

concern over residues, particularly in the diets of infants and children, is leading to new tolerances of the maximum residue level (MRL).

EUROPEAN PROGRAMS

The basis for minor use in the European Communities is Council Directive 91/414/EEC (1). Article 9 states that official or scientific bodies involved in agricultural activities or professional agricultural organizations and professional users may request that the field of application of a plant protection product, already authorized in the member state in question, is extended to purposes other than those covered by this authorization. Member states shall grant an extension of the field of application of an authorized plant protection product and shall be obliged to grant such an extension when it is in the public interest to the extent that

documentation and information have been submitted by the applicant,
it has been established that it does not cause unnecessary suffering and pain to vertebrates to be controlled, it has no harmful effect on human or animal health, directly or indirectly or on ground water and it has no unacceptable influence on the environment,
the intended use is minor in nature, and
users are fully and specifically informed about instructions for use.

Article 10 states that a member state must refrain from requiring the repetition of tests and analyses already carried out in connection with the authorization of the product in another member state, to the extent that agricultural, plant health, and environmental (including climatic) conditions relevant to the use of the product are comparable in the regions concerned.

The following criteria are used in the EU for classifying crops:

	Major crops	Minor crops	Very minor crops
Daily dietary intake	>7.5 g	1.5–7.5 g	<1.5 g
Cultivation area	>10,000 ha	600–10,000 ha	<600 ha
Production	>200,000 t per year	<200,000 t per year	—

The rules for extrapolation are laid down in "Guideline on Comparability, Extrapolation, Group Tolerances and Data Requirements for setting MRL's" (2), which states that the principle is to choose the trial conditions that, under realistic circumstances, would be the least favorable. This means the highest residue situation according to the intended uses, for example, maximum number of applications, highest prescribed dosage, and shortest post harvest interval. For minor and very minor crops, four trials representative of the proposed growing areas are normally required. However, the systematic evaluation of existing information and experience often make it possible to reduce the number of trials needed. Different application methods such as spraying, drenching, dusting, misting, and granule spreading will as a rule not produce comparable residue results and must therefore be documented separately. It is assumed that for the implementation of residue trials, the climatic conditions and weather influences in each of the two regions in Europe, divided by a line through the middle of France, are comparable:

Northern and Central Europe
Southern Europe and the Mediterranean

Data from different countries within the same region may reflect different cultural practices and might therefore be rejected. The evaluation of intended uses within the European Union should be based on residue data mainly generated within the European Union. Trials for residues must be carried out according to GLP rules and efficacy trials according to good experimental practice (GEP).

REMEDIES AGAINST MINOR USE PROBLEMS

The European countries have the same problems as the United States, but governmental funds for research to help acquire information on off-label use are very limited in contrast to the United States. To help solve the problems

M

it is first of all necessary to do what is possible with regard to agricultural practices in order to reduce the need for pesticides.

There are different ways of solving the problems. Expansion of registration by the company is preferred because the safety instructions and directions for use then appear directly from the label. Exemption can be given for a shorter period (usually for no more than 3 years, when withdrawal of registration causes immediate problems) under the following conditions: no alternative pesticides exist, the pest is an essential threat, and growing is performed according to good agricultural practice (GAP). If a contract has been established between a grower and for example, a seed company that buys the crop from the area, the grower can according to the contract be allowed to get the necessary pesticides. When an advisor has inspected the crop, established that the pest is present and a pesticide is needed, the advisor can make a prescription that allows the grower to buy and use a defined amount of a pesticide.

Approved off-label use makes it possible to use a pesticide in a crop that is not on the label and this is a more permanent solution than exemption. For off-label use the product must be registered in the country and be commercially available, and it must be possible to extrapolate residue data. The evidence submitted with the application (safety precautions and statutory conditions) must satisfy the registration authority to the extent that there is a reasonably high probability that the claims of the pesticide's effectiveness and crop safety (where applicable) are justified. Approved off-label use is entirely at the user's own commercial risk and the grower must be in possession of special instructions for use.

The present activities have solved many problems of minor use but lack of finances is still a major problem in getting the necessary information. Ongoing work producing a central, EU-wide compatible database with free on-line research for every member of the database will be of great help in the future. This database for minor crops and off-label uses requires an EU-wide coordination of trials and the supply of trial data for efficacy, phytotoxicity, and residue behavior. For the time being, the database is being compiled in Germany. Once it is finished, consideration should be given to moving it to Brussels to DG VI of the EU Commission.

REFERENCES

1. Anonymous. Concerning the placing of plant protection products on the market. Official Journal of the European Communities, Luxembourg **1991**, *L230*, 32. Council Directive 91/414/EEC.
2. Anonymous. *Guideline on Comparability, Extrapolation, Group Tolerances and Data Requirements for Setting MRL's*; Commission of the European Communities: Luxembourg, 1999; 30. Doc.7525/VI/95–rev. 5.

FURTHER READING

Pallut, W.; Schmidt, H.-H. *Mitteilungen aus der Biologischen Bundesanstalt für Land- und Forstwirtschaft*; Second International Symposium on Minor Uses; Parey Buchverlag: Berlin, 1996; 145. Heft 324.

Schmidt, H.-H.; Holzman, A. *Mitteilungen aus der Biologischen Bundesanstalt für Land- und Forstwirtschaft*; International Symposium on Minor Uses; Parey Buchverlag: Berlin, 1993; 196. Heft 291.

MITE-BORNE DISEASES (SUCH AS SCRUB TYPHUS)

William John Hannan McBride
Cairns Base Hospital, Queensland, Australia

INTRODUCTION

Mites are a common cause of skin irritation. Some species of mite can transmit disease to humans. Scrub typhus and Rickettsialpox are two diseases spread by this means. Scrub typhus occurs in areas of transitional vegetation throughout Asia. Infection is acquired through the bite of the larval stage of trombiculid mites. An eschar develops at the site of the bite. Headache, fevers, and generalized aching are typical symptoms. The illness resolves quickly with appropriate antibiotics. The ecology of the mite, its mammalian hosts, and the vegetation in which it exists has resulted in areas in which intense disease transmission can occur. Knowledge of these areas and the use of methods to prevent disease could result in a reduction of the morbidity caused by this disease. Rickettsialpox is primarily a disease of inner city slums and is transmitted by a mite whose normal host is the common house mouse.

SCRUB TYPHUS

Ecology and Epidemiology

Orientia tsutsugamushi is a small gram-negative intracellular bacillus. It is about 0.8–2.0 μm long and 0.3–0.5 μm in diameter. The bacteria are not thought to elaborate toxins. Pathogenicity is related to the invasion of vascular endothelial cells or the consequent cellular response to infection. There are three major serotypes; Karp, Gilliam, and Kato. Other strains have been identified using traditional and molecular techniques.

The bacterium infects trombiculid mites and establishes a lifelong symbiosis with its host. The infection is passed efficiently from one generation of mite to the next via the ova. The larval stage (chigger) of the trombiculid mite is responsible for the transmission of disease to other animals. The chigger normally feeds only once during its lifetime and so normally there is no opportunity for the mite to pass infection from one host to another. After hatching, the larvae tend to congregate on debris several centimeters above the ground and attach to any passing animal. After attachment the chigger feeds on serous exudate from the host. Feeding may take anywhere from 2 to 12 days and is species dependent. The chigger detaches and progresses through pupal and nymph stages to reach adulthood up to 4 weeks later. The males deposit spermatophores that are collected by the female and egg laying commences. Four hundred eggs may be deposited over a five-month period and the total lifespan of the mite can be up to 15 months.

A number of species of trombiculid mite have been implicated in the transmission of scrub typhus. *Leptotrombium deliense, L. fletcheri, L. akamushi, L. arenicola, L. pallidum*, and *L. scutellare* are the most important vectors for the disease but other species have been implicated also. Animal hosts include many rodent species and other small mammals. The close association of animals with the ground and a few centimetres above the ground is the factor that most determines whether an animal is a host. Thus, most records of natural infection occur in *Rattus* and other murines. Voles and tree shrews are also frequent hosts (1). Bird species can be infected experimentally but the organism has not been isolated from this source in field conditions. Humans are accidental hosts. In endemic areas the percentage of infected chiggers is up to 3% (2). Most rodents in the same areas have evidence of infection, either past or current (3). Transmission of infection from infected animals to uninfected chiggers has been demonstrated (4).

Scrub typhus occurs throughout eastern Asia and the western pacific region (Fig. 1). The disease is well described in Japan, Korea, China, India, the Southeast Asian countries, northern Australia, Vanuatu, and the Solomon Islands. Infection is unknown in the New World, Europe, and northern Russia. There is no definite proof of its existence in Africa. The concurrent presence of murines and *Leptotrombidium* may be indicative of Scrub typhus endemicity.

The term scrub typhus infers that transmission occurs in areas of bushy vegetation. These areas can include transitional vegetation, such as on the edge of rainforests or in recently cleared areas or along watercourses. Areas

Fig. 1 The worldwide distribution of scrub typhus.

including tropical rainforests, alpine meadows, seashore areas, and semidesert regions, however, have all been implicated in transmission (1). Within endemic countries, transmission tends to occur in certain locations. Even within these locations there are smaller areas, as small as several hundred square meters, where transmission is intense. These areas tend to remain stable over time and may represent ecological niches that favor the persistence of infection in the mite population. The encounter between a large group of humans and one of these locations can result in a significant cluster of scrub typhus cases.

The bite of the chiggers responsible for the transmission of scrub typhus often is not associated with itching. Other genera such as *Eutrombicula*, *Neotrombicula*, or *Schoengastia*, which normally infest birds and reptiles, are classically associated with intense itch in humans.

Clinical Features

The incubation period is typically 10–14 days after the bite of an infected chigger but can range from 7 to 21 days. The onset of the illness may be sudden or develop over several days with the sufferer feeling vaguely unwell and having symptoms of fever at night (5). The most common patient complaints are fever, myalgias, and headache.

An eschar (Fig. 2) has been observed in the order of 60–75% of patients with scrub typhus and can be located anywhere on the body. The eschar is generally painless and the patient may be unaware of its presence. Lymph node enlargement occurs particularly in association with the eschar but can occur in lymph node groups elsewhere in the body.

A rash is commonly observed. It is macular and begins, during the first week of the illness, on the trunk and spreads to the arms and legs. Evidence of respiratory tract involvement, manifest as a dry cough, is found in about 25% of cases. Radiological abnormalities have been, however, observed in up to 72% of cases (6). Severe lung involvement manifesting as adult respiratory distress syndrome (ARDS) is occasionally seen. Cardiac involvement is uncommon although myocarditis can be severe. A relative bradycardia is occasionally observed. Neurological involvement is common but rarely severe. Cerebrospinal pleocytosis is seen in about half the sufferers and there

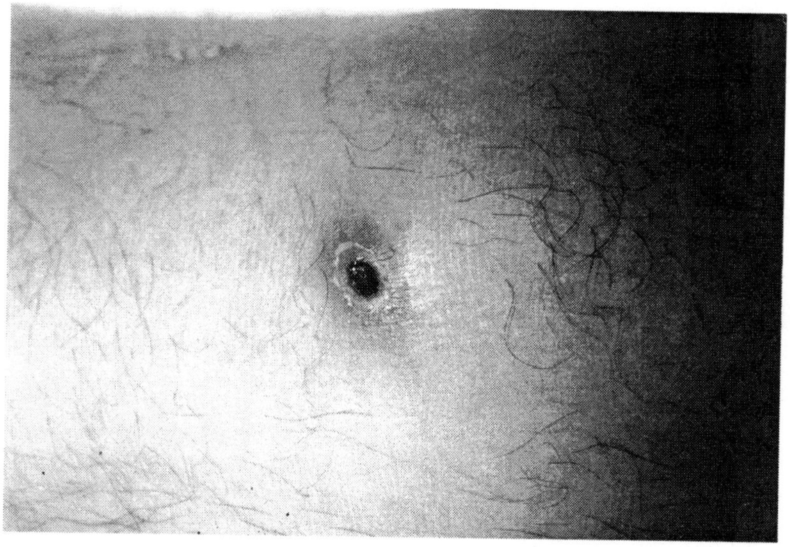

Fig. 2 An eschar found behind the knee in a person with scrub typhus.

is PCR evidence of organism in at least some of these patients (7).

Untreated the illness usually lasts about 2 weeks. In the preantibiotic era mortality rates were in the order of 10–30%.

Diagnostic Considerations

The current standard test for the diagnosis of rickettsial diseases is the immunofluoresence test. The test can be used to detect IgM or IgG subclasses of antibody and is suitable for quantifying antibody. IgM antibody may appear at the end of the first week of illness. A rise in the level of IgG often appears before the end of the second week but may be delayed up to four weeks. A definite diagnosis is best achieved by demonstrating a fourfold rise in antibody titre. More convenient testing methods have been recently been developed. One test, a single-use dot blot assay is sensitive and specific (8). An ELISA test is now available for scrub typhus and may prove useful as a screening test. All these tests have superseded the old Weil-Felix test.

In the early stages of illness, before antibodies develop, the organism may be cultured from the blood. Traditionally, laboratory animals, especially mice, have been used for this purpose.Cell culture is now more widely used. PCR detection based on the amplification of a gene encoding a 56-kDa antigen has been reported (9).

Treatment and Prevention

Tetracycline (doxycycline) and chloramphenicol are both effective in the treatment of scrub typhus. A rapid response (24–48 h) to treatment is a characteristic finding. The time to fever defervesence is shorter with doxycycline than for chloramphenicol (10). Strains of *O. tsutsugamushi* resistant to both antibiotics have been reported from northern Thailand and are associated with poor response to treatment (11). Given the ecology of scrub typhus, antibiotic resistance is not likely to become a widespread problem. The duration of treatment can be as short as three days, which is as effective as 7 days of therapy (12).

The use of Tetracycline should be avoided in children less than 8 years old and in pregnancy.Chloramphenicol is difficult to obtain in some countries because of occurrence of rare but serious side effects. Alternative drugs such as Ciprofloxacin and Azithromycin are active in vitro (13, 14). The use of Azithromycin has been reported in pregnancy (15) and Ciprofloxacin has been used in children (16).

Prevention of scrub typhus involves avoidance of exposure to infected chiggers. This requires detailed knowledge of infested locations and avoidance of these areas, or use of repellants such as a pyrethroid (17).Chemoprophylaxis, using doxycycline, 200 mg weekly, is effective (18); its use should be considered if exposure to infected chiggers is likely to be intense.

OTHER MITE-BORNE DISEASES

Rickettsialpox is a disease caused by *Rickettsia akari*. The organism is related to the spotted fever group rickettsiae, which are normally transmitted via ticks. The mite *Liponyssoides sanguineus*, an ectoparasite of the domestic mouse *Mus musculus* transmits the bacterium (19). It is predominantly a disease seen in urban areas and is associated with overcrowding or poor housing. It has been reported from major cities in the United States, South Africa, Korea, and parts of Russia. After an incubation period of 1–3 weeks, an eschar develops, associated with regional lymphadenopathy. Several days later fever and headache, photophobia, and myalgias develop. A generalized papulovesicular rash develops another 2–3 days later. The disease may be confused with chickenpox. Overall the disease is mild and without treatment, resolves in 2–3 weeks. Treatment with tetracycline brings about a resolution of symptoms. Diagnosis is best achieved using a direct fluorescent antibody test. There is a serological cross-reaction with *R. rickettsia* antigen. Early treatment may delay the development of detectable antibodies. (20).

REFERENCES

1. Traub, R.; Wisseman, C.L., Jr. The ecology of chigger-borne rickettsiosis (Scrub Typhus). J. Med. Entomol. **1974**, *11*, 237–303.
2. Tanskul, P.; Strickman, D.; Eamsila, C.; Kelly, D.J. Rickettsia tsutsugamushi in chiggers (Acari: Trombiculidae) associated with rodents in central Thailand. J. Med. Entomol. **1994**, *31*, 225–230.
3. Frances, S.P.; Watcharapichat, P.; Phulsuksombati, D.; Tanskul, P. Occurrence of orientia tsutsugamushi in chiggers (Acari: Trombiculidae) and small animals in an Orchard near Bangkok, Thailand. J. Med. Entomol. **1999**, *36*, 449–453.
4. Takahashi, M.; Murata, M.; Hori, E.; Tanaka, H.; Kawamura, A., Jr. Transmission of rickettsia tsutsugamushi from Apodemus speciosus, a wild rodent, to larval trombiculid mites during the feeding process. Jpn. J. Expt. Med. **1990**, *60*, 203–208.
5. Doherty, R.L. A Clinical study of scrub typhus in North Queensland. Med. J. Aust. **1956**, *2*, 212–220.
6. Choi, Y.H.; Kim, S.J.; Lee, J.Y.; Pai, H.J.; Lee, K.Y.; et al.

M

Scrub typhus: radiological and clinical findings. Clin. Radiol. **2000**, *55*, 140–144.
7. Pai, H.; Sohn, S.; Seong, Y.; Kee, S.; Chang, W.H.; et al. Central nervous system involvement in patients with scrub typhus. Clin. Infect. Dis. **1997**, *24*, 436–440.
8. Weddle, J.R.; Chan, T.C.; Thompson, K.; Paxton, H.; Kelly, D.J.; et al. Effectiveness of a dot-blot immunoassay of anti-rickettsia tsutsugamushi antibodies for serologic analysis of scrub typhus. Am. J. Trop. Med. Hyg. **1995**, *53*, 43–46.
9. La Scola, B.; Raoult, D. Laboratory diagnosis of rickettsioses: current approaches to diagnosis of old and new rickettsial diseases. J. Clin. Microbiol. **1997**, *35*, 2715–2727.
10. Sheehy, T.W.; Hazlett, D.; Turk, R.E. Scrub typhus. A comparison of chloramphenicol and tetracycline in its treatment. Arch. Intern. Med. **1973**, *132*, 77–80.
11. Watt, G.; Chouriyagune, C.; Ruangweerayud, R.; Watcharapichat, P.; Phulsuksombati, D.; et al. Scrub typhus infections poorly responsive to antibiotics in northern Thailand. Lancet **1996**, *348*, 86–89.
12. Song, J.H.; Lee, C.; Chang, W.H.; Choi, S.W.; Choi, J.E.; et al. Short-course doxycycline treatment versus conventional tetracycline therapy for scrub typhus: a multicenter randomized trial. Clin. Infect. Dis. **1995**, *21*, 506–510.
13. McClain, J.B.; Joshi, B.; Rice, R. Chloramphenicol, gentamicin, and ciprofloxacin against murine scrub typhus. Antimicrob. Agents Chemother. **1988**, *32*, 285–286.
14. Strickman, D.; Sheer, T.; Salata, K.; Hershey, J.; Dasch, G.; et al. In vitro effectiveness of azithromycin against doxycycline-resistant and-susceptible strains of rickettsia tsutsugamushi, etiologic agent of scrub typhus. Antimicrob. Agents Chemother. **1995**, *39*, 2406–2410.
15. Watt, G.; Kantipong, P.; Jongsakul, K.; Watcharapichat, P.; Phulsuksombati, D. Azithromycin activities against orientia tsutsugamushi strains isolated in cases of scrub typhus in northern Thailand. Antimicrob. Agents Chemother. **1999**, *43*, 2817–2818.
16. McBride, W.J.; Taylor, C.T.; Pryor, J.A.; Simpson, J.D. Scrub typhus in north Queensland. Med. J. Aust. **1999**, *170*, 318–320.
17. Ho, T.M.; Ismail, S. Efficacy of three pyrethroids against leptotrombidium fletcheri (Acari: Trombiculidae) infected and noninfected with scrub typhus. J. Med. Entomol. **1991**, *28*, 776–779.
18. Twartz, J.C.; Shirai, A.; Selvaraju, G.; Saunders, J.P.; Huxsoll, D.L.; et al. Doxycycline propylaxis for human scrub typhus. J. Infect. Dis. **1982**, *146*, 811–818.
19. McDade, J. Rickettsial Diseases. In *Topley and Wilson's Microbiology and Microbial Infections*; Oxford University Press: London, 1998; 995–1011.
20. Saah, A.J. Rickettsia Akari (Rickettsialpox). In *Mandell, Douglas, and Bennett's Principles and Practice of Infectious Diseases*; Churchhill Livingstone: Philadelphia, 2000; 2042–2043.

MODELING PEST MANAGEMENT

Andrew Paul Gutierrez
University of California, Berkeley, California, U.S.A.

INTRODUCTION

Pest management is a key component of sustainable agriculture, and may be defined as applied population ecology wherein humans manage populations of plants and/or domesticated animals and their pests and natural enemies in environments modified by weather and agrotechnical inputs. The complexity of managing pests in an agroecosytem requires the development of models that enable specialist to separate losses due to pests from yield variations due to weather and agronomic practice (1). Such models provide the bases for evaluating the dual objectives in modern agriculture of *minimizing* inputs that cause adverse environmental, human, and animal health effects and *maximizing* profits. In modern societies, these goals may result in conflicts between public and private interests, but this is not the case in subsistence ones (2) where the goal is often yield stability.

Some components of crop pest management research are illustrated in Fig. 1, but the same approach applies to medical and veterinary pest problems. The methods fall under the ambit of agroecosystems analysis (3). The analysis must be tritrophic in scope because natural processes such as biological control when correctly managed replace disruptive pesticide inputs that may induce resurgence of target pests, outbreaks of secondary pests, and pesticide resistance. Other agronomic inputs (e.g., fertilizers) may also exacerbate pest levels and must be considered in the pest management system. Several modeling approaches have been used in pest management, and they may be broadly classed as empirical, statistical, operations research, analytical, and simulation (1–6).

MODELING APPROACHES

Empirical models based on trial and error have a long history and fall under the rubric of the common wisdom. Traditional societies worldwide have developed pest management strategies that reduce pest damage and lead to sustainable crop production. These systems may fail when modern agro-technical inputs are introduced, and may require the accumulation of new experience to resolve the problem.

Population sampling and agronomic trials have been used to develop statistical models to assess the costs and benefits of levels and timing of pest abundance and agronomic inputs (i.e., economic threshold studies). These models tend to be static and hence are time and place specific. They may not be able to evaluate factors beyond the range of the data used to develop the model. Such models have yielded useful pest management decision rules.

Attempts have been made to use operations research methods from engineering and economics to determine optimal pest management strategies. Such methods require sufficient knowledge of the system to formulate the optimization model. Often this information is lacking, and/or the biology may be too complicated, requiring reduction in the dimensions of the model. This simplification may obviate their general utility except for strategic purposes.

Pest management is applied population ecology and hence lends itself to demographic modeling methods. Mathematical models of the population dynamics may be simple descriptors of the system having analytical solutions or they may include considerable biological detail requiring numerical simulation. Analytical models may give strategic insights about the system while richer models may yield tactical recommendations for specific abiotic and biotic conditions. Modeling for tactical decision making requires a detailed understanding of the biology of the interacting species as modified by weather and edaphic and agronomic factors. Biologically rich mathematical models are usually implemented as computer simulation models.

There are many approaches to simulation modeling, but increasingly, physiologically based models of the energy dynamics of cropping systems are being developed

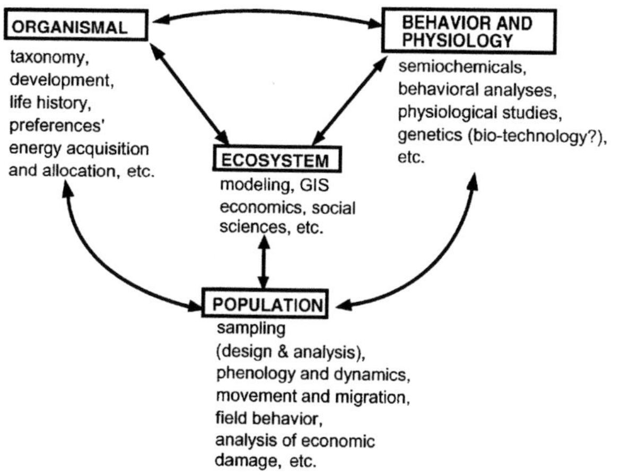

Fig. 1 Components of pest management research using systems analysis as a unifying tool. From a report to the International Centre for Insect Physiology and Ecology (ICIPE), Nairobi, Kenya. (A. P. Gutierrez and J. Baumgartner, personal communication.)

because the importance of the plant as the integrator of all factors is being increasingly recognized. Energy is the currency of biology that when multiplied by price leads directly to economic models (see later). Modeling the energy dynamics within a crop system facilitates the assessment of the relative losses due to pests, weather, and agronomic and edaphic factors.

SIMULATION MODELS OF CROP SYSTEMS

Crops are age-structured populations of plants each with time varying mass–age-structured populations of subunits (i.e., of fruits, leaves, stems, and roots). The number in each age–mass class may vary overtime in response to factors that affect birth–death rates of whole plants and/or their subunits as well as those of higher trophic level populations. The crop may be viewed as a canopy of average plants with populations of pests and natural enemies within (canopy models), or as individual plants each with its own set of pest and natural enemy populations with migration between plants (i.e., a metapopulation). Modeling of individuals in pest and natural enemy populations (individual based models) is not recommended in pest management.

A basic premise of the physiological approach is that all organisms, including the economic one, face the same problems of resource (energy) acquisition and allocation (1, 7). Physiological models assume energy allocation priority first to respiration (maintenance costs in economics), then reproduction (profit) and if assimilate (revenues) remains to growth (infrastructure costs). The shapes of the acquisition functions are convex and those for the maintenance costs are positive exponential (the Q10 rule) with the net being the amount of resources available for allocation. These analogies (3) allow the use of the same model to describe the dynamics of all species including

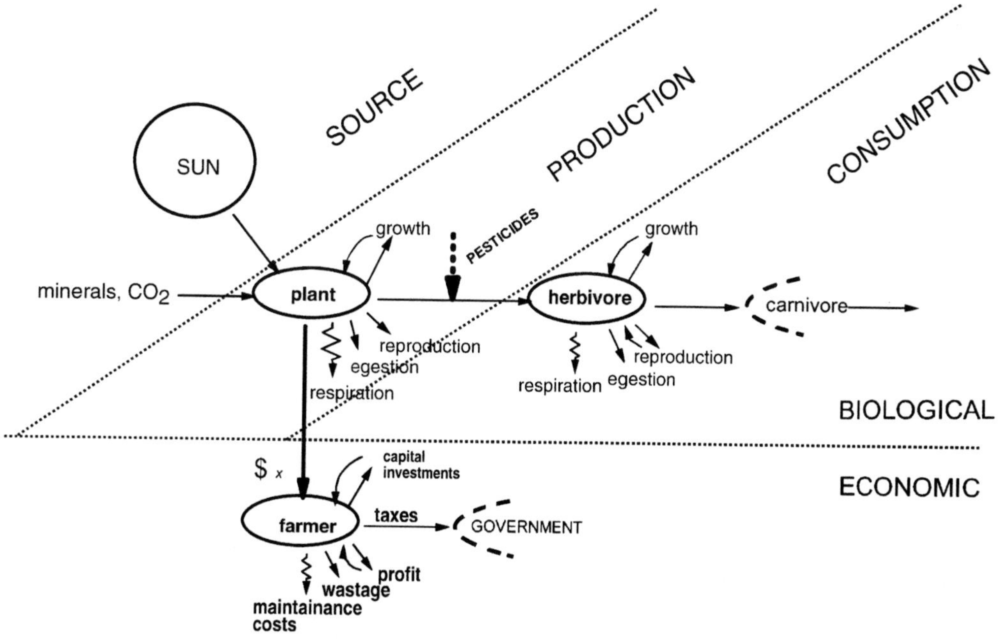

Fig. 2 Analogies between trophic levels including the economic one. (From John Wiley & Sons, New York.)

the economic one (Fig. 2). Each organism is assumed to try to satisfy a genetic (economic) demand for resources, but the process involves imperfect search, and the supply of a resource obtained is always less than or equal to the demand, causing reductions in growth, reproduction, and survival rates from the maximum. In the model, all vital rates are controlled by the supply/demand ratio with only the units and interpretation of the flow rates differing among species.

How biotic and abiotic factors affect plant growth and development is central to developing such models. These factors may affect either the supply (production) or the demand (sinks, e.g., fruits) side of the supply/demand ratio. Occasionally, both sides may be affected and in other cases the pests may attack the standing crop. The supply–demand paradigm simplifies model development allowing assessment of yield loss and in some cases yield compensation in the face of pest damage. Such models facilitate evaluation of the costs and benefits of pest management options basic to the development of dynamic economic thresholds.

SUPPLY–SIDE PESTS

Supply-side pests reduce the photosynthetic rate in various ways. Important supply-side pests are defoliators, sapsuckers, spidermites, nematodes, diseases, and others. Defoliation attacks leaves and may cause wound healing losses, but the effects on yield depend on the age of leaves attacked, the loss rate, and compensation due to increased light penetration. In contrast, spidermites kill leaf cells, reducing photosynthesis in damaged leaves that are not shed, reducing light penetration to lower leaves. Stem borers and vascular plant diseases may slow the photosynthetic rate by reducing the translocation of water and nutrients, and some may kill whole plants. Thrips and armyworms may damage the terminal, inducing developmental delays and reducing yield.

PLANT DEMAND-SIDE PESTS

Pests may attack fruit (e.g., sinks) causing premature abscission altering present and future demands for photosynthate. High abscission rates may cause rank growth as the excess photosynthate is allocated to vegetative growth. Most plant species have a reproductive surplus that allows for varying degrees of compensation. Of crucial importance in compensation is the time and energy lost in abscised fruits. Little time and energy is lost in abscised buds and small fruit, and replacement buds may be produced in many species at rates sufficient for compensation. Attacks on older fruits may involve considerable losses in time and energy often precluding compensation.

The ratio of the cumulative buds initiated to the cumulative numbers abscised may provide the basis for determining whether compensation is possible. Such data yield a convex function that estimates the *compensation point* and may provides a rule of thumb for estimating the economic threshold. For example, loss of 30% of fruit bud in many cotton varieties does not affect yield.

OTHER KINDS OF PLANT PESTS

Some pests affect both sides of the ratio. For example, pests attacking the apical meristem of a plant or branch kill primordial tissues that introduce time delays and alter future fruit and vegetative dynamics. Other pests may attack the standing crop of fruits without appreciably reducing sink demands, making plant compensation unlikely. In such cases, the damage accrues over time.

CONCLUSIONS

Interactions among pests and higher trophic levels (top–down effects) are complicated, time varying, and are always integrated via the plant's physiology and dynamics as modified by abiotic factors (bottom–up effects). Tritrophic models of crop systems enable assembling and understanding these complicated biological relationships (8–10). These models provide the bases for developing ecologically sound pest management strategies that may be implemented as part of total farm expert systems or regionally using geographic information systems technology. These methods fall under the ambit of agroecosystems analysis.

REFERENCES

1. Gutierrez, A.P. *Applied Population Ecology: A Supply-demand Approach*; John Wiley & Sons, Inc.: New York, NY, 1996.
2. Altieri, M.A. Sustainable agriculture. Encyclopedia Agric. Sci. **1994**, *4*, 239–247.
3. Gutierrez, A.P.; Curry, G.L. Conceptual Framework for Studying Crop-Pest Systems. In *Integrated Pest Management Systems and Cotton Production*; Frisbie, R.E., El-Zik,

K.M., Wilson, L.T., Eds.; John Wiley & Sons: New York, 1989; 37–64.
4. Curry, G.L.; Feldman, R.M. *Mathematical Foundations of Population Dynamics*; Series No. 3, TEES Monograph, 1987.
5. deWit, C.T.; Goudriaan, J. *Simulation of Ecological Processes*, 2nd Ed.; PUDOC Publishers: The Netherlands, 1978.
6. DiCola, G.; Gilioli, G.; Baumgärtner, J. Mathematical Models for Age-Structured Population Dynamics. In *Ecological Entomology,* 2nd Ed; Hufaker, C.B., Gutierrez, A. P., Eds.; Wiley: New York, 1998.
7. Regev, U.; Gutierrez, A.P.; Schreiber, A.P.; Zilberman, D. Biological and economic foundations of renewable resource exploitation. Ecolog. Eco. **1998**, *26*, 227–242.
8. Gutierrez, A.P.; Neuenschwander, P.; van Alphen, J.J.M. Factors affecting the establishment of natural enemies: biological control of the cassava mealybug in West Africa by introduced parasitoids: a ratio dependent supply-demand driven model. J. Appl. Ecol. **1993**, *30*, 706–721.
9. Graf, B.; Gutierrez, A.P.; Rakotobe, O.; Zahner, P.; Delucchi, V. A simulation model for the dynamics of rice growth and development. II. Competition with weeds for nitrogen and light. Agric. Sys. **1990**, *32*, 367–392.
10. Wermelinger, B.; Oertli, J.J.; Baumgärtner, J. Environmental factors affecting the life table statistics of *Tetranychus urticae* (Acari: Tetranychidae). III. Host plant nutrition. Exp. & Appl. Acar. **1991**, *12*, 259–274.

MODIFIED ATMOSPHERIC STORAGE

Adel A. Kader
University of California, Davis, California, U.S.A.

INTRODUCTION

Modified atmospheres (MA) or controlled atmospheres (CA) involve the removal or addition of gases resulting in an atmospheric composition around a commodity that is different from that of air (78.08% N_2 20.95% O_2, 0.03% CO_2). Usually this entails reduction of oxygen (O_2) and/or elevation of carbon dioxide (CO_2) concentrations. MA and CA differ only in the degree of control; CA is more exact. The use of MA or CA should be considered a supplement to proper temperature and relative humidity management to maintain the quality and safety of fresh fruits and vegetables and their products throughout the postharvest handling system.

OBJECTIVES

The primary objectives of CA storage of fresh produce are extension of postharvest life by maintenance of appearance; textural, flavor and nutritional quality; and by control of postharvest pathogens and insects. Beneficial effects include the following:

1. Retardation of senescence (ripening) and associated biochemical and physiological changes, i.e., slowing down respiration and ethylene production rates, softening, and compositional changes.
2. Reduction of fruit sensitivity to ethylene action at O_2 levels below 8% and/or CO_2 levels above 1%.
3. Alleviation of certain physiological disorders such as chilling injury of avocado and some storage disorders of apples.
4. CA can have a direct or indirect effect on postharvest pathogens and consequently decay incidence and severity. For example, elevated CO_2 levels (10–15%) significantly inhibit development of botrytis rot on strawberries, cherries, and other horticultural perishables.
5. Low O_2 (<1%) and/or elevated CO_2 (40–60%) can be a useful tool for insect control in some fresh and dried fruits and vegetables, nuts, and grains.

The extent to which CA is beneficial depends on the commodity; cultivar; maturity stage; initial quality; concentrations of O_2, CO_2 and ethylene; temperature; and duration of exposure to these conditions.

In most cases, the difference between beneficial and harmful CA combinations is relatively small. Also, CA combinations that are necessary to control decay or insects, for example, cannot always be tolerated by the commodity and may result in faster deterioration. Potential hazards of CA to the fresh produce include the following:

1. Initiation and/or aggravation of certain physiological disorders such as brown heart in apples and pears.
2. Irregular ripening of fruits such as banana, mango, pear, and tomato can result from exposure to O_2 levels below 2% and/or CO_2 levels above 5% for longer than one month.
3. Development of off-flavors and off-odors at very low O_2 concentrations (as a result of anaerobic respiration) and very high CO_2 levels (as a result of fermentative metabolism).
4. Increased susceptibility to decay pathogens when the fruit is physiologically injured by too-low O_2 or too-high CO_2 concentrations.

TECHNOLOGY IMPROVEMENTS

Several improvements in CA technology have been made in recent years, these include creating nitrogen by separation from compressed air using molecular sieve beds or membrane systems, low O_2 (1.0–1.5%) storage, low ethylene CA storage, rapid CA (rapid establishment of the optimum levels of O_2 and CO_2, and programmed (or sequential) CA storage (e.g., storage in 1% O_2 for two to six weeks followed by storage in 2–3% O_2 for the remainder of the storage period). Other developments, which may expand use of MA during transport and distribution, include using edible coatings or polymeric

M

Table 1 CA storage potential of fruits and vegetables

Duration[a] (months)	Commodities
More than 12	Almond, Brazil nut, cashew, filbert, macadamia, pecan, pistachio, walnut, dried fruits and vegetables
6–12	Some cultivars of apples and European pears
3–6	Cabbage, Chinese cabbage, kiwifruit, persimmon, pomegranate, some cultivars of Asian pears
1–3	Avocado, banana, cherry, grape (no SO_2), mango, olive, onion (sweet cultivars), some cultivars of nectarine, peach and plum, tomato (mature-green)
< 1	Asparagus, broccoli, cane berries, fig, lettuce, muskmelons, papaya, pineapple, strawberry, sweet corn

[a]Storage duration in air is about half of that in CA.

films with appropriate gas permeabilities to create a desired MA around and within the commodity. MA packaging is widely used in marketing fresh-cut fruits and vegetables.

Based on responses of fruits and vegetables to CA at optimum temperatures and relative humidities, we can divide them into six groups according to their storage potential as given in Table 1.

Successful application of atmospheric modification depends upon the commodity, cultivar, maturity stage at harvest, and a positive return on investment (benefit/cost ratio). Many advances in the technology of establishing, monitoring, and maintaining controlled atmospheres have facilitated their expanded use on several commodities, such as apple, avocado, banana, kiwifruit, and mango and vegetables such as asparagus, broccoli, and lettuce during transport. Advances in controlled atmosphere storage technology include faster establishment of desired atmospheres, less fluctuation in O_2 and CO_2 levels, ability to change atmospheric composition as needed during storage, and ability to scrub ethylene from the storage environment. Commercial use of controlled atmosphere storage is greatest on apples and pears worldwide; less on kiwifruits, avocados, persimmons, and pomegranates, nuts, and dried fruits and vegetables. Atmospheric modification during long-distance transport is used on apples, asparagus, avocados, bananas, broccoli, cane berries, cherries, figs, kiwifruits, mangos, melons, nectarines, peaches, pears, plums, and strawberries. Continued technological developments in the future to provide controlled atmospheres during transport and storage at a reasonable cost are essential to greater applications on fresh fruits and vegetables.

BIBLIOGRAPHY

Food Preservation by Modified Atmospheres; Calderon, M., Barkai-Golan, R., Eds.; CRC Press: Boca Raton, FL, 1990; 1–402.

El-Goorani, M.A.; Sommer, N.F. Effects of modified atmospheres on postharvest pathogens of fruits and vegetables. Hort. Rev. **1981**, *3*, 412–461.

Kader, A.A. Biochemical and physiological basis for effects of controlled and modified atmospheres on fruits and vegetables. Food Technol. **1986**, *40* (5), 99–100, 102–104.

Kader, A.A.; Zagory, D.; Kerbel, E.L. Modified atmosphere packaging of fruits and vegetables. CRC Crit. Rev. Food Sci. Nutr. **1989**, *28*, 1–30.

Mitcham, E.J.; Zhou, S.; Kader, A.A. Potential for CA for Postharvest Insect Control in Fresh Horticultural Perishables: An Update of Summary Tables Compiled by Ke and Kader. In *CA 97 Proceedings Volume 1*; Thompson, J.F., Mitcham, E.J., Eds.; Postharvest Horticulture Series No. 15, Department of Pomology, University of California: Davis, 1997; 78–90.

Thompson, A.K. *Controlled Atmosphere Storage of Fruits and Vegetables*; CAB International: Wallingford, UK, 1998; 1–278.

MOLLUSCICIDES

Bernhard Speiser
Research Institute of Organic Agriculture (FiBL), Frick, Switzerland

INTRODUCTION

Molluscicides are used to control pests from the class of gastropods, which belong to the phylum of the molluscs (along with mussels, squids, and other groups). Practitioners distinguish between "slugs" (gastropods without external shells) and "snails" (gastropods carrying an external shell), but slugs are not a taxonomic group. On a worldwide scale, gastropods are pests of lesser economic importance than insects, mites, or nematodes. In certain areas, however, they may reduce yields substantially, severely affect the quality of the harvested products, or transmit human or animal diseases. Although relatively few gastropod species are major pests, they attack a wide range of different crops.

PEST SPECIES AND THEIR DAMAGE

Out of more than 100,000 species of gastropods, only a few dozen are serious regional pests, and only a handful are important pests on a worldwide scale. There are three ecological groups of gastropod pests: terrestrial slugs, terrestrial snails, and freshwater snails. Terrestrial snails can be pestiferous in relatively dry habitats because their shell protects them from desiccation, while slugs become pests in regions with regular rainfalls. Gastropods affect a wide variety of agricultural and horticultural crops such as arable, pastural, and fiber crops; vegetables; bush and tree fruits; herbs and ornamentals; and slugs cause significant damage in gardens of homeowners in industrialized countries. In most crops, the extent of gastropod damage is subject to high regional and year-to-year variation or affects only a small proportion of a field. In addition, gastropods can interfere with aquaculture and can be intermediate hosts of human, animal, or plant diseases.

Terrestrial Pest Slugs

In areas with a temperate climate and plenty of rainfall, slugs can be important pests in arable and horticultural crops and private gardens. During the past decades, slug damage in agriculture has drastically increased. The main reasons for this are reduced tillage, direct seeding, and the spread of fallows as a consequence of set-aside programs (1). Seeds or seedlings of winter wheat, winter barley, oilseed rape, maize, sugar beet, sunflower, soybean, and other field crops can be destroyed completely by slugs. Potatoes, vegetables, strawberries, and ornamentals are often devalued by holes or feeding marks, or by the presence of slugs, their feces, or their slime trails (1–4). Slug activity highly depends on temperature and humidity and may vary greatly between regions, as well as between years.

The gray field slug, *Deroceras reticulatum* (Agriolimacidae), is the most important slug species worldwide. It occurs mainly in arable fields, and can be found in Europe, North and South America, Asia, and Oceania. Other *Deroceras* spp., *Milax* spp. (Milacidae), *Tandonia* spp. (Milacidae), *Limax* spp. (Limacidae), and *Arion* spp. (Arionidae) are important pests in certain areas, but are less widely distributed than *D. reticulatum*. Veronicellid slugs (Veronicellidae) cause severe crop damage and act as vectors of human disease in Central America.

Terrestrial Pest Snails

Californian citrus fruit are regularly damaged by the brown garden snail *Cryptomphalus aspersus* (formerly *Helix aspersa*; Helicidae). Fruit with a nibbled skin suffers mold growth during overseas transport, thus spoiling the entire shipment. In South Australia, a number of snails originating from the Mediterranean area (*Cernuella virgata, Theba pisana*, and *Cochlicella* spp., Helicidae) cause problems in cereals and pastures. In cereal fields, the snails aestivate on the heads, pods, and stalks of cereals and clog the combines. Snail remains can lead to down-grading or rejection of the grain. The quality of pastures deteriorates if snails are present in large numbers. In the Pacific region and particularly in Hawaii, the giant African snail (*Achatina fulica*; Achatinidae) attacks vegetables and ornamentals. This very large (up to 30 cm long) and voracious snail originates from Africa and was introduced either by accident or as a source of food.

M

Aquatic Pest Snails

In Southeast Asia, rice fields are increasingly threatened by the golden apple snail (*Pomacea canaliculata*, Ampullariidae), which was introduced from South America. Several aquatic snails, namely *Bulinus* spp. (Planorbidae), *Biomphalaria* spp. (Planorbidae), and *Oncomelania* spp. (Pomatiopsidae), are intermediate hosts of human schistosomiasis in tropical regions of America, Asia, and Africa. Eliminating these vector snails with molluscicides is one part of a series of measures to reduce the prevalence of schistosomiasis. *Lymnaea* spp. (Lymnaeidae) are vector snails for fasciolosis, a related disease in livestock.

CONTROL OF GASTROPODS

Molluscicide Use

The worldwide molluscicide market is estimated to be $550 million per year. Molluscicide sales reach the highest percentage in Great Britain, followed by France (4% and 2.5% of the total agricultural pesticide market, respectively). In most other countries, molluscicide sales account for less than 1% of the total pesticide market. In Great Britain, the area treated with molluscicides increased 67-fold from 1970 to mid '90s (5). Arable crops account for 90% of molluscicide usage, fodder crops for 7%, vegetables for 2% and soft fruits for 1%. The main crops treated are winter wheat, winter barley, oilseed rape, potatoes, newly sown leys, Brussels sprouts, turnips, swedes, and strawberries. Pellets containing me-taldehyde treat 55% of the surface. Pellets containing methiocarb treat 40% of the surface. For the year 1995, the cost of molluscicides was estimated at a total of $10 million, or 12 $/ha (excluding costs for application). In private gardens, slug pellets contribute significantly to the total pesticide use.

Terrestrial slugs and snails normally are controlled by baiting. Pellets mainly are broadcast. Subsequent pellet applications may be necessary to obtain sufficient crop protection depending on climatic conditions and on the duration of the susceptible life stage of a crop. Ingestion of pellets by slugs results in feeding inhibition and/or death of the slug. Slug pellets consist of an edible matrix (e.g., bran, flour), an active ingredient and additives. Either metaldehyde or insecticides (mainly methiocarb, but also thiodicarb and bensultap) are used as the active ingredient at concentrations of 2–7.5%. Common additives are slug and snail attractants; bittering agents or repellents to prevent ingestion by wildlife, pets, and children; dyes; and preservatives against molding. A vast variety of formulations that differ in composition or manufacturing process is available. This may affect the coverage in the field, the attractivity and palatability to slugs or snails, and the rainfastness. Formulations that provide a good coverage and sustainable pellet integrity give best results.

Aquatic snails have been controlled chemically since the 1950s, as part of programs for control of schistosomiasis. More recently, molluscicides have also been used to protect rice fields from the golden apple snail and to control snails in fish ponds. Niclosamide, metaldehyde, and organic tin compounds are frequently used to control aquatic snails. Trifenmorph, endosulfane, nicotinanilides, and some copper or amide compounds are also active against aquatic snails, but are less frequently used (6).

Until now, resistance of field populations to molluscicides has been observed neither in terrestrial nor in aquatic gastropods.

Side Effects

Metaldehyde has no negative side effects on earthworms, bees, or carabid beetles, while methiocarb and other insecticides may affect these organisms. The sporadic cases of poisoning of dogs or other pets are due to accessible storage or false application in piles, and should not occur if pellets are spread as recommended.

Niclosamide is harmless for humans and cattle, but may be hazardous for fish. If used according to instructions, metaldehyde is harmless for fish and daphnias. Some organotins affect the growth of oysters and other marine wildlife, and their use against snails has been banned in several countries.

Cultural Control

Slug populations are significantly reduced by plowing and other methods of soil cultivation. Slug damage to cereals may be reduced by preparing a fine seed-bed and by sowing at greater depth than usual (3). Adjustment of sowing or planting date may limit damage. Periodic draining protects rice fields against the golden apple snail. Other methods of cultural control of slugs and snails are highly crop-specific and are not listed here.

FUTURE DEVELOPMENTS

In recent years, attempts for soil conservation have promoted reduced tillage and direct seeding. Government sponsored set-aside programs and socioeconomic trends produce extended areas under fallow. These trends, which favor slug populations considerably, will be continued. In addition, higher quality demands may lower the tolerance for slug presence and slug damage on produce. As a result, control of pest slugs and snails will gain importance in the future. In arable crops, trends in molluscicide usage are 1) shift from curative to the regular preventive application in most severely threatened crops and regions; 2) reducing application rates; and 3) application according to climate-based forecasting models of slug activity.

During the past few years, manufacturers of slug pellets have improved formulations resulting in more even coverage, increased palatability, optimal concentration of active ingredients, and increased resistance to physical degradation. Existing molluscicidal compounds will continue to play a major role in the future, but in certain niche markets new molluscicides may be successful. A range of iron, aluminium, and copper compounds are known to be molluscicidal. A slug pellet containing iron(III)phosphate has recently been introduced in home and garden markets in Europe and North America. The nematode *Phasmarhabditis hermaphrodita* (Rhabditidae) is commercialized in several European countries as a biocontrol agent for slugs.

In Southeast Asia, the golden apple snail continues to enlarge its range thus threatening a growing surface of rice fields. For the control of aquatic snails, there is a trend to use environmentally more harmless molluscicides instead of organic tin compounds. Extracts of several tropical plants, particularly *Ambrosia maritima* (Compositae), *Anacardium occidentale* (Anacardiaceae), and *Phytolacca dodecandra* (Phytolaccaceae) might be used in the future as locally produced molluscicides. However, this method is not yet developed sufficiently to allow practical application.

REFERENCES

1. Hammond, R.B. Conservation Tillage and Slugs in the U.S. Corn Belt. In *Slug and Snail Pests in Agriculture*; Henderson, I.F., Ed.; BCPC Symposium Proceedings No. 66; British Crop Protection Council: Farnham, England, 1996; 31–38.
2. Godan, D. Pest Slugs and Snails, Biology and Control. Gruber, S. Trans., Ed.; Springer: Berlin, 1983; 445.
3. Port, C.M.; Port, G.R. The biology and behaviour of slugs in relation to crop damage and control. Agric. Zool.Rev **1986**, *1*, 255–299.
4. South, A. *Terrestrial Slugs. Biology, Ecology and Control*; Chapman & Hall: London, 1992; 428.
5. Garthwaite, D.G.; Thomas, M.R. The Usage of Molluscicides in Agriculture and Horticulture in Great Britain over the Last 30 Years. In *Slug and Snail Pests in Agriculture*; Henderson, I.F., Ed.; BCPC Symposium Proceedings No. 66, British Crop Protection Council: Farnham, England, 1996; 39–46.
6. Webbe, G. The Toxicology of Molluscicides. In *International Encyclopedia of Pharmacology and Therapeutics*; Sec. 125, Pergamon Press: Oxford, England, 1987; 167.

FURTHER READING

Henderson, I.F. In *Slug & Snail Pests in Agriculture*, BCPC Symposium Proceedings No. 66, British Crop Protection Council: Farnham, England, 1996; 450.

Henderson, I.F. In *Slugs and Snails in World Agriculture*, BCPC Monograph No. 41, British Crop Protection Council: Thornton Heath, England, 1989; 422.

MONOCULTURES AND PESTS

Michael J. Weiss
University of Idaho, Moscow, Idaho, U.S.A.

M

INTRODUCTION

The extensive use of monocultures in agriculture production has resulted in some insects and pathogens becoming major pests (1). Ecologists have attempted to explain this phenomenon with two main theories: resource concentration and natural enemies hypothesis. The resource concentration hypothesis is based on the relationship between plant density; as it increases, herbivore density also increases, because the resource (i.e., plant) is more concentrated. This is particularly true for specialist herbivores. The natural enemy hypothesis states that as the plant community become less diverse, it results in less faunal diversity, resulting in reduced food sources for natural enemies. Therefore, the herbivore population increases as a result of reduced predation and parasitism by natural enemies. There is experimental evidence that in some cases, these two theories are synergistic and in others, antagonistic.

THE PROBLEM

A monoculture is the cultivation of single species of a plant, usually within the same cropping season over an extensive area. Monocultures can occur at field, farm, or landscape spatial scales. In industrial agriculture, it is very common for only a single species of plant to be produced in the same field; however, there are exceptions (e.g., oats and alfalfa). Additionally, it is not unusual in the industrial model that within the same ecosystem, the vast areas are planted to a single species of plant (e.g., the Corn Belt).

The vast monocultures in modern industrialized agriculture are often genetically uniform. Extensive monocultures have allowed some insect pests, pathogens, and weeds to exploit these unnatural systems. The extensive use of monocultures as the primary cropping system either on an individual farm or over a larger area has resulted in increased pest densities of some pest species, particularly those that are specialists on the crop plant.

Mechanisms

The ''resource concentration'' versus the ''natural enemies'' hypothesis proposes that more herbivores will accumulate on the host plant when it is grown in a single species stand as compared with the host plant in a mosaic with other species of plants. Therefore, the abundance of suitable food for specialist insects results in an increase of in abundance. Insect density may increase, decrease, or remain constant as host plant density changes; therefore, individual species of insect have different responses to changes in host density. Since the hypothesis was proposed in 1973, it has been refined to include not only plant density, but also plant quality and field size as determinates of herbivore density in relation to a monoculture. Tests of the resource concentration hypothesis have suggested, however, that the size of the area in a monoculture rather then the plant density within the area has a more consistent relationship to herbivore density. In addition to abundance of food resources, scale and plant density in monocultures may influence other intrinsic factors, such as survival of overwintering populations. Plant size, shape, and phenology are also documented as having an influence on insect-plant interactions.

The lack of plant diversity in a monoculture may also reduce the density of natural enemies (the natural enemies hypothesis) hence increasing survival and density of pest herbivores. The premise of this hypothesis is as crop cultures become more diverse, the increase in fauna numbers and diversity increases the opportunity of natural enemies by providing alternative food sources. Additionally, the increase in natural enemies will reduce the density of the herbivore species. Predators in diverse plant communities, in particular, are more polyphagous than predators in monocultures, which tend to be more specialized.

Encyclopedia of Pest Management

Monocultures generally experience greater pest pressure, but the reverse case does occur. Mechanisms are diverse, with experimental evidence supporting both the resource concentration and natural enemy hypothesis. There is experimental evidence that in some polycultures, these two hypotheses are complementary, and in others they appear antagonistic.

The Corn Cropping System

In the United States, corn, *Zea mays* L., currently is planted on more than 80 million acres, with the acreage concentrated in the Corn Belt states of Iowa, Illinois, Indiana, Nebraska, South Dakota, Minnesota, and Wisconsin. In the central Corn Belt corn is typically produced in annual rotation with soybeans, *Glycine max* (L.) Merr. However, in the more arid and irrigated area of the western Corn Belt, corn is often produced every year on the same field. There are more than 30 species of insects and 40 different pathogens that are reported to attack corn in the United States.

Western corn rootworm

The western corn rootworm, *Diabrotica virgifera virgifera* LeConte, is the major pest of corn in the United States. The genus *Diabrotica* is neotropical and coevolved with corn. The biology of the western corn rootworm, univoltine (single generation per year) with an obligatory egg diapause evolved as the selection process to the wet-dry climate of central Mexico. This resulted in the larva stage and the grass host plants being in synchrony, by having the egg diapause being broken by the wet cycle that also allowed the obligate grass hosts to begin growth. Because the larvae of the western corn rootworm cannot travel far in the soil in search of host plant roots, larval success is dictated by the ovipositional site selection by the female. The western corn rootworm is very a specialized insect and has been able to exploit the use of monocultures of corn to become a major pest. The movement away from multiyear crop rotations to continuous corn production on the same field, has ''set the table'' for the western corn rootworm. Recently, a subpopulation of the western corn rootworm has been found in Illinois and Indiana that the females will forage for pollen and oviposit in soybean fields. In the next cropping season, when corn is rotated in to the previous soybean field, significant root injury can occur. Thus, it appears that there is a subpopulation of the western corn rootworm that is adapting to the biennial cropping systems of the central Corn Belt.

European corn borer

The European corn borer, *Ostrinia nubilais* (Hübner) is a nonneartic species, inadvertently introduced into the United States in the early 1900s. This insect has a large host range of more than 300 recorded host plants, including *Zea mays*. The biology of this insect, environmentally voltine dependent, has allowed it become a major pest of corn from the east coast to the northern great plains of the United States. In the majority of the Corn Belt, the European corn borer has two generations per year. The females from the first generation prefer to oviposit on the tallest corn plant; if the plants are not of sufficient height, females will deposit eggs on other host plants. Therefore, large areas of early planted corn will be of sufficient height to recruit females for oviposition. Females of the second generation are attracted to plants that are in the early stages of silking. Conversely, early generation females are attracted to early planted corn and second generation females are attracted to later planted corn. Although this insect is polyphagous, there is experimental evidence that corn grown in a polyculture has lower densities, although the mechanisms have not been elucidated.

Southern corn leaf blight

Southern corn leaf blight caused extensive yield losses in 1970 through the majority of the central and eastern Corn Belt with damage estimates of more than $1 billion. Although this fungal disease is usually only a minor limitation in corn yield, the widespread use of Texas male-sterile cytoplasm, cms-T, in the production of corn hybrids that were extremely susceptible to a new race of *Cochiliobolus heterostrophus*, race T, combined with favorable environmental conditions resulted in significant yield losses. The extensive use of a monoculture cropping in conjunction with the majority of corn hybrids sharing the Texas male-sterile cytoplasm resulted in the epidemic.

REFERENCE

1. Andow, D.A. Vegetational diversity and arthropod population response. Ann. Rev. Entomol. **1994**, *36*, 561–586.

MOSQUITO-BORNE ARBOVIRAL ENCEPHALITIDES

Carl J. Mitchell
Centers for Disease Control and Prevention, Fort Collins, Colorado, U.S.A.

INTRODUCTION

Encephalomyelitis, defined as inflammatory disease of the brain and spinal cord, but generally also involving the meninges, can be caused by a number of arboviruses (arthropod-borne viruses) transmitted by mosquitoes or other arthropods (1–4). This discussion is restricted to those viruses that have a mosquito vector. One or more of these viruses is present on all continents and on many islands that have climates suitable for sustaining mosquitoes (5). Mosquitoes are the principal invertebrate hosts of these viruses and serve as primary vectors. A variety of bird and mammal species serve as enzootic reservoir hosts. Humans and most species of domesticated animals do not serve as virus-amplification hosts, but infections may result in varying degrees of morbidity and mortality.

ARBOVIRUSES

At least 17 mosquito-borne arboviruses are known or suspected to cause encephalomyelitis in humans or domesticated animals, or both; these belong to three families and four genera (Table 1) (1, 5). The viruses, and the standard abbreviations for them, follow: Genus *Alphavirus*: Aura (AURA), eastern equine encephalitis (EEE), Ross River (RR), Semliki Forest (SF), Una (UNA), Venezuelan equine encephalitis (VEE), western equine encephalitis (WEE); genus *Flavivirus*: Japanese encephalitis (JE), Murray Valley encephalitis (MVE), Rocio (ROC), St. Louis encephalitis (SLE), West Nile (WN); genus *Bunyavirus*: California encephalitis (CE), Cache Valley (CV), Jamestown Canyon (JC), La Crosse (LAC); and genus *Phlebovirus*: Rift Valley fever (RVF).

TRANSMISSION CYCLES

There are two types of horizontal virus transmission, i.e., arthropod to vertebrate host to arthropod (1). The basic biologic transmission cycle consists of an infected mosquito feeding on a susceptible vertebrate host, usually bird or mammal or both, depending on the virus and vector involved. The vertebrate host responds by circulating virus in its blood for a variable period. More mosquitoes feed and take up the virus. Following a suitable extrinsic incubation period, when the virus replicates and reaches the salivary glands, infected mosquitoes feed again and may transmit the virus. Infected mosquitoes remain infected for life and have the potential to transmit the virus each time they feed, which generally precedes each gonotrophic cycle and can range from none to a few or many times. In some cases mechanical transmission can occur, usually following an interrupted feeding on a viremic host, by transfer of virus from contaminated mouthparts of mosquitoes in the absence of virus replication (1). Transovarial, or other means of vertical transmission, may occur between mother and offspring via infected eggs. This is an important maintenance mechanism for many bunyaviruses such as LAC virus (1). Finally, venereal transmission of some arboviruses can take place between adult male and female mosquitoes, but this probably is of minor importance for virus maintenance or epidemiology (1).

Different mosquito species vary in their susceptibility to infection and ability to transmit different viruses (1, 2). Also, vertebrate species vary in their susceptibility and in their capacity to serve as virus-amplification hosts by circulating viruses in the blood at concentrations sufficient to infect vector mosquitoes (1, 2). These concepts are referred to as vector competence and vertebrate host competence, respectively. Most virus/vector relationships are fairly specific, and each virus considered here generally is transmitted mainly by one or a few mosquito species. There are exceptions, e.g., VEE and RVF viruses. Virus/host relationships are more variable, and viruses such as EEE, WEE, JE and SLE may be amplified by a number of avian species, depending on the geographic area. Domestic swine also are important amplifying hosts of JE virus.

EPIDEMIOLOGY/ EPIZOOTIOLOGY

In cool-temperate areas encephalomyelitis cases are correlated with vector mosquito population fluctuations

Table 1 Mosquito-borne arboviruses known or suspected to cause encephalomyelitis in humans or domestic animals

Arboviral etiology: Virus[a]	Proven		Suspected		Major distribution
	Humans	Domestic animals	Humans	Domestic animals	
Alphavirus					
AURA				+	S. America
EEE	+	+			N. and S. America
RR				+	Australia
SF				+	Africa, Asia
UNA				+	S. America
VEE	+	+			N. and S. America
WEE	+	+			N. and S. America
Flavivirus					
JE	+	+			Asia
MVE	+			+	Australia
ROC	+			+	S. America
SLE	+				N. and S. America
WN	+	+			Africa, Asia, Europe, N. America
Bunyavirus					
CE	+				N. America
CV			+	+	N. and S. America
JC			+		N. America
LAC	+				N. America
Phlebovirus					
RVF	+				Africa

[a]See text for full names of arboviruses.

and generally peak during summer and early autumn (1, 2). Peak transmission to humans frequently occurs as vector populations are declining. Such populations contain a higher proportion of older mosquitoes that have fed more than once, thus increasing their chances of becoming infected. In tropical and subtropical areas, where mosquito production continues year-round, virus transmission may occur at any time of the year (6).

Disease incidence correlated with age or sex varies depending on the virus (1–3, 6, 7). For example, LAC encephalitis is mainly a pediatric problem with high morbidity and low mortality. In contrast, WEE may cause high mortality in infants, and SLE causes more severe disease in older segments of the population (2). VEE generally causes a mild influenza-like disease in humans, but may have devastating effects on equine populations over wide areas. Other viruses such as EEE can be highly lethal in humans; 30% or more of clinical cases have fatal outcomes. Also, epizootics of EEE cause significant equine mortality during most years, especially along the Eastern Seaboard and the Gulf Coast of the United States. WN virus may cause relatively mild disease in parts of its range (Africa, Asia) and more severe disease in areas where it recently has been introduced (United States) (8). Among the mosquito-borne encephalitides, JE virus is the most important in terms of human morbidity and mortality. Epidemics involving several thousand people occur frequently in Asia. This virus also is of great economic importance since it causes stillbirth in swine infected *in utero*. Virus attack rates for most of the viruses discussed here are highest among individuals who are out-of-doors when vector mosquitoes are most likely to be biting, e.g., during major outbreaks of ROC in Brazil encephalitis cases occurred mainly among forestry and agricultural workers and fishermen 15–30 years of age (6). Acute encephalomyelitis caused by several arboviruses may be followed by neurologic sequelae, of greater or lesser severity, which may last throughout the patient's lifetime (1, 2).

ECOLOGY

Some classic vector/virus associations in the United States are *Culex tarsalis*/WEE and also SLE in the western United States; *Cx. pipiens* complex/SLE throughout the midwest and much of the eastern and southeastern United States; *Cx. nigripalpus*/SLE in Florida; *Ae. triseriatus*/LAC, and *Culiseta melanura*/EEE (enzootic cycle) throughout their ranges; elsewhere, *Cx. tritaeniorhynchus*/JE in parts of Asia; and *Cx. annulirostris*/MVE in Australia (1, 2, 5). For viruses such as VEE and RVF, many mosquito species may serve as vectors. In other instances, e.g., ROC virus, vector relationships are not well defined (6). Given the requirements of the immature stages of mosquitoes for aquatic habitats, seasonal abundance of vector mosquitoes is affected not only by temperature but also by rainfall patterns and agricultural practices. For example, *Cx. tarsalis*-transmitted WEE is closely correlated with the distribution of irrigated agriculture in the United States and *Cx. tritaeniorhynchus*-transmitted JE is associated with rice culture in many parts of Asia (1, 2). The feeding habits of vector mosquitoes are very important in determining their associations with particular viruses. In general, *Culex* mosquitoes feed most frequently on avian hosts and *Aedes* mosquitoes feed more on mammals; there are important exceptions, and seasonal changes in host feeding patterns have been noted and correlated with host availability (2).

PREVENTION AND CONTROL

Commercial vaccines are available for JE, WEE, EEE, and VEE viruses. However, only JE vaccine is approved for use in humans, and it has been very effective in reducing the number of human cases in Japan, Korea, Taiwan, People's Republic of China, and Thailand (1). Monovalent, bivalent, and trivalent vaccines for WEE, EEE, and VEE are widely used for protecting equines from infection in the Americas. Also, in areas where EEE is enzootic, EEE vaccine has been given "off label" (unlicensed) use to susceptible birds reared for commercial purposes such as pheasants, quail, and emu. JE vaccine has proved effective in reducing economic losses in swine herds in Asia.

Vector control also effectively may reduce or prevent virus transmission by mosquitoes (9–11). The greatest successes have occurred in areas with tax-supported organized mosquito control agencies, e.g., California, Florida, and New Jersey in the United States. Source reduction and treatment of aquatic habitats with biological control agents and/or chemicals, accompanied by surveillance of adult mosquito populations and appropriately timed adulticide treatments, yield the best results. Commonly used biological control agents include *Bacillus thuringiensis israelensis, Bacillus sphaericus*, and larvivorous fish such as *Gambusia affinis*. A variety of chemicals including organophosphates, carbamates, pyrethroids, and insect growth regulators are used for mosquito control (11). Organochlorine compounds such as DDT are still manufactured and used in some countries for controlling the anopheline vectors of malaria, but are rarely used for controlling arbovirus vectors. In the event of an arbovirus outbreak, emergency measures often are implemented to quickly reduce adult mosquito populations. These measures include application of ultra-low volume (ULV) space sprays by ground-operated or aerial equipment. In some cases, such treatments may rapidly reduce the number of vector mosquitoes (9). However, beginning such treatments early enough in the course of an epidemic to significantly reduce virus transmission is more problematic. Consequently, since emergency measures often are implemented when epidemics are waning, controversy exists about the usefulness of such measures in preventing or reducing the number of human cases.

REFERENCES

1. *The Arboviruses: Epidemiology and Ecology*; Monath, T.P., Ed.; CRC Press: Boca Raton, FL, 1988–1989; 1–5, 1319.
2. Reeves, W.C.; Asman, S.M.; Hardy, J.L.; Milby, M.M.; Reisen, W.K. *Epidemiology and Control of Mosquito-Borne Arboviruses in California, 1943–1987*; California Mosquito Vector Control Association: Sacramento, CA, 1990; 508.
3. Lundstrom, J.O. Mosquito-borne viruses in Western Europe: a review. J. Vector Ecol. **1999**, *24*, 1–39.
4. Moore, C.G.; McLean, R.G.; Mitchell, C.J.; Nasci, R.S.; Tsai, T.F.; Calisher, C.H.; Marfin, A.A.; Moore, P.S.; Gubler, D.G. *Guidelines for Arbovirus Surveillance in the United States*; DHHS, PHS, Centers for Disease Control and Prevention, National Center for Infectious Diseases, Division of Vector-Borne Infectious Diseases: Ft. Collins, CO, 1993; 83.
5. Karabatsos, N. *International Catalog of Arboviruses*, 3rd Ed.; Centers for Disease Control, U.S. Public Health Service: Ft. Collins, CO, 1985; 1147.
6. *An Overview of Arbovirology in Brazil and Neighbouring Countries*; Travassos da Rosa, A.P.A., Vasconcelos, P.F.C., Travasos da Rosa, J.F.S., Eds.; Instituto Evandro Chagas: Belem, Brazil, 1998; 296.
7. Hubalek, Z.; Halouzka, J. Arthropod-borne viruses of vertebrates in Europe. Acta Scientiarum Naturalium Brno. **1996**, *30*, 1–95.
8. CDC. Outbreak of west Nile-like viral encephalitis—New York. Mor. Mortal Wkly. Rep. **1999**, *48*, 845–849.
9. Mitchell, C.J.; Hayes, R.O.; Holden, P.; Hill, H.R.; Hughes, T.B., Jr. Effects of ultra-low volume applications of malathion in hale county, Texas. I. Western encephalitis virus activity in treated and untreated towns. J. Med. Entomol. **1969**, *6*, 155–162.
10. Mitchell, C.J. Environmental Management for Vector Control. In *The Biology of Disease Vectors*; Beaty, B.J., Marquardt, W.C., Eds.; University Press Colorado: Niwot, CO, 1996; 492–501.
11. Sarhan, M.E.; Howitt, R.E.; Moore, C.V.; Mitchell, C.J. Economic evaluation of mosquito control and narrow-spectrum mosquitocide development in California. Giannini Foundation Research Report **1981**, *330*, 90.

MYCORRHIZAE

James A. Traquair
Agriculture and Agri-Food Canada, London, Ontario, Canada

INTRODUCTION

Mycorrhizae are mutualistic symbioses between plants and fungi located in roots or root-like organs in which there is an exchange of organic carbon and minerals, particularly phosphorus. This mutualistic lifestyle benefits both partners in terms of growth and ecological fitness and has evolved in all the classes of true fungi associated with a wide range of seed plants, ferns, bryophytes, and lycopods (1–3). Fossil roots from the Devonian flora of the Rhynie chert containing intracellular, arbuscular-type fungi support the contention that mycorrhizal associations facilitated the colonization of land by plants about 400 million years ago (2). Mycorrhizal associations are ubiquitous today in widespread geographic localities and in diverse habitats ranging from wetlands to deserts, mineral grassland and agricultural soils to organic heathland soils, and roots of herbs and grasses to roots of trees and shrubs in tropical, temperate, and arctic/alpine environments (1, 2). The details of their structure, development, functioning, and ecology are not covered here and this is not an exhaustive review of the current literature. These topics are described elsewhere in many excellent journal articles, reviews, textbooks, and chapters in textbooks cited for further reading. Only highlights of the general nature and function of mycorrhizae in sustaining environmental quality and in improving plant health and production in the context of agriculture, horticulture, and forestry are given here.

STRUCTURE OF MYCORRHIZAE

Considerable morphological diversity is now recognized in the types of mycorrhizal associations (2, 3). This variation depends on the fungal partner, the plant and macro and microenvironmental conditions under which the association develops. More than one type of mycorrhizae can be identified on the same root system (2). Moreover, the same fungus can produce different types of mycorrhizae on different hosts (2). The common feature and the determinant of function in these diverse mycorrhizae is the presence of a network of pervasive extraradical hyphae that extend from the root into the soil beyond the nutrient depletion zone of fine roots and root hairs (2).

Mycorrhizae are classified morphologically as endotrophic, ectotrophic, or ectendotrophic according to the position of the fungus at the site of nutrient exchange between the symbionts. These transfer sites are marked by highly branched hyphae that increase the surface area of cellular interaction between fungus and plant (2). The exchange sites are generally intracellular arbuscules and coils and various kinds of intercellular hyphae and hyphal networks that are closely appressed to root cell walls (Fig. 1). The intracellular arbuscules and coils are in living cells in which the plasma membrane is not penetrated but is highly invaginated.

BENEFITS OF MYCORRHIZAE

The preservation of existing mycorrhizal relationships and the establishment of new ones in agriculture, horticulture, and forestry settings are advantageous especially in stressful, low fertility, and low moisture environments. Benefits of mycorrhizae to the plant usually are measured in terms of increased water and mineral uptake leading to improved plant growth and improved ecological "fitness" for any particular environment (1–3). The negative effects of tillage and the disruption of mycorrhizal networks in agricultural soils, is reflected in the delayed uptake of nutrients and delayed growth improvements in crop plants such as corn (2, 4). The extraradical mycelial network enhances the uptake of water and dissolved nutrients, particularly phosphorus and minor elements for the plant (2). In the unusual case of orchid endomycorrhizae sugars are transferred to host plant from the saprophytic *Rhizoctonia*-like fungus, and in the case of monotropoid ectomycorrhizae sugars are transferred to the herbaceous plant by way of fungal interconnections to ecto- or ectendomycorrhizal roots of pine and spruce trees (2).

Ecological fitness is also reflected in the increased resistance of plants to diseases and soilborne environmental pollutants (1, 2). Mechanisms of disease resistance vary

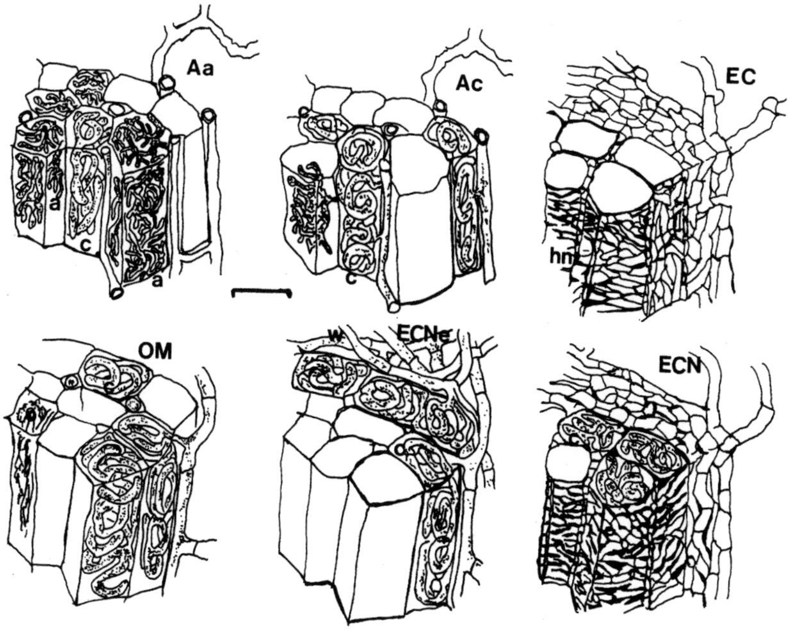

Fig. 1 Root diagrams showing different kinds of mycorrhizae based on morphology of the fungal hyphae and the position of the fungus in relationship to cortical root cells: arbuscular endomycorrhiza (Aa) with many intracellular arbuscules (a) relative to coils (c); arbuscular endomycorrhiza (Ac) with many intracellular coils (c) relative to arbuscules; ectomycorrhiza (EC) with an external mantle of densely packed hyphae and intercellular hyphal network (hn) between cortical cells; ectendomycorrhiza (ECN) with external mantle and intracellular coils (c); ericoid ectendomycorrhiza (ECNe) with weft-like external hyphae (w) and intracellular coils (c); orchid endomycorrhiza (OM) without external mantle and with intracellular coils (c). Bar scale = 20 μm.

with the type of mycorrhizae (5–11). Endomycorrhizal fungi are known to compete with root pathogens for infection sites and to trigger biochemical defenses against plant pathogens (6, 8, 10, 11). Ectomycorrhizal fungi physically exclude pathogens and antagonize them directly and indirectly through antibiosis and mycoparasitic mechanisms (7, 9). The indirect mycorrhizosphere effect is based on the encouragement of naturally occurring microbial antagonists (8, 12). Furthermore, plant ecologists are now recognizing the regulatory role of mycorrhizae in plant community dynamics in grassland, woodland, and heathland soils (2). Mycorrhizae are beneficial also in the revegetation of disturbed forest or mining sites and in the biological remediation of soils polluted by heavy metals and xenobiotic compounds (3, 13). Ectomycorrhizal fungi in particular, are known to chelate heavy metals and to degrade xenobiotic organic pollutants (2).

Benefits to the fungal partner are less obvious. They include access to organic carbon (particularly in the case of biotrophic mycobionts), buffering against environmental perturbation, creation of microenvironment for fungal growth through soil aggregation and the stabilization of supportive plant communities and associative ''helper'' microorganisms in the mycosphere and mycorrhizosphere that solubilize phosphorus, antagonize root pathogens including fungi and nematodes, and suppress competition by soilborne fungal saprophytes (2, 5).

RISKS AND CHALLENGES

The benefits of mycorrhizal colonization in crop production and pest management far outweigh the risks. The novel exploitation of mycorrhizal associations to promote growth and at the same time protect plants is an environmentally friendly biotechnology that is complex yet worthy of more attention (2). Records of mycorrhizal fungi as disease-causing agents are rare. A stunting related to carbon drain on the plant is a problem in tobacco colonized by the arbuscular endomycorrhizal fungus, *Glomus macrocarpum*, in certain soils. The stress of seedling colonization by mycorrhizal biotrophs is usually rapidly nullified in older plants by the advantages of established mycorrhizal associations (2). Control of the delicate balance between parasitism and mutualism is a greater problem in orchid endomycorrhizae where potentially pathogenic Rhizoctonia-like fungal partners are checked by antifungal phenolic compounds produced by the plant (2).

We need to learn more about the ecophysiology of nutrient transfer, the competition between mycorrhizal fungi and other microorganisms, the effectivity of different mycorrhizal fungi, and the mycorrhizal dependency of plants species before we can fully exploit mycorrhizae as a practical biotechnology (2). Molecular biology tools offer encouraging prospects for fundamental and applied studies of the ecology, distribution, activity, and genetic improvement of mycorrhizal fungi (2, 14). In agriculture, the best approach is to conserve and enhance existing endomycorrhizal associations for field crops (1, 15). The introduction of mycorrhizal inoculants in field crops continues to be a challenge because of problems in economical production of sufficient inoculum of endomycorrhizal biotrophs (14, 16). Recent discoveries in the genetics of the host relative to endomycorrhizal colonization, the gene regulation, and biochemistry of plant-fungus interactions will lead to breakthroughs in the artificial cultivation of endomycorrhizal fungi (2). To date, commercial production of this inoculum is based on expensive root aeroponics, root organ culture, and hydroponics technologies (2, 16). In horticulture and forestry there are opportunities to introduce new and better mycorrhizal symbionts because these fungi can be propagated in artificial culture (17). Micropropagation and transplant technologies are exciting ways of delivering mycorrhizal fungi in nursery systems producing tree fruits, high value ornamentals, and vegetable crops (2). The technology for inoculum production for ectomycorrhizal fungi for forest trees based on solid-state and liquid fermentation is better developed than for endomycorrhizal fungi normally associated with horticultural crops (2, 16). Even so, clay-, gel- or peat-based granular (pelletized) and liquid inocula for both types of mycorrhizal fungi are commercially available (16).

REFERENCES

1. Allen, M.F. The Ecology of Mycorrhizae. In *Cambridge University Press*; Cambridge, UK, 1991; 1–184.
2. Smith, S.E.; Read, D.J. *Mycorrhizal Symbiosis, 2nd Ed.; Academic Press: New York,* 1997; 1–605.
3. Wilcox; Mycorrhizae, H.E. *Plant Roots: The Hidden Half,* 2nd Ed.; Waisel, Y., Eshel, A., Kafkafi, U., Eds.; Marcel Dekker, Inc.: New York, 1996; 689–721.
4. McGonigle, T.P.; Miller, M.H. Responses of mycorrhizae and shoot phosphorus of maize to the frequency and timing of soil disturbance. Mycorrhizae **1993**, *4*, 63–68.
5. Barea, J.M.; Azcòn, R.; Azcòn-Aguilar, C. Mycorrhiza and Crops. In *Mycorrhiza Synthesis*; Tommerup, I., Ed.; Academic Press: New York, 1993; 166–189.
6. Benhamou, N.; Fortin, J.A.; Hamel, C.; St. Arnaud, M.; Shatilla, A. Resistance responses of mycorrhizal Ri T-DNA transformed carrot roots to infection by *Fusarium oxysporum f.sp. chrysanthemi.* Phytopathology **1994**, *84*, 958–68.
7. Duchesne, L.C. Role of Ectomycorrhizal Fungi in Biocontrol. In *Mycorrhizae and Plant Health*; Pfleger, F.L., Linderman, R.G., Eds.; The American Phytopathological Society Press: St. Paul, MN, 1994; 27–45.
8. Linderman, R.G. Role of VAM Fungi in Biocontrol. In *Mycorrhizae and Plant Health*; Pfleger, F.L., Linderman, R.G., Eds.; The American Phytopathological Society Press: St. Paul, MN, 1994; 1–25.
9. Morin, C.; Samson, J.; Dessureault, M. Protection of black spruce seedlings against cylindocladium root rot with Ectomycorrhizal Fungi. Can. J. Bot. **1999**, *77*, 169–174.
10. Traquair, J.A. Fungal biocontrol of root diseases: endomycorrhizal suppression of cylindrocarpon root rot. Can. J. Bot. **1995**, *73* (Supplement 1), S89–S95.
11. Slezak, S.; Dumgasgaudot, E.; Rosendahl, S.; Kjoller, R.; Paynot, M.; Negrland, J.; Gianinazzi, S. Endoproteolytic activities in pea roots inoculated with arbuscular mycorrhizal fungus, *Glomus mosseae and/or Aphanomyces euteiches* in relation to bioprotection. New Phytologist **1999**, *142*, 517–529.
12. Azcòn-Aguilar, C.; Barea, J.M. Interactions Between Mycorrhizal Fungi and Other Rhizosphere Microorganisms. In *Mycorrhizal Functioning: An Integrative Plant-Fungus Process*; Allen, M.F., Ed.; Chapman & Hall: New York, 1992; 163–198.
13. Pfleger, F.L.; Stewart, E.L.; Noyd, R.K. Role of VAM Fungi in Mine Land Revegetation. In *Mycorrhizae and Plant Health*; Pfleger, F.L., Linderman, R.G., Eds.; The American Phytopathological Society Press: St. Paul, MN, 1994; 47–81.
14. Wood, T.; Cummings, B. Biotechnology and the Future of VAM Commercialization. In *Mycorrhizal Functioning: An Integrative Plant-Fungus Process*; Allen, M.F., Ed.; Chapman & Hall: New York, 1992; 468–488.
15. Thompson, J.P. What Is the Potential for Management of Mycorrhizas in Agriculture. In *Management of Mycorrhizas in Agriculture, Horticulture and Forestry*; Robson, A.D., Abbott, L.K., Malajczuk, N., Eds.; Kluwer Academic Publishers: Dordrecht, The Netherlands, 1994; 191–200.
16. Mitchell, D.T. Mycorrhizal Associations. In *Exploitation of Microorganisms*; Jones, D.G., Ed.; Chapman & Hall: London, 1993; 169–196.
17. Grove, T.S.; Malajczuk, N. The Potential for Management of Ectomycorrhiza in Forestry. In *Management of Mycorrhizas in Agriculture, Horticulture and Forestry*; Robson, A.D., Abbott, L.K., Malajczuk, N., Eds.; Kluwer Academic Publishers: Dordrecht, The Netherlands, 1994; 201–210.

MYCOTOXINS

J. David Miller
Carleton University, Ottawa, Ontario, Canada

M

INTRODUCTION

Mycotoxins are chemicals that are produced by filamentous fungi that affect human or animal health. By convention, this excludes mushroom poisons. All of these species are Deuteromycetes some of which have a known Ascomycetous stage. The occurrence of mycotoxins is entirely governed by the existence of conditions that favor the growth of the fungi concerned. Under environmental conditions, different fungal species are favored as diseases of crop plants or as saprophytes on stored crops. When the conditions favor the growth of toxigenic species it is an invariable rule that one or more of the compounds for which the fungus has the genetic potential are produced.

FIVE IMPORTANT MYCOTOXINS

Although there are hundreds of fungal metabolites that are toxic in experimental systems, there are only five that are of major agricultural importance: deoxynivalenol, aflatoxin, fumonisin, zearalenone, and ochratoxin (1). Deoxynivalenol occurs when wheat, barley, corn, and sometimes oats and rye are infected by *Fusarium graminearum* and *F. culmorum*. These species cause fusarium head blight in small grains, a major agricultural problem worldwide, as well as a similar disease in corn called gibberella ear rot. Disease incidence is most affected by moisture at flowering and most cultivars and hybrids used today lack genetic resistance to the disease. *F. graminearum* is common in wheat from North America and China. *F. culmorum* is the dominant species in cooler wheat growing areas such as Finland, France, Poland, and The Netherlands. A third species, *F. crookwellense* can also cause head blight or corn ear rot and produces nivalenol and zearalenone (2).

Humans are not less and are probably more sensitive to deoxynivalenol than swine (3). Swine are the domestic animal species most sensitive to the effects of deoxynivalenol because of its neurotoxicity (4). In addition, deoxynivalenol causes changes in immune system function in male mice, occurring at dietary concentrations often encountered by humans. As with other trichothecenes, high exposures increase susceptibility to facultative pathogens such as *Listeria* (5). Cattle, cows, and poultry species are tolerant to deoxynivalenol and milk production is not affected at typical field concentrations (4). Deoxynivalenol is not a carcinogen (6).

Crops that are contaminated by deoxynivalenol can often contain zearalenone albeit at a lower frequency. Zearalenone is more common in maize than small grains (6). Zearalenone is an estrogen analogue and causes hyperestrocism in female pigs at low levels; the dietary no-effect level is less than 1 mg/kg. Cows and sheep are also sensitive to the estrogenic effects of this toxin with depressed ovulation and lower lambing percentages (4). Non-human primates and humans are also very sensitive to the estrogenic effects of zearalenone (3).

Aflatoxin is mainly produced by *Aspergillus flavus* which is a problem in many commodities but most human exposure comes from contaminated corn, groundnuts, and rice. *A. parasiticus* is uncommon outside North and South America and there is more associated with peanuts. *A. flavus* contamination of corn or peanuts occurs in two basic ways. Either airborne or insect-transmitted conidia contaminate the silks and grow into the ear when the maize is under high temperature stress *or* (more commonly) insect- or bird-damaged kernels become colonized with the fungus and accumulate aflatoxin. In either plant, drought-, nutrient-, or temperature-stressed plants are more susceptible to colonization by *A. flavus* or *A. parasiticus* (1).

In poultry, aflatoxin exposure results in liver damage, impaired productivity and reproductive efficiency, decreased egg production in hens, inferior egg-shell quality, inferior carcass quality, and increased susceptibility to disease. The effects of acute and chronic exposure in swine are largely attributable to liver damage. In cattle, the primary symptom is reduced weight gain as well as liver and kidney damage and reduced milk production (7). Aflatoxin is also immunotoxic in domestic and laboratory animals with oral exposures in the μg/g range. Cell-mediated immunity (lymphocytes, phagocytes, mast cells, and basophils) is more affected than humoral immunity (antibodies and complement; (5)). Naturally occurring mixtures of aflatoxins were classified as class 1 human carcinogens and aflatoxin B1 is also a class 1 human carcinogen. (6).

Fumonsisins are produced by *F. verticilloides* (formerly *moniliforme*), *F. proliferatum* and several uncommon fusaria (1). Fumonisins have been found as a very common contaminant of corn-based food and feed in the United States, China, Europe, southern Africa, South America, and southeast Asia (8).

F. verticilloides and *F. proliferatum* can be recovered from virtually all corn kernels including those that are healthy, which suggests that it may be an endophyte i.e., a mutualistic relationship (1). *F. verticillioides* and *F. proliferatum* cause a "disease" called fusarium kernel rot. In parts of the United States and lowland tropics, this is one of the most important ear diseases and is associated with warm, dry years and insect damage. Corn plant disease stress also promotes the growth of *F. verticillioides* and fumonisin formation (1, 2).

Fumonisin causes Equine Leucoencephalamalacia (ELEM), which involves a liquefactive necrosis of the cerebral hemispheres. Clinical manifestations include abnormal movements, aimless circling, lameness, etc., followed by death (4). In swine, high exposures of fumonisin results in porcine pulmonary edema (PPE) caused by fumonisin-induced heart failure. At lower exposures, both liver and kidney damage has been reported in swine. Fumonisin causes feed refusal and changes in carcass quality at dietary concentrations in the low mg/kg range (8).

Exposure to *F. verticillioides*-contaminated maize has been linked to the elevated rates of esophageal cancer in the Transkei for 25 years and this has since been linked to fumonisin exposure (6, 8). Fumonisin caused tumors in mice and rats in the U.S. National Toxicology Program study (NTP TR 496) and is currently an IARC class 2B carinogen (possible human carcinogen).

Ochratoxin is known to be produced by only one species of *Penicillium, P. verrucosum. Aspergillus ochraceous* and several related species also produce ochratoxin on grapes and coffee (6). A few percent of surface-disinfected wheat and barley kernels collected at harvest in the United Kingdom and Denmark were contaminated by *P. aurantiogriseum* and *P. verrucosum* and this was similar to studies from Canada (1).

Ochratoxin is a potent nephratoxin in swine and causes kidney cancer in male Fisher 344 rats. Low exposures result in kidney damage in swine but typically there are no overt signs. At higher concentrations (>2 μg/g), decreased weight gains occur. Poultry are affected showing reduced growth rate and egg production at low ochratoxin concentrations >2 μg/g. Cattle are resistant to ochratoxin concentrations found in naturally contaminated grain (4).

Ochratoxin is suspected as the cause of urinary tract cancers and kidney damage in areas of chronic exposure in parts of eastern Europe. Despite considerable effort, no satisfactory conclusion has been reached regarding the linkage of ochratoxin with urinary tract cancers in humans (6).

MYCOTOXINS AND PESTICIDES

Insect damage is known to promote accumulation of dexoynivalenol, fumonisin, aflatoxin, and probably zearalenone. Steps taken to reduce insect herbivory or drought stress on crops have the general effect of reducing mycotoxin contamination. Bt corn has lowered fumonisin content compared to similar non-Bt hybrids.

There is evidence that application of pesticides can have an undesirable effect on mycotoxin production. There is some direct evidence of this for the trichothecene mycotoxins including deoxynivalenol. Applications of foliar fungicides to wheat reduced disease symptoms compared to controls but deoxynivalenol content was not reduced. Certain concentrations of the fungicide tridemorph resulted in greater production of T-2 toxin per unit dry weight by *F. sporotrichioides*. Barley grown in field plots, inoculated with *F. sporotrichioides* and treated with tridemorph had higher trichothecene concentrations compared to controls (9).

It is also now known that some genotypes of corn and peanuts produce compounds that interfere with mycotoxin production as opposed to fungal growth per se (10). The first pure compound to be isolated with such activity was reported from corn. Such activity was recognized in *Myrothecium*-resistant cultivars of melon but was also seen in a variety of pure plant allelochemicals. At least in melons and corn, such chemicals related to a plant defense system in relation to phytotoxicity of trichothecenes. The role of compounds in peanuts and soybeans that affect aflatoxin production is not clear.

OCCUPATIONAL EXPOSURES

Grain dusts represent an occupational hazard for a number of reasons including the possibility of the allergic disease hypersensitivity pneumonitis and organic dust toxic syndrome, a poorly understood disease. Inhalation of mycotoxins contained in airborne dusts is also a potential health risk. Inhalation of mycotoxins in spores and dusts affects macrophage function and other aspects of lung biology. Inhalation of mycotoxins is a more potent route of exposure for systemic toxicity for aflatoxin, trichothecenes, and ochratoxin (11). Workplace exposure to airborne aflatoxin results in increased relative risk of liver cancer (6). Inhalation exposure to deoxynivalenol in the workplace has been subjected to a formal risk estimate. Workplace exposure to ochratoxin resulted in severe kidney damage (11).

REFERENCES

1. Miller, J.D. Fungi and mycotoxins in grain: implications for stored product research. J. Stored Product Research **1995**, *31*, 1–6.
2. Miller, J.D. Epidemiology of *Fusarium* Ear Diseases. In *Mycotoxins in Grain: Compounds Other Than Aflatoxin*; Miller, J.D., Trenholm, H.L., Eds.; Eagan Press: St. Paul, MN, 1994; 19–36.
3. Kuiper-Goodman, T. Prevention of Human Mycotoxicosis through Risk Assessment and Risk Management. In *Mycotoxins in Grain: Compounds Other Than Aflatoxin*; Miller, J.D., Trenholm, H.L., Eds.; Eagan Press: St. Paul, MN, 1994; 439–470.
4. Prelusky, D.B.; Rotter, B.A.; Rotter, R.G. Toxicology of Mycotoxins. In *Mycotoxins in Grain: Compounds Other Than Aflatoxin*; Miller, J.D., Trenholm, H.L., Eds.; Eagan Press: St. Paul, MN, 1994; 359–404.
5. Pestka, J.J.; Bondy, G.S. Mycotoxin-induced Immune Modulation. In *Immunotoxicology and Immunopharmacology*; Dean, J.H., Luster, M.I., Munson, A.E., Kimber, I., Eds.; Raven Press: New York, 1994; 163–182.
6. IARC. *Monograph 56. Some Naturally Occurring Substances: Food Items and Constituents, Heterocyclic Aromatic Amines and Mycotoxins*; International Agency for Research on Cancer: Lyon, France, 1993.
7. *Mycotoxins and Animal Foods*; Smith, J.E., Henderson, R. S., Eds.; CRC Press: Boca Raton, FL, 1991.
8. International Program for Chemical Safety. *Fumonisins*; World Health Organization: Geneva.
9. D'Mello, J.P.F.; Macdonald, A.M.C.; Postel, D.; Dijksma, W.T.P.; Dujardin, A.; Placinta, C.M. Pesticide use and mycotoxin production in *Fusarium* and *Aspergillus* phytopathogens. European J. Plant Pathology **1998**, *104*, 741–751.
10. Miller, J.D.; Miles, M.; Fielder, D.A. Kernel concentrations of 4-acetylbenzoxazolin-2-one and diferuloylputrescine in maize genotypes and gibberella ear rot. J. Agric. Food Chem. **1997**, *45*, 4456–4459.
11. Miller, J.D. Mycotoxins. In *Handbook of Organic Dusts*; Rylander, R., Pettersen, Y., Eds.; CRC Press: Boca Raton, FL, 1994; 87–92.

NATURAL PESTICIDES

Franck E. Dayan
United States Department of Agriculture, University, Mississippi, U.S.A.

N

INTRODUCTION

The chemical pest control paradigm has dominated modern agricultural practices since its introduction about 60 years ago. It has been one of the key components accompanying the "green" revolution that resulted in tremendous increases in crop yield during the past 50 years. However, newer and more stringent pesticide registration procedures, such as the Food Quality Protection Act in the United States, are affecting the number of chemical tools, especially insecticides, available to farmers. Natural product-based pesticides, along with novel chemistry, are being developed to replace the compounds lost due to the new registration requirements. This review highlights the historical use of natural products in agricultural practices and the impact of natural products on the development of new pesticides. Readers interested in more detail are referred to the following reviews (1, 2).

HISTORICAL USE OF NATURAL PRODUCTS FOR PEST MANAGEMENT

Early agricultural practices relied heavily on crop rotation or mixed crop planting to optimize natural pest control (such as predation, parasitism, and competition) and required a significant amount of manual pest management. The concept of "natural pesticides" is not new. Control of caterpillar infestations in citrus orchards by introducing colonies of ants is reported around 300 BC in ancient Chinese literature, and the use of beneficial insects to control other insect pests was mentioned by Linnaeus in 1752.

The advent of monoculture and intensive agricultural practices, though beneficial in that it allows great increases in yields, is often accompanied by a more pernicious pest problem that has mostly been addressed by the use of synthetic pesticides. In the past few decades, the concept of integrated pest management has become more widely accepted, combining pest population monitoring, mechanical agricultural practices, biocontrol, and natural bioactive compounds to rely less heavily on the indiscriminate use of synthetic pesticides.

STRUCTURAL DIVERSITY IN NATURE

The usually complex carbon skeleton of natural products derived from secondary metabolism is the result of natural selection of molecules that provided some protection against specific biotic challenges. Nature has, in a sense, performed a "high throughput" screen over long period of time to select particularly suitable biologically active compounds. The "high throughput" refers not to the rapidity of the selection, but rather to the innumerable permutations of relatively complex structures that have been made. Structural diversity that resulted from this has been, and still remains, an invaluable source of lead compounds in developing novel agrochemical (and pharmaceutical) products.

Important benefits of natural product-based pesticides, such as the absence of "unnatural" ring structures and the presence of few heavy atoms, are their short environmental half-lives and their tendency to affect novel target sites. This latter fact is particularly important since the need for new modes of action is so pressing. Agrochemical companies are actively seeking novel mechanisms of action for which they can develop new chemistry.

INSECTICIDES DERIVED FROM NATURE

Pyrethrins

Historically, the greatest impact of natural products on pesticide development has been in the area of insecticides. The insecticidal properties of *Chrysanthemum* flower head powder was known for a long time and the discovery of the bioactive pyrethrin component led to the successful development of numerous pyrethroid insecticides (Fig. 1). Natural pyrethrins were quite unstable, being rapidly degraded in the presence of light and oxygen. Several synthetic programs were initiated to generate more stable commercial products. Initial progress was slow but key

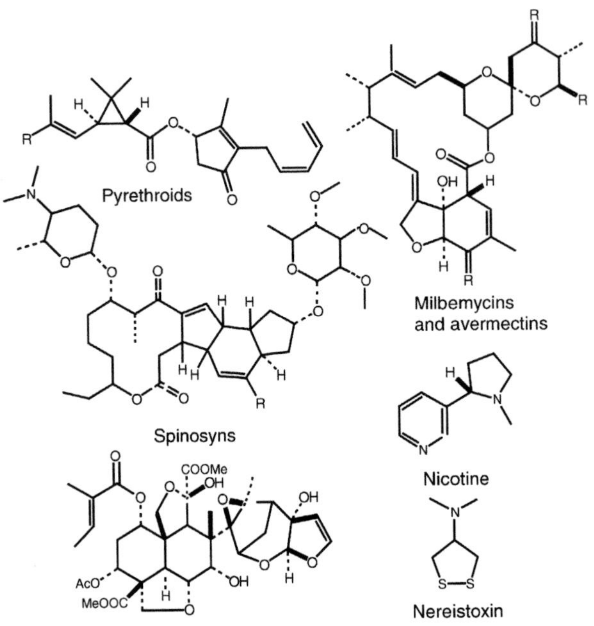

Fig. 1 General structures of natural products with insecticidal activity mentioned in the text.

synthetic analogs with improved stability and selectivity were eventually discovered. The insecticidal activity of pyrethroids is associated with their ability to cause prolonged opening of sodium-ion channels in nerve membranes resulting in a lethal blockage of the nerve signal system in insects.

Nicotine and Nereistoxin Derivatives

Crude extracts of tobacco leaves rich in nicotine (Fig. 1) have been used as insecticides for many years. This alkaloid acts as an acetylcholine receptor agonist and is highly toxic to insects and mammals. The search for insect-specific analogs, such as nithiazin, imidacloprid, nitenpyram, and acetamiprid, led to the discovery of neonicotinoid analogs that specifically bind to insect acetylcholine receptors.

Compounds derived from the general cytotoxin nereistoxin (Fig. 1) with much improved toxicity profiles and strong insecticidal activity have been developed. These compounds have the same mode of action as nicotinoid insecticides.

Milbemycins/Avermectins

The discovery of these complex molecules is fairly recent. Most of these structures were discovered in *Streptomyces* cultures and have shown broad acaricidal, insecticidal, and anthelmintic activities. Both milbemycins and avermectins (Fig. 1) interfere with the opening of glutamate and GABA-gated chloride-channels. While studies aiming at producing simpler synthetic molecules yielded compounds with much lower activities, suitable natural product analogs can be synthesized without great difficulty. These molecules are now the primary tools used against parasites of farm animals and are used as insecticides in crop protection.

Spinosyns

Spinosyns (Fig. 1) are marketed for lepidoptera control in cotton as a mixture obtained by a fermentation process. Their mode of action is unknown but there are no known cases of cross-resistance to this class of inhibitors, suggesting that its target site is different from other known insecticides.

Azadirachtin

Seeds of the neem tree contain the highly insecticidal azadirachtin (Fig. 1). In addition to being a feeding deterrent, this complex molecule is also lethal to insects. It has multiple effects on insects, including disruption of metamorphosis by interfering with ecdysteroid synthesis, but the precise mode of action of azadirachtin is still unknown. Commercial products enriched with azadirachtin and other limonoids are available for controlling insects.

Juvenile-Hormone Mimics and Pheromones

Juvenile hormones are used to prevent certain insects from developing into fertile adults, therefore preventing them from reproducing. Other hormone-type insecticides, such as pheromones, have been used to control insect population in many ways. Initial approaches included using high levels of pheromones to confuse insects and prevent them from mating. The most common use is as an attractant placed in traps coated with either sticky substances that prevent the insects from leaving or insecticides that kill the organisms.

Bacillus thuringiensis

Bacillus thuringiensis produces the insecticidal protein δ-endotoxin during sporulation. When ingested by an insect, δ-endotoxin binds to the epithelial cells of the gut and causes those cells to lyse, leading to the death of the insect. Several commercial products containing various forms of δ-endotoxin are available. Transgenic crops engineered to produce the δ-endotoxin have successfully been commercialized. These crops are insect-resistant since feeding on the foliage is lethal to insects.

FUNGICIDES AND BACTERICIDES DERIVED FROM NATURE

Numerous natural products have been identified and either used directly as fungicides or as templates to develop commercial products. While some of these products have narrow commercial use, others have been used as broad-spectrum fungicides. There are many natural products with excellent antifungal activity. However, few of them have been commercialized because of toxicological problems that often arise during the early stages of development, the excessive cost of large-scale production, and/or a problem with instability in the environment.

Strobilurins

Strobilurins (Fig. 2) are the most recent example of a successful use of natural products to develop an entire new class of fungicides. Their discovery and development has recently been reviewed (3). Their fungicidal activity was first reported 30 years ago. Strobilurins have been identified primarily in basidiomycetes species throughout the world. These compounds are reversible inhibitors of the ubihydroquinone oxidation center of the bc1 complex in the mitochondrial respiratory chain. The discovery and development of strobilurin-derived fungicides has opened an extremely important area of research, with hundreds of international patent applications filed by more than 20 companies and research groups and thousands of analogs generated.

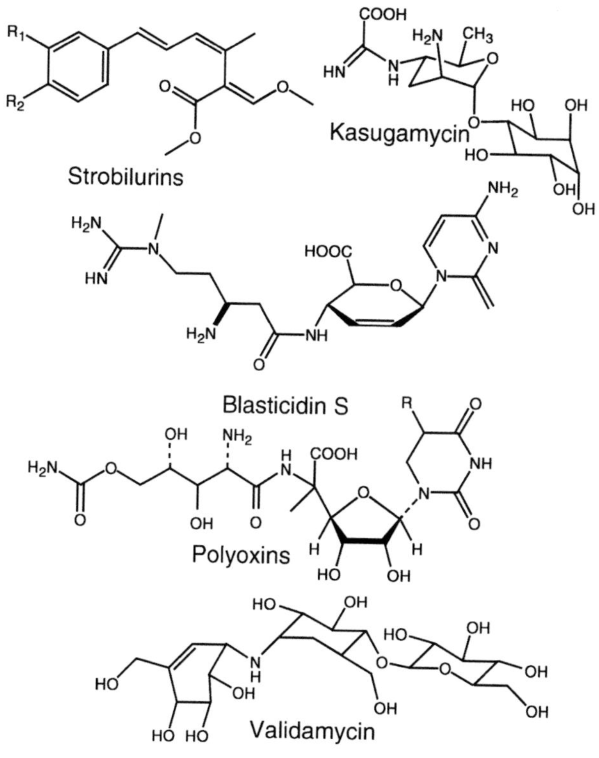

Fig. 2 General structures of natural products with fungicidal activity mentioned in the text.

Kasugamycin

Kasugamycin (Fig. 2), isolated from *Streptomyces kasugaensis*, is a general inhibitor of protein biosynthesis in microorganisms, but does not affect protein synthesis in mammals. It is used as a commercial product for the control of rice blast (*Pyricularia oryzae*) and bacterial diseases caused by *Pseudomonas* spp. in several crops. The use of kasugamycin for fighting fungal diseases in rice paddies is now preferred over blasticidin S, which exhibited slight phytotoxic damage to the crop.

Polyoxins

Several polyoxin metabolites with antifungal activity have been identified. Polyoxin B and polyoxin D (Fig. 2) interfere with chitin synthase activity, preventing cell wall formation. These fungicides are commercially produced by fermentation for the control of rice sheath blight (*Rhizoctonia solani*) and other diseases in fruit and ornamental crops.

Validamycin

Another group of natural fungicides derived from the validamycin (Fig. 2) class has also been developed to control rice sheath blight. These fungicides are considered quite safe because their target site, the enzyme trehalase responsible for the hydrolysis of the disaccharide trehalose, is not found in mammalian systems.

HERBICIDES DERIVED FROM NATURE

Few herbicides have been derived from the backbone of natural compounds, relative to the number of insecticides and fungicides that have been successfully developed from natural products. This is somewhat unexpected since, from a chemical ecology standpoint, fungal/microbial plant pathogens and certain plants, in particular those involved in allelopathic interactions, produce highly phytotoxic molecules. In fact, hundreds of microbial natural products have been patented for potential use as herbicides, but bialaphos and its analog phosphinothricin are the only successful commercialized herbicides from natural products (4). There is no plant product used directly as a herbicide, but striking similarities exists between natural phytotoxins and certain synthetic herbicides (5).

Bialaphos and Phosphinothricin

Bialaphos (Fig. 3) has been isolated from the fermentation of *Streptomyces hygroscopis* and has been commercialized in eastern Asia. When applied directly to plants, it is metabolically converted by the plant to its bioactive form phosphinothricin (Fig. 3). The synthetic version of phosphinothricin has been commercialized as glufosinate (Basta®, Liberty®), and genetically engineered crops resistant to phosphinothricin have been marketed. This natural product-based herbicide has a unique mode of action, targeting glutamine synthetase. Other natural products such as phosalacin, oxetin, and tabtoxin also inhibit glutamine synthetase but have not been developed as commercial herbicides (6).

Monoterpene Cineoles

The commercial herbicide cinmethylin is a 2-benzyl ether substituted analog of the monoterpene 1,4-cineole (Fig. 3). This compound was discovered and partially developed by Shell Chemicals for monocot weed control. Phytotoxic monoterpene cineoles are commonly found as components of essential oils from aromatic plants. While both cinmethylin and 1,4-cineole cause similar symptoms on treated plants, it was reported recently that only the natural product inhibits asparagine synthetase, the key enzyme in asparagine synthesis in plants. Thus, cinmethylin is apparently a proherbicide that requires metabolic bioactivation via cleavage of the benzyl-ether side chain.

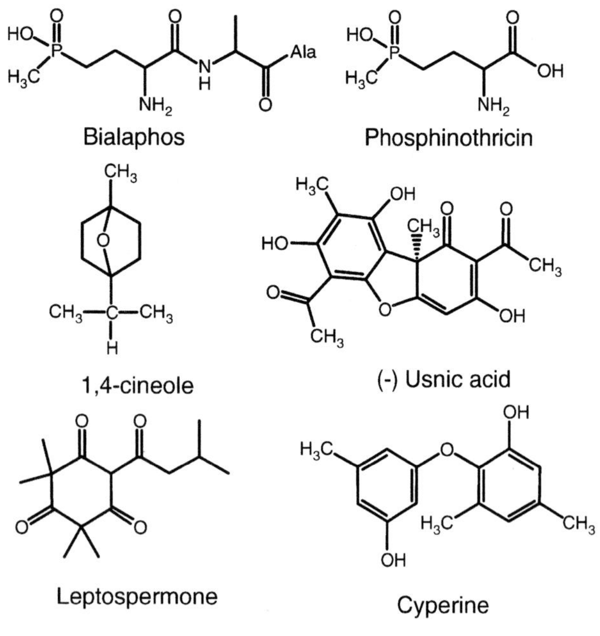

Fig. 3 General structures of natural products with herbicidal activity mentioned in the text.

Triketones

Leptospermone (Fig. 3), a natural triketone isolated from bottlebrush plant (*Calispermon* spp.), is herbicidal and causes bleaching of the foliage. The herbicide sulcotrione, a structural analog of leptospermone, has been shown to inhibit the relatively novel herbicide target site hydroxyphenylpyruvate dioxygenase (HPPD) (7). Inhibition of this enzyme disrupts the biosynthesis of carotenoids and causes bleaching of the foliage. We recently found that HPPD is sensitive to numerous classes of natural products such as the *p*-benzoquinones sorgoleone and maesanin, the *p*-naphthoquinone juglone, and the triketone usnic acid (Fig. 3) (unpublished data).

Cyperine

Cyperine (Fig. 3) is a phytotoxic natural product that has been isolated from several fungal pathogen sources. This phytotoxin has high activity on several plant species, including the problematic weed purple nutsedge (*Cyperus rotundus*) and pokeweed (*Phytolacca americana*). Cyperine is structurally similar to the synthetic diphenyl ether herbicides such as acifluorfen and oxyfluorfen but is a poor inhibitor of protoporphyrinogen oxidase.

Allelopathy

Allelopathy, the ability of a plant to suppress the growth of competing weeds in their immediate surroundings, is being investigated as an alternative approach to weed management. While certain plant species strongly repress the development of other plants, these allelopathic traits have, for the most part, been eliminated from crops by breeding programs selecting for higher yields. For example, certain native rice lines repress the growth of barnyard grass (*Echinochloa crus-galli*), whereas commercial varieties do not affect the growth of this weed. Introducing the biosynthetic pathway of an allelochemical into a crop has now become possible using genetic engineering tools, though it has not yet been achieved successfully (8).

PROSPECT

Traditional pest management has already been deeply influenced by bioactive natural products that are used directly, or in a derived form, as pesticides. These past successes and the current public concern over the impact of synthetic pesticides on the environment ensures a continued, if not increased, interest in searching nature for environmentally friendlier pest management tools.

REFERENCES

1. Pachlatko, J.P. Natural products in crop protection. Chimia **1998**, *52*, 29–47.
2. Copping, L.G.; Menn, J.J. Biopesticides: A review of their action, applications and efficacy. Pest. Manag. Sci. **2000**, *56*, 651–676.
3. Sauter, H.; Steglich, W.; Anke, T. Strobilurins: evolution of a new class and active substances. Angew. Chem. Int. Ed. **1999**, *38*, 1328–1349.
4. Duke, S.O.; Abbas, H.K.; Amagasa, T.; Tanaka, T. Phytotoxins of Microbial Origin with Potential for Use as Herbicides. In *Crop Protection Agents from Nature: Natural Products and Analogues*; Copping, L.G., Ed.; Royal Society of Chemistry: Cambridge, UK, 1996; 82–113, SCI Critical Reviews on Applied Chemistry.
5. Dayan, F.E.; Romagni, J.G.; Tellez, M.R.; Rimando, A.M.; Duke, S.O. Managing weeds with natural products. Pestic. Outlook **1999**, *5*, 185–188.
6. Lydon, J.; Duke, S.O. Inhibitors of Glutamine Biosynthesis. In *Plant Amino Acids: Biochemistry and Biotechnology*; Singh, B.K., Ed.; Marcel Dekker, Inc.: New York, 1999; 445–464.
7. Lee, D.L.; Prisbylla, M.P.; Cromartie, T.H.; Dagarin, D.P.; Howard, S.W.; Provan, W.M.; Ellis, M.K.; Fraser, T.; Mutter, L.C. The discovery and structural requirements of inhibitors of *p*-hydroxyphenylpyruvate dioxygenase. Weed Sci. **1997**, *45*, 601–609.
8. Scheffler, B.E.; Duke, S.O.; Dayan, F.E.; Ota, E. Crop Allelopathy: Enhancement through Biotechnology. In *Recent Advances in Phytochemistry*; Romeo, J., Ed.; Elsevier: Amsterdam, *in press*.

NEMATODE CONTROL OF PESTS

Simon R. Gowen
N.G.M. Hague
The University of Reading, Earley Gate, Reading, United Kingdom

INTRODUCTION

Insect nematology from the mid-nineteenth century to 1930 was primarily descriptive. Between 1930 and the early 1940s an ambitious program was initiated to control the Japanese beetle, *Popillia japonica*, in the United States with *Steinernema glaseri*, the research being curtailed by the onset of the Second World War and the development of the milky disease bacterium, *Bacillus popilliae*, for biocontrol. The next major development was the study by R.A. Bedding of the nematode, *Deladenus siricidicola*, for the control of the wood wasp, *Sirex noctilo*, in Australia and New Zealand. Another nematode that gained recognition for biocontrol was *Romanomermis cucilivorax* for the control of mosquitoes, but the commercial development of this nematode failed because of poor nematode storage and transport capabilities and also because of the success of *Bacillus thuringiensis* subspecies *israelensis*. Of all the nematodes studied for the biological control of insects, the entomopathogenic nematodes have aroused the most interest.

ENTOMOPATHOGENIC NEMATODES

These nematodes, which belong to the families Steinernematidae and Heterorhabditidae, are mutualistically associated with bacteria and their actions against insect larvae and pupae are similar. The free-living, nonfeeding infective juveniles (known as dauer juveniles [DJs]) of these nematodes possess attributes of insect parasitoids or predators and microbial pathogens. Like parasitoids/predators they have chemoreceptors and are motile; like pathogens they are highly virulent, killing their hosts quickly and can be cultured easily in vitro with a high reproductive potential. They have a broad host range; are safe to vertebrates, plants, and other nontarget organisms; and are exempt from registration in most countries including the United States and Europe. The DJs are easily applied using standard spraying equipment, are compatible with many pesticides (not nematicides), and are amenable to genetic selection.

BIOLOGY

The DJ, ensheathed by the second-stage cuticle, carries the symbiotic bacterial cells in its intestinal tract. Once a suitable host is found, the DJ enters the host through natural openings (mouth, anus, or spiracles) or possibly wounds and penetrates into the hemocoele. The DJs of heterorhabditids also possess a dorsal tooth that may assist in the direct penetration through the host's integument. The bacterial cells, voided from the nematode's intestine into the haemolymph, propagate and kill the host by septicemia within 48 hours. The nematodes feed on the bacterial cells and host tissues, produce two or three generations, and emerge from the host cadaver as a new generation of DJs that search for new hosts.

The relationship between the nematode and the bacteria is known as mutualism. These nematodes normally are associated with only one species of bacterium. Some species of bacterium are associated with more than one nematode species (Table 1), but it is possible that there may be several isolates of a particular bacterial species that differ in their genetic make-up. In this association, the nematode relies on the bacterium (symbiont) for killing the insect host, creating a suitable environment for its development by producing antibiotics that suppress competing secondary microorganisms, breaking down the host tissues into suitable nutrients, and serving as a food source.

Host Range, Behavior, and Survival

Entomopathogenic nematodes and their bacterial symbionts kill insects very quickly (12–48 h) and do not form intimate, highly adapted host-parasite relationships characteristic of other insect-nematode infections (e.g., mermithids and allantonematids). These nematodes are pathogenic to hosts in all insect orders, a spectrum well

Encyclopedia of Pest Management

N

Table 1 Some of the more important species of *Steinernema*[a] and *Heterorhabditis*[a] and their respective bacterial symbiont

Nematode genus	Nematode species	Bacterial symbiont
Steinernema	*abbasi*	*Pseudomonas oryzihabitans*[b]
	arenarium (= anomali)	undescribed
	affine (- affinis)	*Xenorhabdus bovienii*
	carpocapsae	*X. nematophilus*
	feltiae	*X. bovienii*
	glaseri	*X. poinarii*
	intermedium (= intermedia)	*X. bovienii*
	karii	undescribed
	kushidae	*X. japonicus*
	rarum (= rara)	*Xenorhabdus sp.*
	riobrave	*Xenorhabdus sp.*
	scapterisci	*Xenorhabdus sp. and other genera*
Heterorhabditis	*bacteriophora*	*Photorhabdus luminescens*
	indica	*P. luminescens*
	megidis	*P. luminescens*
	zealandica	*P. luminescens*

[a] Several other species in both genera have been described (1).
[b] The only symbiont isolated so far from *S. abbasi* (1).

beyond that of any other microbial agent, which is the main reason for the current interest in this group of nematodes. The broad host range of these nematodes can be a two-edged sword, that is, the prospect of the large-scale release of such extreme generalists, lethal to almost every insect they have been tested against, generates images of indiscriminate biocides that may have negative environmental impacts. However, like other biological control agents, steinernematid and heterorhabditid nematodes attack a far wider spectrum of insects in the laboratory, where host contact is assured, environmental conditions are optimal, and no ecological or behavioral barriers to infection exist. Under field conditions these nematodes, although highly infective to a wide range of lepidopterous larvae in the laboratory, are quickly inactivated by environmental extremes (i.e., desiccation, radiation, and temperature) characteristic of exposed foliage. There is considerable evidence that entomopathogenic nematodes possess a restricted host range in natural systems and are therefore not as threatening as their experimental host range might suggest, most of the target hosts being found in the soil.

Field efficacy depends on understanding how nematodes locate, identify, and assess potential hosts. Firstly the nematode DJ must search for the host. Some species of nematode (e.g., *S. carpocapsae*) are found near the soil surface and are known as ''ambushers,'' while species of *Heterorhabditis* are found deeper in the soil where they actively search for hosts, that is, they are ''cruisers.'' The factors that affect active nematode dispersal and host finding in soil include pore space, moisture, plant roots, and temperature, the latter being the most important. Some temperate nematodes, such as *S. feltiae* and *H. megidis* are active at temperatures between 15 and 25°C and may infect hosts as low as 5°C, but other species isolated from the semiarid tropics, that is, *S. riobrave* and *S. abbasi*, are active between 25 and 35°C, while the activity of *S. carpocapsae* lies between 18 and 28°C.

Nematode survival is enhanced in soil where they are buffered from environmental extremes. Most nematodes isolated in temperate regions can survive low temperatures, and slow desiccation at high relative humidity enhances their survival but nematodes isolated in the tropics do not survive below 10°C. Not all DJs are infectious at the same time suggesting that these nematodes may have adopted a survival strategy by staggering their period of infectivity. The presence of the second-stage juvenile (J2) cuticle surrounding the DJs of steinernematids and heterorhabditids enhances survival. The J2 cuticle, easily lost in steinernematids but more difficult to remove in heterorhabditids, improves desiccation tolerance particularly in heterorhabditids.

Field Efficacy

Successful market penetration of nematode-based products depends on providing predictable control that can only be achieved by studying the interplay of abiotic and biotic factors. Nematodes can successfully develop in many different host species but optimal development differs markedly with nematode species or isolates, therefore screening of several nematode species or isolates against a particular target host is essential in any control program. The biology of the nematode, the target host, and the environment in which the nematode will be applied must be considered carefully when designing the control strategy. A large number of field trials is necessary to optimize protocols to achieve predictable control, particularly with respect to moisture (irrigation frequency, and rainfall), soil type, season (temperature), nematode isolate, and method of application. Many insects have been controlled successfully by nematodes (Table 2). Most of these insects live in the soil where the cryptic habitat protects nematodes from desiccation and UV light, buffering the nematodes from day-night temperature extremes and promoting contact between nematodes and insects living there.

Table 2 Some insect groups successfully controlled with steinernematid and/or heterorhabditid nematodes

Insect group	Common name	Major host plants/environment
Curculionidae	Root weevils	Banana[a], berries, citrus, forest seedlings, hops, mint, ornamentals, sweet potato, sugar beet
Chrysomelidae	Flea beetles	Mint, potato, sweet potato, sugar beet, vegetables
Scarabaeidae	White grubs	Berries, field crops, ornamentals, turf
Diptera		
Agromyzidae[b]	Leaf miners	Chrysanthemum, vegetables
Ephydridae	Shore flies	Ornamentals, vegetables
Phoridae	Phorid flies	Mushrooms
Sciaridae	Sciarid flies	Mushrooms, ornamentals, vegetables
Tipulidae	Crane flies	Turf, ornamentals
Muscidae	Flies	Animal-rearing facilities
Lepidoptera		
Noctuidae	Cutworms, army worms	Corn, cotton, peanuts, turf, vegetables
Pterophoridae	Plume moths	Artichokes
Pyralitae	Webworms	Cranberries, ornamentals, turf
Sesiidae	Crown/stem borers	Berries, cucurbits, ornamentals, shrubs
Cossidae	Leopard moth	Apple, pear
Orthoptera		
Gryllotalpidae[c]	Mole crickets	Turf, vegetables
Blattoidea		
Blattellidae	German cockroach	Apartments
Siphonaptera		
Pulicidae		Soil

[a] Target stage is adult using a baited trap.
[b] Above ground application in greenhouses.
[c] Target is nymphs and adults.

COMMERCIAL DEVELOPMENT

Mass Production

Until Rudolf Glaser first devised a method to culture *S. glaseri* on an artificial medium in 1940, entomopathogenic nematodes could be reared only in living insects. Larvae of the greater wax moth are commonly used today for in vivo culture but the costs are too high for large-scale production, but may be useful for cottage industries in developing countries. A monoxenic culture technique (known as the Bedding solid-media technique) achieves a much higher and more consistent yield and production costs can be reduced further by a semi-automated harvesting process. This technique was the method of choice in a decentralized society such as China for the control of the apple borer, *Carposina nipponensis*, and the tree-boring cossid moth, *Holcocercus insularis*.

The Bedding technique can maintain an output level of about 10×10^{12} nematodes per month. Above this point labor costs remain constant and significant. Production by monoxenic liquid fermentation costs less and it has been possible to produce *S. carpocapsae* and *S. feltiae* consistently and efficiently in 7500–80,000 liter fermenters with a yield capacity as high as 100,000 nematodes per ml. Of primary importance for achieving consistently high quality yields of nematodes are optimum aeration, sheer sensitivity, and stability of xenorhabdid bacterium from steinernematids. *Xenorhabdus* bacteria exist as two phases, primary and secondary; only the primary form produces antibiotics and supports nematode reproduction both in vivo and in vitro.

Formulation and Storage

Formulation of nematodes into a stable product has played a significant role in the commercialization of entomopathogenic nematodes. Active DJs must be immobilized to prevent depletion of their lipid and glycogen reserves. Although DJs can be immobilized at low temperatures (5–15°C) this is not necessarily commercially feasible and so nematodes were immobilized onto carriers such as polyacrylimide and alginate gels or by partially desiccating them on clay. Other techniques such as the development of flowable concentrates and immobilization on granules have had some success. Steinernematids, especially *S. carpocapsae*, can be maintained for up to 6 months on gels at room temperature and for up to 12 months under refrigeration. Most heterorhabditids do not have a long shelf life and are difficult to produce and formulate.

Application Methods

Control strategies involve the inundative application of DJs against a particular insect pest with short-term control being the primary goal. As with chemical insecticides, spraying nematodes directly onto the soil surface is the most commonly used method of application but nematodes can also be applied by drip or sprinkler irrigation systems. Field concentrations of approximately 2.5 billion nematodes per ha usually are applied to ensure that sufficient DJs will come in contact with target insects for adequate control. This high application rate is needed to overcome the negative impacts of abiotic and biotic factors in the soil environment. High nematode concentrations are also needed against certain insects that only remain in the soil for a few days and are small in size, for example, the cabbage root fly, *Delia radicum.* Large differences in field efficacy have been reported for different species and isolates. Thus, heterorhabditids are very effective against the Japanese beetle but not against the mole cricket. In contrast, the steinernematid, *S. scapterisci* will control mole crickets.

CONCLUSIONS

Entomopathogenic nematology has a relatively short history dating back to the pioneering research of R.W. Glaser and his colleagues in the 1930s. In the past 20 years there has been considerable advances in the knowledge about these nematodes resulting in the successful commercialization of products containing the DJs of these nematodes. Currently products containing *Steinernema riobrave* for the control of the citrus weevil complex, *S. feltiae* for the control of sciarid and phorid flies in mushrooms, and *S. carpocapsae* and *Heterorhabditis megidis* for the control of the black vine weevil, *Otiorhynchus sulcatus*, in various crops, are available. The current political atmosphere that favors the reduction in the use of chemical pesticides and promotes alternative methods is conducive to the introduction and widespread use of nematode-based insect control products where these are cost effective. The nematode-bacterium complex, unlike other insect pathogens, has been exempt from registration in the United States and elsewhere in the world. In recent years there has been considerable research on the bacteria in this complex to evaluate the status of the antibiotics produced by these bacteria. It remains to be seen whether these bacteria and their metabolites from ''biologically safe'' biological control agents will remain free of regulation.

REFERENCES

1. Boemare, N.; Richardson, P.; Coudert, F. In *Entomopathogenic Nematodes: Taxonomy, Phylogeny and Gnotobiological Studies of Entomopathogenic Nematode Bacterium Complexes*, Proceedings of a Workshop Held at HRI, Wellesbourne, Warwick, U.K., 1999; COST 819 (EUR 18832), 111.

FURTHER READING

Bedding, R.A.; Akhurst, R.; Kaya, H.K. *Nematodes and the Biological Control of Insect Pests*; CSIRO Publications: 1993; 178.

Gaugler, R.; Kaya, H.K. *Entomopathogenic Nematodes in Biological Control*; CRC Press: Boca Raton, FL, 1990; 365.

Georgis, R.; Manweiler, S.A. Entomopathogenic nematodes: a developing biological control technology. Agric. Zool. Rev. **1993**, *6*, 63–92.

Kaya, H.K.; Gaugler, R. Entomopathogenic nematodes. Annu. Rev. Entomol. **1993**, *38*, 181–206.

N

NEMATODE MANAGEMENT

Graham Stirling
Biological Crop Protection Pty. Ltd., Moggill, Queensland, Australia

INTRODUCTION

Nematodes are wormlike invertebrates that are found in a diverse range of habitats, including soil, freshwater, saltwater, and the tissues of plants and animals. There may be as many as a million nematode species, but only a small fraction of these have been described. Plant-parasitic species are equipped with a slender feeding apparatus known as a stylet that enables them to feed on the roots or foliage of plants. Some cause enough damage to be considered economically important pests. They are either ectoparasitic (i.e., they remain largely outside the plant and feed externally by inserting their stylet into plant cells), or endoparasitic (i.e., they enter plant tissue and feed internally). Some of the endoparasites remain migratory throughout their life cycle while others establish specialized feeding sites within roots, lose their wormlike shape and become sedentary. Economically important genera are listed (Table 1) according to their form of parasitism. Some of these nematodes feed above ground but most feed on roots, reducing the size and effectiveness of root systems or affecting the marketability of tubers, rhizomes, or other underground parts of plants. Nematodes reduce yields of most crops by 5–15%, with the sedentary endoparasites, particularly the cyst and root-knot nematodes, being the most important pests worldwide (1, 2).

EXCLUSION VERSUS MANAGEMENT

Although most economically important plant-parasitic nematodes are widely distributed, there are regions where certain species do not occur. The first form of defense against such pests is to limit their spread to new areas. Quarantine measures are therefore used at both national and regional levels to prevent the movement of infested soil or plant material into areas known to be free of particular nematodes. At a local level, dissemination of nematodes can be reduced by ensuring that nursery material is free of endoparasitic species and that nematodes are not transported from place to place in irrigation water, on soil adhering to farm machinery, or by other means.

Once an area is infested with pest nematodes, eradication is rarely an option and the cropping system must be managed to minimize losses from nematodes. This can be achieved by manipulating either the nematode population density or the resistance/tolerance status of the crop. Since the cropping system, economic and environmental considerations, and the presence of other soilborne pests and pathogens influence the range of control options, a systems approach must be adopted when developing a nematode management program. Nematode management is therefore an integral component of total crop management.

INTEGRATED NEMATODE MANAGEMENT

The concept of using a variety of techniques in a compatible manner to reduce pest damage to tolerable levels (usually referred to as integrated pest management or IPM) is as appropriate for nematodes as it is for other pests. The appeal of IPM is that it provides acceptable procedures for managing pests in sustainable agricultural systems. In the case of nematodes, population densities are measured and control tactics are implemented only when infestation levels are above the economic threshold. When measures are required to reduce nematode populations, the economic and environmental impacts of possible control options are considered and the safest and most effective tactics are chosen (3, 4).

Nematode Monitoring

Monitoring involves regularly collecting soil and root samples from a field and processing the samples to determine the nematode species that are present and to estimate their population densities. Because it is a labor-intensive process that can be relatively expensive, moni-

N

Table 1 Some economically important genera of plant-parasitic nematodes

Mode of parasitism	Common name (genus)
Sedentary endoparasite	Root-knot (*Meloidogyne*), cyst (*Heterodera, Globodera*), reniform (*Rotylenchulus*), citrus (*Tylenchulus*), false root-knot (*Nacobbus*)
Migratory endoparasite	Lesion (*Pratylenchus*), burrowing (*Radopholus*)
Ectoparasite	Dagger (*Xiphinema*), stubby (*Trichodorus, Paratrichodorus*), needle (*Paralongidorus*), sting (*Belonolaimus*), spiral (*Helicotylenchus, Rotylenchus*), lance (*Hoplolaimus*), sheath (*Hemicycliophora*), ring (*Criconemella*), stunt (*Tylenchorhynchus*), pin (*Paratylenchus*)
Above-ground parasite	Stem and bulb (*Ditylenchus*), seed and leaf gall (*Anguina*), leaf (*Aphelenchoides*)

toring is not possible in all situations where nematode management decisions are made. It is probably most useful in moderate to high value crops, where decisions are often made about whether a nematicide should be applied.

In most monitoring programs, 20–50 soil or root samples collected from a defined area of a field are mixed together and a single subsample is processed to limit costs. To control obvious sources of variability in count data, systematic sampling patterns are generally used within areas that are selected for soil uniformity or symptom expression. Samples are generally collected before planting when most nematode management decisions must be made. However, samples from perennial crops will be collected at regular intervals during the life of the crop.

Samples are processed in laboratories equipped to work with nematodes. In situations where the key pest (i.e., a nematode species that will cause a significant reduction in crop yield unless some control action is taken) produces distinctive root symptoms, bioassays are sometimes the best way of processing samples.However, it is more common to extract nematodes from roots or soil and count them. A range of extraction techniques is available but whatever procedure is employed, quality control procedures must be adequate to ensure reliability. Several plant-

Table 2 Nematode and crop management strategies to reduce losses from plant-parasitic nematodes

Strategy/tactic	Comments
Exclusion-avoidance	
Quarantine	Effective for many serious pests
Use clean planting material	Useful in new areas for crop-specific pests
Avoid transfer of infested soil/water	Limits dissemination at a local level
Reduce initial population density	
Weed-free fallows	Cropping with a nonhost crop is preferable to bare fallow because of erosion
Crop rotation	Widely used and effective
Organic amendments	High application rates are needed
Interrow crops, antagonistic plants	Limited efficacy
Tillage	Effects limited to upper soil layers
Heat	Effective for planting material and glasshouses
Solarization	Limited by climate and depth of control
Flooding	Impractical in most cropping systems
Resistant cultivars	Effective against cyst and root-knot nematodes
Fumigants and nematicides	Effective; limited by high mammalian toxicity and off-target effects
Biological nematicides	Few products available; unreliable
Suppress nematode multiplication	
Resistant cultivars	Effective if suitable cultivars are available
Organic amendments	Relies on maintaining a biologically active soil
Nematicides (postplant)	Limited by cost and residue problems
Biological control	Most useful in perennial crops
Improve crop tolerance to nematodes	
Tolerant cultivars	Useful, but carryover high nematode densities
Optimum crop management	Water and nutrient management most important

parasitic nematodes are usually present in most samples but it may be necessary only to identify and enumerate species that are key pests.

Prediction of Crop Loss

Before data on nematode densities in particular fields can be used to make nematode management decisions, information must be available on the general relationship between nematode population density and yield. Models that describe such relationships have been developed for many nematode-crop systems by fitting preplant nematode densities to yield data obtained from pot, microplot and field experiments. The nematode population density above which some damage could be expected can be determined from such models and provides a rational basis for forecasting whether losses due to nematodes are likely in a future crop.Because environmental factors and the standard of crop management influence the damage threshold, the threshold levels that have been published in the literature should be used conservatively until their applicability to a particular farming system is confirmed.

Management Tactics

In situations where the likelihood of nematode damage is high enough to warrant taking action, a wide range of management tactics can be used to reduce crop losses from nematodes (5–7). Some of the most useful are listed in Table 2. Although tactics for reducing the number of nematodes feeding on a crop are usually seen as the primary means of minimizing damage caused by an existing nematode infestation, improving the crops tolerance to nematodes can also reduce losses. Since crops that are not stressed for water or nutrients are better able to withstand attack by nematodes, any action taken to improve the standard of crop husbandry will improve the yield of nematode-infested crops.

The range of tactics available in a particular situation will depend on the value of the crop and the intensity of the cropping system. Thus, in low value crops that are grown over large areas with relatively low inputs of fertilizer and pesticides, the main options are crop rotation, resistant cultivars, and a variety of cultural controls. A greater range of options is available in horticultural, ornamental, and vegetable crops but soil fumigation or routine application of nematicides has dominated control practices in high value crops for many years. However, a continuing decline in the availability of chemicals has resulted in a move toward nonchemical controls. Since they are not always as effective as chemicals, several potentially compatible nonchemical control measures may have to be used together to achieve the level of control previously obtained with soil fumigants and nematicides.

THE FUTURE

Despite the fact that plant-parasitic nematodes are part of a complex soil food web, organisms that occupy the same ecological niche as nematodes have played little or no role in nematode management in the past. However, this situation is changing. There is now greater awareness of the benefits of enhancing microbial activity and diversity in soil, the plant growth-promoting effects of some soil microorganisms are better understood, interactions between populations of free-living and plant-parasitic nematodes are being studied, and attempts are being made to better utilize the suppressive effects of the soil microflora and microfauna on nematodes. New strategies for managing nematodes are likely to emerge from research in these and related areas.

Developments in molecular biology are also likely to produce new tools for nematode management. Progress in molecular technologies for identifying and quantifying nematodes in soil will improve IPM systems by decreasing the cost of processing nematode samples and by facilitating a more complete characterization of the diverse range of nematode trophic groups that occur in soil. Because it is now technically possible to transfer genes encoding products detrimental to a target nematode, and to identify, clone and then introduce resistance genes into susceptible crop plants, nematode resistance eventually should be available in a greater range of crops. Provided genetically manipulated food crops are accepted by consumers, such technologies will play an important role in the nematode management programs of the future.

REFERENCES

1. *Plant Parasitic Nematodes in Temperate Agriculture*; Trudgill, D.L., Webster, J.M., Evans, K., Eds.; CAB International: Wallingford, UK, 1993; 1–648.
2. *Plant Parasitic Nematodes in Subtropical and Tropical Agriculture*; Sikora, R.A., Bridge, J., Luc, M., Eds.; CAB International: Wallingford, UK, 1990; 1–629.
3. Duncan, L.W. Current options for nematode management. Annu. Rev. Phytopathol. **1991**, *29*, 469–490.
4. Barker, K.R.; Koenning, S.R. Developing sustainable systems for nematode management. Annu. Rev. Phytopathol. **1998**, *36*, 165–205.
5. *Principles and Practice of Nematode Control in Crops*; Kerry, B.R., Brown, R.H., Eds.; Academic Press: Sydney, 1987; 1–447.
6. Whitehead, A.G. *Plant Nematode Control*; CAB International: Wallingford, UK, 1998; 1–384.
7. Stirling, G.R. *Biological Control of Plant-Parasitic Nematodes*; CAB International: Wallingford, UK, 1991; 1–282.

NONCHEMICAL OR PESTICIDE-FREE FARMING

N

Inger Källander
Swedish Ecological Farmers Association, Katrineholm, Sweden

Gunnar Rundgren
GroLink, Höje, Sweden

INTRODUCTION

Agriculture methods that exclude pesticides and other chemical inputs are spreading rapidly all over the world. Commonly known as organic farming, this agriculture model is increasingly recognized by farmers, consumers, environmentalists and policy-makers as a way to improve environmental, social, and economical sustainability in food production. This entry explains the concept of Organic Farming and gives an overview of the current situation, growth factors and dynamics, and basic practices.

DEFINITION OF CONCEPTS

Nonchemical or Pesticide-Free Farming as a defined agriculture concept and method is known by different names. Organic farming is the term most used worldwide (United States, England, Africa, and Asia) and will be the term used in this article. Other names for the same concept are ecological agriculture (Nordic countries), Agricultura Ecológica (Latin America), and agriculture biologique (France). The International Federation of Organic Agriculture Movements (IFOAM) consists of 750 member organizations in more than 100 countries. The organic farming objectives and basic principles are defined by the worldwide membership in the following way:

Organic Production and Processing is based on a number of principles and ideas. They all are important and are not necessarily listed here in order of importance.

To produce food of high quality in sufficient quantity.
To interact in a constructive and life-enhancing way with natural systems and cycles.
To consider the wider social and ecological impact of the organic production and processing system.
To encourage and enhance biological cycles within the farming system, involving microorganisms, soil flora and fauna, plants and animals.
To develop a valuable and sustainable aquatic ecosystem.
To maintain and increase long term fertility of soils.
To maintain the genetic diversity of the production system and its surroundings, including the protection of plant and wildlife habitats.
To promote the healthy use and proper care of water, water resources, and all life therein.
To use, as far as possible, renewable resources in locally organized production systems.
To create a harmonious balance between crop production and animal husbandry.
To give all livestock conditions of life with due consideration for the basic aspects of their innate behaviour.
To minimize all forms of pollution.
To process organic products using renewable resources.
To produce fully biodegradable organic products.
To produce textiles which are long lasting and of good quality.
To allow everyone involved in organic production and processing a quality of life which meets their basic needs and allows an adequate return and satisfaction from their work, including a safe working environment.
To progress toward an entire production, processing and distribution chain which is both socially just and ecologically responsible.

—*From IFOAM Basic Standards 1998*

RAPID GROWTH WORLDWIDE

Organic farming is rapidly increasing in all parts of the world. In 1998 the acreage covered by organic farming in Europe was approximately 2% with a great variability between the EU member countries, from less than 1% to more than 10% (1). Other countries with a well-developed

Table 1 Certifield and policy-supported organic and in-conversion land area in Europe[a]

	Year end								
	1985	1986	1987	1988	1989	1990	1991	1992	1993[b]
European Union									
Austria	5880	7000	8400	12,320	16,674	21,546	27,580	84,000	**135,982**
Belgium	500	700	**972**	1000	1200	1300	1400	1700	**2179**
Denmark	4500	4800	5035	**5881**	**9553**	**11,581**	**17,963**	**18,653**	**20,090**
Finland	1000	1200	1400	**1500**	**2300**	**6726**	**13,281**	**15,859**	**20,340**
France	45,000	50,000	55,000	60,000	65,000	72,000	**81,225**	85,000	**87,829**
Germany-cert	**24,940**	**27,160**	**33,047**	**42,393**	**54,295**	**90,021**	**158,477**	**202,379**	**246,461**
Germany-other	0	0	0	0	0	15,000	30,000	96,742	126,382
Greece	0	0	0	50	100	150	200	250	**591**
Ireland	1000	1100	**1300**	1500	3700	3800	**3823**	5101	**5460**
Italy	5000	5500	**6000**	9000	11,000	**13,218**	**16,850**	30,000	**88,437**
Luxembourg	350	400	**412**	450	550	600	**634**	500	**497**
Netherlands	**2450**	**2724**	**3384**	5000	**6544**	**7469**	**9227**	**10,053**	**10,354**
Portugal	50	200	**320**	420	550	1000	2000	2000	**3060**
Spain	**2140**	2500	**2714**	3000	3300	**3650**	**4235**	**7859**	**11,675**
Sweden-cert	**1500**	**2500**	**4870**	**8598**	**23,600**	**28,500**	**31,968**	**33,267**	**36,627**
Sweden-other	0	0	0	0	**5092**	**4890**	**5775**	**7161**	**7916**
United Kingdom	6000	7000	**8500**	11,000	**18,500**	31,000	34,000	35,000	**30,992**
EU 15	100,310	112,784	131,354	162,112	221,1958	312,451	438,638	635,524	834,872
EU 15 (% change)		12%	16%	23%	37%	41%	40%	45%	31%
EU Agenda 2000 accession countries									
Cyprus									
Czech Rep					**260**	**3480**	**17,507**	**15,371**	**15,667**
Estonia							**500**	**2350**	1600
Hungary							**2500**	**5400**	6400
Poland					**300**	**550**	**1240**	**2170**	**3540**
Slovenia								**70**	70
Ag 2000	0	0	0	0	560	4030	21,747	25,361	27,277
European Free Trade Association									
Iceland									
Liechtenstein									
Norway	90	114	246	312	534	1578	**2444**	**3225**	**3778**
Switzerland	4830	5520	6630	7275	10,080	12,045	14,100	**17,300**	**20,784**
EFTA	4920	5634	6876	7587	10,614	13,623	16,544	20,525	24,562
EU 21/EFTA	105,230	118,418	138,230	169,699	233,132	330,104	476,929	681,410	886,711
% change		13	17	23	37	42	44	43	30
Other Europe									
Albania									
Andorra									
Belarus									
Bosnia-H									
Bulgaria									
Croatia								**100**	50
Georgia								**100**	100
Latvia								**1240**	1250

1994	1995	1996	1997	1998 est	Utilized agric area		Average annual growth rate		
					000 ha (1995)	% organic (1998)	1 year (1997-98)	5 year (1993-98)	10 year (1988-98)
192,337	**335,865**	**309,089**	**345,375**	350,000	3449	10.15	1.3	24.2	48.0
2683	**3385**	**4261**	6418	7000	1336	0.52	9.1	27.0	22.0
21,145	**40,884**	**46,171**	**64,329**	96,979	2715	3.57	50.8	40.3	35.2
25,822	**44,695**	**84,555**	**102,335**	**126,175**	2605	4.84	23.3	46.7	62.4
94,806	**118,393**	**137,084**	**165,405**	**218,790**	30,277	0.72	32.3	20.3	14.1
272,139	**309,487**	**354,171**	**389,693**	416,518	17,344	2.40	6.9	11.1	27.5
173,128	152,062	121,575	60,307	0	17,344	0.00	– 100.0	– 29.1	#DIV/0!
1188	**2401**	**5269**	**10,000**	10,000	5741	0.17	0.0	82.5	75.7
5390	**12,634**	**20,496**	**23,591**	34,696	4444	0.78	47.1	51.5	44.8
154,120	**204,494**	**334,176**	641,149	830,000	17,294	4.80	29.5	58.3	63.4
538	**571**	**594**	**618**	**785**	127	0.62	27.0	9.9	6.5
10,975	**11,486**	**12,385**	**16,660**	20,000	1981	1.01	20.0	14.6	15.4
7267	**10,719**	**9191**	**12,193**	**24,902**	3981	0.63	104.2	61.5	57.3
17,209	**24,079**	**103,735**	**152,105**	269,465	25,092	1.07	77.2	108.4	71.3
47,921	**83,326**	**113,571**	**117,669**	127,000	3438	3.69	7.9	30.5	37.4
6930	3498	48,741	87,516	130,000	3438	3.78	48.5	271.9	#DIV/0!
32,476	**48,448**	**49,535**	106,000	250,000	15,852	1.58	135.8	61.2	44.3
1,066,074	1,406,427	1,754,599	2,301,363	2,912,310	135,676	2.15	26.5	28.4	33.6
28%	32%	25%	31%	27%	29.9%				
15,818	**14,127**	**17,021**	**20,239**	71,746	**4166**	0.49	254.5	56.8	#DIV/0!
1600	3000	3000	3000	3000	1380	0.22	0.0	17.5	#DIV/0!
8630	**12,325**	9300	16,687	17,000	6136	0.27	1.9	26.9	#DIV/0!
5000	**6855**	8000	9000	10,000	18,743	0.05	11.1	23.7	#DIV/0!
70	70	70	70	70	864	0.01	0.0	0.0	#DIV/0!
31,118	36,377	37,391	48,996	101,816	31,289	0.16	107.8	34.5	#DIV/0!
146	717	1082	1288	1300	2280	0.06	0.9	#DIV/0!	#DIV/0!
		630	630	650	3.5	18.00	3.2	#DIV/0!	#DIV/0!
4520	**5768**	**7897**	**11,706**	12,000	**1005**	1.16	2.5	27.0	50.6
25,230	**31,815**	**58,741**	**71,790**	79,680	**1083**	6.63	11.0	33.1	28.3
29,896	38,300	68,350	85,414	93,630	4371.5	1.95	9.6	32.6	29.6
1,127,088	1,481,104	1,860,340	2,435,773	3,107,756	171,337	1.42	27.6	28.5	33.9
27	31	26	31	28	30.1				
					1100	0.00			
					26	0.00			
					9391	0.00			
					1940	0.00			
	15	15	15		6154	0.00			
50	**120**	120	120		2233	0.01			
100	100	100	100			#DIV/0!			
1250	**1147**	1200	1200		2530	0.05			

(Continued on next page)

Table 1 Certifield and policy-supported organic and in-conversion land area in Europe[a] (*Continued*)

	Year end								
	1985	**1986**	**1987**	**1988**	**1989**	**1990**	**1991**	**1992**	**1993**[b]
Lithuania									**148**
Macedonia									
Malta									
Moldova								**600**	600
Monaco									
Romania									
Russia									3610
San Marino									
Slovakia						**15,140**	**14,773**	**14,700**	**14,724**
Turkey						1037		6077	**5216**
Ukraine									
Yugoslavia									**360**
Other Europe	0	0	0	0	0	16,177	14,773	22,817	26,058
Europe	**105,230**	**118,418**	**138,230**	**169,699**	**232,572**	**342,251**	**469,955**	**678,866**	**885,492**

[a]Based on data supplied direct or published up to 28/5/99 (bold) and provisional estimates subject to confirmation. Publication in graphical format preferred because accuracy of data varies between countries. Data are frequently updated—please request current version before publication. Copyright: Nicolas Lampkin, Welsh Institute of Rural Studies, University of Wales, Aberystwyth, GB-SY23 3AL.

organic sector are the United States (approx. 500,000 ha), Argentina (280,000 ha) and Australia (350,000 ha) (2). The retail sales value of the European market for organic food in 1998 was estimated to be EUR 5—7 billion and in the United States 4.5 billion with an annual growth by 20–30%, a development trend that is expected to continue (2). The growth is clearly the strongest in countries with efficient policy support programs and a dynamic market development (Table 1).

The reason for the growing interest in organic farming as a production model is its recognized positive environmental, social, and economical effects that enhance food security and food safety:

- increased biodiversity due to diversified production systems, the omission of pesticides, and a greater interest to preserve and use local crops and varieties
- reduced nutrient losses and leakage due to cover crops, recirculation, the integration of animals and crops, the omission of chemical fertilizers, a high degree of self-sufficiency in animal fodder, and high nutrient efficiency
- reduced erosion and better water management due to good soil management practices such as mulching, green manuring, and crop diversification
- reduced use of nonrenewable resources due to the omission of chemical fertilizers and the use of onfarm-produced fodder and feedstuffs replacing imported feed components with long transportation
- safer working conditions for farmers and farm workers, and
- better food quality taking into account not only the reduced risk of contamination of food and water but also environmental characteristics of the production system, working conditions of the people involved, and an ethical dimension.

While developed countries put a strong focus on the environmental and health effects, many third world countries increasingly recognize organic farming as a tool to combat poverty, food insecurity, and migration from the rural areas. Numerous examples show how the social and economical situation for the farmers and farm workers can be improved by an optimum use of local resources and cheap technology to reduce the depletion of natural resources and health problems caused by pesticides. In general, organic farming and marketing tend to show new models for economical and social development in agriculture and the rural areas.

ORGANIC FARMING—A HOLISTIC APPROACH

Organic farming can be described as a holistic and systematic approach to agriculture based on two founding principles: reduced use of nonrenewable resources and precaution. The first principle implies optimum use and

1994	1995	1996	1997	1998 est	Utilized agric area	Average annual growth rate
267	**582**	**1118**	**1568**	**4006**	3524	0.04
					1306	0.00
					13	0.00
600	600	600	600		2560	0.02
						#DIV/0!
	980	1000	1001		14,790	0.01
11,941	20,000	20,000	20,000		210,303	0.01
						#DIV/0!
14,762	**18,813**	**27,661**	**27,800**	**50,695**	2446	1.14
5196		**15,250**				#DIV/0!
					41,929	0.00
350	350	350	350		5850	0.01
34,516	42,707	67,414	52,754			
1,130,486	**1,487,434**	**1,890,363**	**2,439,531**			

recirculation of local renewable resources, local production-consumption systems, and enhancing soil fertility and biodiversity. The latter implies exclusion of pesticides and other pollutants as well as genetically modified organisms (GMOs). The common basic principles for organic production are the following:

Diversity is the foundation of organic agriculture. Diversification is designed according to local climatic conditions and available natural resources as well as the social and cultural situation. The size of the farm, available labor, buildings, machinery, and the market access are factors that effect the planning of the production. Crop rotation or succession, intercropping, and agroforestry are examples of diversification methods. The diversity provides balanced nutrition for plants and animals and a natural crop protection. It also makes the total production less vulnerable to falling prices or reduced yields.

Soil management is crucial for all the activities on the farm. Measures to improve microlife activity and soil structure are prerequisites to achieve a balanced nutrient supply, resistance against pests, and a high product quality. Careful soil labor, coverage with living plants, and a supply of organic matter through green manure, mulch, and compost are important measures to activate and increase the microorganisms and build up soil fertility.

Nutrient supply relies on recirculation of organic matter and efficient use of available nutrients. In addition nitrogen-fixing plants (legumes) are used. Farmyard manure is stored and distributed in such ways that it is efficiently used without leaking to the surrounding environment. A certain addition of minerals and lime sometimes may be needed. In those cases the certification body provides a list of allowed inputs.

Green manure, living plants that are used to fertilize the soil, has many positive effects such as nitrogen fixation, improvement of soil structure, organic matter supply, animal fodder, crop protection, erosion reduction, water holding capacity, and weed control. Green manure can be grown as a main crop (clover, beans, peas), intercropped with various species (clover-grain, beans-corn), or grown as a cover crop after harvest of a main crop (rye, mustard).

Weed management consists of preventive measures (crop rotation, intercropping, green manure, and the competitiveness of strong crops) and technological measures (cultivation, delayed sowing, weed harrowing, hoeing, brushing, flame-weeding, and hand-weeding). Several new methods are under development.

Pest management consists, to a great extent, of planning and preventive measures. Strategies have to be elaborated according to the biology of each specific crop and its pests. Polyculture and diversity benefit predators. Balanced fertilization, crop rotation, and intercropping, resistant varieties, and green manure are important measures. Some direct control measure include mulching and biological control with soap, plant extracts, or living organisms (like predators, beneficial bacteria, fungi, and nematodes).

Animal husbandry is planned in a way that grants the animals a life according to their natural needs and behavior. The well-being of the animal is the basis for good animal health and good economy. Basically animals are kept free-range outdoors and babies of mammals suckle their mothers. The animals are allowed to live in a herd and to make up their own ranking order. They should be fed the kind of feed they are naturally adapted to digest. Mixing of animal species reduces parasites and adds to the well being of the animals.

A PRODUCER–CONSUMER DRIVEN MOVEMENT

The objectives of organic farming are shared by an increasing number of consumers. In the consumer's choice of an organic product lies not only the hope of healthy food, but also the satisfaction of being able to encourage sustainable and fair food production and trade. The concept of organic farming is communicated between producers and consumers through certification and labeling and the consumer trust in organic products is built on the transparency and guarantee of the certification system. In addition to private certification and labeling systems, governments have increasingly started to regulate the marketing of organic products. The European Union passed regulation 2092/91 in 1991 and the United States passed the Organic Foods Production Act (OFPA) in 1990. Japan is introducing marketing regulations in the year 2000. International standards are developed by Codex Alimentarius and by IFOAM. IFOAM has had international standards for organic agriculture since 1980.

In Europe organic farming is becoming a key policy issue. The present EU agri-environmental policy which contains special development programs to encourage organic farming. One reason to support development is to respond to increasing consumer demand. Another reason is a gradual convergence of EU policy goals with the objectives of organic farming, including environmental protection, animal welfare, resource use sustainability, food quality, and safety, financial viability, and social justice. The EC Regulation 2078/92 provides financial security for farmers under and after conversion. The United Nations FAO has also identified the potential of organic agriculture, both as a market opportunity and as a sustainable production method (3).

ORGANIC FARMING—A MODEL FOR SUSTAINABLE DEVELOPMENT

As a model for agriculture the organic farming concept has several advantages; it addresses all the mentioned producer and consumer objectives simultaneously, it uses the market mechanisms to achieve the goals, and it is defined and recognized globally. It must however be stressed that organic farming should not be seen as a ready package of solutions to different problems in agriculture. Organic agriculture is not primarily a farming technology. It should rather be understood as a model for development with democratically decided objectives, principles, and guidelines and a political tool for development based on theory and practice. Given serious efforts in research, development, extension, and education organic farming has a great potential to improve and increase the food situation in the world today and in future.

REFERENCES

1. Lampkin, N.; Foster, C.; Padel, S.; Midmore, P. *The Policy and Regulatory Environment for Organic Farming in Europe*, 1999.
2. Rundgren, G. *Development of Organic Agriculture*; International Training Programme, 1999.
3. FAO. Organic Agriculture, Report to the FAO Committee on Agriculture, Jan. 25–26, 1999.

NORTH AMERICAN FOREST LOSSES DUE TO INSECTS AND PLANT PATHOGENS

N

Elizabeth Harausz
David Pimentel
Cornell University, Ithaca, New York, U.S.A.

INTRODUCTION

Insect and pathogen losses in North American forests are a serious problem. In the United States insects and plant pathogens destroy 59.5 million ha annually, a volume loss of 158 million cubic meters (1). The total forested land in the United States is 298.1 million ha, so this is a loss of 20% of total forest production (2). In 1995 this corresponded to a $1.5 billion loss of timber and timber-related products (1). Canada loses an estimated 108 million cubic meters of timber annually out of a total of 284 million cubic meters, which is about 38% loss of total forest production (3).

LOSSES DUE TO INSECTS

The annual U.S. loss to insects is 42.1 million ha, or 14% of forested lands (1). Of this, 6.1 million ha is due to major defoliators (Douglas-fir tussock moth, gypsy moth, spruce budworm, and the western spruce budworm) and 4 million ha to the major bark beetles (the Montana bark beetle and the southern pine beetle) (1). In Canada during 1982–1986, average annual loss to insects was 63 million cubic meters, or 22% of total forested lands (3).

There are 60.7 million ha of spruce and fir forest in the United States and Canada, prime targets of spruce budworm (4). These insects prefer mature and overmature balsam fir, but during epidemics they will infect many types of conifers and even hemlock, tamarack, and pine (4). In the United States there are approximately 4.9 million ha of spruce and fir forests (4). From 1979 to 1983 an average of 1.6 million ha were defoliated each year (4). Canada also suffers from severe spruce budworm infestations. In 1992, over 10 million ha forests sustained tree mortality or moderate (30–69%) to severe (70–100%) defoliation (5).

Other budworms such as the western spruce budworm, the jackpine budworm, and the eastern blackhead budworm also cause significant damage. The western spruce budworm attacks conifers in western North America. However in British Columbia, Douglas fir is the primary target (5). In a peak outbreak in 1987, 800,000 ha of British Columbia forests were affected (5). The jackpine budworm caused 294,000 ha of moderate to severe defoliation (greater than 30% defoliation trees) in Canadian forests in 1995 (6). A 1987 outbreak in Canada of eastern blackhead budworm affected 35,000 ha of balsam fir forests. In 1991, 12,400 ha of trees sustained 30% or greater defoliation, including mortality (5).

The gypsy moth is another major defoliator. In the northeastern U.S., this moth prefers oaks but in outbreaks it will feed on many tree species (4). Insects weaken trees and leave them susceptible to attacks by other insect pests and diseases (4). A tree's radial growth may be reduced as much as 30–50% because of serious defoliation (4). During 1981 in the northeastern U.S., 5.2 million ha were defoliated, the worst gypsy moth outbreak on record (4). Ontario sustained significant damage in 1991, when the gypsy moth defoliated trees on 347,415 ha forest (5).

The forest tent caterpillar is the most prominent aspen defoliator in Canada. It defoliated at least 30% of the foliage on 16.3 million ha of trees in 1992 (5). In 1995 this decreased to 294,000 ha (6). However, by 1996 this damaged area increased again to 854,269 ha of moderate to severe defoliation (7).

The mountain pine beetle is one of the most damaging bark beetle species (4). In the United States this beetle occurs largely in the west. It mostly attacks mature and overmature ponderosa pines, sugar pines, and western white pines, but white bark pines and limber pines are also at risk (4). Since 1980 the mountain pine beetle has killed 30 million U.S. trees (4). In the western U.S. the area infested in 1983 totaled 1.5 million ha and the number of trees killed was about 10 million (4). The area infested within these states was 2% of the total forested land (1). In Canada lodgepole pine are mostly affected by the mountain pine beetle. In 1992, 44,750 ha of pines suffered 30% or greater defoliation, including mortality (5). This beetle was the most damaging pest in 1992, in British Columbia (5). In this providence alone, the beetle killed about 220 million mature pines during a 20-year period (5).

The southern pine beetle is another highly destructive pest. It attacks southern yellow pines, with shortleaf and loblolly pines as principal hosts (4). In 1983, in the southern U.S. states that had outbreak level counties, 4.6 million ha were infested, destroying 3.1 million cubic meters of trees (4). The infested area covered 6.6% of the total forested land in these states (1). A record setting infestation of southern pine beetle occurred in 1995, in South Carolina. An estimated $107 million of timber was destroyed by the beetle (8). This was a greater loss than the beetle had created cumulatively from 1960 to 1994 (8).

The spruce beetle infects conifers, mostly white and Engelmann spruce. Damage to trees, such as blowdowns or cuttings, creates opportunities for beetle infestations. In 1992, 87,000 ha of Canadian spruce were infected with the spruce beetle (5).

LOSSES DUE TO PLANT PATHOGENS

The United States loses 17.4 million ha of timber to plant pathogens each year (1). This loss is out of 298.1 million ha of total U.S. forested lands, or 6% of total forest production (2). From 1982 to 1986 Canada lost an estimated 45 million cubic meters annually (or about 16% of the potential production) to major diseases (3). The three major diseases in the U.S. are root diseases, dwarf mistletoe, and fusiform rust (1). Other major North American pathogens include hypoxylon cankers and white pine blister rust (3). Diseases cause timber losses through tree death, growth reduction, decay, deformations, cull, predisposition to windthrow, and vulnerability to pest insects and other diseases (3). Indirect losses include slow regeneration, reduced stock, stand composition changes, loss of site quality, reduced management options, monetary loss from having to replace stock, and the need to institute special management practices (3).

Root and stem rots are arguably the most damaging forest diseases, causing decay in the roots, butts, and stems of trees (3). These fungi reduce timber volume by killing trees, reducing growth, causing butt and stem cull, and making the trees vulnerable to other pests and to windthrow by weakening the root system (3). It is difficult to determine the damage done by root diseases because they often kill trees that are already weakened by other factors such as environmental conditions (e.g., water shortage) and other insects and diseases (3). However, losses due to growth retardation, mortality, and wood destruction are estimated to be 24 million cubic meters of timber per year in Canada (3). The annual loss due to tree mortality in the Pacific, Rocky Mountain, and southeastern states in the United States from 1979 to 1983 was estimated to be approximately 79 million cubic meters (4).

Stem rots and decays destroyed 25 million cubic meters of timber from Canadian forests in 1985 (3). Similar to root rot, stem rot is often found in trees with multiple infections including other pathogens or insects (3). These rots decrease trunk strength and pulp yield and quality (3). It is difficult to determine the full effects of stem rot because the disease is often hidden inside the trees and the external indicators can be unreliable (3).

Dwarf mistletoe is a parasitic flowering plant that grows on many economically valuable conifers (3). Dwarf mistletoe attacks trees of all ages, decreasing growth, wood quality, tree vigor, and causing tree mortality (3). The toal growth loss resulting from dwarf mistletoe in Canada for 1987 equaled 3.5 million cubic meters (3). In the U.S. western states, from 1979 to 1983, 9.3 million ha were infected (4). This is 6.4% of the total forested land (1). The annual loss due to this infestation was 11 million cubic meters of timber (4). These numbers for the United States are conservative because not all lands are included, only National Forest lands were tallied for Arizona, New Mexico, Colorado, and eastern Wyoming (4).

Fusiform rust is most prevalent in the southeast United States among slash and loblolly pines, being particularly severe on plantations (4). Mortality occurs primarily in young stands with the older stands suffering decreases in the quality of timber (4). In the southeastern states during 1983, fusiform rust caused a loss of 3.1 million cubic meters of timber, a 1983 monetary loss of $100 million (4).

White pine blister rust is a perennial fungal disease of the stem and branches of eastern white pine and western white pine (3). It kills approximately 0.6 million cubic meters of white pine a year in Canada (3). It is believed to be partly responsible for the elimination of white pine in some parts of eastern Canada (3). Hypoxylon cankers are also a significant threat to forests. Canada loses 11.2 million cubic meters of timber due to mortality each year from these cankers (3).

Dutch elm disease is another serious disease. In Manitoba, Canada from 1991 to 1992 16,000 diseased elms were removed (5). At this time it was forecasted that an additional 13,400 trees would be removed from 1992 to 1993 (5).

N

CONCLUSION

Clearly, pathogen and insect damage cause substantial losses throughout North American. To recap, 59.5 million ha are destroyed annually in the United States by pathogens and insects, a volume of 158 million cubic meters (1). This is a loss of $1.5 billion (1995) of timber and timber-related products (1). Canada loses an estimated 108 million cubic meters of timber annually (3). It should also be pointed out that the actual losses caused by these pests are much greater than the value of timber products. Tourism and recreation suffer as well as aesthetics, ecological balance, and animal habitat, to mention a few other effects. There are a variety of methods to reduce pest losses that can be used in isolation or in combination with one another. Direct forms of control include the use of biological controls and/or chemical controls and/or improved silvicultural methods (3). Indirect controls include the elimination of diseased and damaged trees through sanitation. In addition, tree losses can be reduced by increasing species diversity, increasing genetic diversity, planting tree species that are resistant to diseases and insect pests, reducing logging injuries, and reducing silvicultural practices that may increase a tree's susceptibility to diseases and insect pests (3).

REFERENCES

1. Harausz, E.P. The effects of insects and pathogens and the use of pesticides in North American forests. Agric. Ecosyst. Environ., *submitted.*
2. U.S. Department of Agriculture. *Agricultural Statistics, 1995-1996*; GPO: Washington, 1995–1996.
3. Singh, P. Research and management strategies for major tree diseases in Canada: synthesis. For. Chron. **1993**, *69* (2), 151–1062.
4. Loomis, R.C.; Tucker, S.; Hofacker, T.H. *Insect and Disease Conditions in the United States 1979–1983*; U. S. Department of Agriculture: Washington, DC, 1985.
5. Hall, P.J.; Bowers, W.W.; Carew, G.C.; Magasi, L.P.; Hurley, J.E.; Lachance, D.; Howse, G.M.; Applejohn, M.J.; Syme, P.D.; Myren, D.T.; Nystrom, K.L.; Hopkin, A.A.; Meating, J.H.; Power, J.H.; Cerezke, H.F.; Brandt, J.P.; Van Sickle, G.A.; Wood, C.S. *Forest Insect and Disease Conditions in Canada 1992*; Canadian Forest Service: Ottawa, Canada, 1994.
6. CCFS. *Compendium of Canadian Forestry Statistics-1996*; National Forestry Database Program, Government of Canada: Ottawa, Canada, 1997.
7. Forest health conditions in Ontario, forest health bulletin, summer 1996. http://www.glfc.forestry.ca/english/onlinep/summer.html.
8. South Carolina Beetles Kill $107 Million in Pine Timber. In *Orlando Sentinel*; Jan. 15, 1998; A10.

ONCHOCERCIASIS

O

Mike W. Service
Liverpool School of Tropical Medicine, Liverpool, England

INTRODUCTION

Commonly referred to as river blindness because the simuliid vectors (blackflies) breed in rivers and the disease can cause blindness, onchocerciasis is endemic in 34 countries. It occurs in much of West and Central Africa, with limited foci in East Africa from Ethiopia to Tanzania and with isolated pockets in Malawi, Sudan, and southern Yemen, possibly extending into Saudi Arabia. About 95% of an estimated 17.7 million people infected live in Africa. In the Americas the disease is found in localized areas of southern Mexico, Guatemala, Brazil, Venezuela, Ecuador, and Colombia. Some 270,000 people are blind and another 500,000 have severely impaired vision.

THE PARASITES

Adult nematode worms, *Onchocerca volvulus*, live in the subcutaneous tissues of humans often forming tangled masses in fibrous nodules. Adult worms can live up to 17 years. The female (30–50 cm) is ovoviviparous and after about 11 months she produces numerous minute sheathless microfilariae that migrate to the skin. The appearance of microfilariae in the skin occurs 15–18 months after an infective bite, and their presence is accompanied by clinical symptoms. Microfilariae live for about 2 years.

In the simuliid vectors skin microfilariae are ingested with the bloodmeal and while most die, some penetrate the stomach wall and migrate to the thoracic flight muscles where they develop into sausage-shaped stages and undergo two molts. A few survive and elongate into thinner worms that pass through the head and down the short proboscis of the fly. The infective third-stage larva in the proboscis penetrates the host's skin when the females alight to bite. The interval between ingestion of microfilariae and the time to when infective larvae are in the proboscis is about 6–13 days.

The disease is rife only in rural areas and is particularly severe up to 10 km on either side of rivers where the simuliid blackfly vectors breed. This results in fertile valleys being uneconomically farmed or even deserted.

CLINICAL SYMPTOMS

There are no clinical symptoms when infections are light. Moreover, there is often a period of 1–3 years between infection and symptoms, but when they occur they are caused by the microfilariae. The adult worms are of secondary importance and present no more than cosmetic blemishes. In chronic infections, nodules containing adult worms can be felt and sometimes seen over bony protuberances. In Africa nodules are found mainly around the pelvic region, the knees and chest. In Central America nodules more commonly occur on upper parts of the body, especially on the head. In light infections there may be few skin reactions, but in most patients pruritis with skin lesions occur, accompanied by an itchy rash. There can be a thickening of the skin, and loss of its elasticity leading to "hanging groin" and a prematurely aged appearance. Mottled depigmentation ("leopard skin") is common, particularly on the shins. The most important event is deteriorating eyesight that can result in complete blindness. Blindness is more common in savanna regions where more than 10% of the population may be blind and another 20% have impaired vision (Fig. 1). Blindness is caused by the damage done to the cornea and retina by the microfilariae. The eyes often have fluffy opacities.

TREATMENT

The drug of choice is ivermectin (1), of which a single dose of 150 μg/kg bodyweight given once or twice a year will reduce skin microfilariae to low levels for a year. Although regarded as a safe drug it should not be given to

Encyclopedia of Pest Management

Fig. 1 Testing the visual acuity of a villager in Sierra Leone, West Africa, who has onchocerciasis, a disease that results in deteriorating eyesight.

those under 5 years or weighing less than 15 kg, pregnant mothers, and a few other categories. Ivermectin does not kill the adult worms, which will continue to produce microfilariae. Suramin given intravenously is the only drug that is currently available that kills the adult worms, but it is not recommended because it is very toxic.

SIMULIID VECTORS

Simuliid blackflies have an almost worldwide distribution. There are nearly 1600 species in 26 genera, medically the most important being *Simulium*, species of which are vectors of onchocerciasis. Larval habitats consist of all types of flowing water, ranging from small streams (Fig. 2) and lake outlets to very large rivers and waterfalls. Larvae do not swim but are sedentary attaching themselves to submerged vegetation, rocks, and stones. The last larval instar spins a slipper-shaped cocoon that encloses the pupae. Only adult females take bloodmeals and can be disease vectors. They bite out of doors during the daytime. Their rasping mouthparts facilitate the uptake of skin microfilariae of onchocercal worms. Adults often fly just a few hundred meters, but some species can be transported 250–500 km on prevailing winds. The most important vectors of onchocerciasis in sub-Saharan Africa are species within the *Simulium damnosum* complex, such as *S. damnosum*, s.str., *S. sirbanum, S. sanctipauli*, and *S. leonense*. Other less important vectors include *S. neavei*, a phoretic species whose immature stages occur on mayflies (Ephemeroptera) and various crustacea, including freshwater crabs. This species is responsible for transmission in the Congo, Zaire, and Uganda. *Simulium ochraceum* is the principal vector in southern Mexico, while *S. metallicum* is an important vector in Venezuela, and *S. exiguum* is a primary vector in Ecuador and secondary one in Venezuela. In Brazil one of the main vectors is *S. guinense*.

CONTROL

Although insecticidal fogging or spraying of vegetation harboring adult flies has occasionally been undertaken, this results in very temporary and localized control. Most control is aimed at the larval breeding sites, that is, flowing waters. Occasionally control has been achieved by modifying water flow rates and removal of aquatic vegetation, but the most effective method is the application of insecticides to larval breeding sites (2). Insecticides need be applied at only a few selected sites for 15–30 min, because as the insecticide is carried downstream it kills the larvae over long stretches of water. In the past DDT has given good control of blackfly vectors in Nigeria and Uganda, but because of its accumulation in food chains DDT is no longer recommended for larval control. Instead organophosphates such as temephos or microbial ones such as *Bacillus thuringiensis* ssp. *israelensis* are usually recommended. In many situations ground applications of insecticides are impractical either because of the vast size of the rivers or because it is difficult to reach networks of streams surrounded by thick vegetation. Under these circumstances aerial applications from fixed-winged aircraft or helicopters (Fig. 3) are used. Because of the severity of river blindness in the Volta River Basin of West Africa and its devastating effects on rural life, the world's most ambitious and largest vector control program was initiated in 1974 by the World Health Organization (WHO). It is called the Onchocerciasis Control Program (OCP). By 1986 there were 11 countries in this control program, namely Benin, Burkina Faso, Côte d'Ivoire, Ghana, Guinea, Guinea Bissau, Mali, Niger, Senegal, Sierra Leone, and Togo, covering a population of more than 30 million. Some 50,000 km of rivers over an area of about

Fig. 2 A typical larval habitat of the simuliid vectors of onchocerciasis in Ghana, West Africa.

1.3 million km^2 were dosed weekly with temephos that was dropped from helicopters or light aircraft (3). Because of the appearance of temephos resistance in 1980 in some populations and species of the *S. damnosum* complex, some rivers were treated with other insect-icides such as chlorphoxim. In 1982, a rotation of seven insecticides belonging to four chemical groups was introduced, but most do not have the carrying capacity of temephos. The most important new insecticide introduced is *B. thuringiensis* ssp. *israelensis*. Larviciding must continue in some parts of the area until the year 2002 (4). Since 1988 the OCP has been undertaking the large-scale distribution of the microfilaricidal drug ivermectin (Mectizan) that is given just once a year. The manufacturers have said that ivermectin will be given free for as long as is needed. However, there are still distribution costs and logistics of getting the drug to rural communities often with poor lines of communication. Results from the OCP have been spectacular, and transmission of onchocerciasis has ceased over most of the OCP area (5). In 1995 the African Program for Onchocerciasis Control (APOC) was created to cover populations at risk in 19 countries outside the OCP area. The objective is to establish within 12 years, a sustainable community-based ivermectin treatment regimen, backed up with focal larviciding. Because ivermectin does not kill adult worms, control needs to continue for 20 years so that the reservoir of infection in the human population (i.e., adult onchocercal worms) dies out. Mass nodulectomy has been practiced routinely in Mexico and Guatemala for many years and although it has reduced the incidence of severe ocular complications and prevented blindness, patients from which all palpable nodules have been removed still have large numbers of microfilariae in their skin.

Fig. 3 A helicopter dosing a larval habitat of onchocerciasis vectors in West Africa with temephos insecticide.

COSTS

From 1974 to 1997 the total cost of the OCP was about $513 million—that means that each person in the area has been protected at an annual cost of less than a dollar. Cost of the APOC over a 12-year period is estimated to be $161 million.

REFERENCES

1. Anon. Mectizan and onchocerciasis: a decade of accomplishment and prospects for the future. The evolution of a drug into a development concept. Ann. Trop. Med. Parasitol **1998**, *92*, 1–179(Suppl.).
2. Davies, J.B. Sixty years of onchocerciasis vector control: a chronological summary with comments on eradication, reinvasion, and insecticide resistance. Ann. Rev. Entomol **1994**, *39*, 23–45.
3. Samba, E.M. The onchocerciasis control programme in West Africa. An example of effective public health management. WHO. Publ. Health Action **1994**, *1*, 1–107.
4. Molyneux, D.H.; Davies, J.B. Onchocerciasis control: moving towards the millennium. Parasitol. Today **1997**, *13*, 418–425.
5. World Health Organization. Onchocerciasis and its control. WHO. Techn. Ser. **1995**, *852*, 1–103.

OPTIMIZING PESTICIDE APPLICATIONS

Miriam Austerweil
Abraham Gamliel
Institute of Agricultural Engineering, Bet Dagan, Israel

O

INTRODUCTION

The application of pesticides for crop protection involves a series of steps: chemical preparation of the formulation followed by its discharge, dispersal, and deposition on the target. Effective control is usually achieved, but the process of pesticide delivery is not efficient and only a part of the applied chemical reaches the target. Therefore, dosages that are recommended by the chemical manufacturers are usually higher than necessary because they include safety margins to compensate for variations in application techniques and application errors. Optimizing the pesticide application has an essential role in reducing the waste of chemicals and minimizing the risk to the operator and the environment. Improved and precise application focuses on matching the design of the equipment to the characteristics of the target, the physical–chemical properties of the applied pesticide, the micrometeoro-logical conditions, and the biological requirements. The collaboration among the application equipment manufacturers, agrochemical companies, research teams, crop advisers, and farmers could result in recommendations for reduced dosages by more efficient pesticide use.

OBJECTIVES IN OPTIMIZING PESTICIDE APPLICATION

Pesticides are applied to protect agricultural crops from damage caused by insect pests, plant pathogens, weeds, etc. A high proportion of the foliar spray is lost mainly to the ground and by drift. An estimated 1–30% of the applied pesticide reaches the target (1, 2). There is a vast difference between the theoretically effective dosage and that which is used in the field. Dosages, which usually are recommended by the chemical manufacturers, include safety margins to compensate for variations in application techniques and application errors. The multidisciplinary nature of the pesticide application process explains the slow development of this field in the past. Increased concerns for health, safety, and environmental standards have been expressed in new legislation in some countries, which require up to 50% reduction in the per-hectare application of active ingredient (a.i./ha). This has been the driving force urging the design of more effective application equipment (2, 3). Optimized pesticide application should achieve "optimum" spray quality on the selected target and should minimize off-target deposition, thus reducing the risk to the operator and to the environment. Efforts are directed toward improving the delivery process to suit target characteristics and micrometeo-rological conditions, by matching equipment operation, formulation, drop size, and density, and the interaction among them.

An interdisciplinary collaboration among biologists, chemists, physicists, and engineers is necessary to optimize each step of the application process and to elucidate the biological requirements.

Approaches to the Optimization of Pesticide Application

Air utilization

The use of air has proved in many cases to be an effective means of optimizing each step of the pesticide application process (4). Spray generation by twin fluid nozzles with internal air mixing and air inclusion nozzles has been developed; depending on design, they can provide greater spray recovery and reduced drift when compared with low pressure fan nozzles and deflector nozzles.

Air assistance is useful in pesticide delivery, dispersion, and deposition. Advantages of this technique—good crop penetration and coverage of problematic areas (e.g., undersides of the leaves, inside dense plant canopies) with reduced water consumption—have led to its application to a wide variety of crops: trees, protected cultivation, and arable crops. The use of airflows often involves an increased energy consumption and the need to adjust and adapt its characteristics to the target size and shape. There are various designs of crop sprayers with air assistance, including tunnel sprayers, modified boom sprayers with air sleeves, laterally directed air streams, vertical droplet tubes (2, 5), and pulsed air stream (6). The automation of this technique is a further contribution to optimization (7).

Mechanical aerosol generators (cold foggers) and automatic dusters represent air assisted, automatic, very low volume application techniques for greenhouse crop protection (8, 9). The functioning of these devices is based on two operations: the release of the liquid aerosol (mechanically generated) or the dust into a turbulent air stream, and the dispersion and deposition of the droplets or particles by the air movement inside the greenhouse. High spray recovery on both leaf surfaces and efficient pest control is obtained, with 30% reduced dosage compared with high volume (HV) application (8). Dosages are often reduced by 20% or more compared with high volume spraying, as there is less waste on the ground. Supplementary air circulation improves the uniformity of pesticide distribution in the greenhouse and the coverage on both the upper and lower sides of the leaves (8). An important advantage of aerosol application is the rapid pesticide dissipation. Accelerated dissipation of pesticide residues in herb crops was observed after cold fogging treatment, compared with high and low volume application techniques. Reducing dosages by 50% further enhanced this process while providing good coverage (9).

The interaction between air assistance and droplet size affects transport and deposition. In most cases, small droplets appear to be more efficient biologically, but the use of small droplets increases the probability of pesticide drift (3). Thus, a number of means are used to control and minimize drift including using coarser droplets, drift-modifying adjuvants, shields, and shrouds.

Modifying physical–chemical properties of the spray liquid

The design of new adjuvants and formulations should improve the dose-transfer process and should minimize the drift. Advanced methods to evaluate droplet size and velocity distribution (such as wind tunnels, laser instruments, and high-speed photography) assist the accurate determination of the effects of formulation on spray nozzle performance and drift formation (10, 11).

Adjustment and control of the sprayer output to crop structure parameters

Crop structure parameters such as crop area index, canopy volume, and row spacing influence the collection and deposition of the applied pesticide (12). Dose adjustment methods have been developed, based on crop row spacing, crop area, and row volume, as well as machine designs with adjustable output angles. Mathematical models and various sensor types are promising tools to achieve this goal (13, 14).

Selective and site specific chemical application

Accurate application using imaging and sensing technologies, global positioning systems (GPS) and global information systems (GIS) are contemporary aids to precision spraying. Computerized vision has made an impact on designing ''spot'' or ''patch'' sprayers for selective herbicide treatment (15). GPS is used in both ground and aerial applications and to provide weed maps (3, 5).

Electrostatic charging of droplets is an attempt to improve deposition, particularly on the underside of the leaves (2, 16). The practical impact of this technique is little, due to the inconclusive results obtained under field conditions and the higher cost of equipment (2).

Computer simulation models

Mathematical models assist in optimizing the sprayer parameters to improve pesticide efficacy and reduce environmental hazard (3). Such models are used in aerial application, to predict spray drift under field conditions for ground sprayers, to design air assisted application equipment, for the dose-transfer process, to predict droplet collection, and for selecting and designing drift samplers.

Integrated pest management strategies and sophisticated decision systems

Decision systems based on integrated pest management increase the efficacy of pesticide application by ensuring more timely and adequate treatment, combined with improved agronomic practices (3). Legislation affecting the approval of the application equipment, training, and certification to apply pesticides are important for optimal and safe application of pesticides (4).

Optimizing pesticide application could pose conflicts of interests between the agrochemical and sprayer industries, but, nevertheless the collaboration among them, the researchers, crop advisers, and farmers could result in more efficient pesticide use and recommendations for reduced dosages.

REFERENCES

1. Pimentel, D.; Levitan, L. Pesticides amounts applied and amounts reaching pests. BioScience **1986**, *36* (2), 86–91.
2. Matthews, G.A. *Pesticides Application Methods*; 2, Longman: London, 1992.
3. Hall, F.R. Spray deposits: opportunities for improved efficiency of utilization via quality, quantity and formulation. Phytoparasitica **1997**, *25* (Suppl.), 39S–52S.
4. Matthews, G.A. A Review of the Use of Air in Atomisation of Sprays, Dispersion of Droplets Down Wind and

O

Collection on Crop Foliage. In *Aspects of Applied Biology 57, Pesticide Application*; Cross, J.V., Gilbert, A.J., Glass, C.R., Taylor, W.A., Walklate, P.J., Western, N.M., Eds.; University of Surrey: Guildford, UK, January 17–18, 2000; 21–27The Association of Applied Biologists, Wellesbourne, UK.

5. Matthews, G.A. Pesticide application: current status and further developments. Phytoparasitica **1997**, *25* (Suppl.), 11S–19S.
6. Gan-Mor, S.; Grinstein, A.; Beres, H.; Riven, Y.; Zur, I. Improved uniformity of spray deposition in a dense plant canopy: methods and equipment. Phytoparasitica **1996**, *24* (1), 57–67.
7. Austerweil, M.; Grinstein, A. Automatic pesticide application in greenhouses. Phytoparasitica **1997**, *25* (Suppl.), 71S–80S.
8. Austerweil, M.; Gamliel, A. Approaches to Evaluating the Performance of Air-Assisted Pesticide Application Equipment in Greenhouses. In *Aspects of Applied Biology 57, Pesticide Application*; Cross, J.V., Gilbert, A.J., Glass, C.R., Taylor, W.A., Walklate, P.J., Western, N.M., Eds.; University of Surrey: Guildford, UK, 2000; 391–398January 17–18, The Association of Applied Biologists, Wellesbourne, UK.
9. Gamliel, A.; Austerweil, M.; Zilberg, V.; Rabinowich, E.; Manor, H. Improved Application to Reduce Pesticide Residues in Herb Crops. In *Aspects of Applied Biology 57, Pesticide Application*; Cross, J.V., Gilbert, A.J., Glass, C.R., Taylor, W.A., Walklate, P.J., Western, N.M., Eds.; University of Surrey: Guildford, UK, 2000; 273–278January 17–18, The Association of Applied Biologists, Wellesbourne, UK.
10. Miller, P.C. Spray Drift and Its Measurement. In *Application Technology for Crop Protection*; Matthew, G.A., Hislop, E.C., Eds.; CAB International: Wallingford, UK, 1993; 101–122.
11. Parkin, C.S. Methods for Measuring Spray Droplets Sizes. In *Application Technology for Crop Protection*; Matthew, G.A., Hislop, E.C., Eds.; CAB International: Wallingford, UK, 1993; 57–84.
12. Holloway, P.J.; Ellis, M.C. Butler; Webb, D.A.; Western, N.M.; Tuck, C.R.; Hayes, A.L.; Miller, P.C.H. Effects of some agricultural tank-mix adjuvants on the deposition efficiency of aqueous sprays on foliage. Crop Prot. **2000**, *19* (1), 27–37.
13. Hall, F.R. Application to Plantation Crops. In *Application Technology for Crop Protection*; Matthew, G.A., Hislop, E.C., Eds.; CAB International: Wallingford, UK, 1993; 187–213.
14. Walklate, P.J.; Richardson, G.M.; Cross, J.V.; Murray, R.A. Relationship between Orchard Tree Crop Structure and Performance Characteristics of an Axial Fan Sprayer. In *Aspects of Applied Biology 57, Pesticide Application*; Cross, J.V., Gilbert, A.J., Glass, C.R., Taylor, W.A., Walklate, P.J., Western, N.M., Eds.; University of Surrey: Guildford, UK, 2000; 285–292January 17–18, The Association of Applied Biologists, Wellesbourne, UK.
15. Koch, H.; Weisser, P. Sensor Equipped Orchard Spraying–Efficacy, Savings and Drift Reduction. In *Aspects of Applied Biology 57, Pesticide Application*; Cross, J.V., Gilbert, A.J., Glass, C.R., Taylor, W.A., Walklate, P.J., Western, N.M., Eds.; University of Surrey: Guildford, UK, 2000; 357–362January 17–18, The Association of Applied Biologists, Wellesbourne, UK.
16. Lindquist, R.K.; Powell, C.C.; Hall, F.R. Glasshouse Treatment. In *Application Technology for Crop Protection*; Matthew, G.A., Hislop, E.C., Eds.; CAB International: Wallingford, UK, 1993; 275–290.

ORGANIC AGRICULTURE

Kathleen Delate
Iowa State University, Ames, Iowa, U.S.A.

INTRODUCTION

The USDA National Organic Standards Board (NOSB) defines organic agriculture as ''an ecological production management system that promotes and enhances biodiversity, biological cycles, and soil biological activity (1). It is based on minimal use of off-farm inputs and on management practices that restore, maintain, or enhance ecological harmony. The primary goal of organic agriculture is to optimize the health and productivity of interdependent communities of soil life, plants, animals and people'' (1). The term ''organic'' is defined by law. The labels ''natural,'' ''eco-friendly,'' and similar statements do not guarantee complete adherence to organic practices as defined by law.

HISTORY

In 1990, the U.S. Congress passed the Organic Food Production Act (OFPA). This law was heralded as the first U.S. law established to regulate a system of farming. The OFPA requires that anyone selling products as ''organic'' must follow a set of prescribed practices that include avoidance of synthetic chemicals in crop and livestock production, and in the manufacturing of processed products. ''Certified organic'' crops must be raised on land to which no synthetic chemical (any fertilizers, herbicides, insecticides, or fungicides) inputs were applied for three years prior to the crops' sale. Organic certification agencies became established in the United States to deal with a required ''third-party certification.'' There are at least 20 private certification agencies and 15 state agencies certifying organic production and processing in the United States. Proposed rules implementing the federal OFPA law were promulgated in 1997, after seven years of revisions. Unfortunately, these rules did not meet private certification agencies' standards, and a record number of complaints (275,000) were issued in the public comment period. Now that the federal rules are established (released in 2001), all certifiers must utilize the federal standards as the minimum standard for the ''certified organic'' label in the United States. European regulation is under the auspices of the International Federation of Organic Agriculture Movements (IFOAM) with national certification agencies in each country (2). Japan currently certifies under the Ministry of Agriculture and Forestry. Certification for the European Union and Japan is extended to several U.S. certifiers that meet international standards.

Organic agriculture is the oldest form of agriculture on earth. Farming without the use of petroleum-based chemicals (fertilizers and pesticides) was the norm for farmers in the developed world until post-World War II. The war era led to technologies that were adapted for agricultural production. Ammonium nitrate used for munitions during WWII evolved into ammonium nitrate fertilizer; organophosphate nerve gas production led to the development of powerful insecticides. These technical advances since WWII have resulted in significant economic benefits, as well as unwanted environmental and social effects. Organic farmers seek to utilize those advances that yield benefits (e.g., new varieties of crops, more efficient machinery) while discarding those methods that have led to negative impacts on society and the environment, such as pesticide pollution and insect pest resistance (3). Instead of using synthetic fertilizers and pesticides, organic farmers utilize crop rotations, cover crops, and naturally based products to maintain or enhance soil fertility (4, 5). These farmers also rely on biological, cultural, and physical methods to limit pest expansion and increase populations of beneficial insects on their farms. By managing their ecological capital through efficient use of on-farm natural resources, organic farmers produce for diverse and specialized markets that provide premium prices.

Because genetically modified organisms (GMOs) constitute synthetic inputs and pose unknown risks, GMOs, such as herbicide-resistant seeds, plants, and product ingredients are disallowed in organic agriculture. Organic livestock, like organic crops, must be fed 100% organic food or feed in their production. Synthetic hormones and antibiotics are disallowed in organic livestock production. Traditional farmers throughout the world have relied on natural production methods for centuries, maintaining consistent yields within their local environment. While "green revolution" technologies have led to increased yields in many less developed countries, many farmers have seen an increase in pest problems with new varieties and high input-based systems.

Motivations for organic production include economic, food safety, and environmental concerns. All organic farmers avoid the use of synthetic chemicals in their farming systems, but philosophies differ among organic farmers regarding methods to achieve the ideal system. Organic farmers span the spectrum from those who completely eschew external inputs, create on-farm sources of compost for fertilization, and encourage the activity of beneficial insects through conservation of food and natural habitats, to those farmers who import their fertility and pest management inputs. A truly sustainable method of organic farming would seek to eliminate, as much as possible, reliance on external inputs.

WORLDWIDE STATISTICS

USDA does not publish systematic reports on organic production in the United States. The most recent census in 1994 identified 1.5 million acres of organic production in the United States with 4050 farmers reporting organic acreage (6). This figure underrepresents current production because many organic farmers opt to sell their products as organic without undergoing certification. The U.S. organic industry continues to grow at a rate of 20% annually. The industry was listed as a $4.5 billion industry in 1998, with predicted future growth to $10 billion by 2003. The organic industry is a consumer-driven market. According to industry surveys, the largest purchasers of organic products are young people and college-educated consumers. Worldwide consumption of organic products has experienced tremendous growth, often surpassing U.S. figures of 20% annual gain. Much of the increase in consumption worldwide has been fueled by consumers' demand for GMO-free products. Because GMOs are disallowed in organic production and processing, organic products are automatically segregated as GMO-free at the marketplace. European consumers have led the demand for organic products, particularly in countries such as the Netherlands and Scandinavia. Two percent of all German farmland, 4% of Italian farmland, and 10% of Austrian farmland is managed organically (2). Prince Charles of England has developed a model organic farm and established a system of government support for transitioning organic farmers. Major supermarket chains and restaurants in Europe offer a wide variety of organic products in their aisles and on their menus. Industry experts predict that the establishment of federal rules will advance organic sales in the United States. Although the organic industry began as a niche market, steady growth has led to its place in a "segment" market since 1997. The organic dairy industry, for example, expanded by 73% from 1996 to 1997, and continues to grow today. Organic markets can be divided into indirect and direct markets. Indirect or wholesale markets include cooperatives, wholesale produce operations, brokers, and local milling operations. Many supermarket chains buy directly from farmers or from wholesalers of organic products. Because meat can now be labeled as "organic," as of 1999, the marketing of organic beef, pork, chicken, and lamb has been significantly simplified. Roadside stands, farmers' markets and community supported agriculture (CSA) farms constitute the direct marketing end of the organic industry. Most consumers relate their willingness to pay premium prices for food that has been raised without synthetic chemicals because of their concern for food safety and the environment. Supporting local family farmers also enters into their purchasing decisions.

CROP AND PEST PERFORMANCE IN ORGANIC SYSTEMS

The basis for all organic farming systems is the health of the soil (4, 7). In addition to maintaining adequate fertility, organic farmers strive for biologically active soil, containing microbial populations required for nutrient cycling (8, 9). Crop rotations (required for all organic operations) provide nutrients such as nitrogen in the case of legume crops (alfalfa and clover) and carbonaceous biomass upon which beneficial soil microorganisms depend for survival (10, 11). A crop rotation plan is required as protection against pest problems and soil deterioration (12–15). Ideally, no more than four out of six years should be in agronomic crops, and the same row crop cannot be grown in consecutive years on the same land. Legumes (alfalfa, clovers, and vetches) alone, or in combination with small grains (wheat, oats, and barley), must be rotated with row crops (corn, soybeans, amaranth, vegetables, and herbs) to ensure a healthy system. A typical six-year rotation in the Midwestern United States would be corn (with a cover of winter rye)–soybeans–oats (with an underseeding of alfalfa)–alfalfa–corn–soybeans (16, 17). Horticultural crops must be rotated with a leguminous cover crop at least once every five years.

Pest management in organic farming systems is based on a healthy plant able to withstand some pest injury and on the inherent equilibrium in nature, as most insect pests have natural enemies that regulate their populations in unperturbed environments (18, 19). Because only naturally occurring materials are allowed in organic production, insect predators, parasites, and pathogens exist without intervention from highly toxic insecticides (20). Most organic farmers rely on naturally occurring beneficial insects on their farms, but some farmers purchase and release lacewings and other natural enemies every season, for example. There are also commercial preparations of natural insect pathogens, such as *Bacillus thuringiensis* (Bt), which are used to manage pestiferous larvae, such as corn borers. Botanical insecticides, such as neem and ryania, are also allowable in organic production, but as with all insecticides, sprays should be used only as a last resort. Although these materials are naturally based, some materials may affect natural enemies. Prevention is a cornerstone of organic farming (21). Pest-free seeds and trans- plants, along with physical and cultural methods, are used to prevent pest infestations. Physical methods include the use of row covers for protection against insects such as cabbage butterflies and aphids. Cultural methods include sanitation and resistant varieties. Plant varieties are used that have been bred traditionally (i.e., no manipulated gene insertion or engineering involved) for insect, disease, and nematode resistance or tolerance.

Most organic farmers rely on multiple tactics for their weed management (22, 23). Allelopathic crops, cultivation, mulching, and flame burning are all methods available for organic farmers. Allelopathic crops, such as rye and oats, produce an exudate that mitigates against small weed seed germination. Depending on the crop, cultivation offers the least labor-intensive method of organic weed management. Timely cultivation is key; without specific schedules, weeds proliferate. Propane flame burning is generally used in conjunction with cultivation, particularly during times of high field moisture. Mulching with straw or wood chips is commonly used in many organic horticultural operations.

Yields comparable to conventional crops have been shown for organic crops in three university long-term experiments in the United States (South Dakota State University (24), Iowa State University (16), and the University of California–Davis (25) and in many European studies (26). Factoring in an organic premium (ranging from 50 to 400%, depending upon crop and season), organic systems consistently out-performed conventional systems in terms of economics (27, 28, 17). Pest problems were not a critical factor in these organic systems. Other studies have shown the benefits of organic practices, such as composting, in mitigating root-borne diseases (8).

KEY ISSUES REQUIRING ADDITIONAL RESEARCH

Continued verification of the long-term benefits of organic versus conventional farming in terms of soil quality (29), pest management, and nutritional benefits (30) is needed. Key issues include the development of management practices to increase nutrient cycling for maintenance of crop yields and optimize biological control of plant pests and diseases (31). Economic analysis, including risks of the three-year transition required for organic certification, will provide useful information for growers interested in alternative systems (26, 32). Appropriate tillage systems, which protect soil quality and provide adequate soil preparation, remain as important issues for organic producers. The improvement of natural parasiticide formulations, such as diatomaceous earth, is required for optimum organic livestock production. Marketing and support needs include the availability of reliable statistics for organic operations and prices. Although many European countries support their farmers in their organic production practices through environmental subsidies (33), the United States has made small gains in this area. Some state agencies (Minnesota Department of Agriculture) and the USDA Natural Resources Conservation Services (NRCS) through the Environmental Quality Indicators Program (EQIP) offer financial incentives to organic farmers during their transitioning years. More of these support services are needed to encourage farmers interested in the conversion to alternative production (34).

REFERENCES

1. National Organic Program USDA Agricultural Marketing Service: Washington, DC, 1. http://www.ams.usda.gov/nop/ (accessed June 5, 2000).
2. Lampkin, N.H. *The Policy and Regulatory Environment for Organic Farming in Europe*; Universität Hohenheim, Institut für Landwirtschaftliche Betriebslehre: Stuttgart, Germany, 1999; 1–379.
3. Altieri, M. *Agroecology*; Second Ed., Westview Press: Boulder, CO, 1995; 1–433.
4. Lampkin, N.H.; Measures, M. *1999 Organic Farm Management Handbook*, 3rd Ed., Welsh Institute of Rural Studies, University of Wales: Hamstead Marshall, Berkshire, 1999; 1–163.
5. Lockeretz, W.; Shearer, G.; Kohl, D. Organic farming in the corn belt. Science **1981**, *211*, 540–547.
6. Lipson, M. *Searching for the "O-Word": Analyzing the USDA Current Research Information System for Pertinence to Organic Farming*; Organic Farming Research Foundation: Santa Cruz, CA, 1997; 1–85.

O

7. Macey, A.; Kramer, D. *Organic Field Crop Handbook*; Canadian Organic Growers, Inc.: Ottawa, Canada, 1992; 1–256.
8. Drinkwater, L.E.; Letourneau, D.K.; Shennan, C. Fundamental differences between conventional and organic tomato agroecosystems in California. Ecological Appl. **1995**, *5* (4), 1098–1112.
9. Wander, M.M.; Traina, S.J.; Stinner, B.R.; Peters, S.E. Organic and conventional management effects on biologically active soil organic matter pools. Soil Sci. Soc. Am. J. **1994**, *58*, 1130–1139.
10. Drinkwater, L.E.; Wagoner, P.; Sarrantonio, M. Legume-based cropping systems have reduced carbon and nitrogen losses. Nature **1998**, *396*, 262–265.
11. Yeates, G.W.; Bardgett, R.D.; Cook, R.; Hobbs, P.J.; Bowling, P.J.; Potter, J.F. Faunal and microbial diversity in three welsh grassland soils under conventional and organic management regimes. J. Appl. Ecol. **1997**, *34* (3), 453–470.
12. Adee, E.A.; Oplinger, E.S.; Grau, C.R. Tillage, rotation sequence, and cultivar influences on brown stem rot and soybean yield. J. Prod. Agric. **1994**, *7* (3), 341–347.
13. Karlen, D.L.; Varvel, G.E.; Bullock, D.G.; Cruse, R.M. Crop rotations for the 21st century. Adv. in Agron. **1994**, *53*, 1–45.
14. Lipps, P.E.; Deep, I.W. Influence of tillage and crop rotation on yield, stalk rot, and recovery of *Fusarium* and *Trichoderma spp.* from corn. Plant Dis. **1991**, *75* (8), 828–833.
15. Stinner, D.H.; Stinner, B.R.; Zaborski, E.R.; Favretto, M. R.; McCartney, D.A. *Ecological Analyses of Ohio Farms under Long-term Sustainable Management*; Ecological Society of America, August: Knoxville, TN, 1994; 7–11.
16. Delate, K. *Comparison of Organic and Conventional Rotations at the Neely-Kinyon Long-Term Agroecological Research (LTAR) Site*; Leopold Center for Sustainable Agriculture Annual Report, Iowa State University: Ames, IA, 1999; 1–12.
17. Welsh, R. *The Economics of Organic Grain and Soybean Production on the Midwestern United States*; Henry, A., Ed.; Wallace Institute for Alternative Agriculture: Greenbelt, MD, 1999.
18. Neher, D.A. Nematode communities in organically and conventionally managed agricultural soils. J. Nematol. **1999**, *31*, 142–154.
19. Pfiffner, L.; Niggli, U. Effects of bio-dynamic, organic and conventional farming on ground beetles and other epigaeic arthropods in winter wheat. Biol. Agri. Hort. **1996**, *12*, 353–364.
20. Kromp, B.; Meindel, P.; Harris, P.J.C. *Entomological Research in Organic Agriculture*; Academic Publishers: Bicester, England, 1999; 1–386.
21. Lockeretz, W. *Environmentally Sound Agriculture: Selected Papers from the Fourth International Conference of the International Federation of Organic Agriculture Movements (IFOAM)*; Praeger: New York, 1983; 1–426, , Cambridge, MA, Aug 18–20, 1982.
22. Lanini, W.T.; Zalom, F.; Marois, J.; Ferris, H. Researchers find short-term insect problems, long-term weed problems. Ca. Agric. **1994**, *48*, 27–33.
23. Liebman, M.; Ohno, T. Crop Rotation and Legume Residue Effects on Weed Emergence and Growth: Applications for Weed Management. In *Integrated Weed and Soil Management*; Hatfield, J.L., Buhler, D.D., Stewart, B.A., Eds.; Ann Arbor Press: Chelsea, MI, 1997; 181–221.
24. Dobbs, T. Report on Organic and Conventional Grain Trials at South Dakota State University. In *USDA-ERS Conference on The Economics of Organic Farming Systems*; USDA: Washington, DC, 1999; 1–4.
25. Klonsky, K.; Livingston, P. Alternative systems aim to reduce inputs, maintain profits. Ca. Agric. **1994**, *48* (5), 34–42.
26. Lampkin, N.H.; Padel, S. *The Economics of Organic Farming: An International Perspective*; CAB International: Wallingford, Oxon, U.K., 1994; 1–468.
27. Hanson, J.C.; Lichtenberg, E.; Peters, S.E. Organic versus conventional grain production in the mid-atlantic: an economic and farming system overview. Am. J. Alt. Agric. **1997**, *12*, 2–9.
28. Stanhill, G. The comparative productivity of organic agriculture. Agric. Ecosys. Env. **1990**, *30*, 1–26.
29. Lockeretz, W.; Shearer, G.; Klepper, R.; Sweeney, S. Field crop production on organic farms in the midwest. J. Soil Water Conserv. **1978**, *33*, 130–134.
30. Woese, K.; Lange, D.; Boess, C.; Bögl, K.W. A comparison of organically and conventionally grown foods—results of a review of the relevant literature. J. Sci. Food Agric. **1997**, *74*, 281–293.
31. Walz, E. *Final Results of the Third Biennial National Organic Farmers' Survey*; Organic Farming Research Foundation: Santa Cruz, CA, 1999; 1–75.
32. Chase, C.; Duffy, M. An economic comparison of conventional and reduced-chemical farming systems in Iowa. Amer. J. Alt. Agric. **1991**, *6* (4), 168–73.
33. Zygmont, J. In *International Issues Pertaining to Organic Agriculture*, Proceedings of the Workshop: Organic Farming and Marketing Research—New Partnerships and Priorities, Lipson, M., Hammer, T., Eds.; Organic Farming Research Foundation: Santa Cruz, CA, 1999; 317–379.
34. D'Souza, G.; Cyphers, D.; Phipps, T. Factors affecting the adoption of sustainable agricultural practices. Agric. and Resource Econ. Rev. **1993**, *22* (2), 159–205.

ORGANIC FARMING

Brenda Frick
University of Saskatchewan, Saskatoon, Saskatchewan, Canada

INTRODUCTION

Organic farming strives to produce healthful food while maintaining or improving the health of the agro-ecosystem. Organic farmers emphasize a systems approach that manages, respects, and encourages natural, biological processes. Pests are managed through good husbandry practices such as crop rotation, residue management, cultivar selection, crop competition, soil fertility management, and, where necessary, through judicious use of biological and mechanical controls. Standards for organic certification focus on the production process, rather than the product. They emphasize natural processes rather than synthetic products in crop and livestock production, a program for soil building, and diverse crop rotations. Organically produced crops are kept separate from others during their journey from producer to consumer. An audit trail tracks the history of a given product, and helps to assure quality and consumer confidence.

ORGANIC FARMING STANDARDS

Certification standards for organic systems are complex. Minimum standards are set by regional certification bodies and international organizations such as the Organic Crop Improvement Association. National regulations for Canada were developed through the Canadian Organic Advisory Board. Standards are being developed in the United States through the U.S. Department of Agriculture National Organic Program Proposed Rule. A number of European countries and the European Union have developed or are developing standards. Different countries vary in the stringency of their regulations, but in general, the standards prohibit synthetically processed fertilizers, pesticides, growth regulators, antibiotics, and genetically modified or engineered organisms. They require a management plan that includes strategies for crop rotation, soil management, monitoring and problem solving for crop protection, and detailed record keeping. Farms must meet the regulations for a minimum period of time, usually three years, to qualify for certified organic status. The International Federation of Organic Agriculture Movements (IFOAM) offers a basic standard and accreditation for certification bodies that is highly regarded by international traders.

The word ''organic'' is protected by law in some countries. Equivalent terms in other countries include ''biological'' and ''ecological.'' The term ''organic'' is more common in English-speaking countries; the latter terms are more common in mainland Europe.

PEST MANAGEMENT PRACTICES

Organic production is a systems approach to farming. Producers strive to understand the ecological relationships that influence the abundance of the various species in their systems, and to avoid outbreaks of those species that harm the crop. Many of the methods used to favor the crop in the ecological community can be summarized by good crop husbandry—appropriate timing, depth, and rate of seeding; management of soil fertility; selection of locally adapted and competitive crops and crop cultivars. Mechanical and biological pest controls are used as necessary. Off-farm inputs are considered only as a last resort.

Crop rotation is one of the strongest tools that the producer uses. A diverse rotation increases microbial and mycorrhizal activity in the soil and improves crop vigour. Populations of crop-specific pests, such as many diseases and insects, are severely reduced by years when a given crop is not grown. Crop rotations alter the timing and competitive relationships of crops and reduce the build-up of weed communities adapted to any given management practice. Crop and cultivar selection is also very important. Matching crops to fields improves their competitive relations with weeds, and helps them to resist both insect and pathogen attacks (Table 1).

Weeds

In organic systems, weeds are considered to be a part of the ecological system. They are often beneficial in moderating the soil environment, providing habitat and food

Table 1 Influence of insects, diseases, and weeds on quality losses in crops

Loss in size and thousand-kernel-weight (grain)
Loss in size of tubers and roots
Increase of moisture content
Changes in chemical composition of the kernels, tubers, and roots
Discoloration of kernels and tubers
Formation of mycotoxins by plant pathogens
Loss in germination of the seeds and tubers
Reduced processing properties (e.g., baking and malting properties)
Reduced feeding properties (nutrition value)

for micro- and macrofauna, moving soil nutrients to the surface, and indicating soil or management problems. Of course, large weed populations are often detrimental. Weeds can harbor pests, reduce crop yield and quality, and interfere with harvest. When weeds are abundant and considered problematic despite prevention and crop rotation, an organic producer has several options.

A strongly competitive crop is an excellent defense against weeds. Competitive crops, such as fall rye and sweet-clover, and perennial crops, such as alfalfa, are particularly effective at reducing the weed community. Less competitive crops such as flax and lentil are best saved for less weedy fields. Crop competition can be increased by appropriate cultivar selection and by crop management techniques such as heavy seeding, narrow row spacing, good seed-bed management, etc.

Tillage is commonly used to reduce weed populations. Tillage can be used after harvest, before seeding, and in fallow years. Concerns over the negative effects of tillage on soil quality, especially on erosion potential, have reduced the frequency of fall and fallow tillages, though these may be cautiously used for perennial weed control. Delayed seeding after spring tillage remains an important tool. Early tillage stimulates the germination of volunteer crop and weed seeds. These are destroyed with a second tillage at or before seeding. This strategy has been especially effective at reducing the abundance of early emerging species such as wild oats and winter annual weeds such as stinkweed.

Harrowing after seeding or even after crop emergence can also be effective. This strategy is most effective for weeds that emerge from shallow depths, such as green foxtail, in crops that have large, deeply placed seeds. For row crops, interrow cultivation can be effective. The combination of early harrowing across the rows and interrow cultivation can offer good weed control. Other mechanical weed control techniques include flaming, burning, and mowing. Chaff collection at harvest can remove significant numbers of weed seeds. It is most effective for volunteer crop seeds, and weed seeds such as lamb's quarters, that are largely retained on the plant at a height above the stubble.

Biocontrol of weeds includes the use of livestock, weed-eating insects, and weed-suppressing diseases. Livestock can be used for grazing or to consume mowed weeds. Livestock can also be used to consume chaff or seed screenings. Biocontrol insects generally target perennial weed species. For instance, the black-dot spurge beetle has been released for control of leafy spurge. Few fungal biocontrol agents are available. Examples include DeVine® for stranglervine, Collego® for northern joint vetch, and Biomal® for round-leaved mallow. Beneficial organisms can be encouraged by practices such as reducing tillage, maintaining shelter belts, and growing crops, such as legumes and some cereals, that encourage mycorrhizae. These can also be effective for weed control.

Insects And Other Invertebrates

A majority of insects found in crop fields are beneficial or are of no economic importance. Diversity of habitat and wildlife encourages positive interactions among species, and reduces the potential for outbreaks of insect pests.

Organically grown crops may be less attractive to insects than crops grown with abundant synthetic fertilizers. Synthetic fertilizers may result in the accumulation of excess nitrate in plant tissue; composted manure and green manures release nitrogen more slowly, reducing the pot-ential for this accumulation. Excess nitrate makes plants more attractive to insect pests and can increase the re-productive rate of some insects, thus increasing the severity of an insect outbreak.

Crop rotations can be effective against insect pests with limited dispersal capabilities, such as the corn rootworm. For these types of insects, reducing crop residues and volunteer crop are also important.

For widely dispersed insects that are attracted to specific crops, large-scale cropping diversity is important. Monocultures favor these insects. Increasing the presence of nonattractive plant species reduces the incidence of attack. Practices such as strip cropping and intercropping reduce the attraction of insect pests by diluting the aroma of the crop or confusing the insect's search image. For instance, under-seeding canola with yellow sweet-clover may reduce the incidence of flea beetles. Even weed populations may function this way. For instance, weeds in alfalfa may increase the habitat for parasitic wasps that control alfalfa caterpillar. Barrier strips of an unattractive crop around the outside of the susceptible crop may prevent insects from crossing into the susceptible crop. For instance, a border strip of peas may reduce the movement of grasshoppers into wheat. Trap strips of an attractive crop may be sown. After insects accumulate there, it can be mowed or cultivated, thus destroying many of the insects. Small-sized fields also limit the problem of insect outbreaks.

The time of seeding can sometimes be altered to avoid insect pests. For instance, late seeding of canola can reduce the severity of flea beetles and early seeding of wheat can reduce the attack of wheat midge.

Cultivar selection may help prevent insect problems. Wheat cultivars differ substantially in their susceptibility to wheat midge, in large part due to their rate of development.

Biocontrol agents are available for some insect pests. Predators such as ladybugs and lacewings can be used to reduce aphid populations. *Bacillus thuringiensis* (Bt) strains have been developed for control of several insects, including caterpillars and beetles. Some of these control organisms can be purchased. Maintaining a varied habitat in and around the crop field can also help to harbor such natural organisms.

Pheromone traps may be used to lure insects away from crops. Sticky traps can also be used. These trapping methods are particularly appropriate to monitoring insect populations. They are less effective at large-scale insect removal. In high-value crops, such as potatoes or strawberries, insect "vacuums" can be used for larger insects such as Colorado potato beetles or lygus bugs.

A few products can be used under organic certification standards. Soaps and oils can be used to suffocate insects such as aphids. Diatomaceous earth can be used to discourage soft-bodied insect larvae, rusty grain beetle, and slugs and snails. Natural products such as pyrethrum and rotenone are also acceptable under most certification standards.

Diseases

As with insects, many fungal and bacterial species are beneficial or benign. Practices that increase biodiversity will likely increase species that compete with or prey on disease species, and thus reduce the outbreak of disease epidemics. The severity of disease can be limited by reducing the population of the pathogen or the susceptibility of the host, or by changing environmental conditions that favor infection.

Rotation is an important key to reducing disease by reducing the inoculum level of the pathogen. Crop rotation is effective when pathogens are obligate and host specific, with low dormancy and poor aerial spread. Leaf blights of cereals and ascochyta blight of lentil can be reduced by rotations that include different crops. Crop rotation alone is not sufficient to eliminate diseases in perennial crops; diseases such as common root rots and seedling blights with a wide host range; diseases that persist in soil, such as fusarium wilt of flax; diseases with rapid spread, such as powdery mildew of peas; or diseases that are widespread in the air or by insect vectors, such as cereal grain rusts and aster yellows.

Straw, residue, and weed management can be important. Straw is a primary inoculum for some disease species. Incorporation of residues speeds their decomposition and thus reduces the pathogen population. The risks to soil quality with excess tillage need to be balanced against the risk of leaving inoculum in the field. Other sources of infection include volunteers of the target crop, and weeds that are closely related to it. These sources of inoculum can reduce the effectiveness of a rotation away from the target crop.

Where diseases spread slowly from adjacent fields, field edges can be treated separately. Barrier strips or early swathing or mowing of the severely affected area may reduce the spread of disease into the crop.

Seed quality can impact disease potential. Reducing seed damage during harvest, storage, and seeding can reduce the susceptibility to disease, especially seedling blights. This is especially important for seeds such as flax, rye, and pulses. Disease-free seed reduces the spread of seedborne diseases into new areas.

Host susceptibility can be reduced by the selection of appropriate cultivars. Resistance or relative tolerance to a number of diseases varies greatly among cultivars. Differences among cultivars in disease susceptibility may reflect differences in their rate of growth; differences in their architecture and thus canopy humidity; or physical, biochemical, or genetic properties that restrict disease entry. Crop timing may also be important. For instance, earlier seeding may reduce the incidence of diseases such as barley yellow dwarf, powdery mildew of pea, and pasmo of flax. Late seeding of fall-seeded crops can reduce spread of disease by reducing their overlap with similar spring-seeded crops. Late seeding into warm soil can increase seedling vigor, and thus reduce crop susceptibility to seedling blights. Reduced seeding rates can reduce the spread of disease by reducing contact among plants, and by altering the environment in the canopy. Intercropping or strip cropping can be effective as well.

Some environmental manipulations can make disease frequency less severe. Selection of an appropriate field for a given crop can be important. Nutrient imbalance can make diseases such as take-all in cereals more severe. The incorporation of manure and green manures in rotations can reduce the severity of disease by encouraging microbes that are antagonistic to crop pathogens.

FUTURE CONCERNS

Currently, most organic standards prohibit the use of genetically modified organisms. There is concern among organic producers that the popularity of biotechnology in crop breeding will result in the abandonment of traditional breeding programs. If all or even most future genetic disease and insect resistance is incorporated into a genetically modified background, it will be unavailable to organic producers. This will greatly reduce their pest management options.

BIBLIOGRAPHY

Altieri, M.A. *Agroecology: The Scientific Basis of Alternative Agriculture*; Westview Press: Boulder, CO, 1987.

Earthcare: Ecological Agriculture in Saskatchewan; Hanley, P., Ed.; Earthcare Group: Regina, SK, 1980.

Macey, A. *Organic Field Crop Handbook*; Canadian Organic Growers, Inc.: Ottawa, Canada, 1992.

Radesovich, S.; Holt, J.; Ghersa, C. *Weed Ecology. Implications for Management*, 2nd Ed.; John Wiley & Sons: Toronto, ON, 1997.

Food and agriculture organization of the United Nations. Special, organic agriculture and sustainability, defining organic agriculture, sustainable development dimensions, environmental policy, planning & management, 1998. http://www.fao.org/WAICENT/faoinfo/sustdev/epdirect/EPRE0056.HTM (accessed April 20, 1999).

Organic crop improvement association, international certification standards, 1996. htttp://www.gks.com/library/standards/ocia/ociain.html (accessed April 21, 1999).

The national standard of Canada for organic agriculture, Canadian organic advisory board, 1999. http://www.coab.ca/standard.htm (accessed May 1, 1999).

USDA National Organic Program Proposed Rule, United States Department of Agriculture, 1999. http://www.ams.usda.gov/nop/rule.htm (accessed May 1, 1999).

ORIGINS OF WEEDS: BENEFITS OF CLEAN SEED

Heike Vibrans
Colegio de Postgraduados en Ciencias Agrícolas, Montecillo, Mexico

INTRODUCTION

Weeds, as a group, are the plants that participate most in the worldwide floristic exchange that started with the advent of agriculture about 10,000 years ago. Weeds have traveled with humans during migration, commerce, agronomic changes, and military campaigns, some intentionally, others accidentally. On the way, they acquired improved genes from other populations or species, thus becoming more successful—and often more noxious from the human point of view.

An important way to avoid future weed problems is to use clean crop seeds. Not only does it impede the spread of weeds to new regions or countries, it also reduces evolution of the weeds by not allowing genetic exchange among previously separate populations.

ECOLOGICAL ORIGINS OF WEEDS

Habitats that contain species preadapted to become weeds are those that are disturbed naturally with a certain regularity (1–4).

Many weeds of arable fields originated in the vicinity of moving water—in a count from central Europe about 44% of the native weed species. In this type of habitat we find several ecological conditions similar to croplands: exposure to the sun; mechanical disturbance of the soil, often in a regular way; accumulation of refuse, and with it nitrogen; sandy soils that dry easily and heat up quickly; and little initial competition. In temperate areas, *Urtica dioica, Rumex obtusifolius*, and various species of *Polygonum* have their natural habitat in river valleys (3). An example of a river valley weed from the tropics is *Asclepias curassavica* (Fig. 1). Various species of *Brassica*—weeds and crops—originated in coastal vegetation.

Many weeds have their origin in open vegetation types (1, 3, 4):

- forest-grassland borders (ecotones) and forest clearings (*Solidago, Epilobium*, and *Oenothera*);
- vegetation of salty soils (*Salsola, Kochia, Chenopodium, Portulaca*, and others);
- open oak forests heavily influenced by fire in the Middle East, Mexico, and East Asia (*Cosmos and Bidens*);
- the low vegetation on calcareous soils in central Europe;
- the tundra-like open grasslands dominating Europe after the last glaciation; and
- deserts with their flora of annuals.

Large animals graze and browse, trample soil along paths, periodically or regularly destroy herbaceous vegetation around their sites of congregation, and enrich the soil with nitrogen near their resting places. Many of the dominant ruderal or weedy grass species in the tropics have evolved in Africa with its large mammals; several have clear adaptations to these animals (1). For example, the kikuyu grass, *Pennisetum clandestinum*, only flowers when it is cut (or grazed), has seeds that can disperse only through the gut of an animal, and a runner system that has clearly evolved to resist trampling and to cover newly opened gaps in the vegetation. It is an intolerant invader and transformer of tropical highland landscapes, but also a good fodder and lawn plant. Its natural habitat is the forest-grassland ecotone in the East African mountains. Small animals can also disturb vegetation: leaf-cutter ants are a strong influence in large parts of the tropics; termite hills are special habitats with open, nutrient-rich soil; burrowing animals may open up ruderal-like sites.

FURTHER EVOLUTION TOWARD BEING A SUCCESSFUL—NOXIOUS—WEED: NEW ECOLOGICAL CONDITIONS

Weed evolution is highly dynamic (1, 2). Once preadapted plants start to grow in human-made habitats, they evolve under a new set of selective pressures. Microevolution of biotypes may be a rapid process (a few generations), if the selective pressure is strong and the initial genetic material variable (5).

For example, regular harvest with nonspecific mowing and threshing in small-seeded crops will promote plants

Fig. 1 *Asclepias curassavica*, today an ornamental and worldwide tropical weed, growing in its natural habitat along a river in southern Mexico.

with a cycle and seeds similar to the cultivated plant, that are harvested and sown with it (''crop mimics''; flax and *Camelina sativa*; small cereals and corncockle, *Agrostemma githago*). Strong weeding pressure will lead to plants that are vegetatively similar to the crop plant (rice and the barnyard grass, *Echinochloa crus-galli*; maize and teosinte, *Zea mays* subsp. *mexicana* and subsp. *parviflora*) (6, 7). Herbicides select for species that circumvent the biochemical pathways interrupted by the chemicals (*Chenopodium, Amaranthus, Senecio*, and *Capsella*).

Weeds soon expand into the areas from which they were previously absent, and associate with close relatives, with whom they may be interfertile. One of the most important ways in which all plants—wild plants, weeds, and crops—improve their fitness is by hybridization and genetic introgression, that is, by exchanging genetic material, recombining it, and selecting the most successful individuals (5). This exchange of genetic material may take place between populations of the same species, between related species, between crops and related weeds, and even between species of different genera. Several of the most noxious weed species are close relatives of crops, which either derive directly from the cultivated plant, or are results of weed ''improvement'' with genes of the crop (1). Seventeen of Holm's eighteen ''Worst Weeds of the World'' are cultivated in some part of the world (8). Also, hybrid species may result, for example the allopolyploid *Stellaria media*.

Weed problems may also change rapidly. Several species of weeds that were important 200 years ago (e.g., the corncockle, or the cornflower or bachelor's button, *Centaurea cyanus*) are rare today.

Other weed species have been favored by modern agriculture.Monoculture creates very specific habitats for weeds (9). A few species that are best adapted to the cycles of the crop and of the weed control measures can grow with little competition, and build up large populations. Herbicide use has favored close relatives of the crop plants (oats and *Avena fatua*; sugar beets and ''weed beets''), and, of course, species that can develop resistant biotypes. Fertilization promotes plants that are able to convert nutrients into biomass rapidly (e.g., *Chenopodium* and *Stellaria media*). Abandonment of crop rotation and initiation of double-cropping in irrigated areas allow seed banks for some weeds to build up to critical levels. High-yielding, but less competitive, crops leave more room for weed growth. Cereals with short stalks permit more light into the lower strata of the field. No-tillage agriculture promotes perennial weeds.

GEOGRAPHICAL ORIGINS OF WEEDS

Weed floras, just as the floras of cultivated plants, generally contain elements from various parts of the world. However, only in some cases have floras of either type been successful as a whole flora in a new region outside of their origin (an example is New Zealand). The percentage of alien plants—which are generally weeds—in a general flora vary widely, from 3–4% in Israel to 7% in Java to 47% in New Zealand. One notable trend is the high percentage of aliens in island and urban floras (often around 50%). Intermediate positions are held by areas without much tradition in agriculture and colonized by Europeans (Australia and Canada have about 28% aliens). European general floras generally contain 10–15% plants of exotic origin (10, 11).

The following hypotheses have been used to explain the differences in the competitive ability of weeds originating in various parts of the world (12):

On the basis of island theory it has been stated that European weeds are more successful because they evolved on the largest landmass of the earth, and therefore under more competitive conditions. This does not explain why East Asian weeds have not been more successful worldwide (13).

Gray first proposed in 1879 (14) that the number of competitive weed species that an area produces may be related to the time agriculture has been practiced there. This might be combined with the type of agriculture practiced: plow agriculture was invented in the Middle East and went on to conquer agriculture around the world. This would explain the low percentage of alien weeds in the Middle East. Another example is the Valley of Mexico, which only has about 8% aliens, despite human disturbance for thousands of years; traditional maize fields of the area contain only about 12% aliens, and some types have none (15).

Type and length of time of commercial relationships of a region certainly appears to have had an influence on the number of aliens.

Unfortunately, there are few data from the old agricultural centers outside of the Middle East. Weed science in the tropics tends to concentrate on dominant plants in modern agriculture. Studies of the weeds in traditional agroecosystems could be used to answer questions on the origin and the nature of weeds.

THE IMPORTANCE OF CLEAN SEEDS

Worldwide commerce is an efficient vehicle for the dispersal of weed species into new areas. While in Europe the rate of establishment of new exotic species has been leveling off since a peak at the end of the nineteenth century (3, 11), other parts of the world have seen a dramatic increase.Introduction of new species is by far the most important source of new weed problems worldwide. A new introduction may be more successful than in its home range, if it is introduced without its own diseases, pests, and predators.

A large part of weed species was introduced intentionally: of 233 noxious weeds of Australia, 31% were introduced as ornamentals, and another 15% for other uses. In South Africa about 60% of 530 alien species probably were introduced as useful plants (16).However, the rest of the noxious weeds were introduced accidentally.Here, it is more difficult to trace the vector, but there are numerous examples of introductions via contaminated seed shipments. Indeed, crop research and breeding centers have been accused of distributing their weeds worldwide through lots of improved crop seeds (though apparently they have now taken appropriate measures). Today, small seeds (turf and pasture grasses, alfalfa, flower mixtures, etc.) are still problematic, because cleaning them with automatic equipment is not as efficient as for larger seeds.

If farmers sow crop seeds contaminated with weed seeds—even if they are only the weeds of the neighbor—they effectively help the weeds improve their fitness, and therefore their noxiousness. This effect is multiplied, if the contaminated seeds come from faraway areas, or from other continents.

Therefore it is of prime importance to sow uncontaminated crop seeds. The farmer not only avoids spreading new species, but also impedes the mixing of different populations of the existing species. Strict rules on seed hygiene and certification are necessary not only in international commerce, but also on national and regional levels. Farmers sowing their own material, or obtaining their seeds through informal channels, should take care not to promote weed spread and genetic improvement of these plants.

REFERENCES

1. Baker, H.G. The evolution of weeds. Annu. Rev. Ecol. Syst. **1974**, *5*, 1–24.
2. Baker, H.G.; Stebbins, G.L. *The Genetics of Colonizing Species*; Academic Press: New York, 1965.
3. Ellenberg, H. *Vegetation Mitteleuropas mit den Alpen,* 5th Ed.; Ulmer: Stuttgart, Germany, 1996.
4. Zohary, M. *The Geobotanical Foundations of the Middle East*; Gustav Fischer: Stuttgart, Germany, 1973; 2, Swets & Zeitlinger, Amsterdam.
5. Parker, C. Prediction of New Weed Problems, Especially in the Developing World. In *Origins of Pest, Parasite, Disease and Weed Problems*; Cherret, J.M., Sagar, G.R., Eds.; Blackwell: Oxford, 1977.
6. Ellstrand, N.C.; Prentice, H.C.; Hancock, J.F. Gene flow and introgression from domesticated plants into their wild Relatives. Annu. Rev. Ecol. Syst. **1999**, *30*, 539–563.
7. Van Raamsdonk, L.W.D.; van der Maesen, L.J.G. Crop-Weed Complexes: The complex relationship between crop plants and their wild relatives. Acta Bot. Neerl. **1996**, *45* (2), 135–155.
8. Holm, L.D.; Plucknett, D.L.; Pancho, J.V.; Herberger, J.P. *The World's Worst Weeds, Distribution and Biology*; University Press: Honolulu, HI, 1977.
9. Pimentel, D. The Ecological Basis of Insect Pest, Pathogen and Weed Problems. In *Origins of Pest, Parasite, Disease and Weed Problems*; Cherret, J.M., Sagar, G.R., Eds.; Blackwell: Oxford, 1977.
10. Heywood, V.H. Patterns, Extents and Modes of Invasions by Terrestial Plants. In *Biological Invasions: A Global Perspective*; Drake, J.A., Mooney, H.A., di Castri, F., Groves, R.H., Kruger, F.J., Rejmánek, M., Williamson, M., Eds.; Wiley: New York, 1989.

11. Jäger, E.J. Möglichkeiten der prognose synanthroper pflanzenausbreitungen. Flora **1988**, *180*, 101–131.
12. Lonsdale, W.M. Global patterns of plant invasions and the concept of invasibility. Ecology **1999**, *80* (5), 1522–1536.
13. Huston, M.A. *Biological Diversity. The Coexistence of Species on Changing Landscapes*; Cambridge University Press: Cambridge, MA, 1994.
14. Gray, A. The pertinacity and predominance of weeds. Amer. J. Sci. Arts **1879**, *118*, 161–167.
15. Vibrans, H. Native maize field weed communities in South-Central Mexico. Weed Res. **1998**, *38*, 153–166.
16. Wells, M.J.; Poynton, R.J.; Balsinhas, A.A.; Musil, K.J.; Joffe, H.; van Hoepen, E.; Abbot, S.K. The History of Introduction of Invasive Alien Plants to Southern Africa. In *The Ecology and Management of Biological Invasions in Southern Africa*; Macdonald, A.N., Krueger, F.J., Ferrar, A.H., Eds.; Oxford University Press: Cape Town, South Africa, 1986.

ORNAMENTAL CROP PEST MANAGEMENT (INSECTS)

Tetsuo Gotoh
Ibaraki University, Ibaraki, Japan

Xiao-Yue Hong
Nanjing Agricultural University, Nanjing, Jiangsu, China

INTRODUCTION

Ornamental crops include cut flowers, potted floral and foliage plants, hanging baskets, and so on. During the past two decades, the areas under ornamental crop cultivation have increased 1.1–1.8 times in floricultural countries of the world. The areas devoted to ornamental crop production in greenhouses and vinyl houses have tended to increase more than those in the outdoors. In 1997, the greenhouse cultivation area, expressed as a percentage of the total area under ornamental crop cultivation, was 69.7% in Holland, 57.5% in Italy, and 49.4% in Japan (AIPH Yearbook). The wholesale value of nonedible horticulture products excluding bulbs, nursery stocks (trees), and grasses is more than $7.1 billion in Holland, 5.4 billion in the United States, 4.7 billion in Italy, and 4.1 billion in Japan.

Although many pests attack ornamental crops, the most severe problems limiting production of certain crops are caused by six insect orders and one mite order. These include Heteroptera (Hemiptera; lace bugs, plant bugs), Homoptera (Hemiptera; aphids, mealybugs, scales, whiteflies), Thysanoptera (thrips), Coleoptera (beetles, chafers, weevils, soil-living larvae (grubs)), Diptera (bulb flies, fungus gnats, leaf miners, midges), Lepidoptera (armyworms, borers, cutworms, leaf miners, leaf tiers, loopers, rollers, woolybears), Hymenoptera (orchid flies, rose slugs), and Acari (tarsonemid mites, spider mites, eriophyid mites, bulb mites) (Tables 1 and 2). Other pests include springtails (Collembola), cockroaches (Blattodea), slugs and snails (Stylommatophora) and nematodes (Nematoda), which occasionally cause severe problems only to certain crops.

More than 3500 alien (exotic) arthropod species have invaded North America during the past 100 years, and account for ca. 3.5% of all arthropods known in the United States (4). In Japan, exotic arthropods exponentially increased after 1868 and 363 species have been recorded so far (5). Some of them, including sweet potato whitefly, western flower thrips, and leaf miners, cause serious damages to many crops. Therefore, special attention has to be paid to exotic pest invasions.

DAMAGE

Insects, mites, slugs, snails, and nematodes directly suck on or chew plants. Light infestations are usually not harmful to plants, but high infestations may result in leaf curl, wilting, stunting of shoot growth, and delay in production of flowers and fruits, as well as a general decline in plant vigor and sometimes plant death.

Some pests are important vectors of plant diseases, transmitting pathogens in the feeding process. These pests include aphids (cucumber mosaic virus, turnip mosaic virus, tomato aspermy virus, gerbera latent virus), whiteflies (tomato yellow leaf curl virus), thrips (tomato spotted wilt virus, melon yellow spot virus, impatiens necrotic spot virus), and mites (orchid fleck virus, fig mosaic virus, cherry mottle leaf virus, garlic mite-borne mosaic virus).

CONTROL MEASURES

Chemical Control

Use of chemicals such as insecticides and acaricides can effectively control pests (1–3). We do not recommend the use of specific agricultural chemicals for two reasons. First, chemicals can vary from one country to another depending on the chemical registration status and economic development. Second, they have frequently led to the development of pesticide resistance and to harmful

O

Table 1 List of major insect and mite pests of flower and foliage crops

Abbr.[a]	Pests	Order-Family	Genera
Cm1	Springtails	Collembola	
B1	Cockroaches	Blattodea-Blattidae	*Periplaneta*
Ht1	Plant bugs	Heteroptera-Miridae	*Lygus, Poecilocapsus*
Ht2	Azalea lace bugs	Heteroptera-Tingidae	*Stephanitis*
Ho1	Whiteflies and greenhouse whiteflies	Homoptera-Aleyrodidae	*Bemisia, Trialeurodes*
Ho2	Aphids	Homoptera-Aphididae	*Aphis, Acyrthosiphon, Brachycaudus, Macrosiphoniella, Macrosiphus, Myzus, Neomyzus*
Ho3	Soft scales and armored scales	Homoptera-Coccidae, Diaspidae	*Pulvinaria, Lepidosaphes, Chrysomphalus*
Ho4	Mealybugs (scale insects)	Homoptera-Pseudococcidae	*Pseudococcus, Ferrisia*
T1	Thrips	Thysanoptera-Thripidae	*Thrips, Heliothrips, Hercinothrips, Frankliniella, Taeniothrips*
Cp1	Banded cucumber beetles	Coleoptera-Chrysomelidae	*Diabrotica*
Cp2	Lily leaf beetles	Coleoptera-Chrysomelidae	*Lilioceris*
Cp3	Beetles and black vine weevils	Coleoptera-Curculionidae	*Otiorhynchus*
Cp4	Fuller rose beetles	Coleoptera-Curculionidae	*Asynonychus*
Cp5	Iris weevils	Coleoptera-Curculionidae	*Mononychus*
Cp6	Lily weevils	Coleoptera-Curculionidae	*Agasphaerops*
Cp7	Japanese beetles	Coleoptera-Scarabaeidae	*Popillia*
Cp8	Rose chafers	Coleoptera-Scarabaeidae	*Macrodactylus*
D1	Rose midges	Diptera-Cecidomyiidae	*Dasineura*
D2	Lesser bulb flies	Diptera-Syrphidae	*Eumerus*
D3	Narcissus bulb flies	Diptera-Syrphidae	*Merodon*
D4	Leaf miners	Diptera-Agromyzidae	*Agromyza, Liriomyza, Phytomyza (= Chromatomyia)*
D5	Seed midges and chrysanthemum gall midges	Diptera-Cecidomyiidae	*Neolasioptera, Rhopalomyia*
D6	Fungus gnats	Diptera-Sciaridae	*Bradysia, Sciara*
L1	Borers	Lepidoptera-Crambidae	*Ostrinia*
L2	Columbine borer and stem borers	Lepidoptera-Noctuidae	*Papaipema*
L3	Iris borers	Lepidoptera-Noctuidae	*Macronoctua*
L4	Rhododendron borers	Lepidoptera-Sesiidae	*Synanthedon*
L5	Azalea leaf miners	Lepidoptera-Gracillariidae	*Caloptilia*
L6	Greenhouse leaftier	Lepidoptera-Crambidae	*Udea*
L7	Loopers	Lepidoptera-Noctuidae	*Trichoplusia, Pseudoplusia*
L8	Leafrollers	Lepidoptera-Tortricidae	*Platynota*
L9	Armyworms	Lepidoptera-Noctuidae	*Spodoptera*
L10	Budworms and tobacco budworms	Lepidoptera-Tortricidae, Noctuidae	*Platynota, Heliothis, Helicoverpa*

(Continued)

Table 1 List of major insect and mite pests of flower and foliage crops (*Continued*)

Abbr.[a]	Pests	Order-Family	Genera
L11	Cutworms and variegated cutworms	Lepidoptera-Noctuidae	*Peridroma, Platynota, Ostrinia, Spodoptera, Trichoplusia, Feltia*
L12	Earworms and bollworms	Lepidoptera-Noctuidae	*Helicoverpa*
L13	Hornworms	Lepidoptera-Sphingidae	
L14	Woollybears	Lepidoptera-Arctiidae	
Hy1	Orchid flies	Hymenoptera-Eurytomidae	*Eurytoma*
Hy2	Bristly roseslugs	Hymenoptera-Tenthredinidae	*Cladius*
A1	Cyclamen mites	Acari-Tarsonemidae	*Phytonemus*
A2	Tarsonemid mites	Acari-Tarsonemidae	*Polyphagotarsonemus*
A3	Spider mites	Acari-Tetranychidae	*Eotetranychus, Oligonychus, Tetranychus*
A4	Eriophyid mites	Acari-Eriophyidae	*Aceria*
A5	Bulb mites	Acari-Acaridae	*Rhizoglyphus*
S1	Slugs	Stylommatophora-Agriolimacidae, Arionidae, Limacidae	*Agriolimax, Deroceras, Arion, Limax*
S2	Snails	Stylommatophora-Helicidae	*Cryptomphalus, Theba*
N1	Nematodes	Nematoda-Tylenchida	

[a] These abbreviations are adopted in Table 2.
(From Refs. 1–3.)

effects on human beings and the environment, often resulting in their withdrawal from the market. To avoid the development of resistance in crop pests, one should read the label directions carefully before buying and using any pesticides, follow all instructions and carry out pesticide rotation of several kinds of chemicals having different modes of action. Some examples of rotating chemicals are organosynthetic pesticides such as acephate and bifenthrin, biorational products such as petroleum oil emulsifiable concentrate, (horticultural oils) insecticidal soaps, and IGR (insect growth regulators) (3).

Treating the soil between crops with chemicals such as chlorpicrin and pesticides is also effective against pests living in and on the soil, such as white grubs, leafminers, nematodes, and soil-borne disease. Fumigants such as fosthiazate and 1,3-dichloropropene can be adopted to greenhouses and vinyl houses for controlling pests and soil-borne disease. If biological control is being actively used for pest management of ornamental plants, it is better to use selective chemicals having no or less harmful effects on beneficial insects and pathogens.

Biological Control

Numerous beneficial insects and pathogens prey on or infest pests. To date more than 40 predators and parasite insects and 60 living systems (baculoviruses, protozoa, bacteria, fungi, and nematodes) are sold commercially for use in pest control (6).

Pearly green lacewings, ladybirds, aphid gall midges, and parasitic wasps play an important role in the suppression of aphids. Whitefly parasitoids are natural enemies of whiteflies. For thrips, minute pirate bugs (*Orius* spp.) and phytoseiid mites are used. Verticillium wilt is used to control aphids, whiteflies, thrips, and mites. White grubs living in soil can be controlled by parasitic nematodes (*Steinernema* spp.). Lepidopteran egg parasites (*Trichogramma* spp.), Bt (*Bacillus thuringiensis*), white muscardine (*Beuveria* spp.), and parasitic nematodes are biological control agents for lepidopteran pests. For mites, mitephagous thrips, cecidomyiid midges (*Feltiella* spp.), minute pirate bugs, ladybirds and phytoseiid mites are used for control (1, 3, 6). It is worth mentioning that gardeners should avoid using insecticides, which are harmful to beneficial organisms in the garden, and should also strive to keep their plants healthy and growing vigorously because some insects such as migrating aphids are attracted to the yellow-green color of unhealthy plants.

Cultural Control

Proper cultural techniques along with maintenance of vigorous plant growth should minimize the impact of pest damage. The use of resistant cultivars and transgenic varieties provides the first step of defenses against pests, and then precautions should be taken to avoid later infection. Weed control in and around the crop production areas is essential because weeds can act as a reservoir for the pests and virus vectors. Weed removal should be done quickly. Nursery stocks should be checked before planting and the infested ones should be eliminated. Crop production areas should always be sanitized, as this can remove or reduce the chance that problems will occur in the following year. Other cultural controls include hand picking of insects or infected branches, using nonchemical insect baits, applying plastic netting, crop rotation, and using organic fertilizers such as bat guano, blood meal, bone meal, cottonseed meal, and soybean meal. However, using a large amount of organic fertilizer may increase the number of fungus gnats.

Physical Control

Various films such as reflective sheets and ultraviolet-absorbing vinyl film may help prevent pests from landing on crop production areas, but the UV film interferes with the flight behaviour of honeybees. Reflective sheets help to keep away aphids, whiteflies, thrips, leaf beetles, diamondback moth, and blues (Lepidoptera). Light traps, using various color lamps and diode-emitting ultra high luminance light are effective against some pests. Yellow fluorescent light keeps noctuid moths such as tobacco budworm away from carnation and chrysanthemum (7, 8). On the other hand, yellow color attracts aphids, whiteflies, and leafminers, and therefore yellow sticky traps and yellow water pan traps have been used to monitor or reduce populations of these pests. Similarly, blue sticky traps can be used to monitor or detect thrips infestation.

Microscreening helps to reduce the number of pests moving into greenhouses and vinyl houses from adjacent fields. But screens will not exclude all pests, and they also reduce ventilation.

Another method is to raise the temperature of wet soil by covering it with vinyl sheets. This method is effective in excluding nematodes and pathogens between crops in the greenhouses. However, it can be used only in summer when there is plenty of solar radiation. For bulbs, hot-water dipping (30–120 min at 44–50°C) is effective in removing bulb flies and nematodes.

Sex Pheromones

Pheromone traps are mostly used for monitoring pest population densities, and sex pheromones help to reduce the population by mating disruption. Sex pheromones are characterized as having ''mild'' and species-specific effects, that is, they take longer than chemicals to achieve sufficient pest control. Therefore, widespread use of pheromone traps to control ornamental crop pests will be increased in the future, because ornamental crops are required to be less damaged and to have a higher quality in both flowers and foliages.

FUTURE CONCERNS

With world trade increasing, it is becoming more important to deal with exotic pests and diseases of ornamental crops. All countries involved should abide by the Sanitation and Phytosanitary Measures (SPS) agreement proposed by the World Trade Organization, and the International Plant Protection Convention (IPPC) proposed by the FAO. Government departments and research institutes should also conduct a Pest Risk Analysis (PRA) on ornamental crop pests and have a list of ornamental crop pests that need to be quarantined to prevent invasions of exotic pests. In the meantime, research on quarantine measures for quarantined pests should be conducted in case an invasion should occur.

IPM (integrated pest management) combines biological, cultural, physical, and chemical tools to regulate pest populations and aims to minimize economic, environmental, and human health risks. It is essential that those working on ornamental crop pests, as well as consumers, should be educated about IPM. Much research is needed to develop control thresholds for insect pests of ornamental crops. New technologies should be applied to assist IPM programs against ornamental crop pests. These technologies would include a Global Positioning System (GPS) to precisely locate areas of infestation, a Geographic Information System (GIS) to display the information in the most useful ways, and Hazard Analysis Critical Control Point (HACCP) and decision support software to help choose the most effective countermeasures.

The internet was designed to share information from a variety of sources among a large number of end-users. The United States, the European countries, Canada, China, and Australia have established IPM WWW sites to provide extensionists and farmers with information on pest management (9, 10), although most of the information is on farm pests. More information is needed on ornamental pest IPM, covering the basics of pest recognition and scouting, ecology, biology, and control strategies. The value of the internet to IPM as both an information system and decision aid is growing.

Table 2 Major insect and mite pests recorded in flower and foliage crops

Plants	Pest orders					
	Collembola (Cm)	Blattodea (B)	Heteroptera (Ht)	Homoptera (Ho)	Thysanoptera (T)	Coleoptera (Cp)
Ageratum			1	1, 2, 4		
Antirrhinum majus (Snapdragon)			1	2	1	
Aquilegia				2		
Aster			1	2		3
Azalea			2	1, 2, 3, 4	1	3
Begonia				1, 2, 3, 4	1	3
Camellia				1, 2, 3, 4	1	3
Cineraria				1, 2, 4	1	
Coleus			1	1, 4		
Cyclamen				2, 3	1	3
Dahlia			1	2	1	
Delphinium				2		
Dendranthema (Chrysanthemum)			1	1, 2, 4	1	8
Dianthus (Carnation)			1	2		8
Digitalis				2, 4		3
Euphorbia pulcherrima (Poinsettia)				1, 2, 3, 4	1	
Freesia				2	1	
Fuchsia				1, 2, 3, 4	1	3
Gerbera				1, 2	1	
Gladiolus			1	2, 4		
Gloxinia				2, 4	1	3
Hibiscus				1, 2		3
Hydrangea			1	1, 2, 3	1	8
Impatiens			1	2	1	1
Iris				2, 3	1	5
Kalanchoe				2, 3, 4	1	
Lantana (Lily)				2		2
Lilium (Easter lilies)				2, 3	1	2, 6
Lupinus (Lupine)			1	2		
Orchidaceae (Orchid family)		1	1	2, 3, 4	1	3
Pelargonium (Geranium)			1	1, 2, 4	1	3
Petunia			1	2	1	
Rosa (Rose)			1	1, 2, 3	1	4, 7, 8
Saintpaulia (African Violet)	1			2, 4	1	
Salvia				1, 2, 3		
Tagetes (Marigold)			1	2	1	7
Tulipa (Tulip)				2		
Verbena				1, 2	1	
Viola (Pansy)				2, 4		
Zinnia			1	1, 2	1	3

See Table 1 for abbreviations.
(From Refs. 1–3.)

REFERENCES

1. *The Ball RedBook*; Ball, G.V., Ed.; 15th Ed.; Geo. J. Ball Publishing: Chicago, 1991; 802.
2. Larson, R.A. *Introduction to Floriculture*, 2nd Ed.; Academic Press: New York, 1992; 636.
3. Powell, C.C.; Lindquist, R.K. *Ball Pest & Disease Manual*, 2nd Ed.; Ball Publishing: Batavia, IL, 1997; 426.
4. Cox, G.W. *Alien Species in North America and Hawaii, Impacts on Natural Ecosystems*; Island Press: Washington, DC, 1999; 387.
5. Kiritani, K.; Morimoto, N. Fauna of exotic insects in Japan

Pest Orders					
Diptera (D)	Lepidoptera (L)	Hymenoptera (Hy)	Acari (A)	Stylommatoph (S)	Nematoda (N)
	6, 7, 10,12		1, 3		
6	7, 11		1, 3		
4	2				
4	1				
	4, 5, 6		3		1
6			1, 3		
	7, 8				
4, 6	7, 11		3		
6			3	1	
6			1, 3		
	1		1, 3		
4	1		2, 3		
4, 5, 6	1, 7, 9, 10, 11		1, 3, 4	1, 2	1
4, 6	7, 11		3		
6	8		3		
			5		
			2, 3		
4			2, 3		
	11		3, 5		1
			1		
	6		3	1, 2	1
			1, 3		
2	3		5		
	7		2		
	1		5		
5, 6	2, 11		5	1	
6	11		3	1, 2	
		1		1, 2	
6	7, 8, 11		2, 3	1	
4	11, 13, 14		1		
1	8, 11	2	3	1	
6			1, 3		1
4					
3			3		
6			5		
	8, 14		2, 3		
	11		3		
4			3		

with special reference to North America. Biological Invasions, *in press*.

6. Copping, L.G. *The BioPesticide Manual*, 1st Ed.; British Crop Protection Council: Surrey, 1998; 333.
7. Masuda, T.; Nakamura, K.; Kumazawa, K. Control of tobacco budworm, *Helicoverpa armigera* (Hübner), by yellow fluorescent light on carnation in greenhouses. Annu. Rep Kanto-Tosan Plant Prot. Soc. **1977**, *44*, 279–281 (in Japanese).
8. Kunimoto, Y.; Inda, K. Control of tobacco budworm, *Helicoverpa armigera* (Hübner), in chrysanthemum field by overnight illumination with yellow fluorescent lamp. Bull. of Nara Agri. Exp. Stn. **1999**, *30*, 30–31 (in Japanese).
9. Natural Resources Institute. IPMEurope Network, 2000; http://www.nri.org/IPMEurope/homepage.htm (accessed April 2001).
10. North Carolina State University. National Integrated Pest Management Network, 2000; http://www.reeusda.gov/nipmn (accessed April 2001).

OTHER MAMMAL DAMAGE AND CONTROL

Rex E. Marsh
University of California, Davis, California, U.S.A.

INTRODUCTION

Nonrodent mammalian pests makeup a highly diverse group of animals and come into conflict with humans in a variety of situations, including urban, suburban, and rural areas (Table 1). They are responsible for economic damage to agriculture, forestry, landscaping, and structures (1). Some species are implicated in the dissemination of diseases, some are predators of other species, and others are simply an annoyance or nuisance. A number of these animals are considered pests only in special situations; in other settings they may be considered highly desirable animals. Control methods and management approaches used to restrict damage or resolve the created problems are as diverse as the pest species themselves. Control methods include but are not limited to trapping, shooting, poison baits, chemical repellents, frightening devices, and fumigants. Physical barriers in the form of exclusion from buildings and fencing out the pest species are major management approaches, along with habitat modification, modifications in agricultural and forestry cultural practices, and changes in livestock husbandry (2).

IMPLICATED ANIMALS

Rodents, their damage and control, are covered in other articles in this encyclopedia, as are deer and their damage and control, and the influence of vertebrate pests, including mammalian predators, on livestock production. This section, therefore, is devoted mostly to miscellaneous nonrodent mammal pests, principally raccoons, opossums, skunks, armadillos, bats, moles, and rabbits. Less widely recognized pests include such nonrodent pests as weasels, badgers, shrews, and even black bears. Still other species are mentioned as a means of illustrating the wide diversity of wildlife/human conflicts and management practices.

Raccoons, opossums, skunks, and armadillos are frequently referred to as nuisance animals as they are more of a nuisance than anything else when roaming around rural farm buildings or when taking up residence in a suburban residential community (3). Such nuisance animals have given rise to a growing service industry, the Nuisance Wildlife Control Operators (NWCOs), who trap and remove the animals as a service. The animals may be relocated or dispatched, depending on the pest species, state laws, and situation. Rodents such as garden-loving woodchucks and attic-living tree squirrels are also often trapped as part of the NWCOs services. Structural Pest Control Operators (PCOs) commonly provide control of insects and rodents, that is, rats and mice, within buildings and they most often also provide bat removal and exclusion services.

DAMAGE AND CONTROL

Structural Pests

A number of bat species take up residence in dwellings or other man-made structures. In such situations, their presence is or may be undesirable because of the resulting noise they create and the odor of guano and the insects it supports. Because bats, along with skunks, raccoons, and foxes, depending on the region of the country, are also the wildlife species most often found infected with the deadly rabies disease, health departments always caution the public against too close an association or contact with these animals.

The common practice for ridding bats from buildings is to seal off all openings except one or two and install one-way bat escape valves, which permit the bats to leave but not return. Once the bats are all gone, the last of the entry points are sealed.

Shrews, most significant as feeders on conifer seed, on rare occasions enter and take up residence in structures and are removed by trapping. It is not uncommon for destructive raccoons to enter and nest in chimneys or to rip off roof shingles to gain access to an attic to secure a place to rear their young. Skunks, raccoons, and opossums will all take up residence and give birth to young beneath dwellings or other buildings. Associated ectoparasites from these and other mammal pests may infest pets and humans alike and some may be vectors of diseases (4).

Table 1 Mammal pests and the nature of their conflicts with human interests

Mammal pests	Structural pests	Threats to humans	Mammals as predators	Pests of gardens and landscaping	Pests of forestry and agricultural crops
Armadillos			X		
Badgers			X	X	X
Bats	X				
Bears		X	X		X
Bobcats			X		
Coyotes		X	X		
Foxes			X		
Moles				X	X
Mountain lions		X	X		
Opossums	X		X	X	
Rabbits				X	X
Raccoons	X		X	X	
Shrews	X			X	X
Skunks	X		X	X	
Weasels			X		

Trapping is the remedy most often used, followed by sealing them out.

Threats to Humans

Mountain lions, black and grizzly bears, and coyotes are native predators that are occasionally involved in attacks on humans in recreational or suburban areas, but fatalities are very rare. These problems are increasing due to the expanding human population and its increasing encroachment into the wildlife's habitat and space. Coping with these particular potential problems focuses on public education. Some temporary protection from bears relies on the use of irritating pepper (capsaicin) sprays. Persistently threatening animals are generally trapped and relocated, while an offending animal is generally dispatched.

Mammals as Predators

In suburban areas, fox, coyotes, bobcats, and mountain lions are frequently implicated in preying upon pets such as domestic cats and small dogs. In these situations, population reduction of the offending species by trapping or shooting may be necessary. These same native mammalian predators that threaten the lives of our pets, and occasionally cause human fatalities, are responsible for serious economic losses to livestock production, including poultry. Weasels, skunks, opossums, raccoons, badgers, and bears may also create predation problems (5). The black bear's love of honey makes it a particular problem to beekeepers. Predators often work to the detriment of efforts to save threatened and endangered wildlife species by decimating the breeding stock and valuable offspring.

Trapping or shooting often accomplishes population control of large predators. Fencing, the use of guard dogs, and livestock husbandry changes are important in resolving many problems. Fencing is the major means of protecting apiaries from bear.

Pests of gardens and landscaping

Moles and rabbits, although their behavioral characteristics are similar to rodents, do not belong to that group. They are of major importance as pest species in a number of situations. Moles, mostly through their burrowing activities, disfigure turf and disturb plantings. Rabbits feed on a wide variety of plants and are capable of debarking and killing numerous trees. Skunks and raccoons dig up lawns while seeking grubs, and raccoons are notorious for raiding garbage cans and for stripping ripening ears from sweet corn. Opossums and armadillos are sometimes nuisance or garden pests regionally. The control of moles is by trapping, while fencing, tree protectors, chemical repellents, trapping, and shooting represent the controls for rabbits. The nuisance animals are usually trapped and relocated or euthanized.

Pests of forestry and agricultural crops

Rabbits (including hares) and black bears can be serious pests to forest trees and reforestation efforts. Rabbits clip off the tops of small seedlings and girdle the trunks of forest saplings. Bears, especially in the northwest, claw the trunks of sizable timber trees, stripping them of bark to feed on the cambium tissue. Providing alternative food to satisfy the bear's nutritional needs at this critical seasonal period when damage usually occurs has been one unique solution to the problem. Trapping and shooting are also used as control measures for both bears and rabbits. Chemical repellents and plastic mesh tree guards are used to protect forest tree seedlings and young orchard trees from rabbits. When their populations are high, rabbits extensively damage hay and forage crops, especially alfalfa, as well as certain vegetable crops. Jackrabbits are the worst of the rabbit offenders, especially in the western part of the country. Because of their vast numbers and the inconsistent results of other control measures, fencing them out may be the best solution to protecting high value crops. In other situations, rabbit drives and shooting or poison baits may be employed to reduce jackrabbit numbers.

REFERENCES

1. Anonymous. *Vertebrate Pests: Problems and Control*; National Academy of Sciences: Washington, 1970; 1–154.
2. Hygnstrom, S.E.; Timm, R.M.; Larson, G.E. *Prevention and Control of Wildlife Damage*; Nebraska Cooperative Extension Service, USDA-APHIS-Animal Damage Control, and Great Plains Agricultural Council, University of Nebraska: Lincoln, 1994; 1–936.
3. Timm, R.M.; Marsh, R.E. Vertebrate Pests. In *Handbook of Pest Control—Mallis,* 8th Ed.; Moreland, D., Ed.; Mallis Handbook & Technical Training Company: Cleveland, 1997; 954–1019.
4. Marsh, R.E. Vertebrate Pest Management. In *Advances in Urban Pest Management*; Bennett, G.W., Owens, J.M., Eds.; Van Nostrand Reinhold: New York, 1986; 253–285.
5. Eadie, W.R. *Animal Control on Field, Farm, and Forest*; Macmillan: New York, 1954; 1–258.

OVIPOSITIONAL DISRUPTION EMPLOYING SEMIOCHEMICALS

O

P. Larry Phelan
The Ohio State University, Wooster, Ohio, U.S.A.

INTRODUCTION

Behavioral disruption capitalizes on the mechanisms used by the pest for finding and selecting mates, food, and hosts for offspring, processes required for survival and reproductive success. Garnering the king's share of recent attention in this area has been the use of sex pheromones for mating disruption of adults, and to a less extent, antifeedants for developmental disruption of immature stages. Ironically, although few examples of disruption of host finding can be seen in modern pest-management programs, this approach has the longest historical precedent, mostly through use of plant volatiles. Long before any behaviorally active components were characterized, dried herbs and aromatic plant oils were used to protect stored foods and clothes. In addition, crops have been protected from colonization by insect pests through intercropping with odorous nonhost plants. Despite the recent lack of attention, there are good arguments for why this avenue has the potential for effective pest control that is not ecologically disruptive.

EXAMPLES OF IMPLEMENTATION

The subject of ovipositional disruption and its implementation is much more varied than that of disruption by sex pheromones, as host finding and egg laying is a concatenate process modulated by a number of different mechanisms and cues. In the chemical modality, ovipositional decision making may be mediated by plant chemicals and/or pheromones.

Plant Chemicals

Plant-produced volatiles can mediate host finding and longer-distance orientation, while nonvolatiles often determine egg deposition once contact is made. Disruption using plant chemicals can be achieved by a number of mechanisms, including confusion, masking, diversion, and repellency. Repellency is self-explanatory and utilizes plant com-pounds that are irritating or otherwise sensorily repugnant to a wide range of arthropods, such as capsaicin or other herbal essences. Diversion uses host-derived compounds to attract the pest to an artificial trap or a trap crop, which can be treated with insecticide. Populations of apple maggot fly (*Rhagoletis pomonella*) can be monitored, and in some applications reduced, by diversion to sticky red spheres baited with apple volatiles (1). In this system, host finding is mediated by two modalities, a visual response to red spheres and an olfactory-mediated attraction to apple odor. Since this strategy uses the same mechanisms of attraction as the fruit, there is the potential problem of deploying enough traps to compete. However, since infestation occurs largely through immigration, competition is reduced by placing traps on the orchard perimeter to intercept flies prior to entering the orchard (2). A large resident population of flies within the orchard would probably make this strategy impractical. In such cases, the diversion approach might be enhanced by a "push-pull" strategy, in which use of an attractive trap or trap crop is combined with treating the crop with repellent or nonhost odors.

Ovipositional disruption through confusion and masking has been demonstrated for the navel orangeworm (*Amyelois transitella*). This major pest of nut crops in California effects its damage by larval feeding directly on the nut meat, and oviposition is mediated through olfactory orientation to fatty acids emitted by the nut (3). A confusion-based disruption of oviposition first was demonstrated by broadcast application of almond oil and presscake (4). Presumably, this approach interferes with the ability of the female to distinguish nuts from the rest of the tree. Larvae from eggs not laid on the nuts have a low probability of finding the nut. This methodology was later refined using the attractive long-chain fatty acids themselves for disruption (5). A "third-generation" disruption of navel orangeworm was most recently developed by masking the attractive odor of almonds with a form of soybean oil. The ratio of fatty acids in soybean triglycerides is different from that of almond oil and is not itself attractive to this pest. Thus, it appears that the odor profile of the host plant is masked, and egg laying is not stimulated. Available as the product Stealth NOW, crop damage is reduced by 70–75%, while beneficial arthropods are preserved. This emulsifiable concentrate apparently is the first of its kind to reach commercial application.

Host-Marking Pheromones

A number of species deposit a marking pheromone during oviposition that deters egg laying by females arriving subsequently (6). These pheromones have the effect of spacing eggs to reduce intraspecific competition that would lead to lower survivorship of all individuals. A number of ecological constraints determines the probability that such pheromones are utilized by a species. The use of marking pheromones to reduce host colonization has been most successful in the cherry fruit fly (*Rhagoletis cerasi*), where broadcast application of synthetic pheromone reduced fruit damage by 90% (7). Although not yet commercially viable, recent developments suggest considerable potential for this strategy.

ECOLOGICAL PARAMETERS AND SUCCESS OF OVIPOSITIONAL DISRUPTION

There has been little research either on the mechanisms underlying ovipositional disruption or on the impact of ecological parameters and behavioral mechanisms on this approach. However, one can predict those parameters that might favor its success and practicality.

Narrow Window of Oviposition or Crop Vulnerability

The success of ovipositional disruption is likely to be inversely related to the duration of the period of pest adult activity. A univoltine species whose egg laying is restricted to a limited period of crop phenology would be the best candidate due to the limited time during which a formulation must remain active and the reduced need for repeat applications.

Single Important Pest or Multiple Species Using Same Mechanism for Host Finding

If additional species present at the same time require insecticidal intervention, control of only one species with host volatiles may be of little use. On the other hand, if secondary pests arise through pesticidal control of the key pest, there would be stronger incentive to use a nontoxic strategy that would not interfere with natural enemies. In addition, pesticide use could be reduced for the crop if different pests are asynchronous in their activity.

Pest Is a Tissue Specialist

Disruption probably would be easier when the pest must search for a particular structure of the plant, such as the fruit. This would be an essential characteristic if disruption is to be achieved through confusion with host attractants. If the pest is a leaf-feeder or if larvae are relatively mobile and can find tissue on their own, application of host attractants could actually increase damage.

Pest Infests Crop via Immigration

In many cases, disruption of species immigrating into the crop will be easier to achieve, either through interception with toxic baits (e.g., apple maggot fly) or trap crops, or by masking with nonhost odors. Resident populations may be able to use other short-range cues for egg laying or the sheer number of potential oviposition sites may outcompete the disruption strategy.

Breadth of Host Range

The effect of this parameter may be context-dependent; however, one might predict successful disruption is more

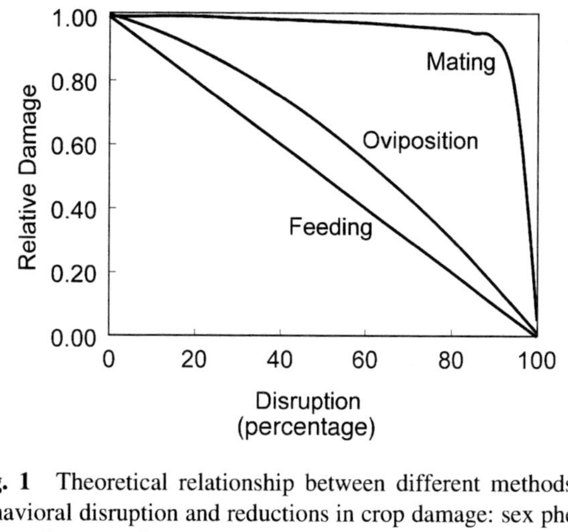

Fig. 1 Theoretical relationship between different methods of behavioral disruption and reductions in crop damage: sex pheromones for mate-finding by males, plant chemicals or marking pheromones for ovipositional host-finding by females, and antifeedants for feeding by larvae.

likely with generalist species. Once mated, females usually exhibit a strong propensity to lay eggs, and often become less selective in their choice of oviposition sites. The presence of alternative hosts will provide an outlet for egg laying. In addition, a principle for managing pesticide resistance is to provide refugia that ensure survival of susceptible genotypes, whose genes can be maintained in the population. A pest with a broad host range is likely to find such refugia in alternative hosts, reducing natural selection for finding the value crop in the face of disruption.

COMPARISON TO MATING DISRUPTION

Ovipositional disruption shares many of the same mechanisms governing mating disruption with pheromones; however, key differences also can be identified. Foremost is the fact that while pheromone disruption targets the male and only indirectly impacts pest damage, ovipositional disruption targets the gravid female, resulting in a more direct impact on damage (Fig. 1). Since males can mate many times and a single female can lay many eggs, pheromone disruption must be highly effective (possibly >95%) to impact crop damage significantly. In contrast, disruption of egg laying directly reduces the population of damage-causing larvae. The two strategies also differ with regard to the effects of pest immigration. If females immigrate into an area after mating has already occurred, pheromone disruption is ineffective. In contrast, ovipositional disruption may be most effective for immigrating pests, as described earlier. Thirdly, while pheromone disruption works best at low population levels, the success of ovipositional disruption by masking should be independent of population density. Finally, both sex and host-marking pheromones are usually novel compounds that require expensive synthetic procedures. Plant-derived compounds, on the other hand, may be readily available as commodities or even waste products of crop production, reducing the costs of this strategy. This differential is offset somewhat by the need for very small quantities of sex pheromones relative to plant-derived chemicals.

FUTURE DIRECTIONS

Ovipositional disruption using host-plant chemistry and marking pheromones is a relatively undeveloped strategy for control. Attention to the ecological considerations noted above should provide guidance for which crop–pest systems are most amenable to this approach. Like many other aspects of IPM, behavioral disruption requires better understanding of pest biology and knowledge of timing for implementation than may be necessary for traditional insecticides. However, when properly developed, this approach can provide selective suppression of crop damage while protecting farm workers, beneficial arthropods, and wildlife.

REFERENCES

1. Duan, J.J.; Prokopy, R.J. Control of apple maggot flies with pesticide-treated red spheres. J. Econ. Entomol. **1995**, *88*, 700–707.
2. Prokopy, R.J.; Mason, J.L.; Christie, M.; Wright, S.E. Arthropod pest and natural enemy abundance under second-level versus first-level integrated pest management practices in apple orchards: a 4-year study. Agric. Ecosys. Environ. **1996**, *57*, 35–47.
3. Phelan, P.L.; Roelofs, C.J.; Youngman, R.R.; Baker, T.C. Chemical characterization of volatiles mediating ovipositional host–plant finding by *Amyelois transitella* females. J. Chem. Ecol. **1991**, *17*, 599–613.
4. Van Steenwyk, R.A.; Barnett, W.W. Disruption of navel orangeworm (Lepidoptera: Pyralidae) oviposition by almond by-products. J. Econ. Entomol. **1987**, *80*, 1291–1296.
5. Baker, T.C.; Phelan, P.L. Ovipositional disruption of the navel orangeworm with fatty acids. US Patent No. 5, 104, 654, 1992.
6. Roitberg, B.D.; Prokopy, R.J. Insects that mark host plants. BioScience **1987**, *37*, 400–406.
7. Aluja, M.; Boller, E.F. Host marking pheromone of *Rhagoletis cerasi*: field deployment of synthetic pheromone as a novel cherry fruit fly management strategy. Entomol. Exp. Appl. **1992**, *65*, 141–147.

PEST CONCEPT

Larry P. Pedigo
Iowa State University, Ames, Iowa, U.S.A.

P

INTRODUCTION

Integrated pest management seeks to reduce the status of pests by following principles of ecology and using the latest advances in technology. Much of the foundation of insect pest management rests on determining whether an organism is truly a pest and, if so, just how serious a problem it is causing. Therefore, the concept of "pest" is rudimentary in defining the principles of integrated pest management.

DISCUSSION

Simply put, a pest is a species that interferes with human activities. According to Australian entomologist P.W. Geier and his colleagues, the quality of being a pest is anthropocentric (considering humans as the central fact or final aim and end of a system) and circumstantial. For example, termites feeding on dead wood in a forest serve an important ecological function, one of degradation in the process of returning nutrients to the soil. Clearly, termites are not pests in this context; they are beneficial to humankind. The same species, performing the same ecological function in the environment of a human home, is an important pest. An understanding of the often significant ecological roles played by pest species in unmanaged environments, as well as agricultural ones, frequently gives insights into how to deal with them. Moreover, this understanding can aid in developing more tolerant attitudes toward pest presence.

PEST TYPES

Organisms that are considered pests mostly include insects, mites, ticks, nematodes, weeds, fungi, bacteria, viruses, rodents, birds, molluscs, and crustaceans. Although some of the microorganisms listed cause diseases of humans and livestock, they are not usually referred to as pests. As a rule, viruses and microorganisms dealt with by physicians and veterinarians are not considered pests and usually are not under the purview of integrated pest management.

A great share of pest problems are encountered in agricultural and forest production systems. Here, problems arise because of significant population densities not simply because of species presence. Most pests of agriculture (including agronomy, animal science, horticulture, and forestry) are those species whose activities, enhanced by population density, cause economic losses.

Other pests include those whose mere presence is objectionable. These sometimes are referred to as aesthetic pests. Economic losses based on aesthetics are difficult to determine but are nonetheless real. In instances where insects enter the household to spend the winter, for example, boxelder bugs (*Leptocoris trivittatus*), some persons may be motivated to spend money on insecticidal control and therefore suffer an economic loss. Others may not be as bothered by insect presence. Presence of insects or insect parts in food, however, is another matter, being regulated by law. Insect presence in food processing and storage requires costly sanitation measures and, often, the disposal of tainted products. Such losses can be measured objectively and are significant.

Another group of pests, mostly insects, can be termed medical pests. Losses caused by medical pests also are very significant but, like some aesthetic pests, are extremely difficult to measure. Loss of work or work efficiency may be measurable in economic terms, but economic loss from discomfort, pain, and even loss of life is difficult to assign. Such problems make application of integrated pest management principles quite difficult.

BIBLIOGRAPHY

Flint, M.L.; van den Bosch, R. *Introduction to Integrated Pest Management*; Plenum Press: New York, 1981; 1–2.

Geier, P.W.; Clark, L.R.; Briese, T. Principles for the control of arthropod pests, I. Elements and functions involved in pest control. Protection Ecol. **1983**, *5*, 1–96.

Matthews, G.A. *Pest Management*; Longman: New York, 1984, 1–6.

Pedigo, L.P. Integrated Pest Management. In *McGraw-Hill Yearbook of Science and Technology*; McGraw-Hill: New York, 1985; 22–31.

Pedigo, L.P. *Entomology and Pest Management*, 3rd Ed.; Prentice-Hall: Englewood Cliffs, NJ, 1999; 31–32.

Encyclopedia of Pest Management

PEST MANAGEMENT IN ECOLOGICAL AGRICULTURE

Barbara Dinham
Pesticide Action Network UK, London, United Kingdom

INTRODUCTION

Strategies to develop ecological farming methods stem from economic, health, environmental, and practical concerns with chemical pesticides. Ecological agriculture, also called sustainable agriculture (1), agroecology, low-external input, regenerative agriculture (2), and farmer-participatory integrated pest management (IPM), has no single definition (3). The Food and Agriculture Organisation (FAO) of the United Nations suggests that such sustainable development (in the agriculture, forestry, and fisheries sectors) conserves land, water, plant, and animal genetic resources, is environmentally nondegrading, technically appropriate, economically viable, and socially acceptable (4).

Ecological agriculture implies approaches to prevent or minimize application of chemical pesticides, which promote local inputs, and where increased farmer knowledge becomes the basis of managing pests and improving yields and sustainability (5). Organic agriculture is distinctive in requiring an approved body to certify that no chemical fertilizers or pesticides have been used. This provides a guarantee for consumers, and can enable farmers to receive a premium for their crops.

PEST MANAGEMENT STRATEGIES

Ecological pest management practices have been adopted in both industrialized and developing countries. Most farming systems still use certain aspects: crop rotation, field clearance to destroy pest refuges, resistant varieties, early or late planting regimes. More specific ecological strategies employed by farmers vary widely, according to the cropping system, whether the farmer is in an industrialized or a developing country, and the locally available inputs.

As interest in ecological pest management grows, it is increasing the demand for biological technologies. The biological control agent in most widespread use, *Bacillus thuringiensis* (Bt) can be easily produced in large quantities, and can be used like a chemical. As a result it is applied in both conventional and organic agriculture in many countries. Bt has proved extremely effective in controlling persistent pests such as the diamondback moth, which plagues cabbages and related crops, and has led to a pesticide treadmill in many areas (6).

Biological pest controls require the development of breeding centers for insect predators and parasitoids. Cuba has the most advanced program globally with a country-wide network of over 300 centers for the Reproduction of Entomopathogens and Entomophages supplying bacteria, fungi, viruses, and insect parasitoids. These include *Lixophaga Diatraege* for cane stem borer, the parasitic wasp *Trichogramma*, and the fungal disease *Beauvaria bassiana* against a total of seven pests (7).

Particularly in developing countries, many indigenous plants and locally adapted technologies are used against a variety of pests. Most widely known is the neem tree *(Azardirachta indica)*, native to India and also found in parts of Africa, whose leaves are effective against many pests. Other common solutions are pyrethrum, chili peppers, wood ash, and castor oil seeds (8, 9).

SPREADING ECOLOGICAL PRACTICES

Ecological strategies for pest management are characterized by a holistic approach and not only the substitution of biological controls for chemical pesticides. Management strategies include soil conservation, seed selection, and maintenance of agricultural biodiversity. Participation of the women and men farmers to ensure cultural and local appropriateness is of central importance in the development of new strategies. Farmer Field Schools (FFS) have proved a successful training approach, where training takes place at field level, focusing on recognition of pests and predators and their life cycles. Designing in-field experiments, examining economic losses from pest damage, and encouraging observation of local plants that may act as a trap crop for pests or have repellent properties, become part of the armory (10).

P

ECOLOGICAL DIVERSITY FOR DIVERSE CROPPING SYSTEMS

Fragile tropical soils may benefit most from ecological strategies, but these farming approaches are not restricted by crop, climate, or continent. Table 1 indicates a range of cropping systems and countries that have benefited from IPM programs. One of the most successful examples is in European glasshouses where 70% of commercial glasshouses have been managed through IPM based on biological controls for more than 15 years. In these intensive production systems, IPM enables growers to control all major pests and avoid most pesticide use.

Cotton is the crop that uses the most chemical insecticides worldwide. Nevertheless, cotton IPM and organic strategies have demonstrated that ecological alternatives are successful. An FFS trainer- and farmer-training program in Pakistan prevented insecticide applications in the first 8–10 weeks after planting, allowing natural enemy populations to build up, and giving higher yields in seven of the 10 demonstration plots (11). Similar successes have been achieved in India, Zimbabwe, and elsewhere.

Adverse environmental impacts of pesticide use in conventional cotton systems and the problems of insect resistance to pesticides have encouraged farmers to invest in organic systems. While still a small proportion of overall cotton production, the growing consumer support for ecological fiber is likely to further encourage producers. The United States is the largest single producer with 32% of the total certified organic cotton fiber production in 1997, followed by Turkey with 22%. Organic cotton production is well suited to small-scale cropping systems and in the same year 15% of certified organic cotton fiber came from India, 19% from Africa, and 11% from Latin America (mainly Peru) (12).

In Indonesia IPM strategies based on improving farmers' knowledge of ecological pest management were highly successful against infestations of brown rice-hopper that decimated the rice crop in the mid-1980s. An IPM rice program supported by the FAO in South and South East Asia that began in 1980 has targeted farmers using high chemical inputs. More than 500,000 farmers have been trained and now save on average U.S. $10 per hectare per season, while maintaining or increasing yields (13).

Farmers develop their own strategic improvements. In an area of low rainfall and high soil erosion in Burkina Faso, local groups and villagers worked with the government and local organizations to develop an ecological approach ranging from tree planting to increased use of manure. Covering more than 200 villages, farmers increased sorghum yields from 870 kg/ha to 1650–2000 kg/ha. In the semi-arid Machakos region of Kenya farmers developed appropriate agricultural techniques, building terracing, selective animal grazing, and manure collection, significantly increasing soil fertility and yields (14).

The major physical constraints to adoption of ecological pest management relate to agricultural production systems rather than crops, regions, and climate. Large-

Table 1 Impact of selected IPM programs on pesticide use, crop yields, and annual savings

Country and crop	Average change in pesticide use (as % of conventional treatments)	Changes in yields (as % of conventional treatments)	Annual savings of program (US $ '000)
Togo, cotton	50	90–108	11–13
Burkina Faso, rice	50	103	No data
Thailand, rice	50	No data	5–10,000
Philippines, rice	62	110	5–10,000
Indonesia, rice	34–42	105	50–100,000
Nicaragua, maize	25	93[a]	No data
United States, nine commodities	No. of applications up, volumes applied down	110–130	578,000
Bangladesh, rice	0–25	113–124	No data
India, groundnuts	0	100	34
China, rice	46–80	110	400
Vietnam, rice	57	107	54
India, rice	33	108	790
Sri Lanka, rice	26	135	1,000

[a]Lower yields, but higher net returns. (From Ref. 2.)

scale monoculture, for example, does not easily lend itself to ecological approaches because of dependence on single varieties, loss of natural soil fertility, and emergence of specialized pests that kill natural enemies. The most sustained improvements can be found when government policies support ecological practices.

ENVIRONMENTAL AND ECONOMIC BENEFITS

Because of the many starting points, farmers have different motivations for adopting ecological approaches. In industrialized countries, where farmers generally rely heavily on external inputs, ecological practices may lower yields, but improve the environment.

In developing countries, some 2.3–2.6 billion people are supported by agriculture using the higher inputs of green revolution technologies and in these areas farmers would generally stabilize or achieve slightly higher yields with ecological pest management, also gaining environmental benefits. The remaining 1.9–2.2 billion people are largely fed by traditional agriculture and farmers. Here, ecological management strategies can substantially improve yields and income (15).

The environmental benefits of ecological agriculture stem from reductions in chemical inputs, which threaten biodiversity, pollute water sources, and kill fish and other nontarget beneficials, often including cattle or domestic animals. Some entomologists believe that a significant proportion of the most serious insect pest problems have been introduced or worsened as pesticide use eliminates local natural enemies.

Economically, dependence on pesticides and poor management strategies can have devastating impacts, and the health and environmental costs of pesticide use are rarely calculated. In India cotton production accounts for more than 50% of pesticides used and poor application practices have resulted in insect pest resistance to chemicals. Farmers lacked the know-how to develop alternatives, and became deeply indebted to money lenders and pesticide dealers. The FAO has estimated that IPM could reduce pesticide use in Asian rice crops by 50% or more, without compromising yields, and maintain or improve net returns to the farmer—savings could amount to U.S. $1 billion (16).

In areas of fragile and problem soil, intensive agriculture can lead to desertification, while resource-conserving systems in arid and semi-arid regions can deliver sustainable and often increasing yields. Farmers in these regions can rarely afford the external inputs and small improvements in yield would have profound impacts on the lives of the largely poorer populations that these agricultural systems support.

RISKS IN ADOPTING ECOLOGICAL STRATEGIES

The risks involved in adopting ecological pest management strategies are lower for small-scale farmers who are not yet using chemical inputs, and for farmers on a pesticide treadmill caused by high dependence and pest resistance to chemical inputs. In these instances the main risks are from short-term projects imposed without involvement of women and men in the farming communities who may swap one kind of dependency for another—the goodwill of donors. Farmers need knowledge to transfer to more ecological pest management strategies, and even successful IPM training can be limited by lack of funds for follow up.

Farming systems in industrialized countries that are effectively managing pesticide usage are likely to lose yields when adopting ecological strategies, particularly during the transitional stage. These farmers are also likely to face higher labor costs, though offset to some extent by lower input costs. The benefits are longer term and less tangible: improvements to the environment and to health, and in maintaining yields over time. For farmers who adopt organic practices and seek certification, the initial lower yields can be offset by a premium for their crops.

REFERENCES

1. In *Promoting Sustainable Agriculture and Rural Development, Agenda 21* Ch. 14, UNCED United Nations Conference on Environment and Development, Rio de Janiero, 1992. http://www.igc.apc.org/habitat/agenda21/ (accessed January 2001).
2. Pretty, J. Local Groups and Institutions for Sustainable Agriculture. In *Regenerating Agriculture: Policies and Practice for Sustainability and Self-Reliance*; Earthscan Publications: London, 1995; 131–162.
3. UNDP. In *Benefits of Diversity: An Incentive Toward Sustainable Agriculture*; UNDP: New York; 1992.
4. In *The den Bosch Declaration and Agenda for Action on Sustainable Agriculture and Rural Development*, Report of the Conference, FAO/Netherlands Conference on Agriculture and the Environment, FAO and Ministry of Agriculture, Nature Management and Fisheries of the Netherlands, The Netherlands, April 15–19, 1991.
5. Reijntjes, C.; Haverkort, B.; Waters-Bayer, A. Low-External-Input and Sustainable Agriculture (LEISA): An Emerging Option. In *Farming for the Future: An Introduction to*

Low-External-Input and Sustainable Agriculture; Macmillan: London, UK, 1992; 2–21.
6. CAB International. In *Global Crop Protection Compendium*; CAB International: Wallingford, UK, 1999; (CD ROM).
7. Management of Insect Pests, Plant Diseases and Weeds and Soil Management: A Key to the New Model. In *The Greening of the Revolution: Cuba's Experiment with Organic Agriculture*; Benjamin, M., Rosset, P., Eds.; Ocean Press: Melbourne, 1994; 35–65.
8. Stoll, G. Methods of Crop and Storage Protection. In *Natural Crop Protection Based on Local Farm Resources in the Tropics and Subtropics*; Agrecol, Verlag Josef Margraf: Langen, Germany, 1986; 80–167.
9. Elwell, H.; Maas, A. *Natural Pest and Disease Control*; Natural Farming Network: Harare, Zimbabwe, 1995; 3–128.
10. www.communityipm.org/ (accessed January 2001).
11. Poswal, A.; Williamson, S. Off the 'Treadmill': Cotton IPM in Pakistan. In *Pesticides News*; 1998 June, *40*, 12–13.
12. Myers, D.; Stolton, S. *Organic Cotton: From Field to Final Product*; Intermediate Technology and the Pesticides Trust: London, 1999; 1–120.
13. Farmer First–Field Schools are a Key to IPM Success. In *Growing Food Security: Challenging the Link between Pesticides and Access to Food*; Dinham, B., Ed.; Pesticide Action Network UK: London, 1996; 87–88.
14. Tiffen, M.; Mortimore, M.; Gichuki, F. Management and Managers. In *More People, Less Erosion—Environmental Recovery in Kenya Part III*; John Wiley & Sons: Chichester, UK, 1994; 131–226.
15. Pretty, J. *Regenerating Agriculture: Policies and Practice for Sustainability and Self-Reliance*; (op.cit 2), Earthscan Publications: London, 1995; 1–25.
16. In *FAO. Rice and the Environment: Production Impact, Economic Costs and Policy Implications*, Committee on Commodity Problems, Intergovernmental Group on Rice, Seville, Spain, May 14–17, 1996. FAO: Rome, 1996; 1–13. CCP:RI 96/CRS1.

P

PEST MANAGEMENT IN ORGANIC FARMING

Ján Gallo
Slovak Agricultural University, Nitra, Slovakia

INTRODUCTION

Organic farming in its essence is defined as a system of efficient production that is expected to create an environmentally beneficial, integrated, human, and economically stable agricultural system. Organic farming changes the philosophy of human consideration of nature from anthropocentric and technocratic arrogance to the gradual return of a holistic view in which humans are a part in the sense of the old trueism: "*We have not received soil, water and landscape from our parents, but we have borrowed them from our children.*" According to this philosophy, nature is the uniform whole with its internal natural value and from this point of view ecological, economic, and social aspects of agricultural production are derived. The organic system of farming is based on a maximally closed flow of energy without any additional outside inputs; therefore it is a sustainable system that uses local and renewable resources. Control of harmful biotic (diseases, pests, and weeds) and abiotic factors is an integral part of the system (1, 2).

STRATEGY OF CONTROL AGAINST ANIMAL PESTS

The strategy of controlling harmful animal organisms in the system of organic farming is based first of all on natural factors of resistance. It includes mainly internal factors (genetic predisposition of organisms) as well as the external ones that act inside the ecosystem. External ecosystem factors are of an abiotic (pedo-hydrometeo factors) and biological character. The biotic resistance of the environment has been applied prevailingly against animal pests using mainly the factor of natural mortality (2, 3).

Basic strategic objectives for pest regulation can be summarized as follows:

A species that seems to be a potential pest can survive under certain acceptable abundance. This means that the population density of an individual species is regulated and those measures that are directed to species eradication at all costs are not realized. The term "pest" is relative, and for this reason, animals in agricultural crops within organic farming cannot be subdivided to harmful, indifferent, and useful.

The unit of regulation is the ecosystem; and for this reason, there is a need to contribute to maintaining its autoregulation capability in maximum range.

Great emphasis is placed upon the knowledge of natural antagonists (parasitoids, predators, and pathogens).

Miscellaneous species must be maintained within the landscape.

Measures against pests have to be in accordance with economic aspects.

Organic farming does not represent the simple substitution of synthetic chemical pesticides by biological ones, it is a system of complex measures that promotes the good health of plants and not the total annihilation of pathogens (2–4).

REGULATION OF ANIMAL PESTS

Animal pests live in all field crops, but within each individual crop they require different conditions for their development. Under our conditions, 69% of field crop pests are being stayed on arable soil during the year. This linkage to arable soil is manifested in the population dynamics of individual species.

This strategy of control against animal pests requires that the regulation of animal pest occurrences is realized mainly through the utilization of

1. the phenomenon of biotic resistance of the environment and
2. anthropogenic measures.

The Phenomenon of Biotic (Natural) Resistance of the Environment

Natural mortality (which is caused first of all by the activity of predators, parasitoids, and parasites) is a significant natural regulator (bioregulator) of plant pest populations in agriculture. This regulation is manifested in the long-term maintenance of pest population density below the level of harmfulness. Bioregulators in the system of organic farming belong mainly to the group of spiders (*Araneida*), predacious mites (*Acarina*), and parasitoidic insects (*Insecta*) predominantly within the order of *Hymenoptera* and *Diptera*. Another group of bioregulators can be found within the group of phytopathogenic microorganisms (for instance fungi, bacteria, and others). Spiders (*Araneida*) occur in all biotopes on soil surfaces and on the above-ground parts of plants and can effectively reduce the population of harmful insects, mainly aphids (*Aphidoidea*), in field crops. Predacious mites (*Acarina*) belong to edaphic fauna and within stands of field crops are effective natural regulators of pests that live in the soil. Predacious polyphagous bugs, (*Heteroptera*) for example, within the family of *Anthocoridae* feed on the larvae of thrips (*Thysanoptera*), aphids (*Aphidoidea*), and mites (*Acarina*). The golden-eyed fly (*Chrysopa perla*) is able to destroy 200 to 500 aphids during its larval stage, but it can also feed on *Diaspididae, Tetranychidae*, and the larvae of flies. Parasitoidic *Hymenoptera*, predominantly the species that belongs to the families of *Ichneumonidae, Braconidae*, and *Chalcidoidae*, in the larval stage feed on the different stages of insect, starting with eggs and finishing with imagos. Imagos of these parasites as well as their larvae frequently injure the body of the host, feed on its body fluid, and finally cause the death of the host. For example, *Apanteles, Angitia*, and *Microgaster* are some of the main regulators of white butterflies (*Pieridae*). *Tersilochus heterocerus* is an important parasitoid of rape beetle (*Meligethes aeneus*). Beetles of the family of *Carabidae* represent a significant group of predacious insects (for instance *Calosoma sycophanta, C. inquisitor*) in field crops. They feed on aphids (*Aphidoidea*), larvae of beetles (*Coleoptera*), caterpillars of butterflies (*Lepidoptera*), and larvae of *Diptera* that they chase on the soil surface. Beetles (*Coleoptera*) within the family of *Staphilinidae* stay on the surface of soil and feed on aphids, eggs, larvae, and cocoons of insects, partially also on fungi (*Aleochara bilineata* significantly regulates *Delia brassicae*). Another very important group of predacious insects in field crops are ladybugs (*Coccinellidae*), the larvae and imagos of which feed on aphids (*Aphidoidea*) and help to reduce their numbers (an imago consumes 40 to 60 aphids per day). *Syrphidae* live in all cultural crops and their larvae feed on aphids and belong to the effective factors through which an occurrence of aphids is regulated in cereals, sugar beet, potatoes, and legumes (a larva consumes about 100 aphids per day) (3, 5).

It is necessary to protect the above-mentioned components of natural resistance in the environment and create optimum conditions for their existence. However, in many cases, the overpopulation of pests cannot be regulated only by means of their natural enemies (bioregulators).

Anthropogenic Measures

In the cases when biotic resistance of environment is less effective and there is a danger of substantial reduction in yields or deterioration of bioproduct quality, anthropogenic measures can be applied. Anthropogenic measures fulfill two essential roles:

1. to prevent pests from attacking healthy plants
2. to improve the health of attacked plants.

These are mainly indirect (preventive or prophylactic) methods for pest regulation. Direct (repressive or therapeutic) methods can be used only in the most necessary cases, and only those that are in accordance with International Federation of Organic Agriculture Movements (IFOAM) rules. Indirect measures include crop rotation, selection of proper species and cultivars, appropriate soil cultivation, suitable fertilization, green fertilizing, and outside and inside quarantine. Crop rotation seems to be the dominant sustainable procedure on the farm that protects, renews, and supports biodiversity and also plays a key role in soil fertility preservation. By means of the above-mentioned indirect principles, overpopulation of pests can be restricted completely, or at least very substantially. *Zabrus gibus* damages winter wheat five to eight times less when winter wheat follows after sunflowers or maize within crop rotation. If cereal crops are grown several years in the same place, they are frequently damaged by *Scotia segetum*, nematodes (*Heterodera*), and frit fly (*Oscinella frit*). Within organic farming, the portion of cereals in crop rotation would not exceed 50%. Early plowing of stubble can eliminate 70 to 80% of thrips (*Haplotrips tritici*). On the other side, late sowings of peas are more likely to be attacked by pea aphids (*Acyrthosiphon onobrychys*) and weevils (*Apion* spp.). Harvesting maize to the shortest possible stubble (0.08 to 0.10 m) substantially decreases the number of overwintering caterpillars of European corn borer (*Ostrinia nubilalis*). If the harvest is done early enough it enables about 80% of pests to be removed together with plant material. Occurrence of bean weevils (*Acanthoscelides obsoletus*) on bean plants can be effectively reduced if rows of beans are alternated with rows of maize, which by its "odor" causes this pest to become disoriented as it is flying toward bean plants. Fertilization within organic farming is based predominantly on own farm organic fertilizer utilization, however, its application must follow the rules of IFOAM (3, 5).

For the future, solutions are needed for problems related to bioproducts from the raw materials grown within organic farming and that have received a certificate according to the rules of IFOAM. We have to face the fact that growing plants are exposed to biotic and abiotic stresses and in some cases they react by forming ''natural pesticides'' that can be dangerous for humans (e.g., carcinogenic psoralens in celery, parsley, and parsnip). While the above-mentioned and similar matters have to be solved, the medicinal advantages of bioproducts resulting from organic farming must not be doubted (1, 5).

In addition to the principles of sustainable development, appropriate laws should be created including legislative treatment of organic farming and biofoods production for each respective country (224/1998 Z.z in the Slovak Republic). By means of the law a new historical paradigm will be defined, within which nature will become a globally respected value.

REFERENCES

1. Klinda, J. *Agenda 21 a ukazovatele trvalo udrzatel'ného rozvoja (Agenda 21 and Indicators of Sustainable Development)*; MZP SR: Bratislava, 1996; 1–517, ISBN 80-88833-03-5.
2. Gallo, J.; Sedivy, J. *Integrovanáochrana rastlín (Integrated Pest Management)*; VES VSP: Nitra, 1992; 1–163, ISBN 80-73137-061-4.
3. Neuerburg, W.; Padel, S. *Organisch-biologischer Landbau in der Praxis*; BRD: Munchen, 1992; 1–460.
4. Lampkin, N. *Organic Farming*; Farming Press Books: Ipswich, United Kingdom, 1990; 214–271, ISBN 0-852 36-191-2.
5. Petr, J.; Dlouhy, J. *Ekologické zemêdêlství (Ecological Agriculture)*; Zemêdêlství nakladatelství Brázda: Praha, 1992; 1–305, ISBN 80-209-0233-3.

PEST MANAGEMENT IN TROPICAL AGRICULTURE

P

Charles J. Muangirwa
Tropical Pesticides Research Institute, Arusha, Tanzania

INTRODUCTION

Integrated pest management has a bright future in tropical agriculture on the basis of adoption of available options of cultural practices, as well as numerous pest management options that are likely to emerge from understanding the interaction of the tropical biological diversity. Pesticides are used for pest management in the tropics, more so on cash crops than on food crops. While the use of pesticides is on a decline elsewhere, it is likely to increase in the tropics even on food crops and threaten basic biological components of pest management, the environment, and safety to people. Generally, there has been an improvement from the use of persistent pesticides over wide areas to application of less persistent pesticides on specified sites based on pest ecology. There is a need to undertake training on safe use of pesticides at all levels, create awareness that pesticide use is only one among other pest management options, and establish poison information centers.

Overall, efforts should continue to promote the concept of integrated pest management as a system. The strategy should be to produce sustainable pest management options based on an understanding of tropical biological diversity and, at the same time, improve decision making for the adoption of existing and emerging options.

BASIS OF PEST MANAGEMENT IN TROPICAL AGRICULTURE

Pest management in tropical agriculture is basically influenced by: 1) high diversity of organisms (biological diversity), 2) diversity of physical features and ecosystems, 3) moderate seasonal weather changes, and 4) the human factor.

The high biological diversity of the tropics includes a wide range of interacting organisms from which have been identified pests that affect health, crops, and livestock, as well as cultural practices for pest management. Today, the interaction is known to include chemical, visual, and audible cues; and has strong implication on known and potential pests, natural enemies of pests, natural products for pest management, and pest/host/crop resistance.

Physical features of the tropics are also diverse, and include high-altitude temperate zones, rain forests, savannah woodlands, grasslands, and desert zones. Each zone has its own biological diversity either existing in isolation or overlapping with others. Moderate seasonal weather changes in the tropics affect populations indirectly, through (modification of) interactions between organisms. An understanding of the dynamics of such interactions would strengthen the contribution of pest management to sustainable tropical agriculture (Fig. 1).

Generally, humans influence pest management through their objectives, perceptions, problems, and adoption of options for attaining those objectives or solving those problems (2). Most tropical countries are involved in subsistence agriculture, which is dominated by rudimentary cultural practices of pest management. Pest management on commercial farms is mainly based on temperate models.

As can be seen above, pest management for sustainable tropical agriculture should be regarded as a system of options that is dominated by tropical biological diversity, and should be processed starting and ending with the consideration of human factors (3, 4).

CURRENT PEST MANAGEMENT PROBLEMS IN TROPICAL AGRICULTURE

Key problems in tropical pest management include: 1) low application of pest management options that are based on tropical pest ecology, 2) lack of information on tropical pest ecology, 3) dependency on synthetic pesticides, 4) exotic pests, and 5) migrant pests.

Low Application of Pest Management Options That Are Based on Tropical Pest Ecology

Subsistence farmers in the tropics have applied rudimentary cultural practices in pest management on their own,

Generations of insect pests (i.e. rice stem borer brown plant hopper) overlap when rice is grown intensively in Southeast Asia, thus making it difficult to time insecticide applications. Besides, insecticides kill natural enemies of these pests and hence create futher room for increases in pest populations.

The rice stem borer and brown plant hopper are better controlled by pesticides in temperate Japan and Korea where pest populations are synchronized and less dependent on natural enemies.

Fig. 1 Difference in the control of insect pests of rice (rice stem borer and brown plant hopper) in tropical Southeast Asia and temperate Japan and Korea. (From Ref. 1.)

perhaps in desperation. Even where there has been a scientific breakthrough, say in biological control, such options have been viewed as risky by commercial farmers, and too costly to be adopted by subsistence farmers (Fig. 2).

Lack of Information on Tropical Pest Ecology

The rudimentary cultural pest management practices used by subsistence farmers and the overall interaction of tropical biological diversity have not been researched to exploit their full potential for formulating sustainable pest management options. Information, expertise, and facilities are lacking from the taxonomy level to the biology, behavior, and ecology of most organisms in the tropics. In effect, this has left room for dependency on pesticides. On the other hand there has been some progress in designing and using novel pest management options (Fig. 3).

Lack of taxanomic expertise and facilitics has often resulted in delay in intervention or misidentification, particularly of exotic pest; for example, coffee mealy bug in Kenya, floating aquatic fern in Lake Kariba (Zimbabwe), larger grain borer in Tanzania, and cassava mealy bug in Uganda.

However; prompt and correct identification can result in successful control, as was the case with the New World scew worm. The worm entered Africa through Libya and was a threat to the continent's livestock and game.

Ironically, in most cases, farmers have been the first to note and report new damaging organisms, sometimes in dispute with authorities.

Fig. 2 Need of prompt identification of pests. (From Refs. 4 and 5.)

Control of tsetse flies, the vectors of trypanosomiasis in Africa, has evolved through various stages:

i. Environment manipulation—cutting tress that constitute tsetse habitat and shooting vertebrate hosts.

ii. Spraying residual insecticides on resting sites on trees or spraying aerosol insecticides to whole vegetation with aircraft.

iii. Use of traps, specifically designed for tsetse flies, baited with tsetse attractants and pesticides.

In the process it has been possible to avoid environment destruction resulting from cutting tress and killing animals. Also the use of pesticides has been reduced from spraying the whole emvironment to specific targets and even total avoidance of pesticides.

Exposure of people to pesticides has also been reduced while participation of affected communities has enhanced. This has happened because of the growing understanding of biology, behavior, and ecology of tsetse files.

Fig. 3 Understanding of biology, behavior, and ecology of tsetse flies leads to enviromentally friendly pest management options. (From Ref. 6.)

Dependency on Synthetic Pesticides

Dependency on synthetic pesticides of today's agriculture has negative effects on: 1) safety of people, 2) pest management, and 3) environment.

Pesticide safety

About 60–85% of the workforce in the tropics is engaged in agriculture. About 2.9 million cases of acute exposure to pesticides are reported annually from the tropics, out of which about 220,000 deaths occur (7). Commonly, exposure occurs during mixing, loading, and applying of pesticides. Exposure during mixing and loading is greatest when pesticides that penetrate the skin are used. Exposure is mainly due to poor knowledge of pesticide safety at farm level among extension workers, and is complicated by lack of skills among health workers to attend pesticide safety related problems. Although some populations have been exposed to pesticides for long periods, there are no direct and concrete records of chronic symptoms. Generally, farm workers are more exposed to pesticides than peasants. Peasants are likely to be exposed to pesticides in activities related to cash crop production than food crop production.

The growing demand for food may contribute to an increase in the use of pesticides on food crops as well. As there is no structured reporting system on pesticide poisoning, the available data is only indicative.

Effects of pesticides on pest management

The effect of pesticides on pest management has mainly been recorded in relation to natural enemies of pests. Decline in natural control mechanisms has resulted in an increase in status of present pests and occurrence of new pests (6). Pesticide resistance has been associated with application of low doses that kill susceptible individuals of a pest population selectively, leaving resistant individuals to reproduce and increase to high proportions.

Pesticides and environment

Pesticide residues have been traced in fish and fish-eating birds (8); human bodily fluids including milk; soil, and water. Most of the residues observed were traces of organochlorine pesticides, which accumulate in the fat of organisms and in the environment. Biological assays based on levels of acetylcholinesterase in blood have indicated exposure to organophosphates and carbamate pesticides as well. There is a growing concern that the breakdown of pesticides in the tropics results in lesser-known compounds that are more toxic than the original pesticides (9).

Migrant Pests and Tropical Pest Management

Major migrant pests affecting tropical agriculture include birds (Queleas) and insects (locusts and armyworms). Migrant pests are a problem because they can travel in large numbers over long distances and denude vast areas of crops and other vegetation. Control of migrant pests is so far dependent on pesticides applied at virtually any cost, so as to prevent disasters. Environmental implications of the control measures are yet to be assessed. Studies on physiology, ecology, and behavior of queleas and locusts in relation to migration gives some hope on the use of nonpesticidal control measures of the pest.

Exotic Pests of Tropical Agriculture

Exotic pests that invade the tropics flourish to become key pests because of: 1) lack of interaction with resident organisms—particularly natural enemies—and 2) favorable conditions throughout all seasons. Invasion of the tropics by exotic pests is often due to inadequate enforcement of quarantine requirements and general ignorance about quarantine. Routes of exotic pests include food aid, germ plasm movement, and traveling. Careful introduction of natural enemies from native localities controls exotic pests in their new homes (9).

TRENDS IN TROPICAL PEST MANAGEMENT

Trends in pest management in tropical agriculture include three eras: 1) before synthetic pesticides era, 2) synthetic pesticides era, and 3) integrated pest management era.

Before Synthetic Pesticides Era

Tropical pest management before the synthetic pesticides era was characterized by the use of cultural practices and the use of plants with pesticidal properties, all based on farmers' experiences on the tropical physical and biological diversity. Experts of the presynthetic pesticides era were involved in taxonomy, biology of pests, and advocacy of cultural practices.

Synthetic Pesticides Era

Synthetic pesticides were used globally for the first time in the mid-1940s. At that time cash crops were introduced to most tropical countries by colonialists. Synthetic pesticides were adopted almost immediately for the management of pests in cash crops. Farmers and peasants were taught about pesticides, but as part of crop agronomy and animal husbandry, and not on personal or environmental safety.

During this era, experts drifted from being naturalists to pesticide applicators. There was also an increase in training of natives at various levels, as a postcolonial activity (10). However, training emphasized the use of pesticides more than a naturalistic approach to pest management, and on the control aspects of pesticide rather than on safety to users and environment. Research on pesticides was initially aimed at testing if a chemical killed a pest and would reduce pest population. Subsequently, such studies evolved to include ecology of pests and related organisms, pesticide application, and pesticide residues. Problems of pesticide poisoning, pest resistance, and emergence of new pests were observed some four decades ago. Over the years there was low use of pesticides on food crops, but the situation changed in response to demand for food to feed the increasing human population. It appears that peasants transferred their experiences from use of pesticides on cash crops to the control of pests on food crops without technical considerations.

Integrated Pest Management Era

The synthetic pesticides era is coming to an end or taking a different outlook in tropical agriculture because of: 1) Negative impact of pesticides on users' health and on the environment. 2) Pesticides failing to control pests. 3) Likelihood of applying pesticides on specified targets, based on pest ecology, and hence reduce environmental impact. 4) Lessons learnt during the before-pesticides era that cultural practice and use of plants with pesticidal properties are effective pest management options. 5) Emerging technologies based on cultural practices and studies on interactions between organisms revealing a wide range of pest management options (1, 5, 7, 11). 6) Increasing interest in pest management by various parties. A number of governments have declared pest management as an official policy, various institutions (governmental, NGOs, and private, etc.) are involved in various aspects of pest management including pest ecology and biology research, safe use of pesticides, and adoption of the use of alternative pesticides. Multinational pesticide companies have also started taking interest in pest management, particularly in areas of biotechnology. 7) Increase in action by governments to undertake pesticides registration and control activities as a move to minimize negative effects of synthetic pesticides.

DISCUSSION

By observing the integrated pest management era in the tropics through the problem-free cultural practices (i.e., item 4 above) and the emerging knowledge on the interaction of organisms (5 above) it is now clear that reliance on pesticides as a single option for achieving desired health, crop, or livestock protection is unnecessary. In effect, interaction between organisms in the tropics represents a wide range of ''suppressing'' factors including natural enemies of pests, natural products, cultural practices, crop, and host and pest resistance, etc. We can add pesticides to this list, but only to play a complimentary role. The various options for achieving the desired protection are presently limited, but would increase depending on the increase in knowledge of interactions between organisms in the tropics. Socioeconomic considerations and training are increasingly becoming important components of pest management. Considering the multiplicity and interaction between the various options in the tropics, it is now increasingly acceptable that pest management be implemented as an integrated system rather than a single act of reducing pest population or decreasing damage. Generally, the systems approach to pest management in tropical agriculture would be sustainable and be optimized if implemented through two complementary strategies: 1) Reduction of constraints to adoption of sustainable pest management options at various levels, i.e., to improve decision making, and 2) research aimed at producing sustainable pest management options, based on understanding of tropical pest ecology.

REFERENCES

1. N.R.I. *A Synopsis of Integrated Pest Management in Developing Countries in the Tropics*; Natural Resources Institute: Chatham, U.K., 1999; 20.
2. Norton, G.A. A Decision-analysis approach to integrated pest control. Crop Prot. **1982**, *1* (2), 196–199.
3. Norton, G.A.; Adamson, D.; Aitken, L.G.; Bilston, L.J.; Foster, J.; Fronk, B. Facilitating IPM: role of participatory workshops. Int. J. Pest Manage. **1999**, *45* (2), 85–90.
4. Castella, J.C.; Jourdain, D.; Trebuil, G.; Napompeth, B.; Jourdain, G.A.; Napompheth, B.A. Systems approach to understanding obstacles to effective implementation of IPM in Thailand: key issues for the cotton industry. Agric. Ecosyst. Environ. **1999**, *72* (1), 16–34.
5. BioNet–International. Taxonomy in the Biological Control of Pests, Diseases, and Weeds. In *BioNET–International: The Business Plan*; 1999; 3–65. http://www.bionet.intl.org.
6. Nyambo, B.; Murphy, S.T.; Barker, P.; Walker, J. In *Biocontrol in Coffee Pest Management in Tropical Entomology*, Proceedings of the 3rd International Conference on Tropical Entomology, Nairobi, Kenya, Oct. 30–Nov. 4, 1994; ICIPE Science Press, 1998; 155–168.
7. Hargrove, J.W. In *Trypanosomiasis Management Using Bait: Some Implications of Tsetse Behavior and Ecology*, Proceedings of the 3rd International Conference of Tropical Entomology, Nairobi, Kenya, Oct. 30–Nov. 4, 1994; ICIPE Science Press, 1998; 155–168.
8. Ijani, A.S.M.; Katondo, J.M.; Malulu, J.M. In *Effects of Orgonochlorine Pesticides in Birds in United Republic of Tanzania, Environmental Behavior of Crop Protection Chemicals,* Proceedings of IAEA/FAO Conference, Viena, Italy, July 1–5, 1996; IAEA/FAO: Rome, Italy, 1996; 460–461.
9. Lehtinen, S. Pesticides. In *African Newsletter on Occupational Health and Safety*; 1999; 9 (1), 23.
10. Paasivirta, J.; Palm, H.; Paukku, R.; Ak'habuhaya, J.; Lodenius, M. Chlorinated insectides residues in Tanzania environment Tanzadrin. Chemosphere **1988**, *17* (10), 2055–2062.
11. Hilder, V.A.; Gatehouse, A.M.R. Biological control in developing countries: towards its wider application in sustainable pest management. Med. Fac. Landb. Rijksaniv. Gent. **1990**, *55* (2a), 216–223.

PEST POPULATION MONITORING

Johann Baumgärtner
International Centre of Insect Physiology and Ecology (ICIPE), Nairobi, Kenya

Cesare Gessler
Institute of Plant Sciences—Pathology, Swiss Federal Institute of Technology, Zürich, Switzerland

P

INTRODUCTION

To monitor pests like weeds, diseases, and many arthropods in integrated pest management (IPM) means to oversee or carefully watch their occurrence and development for decision-making purposes. IPM is a well-established system of compatible control methods, which are applied on the basis of specific monitoring information combined with ecological and economic consideration (1, 2).

Since assessment of the total number of pests or their damage symptoms is generally impractical, sampling procedures are employed to estimate densities per sampling unit or proportions of occupied sampling units in a sampling universe (2–4). Some IPM programs require classification of target organisms or damage symptoms, while others rely on presence or absence information. Monitoring information is collected once or repeatedly during a cropping period or season and related to a defined action threshold. Less often forecasting systems are developed, which predict pest development and final crop yields. Control methods are applied if the action threshold is exceeded (2–6).

Individuals of target populations generally pass through various life stages with different lifestyles. Their development is driven by environmental factors with often highly stage-specific effects. Sometimes the individuals respond to an anticipated or perceived deterioration of the habitat by migration or entering a resting stage. A cost-efficient monitoring program generally focuses on easily detectable life stages, and activities might be suspended if monitoring becomes inefficient. Moreover, preference is given to life stages that allow the best prediction of yield loss.

If the number of generations is less than one per cropping period or season, a single well-timed estimate of density may suffice. Few weeds, many arthropod pests, and most diseases have overlapping generations, and sampling procedures must take into account the likely presence of more than one life stage at any point in time. In general, multigeneration systems require repeated monitoring because of uncertainties in predicting target population densities and phenologies.

A wide range of sampling techniques is available (3, 6). The sampling universe may be a region, an assemblage of farms or fields, but a field usually provides the appropriate level of spatial resolution. In general, weeds or disease symptoms can be counted directly in the field, while in the case of many arthropods the sampled material is often taken from the field and inspected elsewhere. The structure of the sampling universe; the spatial distribution patterns of target population; or their damage symptoms, sampling costs, and reliability of the estimate are among the most important factors in the design of a sampling strategy (3, 4).

PURPOSE AND SCOPE

Monitoring of pest populations occurs for many reasons. On a large geographical scale, quarantine services monitor presence and absence of arthropod pests or diseases on imported goods and initiate strategies of containment and eradication whenever their presence is recorded. The relevant information is disseminated via various channels to a wide range of decision makers. Details of such operations are given, for example, by the European and Mediterranean Plant Protection Organization (EPPO). On a small geographical scale, plant protection services may provide information on the appearance of a pest and encourage farmers to initiate monitoring activities in their fields.

Monitoring is often done to assess the genetics of a target organism. For example, genotypes of cereal rust diseases are continuously monitored to assure maintenance of current resistance levels. The monitoring of target organism genetics is a prerequisite for management of host plant resistance and maintenance of control measure efficacy.

Pest control methods generally require information on population densities or proportions of infested units with reference to crop growth stages (2, 3). In some cases, consecutive monitoring data are summed to provide information on the dynamics of target population development and crop yield formation (6–8). For example, the density of heteropterans is summed over time to calculate heteroptera-days and this is used as a measure for an arthropod pest load on cowpea yield formation (9). In other cases one or more observations are used to predict future pest densities and their impact on yields (10). For example, the EPIPRE (Epidemic prevision) forecasting system relies on monitoring information to estimate maximum aphid density and disease incidence, and these are related to future potential yield losses. This estimate is used to justify chemical control (2).

A single point estimate may be sufficient to estimate densities or damage symptoms. Such methods are appropriate for diseases, which have a small potential for high yield loss, and which develop relatively slowly under the influence of environmental factors that vary little over years. Fast-developing diseases such as potato blight, that have a high yield loss potential, require continuous monitoring. Detection of disease presence is often sufficient to initiate chemical control measures (2, 4).

ECOLOGY

A cost-efficient monitoring program must take into account the life history and the phenology of the target organism (2, 3). An appropriate sampling strategy must anticipate the effects of unfavorable environmental conditions or deteriorating habitats that often cause pests to migrate or to enter a resting stage. Under such circumstances, the target organism may not easily be detected and monitoring activities may be suspended. For example, the monitoring of codling moths by means of pheromone traps may start only once a period of physiological time has elapsed after a specific biofix date. During most of this period the individuals are difficult to detect.

The monitoring of diseases such as grape downy mildew can be delayed until the overwintering oospores are mature, microclimatic conditions permit infection, heavy rains allow dispersal, and the first incubation period has passed.

SAMPLING TECHNIQUES AND PLANS

While research programs generally require intensive sampling programs for estimation of population parameters, extensive and cost-efficient sampling procedures are needed for IPM-related monitoring (2, 3). On a spatial scale, field-specific information is more frequently required than information specific to farms or regions. On a temporal scale, as mentioned above, a single-point estimate may be possible, but repeated sampling is often recommended.

A wide range of sampling techniques is used in monitoring programs (2, 3, 6). Remote sensing technology may allow the detection of target organisms or damage symptoms. Some target organisms can be sampled from the air, from on or within the plant, or from the ground to assess relative and absolute pest population densities. Spore traps and pheromone-baited traps allow sampling of pathogens and pests from the air, plant inhabiting target organisms are monitored by visual methods, and soil samples processed in special apparatus are used to sample target organisms from the soil.

Some techniques allow the estimation of absolute densities recorded per unit surface or volume, while others provide estimates of relative densities. Both measurements can be related to adequately defined action thresholds. Often, relative estimates are poorly correlated to yield losses, as in the case of codling moth flight intensities or apple scab spore counts. In the former case, monitoring yields useful information on the occurrence of economically relevant life stages for the optimum timing for undertaking chemical or biological control. Traps are used to monitor apple scab spores. The spore counts, however, are not directly related to damage. However, supplemented with microclimatic data and ascospore occurrence, the counts trigger control recommendations.

The reliability of population estimate is affected by systematic and random errors (6). Stratified and multistage sampling procedures consider the homogeneity and the structural complexity of the sampling universe (2, 3). Sampling plans are derived from statistical models describing the spatial distribution of target organisms or damage symptoms. In general, sequential estimation in general and sequential binomial sampling plans relating the proportion of infested sampling units to adequately defined action threshold are particularly cost efficient. The design of cost-effective sampling strategies is important since it would lead to a more widespread use of reliable monitoring information.

FUTURE DEVELOPMENT

Monitoring programs will become more cost efficient because improved understanding of target organism dynamics and yield forming help to place sampling activities optimally in time and space and reduce the numbers of samples. Sequential sampling programs make efficient use of already obtained sampling information. Currently, monitoring in relation to action thresholds is done primarily for rationalizing chemical control, but in the future monitoring methods will be developed and adapted to a wide range of control methods including biological control (2, 3). Rather than focusing on individual pest populations, future monitoring systems will increasingly target multi-species population systems.

Emphasis will be given to monitoring the genotypes of target organisms and host plants. For example, molecular markers prone to automatization will replace the cumbersome search for rare target organisms and facilitate detection of resistance levels.

The improvement of existing and development of new monitoring techniques will further improve decision-making in IPM systems. Refined remote-sensing techniques will replace location-specific measurements. Forest and urban diseases, for example, are already detected by using airborne digital imagery (11). The proposed use of trained neural networks and fuzzy inference systems may improve the understanding of pest population dynamics and facilitate the design of programs for regional and local population monitoring.

The development of computer based decision-support systems is a strong incentive for further development of monitoring systems. Improved access to information systems will assist the decision makers to identify pests as well as damage symptoms, provide guidance for monitoring activities, and facilitate the selection of adequate control measures. Computer based decision-support systems will facilitate communication and efficiently link automated monitoring devices to remote sensed data. The quality of forecasts will be improved by combining geo-referenced monitoring information with geographical information including remote sensing data.

P

REFERENCES

1. Dent, D. *Integrated Pest Management*; Chapman & Hall: London, 1995; 56.
2. *Techniques for Reducing Pesticide Use*; Pimentel, D., Ed.; John Wiley & Sons: New York, 1997; 444.
3. *Handbook of Sampling Methods for Arthropods in Agriculture*; Pedigo, L.P., Buntin, G.D., Eds.; CRC Press: Boca Raton, FL, 1995; 714.
4. Madden, L.V.; Hughes, G.; Munkvold, G.P. Plant disease incidence: inverse sampling, sequential sampling, and confidence intervals when observed mean incidence is zero. Crop Prot. **1996**, *15*, 621–632.
5. Roux, O.; Baumgärtner, J. Evaluation of mortality factors and risk analysis for the design of an integrated pest management system. Ecological Modeling **1998**, *109*, 61–75.
6. Schaub, L.; Stahel, W.A.; Baumgärtner, J.; Delucchi, V. Elements for assessing mirid (Heteroptera, Miridae) damage threshold on apple fruits. Crop Prot. **1998**, *7*, 118–124.
7. Gaunt, R.E. New technologies in disease measurement and yield loss appraisal. Can. J. Plant Pathol. **1995**, *17*, 185–189.
8. *Spatial Components of Plant Disease Epidemics*; Jaeger, M.J., Ed.; Prentice Hall: Englewood Cliffs, NJ, 1989; 243.
9. *Experimental Techniques in Plant Disease Epidemiology*; Kranz, J., Rotem, J., Eds.; Springer: Berlin, 1988; 299.
10. Dreyer, H.; Baumgärtner, J. The influence of post-flowering pests on cowpea seed yields with particular reference to damage by heteroptera in Southern Benin. Agr., Ecosys. Environ. **1995**, *53*, 137–149.
11. Nutter, F.W., Jr. Quantifying the temporal dynamics of plant virus epidemics: a review. Crop Prot. **1997**, *16*, 603–618.

PEST STATUS

Joe Funderburk
University of Florida, Quincy, Florida, U.S.A.

INTRODUCTION

Pest status is the ranking of a pest relative to the economics of dealing with the species (1). The importance of a pest to producing a crop or raising livestock varies with a number of factors. The crop and its market value plus its susceptibility to damage contribute to pest status. Population numbers of the pest, feeding and ovipositional characteristics resulting in injury, and management costs also affect pest status. The environment in which the pest–crop interaction occurs will ultimately mediate the importance of a pest to a crop and determine the measure and constraints of pest management programs. In most agricultural situations, crops and livestock are able to tolerate some level of injury before economic loss occurs. The same is true for ornamental plants grown for aesthetic value. Pest types are noneconomic, occasional, perennial, and severe. Perennial and severe pests pose serious challenges to a production system because of the intensity of management needed to prevent economic losses. Pest status of exotic pests is usually greater when the species has been accidentally introduced into new ecological areas because the normal complement of natural enemies is not present. Most of the key pests in the United States are the result of accidental introductions. For food and other products exported to overseas markets, the mere presence or detection of a pest can render it unacceptable.

FACTORS AFFECTING PEST STATUS

Producers must evaluate the importance or potential importance of many pests attacking crops. The economic injury level is used to define pest status and make objective pest management decisions (2). It incorporates biological information about pest and host, and economic criteria including crop value and management costs. Utilizing the economic injury level is an objective way to measure the importance of a pest insect, disease, or weed whether therapeutic or preventive management tactics are considered. Novel approaches for defining pest status, such as the aesthetic injury level (3), the environmental economic injury level (4), and the multiple-species economic injury level (5), are based on modifications of the components of the economic injury level model. By using these approaches for making objective decisions, the producer must predict market value, which is one of the most variable of the factors determining pest status. A pest automatically assumes higher or lower importance because of changing economics. The cost of using a management tactic or of using multiple tactics must be estimated before profitability of action can be assessed. These costs are usually more stable than crop market values.

Injury is another factor influencing pest status. Injury is governed by pest and by host populations. The pest impairs the ability of the host to survive, grow, and reproduce. The type of host and the production situation play a role in determining the kind and degree of injury. The relationship between injury and loss of utility (damage) is the biological component in determining the economic injury level (6). This relationship provides the basis for assessing how many pests can be tolerated economically before management tactics (preventive or therapeutic) are employed.

Factors influencing the relationship between injury and damage are the time of injury, the part of a plant injured, the type of injury, the intensity of injury, and environmental influences on a host's ability to withstand injury (6). Categories of injury to plants from insect pests include, leaf feeders, assimilate sappers, stand reducers, turgor reducers, and fruit feeders (7). Injury from insects tends to occur over a short duration (acute) whereas injury from weeds and disease occurs usually over a long period of time (chronic). Direct injury to yield producing plant structures is less tolerable than injury to nonyield producing structures.

The environment is an important factor affecting damage and pest status. Weather factors such as temperature and rainfall, affect the ecology of pest species and therefore pest numbers. Crops previously stressed by biotic or abiotic factors may have different responses to subsequent stressors. The impact of injury from pests is frequently more severe for plants under drought conditions than those growing under optimal conditions. Further, damage to crops is greater from some pests, especially weeds, if the plants are suffering from inadequate nutrition.

The social/political environment also can greatly affect determination of pest status. Crop value and pest management costs are affected by government laws and regulations (1, 3, 7–9). These laws and regulations are established based on scientific information of food safety and environmental protection, as well as public perception of the issues. Standards of grade and quality of food are affected by consumer acceptance of appearance. The first line of defense against pests is to prevent their entry into areas where they may become pests. Quarantine laws require that imported agricultural products be completely free of pest organisms, which places restrictions on where the crop is grown and on pest control programs. The spread of serious pests worldwide and the potential for additional spread has had great effects on pest status of these pests.

The pest management strategy and intensity of management activities are determined by the status of a pest in the production system (1). The status of a pest is categorized based on the long-term average density of the pest (general equilibrium position) compared to the economic injury level (Fig. 1). The general equilibrium position of a noneconomic pest is well below the economic injury level and the greatest population density never reaches the economic injury level. The cost of reducing injury is greater than the losses the pest inflicts. The most common pest type is the occasional pest. The general equilibrium position is far below the economic injury level, but populations sporadically result in economic damage. Occasional pests are usually managed by detecting an impending outbreak and therapeutically applying a management tactic before economic damage occurs. Most insect pests of soybean in the United States are either noneconomic or occasional pests. The soybean nodule fly, *Rivellia quadrifasciata* (Macquart), is an example of a noneconomic pest, and the soybean looper, *Pseudoplusia includens* (Walker), is an example of an occasional pest.

The general equilibrium position of a perennial pest is below but very close to the economic injury level (Fig. 1). Without management, populations of such a pest are frequently above the economic injury level. Severe pests are a constant problem. The general equilibrium position is above the economic injury level. Severe and perennial pests are considered key pests because of the intensity of management needed to prevent damage. Such pests usually damage the harvested product, such as the fruits, or they occur in very dense populations. The pest management strategy for severe and perennial pests is to reduce the general equilibrium position of the population by combining several management tactics. The western flower thrips, *Frankliniella occidentalis* (Pergande), is

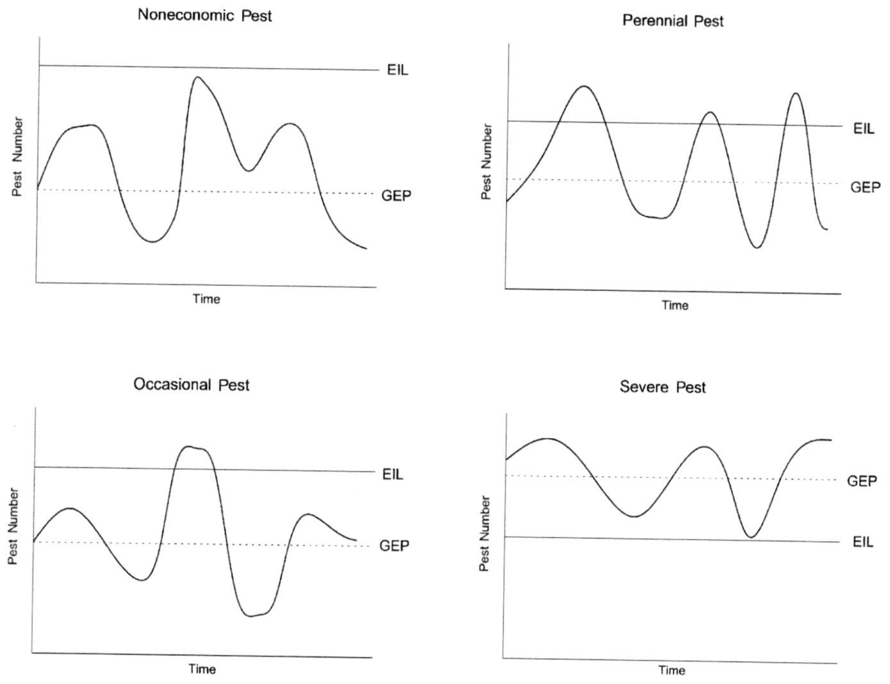

Fig. 1 Graphs representing the four pest types. EIL—economic injury level; GEP—general equilibrium position.

native to the southwestern United States, but the pest has been accidentally introduced into many areas worldwide where it is a key pest of many vegetable, ornamental, and agronomic crops grown in the field or greenhouse. The species is also one of seven known species of thrips that are vectors of tospoviruses. Tospoviruses are key pests of many crops in many parts of the world (10).

FUTURE CONCERNS

The increase in international trade of agricultural products has resulted in the spread of many pests worldwide (7). Many of these alien pests have greater pest status in the region where they have been introduced because the normal complement of natural enemies is not present. Exotic pests can cause severe economic losses to producers and are responsible for suppressive measures that are detrimental to the environment. Current policy is for further expansions in international trade including agricultural products. This undoubtedly will result in continued accidental introductions of pest organisms into new ecological areas. Management programs for pests will be intensified due to the change in the status of pests in these ecological areas. At the same time, strict quarantine laws will continue to require that imported products be free of pests. This further influences the status of pests for exported agricultural products.

REFERENCES

1. Pedigo, L.P. *Entomology and Pest Management,* 3rd Ed.; Prentice-Hall: Upper Saddle River, NJ, 1999; 691.
2. *Economic Thresholds for Integrated Pest Management*; Pedigo, L.P., Higley, L.G., Eds.; University of Nebraska Press: Lincoln and London, 1996; 327.
3. Raupp, M.J.; Davidson, J.A.; Koehler, C.S.; Sadof, C.S.; Reichlderfer, K. Decision-making considerations for aesthetic damage caused by pests. Bull. Entomol. Soc. Am. **1988**, *34*, 27–32.
4. Pedigo, L.P.; Higley, L.G. A new perspective on the economic injury level concept and environmental quality. Am. Entomol. **1992**, *38*, 12–21.
5. Hutchins, S.H.; Funderburk, J.E. Injury guilds: a practical approach for managing pest losses to soybean. Agric. Zool. Rev. **1991**, *4*, 1–21.
6. Pedigo, L.P.; Hutchins, S.H.; Higley, L.G. Economic injury levels in theory and practice. Annu. Rev. Entomol. **1986**, *31*, 341–68.
7. Funderburk, J.; Higley, L.; Buntin, G.D. Concepts and directions in arthropod pest management. Adv. Agron. **1993**, *51*, 125–72.
8. *Introduction to Insect Pest Management*; Luckman, W.H., Metcalf, R.L., Eds.; John Wiley: New York, 1994; 651.
9. *Decision Tools for Pest Management*; Mumford, J.D., Norton, G.A., Eds.; CAB International: London, 1993; 297.
10. Ullman, D.E.; Sherwood, J.L.; German, T.G. Thrips as Vectors of Plant Pathogens. In *Thrips as Crop Pests*; Lewis, T., Ed.; CAB International: London, 1997; 539–65.

PEST/HOST-PLANT RELATIONSHIPS

Zamir K. Punja
Simon Fraser University, Burnaby, British Columbia, Canada

INTRODUCTION

Through evolutionary time, most pests have developed specific types of relationships or interactions with the plant hosts they attack. These interactions may start out with appropriate recognition of the signals or cues that identify a susceptible host, proceed to the establishment of a parasitic/herbivorous/competitive interaction, and end up with the ultimate visible crop damage that results in economic yield loss. An understanding of these various relationships is pivotal to the development of appropriate pest control strategies. Most control methods generally are based on attempts to disrupt or alter aspects of the establishment/development of these pest–host plant relationships. The success of these pest control measures in many instances is dependent upon how well characterized these various evolved relationships are, and the degree to which they can be effectively disrupted. These relationships will be discussed in the context of plant pathogens, insects, and weed pests.

PLANT PATHOGENS

The successful development of a plant pathogen is dependent upon the availability of a plant in which defense responses are absent (a susceptible host), coupled with a physical/chemical/biological environment that is conducive to the growth, development, and reproduction of the pathogenic organism (1). Plants in which there is no recognition between the pathogen and the plant, thereby not resulting in the establishment of any type of relationship, are termed nonhosts. Plants in which there is an early recognition, followed by induction of defense mechanisms against the pathogen that circumvent the establishment of a relationship, are termed resistant plants (1).

Once successful entry of a pathogen into a host is achieved, the ensuing relationship may traverse a broad spectrum, ranging from obligatory for the pathogen (as in obligate parasites such as viruses, viroids, phytoplasmas, biotrophic fungi, most nematode species, and parasitic plants), to semiobligatory or facultative saprophytism (as in hemibiotrophic fungi, some bacteria, and a few nematode species), to nonobligatory or facultative parasitism (as in many fungi and bacteria). The degree of dependency of the pathogen on its host plant for growth and reproduction, i.e., obligatory versus facultative parasitism, has no bearing on the amount of resulting crop damage or the ease with which disease control is attained. Many obligate parasites, e.g., viruses, parasitic plants, rust and mildew fungi, significantly reduce plant growth without killing the host, while others, e.g., soilborne fungi, bacteria, can decimate all of the plants in an area under the optimal environmental conditions (1).

In the obligate parasite host–plant relationship, nutrients and water are diverted by the pathogen from its host without destruction of a large number of host cells. This specialized relationship may be restricted to only a few plant species or may be much broader. In the nonobligatory relationship, nutrients are obtained from host tissues that are destroyed by enzymes and/or toxins of the pathogen. Such a relationship generally is nonhost specific and involves a wide range of plant species.

The strategies developed to successfully control these plant pathogens rely to some degree upon their host relationships. For obligate parasites that are totally dependent upon their host, strategies to prevent entry into the plant and further establishment (through, for example, the development of resistant cultivars or transgenic plants expressing antifungal proteins) have the highest chance of success. Since these obligatory pathogens cannot survive outside of a living host, prevention of infection is an effective means of control. On the other hand, for nonobligatory pathogens (such as many soil fungi), attempts to reduce saprophytic development in the absence of a host (through, for example, addition of organic amendments) rely on reducing pathogen survival and inoculum levels (1).

INSECT PESTS

The trophic relations of phytophagous insects with host plants are intriguing and an understanding of the mechanisms involved can lead to the development of innovative

Encyclopedia of Pest Management

pest control strategies. Research has demonstrated that the interactions between the host plant and insect pest populations are complex and several may operate simultaneously. Recognition of a potential host by herbivorous insects is the first step in a sequence of events leading up to feeding injury on crop plants. Most insects are associated with a specific range of host plants, which may range from a single species (monophagous) to several species in different plant families (polyphagous). Host selection involves a recognition of plant and environmental cues by the insect, followed by a change in behavioral pattern leading to arrival at the host and subsequent feeding and/or oviposition. The ability of an insect to find and assess a plant depends on an appropriate sensory system to convey the necessary amount of information about the environment (2–4).

The sensory systems most involved are the visual and the chemical. It has been demonstrated that many phytochemicals (volatile and nonvolatile) can act as attractants and feeding stimulants. For many insects, host–plant-specific secondary plant chemicals, e.g., terpenes, can serve as signals detected by chemoreceptors or olfactory receptors that attract them to a potential host. In other cases, chemical constituents in the plant can serve as deterrents or antifeedants, e.g., alkaloids and phenolics. The balance between the phagostimulatory components, i.e. nutrients, and the deterrent compounds within a plant can have an impact on the acceptability of the tissues to a particular species of insect. Physical characteristics of the plant, such as the presence of trichomes or a thick waxy cuticle, can deter insects from feeding or ovipositing, thereby minimizing damage to the host. Many of these observations have been the outcome of studies comparing resistant versus susceptible plant species to insect attack. Plant shape, size, and color are also important cues for host selection and oviposition in many insect species (2–4).

The injury caused to crop plants by insects can be the direct result of reduced photosynthate available to the plant due to feeding injury and uptake of nutrients (such as carbohydrates, proteins, and lipids) by the insect. In addition, injury to plants can also result from secretion of toxins into tissues by certain insects and predisposition of affected plants to other biotic and abiotic stresses. Transmission of pathogens, e.g., viruses, fungi, and bacteria, can also result from insect feeding in plants.

Many of the methods developed for control of insects attempt to disrupt the initial visual or chemical recognition responses to a potential host, e.g., by use of physical control methods such as mulches or barrier crops, or by chemical methods such as applications of pheromones. In addition, alteration of chemical components in a plant to deter feeding by selection and breeding can potentially reduce damage.

WEEDS

With the exception of the parasitic plants, weeds do not establish any specific types of relationships with crop plants when compared to plant pathogens and insects. However, the characteristics of weedy plant species allow them to compete effectively with crop plants for space, light, water, and nutrients. Weedy plants grow rapidly under a broad range of environments and are well adapted for long-term survival. During the ensuing crop–weed interactions, crop yield is reduced significantly, particularly if the competition is established early and is allowed to continue. Uptake of water and nutrients, such as nitrogen and phosphorous, by the weed and competition for space and light occurs to the detriment of the crop plant. In addition, weeds can affect crop yield by reducing quality of the harvested product by acting as hosts to enhance populations of pathogens and insects, and by interfering with crop harvest practices (5, 6).

In some cases, weeds establish themselves in close association with a crop plant as a "crop mimic," i.e., close rssemblance to the crop, making control difficult (e.g., barnyard grass in rice cultivation). An indirect form of interaction in some weeds with crop plants is allelopathy—the production and release of toxic substances (allelochemicals) into the immediate environment of adjacent plants that negatively affects their growth (e.g., knapweed and quackgrass). In some instances, the toxic effects may be due to decomposing residues and microbial activities.

Disruption of the interactions between weeds and crop plants relies on the use of control methods that place the crop in a competitive position. The use of tillage practices, herbicide application, cultural methods, and biological control can all have an impact on suppressing weed populations (5, 6).

REFERENCES

1. Agrios, G.N. *Plant Pathology*, 4th Ed.; Academic Press: San Diego, CA, 1997; 635.
2. *Herbivorous Insects. Host-Seeking Behavior and Mechanisms*; Ahmad, S., Ed.; Academic Press: New York, NY, 1983; 257.
3. Bernays, E.A.; Chapman, R.F. *Host-Plant Selection by Phytophagous Insects*; Chapman and Hall: New York, NY, 1994; 312.
4. *The Host-Plant in Relation to Insect Behaviour and Reproduction;* Jermy, T., Ed.; Plenum Press: New York, NY, 1996; 322.
5. Radosevich, S.; Holt, J.; Ghersa, C. *Weed Ecology—Implications for Management*, 2nd Ed.; John Wiley & Sons: New York, NY, 1997; 589.
6. Zimdahl, R.L. *Fundamentals of Weed Science*; Academic Press: San Diego, CA, 1993; 450.

PESTICIDE MUTAGENESIS

Geraldo Stachetti Rodrigues
Embrapa Environment, Jaguariúna, Sao Paolo, Brazil

INTRODUCTION

Due to their very role as bioactive chemicals, pesticides tend to form electrophilic metabolites capable of reacting and combining with biologic macromolecules. Preferred sites of action include nucleophilic oxygen as well as nitrogen atoms, both of which are abundant in DNA predisposing the genetic material to mutagenic covalent binding. The main deleterious effect resulting from exposure to environmental mutagens is the possible initiation of cancer. Other manifestations of genotoxic environmental pollutants such as pesticides include heritable genetic diseases, reproductive dysfunction, and birth defects (1). Epidemiological studies provide evidence that several types of tumors and other carcinogenic manifestations are in excess among farmers and other occupational groups associated with pesticide handling. In addition to epidemiological evidence, laboratory and field monitoring data indicate that increases in chromosome aberration, recombination, sister-chromatid exchange, and other genotoxic events are in excess for pesticide-exposed groups, pointing out a generic genetic activity of pesticides in humans (2).

GENOTOXIC PESTICIDE PREVALENCE

A large proportion of the pesticides presently available in the market both for agricultural and home use show mutagenic activity in one or the other mutation assay. An official evaluation of pesticide genotoxicity once performed under the so-called Substitute Chemicals Program came to conclude that of varied classes of pesticide chemicals assayed for genetic toxicity with diverse test systems, 54% presented definite positive results in at least one test system (1). There seems to be no association between in vivo toxicity of the pesticides studied and their potential for genetic damage in short-term mutation assays. Most pesticide molecules will produce genotoxic nitrosamines or N-nitroso metabolites in the environment or in vivo, and direct-acting mutagenic compounds are formed and accumulated as conjugates in plants after metabolic activation of promutagenic pesticide molecules, being made available for transfer to fauna and humans through the food chain (3). The prevalence of market-available pesticides with genotoxic potential is illustrated in Table 1, which shows that most compounds induce a definite positive response in at least one of the sensitive bioassays studied.

ECOLOGICAL IMPACTS OF PESTICIDE MUTAGENESIS

In addition to the risks to human health posed by mutagenic pesticides, there are important ecological risks as well, such as the threat posed by pesticides to the stability of the ecosystems through the cumulative introduction of deleterious mutations into the genetic pool. Indeed, it has been demonstrated that pesticides and other environmental genotoxicants are capable of altering the genetic makeup of some natural populations. The frequency of mitotic chromosome aberrations has been shown to increase in several weed species growing in herbicide-sprayed fields. These effects may result in interference with the evolutionary trends of natural fauna and flora. Three main mechanisms could come into play: 1) an increased pesticide resistance in certain species (due to selection), 2) the elimination of certain susceptible species (influencing the course of selection), and 3) the origination of morphological differences of a nature sufficient for taxonomic recognition (for example, through polyploidy). The pace of genetic change, for instance development of resistance to pesticides, is directly proportional to the effectiveness of the selection pressure, given that the appropriate adaptive character is present. It has been hypothesized, however, that the adaptive challenges posed by environmental stresses dramatically enhance the mutation rate, by inducing de novo adaptive variation involving both already known (such as

Table 1 Genetic hazard potential of pesticides as detected in plant genotoxicity bioassays[a]

	Genetic toxicity potential[b]		
Pesticide group	Developmental flaws	Reproduction impairment	Hereditary effects
Insecticides	84	73	83
Herbicides	75	72	78
Fungicides	68	74	58

[a]Plant bioassays are considered risk-averse systems, due to their high sensitivity and low specificity.
[b]Percentage positive in at least one bioassay system.
(Adapted from Ref. 9.)

transoposable elements) as well as possibly unrecognized genetic mechanisms.

Pesticide Mutagenesis and Pesticide Resistance

Actually, the utilization of physical and chemical mutagens (including pesticides such as the fungicide benomyl) to cause an increase in mutation rate for the induction and subsequent selection of desirable traits is not a new concept in plant breeding, especially in relation to development of resistance to herbicides. In such breeding programs, seeds of crop plants are soaked in mutagenic agents (such as the alkylating agent ethyl methanesulfonate [EMS]) and the subsequent progeny are grown under herbicide pressure. The mutagen-treated progeny lines frequently provide many resistant seedlings, while the progeny of untreated plants commonly provide none. Similarly, it has been shown that a single nuclear mutation can be responsible for an increase in resistance, denoting that resistance induction by mutation is probably a common phenomenon in nature.

The potential increase in mutation rate and selection pressure to which plants, animals, and microbes are subjected in pesticide-sprayed fields, coupled with the possibility that single mutations may confer resistance to pesticides, may have direct implications on the very effectiveness of pesticides themselves, which can be drastically impaired by the introduction of adaptive variation into the genetic pool of pest populations, possibly increasing the rate of pesticide resistance development (4, 5).

PESTICIDE MUTAGENESIS ASSESSMENT

Risks associated with pesticide mutagenesis to humans and wildlife can be assessed by using screening systems that are sensitive and able to detect the whole mutagenic spectrum. There are more than 100 short-term bioassays for evaluating the potential genotoxic effects of pesticides, but since no single system encompasses the whole spectrum of possible genetic toxic effects, a combination of evaluation procedures is recommended for the assessment of pesticide mutagenesis.

Assessment Methodology

As a consequence of the very large number of genetic toxicity evaluation techniques described in the specialized literature most assays can be considered ancillary and will be employed only for specific ends. The assays accepted for routine evaluation of pesticides fall in one of six testing categories: 1) microbial assays, either a) prokaryotic (bacterial, such as the *Salmonella typhimurium, Escherichia coli, Bacillus subtilis*) or b) eukaryotic assays (fungi, such as *Saccharomyces cerevisae, Neurospora crassa*); 2) in vitro isolated eukaryotic cell lines; 3) Host-mediated assays; 4) in vivo animal; and 5) in vivo plant bioassays (6). Most microbial assays rely on auxotrophic cell lines, that is, populations incapable of synthesizing specific amino acids or other growth factors. Colonies that become capable of growing in factor-deficient media after treatment indicate mutational reversion to normal condition, and the frequency of reversion is correlated to mutation potency of the chemical being tested (7). In order to approximate the studies to higher organism subjects, several assays involve in vitro exposure of mammalian cell lines. Mouse bone marrow and erythrocytes, hamster ovary cells, and human lung fibroblast are some examples, and several genetic endpoints can be evaluated, such as chromatid exchanges, unscheduled DNA synthesis, and micronucleus frequency (8). Aside from the practical advantages offered by the microbial and isolated cell lines bioassays (ease of manipulation, asepticism, small space and large population assayed, and low cost), the tests often depend on an external metabolic activation complement, for the microbial and isolated cell lines may be incapable of responding to promutagenic compounds if these are not partially metabolized. The host-mediated assays are devised to circumvent this deficiency. Mammalian subjects are exposed to the pesticide, the microbial cell line is injected into this treated subject, and the cell line is later recovered and evaluated for mutation induction. Alternatively, for plant metabolic activation, whole plants are treated with the pesticide and their extracts are applied directly into the microbial assay. In vivo animal assays fall mostly into two categories: 1) micronucleus assays, which involve the encapsulation of chromosome fragments resulting from pesticide-induced aberrations and breaks in small nuclei easily seen inside the cells shortly after division, normally evaluated in peripheral blood cells of mouse and other mammals; and 2) sex-linked lethal assays, which involve the elimination of marker characters in interbred organisms, normally the fruit fly *Drosophila melanogaster* (6). Finally, plant bioassays involve a variety of endpoints, from micronuclei in pollen mother cells (*Tradescantia*) and root tip meristematic cells (*Allium, Vicia*), reversion and crossing-over in chlorophyll deficient lines (*Hordeum, Glycine*), sugar-specific starch production in pollen (*Zea*), flower pigmentation alteration (*Tradescantia*), and many other endpoints in many different species (9). An important advantage of the plant systems is the capability of in situ exposure, eliminating most of the sample manipulation and allowing evaluation in ''real world'' situations (10).

FUTURE CONCERNS

The evidences of metabolic activation of pesticides into direct acting mutagens, especially in plants, and the complex interactions that may occur by the synergistic action of the multitude of chemicals presently under common usage warrant devoted attention to the evaluation of the potential genetic consequences of pesticide mixtures and derived metabolites accumulated through food chains. This complexity is further deepened by the unpredictability of mutational effects as a result of pleiotropism and other unforeseeable events. Studies of this nature should attempt to: 1) point out the impossibility of predicting all genetic effects of pesticides, and 2) recognize that small, improbable effects may have important consequences when imposed onto the very large populations exposed to pesticides in the environment (Table 1).

REFERENCES

1. Waters, M.; Sandhu, S.S.; Simmon, V.F.; Mortelmans, K.E.; Mitchell, A.D.; Jorgenson, T.A.; Jones, D.C.L.; Valencia, R.; Garrett, N.E. Study of Pesticide Genotoxicity. In *Genetic Toxicology: An Agricultural Perspective*; Fleck, R.A., Hollaender, A., Eds.; Plenum Press: New York, 1982; 21, 275–326 Basic Life Sciences.
2. Maroni, M.; Fait, A. Health effects in man from long-term exposure to pesticides—a review of the 1975–1991 literature. Toxicology **1993**, *78* (1–3), 1–180.
3. Gentile, J.M.; Gentile, G.J.; Bultman, J.; Sechriest, R.; Wagner, E.D.; Plewa, M.J. An evaluation of the genotoxic properties of insecticides following plant and animal activation. Mutation Res. **1982**, *101*, 19–29.
4. Rodrigues, G.S.; Pimentel, D.; Weinstein, L.H. In situ assessment of pesticide mutagenicity in an integrated pest management program I—tradescantia micronucleus assay. Mutation Res. **1998**, *412*, 235–244.
5. Rodrigues, G.S.; Pimentel, D.; Weinstein, L.H. In situ assessment of pesticide mutagenicity in an integrated pest management program II—maize waxy pollen mutation assay. Mutation Res. **1998**, *412*, 245–250.
6. Epstein, S.S.; Legator, M.S. *The Mutagenicity of Pesticides: Concepts and Evaluation*; MIT Press: Cambridge, 1971; 220.
7. Laborda, E.; De La Pena, E.; Barrueco, C.; Valcarce, E.; Canga, C. Mutagenic evaluation of pesticides. Revista de Sanidad e Higiene Pública **1985**, *59* (9/10), 1201–1214.
8. Hrelia, P.; Vigagni, F.; Maffei, F.; Morotti, M.; Colacci, A.; Perocco, P.; Grilli, S.; Cantelli-Forti, G. Genetic safety evaluation of pesticides in different short-term tests. Mutation Res. **1994**, *321* (4), 219–228.
9. Sharma, C.B.S.R.; Panneerselvan, N. Genetic toxicology of pesticides in higher plant systems. Critical Rev. in Plant Sci. **1990**, *9* (5), 409–442.
10. Rodrigues, G.S.; Ma, T.H.; Pimentel, D.; Weinstein, L.H. Tradescantia bioassays as monitoring systems for environmental mutagenesis—a review. Critical Rev. Plant Sci. **1997**, *16* (4), 325–59.

PESTICIDE REDUCTION IN DEVELOPING COUNTRIES

George Ekstrom
National (Swedish) Chemicals Inspectorate, Solna, Sweden

INTRODUCTION

In developing countries agricultural population groups may handle pesticides without being aware of the health hazards, without being able to read any safety advice (if available), and without being able to protect themselves properly. The resulting public health impact includes 3 million hospitalized with severe pesticide poisonings and 220,000 deaths annually, predominantly in the developing world. The environmental effects include air, soil, food, and water pollution; loss of biodiversity; and interference with natural pest control processes. Pesticide reduction measures aim at a reduction of these and other pesticide risks to human and animal health and the environment, a reduced reliance on chemical pest control, and a reduction in quantities of chemical pesticides used.

RELIANCE AND USE (1)

Pesticide reduction activities, as presently seen in many developing as well as industrialized countries, may be looked on as a reversal of the chemical component of agricultural intensification experienced since the introduction of chemical pesticides in the late 1940s (Fig. 1). The annual global use of chemical pesticides amounts to 2.5 million metric tons calculated as biologically active substances. The four largest country markets for agrochemicals are the United States (20% of the world market), Japan, France, and Brazil. Developing countries with large or fast growing pesticide markets are Argentina, China (possibly already among the two to five largest markets), India, Indonesia, Mexico, Pakistan, Philippines, and Vietnam. The global market value of pesticides amounts to more than USD 30,000 million per year. The developing countries' total share of the market is estimated at 24% on a weight basis; Asia 12%, Latin America 8%, and Africa 4%. In value terms, the market share for pesticides used in developing countries has increased from 22% in 1978 to 31% in 1997.

Decreasing markets for pesticides in several European and other markets in combination with a potential for increased pesticide use in developing countries put severe pressures on the latter. Traditional crop protection practices fall into oblivion or are looked upon as unmodern or obsolete. Overreliance on the effectiveness of pesticides has led to loss of traditional nonchemical methods of pest control.

CONTROL PROBLEMS (2)

In addition to severe health and environmental problems resulting from current pesticide use practices, developing countries also have infrastructural problems in controlling pesticides in a life-cycle perspective including import, local production, distribution, marketing, extension services, proper use, and waste disposal (Table 1). The agrochemicals market in a developing country may consist of three separate entities: the "white market" with locally approved and registered pesticides; the "gray market" with pesticides imported by or on behalf of a ministry without having been registered by the competent authority, or, if registered, often without prescribed labeling; and the "black market" with unregistered pesticides being brought into the country illegally across unguarded borders.

A recent FAO questionnaire to member states indicates several continuing serious deficiencies in critical areas of pesticide regulation and control in many countries, particularly in Africa and Latin America. At least half of those responding indicated the need for technical assistance and increased government support to strengthen their national capabilities and infrastructures necessary to operate their pesticide control schemes effectively.

GAINING CONTROL: GOVERNMENT RESPONSIBILITIES (3)

Governments have the overall responsibility and should take the specific powers to regulate the distribution and use of pesticides in their countries. The international community has created a number of pesticide reduction in-

Encyclopedia of Pest Management

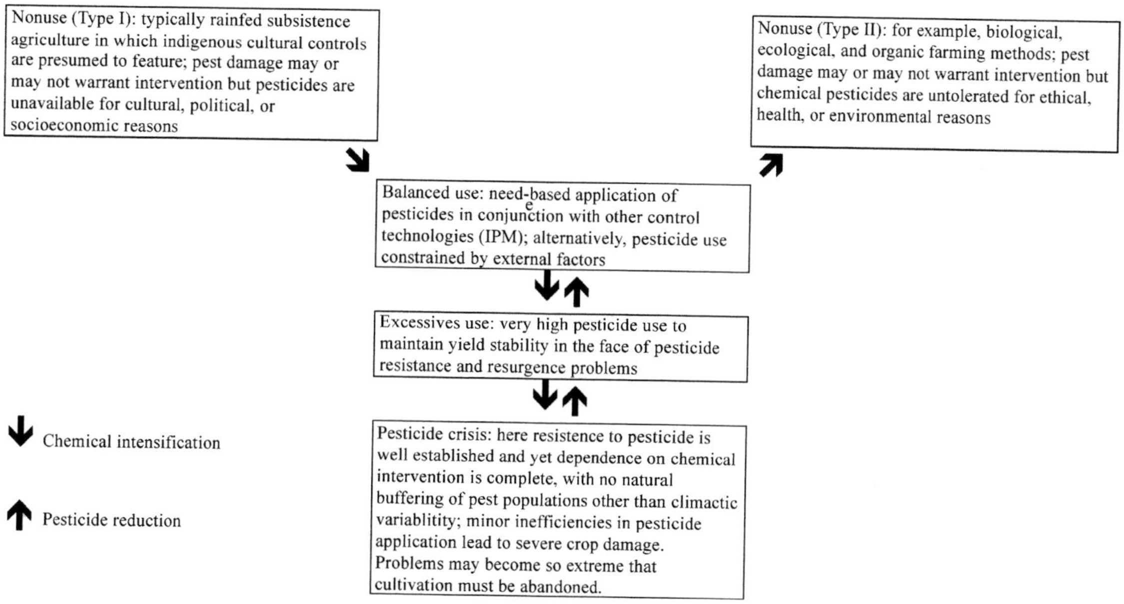

Fig. 1 Pesticide reduction versus chemical intensification. Adapted from NRI, *A Synopsis of Integrated Pest Management in Developing Countries in the Tropics;* Natural Resources Institute, Chatham (U.K.), 1992.

struments to assist governments (Table 2) such as the International Code of Conduct on the Distribution and Use of Pesticides. The Code was developed by the Food and Agriculture Organization of the United Nations to define governments' and other stakeholders' responsibilities and to serve as a point of reference until governments have established adequate regulatory infrastructures for pesticide control.

The Code recommends that governments should adopt and implement national pesticide risk reduction programs; amend agricultural policy instruments to encourage and facilitate use of biologically based farming methods; help farmers adopt biologically based pest management systems; expand and encourage farmer-driven research and education; encourage research, development, and commercialization of safer technologies; support information transfer systems; establish training and education programs for users; initiate farmer exchange programs; and monitor progress in reducing pesticide risks. Many developing countries have appointed national focal points to

Table 1 Some pesticide control problems faced by developing countries

1. Lack of facilities for disposal of waste pesticides (66% of all responding developing countries)
2. Export difficulties caused by pesticide residues in food (65%)
3. Pesticides available through distribution outlets that also deal with food, medicines, and other products for internal consumption (65%)
4. Unavailability of an appropriate size range of pesticide packages, suitable for end use (to reduce handling and other hazards) (62%)
5. Pesticide labels sometimes or generally not clear and concise (44%)
6. No regulations in force to restrict the availability of pesticides (43%)
7. Lack of a national pesticide registration and control scheme (20%)
8. The government and responsible authorities are not in a position to effectively enforce prohibition of importation, sale, and purchase of an extremely toxic product in their territory (19%)

(From FAO. 1996. *Analysis of Government Responses to the Second Questionnaire on the State of Implementation of the International Code of Conduct on the Distribution and Use of Pesticides.* Food and Agriculture Organization of the United Nations, Rome, 1996. http://www.fao.org/waicent/faoinfo/agricult/agp/agpp/pesticid/default.htm; accessed March 2001.)

Table 2 Selected international instruments relevant to pesticide reduction

1. Convention Concerning Safety in the Use of Chemicals at Work, Convention 170, International Labour Conference. 1990; http://www.ilo.org/public/english/protection/safework/cis/products/safetytm/c170.htm;
2. Rotterdam Convention: Prior Informed Consent Procedure for Certain Hazardous Chemicals and Pesticides in International Trade. UNEP/FAO. 1999; http://www.fao.org/waicent/faoinfo/agricult/agp/agpp/pesticid/pic/picho me.htm (accessed March 7, 2001).
3. International Code of Conduct on the Distribution and Use of Pesticides. FAO. 1989; http://www.fao.org/waicent/faoinfo/agricult/agp/agpp/pesticid/ (accessed March 7, 2001).
4. Agenda 21, Chapter 14, Promoting Sustainable Agriculture and Rural Development. United Nations Conference on Environment and Development. Rio de Janeiro. 1992; http://www.igc.apc.org/habitat/agenda21/ch_14.html (accessed March 7, 2001).
5. Agenda 21, Chapter 19, Environmentally Sound Management of Toxic Chemicals Including Prevention of Illegal International Traffic in Toxic and Dangerous Products. United Nations Conference on Environment and Development. Rio de Janeiro. 1992; http://www.igc.apc.org/habitat/agenda21/ch_19.html (accessed March 7, 2001).
6. Guidelines for Aid Agencies on Pest and Pesticide Management, Guidelines on Aid and Environment No. 6. OECD. Paris. 1995; http://www.oecd.org/dac/pdf/guid6.pdf (accessed March 7, 2001).
7. Cultivating our Future: Crop Protection Industry's Contribution to Achieving Sustainable Agriculture; http://www.gcpf.org/industryframes.html (accessed March 7, 2001).
8. Responsible Care: A Voluntary Chemical Industry Action Program. European Chemical Industry Council. Brussels; http://www.cefic.be/activities/hse/rc/ (accessed March 7, 2001).
9. Product Stewardship — Managing the Safe Use of Chemical Products Along Their Life Cycle; http://www.cefic.be/activities/hse/rc/ (accessed March 7, 2001).

function as liaison centers with relevant United Nations specialized agencies and programs and the Intergovernmental Forum on Chemical Safety (IFCS) (Table 3).

REDUCING RISKS (4–6)

Solutions for reducing risks include education and training of pesticide suppliers, agricultural extension workers, farmers, farmworkers, and plantation workers; promotion of hazard awareness across all sectors of society (including schools); removal of hazardous and persistent pesticides; restrictions in availability to the general public of highly toxic pesticides; and introduction of waste disposal programs. Recommendations on restrictions in availability of pesticides, based on the WHO classification of pesticides by acute health hazard, have been put forward by WHO and FAO jointly. In the absence of a globally harmonized

Table 3 National focal points and information centers and their roles in pesticide reduction

Focal point/information center	Number of developing countries[a] with focal point/information center	Role in pesticide reduction
National Focal Point for UNEP/Infoterra	114 (86%)	Hazard communication (environmental impact)
National Contact Point for FAO/WHO Food Standards Program: Codex Alimentarius	111 (84%)	Hazard communication (food residues); potential for risk reduction
Designated National Authority for the FAO/UNEP Prior Informed Consent (PIC) procedure	107 (81%)	Hazard communication (banned and severely restricted pesticides); potential for risk reduction
National Focal Point for the Intergovernmental Forum on Chemical Safety	73 (55%)	Promotion of environmentally sound management of pesticides
National Association of Manufacturers and Importers of Agrochemicals	41 (31%)	Promotion of "Responsible Care"; and "Product Stewardship"; education of pesticide suppliers and training of end-users
National Poison Center	32 (24%)	Hazard communication (pesticide accidents and poisonings)

[a] G77 member states

(From Ekstrom, G.; Ed. *World Directory of Pesticide Control Organizations*. 3rd Ed.; Royal Society of Chemistry, Cambridge, U.K., 1996.)

and accepted labeling system for pesticide products, the end-user may encounter labels of many different layouts and degrees of complexity, including texts in foreign languages, various color codes, and unfamiliar symbols. A system with easily understood ''pictograms'' has been introduced and recommended jointly by FAO and the pesticide industry.

Pesticides have been selected by governments for substitution either on a case-by-case basis or grouped by particular adverse properties as reflected by, for example, a WHO classification as extremely or highly hazardous (Classes Ia and Ib, respectively); an IARC (International Agency for Research on Cancer) or U.S. EPA classification as carcinogenic, probably carcinogenic, or possibly carcinogenic; or a UNEP classification as persistent in the environment. Candidates for national substitution have also been found among pesticides with particular problems highlighted by national, regional, or international public interest nongovernmental organizations such as the Pesticides Action Network (PAN).

REDUCING RELIANCE AND USE: CHANGEOVER TO IPM (4, 7–10)

Integrated Pest Management

Key elements of integrated pest management (IPM)[a] include recognition of farmers as experts; establishment of farmer field schools; integration of research and field staff in IPM implementation; and public education programs to change the perception of pests and pesticides. Objectives include self-sufficiency in food, increased productivity, sustainable production, community development, environmental protection, and prevention of sales of highly toxic pesticides. Crops under IPM include rice in most countries but also apples, potatoes, chili, mandarin oranges, coconut, crucifers, mung beans, maize, mango, tangerines, pomello, pulses, cotton, sugarcane, durian, cabbage, and tomatoes. Crops reported to be promising for IPM include peanuts, shallot, soya, tea, cacao, abaca, bananas, and cut flowers. Constraints to IPM have been found to be poor institutional capacity to integrate stakeholders, no uniform understanding of IPM, poor understanding of farmers' indigenous knowledge, misuse of pesticides, poor implementation of laws and regulations, inadequately trained crop protection specialists, and poor linkage between research and extension.

National Initiatives

Several developing countries have launched national pesticide reduction initiatives based on or including IPM. In Bangladesh results obtained include 90% reduction in pesticide use in rice cultivation and reintroduction of fish cultivation in rice fields. In China a Green Certificate program was introduced and highly toxic pesticides were banned for use in vegetables. In Cuba biological pest control has been defined as a national priority, and a National Food Program has been established. An ''Alternative Model'' for agricultural production was introduced as the official government policy with emphasis on crop diversity, organic fertilizers, biological control, biopesticides, and microbial antagonists with 80% of all pest control biologically based. Another goal was for the country to be self-sufficient in basic fruits and vegetables, all grown organically. Systems were introduced to facilitate information exchange between farmers, extension agents, and scientists. Selected pesticides known to cause harm to farmers and rural communities were banned.

In India training of trainers was introduced through season-long training programs, training of agricultural officers and farmers through field schools, IPM demonstrations in various crops, and distribution of pamphlets. In Indonesia a national IPM program was launched. Iran has established a High Council of Policy and Planning for Reduction of Agricultural Pesticides. In Malaysia a National IPM Committee was established. Results obtained include reduced pesticide use, increased farmers' knowledge of biological and ecological control methods, increased farmer confidence, and increased cooperative spirit in the concerned villages. In Nepal IPM was appointed official policy for crop protection and a national focal point for IPM was established. In the Philippines, a national IPM program was established aiming at IPM as a standards approach to crop husbandry. In Sri Lanka, IPM was the official recommendation for national food production. Pesticides in WHO Classes 1a and 1b were prohibited.

Nongovernmental Organizations

In a developing country situation with weak or no pesticide control infrastructure, including a poorly enforced pesticide registration scheme, nongovernmental stakeholders have shown that substantial contributions can be made aiming at an informed (safe and environmentally

[a]Integrated pest management (IPM) means the careful integration of a number of available pest control techniques that discourage the development of pest populations and keep pesticides and other interventions to levels that are economically justified and reduce/minimize risks to human health and the environment. IPM emphasizes the growth of healthy crops with the least possible disruption of agroecosystems, thereby encouraging natural pest control mechanisms.

sound) and reduced use of pesticides, for example, within the context of an IPM regime. Nongovernmental organizations, therefore, play an important role in a country's pesticide reduction efforts. National farmers' and farmworkers' associations, agricultural and other academic training and research centers, poison centers, mass media, and environmental and consumer groups in collaboration with their regional and global partner organizations; e.g., Consumers International, The International Union of Food and Allied Workers Associations (IUF), and the Pesticides Action Network, have substantial potential to influence policies and practices.

REFERENCES

1. Chemical Hazards in Developing Countries. In *The Science of the Total Environment*; Ramel, C., Dardozzi, R., Eds.; Supplement 1, 1996, 188.
2. FAO. FAO/OECD Pesticide Risk Reduction Survey: Activities in Selected Non-OECD Countries. In *Food and Agriculture Organization of the United Nations: Rome.* http://www.fao.org/ag/agp/agpp/pesticid /manage/survey/ risk.htm (accessed April 1999).
3. Pesticides Trust. In *Control of Pesticides and IPM: Implementation of Farmer Participatory Integrated Pest Management and Better Chemical Management in ACP States*; Commission of the European Union, Directorate General VIII: Brussels, Belgium, 1998.
4. Geier, B. *Organic Agriculture Worldwide—A Fast Growing Reality for 100% Pesticide Risk Reduction*, International Conference on Pesticide Use in Developing Countries, San José, Costa Rica, February 23–28, 1998; International Federation of Organic Agriculture Movements: Tholey–Theley, Germany, 1998.
5. Gips, T. Breaking the Pesticide Habit: Alternatives to 12 Hazardous Pesticides. In *International Alliance for Sustainable Agriculture*; University of Minnesota: Minneapolis, MN, 1987.
6. Morner, J. *Replacing Persistent Organic Pesticides: Guidance on Strategies for Sustainable Pest and Vector Control*, Final Draft; UNEP Chemicals, United Nations Environment Program; Geneva, Switzerland: October 1999.
7. Scialabba, N. *Factors Influencing Organic Agriculture Policies with a Focus on Developing Countries*, IFOAM 2000 Scientific Conference, Basel, Switzerland, August 2000; 28–31. http://www.fao.org/organicag/doc/BaselSum-final.doc (accessed March 8, 2001).
8. *Experience Suggests Countries Can Significantly Reduce Pesticide Use*, Ecological Agriculture Projects; McGill University: Ste-Anne-de-Bellevue, Canada. http://www.eap.mcgill.ca/MagRack/EC/ec1_4_1.htm (accessed March 8, 2001).
9. Myers, D.; Stolton, S. Organic Cotton: From Field to Final Product. In *The Pesticides Trust*; London, 1999.
10. Watts, M.; Macfarlane, R. Reducing Reliance: A Review of Pesticide Reduction Initiatives. In *Pesticide Action Network Asia and the Pacific*; Penang: Malaysia, 1997.

PESTICIDE RISK TO BEE POLLINATORS

P

Gavin Lewis
JSC International Ltd., Harrogate, Yorkshire, United Kingdom

John H. Stevenson
Harpenden, Hertfordshire, United Kingdom

Ingrid H. Williams
IACR-Rothamsted, Harpenden, Hertfordshire, United Kingdom

INTRODUCTION

Pesticides and bee pollinators are both essential for the efficient production of many crops (1). However, pesticides have potential to kill not only target pests but also nontarget insects, such as bee pollinators. To minimize the risk of poisoning bees with pesticides, systems of carefully controlled pesticide application, overseen by government regulation, have been developed in many parts of the world.

An estimated one-third of the world's crops require pollination by insects to set fruits and seeds; about 70% of these are pollinated by bees (Fig. 1), an activity valued worldwide at \$65–70 billion (1). The honey bee is the most widespread and widely used bee pollinator and many beekeepers rent their colonies to growers to supplement the pollination service provided by naturally occurring bees. Other bee species, such as the alfalfa leafcutting bee (*Megachile rotundata*), the alkali bee (*Nomia melanderi*), and mason bees (*Osmia spp.*) are also managed in some countries for supplementary field crop pollination, and commercially reared bumble bees (*Bombus spp.*) are extensively used to pollinate glasshouse crops, such as tomatoes (2).

FACTORS AFFECTING THE RISK TO BEES FROM PESTICIDES

Bees may be exposed to pesticides in several ways (3). During foraging on a crop or nearby, they may be subject to direct contact from overspray or spray drift; they may contact residues on plant surfaces, or ingest them in contaminated pollen, nectar, or water. Juvenile stages or nurse bees may be exposed to residues in food/water collected by foragers. Bees may also be exposed to pesticides used for nonagricultural purposes, e.g., wood preservatives in potential nesting sites, or by deliberate poisoning, e.g., where they are perceived as a "nuisance" or threat to people.

Each pesticide has an inherent level of toxicity to bees, which may vary according to the route of exposure (contact or oral), e.g., as a result of its mode of action. However, the application rate of a pesticide varies according to its biological activity and usage and so the degree of exposure varies correspondingly. This aspect has been formalized in the hazard ratio which divides the application rate of a pesticide by its toxicity (LD_{50} value). The method of application also has an influence. Aerial applications appear to be more hazardous than ground-based applications, possibly because the bees have less time to escape. Formulation type may affect the degree to which inherent toxicity is expressed, for example, microencapsulated formulations of pollen grain size may be collected together with pollen. The duration of toxicity, i.e., the extent of residual contact, varies with the rate of pesticide breakdown which, in turn, depends on the compound concerned and environmental factors, e.g., temperature, light intensity, rainfall, etc.

Bee species vary in their susceptibility to pesticides and this may relate to specific compounds or groups of compounds. Thus, alkali and leafcutter bees are generally more tolerant than the honey bee, although their greater surface-to-volume ratio results in higher susceptibility to field-weathered residues. Different races within a species may also vary in susceptibility, e.g., Africanized honey bees are more tolerant than other races to most chemicals tested.

Plant species vary in their attraction to bees thus affecting potential exposure, for example, oilseed rape and mustard are generally attractive while alfalfa is favored more by alkali and alfalfa bees than by the honey bee. Other plants not normally visited by bees, e.g., cereals and sugar

Fig. 1 Honeybee on *Phacelia tanacetifolia.*

cane, may become a food source for them when infested with aphids producing honeydew.

Environmental factors may affect the risk from pesticides. The availability of alternative forage, which will vary through the season, will determine the degree of foraging activity and the extent of exposure on a particular crop. Seasonal patterns may also be found in relation to the life history strategy of different bee species. Climatic conditions influence foraging activity and temperature may also have specific effects, e.g., pyrethroids show a negative temperature coefficient. Diurnal activity patterns may affect exposure, particularly in relation to application timing, and this may in turn be influenced by temperature, nectar flow, etc.

REGULATION OF PESTICIDE APPLICATION

An assessment of the risk to honey bees from the use of pesticides has been a regulatory requirement in N. America and many European countries for more than 20 years. U.S. EPA guidelines comprise three sequential tests: honey bee acute contact (OPPTS 850.3020); honey bee toxicity of residues on foliage (OPPTS 850.3030); field testing for pollinators (OPPTS 850.3040) (4).

In Europe, several countries have their own honey bee data requirements, e.g., Germany, the UK, and the Netherlands. The International Commission for Plant-Bee Relationships (ICPBR) ''Bee Protection Group,'' established a European forum for the harmonization of methods for assessing the toxicity of pesticides to bees in 1980. This brings together experts from industry, government, and academia to discuss testing methodology and risk assessment to bees. The group has provided a consensus approach formalized in the European Plant Protection Organisation (EPPO) and Council of Europe (CoE) scheme which now forms the basis of regulatory requirements throughout the European Union under Directive 91/414/EEC (5).

The European pesticide risk assessment requirements for honey bees are also based on a sequential testing scheme (6). Intrinsic toxicity is assessed with two tests for the main routes of exposure—contact and oral. A preliminary risk assessment is made using the hazard ratio. If further testing is considered necessary, semifield or cage testing using free-flying bees contained on treated plots of flowering plants and/or full field trials conducted under normal agricultural conditions are undertaken (7). All pesticides have to pass through this assessment before they can be used and, if necessary, specific restrictions may be imposed to deal with any risk identified, e.g., no use on flowering crops.

IN-USE MONITORING AND GOOD AGRICULTURAL PRACTICE

Honey bees are a particularly useful indicator species (representative of pollinating insects) as beekeepers can monitor field incidents that might be related to pesticide use. Several countries, e.g., Germany, UK, and the Netherlands have monitoring schemes, which help to verify the effectiveness of the regulatory scheme, to validate specific elements within it, and to identify existing or new problems.

Never apply pesticides to flowering crops being foraged by bees UNLESS the pesticide AND/OR its mode of action are known to be safe to bees.

REFERENCES

1. Free, J.B. *Insect Pollination of Crops*; Academic Press: London, 1993; 684.

2. Matheson, A.; Buchmann, S.L.; O'Toole, C.; Westrich, P.; Williams, I.H. *The Conservation of Bees*; Academic Press for the International Bee Research Association and the Linnean Society: London, 1996; 254.
3. Adey, M.; Walker, P.; Walker, P.T. *Pest Control Safe for Bees*; International Bee Research Association: London, 1986; 224.
4. Johansen, C.A.; Mayer, D.F.; Eves, J.D.; Kious, C.W. Pesticides and bees. Environ. Entomol. **1983**, *12*, 1513–1518.
5. Aldridge, C.A.; Hart, A.D.M. *Validation of the EPPO/CoE Risk Assessment Scheme for Honeybees,* Proceedings of the 5th International Symposium on the Hazards of Pesticides to Bees, Wageningen, The Netherlands, October 2–28, 1993; Plant Protection Service, 1993; 37–42.
6. OEPP/EPPO. Honeybees. *Decision-making Scheme for the Environmental Risk Assessment of Plant Protection Products*; Ch. 10, OEPP/EPPO Bulletin, 1993; 23, 151–159.
7. Smart, L.E.; Stevenson, J.H. Laboratory estimation of toxicity of pyrethroid insecticides to honeybees: relevance to hazard in the field. Bee World **1982**; *63*, 150–152.

P

PESTICIDE SENSITIVITIES

Ann McCampbell
Multiple Chemical Sensitivities Task Force of New Mexico, Santa Fe, New Mexico, U.S.A.

INTRODUCTION

Almost all pesticides are toxic and pose some health risks to everyone. There are, however, a growing number of people who react adversely to pesticides at levels far below those that cause symptoms in the average person. For these pesticide-sensitive people, exposures to even minute amounts of insecticides, herbicides, and other pesticides can trigger severe symptoms and prolonged ill health. Some are so sensitive to pesticides that they can not tolerate the pesticide residues on conventionally grown food and must eat only organic food. Many people report that they developed pesticide and other chemical sensitivities after a pesticide exposure, such as having their home or office sprayed with pesticides, being exposed to aerial drift of pesticides, or through an occupational exposure. Some farmers, ranchers, and pest control operators have developed pesticide sensitivities as a result of their work-related pesticide exposures. Reports from around the world indicate that pesticide and chemical sensitivities are a global problem.

HOW PREVALENT ARE PESTICIDE SENSITIVITIES?

Two random population-based surveys conducted by the California Department of Health Services (1) and New Mexico Department of Health (2) found that 16% of the respondents reported being unusually sensitive to common chemicals such as pesticides. In both studies, women were twice as likely as men to report being chemically sensitive. Otherwise, chemically sensitive respondents were evenly distributed among age groups, educational and income levels, and geographic locations. The percentage of respondents reporting chemical sensitivity were also similar in ethnic and racial groups except for Native Americans, who reported a higher prevalence.

Many people who are sensitive to pesticides also have adverse reactions to other common chemicals, such as those found in perfumes, tobacco smoke, new carpets, air ''fresheners,'' fresh ink, new paint and building materials, gasoline, vehicle exhaust, and many cleaning and laundry products. However, exposures to pesticides typically induce their worst reactions. Some people are only mildly affected while others have a more severe form of the illness called multiple chemical sensitivities (MCS). In the New Mexico study (2), 2% of the respondents reported they had been diagnosed with MCS. In California (1), 3.5% of the respondents reported that they had been diagnosed with MCS and were chemically sensitive.

WHAT CAUSES PESTICIDE/CHEMICAL SENSITIVITIES AND MCS?

Many people develop pesticide and chemical sensitivities after moving into a new home, working in a recently remodeled office, spending time in a sick building, or after being exposed to pesticides or solvents. Others slowly become ill over a period of years, seemingly as the result of the cumulative exposures of everyday life. In a survey of 6800 chemically sensitive people (3), 80% stated they knew ''when, where, with what, and how they were made ill.'' Of this group, 60% blamed pesticides for causing their illness.

Exposure to organophosphate pesticides is most implicated in triggering pesticide/chemical sensitivities and MCS. The 1997 EPA Review of Chlorpyrifos Poisoning Data (4) stated that MCS was the most commonly reported long-term health effect of chlorpyrifos poisoning. Though rarely acknowledged by pesticide manufacturers, at least one label for a chlorpyrifos-containing pesticide product warns that ''repeated exposure to cholinesterase inhibitors may without warning cause prolonged susceptibility to very small doses of any cholnesterase inhibitor'' (5). In addition, Miller and Mitzel (6) found that individuals who developed chemical sensitivities after a cholinesterase-inhibiting pesticide exposure (e.g., exposure to an organophosphate or carbamate pesticide) were found to have

Encyclopedia of Pest Management

significantly greater symptom severity than those who developed sensitivities following exposures to building remodeling chemicals.

Farmers and ranchers are known to develop pesticide and chemical sensitivities. Sheep dippers in the United Kingdom exposed to organophosphate insecticides have reported MCS-like illnesses (7). Tabershaw and Cooper studied 114 agricultural workers who experienced acute organophosphate poisoning, some of whom developed persistent MCS-like symptoms. Three years after their acute exposure, 22 workers (19%) reported that even a ''whiff'' of pesticide made them feel ill (8).

Besides those who work with pesticides, individuals who are less able to detoxify pesticides are at increased risk of pesticide injury, including the development of pesticide sensitivities. This group includes young children, the elderly, and those with acquired or genetic deficiencies in detoxifying enzymes. For example, a number of organophosphate insecticides are detoxified by serum paraoxonases (PON1). There is wide variation in subtypes and levels of these enzymes in the population due to genetic polymorphisms (9). Regardless of genetic make-up, all newborns have very low levels of PON1. These and other enzyme differences contribute to a very wide variation in individual susceptibility to organophosphate poisoning.

Haley et al. (10) found an association between low levels of a type of serum paraoxonase and neurologic symptoms in Gulf War Veterans. Veterans who served in the Gulf War were exposed to numerous cholinesterase-inhibiting substances including insecticides and antinerve gas pills (pyridostigmine bromide). These results suggest that certain soldiers may have had an impaired ability to metabolize cholinesterase-inhibiting chemicals that led to their increased susceptibility to becoming ill. Of note is that many ill Gulf War veterans also report being chemically sensitive. Kipen et al. (11) found that 36% of ill veterans on the Gulf War Registry reported having developed chemical sensitivities.

WHAT ARE THE SYMPTOMS OF PESTICIDE SENSITIVITIES?

Chemically sensitive people react to a wide variety of pesticides, including insecticides (organophosphates, pyrethroids, and carbamates), herbicides, fungicides, and fumigants. Exposures may occur via inhalation, ingestion, or skin contact. Exposures to even minute amounts of pesticides from, for example, pesticide drift or volatization from neighborhood lawn treatments, driving on a street whose roadside has been sprayed with herbicides, being in a house that was treated with pesticides years earlier, or eating food with pesticide residues can cause serious health problems for someone who is pesticide sensitive. Symptoms can range from mild to life-threatening and include, but are not limited to, headache, trouble concentrating, nausea and vomiting, fatigue, dizziness, tremulousness, rashes, asthma, irregular heartbeat, and joint and muscle pain. Neurologic and gastrointestinal symptoms often predominate following pesticide exposures.

Case example 1 A 35-year-old woman with pesticide sensitivities felt ill after a glyphosate-containing herbicide being used in a neighbor's yard wafted through her open window. She became progressively more nauseated over the next few days and began vomiting.By the fifth day after the exposure, she began to vomit blood and was admitted to the

1) The symptoms are reproducible with repeated chemical exposure.

2) The condition is chronic.

3) Low levels of exposure (lower than previously or commonly tolerated) result in manifestations of the syndrome.

4) The symptoms improve or resolve when the incitants are removed.

5) Responses occur to multiple chemically unrelated substances.

6) Symptoms involve multiple organ systems.

Fig. 1 Concensus criteria for diagnosing multiple chemical sensitivities (MCS).

hospital with gastrointestinal bleeding. Endoscopy revealed erosive lesions in her esophagus.

Case example 2 A 42-year-old woman with severe pesticide sensitivities took a bite of steamed zucchini that she had been told had been organically grown. She immediately developed nausea, abdominal pain, dizziness, irregular heartbeat, and numbness and tingling on her face.When she called the store where the zucchini had been purchased, she was told that the zucchini was NOT certified organic.

Physicians diagnose pesticide and chemical sensitivities by taking a history, performing a physical exam, and determining whether a person's symptoms come and go in relation to chemical exposures. Though not diagnostic in themselves, chemically sensitive people often have a variety of abnormal laboratory tests such as single photon emission computed tomography (SPECT) brains scans (12), electroencephalograms (following chemical challenge) (13), immune studies (14), tests of porphyrin metabolism (15), and neuropsychological evaluations. Consensus criteria for diagnosing MCS were published in 1999 (15) (see Fig. 1).

WHAT IS THE MECHANISM OF PESTICIDE AND CHEMICAL SENSITIVITIES?

A small percentage of pesticide-sensitive people may have traditional IgE (immediate) or IgG (delayed hypersensitivity) allergic reactions, such as immune-mediated rashes or asthma. Others have typical toxic reactions to pesticides, but at doses far below the norm. The mechanism for the sensitivities in the vast majority of pesticide-sensitive people, whose symptoms may vary from person to person, has not been well elucidated.The leading view, however, is that chemical sensitivities result primarily from a neurotoxic injury. Besides evidence of neurologic dysfunction, most people who are sensitive to pesticides and chemicals also have evidence of endocrine, detoxification, and immune system abnormalities.

WHAT IS THE IMPACT OF PESTICIDE AND CHEMICAL SENSITIVITIES?

According to Ashford and Miller (16), ''existing evidence does suggest that chemical sensitivity is on the rise and could become a large problem with significant economic consequences related to the disablement of productive members of society.'' The impact on people who develop pesticide and chemical sensitivities can be enormous. It is not uncommon for people to lose their jobs, careers, homes, savings, friends, and families, besides losing their health. People are often too sick to work and cannot tolerate most workplaces and homes. Their illness can also be confusing and overwhelming to friends and family members who may distance themselves rather than offer support.

The cost to society is also high when one considers the lost productivity of skilled professionals and employees and the cost of supporting disabled workers. Chemical sensitivities is considered a potentially disabling condition by the U.S. Social Security Administration (17) and U.S. Department of Housing and Urban Development (HUD) (18), and is covered under the Americans with Disabilities Act on a case-by-case basis (19). Workers in some states have also been awarded workers' compensation benefits.

In summary, pesticide and chemical sensitivity is a large and growing public health problem that warrants greater attention by the private and public sectors. The agricultural community, in particular, would benefit from a better understanding of the risks to agricultural workers of developing pesticide and chemical sensitivities, the need to notify pesticide-sensitive people of anticipated pesticide applications, and the influence of people who are intolerant of conventionally grown food on the increasing sales of organic food.

REFERENCES

1. Kreutzer, R.; Neutra, R.R.; Lashuay, N. Prevalence of people reporting sensitivities to chemicals in a population-based survey. Am. J. Epidemiol. **1999**, *150* (1), 1–12.
2. Voorhees, R. Results of Analyses of Multiple Chemical Sensitivities Questions. In *1997 Behavioral Risk Factor Surveillance System*; New Mexico Department of Health, 1999; 1–45.
3. National foundation for the chemically hypersensitive. Cheers **1989**, *1* (6).
4. Blondell, J.; Dobozy, V.A. *Review of Chlorpyrifos Poisoning Data Memorandum*; Office of Prevention, Pesticides and Toxic Substances, U.S. Environmental Protection Agency Washington, DC, 1997; 1–70.
5. *Orkinban Plus Concentrate label*; EPA Reg. No. 6754-48, Dettelbach Pesticide Corporation, 1985.
6. Miller, C.S.; Mitzel, H.C. Chemical sensitivity attributed to pesticide exposure versus remodeling. Arch. Environ. Health **1995**, *50* (2), 119–29.
7. Monk, J. Farmers fight chemical war. Chem. Ind. **1996**, *108*.
8. Tabershaw, I.; Cooper, C. Sequelae of acute organic phosphate poisoning. J. Occup. Med. **1966**, *8*, 5–20.
9. Furlong, C.E.; Richter, R.J.; Shih, D.M.; Lusis, A.J.; Alleva, E.; Costa, L. Genetic and temporal determinants of

pesticide sensitivity: role of paraoxonase (PON1). Neurotoxicology **2000**, *21* (1–2), 91–100.
10. Haley, R.W.; Billecke, S.; La Du, B.N. Association of low PON1 Type Q (Type A) arylesterase activity with neurologic symptom complexes in Gulf War Veterans. Tox. Appl. Pharm. **1999**, *157* (3), 227–233.
11. Kipen, H.M.; Hallman, W.; Kang, H.; Fiedler, N.; Batelson, B.H. Prevalence of chronic fatigue syndrome and chemical sensitivities in Gulf Registry Veterans. Arch. Env. Health **1999**, *54* (5), 313–318.
12. Ross, G.H.; Rea, W.J.; Johnson, A.R.; Hickey, D.C.; Simon, T.R. Neurotoxicity in single photon emission computed tomography brain scans of patients reporting chemical sensitivities. Toxicol. Ind. Health **1999**, *15* (3–4), 415–420.
13. Fernandez, M.; Bell, I.R.; Schwartz, G.E. EEG sensitization during chemical exposure in women with and without chemical sensitivity of unknown etiology. Toxicol. Ind. Health **1999**, *15* (3–4), 305–312.
14. Ziem, G.; McTamney, J. Profile of patients with chemical injury and sensitivity. Env. Health Perspect **1997**, *105* (Supplement 2), 417–436.
15. Bartha, L.; et al. Multiple chemical sensitivity: a 1999 consensus. Arch. Env. Health **1999**, *54* (3), 147–149.
16. Ashford, N.A.; Miller, C.S. Chemical Exposures, Low Levels and High Stakes. 2nd Ed.; Van Nostrand Reinhold New York, 1998; 26.
17. Department of Health and Human Services. *Medical Evaluation of Specific Issues-Environmental Illness, Program Operations Manual System*, SSA Publication No. 68–0424500; Part 04, Chapter 245, Section 24515.065; Social Security Administration: Baltimore, 1988.
18. U.S. Department of Housing and Urban Development. In *Multiple Chemical Sensitivity Disorder and Environmental Illness as Handicaps*; Washington, DC, 1992; 1.
19. Equal Employment Opportunity Commission. In *Americans with Disabilities Handbook*; EEOC–BK–19, U.S. Department of Justice Washington, DC, 1991; III–21.

P

PESTICIDE TERATOGENESIS

Geraldo Stachetti Rodrigues
Embrapa Environment, Jaguariúna, Sao Paolo, Brazil

INTRODUCTION

Whether a chemical compound may or may not be released to the environment for any specified reason depends on the balanced consideration of the relative benefits and drawbacks associated with its use. Often, clear economic or convenience benefits foist specified uses for substances that health or environmental drawbacks might justify restraint. The widespread usage of pesticides is an example of societal choice to endure a certain risk from exposure in exchange for more abundant produce with improved cosmetic standards. However, when exposure comes to imply possible harm to our children, either born or unborn, even the smallest risk is deemed unacceptable and the sources of potential exposure are sought out and painstakingly eliminated. The drastic restriction on organochlorine pesticides followed swiftly the reports of maternal milk contamination, but almost too late for a great many vulnerable wildlife species that were reportedly submitted to severe impact for decades of crop and forest spraying with these chemicals.

The study of the potential deleterious effects of chemical agents on the reproductive cycle of mating organisms, their generated conceptus, and their offspring is the subject matter of the branch of toxicology called teratogenesis or reproductive toxicology.

DEFINITION AND SCOPE OF REPRODUCTIVE TOXICOLOGY

Teratogenesis is a general denomination for the ability of chemical and physical agents (e.g., pesticides) to induce reproductive dysfunction in mating organisms with consequential (or independent) malformation of the eventual conceptus or abnormal development or behavior of the offspring. The study of reproductive toxicity involves the understanding of at least three distinct segments of the reproductive cycle, from 1) fertility and general reproductive performance of the mating organisms, through 2) the evaluation of teratogenesis proper and embryotoxicity, to 3) the observation of peri- and postnatal structural and functional development of the offspring. In order to encompass such broad scope, reproductive toxicity evaluation requires the appraisal of exposure through multiple generations and assessment of several endpoints. Important developmental steps of the offspring's maturation process must be taken into consideration, from postnatal organ maturation (i.e., the central nervous system and reproductive tract) through complex behavioral development, intelligence, and sensorial performance to the success in coping with the stresses of life such as sexual maturation, parenthood, and the aging process (1).

Reproductive Toxicity and the Environment

The contribution of the environment and its contaminants for the occurrence of birth defects was brought to public attention in the 1940s, as regarded by excess or deficiency of vitamin A. Prior to 1960, governmental recommendations for the assessment of chemical effects on the reproductive cycle involved only conventional six-week chronic-toxicity testing in male and female rodents followed during two pregnancies. Fetal survival was the main evaluation parameter and a general belief in the imperviousness of the placenta still prevailed.

During the early 1960s the thalidomide disaster evidenced, on the one hand, the greater vulnerability of the embryo and fetus, and on the other, that the complexity of the mother–child unity warranted special consideration. A variety of chemical agents, from medicines to natural products, to drugs of abuse, to pesticides were identified as teratogenic in the 1970s. The classical example of inhibited spermatogenesis and loss of libido in men exposed to the pesticide fumigant dibromochloropropane (DBCP) illustrates the case of one specific substance inducing a distinctive response on the male organism only. On the other hand, estrogen-like substances, such as several organochlorine pesticides (e.g., DDT, chlordecone, methoxychlor, and mirex), have been demonstrated to induce defeminization, miscarriages, malformations, and transplacental carcinogenesis. This variety of effects results

Table 1 Reproductive toxicity of selected pesticides

Pesticide group	Chemical class	Compound	Teratogenicity experimental outcome				Comments
			Positive	Negative	Controversial	No data	
Insecticides	Inorganic	Arsenicals	X				Embryolethal and teratogenic effects in several mammalian species.
		Mercurials	X				Serious neurologic disorders in children born to exposed mothers.
	Haloalcane	Dibromochloropropane	X				Impairment of spermatogenesis and loss of libido in exposed men.
		Ethylene dibromide	X				Causes oligospermia and spermatozoa degeneration.
	Organochlorine	Aldrin/Dieldrin	X				Increases fetotoxicity and teratogenic abnormalities in mice and hamsters.
		Camphechlor (Toxaphene)		X			
		Chlordane			X		At high dosages decreases fertility and viability of offspring. Affects reproductive system and causes defeminization in rats and mice.
		Chlordimeform			X		Data inadequate to evaluate teratogenicity to experimental animals.
		Chlorobenzilate				X	No adverse effects observed, but teratogenic potential has not been fully determined.
		DDT			X		Impairs reproduction and/or development in mice, rats, rabbits, dogs, and avian species.
		Dicofol			X		High doses appear to have adverse effect on preimplantation stages of embryonic development. Causes feminization of avian male embryos.
		Endosulfan				X	
		Endrin		X			Not teratogenic even at doses causing maternal or fetotoxicity.
		Heptachlor		X			Can interfere with offspring viability.
		Hexachlorobenzene	X				Caused cataracts in both parents and progeny in rats.
		Kelevan				X	Produces fetotoxicity and some teratogenic effects.

(Continued)

P

Table 1 Reproductive toxicity of selected pesticides (*Continued*)

Pesticide group	Chemical class	Compound	Teratogenicity experimental outcome: Positive	Negative	Controversial	No data	Comments
		Lindane	X				Causes steroid hormone deficiency resulting in reproductive and developmental failure.
		Mirex	X				And chlordecone/kepone (degradation products). Causes defeminization and persistent aberrant sexual behavior.
	Organophosphate	Chlorpyrifos			X		Negative in experimental animals, induced birth defects in children exposed in uterus.
		Dichlorvos			X		Not teratogenic at nontoxic doses. At high doses causes disturbances in spermatogenesis in mice and rats. Active alkylating agent.
		Dimethoate		X			Affects mating.
		Fenitrothion	X				Teratogenic in fish assays.
		Malathion		X			
		Parathion			X		Teratogenic to mice only when administered at maternally lethal dose.
		Tetrachlorvinphos				X	
		Trichlorfon	X				Reported to have alkylating property. Teratogenic to rats, mice, hamsters, guinea pigs and pigs.
	Carbamate	Aldicarb		X			No indication of teratogenicity.
		Carbaryl	X				Causes visceral abnormalities and reduced ossification.
		Thiodicarb	X				Causes visceral abnormalities and reduced ossification. Decreases sperm count, motility, and live/dead ratio.
	Pyrethroid	D-phenothrin		X			
Herbicides	Acrylaldehyde	Acrolein		X			
	Bipyridilium	Diquat and Paraquat			X		Causes adverse effects on sperm development, negative results in long-term animal experiments.
	Chlorophenoxyacetic	MCPA	X				Contaminants (phenol and chlorocresol) may be implicated.

(Continued)

Table 1 Reproductive toxicity of selected pesticides (*Continued*)

Pesticide group	Chemical class	Compound	Teratogenicity experimental outcome: Positive	Negative	Controversial	No data	Comments
		2,4-D			X		Inconsistent results attributed to absence of contaminants.
		2,4,5-T	X				Contaminants (TCDD) may be the causal agent.
	Dinitroaniline	Trifluralin	X				Embryolethal, causes skeletal variants in mice.
	Glycine derivative	Glyphosate		X			
	Nitrobenzene	Dinoseb	X				Causes sperm morphological alterations and sperm oligospermia.
	Phenylurea	Fluometuron				X	
	s-triazines	Ametryn			X		Causes delayed ossification, abortions, and resorptions.
		Atrazine	X				
		Cyanazine	X				Causes anophtalmia and microphtalmia.
	Thiocarbamate	Diallate				X	
Fungicides	Benzimidazols	Benomyl	X				Spindle poison.
		Carbendazin	X				
	Chloroaromatic and phenolic	Dichlorobenzene		X			
		Tetrachloronitrobenzene or Tecnazene		X			
		Pentachloronitrobenzene or Quintozene		X			Positive results obtained with Quintozene involve Hexachlorobenzene.
		Pentachlorophenol	X				Metabolites cause teratogenesis in fish.
		Phenylphenol				X	
	Dicarboximide	Vinclozolin	X				Causes hormone disruption and late abnormal sexual behavior.
	Dithiocarbamate	Maneb	X				Ethylenethiourea is the causal agent.
		Ziram	X				Ethylenethiourea is the causal agent.
	Phtalimide	Captafol	X				
		Captan			X		Little if any effect at maternally tolerated doses. Inconclusive.
	Phtalonitrile	Chlorothalonil				X	
Proportion of total (%)			45	21	18	16	

From the Pesticide Datasheet Series of the World Health Organization, the International Association for Research on Cancer Working Groups, and selected references.

from the interference of these substances with the metabolism of steroid and protein hormones, therefore altering a whole spectrum of complex developmental functions (2).

REPRODUCTIVE TOXICITY PREVALENCE

Reproductive Toxicity and Public Health

In the United States more than 560,000 miscarriages, stillbirths, and infant deaths and 200,000 birth defects occur each year. Prevalence figures are often underestimated because many congenital defects and disorders are not detectable during early postnatal life, averaging just 1.5–3.5% at birth but up to 16.0% at age 6. If one considers other less evident behavioral, sensorial, intellectual, and psychological disorders, these figures could be even higher. Estimates are that from 1 to 7% of congenital diseases in humans is drug or chemical related (3).

Chemical Insults and Reproductive Toxicity

The capability of a chemical agent to produce teratogenesis may not have any relationship with its ability to induce toxicity in males or nonpregnant females, and no extrapolation should be presumed. A chemical may cause harm anywhere in the body, often resulting in death of the particular cell or group of cells exposed. This case normally involves just transient and localized damage, for most cells are quite dispensable. However, when the change affects the genetic functioning of the cell while still permitting it to divide, the change may be inherited by all descendant cells, resulting in less localized damage. The effect may then be carcinogenic; or if occurred in the embryo, it may be teratogenic, impairing the further normal development of the unborn organism.

A toxic substance (e.g., a pesticide) may affect the reproductive cycle in one or more of the following ways: 1) Alteration of DNA or RNA synthesis and replication, chromosome number or structure, or gene mutation, translocation, deletion, or insertion. Hence, any chemical that is, or may be activated to, an electrophile (e.g., alkylating agents such as the *s*-triazine herbicides atrazine and simazine, or several organophosphorous insecticides including methyl-parathion, trichlorfon, and dichlorvos) must be considered a suspect teratogen. 2) Impairment of spermatogenesis and normal sperm viability, motility, or morphology (for example, due to the fumigants ethylene dibromide and DBCP), be it a result of direct toxicity on spermatogonial or sperm cells (as caused by organochlorine compounds such as DDT, methoxychlor, heptachlor, and chlordane; and nitrobenzenes such as the herbicide dinoseb); by interference with meiosis (spindle poisons such as benomyl and carbendazin); or by activated electrophiles (such as the organophosphates trichlorfon and dichlorvos, among others). 3) Deviations in sexual behavior, as induced by the insecticide kepone (chlordecone, a degradation product of mirex), and by estrogenic organochlorines. 4) Disruption of endocrine function, which in turn affects several sites as caused by estrogen-like (i.e., several organochlorines) and antiandrogen compounds (such as the fungicide vinclosolin), or by induction of the mixed-function oxidases, which increase the clearance of endogenous steroid hormones (for example as caused by the insecticides mirex, heptachlor, and endosulfan, among many others). 5) By direct toxicity to the mother or embryo/fetus (4–6).

Ecological Consequences of Teratogenic Pesticides

In addition to their potential role as human reproductive toxicants, pesticides are also implicated in reproductive failure of wildlife species exposed to pesticide sprays and residues. One main example of the ecological consequences of teratogenic pesticides is related to the organochlorines, which besides inducing malformations in embryos caused calcification problems in eggshells and impaired reproduction of several wild bird species. This problem still remains in numerous areas, due to organochlorine residue accumulation through food chains reaching toxic levels mostly in birds of prey; and to wild populations exposed to organochlorine contaminated sites.

Several reproductive toxicity assays are available to evaluate pesticide impact on wildlife or their surrogates, and many compounds have been shown to cause teratogenesis in fish, amphibian, avian, and mammalian species. Most of these are standardized two-generation assays, and realistic extrapolation regarding the impact on wild populations is difficult.

PESTICIDE TERATOGENESIS ASSESSMENT

Assessment Protocol

There are both in vitro and in vivo testing protocols designed for reproductive toxicity evaluation. In vitro assays involve many different cell and tissue culture techniques utilizing either mammalian or nonmammalian tissues. The officially accepted reproductive toxicity and teratogenicity testing sequence presently involves in vivo multigeneration exposure of at least two species, a rodent (usually rat or mouse) and a nonrodent (usually rabbit), in three experimental segments. Segment I studies are the most comprehensive, and aim at assessing the reproductive process from gonadal function through mating behavior; pregnancy loss; parturition; incidence of teratogenesis; and postnatal survival, growth, and development through lactation. Male animals (usually rats, in groups of 10) are exposed for 60 days (to encompass the whole spermatogenesis cycle), and females (usually 20) for 14 days followed by mating. Treatment is continued during gestation and weaning. Three exposure doses, the higher sufficient to produce some maternal toxicity, and a control are included. Half of the pregnancies are terminated by the middle of gestation for embryonic evaluation, and the remaining females are allowed to deliver and wean their offspring, which are then euthanized and evaluated for gross, visceral, and skeletal abnormalities. Segment II is specific for embryotoxic and teratogenic evaluation. It involves treatment of pregnant females of two species (a rodent and a nonrodent) during the period of organogenesis. Fetuses are delivered by cesarean section one day prior to parturition and all parameters are examined. Segment III is the perinatal–postnatal testing, emphasizing fetal development/delivery, lactation, and neonatal survival. Pregnant female rats are exposed during the last third of gestation through weaning, and variations in this segment have included key evaluations in the adult animals, such as sensorial, behavioral, and reproductive performance. There is growing interest in the evaluation of transplacental carcinogenesis, be it of juvenile or adult onset (1).

PESTICIDES AND REPRODUCTIVE TOXICITY

A significant proportion of a list of pesticide compounds evaluated for reproductive toxicity under the World Health Organization (7) and the International Association for Research on Cancer (8) Working Groups yielded positive or controversial results, while information is lacking for some 16% of the compounds (Table 1). Most results of animal experimentation on pesticide teratogenesis involve testing at maternally toxic doses, and caution must be exercised in drawing conclusions from these studies. However, as a precautionary principle, any significant exposure of humans of childbearing age to pesticides and pesticide residues should be avoided.

REFERENCES

1. Miller, R.K.; Kellogg, C.K.; Saltzman, R.A. Reproductive and Perinatal Toxicology. In *Handbook of Toxicology*; Haley, T.J., Berndt, W.O., Eds.; Hemisphere Publishing Co.: New York, 1987; 195–309.
2. Chernoff, N. The Reproductive Toxicology of Pesticides. In *Toxicology of Pesticides: Experimental, Clinical and Regulatory Perspectives*; Costa, L.G., Galli, C.L., Murphy, S.D., Eds.; NATO ASI-Cell Biology, Springer-Verlag: Berlin, 1987; 109–123, H13.
3. Harbison, R.D. Teratogens. In *Toxicology: The Basic Science of Poisons*; Doull, J., Klaassen, C.D., Amdur, M.O., Eds.; Macmillan Publ. Co. Inc.: New York, 1980; 158–175.
4. Durham, W.F. Toxicology of Insecticides, Rodenticides, Herbicides, and Fungicides. In *Handbook of Toxicology*; Haley, T.J., Berndt, W.O., Eds.; Hemisphere Publishing Co.: New York, 1987; 364–383.
5. Marrs, T. Toxicology of Pesticides. In *General and Applied Toxicology*; Ballantyne, B., Marrs, T., Turner, P., Eds.; Stockton Press: New York, 1993; 2, 1329–1341.
6. Ratcliffe, J.M.; McElhatton, P.R.; Sullivan, F.M. Reproductive Toxicity. In *General and Applied Toxicology*; Ballantyne, B., Marrs, T., Turner, P., Eds.; Stockton Press: New York, 1993; 2, 989–1020.
7. WHO Working Group. In *Environmental Health Criteria*; Series on Pesticide Datasheets.
8. International Association for Research on Cancer Working Group. *IARC Monographs on the Evaluation of the Carcinogenic Risk of Chemicals to Humans*; Series on Pesticide Datasheets.

PHYSICAL BARRIERS FOR THE CONTROL OF INSECT PESTS

Gilles Boiteau
Agriculture and Agri-Food Canada, Fredericton, New Brunswick, Canada

INTRODUCTION

Historically prominent among insect control methods, physical barriers continue to play a significant supporting role today and are likely to have a significant role in the integrated pest management (IPM) of the future. The adoption rate of physical barriers against insects by crop managers will depend on their cost, efficacy, ease of use, public incentives, consumer demand, and availability of other control methods (desperate need for control). Physical barriers have been defined as the alteration of the environment by physical means to make it hostile or inaccessible to insect pests (1, 2). Examples surround us wherever we live in the world: sealed packaging protects our food from insects and other pests, screen doors and windows protect our houses from houseflies, and protective clothing and nets protect us from biting insects that can in some locations vector deadly human diseases.

ATTRIBUTES OF PHYSICAL BARRIERS

A physical barrier is a structure made up of wood, metal, plastic, or other materials used to obstruct or close passage or to fence in a space (1). Physical barriers have many positive attributes: high compatibility with other control methods, low operation costs, encouragement of native populations of biological control agents, and low environmental impact.

Although barriers tend to be most effective against flightless insect species or stages, they can be used against flight-capable species or stages to modify flight direction or landing location and frequency.

In Peru, crop damage caused by the Andean potato weevil, *Premnotrypes latithorax* (Pierce), has been reduced significantly using trenches (3). The peripheral physical barriers effectively halt the migration of the flightless weevils from field to field (3). Weevils are captured or killed by insecticides applied in the trench. Similar earthworks have been used in North America to stop the movement of chinch bugs, *Blissus leucopterus leucopterus* (Say), between crops (2). The lining of the trench with a thin layer of plastic that rapidly becomes covered with a fine layer of fine soil particles makes it effective at catching and retaining insect pests such as the Colorado potato beetle, *Leptinotarsa decemlineata* (Say) (Fig. 1) and does away with the need for insecticides (4). Plastic extrusion was used to develop an above-ground reusable barrier against the Colorado potato beetle. The barrier and trap are based on the principles of the plastic-lined trench (5).

In the management of plant diseases, control methods can be directed to the disease itself or to the insect vector. Organic and synthetic mulches, nets, and row covers can be used to reduce the landing of insect vectors in crops. Natural windbreaks made up of trees and shrubs and artificial windbreaks made up of different materials can be used to modify the air movement around the fields to change the distribution of small insects (6). The effectiveness of these techniques is specific to the pest and the crop (7, 8). In forestry, windbreaks consisting of deciduous trees can mask spruce or balsam fir seed orchards from insects. In the urban environment, barriers of burlap covered with sticky materials that are environmentally safe prevent tent caterpillars and gypsy moth from climbing up and destroying important specimen trees.

Effectiveness and Cost

The physical barriers tend to produce a progressive reduction of insect pest numbers whereas insecticides provide a sharp reduction in the abundance. As a result, the effectiveness (%) of a physical barrier at any given time may seem low compared to an insecticide or a mechanical control such as a propane flamer for example. However, the effectiveness of the barrier is ongoing throughout the crop season whereas the effectiveness of the insecticide is short lived. The cost of registering, marketing, and purchasing insecticides are continually increasing. The cost of developing physical barriers is generally lower and generally bypasses strict registration procedures that insecticides require.

Encyclopedia of Pest Management

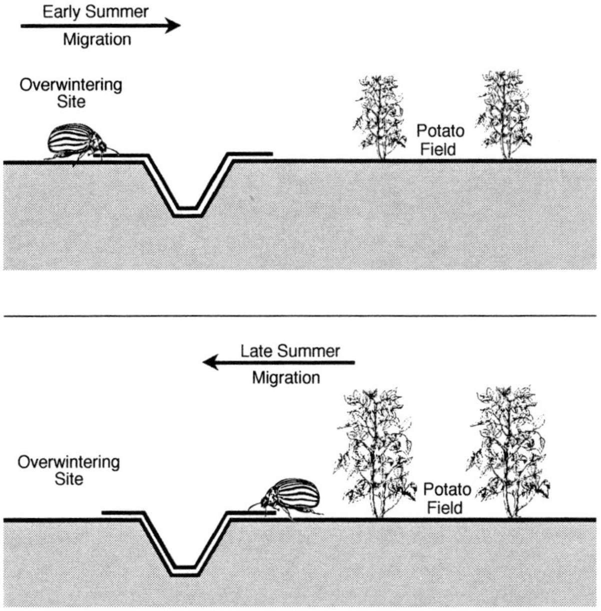

Fig. 1 Plastic-lined trench barrier for the protection of the potato crop from early summer migrating Colorado potato beetles and the trapping in late summer of potato beetles emigrating to the overwintering sites. The dust adhering to the plastic prevents the beetles from climbing out of the trench.

The combined cost of materials, installation, and operation for barriers covering a field crop is often high and at best similar to that for synthetic insecticides. Barriers spread over a crop, such as row covers, can be very effective but are rarely economical except for high value crops (9, 10). As a result, barriers tend to be recommended for small fields, especially organic production where the value of the crop is high. Cost has been a major impediment to their use in large commercial crops. However, it is important to distinguish barriers placed around fields from barriers or other control materials spread over the field surface. The cost per hectare of barriers installed around the perimeter of fields to intercept incoming and outgoing insect pests actually decreases as the area of the field increases (Fig. 2).

The perimeter of fields increases linearly while their area increases exponentially as their size increases. Thus the cost per unit area of perimeter-based control techniques, such as plastic lined trenches against potato beetles, decreases exponentially with increasing field size, while the cost of surface-based controls such as chemical insecticides remains constant no matter the field size.

For example, plastic-lined trenches against Colorado potato beetles are cost competitive with insecticides for fields of 10 ha or more (Fig. 2) (4). The inclusion of physical barriers and other perimeter-based control methods ensures that the IPM strategy is not restricted to small organic farms since it can in fact be most economical for large commercial farms. The higher cost of perimeter barriers results from the expensive price of designs that are often still experimental, not fully optimized or mechanized (11) and from the lack of reliable insect abundance forecasting methods for this preventative control method.

The cost of permanent or semipermanent barriers can be amortized over a period of time (8). A system of permanent barriers around small fields of high value crops rotated among these nested blocks allows the sharing of barriers between blocks and therefore a reduction in cost (8). Such a system has been proposed for high value vegetable crops infested by root feeding insects for which the inventory of effective insecticides is low and decreasing. Studies have shown that exclusion fences will reduce infestation levels and damage by key root maggot species in several vegetable crops where effective insecticides are becoming scarce.

THE FUTURE

The future of physical barriers seems to be linked to that of insecticides in crop and forest production.

Insecticide technology is effective and easy to use. With physical barriers, more handling is required including initial placement; removal for weeding, fertilizing, and harvesting; and final storage or disposal of the barriers. Most barriers must be used in combination with other control methods, thus increasing the complexity of the management system. However, new materials such as

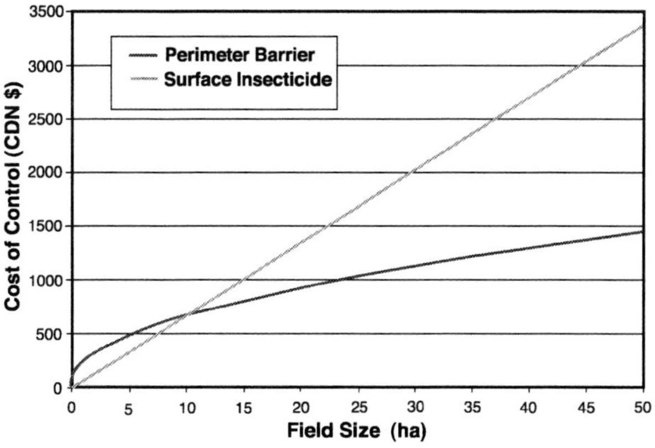

Fig. 2 Changes in the cost of Colorado potato beetle control as a function of field size for perimeter (trench barrier) and surface-based (insecticide) control methods.

spun-bonded polyester row covers and processes for manufacturing and installing these materials and devices are making previously impractical methods economically feasible. The use of physical barriers will first increase in crops for which there are no insecticides available or for which the consumers do not accept the use of insecticides.

REFERENCES

1. Banks, H.J. Physical control of insects—recent developments. J. Aust. Entomol. Soc. **1976**, *15*, 89–100.
2. Metcalf, R.L.; Metcalf, R.A. Destructive and Useful Insects. In *Their Habits and Control*, 5th Ed.; McGraw-Hill, Inc.: New York, 1993; 19–23, Ch. 9.
3. Alcazar, J.; Cisneros, F. Integrated Management for Andean Potato Weevils in Pilot Units. In *Program Report 1995–96*; The International Potato Centre: Lima, 1997; 169–176.
4. Boiteau, G.; Pelletier, Y.; Misener, G.C.; Bernard, G. Development and evaluation of a plastic trench barrier for protection of potato from walking adult Colorado potato beetles (Coleoptera: Chrysomelidae). J. Econ. Entomol. **1994**, *87*, 1325–1331.
5. Boiteau, G.; Osborn, W.P.L. Comparison of plastic-lined trenches and extruded plastic traps for controlling *Leptinotarsa decemlineata* (Coleoptera: Chrysomelidae). Can. Entomol. **1999**, *131*, 567–572.
6. Pedgley, D.E. *Windborne Pests and Diseases: Meteorology of Airborne Organisms*; Ellis Horwood Limited: Chichester, England, 1982; 250.
7. Raccah, B. Nonpersistent Viruses and Control: Epidemiology and Control. In *Advance in Virus Research*; Maramorosch, K., Murphy, F.A., Shatkin, A.J., Eds.; Academic Press: Orlando, FL, 1986; 31, 387–429.
8. Boiteau, G.; Vernon, R. Barrières Physiques contre les Insectes Nuisibles. In *La Lutte Physique en Phytoprotection*; Vincent, C., Panneton, B., Fleurat-Lessard, F., Eds.; Editions INRA: Paris, 2000; 237–61.
9. Cohen, S. Reducing the spread of aphid-transmitted viruses in peppers by coarse-net cover cucumber mosaic virus and potato virus Y. Phytoparasitica **1981**, *9*, 69–76.
10. Ferro, D. Mechanical and Physical Control of the Colorado Potato Beetle and Aphids. In *Potato Insect Control: Development of a Sustainable Approach*; Duchesne, R.-M., Boiteau, G., Eds.; Ministère de l' Agriculture, des Pêcheries et de l' Alimentation du Québec: Sainte-Foy, Québec, 1995; 53–68.
11. Misener, G.C.; Boiteau, G.; McMillan, L.P. A plastic-lining trenching device for the control of Colorado potato beetle: beetle excluder. Am. Potato J. **1993**, *70*, 903–908.

PHYTOTOXICITY

Céline Boutin
Canadian Wildlife Service, Hull, Québec, Canada

P

INTRODUCTION

Pesticides are substances used to kill or suppress pests, especially insects (insecticides), pathogens (fungicides), and weeds (herbicides). They are widely used in agriculture; horticulture; forestry including plantations; and in other areas such as rights of ways, golf courses, and greenhouses; and for domestic purposes (e.g., lawns). The use of pesticides worldwide amounts to 2.5 million metric tons. The advantage of using pesticides from an agronomic point of view has been amply considered. What is frequently not taken into account in the cost/benefit equation is the unwanted phytotoxicity that frequently occurs following pesticide application, for example, injury to crops and natural vegetation including trees and shrubs, as well as repercussions of these injuries at various trophic levels.

PESTICIDES MOVING OUTSIDE THEIR AREA OF INTENDED USE

Regardless of the method of application, it is generally accepted that there will always be some misplacement of pesticides off-target, at the time of application through mechanisms such as drift, or after the application through volatilization, runoff, leaching or soil particles.

Pesticide drift off fields can be more than 50% of the applied amount with aerial application, up to 35% with mist-blower sprayers, and up to 20% with ground-mounted tractors. Occurrences of off-target damage through volatilization (vapor drift), with soil particles, or via runoff have been documented. Pesticides can move several if not thousands of kilometers from the site of application (e.g., residues of volatile pesticides have been detected in the Canadian tundra). Thus, given the amount of pesticide used and the movement of pesticide upon and after application, there is considerable potential for pesticides to affect nontarget organisms in the environment.

DAMAGE TO CROPS

Pesticides can occasionally cause injury to crops to which they are applied. Under less than optimal soil or weather conditions, spray overlap and ill-tied spraying may lead to important crop yield reduction. Most likely however, damage will take place to crops growing in the vicinity of the intended field of application or through persistence of pesticides over a period of two or more growing seasons.

Damage to neighboring crops has been reported with 2,4-D, MCPA, atrazine, glyphosate, mecoprop, and others. More recently, ALS inhibitors (acetolactate synthase enzyme inhibiting herbicides) have been the cause of growing concern. Among these, sulfonylurea and imidazolinone herbicides are especially trouble-prone due to their potentially high persistence, mobility, broad spectrum of activity, and high phytotoxic ability at below analytical detection levels. Recent experiments have demonstrated that a dose of 0.2% of the field application rate of chlorsulfuron caused a substantial reduction in the reproduction of cherry trees (*Prunus avium*), and that yield and quality were dependent on the concentration of spray, frequency of herbicide exposure, and susceptibility linked with the growth stage. Numerous other species are known to be sensitive to chlorsulfuron at very low doses, including peas (*Pisum sativa*), alfalfa (*Medicago sativa*), wine grape (*Vitis spp.*). The sulfonylurea metsulfuron methyl can affect beans (*Phaseolus vulgaris*) at 1% label rate (0.045 g-ai/ha). Imidazolinone herbicides are known to alter potato (*Solanum tuber*) size and quality at 0.1 time recommended field rates. While reductions in the weight and the overall yield of sensitive species is problematic, another major concern is the potential lowering of fruit quality that can lead to important economic loss.

Most importantly with ALS inhibitors, significant reproductive damage can arise without visible symptoms on the vegetative parts. Furthermore, because sulfonylureas exhibit biological effects at levels lower than those that can be detected by standard analytical methods, the cause–effect relationship cannot be established with certainty under field situations.

Soil persistence may cause problems at the location of initial herbicide application, but may also impact plants outside the field through erosion and movement of soil particles. This was noted for a particularly persistent sulfonylurea, sulfometuron methyl, which damaged crops when contaminated soil dust was blown onto adjacent fields from the roadsides where the herbicide had been applied. Injury to crops can also arise when pesticide residues persist in soil thus damaging a sensitive crop planted the following season. A most interesting case occurred in Britain in 1989 where the herbicide clopyralid used on sugar beet later fed to cattle (pulp), injured chrysanthemums through the contaminated cattle manure.

Nonherbicides may exert secondary effects on plants. Carbaryl, a carbamate insecticide widely used in agriculture and noncrop situations is known to affect several tree and crop species. Fungicides can also be toxic to crops.

DAMAGE TO FORESTS AND TREES

Scorched and distorted foliage, partial or complete transient defoliation, and dead trees and shrubs are common in hedgerows, windbreaks, shelterbelts, and forest edges abutting cropfields regularly sprayed with herbicides. Sulfonylureas sprayed on grain crops growing in the vicinity of some orchards and almond trees in the United States, alluded to above, were likely connected with low yields. Although pesticide injury may not kill woody plants, they may become more susceptible to attack by diseases or insects.

DAMAGE TO NATURAL HERB VEGETATION

Crops and trees are species of economic importance. The full extent of pesticide impact on noneconomical species, populations, and ecosystems is naturally much less known and complex to quantify. A few examples will suffice to illustrate possible impacts.

In Britain it was observed that lethal effects of low doses of herbicides can occur on terrestrial plants of conservation interest at 2–6 m from field edges (20 m for seedlings). Damage at greater distances was observed although plants could recover within the growing season. The long-term effects of sublethal injury remain largely unknown. Yet it was recently demonstrated in Canada that recurrent applications of herbicides, combined with other agricultural practices such as use of fertilizers and tillage, had a long-term effect on plant populations inhabiting hedgerows and woodlot edges adjacent to cropfields.

A few studies attest to the toxicity and effect of herbicides to wetland plants. In western North America, the potential for contamination of prairie wetlands is high due to their interspersion among croplands. Metsulfuron methyl, widely used in these areas, is known to affect some wetland species at 1% of the recommended rate in agriculture (0.045 g-ai/ha). This herbicide can be applied at 560 g-ai/ha in forestry. Plants submerged in water, or with floating leaves, and algae have shown sensitivity to several herbicides.

INDIRECT EFFECTS—OTHER TROPHIC LEVELS

Phytotoxic effects of pesticides, particularly herbicides, have repercussions at other trophic levels, by modifying plant phenology, morphology, and physiology or by changing the species composition and diversity.

Application of 2,4-D has been shown to cause a delay in the germination of seeds and growth of wheat sufficient to favor an increase of wireworm (*Agriotes spp.*) damage. The same herbicide also enhanced the protein content in wheat resulting in a proliferation of aphids. The augmentation within crops of nitrogenous compound concentrations induced by 2,4-D led to an increase of pests and pathogens in maize, rice, and broad beans. Conversely, an increase in nitrate content can render forage toxic to livestock. Nodulation and nitrogen fixation can be disrupted by herbicides resulting in adverse effects on growth and reproduction of legume crops. The total removal of weeds associated with some crops can alter the subtle plant/insect/pathogen interaction. In maize, the removal of Johnsongrass (*Sorghum halepense*) weeds enhanced the infection severity caused by two pest viruses when their insect vectors were forced to move to maize rather than feed on the preferred weed species.

By reducing plant diversity at the margins of fields and in adjacent habitats, the diversity and abundance of invertebrates is reduced, which in turn will affect other trophic levels. One of the best documented studies illustrating the consequences of alterations of plant species composition and habitat quality was performed in Britain. It was found that grey partridge (*Perdix perdix*) numbers declined by 80% between 1952 and the mid-1980s. A review of several studies conducted since the 1960s has led to the conclusion that the use of herbicides, and to a lesser extent of insecticides, precipitated the decline of grey partridge populations. By and large the dramatic reduction in the numbers of this species in agricultural land was attributed to declining chick survival early in the season due to weed removal by herbicides, accompanied with a shortage of insects at this very crucial period of the year. The alteration of preferred nesting sites due to the removal of hedgerows and reduction of margins typical of modern agricultural practices was also a contributing factor.

ECONOMIC LOSSES

In terms of cost, effects of pesticide damage to crops cannot be accurately measured. Between 0.1 and 0.25% of annual crop production in the United States was estimated to be lost to pesticide drift and residue persistence. Additional hidden costs related to the phytotoxicity of crops include the development of resistance to pesticides, especially of weeds to herbicides, which may induce additional control to be made and extra research for developing new products. Furthermore, changes in the weed composition may follow the control or elimination of a weed, with new troublesome species emerging. A hidden cost may also include the deterioration of woodlots and hedgerows interspersed among agricultural land, and used by farmers for wood supply. In general, any attempt to quantify costs that arise from damage caused by pesticide use is more difficult for wildlife due to the inherent problems of placing value on natural vegetation and the wildlife that depends on it, not to mention the fact that such damage goes largely unreported.

BIBLIOGRAPHY

Boutin, C.; Freemark, K.E.; Keddy, C.J. Overview and rationale for developing regulatory guidelines for non-target plant testing with chemical pesticides. Environ. Toxicol. Chem. **1995**, *14* (9), 1465–1475.

Boutin, C.; Jobin, B. Intensity of agricultural practices and effects on adjacent habitats. Ecol. Appl. **1998**, *8* (2), 544–557.

Boutin, C.; Lee, H.B.; Peart, T.E.; Batchelor, S.P.; Maguire, R.J. Effects of the sulfonylurea herbicide metsulfuron methyl on growth and reproduction of five wetland and terrestrial plant species. Environ. Toxicol. Chem. **2000**, *19* (10), 2532–2541.

Conacher, J.; Conacher, A. *Herbicides in Agriculture: Minimum Tillage, Science and Society*; Geowest No. 22, University of Western Australia: Nedlands, WA, 1986; 169.

Eagle, D.J. Agrochemical damage to UK crops. Pestic. Outlook **1990**, *1* (2), 14–16.

Freemark, K.E.; Boutin, C. Impacts of agricultural herbicides on terrestrial wildlife in temperate landscape: a review with special reference to North America. Agric., Ecosys. and Environ. **1995**, *52*, 67–91.

Heap, I.M. The occurrence of herbicide-resistant weeds worldwide. Pestic. Sci. **1997**, *51*, 235–243.

Peterson, H.G.; Boutin, C.; Martin, P.A.; Freemark, K.E.; Ruecker, N.J.; Moody, M.J. Aquatic phytotoxicity of 23 pesticides applied to expected environmental concentrations. Aquat. Toxicol. **1994**, *28*, 275–292.

Pimentel, D.; Acquay, H.; Biltonen, M.; Rice, P.; Silva, M.; Nelson, J.; Lipner, V.; Giordano, S.; Horowitz, A.; D'Amore, M. Environmental and economic costs of pesticide use. BioScience **1992**, *42* (10), 750–760.

Sheehan, P.J.; Baril, A.; Mineau, P.; Smith, D.K.; Harfenist, A.; Marshall, W.K. *The Impact of Pesticides on the Ecology of Prairie Nesting Ducks*; Technical Report Series No. 19, Canadian Wildlife Service, Environment Canada: Ottawa, Canada, 1987; 1–641.

PLANT DENSITY

Kohji Yamamura
National Institute for Agro-Environmental Sciences, Tsukuba, Japan

INTRODUCTION

Plant density and row spacing primarily influence crop yield by changing the growth of the crop. If the growth conditions are optimal, increase in plant density generally increases the yield by increasing the light interception by the crop (1). However, dense planting does not always increase the yield in actual fields, because the plant density also influences the interaction among crop, insect pests, phytopathogens, and weeds. The number of insect pests per plant usually decreases with increasing plant density, as a result of dilution of colonists among the plants. Incidence of several insect-borne virus disease decreases with increasing plant density accordingly. Conversely, incidence of most fungal disease increases with increasing plant density, because of the faster rate of transmission between adjacent plants in the denser stand. Weeds are sometimes suppressed in densely planted fields as a result of competition between crop and weeds. The relative importance of these interactions changes depending on the crop systems.

INSECT–CROP INTERACTION

Root (2) proposed the "resource concentration hypothesis" whereby "many herbivores, especially those with a narrow host range, are more likely to find hosts that are concentrated (i.e., occur in dense or nearly pure stands)." This hypothesis predicts that the density of herbivores per host plant (or per leaf area) is higher in dense stands of their host plants (Type I in Fig. 1). Several experiments such as that on the abundance of the Mexican bean beetle *Epilachna varivestis* (Mulsant) supported this hypothesis. However, most of the experimental results examining the effect of plant density contradict this hypothesis if the size of the experimental plot is kept constant. Thus, we can adopt the opposite hypothesis, which may be called the "resource diffusion hypothesis": herbivores use more efficiently hosts that are diffused (i.e., sparsely distributed) (Type II or III in Fig. 1). For example, increasing the number of collard plants per hectare 16-fold reduced the number of pest insects per leaf surface area by about one-half and thereby reduced the total damage to the crop (3). The number of eggs of rice water weevil *Lissorhoptrus oryzophilus* (Kuschel) per plant decreased with increasing density of rice plants (4). The number of eggs of the small white butterfly *Pieris rapae crucivora* (Boisduval), the diamondback moth *Plutella xylostella* (Linnaeus), the beet semilooper *Autographa nigrisigna* (Walker) per plant decreased with increasing cabbage density (5). In this case, population per unit ground area curvilinearly increased, showing the Type II response in Fig. 1. This response is partly due to the characteristics of oviposition behavior of females. Females of *P. rapae crucivora*, as well as the subspecies *P. rapae* (L.), lay eggs sparsely even in a dense stand of cabbages (5, 6). Hence, the number of eggs per ground area increases curvilinearly, approaching an upper limit. The Type II curve is also partly attributable to a "dilution effect" caused by a limited total number of pests. In contrast, the density of several aphids shows a Type III curve. For example, increase in the seeding rate of spring tick bean from 40 to 647 kg per hectare decreased the peak density of *Aphis fabae* (Scop.) per plant. In this case, the peak density of *A. fabae* per unit ground area became maximum in the mid range of seeding rate (between 40 and 162 kg per hectare) (7). An optomotor response of aphids appears to be a cause of such a Type III response; they are attracted to sparse planting, because the contrast between leaf and soil in a sparsely planted field generates the appropriate long-wave emission that induces alighting behavior (8).

Plant density sometimes alters the plant shape and growth duration, which indirectly influences the damage caused by the pest. For example, damage to corn by larvae of the southwestern corn borer *Diatraea grandiosella* (Dyar) increases with increasing plant density, because the sparse planting ensures large healthy stalks that prevent lodging. Narrow plant spacing of certain cotton varieties enables earlier fruiting. Such a short growth duration en-

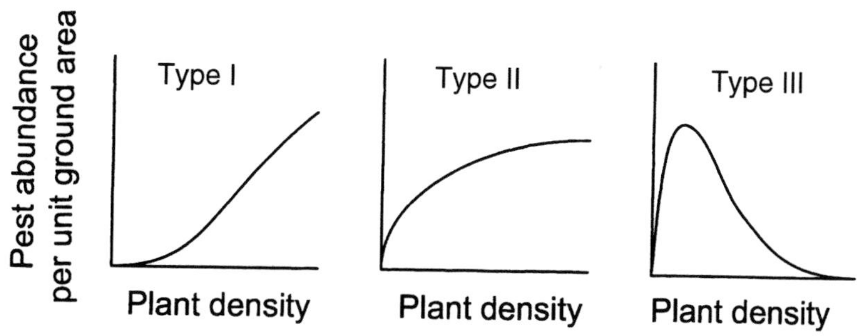

Fig. 1 Three types of the effects of plant density on the abundance of insect pests and phytopathogens.

hances escape from the damage caused by boll weevil *Anthonomus grandis* (Boheman) in Texas. Survival rate of several insect pests may also be influenced by plant density. For example, the survival rate of larvae of lepidopterous pests in cabbage fields was higher in dense stands (9). A higher survival rate enhances the abundance of pests in the next generation by enhancing the overall reproduction rate. When pests reproduce several generations during the growth period of the crop, the density of pests may become larger in dense stands, generating a Type I response in the long run.

PATHOGEN–CROP INTERACTION

Plant density causes both positive and negative effects on plant disease incidence. Burdon and Chilvers provided a summary of 69 studies about the effect of plant density on disease incidence (10). Of the total, 39 references provided evidence for a positive correlation between host density and disease incidence (Type I in Fig. 1), and 24 references provided evidence for a negative correlation (Type II or III). A majority of the positive correlations between disease incidence and plant density related to fungal diseases. For example, increase in celery density from 3.4 to 15 per m^2 doubled the percentage of infection by celery blight caused by *Cercospora apii* (Fres). This was probably due to facilitated splash dispersal between adjacent plants. Higher survival rate of inoculum during dispersal between plants enhanced the overall reproduction rate of pathogens in denser stands, generating a Type I response. On the other hand, most of the negative correlations between disease incidence and plant density were related to viral diseases transmitted by aphids. For example, increase in sugar-beet density (from 17.5 to 126.5 thousand per hectare) decreased the proportion of plants infected by beet mild yellowing virus from 51 to 12% (11). In this case, the number of infected plants per unit ground area (i.e., absolute number of diseased plants) increased from 9.9 to 19.6 thousand per hectare (Type II). In the other cases, such as groundnut rosette disease transmitted by *Aphis craccivora* (Koch), both the proportion and absolute number of infected plants decreased with increasing plant density (Type III).

WEED–CROP COMPETITION

Reduced row spacing increases light interception by the crop canopy and lessens the quantity of light incident on leaves of weeds. Accordingly, narrow row spacing will increase the ability of the crop to compete with weeds (12). Hence, decreased row width often reduces the requirement of herbicides and also enables a reduction in the frequency of postemergence tillage for weed control. For example, reducing soybean row width from the conventional 76-cm spacing to 25 cm reduced the biomass of a mixture of several common annual weeds by 52%. Weed control and crop yields of both narrow row sunflower (25 cm) and corn (38 cm) receiving 25–33% of their recommended herbicide application rates were equivalent to that of their conventional (76-cm row) counterparts receiving full herbicide rates plus interrow cultivation.

INTEGRATION

The plant density that maximizes the light interception by crop will sometimes minimize the light interception by weeds. Hence, the plant density to suppress weeds may not significantly contradict to the optimal plant density yielding highest crop yield. As for insect pests and plant diseases, if they have Type III response in Fig. 1, we can also readily exploit the plant density to enhance the yield. We are confronted with a dilemma when the pest species shows Type I response. Attempts to maximize crop yield through high planting density will tend to increase pest incidence. Yet the converse, reducing plant density, might appear unlikely to provide an economically acceptable approach even when it can reduce pest incidence. Combination with other control tactics might be necessary in such a case.

REFERENCES

1. Clegg, M.D.; Francis, C.A. Crop Management. In *Sustainable Agriculture Systems*; Hatfield, J.L., Karlen, D.L., Eds.; Lewis: Boca Raton, FL, 1994; 135–156.
2. Root, R.B. Organization of a plant-arthropod association in simple and diverse habitats: the fauna of collards (*Brassica oleracea*). Ecol. Monogr. **1973**, *43*, 95–124.
3. Pimentel, D. The influence of plant spatial patterns on insect populations. Ann. Entomol. Soc. Am. **1961**, *54*, 61–69.
4. Thompson, R.A.; Quisenberry, S.S. Rice plant density effect on rice water weevil (Coleoptera: Curculionidae) infestation. Environ. Entomol. **1995**, *24* (1), 19–23.
5. Yamamura, K. Relation between plant density and the arthropod density in cabbage fields. Res. Population Ecol. **1999**, *41* (2), 177–182.
6. Root, R.B.; Kareiva, P.M. The search for resources by cabbage butterflies (*Pieris rapae*): ecological consequences and adaptive significance of markovian movements in a patchy environment. Ecology **1984**, *65* (1), 147–165.
7. Way, M.J.; Heathcote, G.D. Interactions of crop density of field beans, abundance of *Aphis fabae* (Scop.), virus incidence and aphid control by chemicals. Ann. Appl. Biol. **1966**, *57*, 409–423.
8. Kennedy, J.S.; Booth, C.O.; Kershaw, W.J.S. Host finding by aphids in the field III: visual attraction. Ann. Appl. Biol. **1961**, *49*, 1–21.
9. Yamamura, K.; Yano, E. Effects of plant density on the survival rate of cabbage pests. Res. Population Ecol. **1999**, *41* (2), 183–188.
10. Burdon, J.J.; Chilvers, G.A. Host density as a factor in plant disease ecology. Annu. Rev. Phytopathol. **1982**, *20*, 143–166.
11. Heathcote, G.D. The effect of plant spacing, nitrogen fertilizer and irrigation on the appearance of symptoms and spread of virus yellows in sugar-beet crops. J. Agric. Sci. **1974**, *82*, 53–60.
12. Forcella, F.; Burnside, O.C. Pest Management—Weeds. In *Sustainable Agriculture Systems*; Hatfield, J.L., Karlen, D.L., Eds.; Lewis: Boca Raton, FL, 1994; 157–197.

PLANT GROWTH PROMOTING RHIZOBACTERIA

Susan M. Boyetchko
Agriculture and Agri-Food Canada, Research Centre, Saskatoon, Saskatchewan, Canada

P

INTRODUCTION

The rhizosphere is a layer of soil surrounding plant roots that has a high level of microbial activity, particularly because of nutrients secreted by plant roots in the form of soluble exudates (e.g., sugars, amino acids, organic acids, and other photosynthates). The term plant growth promoting rhizobacteria (PGPR) was first used to describe a group of bacteria inhabiting the rhizosphere that exhibited beneficial effects on plant development, primarily by suppressing the activity of plant pathogens through biological control (1, 2). Biological control is the use of biotic agents to control or suppress plant diseases while improving plant health. This often involves the reduction of pathogen inoculum or disease-producing activity by antagonistic organisms. Various strains of PGPR identified as potential biocontrol agents include species of *Bacillus, Pseudomonas, Burkholderia, Enterobacter*, and *Serratia*. There is a proposal under consideration to change the term PGPR into two new groups: 1) biocontrol plant growth promoting bacteria (biocontrol-PGPB), which refers to bacteria that enhance plant growth through biological control activity; and 2) plant growth promoting bacteria (PGPB) referring to bacteria that enhance plant growth directly. This distinguishes the two groups of bacteria based on their functional features (3).

DEVELOPMENT AND ROLE OF PGPR IN CROP IMPROVEMENT

There is tremendous potential for exploiting PGPR for biological control of plant diseases to improve crop yields (4). Some factors that justify the selection of biological control as a pest management strategy include lack of chemical fungicides for control of a variety of plant pathogens, particularly soilborne diseases, increased regulation of existing chemical fungicides, and government initiatives to restrict agricultural pesticide use. Environmental impacts such as pesticide residues in the soil, water, and food; pesticide spray drift; and development of fungicide-resistant pathogens are additional issues that support the choice of biological control as a disease control strategy. Although biological control should not be viewed as a replacement for chemicals or other forms of disease management, it should be considered as a tool for farmers to implement in an integrated pest management program.

Biological Control Using PGPR

PGPR strains enhance plant growth indirectly by decreasing the natural populations of disease causing organisms or by reducing the populations of deleterious rhizobacteria that cause plant growth inhibition and root development (1). There are several commercially available products for biological control of plant pathogens (Table 1). *Agrobacterium radiobacter* is marketed under a variety of trade names (Galltrol-A, Norgall, Diegall, and Norbac 84C) for control of crown gall on stone fruits. Products such as Epic, Kodiak, and Kodiak HB are formulations of *Bacillus subtilis* for control of diseases caused by *Rhizoctonia* and *Fusarium* species in cotton, legumes, vegetables, and ornamentals. *Pseudomonas fluorescens* is being sold in Australia under the trade name Conquer, while a granular peat formulation of *P. fluorescens*, under the trade name Dagger G™, looks promising as a biocontrol agent for cotton diseases. *Streptomyces griseoviridis* is commercially available as Mycostop, a biofungicide for control of damping-off, root and stem rot, and wilt disease (5, 6).

Several modes of action that have been attributed to the disease suppressing properties of PGPR include antibiosis, competition, parasitism and lysis, and induced resistance (6, 7). Antibiosis, one of the most commonly cited modes of action, is the production of one or more antibiotics, such as pyoluteorin, pyrrolnitrin, phenazine 1-carboxylate, kanosamine, zwittermicin A, and hydrogen cyanide that have antagonistic activity against the target pathogen. Competition involves the pathogen and PGPR competing for food and essential nutrients in the same infection court. One classic example is the production of bacterial-derived siderophores that chelate iron to make it unavailable to other

Table 1 Examples of commercially developed bacterial agents for biological control of plant pathogens

Bacterial agent	Product name	Target pathogen/disease	Company
Agrobacterium radiobacter strain 84	Galltrol-A	Crown gall	AgBioChem, Inc., Orinda, CA
Bacillus subtilis	Epic, Kodiak, Kodiak HB, Kodiak AT	*R. solani, Fusarium* spp., *Alternaria* spp., and *Aspergillus* spp. that attack roots	Gustafson, Inc., Dallas, TX
B. subtilis	Serenade	Powdery mildew, downy mildew, Cercospora leaf spot, early blight, late blight, brown rot, fire blight, others	AgraQuest, Inc., Davis, CA
Burkholderia cepacia	Deny (formerly Blue Circle, Precept)	*Rhizoctonia, Pythium, Fusarium*	Stine Microbial Products, Shawnee, KS
Pseudomonas cepacia	Intercept	*R. solani, Fusarium* spp., *Pythium* sp.	Soil Technologies Corp, Fairfield, IA
Pseudomonas fluo rescens A506	BlightBan A506	*Erwinia amylovora*, frost damage, russet-inducing bacteria	Plant Health Tech., Fresno, CA
P. fluorescens	Conquer	*Pseudomonas tolaasii*	Mauri Foods, North Ryde, Australia
P. fluorescens NCIB 12089	Victus	*P. tolaasii* (bacterial blotch)	Sylvan Spawn, Kittanning, PA
P. solanacearum (nonpathogenic)	PSSOL	*P. solanacearum*	Natural Plant Protection, Nogueres, France
Pseudomonas syringae ESC-10	Bio-save 100/1000	*Botrytis cinerea, Mucor pyroformis, Geotrichum candidum, Penicillium* spp.	EcoScience Corp., Orlando, FL
Streptomyces griseoviridis K61	Mycostop	*Fusarium* spp., *Alternaria brassicola, Phomopsis* spp., *Phytophthora* spp.	Kemira Agro Oy, Porkkalankatu, Finland

rhizosphere-inhabiting microbes (e.g., plant pathogens). Although not fully understood with bacteria, parasitism and lysis refers to the digestion of fungal cell walls of the pathogen that may occur through the secretion of lytic enzymes. Finally, induced resistance is associated with changes in the plant's defense mechanisms that can lead to a reduction in disease levels or increased tolerance of the plant to a pathogen. PGPR that are preinoculated on the plant may cause changes in the plant's ability to defend itself against a variety of plant pests.

CHALLENGES AND RISKS

Biological

One of the major criticisms of biological control agents is their inconsistent performance, particularly under field conditions. Erratic performance in the field may be overcome through the selection of strains that have better survival and competitiveness with other indigenous microorganisms, greater compatibility with crop species, and greater efficacy under a broad range of environmental conditions. Other factors that can contribute to inconsistent performance are the absence of disease pressure from the target pathogen(s) and the presence of pathogens that the biocontrol agent is not capable of controlling. Several biocontrol agents have a narrow spectrum of activity that can limit the extent of crop enhancement. These challenges may be overcome by developing a combination of strains that: 1) suppress a variety of plant diseases, and 2) have different levels of activity under a variety of environmental and soil conditions. Strain improvement can also be accomplished through genetic manipulation. Selection of genes for enhanced antibiotic production, or combining a variety of modes of action into one bacterial strain through genetic engineering may be possible to improve field performance. However, a better understanding of the underlying modes of action for the various PGPR strains will be required.

Technological

One of the major challenges for development of biocontrol products is the selection and availability of suitable liquid culture for mass production with cost-effective media components and formulation technologies. Liquid fermentation is easily achieved with bacteria; however, understanding physiological changes, including metabolites associated with mode of action, in addition to their survival during drying and storage requires greater evaluation.

It is important to monitor the growth rate, propagule yield, and metabolite production as well as performance of the fermented and formulated bacteria. Selection of suitable formulations should assure a high level of efficacy and shelf life (5, 8).

Environmental

Biological control offers an additional disease control measure that is environmentally safe when compared to some chemical control products. Public concerns continue to increase about the adverse effects of chemical residues in the environment and on human health. However, environmental groups argue that applying high doses of biocontrol agents into the environment may have detrimental effects on nontarget organisms in the agroecosytem and that they may displace indigenous microflora. Researchers have demonstrated that introduced organisms, such as these biocontrol agents, do not persist at high levels in the environment and that their populations comprise a very small fraction of the microbial community after one season. DNA technologies also allow the tracking of biocontrol agents to conduct environmental risk assessments in order to assure that introduction of microbial-based products do not have detrimental effects on the environment.

Economical

One of the most attractive reasons for pursuing biological control agents compared to chemical pesticides is the deregistration of currently available chemical fungicides and legislation-driven mandates to reduce chemical pesticide use worldwide. The costs associated with developing and bringing a new chemical to the marketplace can be as high as $80 million and take 8–10 years (9). However, the reduced costs associated with discovery, development, and registration of microbial biocontrol agents ($0.8–1.6 million) may prompt industry to invest in the development of these products. Some of the major issues facing biocontrol products include the availability of cost-effective large-scale fermentation and formulation technologies. The narrow spectrum of pests controlled by many biological control agents may necessitate combinations of microbially derived products with chemical pesticides. Compatibility between these products must be established prior to use in an integrated program. Unfortunately, the level and speed of disease control expected with biocontrol agents is often similar to that expected for chemical fungicides, which may not be realistic. The degree of control is often affected by environmental conditions prevalent at the time prior, during, and after application. It will take a great deal of public education and consumer awareness to change our perception and expectations of biological control for it to be generally accepted.

FUTURE VALUE OF PGPR TECHNOLOGY

The potential markets of PGPR as biocontrol products include: 1) diseases where chemical fungicide applications are cost-prohibitive, 2) diseases where specific chemicals have restricted use or have been deregistered, 3) control of pathogens that have developed fungicide resistance, and 4) crop production systems that are in contained or controlled environments (e.g., greenhouse operations). Biological control strategies should be considered as one component in an integrated pest management program, rather than as a replacement for other disease control measures or a silver bullet method targeted to a specific pathogen.

REFERENCES

1. Kloepper, J.W. Plant Growth-Promoting Rhizobacteria as Biological Control Agents. In *Soil Microbial Ecology*; Metting, F.B, Jr., Ed.; Marcel Dekker, Inc.: New York, 1993; 255–274.
2. Lazarovits, G.; Nowak, J. Rhizobacteria for Improvement of Plant Growth and Establishment. HortScience **1997**, *32* (2), 188–192.
3. Bashan, Y.; Holguin, G. Proposal for the division of plant growth-promoting rhizobacteria into two classifications: biocontrol-PGPB (plant growth-promoting bacteria) and PGPB. Soil Biol. Biochem. **1998**, *30*, 1225–1228.
4. Weller, D.M.; Thomashow, L.S.; Cook, R.J. Biological Control of Soil-Borne Pathogens of Wheat: Benefits, Risks and Current Challenges. In *Biological Control: Benefits and Risks*; Hokkanen, H.M.T., Lynch, J.M., Eds.; Cambridge University Press: New York, 1995; 149–160.
5. Elad, Y.; Chet, I. Practical Approaches for Biocontrol Implementation. In *Novel Approaches to Integrated Pest Management*; Reuveni, R., Ed.; Lewis Publishers: London, 1995; 323–338.
6. Boyetchko, S.M. Biological Control Agents of Canola and Rapeseed Diseases—Status and Practical Approaches. In *Biotechnological Approaches in Biocontrol of Plant Pathogens*; Mukerji, K.G., Chamola, B.P., Upadhyay, Eds.; Kluwer Academic/Plenum Publishers: New York, 1999; 51–71.
7. Handelsman, J.; Stabb, E.V. Biocontrol of soilborne plant pathogens. The Plant Cell **1996**, *8*, 1855–1869.
8. Fravel, D.R.; Connick, W.J. Jr.; Lewis, J.A. Formulation of Microorganisms to Control Plant Diseases. In *Formulation of Microbial Biopesticides*; Burges, H.D., Ed.; Kluwer Academic Publishers: Amsterdam, 1998; 187–202.
9. Utkhede, R.S. Potential and problems of developing bacterial biocontrol agents. Can. J. Plant Pathol. **1996**, *18*, 455–462.

PLANT PATHOGEN PEST MANAGEMENT

Jalpa P. Tewari
University of Alberta, Edmonton, Alberta, Canada

INTRODUCTION

During ancient times, plant diseases were believed to be caused by the wrath of God. Hence, ceremonies were held to appease Gods, thereby reducing the scourge of plant diseases. A trial-and-error era followed this fatalistic period, leading to recognition of certain agricultural management practices or treatments that would reduce disease severity. Development of the fungicide formulation Bordeaux mixture was based on a chance discovery in 1882 (1). However, as more information was gathered on the causal agents of plant diseases and on the mechanisms of pathogenicity, more targeted and scientifically planned approaches came about.

The expression disease control has the connotation of eliminating the disease completely. However, elimination is neither practically feasible, necessary, nor desirable. Therefore, the concept of disease management is preferable over disease control. This is one of the important principles of integrated pest management, in addition to using environment-friendly approaches as far as possible, adequate pathogen/disease diagnosis and monitoring, and economic threshold considerations for the judicious application of all disease management strategies in concert.

PATHOGEN/DISEASE DIAGNOSIS

Timely pathogen/disease diagnosis and monitoring are essential for effective disease management. Depending on the kind of pathogen, identification may be required at the genus, species, and race level. Pathogen identification allows access to the knowledge base relating many attributes of the disease, such as the host range, symptoms, signs, pathogen life cycle, modes of dispersal/transmission, speed of disease progression, biochemical features of the pathogen, survival strategies, epidemiology, and the disease cycle. The aforesaid can help in identification of vulnerable phases of the pathogen that can be targeted in disease management. Pathogen identification is based on morphological, pathological, biochemical, serological, and molecular criteria. Molecular probing of plant pathogens based on identification of unique sequences in the DNA is currently gaining increasing significance and application for disease diagnosis.

DISEASE FORECASTING

For important and recurring plant diseases, it is desirable to have the disease forecasting programs in place. Such programs are based on assessing/monitoring primary and/or secondary inocula, host condition, and forecasting weather conditions, or a combination of all these. Based on this and other detailed information on plant diseases, computer simulation of epidemics and expert systems allowing rational disease management decisions have been developed. The latter attempt to emulate human decision making expertise for use in disease management (2).

STRATEGIES OF PATHOGEN MANAGEMENT

Plant disease management practices operate by affecting components such as the host, pathogen, and environment. More specifically, they reduce the introduction of pathogens or reduce the initial amount of inoculum and retard their development. Management practices can be applied once windows of opportunities have been identified through detailed research on pathogens and diseases. The choice of management technique(s) will depend on the value and type of crop, such as field and horticultural crops grown in extensive acreages, crops grown in smaller land parcels or gardens, and greenhouse-grown crops. Greenhouse ecosystems allow modulation of the environment that is hard if not impossible to do under field conditions. Human cultural and economic factors in various parts of the world also affect the kinds of management techniques used. The four basic methods of plant disease management—exclusion, eradication, host resistance, and protection (3)—are briefly described below.

Exclusion

This disease management technique is based on the premise that some pathogens have a restricted distribution. The restricted distribution may extend locally or to major geographic regions such as continents and countries or to provinces thereof. The international traveler routinely has to undergo quarantine inspections when entering a country through a port of entry. However, the magnitude, ease, and speed of present-day travel by humans has made the practical application of quarantine regulations very difficult. Various countries require International Phytosanitory Certificates for importing seed, vegetative propagating material, and plant produce. Such measures have greater chances of success in excluding pathogens from certain regions if these regions are secluded by extensive natural barriers, such as oceans or lofty mountain ranges that may restrict natural dispersal of inoculum. Political boundaries alone are unable to stop the ''march'' of plant pathogens. In spite of precautions, the Karnal bunt disease of wheat caused by *Tilletia indica* was recently introduced into the United States from Mexico.

Certification of seed and vegetatively propagated plant parts is done in many parts of the world to raise and make the pathogen-free germplasm and propagating material available to growers. Certification is often combined with indexing, especially for vegetative propagating materials, where disease-free status is ascertained after testing the plant materials in the laboratory and/or growing the plants in the greenhouse. Numerous culturing, indicator host inoculation, serological, and DNA-based diagnostic techniques are in use for indexing. Many seed-borne pathogens such as *Ustilago nuda* on barley, *Leptosphaeria maculans* on canola (*Brassica* spp.), and *Alternaria brassicae* on canola (*Brassica* spp.) are also monitored by various techniques.

Eradication

Numerous kinds of management practices in this category attempt to eradicate the pathogens. However, in actual practice these measures result only in reduction and not eradication of pathogen populations.

The farmer is instrumental in initiating many cultural techniques such as crop rotation, management of crop stubble, host eradication, and many others. From a disease management standpoint, crop rotation is really a technique whereby the soil is sanitized of the multiplication and perennation habitats of certain plant pathogens. Many factors such as saprophytic abilities, presence of survival structures, host range, and the presence of long-range inoculum dispersal systems in pathogens will affect the success of crop rotation. The types of tillage practices used also affect the rate of decomposition of crop stubble in the soil. Basic studies that may lead to development and formulation of a microbial inoculant containing the white wood-rotting bird's nest fungus, *Cyathus olla*, are currently in progress. This inoculant may accelerate stubble decomposition in the soil and thereby reduce pathogen populations. Host eradication has been effective in managing citrus canker caused by *Xanthomonas axonopodis* in Florida. Similarly, eradication of common barberry (*Berberis vulgaris*), the alternate host of *Puccinia graminis tritici*, has resulted in elimination of early infections and reduction in new pathogen-race evolution in Canada and the United States.

Currently, there is a great deal of interest in biological control methods that involve reduction in pathogen populations through other organisms (4). These kinds of interactions occur routinely in nature and such disease management strategies are considered to be environment-friendly. Management of the crown gall disease caused by *Agrobacterium tumefaciens* and the closely related bacterium *A. radiobacter* is one of the oldest and most well-known success stories of biocontrol systems. Some biocontrol agents are being genetically modified to over-express the gene active in disease management. Natural and introduced fluorescent Pseudomonads in the rhizosphere are responsible for invoking systemic resistance and some other effects, leading to reduction in disease (5). Similar effects caused by the use of compost are also considered to be mediated through microbial action. Public concerns with synthetic chemicals have also led to research and development of biocontrol products for reducing postharvest decay of fruits (6). The efficacy of biocontrol agents is often lower than that shown by chemical agents. For this reason, the fungicidal sugar analog 2-deoxy-D-glucose is being tested in combination with a biocontrol agent for reducing postharvest fruit spoilage. Also, the biocontrol agents are often thought to be less reliable and less persistent in nature. These and other dogmas and myths are now being challenged, at least in some systems. Clearly, more biocontrol agents will come into commercial use as more and more research is done.

Bioelicitors of host defense offer new environment-friendly opportunities for disease control and are thought to have potential as a new generation of crop protectants (7). A number of recent studies have shown induction of host defense system(s) and disease reduction through application of saprophytes, avirulent pathogens, and pathogens of related plants or fractions thereof. Recently, the compound salicylic acid, which is closely related to aspirin (acetosalicylic acid), has also been identified as an activator of plant defenses (8).

Many pathogens and diseases are managed by a variety of physical treatments such as heat, hot air, hot water, drying, refrigeration, and radiation. Some of these methods have been in use for a long time.

Host Resistance

Through resistance, a plant is able to retard or prevent entry or subsequent progress of a pathogen. Resistance is the preferred way of disease management as it is environment-friendly and reduces input costs. There are two types of true resistance: vertical and horizontal (1). Vertical resistance is high level, race-specific, and conditioned by one or a few major genes. A race change in the pathogen is able to overcome this type of resistance. Hence, vertical resistance in not durable. Horizontal resistance, on the other hand, is of a lower order. It is conditioned by many minor genes. Hence, it is durable and effective against all races of the pathogen. Due to ease of manipulation of one or a few genes, plant breeders have made extensive use of vertical resistance in crop improvement. Similarly, several genes encoding for plant pathogenesis-related proteins, such as β-1,3-glucanase, chitinase, and peroxidase have been cloned and transferred to plants making them resistant. Another type of resistance is known as ''slow diseasing.'' Slow diseasing is partial resistance partitioned in different proportions between horizontal and vertical resistance. Recently, there has been considerable interest in slow diseasing due to the durable nature of its horizontal component.

Availability of plant genetic variability in terms of resistance genes is required for any crop improvement program. Some sources of genetic variability include closely related wild plants, breeding lines, and land races of crop plants. Mutagenesis, somaclonal variation, and selection against pathogenicity-determinants of disease-causing agents can also be used for generating variability. Until recently, emphasis has been on developing crops with resistance to single diseases, but the future will see more research on multiple disease resistance.

Protection

Plants can be protected from exposure to the inoculum and infection by many methods. The most common methods entail application of chemicals such as the fungicides, bactericides, nematicides, viricides, and some others in the infection court. Herbicides are used for the control of parasitic flowering plants. Chemicals can be applied as seed or soil treatment, foliage sprays or dusts, wound dressing, and fumigants. Many chemicals do not penetrate plant tissues, remain on the plant surface, and act through contact action. Another group of chemicals are systemic, being absorbed readily by plant tissues. The majority of the systemic chemicals move upward in the plant along with the transpiration stream. The chemicals may function as protectants, in which case they must be applied in the infection court in advance of the arrival of the pathogen to prevent infection. Alternatively, they may function as eradicants and kill or suppress the pathogen already invading the host tissue (1).

There are increasing environmental and health concerns, some real and others perceived, from the public at large regarding the use of plant protection chemicals. Some of these concerns extend even in relation to atmospheric components. Methyl bromide, one of the five most widely used chemicals in the United States, is scheduled to be phased out in 2001 due to its role as an ozone depleter (9) and all other chemicals are being reexamined for safety to public. There is interest in chemicals perceived by public to be ''friendly'' such as sodium carbonate or bicarbonate, calcium and silicon compounds, and protectants tailored after naturally occurring chemicals (10).

REFERENCES

1. Agrios, G.N. *Plant Pathology*, 4th Ed.; Academic Press: San Diego, 1997; 635.
2. Robinson, B. Expert systems in agriculture and long-term research. Can. J. Plant Sci. **1996**, *76* (4), 611–617.
3. Lamey, H.A. *Plant Diseases. Their Development and Control*; Publication No. EB–31, Cooperative Extension Service, North Dakota State University: Fargo, ND, 1982; 3–14.
4. Harman, G.E. Myths and dogmas of biocontrol: changes in perceptions derived from research on *Trichoderma harzianum* T–22. Plant Dis. **2000**, *84* (4), 377–393.
5. Raupach, G.S.; Kloepper, J.W. Mixtures of plant growth-promoting rhizobacteria enhance biological control of multiple cucumber pathogens. Phytopathology **1998**, *88* (11), 1158–1164.
6. Wilson, C.L.; El Ghaouth, A.; Chalutz, E.; Droby, S.; Stevens, C.; Lu, J.Y.; Khan, V.; Arul, J. Potential of induced resistance to control post-harvest diseases of fruits and vegetables. Plant Dis. **1994**, *78* (9), 837–844.
7. Lyon, G.D.; Newton, A.C. Do resistance elicitors offer new opportunities in integrated disease control strategies? Plant Pathol. **1997**, *46* (5), 636–641.
8. Ryals, J.; Uknes, S.; Ward, E. Systemic acquired resistance. Plant Physiol. **1994**, *104* (4), 1109–1112.
9. Ristaino, J.B.; Thomas, W. Agriculture, methyl bromide, and the ozone hole. Can we fill the gaps? Plant Dis. **1997**, *81* (9), 964–977.
10. Ypema, H.L.; Gold, R.E. Kresoxim-methyl: modification of a naturally occurring compound to produce a new fungicide. Plant Dis. **1999**, *83* (1), 4–19.

PLANT QUARANTINE

Guy J. Hallman
Agricultural Research Service, Weslaco, Texas, U.S.A.

P

INTRODUCTION

Introduced pests cause considerable losses to plants; damages to crops caused by exotic arthropods in the United States alone have been estimated at $14.4 billion annually, and the number of introduced species worldwide is increasing to unprecedented levels (1). While exotic insects constitute about 1% of the extant insect fauna of the United States, more than 40% of the serious insect pests in the United States are of foreign origin (2). In some areas introduced species account for the vast majority of pests, for example, 98% in Hawaii. Plant-feeding organisms also cause major environmental harm besides damaging agriculturally important plants. Plant quarantines are established in an attempt to prevent pests from spreading to new areas.

This article concentrates on the plant quarantine situation in the United States; most countries have similar legislation aimed at preventing the geographic spread of plant pests (3). Nine regional plant protection organizations exist (Table 1) and on a global level the International Plant Protection Convention (IPPC) (1952), ratified by 111 countries, is devoted to preventing the spread of plant pests.

HISTORY OF PLANT QUARANTINE IN THE UNITED STATES

The United States has a history of regulating plant quarantine issues that dates back a century and can be used to illustrate the development of a national phytosanitary policy. In 1905 the first pest regulatory law in the United States, the Insect Pest Act, was passed; it provided authority to regulate the importation and interstate movement of live injurious insects. But the Plant Quarantine Act of 1912 is considered the first significant regulatory legislation in the United States to seriously restrict plant pest movement. It authorized the federal government under the U.S. Department of Agriculture (USDA) to impose plant quarantines and prevented the entry or interstate movement of quarantined plants without an appropriate permit or certificate of inspection deeming the transported plants to be free of quarantined pests. By 1930 the exponential increase in pest establishment into the United States, evident since 1880, was reduced to a much milder (although still threatening) linear increase due in large part to pest regulatory legislation and enforcement. In 1928 an amendment to the Plant Quarantine Act gave authorized inspectors the right to stop and search, without a warrant, people and their conveyances and seize and destroy articles in violation of quarantine legislation. The Organic Act (1944) expanded the USDA's role to provide certification of quarantine pest-free status to agricultural products exported from the United States to meet phytosanitary requirements of foreign countries and enhance U.S. export capabilities. The act also gave the USDA authority to cooperate in plant pest regulatory actions with a foreign country (Mexico). In 1976 amendments to the Organic Act expanded cooperative authority to all countries of the Western Hemisphere. The Federal Plant Pest Act (1957) replaced the Insect Pest Act (1905) and broadly defined "pest" as insects, rodents, nematodes, fungi, weeds, or any other forms of terrestrial or aquatic plants, animals, viruses, bacteria, or other microorganisms that threaten U.S. crops or other plant life. It also reinforced the authority of inspectors to seize, treat, or destroy products containing pests that threaten U.S. agriculture. The Plant Protection Act (2000) increases penalties for smuggling agricultural products up to US $50,000 for individuals and $250,000 for businesses and gives the Secretary of Agriculture authority to subpoena evidence and witnesses. Today regulatory agencies worldwide at national and provincial levels have broad authority to place items and areas under varying levels of quarantine and perform the necessary duties including inspection without a warrant, confiscation, and destruction of commodities suspected of harboring quarantined pests.

QUARANTINE METHODS

In the United States, the Plant Quarantine Act and later acts delegate legal authority to the Secretary of Agricul-

Table 1 Regional plant protection organizations

Acronym	Name	Members
APPPC	Asian and Pacific Plant Protection Commission	Australia, Bangladesh, Cambodia, China, Fiji, France (for French Polynesia), India, Indonesia, Laos, Malaysia, Myanmar, Nepal, New Zealand, Pakistan, Papua New Guinea, Philippines, Portugal (for Macau), Republic of Korea, Solomon Islands, Sri Lanka, Thailand, Tonga, Viet Nam, Western Samoa
CPPC	Caribbean Plant Protection Commission	Barbados, Colombia, Costa Rica, Cuba, Dominica, Dominican Republic, France (for Guadeloupe, French Guiana, Martinique), Grenada, Guyana, Haiti, Jamaica, Mexico, Netherlands (for Aruba, Netherlands Antilles), Nicaragua, Panama, Saint Kitts and Nevis, Saint Lucia, Suriname, Trinidad and Tobago, United Kingdom (for British Virgin Islands), United States of America (for United States Virgin Islands, Puerto Rico), Venezuela
COSAVE	Comité Regional de Sanidad Vegetal para el Cono Sur	Argentina, Brazil, Chile, Paraguay, Uruguay
CA	Comunidad Andina	Bolivia, Colombia, Ecuador, Peru, Venezuela
EPPO	European and Mediterranean Plant Protection Organization	Albania, Algeria, Austria, Belgium, Bulgaria, Croatia, Cyprus, Czechia, Denmark, Estonia, Finland, France, Germany, Greece, Guernsey, Hungary, Ireland, Israel, Italy, Jersey, Jordan, Kyrgyzstan, Latvia, Lithuania, Luxembourg, Macedonia, Malta, Morocco, Netherlands, Norway, Poland, Portugal, Romania, Russia, Slovakia, Slovenia, Spain, Sweden, Switzerland, Tunisia, Turkey, Ukraine, United Kingdom
IAPSC	Inter-African Phytosanitary Council	Algeria, Angola, Benin, Botswana, Burkina Faso, Burundi, Cameroon, Cape Verde, Central African Republic, Chad, Comoros, Congo, Côte d'Ivoire, Djibouti, Egypt, Equatorial Guinea, Ethiopia, Gabon, Gambia, Ghana, Guinea, Guinea-Bissau, Kenya, Lesotho, Liberia, Libya, Madagascar, Malawi, Mali, Mauritania, Mauritius, Mozambique, Namibia, Niger, Nigeria, Rwanda, São Tomé and Principe, Senegal, Seychelles, Sierra Leone, Somalia, South Africa, Sudan, Swaziland, Togo, Tunisia, Uganda, United Republic of Tanzania, Zambia, Zimbabwe
NAPPO	North American Plant Protection Organization	Canada, Mexico, United States
OIRSA	Organismo Internacional Regional de Sanidad Agropecuario	Belize, Costa Rica, El Salvador, Guatemala, Honduras, Mexico, Nicaragua, Panama
PPPO	Pacific Plant Protection Organization	Australia, Cook Islands, Federated States of Micronesia, Fiji, France (for French Polynesia, New Caledonia, Wallis and Futuna Islands), Kiribati, Marshall Islands, Nauru, New Zealand, Niue, Northern Mariana Islands, Palau, Papua New Guinea, Solomon Islands, Tokelau, Tonga, Tuvalu, United Kingdom (for Pitcairn), United States (for American Samoa and Guam), Vanuatu, Western Samoa

ture to impose quarantines and certify plants and their products for exportation. This authority includes taking various forms of emergency action without prior court approval to stop the spread of invasive plant pests; restrict or prohibit imports that may carry exotic pests, promulgate regulations for cooperative federal-state programs; conduct cooperative federal-state-international suppression, control, or eradication measures; and provide export certification for domestic plants and plant products on request. Pest risk analyses are conducted to determine if and to what degree a proposed importation represents a risk to U.S. agriculture and plant life. The analysis considers pests' potential to cause economic, social, and environmental damage and what safeguards could be used to reduce that risk. The owners of the product may be ordered to destroy, treat, or return to the point of origin an infested product at their expense.

When a quarantine is established the regulations include a geographic definition of the quarantined area, a list of regulated or prohibited items from that area, the conditions under which regulated items may be moved out of that area, and provisions under which certificates or permits may be issued to allow movement of regulated items out of the area. The USDA is exhorted to use the least disruptive and costly means to prevent the spread of exotic pests without unnecessarily restricting commerce. Imprisonment of up to one year as well as the fines previously noted may be imposed against individuals who violate plant quarantine regulations or forge or alter phytosanitary documents. Compensation sometimes is made to reduce economic impact to affected persons when plants or their products must be destroyed through no fault of their own; for example, this has happened in recent years related to Karnal bunt, *Tilletia indica*, (a smut fungi) eradication. If unwarranted destruction or other economic loss is incurred by interested parties related to a regulatory decision, they may pursue legal action against the United States within one year of the regulatory action being taken.

FUTURE CONCERNS

The world has entered a period of accelerated pest introductions unmatched since the early years of regulatory legislation. Several reasons contribute to this increase in the establishment of exotic pests: 1) Increased concern over invasive species that do not attack economic plants but plants in natural ecosystems, especially endangered plants, results in more encounters of them than may have been noticed in the past. 2) Increased and more rapid travel to all corners of the earth has increased the probability that exotic pests will be transported. 3) Increases in amounts and varieties of commodities traded have increased the taxonomic range and risk of pest importations. 4) Reduced resources dedicated to regulatory efforts in many parts of the world during economic, social, and/or political downturns in Asia, Latin America, and the former Soviet Union has resulted in reduced vigilance toward exotic pests. 5) Global ecological changes including "global warming," loss of habitat and biodiversity, increased nitrogen deposition in nitrogen-deficient ecosystems, and increases in atmospheric carbon dioxide are expected to elevate the chances of invasive species, which are often quite adept at filling vacant niches and invading changing habitats, becoming established. 6) The low penalties applied to most cases of nonnarcotic smuggling are serving as less of a deterrent given the relatively increased value of some of the items being smuggled.

Several other problems plague plant pest regulatory agencies worldwide. An increasingly large backlog of requests to export agricultural commodities to affluent countries results in rulings that may be delayed by years. Some countries, such as Australia, have responded to this increased demand by delegating part of the data collecting for the risk analyses to the interested parties. The unification of regional markets causes phytosanitary quarantines to become spread among countries that did not have these quarantines before. Progress in achieving consistency within and among countries for phytosanitary regulations has been made by the IPPC, although problems persist. There is inconsistency in assigning risk to different pest situations (4). The meager data on the thousands of potential pests, their possible importance, and ability to become established make risk analyses very difficult and imprecise. The use of dubious quarantines as artificial trade barriers has been successfully challenged in the World Trade Organization, although enforcement is not very effective and accomplished largely by retaliatory threats.

There are still perhaps 20,000 foreign pests that could have a negative impact on commercial plants in the United States, with like numbers for many other countries. Now that concern has spread to exotic organisms negatively affecting ecosystems, the number of unwanted species and the effort required to protect agricultural and natural ecosystems throughout the world increases several fold.

REFERENCES

1. Pimentel, D.; Lach, L.; Zuniga, R.; Morrison, D. Environmental and economic costs of nonindigenous species in the United States. BioScience **2000**, *50* (1), 53–65.
2. Schwalbe, C.P. History of subsection eb: regulatory. Bull. Entomol. Soc. America **1989**, *35* (3), 140–143.
3. Wilson, C.L.; Graham, C.L. *Exotic Plant Pests and North American Agriculture*; Academic Press: New York, 1983; 522.
4. Liquido, N.J.; Griffin, R.L.; Vick, K.W. *Quarantine Security for Commodities: Current Approaches and Potential Strategies*; USDA, Agric. Res. Service: Beltsville, MD, 1997; 50.

P

PLANTING TIMES

R.A. Balikai
University of Agricultural Sciences, Dharwad, Karnataka, India

INTRODUCTION

The main objective of manipulating planting time is largely to avoid a pest. Careful selection of planting dates makes it possible to avoid the egg-laying period of a particular insect, and allow young plants to establish to a tolerant stage before the attack occurs. This practice allows a shorter period of susceptibility during which the insect may attack or the crop may reach a mature stage before a certain pest becomes abundant. It also helps to minimize pest damage by causing asynchrony between the development of the host plant and its pest, or by synchronizing insect pests with their natural enemies, or by crop production with available preferred alternate host plants of the pest.

FACTORS INFLUENCING PLANTING TIME

The planting time in various field crops is governed not only by the necessity of evading the ravages of insect pests and diseases, but also by the environmental requirements of the crop for exploiting its full yield potential, since planting time certainly results in a conflict between pests and agronomic practices, as number of factors influence dates of planting in relation to yield maximization. Some of the important factors are listed below.

1. In dryland tracts, seasonal rainfall pattern and soil moisture balance should be taken advantage of to decide the planting dates in order to obtain higher yields.
2. Late maturing genotypes having higher yield potential need to be planted earlier than medium– or early–maturing ones.
3. Exposure of crop to extremely high air temperature results in lower yields.
4. Nonoptimum time of planting may result in winter injury to wheat crop.
5. Early planting helps in effective weed control.
6. Delayed planting results in delayed harvesting, resulting in losses due to lodging, weathered grain, grain sprouting in the head, and green heads from late tillering.

MANIPULATION OF PLANTING TIMES

Varying the planting time of crops works as a means of pest control by creating asynchrony between crop phenology and pest phenology resulting in reduced rate of colonization. By adjusting planting times, the pest or its damage can be avoided by three basic means: 1) early planting 2) late (delayed) planting, and 3) optimum planting time that provides injury-free periods during the course of crop growth and development (1). The principles under consideration are discussed below with suitable examples.

Early Planting

There are several examples to illustrate the use of early crop planting to avoid a pest. In most of the cases the approach allows the production of a crop before the insect pests reach damaging numbers. For example, in the United States, sorghum midge populations increase on Johnson grass and a few grassy hosts in spring and early summer, then migrate to sorghum fields as the crop begins to flower (2). Thus early-planted sorghum escapes economic damage while injury intensifies with later planting dates.

Late Planting

There are several classic examples of delayed planting to illustrate this principle. Delayed fall planting of winter wheat is an effective and widely used method of escaping injury by the Hessian fly, *Mayetiola destructor* (Say). The idea behind this practice is to see that wheat will not be made available for fly attack until most of the oversummering adult flies have emerged and died.

Table 1 Effect of crop planting times on insect abundance and/or potential damage

	Corn planting time			Sorghum planting time			Cotton planting time			Wheat planting time	
Insect	Early	Late	Insect	Early	Late	Insect	Early	Late	Insect	Early	Late
Seed-corn beetle	+	–	Wireworm	+	–	Thrips	+	–	Wireworm	+	–
Seed-corn maggot	+	–	Whitegrub	+	–	Cotton aphid	+	–	White grub	+	–
Wireworm	+	–	Flea beetles	–	+	Cotton fleahopper	+	–	Grasshopper	+	–
Corn root aphid	–	+	Chinch bug	–	+	Boll weevil	–	+	Hessian fly	+	–
N. corn rootworm	–	+	Yellow sugarcane aphid	+	–	Boll worm	–	+	Oat bird-cherry aphid	+	–
W. corn rootworm	–	+	Cutworms	+	–	Tobacco budworm	–	+	Corn leaf aphid	+	
S. corn rootworm	–	+	S. corn rootworm	–	+	Pink bollworm	–	+	Green bug	+	–
Thrips	–	+	Greenbug	+	–	Cotton leafworm	–	+	Cutworms	+	–
Mite	–	+	Corn leaf aphid	–	+	Cabbage looper	–	+	Armyworm	+	–
European corn borer	+	–	Fall armyworm	–	+	Lygus bug	–	+	Lesser cornstalk borer	+	–
S.W. corn borer	–	+	Beet armyworm	–	+	Stink bug	–	+	Flea beetle	+	–
Corn earworm	–	+	Corn earworm	–	+	Spider mite	–	+	Leaf hopper	–	+
Fall armyworm	–	+	Sorghum midge	–	+	Cotton square borer	+	–	Thrips	+	–
Armyworm	+	–	Sorghum webworm	–	+	Garden webworm	–	+	Wheat curl mite	+	–
Chinch bug	–	+	S.W. corn borer	–	+	Cutworm	+	–			
Corn leaf aphid	–	+	Sugarcane borer	–	+	Leaf miner	–	+			

Note: + = Increase in population density or damage; – = reduction in population density or damage.
(Adapted from Ref. 1.)

Planting to Attain Injury-Free Periods

In this principle, the crop plantings may be adjusted in time to allow production during an injury-free period. This practice is applicable for pests whose numbers decline drastically for a period of time. This often occurs when the pest has discrete, nonoverlapping generations. Adjusting sunflower planting to escape damage by the sunflower moth, *Homoeosoma electellum* (Hulst.), provides a good example (3).

ROLE OF PLANTING TIMES IN DIFFERENT FIELD CROPS

The importance of planting times in different field crops is discussed below with various practical examples. The reaction of various pests to altered plantings is also provided in Tables 1 and 2.

Cereals

Early planting of corn in the United States would appear to be an important management tactic for corn root aphid [*Anuraphis maidiradicis* (Forbes)], northern [*Diabrotica longicornis* (Say)], western [*D. virgifera* (Leconte)], and southern corn root worm [*D. undecimpunctata howardi* (Barber)], thrips, southwestern corn borer [*Diatraea grandiosella* (Dyar)], corn earworm [*Helicoverpa zea* (Boddie)], fall armyworm [*Spodoptera frugiperda* (J.E. Smith)], chinch bug [*Blissus leucopterus* (Say)], corn leaf aphid [*Rhopalosiphum maidis* (Fitch)], mites [*Oligonychus* spp.], and different species of grasshopper (1) (Table 1).

Date of planting as a method of downy mildew control allows seedlings and young plant growth to occur during a time that is not favorable for sporangial development, formation, and infection, or during a time when the inoculum source of the crop has been harvested (4). Though large-scale planting of cereals with the first rains consi-

Table 2 Effect of crop planting times on insect abundance and/or potential damage in India

		Planting time	
Crop	Insect	Early	Late
Cereals			
Barley	*Rhopalosiphum maidis* (Fitch.)	−	+
Common, millet	*Atherigona* spp.	+	−
Corn	*Atherigona naquvi* (Steyskal)	−	+
Rice	*Cnaphalocrosis medinalis* (Guenee)	−	+
	Nilaparvata lugens (Stal.)	−	+
	Orseolia oryzae (Wood Mason)	−	+
Wheat	*Mythimna separata* (Walker)	−	+
Sorghum	*Chilo partellus* (Swinhoe)	+	−
	Contarinia sorghicola Coq.	+	−
	Atherigona soccata (Rondani)	−	+
Commercial crop			
Sugarcane	*Chilo infuscatellus* (Snell.)	−	+
Oil seeds			
Linseed	*Dasyneura lini* (Barnes)	−	+
Rape seed & mustard	*Lipaphis erysimi* (Kalt.)	−	+
Safflower	*Uroleucon carthemi* (H.R.L.)	−	+
	Uroleucon compositae (Theobald)	−	+
Soybean	*Rivula* sp.	+	−
	Obereopsis brevis (Swedenbord)	+	−
Peanut	*Holotrichia serrata* (Fab.)	−	+
Sesame	*Antigastra catalaunalis* (Duponchel)	−	+
Pulses			
Chickpea	*Helicoverpa armigera* (Hub.)	−	+
Cowpea	*Maruca testulalis* (Geyer)	−	+
	Cydia ptychora (Mayrick)	−	+
Pigeonpea	*Helicoverpa armigera* (Hub.)	−	+

Note: + = increase in population density or damage; − = reduction in population density or damage.
(From Ref. 6.)

derably reduces downy mildew incidence, delayed planting generally exposes young plants to higher amounts of inoculum (4, 5).

When fields are planted synchronously, the length of crop period in the area is reduced, thus fewer generations of insects are completed to damage the crop. In Asia and Africa early and uniform planting reduces damage by the sorghum shootfly and various grasshoppers. For many insect pests of sorghum, early planting would appear to be an effective management tactic, since population densities and damage increase with the lateness of the crop with certain exceptions (Table 1). In India, early planting has been found to reduce gall midge, leaf folder (6) and planthopper (7) damage in rice. But there are reports that if there are premonsoon rains, early planted crop suffers due to gall midge much more than that of optimum planting. This is probably due to the development of midges on alternate hosts that come up after rains. Early planting of *rabi* sorghum reduces shootfly incidence with increased grain yield (8, 9).

Medium season seeding of winter wheat is usually most favorable for any locality. However, planting earlier than the optimum seeding time is a common practice in many areas probably to take advantage of favorable soil moisture conditions. But too early seeding may lead to injury from fungus disease that develops under warm conditions. Early planting of wheat also generally increases the severity of insect pests (Table 1). Timely seeding of wheat (mid-November) in most parts of India helps to reduce the incidence of root diseases, Karnal bunt, and shootfly. The late sown crop is more liable to the attack of Karnal bunt and shootfly whereas the early sown crop suffers more from root diseases. It is also helpful in keeping weeds under check (10).

Oil seeds

Early planting of peanut at the onset of the rainy season decreases the bud necrosis incidence and losses due to the disease can be avoided. A crop planted late in rainy season is exposed to high levels of vector influx during the seedling stage and thus losses are more. It is necessary to monitor the population of thrips in groundnut to determine the optimum time of planting by avoiding the peak population influx period (11). Early planting of peanut reduces incidence of rust, leaf spots, and tikka diseases (12, 13). Leaf spots and rusts, being mainly airborne diseases, spread quickly where there is continuity of host plants over large areas. In the case of safflower, early planting reduces the damage by aphid (Table 2) while too early planting leads to heavy attack by alternaria leaf spot disease. Soybean under optimum seeding suffers less due to stemfly and leaf rollers. Similarly, adjusting seeding time in the case leaf of sunflowers in such a way that flowering takes place during December (a month that not only harbors less head borers but also there is a maximum activity of pollinators) for effective pollination of the crop.

Cotton

Timely and synchronous planting has been found to reduce bollworm damage in India. The early- and late-planted cotton suffered higher damage due to pink and spotted bollworms (14). Ideally, cotton planting should be delayed until the soil temperature reaches at least 16°C. This occurs in early March along the Gulf Coast and in Arizona, after mid-April in most of the cotton belt and in California. The late planted cotton is exposed to a large number of pests such as the boll weevil, tobacco budworm, bollworm, and pink bollworm (Table 1). However, delayed planting of cotton to cause suicidal emergence of overwintering pink bollworms and bollweevils has been shown to constitute promise as a management tactic. The damage by sucking pests is sometimes more severe in early-planted cotton as lower temperature retarded growth, resulting in an intensification of pest-induced plant damage (1).

Pulses

Early planting can be used to minimize pod borer damage to chickpeas in Northern India (Table 2). Two peaks of pod borer occur during December and March. During second peak, the pest inflicts severe damage to the chickpea crop. An early (October) planted crop escapes with the least damage. Late planting (December–January) matures during late March to April and suffers heavy damage. November planting also suffers moderate damage (15).

FUTURE CONCERNS

The monetary benefits from these practices, though not easily definable, are tremendous. The role of planting time has enabled avoidance of losses of millions of dollars annually, otherwise caused by Hessian fly in wheat, by midge in sorghum, by corn earworm in corn, and by cotton bollworm and tobacco budworm in cotton. However, since the planting dates depend on the prevailing environmental situation, geographic area, and pest under consideration, the planting dates for various crops and pests need to be identified for different areas/regions and should be incorporated into the IPM packages. This would serve as a best and long-lasting noncash input.

REFERENCES

1. Teetes, G.L. The Environmental Control of Insects Using Planting Times and Plant Spacing. In *CRC Handbook of Pest Management in Agriculture*; Pimentel, D., Ed.; CRC Press, Inc.: Boca Raton, FL, 1981; 1, 209–221.
2. Thomas, J.G. The sorghum midge and its control. Tex. Agric. Ext. Serv. Leafl. **1969**, 942.
3. Mitchell, T.L.; Ward, C.R.; Teetes, G.L.; Schaefer, C.A.; Bynum, E.D.; Brigham, R.D. Sunflower pest population levels in relation to date of planting on the Texas high plains. Southwest. Entomol. **1978**, *3*, 279.
4. Tantera, D.M. Cultural practices to decrease losses due to corn downy mildew diseases. Trop. Agric. Res. Series **1975**, *8*, 165–173.
5. Siradhana, B.S.; Dange, S.R.S.; Jain, K.L.; Rathore, R.S. Effect of sowing dates on the natural incidence of sorghum downy mildew of maize in Udaipur, Rajsthan. Indian Phytopathol. **1975**, *28*, 140–141.
6. Rathore, Y.S.; Lal, S. *Pest Management Through Cultural Practices in India*; Indian Institute of Pulses Research: Kanpur, India, 1994; 9–14.
7. Sontakke, B.K.; Naik, R.P. Managing plant hoppers for more rice. Indian Farming **1990**, *40* (2), 21–22.
8. Balikai, R.A. Effect of different dates of sowing on shootfly incidence and grain yield of sorghum. Insect Env. **1999**, *5* (2), 57–58.
9. Balikai, R.A.; Yelshetty, S.; Kullaiswamy, B.Y. Effect of dates of sowing in combination with different insecticides and bioagent on sorghum shootfly. Agric. Sci. Digest **1998**, *18* (4), 261–263.
10. Tandon, J.P. Research, Development and Management for Production of Wheat. In *Integrated Pest and Disease Management*; Upadhyay, R.K., Mukerji, K.G., Chamola, B.P., Dubey, O..P., Eds.; A.P.H. Publishing Corporation: New Delhi, India, 1998; 131–144.
11. Mayee, C.D. Enhancing Productivity of Groundnut in India Through IDM. In *Integrated Pest and Disease Management*; Upadhyay, R.K., Mukerji, K.G., Chamola, B.P., Dubey, O.P., Eds.; A.P.H. Publishing Corporation: New Delhi, India, 1998; 343–360.
12. Ghewande, M.P.; Sukla, A.K.; Pandey, R.N. Management of foliar diseases of groundnut through agronomic practices. Indian Bot. Reporter **1986**, *5*, 179–181.
13. Kodmelwar, R.W.; Ingle, A.Y. Effect of sowing dates, spacing and meteorological factors on the development of tikka and rust of groundnut. Indian Phytopathol. **1989**, *42*, 274.
14. Dhaliwal, G.S.; Arora, R. *Principles of Insect Pest Management*, 1st Ed.; National Agricultural Technology Information Centre: Ludhiana, India, 1996; 51–52.
15. Rathore, Y.S.; Nwanze, K.F. Pest Management Through Cultural Practices: Innovative Methods. In *Pests and Pest Management in India—The Changing Scenario*; Sharma, H.C., Veerabhadra, Rao M., Eds.; Plant Protection Association of India: Hyderabad, India, 1993; 216–228.

PNEUMATIC CONTROL OF AGRICULTURAL INSECT PESTS

Charles Vincent
Agriculture and Agri-Food Canada, Saint-Jean-sur-Richelieu, Quebec, Canada

INTRODUCTION

Pneumatic control essentially involves the use of an airstream to dislodge and kill insect pests. Insects can be dislodged from plants with negative (aspiration) or positive (blowing) air pressure. Once the insects have been dislodged, they are killed by a system of turbines or are collected and killed in a dedicated system upstream of the blower.

The use of vacuum is not new. For example Crosby and Leonard (1) studied the effect of a vacuum machine to control insect pests of vineyards in New York State. The advent of synthetic insecticides offered efficient and cheap control tactics to the growers and eclipsed the appeal of physical control methods. However, there has been, in the past few years, a renewed interest in pneumatic control methods, as evidenced by the numerous articles published in agricultural magazines (2–4). The quantitative value of pneumatic control is often poorly documented in these articles. Our objective is to summarize key published information on pneumatic control with special reference to two agricultural systems, that is, the tarnished plant bug on strawberries and the Colorado potato beetle on potatoes, as treated by Vincent and Boiteau (5). Readers interested in researches on pneumatic control of various insects attacking celery plantations should consult Weintraub and Horowitz (6) and those interested in engineering aspects are referred to Khlelifi et al. (7) and Lacasse et al. (8).

MANAGEMENT OF TARNISHED PLANT BUG ON STRAWBERRIES

Plant bugs (Miridae) are key pests of strawberry in North America and in commercial fields; their control relies mainly on the use of synthetic insecticides. As few natural enemies are available for the control of plant bugs, pneumatic control has been researched to allow strawberry production with a minimum of insecticides.

In California, Driscoll Associates used the Bugvac®, a tractor-mounted device to control *L. hesperus* Knight (Miridae). When passed at 5 km per hour, another vacuum device, the Ag-Vac®, removed up to 80% of western plant bugs (*Lygus hesperus* Knight) under field conditions (9). According to Grossman (10), satisfactory results were attained by treating California strawberry fields twice a week (no data given).

Pickel et al. (11) compared three types of vacuum devices operated at speeds ranging from 4 to 8 km per hour. Passes were done once or twice a week, regardless of the size of the plant bug populations. In plots treated once a week, adult and nymph populations decreased respectively by 74% and 43%. However, the decrease in damage to the fruit was not sufficient for the treatment to be commercially viable in California. The authors concluded that a higher airflow speed could increase the effectiveness of vacuum treatments.

A tractor-mounted vacuum collector (the Biovac®) was tested on a commercial strawberry farm in Quebec, Canada (12), where the tarnished plant bug (*Lygus lineolaris* Palisot de Beauvois) reduces both the quantity and quality of the harvest by feeding on the fruit (13). The authors concluded that this prototype of pneumatic machine was of limited use in a commercial situation.

To achieve a better understanding of the effects of Biovac treatments on *L. lineolaris* populations and to improve the efficiency of the machine, Vincent and Chagnon (14) set up a test bench to study the influence of relative height of the hood and forward velocity and bug behavior on removal effectiveness. Individuals (nymphs or adults) were marked on the back with fluorescent powder and 99% of them were found in less than 5 min. by illuminating the test area with an ultraviolet lamp. Optimal efficiency was achieved when the forward velocity was 4 km per hour, for airflow speeds of 30.7 m per sec and when the hood was

Encyclopedia of Pest Management

at two/three of the strawberry canopy (15). Differences in the proportion of insects vacuumed can be explained by the relative distance of the insects from the hood shortly before the machine passes; the percentage of success being inversely proportional to this distance.

Effects on Pollination and Pollinators

The Biovac disperses pollen by vacuuming it up and then ejecting it. Neither the distance of the recipient plant from the originating plant nor the time of day of treatment had a significant effect the pattern of pollen dispersal (16). Although the viability of pollen dispersed by the Biovac was half that of pollen collected directly from flowers, treatment with the Biovac increased the rate of pollination marginally.

When the Biovac was passed 19% of pollinators flew away before it arrived (17). Of those remaining on the flowers upon Biovac passage, 61% of individuals were vacuumed and the rest remained clinging to the flowers after the machine had passed.

MANAGEMENT OF COLORADO POTATO BEETLE ON POTATOES

Boiteau et al. (18) conducted field tests of a vacuum collector for Colorado potato beetles that used a combined positive/negative airflow system. The advantage of this design is that it takes less energy to dislodge an insect with a positive airstream and then aspirate it than it does to dislodge an insect by aspiration alone

The Beetle Eater®, tested by Boiteau et al. (18) and Boiteau and Misener (19), directs a positive airstream at the base of plants from above the row, shaking the plants and causing the beetles to fall off. Laboratory tests have shown that more than 90% of adults (male and female) and larvae fell off potato plant leaves when subjected to vibrations greater than 20 Hz with an amplitude greater than 0.6 mm (19). Because the combined frequency and amplitude create an acceleration force that is lower than the beetles' ability to cling to the leaves, the beetles' drop-off behavior cannot be attributed to the acceleration force produced by shaking the plants. Instead, it should be an escape behavior. Interestingly, the drop-off frequency was lower in insects holding onto leaf margins than for those on the leaf surface. More than 82% of the beetles dropped off within the first two seconds of the five-second test period. This suggests that they are reacting spontaneously to vibrations of a certain frequency and amplitude by dropping off the plant, a behavior known as thanatosis. Consequently, pneumatic machines for Colorado potato beetles can be operated at a limited range of travel speeds, particularly if their effectiveness is strongly linked to the aspiration of free-falling insects.

In the field, the Beetle Eater dislodged about 61% of adults. However, 13% of the dislodged adults fall to the ground, resulting in a real harvest of 48%. Among the first and second larval instars, 42% of the beetles are collected and 3% fall to the ground, while in third and fourth instars, 50% of the beetles are collected and 23% fall to the ground. The larvae on the ground can climb back up the plant and cause additional damage. Terminal velocities of free falling individuals for adults and fourth, third, and second instars were 9.4, 9.5, 7.3, and 5.9 m per sec respectively (20).

The force required to remove an individual from a leaf was as high as 40 mN for adult beetles, 30 mN for third and fourth instar larvae, and 10 mN for second instar larvae (20). For all life stages, the ratio of insect weight to removal force was less than one and that explains why individuals of all stages were found on the ground during the field tests after the machine passed (18). Adults are much more mobile than larvae. They can respond to stimuli such as shade, vibrations, and probably also to aspiration with a thanatosis reflex. This may explain, in part, why, in field tests, a significant proportion of adults fell to the ground despite the high removal forces produced by aspiration.

As the maximum (theoretical) efficiency of a collector is 98% for the small larval instars, 97% for adults, and 77% for the large larval instars (18), the efficiency of vacuum collection in an individual field depends on the relative abundance of the different insect life stages and the variations in weight within each stage. The effectiveness of vacuum collection systems themselves is highly variable, and varies also from one potato field to the next. Vacuum collector efficiency can probably be improved by increasing the suction force and the intensity of vibrations transmitted to the plant.

Effects on Nontarget Arthropods and on Disease Transmission

Capture rates of 62% for a number of predator species belonging to the taxa Arachnida, Chrysopidae, and Coccinellidae were reported by Boiteau et al. (18). Because native insects can play a role in controlling the Colorado potato beetle in potato fields (22), these species must be taken into account when using the collector. If an inundative release of predators is planned for a potato field, it should be coordinated with vacuum collector treatments (21). The Beetle Eater was found to remove potato aphids [*Macrosiphum euphorbiae* (Thomas)] at a rate of at least 56%, showing that the collector has potential for controlling other potato pests.

Field trials showed that the vacuum collector did not spread the PSTVd viroid or the PVX virus, even with a large number of infected plants and multiple passes (18).

CONCLUSION

In conclusion, at the present time pneumatic control cannot be used as sole control method in agricultural systems. The technology of pneumatic control of insect pests needs to be refined in order to be acceptable to farmers growing crops for extremely competitive markets. Several avenues of research should be explored further to make pneumatic control devices suitable for commercial operations, that is, field behavior of the insects, design of efficient machines, and fine-tuning of operational parameters.

ACKNOWLEDGEMENTS

Giles Boiteau (Agriculture and Agri-Food Canada, Fredericton, N.B.) commented on the manuscript.

REFERENCES

1. Crosby, C.R.; Leonard, M.D. The tarnished plant-bug. Cornell U. Bull. **1914**, 364.
2. Hillsman, K. Pest vacuums: innovative equipment sweeps up bugs. The Grower **1988**, *21* (12), 30–31.
3. Stockwin, W. Sweeping away pests with bugvac. Am. Vegetable Grower **1988**, *36* (11), 34–38.
4. Street, R.S. Is vacuum pest control for real? Agrichem. Age **1990**, *34* (2), 22–23, 26.
5. Vincent, C.; Boiteau, G. La Lutte contre les Insectes Nuisibles par Aspiration. In *La Lutte Physique en Phytoprotection*; Vincent, C., Panneton, B., Fleurat-Lessard, F., Eds.; INRA Editions: Paris, 2000; 293–306.
6. Weintraub, P.; Horowitz, A.R. La Lutte contre les Insectes Ravageurs à l'Aide d'Aspirateurs: l'Expérience Israélienne. In *La Lutte Physique en Phytoprotection*; Vincent, C., Panneton, B., Fleurat-Lessard, F., Eds.; INRA Editions: Paris, 2000; 321–332.
7. Khelifi, M.; Laguë, C.; Lacasse, B. Lutte Pneumatique contre les Insectes en Phytoprotection. In *La Lutte Physique en Phytoprotection*; Vincent, C., Panneton, B., Fleurat-Lessard, F., Eds.; INRA Editions: Paris, 2000; 283–292.
8. Lacasse, B.; Laguë, C.; Roy, P.-M.; Khelifi, M. Contrôle Pneumatique du Doryphore de la Pomme de Terre. In *La Lutte Physique en Phytoprotection*; Vincent, C., Panneton, B., Fleurat-Lessard, F., Eds.; INRA Editions: Paris, 2000; 307–320.
9. Southam, W.J. *Ag-Vac Agricultural Insect Vacuums*; Watsonville, CA, 1990; 3, (mimeo).
10. Grossman, J. Update: strawberry IPM features biological and mechanical control. IPM Practitioner **1989**, *11* (5), 1–4.
11. Pickel, C.; Zalom, F.G.; Walsh, D.B.; Welch, N.C. Efficacy of vacuum machines for *Lygus hesperus* (Hemiptera: Miridae) control on coastal California strawberries. J. Econ. Entomol. **1994**, *87* (6), 1636–1640.
12. Vincent, C.; Lachance, P. Evaluation of a tractor-propelled vacuum device for the management of tarnished plant bug populations in strawberry plantations. Environ. Entomol. **1993**, *22* (5), 1103–1107.
13. Vincent, C.; de Oliveira, D.; Bélanger, A. The Management of Insect Pollinators and Pests in Quebec Strawberry Plantations. In *Monitoring and Integrated Management of Arthropod Pests of Small Fruit Crops*; Bostanian, N.J., Wilson, L.T., Dennehy, T.J., Eds.; Intercept Ltd.: Andover, Hampshire, U.K., 1990; 177–192.
14. Chagnon, R.; Vincent, C. A test bench for vacuuming insects from plants. Can. Agric. Eng. **1996**, *38* (3), 167–172.
15. Vincent, C.; Chagnon, R. Vacuuming tarnished plant bug on strawberry: a bench study of operational parameters versus insect behavior. Entomologia Experimentalis et Applicata **2000**, *97* (3), 387–2000.
16. Chiasson, H.; de Oliveira, D.; Vincent, C. Effects of an insect vacuum device on strawberry pollination. Can. J. Plant Sci. **1995**, *75* (4), 917–921.
17. Chiasson, H.; Vincent, C.; de Oliveira, D. Effect of an insect vacuum device on strawberry pollinators. Acta Hortic. **1997**, *437*, 373–377.
18. Boiteau, G.; Misener, C.; Singh, R.P.; Bernard, G. Evaluation of a vacuum collector for insect pest control in potato. Am. Potato J. **1992**, *69* (3), 157–166.
19. Boiteau, G.; Misener, G.C. Response of colorado potato beetles on potato leaves to mechanical vibrations. Can. Agric. Eng. **1996**, *38* (3), 223–227.
20. Misener, G.C.; Boiteau, G. Suspension velocity of the colorado potato beetle in free fall. Am. Potato J. **1993a**, *70* (4), 309–316.
21. Misener, G.C.; Boiteau, G. Holding capacity of the colorado potato beetle to potato leaves and plastic surfaces. Can. Agric. Eng. **1993b**, *35* (1), 27–31.
22. Cloutier, C.; Jean, C.; Bauduin, F. In *More Biological Control for a Sustainable Potato Pest Management Strategy, Potato Insect Pest Control*, Proceedings of a Symposium, Quebec City, Canada, 31 July–1 August 1995; Duchesne, R.-M., Boiteau, G., Eds.; Quebec Ministry of Agriculture: Quebec City, 1995; 15–52.

P

POPULATION THEORY AND PEST MANAGEMENT

Alan A. Berryman
Washington State University, Pullman, Washington, U.S.A.

INTRODUCTION

Pest managers tend to be practical people. The reason is simple: their clients need immediate solutions to their pest problems, for without them their crops can be destroyed and their profits wiped out in a single growing season. Practical people tend to look for quick and effective solutions to their problems—the proverbial "silver bullet." Thus it was when DDT was discovered in 1939 and hailed as the miracle insecticide—the final answer to the insect problem (1). Today there is the tendency to look to genetic engineering or biotechnology for the silver bullet.

Academics, on the other hand, are often theoretically minded. They are mainly interested in understanding how nature works rather than in the realities of crop production. Nature is often complex, and its understanding and explanation may require complex reasoning, complex jargon and, sometimes, complex mathematical models. It is not surprising that practical pest managers and theoretical academics often find it difficult to communicate.

WHY THEORY IS IMPORTANT TO PEST MANAGERS

IPM students sometimes ask me why they should learn esoteric theory when their job is to solve practical problems. My usual answer is to draw on analogy: Engineers are also practical people but their education is burdened with mathematics, mechanics, and physics, much of it at the theoretical level. Engineers realize that they cannot build bridges and rockets without considering the laws of nature. How can one put a man on the moon without understanding the theories of gravity and planetary motion? Impossible! When engineers ignore or overlook theory, they sometimes reap the dire consequences. And so it was with the Tacoma Narrows Bridge, destroyed by harmonic motion during a violent windstorm. If the engineers had paid more attention to the theories of complex dynamics and harmonics, they might have avoided this disaster.

When DDT was first employed to control insects, some academics (those who understood the theory of evolution and took it seriously) predicted that insects would soon become resistant to the new pesticide. Although most pest managers now understand the theory of evolution and use it in their everyday operation (e.g., resistance management), they generally ignored or ridiculed the pessimistic academics when DDT first came on the scene. They had their silver bullet and it worked—for a time.

Theory provides us with an understanding of nature, and this understanding enables us to anticipate the consequences of our actions. Theory, therefore, is an essential part of any thoughtful and responsible human activity, including pest management.

WHY POPULATION THEORY?

IPM is concerned with the management of pest *populations*. It seems reasonable, therefore, to expect pest managers to understand what causes populations to change in time and space—what is generally known as *population dynamics* (2). This is not to say that other theoretical knowledge is not important. We have already seen that an understanding of evolutionary theory is essential if we are to anticipate and manage pesticide resistance. I could also argue that a theoretical understanding of community ecology, food webs, and predator–prey dynamics is also necessary. What I am saying is that pest managers should, at the very least, have a good understanding of population dynamics theory.

Population theory concerns itself with the general explanation and understanding of changes in numbers, density, age-classes, and so on in populations of living organisms. However, because it is the density of pests per unit area of crop or cropland that determines the level of economic damage, pest managers are usually concerned more with changes in population density than with other population variables.

A theory can be viewed as a systematic statement of the principles and relationships underlying an observed natural phenomenon. Population dynamics theory involves two sets of principles and relationships, some pertaining to dynamic systems in general and others to populations of living organisms in particular.

The first general principle recognizes that changes in any dynamic system can be caused by either outside (*exogenous*) or internal (*endogenous*) effects. Exogenous effects cause changes in the system but are themselves unaffected by those changes. For example, weather can affect insect population dynamics, say, by killing insects, but is not influenced itself by insect numbers. In ecological parlay, exogenous effects are called *density-independent* because they act independently of pest density.

On the other hand, endogenous effects are caused by feedback loops within the dynamic system (notice that a dynamic system is thus defined as a group of variables linked together by feedback loops) (2). For example, suppose that a pest population increases for some reason or another. This provides more food for predators, causing their populations to increase. More predators eat more prey, and so the pest population is eventually suppressed. Notice that an original *increase* in pest density is followed by an eventual *decrease* in density through the action of predators. This is called a *negative* feedback loop because the initial change results in an opposing effect (increase in some quantity results in an eventual decrease in that quantity, or vice versa). Notice that the negative feedback action causes the density of the pest population to return towards its original level and, therefore, opposes changes in the characteristic state of the system. In other words, negative feedback loops tend to stabilize dynamic systems around what are called *equilibrium points* (Fig. 1A). Engineers would say that negative feedback is necessary to stabilize the dynamics of rockets and airplanes, or to keep them on track (the track being the equilibrium point). However, although negative feedback is necessary for stable dynamics, it is not sufficient. To be stable, negative feedback must act *rapidly*, otherwise the variables may oscillate around their equilibrium points, with the degree of oscillation being directly dependent on the length of the feedback time-delay (Fig. 1B). The speed of action of a negative feedback loop depends, to a large degree, on the number of variables involved in the loop because each variable needs time to change, quantitatively, in response to the other; e.g., predators require time to produce more offspring after being confronted with more prey.

Endogenous effects can also be created by *positive feedback*. However, unlike the stabilizing effects of negative feedback, positive feedback induces instability by accentuating or amplifying changes in the system. Positive feedback is the force behind population explosions,

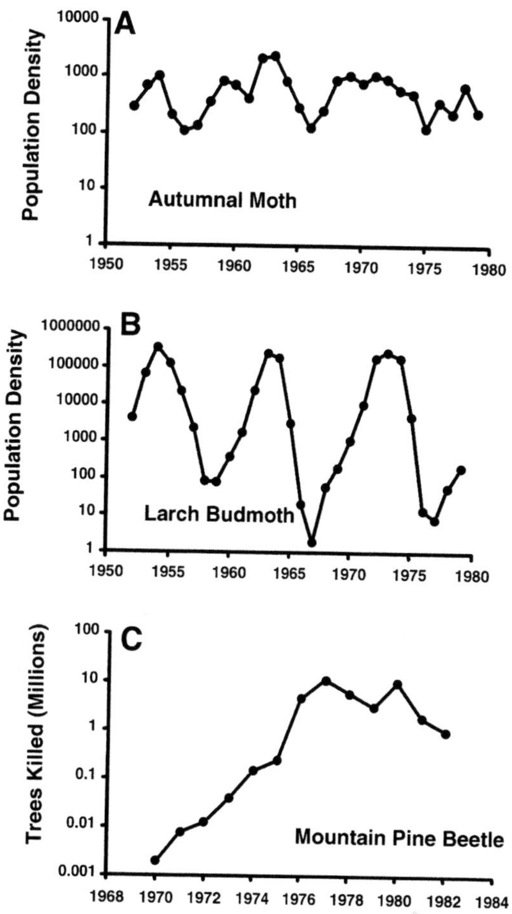

Fig. 1 Three major types of population dynamics: (**A**) Relatively stable population of the autumnal moth in the Swiss Alps. Notice that numbers fluctuate around a characteristic, or average, level of abundance. (**B**) Strongly oscillatory (cyclic) population of the larch budmoth in the Swiss Alps. Notice that densities oscillate by several orders of magnitude in a regular 9-year cycle. (**C**) Unstable outbreak of the mountain pine beetle in Glacier National Park. Notice that annual tree killing increased from a few thousand in 1970 to more than 10 million in 1977 as the beetle population erupted through the park.

inflation spirals, arms races and, interestingly, the process of evolution. Positive feedback can also create unstable breakpoints or thresholds in dynamic systems and are important in the theory of insect outbreaks (Fig. 1C) (3).

In ecology, endogenous feedback loops are usually referred to as *density-dependent* because the feedback is induced by, and effects, population density. Negative feedback loops are then identified as *direct* density-dependent (or just plain density-dependent) and positive feedback *inverse* density-dependent. *Delayed* density-dependent refers to negative feedback loops that operate slowly and, thereby, create the conditions for oscillatory dynamics. Direct density-dependence usually results from competition between individuals for food, hiding places, or other resources (2). It can also result from the feeding of general predators that switch to the most abundant prey species and/or aggregate in areas where prey are abundant. Delayed density-dependence often arises from feedback between a species and its food supply or natural enemies. Finally, inverse density-dependence usually results from cooperation between individuals during defense or hunting activities.

These general principles and relationships outlined above are important to the pest manager. The overall goal of IPM is to maintain pest populations at low, nondamaging, densities. In general systems parlay, this means creating a stable *and* low equilibrium. In order to do this we must either create our own negative feedback loop or take advantage of natural negative feedback loops. For example, the development of an economic action level, a pest density above which the grower initiates control actions, creates a negative feedback loop because the decision to control depends on pest density (direct density-dependent). Biological control, of course, is the use of natural feedback loops, as between the pest and its enemies, to control the pest population at subeconomic levels. A thorough understanding of this, however, may require one to delve into what is called predator-prey theory (4, 5). In contrast, the pest manager usually tries to avoid instabilities, such as pest population explosions (outbreaks) created by positive feedback. The theory of outbreaks (3) is yet another theoretical area important to the pest manager. Of course, there is much more I could say about the theory of population dynamics, but space is limited. Nevertheless, I hope to have convinced the reader that theoretical knowledge is important for the intelligent management of pest problems, and that population theory is perhaps of prime importance. Those who wish to dig deeper can consult the following references.

REFERENCES

1. Berryman, A.A. Population theory: an essential ingredient in pest prediction, management, and policy-making. Am. Entomol. **1991**, *37*, 138–142.
2. Berryman, A.A. *Principles of Population Dynamics and Their Application*; Stanley Thornes: Cheltenham, UK, 1999.
3. Barbosa, P.; Schultz, J.C. *Insect Outbreaks*; Academic Press: New York, 1987; 202–204.
4. Huffaker, C.B.; Gutierrez, A.P. *Ecological Entomology*, 2nd Ed.; John Wiley: New York, 1999.
5. Hawkins, B.A.; Cornell, H.V. *Theoretical Approaches to Biological Control*; Cambridge University Press: Cambridge, UK, 1999.

POSTHARVEST CROP LOSSES (INSECTS AND MITES)

P

Dongfeng Cao
David Pimentel
Kelsey Hart
Cornell University, Ithaca, New York, U.S.A.

INTRODUCTION

Postharvest crop losses have been reduced significantly in the past 50 years, from 49% in 1949 to 25% in 1998. These changes followed several international conventions on the subject of Postharvest Food Losses, since the 1974 World Food and the 7th Special Session of the United Nations General in 1975. The discussions generated at the conferences led to the development of grain storage facilities. However, the environmental and ecological problems and global food shortage associated with the rapidly growing world population—that has doubled during the past 40 years to 6 billion—has directed attention once again toward the potential seriousness of postharvest food losses. Annually, about 25% of total food products (about 1225–2300 million tons, including 600–800 million tons of cereal products, 250–500 million tons of root and tuber crops, and 375–1000 millions tons of fruits and vegetables) are lost worldwide during the postharvest period. Approximately 20% (5%, or about 500 million tons, of the total food product) of these losses are due to destruction by insects and mites in storage, despite the widespread use of chemical insecticides on these crops. The economic costs of these losses total about $25.8 billion worldwide each year. (Table 1).

WORLDWIDE IMPACT

Insects and mites are one of the major causes of postharvest crop losses during storage. In Pakistan, for example, insects cause about 40–70% of the total postharvest crop loss in food grains despite the employment of chemical insecticides and protection methods. Without the employment of these practices, pest insects (including mites) would damage from 50% to 90% of postharvest crops.

Exact figures for the postharvest crop losses caused by insects and mites are difficult to estimate, as these losses can range from 0% to 90%, depending on the type of food stored, the pest or pests involved, and the location and type of the storage system. The total global postharvest crop losses caused by pest insects and mites are conservatively estimated at about 5% of all harvested crops worldwide.

In developed countries, the loss is lower than this global average level of 5% for pest insects and mites (Table 1). For example, the postharvest food losses caused by pest insects and mites in the United States total approximately 3% of harvested foods. In most developing counties, however, the postharvest losses caused by pest insects and mites range from 4% to 8% of the total harvested crops. This figure can be much higher in specific regions, depending again on the type of insect pests involved, the specific foods being stored, and the storage systems and length of time the crops are stored. In India, the average loss to insect pests during storage in two-year-old wheat was 8.9%, with losses in individual cases reaching as high as 15%.

In addition to losses in the amount/weight of stored crops, insect and mite pest infestations of stored food diminishes food quality. Postharvest pest infestation results in contamination of the stored food with impurities such as droppings, cocoons and insect fragments, reduced nutritional value and seed germination in stored products, and increased microorganism contamination caused by an increase of temperature and moisture in the storage system.

At present, more than 3 billion people worldwide—greater than half the total world population—are malnourished. As the population continues to increase and global food shortages continues to grow, it is vital to limit food losses by preventing and controlling postharvest crop losses to insects and mites. This need is especially pressing in developing countries, where food storage and postproduction systems are limited and food shortages can be significant due to rapid population growth rates and limited resources.

Table 1 Postharvest crop loss caused by insects and mites

Country/Region	Total losses due to all pests (%)	Source	Losses due to insects & mites (%)
World	25	FAO, 1998	5
United States	5–15	Pimentel, 1998	1–3
Developing countries	20–40	He, 1996	4–8
Tropics	30	Milner, 1978	6
India	30	Pariser, 1984	6

DEVELOPMENT OF STORAGE PEST INSECT CONTROL

Despite the sometimes significant postharvest crop losses to pests, food storage systems are typically designed to limit pest losses. Several different methods including physical, chemical, biological, and biotechnological measures are employed for pest control in stored food products (Table 2). An examination of postharvest crop protection in the past 30 years shows that many agencies have managed to reduce food losses caused by insects and mites during storage by approximately 5%, primarily with careful design of storage facilities and systems. Generally, these advanced storage facilities can reduce the losses common to traditional storage systems by half by implementing modern storage technologies. These modern methods include cold, controlled-atmosphere, airtight, and forced ventilation storage. These storage systems make the environment unsuitable for insects and other pests. In addition to the above methods, 40 countries have registered to use short-wave radioactive techniques to combat food-grain insect pest infestations.

However, while the above food storage methods have been effective in reducing food storage losses to pests, the development and implementation of these methods require significant financial investments. Most developing countries lack the funds to invest in these advanced control technologies. In some developing countries, about 80% of food grain is stored at the farm level; farmers still primarily use traditional methods to protect stored products from pest infestations. Traditional methods include storage facilities that are not 100% pest proof. Many farmers in developing countries are turning toward increased use of insecticides and fumigants, which are popular because they are convenient and fairly inexpensive. Adverse health and environmental consequences can result if traditional farm storage sites are not sealed properly when insecticides and/or fumigants are used.

Though fumigation and insecticide applications are popular methods of controlling insect and mite pests in postharvest crops throughout the world, nonchemical measures are growing increasingly important because of the adverse effects of the addition of insecticides to foods. These negative consequences can include human poisonings and the development of insect resistance to insecticide. Integrated pest management programs aim to use nonchemical pest control methods, and only as a last resort use small amounts pesticides. Some natural insecti-

Table 2 Measures of storage pest control

1. **Preparation of stores and hygiene measures**: Choose cool, dry, and clean storage sites a convenient distance from field; design new storage facilities to minimize the introduction or entrance of insects and mites from the outside by eliminating any cracks or holes, minimizing the number of corners, and allowing for adequate ventilation; clean and sterilize old stores with lime chalk, and safe pesticides prior to reusing; and clean and treat all change tools for transportation to equipment for loading and unloading of food products.
2. **Preparation of postharvest crops for storage**: Choose crop varieties that are resistant to insect and mite infestation; harvest on time to avoid preventable post infections; keep clean, dry and cool; process, package, sterilize and disinfest necessarily; apply safe insecticides to control insects and mites and store in the proper system for the specific crop.
3. **Control of storage insects and mites during storage**: Use safe insecticides as necessary by spraying, dusting, and/or fumigating; utilize physical measures such as airtight storage, vacuum packaging and storage, cold storage, controlled atmosphere storage, dry storage, high temperature storage radiation, inert gases, inert dusts, physical barriers (oil and wax), physical controls implement biological pest control methods by introducing natural predators, parasitoids, pathogenic agents, and biological insecticides like Bt.; use biotechnical measures of pest control such as baiting, pheromones, attractants, repellents, growth regulation, resistant varieties, and safe insecticides from plant products.

P

cides, such as neem (*Azadrichita indica*), can be used to control stored pest insects and mites.

FUTURE CONCERNS

The development of safe and effective nonchemical pest-controls in food storage systems is easier than the development of those technologies for use in preharvest pest control. Storage systems are usually simple and closed, and are therefore easy to manipulate. Given the current increasing land and energy resource shortages and environmental and ecological problems associated with pesticides, sound nonchemical means of reducing post-harvest losses to pests should receive priority for implementation. The grain and other food products protected by sound postharvest pest control can help reduce the world food shortages.

BIBLIOGRAPHY

Adams, J.M. A review of the literature concerning losses in stored cereals and pulses published since 1964. Tropical Sci. **1977**, *19* (1), 1–28.

Baloch, U.K. *Integrated Pest Management in Foodgrains*, Food & Agriculture Organization of the United Nations and Pakistan Agricultural Research Council, Islamabad, Pakistan, 1992; 8–43.

Bourne, M.C. The world problem of post-harvest food losses. Nutrition Hlth. **1982**, *1* (1), 24–28.

FAO. http://www.fao.org/inpho/vlibrary (accessed Aug 1999).

Greer, T.V. Decreasing post-harvest food loss in the third world: the storage factor. J. Int. Food Agribusiness Marketing **1990**, *2* (2), 7–21.

He, H.; He, Y. Estimation of grain post-production operating methods. Agric. Mechanization in Asia, Africa and Latin America **1996**, *27* (1), 54–58.

Kantor, L.S.; Lipton, K.; Manchester, A.; Oliveira, V. Estimating and addressing america's food losses. Food Rev. **1997**, *20* (1), 2–26.

Milner, M.; Scrimshaw, N.S.; Parpia, H.A.B. Post-Harvest Losses: A Priority of the V.N. University. In *World Food, Pest Losses, and the Environment*; Pimentel, D., Ed.; An AAAS Selected Symposium, American Association for the Advancement of Science: Washington, DC, 1976; 13, 105–196.

Okezie, B.O. World Food Security—The role of post-harvest technology. Food Technology **1998**, *52* (1), 64–69.

Wright, M.A.P. Loss assessment methods for durable stored products in the tropics: appropriateness and outstanding needs. Tropical Sci. **1995**, *35*, 171–85.

POSTHARVEST FOOD LOSSES (VERTEBRATES)

Dongfeng Cao
David Pimentel
Kelsey Hart
Cornell University, Ithaca, New York, U.S.A.

INTRODUCTION

Several vertebrates, such as rodents and birds, are important pests of postharvest crops, causing important losses of stored food worldwide. Vertebrate pests cause 5% (approximately 500 million tons) of postharvest food losses. In India, for example, an estimated storage loss of food grain totaled about 9%. These losses were mostly due to rodents. Conservative estimates suggest that each year vertebrate pests cause $25–30 billion of storage food losses worldwide, despite the use of some effective pest controls measures.

POSTHARVEST LOSSES CAUSED BY VERTEBRATE PESTS

The estimated 5% loss of stored food products to vertebrate pests (mostly rodents) is conservative. Postharvest food losses due to vertebrate pests are difficult to estimate because the damage assessments vary widely. Losses range from 1 to 90%, depending on the location, storage facility, food, and pests investigated.

Storage facility construction and management are important for vertebrate pest control. In developed countries, the losses caused by vertebrate pests are typically lower than the global average of 5%, due to their well-designed food storage and processing systems. In developing countries, however, losses to vertebrate pests are much higher (Table 1). For instance, it has been estimated that the overall losses of grain to rodents in India are about 25% in the field before harvest, and 25–30% postharvest. Storage losses to rodents alone cost at least $5 billion annually in stored food and seed grain in India.

According to some reports of the Food and Agriculture Organization (FAO) of the United Nations (1999), rodents eat an amount of food equivalent to 7% (rats) to 15% (mice) of their body weight daily, consuming about 6.5 kg (rats), 1.5 kg (mice) of grain individually each year. In the United States, there are more than 1.25 billion rats; India alone has 2.5 billion rats that cause from $10 to $15 per rat in damages each year. When additional losses to mice, birds, rabbits, and other animals are considered, the significance of postharvest crop losses caused by vertebrate pests is enormous.

In addition, the actual losses to vertebrate pests are even higher than simply the cost of the stored products eaten by the animals, since vertebrate pests also contaminate uneaten food with urine, feces, hair, and pathogenic agents. Rodents can transfer about 50 diseases to humans, including typhoid, paratyphoid, trichinosis, scabies, plague, and hemorrhagic fevers like ebola. Birds can also spread several diseases. Rock pigeons, for examples, are serious pests because they can spread histoplasmosis, parrot fever, ornithosis, and encephalitis. The costs resulting from the spread of disease are not normally considered when assessments of storage losses to vertebrate pests are made, but should be taken into account. Finally, because rodents prefer to feed on protein- and vitamin-rich food—like plant embryos—they can cause significant losses in nutrition and germination ability of the seeds in storage.

CONTROL OF POSTHARVEST VERTEBRATE PESTS

The control of postharvest vertebrate pests is specific because of the unique characteristics of these pests. Vertebrate pests are easily attracted to grain and other stored produce. A few simple but important ways to keep these pests from infesting stores are cleaning up spilled grain in the surroundings of the stores, and keeping storage areas tight to prevent rodents from entering them.

It is difficult to reduce vertebrate pest populations effectively because these animals can typically find their way into warehouses and similar storage facilities. Rodent pests, such as the black rat (*Rattus rattus*), the Norway rat (*Rattus norvegicus*), and the house mouse (*Mus musculus*), are widespread worldwide.

Many traditional vertebrate-pest control measures are still used today (Table 2). Some of these controls include traps and poison baits. Overall, tight storage areas are necessary to prevent the invasions of vertebrate pests. Developing countries and farmers worldwide suffer heavy losses from vertebrate pests.

Rodenticides, while effective at controlling pests, pose risks to humans and livestock. Rodents have evolved resistance to many common rodenticides, resulting in the increased use of increasingly toxic rodenticides. Due to these problems, effective nonchemical vertebrate pest control measures need to be investigated and developed. Physical control, biological control, and biotechnological control could all help protect postharvest food resources from vertebrate pest losses. Integrated pest management in food storage and processing systems offer safe control measures by comprehensively considering storage pests, postharvest safety, and potential environmental problems.

Postharvest vertebrate pest control should be more effective and economical than vertebrate pest control in the field because tight storage systems make it easier to exclude vertebrate pests.

Table 2 Postharvest vertebrate pest control measures

1. Preventive measures to keep vertebrate pests out of storage: Choose dry, tight, storage facilities. The tight storage units require solid doors, ventilation, and construction.
2. Vertebrate control measures to keep product damage down to a minimum: Control measures include mechanical traps and chemical control measures such as acute poisons (zinc phosphide) and chronic poisons (anticoagulants).
3. Integrated vertebrate pest control: Use available preventive and control measures that consider environmental, ecological, and economic factors to effectively control vertebrate pests.

FUTURE CONCERNS

With more than 3 billion people (half of the world's population) currently malnourished worldwide, and with preharvest and postharvest food losses amounting to more than 50% of the total potential world food, there is a critical need to reduce food losses to pests including vertebrate pests like rodents. Estimates suggest that more than 5% of the world's postharvest food is lost to vertebrate pests, primarily rodents. Working to prevent these food losses is a step toward curtailing the global food security problem.

Table 1 Postharvest food losses caused by rodents

Country	Losses (%)	Country	Losses (%)
Brazil	4–8	Nepal	3–5
Bangladesh	2–5	New Hebrides	10
China	5–20	Nigeria	3–5
Egypt	0.5–1	Pakistan	5–10
Ghana	2–3	Philippine	2–5
India	5–15	Qatar	1
Iran	0.2	Sierra Leone	2–3
Iraq	3	Solomon Islands	5
Israel	5	Syria	5–25
Laos	5–10	Thailand	5
Republic of Korea	20	Turkey	5–15
Tunisia	6–8	United Arab Emirates	5–10

Sources: Buckie and Smith 1994; Prakash and Mathur, 1988.

BIBLIOGRAPHY

Bell, A. Integrated Rodent Management in Post-harvest Systems. http://www.fao.org/inpho/vlibrary/gtzhtml/x0280e/x0280e00.htm (accessed Oct 1999).

Baloch, U.K. *Integrated Pest Management in Foodgrains*; Food and Agriculture Organization of the United Nations and Pakistan Agricultural Research Council: Islamabad, Pakistan, 1992; 8–43.

Bourne, M.C. The world problem of post-harvest food losses. Nutr. Health **1982**, *1* (1), 24–28.

FAO. Manual of the Prevention of Post-harvest Grain Losses-Rodent Pests http://www.fao.org/inpho/vlibrary/gtzhml/x0065e/x0065E0j.htm (accessed Oct 1999).

Hone, J. *Analysis of Vertebrate Pest Control*; Cambridge University Press: Cambridge, UK, 1994.

Prakash, I. *Rodent Pest Management*; CRC Press: Boca Raton, FL, 1988; 459–464.

Prakash, I.; Mathur, R.P. *Management of Rodent Pests*; Kapoor Art Press: New Delhi, India, 1987; 20–23.

Richards, C.G.J.; Ku, T.Y. *Control of Mammal Pests*; Taylor & Francis: Basingstoke, 1987; 293–312.

Wright, M.A.P. Loss assessment methods for durable stored products in the tropics: appropriateness and outstanding needs. Tropical Sci. **1995**, *35*, 171–185.

POSTHARVEST INSECT CONTROL WITH INERT DUSTS

Paul Fields
Cereal Research Centre, Agriculture and Agri-Food Canada, Winnipeg, Manitoba, Canada

Zlatko Korunic
Diatom Research and Consulting, Guelph, Ontario, Canada

INTRODUCTION

Inert dusts such as clay, sand, rock phosphate, ashes, diatomaceous earth, and synthetic silica have been used as insecticides for thousands of years by aboriginal peoples in North America and Africa, and are also used in modern grain storage facilities (1). Modern-day research on inert dusts as a stored-grain protectant began in the 1920s (1–4). Some inert dusts work by damaging the insect cuticle causing death by desiccation. Other dusts such as clay, sand, and ashes are used at higher doses (above 10 wt%), and they work by providing a physical barrier against insects (3). The main advantage of inert dusts is their low mammalian toxicity. In Canada and the United States, diatomaceous earth is registered as an animal-feed additive and silicon dioxide is registered as a human-food additive (1). Also, inert dusts are effective for long durations and they do not affect end-use quality (2). Their main disadvantages are that they are dusty to apply, do not work at high relative humidities, and impede the flow of grain (5, 6).

TYPES OF INERT DUSTS

Mineral Dusts

Rock phosphate, ground sulphur, lime (calcium hydroxide), limestone (calcium carbonate), and salt (sodium chloride) have shown some activity against stored-product insects (3).

Earths and Ashes

Powdered clay, sand, and earth have been traditionally used as a control measure by applying a thick layer of dust on the top surface of a grain bulk. These are less effective than diatomaceous earth and synthetic silica, and like the mineral dusts need to be used at much higher doses.

Diatomaceous Earth

Diatomaceous earth is made up of the fossilized skeletal remains of diatoms (Fig. 1), single-celled algae that are found in fresh and salt water. Diatoms are microscopic and have a fine skeleton made up of amorphous silica ($SiO_2 + nH_2O$). The accumulation of diatom skeletons over thousands of years produces the soft sedimentary rock, diatomaceous earth. The diatomaceous earth deposits currently mined are millions of years old, and some deposits are hundreds of meters thick. The major constituent of diatomaceous earth is amorphous silicon dioxide (SiO_2) with minor amounts of other minerals (aluminum, iron oxide, calcium hydroxide, magnesium, and sodium). The insecticidal activity can vary by 20 times, depending on the geological origin of the diatomaceous earth (3, 4). Effective diatomaceous earths have SiO_2 content above 80%, a pH below 8.5, and a tapped density of below 300 gL^{-1} (5).

Synthetic Silica

Synthetic silica are manufactured by various methods, and all have the common formula SiO_2. The different types of silica have different specific surface area, particle size, drying loss, ignition loss, and structure, which may effect their insecticidal activity. They are very light powders, are the most effective of all inert dusts, and have an acute rat LD_{50} of 3160 $mgkg^{-1}$ (1).

FACTORS AFFECTING EFFICACY

For the most part, inert dusts are effective because they damage insect cuticle and insects die from desiccation.

Encyclopedia of Pest Management

P

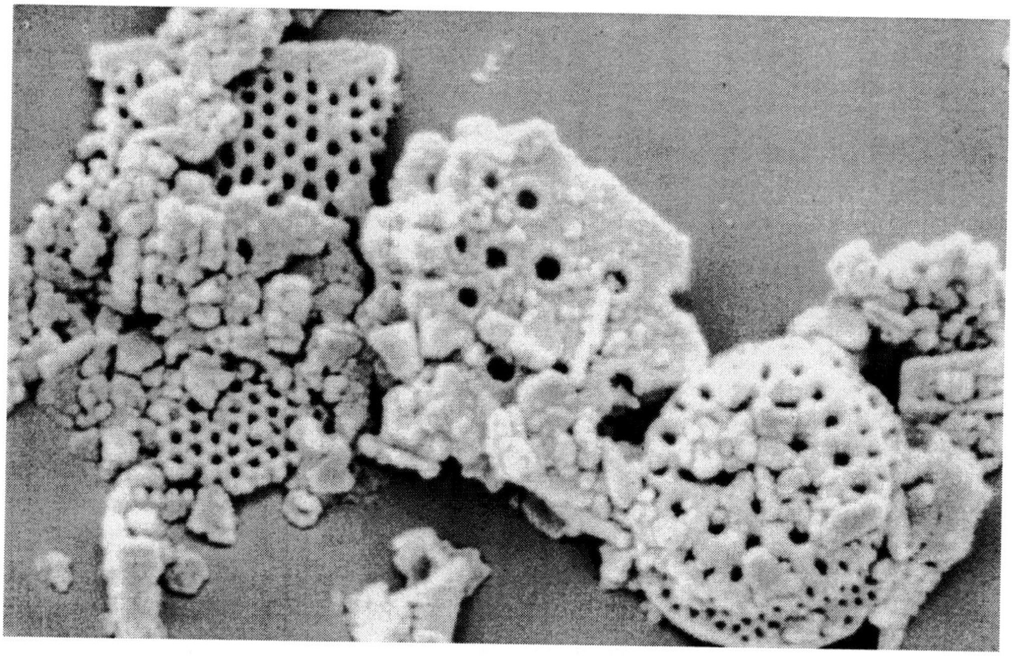

Fig. 1 Scanning electron micrograph of a saltwater diatomaceous earth.

As insects move through the grain or across a treated surface, they pick up dust particles on their cuticles. The silicon-dioxide-based inert dusts are thought to absorb the cuticular waxes. Abrasion of the cuticle is thought to be less important as a mode of action for the silica-based inert dusts, but may be the main mode of action for the other inert dusts. Insects die when they have lost approximately 60% of their water or about 30% of their body weight (1). Hence the lower the relative humidity, the higher the efficacy of the inert dusts. For example,

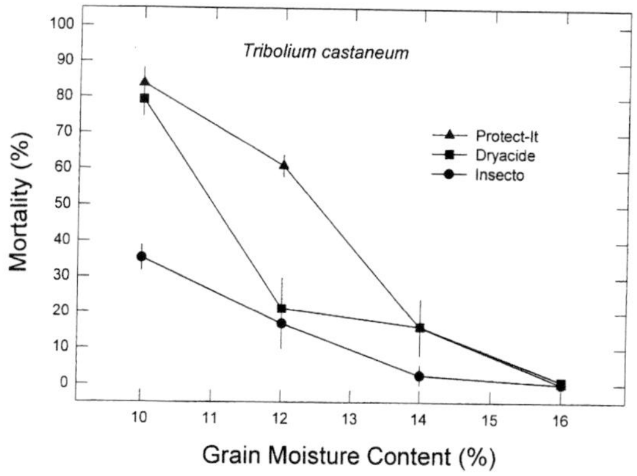

Fig. 2 The mortality of *Tribolium castaneum* held on wheat for 14 days at 30°C treated with three diatomaceous earths at 300 ppm. (Modified from Ref. 7.)

Tribolium castaneum has 35–85% mortality at 300 ppm at 10% moisture content in wheat, but there is no mortality at 16% moisture content (7) (Fig. 2).

Not all insects have the same sensitivity to inert dusts (Table 1). The type of grain also affects efficacy, in de-

Table 1 The lethal dose for 50 and 90% of the population (LD_{50} and LD_{90}) of various stored-product insects in wheat at 10% moisture content treated with Protect-It® diatomaceous earth after seven days

Insect	LD_{50} (ppm) (90% confidence intervals)	LD_{90} (ppm) (90% confidence intervals)
Cryptolestes ferrugineus	52 (45–61)	96 (89–112)
Oryzaephilus surinamensis	50 (24–72)	158 (136–185)
Sitophilus granarius	204 (188–221)	373 (347–406)
Sitophilus oryzae	260 (242–281)	436 (393–496)
Tribolium castaneum	325 (303–347)	421 (397–449)
Rhyzopertha dominica	340 (313–367)	596 (560–637)

(From Korunic and Fields, unpublished data.)

creasing order for the amount of diatomaceous earth needed to obtain control: milled rice < sunflower < maize < paddy rice < oats < barley < wheat < durum (2).

METHODS OF APPLICATION

Diatomaceous earth is registered (as of spring 2000) as a grain protectant or for treating grain storage structures in Australia, Canada, China, Croatia, Denmark, Germany, Indonesia, Japan, Saudi Arabia, and the United States. In Australia, diatomaceous earth is used principally as a treatment for empty silos, and the entire grain mass cannot be treated for grain destined for the bulk handling companies because of the adverse effects on the handling properties of the grain treated with diatomaceous earth. In most other countries, there are no restrictions on the destinations of diatomaceous-earth-treated grain. In India, during the 1960s, 70% of the grain was treated with activated kaolin clay. Egypt used rock phosphate and sulphur (1), and in the Philippines and Honduras lime is still used (3).

For grain treatment, the diatomaceous earth is applied as a dust as the grain is augered into the granary. In the United States, often only the grain surface or the top 10–20% of the grain is treated. This reduces the amount of diatomaceous-earth needed to protect the grain mass, hence limiting the negative effects of diatomaceous earth treatment. One solution to the airborne dust problem is to apply the diatomaceous earth in a water suspension. This method of application is widely used in Australia to treat empty structures, but slurry application slightly reduces efficacy. On the farm, it is possible to treat empty granaries by blowing the diatomaceous earth in through the aeration ducts.

Another way to reduce the problems associated with diatomaceous earth is to lower the concentration needed to achieve control. A mixture of diatomaceous earth (90%) and silica gel (10%) doubles the efficacy compared to diatomaceous earth used alone (2).

Diatomaceous earth can be combined with other treatments to increase efficacy. In Australia, diatomaceous earth is used as a top dressing in conjunction with long duration, low-dose phosphine fumigation (SiroFlo®). The major limitation of SiroFlo® used alone is that phosphine concentrations are too low to obtain control at the surface of the grain bulk. Diatomaceous earth serves a dual function in providing a physical barrier to retain the phosphine and controlling insects directly. In a similar role, diatomaceous earth is used as a top dressing in conjunction with cooling the grain mass by ambient air or refrigerated aeration. Diatomaceous earth can also be used in conjunction with heat to provide a more complete control of stored-product insects in food-processing facilities (2).

LIMITATIONS

Despite the numerous advantages of inert dusts, their use to control stored-product insects remains limited. The main problem with the use of diatomaceous earth is that it reduces grain bulk density and grain fluidity (2, 6). Adding diatomaceous earth at 50 ppm reduces grain bulk density by approximately 2 $kghL^{-1}$ for wheat (5). As bulk density is a grading factor, the addition of diatomaceous earth can reduce the grain grade. Another limitation is that grain must be dry for the diatomaceous earth to cause enough desiccation to control insect populations. Application of inert dusts can be undesirable because of the dust generated. Diatomaceous earth can be used as a mild abrasive, and there is concern over increased wear on grain handling machinery. However, tests need to be conducted to determine if diatomaceous earth does increase the wear on grain handling and milling equipment.

Depending on the source and processing, diatomaceous earth can contain 0.1 to 60% crystalline silica. The diatomaceous earths registered as insecticides generally have less than 7% crystalline silica. Crystalline silica has been shown to be carcinogenic if inhaled (8). However, the use of proper dust masks and diatomaceous earth with low crystalline silica can effectively protect against this health risk.

FUTURE OF INERT DUSTS

In the last decade there has been an increase in the use of diatomaceous earth because its low mammalian toxicity responds to the concerns of worker's safety, food residues, and resistant insect populations associated with the use of chemical insecticides. If the newer formulations can respond to the limitations of diatomaceous earth, there will be an even wider adoption of diatomaceous earth to control stored-product insect pests. We expect that to address the respiratory health concerns associated with crystalline silica, these new formulations will have less than 1% crystalline silica and only a minor fraction of their particles in the respirable range. The diatomaceous earths will come from deposits that have been rigorously tested to ensure high efficacy and may be combined with additives to enhance activity.

Resistance to residual insecticides has been one of the reasons to search for alternatives to chemical insecticides. Laboratory experiments have shown that several stored-product pests can have up to a twofold reduction in susceptibility when exposed to diatomaceous earth for five to seven generations (4). Although there are no reported cases of insects developing resistance to diatomaceous-earth in commercial stores, these results suggest that it will be necessary to use resistance management strategies to prevent widespread resistance to diatomaceous earth products.

REFERENCES

1. Ebeling, W. Sorptive dusts for pest control. Ann. Rev. Entomol. **1971**, *16*, 123–58.
2. Fields, P.G.; Korunic, Z.; Fleurat-Lessard, F. Control of Insects in Postharvest: Inert Dusts and Mechancial Means. In *Physical Control Methods in Plant Protection*; Vincent, C., Panneton, B., Fleurat-Lessard, F., Eds.; Springer-Verlag: Berlin, *in press*.
3. Golob, P. Current status and future perspectives for inert dusts for control of stored product insects. J. Stored Prod. Res. **1997**, *33*, 69–79.
4. Korunic, Z. Diatomaceous earths, a group of natural insecticides. J. Stored Prod. Res. **1998**, *34*, 87–97.
5. Korunic, Z. Rapid assessment of the insecticidal value of diatomaceous earths without conducting bioassays. J. Stored Prod. Res. **1997**, *33*, 219–229.
6. Fields, P.G. In *Diatomaceous Earth: Advantages and Limitations*, Proceedings of the 7th International Working Conference on Stored Products Protection; Jin, Z., Liang, Q., Liang, Y., Tan, X., Guan, L., Eds.; Sichuan Publishing House of Science & Technology: Chengdu, 1999; 1, 781–784.
7. Fields, P.G.; Korunic, Z. The effect of grain moisture content and temperature on the efficacy of diatomaceous earths from different geographical locations against stored-product beetles. J. Stored. Prod. Res. **2000**, *36*, 1–13.
8. International Agency for Research on Cancer (IARC). *Silica, Some Silicates, Coal Dust and Para-aramid Fibrils*; IARC Working Group on the Evaluation of Carcinogenic Risks to Humans: Lyon, 1997; 68, 506.

POSTHARVEST PEST LOSSES

Somiahnadar Rajendran
Central Food Technological Research Institute, Mysore, India

INTRODUCTION

Postharvest pest losses, the measurable quantitative and qualitative losses caused by the attack of biological agents such as insects, mites, rodents, and microbes on stored products, have been estimated at 10% world over (1). More than 600 species of beetles and 70 of moths among insects, 355 species of mites, 40 of rodent, and 150 of fungi have been reported to be associated with stored products. The pests consume the food commodities and contaminate them with their filth comprising fragments, rodent hairs, pellets, and toxins, thereby affecting the market value of the produce (Figs. 1 and 2). Substantial expenditure is involved in controlling the depredators using pesticides and by other means. Prevention of losses is a challenge for the technologists, food industries, and governments around the world.

WORLDWIDE LOSSES

Worldwide average foodgrains losses have been put at 5% due to insects, 2% for rodents, and 5–30% for molds and mycotoxins (2). Losses due to pests are invariably more with on-farm than off-farm (central and commercial) storages. In the developed countries, losses have been generally at 1% while 10–30% losses have been reported in the developing countries (3). Meteorological conditions favorable for pest proliferation, absence of incentives for quality, and lack of resources and expertise are some of the factors contributing to the higher loss (4). Postharvest losses in foodgrains due to pests varied from 0.2 to 8% in China, <3% in India, and 0.5 to 30% in some countries in Africa and Asia. Postharvest spoilage of grains due to fungi and aflatoxins has been estimated to be 3.6% in China and about 5% each in Indonesia, Philippines, and Thailand (5).

In Canada, foodgrains losses due to insect, mites, and fungi have been reported as 0.5% valued at US $17 million. However, the estimated annual financial losses (direct and indirect losses) ranged from $135 to 380 million (6). In Australia, the annual crop losses due to mice in some states since 1990 have been estimated at AU $26 million (7).

LOSSES IN THE UNITED STATES

In the United States, the annual losses of postharvest foods and food products due to insects ranged from $2 to 6 billion and that caused by rodents and other vertebrates were noted at $2.5 billion (8). Economic losses due to aflatoxin in food commodities in the United States and Canada have been put at $5 billion per year (9). According to a 1990 survey, the estimated stored grain losses due to insects and fungi exceed $500 million per year. Losses of wheat alone have been reported to exceed $300 million in the United States (10). Postharvest losses due to rodents are known to exceed $900 million per year and about $100 million are spent for control (11).

PREVENTION OF LOSSES

Various control/management techniques are adopted in the postharvest sector to check the losses due to pests (Table 1). The control strategies are undergoing changes in the context of consumer perception about pesticide contaminants, pest resistance, and new regulations on conventional pesticides. Accordingly, the frequency of application and dosage rates of chemical protectants and fumigants are decreasing and physical and biological methods are getting priority. Pest resistance has been a major impediment to achieve 100% insect control. To address the resistance crisis, alternate chemicals, insecticide cocktails, and improved fumigation technologies (Siroflo, Closed Loop Fumigation, and on-site generation) are being used. Automation in monitoring temperature and insect density in bulk grain storages as early warning systems will further improve the efficiency of pest con-

Fig. 1 Wheat damaged by the Khapra beetle, *Trogoderma granarium.*

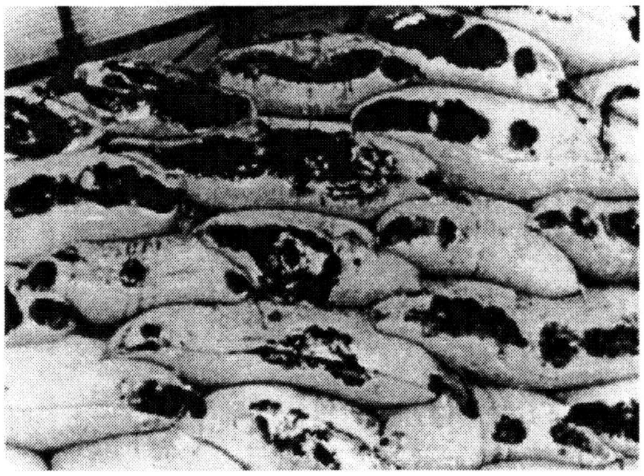

Fig. 2 Mold attack in bagged milled rice.

trol operations. Of late, computer-assisted expert systems have been put to use as a training tool for forecasting pest activity and to make sound management decisions (12). Integrated pest management (IPM) involving compatible control techniques that are cost-effective and also satisfy the demands of the consumer and regulatory authorities have to be implemented widely to increase profits and reduce economic loss. IPM is practiced in the United States, Europe, and in a few developing countries (4, 12).

Aeration with ambient or chilled air is important for mold control in grains in addition to controlling insects. Development of plant cultivars that are resistant to fungal invasion and toxin production is necessary. As regards rodents, effective control techniques are already available. Nevertheless, an integrated approach comprising good housekeeping, use of physical traps, poison baiting, and burrow fumigation is essential. Efficient control requires continuous and concerted effort of all personnel involved in storage of food grains (13).

Table 1 Measures to reduce postharvest losses

Control method	Impediments	Future
Insects		
Fumigants	Resistance, hazard to environment	Modified formulations and application techniques, new fumigants
Contact insecticides	Resistance, new regulations, consumer aversion	Mixtures, new compounds, combination with aeration or manipulated temperatures
Heat	Critical temperatures	Hot air in fluidized bed, integrate with inert dusts or fumigants
Aeration	Alone not effective	Integrate with inert dusts or insecticide capping or fumigants
Controlled atmospheres	High gas-tightness, cost	On-site generators, combined with heat
Insect growth regulators	Slow acting, cost	In mixture with conventional insecticides
Inert dusts	Worker safety, affects grain flow rate	Slurry treatment, combined with insecticides/ fumigants/ aeration/ heat
Trapping/monitoring	Species-specific, alone not effective	In combination with insecticides and as insect pathogen dispenser
Biological agents	Species-specific	Role in IPM
Rodents		
Acute poisons	Bait shyness, risks to nontarget animals	Improved bait formulations and bait stations
Anticoagulants	Resistance	Pulsed baiting, chemical sterilants
Microbes		
Drying to safe moisture	Energy intensive, cost	Affordable dryers, grain chillers, development of resistant cultivars

THE FUTURE

The world's food supply can be increased by at least 5% by preventing postharvest pest losses. There is a general tendency towards achieving reduced food losses over the years due to the increasing demand for foodstuffs free from pest contaminants. This trend is likely to continue, provided that we improve the pest management regimes. In this context, aeration, heat treatment, inert dusts, automated pest and temperature monitoring, and expert systems will play significant roles. A holistic approach deploying compatible physical, biological, and chemical control techniques is essential to reduce the postharvest pest losses. This is a continuous process and any lapse at any time will lead to pest outbreaks causing huge financial losses.

REFERENCES

1. Boxall, R.A. PostHarvest Losses to Insects—A World Overview. In *Biodeterioration and Biodegradation*; Rossmoore, H.W., Ed.; Elsevier: Barking, 1991; 8, 160–175.
2. Tipples, K.H. Quality and Nutritional Changes in Stored Grain. In *Stored Grain Ecosystems*; Jayas, D.S., White, N.D.G., Muir, W.E., Eds.; Marcel Dekker, Inc.: New York, 1995; 325–351.
3. Scholler, M.; Prozell, S.; Al-Kirshi, A.G.; Reichmuth, Ch. Towards biological control as a major component of integrated pest management in stored product protection. J. Stored Prod. Res. **1997**, *33* (1), 81–97.
4. Schulton, G.G.M. In *Stored-Product Protection in Warm Climates*, Proceedings of the 5th International Working Conference on Stored Product Protection, Bordeaux, France, Sept 9–14, 1990; Fleurat-Lessard, F., Ducom, P., Eds.; INRA/SDPV: Bordeaux, 1991; 1557–1570.
5. Lubulwa, A.S.G.; Davis, J.S. In *Estimating the Social Costs of the Impacts of Fungi and Aflatoxins in Maize and Peanuts*, Stored Product Protection, Proceedings of the 6th International Working Conference on Stored Product Protection, Canberra, April 17–23, 1994; Highley, E., Wright, E.J., Banks, H.J., Champ, B.R., Eds.; CAB International: Wallingford, 1994; 1017–1042.
6. Anonymous. *Stored Products Research Programme: Strategic Focus*; Agriculture Canada Research Branch: Winnipeg, 1993; 8.
7. www.brs.gov.au/agrifood/pests/rodent.html (accessed May 2000).
8. Scott, H.G. Nutrition Changes Caused by Pests in Food. In *Ecology and Management of Food-Industry Pests*; Gorham, J.R., Ed.; Association of Official Analytical Chemists: Arlington, 1991; 436–467.
9. Miller, J.D. Agriculturally-important mycotoxins—a review. ASEAN Food Handling Newsl. **1994**, *44*, 5–7.
10. Kankel, P.; Criswell, J.T.; Cuperus, G.W.; Noyes, R.T.; Anderson, K.; Fargo, W.S. Stored product integrated pest management. Food Reviews Int. **1994**, *10* (2), 177–193.
11. Maheen, A.P. *Rats and Mice: Their Biology and Control*; Rentokil Limited: East Grinstead, UK, 1984; 383.
12. Hagstrum, D.W.; Flinn, P.W. Integrated Pest Management of Stored Grain Insects. In *Storage of Cereal Grains and Their Products*; Sauer, D.B., Ed.; American Association of Cereal Chemists, Inc.: St. Paul, MN, 1992; 535–562.
13. Smith, R.H. Rodents and Birds as Invaders of Stored-grain Ecosystems. In *Stored-Grain Ecosystems*; Jayas, D.S., White, N.D.G., Muir, W.E., Eds.; Marcel Dekker, Inc.: New York, 1995; 289–323.

POULTRY PEST MANAGEMENT (ARTHROPODS)

P

Nancy C. Hinkle
University of California, Riverside, California, U.S.A.

INTRODUCTION

Poultry pests include both on-animal (ectoparasites) and environmental pests (Table 1). Ectoparasites are harmful to animal health, reduce production, and may affect poultry workers. Pests such as flies and beetles damage facilities, transmit pathogens, and migrate from the premises, resulting in potential social and legal conflicts (1). They are major concerns to producers, and significant effort and money are expended on their control. Understanding pest biology and behavior is essential to developing effective integrated suppression programs for poultry pests.

ECTOPARASITES

Lice and mites are the most important external parasites of poultry (2). Infestation may result in weight loss in birds and decreased egg production.

Lice

Lice are small insects whose feeding causes irritation and itching so that infested birds spend considerable time scratching in attempts to rid themselves of the pests (Fig. 1). All lice on poultry are chewing lice, with mouthparts adapted for feeding on feathers and skin scales (Fig. 2). They are permanent ectoparasites, spending their entire life on the bird. Injury is due to irritation or itching caused by lice crawling and gnawing on skin and feathers. Birds become restless and do not feed or digest their food properly. The adult female lays its eggs on the bird, gluing them to feathers. The eggs (nits) hatch in a few days and the young lice mature in about two weeks. Because lice spend their entire life cycle on the bird, control strategies must be directed at the host. On-bird biological control options are extremely limited, as are mechanical and cultural possibilities.

Mites

There are several mites that are ectoparasites of birds. The depluming mite attacks skin near the base of feathers and the intense itching associated with their activity impels the host to pluck out its feathers. The scalyleg mite burrows in the skin, producing a lifting of skin scales and a thickened shank condition with deformity and encrustation, causing the leg to look swollen to more than double its normal size.

The red chicken mite has been reported capable of transmitting fowl pox virus. Feeding by red chicken mites greatly reduces egg production, forces setting hens to leave their nests, results in exsanguination of young chicks, and leads to severe loss of blood and vitality, predisposing birds to disease. These mites feed on birds at night, retreating during the day to crevices in the poultry house where they lay their eggs. Thus, acaricide treatment requires thorough application to penetrate all available hiding places. The red chicken mite can survive starvation for more than eight months, so even a thorough cleanout may not be adequate to eliminate an infestation.

The northern fowl mite is one of the most important external parasites of poultry, parasitizing both domestic fowl and wild birds (Fig. 3) (2). Mite levels and their effects on egg production seem to be related to bird strain and age, with young birds being more susceptible to high infestations and antibodies producing some degree of resistance in older birds. They have been shown capable of

Table 1 Common poultry pests

Chicken body louse (*Menacanthus stramineus*)
Shaft louse (*Menopon gallinae*)
Fluff louse (*Goniocotes gallinae*)
Wing louse (*Lipeurus caponis*)
Head louse (*Cuclotogaster heterographus*)
Scaly leg mite (*Knemidokoptes mutans*)
Depluming mite (*Knemidokoptes gallinae*)
Red chicken mite (*Dermanyssus gallinae*)
Northern fowl mite (*Ornithonyssus sylviarum*)
Fowl tick (*Argas persicus*)
Bedbug (*Cimex lectularius*)
Sticktight flea (*Echidnophaga gallinacea*)
House fly (*Musca domestica*)
Little house fly (*Fannia canicularis*)
Darkling beetle, lesser mealworm (*Alphitobius diaperinus*)
Hide beetle (*Dermestes maculatus*)

persisting in the absence of a host for up to six weeks, although most starve within a couple of weeks. Mites are frequently transmitted from bird to bird by contact or by crawling onto new hosts from secluded places in the environment. Mites may be introduced into a "clean" house by bringing in infested birds or from wild birds. Northern fowl mite control is primarily based on chemical suppression, although prevention of mite introduction into the bird population is the only dependable means to keep birds mite-free (3).

Ticks

Poultry ticks are active at night, traveling several feet to the host and back to hiding places where they spend the day. Females digest their blood meals and lay egg batches in crevices. Chemical control involves removing birds

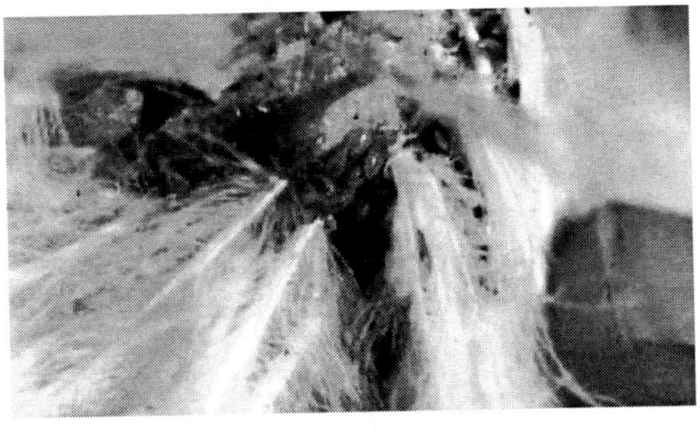

Fig. 1 Chicken body lice occur in large numbers, and their activity abrades the skin, producing scabs.

Fig. 2 *Menacanthus stramineus*, the chicken body louse, feeds on feathers and skin scales.

and applying acaricides to thoroughly cover walls, ceilings, floors, cracks, and crevices using high-pressure spray (4).

Bedbugs

Bedbugs are bloodsucking nocturnal parasites that have a wide host range, including humans and other mammals as well as poultry (3). Unfed bedbugs are about 4 mm long, oval, and flattened. These insects feed on birds at night, sucking blood until they swell to double their size. During the day, they hide in cracks and crevices within the house, where the females lay egg batches. Both nymphal and adult bedbugs can live for several months without feeding, so facility spelling is ineffectual.

Fleas

Sticktight fleas typically attach to the bird's head, often in clusters numbering 100 or more. They embed their mouthparts into fleshy areas and may remain attached for up to three weeks. Females lay their eggs while attached and the eggs roll off into the nest or litter, where larvae develop. Following pupation, adult fleas seek hosts, attach, and repeat the cycle. Adult feeding causes irritation,

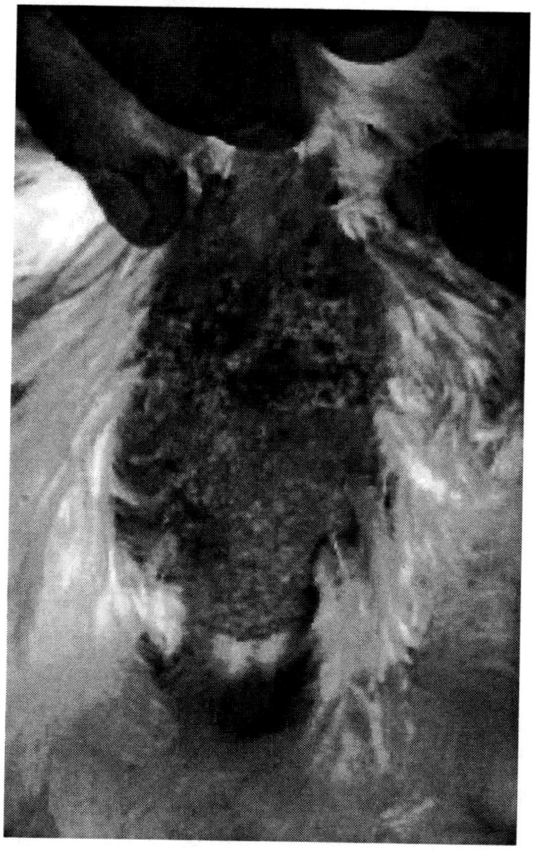

Fig. 3 Northern fowl mites (*Ornithonyssus sylviarum*) infest the vent; their blood-sucking results in scabbing and feather loss.

which can kill young birds and reduce growth and egg production. Clustered sticktight fleas can cause the eyes to swell shut. Sticktight fleas may be found infesting a variety of other hosts including wild birds, dogs, cats, and horses. Because they are found on such a wide host range, carrier animal exclusion is a critical aspect of flea elimination.

Potential Ectoparasite Control Strategies

Poultry ectoparasites present a control challenge because the birds' water-repellent feathers protect the arthropods from insecticide sprays. While many chemicals may be active against mites and lice, they must contact the pests to be effective, so insecticidal spray penetration through the feathers is critical (3).

Because birds demonstrate increasing tolerance/resistance to some ectoparasites as they age, possibilities exist for employing host animal resistance to suppress these pests (2). Using either traditional breeding techniques or molecular manipulations, poultry strains that do not support ectoparasite populations may be developed. Similarly, vaccines may be produced that confer protection against ectoparasitic arthropods.

NUISANCE PESTS

Flies

A variety of flies are associated with poultry, with house flies and lesser house flies constituting the primary control challenges (4). House fly populations are typically highest in late spring and summer, while lesser house fly populations peak in early spring. House flies are pestiferous because they enter homes and land on food and eating surfaces, producing fecal and regurgitant spots. Lesser house flies are bothersome in that they congregate and hover for hours at a time. Both flies are capable of mechanically transmitting the causative agents of some diseases.

Control of filth-breeding flies is dependent on manure and moisture management. Moist poultry manure is highly attractive to adult flies and provides ideal conditions for immature development. Rapidly dried manure loses its attractiveness as well as its ability to support maggot development. Good ventilation is supportive in fly suppression; not only does it help to dry manure rapidly but moving air disturbs adult flies, driving them from the area and hampering oviposition.

In addition to cultural and mechanical fly suppression, chemical controls are frequently employed either as adulticidal sprays, toxic baits, or larvicides. However, residual and space spraying often has undesirable ancillary effects on extant beneficial arthropods. One insect development inhibitor has been formulated as a spray and a poultry feed-through material to suppress maggot development in the manure. Unfortunately, however, flies have developed high resistance levels to many of these chemical controls (5).

In stable ecosystems, biological control results from a variety of parasitoids, predators, pathogens, and competitors (6). Long-term manure accumulation permits establishment by a diverse fauna of competitors and natural enemies of pest flies. Techniques for manipulating these resident beneficial arthropods are being developed. Commercially available fly biocontrol options are currently limited to parasitoids, including those in the genera *Muscidifurax* and *Spalangia*. The challenges of biological fly suppression include operational factors as well as technical and biological constraints. Using biological control effectively requires much more thought than applying pesticides. Use of parasitoids is time-sensitive, requiring that releases be made when the proper host stages are present. Commercially produced parasitoids may exhibit qualities that are maladaptive for specific environments or inadequate quality control may yield a less competitive or incompatible product. Because parasitoids are themselves biological entities, they are subject to disease, predation, starvation, desiccation, etc., and are functional only within specific environmental parameters.

The opposite strategy of removing manure at frequent intervals to break the fly developmental cycle is time and labor intensive, and obviates use of biological control. If frequent cleanout is employed, it is essential that it be thorough; even minimal manure residue provides adequate habitat for maggot development.

Darkling Beetles

Adult darkling beetles are about a quarter inch long and shiny black. Larvae may reach 1/3 inch in length and resemble small wireworms. Both larvae and adults are very active, burrowing in the litter and feeding on spilled feed and other organic material. Prior to pupation, larvae tunnel into the soil or into wooden structures and, in so doing, often produce damage to the wood. They can create even more damage in foam insulation. Darkling beetle larvae have been found to harbor a number of disease organisms including fowl pox, Newcastle disease, avian leukosis, *E. coli*, and *Salmonella*.

Although habitat removal is the most effective darkling beetle control, chemical suppression remains an integral strategy. Birds cannot be present when insecticides are applied, so treatments must be made when the house has been cleaned out and before birds are brought back in. Following clean-out, litter must be removed from the premises to prevent beetles from crawling or flying back into the house (4).

FUTURE DIRECTIONS

Poultry pest challenges and available suppression options depend on production systems, climate, and action thresholds. Control strategies must be socially and environmentally compatible, economically viable, and operationally tenable.

Integrated pest management employs physical, mechanical, cultural, biological, and chemical suppression techniques (4). Use of each component is dependent on compatibility with environmental conditions, agronomic practices, and production systems. Engineering the facility to incorporate management techniques that contribute to fly, beetle, and ectoparasite suppression must begin before construction. Mechanical controls include exclusion, habitat disruption, trapping, and electrocutor devices.

Control strategies will be quite different for ectoparasites that spend their entire life cycle on the host (lice, most mites) compared with those that are temporary parasites (bedbugs, ticks). Approaches to suppression of nuisance pests such as flies and beetles are entirely different (7).

Considerable research remains to be done before biocontrol organisms can be considered as practical poultry pest management tools. While use of parasitoids and predators for ectoparasite control presents some unique challenges, pathogens may be more feasible for on-host suppression. Bacteria and fungi have been investigated as biopesticides, but development is constrained by markets and operational considerations. Flies and beetles, on the other hand, are logical targets for use of parasitoids, pathogens, predators, and competitors.

REFERENCES

1. Geden, C.J.; Arends, J.J.; Axtell, R.C.; Barnard, D.R.; Gaydon, D.M.; Hickle, L.A.; Hogsette, J.A.; Jones, W.F.; Mullens, B.A.; Nolan, M.P., Jr.; Nolan, M.P., III; Petersen, J.J.; Sheppard, D.C. Poultry. In *Research and Extension Needs for Integrated Pest Management for Arthropods of Veterinary Importance*; Hogsette, J.A., Geden, C.J., Eds.; 1999. http://www.ars-grin.gov/ars/SoAtlantic/Gainesville/cm_fly/ Lincoln.html (accessed Sept 2000).
2. DeVaney, J.A. Symposium: Arthropods of economic importance to the poultry industry: ectoparasites. Poultry Sci. **1986**, *65*, 649–656.
3. Axtell, R.C.; Arends, J.J. Ecology and management of arthropod pests of poultry. Ann. Rev. Entomol. **1990**, *35*, 101–126.
4. Axtell, R.C. Poultry integrated pest management: status and future. Integ. Pest Manag. Rev. **1999**, *4*, 53–73.
5. Scott, J.G.; Alefantis, T.G.; Kaufman, P.E.; Rutz, D.A. Insecticide resistance in house flies from caged-layer poultry facilities. Pest. Manag. Sci. **2000**, *56*, 147–153.
6. Rutz, D.A.; Patterson, R.S. *Biocontrol of Arthropods Affecting Livestock and Poultry*; Westview Press: Boulder, CO, 1990; 174.
7. Turner, E.C., Jr. Structural and litter pests. Poultry Sci. **1986**, *65*, 644–648.

PREDACIDES

Peter J. Savarie
National Wildlife Research Center, Fort Collins, Colorado, U.S.A.

P

INTRODUCTION

Predacides are lethal agents that remove offending predators, and their use is the one of several techniques of an integrated program for control of predation (1, 2). The use of predacides in the United States is a controversial issue for livestock producers, resource managers, biologists, and the general public. Historically, the predominant application of predacides has been for the control of mammalian carnivores such as red foxes (*Vulpes vulpes*), coyotes (*Canis latrans*), and wolves (*C. lupus*) that prey on livestock and poultry. Currently, there are no predacides for wolves because they are endangered in most of the United States. In addition to their use for reducing predation on livestock, predacides have been used in conservation programs for control of predators and other animals such as exotic invasive species that prey upon federally designated threatened or endangered species, or are vectors of communicable disease. Four predacide products are registered by the United States Environmental Protection Agency (EPA) (Table 1) for mammalian predator control and each has stringent regulations and directions for use. Three products have specific use restrictions and are classified as Restricted Use Pesticides, requiring that applicators be trained and certified. The use of any predacide must conform to all applicable federal, state, and local laws and regulations. Taking only offending animals that cause predation is a major goal in lethal predator management and is accomplished through the use of specific delivery systems and placement of predacides. Through unique delivery devices and restrictive requirements on their use, exposure of the predacide in the environment and to nontargets is reduced.

PREDACIDE FOR DEPREDATING COYOTES

One of the most selective techniques for depredating coyotes is the sodium fluoroacetate (Compound 1080) Livestock Protection Collar (3, 4). The Livestock Protection Collar is a Restricted Use Pesticide for use on sheep or goats and takes advantage of the coyote's attack behavior of biting on the throat to kill the animal. The distinct advantage of using the collar is that only the coyote that is actually killing the sheep or goat is affected; coyotes that may not be killing livestock are unaffected. The collar has two rubber reservoirs containing a solution of sodium fluoroacetate and is attached on the throat of a sheep or goat with straps around the neck. A coyote that attacks the throat of a collared animal bites into the reservoir and receives an oral lethal dose of sodium fluoroacetate. Sodium fluoroacetate is highly toxic (5) and strict state and federal regulatory requirements, and 18 use restrictions govern its use in the collar. One restriction limits use of collars in fenced pastures only; collars cannot be used on open range. Training and certification of applicators is required on a state-by-state basis and currently seven states have EPA-approved programs. Although the collar has high selectivity, it has not been widely used because of the high level of management resources required to direct the coyote's attack to the collared livestock. To be most effective the collar must be used with specific sheep and goat husbandry practices where predation is high and regular. The usual procedure is to place collars on a small group of penned animals from a flock and move the other livestock to another vicinity.

FUMIGANT FOR DENS

The Large Gas Cartridge is a fumigant for control of coyotes, red foxes, and striped skunks (*Mephitis mephitis*) in dens only and is not a Restricted Use Pesticide. The active ingredients in gas cartridges, sodium nitrate and charcoal, produce carbon monoxide when ignited (6). The gas cartridge is also very selective because the dens of the target animals can be identified by size, tracks, remains of prey, scat, and observation of animals at the site. Coyote and fox predation on livestock is high during the spring of the year and gas cartridges are used primarily at this time when the adults are providing food for their young. The biological basis for fumigating dens was demonstrated in

Table 1 Predacides registered by the United States Environmental Protection Agency (Registrant for these predacides is the U.S. Department of Agriculture, Animal, and Plant Health Inspection Service.)

Product label name/EPA registration number	Active ingredient(s)	Target species
[a]Sodium fluoroacetate (Compound 1080) livestock protection collar/56228-22	Sodium fluoroacetate	Coyote
[b]Large gas cartridge/56228-21	Sodium nitrate, charcoal	Coyote, red fox, striped skunk
[a]M-44 Cyanide Capsules/56228-15	Sodium cyanide	Coyote, red fox, gray fox, wild dog
[a]M-44 Cyanide Capsules/56228-32	Sodium cyanide	Arctic fox

[a]Restricted use pesticide
[b]General use pesticide

a study that showed that killing of sheep by coyotes was sharply reduced when the pups in dens were removed (7). Sheep kills were reduced by 92% when pups were eliminated, 99% when both pups and adults were removed, and only 4% when no removal was conducted, suggesting that provisioning pups motivates the killing of larger prey animals. Dens may be treated if skunks are creating health and safety hazards (i.e., rabid animals) or are depredating threatened or endangered species.

CYANIDE EJECTOR DEVICE

M-44 Cyanide Capsules is another Restricted Use Pesticide; it has 26 use restrictions and requires that applicators be trained and certified. The M-44 has two product labels. One label is for controlling coyotes, red foxes, gray foxes (*Urocyon cinereoargenteus*), and wild dogs (*Canis familiaris*) that prey upon livestock and poultry, threatened or endangered species, or are vectors of communicable disease. The principal use is for coyote damage management in pastures, rangeland, and forest land (8). The other label is for control of arctic foxes (*Alopex lagopus*) that deprecate threatened or endangered species in the Aleutian Islands, Alaska. The M-44 ejector is a spring-powered device that consists of an ejector mechanism, a base that is placed in the ground to contain the ejector, capsule holder, and capsule containing sodium cyanide. The capsule holder is wrapped with absorbent material that contains a lure bait and protrudes above the ground when set. The ejector is triggered when a canid pulls hard on the baited M-44 unit and sodium cyanide is ejected into the animal's mouth; smaller animals cannot activate the device. Moisture in the animal's mouth reacts rapidly with sodium cyanide releasing hydrogen cyanide gas and the animal succumbs within a few seconds. When M-44s are placed in the field, applicators use their expertise in animal behavioral patterns to minimize the risk of attracting nontarget animals to the device. The risk to nontarget animals is also reduced through the use of specialized lures and attractants designed for the species of concern.

FUTURE PROSPECTS

In recent years, public attitudes have not supported techniques of lethal control of carnivores and with increased regulations, the use of predacides for depredation management will most likely be reduced as reliable efficacious nonlethal techniques are developed. Several nonlethal control methods such as frightening devices (e.g., lights, sirens, pyrotechnics) and guard animals such as dogs, llamas, and donkeys have had various degrees of success in reducing predation in some situations but are not practical for all livestock husbandry situations. It is appealing to think that repellents or substances that produce conditioned taste aversion could be developed to deter predation, but investigations to date have not shown much promise (9). Fertility control, with the presumption that reductions in numbers of offspring would result in less livestock predation, is a promising area of investigation that is receiving renewed attention (10). However, a useable product, if successful, will still take years of development.

REFERENCES

1. Andelt, W.F. Carnivores. In *Rangeland Wildlife*; Krausman, P.R., Ed.; Society for Range Management: Denver, CO, 1996; 133–155.
2. Knowlton, F.F.; Gese, E.M.; Jaeger, M.M. Coyote depredation control: an interface between biology and management. J. Range Manage. **1999**, *52*, 398–412.
3. Connolly, G.; Burns, R.J. In *Efficacy of Compound 1080*

Livestock Protection Collars for Killing Coyotes That Attack Sheep, Proceedings of the Vertebrate Pest Conference, 1990; 14, 269–276.
4. Burns, R.J.; Zemlicka, D.E.; Savarie, P.J. Effectiveness of large livestock protection collars against depredating coyotes. Wildl. Soc. Bull. **1996**, *24* (1), 123–127.
5. Savarie, P.J.; Connolly, G.E. Criteria for the Selection and Development of Predacides. In *Vertebrate Pest Control and Management Materials: Fourth Symposium*; Kaukeinen, D.E., Ed.; ASTM Special Technical Publication 817: Philadelphia, PA, 1983; 278–284.
6. Savarie, P.J.; Tigner, J.R.; Elias, D.J.; Hayes, D.J. In *Development of a Simple Two-Ingredient Pyrotechnic Fumigant*, Proceedings of the Vertebrate Pest Conference, 1980; 9, 212–221.
7. Till, J.A.; Knowlton, F.F. Efficacy of denning in alleviating coyote depredations upon domestic sheep. J. Wildl. Manage. **1983**, *47* (4), 1018–1025.
8. Connolly, G.E. In *M-44 Sodium Cyanide Ejectors in the Animal Damage Control Program, 1976–1986*, Proceedings of the Vertebrate Pest Conference, 1988; 13, 220–225.
9. Conover, M.R.; Kessler, K.K. Diminished producer participation in an aversive conditioning program to reduce coyote depredation on sheep. Wildl. Soc. Bull. **1994**, *22*, 229–233.
10. DeLiberto, T.J.; Conover, M.R.; Gese, E.M.; Knowlton, F.F.; Mason, R.J.; Miller, L.; Schmidt, R.H.; Holland, M. *Fertility Control in Coyotes: Is It a Potential Management Tool?* Proceedings of the Vertebrate Pest Conference, 1998; 18, 144–149.

PREVENTION

Larry P. Pedigo
Iowa State University, Ames, Iowa, U.S.A.

INTRODUCTION

Integrated Pest Management (IPM) prevention has been called Preventive Pest Management (Preventive PM) and has been a major concern in IPM development. Preventive PM is, or at least should be, the ultimate form of IPM.

Preventive PM implies that action should be taken against a pest before injury occurs. Therefore, the tactics are often employed without knowledge of pest presence or status at a particular point in time. For example, seeds of resistant plants are chosen and plans made in advance of the growing season, often long before any knowledge of a pest's economic importance in a particular growing season is known.

LOWERING THE PEST'S GENERAL EQUILIBRIUM POSITION

Prevention can be achieved by focusing on either the pest or the host. Very often it focuses on the pest. The ecological objective is to lower the pest's general equilibrium position (average density) and, subsequently, the average level of crop damage to a position below the economic damage level, which is where costs equal benefits.

Lowering the general equilibrium position can be accomplished either by prohibiting pest establishment or quickly limiting pest population growth when pest establishment occurs. Tactics for this purpose include many biological controls, crop rotations, sanitation and tillage, altering planting dates to prevent or reduce colonization, trap cropping, spatial arrangements of plants (e.g., row widths and adjacent crops), and cultivar selection for insect resistance (e.g., antibiosis, antixenosis).

Most of these tactics are compatible for integrating into the overall Preventive PM program. It would seem that one of the least compatible tactics is the use of insects as biological control agents. Although the prevailing viewpoint is that these biological controls are compatible with other tactics, it would not seem so. This is because tactics used to suppress the pest also usually suppress natural-enemy populations, if in no other way than to reduce the natural enemy's food source. This should not be taken to imply that insects not be used in biological control but rather that these agents are some of the most difficult to integrate into the management program. Therefore, special emphasis is needed for this integration.

Seemingly, if tactics work in a density-independent manner they tend to be compatible for inclusion in the pest management program. Conversely, if effectiveness of the tactic is promoted by pest density, as with many insect natural enemies and disease-causing organisms, then the tactic becomes less compatible when used in conjunction with other density-reducing tactics.

Prospects for natural enemy compatibility have improved in recent years with more efficient rearing procedures and methods of augmentation. Consequently, there will be a greater potential for augmentation, thereby making the natural enemies less density dependent for effectiveness in management. Genetic improvements through conventional selection and bioengineering to produce natural enemies resistant to insecticides and other tactics also will enhance compatibility.

RAISING THE LEVEL AT WHICH ECONOMIC DAMAGE OCCURS

The other area of prevention focuses on the crop itself. In this instance, attempts are made to manage the loss from the pest rather than managing pest density. In other words, the goal here is to reduce the level of injury per pest.

The ecological objective of this approach is to raise the level at which economic damage occurs. Here, the manager looks for ways to make the crop less vulnerable to a developing pest population. For most crops, this could mean irrigating or increasing nutrient inputs to increase plant vigor, changing planting dates to present a less vulnerable stage to a pest, and/or selecting tolerant cultivars.

Tolerance to pest injury is a tactic often ignored by plant breeders, pest managers, and others, though it may be the most sustainable tactic that can be used. Advantages of tolerance are twofold. First, tolerance is completely compatible with insect natural enemies. It allows higher pest densities without significant yield loss and, therefore, is conducive to factors acting in a density-dependent manner. Second, because no mortality pressure is placed on the pest population, there is little chance for development of resistance to the tactic. Indeed, because of the latter, tolerance may be one of the premier IPM tactics. However, much more must be known about tolerance mechanisms in plants and modes of inheritance of these before this approach can be more widely adopted.

PESTICIDES NOT RECOMMENDED AS A PREVENTIVE TACTIC

In seeking preventive tactics, pesticides are not recommended for Preventive PM. They have been used widely for prevention, and many growers still use them in this way today. However, the use of pesticides in regularly scheduled applications frequently has resulted in pesticide resistance and other forms of ecological backlash. Therefore, this approach cannot be considered part of a sustainable IPM program.

BIBLIOGRAPHY

All, J.N. In *Importance of Designating Prevention and Suppression Control Strategies for Insect Pest Management Programs in Conservation Tillage*, Proceedings of the 1989 Southern Conservation Tillage Conference, University of Florida: Tallahassee, FL, 1989; 1–3.

Cate, J.R.; Hinkle, M.K. *Integrated Pest Management: The Path of a Paradigm*; Special Rept., National Audubon Soc.: Washington, DC, 1993; 43.

Pedigo, L.P. Closing the gap between IPM theory and practice. J. Agric. Entomol. **1995**, *12*, 171–181.

Pedigo, L.P. *Entomology and Pest Management*, 3rd Ed.; Prentice-Hall: Englewood Cliffs, NJ, 1999; 691.

Quisenberry, S.S.; Schotzko, D.J. Integration of plant resistance with pest management methods in crop production systems. J. Agric. Entomol. **1994**, *11*, 279–290.

Rabb, R.L. A sharp focus on insect populations and pest management from a wide-area view. Bull. Entomol. Soc. Am. **1978**, *24*, 55–61.

Stern, V.M.; Smith, R.F.; van den Bosch, R.; Hagen, K.S. The integrated control concept. Hilgardia **1959**, *29*, 81–101.

P

PRINCIPLES OF PEST MANAGEMENT WITH EMPHASIS ON PLANT PATHOGENS

Kitty F. Cardwell
International Institute of Tropical Agriculture, Cotonou, Bénin

Chris J. Lomer
Royal Veterinary and Agricultural University (KVL)
Copenhagen, Denmark

INTRODUCTION

Biotic plant diseases are caused by pathogenic microbial agents: fungi, bacteria, nematodes, viruses, and viroids. Plant diseases reduce global food and fiber production by an estimated 20% annually, but during severe epidemics, localized losses can reach 100%. From the 1930s to the 1960s, plant pathologists were confident that disease could not only be controlled, but perhaps even eliminated using chemicals and host plant resistance. Limitations to this approach emerged as host plant resistance was not durable if the genetic basis for it was too narrow, and chemicals were perceived as expensive, deleterious to the environment, or not durable if the target microbe could develop resistance. Thus, the concept of eradicating disease has given way to the concept of integrated disease management to hold damage to economically acceptable levels (1). There are numerous control methods that may be used singly or in combination (Table 1).

HOST PLANT RESISTANCE

Host plant resistance to disease is a physiological response that limits the growth and development of a pathogen. The response to an invading pathogen may be highly specific so that the reaction is immunity (no disease development at all), or general, so that disease impact is reduced (2, 3).

Specific Host-pathogen Interactions: Vertical Resistance

Many plant pathogens are specifically coevolved with their plant hosts; a specific gene within both the plant host and the pathogen will code for mutual recognition and response (gene for gene relationship). When a plant has resistance to a pathogen based on a single specific gene this is described as vertical resistance. In natural ecosystems, a plant population consists of resistant and susceptible individuals, while the pathogen population consists of individuals (races) that are either virulent or avirulent to the host (4). Thus, some proportion of the plant population may be diseased, but the plant and pathogen populations are in equilibrium over time, allowing both to survive. In more intensive plant production systems, the aim is to minimize losses due to diseases. Through plant breeding, scientists strive to select only the resistant genotype and distribute this as a variety or hybrid. Unfortunately, such vertical resistance may be broken down when a pathogen gene locus mutates to virulence, or more frequently, when virulent strains already exist in the pathogen population. The virulent strain or mutant is specifically selected for as it is the only one that can infect and reproduce on the host. If the population of plants has been bred for uniformity in the resistance gene locus, then an epidemic can occur so that every plant becomes diseased (4). There are a number of strategies to avoid ''breakdown'' of a single vertical resistance gene: 1) numerous resistance genes bred into a single variety; 2) several genes bred into isolines and the isolines mixed into multigenic varieties; and 3) geographic or temporal deployment of different resistant cultivars so that spread of a virulent race of pathogen will be checked.

General Resistance: Horizontal Resistance

Horizontal or general resistance to pathogens can occur in the absence of any specific genetic recognition/response interaction (3). Disease may still occur, but the plant's

Table 1 General methods of disease control (1)

A. Avoidance of the pathogen
   (1) Choice of geographic area
   (2) Planting and harvesting date decisions
   (3) Use of disease-free planting stock
   (4) Modification of cultural practices
B. Exclusion of the pathogen
   (1) Treatment of seeds/planting material to kill resident pathogens
   (2) Meristem excision and tissue culture
   (3) Inspection, certification, and plant quarantine
   (4) Control of insect vectors
C. Management of the pathogen population
   (1) Biological control and microbial niche management
   (2) Crop rotation with nonhost plants or trap plants, and field fallowing
   (3) Intercropping with nonhosts
   (4) Removal of diseased plants and plant parts from the field
   (5) Understanding the role of weeds and weed hosts in and around the field
   (6) Sanitation and management of crop residues
   (7) Soil microbe management with chemical drenches, solarization, or mulching
   (8) Fungicide sprays
D. Protection of the plant
   (1) Application of chemical or biological control agents as prophylaxis against infection
   (2) Inoculation for induction of SAR
   (3) Control of insect vectors
   (4) Modification of plant nutrition
E. Development and deployment of host plant resistance
   (1) Selection and breeding for resistance (vertical, horizontal, and population resistance)
   (2) Genetically modifying plant with nonhost genes to induce a "nonhost" response
   (3) Genetically modify plant to contain viral coat protein to induce resistance to same virus
   (4) Use of external chemical or biological resistance inducers
   (5) Temporal and spatial mixing of different resistance genotypes
F. Therapy applied to diseased plant l
   (1) Application of "curative" chemicals
   (2) Heat treatment (i.e., seed sanitation)
   (3) Surgery

nonspecific responses to external stresses reduce pathogen growth and development. The mechanism of response may be chemical or mechanical. A chemical response is a cascade of genetic signaling elicited by a pathogen (or other stresses) leading to the production and mobilization of pathogen suppressive metabolites in and around the plant i.e., phytoalexins, phenolics, and chitinases. This is sometimes referred to as "systemically acquired resistance" (SAR). SAR can also be induced in response to chemicals, nonpathogenic microbes, and environmental stress.

Physical characteristics that give broad-based resistance such as waxy cuticle, quick growth habit, fewer stomata, etc., make it harder for the pathogen to gain entrance. This type of resistance, horizontal resistance, usually involves more than one gene and is considered to be more stable than vertical resistance.

CHEMICAL CONTROL

Fungicides

At the end of the twentieth century, in the Americas and Japan, fungicides comprise about 15% of the total pesticide market share, while in Europe, they are used more than other chemical pest control measures. In Africa, seed protectant and seed dressing fungicides are often used because of low cost. In China, antibiotic fungicides are produced regionally in small-scale factories. Modern fungicides are nonphytotoxic and have low toxicity for other organisms in the ecosystem (5). The modes of action are specific to fungal metabolism such as inhibition of motility of swimming spore flagella, inhibition of respiration, thickening of cell walls, interference with cell division, or block age of fungal lipid or protein synthesis. Nevertheless, anywhere that a fungicide is relied upon too heavily, and multiple sprays per cropping period are applied, it is possible to have problems of residual build-up. High levels of residue of any chemical can have unforeseeable deleterious effects on nontarget organisms.

Another unwanted effect of overreliance upon a chemical control is that loss of sensitivity to the pesticide may occur. Resistance to fungicides occurs through the development of and selection for strains that are insensitive to the mode of action of the chemical. Thus, useful, ecologically benign fungicides have to be managed carefully if they are to remain effective. The development of pathogen resistance or insensitivity may be delayed by avoiding repeated deployment of the fungicide as the sole control measure. Fungicides can be indispensable and sustainably effective components of integrated disease management systems when combined with other control measures.

Other Compounds

Antibiotics for control of fungal and bacterial diseases are rarely considered economical. Plant volatiles and oils with fungistatic effects have been reported as traditional or indigenous knowledge from Africa, Southeast Asia, and South America. The efficacy and consistency of effect has not been tested rigorously, but these local remedies may provide some options in small-scale agricultural systems. Chemical induction of systemically acquired resistance is an approach that has led to the development of commercial products.

CULTURAL CONTROL

All cropping systems involve different management practices, dependent on the agroeconomic scale of operation, which can be manipulated to influence disease levels. These may involve pathogen avoidance, for example, managed time of planting, and pathogen population management via interplanting, weeding, tillage, clearing plant litter, roguing, mulching, and the addition of organic matter to soil. For a more complete list see Table 1 (1, 2, 6).

BIOLOGICAL CONTROL

Although management of plant diseases is currently dominated by the use of resistant plant varieties, fungicides and cultural practices, other ecologically sustainable interventions are being developed (7, 8). Microorganisms inhabit soil, rhizosphere, or phyllosphere communities where interactions range from synergistic to antagonistic. Plant disease occurs on a susceptible host when a pathogen population finds little competition or challenge from the niche community and environmental factors favor the pathogen's growth and development. Conversely, disease development may be constrained by biological processes in the niche environment. We may be able to alter the balance in favor of the plant by the addition or augmentation of various beneficial microorganisms as biological or microbial pesticides.

Hyperparasitic organisms are pathogenic to the pathogen (fungi on other fungi, fungi on nematodes, viruses on fungi, bacterial and viral pathogens of nematodes, etc.).

Suppressive microorganisms function by crowding or direct antagonism (e.g., suppression of *Geaumannomyces*, a fungal pathogen that causes severe disease in wheat, by *Pseudomonas fluorescens*, a soil-inhabiting bacterium).

Competitors—some strains or species of microbe—outcompete or displace another from the niche. A virulent plant pathogen can be displaced by an avirulent or ''hypovirulent'' strain of the same species. Temporal displacement can occur when one microbe occupies a niche before the other one arrives.

Mass production, formulation, and application techniques of microbial biological control agents have been developed, including various methods of liquid and solid-state culture for fungi and bacteria. Many fungi are known to store well in dry powder form. Carrier substances ranging from water, oil, kerosene, dust, and clay to agricultural wastes such as cereal hulls have been formulated for delivery of biological control agents. Globally, a serious constraint to development of this disease control technology is the economy of scale. Another constraint is the lack of uniformity in regulatory requirements and registration procedures for biological agents (9).

BIOTECHNOLOGICAL CONTROL OPTIONS

Advances in biotechnology offer several control options.

Cross-protection—when a plant cell is invaded by a virus, it is generally not susceptible to infection by another. Thus, an option for control of viral diseases is to genetically modify the plant so that each cell contains a piece of the viral coat protein, inducing permanent cross-protection (10).

Some pathogens pick up small pieces of satellite RNA or DNA, which can reduce their fitness and reduce disease. Vectors can be used to introduce these viral sequences into the host plant.

Nonhost resistance—many plant pathogens can cause disease only on specific host plants. The relationship between a parasite and its host involves "recognition" of the host by the parasite. A gene from a nonhost plant placed in the genome of a host plant may render the modified plant a nonhost by changing some aspect of recognition.

Finally, resistance genes transfer within and between species, and resistance gene amplification within a species are possible.

REFERENCES

1. Zadocks, J.C.; Schein, R.D. *Epidemiology and Plant Disease Management*; Oxford University Press: New York, 1979; 427.
2. Chaube, H.S.; Singh, U.S. *Plant Disease Management: Principles and Practices*; CRC Press: Boca Raton, USA, 1991; 319.
3. Vanderplank, E. *Disease Resistance in Plants*; Academic Press, Inc.: London, 1968; 194.
4. Wolfe, M.S.; Caten, C.E. *Populations of Plant Pathogens, Their Dynamics and Genetics*; Blackwell Scientific Publications: Oxford, 1987; 280.
5. Lyr, H. *Modern Selective Fungicides: Properties, Applications, Mechanisms of Action*; Gustav Fischer Verlag: Jena, New York, 1995; 595.
6. Parker, C.A.; Rovira, A.D.; Moore, K.J.; Wong, P.T.W.; Kollmorgen, J.F. *Ecology and Management of Soilborne Plant Pathogens*; APS Press: St. Paul, MN, 1985; 358.
7. Cook, R.J.; Baker, K.F. *The Nature and Practice of Biological Control of Plant Pathogens*; APS Press: St. Paul, MN, 1989; 539.
8. Hornby, D. *Biological Control of Soil-borne Plant Pathogens*; CAB International: Wallingford, Oxon, UK, 1990; 479.
9. Burges, H.D. *Formulation of Microbial Biopesticides, Beneficial Microorganisms, Nematodes and Seed Treatments*; Kluwer Academic Publishers: Dordrecht, 1998; 412.
10. Kaniewski, W.K.; Thomas, P.E. *Molecular Biology 12*; 1999. www.biotech-info.net (accessed April 18, 2000).

P

PROTECTIVE CLOTHING

Andrew Gilbert
Ministry of Agriculture, Fisheries, & Food, Sand Hutton, York, United Kingdom

INTRODUCTION

Protective clothing is normally associated with operations involving the handling and application of pesticides (1). This is because these tasks carry an inherent risk of personal contamination by the pesticides being used and it is important to reduce or eliminate the consequent risk of personal exposure to them, which may in turn lead to detrimental health effects. To function effectively, protective clothing must resist ingress by contaminants that deposit upon its outside surface, and be capable of removal by the wearer after use without transfer of contaminants to their own clothing or skin. Furthermore, the items worn should also be comfortable to the wearer and not impede the conduct of the work concerned. Items should be readily available and affordable, to suit the practical and economic needs of the user in the context of their working environment. Thus, protective clothing forms part of the risk-benefit equation that should accompany (and justify) rational pesticide use.

WHO SHOULD WEAR PROTECTIVE CLOTHING?

It can be assumed that protective clothing will be necessary to be worn by professional users of pesticides. This is because of the relatively high level of usage and long duration of work involved at the commercial scale, each of which factors increases the likelihood of routine personal contamination, as well as the relatively hazardous properties of some pesticide products (e.g., fumigants) that are approved for use only by professional operators. However, the principles that underlie the wearing of protective clothing as a component of good occupational hygiene may also apply to amateur pesticide users. The inherent hazard from pesticide products approved for "amateur" use is lower and likely usage patterns involve less product, used over a shorter time, but there is possibly a greater tendency toward incidental contamination during preparation and application than is expected from trained and experienced operators.

WHAT PROTECTIVE CLOTHING IS AVAILABLE?

Protective clothing comprises a very wide range of many sorts of items of personal protective equipment (PPE) that can be worn on the body (2). This includes coveralls (as single- or multiple-piece garments) providing a barrier enclosing the body and major limbs, aprons giving supplementary protection to coveralls worn underneath, hats (or coverall hood) to cover the head, face shield or eye protectors to cover the face, respirators that filter inhaled air and cover the lower face, as well as boots and gloves to protect feet and hands.

Most sorts of PPE come as different types, depending on the performance expected from their design and materials of construction. Materials essentially fall into two groups: air-permeable (or "breathable") and air-impermeable. Breathable fabrics (e.g., woven cotton) are needed for comfort and wearability in most situations, but these may have limited resistance to penetration[a] by external contaminants and cannot resist permeation[b] due to their porosity. Air-impermeable fabrics (e.g., rubber) offer a physical barrier preventing penetration by external contaminants and provide varying degrees of resistance to their permeation, so are suitable for construction of garments that are expected to become routinely subject to contamination in use (e.g., gloves). Permeation processes depend greatly on the composition of contaminants, thus water-based solutions tend to permeate rubber far less than those based on organic solvents, hence materials must be

[a]The process of movement of substance through porous materials, seams, pinholes, or other imperfections in a material on a nonmolecular level.

[b]A process by which a substance moves through an unbroken material on a molecular level.

chosen to provide sufficient resistance to breakthrough and subsequent passage of contaminant, to protect the wearer. Respirators must allow sufficient passage of air so as not to inhibit breathing yet be capable of filtering out airborne contaminants, be they particulates, vapors/gases, or both. Alternative types and grades of filter are available and must be selected to withhold the size of particles, or resist breakthrough, by the gas or vapor concerned.

WHERE PROTECTIVE CLOTHING IS NEEDED

An operator's PPE ensemble should cover any body part where contamination is possible (generally everywhere to some degree), but especially those parts of the body where deposits are likely to be greatest. The likelihood of contamination depends on the type of pesticide being used and the way in which it is to be handled or applied. If a high concentration of airborne contamination is expected (e.g., fumigants, high vapor pressure active ingredients, aerosol, fog, or mist applications) then respiratory protection will be needed, and probably whole body protection as well. Sprays (and fine dusts or powders) that may impinge on the body either directly or by drift, can deposit contaminants widely so will require a coverall to be worn. If deposits tend to become localized on certain parts of the body (e.g., those that are closest to the sprayed target) then more localized protection may be appropriate (e.g., waterproof leggings). Direct transfer of contamination to the body from pesticide treated materials or from application equipment (e.g., knapsack sprayers) can be prevented by wearing appropriate localized PPE, such as an apron or a waterproof cape.

Table 1 Hierarchy for consideration of possible exposure control measures

Order to consider	Control measure
First	Elimination of hazardous substances or their substitution with less hazardous alternatives
Second	Use of engineering control measures to improve containment of hazardous substances, so reducing the likely level of contamination leading to exposure
Third	Implementation of operational controls and safe systems of work to lessen the chances of incidental events leading to exposure
Fourth	Use of PPE, which may include items of chemical protective clothing
Fifth	Conduct of exposure and/or health monitoring strategies

Note: The above measures do not obviate the need to employ trained operatives.

WHEN PROTECTIVE CLOTHING IS NEEDED

PPE should always be worn when there is any chance of personal contamination by pesticides, simply as a measure of good occupational hygiene. However, if relatively hazardous pesticides are being used (3) so personal exposure to pesticide could cause adverse health effects, PPE has a key role to play in controlling occupational exposure. In all cases the decision whether to use PPE and the type to be worn must be made *only* in the context of a proper assessment of the hazard (from the pesticide), the potential risk to health (from likely personal exposure), and all aspects of available alternative strategy for controlling exposure to the hazardous substance (see Table 1).

WHO DECIDES WHICH PPE?

There are several parties who normally should contribute information necessary to make the right decision. Primarily, it is for the regulatory authority and supplier of any approved pesticide to address the requirement for PPE to be worn depending on the hazardous properties of the pesticide product, the probable level of occupational exposure, and so the potential risk to health arising from its use. The label of the pesticide (or associated advice supplied with the product) will give recommendations for minimal PPE necessary to be worn for the safe use of the product. Advice normally covers the requirements during handling and preparation for use of the concentrated product separately from those during use or application of diluted pesticide. Handling of concentrated liquids normally will require boots, gloves, coverall (with hood), apron, and face-shield (plus respirator, depending on the nature of the pesticide). Use of diluted pesticides normally requires less stringent precautions, but boots, coverall, and head protection are recognized as sensible basic work wear, with gloves and face-shield or eye protection being options available to prevent exposure via hands and face in circumstances where these are likely to become contaminated. Some pesticide labels may specify the design and materials of necessary PPE, but in many cases the choice of these is left to the wearers (or their advisor) and will depend on how they intend to use the pesticide. Officially approved label guidance should be read in the context of any relevant code of practice, for example the ''Green''

Code in the UK (4), which will give advice on how a user's individual circumstances are to be assessed, and risks addressed. A responsibility also falls on the employer of any pesticide user (including the self-employed) to have carried out an adequate safety assessment and to have taken the steps necessary to ensure the adequate control of potential risks to the health of employees arising from occupational exposure to pesticides. In this regard there are sources of guidance available (5), which assist them, or their professional consultant, to conduct and implement the respective safety assessment. Those who manufacture PPE must ensure that it is appropriately tested and correctly labeled according to its performance limitations, and those who supply or procure PPE must ensure that they have assessed the purpose that the PPE must serve so that it will be up to the job. Well-established traditions of "best practice" (6) associated with professional usage of pesticides are also reflected in formal national standards (7) that encompass the broader range of industrial and professional use of hazardous chemicals.

Ultimately the wearer of any PPE has to play a vital part, insofar as they should never ignore the need to select the right PPE; to wear it correctly; and to ensure its proper storage, maintenance, and serviceability. Only the wearer of any PPE can ensure that it is functioning as it should in the field during work, and that any obvious damage, suspected failure, or occurrence of ingress of contamination (i.e., leakage) is promptly detected and the fault rectified. The importance of avoiding transfer of external contaminants to the inside surfaces of protective clothing should be borne in mind. Particularly, routine and regular renewal of protective gloves is beneficial in reducing the potential for exposure via the hands, where occluded contact with internal contaminants may otherwise promote their enhanced absorption through the skin. Exposure inside gloves, even when correctly selected, may be determined by the number of times the gloves are removed and ingress of contaminant cannot be avoided.

REFERENCES

1. Matthews, G.A. *Pesticide Application Methods*, 3rd Ed., Blackwell Scientific: Oxford, 2000.
2. Oklahoma State University On-line State Library, Personal Protective Equipment on the Internet. http://www.pp.okstate.edu/ehs/links/ppe.htm (accessed Aug 2000).
3. Bates, J.A.R. Health and environmental hazard classification of pesticides. Pestic. Outlook (**2000, June**), 109–115.
4. Ministry of Agriculture (MAFF)/Health and Safety Commission (HSC). In *1998 Code of Practice for the Safe Use of Pesticides on Farms and Holdings (Green Code)*; MAFF Publications: London, 1998.
5. British Crop Protection Council/ATB Landbase. In *Using Pesticides: A Complete Guide to Safe, Effective Spraying*; BCPC Publication Sales: Bracknell, UK, 1999. http://www.bcpc.org (accessed Aug 2000).
6. Gilbert, A.J.; Bell, G.J. Test methods and criteria for selection of types of coveralls suitable for certain operations involving handling or applying pesticides J. Occupational Accidents **1990**, *11*, 255–268.
7. British Standards. In *British Standard Recommendations for the Selection, Use and Maintenance of Chemical Protective Clothing, BS7184*; British Standards Institution: London, 1989. http://www.bsi-global.com (accessed Oct 2000).

PROTOZOAN CONTROL OF PESTS

Leslie C. Lewis
United States Department of Agriculture, Ames, Iowa, U.S.A.

P

INTRODUCTION

Protozoa is a subkingdom of eukaryotic organisms that is taxonomically subdivided into several phyla, some of which contain entomogenous species. Species in the phylum Microspora have been investigated extensively as possible components of integrated pest management programs. Microsporidia are used herein to exemplify how Microspora suppress or control pest insects. Microsporidia are ubiquitous, obligatory intracellular parasites that are disease agents for several insect species; however, they are nonconventional because they are not an apply-and-kill insecticide. Microsporidia are naturally maintained in an insect population thereby causing a chronic, low-level suppression sometimes referred to as the ''common cold'' of the insect world. An infection with microsporidia can be relatively benign unless insects are stressed by other biotic and abiotic factors. Microsporidia are maintained in a population by both horizontal and vertical transmission. *Nosema pyrausta*, a microsporidium infecting the European corn borer, *Ostrinia nubilalis*, is used to illustrate these methods of transmission.

In horizontal transmission a spore is eaten by a European corn borer larva. In the midgut, it germinates, extrudes a polar filament, and injects sporaplasm into a midgut cell. The inoculum passes through a developmental cycle in which spores are produced that infect other tissues. Spores in infected midgut cells are sloughed into the gut lumen and are passed in the feces to the corn (*Zea mays*) plant. These spores remain viable and are consumed during larval feeding so that the infection cycle is repeated in midgut cells of the new host. If a female larva is infected, *Nosema* is passed to the filial generation by vertical transmission. As the infected larva develops through to an adult the ovarial tissue and developing oocytes become infected with *N. pyrausta*. The embryo is infected within the yolk. When larvae hatch they are infected with *N. pyrausta*. Both horizontal and vertical transmission maintain *N. pyrausta* in natural populations of European corn borer (1). *N. pyrausta* suppresses populations of European corn borer by reducing oviposition, percentage hatch, and survival of infected neonate larvae (2, 3).

In horizontal transmission, microsporidia are usually passed from larva to larva within a generation; however, horizontal transmission also can occur between generations as in European corn borer. Egg masses infected with *N. pyrausta* were placed in the whorl of V7 (4) stage corn plants. When the larvae hatched, they fed and defecated on the corn plants, passing viable spores to protected parts of the plants. European corn borers from the subsequent generation laid eggs on the same plants. These larvae were infected (80%) from the *N. pyrausta* inoculum left on the plant from the previous generation's larvae (5).

CONVENTIONAL APPLICATIONS

Microsporidia have been used experimentally as microbial insecticides, but such use is limited because microsporidia must be produced in vivo and commercial availability is rare. *N. pyrausta* applied to V7-stage corn with a backpack sprayer (22.56×10^9 spores/plant) reduced the number of European corn borer larvae per plant by 48% and produced an *N. pyrausta* infection in 62% of the recovered larvae. When similar amounts of *N. pyrausta* were applied to R1-stage corn the number of larvae per plant was reduced by 17 to 43% (6).

Nosema locustae produced in laboratory-reared *Melanoplus bivitattus* was applied to rangeland at rates of 0, 50, 100, 200, 400, and 800 spores/6.5 cm^2 to suppress several grasshopper species. One *M. bivittatus* can produce in excess of 1×10^9 spores; enough inoculant to treat 3 acres of rangeland at 50 spores/6.5 cm^2 (7). A rate of 50 spores/6.5 cm^2 was not statistically different from an application of 800 spores/6.5 cm^2 in reducing numbers of grasshoppers. Additional studies concluded that 6.3 to 9.4×10^9 spores formulated on 0.5–0.7 kg of wheat bran reduced density in all grasshopper species by 50–60% (8).

Other pest insects have been experimentally suppressed with microsporidia, e.g., Mexican bean beetle, *Epilachna varivestis,* with *Nosema varivestis* and *Nosema epilachnae* (9); black cutworm, *Agrotis ipsilon*, with *Vairimorpha nectarix* (10, 11); mormon crickett, *Anabrus simplex*, with *N. locustae* (12); and stored-product Coleoptera with *Nosema spp.* (13).

NATURAL SUPPRESSION

The impact of natural infections of microsporidia has been extensively documented. *Burnella dimorpha* infects the tropical fire ant, *Solenopsis geminata*, and is instrumental in its suppression. Pupae within an ant colony infected with *B. dimorpha* rupture and are eaten by adult worker ants. The workers filter the spores, and hold them in the infrabuccal cavity from which they are expelled and fed along with food to 4th instars. Thus, the workers are vectors of *B. dimorpha* within the colony (14).

Nosema fumiferanae infects western spruce budworm, *Christoneura fumiferanae*. Although the infection is usually not lethal, it can reduce pupal weight, and adult longevity and fecundity (15–17). The gypsy moth, *Lymantria dispar*, a serious pest of woodlands in the eastern United States is not infected by any microsporidia. In eastern Europe, however, *L. dispar* is infected with several species of microsporidia. Currently, methodologies are being developed and risk assessment analyses are being conducted to introduce microsporidia from Europe into the U.S. gypsy moth population (18).

Lucilia cuprina, the Australia sheep blowfly, is infected with the microsporidium *Octosporea muscaedomesticae*. Investigations are ongoing to use this microsporidium as a control agent against this livestock pest (19, 20).

INFLUENCE OF MICROSPORIDIA COMBINED WITH ABIOTIC AND BIOTIC FACTORS

Chemical and microbial insecticides, beneficial insects, host plant resistance, and temperature have been used with microsporidia to suppress insects. Combinations of *N. pyrausta*, *Bacillus thuringiensis*, carbaryl, and carbofuran were applied to V7 stage corn to suppress larval populations of the European corn borer. All entities acted independently in suppressing larvae (21). In additional studies, *B. thuringiensis*, carbaryl, and carbofuran acted independently when applied to corn plants infested with *N. pyrausta*-infected larvae (22).

Reduced rates of carbaryl and *N. locustae* were effective in decreasing populations of several species of rangeland grasshoppers. Carbaryl caused immediate knock-down of some grasshoppers; the survivors were infected with *N. locustae*, which debilitated them and exposed them to predation (23).

Nosema pyrausta and larval and egg parasitoid combinations also suppress European corn borer larvae. *Trichogramma brassicae*, an egg parasitoid, develops in *N. pyrausta*-infected eggs of European corn borer. The viability of *T. brassicae* was reduced slightly, however, both organisms contributed to suppression (24, 25).

Macrocentrus grandii, a parasitoid of European corn borer larvae, co-exists with *N. pyrausta*. *N. pyrausta* has a negative impact on *M. grandii* but the two organisms do survive in the ecosystem and contribute to European corn borer population suppression (26–28).

Some cultivars of corn are naturally resistant to feeding by European corn borer. Natural resistance ranges from fully susceptible to fully resistant to larval feeding in many hybrid corn populations. Survival of European corn borer on plants with medium-to-full resistance is greatly reduced if the larvae are infected with *N. pyrausta* (29, 30).

Cold, wet weather restricts activity of some adult insects, especially gravid females. Oviposition by European corn borers infected with *N. pyrausta* is dramatically reduced when the insects are maintained in an environment not favorable for oviposition. A noninfected European corn borer population held under favorable conditions produced 35 times as many eggs as an infected population held under adverse conditions (31). Most likely this same scenario occurs in nature.

SUMMARY

Protozoa are instrumental in suppressing many insect pests. The phylum Microspora, one of several phyla of Protozoa, contains several species of microsporidia indigenous to insect populations. Some of these microsporidia have been used to manage crop insect pests; however, equal success in insect management is achieved when an additional management component, e.g., microbial and chemical insecticides, host plant resistance, or parasitoids is combined with the microsporidia. Yes, protozoa suppress insect populations, however, the effect can be maximized by the addition of another management component.

REFERENCES

1. Lewis, L.C. Biological Control. In *Ullman's Encyclopedia of Industrial Chemistry,* 5th Ed.; VCH Publishing: Florida, 1985; A4, 77–97.
2. Kramer, J.P. Some relationships between *perezia pyraustae* (Paillot) (Sporozoa, Nosematidae) and *pyrausta nubilalis*

(Hübner) (Lepidoptera, Pyralidae). J. Insect Pathol. **1959**, *1*, 25–33.

3. Lewis, L.C.; Lynch, R.E.; Guthrie, W.D. Biology of European corn borers reared continuously on a diet containing fumidil B. Ann. Entomol. Soc. Am. **1971**, *64*, 1264 –1269.
4. Ritchie, S.W.; Hanway, J.J.; Benson, G.O. How a corn plant develops. Iowa State Univ. Sci. Tech. Spec. Rep. **1997**, *48*.
5. Lewis, L.C.; Cossentine, J.E. Season long intraplant epizootics of entomopathogens, *Beauveria bassiana* and *Nosema pyrausta*, in a corn agroecosystem. Entomophaga **1986**, *31* (4), 363–369.
6. Lewis, L.C.; Lynch, R.E. Foliar application of *Nosema pyrausta* for suppression of populations of European corn borer. Entomophaga **1978**, *23* (1), 83–88.
7. Henry, J.E. Experimental application of *Nosema locustae* for control of grasshoppers. J. Invertebr. Pathol. **1971**, *18*, 380–394.
8. Henry, J.E.; Tiahrt, K.; Oma, E.A. Importance of timing, spore concentrations, and levels of spore carrier in applications of *Nosema locustae* (Microsporida: Nosematidae) for control of grasshoppers. J. Invertebr. Pathol. **1973**, *21*, 263–272.
9. Brooks, W.M. Comparative effects of *Nosema epilachnae* and *Nosema varivestis* on the Mexican bean beetle, *Epilachna varivestis*. J. Invertebr. Pathol. **1986**, *48*, 344–354.
10. Cossentine, J.E.; Lewis, L.C. Impact of *Vairimorpha necatrix* and *Vairimorpha* sp. (Microsporida: Microsporida) on *Bonnetia comta* (Diptera: Tachinidae) within *Agrotis ipsilon* (Lepidoptera: Noctuidae) hosts. J. Invertebr. Pathol. **1986**, *47*, 303–309.
11. Grundler, J.A.; Hostetter, D.L.; Keaster, A.J. Laboratory evaluation of *Vairimorpha necatrix* (Microspora: Microsporidia) as a control agent for the black cutworm (Lepidoptera: Noctuidae). Environ. Entomol. **1987**, *16* (6), 1228–1230.
12. Henry, J.E. Experimental control of the mormon cricket, *Anabrus simplex*, by *Noseam locustae* (Microspora: Microspora), a protozoan parasite of grasshoppers (Orthoptera: Acrididae). Entomophaga **1982**, *27* (2), 197–201.
13. Khan, A.R.; Selman, B.J. *Nosema* spp. (Microspora: nosematidae) of Stored-product Coleoptera and Their Potential as Microbial Control Agents. In *Management and Control of Invertebrate Crop Pests*; Russel, G., Ed.; Andover: Hampshire, UK, 1989; 133–163.
14. Jouvenaz, D.P.; Ellis, E.A.; Lofgren, C.S. Histopathology of the tropical fire ant, *Solenopsis geminata*, infected with *Burenella dimporpha* (Microspora: Microsporida). J. Invertebr. Pathol. **1984**, *43*, 324–332.
15. Wilson, G.G. Effects of occurrence of *Pleistophora schubergi* (Microsporida) on the spruce budworm, *Choristoneura fumiferana* (Lepidoptera: Tortricidae). Can. Entomol. **1982**, *114*, 81–83.
16. Wilson, G.G. In *A Dosing Technique and the Effects of Sublethal Doses of* Nosema fumiferanae *(Microsporidia) in a Spruce Budworm* (Choristoneura fumiferanae) *(Lepidoptera: Tortricidae) Population*, Proceedings of the Entomological Society of Ontario, 1983; 108, 144–145.
17. Wilson, G.G. In *The Transmission and Effects of* Nosema fumiferanae *and* Pleistophora schubergi *(Microsporidia) on* Choristoneura fumiferanae *(Lepidoptera: Tortricidae),* Proceedings of the Entomological Society of Ontario, 1984; 115, 71–75.
18. Solter, L.F.; Maddox, J.V.; McManus, M.L. Host specificity of microsporidia (Protista: Microspora) from European populations of *Lymantria dispar* (Lepidoptera: Lymantridae) to indigenous North American lepidoptera. J. Invertebr. Pathol. **1997**, *69*, 135–150.
19. Cooper, D.J.; Pinnock, D.E.; Bateman, S.M. Susceptibility of *Lucilla cuprina* (Wiedemann) (Diptera: Calliphoridae), to *Octosporea muscaedomesticae* flu. J. Aust. Entomol. Soc. **1983**, *22*, 292.
20. Smallridge, C.J.; Cooper, D.J.; Pinnock, D.E. The effect of the microsporidium *Octosporea muscaedomesticae* on adult *Lucilia cuprina* (Diptera: Calliphoridae). J. Invertebr. Pathol. **1995**, *66*, 196–197.
21. Lublinkhof, J.; Lewis, L.C.; Berry, E.C. Effectiveness of integrating insecticides with *Nosema pyrausta* for suppressing populations of the European corn borer. J. Econ. Entomol. **1979**, *72*, 880–883.
22. Lewis, L.C.; Lublinkhof, J.; Berry, E.C.; Gunnarson, R.D. Response of *Ostrinia nubilalis* (Lep.: Pyralidae) infected with *Nosema pyrausta* (Microsporida: Nosematidae) to insecticides. Entomophaga **1982**, *27* (2), 211–218.
23. Onsager, J.A.; Rees, N.E.; Henry, J.E.; Nelson, F.R. Integration of bait formulations of *Nosema locustae* and carbaryl for control of rangeland grasshoppers. J. Econ. Entomol. **1981**, *74*, 183–187.
24. Nelson, F.R.; Sajap, A.S.; Lewis, L.C. Effects of the microsporidium *Nosema pyrausta* (Microsporida: Nosematidae) on the egg parasitoid, *Trichogramma nubilale* (Hymenoptera: Trichogrammatidae). J. Invertebr. Pathol. **1988**, *52*, 294–300.
25. Saleh, M.; Lewis, L.C.; Obrycki, J.J. Selection of *Nosema pyrausta* (Microsporidia: Nosematidae)-infected *Ostrinia nubilalis* (Lepidoptera: Pyralidae) eggs for parasitization by *trichogramma nubilale* (Hymenoptera: Trichogrammatidae). Crop Protection **1995**, *14*, 327–330.
26. Andreadis, T.G. *Nosema pyrausta* infection in *Macrocentrus grandii*, a braconid parasite of the European corn borer, *Ostrinia nubilalis*. J. Invertebr. Pathol. **1980**, *3*, 229–233.
27. Seigel, J.P.; Maddox, J.V.; Ruesink, W.G. Impact of *Nosema pyrausta* on a braconid, *Macrocentrus grandii,* in central Illinois. J. Invertebr. Pathol. **1986**, *47*, 271–276.
28. Cossentine, J.E.; Lewis, L.C. Development of *Macrocentrus grandii* (Goidanich), within microsporidian-infected *Ostrinia nubilalis* (Hübner) host larvae. Can. J. Zool. **1987**, *65*, 2532–2535.
29. Lewis, L.C.; Lynch, R.E. Influence on the European corn borer of *Nosema pyrausta* and resistance in maize to leaf feeding. Environ. Entomol. **1976**, *5*, 139–142.

P

30. Lynch, R.E.; Lewis, L.C. Influence on the European corn borer of *Nosema pyrausta* and resistance in maize to sheath-collar feeding. Environ. Entomol. **1976**, *5*, 143–146.
31. Lewis, L.C. *Effect of Nosema pyrausta and Temperature on Egg Production by Ostrinia nubilalis*, Proceedings of the XVIII Conference of the International Working Group on *Ostrinia*; Turda, Romania, Sept 11–16, 1995.

BIBLIOGRAPHY

Brooks, W.M. Entomogenous Protozoa. *CRC Handbook of Natural Pesticides, Vol. 5: Microbial Insecticides Part A Entomogenous Protozoa and Fungi*; Ignoffo, C.M., Ed.; CRC Press: Boca Raton, Florida, 1988.

Canning, E.U. An Evaluation Protozoal Characteristics in Relation to Biological Control of Pests. In *Parasitology Vol. 19: Parasites as Biocontrol*; Anderson and Canning, Eds.; Symposia of the British Society for Parasitology: London, 1982; 119–149.

Goettel, M.S.; Johnson, D.L. Microbial Control of Grasshoppers and Locusts. In *Memoirs of the Entomological Society of Canada*; Entomological Society of Canada: Ontario, Canada, 1997.

Wilson, G.G. Protozoans for Insect Control. In *Microbial and Viral Pesticides*; Kurstak, E., Ed.; Marcel Dekker, Inc.: New York, 1982; 587–600.

PUBLIC HEALTH AND COSTS OF PESTICIDES

Kelsey Hart
David Pimentel
Cornell University, Ithaca, New York, U.S.A.

P

INTRODUCTION

Since the first use of DDT for crop protection in 1945, the total amount of pesticides used in agriculture worldwide has been staggering. In 1945, about 50 million kg of pesticides were applied worldwide. Today, global usage is currently at about 2.5 billion kg per year, an approximate 50-fold increase since 1945 (1). Unfortunately, the toxicity of most modern pesticides is more than 10-fold greater than those used in the early 1950s (1), so the potential hazards have increased as well. In fact, studies have linked exposure to pesticides with a variety of human health problems, from asthma to cancer (2). High levels of exposure to toxic pesticides can even result in fatal poisonings. In addition, we are discovering that we can be unknowingly exposed to pesticides and pesticide residues through the food we eat, water we drink, and air we breathe (3), and that both short-and long-term exposure to pesticides can lead to chronic health effects. Based on the available data, estimates are that human pesticide poisonings and related illnesses in the United States cost about $933 million each year (3).

EXPOSURE TO PESTICIDES

While farmers and pesticide applicators typically are exposed to higher levels and more kinds of pesticides, the general public is also exposed—often unwittingly—to pesticides and pesticides residues in their daily lives. For example, about 35% of the foods purchased by American consumers have detectable levels of pesticide residues (4). This estimate is, in fact, conservative because we currently test for only about one-third of the pesticides in use.

In addition to food contamination, the public can also be exposed to pesticides in other ways. The principle exposure of the general public in the United States occurs in the home; about 90% of all U.S. households use pesticides on their lawn, garden, and/or the inside of the home (5). Drinking water can also contain significant chemical and pesticide residues; at present, more than 10% of U.S. rivers and 5% of U.S. lakes are measurably polluted with pesticides (5). Ground water supplies can also be polluted when pesticides seep into aquifers or wells (6). Finally, public exposure to pesticides can occur through accidents or spills, or even through the air during application, when pesticides drift from the target area into more populated towns and cities. In fact, only 25% to 50% of pesticides applied by aircraft under ideal weather conditions actually reach the target area (1).

ACUTE EFFECTS: PESTICIDE POISONINGS

In 1945, when synthetic pesticides were first used, few pesticide poisonings were reported. But by the late 1960s, both pesticide usage and toxicity had increased so dramatically that the number of human pesticide poisonings was substantial (1). Unfortunately, this trend has continued into the present (Fig. 1). Just in the last decade, the total number of pesticide poisonings in the United States has increased from 67,000 in 1989, to the current level of 110,000 per year (2). Worldwide, the increased use of pesticides results in approximately 26.5 million cases of occupational pesticide poisonings each year, and an unknown number of nonoccupational pesticide poisonings (7). Of all these estimated poisoning episodes, about 3 million cases are hospitalized, resulting in approximately 220,000 fatalities and about 750,000 cases of chronic illness every year (8).

Poisonings can occur when pesticides contact the skin or eyes, are inhaled, or are ingested. Typically, pesticides can have acute local effects on the area they directly contact—skin, eye, or respiratory tract irritation—in addition to acute systemic effects. Different pesticides act on different systems in the body in a variety of ways, but most common insecticides—organophosphates and organochlorines (i.e., Parathion)—have acute neurotoxic effects. This means that they impair normal functioning of the brain and/or the spinal cord, which can result in

Encyclopedia of Pest Management

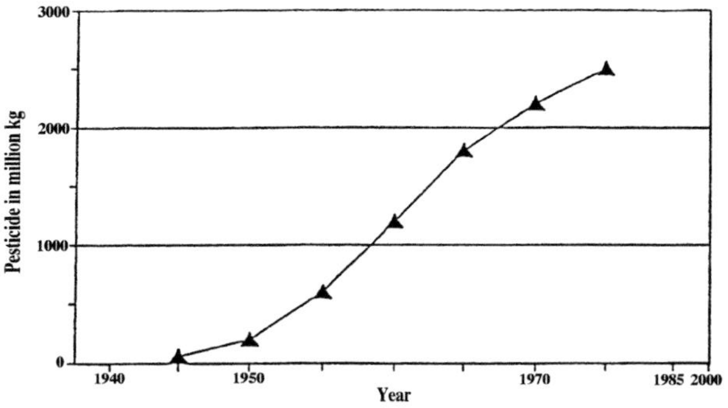

Fig. 1 Trend in annual world pesticide use (2).

tremors, paralysis, seizures, and other systemic effects. Large degrees of exposure or exposure to highly toxic pesticides can be fatal.

CHRONIC EFFECTS: CANCER AND OTHER HEALTH CONCERNS

Exposure to pesticides, though, does not always occur at a level sufficient to produce these acute symptoms. Many people are exposed to low levels of pesticides over a long period of time through their occupations or in the food they eat or water they drink. If no overt symptoms occur close to the time of exposure, we often assume that no damage is being done. However, pesticides have been associated with numerous chronic illnesses and health problems, especially in some highly sensitive individuals.

Chronic effects of pesticides are diverse and can affect most systems of the human body. The major types of chronic health effects that pesticides can have are neurological effects, respiratory and reproductive effects, and carcinogenic effects. The chronic neurotoxic effects of pesticide exposure are not well understood, but there is some evidence linking pesticide exposure with symptoms like fatigue, muscle weakness, and sensory disturbances as well as cognitive effects such as memory loss, language problems, and learning impairment. The malady organophosphate induced delayed poly-neuropathy (OPIDP) is well documented and includes irreversible neurological effects.

In addition to their neurotoxic effects, pesticides can have adverse effects on the respiratory and reproductive systems. For example, 15% of a group of professional pesticide applicators suffered asthma, chronic sinusitis, or chronic bronchitis as compared with 2% of people who used pesticides infrequently (9). Studies have also linked pesticides with reproductive effects such as infertility and fetal deformities, but the data are still inconclusive (10). Some pesticides have been shown to cause testicular dysfunction or sterility in animals, and similar effects are suspected in humans, but are less well understood at present (10). Sperm counts in males in Europe and the United States, for example, declined by about 50% between 1938 and 1990 (11)—a time period during which the use of synthetic pesticides increased about 30-fold and toxicity per pound increased about 10-fold (1). At present, there is evidence that human sperm counts continue to decrease by about 2% per year (3).

U.S. data indicate that 18% of all insecticides, and about 90% of all fungicides, are carcinogenic (12), and many studies have shown that risks for certain types of cancers are higher in people—such as farmworkers and pesticide applicators—who are more frequently exposed to certain pesticides. Certain pesticides have been shown to induce tumors in lab animals; there is some evidence to suggest that they may have similar effects in humans (10).

Many pesticides are also estrogenic—they mimic or interact with the hormone estrogen—linking them to the increased breast cancer rate among some groups of women in the United States. The breast cancer rate rose from 1 in 20 in 1960 to 1 in 8 in 1995 (13). There was a concurrent increase in pesticide use during that time period, and, although it has not been concretely linked to the increase in breast cancer rates, some studies suggest that exposure to pesticides is related to breast cancer incidence. Pesticides that interfere with the body's endocrine—hormonal—system can also have reproductive, immulogic, or developmental effects; these effects are well documented in animals and just starting to be understood in humans (10). While endocrine disrupting chemicals may appear less dangerous at first glance—hormonal effects rarely result in acute poisonings or sudden death—their effects on reproductive and developmental processes may prove to have far-reaching and even more sinister consequences (10).

EFFECTS ON CHILDREN

The negative health effects that pesticides can have—both acute and chronic—can be more significant in children than in adults, for several important reasons. First, children have much higher metabolic rates than adults, and their ability to activate, detoxify, and excrete toxic compounds is different from that of adults. Also, because of their smaller physical size, children are exposed to higher levels of pesticides per unit of body weight. In addition, certain types of pesticides are inherently more dangerous for children than for adults (2, 10). For example, the organophosphate and carbamate classes of pesticides adversely affect the nervous system by inhibiting cholinesterase, a critical enzyme, and can produce difficulty breathing, partial or total paralysis, convulsions, coma, or death. This problem is particularly significant for children since their brains are more than five times larger in proportion to their body weight than an adult's brain, making cholinesterase even more vital. In a California study, 40% of the children working in agricultural fields had blood cholinesterase levels below normal, a strong indication of organophosphate and carbamate pesticide poisoning (14). In addition, a study in England and Wales has shown that 50% of all pesticide poisoning incidents in those countries involved children less than 10 years of age (15). Use of pesticides in the home is also linked to childhood cancer (16).

Because these studies and others have demonstrated that children's increased sensitivities to toxicants is a significant concern, some pesticide regulations have recently been reevaluated to provide special protections for children and infants. The Food Quality Protection Act of 1996 requires the EPA to specifically address risks that pesticides pose to infants and children and provides for an additional safety factors to take into account the potentially greater exposure and/or sensitivity to pesticide effects on infants and children.

ECONOMIC COSTS AND CONCLUSIONS

Although no one can place a precise monetary value on a human life, the economic "costs" of human pesticide poisonings have been estimated. For our assessment, we use the conservative estimate of $2.2 million per human life—the average value that the surviving spouse of a slain New York City policeman receives (17). Available estimates suggest that human pesticide poisonings and related illnesses in the United States cost about $933 million per year (3). Pesticide use, though, provides a substantial net agricultural return of $12 billion/year (2). Are the public health risks associated with pesticide use a great enough concern to warrant a reduction in pesticide use?

Given the known—and suspected—adverse affects of pesticide use on public health discussed in previous sections, it seems fairly clear that our present levels and methods of pesticide use present significant public health dangers and concerns. However, the abrupt and complete cessation of synthetic pesticide use is not without substantial costs of its own. Termination of pesticide use would result in huge crop losses—the food supply would be severely reduced and a huge number of people would starve in a short time (18), an especially dire consequence given the 3 billion people—half the world's population—currently malnourished worldwide (19).

Clearly, it is essential that all the costs and benefits—economic, environmental, and social/health—of pesticide use be considered when current and future pest control programs are being developed and evaluated. A recent study estimated that the environmental and social costs related to U.S. pesticide use—crop losses, public health effects, pesticide resistance, water pollution, and other environmental effects—total $8.3 billion each year (3). Furthermore, another study has shown that U.S. pesticide use can be reduced up to one-half without any reduction in crop yields and cosmetic standards, and only a minimal 0.6% increase in food costs (20). Therefore, it is clear that our current methods of chemical pesticide use need to be reconsidered and evaluated with the above-outlined public health effects and the related economic costs in mind, toward the goal of the development of sound, sustainable pest management practices that maximize the benefits of pesticide use while at the same time minimizing the adverse effects that pesticides can have on human health.

REFERENCES

1. Pimentel, D. Protecting Crops. In *The Literature of Crop Science*; Olsen, W.C., Ed.; Cornell University Press: Ithaca, NY, 1995; 49–66.
2. Pimentel, D.; Hart, K.A. Ethical, Environmental, and Public Health Implications of Pesticide Use. In *Perspectives in Bioethics*; Galston, A., Ed.; Johns Hopkins University Press: Baltimore, MD, 2000, *in press*.
3. Pimentel, D.; Greiner, A. Environmental and Socio-Economic Costs of Pesticide Use. In *Techniques for Reducing Pesticide Use*; Pimentel, D., Ed.; John Wiley & Sons: Chichester, UK, 1997.
4. FDA. Food and drug administration pesticide program residues in foods—1989. J. Assoco. Anal. Chem. **1990**, *73*, 127A–146A.
5. Pimentel, D.; Wilson, C.; McCullum, C.; Huang, R.; Dwen, P.; Flack, J.; Tran, Q.; Saltman, T.; Cliff, B. Economic and environmental benefits of biodiversity. BioSci **1997**, *47* (11), 747–757.
6. Pimentel, D.; Acquay, H.; Biltonen, M.; Rice, P.; Silva, M.; Nelson, J.; Lipner, V.; Giordano, S.; Horowitz, A.; D'Amore, M. Assessment of Environmental and Economic Impacts of Pesticide Use. In *The Pesticide Question: Environment, Economics and Ethics*; Pimentel, D., Lehman, H., Eds.; Chapman & Hall: New York, 1993; 47–84.
7. UNEP. *Global Environmental Outlook*. United Nations Environment Program; Nairobi, 1997.
8. WHO. *Our Planet, Our Health: Report of the WHO Commission on Health and Environment*; World Health Organization; Geneva, 1992.

P

9. Weiner, B.P.; Worth, R.M. Insecticides: Household Use and Respiratory Impairment. In *Adverse Effects of Common Environmental Pollutants*; MSS Information Corporation: New York, 1972; 149–151.
10. Colborn, T.; Dumanoski, D.; Myers, J.P. *Our Stolen Future: Are We Threatening Our Fertility, Intelligence, and Survival? A Scientific Detective Story*; Dutton: New York, 1996.
11. Carlsen, E.; Giwercman, A.; Keiding, N.; Skakkebaek, N.E. Evidence for decreasing quality of semen during past 50 years. Brit. Med. J. **1992**, *305*, 609–613.
12. NAS. *Regulating Pesticides in Food*; National Academy of Sciences: Washington, DC, 1987.
13. McCarthy, S. Congress takes a look at estrogenic pesticides and breast cancer. J. Pestic. Reform **1993**, *13* (4), 25.
14. Repetto, R.; Baliga, S.S. *Pesticides and the Immune System: The Public Health Risks*; World Resources Institute: Washington, DC, 1996.
15. Casey, P.; Vale, J.A. Deaths from pesticide poisoning in England and Wales: 1945–1989. Human Expt. Toxicol. **1994**, *13*, 95–101.
16. Leiss, J.K.; Savitz, D.A. Home pesticide use and childhood cancer: a case-control study. Am. J. Public Health **1995**, *85*, 249–252.
17. Nash, E.P. *What's a Life Worth*; New York Times, 1994.
18. Lehman, H. Values, Ethics, and the Use of Synthetic Pesticides in Agriculture. In *The Pesticide Question: Environment, Economics and Ethics*; Pimentel, D., Lehman, H., Eds.; Chapman & Hall: New York, 1993; 347–379.
19. WHO. *Micronutrient Malnutrition—Half the World's Population Affected*, World Health Organization; Geneva, 13 November 1996; 78, 1–4.
20. Pimentel, D.; McLaughlin, L.; Zepp, A.; Lakitan, B.; Kraus, T.; Kleinman, P.; Vancini, T.; Roach, W.J.; Graap, E.; Keeton, W.S.; Selig, G. Environmental and Economic Impacts of Reducing U.S. Agricultural Pesticide Use. In *Handbook on Pest Management in Agriculture*; Pimentel, D., Ed.; CRC Press: Boca Raton, FL, 1991; 679–718.

QUALITY AND YIELD LOSSES

Elke Pawelzik
Georg-August-University Goettingen, Goettingen, Germany

Q

INTRODUCTION

Agronomy and environment as abiotic factors have a significant influence on the occurrence of pests and weeds. There exist various relationships between several agronomic factors (e.g., nutrient supply), environmental factors (e.g., rainfall), and pests and weeds. On the other hand, the kind of crop and its cultivar is also important for the occurrence and extension of pests. Depending on the different interactions between abiotic and biotic factors, and insect pests and mites, plant pathogens and weeds cause both yield and quality losses. The results are often considerable economic losses. Furthermore, crops with reduced quality cannot be used or can be used only with restrictions for several food or feed purposes.

QUALITY LOSSES

Insects, plant pathogens, and weeds damage quality in different ways. Insects and disease attack the plant in various parts and at different stage of development, which may lead to considerable changes in physical characteristics, chemical composition, and processing properties of grains, tubers, etc. (Table 1). Several pests attack plants in early development stages, which may reduce the seeding rate and lead to larger kernels, but there is little effect on their quality. Pests attacking the roots influence quality of crops interfering with nutrient and water uptake. In some cases lodging will be induced, and the main symptoms shown in grain are reduced kernel size and test weight and shriveling, respectively. Pests damaging the stem restrict nutrient and water transport, which may lead to reduced kernel size, test weight, and protein quality, whereas protein concentration is mostly unchanged or even higher. The influence of pests, which damage foliage on quality parameters, depends on the stage of development of the crop and when the pests attack it. If for example pests attack during tillering and early stem elongation there is only little effect on quality. Pest attack in later stages of plant development could cause reduced accumulation of stem carbohydrate reserves, which may lead to shriveling of the kernels. Reduced kernel size occurs also if pests attack during the endosperm cell division phase of grain growth. Pests directly affecting grains during their filling period cause negative changes in quality. Other pests feed on unripe kernels and cause shriveling as well as increase of α-amylase activity (1). Pathogens have different effects on grain protein accumulation (2). *Septoria* spp. may have only little influence on nitrogen translocation and may raise protein concentration in the kernel. On the other hand, mildew may decrease protein concentration by reducing nitrogen retranslocation from leaves. *Fusarium* spp., which became more and more important in Europe and North America in the past 20 years, infects developing wheat ear, which may result in reduced test weight of the kernels and cause significant changes in the composition of proteins, polysaccharides, and lipids. *Fusarium*-damaged kernels have a poor baking performance due to the degradation of wheat gluten proteins (3). Furthermore, the development of *Fusarium* is often associated with the formation of mycotoxins, particularly the trichothecene deoxynivalenol. *Fusarium*-damage of barley may lead to poor malting properties.

The influence of weeds on quality losses depends on the competition between weeds and crops. When weeds establish themselves before grain growth of crops, the result may be a reduction in ear number per area and kernel number per ear. Weeds growing strong in later stages of development of crops may decrease kernel size and specific weight through a reduction in photosynthesis by shading and competing for soil water and nutrients. In that way they may reduce grain protein concentration, leading to injured quality. Moreover, some weeds often mature later than the crop and therefore they have higher moisture content and may cause deterioration of the crops after harvest and during storage (e.g., *Ranunculus arvensis, Sonchus arvensis*). Plants growing from seeds shed by a previous crop are also weeds and can affect the grain from the current crop. In wheat they may have an influence on the falling number of the kernels (4). Self-sown cereal plants often mature earlier than the current crop and therefore they may cause preharvest sprouting in wet weather. Preharvest sprouting of wheat leads to high α-amylase activity, which results in poor-quality bread with low loaf volume and sticky crumb. When weed seeds

Table 1 Influence of insects, diseases, and weeds on quality losses in crops

Loss in size and thousand-kernel-weight (grain)
Loss in size of tubers and roots
Increase in moisture content
Changes in chemical composition of the kernels, tubers, and roots
Discoloration of kernels and tubers
Formation of mycotoxins by plant pathogens
Loss in germination of the seeds and tubers
Reduced processing properties (e.g., baking and malting properties)
Reduced feeding properties (nutrition value)

get into grain for milling and processing, several of them (e.g., *Sinapis arvensis*, *Trigonella folnum graecum*) may cause discoloration of grain products like flour and bread. Furthermore, other weeds may influence flavor and odor of flour, bread, baked goods, and noodles (e.g., *Coronilla varia*, *Thlaspi arvense*). Some weed varieties are toxic and their seeds can contaminate grains and their products (e.g., *Polygonum convolvulus*, *Sinapis arvensis*) (5). Moreover, weed plants may indirectly influence the quality of crops. Frequently they are hosts or interim hosts for insects and plant pathogens and in that way they are responsible for further extension of pests.

YIELD LOSSES

Yield losses caused by insects, pathogens, and weeds occur as the result of reduced growth as well as the dying off of whole plants.

Losses are caused mainly by insects, mites, plant pathogens, and weeds, whereas mammal and bird pests cause usually losses to a lesser extent. Injuries and losses result from the attack on the growing plant by pests, which use different parts of the plant (seedling, root, stem, foliage, flower, kernel) as nutrition source. Weeds cause losses because they compete with crops for nutrients, water, and light. The period between the emergence of crop and weed strongly affects weed competitiveness and has been found, e.g., for rice yield losses, more important than weed density (6). Between cropping practice, biotic constraints, and yield losses different relationships exist. In a continuously cropping experiment for corn it has been reported that without sufficient fertilization and weeding, yields declined to less than 30% of the original in the second year, whereas with fertilization only, yields began to decrease in the third year (7).

In several regions damages caused by insects, pathogens, and weeds might be associated with increased agricultural intensification, if there have not been achieved adequate plant control and protection (8).

In the last 50 years yield losses from plant pathogens have rapidly increased and amount to 12% of harvest (9). Reasons for this development are increasing cropping frequency (reduced crop diversity, monoculture) and crop inputs (nonorganic fertilizers, pesticides), as well as reduced soil cultivation.

FUTURE CONCERNS

Increasing yield and quality losses by pests and weeds are associated with extended agricultural intensification. Control of pest and weed development especially by agronomical or chemical means is necessary in minimizing yield losses and preventing poor quality. Thus, there will be a requirement for better biological and chemical control strategies in good agriculture practice or integrated systems of plant protection. Efforts of plant protection science in cooperation with other disciplines of agriculture science should find answers to many of the unsolved problems of preventing crop losses.

REFERENCES

1. Helenius, J.; Kurppa, S. Quality losses in wheat caused by the orange wheat blossom midge *Sitodiplosis mosellana*. Ann. Appl. Biol. **1989**, *114*, 409–417.
2. Gooding, M.J.; Smith, S.P.; Davies, W.P.; Kettlewell, P.S. Effects of late-season applications of propioconazole and tridemorph on disease, senescence, grain development and the breadmaking quality of winter wheat. Crop Prot. **1994**, *13*, 362–370.
3. Pawelzik, E.; Permady, H.H.; Weinert, J.; Wolf, G.A. Untersuchungen zum einflussβ einer fusarienkontamination auf ausgewählte qualitätsmerkmale von weizen. Getreide, Mehl und Brot **1998**, *52*, 264–266.
4. Garstang, J.R. The Effect of Volunteers on Cereal Quality and Profitability, Volunteer Crops as Weeds. In *Aspects of Appl. Biol.*; Froud-Williams, R.J., Knott, C.M., Lutman, P.J.W., Eds.; 1993; *35*, 67–74.
5. Wolff, G. Schädliche unkraut- und grassamen. Mühle u. Mischfuttertechnik **1987**, *124*, 579–584.
6. Florez, J.A.; Fischer, A.J.; Ramirez, H.; Duque, M.C. Predicting rice yield losses caused by multispecies weed composition. Agron. J. **1999**, *91*, 87–92.
7. Sanchez, P.A. Soil Management in Shifting Cultivation Areas. In *Properties and Management of Soils in the Tropics*; John Wiley & Sons, Inc.: New York, 1976; 398–399.
8. Savary, S.; Srivastava, R.K.; Singh, H.M.; Elazegui, F.A. A characterisation of rice pests and quantification of yield losses in the rice-wheat system of India. Crop Prot. **1997**, *16*, 387–398.
9. Hoffmann, G.M.; Nienhaus, F.; Schönbeck, F.; Weltzien, H.C.; Wilbert, H. Phytomedizin als Wirtschaftsfaktor. In *Lehrbuch der Phytomedizin*, 3rd Ed.; Blackwell: Oxford, 1994; 15–18.

REFLECTIVE MULCHES

Charles G. Summers
University of California, Davis, California, U.S.A.

INTRODUCTION

Many insects are attracted to or repelled by light of particular wavelengths. This behavior can be exploited to reduce the landing frequency of flying aphids and whiteflies by repelling them with mulches that reflect ultraviolet light. The result is that both insect infestation levels and the incidence of many aphid- and whitefly-transmitted viruses in crops growing over such mulches are reduced. The most commonly used reflective mulch is either polyethylene film to which a thin layer of aluminum has been applied, known as ''metalized mulch,'' or extruded gray colored polyethylene film. The former resembles aluminum foil while the latter appears a shiny gray color to the human eye.

DAMAGE CAUSED BY APHIDS AND WHITEFLIES

Aphids and whiteflies cause a multitude of physical injuries to plants. Direct feeding removes large quantities of photosynthates and fluids causing stunting, wilting, or death. While the severity of injury is usually proportional to insect density, some species also inject a salivary toxin into the plant as a part of the feeding process resulting in damage disproportional to their numbers. Their feeding may also cause growth distortions involving either stimulation or inhibition of plant tissues. Both aphids and whiteflies secrete copious amounts of honeydew, a sugary excretion product, on which several species of fungi, collectively called ''sooty molds,'' grow. The growth of sooty mold on the leaf surface blocks sunlight from reaching the chloroplasts and significantly decreases photosynthesis.

Among the most destructive attributes of aphids and whiteflies is their ability to transmit plant virus diseases. While high aphid or whitefly densities are usually required to cause feeding injury, a single viruliferous individual may infect a plant or several plants with a debilitating disease. Together with leafhoppers, aphids and whiteflies transmit more than 90% of the arthropod-borne plant viruses. Control of these viruses, particularly the nonpersistent aphid-borne viruses, is extremely difficult since they can be both acquired and transmitted in less than a minute. Attempts to limit nonpersistently transmitted virus spread by controlling the insect vectors with insecticides have proven unsatisfactory. In some cases, spraying insecticides actually increases virus spread by increasing vector movement. Resistance to many of the aphid and whitefly-borne viruses, particularly those infecting annual vegetable crops, is not available.

PREVENTING AND DELAYING APHID AND WHITEFLY COLONIZATION OR VIRUS TRANSMISSION

Flying aphids are repelled by light wavelength in the range 3000–4000 Å. While less well understood, whiteflies appear to respond similarly to these same wavelengths. This knowledge can be exploited to reduce or delay both aphid and whitefly colonization and virus transmission by decreasing their landing occurrence on susceptible hosts. Reflective material, usually a plastic film, is placed on the soil surface as a mulch. Holes are cut or burned in the plastic through which transplants or seeds can be placed. Some reflective mulches have a lattice cut down the center through which seedlings can emerge. While numerous types of reflective plastics are available, the two most common are gray extruded polyethylene and polyethylene coated with a thin layer of aluminum. There appears to be no significant differences in the effectiveness of the two in repelling aphids, but the metalized mulches are more effective in repelling whiteflies. Some metalized mulches utilize polyester or polypropylene as the base film, but these are more expensive and less durable under field conditions than polyethylene. Occasionally white polyethylene is used, but it is significantly less effective than either gray or metalized mulch in repelling insects. Another variation is black polyethylene with a stripe of aluminum paint down the center. It too is less effective than solid reflective mulch. Reflective mulches have been shown to decrease the incidence of nonpersistent aphid-borne viruses including cucumber mosaic, zucchini yellows mosaic, and watermelon mosaic-II in cantaloupe, squash, cucumber, and pumpkin by 75–85%. In addition, the initiation of disease infection is delayed by 2–4 weeks (1). The buildup of whitefly numbers has been significantly delayed in these crops as well as in tomatoes, broccoli, and eggplant. In pumpkin and cucumbers, whitefly density on plants growing over metalized mulch was reduced 10- to 14-fold compared to plants growing over bare soil. In pumpkin, the incidence of squash silverleaf, a disorder caused by silverleaf whitefly feeding, was 12% in plants growing over metalized mulches compared to 100% in plants growing over bare soil (2).

The effectiveness of reflective mulches is proportional to the surface area available to reflect ultraviolet light. If alternate rows are mulched, the incidence of virus infection in the mulched rows remains low while that in the unmulched rows is nearly as high as in portions of the field without mulch. When the plant canopy covers ca. 60–70% of the mulch surface, mulches cease to be effective. At this point, the plants become increasingly susceptible to insect colonization or virus infection. However, many plants have reached maturity by this time and can tolerate disease better than do younger plants. In most situations, nearly normal yields are obtained. In order to provide maximum protection, reflective mulches must be in place when plants emerge. Aphids can transmit viruses to plants in the cotyledon stage and whitefly adults can lay enough eggs on the first true leaf to cause severe stunting when the eggs hatch.

In addition to reducing aphid and whitefly numbers and the incidence of virus diseases, plastic mulches retard weed growth, conserve soil moisture, and promote improved soil nutrient absorption. Reflective mulches may also reduce the incidence of some leafhoppers, leafminers, and thrips. Reflective mulches are best suited for use in high value vegetable crops.

REFERENCES

1. Summers, C.G.; Stapleton, J.J.; Newton, A.S.; Duncan, R.A.; Hart, D. Comparison of sprayable and film mulches in delaying the onset of aphid-transmitted virus diseases in zucchini squash. Plant Dis. **1995**, *79* (11), 1126–1131.
2. Summers, C.G.; Stapleton, J.J. Management of vegetable insects using plastic mulch, 1997 season review. U.C. Plant Prot. Q. **1998**, *8* (1/2), 9–11.

REFUGIA FOR PESTS AND NATURAL ENEMIES

R

Robert Verkerk
Imperial College of Science, Technology, and Medicine, Berkshire, United Kingdom

INTRODUCTION

The adoption of within-field monocultures that now typify modern agriculture is known to discriminate against natural enemies so reducing their impact compared with more diverse systems, as found in uncultivated land. The term "refugium" has historically been most widely used in zoogeography and it refers to a place to which a species or group of animals has retreated for various reasons. In this zoogeographical context, refugia are generally considered to be restricted habitats in uncultivated or virgin lands. However, the term refugia has more recently been adopted in the pest management literature and is used to refer to particular areas within or outside cultivated land, where some degree of protection of pests, natural enemies or both, is conferred. The protection can be in many different forms and these have both spatial and temporal components. Examples of refugia include the provision of uncultivated field boundaries, within-field alternative hosts, floral resources, overwintering sites, unsprayed crop areas, and toxin-free crop areas within toxin-containing (transgenic) crop monocultures.

Refugia in relation to pest management may be naturally occurring or artificially created and can be categorized broadly into two main groups: those for pests and those for natural enemies. These will be considered in turn.

REFUGIA FOR PESTS

The two most important types of pest refugia, which may be present within or outside a crop, are: 1) refugia that intentionally prevent or reduce exposure of pests to pesticides, and 2) those which unintentionally provide protected areas that act as obstacles to the delivery of effective pest management, be this in the form of biological, chemical, or any other means of control.

Intentional Pest Refugia

Protection in such refugia is mostly from exposure to pesticides that may either be applied conventionally or be present as toxins within transgenic crops. These refugia are used by pest managers for two distinct purposes: firstly, as a means of maintaining a low abundance of the pest in or near the crop as a food source for natural enemies that, if deprived of food and allowed to die or emigrate, might otherwise be unable to restrain pest development following the pest's re-immigration; and secondly, they are used increasingly for the purpose of delaying the potential development of pesticide resistance that is regarded widely as a major constraint to the intensive use of pesticides in pest management programs (1).

Since the 1960s, a considerable amount of ecological theory and practical experimentation has been used to develop optimum strategies for maintaining stability of pest-natural enemy population dynamics. A consensus of opinion has emerged in support of the view that control measures that attempt to completely eliminate a pest organism are often counter-productive to long-term and sustainable pest management. Refugia for pests (and their natural enemies) are recognized widely as being of crucial importance in maintaining low and stable pest (host) densities (2, 3).

Apart from the value of refugia for pests, it is important to distinguish refugia for alternate hosts of natural enemies that do not necessarily achieve pest status. These alternate hosts may be insects that are closely related to the pest species (e.g., same family or genus, different species). For example, in Central Europe, maintaining broom (*Cytisus scoparius*), often regarded as a weed, has been shown to be beneficial since it helps to perpetuate the braconid wasp parasitoid *Aphidius ervi* via economically unimportant aphid hosts. The increased abundance of this parasitoid in turn provides improved biological control of the economically important pea aphid (*Acyrthosiphon pisum*) on alfalfa in the same region (4).

The use of refugia has formed one of the principle strategies to reduce the risk of development of resistance to transgenic crops that incorporate genes from the bacterium *Bacillus thuringiensis* (*Bt*). Such transgenic crops (e.g., so-called *Bt* cotton) have been widely adopted in large-scale monocultures as a means of combating lepidopterous pests. Refuges have generally taken the form of conventional crops (i.e., the same crop without the transgenes), although controversies abound as to the optimum size of such refuges, the optimum dose of *B. thuringiensis* toxins, and the extent to which unstructured refuges can slow down the development of resistance (5). In the case of transgenic (*Bt*) cotton, refuge areas of conventional cotton comprising 20% of the total crop area are now widely adopted.

Refugia can also be comprised of a crop of a different species to that of the principal crop; for example, pyrethroid resistance in the cotton bollworm (*Helicoverpa armigera*) was shown to be delayed when pyrethroid-susceptible bollworm adults immigrated into cotton fields from surrounding unsprayed corn fields that acted as refugia for this pest (6).

Owing to the complexity of interactions between pests, natural enemies, the crop, and the abiotic environment, as well as the influence of pesticides where they are used, it is very difficult to predict the benefits of using refugia, or to decide precisely on what type and size of refugia would provide the greatest benefits in terms of crop yield and sustainable pest management. If the refugia occupy crop space and do not have direct agricultural value themselves, it is crucial that an optimum balance is struck between the relative scales of the crop and the refugia.

Unintentional Pest Refugia

The presence of pest refugia does not always confer benefits for pest management. The presence of natural or cultivated habitats close to a crop can present problems if natural enemies are not able to function effectively in the crop environment. This is the case with two species of coreid fruitspotting bug (*Amblypelta nitida and A. lutescens lutescens*) in Australia, which tend to be particularly problematic in orchards close to rainforests (7). The rainforests are suspected of containing alternate hosts for the bugs, which then emigrate to adjacent fruit crops where they are not controlled effectively by natural regulatory processes (e.g., natural enemies, partial plant resistance) (7).

REFUGIA FOR NATURAL ENEMIES

Knowledge that uncultivated land can support a diverse range of natural enemies that can help to control pests on agricultural crops has long been known. However, in many industrialised countries, the mechanisation of agriculture, as well as the availability of high-yielding crop varieties and synthetic pesticides has meant that pest management programs have often failed to emphasize the importance of natural enemy refugia. In Britain, for example, hedgerows and uncultivated field margins were destroyed as a result of these modern technologies and only recently have there been attempts to reverse this trend in the wake of increasing awareness of the problems associated with large-scale monocultures and over-reliance on pesticides (8).

Within the crop, natural enemy refugia can take the form of unsprayed crop areas, protected plant parts of the crop itself, noncrop plants that favor natural enemies, or alternate hosts that encourage economically unimportant arthropods, which in turn provide a food source for natural enemies. Outside the crop, uncultivated field margins, hedgerows, "conservation headlands," and "live fences" have all been used, at least in part, to provide refugia for natural enemies that can subsequently move into the crop environment.

To some extent, the pest management benefits associated with strip farming, mixed cropping, and intercropping, common to many forms of peasant agriculture, are caused by parts of the cropping system acting as refugia for natural enemies. In such cropping systems, generalist (nonspecialized) natural enemies can move between crops, strips or even "weeds", depending on the build-up of pests (hosts). For example, the abandonment of strip farming in Peru during the 1950s is considered one of the primary causes of the bollworm outbreaks that developed in cotton, the incidence of which has been greatly reduced following the rediversification of the agro-ecosystem in the region (9).

The effective use of refugia generally requires a thorough knowledge of crop–pest–natural enemy interactions. Such interactions will vary considerably in both space and time and are often referred to as tritrophic interactions as they involve three distinct feeding levels. However, to date, there have been relatively few tritrophic studies on crop-based systems and further studies of this type will be important for the development of more scientific approaches to the use of refugia (10). It is often not sufficient to simply leave uncultivated sections within or outside the crop. Floral composition of refuges, their location and their dimensions should be carefully considered in relation to the specific crop–pest–natural enemy situation in question.

There are many examples in the literature indicating the effectiveness of natural enemy refugia across a wide range of agro-ecosystems. Some examples are considered ahead.

The planting of lucerne (*Medicago sativa*) strips in cotton fields in Australia has been shown to increase the abundance of important predators (11). Predatory beetles, bugs, and lacewings were five–sevenfold more abundant in the lucerne strips than in the cotton crop itself and these predators in turn migrated into the cotton and were on average often twice as abundant in cotton with lucerne strips compared with cotton without (11).

Wild brassicas (e.g., *Barbarea vulgaris* and *Brassica* kaber) planted in the vicinity of cultivated brassicas provide floral nectar reserves for parasitoids of the diamondback moth, so improving the potential for biological control (12). Such noncrop plants can sometimes provide a dual function in pest management, acting both as refuge sites for natural enemies as well as trap crops for the pest. However, there is also the possibility that some plants known to favor natural enemies will also attract pests to the vicinity.

Border-planting of *Phacelia tanacetifolia* (Hydrophyl laceae) as a floral (pollen) resource has also been shown to be effective in increasing the abundance of aphid-eating hover flies, the larvae of which are often key natural enemies of aphids (13). Adult female hover flies generally need to consume pollen before they can lay fertile eggs so it is important that such floral reserves are present close to the crop.

The common farming practice of using prune trees as parasitoid refuges has been shown to improve biological control of the western grape leafhopper, *Erythroneura elegantula* in vineyards in the western United States (14). Almost twice the abundance of the egg parasite *Anagrus epos* was found in vineyards with prune tree refuges compared with those without, the trees acting both as overwintering sites and windbreaks, which aid parasitoid flight and dispersal (14).

Protected parts of certain crops that are relatively free from the disturbances associated with agronomic practices can also provide important refugia for natural enemies. For example, the area beneath the plucking surface of tea plants is well known as a refuge for a range of natural enemies of important tea pests. Research has shown that domatia, which are small, sometimes elaborate pits or shelters at the junctures of veins on the lower surface of leaves of some plants, are inhabited primarily by predatory rather than herbivorous arthropods. A group of researchers has shown that by adding artificial domatia to cotton plants, fruit production could be increased by 30%, compared with control plants (15). This increase was caused by enhanced predation of herbivorous mites by predatory ones that were able to harbor in the artificial domatia (15).

Some natural enemies overwinter within plant parts so the removal of stubble, fruits, or other crop litter may be counter-productive as these plant parts act as refugia for the natural enemies. For example, in Australia, several species of predatory mite (*Typhlodromus spp.*) have been shown to overwinter in the calyx cavities of apple fruits so that early season phytophagous mite control can be improved if apples are left on the ground through winter (16).

REFERENCES

1. Midgarden, D.; Fleischer, S.J.; Weisz, R.; Smilowitz, Z. Site-specific integrated pest management impact on development of esfenvalerate resistance in Colorado potato beetle (Coleoptera: Chrysomelidae) and on densities of natural enemies. J. Econ. Entomol. **1997**, *90*, 855–867.
2. Murdoch, W.W.; Briggs, C.J. Theory for biological control: recent developments. Ecology **1996**, *77*, 2001–2013.
3. Takagi, M. Perspective of practical biological control and population theories. Res. Pop. Ecol. **1999**, *41*, 121–126.
4. Starý, P. *Biology of Aphid Parasites (Hymenoptera: Aphidiidae) with Respect to Integrated Control*; Junk, W., Dr., N.V., Eds.; The Hague: The Netherlands, 1970; 643.
5. Tabashnik, B.E. In *Delaying Insect Adaptation to Transgenic Plants—Seed Mixtures and Refugia Reconsidered*, Proceedings of the Royal Society of London, Ser. B, 1994, 255, 7–12.
6. Han, Z.J.; Wang, Y.C.; Zhang, Q.S.; Li, X.C.; Li, G.Q. Dynamics of pyrethroid resistance in a field population of *Helicoverpa armigera* (Hubner) in China. Pestic. Sci. **1999**, *55*, 462–466.
7. Waite, G.K.; Huwer, R.K. Host plants and their role in the ecology of the fruitspotting bugs *Amblypelta nitida* (Stal) and *Amblypelta lutescens lutescens* (Distant) (Hemiptera: Coreidae). Aust. J. Entomol. **1998**, *37*, 340–349.
8. van Emden, H.F. Plant Diversity and Natural Enemy Efficiency in Agroecosystems. In *Critical Issues in Biological Control*; Mackauer, M., Ehler, L.E., Roland, J., Eds.; Intercept: Andover, England, 1990; 63–80.
9. van Emden, H.F. *Pest Control*, 2nd Ed.; Edward Arnold: London, 1989; 117.
10. Verkerk, R.H.J.; Leather, S.R.; Wright, D.J. Review article: the potential for manipulating crop-pest-natural enemy interactions for improved insect pest management. Bull. Ent. Res. **1998**, *88*, 493–501.
11. Mensah, R.K. Habitat diversity: implications for the conservation and use of predatory insects of *Helicoverpa* spp. in cotton systems in Australia. Int. J. Pest. Manage. **1999**, *45*, 91–100.

12. Idris, A.B.; Grafius, E. Effects of wild and cultivated host plants on oviposition, survival, and development of diamondback moth (Lepidoptera: Plutellidae) and its parasitoid diadegma insulare (*Hymenoptera: Ichneumonidae*). Environ. Entomol. **1996**, *25*, 825–833.
13. White, A.J.; Wratten, S.D.; Berry, N.A.; Weigmann, U. Habitat manipulation to enhance biological control of brassica pests by hover flies (Diptera: Syrphidae). J. Econ. Entomol. **1995**, *88*, 1171–1176.
14. Murphy, B.C.; Rosenheim, J.A.; Granett, J. Habitat diversification for improving biological control: abundance of *Anagrus epos* (Hymenoptera: Mymaridae) in grape vineyards. Environ. Entomol. **1996**, *25*, 495–504.
15. Agrawal, A.A.; Karban, R. Domatia mediate plant-arthropod mutualism. Nature **1997**, *387*, 562–563.
16. Gurr, G.M.; Thwaite, W.G.; Valentine, B.J.; Nicol, H.I. Factors affecting the presence of typhlodromus spp. (Acarina: Phytoseiidae) in the calyx cavities of apple fruits and implications for integrated pest management. Exp. Appl. Acarol. **1997**, *21*, 357–364.

REGULATING PESTICIDES

Matthias Kern
Deutsche Gesellschaft für Technische Zusammenarbeit (GTZ) GmbH, Bonn, Germany

R

INTRODUCTION

Pesticides are introduced directly into the environment, e.g., they are applied to fields as plant protectants, to control vectors, etc. One of the main requirements for any pesticide is that it effectively controls the target organism. Unfortunately, the specificity of many pesticides is rather low, making them toxic for nontarget organisms, too. This also makes them a potential hazard for their users, for the consumers of agricultural products that have been treated with them, and for the environment. Consequently, it is of particular importance that measures be taken to protect people and the environment against the harmful effects of pesticides. For this reason, pesticides belong to that group of chemicals that must be most carefully examined and most stringently regulated.

The purpose of pesticide control legislation is to establish rules and principles for the management of pesticides such as to maximize the benefits from their application while minimizing or ruling out negative side effects (1).

ELEMENTS

The exercise of control over pesticides applies not only to their application. As with any other hazardous type of chemical, due allowance also must be made for their manufacture, transport, national and international distribution, storage, further processing and, as the case may be, disposal (2).

To put these control measures on a legal footing, the fundamental conditions for the management of pesticides must be embodied in the law. Depending on the circumstances prevailing within a given country, the relevant legislation could address pesticides or plant protection, chemicals or environmental protection. In any case, the law must stipulate that no person, business, or organization shall be allowed to distribute or apply any pesticide that has not been registered with, or at least provisionally approved by, the responsible authority. The details of the registration process and of pesticide management must be governed by various subsidiary implementing regulations that can be adjusted to accommodate technical progress without necessitating alteration of the law itself.

Usually, a registration process regulates the distribution of pesticides. The registration requirements for pesticides define their quality, range of application, labeling, packaging and applicable safety measures. In this context, protection of the users and the environment, as well as the proprietary rights of the manufacturers must be taken into account.

For the postregistration scope, legal directives must be in place to govern the sale of pesticides, the training of retailers and users, the award of licenses, and the control activities at various levels.

The areas of production, repackaging, and transport must not necessarily be covered by the pesticide-specific legislation if they already are adequately governed by laws applicable to chemicals in general.

The stipulation of responsibilities is of major importance for the legal application. As a rule, responsibility for the registration and control of pesticides lies with the ministry of agriculture, health or the environment, with decisions concerning the registration and control normally being coordinated between several ministries or subsidiary authorities according to a prescribed procedure. Customs must be involved in the control of importing and exporting activities.

IMPLEMENTATION

In 1985 the United Nations Food and Agriculture Organization (FAO) conference adopted a voluntary "International Code of Conduct on the Distribution and Use of Pesticides" in which the principles of pertinent legislation and of the environmentally sound management of pesticides are compiled for international reference purposes (3).

One of the main prerequisites for the environmentally sound management of pesticides is that an appropriate legal framework be in place and amenable to enforcement of its various laws and implementing regulations. Chapter 19 of Agenda 21, which deals with the environmentally sound management of toxic chemicals, postulates that this be accomplished on the maximum possible scale in all countries by the year 2000 (4).

Surveys conducted by FAO in 1994 show that all industrialized countries and nearly all developing countries have laws regulating the control of pesticides. In many of the emerging countries, however, effective enforcement of their pesticide laws is rendered impossible by a lack of appropriate implementing regulations and/or means of monitoring compliance with the laws. Of the developing countries, 87% stated that their governments were providing little or no resources for the control of pesticide (5).

PROBLEMS

The management of pesticides in the absence of appropriate safety measures can cause problems, the gravity of which will differ according to the prevailing framework conditions.

While developing countries consume only about 20% of all agrochemicals, they account for 70% of all acute pesticide poisonings according to International Labour Organization estimates (6). That corresponds to some 1.1 million cases of pesticide poisoning each year. Most of these cases are attributable to inadequately trained users, improper application techniques, defective application equipment, and/or inadequate safety practices. Implementation of effective work safety regulations should help alter the situation.

In countries with no means of control, improper use is often compounded by problems attributable to the quality of the marketed products. A product containing the wrong active ingredient or having the wrong concentration will not have the envisaged effect on the target organism and will not secure the envisaged benefits for its user. Some 30% of all pesticides sold in developing countries do not comply with international standards regarding the quality of their active ingredients (7). If the quality of the labeling and packaging is also taken into consideration, the percentage of deficient products will be even higher. Considering that the label is the foremost source of information for the user with regard to indication, dosing and safety practices, proper labeling is of central importance for the proper use of the pesticides. Effective supervision of the market and of compliance with registration procedure specifications is a decisive prerequisite for actual compliance with quality standards.

Maximum residue limits must be specified in order to avoid problems with food contamination by pesticides. At the international level, those defined by the Codex Alimentarius Commission serve as guideline data in cases where no national values are specified. Here, too, corresponding target values can be effective only if the residue levels are monitored. In a number of cases, agricultural products from countries in which check analysis is either inadequate or nonexistent were rejected by importing countries on the grounds of maximum-limit transgression. For the exporting country, that means a substantial economic loss. In order to succeed in the international marketplace, more and more countries are beginning to monitor both the pesticide residue levels in food and the use of pesticides on crops. In this instance, economic pressure is accelerating the implementation of legal prescriptions.

INTERNATIONAL CONVENTIONS

Comprehensive national pesticide legislation and control are justifiable for countries with relatively large pesticides markets and corresponding infrastructures. For many smaller countries, however, such structures remain beyond their financial and administrative capacities. Due to the fact that international standards capable of adaptation to national and regional circumstances are already in place, many countries work together at the regional level with regard to the control of pesticides.

As trade flows become increasingly internationalized, it is sensible that the authorities in industrialized and developing countries engage in joint supervisory activities, and such action is increasingly being called for.

In September 1998 the "Convention on the Prior Informed Consent (PIC) Procedure for Certain Hazardous Chemicals and Pesticides in International Trade" was signed in Rotterdam (8). The Rotterdam Convention serves chiefly to provide protection for developing countries that have not yet instituted adequate import controls of their own. The exporting countries pledged to inform the importing countries regarding any plans to transport certain hazardous chemicals which are either banned for reasons of health or environmental protection or are subject to severe restrictions—and to export such chemicals only with the express prior (informed) consent of the importing country.

The FAO and the United Nations Environment Programme (UNEP) introduced the PIC procedure in 1989 on an international voluntary basis. By 1993, 106 countries were participating, and that number grew to 155 by 1999.

While the cooperative principle enjoys widespread approval, it will not become legally binding until the Rotterdam Convention enters into force. What makes this convention so special is that the authorities in the exporting country are bound by importing decisions made by recipient countries, i.e., outside the territory of the former. Hence, the ''borders'' of national legislature are beginning to blur. Now, many exporting countries are having to take measures to ensure that their customs authorities become active in the control of pesticide imports and exports.

The uncontrolled international distribution of persistent organic pollutants (POPs) also presents problems (9). This category includes a number of pesticides, e.g., aldrin, dieldrin, DDT, and mirex. POPs are slow to break down in nature, and the wind can spread them far beyond a country's boundaries. Eventually, they are deposited by precipitation in cooler climate zones, where they accumulate in the food chain. Thus, countries of the north like Canada and Sweden are keenly interested in the global restriction of such substances. International negotiations on the drafting of a convention that would globally restrict the manufacture and use of certain POPs were entered into in 1998. Today, POPs are chiefly produced and applied in developing countries. Those countries, however, are only willing to discontinue the relatively cheap production and use of such products if they see attractive economic alternatives and if financing of the conversion process is assured. Thus, the conclusion of an internationally binding convention on the reduction of global contamination by persistent chemicals is heavily dependent on the economic interests of individual countries, as well as on how much importance each country attaches to the attendant risks.

CONCLUSIONS

Pesticide legislation is part of general chemicals legislation, because many quality standards and safety practices apply in like manner to pesticide, household and industrial chemicals. The respective separate provisions apply only to product-specific characteristics. National legislation in this area is becoming increasingly dependent on global cooperation in order to accommodate expanding international trade in chemicals and take account of supra-regional pollution. The Rotterdam Convention pertaining to international trade in certain hazardous products constitutes an important step in that direction. Likewise, the phasing out of certain pesticides, e.g., POPs, can be achieved only through an international approach. However, as demonstrated by the successful efforts to phase out ozone-depleting substances by way of the Montreal Protocol, this is possible.

The present trend indicates that, in the future, the sale, handling, transportation, and application of chemicals will be controlled within a global context, and the responsibility for its implementation will be divided equally among the industrialized and the developing countries.

REFERENCES

1. FAO. *Pesticide Management Guidelines*; Food and Agriculture Organization of the United Nations: Rome, 1998. http://www.fao.org/WAICENT/FaoInfo/Agricult/AGP/AGPP/Pesticid/Code/Guide.htm (accessed Dec. 1999).
2. Lönngren, R. *International Approaches to Chemicals Control*; The National Chemicals Inspectorate KEMI, Sweden, 1992; 512.
3. FAO. *International Code of Conduct on the Distribution and Use of Pesticides*; Food and Agriculture Organization of the United Nations: Rome, 1990. http://www.fao.org/WAICENT/FaoInfo/Agricult/AGP/AGPP/Pesticid/Code/PM_Code.htm (accessed Dec. 1999).
4. UNCED. *Environmentally Sound Management of Toxic Chemicals, Including Prevention of Illegal International Traffic in Toxic and Dangerous Products;* Agenda 21, Chapter 19, United Nations Conference on Environment and Development, Rio de Janeiro; 1992. http://www.igc.apc.org/habitat/agenda21/ch-19.html (accessed Dec. 1999).
5. FAO. *Analysis of Government Responses to the Second Questionnaire on the State of Implementation of the International Code of Conduct on the Distribution and Use of Pesticides*; Food and Agriculture Organization of the United Nations: Rome, 1996; 101.
6. ILO. *The ILO Programme on Occupational Safety and Health in Agriculture*; International Labour Organization: Geneva, 1998. http://www.ilo.org/public/english/90travai/sechyg/agrivf02.htm (accessed Dec. 1999).
7. Kern, M.; Vaagt, G. Pesticide quality in developing countries. Pesticide Outlook **1996**, (Oct), 7–10.
8. FAO. *Interim Joint FAO/UNEP Secretariat for the Operation of the PIC Procedure*; Food and Agriculture Organization of the United Nations: Rome, 1999. http://www.fao.org/WAICENT/FaoInfo/Agricult/AGP/AGPP/Pesticid/PIC/pichome.htm (accessed Dec. 1999).
9. UNEP. Chemicals. In *Persistent Organic Pollutants*; United Nations Environment Programme: Geneva, 1999. http://irptc.unep.ch/pops/ (accessed Dec. 1999).

REGULATING PESTICIDES (LAWS AND REGULATIONS)

Praful Suchak
Suchak's Consultancy Services, Mumbai, India

INTRODUCTION

Why Regulate Pesticides?

Chemical or biological pesticides have target specific toxicity that controls or eradicates pests falling under different groups. These products, though developed for specific usage, could have adverse effects on living beings and the environment and unchecked use can cause havoc. Regulating pesticides, therefore, would assure reasonable safety in use of these toxic substances and ensure that risks from pesticides to humans and their environment are minimized and are consistent with the benefits achieved by their use in terms of reduced losses.

Regulating pesticides at the international and national level should consider social costs in line with social benefits. Pesticides impose costs on society, such as health risks and environmental degradation, which are not borne by the user. The available policy remedies include bans on individual or classes of chemicals that prohibit the introduction of hazardous compounds into the environment, and economic instruments such as taxes, registration fees, and import duties that work to redistribute and adjust the social costs occurring for pesticide use and also provide the government with revenues that can be used to cover health costs and environmental clean-up activities.

HISTORY

The United States in 1910 introduced the Federal Pesticide Act that underwent complete metamorphosis to become the Federal Insecticide, Fungicide & Rodenticide Act (FIFRA) in 1947, which since 1970 is under the auspices of the Environmental Protection Agency.

Australia initiated pesticide legislation with one state in 1925 and by 1945 all states had their individual laws. The Industry Association brought law common to all states in 1995. By the end of 1999 about 95% of the countries in the world had adopted full/partial regulatory systems.

Early in-depth studies were not carried out on the long-term effects of: 1) repeated exposures, 2) residual toxicity, 3) accumulated toxicity, and 4) the impact on environment. With additional knowledge on the cumulative toxicity of chlorinated hydrocarbons such as DDT having come to light, the regulating authorities have started demanding the generation of additional critical toxicological data to assess short-term, long-term, and environmental toxicity of earlier registered pesticides. The European Union has already undertaken reviews of 90 molecules in the first phase by a Commission regulation dated December 11, 1992, to be completed in 12 years, and a further 148 molecules in the second phase effective March 1, 2000. The remaining substances in the European Union would be included in third phase.

Regulatory requirements for pesticides have undergone a change over the past half a century. With the advent of highly sophisticated testing equipment, more knowledge about harmful effects of the toxic chemicals has come to light. Consistent watch by environmentalists and organizations like the Pesticides Action Network (PAN), Greenpeace, Save the Planet groups, and other nongovernmental organizations has resulted in added awareness resulting in hosts of data requirements for registration/reregistration of pesticides.

Although all developed countries and most of the developing countries have their own legislation to regulate pesticides, there have been vast variations in data requirements for registrations between these countries. With globalization it has became imperative to have harmonized data requirements so that the registrant can hope for faster registration in different (pesticide consuming) countries.

AVAILABLE INTERNATIONAL GUIDELINES

1. Agenda 21 of the United Nations Conference on Environment and Development (UNCED)
2. The Codex Alimentarius
3. The FAO International Code of Conduct and Prior Informed Consent (PIC)
4. WTO and International Trade with respect to pesticides
5. Agreement on Persistent Organic Pollution (POP)

6. Guidelines of Minor Donor Institutions on the purchase of pesticides

IMPLEMENTATION PROBLEMS

Although FAO took the lead to harmonize data requirements in participating nations for registrations of pesticides, certain problems and practical difficulties have occurred such as

1. The original registrant, having invested huge amounts in data generation, is unable to protect the data
2. Absence of confidentiality assurance by the registering country, creating difficulties in multiple country registrations
3. Recommended uses differ from country to country, resulting in difficulties
4. Unchecked dumping of unsafe or banned pesticides in less-developed countries
5. New registrations by a company other than the original registrant by providing data generated by such a company could not be checked

STEPS UNDERTAKEN

Although though PIC entry of banned pesticides could be prevented, this instrument has not been fully effective. Once it becomes fully operational legally things should improve.

With the United States implementing the Food Quality Protection Act and fixing maximum residue limits for 3000 toxic compounds, countries worldwide would need to harmonize their registrations on toxic chemicals so as to meet the residue levels in food.

The formation of the European Union with 15 member countries, OECD with 29 members, and the Technical Working Group having EPA, Canada, and Mexico, has accelerated the pace towards harmonization. However, since a vast disparity exists between developed countries on one side and developing countries on the other side, it is rather difficult to have a unified data requirement, particularly in case of risk assessment.

Acceptance of electronic data submission and dossier/monograph submissions and joint reviews by EU would also pave the way toward harmonization and would address questions in the nondietary exposure area.

Apart from studies related to bioefficiency of the product, the toxicological studies of the toxicant, its analogues, impurities and breakdown products, residual toxicity, etc., as listed in Appendix would help understanding and regulating pesticides.

PRESENT SCENARIO AND PROBABLE REMEDIES

Substantial evidence exists that pesticides are being applied in a technically and economically inefficient manner. Many developing countries subsidize pesticides and equipment, resulting in excessive use of pesticides.

Also in developing countries, the current legal environment and enforcement capabilities have been inadequate and dysfunctional, thus exerting a significant impact on current levels of pesticide use. This is partly due to lack of resources and partly due to manipulation by vested interests.

The inadequacies of the existing regulatory framework, institutional rigidities, and a bias in favor of pesticide-dependent paths also contribute to improper use of pesticides.

A major problem confronting many countries is the absence of well-established procedural mechanisms for public involvement in the decision making process including crop protection policy. Competing interests with a stake in the process, including farmers, the pesticide industry, and policy makers responsible for food security, argue for a more liberal regulatory stance. On the other hand, environmentalists, public health workers, and consumers demand strict regulation and reduced pesticide volumes.

To be more effective, pesticide regulation and implementation should be handled by a neutral agency like the Ministry of Environment or similar organization and not the Ministry of Agriculture or other interested ministry.

Pesticide policy needs to be integrated into the broader public policy debate concerning the nations' agricultural, environmental, and health strategies.

Nevertheless, two general principles should apply. First, dispassionate analysis of the costs and benefits of pesticide use would provide a useful tool for the formulation of normal policies; and second, the broader and more inclusive the debate, the more likely it is that the outcome will serve the public rather than specific private interests.

FUTURE GLOBAL POLICY

A uniform global regulatory system needs to ensure

1. Agricultural chemical use increases agricultural output

2. Food supplies are safe from harmful toxicants/residues
3. Reduced-risk chemical pesticides, biopesticides, and nonchemical alternatives are encouraged
4. Uniform MRLs to eliminate trade barriers
5. Uniform health-based safety standards for pesticide residues
6. Special provisions for certain groups of the population including infants and children

APPENDIX 1

Toxicological and Other Data Requirements for Pesticide Registration

1. Identity of active substance
   Chemical name
   Empirical and structural formula
   Molecular mass
   Method of manufacture (synthesis pathways)
   Purity
   Identity and content of isomers
   Impurity and additives

2. Physical and chemical properties
   Melting point
   Boiling point and relative density
   Vapor pressure
   Volatility
   Appearance
   Absorption spectra-molecular extinction at relevant wavelength
   Solubility in water/organic solvents
   Partitioning coefficient N-octanol/water
   Stability and hydrolysis rate in water
   Photochemical degradation on surface, in water, and in air
   Thermal stability and stability in air

3. Analytical method
   Analytical method for the determination of the pure active substance in the technical grade.
   For breakdown products and additives in plant products, soil, water, animal body fluids, and tissues.

4. Toxicological and metabolism studies
   Studies on acute toxicity—oral, percutaneous, inhalation, intraperitoneal, skin and, where appropriate, eye irritation, and skin sensitization.
   Short-term toxicity—oral, cumulative toxicity, and other routes inhalation or dermal.
   Chronic toxicity—oral, long-term toxicity, and carcinogenicity.
   Mutagenicity—reproductive toxicity-teratogenicity and multigeneration studies in mammals.
   Metabolism studies in mammals—absorption, distribution, and excretion studies, elucidation of metabolic pathways.
   Supplementary studies—neurotoxicity studies—toxic effects of metabolites from treated plants and toxic effects on livestock and pests.
   Medical data—medical surveillance on manufacturing plant personnel, clinical cases, poisoning incidents from industry and agriculture sensitization/allergenicity observations, observations on exposure of the general population, and epidemiological studies if appropriate. Diagnosis and specific signs of poisoning, clinical tests, and prognosis of expected effects of poisoning.
   Proposed treatment: first aid measures, antidotes, and medical treatment.
   Summary of toxicological studies and conclusions, critical scientific evaluation with regard to all toxicological data, and other information concerning the active substance.

5. Residues in or on treated products, food and feed metabolism in plants and livestock
   In treated plants (distribution, metabolism, binding constituents, etc.).
   In livestock (uptake, distribution, metabolism, binding constituents, etc.).

6. Fate and behavior in the environment
   Studies on aerobic and anaerobic degradation under laboratory conditions in different soil types.
   Adsorption and desorption in different soil types including metabolites.

Mobility of the active ingredients in different soil types.

Behavior in water and air, rate and route of degradation.

7. Ecotoxicological studies

   Effects on birds, fish, aquatic organisms such as Daphnia magna, algae, honeybees, earthworms, other nontarget macroorganisms and microorganisms.

8. Information concerning the labeling including indication of danger and safety measures.

BIBLIOGRAPHY

Pesticides Policies in Developing Countries—Do They Encourage Excessive Use? In *World Bank Discussions Paper No. 238*; 1994.

Asian Development Bank. In *Handbook on the Use of Pesticides in Asia Pacific Region*; ADB: Manila, Philippines, 1987.

Pesticide Policy Project Hannover; Publication Serial No. 1, January 1995; No. 2, November 1995; No. 3, December 1995; No. 4, December 1996; No. 5, December 1996; No. 6, 1998; No. 7, April 1999; No. 8, April 1999.

EC Directives 91/414/EEC and Subsequent Directives Including 1999/80/EC.

Proceedings of Asia Pacific Crop Protection Conference 1997 and 1999, PMFAI: Mumbai, India.

Global Pesticides Directory, 2nd Ed.; Suchak's Consultancy Services: Mumbai, India, 1997. suchakgr@vsnl.net

Pesticides News; No. 20–47, Pesticides Action Network (PAN): London, 1993 to 1999.

Guidelines on the Operation of Prior Informed Consent (PIC) Rome FAO 1990, Guidance to Government in PIC Rome 1991, and Other FAO Publications.

U.S. EPA Pesticides Information Network. www.cdpr.ca.govt/docs/epa/epachim.htm

REPELLENTS

John A. Pickett
T.O. Olagbemiro
IACR-Rothamsted, Hertfordshire, United Kingdom

INTRODUCTION

Repellents are compounds that cause avoidance behavior in organisms, principally animals. As natural products, they represent a group of semiochemicals (a term derived from the Greek root semeion, meaning sign- or signal-chemical) that have a behavioral or developmental role without the mechanism involving direct physiological interactions. Many commercial repellents are purely synthetic compounds that do not occur in nature and are therefore not normally incorporated under the term semiochemical.

ACTION AND TYPES

Repellents act prior to direct contact and so do not include the antifeedants, which interfere with feeding and colonization by animals through gustatory or contact chemosensory interactions. Repellents are required to have a vapor pressure commensurate with sufficient volatility to be detectable by olfactory or aerial chemosensory interactions. Thus, they fall into a molecular weight range of up to the low hundreds, with the higher molecular weight compounds tending to be hydrocarbons. Indeed, only very low molecular weight compounds are hydrophilic and the larger the molecule, the more lipophilic it will be.

Repellents can also be waterborne. However, to comply with the definition, such compounds would generally fall into the low molecular weight lipophilic class and be detected by the aquatic equivalent of the olfactory process, rather than through gustation.

Nonnatural product repellents include a group of commercially important compounds which are active against mosquitoes (Diptera: Culicidae) and other nuisance or biting flies that may act as vectors of pathogens causing important diseases. The most extensively used is *N,N*-diethyl-*m*-toluamide (DEET, Fig. 1) discovered in 1954 (1). Synthetic repellents active against carnivorous fish, particularly sharks (e.g., Carcharhinidae) have been developed but tend to involve inorganic compounds, for example those releasing copper ions, which do not fall into the definition of repellents given here. Recently, a new repellent (Fig. 2) has been developed from a synthetic structure-activity study based on DEET (2).

Many natural products have been investigated for fly repellency and include a number of isoprenoids (3). A specific mosquito repellent identified from the essential oil of *Eucalyptus camaldulensis*, eucamalol, 3-formyl-6-isopropyl-2-cyclohexen-1-ol (Fig. 3), is reported to exhibit potent activity against the mosquito *Aedes aegypti* (4). The role of these natural products seems to be in conveying to the carnivorous, or haematophagous, fly the presence of plant material having inherent defenses based on secondary metabolites to which they would not be adapted. Such repellency can be seen against other carnivorous animals, as is the case with citrus (Rutaceae) essential oils against domestic cats and other members of the Felidae. However, with higher animals such as mammals, the complicated behavioral psychology involved in learning means that such repellents are readily overcome and thereby rendered ineffective.

SEMIOCHEMICALS

In extension of the theory of repellency based on the involvement of nonhost signaling, studies are being directed towards the identification of semiochemicals, produced by plants, that may indicate unsuitability. For example, host plants of pest insects such as aphids (Homoptera: Aphididae) can, at various stages in their life cycles, change from being attractant to being repellent, thereby providing semiochemicals of potential use in crop protection (5). Furthermore, plants that are not acceptable as hosts, either

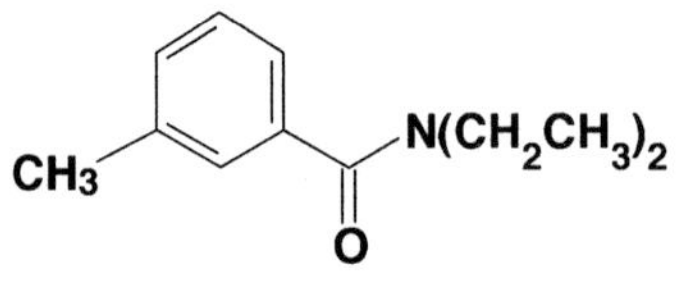

Fig. 1 *N,N*-Diethyl-*m*-toluamide (DEET).

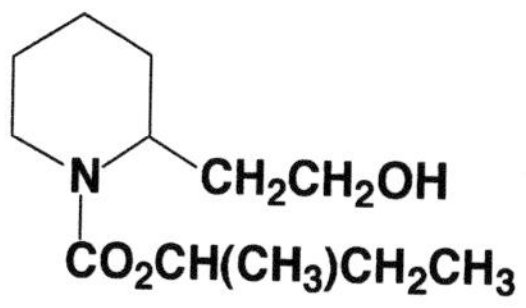

Fig. 2 1-(2-Methylpropoxycarbonyl)-2-(2-hydroxyethyl)piperidine [Bayrepel®].

through inherent secondary metabolism to which the herbivore is not adapted, or by having such secondary metabolism-based defences induced by earlier colonization attempts, can also give rise to useful repellents (6). The discovery that plant compounds having such roles are detected by interactions with highly specific olfactory neurons has led to a search for similar mechanisms in interactions between pest flies, discussed above, and vertebrates unsuitable as hosts. It is now known that cattle showing relatively low loadings of flies such as the horn fly, *Haematobia irritans*, produce compounds that cause this reduction in colonization (7). Evidence is also mounting that the same is true of human beings, specifically in terms of selection by the main sub-Saharan African malaria vector, *Anopheles gambiae*, and the Scottish biting midge, *Culicoides impunctatus* (8).

CONCLUSION

The commercially available range of synthetic repellents is now seeing inroads from natural products that, although not as effective, carry greater public acceptance. Nonetheless, for the future, extended studies on mechanisms of nonhost avoidance should provide new and more robust repellents for protection against human disease vectors, and for the control of pests of livestock and of agricultural and horticultural crops.

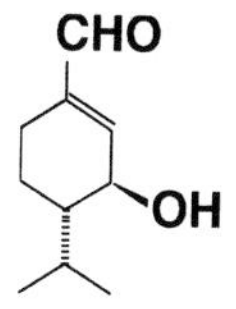

Fig. 3 3-Formyl-6-isopropyl-2-cyclohexen-1-ol (eucamalol).

REFERENCES

1. McCabe, E.T.; Barthel, W.F.; Gertler, S.L.; Hall, S.A. Insect repellent III. *N,N*-diethylamides. J. Org. Chem. **1954**, *19*, 493–498.
2. Boeckh, J.; Breer, H.; Geier, M.; Hoever, F.-P.; Kruger, H.-W.; Nentwig, G.; Sass, H. Acylated 1,3-aminopropanols as repellents against bloodsucking arthropods. Pestic. Sci. **1996**, *48*, 359–373.
3. Pickett, J.A. In *Lower Terpenoids as Natural Insect Control Agents, Ecological Chemistry and Biochemistry of Plant Terpenoids*, Proceedings of the Phytochemical Society of Europe; Harborne, J.B., Thomas-Barberan, F.A., Eds.; Clarendon Press: Oxford, 1991; 297–313.
4. Wantanabe, K.; Shona, Y.; Kakimizu, A.; Okada, A.; Matsuo, N.; Satoh, A.; Nishimura, H. New mosquito repellent from *Eucalyptus camaldulensis*. J. Agri. Food Chem. **1993**, *41*, 2164–2166.
5. Pettersson, J.; Pickett, J.A.; Pye, B.J.; Quiroz, A.; Smart, L.E.; Wadhams, L.J.; Woodcock, C.M. Winter host component reduces colonization by bird-cherry-oat aphid *Rhopalosiphum padi* (L) (Homoptera, Aphididae) and other aphids in cereal fields. J. Chem. Ecol. **1994**, *20* (10), 2565–2573.
6. Hardie, J.; Isaacs, R.; Pickett, J.A.; Wadhams, L.J.; Woodcock, C.M. Methyl salicylate and (−)-(1*R*,5*S*)-myrtenal are plant-derived repellents for black bean aphid, *Aphis fabae* (Scop.) (Homoptera: Aphididae). J. Chem. Ecol. **1994**, *20* (11), 2847–2855.
7. Birkett, M.A.; Pickett, J.A.; Wadhams, L.J.; Woodcock, C.M.; Prijs, H.J.; Thomas, G.; Jespersen, J.; Van-Jensen, K.M. In *Control of Dipteran Pest of Cattle: Development of a Push–Pull Strategy*, 9th International Congress of Pesticide Chemistry: London, UK, 1998.
8. Pickett, J.A.; Woodcock, C.M. The Role of Mosquito Olfaction in Oviposition Site Location and in the Avoidance of Unsuitable Hosts. In *Olfaction in Mosquito-host interactions*; Ciba Foundation Symposium No. 200, Wiley: Chichester, UK, 1996; 109–123.

REPRODUCTIVE AND DEVELOPMENTAL EFFECTS FROM OCCUPATIONAL PESTICIDES EXPOSURE

Shao Lin
State University of New York at Albany, Rensselaer, New York, U.S.A.

INTRODUCTION

Increased public health concern has been directed at occupational exposure to pesticides and reproductive hazards for workers, their spouses, as well as developmental effects on their children. Reproductive effects are defined as occurrences of adverse effects on reproductive organs and related endocrine system. Common reproductive outcomes include changes in sexual behavior, infertility, pregnancy outcomes, and functional change (1). Developmental effects mean the occurrence of adverse effects on the developing organism including death of the developing organism (which can also be considered as reproductive effects), structural abnormality, growth retardation, and functional deficiency (2).

WHAT ARE POTENTIAL PESTICIDE EXPOSURE PATHWAYS FOR WORKERS' FAMILY MEMBERS?

Pesticides used at work can be brought home inadvertently through workers' clothing, hair, hands or nails, and body fluids. Poor hygienic behaviors among pesticide handlers have been found to be associated with measurable blood levels of pesticides (chlordecone) in family members (3). Higher levels of urinary biomarkers of organophosphorous pesticides were reported among the children of pesticide applicators than in the reference group (4). Studies have found organophosphorous concentrations in the soil and dust of the homes of exposed workers to be significantly higher than in control groups (4).

Several factors affect family members' exposure including age, workers' exposure status, and residence near farmland. Younger children usually receive greater exposure than their older siblings because of behavioral differences (more hand-to-mouth behavior) and higher susceptibility. The exposure potential for families of agricultural workers may be greater than for families of other workers.

WHAT ARE THE REPRODUCTIVE EFFECTS OF PESTICIDES ON MALES?

The reproductive toxicity of some pesticides on men is well established. Infertility and reduction of sperm counts in men who were occupationally exposed to dibromochloropropane (DBCP), a nematicide, has been documented (5). Two other pesticides that have been found to affect male reproductive systems are chlordecone and ethylene dibromide (EDB). Chlordecone was shown to affect sperm parameters among production workers and there was evidence of suggestive effects on spermatogenesis for EDB (6). These three pesticides have not been considered as occupational hazards since their uses have been suspended in the United States due to known toxicities of DBCP and chlordecone, and public health concerns relating to EDB.

There is increasing research interest in the use of time to pregnancy as an indicator of the reproductive toxicity of pesticide exposure. de Coke et al. (7) found that spouses of Dutch fruit growers with high exposure to pesticides took twice as long to conceive and had a reduced fecundability ratio compared to spouses of the low-exposure group during the spraying season. A significantly higher proportion of couples consulted physicians for fertility problems among the high-exposure group (28%) than the low-exposure group (8%).

WHAT ARE THE REPRODUCTIVE EFFECTS OF PESTICIDES ON FEMALES?

Common endpoints of reproductive toxicity on females include infertility, time to pregnancy, early pregnancy loss, spontaneous abortion, and fetal death. Infertility is clinically defined as failure to conceive after 12 months of unprotected intercourse. Early pregnancy loss means pregnancies that have been detected by highly specific pregnancy tests but are lost prior to clinical recognition of the pregnancy. Spontaneous abortion refers to pregnancies that terminate naturally prior to 20 weeks of completed gestation. Fetal deaths or stillbirths are pregnancies that terminate naturally at, or after, 20 weeks.

Several studies have demonstrated positive associations between female occupational exposure to pesticides and spontaneous abortions or fetal deaths (8–11). In a large study conducted in China (8), spontaneous abortion was reported to be positively associated with occupational exposure to baumenfon, dikushuang, iprobenfos, chlordimeform, and carobofuran [Risk Ratio (RR) ranged from 1.90 to 4.00]. In addition, women exposed to chlordimeform, dikushuang, and iprobenfos were found to have increased risks of stillbirths. Another study conducted among 56,067 women in Montreal hospitals (9) showed positive effects of pesticide or germicide exposures on stillbirths [Odds Ratio (OR): 3.1, 95% Confidence Interval (CI): 1.1–8.6]. Similarly, spontaneous abortions were more frequent in ornamental plant growers who used several pesticides in a study conducted in Argentina (OR = 1.96) (10). However, the data from the 1980 U.S. National Natality and Fetal Mortality Survey (11) found no association between stillbirths and exposure to three organic compounds including alicyclic halogens, other alicyclic hydrocarbons, and chlorinated hydrocarbons.

In terms of environmental exposure, the reproductive effects of aerial spraying of the organophosphate insecticide malathion was reported in a study conducted in California (12). By using residence in or near a spray corridor as an exposure surrogate, the study did not find statistically significant associations for spontaneous abortions and stillbirths.

WHAT ARE EMBRYONIC OR DEVELOPMENTAL EFFECTS DUE TO PARENTAL EXPOSURE?

The greatest public health concern has focused on embryonic and developmental effects from parental pesticide exposure including spontaneous abortions and fetal deaths, congenital malformations, and low birthweight or prematurity.

Karmaus et al. (13) investigated embryonic effects of maternal occupational exposure to wood preservatives used in daycare centers such as pentachlorophenol (PCP) and lindane in Germany, and found significantly reduced birthweights and birth lengths in exposed pregnancies. By using self-reported information in Australia, Kricker et al. (14) found an increased risk of limb defects among children whose mothers were exposed to pesticides from garden/workplace (OR: 3.4, 95% CI: 1.9–5.9) and extermination of household infestations (OR: 2.4, 95% CI: 1.3–4.5). Similarly, Lin's study in New York State (15) showed consistently increasing but not statistically significant associations between parental exposure to every type of pesticide and associated limb defects. A large study in China reported a positive relationship between maternal exposure to pesticides in the first trimester and central nervous system malformations (16). However, a similar result was not found between all birth defects combined and pesticide exposure.

In the Ontario Farm Family Health Study (17), miscarriage increased in combination with paternal use of thiocarbanates, carbaryl, and unclassified pesticides on the farm. Preterm delivery was also associated with mixing or applying yard herbicides (OR: 2.1, 95% CI: 1.0–4.4). A study among 9512 fathers who worked in British Columbia sawmills found elevated risks for developing anomalies of cataracts, anencephaly or spina bifida, and genital organ defects among the offspring of workers with exposure to dioxin-contaminated chlorophenols (18). However, no significant associations were found for birthweight, prematurity, or fetal deaths.

Perhaps one of the most widely debated questions is the developmental effect among offspring of veterans who were exposed to Agent Orange, a widely used herbicide, during the Vietnam War. During the manufacturing process, a component of Agent Orange was contaminated by small but highly toxic amounts of 2,3,7,8-tetrachlorodibenzo-*p*-dioxin (TCDD), a potential teratogen in animal studies. By comparing the children of veterans who served in Southeast Asia during the same period but did not spray herbicides, Wolfe et al. (19) found no meaningful increase in the risk of spontaneous abortion or stillbirth among the offspring of veterans of Operation Ranch Hand, the unit responsible for aerial spraying of herbicides in Vietnam War. An elevated risk of nervous system defects was reported among the children with paternal dioxin exposure, but it was based on sparse data. In 1992, the United States Air Force (20) released the results of the first study examining individual serum TCDD levels on reproductive outcomes and indicated a significantly higher median value of TCDD among the Ranch Hand personnel (12.8 pg/g) than in the controls (4.2 pg/g). Neonatal death was also found to be associated with TCDD levels (OR: 5.5, 95% CI: 1.5–20.7). No clear association was detected between TCDD level and birth defects, sperm count or percentage of abnormal sperm.

FUTURE DIRECTION AND CONCLUSION

A major problem in prior studies was the lack of accuracy of exposure assessment and classification since many studies used either self-reported data or surrogate information including job title, industrial and occupation, and even rural residence rather than actual measurements. Exposure, defined through job title or occupation, can vary greatly across studies or even within a study and geographic locations. Another important issue is poor specificity of exposure when ''exposure'' was defined by occupation only. Pesticides include wide varieties of chemicals, and the specific substances most responsible for adverse reproductive effects are still unknown. Exposure timing windows and their relationship to the critical period of adverse reproductive outcomes has not been clarified. In addition, the low prevalence of most reproductive outcomes and the infrequency of pesticide exposures require a large study population, making study design challenging. Confounding or effect modification by some variables such as age, training, and use of protective equipment should be considered in future studies.

In conclusion, as Sever et al. (6) indicated, evidence is accumulating on reproductive and developmental toxicity due to parental pesticides exposure. The reproductive outcomes of particular concern include infertility, spontaneous abortion, neonatal death, and some specific congenital malformations. The specific types of pesticides currently in use that are associated with these outcomes are still not clear. Therefore, the evidence relating pesticides to reproductive and developmental effects is far from complete.

REFERENCES

1. Environmental Protection Agency, proposed guidelines for assessing female reproductive risks notice. Fed. Regist. **1988**, *53*, 24834–24847.
2. Environmental Protection Agency, guidelines for developmental toxicity risk assessment. Fed. Regist. **1991**, *56*, 63798–63826.
3. Cannon, S.B.; Veazey, J.M.; Jackson, R.S. Epidemic kepone poisoning in chemical workers. Am. J. Epidemiol. **1978**, *107*, 529–537.
4. Fenske, R.A. Pesticide Exposure Assessment of Workers and Their Families. In *Human Health Effects of Pesticides*; Keifer, M.C., Ed.; Hanley & Belfus, Inc.: Philadelphia, 1997; 221–237.
5. Whorton, D.; Krauss, R.M.; Marshall, R.M.; Milby, T.H. Infertility in male pesticide workers. Lancet **1977**, *2*, 1259–1261.
6. Sever, L.E.; Arbuckle, T.E.; Sweeney, A. Reproductive and Developmental Effects of Occupational Pesticide Exposure: The Epidemiologic Evidence. In *Human Health Effects of Pesticides*; Keifer, M.C., Ed.; Hanley & Belfus, Inc.: Philadelphia, 1997; 305–325.
7. de Coke, J.; Westveer, K.; Heederik, D. Time to pregnancy and occupational exposure to pesticides in fruit growers in the Netherlands. Occup. Environ. Med. **1994**, *51*, 693–699.
8. Pan, X.; Wang, J.; Wu, Z. A prospective study on the relationship between environmental exposure to pesticides and adverse pregnancy outcomes. China Environ. Sci. **1993**, *4*, 91–96.
9. Goulet, L.; Theriault, G. Stillbirth and chemical exposure of pregnant workers. Scand. J. Work Environ. Health **1991**, *17*, 25–31.
10. Matos, E.L.; Loria, D.I.; Albiano, N. Pesticides in intensive cultivation: effects on working conditions and workers' health. PAHO Bull. **1987**, *21*, 405–416.
11. Savitz, D.A.; Whelan, E.A.; Kleckner, R.C. Self-reported exposure to pesticides and radiation related to pregnancy outcome-results from national natality and fetal mortality surveys. Public Health Rep. **1989**, *104*, 473–477.
12. Thomas, D.C.; Pettiti, D.B.; Goldhaber, M. Reproductive outcomes in relation to malathion spraying in the San Francisco bay area, 1981–1982. Epidemiology **1992**, *3*, 32–39.
13. Karmaus, W.; Wolf, N. Reduced birthweight and length in the offspring of females exposed to PCDFs, PCP, and Lindane. Environ Health Perspect. **1995**, *103* (12), 1120–1125.
14. Kricker, A.; McCredie, J.; Elliott, J.; Forrest, J. Women and the environment: a study of congenital limb anomalies. Community Health Studies **1986**, *10*, 1–11.
15. Lin, S.; Marshall, E.G.; Davidson, G.K. Potential parental exposure to pesticides and limb reduction defects. Scand. J. Work Environ. Health **1994**, *20*, 166–179.
16. Zhang, J.; Cai, W.-W.; Lee, D.J. Occupational hazards and pregnancy outcomes. Am. J. Ind. Med. **1992**, *21*, 397–408.
17. Savitz, D.A.; Arbuckle, T.; Kaczor, D.; Curtis, K.M. Male pesticide exposure and pregnancy outcome. Am. J. Epidemiol. **1997**, *146* (12), 1025–1036.
18. Dimich-Ward, H.; Hertzman, C.; Teschke, K.; Hershler, R.; Marion, S.A.; Ostry, A.; Kelly, S. Reproductive effects of paternal exposure to chlorophenate wood preservatives in the sawmill industry. Scand. J. Work Environ. Health **1996**, *22* (4), 267–273.
19. Wolfe, W.H.; Michalek, J.E.; Miner, J.C.; Rahe, A.J.; Moore, C.A.; Needham, L.L.; Patterson, D.G., Jr. Paternal serum dioxin and reproductive outcomes among veterans of operation ranch hand. Epidemiology **1995**, *6* (1), 17–22.
20. Wolfe, W.H.; Michalek, J.E.; Miner, J.C.; Rahe, A.J. Air Force Health Study. In *An Epidemiologic Investigation of Health Effects in Air Force Personnel Following Exposure to Herbicides: Reproductive Outcomes*; U.S. Air Force, Brooks Air Force Base, TX, 1992.

REPRODUCTIVE CONTROL METHODS

R

Lowell A. Miller
USDA/National Wildlife Research Center, Fort Collins, Colorado, U.S.A.

INTRODUCTION

In the United States, conflicts between humans and animals continue to increase. Overpopulation of white-tailed deer is an increasing problem in some sections of the United States. Rats cause major damage worldwide, yet attempts to control them by contemporary means (e.g., poisons or traps) are often less than satisfactory. Avian species, such as blackbirds, cause economic damage to sunflower growers; and starlings, pigeons, and Canada geese are becoming an increasing problem in cities. A growing interest in nonlethal methods for population control of nuisance or damaging species of wildlife has fostered research in reducing fertility of these overabundant wildlife species. Fertility may be reduced by interfering with the fertilization of the egg (contraception) or interfering with the implantation or development of the fertilized egg (contragestion).

The most important fundamentals for success in inducing infertility in a particular species are an understanding of the reproductive behavior and physiology of that species and selecting the most suitable infertility agent. Examples of reproductive behaviors that need to be considered are seasonal versus year-round breeding, monogamous or polygamous mating, multiestrus or monestrus; and does the species need a specific vegetation, temperature, or landscape to be successful in reproduction. If one could change a critical factor needed for successful reproduction, one could effectively reduce the reproductive rate of the target species. A common reason some species are overabundant is that they are adaptable to multiple or changing environments, thus their populations increase in spite of a rapidly changing landscape. Less adaptive species many times become extinct.

Recent infertility research has centered around immunocontraceptive vaccines, which control fertility by stimulating the production of antibodies against gamete proteins, reproductive hormones, or other proteins essential for reproduction. These antibodies interfere with the normal physiological activity of the reproductive system. Two common targets for immunocontraception are gonadotropin releasing hormone (GnRH) and zona pellucida (ZP).

GONADOTROPIN RELEASING HORMONE

The use of gonadotropin releasing hormone (GnRH) immunocontraceptive vaccine can shut down the reproductive activity of both sexes by causing development of antibodies blocking GnRH thus preventing the release of other essential reproductive hormones.

GnRH is produced in the brain by the hypothalamus. It controls the release of the pituitary reproductive hormones, follicle stimulating hormone, and luteinizing hormone, which in turn control the functions of the ovaries and testes. Antibodies to the hypothalamic hormone will reduce the circulating level of biologically active GnRH, thereby reducing the subsequent release of reproductive hormones. The reduction or absence of these hormones leads to atrophy of the gonads and concomitant infertility of both sexes.

GnRH contraceptive vaccines have been evaluated as immunocastration agents in pets, cattle, horses, sheep, and swine for more than 10 years, but little research has been done in wildlife species. Recently, in studies with Norway rats it was found that both males and females immunized with a GnRH vaccine were 100% infertile. The National Wildlife Research Center (NWRC) just finished a long-term study on the effect of GnRH on white-tailed deer in which we achieved an 86% reduction in fawning during active immunization and a 74% reduction over 5 years (Table 1).

ZONA PELLUCIDA

Zona pellucida (ZP) immunocontraceptives were the first to receive widespread publicity as contraceptives in deer and feral horses. ZP is an acellular glycoprotein layer located between the oocyte and the granulosa cells on the outer surface of the egg. For a sperm to fertilize the egg, it must first bind to a receptor on the ZP. An enzyme in the sperm then breaks down the ZP, allowing the sperm passage into the ovum. Antibodies to this glycoprotein layer result in infertility either by blocking the sperm from binding to and penetrating the ZP layer or by interference with oocyte maturation, leading to the death of the deve-

Encyclopedia of Pest Management

Table 1 Summary of fawns born during 5-year GnRH vaccine study

Year	Treatment	Fawns/does
1994–95	Primed and boosted	3/4
1995–96	Primed and boosted	0/6
1996–97	Boosted	1/8
1997–98	No boost	3/8
1998–99	No boost	9/9
Breeding herd		156/90 (x = 1.7)
Sham controls		35/19 (x = 1.8)
GnRH treated		16/35 (x = .46)
= 74% reduction in fawns in GnRH group		

x = Average number of fawns/does.

loping oocyte. The ZP vaccine in use today comes from the pig ovary and is called porcine zona pellucida (PZP). PZP vaccine has been used to produce immunosterilization in dogs, baboons, horses, burros, coyotes, and white-tailed deer. In a white-tailed deer study at Pennsylvania State University, we achieved 89% reduction in fawning during the 2 years of active immunization. A 76% reduction in fawning was observed over the entire 7-year study (Table 2).

STATUS OF ORAL IMMUNOCONTRACEPTION

Fertility control as a technology is available today, but only in laboratory studies, pen studies, and in limited field situations with small numbers of animals. Immunocontraceptive and contragestive vaccines are being produced in limited quantities and animals injected with these vaccines become infertile for 1–3 years. However, to be practical for controlling free-ranging animal populations, these agents will have to be given orally. The technology for developing oral vaccines is in its infancy. Oral delivery is a very difficult technology to develop and may increase USDA and FDA regulatory involvement because it is a new and unproven technology.

Warren discussed a number of the factors relevant to the practical and logistical implementation of contraceptives for controlling wildlife. He correctly pointed out that development will take a team approach involving the laboratory scientists (e.g., immunologists, molecular biologists, and reproductive physiologists) who develop the contraceptive vaccines and associated technologies, and the wildlife biologists who will need to contribute to the development of delivery systems and the means to measure field efficacy and safety.

INDUCED INFERTILITY AS A MANAGEMENT TOOL IN AVIAN SPECIES

Interfering with egg laying or the hatchability of the egg appears to be the best approach to reducing the reproductive capacity in birds. Egg addling, including shaking,

Table 2 White-tailed deer PZP immunocontraception

Year	Treatment	Fawns/does	Yearly average	Cumulative average
1992–93	Primed and boosted	4/11	.36	.36
1993–94	Boosted	1/11	.09	.23
1994–95	4/9 boosted	1/9	.11	.19
1995–96	None boosted	3/9	.33	.23
1996–97	None boosted	7/9	.78	.33
1997–98	None boosted	9/8	1.13	.44
1998–99	None boosted	8/7	1.14	.52
Breeding herd		156/90		1.73
Sham controls		35/19		1.84
PZP treated		33/64		.52
		0.72% reduction in fawns in PZP group		

freezing, or oiling eggs in a nest effectively reduces egg hatchability. However, this method is labor intensive and may be useful only in small-scale operations. We are currently studying Nicarbazin (NCZ), a compound that interferes with egg laying and egg hatchability. Research is being conducted in collaboration with Koffolk Inc., the manufacturer of NCZ. Koffolk has an FDA-approved use of NCZ in broiler chickens for control of coccidiosis caused by *Eimeria* (protozoan parasites that are passed from chicken to chicken in the feces). Large-scale broiler chicken production is not possible without effective control of this parasite.

One of NCZ's side effects is that if fed to breeder or layer chickens, their eggs, although fertilized, often do not hatch. NCZ causes bleaching of the brown color in the eggshell and yolk-mottling (white spots occurring on the yolk) due to fluid transfer from the albumin into the yolk via increased yolk membrane permeability. Severe yolk membrane breakdown causes the yolk and the albumin (white) to blend together, destroying the conditions necessary for viable development of the embryo. When NCZ is withdrawn from the diet, egg production resumes within a few days. Koffolk, Inc. is interested in development of NCZ as an oral goose fertility control agent (Fig. 1).

Another oral contraceptive being studied is 20,25-diazacholesterol, an extremely potent inhibitor of cholesterol synthesis in laboratory animals. The 20,25-diazacholesterol is structurally identical to cholesterol except for the replacement of the carbons with nitrogens at positions 20 and 25. 20,25-diazacholesterol (which we call DiazaCon) acts to reduce the production of reproductive hormones in two ways: by reducing the cholesterol in the endocrine cell and reducing the cholesterol side chain cleavage needed for the production of the reproductive hormones. Conjugated linoleic acid (CLA) might have potential as an oral contraceptive in avian species. Linoleic acid is a 18 carbon fatty acid found in many plant seed oils with double bonds in the 9, 12 position. CLA has double bonds in either the 9, 11 or the 10, 12 positions. This change in the double bond position makes the molecule ineffective as a linoleic acid molecule. When CLA is added to the chicken feed at 0.5% by weight concentration and fed to laying chickens, their egg yolks will solidify when chilled in the refrigerator. When the clutch is being laid, the bird does not sit on the

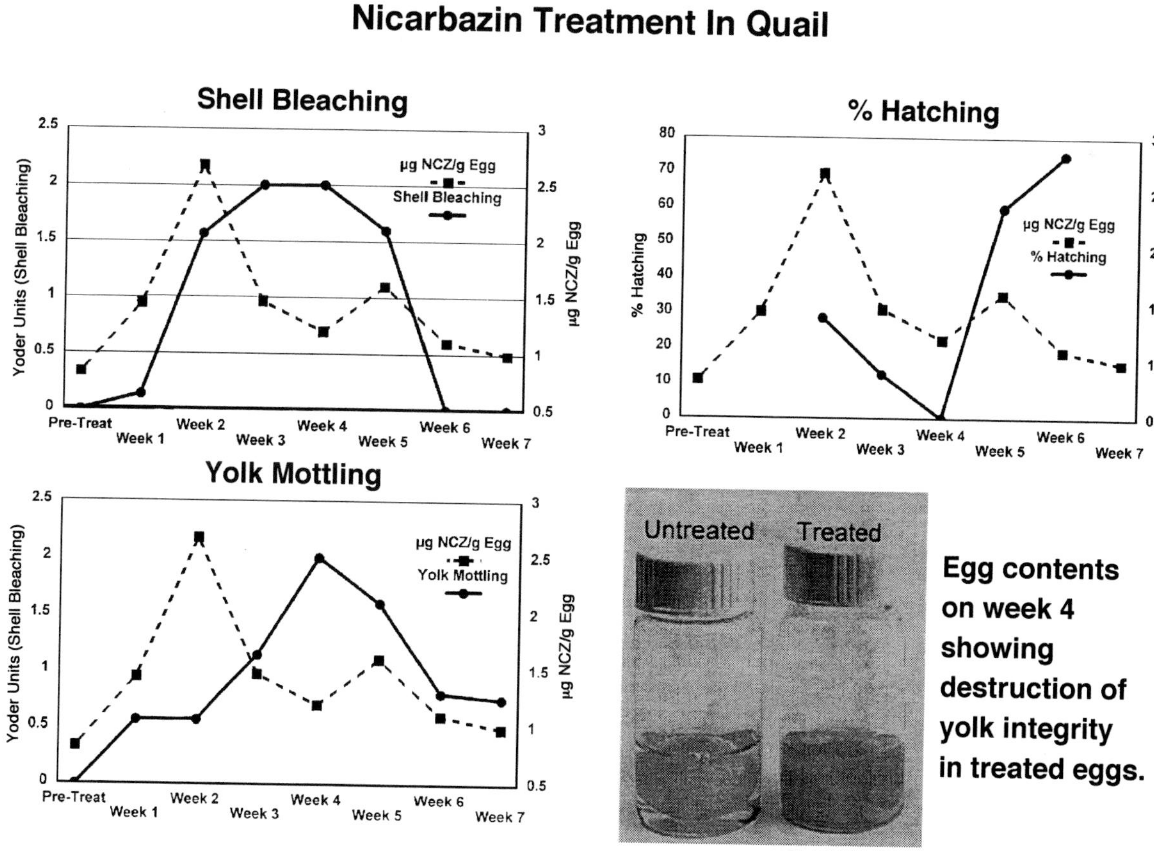

Fig. 1 Nicarbazin treatment in quail.

nest until the clutch is complete. As the temperature drops in the night, the unprotected yolks of CLA-fed birds will solidify. This change in density apparently is not completely reversible and interferes with the hatchability of the eggs.

NEED FOR MORE RESEARCH IN INDUCED INFERTILITY

As human and animal populations increase and as public opinion against hunting and lethal control increases, the need for fertility control becomes extremely important. Wildlife contraceptive technologies are potentially valuable as a new tool to be integrated with more traditional methods of wildlife population management: however, there is relatively little funding for this type of research. Because of the variety of animals that have become overabundant there is a large variation in their reproductive physiology that needs to be understood before their reproductive capacity can be curtailed effectively.

BIBILOGRAPHY

Adams, T.E.; Adams, B.M. Feedlot performance of steers and bulls actively immunized against gonadotropin-releasing hormone. J. Anim. Sci. **1992**, *70*, 691–698.

Chin, S.F.; Strokson, J.M.; Albright, K.J.; Cook, M.E.; Pariza, M.W. Conjugated linoleic acid is a growth factor for rats as shown by enhanced weight gain and improved feed efficiency. J. Nutr. **1994**, *124*, 2344–2349.

Jones, J.E.; Solis, J.; Hughes, B.L.; Castaldo, D.J.; Toler, J.E. Production and egg quality responses of white leghorn layers to anticoccidial agents. Poultry Sci. **1990**, *69*, 378–387.

Miller, L.A.; Fagerstone, K.A. In *Induced Infertility as a Wildlife Management Tool*, Proceedings of the Nineteenth Vertebrate Pest Conference; Salmon, T.P., Crabb, A.C., Eds.; University of California–Davis: Davis, CA, *in press*.

Miller, L.A.; Johns, B.E.; Elias, D.J. Immunocontraception as a wildlife management tool: some perspectives. Wildl. Soc. Bull. **1998**, *26*, 237–243.

Miller, L.A.; Johns, B.E.; Elias, D.J.; Crane, K.A. Comparative efficacy of two immunocontraceptive vaccines. Vaccine **1997**, *15*, 1858–1862.

Miller, L.A.; Johns, B.E.; Killian, G.J. Immunocontraception of white-tailed deer with GnRH vaccine. Am. J. Reprod. Immunol. *44* (5), 266–274.

Miller, L.A.; Johns, B.E.; Killian, G.J. Long-term effects of PZP immunization on reproduction in white-tailed deer. Vaccine **1999**, *18*, 568–574.

Turner, J.W.; Kirkpatrick, J.F.; Liu, I.K.M. Effectiveness, reversibility, and serum antibody titers associated with immunocontraception in captive white-tailed deer. J. Wildl. Manage. **1996**, *60*, 45–51.

Warren, R.J. Should wildlife biologists be involved in wildlife contraception research and management? Wildl. Soc. Bull. **1995**, *23*, 441–444.

RESISTANCE MANAGEMENT

Kun Yan Zhu
Kansas State University, Manhattan, Kansas, U.S.A.

R

INTRODUCTION

Resistance management refers to the use of every possible integrated pest management (IPM)-compatible technique and method to avoid or delay the development of pest resistance to pesticides, or to revert the resistance to less resistant levels in pest populations (1). Resistance management can be involved in dealing with pest resistance to either conventional chemical pesticides or transgenic plants containing toxin genes from other organisms such as the bacterium *Bacillus thuringiensis* (Bt). It also may include the selection of pesticide resistance of natural enemies of pests. However, the objectives and tactics of resistance management are often the opposite of those used for pests (2).

OVERVIEW

Development of pesticide resistance is a natural response of a pest to selection by pesticides. Resistance will remain an ongoing dilemma as long as pesticides are used in pest management programs. In this sense, resistance can be only mitigated rather than managed (3). However, the speed of resistance development in a pest population is dependent on genetic (e.g., frequency, number and dominance of genes for resistance, and fitness) and biological (e.g., generation turnover, number of progeny, and migration) factors of the pest (4) and operational factors (e.g., chemical nature, persistency, mode of action, number of applications per season, and application timing) of pesticides (5, 6). Genetic and biological factors represent inherent characteristics of a pest species that are mostly beyond our control, whereas operational factors can and must be considered for preventing or delaying resistance development (1). The primary goals of using operational tactics for resistance management are to reduce resistance selection pressure by pesticides in pest populations (Table 1). Although several operational factors need to be considered, implementation of the operational tactics can be varied to achieve the best practice of resistance management.

Resistance management is also highly dependent on nonchemical tactics, which include: 1) planting pest-resistant crops or varieties, 2) employing all possible cultural practices to enhance host-plant resistance to pests, 3) deploying and conserving biological control agents, and 4) monitoring pest densities and evaluating economic injury levels so that pesticides are applied only when necessary. Again, these approaches are used to reduce the resistance selection pressure to pest populations by pesticides. Because resistance management is an integral part of IPM, it should be regarded as a subcomponent of IPM.

Resistance management can be greatly facilitated by enlightened policies governing pesticide development, registration, marketing, regulation, and education. For example, transgenic Bt corn has been commercialized in the United States with the European corn borer (*Ostrinia nubilalis*) as the primary target insect pest. The U.S. Environmental Protection Agency has mandated that certain percentages of non-Bt corn be planted along with Bt corn as a resistance management strategy. The non-Bt corn in or adjacent to a Bt corn field serves as a ''refugium'' that allows some corn borers to avoid resistance selection by transgenic Bt corn. Reduction of selection pressure on European corn borer by Bt corn is expected to mitigate resistance development in this pest (7).

EXAMPLES OF RESISTANCE MANAGEMENT

Although both the models and tactics of resistance management have been examined and discussed in some detail for several pest species, deployment and validation of resistance management in a pest management program are relatively difficult because of the complexity of the program under field conditions. One of the best known examples is management of resistance synthetic pyrethroid resistance in the cotton bollworm (*Heliothis armigera*) in Australia (8). Pyrethroid resistance was detected

Table 1 Main operational tactics for resistance management

1. Use of pesticides only when pest populations reach treatment thresholds.
2. Choice of pesticides that are not closely related to previously used ones.
3. Choice of pesticides and formulations that are not very persistent and do not result in prolonged release.
4. Rotation or alternation of different pesticides with different natures or modes of action.
5. Avoidance of areawide applications with a single or few pesticides.
6. Use of pesticides only at certain life stages of the pest.
7. Reduction of treatments with pesticides to the lowest possible number compatible with acceptable pest control.
8. Implementation of pest refugia when pesticides are applied in the field.

first in this pest in eastern Australia in 1983. Control failures were reported in January and February of the same year on 15 different farms in the Emerald Irrigation Area of central Queensland. In response to the outbreak of resistance to synthetic pyrethroids, a large-scale strategy to manage this resistance was implemented within 6 months of the original failures.

The strategy involved restricting the use of synthetic pyrethroids to one 6-week period each summer and limiting the number of pyrethroid sprays to three per grower. The 6-week period during which synthetic pyrethroid sprays were allowed approximates the time needed for one generation of the cotton bollworm. During this time, mixtures with other chemicals, such as an ovicide, were encouraged. The strategy applied to all hosts of the cotton bollworm, including cotton, cereal, oil seed, grain legume, tomato, and tobacco crops, and to the control of other insect pests in these crops.

The strategy was implemented by voluntary cooperation among government departments, growers likely to use synthetic pyrethroids in the designated areas, grower organizations, entomological consultants, representatives of insecticide companies, research scientists, and extension agents. All parties involved were consulted in the design of the strategy and in its annual appraisal by government authorities and scientists from the Commonwealth Scientific and Industrial Research Organization (CSIRO). Growers were educated about the strategy at field days, at meetings, and through the press. The resistance management strategy was supported further by the pesticide companies through the auspices of the Pyrethroids Efficacy Group (PEG), the Insecticides Resistance Action Committee (IRAC), and the Australian Insecticide Resistance Action Committee (AIRAC).

Although direct evaluation of the program was not possible because it did not include a control area where the strategy had not been implemented, the program was judged successful. Not further major field-control failures with synthetic pyrethroids occurred in the first three seasons. The key to success in this example was the voluntary cooperation among state authorities, pesticide companies, research scientists, and growers in implementing the resistance management strategy.

Other examples of resistance management implemented in the field include pyrethroid resistance in the housefly (*Musca domestica*) in Denmark and in the pear psylla (*Psylla pyricola*) and the tobacco budworm (*Heliothis virescens*) in the United States, organotin resistance in the twospotted spider mite (*Tetranychus urticae*) in the United States (9), and triazine herbicide resistance in weeds in several different countries (10). However, the degree of success in achieving resistance management could differ from pest to pest and from area to area because of a wide variety of biological, management, and policy-related factors.

FUTURE CONCERNS

The use of pesticides will continue to be an indispensable component of pest management systems for the foreseeable future. However, the availability of pesticides is decreasing because of increased difficulties in discovering new pesticides, increases in stringency and costs for registration and reregistration of pesticides through federal and state agencies, cancellations of many pesticide uses, and rapid development of pest resistance to pesticides. Resistance management offers a realistic opportunity to forestall resistance development. Therefore, every effort should be made to promote the long-term effectiveness of the currently available or new pesticides for pest management.

See also *Integrated Pest Management*, pages 413–415; *Mechanisms of Resistance: Antibiosis, Antixenosis, Tolerance, Nutrition*, pages 483–486; *Principles of Pest Management with Emphasis on Plant Pathogens*, pages 666–669; *Resistance to Pesticides*, pages 715–717; *Transgenic Crops (Annuals)*, pages 846–849.

REFERENCES

1. Graves, J.B.; Leonard, B.R.; Ottea, J.A. Chemical Approaches to Managing Arthropod Pests. In *Handbook of Pest Management*; Ruberson, J.R., Ed.; Marcel Dekker, Inc.: New York, 1999; 449–486.
2. Croft, B.A. Developing a Philosophy and Program of

Pesticide Resistance Management. In *Pesticide Resistance in Arthropods*; Roush, R.T., Tabashnik, B.E., Eds.; Chapman & Hall: New York, 1990; 277–296.
3. Hoy, M.A. Myths, models and mitigation of resistance to pesticides. Phil. Trans. R. Soc. Lond. B **1998;** *353* (1376), 1787–1795.
4. Georghiou, G.P.; Taylor, C.E. Genetic and biological influences in the evolution of insecticide resistance. J. Econ. Entomol. **1977**, *70* (3), 319–323.
5. Roush, R.T. Designing resistance management programs: how can you choose? Pestic. Sci. **1989**, *26* (4), 423–441.
6. Tabashnik, B.E. Managing resistance with multiple pesticide tactics: theory, evidence, and recommendations. J. Econ. Entomol. **1989**, *82* (5), 1263–1269.
7. *Bt-Corn and European Corn Borer: Long-Term Success through Resistance Management*; Ostlie, K.R., Hutchison, W.D., Hellmich, R.L., Eds.; N.C. Regional Publication (NCR–602), University of Minnesota: St. Paul, MN, 1997.
8. Daly, J.C. Insecticide resistance in *Heliothis armigera* in Australia. Pestic. Sci. **1988**, *23* (2), 165–176.
9. Plapp, F.W., Jr.; Jackson, J.A.; Campanhola, C.; Frisbie, R.E.; Graves, J.B.; Luttrell, R.G.; Kitten, W.F.; Wall, M. Monitoring and management of pyrethroid resistance in the tobacco budworm (lepidopetra: noctuidae) in Texas, Mississippi, Louisiana, Arkansas and Oklahoma. J. Econ. Entomol. **1990**, *83* (2), 335–341.
10. Heap, I.M. The occurrence of herbicide-resistant weeds worldwide. Pestic. Sci. **1997**, *51* (3), 235–243.

RESISTANCE MANAGEMENT IN HOST-PLANT RESISTANCE

Sanford D. Eigenbrode
University of Idaho, Moscow, Idaho, U.S.A.

INTRODUCTION

Crop varieties resistant to pathogens or arthropod pests can lose their resistance as a result of pest adaptation. The typical solution to this problem is continual, sequential development and release of crop varieties with new genes that confer resistance to the adapted pests. For example, in wheat eight genes for resistance to the Hessian fly, *Mayetiola destructor* (1), and numerous genes for resistance to stem rust, *Puccinia graminis*, have been deployed sequentially to counter adaptation by these pests. Although effective in the short term, sequential release is costly and probably not sustainable because the number of available pest resistance genes is finite.

As a solution, methods have been devised to increase the ''durability'' or effective lifetime of pest-resistant crop varieties (2–4). These methods include breeding for inherently durable kinds of resistance and deploying resistance in ways designed to slow pest adaptation. Although implementation of the methods has been slow, increasing incentives to improve the durability of resistance, especially of genetically engineered pest-resistant varieties (3), may lead to their wider adoption.

THE PROBLEM AND GENERAL APPROACH

Pest-resistant crop varieties typically reduce survival and reproduction of target pests. If the pest population is genetically variable for traits affecting pest survival and reproduction on the resistant crop, if survival among genotypes differs sufficiently, and if enough of the pest population is exposed to the resistance, adapted or ''virulent'' pest genotypes will become predominant due to natural selection. For a more complete discussion of this process with examples, see ''Resistance to Host-Plant Resistance'' in this encyclopedia.

Strategies to improve the durability of resistance aim primarily to reduce the rate at which virulent genotypes or genes increase in a pest population exposed to the resistant variety. This can be achieved by reducing selection pressure, that is, the relative reproductive advantage for virulent pest genotypes, or by reducing the probability that such virulent alleles or genotypes exist in the first place. The methods employed include aspects of cultivar breeding as well as management strategies that are in the hands of the producer.

BREEDING FOR DURABLE RESISTANCE

Horizontal Versus Vertical Resistance

One of the most influential concepts in plant pathology is the gene-for-gene model of plant disease first proposed by Flor (5). Flor observed that for every resistance gene he examined in flax, there was a corresponding virulent gene in the pathogen, flax rust, *Melamspora lini.* Flor proposed that a one-to-one correspondence between resistance genes and virulence genes was widespread. Indeed, gene-for-gene relationships have been verified between crops and pestiferous fungi, bacteria, viruses, parasitic plants, nematodes (6), and at least one insect species, the Hessian fly (1). Where a gene-for-gene relationship exists, pest resistance based on a single gene will be effective only against those pest genotypes lacking the corresponding virulent gene. This situation, termed ''vertical resistance,'' should have low durability because just a single-gene change in the pest will overcome it. In contrast, so-called ''horizontal resistance,'' dependent on many genes, each with minor effects, should be more durable because virulence will require many genetic changes in the pest (7). Despite the advantage for durability, breeding for horizontal resistance is uncommon, partly because it is laborious and time consuming. In addition, the gene-for-gene model for predicting durability does not hold for all pests; there are examples of single-gene resistance with good durability. Interactions

R

between major and minor genes and the mechanisms of pathogenicity all influence durability in complex ways. Thus, deliberate breeding for polygenic resistance may not always be the most efficient route to durability.

Tolerance

Crop varieties with a type of resistance known as "tolerance" are less damaged by pests, but do not reduce pest survival and reproduction, do not exert selection on the pest (8), and should be durable. Tolerance contributes to disease and insect resistance in many released crop varieties, but the comparative durability of crops with tolerance has not been documented systematically. Breeding specifically for tolerance is uncommon primarily because it is hard to measure (9) requiring measuring plant damage or injury in response to a range of known pest infestation levels. Tolerance, which can be caused by a variety of complex mechanisms, is almost certainly polygenically inherited, complicating the breeding effort. Tolerance alone may have limited effectiveness because unchecked pest populations eventually can reach damaging levels, lowering incentives for breeding specifically for tolerance.

Antixenosis

A type of insect resistance known as "antixenosis" deters or repels insects that otherwise would attack the crop, but does not necessarily kill these insects (8). Antixenosis cannot occur toward pathogens, which do not actively select their host plants. Antixenosis should be durable, if alternative host plants are available on which pests that reject the crop can survive and reproduce. Incorporating antibiosis (resistance that kills the pest) or using insecticides on a crop with antixenosis can theoretically select for insects that avoid the crop, providing extremely stable resistance. In practice, alternative hosts may not be abundantly available, weakening the effectiveness and potential durability of antixenosis.

Tissue-Specific Resistance

An approach made feasible by transgenic technology is expression of resistance genes only in economically essential plant tissues (3). Some crops can tolerate substantial attack to nonharvested tissues without sustaining yield or quality reductions. For example, the flag leaf provides most of the photosynthate to fill the head of cereal crops, whereas other leaves are less important. If resistance genes protect the essential plant tissues, the rest of the plant can support pest survival and reproduction, reducing selection for virulence. Tissue-specific expression has not yet been applied to enhance durability of transgenic pest-resistant crops.

Partial Resistance

Low levels or "partial resistance" to pests coupled with other types of controls, including biological control and judiciously applied pesticides, can provide more sustainable suppression of pests (10). Because the combination of methods exerts low selection pressure for pest adaptation to any of the component control measures, each should be more durable. The effect is similar to horizontal resistance in which several genes with minor effects are combined in a single resistant variety. Insecticides are sometimes used to control target insects on resistant crop plants, but whether durability of the resistance is increased by this practice has not been documented.

DEPLOYING PEST RESISTANCE TO INCREASE DURABILITY

Rotating Resistant Varieties and Genes

If the frequency of virulence genes declines in a pest population after the corresponding resistant variety is withdrawn from production for a period, it may become possible to deploy that resistance effectively at a later date. Several genes might be deployed in a rotation extending their collective. The durability of such a strategy will depend on how rapidly virulence declines in the absence of selection.

Mixtures or "Multilines"

Resistant and susceptible varieties planted together in mixtures will exert reduced selection for virulence because avirulent pests will survive on the susceptible plants. Similarly, mixtures ("multilines") of varieties with different resistance genes will reduce the relative selective advantage of each virulent allele (11). In addition, successful pest movement or spread between adjacent plants will be impeded in a multiline within a single season. The spatial scale of varietal mixtures may have to be large for mobile insects to ensure individuals cannot survive by attacking multiple resistant genotypes, thus diluting the resistance effect. Mixtures and multilines have several limitations. First, all the components of a multiline must be similar in performance, time to harvest, and quality because they will be harvested together. This is a challenge to the breeder. Second, although mixtures at larger spatial scales do not have the first difficulty, planting and harvest can be complicated for the grower. Third, together a multiline will have only partial resistance, because some of the pest population will be adapted to each variety.

Stacking Resistance Genes

More than one gene for resistance to a particular pest can be bred into a single crop variety in a process called gene stacking or "pyramiding." This increases durability because to overcome the resistance, pests must possess genes adapted to each resistance gene (4). Stacking resistance genes is difficult because different resistant genotypes typically have similar phenotypes, for example, no disease symptoms or low pest survival. As a result, the breeder can not easily distinguish plants that have multiple genes from those with only one gene. Additional steps such as test crosses, or challenging each plant or line with multiple pest races or biotypes become necessary. The recent development of molecular markers potentially simplifies gene stacking because plants with multiple genes can be quickly identified by the presence of each of their associated markers. With this powerful tool, gene stacking for pest resistance and other characteristics, including yield, is being pursued in many breeding programs.

THE HIGH-DOSE/REFUGE STRATEGY

The recent development and deployment of commercial crop varieties engineered with genes for *Bacillus thuringiensis* (Bt) insecticidal proteins has been accompanied by concerns about the durability of such resistance. Pest adaptation to these varieties threatens their viability, but also could accelerate insect pest adaptation to Bt-based insecticides, which are among the best microbial insecticides available. Accordingly, the EPA, in consultation with scientists and industry, has promulgated a set of guidelines for growing Bt-engineered crops to increase their durability. These guidelines are based on what is known as the "high dose/refuge" strategy (3). This strategy exploits the ability to engineer crops with extremely high levels of Bt toxins that can kill 99% of target pests. In theory, the surviving pests should be only those rare individuals homozygous for alleles that provide resistance to Bt toxins. If these few survivors mate predominately with pests not exposed to the toxin, the succeeding generation will again have few homozygous virulent individuals, thus extending the durability of the resistance. To ensure sufficient numbers of pests are not exposed to the toxin, a proportion of the crop acreage must be planted with untransformed susceptible plants, as a so-called "refuge." The EPA guidelines now recommend that refugia comprise 4% of the cropped area. Theory predicts this strategy will be effective, depending on some assumptions: 1) that there are low natural frequencies of alleles for adaptation to Bt in the pest population, 2) that the crop kills 99% or more of target pests feeding on the Bt crop, and 3) that pest population dynamics and movement patterns ensure mating between pests surviving on the Bt crop and pests from the refugia (12). Although based on the best available data, many of these assumptions have not yet been tested rigorously. Each target insect-crop system will differ in specific requirements for success of the high-dose/refuge strategy. Ongoing research is aimed to provide the necessary data. Success of this strategy will also depend on growers adhering to EPA guidelines, which are not mandatory.

CONCLUSION

Sequential release of resistance genes remains the principal response to pest adaptation. The limited sustainability of this approach will eventually require more concerted efforts to enhance resistance durability. Known strategies to increase the durability of pest-resistant crops have been applied only to a limited extent. Implementing these strategies will require more resources because of the increased costs of breeding for durability and ensuring that deployment strategies effectively increase durability. Optimal deployment requires adequate knowledge of pest biology and genetics and the effects of other natural and artificial controls on pest populations. It is to be hoped that the ongoing effort to increase the durability of Bt-transgenic insecticidal crop varieties will succeed, and encourage similar efforts to enhance durability of both transgenic and conventional pest resistance in the future.

REFERENCES

1. Ratcliffe, R.H.; Hatchett, J.H. Biology and Genetics of the Hessian Fly and Resistance in Wheat. In *New Developments in Entomology*; Bondari, K., Ed.; Research Signpost, Scientific Information Guild: Trivandrum, India, 1997; 47–56.
2. *Durable Resistance in Crops*; Waller, F.J.M., Van der Graaff, N.A., Lamberti, F., Eds.; NATO Advanced Science Institutes Series A; Plenum: New York, 1983; 55, 454.
3. Gould, F. Sustainability of transgenic insecticidal cultivars: integrating pest genetics and ecology. Annu. Rev. Entomol. **1998**, *43*, 701–726.
4. Gould, F. Simulation models for predicting durability of insect-resistant germplasm: a deterministic diploid, two-locus model environ. Entomol. **1986**, *15* (1), 1–10.
5. Flor, H.H. The complementary genic systems in flax and flax rust. Adv. Genet. **1956**, *8*, 29–54.
6. Crute, I.R.; Holub, E.B. *The Gene-for-Gene Relationship*

in Plant-Parasite Interactions; CAB International: Oxon, U.K., 1997; 427.

7. Person, C.; Fleming, R.; Cargeeg, L.; Christ, B. Present Knowledge and Theories Concerning Durable Resistance. In *Durable Resistance in Crops*; Lamberti, F., Waller, J.M., Van der Graff, V.F., Eds.; NATO Advanced Science Institutes Series A; Plenum: New York, 1983; 55, 27–40.
8. Painter, R.F. *Insect Resistance in Crop Plants*; Macmillan: New York, 1951; 520.
9. Smith, C.M. *Plant Resistance to Insects: A Fundamental Approach*; John Wiley & Sons: New York, 1989; 286.
10. Thomas, M.; Waage, J. *Integration of Biological Control and Host Plant Resistance Breeding*; Technical Centre for Agricultural and Rural Cooperation (ACP–EU), The International Institute of Biological Control of CAB International: Wageningen, The Netherlands, 1996; 99.
11. Mundt, C.C.; Brophy, L.S. Influence of number of host genotype units on the effectiveness of host mixtures for disease control: a modeling approach. Phytopathology **1988**, *78*, 1087–1094.
12. Alstad, D.N.; Andow, D.A. Managing the evolution of insect resistance to transgenic plants. Science **1996**, *268*, 1894–1896.

RESISTANCE TO HOST-PLANT RESISTANCE

Sanford D. Eigenbrode
University of Idaho, Moscow, Idaho, U.S.A.

INTRODUCTION

Most insect and pathogen populations have a large amount of genetic variability for traits, including those governing responses to host defenses. If a defense mechanism predominates in hosts, as occurs when a resistant crop variety is used over a wide area, pest individuals genetically adapted to the defense will leave more progeny. Over time, adapted types become predominant and resistance to the pest fails. Adaptation by pests can involve a single genetic change, or the accumulation and recombination of multiple genes (more properly, forms of genes or alleles) for adaptation. Pests that have overcome resistance (or the associated genes for adaptation) are typically termed ''virulent'' to a specific type of resistance (1, 2). Whether host-plant resistance (HPR) breakdown occurs will depend on: 1) whether there is sufficient genetic variation in the pest population and 2) whether survival and reproduction of genetically virulent pests is sufficiently greater than survival and reproduction of nonadapted pests. This second requirement depend upon the area planted to the resistant variety and how consistently it is planted over multiple seasons (3). Because these conditions are restrictive, not all deployed HPR has been overcome by pest evolution. Where it has occurred, the consequences have been costly and problematic.

INSECT EXAMPLES

Genetic variants of insect pests that have overcome HPR (and occasionally other control measures) are termed ''biotypes.'' Examples are shown in Table 1. The term biotype implies that these genetic variants are somehow discrete and unique without intermediate forms. In fact, this rarely occurs because mating among insects can produce various combinations of virulent genes (5). For example, different Philippine populations of the rice brown planthopper, *Nilaparvata lugens*, can be selected by rearing them for several generations on resistant rice varieties to produce one of several so-called biotypes of this pest. Nevertheless, the term biotype remains in use to describe resistance-adapted insect pests. The term biotype may be most applicable for virulence based on single genes in the insect pest; populations in which these single genes are fixed are clearly virulent biotypes. The only certain occurrence of this is in the Hessian fly, *Mayetiola destructor*. For every resistance gene available in wheat, there are corresponding genes in the fly that can overcome the resistance, in a classic ''the gene-for-gene'' relationship. Presently, there are 25 genes for resistance to the fly and 16 genetically based biotypes of the fly with patterns of virulence to each resistance gene (6). Even this becomes complicated by the presence of multiple biotypes of Hessian fly within geographic areas small enough to permit interbreeding among the flies.

The majority of reported insect biotypes are aphids (Table 1). This might be because the parthenogenic reproduction and high reproductive capacity of aphids favors adaptation to resistance. It may also reflect the fact that HPR against aphids is more likely to be due to ''antibiosis'' that reduces pest survival, thus selecting for adaptation.

PATHOGEN EXAMPLES

Adaptation to resistance by pathogens occurs more frequently than by insects. For practical purposes pathologists and plant breeders generally assume that plant pathogens occur in multiple, genetically distinct ''races'' (the term typically used for fungal pathogens) or ''pathovars'' (the term typically used for bacteria), capable of causing disease in specific plant genotypes. This problem was discovered early in the effort to develop pathogen-resistant crops. Within 9 years of the first reported incorporation of a single gene for resistance to a pathogen, resistance-breaking races of that pathogen had been discovered (races of *Puccinia graminis tritici*, stem rust in wheat, reported in 1916). As a result, it has become necessary to discover and release multiple genes for resistance to many pathogens to overcome these virulent races. Rusts, in the genus

R

Table 1 Examples of insect biotypes that have overcome host-plant resistance

Crop	Pest	Reported number of biotypes
Alfalfa	Pea aphid	4
	Spotted alfalfa aphid	6
Apple	Wooly apple aphid	2
	Rosy leaf-curling aphid	3
Corn	Corn leaf aphid	3
Raspberry	Raspberry aphid	4
Rice	Green rice leafhopper	2
	Green leafhopper	3
	Brown planthopper	4
	Rice gall midge	4
Brassica spp.	Cabbage aphid	2–4
Wheat	English grain aphid	3
	Hessian fly	16
	Greenbug	8

(From Refs. 1, 2, 4.)

Puccinia, are a good example. In the northern United States and Canada, as many as six races of wheat stem rust can be isolated from wheat during a season. In the same geographic area during a 7-year period, 19 were detected in wheat leaf rust (*P. recondita*) and 35 in oat crown rust (*P. coronata*) respectively. Of these races some are virulent (cause disease) in wheat varieties resistant to other races. Whether virulent races become predominant in the field depends on the conditions for evolution of pest resistance outlined above (1, 3, 7). The number of races of any given pathogen rarely is known accurately and, at any rate, can change quickly. However, an idea of the extent of variation in pathogens can be appreciated on the basis of how many examples of documented gene-for-gene relationships are known involving pathogens. Whereas only perhaps one such relationship is known for insects, at least 24 are known for pathogens (Table 2).

WHAT CAN BE DONE ABOUT PESTS OVERCOMING HOST-PLANT RESISTANCE?

Pest adaptation to host plant resistance is serious and widespread. Because the crops affected include the major staple cereals (Tables 1 and 2), the threat to food security is substantial. Although some resistance apparently is stable, adaptation has occurred by principal pests of important food crops worldwide. Sustainable production of these crops depends on HPR. Because genetic variation in plant pathogens and pest insects is the rule rather than the exception, theory predicts that the potential for evolution of resistance-breaking types is always there and that vigilance is prudent.

The typical response to pests overcoming HPR has been sequential release of additional resistance genes. This is costly because of the time and effort required to screen for this resistance and to incorporate it into varieties. It also may not be sustainable because the number of genes for resistance to each pest is finite. Sequential and cyclical release of a limited number of resistance genes is potentially durable, but requires that between release of each resistance gene the virulence genes become rare in the pest or pathogen population, which may or may not occur. Recognizing these limitations, scientists have devised alternative strategies for developing and releasing resistant varieties to improve durability.

One class of strategies for increasing durability is simultaneous release of multiple genes for resistance, either in varietal mixtures, or combined or "stacked" within the same variety. By not favoring one race or biotype, these strategies can slow selection for virulence. Stacked genes additionally can be effective because pests without virulence genes for all resistance genes in the crop should

Table 2 Cases in which host-plant resistance based on a single gene has been overcome by a single gene in a pathogen, producing gene-for-gene relationships

Host	Pathogen
Triticum	*Puccinia striiformis*
	Puccinia graminis tritici
	Puccinia recondita
	Erysiphe gramnis tritici
	Ustilago tritici
	Tilletia caries
	Tilletia foetida
Zea	*Puccinia sorghi*
	Helminthosporium turicicum
Coffea	*Hemileia vestatrix*
Avena	*Puccinia graminis avenae*
	Ustilago avenae
Linum	*Melampsora lini*
Helianthus	*Puccinia helianthi*
Hordeum	*Erysiphe graminis hordei*
	Ustilago hordei
Solanum	*Phtyophthora infestans*
	Snynchytrium endobioticum
Lycopersicon	*Cladosporium fulvum*
Malus	*Venturia inaequalis*
Phaseolus	*Colletotrichum lindemutheanum*
Oryza	*Pyricularia oryzae*
Gossypium	*Xanthomonas malvacearum*

(From Ref. 2.)

have suppressed or no reproduction. Only pests with all virulence genes will overcome the resistance. Recombination of two or more specific virulence genes in a single pest individual will be extremely rare compared with the frequency of each gene individually (generally it will be the product of their individual frequencies). Stacking resistance genes is laborious and has not been practical for most types of resistance. Molecular markers now being developed for many pest resistance genes will make this strategy more feasible.

Another way to obtain multiple genes for resistance is to select for resistance based on screening with a genetically variable pest population and slowly selecting for resistant crops types over multiple generations. Theoretically, this resistance will be polygenic, that is dependent on many minor genes, and will be durable because no one genetic adaptation by the pest will be able to overcome it. This type of resistance is termed "horizontal," and may be contrasted with "vertical" resistance that is conditioned by a few genes but effective against only few pest genotypes (8). Some theorists strongly advocate reliance on horizontal resistance. Unfortunately, developing horizontal resistance requires selecting with diverse pest populations throughout the breeding process, which can be difficult to achieve.

Another approach to improving durability of resistance is to develop resistance that does not select for pest adaptation. Examples are resistance that deters insects from feeding or ovipositing on a crop or resistance that confers plant tolerance to insect or pathogen attack. Because the survival and reproduction of pests attacking these plants are not affected, there should be no selection for virulent pests. This effect probably does contribute to durability of some existing deployed varieties but this has not been clearly documented. A problem is that mechanisms of many resistant varieties, and how much of the resistance is due to tolerance or deterrence of pests, is not always known.

Durable resistance, when it occurs, is probably due to mechanisms like those described above. Specific aspects of the interactions between pests and crops also may influence durability. Study of existing durable resistance will help understand these mechanisms and may help develop more durable pest resistant crops in the future. Typically, durable resistance is less well studied because the imperative to understand it is less intense than the imperative to address the breakdown of resistance. For a more complete review of the strategies that can be used to increase the durability of resistance to insects and pathogens, see "Resistance Management in Host-Plant Resistance" (pages 708–711) for a more complete description of this high-dose strategy. A recent concern is the development of insect-resistant crops based on transgenic technology. Strategies to increase the durability of this resistance are similar to those described above, but include taking advantage of the technology's ability to produce extremely resistant plants capable of killing most of the insects that feed on them (8).

REFERENCES

1. Crute, I.R.; Holub, E.B. *The Gene-for-Gene Relationship in Plant–Parasite Interactions*; CAB International: Oxon, UK, 1997; 427.
2. Fry, W.E. *Principles of Plant Disease Management*; Academic Press, Inc.: London, 1982; 378.
3. Kim, K.C.; McPheron, B.A. *Evolution of Insect Pests: Patterns of Variation*; John Wiley & Sons: New York, 1993; 479.
4. Smith, C.M. *Plant Resistance to Insects: A Fundamental Approach*; John Wiley & Sons: New York, 1989; 226.
5. Claridge, M.F.; Den Hollander, J.D. The biotype concept and its application to insect pests of agriculture. Crop Prot. **1983**, *2*, 85–95.
6. Ratcliffe, R.H.; Hatchett, J.H. Biology and Genetics of the Hessian Fly and Resistance Wheat. In *New Developments in Entomology*; Bondari, K., Ed.; Research Signpost, Scientific Information Guild: Trivandrum, India, 1997; 47–56.
7. Person, C.; Fleming, R.; Cargeeg, L.; Christ, B. Present Knowledge and Theories Concerning Durable Resistance. In *Durable Resistance in Crops*; Lamberti, F., Waller, M.J., Van der Graaff, V.F., Eds.; NATO Advanced Science Institutes Series A; Plenum: New York, 1983; 55, 27–40.
8. Gould, F. Sustainability of transgenic insecticidal cultivars: integrating pest genetics and ecology. Annu. Rev. Entomol. **1998**, *43*, 701–726.

RESISTANCE TO PESTICIDES

Steven J. Castle
USDA-ARS, Phoenix, Arizona, U.S.A.

R

INTRODUCTION

The occurrence of pesticide resistance stands as one of the most chronic and formidable problems for crop production and public health worldwide. The evolution of different resistance mechanisms by pestiferous organisms has diminished the lethal action of pesticides and led to greater difficulties in controlling pest populations. Resistance as a chronic problem is a result of its occurrence in an increasing number of species, but also its recurrence in resistant pest species that evolve more than one resistance mechanism or employ existing ones to counter novel pesticides. As a formidable problem in pest management, resistance often diminishes control of pest populations and contributes to a pesticide treadmill, i.e., more applications to control a resistant pest leading to more intractable resistance leading to more pesticide applications. Greater awareness of the conditions that produce resistant pest populations combined with counter-resistance tactics have begun to improve the outlook for managing and curtailing resistance.

PESTICIDES AND THE AGELESS BATTLE AGAINST PESTS

From the time that the first human settlements cultivated the earth's resources into a dependable food supply, people have struggled against the natural order using whatever means necessary to maximize productivity. But competitor species of insects and mites, plant pathogens, and weeds have been among the strongest forces working against humankind's efforts. On a different front, insect and mite parasites and disease vectors have long impacted the human condition. The cumulative toll from both agricultural and medical fronts on millennial generations of humans is incalculable. What can be ascertained, however, is that there has been no letup in the challenges by pests to undermine human health and well-being, nor in our resolve to prevent it.

Viewed in this context, the incredulous reaction to the phenomenal pest control attained when DDT was first introduced in the 1940s can perhaps be better understood. Never before had there been such excellent control over diverse insect pests both destructive to agriculture or dangerous to public health. This early euphoria helped spawn what was termed the era of optimism within the Age of Pesticides (1). There was a rush to develop additional synthetic organic compounds and to incorporate them into crop and public health protection programs. It wasn't long, however, before the first incidence of resistance to DDT in 1946 was observed in houseflies in northern Sweden, just two years after DDT had been introduced to that part of the country. By 1957, resistance to DDT had been documented in 27 species, increasing to 37 species by 1962 (2). What had initially been hailed as an infallible pest control product was now being diminished by the evolutionary phenomenon of pesticide resistance.

Difficulties in controlling certain agricultural insect pests with inorganic insecticides had already been acknowledged in the early part of the 20th century, and by 1937 at least seven species were known to resist one or more inorganic treatments (2). These early cases were indicative of an innate capacity of insects to defend against the toxic activities of chemicals, whether they were inorganic compounds or the yet to be introduced organic compounds.

THE CAUSES OF RESISTANCE

Variability in the genetic makeup of a population allows some individuals to survive specific mortality factors such as those exerted by pesticides. The particular trait, or traits, that enable these individuals to survive may be rare within a population initially. But with each additional pesticide application, resistant individuals that survive pesticide exposure gain selective advantage over susceptible individuals that do not survive (Fig. 1). Resistant genotypes multiply in an environment with continuous pesticide pressure so long as their particular resistance mechanism(s) are effective against a prevailing pesticide or pesticide spectrum. The level of resistance attained to a particular pesticide is a function of the mode of inheritance

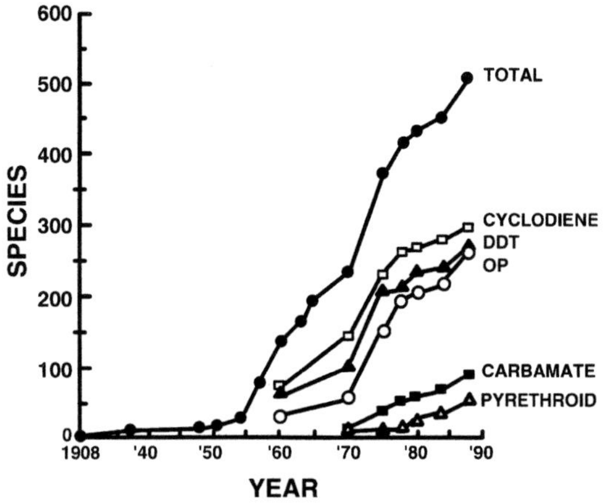

Fig. 1 Chronological increase in numbers of arthropod species resistant to insecticides (total), and those resistant to each of the five principal classes of insecticides. (From Ref. 3.)

at a gene locus and the resulting quantitative expression of that gene. The type of resistance expressed depends on what behavioral or physiological mechanisms are controlled by the resistance gene(s). Multiple resistance mechanisms can act synergistically within certain individuals to resist extreme levels of exposure to one type of pesticide, or act independently against pesticides with different modes of action. Quite often, a single metabolic mechanism or change in a target site will confer cross-resistance to two or more pesticides related by class or not. The character of resistance can vary drastically or subtly between populations due to many possible combinations of resistance genes. But in general, resistance is expressed according to the intensity of selection by a pesticide, the mechanism(s) involved, and the frequency that genes conferring resistance occur in a population.

INCREASED DEPENDENCY ON PESTICIDES

The broad-spectrum activity and greater toxicity of DDT quickly changed the level of dependency on insecticides. Knowledge of the insecticidal activity of DDT and related structures opened up the synthetic pathways that ultimately gave rise to organophosphorous and methyl carbamate compounds, two of the three principal classes of insecticides along with the chlorinated hydrocarbons that would dominate the insecticide market between 1950 and 1980. As new insecticides continued to be developed, the first synthetic organic fungicides and then herbicides became available during the 1950s and 1960s, respectively. The development and implementation of synthetic pesticides occurred at a time when the mode of agriculture was changing to one of greater mechanization, increased fertilization, irrigation, and cultivation that together with new genetic varieties enabled more intensive production on larger areas of land. But these changes also improved the quality of resources that insect pests had already been exploiting, only now with net increases in their fitness levels on agronomically and genetically improved crops. Larger areas under cultivation worldwide also meant more opportunities for weeds to colonize disturbed habitat and for epidemics of plant pathogens to erupt in monocultured crops. This confluence of independent, yet interrelated events involving the development of pesticide technology at a time of agricultural intensification and agronomic improvements resulted in more intensive pesticide usage during a period of increasing pest pressure. The chemical control paradigm had been set with the early successes of DDT against insects. Farmers continued to find recourse for controlling damaging pest populations through an almost interminable production of new pesticides by a growing agrochemical industry.

THE CONSEQUENCES OF RESISTANCE

Attention soon focused on the negative aspects of pesticide use with increasing reports of resistant species. Moreover, there was a growing awareness that in addition to pest species becoming resistant and more difficult to control, pesticide use brought other undesirable consequences such as secondary pest outbreaks and harm to nontarget organisms including vertebrate species. Examples accumulated where entire agricultural systems that had become overly reliant on pesticides were threatened with collapse due to insecticide-resistant pests and decimation of beneficial insect populations (Table 1). The pattern of destruction in cotton-intensive systems worldwide became so familiar that five phases of agricultural production that progressed from simple farm systems to pesticide-induced catastrophes were identified: 1) subsistence, 2) exploitation, 3) crisis, 4) disaster, and 5) recovery. Each of the examples presented in Table 1 experienced these five phases, although the recovery phase was protracted in some due to severely destabilized production conditions.

Aside from the economic disasters that have taken place through misuse of pesticides, resistance has a persistent and insidious cost that directly affects farmers through increased applications to establish control, while pesticide manufacturers lose revenue when particular products are no longer effective against one or more pests. One estimate suggests that resistance adds an extra $40 million to the annual insecticide bill in the United States

R

Table 1 Examples of agricultural regions severely affected by pest outbreaks following the onset of insecticide resistance

Region and country	Time period	Resistant pest(s)	Resistance-tarnished chemicals	Damage
Cañete Valley, Peru, Nicaragua	1950–56	*Heliothis virescens*	BHC	Severe economic losses
	1965–72	*Helicoverpa zea, Spodoptera sunia, Bemisia tabaci*	Toxaphene-DDT-Methyl parathion	Loss of production area; high treatment costs; insecticide poisonings; malaria resurgence
Matamoros-Reynosa, Mexico	1962–70	*H. virescens*	Methyl parathion	Cotton production closed out
The Gezira, Sudan	1976–85	*Bemisia tabaci H. armigera*	Monocrotophos, DDT	Reduced yields; high treatment costs; decreased production
Ord Valley, Australia	1974–78	*H. armigera*	DDT	High treatment costs; cotton production closed out

due to additional treatment costs or alternative controls. The external costs of resistance are more difficult to tabulate, but environmental degradation and public health hazards including death and chronic illness are too often known wherever pesticides have been used intensively.

The increasing numbers of resistant pests have forced many pesticides out of the marketplace and at the same time limited the control options available for farmers. The most recent tally of insects and mites resistant to insecticides/acaracides placed the total at least at 504 species (3). Additionally, more than 150 species of plant pathogenic fungi are resistant to fungicides (4) and 153 weedy plants are resistant to herbicides (5). It has been argued that each of these values understates the true magnitude of the resistance problem because they do not account for multiple cases of resistance within individual species. If, for example, every insecticide/acaracide for which resistance has occurred was summed for each of the 504 resistant species of insects and mites, then a total of 1640 cases of species/insecticide resistance are known. Similarly, there are 216 cases of herbicide resistance when all forms of resistant biotypes within multiple-resistant weed species are included. It should be recognized, however, that species tallies also tend to exaggerate the resistance problem because only some populations develop resistance.

RESISTANCE PROGNOSIS

Pesticide resistance remains an ever-present challenge to agricultural production and public health. However, many positive changes in pest management have greatly improved the prognosis for minimizing the damaging effects of pesticide resistance. Resistance management has developed into an applied discipline with an extensive theoretical foundation and a growing number of practical examples from which to critically evaluate alternative approaches. Realization that active ingredients are finite and have a certain preciousness is helping to shift attitudes concerning the importance of conserving pesticides as essential tools in the endless battle against pests. At the same time, a trend toward greater specificity and toxic activity of pesticides to targeted pests, but decreased toxicity to nontarget organisms, has greatly improved the quality of both pest and resistance management in numerous areas. The bioengineering of crop plants to express toxins at high concentrations without exposing nontarget organisms represents a monumental advancement in pest control. Vigilant anti-resistance tactics will have to be implemented to protect new technological innovations and active ingredients to avoid the mistakes of the past.

REFERENCES

1. Metcalf, R.L. Changing role of insecticides in crop protection. Ann. Rev. Entomol. **1980**, *25*, 219–256.
2. Brown, A.W.A.; Pal, R. *Insecticide Resistance in Arthropods*; WHO Monograph Series No. 38, WHO: Geneva, 1–491.
3. Georghiou, G.P.; Lagunes-Tejeda, A. *The Occurrence of Resistance to Pesticides in Arthropods*; FAO: Rome, 1991, 1–318.
4. Holt, J.S.; LeBaron, H.M. Significance and distribution of herbicide resistance. Weed Technol. **1990**, *4*, 141–149.
5. Heap, I. *International Survey of Herbicide Resistant Weeds*. www.weedscience.com (accessed July 27, 2000).

RHIZOBIA

F.L. Walley
University of Saskatchewan, Saskatoon, Saskatchewan, Canada

J.H. Stephens
MicroBio RhizoGen. Corp., Saskatoon, Saskatchewan, Canada

INTRODUCTION

Rhizobia is a general term referring to diazotrophic (N_2-fixing) soil bacteria belonging to the family Rhizobiaceae, which include *Rhizobium, Bradyrhizobium, Sinorhizobium*, and *Azorhizobium* (1). These soil bacteria are capable of establishing an endocytobiotic association (an association between two organisms in which one organism lives within the cells of the second organism) with certain plants, resulting in the formation of nodules on the plant roots. Within these nodules the bacteria produce an oxygen sensitive nitrogenase enzyme that converts, or fixes, atmospheric N_2 to NH_4^+. The fixed N is subsequently exchanged for an energy supply in the form of plant derived carbohydrates. Thus, the association is symbiotic and one in which the host plant supplies the bacteria with energy, and the bacteria supplies the plant with a source of fixed N. Agriculturally, the most important rhizobia are those that form symbiotic associations with legumes.

SYMBIOTIC RELATIONSHIP AND FUNCTION

Typically, the association between a legume host and the *Rhizobium* is highly specific—no single rhizobial strain can nodulate all legumes, and every rhizobial species has a definable host range (Table 1) (2). The legume host range is, for some rhizobial strains, very limited whereas other strains are characterized by a broad legume host range—a characteristic termed ''symbiotic promiscuity'' (3).

The controlled and coordinated expression of a large number of genes of both legume host and rhizobial partner contributes to the expression of host specificity. In the early stages of host recognition legume roots secrete chemicals known as ''flavonoids'' into the rhizosphere. The flavonoids, which may act as chemical attractants, induce the *Rhizobium* to produce a suite of biochemicals required for root infection. The ability to produce flavonoids apparently is an evolutionary preserve of leguminous plants and consequently, only legumes can develop nodules in the presence of *Rhizobium* (2).

Having received the flavonoid message, the rhizobia attach to host root hairs (the point of infection) behind the root apical meristem. Specific rhizobial nodulation genes are switched on, and the *Rhizobium* begin to synthesize ''Nod'' factors that subsequently set off a series of events in the legume host root that ultimately result in nodulation.

Root ''lectins,'' also released by the legume host, may have a similar role in the signaling process. Lectins are glycoproteins, glycolipids, or protein of nonimmune origin capable of recognizing and binding (reversibly) to carbohydrates. Lectins may play a role as Nod factor receptors, or may mask other surface polysaccharides that otherwise could elicit a host defense mechanism antagonistic to rhizobial infection factors (4).

STAGES OF INFECTION

The initial infection stages have been the subject of many extensive reviews (5, 6). Soon after bacterial binding, the host root hair responds to the presence of the rhizobia by curling. The tightly curled root hair entraps the bacteria and a local lesion is formed by hydrolysis of the root hair cell wall. The rhizobia penetrate the root hair and the development of an infection thread begins. Infection threads are tubular structures that carry *Rhizobium* cells from the root surface into the root cortex. When the infection thread is initiated, cortical cell division leading to nodule development begins. As the thread approaches the region of newly dividing meristematic cells, it branches several times thus carrying the multiplying bacteria into many host cells.

Within the root cortex, enhanced cell division results in the formation of a ''nodule primordium'' that ultimately gives rise to the nodule tissue. In temperate legumes, such as pea, nodule primordia are formed from the root inner cortex, resulting in persistently active

Table 1 Nodulation host range among rhizobia

Rhizobium spp.	Host legume
R. meliloti	*Melilotus, Medicago, Trigonella*
R. leguminosarum bv. *viciae*	*Pisum, Lens, Vicea, Lathyrus*
R. leguminosarum bv. *trifolii*	*Trifolium*
R. leguminosarum bv. *phaseoli*	*Phaseolus*
R. lupini	*Lupinus, Ornithopus, Lotus*
R. loti	*Lotus, Lupinus, Anthyllis, Dorycnium*
R. fredii	*Glycine soja, G. max*
Bradyrhizobium japonicum	*G. max, Siratro,* cowpea, etc.
B. elkanii	*G. max*
Bradyrhizobium sp. Cowpea	*Cicer arietinum*

(Adapted from Ref. 2)

growing cells that give rise to indeterminate nodules (nodules that continue to grow by cell division). In tropical legumes such as soybean [*Glycine max* (L.) Merr.] and common beans (*Phaseolus vulgaris*), nodule primordia are formed from the outer root cortex and give rise to determinate nodules that increase in size by expansion rather than cell division.

Following release into nodule cells, the bacteria alter shape, increase in size, and differentiate into enlarged forms called ''bacteroids,'' which synthesize the nitrogenase enzyme. The term ''symbiosome'' now is used widely to describe the N_2 fixing unit comprising the bacteroid and associated membrane structures. The number of symbiosomes within each infected cell can well exceed 100, and in mature legume nodules the enlarged infected cells may house thousands of symbiosomes.

The host cell is the only source of energy and nutrients for the bacteroids. Under normal physiological conditions, the N_2 fixation capacity of a symbiotic association is limited by the supply of photosynthates from the host plant. Thus, any factor that limits photosynthesis or general crop vigor, such as light intensity, plant density, disease, insect infestation, or CO_2 partial pressure will reduce N_2 fixation.

That the maintenance of plant health, and hence biological N_2 fixation, is of significant importance in agriculture is evident by the estimate that total N fixed by legume crops is 70×10^6 m ton yr^{-1} (7). Furthermore, many agriculturally important legumes can derive as much as 80–90% of their N requirements through the process of N_2 fixation. Indeed, estimates of the quantity of N fixed for soy, lucerne, clover, and lupin are 234, 208, 153, and 169 $ha^{-1}yr^{-1}$, respectively.

INOCULANTS

Legume inoculants, comprised of living rhizobial cultures mixed with, and incubated in, a carrier material suitable for application to legume seeds or the seed environment, have been produced in North America since 1896 when a U.S.patent relating to the inoculation of soil during the cultivation of legume crops was issued to Nobbe and Hiltner (8). This process exploits the N_2 fixation phenomena through enhanced soil target organism populations and has been adopted virtually worldwide. Proculants, in most instances, fundamentally follows the processes developed by Nobbe and Hiltner and only in recent years have acceptable alternatives to peat-based products (or peat equivalents) including liquid (seed applied) and granular (soil applied) formulations, been evident (9, 10). With the exception of the soil applied granular products, a broad range of seed applied protectant compounds, used in many agricultural systems may result in diminished populations of applied inoculants with this diminution resulting in decreased probabilities that any given legume crop will reach its N_2 fixing potential.

REFERENCES

1. Udvardi, M.K.; Kahn, M.L. Evolution of the (Brady) rhizobium-legume symbiosis: why do bacteroids fix nitrogen? Symbiosis **1992**, *14*, 87–101.
2. Young, J.P.W.; Johnston, A.B.W. The evolution of specificity in the legume-rhizobium symbiosis. TREE **1989**, *4* (11), 341–349.
3. Sharma, P.K.; Kundu, B.S.; Dogra, R.C. Molecular mechanism of host specificity in legume-rhizobium symbiosis. Biotechnol. Adv. **1993**, *11*, 741–779.
4. Ridge, R.W.; Kim, R.A.; Yoshida, F. The diversity of lectin-detectable sugar residues on root hair tips of selected legumes correlates with the diversity of their host ranges for rhizobia. Protoplasma **1998**, *202*, 84–90.
5. Sprent, J.I. Which steps are essential for the formation of functional legume nodules? New Phytol. **1989**, *111*, 129–153.
6. Roth, L.E.; Stacey, G. Rhizobium-legume Symbiosis. In *Microbial Cell–Cell Interactions*; Dworkin, M., Ed.; American Society for Microbiology: Washington, DC, 1991; 255–301.
7. Brockwell, J.; Bottomley, P.J. Recent advances in inoculant technology and prospects for the future. Soil Biol. Bioch. **1995**, *27*, 683–697.
8. Smith, R.S. Legume inoculant formulation and application. Can. J. Microbiol. **1992**, *38*, 485–492.
9. Stephens, J.H.G.; Rask, H.M. Inoculant production and formulation. Field Crops Res. **2000**, *65*, 249–258.
10. Fouilleux, G.; Revellin, C.; Catroux, G. Short-term recovery of *Bradyrhizobium japonicum* during an inoculant process using mineral microgranules. Can. J. Microbiol. **1994**, *40*, 322–325.

RISKS OF BIOLOGICAL CONTROL

Barbara I.P. Barratt
AgResearch, Ltd., Invermay Agriculture Center, Mosgiel, New Zealand

INTRODUCTION

Pests often establish in a new country without the suite of predators, parasites, and pathogens that keep them in check in their natural range. Classical biological control relies upon reestablishing natural enemies with their hosts, and so has generally been accepted as an environmentally risk-free form of weed and pest management, avoiding the human health and environmental problems associated with pesticides. However, unlike pesticides, the release of living organisms as biological control agents, which are able to reproduce and actively disperse, has to be considered permanent and irreversible. There has been very little evidence for adverse environmental impacts of biological control introductions, partly attributable to the lack of study of nontarget impacts, however, recent case studies carried out after the release of biological control agents has provided some examples. While prerelease risk assessments are highly desirable, accurate prediction of host range and possible ecological impacts post-release are difficult to determine. These uncertainties have given rise to considerable debate between biological control practitioners and conservationists about the safety of biological control.

THE NATURE OF THE RISK

Adding a new species to an ecosystem is bound to impact in some way, and because of the complex nature of communities, it is almost impossible with current knowledge to predict the ramifications that might occur through the trophic levels within a biological system. The potential risks from biological control introductions come from either direct effects on nontarget species by host range expansion, or indirect effects on the community into which the new species has arrived. Novel hosts might include indigenous species that are taxonomically rela ted to the intended host including beneficial or valued species, and species that occur in or near the habitats where the target host is found, or to where it disperses.

Until recently, there has been little research undertaken to demonstrate the impact of biological control agents on nontarget species. It has been shown that the weevil *Rhinocyllus conicus* (Froehlich), introduced into the United States to control exotic thistles, has also adopted native thistle species as hosts and is significantly reducing seed production (1). This is one of few examples where the impact of a biological control agent on nontarget species has been quantified. In New Zealand, the braconid parasitoid *Microctonus aethiopoides* Loan, introduced to control the weevil lucerne pest *Sitona discoideus* Gyllenhal, has been found to parasitize a number of nontarget weevil species in the field, with parasitism levels of up to 70%. Ironically, one of the species attacked by *M. aethiopoides* in New Zealand is the weed biological control agent, *R. conicus,* although the impact of this on thistle control has not been assessed.

Most well-documented examples of biological control programs that have had severe negative environmental impacts have involved the introduction of vertebrates. The Indian mongoose (*Herpestes auropunctatus*) was introduced to the West Indies and Hawaii to control rats in sugarcane, but being unable to climb trees, it was only able to control the Norway rat and not the less numerous tree rat. The latter, having been suppressed by the former, then became more abundant. The mongoose also attacked domestic and native birds and lizards, and the reduction in numbers of lizards resulted in an increase in numbers of the sugarcane beetle (2). There are examples also of generalist predators that have been very damaging to nontarget organisms, for example the predatory snail, *Euglandina rosea.* This was released in many countries, often contrary to the advice of scientists, to control the giant African snail, *Achatina fulica,* an agricultural pest. However, *E. rosea* has had a devastating effect on native snails, in some instances causing extinctions (3).

Given the regulatory safeguards that we have in place today, such poorly conceived deliberate introductions are unlikely. However, impacts of insects are less obvious and poorly studied. While there are few documented cases, there is circumstantial evidence to suggest that insect biological control agents have the capacity to cause species extinctions, particularly in island communities. In Fiji, the introduction of the tachinid *Bessa remota* Aldrich to control the coconut moth, *Levuana iridescens* Bethune-Baker is thought to have caused the extinction of its host (4). Similarly, the citrus psylla, *Trioza erytreae* Del Guericio was apparently exterminated from Reunion Island by the eulophid parasitoid *Tetrastichus dryi* Waterston (5). In both cases, the parasitoids maintained their populations on alternative species, despite the decline of the target hosts.

Nontarget impacts can be very complex and unpredictable. A good example of this is the extinction of the large blue butterfly, *Maculinea arion* (L.) partly attributable to the biological control of rabbits in the United Kingdom using the myxoma virus (6). The butterfly larvae are unusual in that they live in the nests and feed on the brood of *Myrmica* ants. Changing land use coupled with the rapid decline of rabbits in the United Kingdom caused many grassland habitats to revert to scrub with loss of suitable habitat for the ants. In New Zealand, the introduced braconid wasp, *M. aethiopoides* attacks several nontarget weevil species including native broad-nosed weevils (7). In native grassland sown with legumes to improve soil fertility, these weevils can cause severe damage to the developing legume seedlings. While parasitism by *M. aethiopoides* could be considered an added benefit, the native weevils are also weed seedling feeders and may play a role in controlling the seedling establishment of *Hieracium* (hawkweed), a serious agricultural and conservation weed in native grassland.

REDUCING THE RISK

The risk of adverse impacts from biological control releases probably can never be eliminated, but minimized by carefully considered prerelease studies. Protocols for host specificity testing of weed biological control agents have been developed that depend upon a system of "centrifugal phylogenetic testing" (8). This means that nontarget species from those most closely related to the target weed to those more distantly related are exposed in sequence to the proposed biological control agent. This enables a profile of the host range of the organism to be developed and facilitates the decision on whether or not to proceed with field release. With insects, assembling a suitable list of potential nontarget species to be tested is more difficult because of the much larger number of species involved, and the lack of precise taxonomic information. However, the principles of the system used for weed biological control agents can to some extent be extended to insects (9).

There has been considerable debate about the desirability of host specificity in proposed biological control agents. While some contend that the presence of alternative hosts can assist the biological control agent through periods when the target host is scarce (10), others maintain that it is irresponsible to release any biological control agent unless it can be demonstrated to be completely host specific. Then again, it is argued that laboratory tests can be unreliable because in confinement in an artificial environment some proposed biological control agents attack species that would not be attacked in the field (11) and vice versa. Clearly there is still a great deal of research that needs to be carried out to underpin the design of reliable and feasible protocols. A possible approach is to verify the prerelease predictions of host range made from laboratory tests with postrelease field studies to determine realized host range in the field (7). This can help to build up a database that can assist decision makers in the future. Clearly, the environmental risks associated with the introduction of a biological control agent have to be weighed against the economic benefits of controlling a pest, or in some cases, the environmental cost of doing nothing.

In many countries, regulation of biological control agent introduction has been tightened in recent years. In 1995, the Food and Agriculture Organization (FAO) Council ratified an international "Code of Conduct for the Import and Release of Biological Control Agents" with the intention of providing a set of standards and guidelines for "best practice" biological control agent introduction (12). It is recommended in the Code that proposed importers of biological control agents provide information on agricultural and environmental nontarget effects. It is accepted that regulation of biological control agent introduction is required in the public interest because of its irreversibility, and the potential for biological control agents to disperse to habitats other than those where they were released (13). In New Zealand environmentally protective legislation has recently been enacted in which there is a clear requirement for biological control introductions to be compatible with safeguarding the life-supporting capacity of air, water, and ecosystems; the sustainability of flora and fauna; and the intrinsic value of ecosystems (14). A "precautionary approach" is adopted in managing adverse effects where there is scientific uncertainty about those effects. The challenge will be to maximize biological control safety without compromising its potential benefits.

CONCLUDING COMMENT

Many biological control practitioners have expressed apprehension at the level of concern shown about potential nontarget impacts of introduced biological control agents. It has been stated that biological control introductions for insect pest control would have to stop if it became mandatory to assess and avoid impact on all insect species, rather than being limited to species that are endangered and threatened. Others have expressed a cautionary view that the benefits to be gained from biological control should not be compromised by overstatement of the negative impacts. However, biological control practitioners and other stakeholders need to acknowledge and appropriately address the potential risks of biological control programs. Ideally, research teams should be assembled to address target and nontarget impacts of biological control agents so that predictions can be made with more confidence, and realistic risk/benefit analyses can be considered by regulatory agencies. It is likely that environmental, social, and regulatory requirements for biological control will become more rigorous in the future, which inevitably will mean that biological control programs will be more expensive and slower to implement, but benefits should accrue from increased success rates and improved environmental safety.

REFERENCES

1. Louda, S.M. Negative Ecological Effects of the Musk Thistle Biological Control Agent, *Rhinocyllus conicus*. In *Nontarget Effects of Biological Control Introductions*; Follett, P.A., Duan, J.J., Eds.; Kluwer Academic Publishers: Norwell, MA, 1999; 213–243.
2. Pimentel, D. Environmental risks associated with biological control. Ecol. Bull. **1980**, *31*, 11–24.
3. Civeyrel, L.; Simberloff, D.A. Tale of two snails: is the cure worse than the disease? Biodiv. Conserv. **1996**, *5* (10), 1231–1252.
4. Howarth, F.G. Environmental impacts of classical biological control. Annu. Rev. Entomol. **1991**, *36*, 489–509.
5. Aubert, B.; Quilici, S. Nouvel equilibre biologique observe à la réunion sur les populations de psyllides après l'introduction et l'establissement d'hymenopteres chalcidiens. Fruits **1983**, *38*, 771–780.
6. *The IUCN Invertebrate Red Data Book*; Pyle, R.M., Collins, N.M., Wells, S.M., Eds.; International Union for Conservation of Nature and Natural Resources: Gland, Switzerland, 1983; 451–457.
7. Barratt, B.I.P.; Goldson, S.L.; Ferguson, C.M.; Phillips, C.B.; Hannah, D.J. Predicting the Risk from Biological Control Agent Introductions: A New Zealand Approach. In *Nontarget Effects of Biological Control Introductions*; Follett, P.A., Duan, J.J., Eds.; Kluwer Academic Publishers: Norwell, MA, 1999; 59–75.
8. Wapshere, A.J. A strategy for evaluating the safety of organisms for biological weed control. Ann. Appl. Biol. **1974**, *77*, 201–211.
9. Barratt, B.I.P.; Ferguson, C.M.; McNeill, M.R.; Goldson, S.L. Parasitoid Host Specificity Testing to Predict Host Range. In *Host Specificity Testing in Australasia: Towards Improved Assays for Biological Control*; Withers, T.M., Barton, Browne L., Stanley, J.N., Eds.; CRC for Tropical Pest Management: Brisbane, Australia, 1999; 70–83.
10. Nechols, J.R.; Kauffman, W.C.; Schaefer, P.W. Significance of Host Specificity in Classical Biological Control. In *Selection Criteria and Ecological Consequences of Importing Natural Enemies*; Kaufmann, W.C., Nechols, J.E., Eds.; Entomological Society of America: Lanham, MD, 1992; 41–52.
11. Sands, D.P.A. Effects of Confinement on Parasitoid-Host Interactions: Interpretation and Assessment for Biological Control of Arthropod Pests. In *Pest Control in Sustainable Agriculture*; Corey, S.A., Dall, D.J., Milne, W.M., Eds.; CSIRO Canberra: Canberra, Australia, 1993; 196–199.
12. Schulten, G.G.M. *The FAO Code of Conduct for the Import and Release of Exotic Biological*, EPPO/CABI Workshop on Safety and Efficacy of Biological Control in Europe; Smith, I.M., Ed.; Blackwell Science Ltd.: Oxford, 1997; 29–36.
13. Van Lenteren, J.C. *Benefits and Risks of Introducing Exotic Macro-Biological Control Agents into Europe*, EPPO/CABI Workshop on Safety and Efficacy of Biological Control in Europe; Smith, I.M., Ed.; Blackwell Science, Ltd.: Oxford, 1997; 15–27.
14. ERMA New Zealand. *Annotated Methodology for the Consideration of Applications for Hazardous Substances and New Organisms Under the HSNO Act 1996*; ERMA New Zealand: Wellington, New Zealand, 1998; 28.

FURTHER READING

Caltagirone, L.E.; Huffaker, C.B. Benefits and risks of using predators and parasites for controlling pests. Ecol. Bull. **1980**, *31*, 103–109.

Nontarget Effects of Biological Control; Duan, J.J., Follett, P.A., Eds.; Kluwer Academic Publishers: Norwell, MA, 2000; 316.

Biological Control: Benefits and Risks; Lynch, J.M., Hokkanen, H.M.T., Eds.; Plant and Microbial Ecology Research Series 4 N, Cambridge University Press: Cambridge, UK, 1995; 304.

Simberloff, D.; Stiling, P. How risky is biological control? Ecology **1996**, *77* (7), 1965–1974.

RISKS OF VERTEBRATE PESTS

Mary Bomford
Bureau of Rural Sciences, Canberra, Australia

R

INTRODUCTION

Many vertebrate species have established exotic populations around the world that have become pests of agriculture and the natural environment (1–4). Hence trading and keeping of exotic birds, mammal, reptiles, and amphibians pose risks that new species will escape or be released from captivity and establish wild pest populations. Wildlife management authorities in many countries are now restricting the import and keeping of species that could become future pests. To do this effectively, risk assessments are required to determine the probability of escapes or releases occurring, the probability that released animals would establish wild populations that harm agriculture or the environment, and the probability that newly established populations could be eradicated. These probabilities can be hard to predict because they are dependent on many factors on which there is often inadequate knowledge.

PROBABILITY OF ESCAPES OR RELEASES

Releases or escapes of a species are more likely to occur if

1. large numbers of individuals are kept at many locations;
2. security requirements for keeping the species are low or poorly enforced;
3. the species has a low commercial value;
4. keepers have low awareness of the risks posed by exotic species;
5. species are kept in areas prone to disasters such as floods, fires, hurricanes, or wars; or
6. people have a motive to release animals, for example, for "animal rights" protests, or to establish a wild population for hunting or fishing.

PROBABILITY OF ESTABLISHMENT

Studies comparing successful species introductions with introductions that have failed indicate that an exotic population is more likely to establish if a species

1. has more individuals released at more sites and on more separate occasions;
2. is released at sites that have a close climate match to locations where the species already occurs;
3. has a widespread global distribution;
4. has a history of establishing exotic populations in other places;
5. produces more young per year; or
6. is nonmigratory in its endemic range.
   Two additional factors that many ecologists believe may make a wild population more likely to establish are if a species
7. can live in disturbed habitats, such as suburban or agricultural environments or
8. has a broad, (nonspecialist) diet.

The release of large numbers of animals enhances the chance of successful establishment (5–8). Chance events such as accidents, fires, and floods are more likely to drive small populations to extinction and they may also be more susceptible to extinction due to such factors as increased risk of predation, not finding a mate, or increased competition (9). The threshold minimum population size for successful invasion is not known for most species, although below a minimum threshold of about 20 individuals, survival of a population is unlikely. For example, Griffith et al. (6) found that for birds, the success of establishment dropped sharply below a release group size of 20 individuals. Below this threshold the population may drift randomly up or down, depending on chance events. Thus, a population may establish even if its initial size is below the threshold, should numbers happen to drift above the threshold. An ability to produce many young per year may help a species establish because it enables a small group of animals to increase rapidly above the threshold where they are at a high risk of extinction.

Repeated releases over an extended period and at different sites will increase the chance of successful invasion because the release experiment is repeated many times, under different biotic and abiotic conditions including different climate and season, condition of released animals, and numbers of natural enemies present. A suitable climate for a species will increase the chance that suitable habitat is available in the release site. The degree of climate match between a species' known geographic range and a nominated release site can be assessed using computer programs designed for this purpose, for example BIOCLIM (10) and Climate (11). A species that has a broad geographic range is likely to be able to tolerate wide habitat and climatic variability and therefore is more likely to be able to adapt to conditions in its release site. Similarly, dietary generalists are more likely to be able to use food resources present in their release site than species with specialist diets.

The above factors have limitations for predicting establishment success. Climate match analyses are often fairly coarse, which limits their predictive power. Some animals can tolerate a wider range of physical conditions than that of their current range. Also, although a history of being a successful invader can be used in risk assessment for future invasions because many invasive species establish as exotics in more than one country, other potential invaders will not have been released into new countries where they have had the opportunity to demonstrate their invasion potential.

PROBABILITY OF DAMAGE

Types of harm that may be inflicted by exotic species include

1. Agricultural damage—Exotic species often reduce agricultural productivity by competition with, or predation or harassment of, stock; physical damage to land, pastures, or drainage and irrigation systems; and pollution of waters (4, 12). Exotic birds and mammals eat or damage grain, stock and poultry food, fruit, timber, vegetables, and stored products. Of the 67 species of introduced mammals reviewed by Lever (2), 50 (75%) were recorded as having inflicted damage to agriculture or forestry. Taxa particularly prone to cause damage to agriculture or forestry were Canidae (foxes and dogs), Mustelidae (stoats and ferrets), Cervidae (deer), Bovidae (cattle, sheep, and goats), and Leporidae (rabbits and hares). Of the 133 species of introduced bird species reviewed by Lever (3), 52 (39%) were reported as having inflicted damage to agriculture, mainly to standing crops of grain, oil seed, fruit, or vegetables. Bird taxa particularly prone to cause agricultural damage included Psittaciformes (parrots), Fringillidae (old-world finches), Ploceidae (sparrows and weavers), Sturnidae (starlings and mynahs), and Corvidae (crows). Many of the bird and mammal species that Lever (2, 3) did not report as inflicting damage were present only in small numbers.
2. Environmental damage—Exotic vertebrates can threaten or cause extinction of native wildlife species by competition for resources such as food, space, water, shelter, and nest sites, and by predation or disturbance. They also can degrade habitats, particularly by changing the species composition of plant communities (causing erosion) and by polluting water bodies (4, 12). Islands are particularly prone to environmental damage and to species extinction. Ebenhard (12) describes the environmental damage inflicted on a global scale by 330 species of introduced birds and mammals. At least 40% of mammal introductions can be linked to some ecological impact on the populations of other plants or animals. Ebenhard (4) considers that many of the ecological impacts of exotic birds go unreported because they have been little studied and are often hard to demonstrate.
3. Spread of parasites or diseases—Exotic animals can harbor and spread many diseases and parasites affecting people, stock, and native wildlife (4). For example, the extinction of several native Hawaiian bird species has been attributed to diseases brought in by exotic birds.
4. Attacking or annoying people—Aggressive behavior plus the possession of organs capable of inflicting harm, such as sharp teeth, claws, spines, a sharp bill, or toxin-delivering apparatus may enable individual animals to harm people. Some species, especially those with communal roosts, may cause unacceptable noise.
5. Harm to buildings and equipment—Species that live around people may damage buildings, vehicles, fences, roads, and equipment by chewing or burrowing or polluting with droppings or nesting material.
6. Damage from control measures—Attempts to control or eradicate exotic vertebrates can have

undesirable effects on nontarget species or the environment.

The possible development of new, unpredictable behavior patterns, and of phenotypic or genotypic shifts, brings a strong element of uncertainty to predicting potential impacts of newly established exotic vertebrates. An example is the raccoon-dog (*Nyctereutes procyonoides*) in the Caucasus, which reportedly changed its diet from the fish and crabs it eats in its native Japan, to birds, hares, and poultry. Further, in a species' original habitat, competitors, parasites, diseases, predators, and food species with coevolved defense mechanisms may prevent it from using resources it might otherwise exploit and may keep its numbers in check. When released from these restraints in a new environment, a species may expand its ecological niche and become far more abundant than in its natural range. Exotic animal populations have been reported to reach densities up to 45 times the densities recorded in their native habitats. Species that are not pests in their natural range then become pests in the new environment.

PROBABILITY OF ERADICATION

No abundant and widespread mainland population of an exotic vertebrate pest has ever been eradicated by control measures in any country (13). The sooner eradication is attempted after establishment the higher will be the chance of success. There is usually a time lag following a release before numbers build up sufficiently for the population to go into the steep increase phase of exponential growth and before widespread dispersal occurs. Young of a species born in a new environment are often better able to survive than the original escapees so chances of eradication are correspondingly higher for escapees than for their offspring.

Early eradication attempts when numbers are still low after release carry no guarantee of success. For example, a major campaign to eradicate the monk parakeet (*Myiopsitta monachus*) in the United States attempted when numbers were still relatively low failed. Bomford and O'Brien (13) identify six criteria that can be used to assess whether eradication of a species is likely to be achievable:

1. Immigration is zero—If animals can immigrate onto the eradication area, or continue to be released from captive populations, then eradication will be unachievable or transient.
2. All animals are at risk—All reproductive and potentially reproductive members of the population must be susceptible to removal for eradication to be feasible. For example, if trap-shyness is inherited, or if animals develop neophobia to poison baits, then a subset of the population is not at risk and will not be eradicated.
3. Rate of removal exceeds rate of increase at all population densities—For eradication to be successful the removal rate must exceed the rate of replacement at all densities. Therefore, species with high intrinsic rates of increase are often harder to eradicate because they can recover rapidly to their previous population size. During eradication or control programs, the last few individuals take the most effort to locate and remove. The adequacy of available techniques to detect and remove a species, particularly at low density, will be crucial to the success of control campaigns. Species with low dispersal ability can be contained readily and eradication can be achieved by intensive culling.
4. Population can be monitored at all densities—If animals cannot be detected at low densities there will be no way of ensuring that control efforts are still causing the population to decline, and no way of determining if eradication has been achieved. This requirement is difficult to meet for many species. The visibility of a species will affect the chances of locating individuals. Attributes affecting visibility include habitat preference; body size; group size; behavior; appearance; and types of tracks, signs, and calls.
5. Suitable sociopolitical environment—Social and economic factors can play an overriding role in determining the prospects for successful eradication. Killing animals is unattractive to many people, particularly if controversial techniques are employed or if people become accustomed to seeing an attractive exotic species. Legislation relating to animal welfare, poisons, or the conservation of nontarget species may restrict the use of control techniques and prevent eradication.
6. Cost-benefit analysis favors eradication over control—Because species usually do little harm when they first establish and the potential for future harm may not be obvious, the costs of eradication may be seen to outweigh the benefits.

CONCLUSION

The above factors may be used to predict the probabilities of release, establishment, damage, and eradication. But at best they can only provide indicative estimates because the theoretical knowledge on species invasion, the available data on individual species, and the processes used to assess them all have limitations.

REFERENCES

1. Long, J.L. *Introduced Birds of the World*; Reed: Sydney, 1981; 528.
2. Lever, C. *Naturalized Mammals of the World*; Longman: London, 1985; 487.
3. Lever, C. *Naturalized Birds of the World*; Longman: London, 1987; 615.
4. Ebenhard, T. Introduced birds and mammals and their ecological effects. Swedish Wildl. Res. Viltrevy **1988**, *13*, 1–107.
5. Newsome, A.E.; Noble, I.R. Ecological and Physiological Characters of Invading Species. In *Ecology of Biological Invasions: An Australian Perspective*; Groves, R.H., Burdon, J.J., Eds.; Australian Academy of Science: Canberra, 1986; 1–20.
6. Griffith, B.; Scott, J.M.; Carpenter, J.W.; Reed, C. Translocation as a species conservation tool: status and strategy. Science **1989**, *245*, 477–480.
7. Bomford, M. *Importing and Keeping Exotic Vertebrates in Australia: Criteria for the Assessment of Risk*; Bureau of Rural Resources Bulletin 12, Australian Government Publishing Service: Canberra, 1991; 92.
8. Veltman, C.J.; Nee, S.; Crawley, M.J. Correlates of introduction success in exotic New Zealand birds. Am. Naturalist **1996**, *147*, 542–557.
9. Williamson, M. Mathematical Models of Invasion. In *Biological Invasions. A Global Perspective*; Drake, J.A., Mooney, H.A., di Castri, F., Groves, R.H., Kruger, F.J., Rejmanek, M., Williamson, M.W., Eds.; John Wiley & Sons: Chichester, 1989; 329–350.
10. Martin, W.K. The current and potential distribution of the common myna acridotheres tristis in Australia. Emu **1996**, *96*, 166–173.
11. Pheloung, P.C. *CLIMATE: A System to Predict the Distribution of an Organism Based on Climate Preferences*; Agriculture Western Australia: Perth, 1996; 19.
12. Olsen, P. *Australia's Pest Animals: New Solutions to Old Problems*; Kangaroo Press: Sydney, 1998; 160.
13. Bomford, M.; O'Brien, P. Eradication or control for vertebrate pests? Wildl. Soc. Bull. **1995**, *23*, 249–255.

RODENT DAMAGE AND CONTROL

R

Thomas J. Mbise
Tropical Pesticides Research Institute, Arusha, Tanzania

INTRODUCTION

Rodents belong to the Mammalia order Rodentia, and are subdivided into three suborders: Sciuromorpha, Myomorpha, and Hystrichormorpha. The most distinctive characteristic of this order is their teeth. They have a constantly growing single pair of incisors on each jaw. Separating these incisors from the molars is a large gap called diastema. Due to the constant growing of the incisors, the animals are obliged to gnaw steadily in order to wear them down. As a result the teeth are always sharp, enabling them to damage even materials such as masonry and electric cables. The most worldwide spread rodents are of the family Muridae (Murids) suborder Myomorpha. This includes the commensal rodents largely due to their ability to adapt to and then exploit new situations rapidly.

RODENT PROBLEMS

A rodent is defined as a pest if it is troublesome locally or over a wide area to one or more people. "Troublesome" refers to health or safety hazard; damage or contamination of human food; damage of agricultural crops, livestock products, forests, pastures, urban and rural structures and other properties; or mere general nuisances.

Rodents of Public Health Importance

Most of those rodents living close to and in human dwellings would generally be considered to be of public health importance. This is because many species of rodents may serve as vectors of diseases to humans through the direct transfer of disease-causing organisms or as reservoirs of disease that are transmitted to people by arthropod vectors. Contamination of foodstuffs by rodents in human dwellings is a common phenomenon. This contamination is by feces, hairs, or urine that leads to declaring foodstuffs unfit for human consumption and as animal feed as well. Urine and feces contamination also leads to disease transmission. While contaminating foodstuffs, rodents consume some, in which case in some areas/regions it may lead to food shortage and hence cases of malnutrition.

The rodents of public health importance are the well-known and widespread commensal species, *Rattus norvegicus, R. rattus*, and *Mus musculus*. In addition, other species are also of public health importance, e.g., *Mastomys natalensis* in Africa, *R. exulans* in the Far East (Burma, Indonesia, Malaysia, Thailand, and Vietnam), where it serves as a reservoir for plague and murine typhus, and *Bandicota bengalensis* in South and Southeast Asia (1).

Diseases commonly associated with rodents include plague, salmonellosis, leptospirosis, murine typhus, trichinosis, rabies, lassa, and many haemorrahagic fevers, etc. (2). The list could be even longer, especially if one considers that in the laboratory rodents have been used as testing animals for many diseases and thus are considered potential carriers.

Rodents of Agricultural Importance

Several crop losses have been experienced globally due to regular rodent outbreaks or eruptions. Figures reflecting these losses are usually based on guesswork resulting from short-term surveys. The geographical distribution of rodent species explains which species would be pest suspects in certain geographical regions.

In Africa south of Sahara, *M. natalensis* is the predominant species responsible for huge crop losses (3). Cereals grains mostly damaged by this species include wheat, sorghum, maize, mealie (Indian corn), and rice. Another species common in the region is the Nile rat *Arvicanthus niloticus* causing damage to most cereals, sugar cane, and cotton, especially in the Nile valley and delta areas. Others, but localized, are *Merione* spp. in North Africa, particularly in sandy areas, attacking cereals, groundnuts, and vegetables; *Oenomys, Stochomys, Funisciurus, Cricetomys*, and *Thryonomys* mainly found in forests, attacking cocoa, oil palm, maize, sugar cane, and rice; Gerbils (*Gerbillus, Tatera, Taterillus* spp.) in semidesert and savannah areas, attacking cereals and groundnuts; and *Rhabdomys pumilis* in the highlands of Eastern and Southern Africa, responsible for cereal losses and conifer damage.

In Australia three principal rodent species are a problem in agriculture (4). These are *M. musculus*, attacking cereals; *Rattus conatus* and *Melomys littoralis*, attacking sugar cane. In South America, the principal rodent pest species in rice fields are *Holochilus trasiliensis, Sigmodon hispidus, S. alstoni, Zygodontomys brevicauda*, and *Oryzomys* sp. (5).

Several rodent pest species are known to affect food production in India (6). These include the larger and lesser Bandicoot rat, *Bandicota indica* and *B. bengalensis*, respectively; the short-tailed mole rat, *Nesokia indica*; the soft-furred field rat, *Rattus meltada*; the Indian desert gerbil, *Meriones hurrinae*; the Indian gerbil, *Tatera indica*; the Indian crested porcupine, *Hystrix indica*; and the five-striped squirrels, *Funambulus pennanti* and *F. palmarum*.

Most of the rodent species explained above appear to have acquired pest status after acquiring predominance status in the region. Some of these species acquired predominance after some environmental changes made by humans in an effort to improve living standards, including food requirements that involved opening of more land for agricultural use, introduction of poultries in the urban areas, etc.

Others acquired predominance from their excellent ability to exploit new habitats, which in some cases one species pushed away another. A good example here is the predominance of *R. rattus* in residential areas in Africa south of Sahara. Previously, *M. natalensis* was the house/hut rat but was pushed away to occupy peridomestic and field habitats after the invasion of *R. rattus*.

Economic Estimates of Damage

Economic losses due to rodents are so large and so widespread that they defy precise estimation. Estimates worldwide put the loss of food annually to rodents at about 11 kg per person and that its value is equivalent to the combined Gross National Product of 25 of the poorest countries in the world (7). Such figures could have been derived from estimating that rats and mice would eat an amount of food equivalent to 7% and 15% respectively of their body weight daily. These figures represent the actual grain consumed by the animals but do not show the damage done to the containers (e.g., bags) or the spilled grains after the container had been damaged, nor the amount of grain contaminated with rodent faeces, urine, or hairs.

In Venezuela it is estimated that rice loss from rodents is equivalent to $40 million per year (5) and in the United States crop losses to rodents and associated commensal rodent problems probably exceeds $1 billion a year (8).

Losses or damage caused by rats in Asia run into millions of dollars annually for many countries. In Bangladesh the overall annual losses have been estimated to exceed $50 million and in many years two to three times that amount (9). In the Comoro Islands and Colombia losses in coconut crops due to rodents is usually more than 30% and 77% respectively (8).

RODENT CONTROL

The usual rodent control practices would involve lethal or non-lethal techniques. Application of these techniques either in combination or in sequence, i.e., integrated pest management, may add up to effective and lasting results.

Chemicals for Rodent Control

Major chemicals used in rodent control fall into two main groups—chronic and acute rodenticides. This classification is mainly based upon the mode of rodenticidal action rather than upon chemical structure or physiological action. Most of the chronic chemicals, the exception of calciferol, are anticoagulants.

Acute rodenticides

Acute rodenticides, referred to as single dose or quick-acting poisons, are classified into three hazard use categories. These are: 1) highly toxic and extremely hazardous to people and animals, e.g., arsenic trioxide and bromethaline; 2) moderately toxic and hazardous to people and animals, e.g., pyrinuron and zinc phosphide; and 3) lower toxicity and least hazardous to people and animals, e.g., norbormide and red squill. Rodents that ingest sublethal doses develop bait shyness behavior leading to ineffective control.

Because of their health and environmental hazards in general, and availability of alternative poisons, use of acute rodenticides is restricted in many countries. For example in Tanzania, East Africa, zinc phosphide is registered under "Restricted use" status, implying that it is to be used only under the supervision of a pest control officer.

When a large area needs to be treated, especially as a result of rodent outbreaks, acute rodenticides have been the products of choice. A good example is the aerial application of strychnine in Australia to control high densities of mice in more than 250,000 ha of crops (10).

Chronic rodenticides

These are referred to as single dose or slow-acting anticoagulants that act on the animal by disrupting the mechanism that controls blood clotting leading to fatal internal hemorrhages.

In 1950, the first anticoagulant compound, warfarin, was developed and was regarded as a panacea to chemical rodent control practices. This panacea did not last long due to development of resistant populations. Resistance is present in a number of countries to warfarin and other anticoagulant rodenticides as well. Since all anticoagulants work in essentially the same fashion, i.e., one particular site of action, cross resistance has developed diffusely among them. To counter the resistance problem, more potent compounds were therefore developed and thus today we have two groups of anticoagulant rodenticides referred to as first and second generations. The first generation requires multiple feeding and the second generation a single feeding for a lethal dose respectively. Basing on this difference, a saturated baiting system is best suited to first generation rodenticides and pulse baiting to second generation rodenticides. However, both groups have one thing in common—the delayed death.

HEALTH AND ENVIRONMENTAL HAZARDS

Hazards caused by rodenticides could be either primary (direct poisoning) or secondary (indirect poisoning). Primary poisoning results from nontarget animals, including humans, ingesting baits intended for rodents. Which animals are at a higher risk depends on the formulation and mode of application. Where whole grains are used, grain-eating animals, especially birds, stand a higher risk of direct poisoning. In situations where aerial application is used, birds and many other animals are at risk because of vast area coverage. In one instance of aerial baiting of strychnine in Australia for the control of house mice, when birds were tested for strychnine, it showed up in 82% of the granivores, 55% of the raptors, and 69% of the omnivores.

Acute rodenticides, being fast acting and lacking an antidote, pose more hazards. Fortunately, they are no longer in extensive use as are the chronic ones, the anticoagulants, which are slow acting and have an antidote in the form of Vitamin K.

Efforts to reduce hazards caused by rodenticides are widely in practice. Apart from aerial baiting the usual practice is to place baits in baiting containers deliberately designed to prevent access by nontarget species as much as possible. Various types of baiting boxes are available; otherwise fabricating one using readily available materials is usually the case. By using baiting boxes or containers, uneaten baits can be collected and disposed of accordingly. Following rodenticide application is the removal of carcasses before raptors or other animals pick them up.

Rodenticide baits are colored deliberately to deter birds and children from picking them up or mistaking them for cakes.

Apart from improving and enforcing regulations regarding the use of pesticides (which differ from country to country) training of farmers and the general public on hazards caused by rodenticides is absolutely important. Updating the extension staff with current information available could do this.

NONCHEMICAL TECHNIQUES OF RODENT CONTROL

Nonchemical techniques are numerous, very effective, and user-friendly. Some of these include: 1) environmental sanitation, that is, removal of source of food, water or cover; 2) rodent exclusion techniques including introduction of barriers; 3) mechnical proofing—that is, having no openings into structures that rats and mice can penetrate; and 4) habitat manipulation involving elimination of rodent harborage by modifying the surroundings (11). Others in use but extremely exclusive are electrical barriers and ultrasound.

Traps, available in various forms, are a novel idea in controlling rodents, but with limitations. They are recommended for use in small or localized infestations and also in places where the use of rodenticides is considered undesirable. Traps are also used to remove the survivors of poison treatment.

Predation is another nonchemical rodent control technique. Rodent predators include birds, foxes, ferrets, weasels, snakes, skunks, mongooses, dogs, and cats. The idea of using predators is questionable as the population may not be reduced to a level accepted economically in given circumstances.

A cat in enclosed premises (e.g., a warehouse) could reduce or eliminate the population of rats and mice just like a barn owl could remove a considerable number of rats in oil palm or coconut plantations. Introduction of a predator for rodent control, however, has to be done with a lot of care; otherwise the predator could become a pest to other desirable wildlife.

REFERENCES

1. Gratz, N.G. In *The Global Public Health Importance of Rodents*, Proceedings of a Conference on The Organization and Practice of Vertebrate Pest Control, Elvetham Hall,

Hampshire, England, Aug. 30–Sept. 3, 1982; Dubock, A.C., Ed.; ICL Plant Protection Division: Fernhurst, Haslemere, Surrey, England, 1984; 413–435.
2. Gratz, N.G. The burden of rodent-borne diseases in Africa south of the Sahara. Belg. J. Zool. **1997**, *127* (suppl. 1), 71–84.
3. Fiedler, L.A. Rodent Problems in Africa. In *Rodent Pest Management*; Prakash, I., Ed.; CRC Press: Boca Raton, FL, 1988; 35–65.
4. Armstrong, D.J. *Rodents in Australia*, Proceedings of a Conference on The Organization and Practice of Vertebrate Pest Control, Elvetham Hall, Hampshire, England, Aug. 30–Sept. 3, 1982; Dubock, A.C., Ed.; ICL Plant Protection Division: Fernhurst, Haslemere, Surrey, England, 1984; 49–52.
5. Williams, J.O. *Rodents and their Problems in South America and the Caribbean*, Proceedings of a Conference on The Organization and Practice of Vertebrate Pest Control, Elvetham Hall, Hampshire, England, Aug. 30–Sept. 3, 1982; Dubock, A.C., Ed.; ICL Plant Protection Division: Fernhurst, Haslemere, Surrey, England, 1984; 53–55.
6. Prakash, I. *Vertebrate Pest Problems in India*, Proceedings of a Conference on The Organization and Practice of Vertebrate Pest Control, Elvetham Hall, Hampshire, England, Aug. 30–Sept. 3, 1982; Dubock, A.C., Ed.; ICL Plant Protection Division: Fernhurst, Haslemere, Surrey, England, 1984; 29–35.
7. Gwinner, J.; Harnisch, R.; Muck, O. Manual of the Prevention of Post-Harvest Grain Losses, Chapter 11. Rodent Pests, 1996. http://www.oneworld.org/globalprojects/humcdrom (accessed February 29, 2000).
8. Elias, D.J. Pests with backbones. CERES **1988**, *21* (2), 122.
9. Posamentier, H. Damage by rats cannot be tolerated any longer. Gates Commun. **1985**, *3*.
10. Brown, P.R.; Lundie-Jenkins, G. Non-target mortalities during aerial strychnine baiting of house mice. Wildl. Res. **1999**, *26* (1), 117–128.
11. Horskins, K.; Wilson, J. Cost effectiveness of habitat manipulation as a method of rodent control in Australia Macadamia Orchards. Crop Prot. **1999**, *18* (6), 379–387.

RODENTICIDE (AND VERTEBRATE PESTICIDE) EFFECTS ON WILDLIFE HEALTH

R

Charles Eason
CENTOX Centre for Environmental Toxicology, Lincoln, New Zealand

INTRODUCTION

Uses, Classes, and Definitions

Worldwide, vertebrate pest control principally targets rats and mice, and is undertaken using baits, usually cereal pellets containing a poison, to reduce losses to agricultural production from damage to crops, or transmission of disease to livestock or people. In addition, vertebrate pesticides are used on a broad scale to mitigate conservation problems caused by the impact of rodents and other introduced species on indigenous plants and animals in unique ecosystems such as Australia, New Zealand, Hawaii, and other island habitats.

The chemicals used for controlling vertebrate pests are classified as: 1) anticoagulants, which include warfarin, pindone, diphacinone, coumatetralyl, difenacoum, bromadiolone, brodifacoum, flocoumafen, and difethalione; or 2) nonanticoagulants, include any substance that does not fall in the former category, such as strychnine, cyanide, zinc phosphide, sodium monofluoroacetate (1080), cholecalciferol, and calciferol. There is some confusion with the use of the terms toxins, toxicants, poisons, vertebrate pesticides, and rodenticides. Toxic substances that are or were originally of natural biological origin derived from microbes, plants, and animals are usually described as toxins [e.g., cholecalciferol (vitamin D_3), cyanide, and 10 80]. Toxicants are toxic substances that do not occur naturally (e.g., brodifacoum and pindone). The term, ''poison,'' ''vertebrate pesticide,'' or ''rodenticide'' can be used to cover both toxins and toxicants. Some compounds have multiple uses; for example, cholecalciferol and warfarin are used as vertebrate pesticides or rodenticides, but the former is commonly regarded as a vitamin and the latter a drug used to treat blood clotting disorders in humans. This should not be surprising since Paracelsus noted in the early 1500s, ''All substances are poisons and it is only the dose that makes a distinction between one which is a poison and one which is a remedy.'' Vertebrate pesticides (most commonly rodenticides) are distinguished from insecticides (toxic to insects), herbicides (toxic to plants), and fungicides (toxic to fungi). In this regard, 1080 is unusual as it is known to be toxic to both invertebrate and vertebrate species and could therefore be classified as an insecticide, a vertebrate pesticide, or a rodenticide.

RISKS TO WILDLIFE

The risk to nontarget wildlife from baits containing vertebrate pesticides will be determined in part by the animal's intrinsic susceptibility, the properties of the poisons used (such as their toxicokinetics), bait design and their deployment, and the site-specific complexities of the food webs in areas in which they are used, which may limit or exacerbate the exposure of nontarget species. The risk of adverse effects to wildlife will therefore depend on the inherent toxicity of the pesticide (i.e., hazard) and the potential exposure of the animal to toxic bait or residues of the pesticide in the environment.

Risks = Hazard (H) Exposure

At present, all vertebrate pesticides used for the control of rodents and other unwanted mammals are potentially hazardous to all vertebrate species (despite some species variation and idiosyncrasies in response), and some such as 1080 are toxic to invertebrates. Primary poisoning refers to poisoning resulting from ingestion of toxic baits, and will vary depending on the properties of the pesticide and how it is used. Secondary poisoning occurs when poison is ingested by eating another animal that has been deliberately poisoned, such as, for example, when an owl eats a rat containing residues of a rodenticide.

The use of toxic bait for commensal rodent control in confined areas such as homes, factories, farmhouses, or grain stores, and their application in bait stations, reduces but does not eliminate (1, 2) the risk of primary poisoning of birds, other mammals, reptiles, or invertebrates. Field or broad-acre usage of vertebrate pesticides potentially put a wider range of nontarget species at risk.

ANTICOAGULANT EFFECTS ON WILDLIFE—BRODIFACOUM—A CASE STUDY

Baits containing first-generation anticoagulant rodenticides (FGARs) (e.g., warfarin, pindone, diphacinone, and coumatetralyl), and second-generation anticoagulants (SGARs) (e.g., brodifacoum, flocoumafen, bromadiolone, and difethalione), may cause lethal or sublethal adverse effects in wildlife if ingested. All these compounds have the same mode of action, i.e., interfering with the synthesis of clotting factors, which results in bleeding and death. FGARs were developed in the 1950s and 1960s, and SGARs in the 1970s and 1980s partly to overcome resistance to FGARs. The SGARs such as brodifacoum are more toxic than FGARs to many nontarget species by both primary and secondary poisoning (3, 4), and present a greater risk to wildlife.

The principal use of anticoagulants worldwide has been for control of commensal rodents, primarily Norway rats (*Rattus norvegicus*), ship rats (*Rattus rattus*), and house mice (*Mus musculus*).

The latent period between the time of ingestion of brodifacoum and the onset of clinical signs varies considerably for different species. In rats the onset of symptoms and death usually occur within a week. However, in a marsupial, the Australian brushtail possum (*Trichosurus vulpecula*), this process is protracted and may take between one and four weeks (5). Clinical signs in wildlife, including mammals, birds, and reptiles, usually reflect some manifestations of hemorrhage. Onset of signs may occur suddenly, especially when hemorrhage of the cerebral vasculature or pericardial sac occurs. Signs of anticoagulant toxicosis in wildlife commonly include anemia and weakness. Hemorrhaging may be visible around the nose, mouth or beak, eyes, and anus of mammals. Swollen, tender joints are common and, if hemorrhage involves the brain or central nervous system, ataxia or convulsions can occur. Poisoned animals die of multiple causes related to anemia or hypovolemic shock (5).

The greater potency of second-generation anticoagulants, such as brodifacoum, and their greater potential to affect wildlife compared to first-generation anticoagulants, such as warfarin and pindone, is related to their greater affinity for vitamin K-epoxide reductase and subsequent accumulation and persistence in the liver and kidneys after absorption (6). All anticoagulants attach to this common binding site, but the second-generation anticoagulants have a greater binding affinity than the first-generation compounds (7). All tissues that contain vitamin K-epoxide reductase (e.g., liver, kidney, and pancreas) are target organs for bioaccumulation of these toxicants.

Nontarget wildlife deaths due to primary and secondary poisoning have been associated with application of anticoagulant baits for agricultural pest control, rat eradication programs on islands, and large-scale eradication and control program for introduced species. The perceived hazards of primary and secondary poisoning to nontarget wildlife have restricted registration for field use in the United States (8) of second-generation anticoagulants such as brodifacoum. The detection of brodifacoum residues in New Zealand (9–11), in birds such as kiwi (*Apteryx* spp.) weka (*Gallirallus australis*), morepork (*Ninox novaeseelandiae*), Australasian harrier (*Circus approximans*), Pū keko (*Porphyrio porphyrio*), and grey duck (*Anes superciliosa*), raises serious concerns about the long-term consequences of field use of brodifacoum.

When baits containing brodifacoum or other SGARs are used in the field, bird species most at risk from feeding directly on cereal-based toxic baits are those that are naturally inquisitive and have an omnivorous diet. In New Zealand these include weka, Pū keko, and kea (*Nestor notabilis*). The risk of secondary poisoning is probably greatest for predatory and scavenging birds (especially the weka, Australasian harrier, morepork, and southern black-backed gull [*Larus dominicanus*]) that feed on target species (e.g., live or dead rodents, rabbits, and possums). Recently, surveys in New Zealand demonstrate widespread wildlife contamination, which extends beyond predatory and scavenging birds that have eaten bait or scavenged carcasses, as well as game species (11–17). This pattern is mirrored globally where there is field use of SGARs or commensal use where wildlife can access carcasses, e.g., around farm buildings or on agricultural lands (18–20). In a global context, predators and scavengers are most likely to be at risk from SGARs because these are persistent toxic substances that will bioaccumulate, even when the intervals between exposure are large (e.g., many months or even years). Wildlife or feral mammalian carnivore losses have been reported from New Zealand, England, France, and the United States (17, 20–24).

When using brodifacoum in species conservation program, detrimental effects on some individuals need to be weighed against potential improved survival and breeding success of the nontarget species in the absence of rodents.

In the short term, establishing whether or not the extent of brodifacoum transfer through the food web can be contained, e.g., by only using brodifacoum intermittently in pulses alongside trapping and other control tools, or switching to less persistent vertebrate pesticides and traps, is a key research priority. Where brodifacoum is used on just one occasion for island eradication of rodent populations to provide sanctuaries for endangered birds and other animals, the risks are likely to be outweighed by the benefits.

When considering "secondary poisoning," most authors consider this to relate to one vertebrate species eating another, which is usually the target species. The only confirmed report of secondary poisoning of insectivorous birds with brodifacoum was in a zoo, where avocets, rufous-throated ant pittas, golden plovers, honey creepers, finches, thrushes, warblers, and crakes died in an aviary after feeding on ants and cockroaches that had eaten brodifacoum baits (3). However, the potential for invertebrates to "carry" poison to birds and other animals following field use has been suggested (17).

FUTURE USE OF VERTEBRATE PESTICIDES

Anticoagulants, initially the FGARs and subsequently the SGARs such as brodifacoum, have been pivotal for rodent control over the last 50 years, and are likely to remain important in preventing disease, damage crops, and threats to endangered species for the foreseeable future.

The current focus of vertebrate pesticide researchers is to design "use patterns" and formulations that are effective at killing pest species but are also less hazardous to other wildlife. Despite these initiatives, there has been some inappropriate SGAR use and associated unwanted wildlife toxicosis. The availability of an antidote (vitamin K_1) and the time lag between exposure and onset of symptoms provide an opportunity for treatment of poisoned pets and wildlife. Nevertheless, safer rodenticides are required and this will be a new challenge for the 21st century.

See also *Rodent Damage and Control*, pages 727–730.

ACKNOWLEDGMENTS

Dr. Phil Cowan and Christine Bezar of Landcare Research are acknowledged for scientific and editorial comment. The New Zealand Foundation for Science and Technology and the Department of Conservation are acknowledged for comments on this manuscript.

REFERENCES

1. McDonald, R.A.; Harris, S.; Turnbull, G.; Brown, P.; Fletcher, M. Anticoagulant rodenticides in stoats (*Mustela erminea*) and weasels (*Mustela nivalis*) in England. Environ. Pollut. **1998**, *103*, 17–23.
2. Hosea, R.C. In *Exposure of Non-target Wildlife to Anticoagulant Rodenticides in California,* Proceedings of the 19th Vertebrate Pest Control Conference, San Diego, California, Univ. of Calif., 2000; 236–244, *in press.*
3. Godfrey, M.E.R. Non-target and secondary poisoning hazards of "second generation" anticoagulants. Acta Zool. Fenn. **1985**, *173*, 209–212.
4. Eason, C.T.; Spurr, E.B. Review of the toxicity and impacts of brodifacoum on non-target wildlife in New Zealand. N.Z. J. Zool. **1995**, *22*, 371–379.
5. Littin, K.; O'Connor, C.; Eason, C.T. In *A Comparison of the Efficacy of Brodifacoum in Rats and Possums,* Proceedings of the New Zealand Plant Protection Conference, 2000; 53, 299–304.
6. Huckle, K.R.; Hutson, D.H.; Warburton, P.A. Elimination and accumulation of the rodenticide flocoumafen in rats following repeated oral administration. Xenobiotica **1988**, *18*, 1465–1479.
7. Parmar, G.; Bratt, H.; Moore, R.; Batten, P.L. Evidence for a common binding site in vivo for the retention of anticoagulants in rat liver. Hum. Toxicol. **1987**, *6*, 431–432.
8. Colvin, B.A.; Jackson, W.B.; Hegdal, P.L. In *Secondary Poisoning Hazards Associated with Rodenticide Use,* Proceedings of the 11th International Congress on Plant Protection, 1991; Magallona, E.D., Ed.; 60–64.
9. Roberston, H.A.; Colbourne, R.M.; Nieuwland, F. Survival of little spotted kiwi and other forest birds exposed to brodifacoum rat poison on Red Mercury Island. Notornis **1993**, *40*, 253–262.
10. Murphy, E.C.; Clapperton, B.K.; Bradfield, P.M.F.; Speed, H.J. Brodifacoum residues in target and non-target animals following large-scale poison operations in New Zealand podocarp—hardwood forests. N.Z. J. Zool. **1998**, *25*, 307–314.
11. Dowding, J.E.; Murphy, E.C.; Veitch, C.R. Brodifacoum residues in target and non-target species following an aerial poisoning operation on Motuihe Island, Hauraki Gulf, New Zealand. N.Z. J. Ecol. **1999**, *23* (2), 207–214.
12. Murphy, E.C.; Robbins, L.; Young, J.B.; Dowding, J.E. Secondary poisoning of stoats after an aerial 1080 poison operation in Pureora Forest, New Zealand. N.Z. J. Ecol. **1999**, *23* (2), 175–182.
13. Gillies, C.A.; Pierce, R.J. Secondary poisoning of mammalian predators during possum and rodent control operations at Trounson Kauri Park, Northland, New Zealand. N.Z. J. Ecol. **1999**, *23* (2), 183–192.

14. Meenken, D.; Wright, K.; Couchman, A. Brodifacoum residues in possums (*Trichosurus vulpecula*) after baiting with brodifacoum cereal bait. N.Z. J. Ecol. **1999**, *23* (2), 215–217.
15. Eason, C.T.; Milne, L.; Potts, M.; Morriss, G.; Wright, G.R.G.; Sutherland, O.R.W. Secondary and tertiary poisoning risks associated with brodifacoum. N.Z. J. Ecol. **1999**, *23* (2), 219–224.
16. Robertson, H.A.; Colbourne, R.M.; Graham, P.J.; Miller, P.J.; Pierce, R.J. Survival of brown kiwi (*Apteryx mantellii*) exposed to brodifacoum poison in Northland, New Zealand. N.Z. J. Ecol. **1999**, *23* (2), 225–231.
17. Stephenson, B.M.; Minot, E.O.; Armstrong, D.P. Fate of moreporks (*Ninox novaeseelandiae*) during a pest control operation on Mokoia Island, Lake Rotorua, North Island, New Zealand. N.Z. J. Ecol. **1999**, *23* (2), 233–240.
18. Young, J.; De Lai, L. Population declines of predatory birds coincident with the introduction of klerat rodenticide in North Queensland. Aust. Bird Watcher **1997**, *17*, 160–167.
19. Shore, R.F.; Birks, J.D.S.; Freestone, P. Exposure of non-target vertebrates to second-generation rodenticides in Britain, with particular reference to the polecat *Mustela putorius*. N.Z. J. Ecol. **1999**, *23* (2), 199–206.
20. Stone, W.B.; Okoniewski, J.C.; Stedlin, J.R. Poisoning of wildlife with anticoagulant rodenticides in New York. J. Wildl. Dis. **1999**, *35* (2), 187–193.
21. Alterio, N. Secondary poisoning of stoats (*Mustela erminea*), feral ferrets (*Mustela furo*), and feral house cats (*Felis catus*) by the anticoagulant poison, brodifacoum. N.Z. J. Zool. **1996**, *23*, 331–338.
22. Alterio, N.; Brown, K.; Moller, H. Secondary poisoning of mustelids in a New Zealand *Nothofagus*. Forest. J. Zool. (Lond.) **1997**, *243*, 863–869.
23. Berny, P.J.; Buronfosse, T.; Buronfosse, F.; Lamarque, F.; Lorgue, G. Field evidence of secondary poisoning of foxes (*Vulpes vulpes*) and buzzards (*Buteo buteo*) by Bromadiolone, a 4-year survey. Chemosphere **1997**, *35* (8), 1817–1829.
24. Birks, J.D.S. Secondary rodenticide poisoning risk arising from winter farmyard use by the European polecat *Mustela putorius*. Biol. Conserv. **1998**, *85*, 233–240.

ROLE OF EPA IN PESTICIDE REGULATION[a]

Janice King Jensen
Lindsay Moose
U.S. Environmental Protection Agency, Washington, D.C., U.S.A.

INTRODUCTION

The mission of the U.S. Environmental Protection Agency (EPA) is to safeguard human health and the natural environment upon which life depends. EPA's Office of Pesticide Programs (OPP), with assistance from its regional offices and state and tribal partners, works to protect human health and the environment from unreasonable risks associated with pesticide use and to ensure that pesticide residues in food are safe.

SCOPE OF PESTICIDE REGULATION

Pesticides subject to EPA regulation include insecticides, herbicides, fungicides, rodenticides, disinfectants, plant growth regulators, and other substances intended to control pests. Pesticides play a role in many aspects of everyday life, from agriculture and greenhouses to lawns, swimming pools, and food service establishments. There are about 20,000 registered pesticide product formulations, containing approximately 675 active ingredients and 1835 other ingredients. About 470 pesticide active ingredients are used in agriculture, and EPA has established more than 9000 tolerances (maximum allowable residue limits) for pesticides that may be present in food.

EPA's regulation of pesticides directly or indirectly affects approximately 30 major pesticide producers, another 100 smaller producers, 2500 formulators, 29,000 distributors and other retail establishments, 40,000 commercial pest control firms, one million farms, three and a half million farm workers, several million industry and government users, and all households. Within OPP, approximately 800 people in nine divisions carry out activities relating to pesticide regulation and management. In addition, a large number of people in other EPA offices, including regional offices, provide administrative, legal, enforcement, and research support.

[a]The material in this article has been subject to Agency technical and policy review. The views expressed in this article are those of the authors and do not necessarily represent policies of the U.S. EPA.

STATUTORY FRAMEWORK

EPA regulates pesticides under two major statutes: the Federal Insecticide, Fungicide, and Rodenticide Act (FIFRA) (1) and the Federal Food, Drug, and Cosmetic Act (FFDCA). FIFRA requires that pesticides be registered (licensed) by EPA before they may be sold or distributed for use in the United States, and that they not cause unreasonable adverse effects to people or the environment when used according to EPA-approved label directions. Under FFDCA, EPA sets tolerances (2) for pesticide residues in food. Tolerance requirements apply equally to domestically produced and imported food, and any food with residues not covered by a tolerance (or in amounts that exceed an established tolerance) may not be legally marketed in the United States.

On August 3, 1996, President Clinton signed the Food Quality Protection Act (FQPA) (3) into law. The new law, passed unanimously by both Houses of Congress, requires major changes in pesticide regulation and affords EPA unprecedented opportunities to provide greater health protection, particularly for infants and children. Major provisions include:

- Establishing a single, health-based standard for all pesticide residues in food, eliminating past inconsistencies in the law that treated residues in some processed foods differently from residues in raw and other processed foods;
- Providing for a more complete assessment of potential risks when establishing or maintaining a tolerance, with special protections for potentially sensitive groups such as infants and children; and

Requiring a reassessment of all existing pesticide residue limits in accordance with the new standard of safety.

PESTICIDE REGISTRATION

Registration (4) is a premarket review and licensing program required for all pesticides sold or used in the United States. Every registered pesticide product is required to bear a label, which includes the producing establishment number, the product registration number, an active ingredients statement, warning or precautionary statements, and directions for use. Deviation from pesticide label directions is a violation of Federal law. Under FIFRA, states have the primary enforcement authority over pesticide misuse.

In a typical year EPA reviews more than 5000 registration submissions that vary from routine label amendments to applications for new active ingredients. Most submissions seek to amend existing product registrations or to register new formulations containing active ingredients already registered with EPA. EPA receives about 20 applications for the registration of new active ingredients each year. Registration of a new active ingredient requires a significant investment in time and money by the registrant. For example, data development for a major agricultural chemical can cost $10 million dollars or more and take several years to complete.

EPA bases registration decisions primarily on evaluation of test data provided by applicants. These studies relate to potential toxic effects to humans (such as skin and eye irritation, cancer, birth defects, or reproductive system disorders); environmental fate (or how the pesticide behaves in the environment); ecological effects (toxicity of the pesticide to birds, fish, and other nontarget organisms), as well as other potentially harmful effects. Post-FQPA, EPA must also consider nondietary exposure, such as residential and drinking water, and cumulative risk from pesticides with a common mode of toxicity.

TOLERANCES

Before a pesticide can be registered for use on food or animal feed crops, EPA must establish a tolerance (5) or grant an exemption from the requirement of a tolerance. EPA establishes tolerances and exemptions under authority of the FFDCA to ensure that consumers are not exposed to unsafe pesticide residues in food. Under FQPA, EPA must review all tolerances existing as of enactment of the law to assure the requirements of the new health-based standard are met. The Agency was given 10 years, until August 3, 2006, to complete this review, giving priority to the review of pesticides that appear to pose the greatest potential risk to human health (including organophosphates and carbamates).

The data required for a tolerance include residue chemistry and short- and long-term toxicity feeding studies in animals. The goal of EPA's review is to determine the possible health effects from aggregate exposure to a chemical, including dietary, residential, and water exposure, and whether such exposure represents an acceptable level of risk. This risk determination must not be made only for an individual pesticide, but for the cumulative effect of groups of pesticides which share a common mechanism of toxicity. Before establishing a tolerance, EPA must reach a conclusion that, under the proposed use conditions, there is a reasonable certainty that no harm will result from exposure to the chemical pesticide residue, and the Agency must specifically outline its determination regarding the safety of infants and children.

While EPA is responsible for setting tolerances, monitoring and enforcement are the responsibility of the U.S. Department of Health and Human Services' Food and Drug Administration (FDA) for most foods, and by the U.S. Department of Agriculture's (USDA) Food Safety and Inspection Service for meat, poultry, and some egg products. Although tolerances are generally established in support of registration, EPA also establishes tolerances for imported commodities for pesticide uses that are not registered domestically. If pesticide residues exceed the tolerance, or no tolerance exists, the crop may be considered adulterated and is subject to seizure regardless of whether the pesticide use is permissible in a foreign country.

TRENDS IN PESTICIDE REGULATION

The introduction of synthetic chemicals in agriculture and other settings began in earnest following World War II. Early successful compounds included organochlorine insecticides, such as DDT, and phenoxy herbicides, such as 2,4-D. By the 1960s, organophosphate and carbamate insecticides and triazine herbicides predominated the U.S. market. In the 1980s, synthetic pyrethroid insecticides were introduced, and biological pesticides became increasingly popular. Recently, many pesticide companies are responding to public concern about pesticide risks by developing and registering pesticides with nontoxic modes of action and genetically engineered plants and microorganisms. EPA policies and activities are helping influence trends in pesticide regulation. For example:

Reduced-Risk Pesticides. EPA's registration process gives the highest priority to lower-risk alternatives to existing pesticides. These alternatives may be synthetic or biopesticides (also known as biological pesticides) that are derived from such natural materials as animals, plants, bacteria, and certain minerals. Biological pesticides now comprise the single fastest growing segment of new pesticide registrations. One study projected that biopesticide sales (not including beneficial insect organisms) will increase by more than 20% annually from 1995 to 2000.

Reduced Use of Pesticides. EPA, USDA, and FDA work together to promote alternative agricultural methods that will reduce the use of pesticides for food production and afford greater protection for consumers. EPA hopes to encourage the replacement of chemical-intensive agricultural practices with informational- and technology-intensive practices. These include greater use of traditional practices such as scouting for pests, as well as newer techniques such as computer metering devices for applying pesticides and satellite monitoring to assess the need for, and timing of, pesticide applications.

Integrated Pest Management (IPM). EPA is also working with fellow Agencies, growers, states, and tribes to promote sustainable agriculture and IPM practices that help reduce use of pesticides that may be less safe. Such changes are expected to improve environmental quality without harming grower profits.

Reregistration of Older Pesticides. Since FIFRA was first enacted in 1947, standards for approval and test data requirements have evolved. To ensure that previously registered pesticides comply with current scientific and regulatory standards, FIFRA requires the review and reregistration of existing pesticides. The reregistration process ensures that adequate data are available to enable EPA to fully assess the risks associated with pesticide use and pesticide tolerances.

Tolerance Reassessment. As a result of reassessment, tolerances may be retained, modified, or revoked to ensure that the safety standard is met. In addition, when all registrations for a pesticide's use on a food commodity are canceled, EPA generally takes action to revoke the established tolerance for that commodity. Reassessment and reregistration activities have led to a significant increase in tolerance revocations.

INTERNATIONAL HARMONIZATION AND COOPERATION

EPA actively promotes international harmonization (6) of pesticide regulation to improve food safety, increase the efficiency of pesticide regulation, and minimize international trade barriers. Improved international harmonization and acceptance of pesticide standards is also important to food safety in countries which lack the resources to set their own national standards. EPA has sponsored workshops in Latin America (7) and other locations to provide information on the toxicity of pesticides, and to provide information about U.S. pesticide laws and standards for growers who wish to export to the United States. In addition, with the assistance of the U.S. Agency for International Development, EPA is working to foster the exchange of pesticide regulatory information with countries such as Indonesia (7).

Under the North American Free Trade Agreement (8), the United States and Canada are working together to evaluate reduced-risk pesticides, allowing a shared review burden and quicker introduction of safer products. To date, four reduced-risk pesticides have been jointly reviewed and simultaneously registered in both Canada and the United States. The two countries have also shared evaluation work on more than 80 additional pesticides. Canada and the United States have developed residue zone maps that allow pesticide residue studies conducted in one country to support registration and tolerances in the other, and these maps are currently being extended to Mexico.

REFERENCES

1. EPA Office of Pesticide Programs web site. http://www.epa.gov/pesticides/fifra.htm (accessed Jan 2001).
2. EPA Office of Pesticide Programs web site. http://www.epa.gov/pesticides/food (accessed Jan 2001).
3. EPA Office of Pesticide Programs web site. http://www.epa.gov/oppfead1/fqpa/ (accessed Jan 2001).
4. EPA Office of Pesticide Programs web site. http://www.epa.gov/pesticides/chemreg.htm (accessed Jan 2001).
5. EPA Office of Pesticide Programs web site. http://www.epa.gov/pesticides/food/viewtols.htm (accessed Jan 2001).
6. EPA Office of Pesticide Programs web site. http://www.epa.gov/oppfead1/international/#A (accessed Jan 2001).
7. EPA Office of Pesticide Programs web site. http://www.epa.gov/oppfead1/international/#E (accessed Jan 2001).
8. EPA Office of Pesticide Programs web site. http://www.epa.gov/oppfead1/international/#A2 (accessed Jan 2001).

ROLE OF FDA IN PESTICIDE LAWS

Michael A. Kamrin
Michigan State University, East Lansing, Michigan, U.S.A.

INTRODUCTION

Responsibilities for the management of the risks of pesticides to human health and the environment are divided among a number of state and federal agencies. At the federal level, the agencies that play the largest roles are the Environmental Protection Agency (EPA), the Food and Drug Administration (FDA), and the U.S. Department of Agriculture (USDA). The major pieces of legislation governing pesticides in foods and feeds are the Federal Insecticide, Fungicide, and Rodenticide Act (FIFRA), the Federal Food, Drug and Cosmetic Act (FFDCA), and the Food Quality Protection Act (FQPA), which amends both FIFRA and FFDCA.

Currently, the EPA is responsible for pesticide registration and for setting tolerance limits for pesticide residues in both raw agricultural commodities and processed foods; the FDA for collecting residue data, enforcing tolerances, and conducting diet surveys; and the USDA for collecting data on dietary exposure, food consumption, and pesticide usage, and enforcing some tolerances. Within the FDA, the Center for Food Safety and Applied Nutrition (CFSAN) has the main responsibility for carrying out pesticide-related programs.

THE DELANEY AMENDMENT

Because the definitions of ''pesticide chemical'' and ''food additive'' were changed under FQPA, the roles of FDA and EPA with respect to regulating pesticides in foods were altered. Of particular importance, carcinogenic pesticides that concentrate in processed foods are no longer regulated by FDA under the Delaney Amendment, a part of FFDCA that required the banning of such pesticides from these foods; instead, they are now regulated by EPA under the general standard of ''reasonable certainty of no harm.''

REGULATORY MONITORING

The FDA carries out an extensive program of pesticide residue monitoring to support its role in enforcing tolerances in imported foods and domestic foods shipped in interstate commerce (1). Individual lots of foreign foods are collected at points of entry and individual lots of domestic foods near the point of production. Samples are analyzed for residues of approximately 400 pesticides at detection limits generally well below tolerance levels.

The samples are collected either for surveillance purposes (i.e., to determine residue levels in the absence of specific evidence of contamination) or for compliance purposes (i.e., to resample shipments that were found to have violative pesticide residues). Violations can occur when tolerance limits are exceeded or when pesticide residues are found in foods that do not have a tolerance established for that pesticide. The FDA is responsible for enforcing tolerances for all foods except those for which the USDA is responsible, i.e., meat, poultry, and some egg products. The results of the monitoring are published annually on the World Wide Web (1). In 1998, there were approximately 5000 domestic and 3600 import samples analyzed.

The FDA also uses pesticide residue data generated under the USDA's Pesticide Data Program (PDP) in its regulatory activities. USDA data on the incidence of particular pesticide residues in specific commodities are particularly useful to the FDA for refining its sampling activities related to enforcement of tolerances. Because of the importance of the PDP to FDA and EPA, representatives of both agencies serve on the Executive Steering Committee that coordinates PDP planning and policy.

INCIDENCE-LEVEL MONITORING

In addition to monitoring in direct support of regulation, FDA also analyzes pesticides in certain foods to: 1) get a better understanding of the prevalence of certain pesticides in particular commodities, and 2) determine how statistically representative regulatory monitoring samples are. For example, a two-year survey of triazine residues was completed in 1997 to gain a better understanding of the prevalence of this class of pesticides in various commodities (1).

TOTAL DIET STUDY (TDS)

TDS, also known as the Market Basket Study, is an annual program to measure pesticide residue levels in foods, data which are then used to estimate pesticide intakes in representative diets of Americans in a number of different age groups (2). The foods that are included in the TDS are purchased by FDA personnel from supermarkets or grocery stores four times a year in four regions of the country. A composite of like foods purchased in three cities in a given region is known as a Market Basket. Currently, 264 different foods in 18 Market Baskets are sampled during the TDS. The 264 foods have been chosen as representative of more than 3500 different foods: e.g., apple pie is considered to represent all fruit pies and pastries.

Foods are prepared for consumption, i.e., table ready, before analysis. Many of these foods are prepared items containing multiple ingredients rather than single foods. Comparable data are collected for 14 age/sex groups ranging from 6–11-month-old infants to adults more than 70 years of age. The results of the TDS are published annually by the FDA.

HAZARD ANALYSIS AND CRITICAL CONTROL POINT (HACCP) SYSTEM

In 1997, the FDA implemented a rule requiring use of the HACCP methodology in the seafood industry. The HACCP system focuses on identifying critical steps in the chain from food production to consumption at which potential hazards can be controlled or eliminated (3). The rule applies to all seafood processors and importers and must address all potential hazards including pesticides. It includes special provisions that importers must follow to verify that overseas suppliers follow HACCP. FDA is currently considering adopting an HACCP-based food safety system on a wider basis. FDA has incorporated HACCP into its Food Code, a document that gives guidance to state and territorial agencies that license and inspect food service establishments, retail food stores, and food vending operations.

ACTION LEVELS FOR POISONOUS OR DELETERIOUS SUBSTANCES IN HUMAN FOOD AND ANIMAL FEED

Under the Federal Food, Drug and Cosmetic Act, the FDA has authority to set action levels and tolerances for substances that are unavoidable contaminants in foods (4). If they are avoidable contaminants, e.g., pesticides applied to crops, then they are covered by FQPA and the EPA sets the tolerances. The action levels and tolerances are limits at or above which FDA will take action to remove products from the market. In addition, if an unavoidable contaminant that does not have an action level or tolerance established is detected, FDA may also take action. Action levels and tolerances are set on a food-specific basis, i.e., there are separate limits that apply to each contaminant in each specific food or feed.

There are a number of pesticides for which action levels have been set in at least one food. They are: aldrin/dieldrin, benzene hexachloride, chlordane, chlordecone, dicofol, DDT/DDE/TDE, ethylene dibromide, heptachlor/heptachlor epoxide, lindane, and mirex. Action levels have been set in a number of foods/feeds for most of these but such values have been established for chlordecone only in crabmeat, fish, and shellfish; for dicofol only in processed animal feed; and mirex only in fish. The currently applicable action levels and tolerances are published by the FDA on the World Wide Web (4).

ANTIMICROBIALS

As indicated earlier, the provisions of the FQPA changed the relative roles of EPA and FDA with respect to pesticides. One particular impact was on antimicrobial substances, i.e., agents that may be used against microbes on or in foods or against microbes in or on substances that may come into contact with food. As a result of FQPA and a subsequent clarifying Act, the Antimicrobial Regulation Technical Corrections Act of 1998 (ARTCA), jurisdiction over antimicrobials is divided between EPA and FDA (5).

For example, while EPA regulates antimicrobials applied to raw agricultural commodities in the field and during transport, the FDA regulates antimicrobials applied where food is prepared, packed, or held for commercial purposes. FDA also regulates all antimicrobials used in or on processed food as food additives, except ethylene oxide and propylene oxide. In addition, FDA regulates antimicrobial substances incorporated into or applied to food packaging materials. However, it does not regulate such substances incorporated into or applied to a semipermanent or permanent food-contact surface.

SUMMARY

The main roles of the FDA with respect to pesticides in foods are to provide input into EPA registration and tolerance decisions and to enforce tolerances that have been established for most foods, except for poultry, meat, and some egg products. In its collaborative role, FDA collects data on pesticide residues in foods and conducts total diet surveys.

REFERENCES

1. Food and Drug Administration. *Food and Drug Administration Pesticide Program—Residue Monitoring—1998*; FDA Center for Food Safety and Applied Nutrition: Washington, DC, 1999, 26. http://vm.cfsan.fda.gov/acrobat/pes98rep.pdf (accessed Nov. 1999).
2. Food and Drug Administration. *Total Diet Study*; FDA Center for Food Safety and Applied Nutrition: Washington, DC, 1999. http://vm.cfsan.fda.gov/~comm/tds-toc.html (accessed Nov. 1999).
3. Food and Drug Administration. *HACCP: A State-of-the-Art Approach to Food Safety*; FDA Center for Food Safety and Applied Nutrition: Washington, DC, 1999, http://vm.cfsan.fda.gov/~lrd/bghaccp.html (accessed Nov. 1999).
4. Food and Drug Administration. *Action Levels for Poisonous or Deleterious Substances in Human Food and Animal Feed*; FDA Center for Food Safety and Applied Nutrition: Washington, DC, 1998; 16. http://vm.cfsan.fda .gov/~lrd/fdaact.html (accessed Nov. 1999).
5. Food and Drug Administration. *Antimicrobial Food Additives—Guidance*; FDA Center for Food Safety and Applied Nutrition: Washington, DC, 1999, 11. http://vm.cfsan.fda.gov/~dms/opa-antg.html (accessed Jan. 2000).

ROLE OF USDA IN PESTICIDE LAWS

Michael A. Kamrin
Michigan State University, East Lansing, Michigan, U.S.A.

R

INTRODUCTION

Responsibilities for the management of the risks of pesticides to human health and the environment are divided among a number of state and federal agencies. At the federal level, the agencies that play the largest roles in protecting human health from pesticides in foods and feeds are the Environmental Protection Agency (EPA), the Food and Drug Administration (FDA), and the Department of Agriculture (USDA). The major pieces of legislation governing pesticides are the Federal Insecticide, Fungicide, and Rodenticide Act (FIFRA), the Federal Food, Drug and Cosmetic Act (FFDCA), and the Food Quality Protection Act (FQPA), which amends both FIFRA and FFDCA.

Under current legislation, the EPA is responsible for pesticide registration and for setting tolerance limits for pesticide residues in foods; the FDA for enforcing EPA tolerances in most foods, collecting residue data and conducting diet surveys; and the USDA for enforcing EPA tolerances in meat, poultry, and some egg products; and collecting data on dietary exposure, food consumption, and pesticide usage. Within the USDA, several units collaborate in collecting and analyzing these data. These include the Agricultural Marketing Service (AMS), the Agricultural Research Service (ARS), the Economic Research Service (ERS), the Food Safety and Inspection Service (FSIS), and the National Agricultural Statistics Service (NASS).

PESTICIDE DATA PROGRAM (PDP)

The Agricultural Marketing Service is responsible for administering the Pesticide Data Program (PDP) which has been in operation since 1991 (1). Pesticide monitoring activities are conducted under a Federal–State partnership with 10 states that collect samples of food in commercial distribution for residue analysis. The states are California, Colorado, Florida, Maryland, Michigan, New York, Ohio, Texas, Washington, and Wisconsin, representing more than half of the U.S. population. Summaries of the results of these residue analyses covering fruits and vegetables, wheat, soybeans, corn syrup, and milk are published annually.

With the passage of the FQPA in 1996, the USDA was given responsibility for improved data collection of pesticide residues, including increased sampling of foods most likely to be consumed by infants and children. These data are of particular significance under FQPA since the Act requires the EPA to include more information about actual residues in its tolerance setting judgements and to explicitly incorporate concerns about possible effects on infants and children in these decisions.

In addition, the data collected as part of the PDP are used by the FDA in tolerance enforcement activities and by the USDA Foreign Agricultural Service in global marketing efforts. Further, the ERS uses these data to evaluate pesticide alternatives and to contribute to the benefit side of the equation used by EPA in making pesticide management decisions.

SURVEY OF FOOD INTAKES BY INDIVIDUALS

The ARS administers the periodic survey of food intakes by individuals (2). National food intake surveys have been administered since the 1930s and the latest comprehensive survey, known as the Continuing Survey of Food Intakes by Individuals (CSFII), was conducted in 1994–1996. In this survey, about 15,000 individuals of all ages were interviewed personally to collect 24-h recall data. Although this sample was comprehensive, increased regulatory concern about infants and children led to a Supplemental Children's Survey focused on this population that was conducted in 1997–1998. This supplemental survey collected data from about 5,000 children aged 0–9 years.

The ARS is working with EPA to convert these intake values into consumption profiles that can be used in performing dietary risk assessments. This requires translating foods "as consumed" into approximately 300 raw agricultural commodities that are ingredients in these foods. As indicated earlier, under the provisions of FQPA, these data have taken on added importance for pesticide regulation.

PESTICIDE USE SURVEYS

NASS conducts surveys of pesticide use that are critical for the analysis of the magnitudes of human exposures, the impacts of regulatory activities, and the potential for risk mitigation (3). Data are collected on pesticide use on field crops (corn, cotton, potatoes, soybeans, and wheat), fruit and vegetable crops, livestock, pasture, and range as well as postharvest pesticide use on potatoes, apples, corn, and wheat.

These data are published at the state level (except for livestock) by commodity and active ingredient and include percentage of acres treated, application rate and frequency, and total active ingredient by crop. Surveys are conducted every year for field crops but those for fruits and vegetables are conducted in alternate years. They are carried out only in the major production states, numbering about 10 to 15, for each survey. The survey covers about 80–90% of the total planted acreage for field crops, fruits, and vegetables.

NATIONAL AGRICULTURAL PESTICIDE IMPACT ASSESSMENT PROGRAM (NAPIAP)

NAPIAP is a cooperative program with land grant universities to promote informed decision making on agricultural pesticides. Since the passage of the FQPA, NAPIAP has focused its efforts on development of state and regional crop–pest profiles that detail pest management practices for each crop including critical pesticide uses. These profiles can provide critical information to EPA with regard to pesticide regulation as well as important information that can be used to develop alternatives to pesticides that may lose registration as a result of FQPA.

IR-4 MINOR USE PROGRAM

The IR-4 Program is designed to promote registration for pesticides that will be used on minor crops; i.e., those crops grown in such small quantities that it is not economical for the private sector to invest heavily in registration activities for pesticides needed on these crops. These include a variety of fruit and vegetable crops, nursery and floral crops, and turf grass. In addition, IR-4 supports registration of biopesticides (biological control agents). With the passage of FQPA, IR-4 has also devoted significant attention to defending existing registrations, and to accelerating registration of reduced risk pesticides important to integrated pest management systems.

FOOD SAFETY AND INSPECTION SERVICE (FSIS) NATIONAL COOPERATIVE RESIDUE PROGRAM

Under the traditional FSIS national residue program, samples of meat and poultry from domestic slaughter establishments and samples of egg products from egg processing plants are collected and tested. Analyses may be performed to monitor the incidence of violations of pesticide tolerances set by EPA or for enforcement purposes so as to limit the amount of violative residues in the food supply. Pesticides chosen for analysis are selected on the basis of their likelihood to produce a residue, their toxicity, and the potential for human exposure to the residue. The FSIS publishes its likelihood rankings of pesticides under this system on the World Wide Web (4).

HAZARD ANALYSIS AND CRITICAL CONTROL POINT (HACCP) SYSTEM

As a result of reforms of USDA regulations in 1996, meat and poultry establishments must implement HACCP systems to ensure the safety of their products. The HACCP system focuses on identifying critical steps in the chain from food production to consumption at which potential hazards can be controlled or eliminated (5). While this hazard analysis focuses primarily on microbial contaminants, it also includes hazard analyses for chemical (including pesticide) and physical residues. The FSIS is responsible for carrying out the HACCP regulations promulgated in 1996.

The National Cooperative Residue Program (NCRP) is designed to assist slaughter establishments to implement HAACP. The program includes identifying those pairs of pesticides and slaughter animals that are most likely to result in violative residues and recommending steps that can be taken to prevent such residues from occurring. In addition, the NCRP alters the way in which enforcement actions can be taken.

PESTICIDE RECORDKEEPING

Under the provisions of the Federal Pesticide Recordkeeping Program, the Agricultural Marketing Service requires all certified private pesticide applicators to keep records of their use of federally restricted pesticides. These records must include, with some exceptions, for each pesticide application: 1) brand or product name; 2) EPA registration number; 3) total quantity of pesticide applied; 4) date of application; 5) location of application; 6) crop, product, or site treated; 7) size of area treated; and 8) name and certification number of applicator. Records are required to be made within 14 days of application and to be kept for two years.

SUMMARY

The major roles of USDA in regulating pesticides in foods and feeds are to collect data for use by EPA in the pesticide registration and tolerance setting processes and to enforce regulations aimed at eliminating contamination of meat, poultry, and milk by pesticide residues.

REFERENCES

1. U.S. Department of Agriculture. *Pesticide Data Program, Annual Summary Calendar Year 1997*; Agricultural Marketing Service: Washington, DC, 1998; 109. http://www.ams.usda.gov:80/science/pdp/97summ.pdf
2. Enns, C.W.; Goldman, J.D.; Cook, A. Trends in food and nutrient intakes by adults: NFCS 1977–78, CSFII 1989–91, and CSFII 1994–95. Family Economics and Nutrition Review **1997**, *10* (4), 1–14.
3. U.S. Department of Agriculture. *Agricultural Chemical Usage Reports*; National Agricultural Statistics Service: Washington, DC, 1999. http://usda.mannlib.cornell.edu/reports/nassr/other/pcu-bb/ (accessed Jan. 2000).
4. U.S. Department of Agriculture. *1998 National Residue Plan for 1998: Appendix III: Compounds Historically Ranked under the Compound Evaluation System*; Food Safety and Inspection Service: Washington, DC, 1998; 7. http://www.fsis.usda.gov:80/OPHS/bluebook/hasect1.htm (accessed Dec. 1999).
5. U.S. Department of Agriculture. *Pathogen Reduction/HACCP & HACCP Implementation*; Food Safety and Inspection Service: Washington, DC, 1999. http://www.fsis.usda.gov/OA/haccp/imphaccp.htm (accessed Jan. 2000).

R

SAMPLING

David E. Legg
University of Wyoming, Laramie, Wyoming

S

INTRODUCTION

Sampling is the only method by which pest managers can efficiently determine the level of pest infestation in target populations. Sampling for pest populations can be done by examining a fixed number of samples or a variable number of samples. Popular plans for examining a fixed number of samples include sampling along a predetermined path and/or sampling at predetermined intervals. Frequently used plans for examining a variable number of samples include the rapid classification of pest populations relative to a critical value. Many sampling plans for pest management have been implemented without first being examined for their accuracy or repeatability (precision). It was suggested that these plans be examined to see if they are providing accurate and/or precise estimates of pest population averages and standard errors (1).

ACCURACY AND PRECISION

Managers need to know the levels of pest infestation in specified fields, rangelands, or herds (target populations) before they can make correct management decisions. Those levels can be determined in one of two ways. The first is to use a population dynamics model. Unfortunately, predictions or forecasts from most models are not reliable. The second is to visit the target populations and personally determine the levels of infestation. This is done in such a way that the determinations are accurate and precise. Accuracy indicates that the determinations are representative of the true levels of infestation. If they are not, they are *biased*. Bias is a consistent under- or overdetermination of an infestation. Precision indicates that, if several determinations were conducted on the same population at the same time, their results would be consistent; pest managers have great confidence in precise determinations. There are two ways in which pest populations can be accurately and precisely determined: through a census or through representative sampling (1).

A census involves counting all pests in a carefully defined target population. Assuming that no counting errors are made, the census provides an exact determination of the level of infestation. Such determinations are both accurate (right on the mark) and precise (very repeatable). Unfortunately, the resources required to conduct a census are often greater than the resources available to pest managers. In lieu of a census, then, a small number of samples from the target population are examined. From these, the average number of pests and the standard deviation are estimated. If those samples are *representative* of the target population, the estimated average and standard deviation are unbiased estimates of the census average and standard deviation (2).

Obtaining a Representative Sample

There are many potential ways to obtain a representative sample from a target population. The most important makes use of *randomly selected* sample units. In principle, selecting a random sample is simple. In practice, however, selecting a random sample is difficult, especially in agricultural settings. Some other ways to obtain a representative sample do not use random selection. These ways, or *methods*, are "suspect" and should be tested to ensure they provide accurate pest population determinations before they are implemented (1).

Accuracy of a Suspect Sampling Method

Pest managers can test to see if a suspect sampling method provides accurate determinations of pest populations in one of two ways. The first involves conducting the census *and* the suspect sampling method on the same population, with the sampling method being used to obtain several estimates of that population's average. Then, those estimates themselves are averaged (m). If the suspect sampling method provides an accurate estimate of the census average (μ), then $\mu - m$ will be less than one-tenth the standard deviation of m. The second way to test a suspect sampling method is to conduct several random sampling efforts on a target population, estimate the population average for each, and calculate the average of those averages; by definition, an average of the averages that are estimated from a random sampling effort is μ. Then, if the suspect sampling method provides an accurate estimate of μ, $\mu - m$ will be less than one-tenth the standard deviation of m. Once it has been shown that the suspect sampling method provides a representative sample, that method may be used to determine the level of pest infestation with either a fixed or variable number of samples (1).

EXAMINING A FIXED NUMBER OF SAMPLES

Several plans may be used for examining a fixed number of samples, including unrestricted random sampling, stratified random sampling, cluster sampling, multistage sampling, systematic sampling, and transect sampling. With any of these, the total samples examined is fixed. The most commonly used fixed-sample number plans involve systematic and/or transect sampling. Systematic sampling is a systematic examination of sample units throughout the target population. Examples include examining every 50th plant in a row crop, or the center tree in each half-hectare of an orchard. Transect sampling may be thought of as randomly examining sample units along a line that is traversed across the target population. Examples include sampling along straight-line, zig-zag, U-shaped, or circular routes that are taken across row crops, orchards, or forests. An unbiased estimate of the pest population average may be obtained from transect sampling, even when a number of transects are systematically arranged across the target population, and more than one randomly selected starting point is used. Be advised, however, that biased estimates of the standard deviation may occur when formulae for unrestricted random sampling are used for its estimation. This warning also applies to data collected from systematic sampling (2).

EXAMINING A VARIABLE NUMBER OF SAMPLES

Estimating the level of pest infestation by examining a variable number of samples can be done either through precision-based or classification-based sampling. Precision-based sampling, also known as *fixed-precision sequential sampling*, requires the pest manager to estimate the precision of a population average, after each sample is examined, by dividing that average by its standard error. Sampling stops when precision is less than or equal to some value. Mathematical models exist for expressing precision as a function of cumulative pests counted and total samples examined. Theoretically, this can be done for any specified level of precision. Users of these models should be aware that most generate plans that are somewhat biased in the sense that the observed precision is not equal to the desired precision. Users of precision-based sampling should also be aware that these plans may require the examination of many samples, especially when the level of infestation is low (3).

Classification sampling, which is often referred to as *fixed-error sequential sampling*, allows pest managers to rapidly classify populations into high or low categories relative to a threshold. This kind of sampling is done so the probability of misclassification is fixed at some maximum-allowable level (error level). Fixed-error sequential sampling plans are also generated through mathematical models. Users of these plans should be aware that most are somewhat biased in that they do not perform at the specified error levels. Users of these plans should also know that they classify populations into high or low categories at the expense of obtaining precise estimates of the pest population (4–6).

FUTURE CONCERNS

There are many sampling plans that were not studied for bias or precision before they were implemented. It may be useful to conduct studies on those plans now. Results from studying one commonly used, fixed-sample size plan for grasshopper assemblages indicated that novices tended to overestimate the level of grasshoppers that inhabit Wyoming rangelands. Results also showed that even experienced samplers had significantly differing estimates of grasshopper assemblages, some of which led to diverging management decisions.

REFERENCES

1. Legg, D.E.; Moon, R.D. Bias and Variability in Statistical Estimates. In *Handbook of Sampling Methods for Arthropods in Agriculture*; Pedigo, L.P., Buntin, D., Eds.; CRC Press: Boca Raton, FL, 1994; 55–69.
2. Thompson, S.K. *Sampling*; John Wiley: New York; 1992.
3. Hutchison, W.D. Sequential Sampling to Determine Population Density. In *Handbook of Sampling Methods for Arthropods in Agriculture*; Pedigo, L.P., Buntin, D., Eds.; CRC Press: Boca Raton, FL, 1994; 207–243.
4. Binns, M.R. Sequential Sampling for Classifying Pest Status. In *Handbook of Sampling Methods for Arthropods in Agriculture*; Pedigo, L.P., Buntin, D., Eds.; CRC Press: Boca Raton, FL, 1994; 137–174.
5. Jones, V.P. Sequential Estimation and Classification Procedures for Binomial Counts. In *Handbook of Sampling Methods for Arthropods in Agriculture*; Pedigo, L.P., Buntin, D., Eds.; CRC Press: Boca Raton, FL, 1994; 175–205.
6. Wald, A. *Sequential Analysis*; John Wiley: New York, 1947.

SCABIES MITES

Baik Kee Cho
The Catholic University of Korea, Seoul, South Korea

INTRODUCTION

The scabies mite is barely visible, broad oval shaped, subglobose above, and flattened ventrally. The mite produces scabies in 40 different hosts including humans and domestic animals such as dog, pig, horse, goat, cat, cattle, etc. (1). The adult female mite makes a tunnel (burrow) in the horny layer of the epidermis to stay and lay eggs. Scabies is a highly contagious pruritic dermatosis and the spread is associated with overcrowding, poor hygiene, and delayed diagnosis. It is also considered a sexually transmitted disease because close skin-to-skin contact is necessary for spreading the disease.

MORPHOLOGY OF THE SCABIES MITE

The scabies mite, *Sarcoptes scabiei*, is classified in the order Acarina, suborder Sarcoptiformes, cohort Acaridae, family Sarcoptidae, subfamily Sarcoptinae (2). The body surface is covered with fine striations and the color is creamy white with brown sclerotized legs and mouthparts. The adult female measures 300–500 μm long by 230–340 μm wide. There are a large number of small triangular scales (squamous dorsal thorns) on the dorsum of the female. Human scabies mites have bare areas without triangular scales, while almost all canine scabies mites do not (1). Three pairs of short cone-like spines (dorsal cones) are arranged symmetrically on the upper dorsal surface. Seven pairs of inverted carpet-tack-like spines (dorsal spines) are arranged in two rows of three and four respectively on either side of the midline posteriorly (Fig. 1A). Each leg has five segments (the tarsus, tibia, genu, femur, and trochanter), which appear as if telescoped into each other. The tarsus of legs I and II terminates in an elongated peduncle tipped with a small disc. The tarsus of legs III and IV ends in a long seta. The male *Sarcoptes* is 213 μm long by 162–210 μm wide. The most obvious differences from the female are legs IV and the genital apparatus (3). Leg IV has an elongated peduncle on the tarsus instead of a seta. In the midline between legs IV there is a chitinous structure housing the male genital organ (Fig. 1B). The larva has only six legs and is about 215 μm long by 156 wide (Fig. 1C). The nymph forms are seldom found clinically except in Norwegian scabies and animal scabies.

BIOLOGY OF THE SCABIES MITE

The human scabies mite is an obligate parasite and it can thrive and multiply only on human skin. The female *Sarcoptes* forces the stout spines on the dorsal side against the roof of the burrow to prevent backward movement (2). The male is the most active of all forms and runs around on the skin surface seeking a female (3). It is believed that copulation is done in the female burrow and the female copulates only once.

The female lives approximately 4–6 weeks and lays on average 40–50 eggs. The eggs hatch after 3–4 days followed by a larval stage for 3–4 days. The emerging protonymph stage lasts for 3–4 days and a second molt may give rise to an adult male or second nymph. The latter has to go through a third molt to produce a young adult female (2).

SCABIES

Clinical Symptoms

Scabies is characterized by severe itching and widespread skin eruption. The itch is characteristically nocturnal and is scarcely noticeable in the daytime. It usually starts after the patient retires to bed and becomes warm. The burrows and the associated papules and vesicles are directly related to the presence of the *Sarcoptes*, while the generalized rash and itching sensation are related to the development of sensitization to the mite. There is a symptomless period of about 3–4 weeks in scabies before sensitization develops, which sometimes could be shortened to 1 week if the initial infestation is very heavy (4).

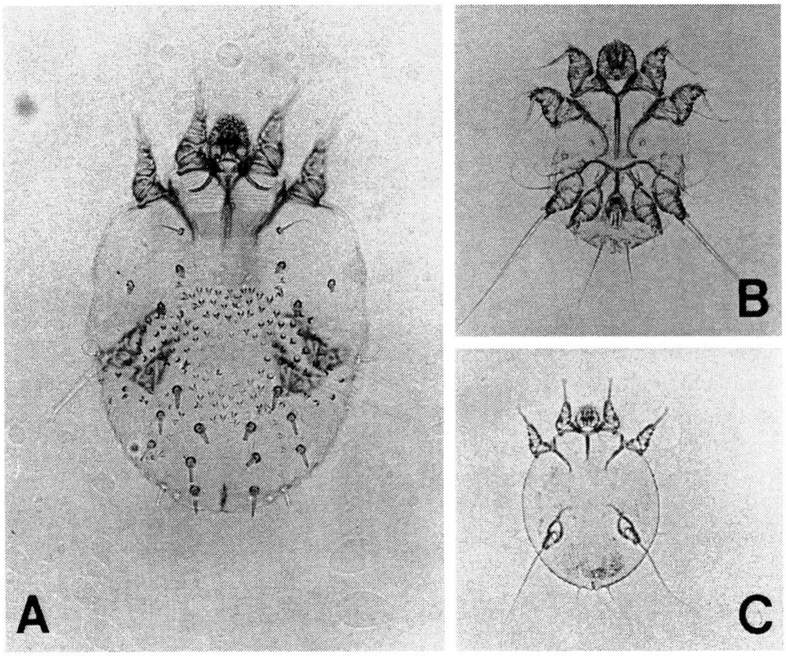

Fig. 1 Human scabies mite: dorsal surface of female (**A**), ventral surface of male (**B**), and ventral surface of hexapod larva (**C**).

The burrow made by the adult female *Sarcoptes* lies in the horny layer of the epidermis and presents clinically as a grayish line resembling a pencil mark about 5 mm long on average (Fig. 2A). The side of the fingers, the finger webs, the wrist, and the male genitalia are the common sites of the burrow. In man, pruritic papules or nodules on the scrotum (Fig. 2B) and penis are very diagnostic features.

In infants, soles and ankles are also common sites (Fig. 2C). Very often, especially on the sides of the fingers and wrists, tiny pinpoint papules or vesicles may be seen close to the burrows.

Norwegian Scabies (Crusted Scabies)

The condition was first described in lepers by two Norwegian doctors. The patients are immunologically deficient, physically debilitated, or mentally retarded. Pruritus generally is absent and crusted lesions are extensive (Fig. 2D). The crust and exfoliative skin scales contain hundreds to thousands of mites per 1 g (5). It can be transmitted explosively to medical personnel and other inpatients by medical instruments or direct contact.

Animal Scabies

Canine scabies in humans

The nonhuman strain of *Sarcoptes* that infests humans most frequently is that from the dog (1). Papular lesions always appear first on the areas of contact with the dog. The areas commonly affected are the forearms, the chest, the anterior abdominal wall, and the front of the thighs. Itching is more pronounced in the evening.

Treatment or removal of the infested dog results in gradual subsidence of the eruption.

Canine scabies in dogs

Skin lesions quickly become excoriated by scratching, leading to crusting. The first lesion may appear on the muzzle, around the eyes (Fig. 3), and edges of the ears. The hair tends to become matted together and often falls out in localized or widespread areas. Hundreds or thousands of all forms of *Sarcoptes* are found per 1 g of scale or crust from the untreated dog (6). If the animal is untreated, secondary infection, cachexia, and death ensues.

Diagnosis of Scabies

For the dermatologist with some experience, scabies with typical symptoms can be one of the easiest diseases to diagnose. Pruritic papules and nodules on the penis and scrotum are a diagnostic feature of scabies as well as the burrows on the predilection sites. Definitive diagnosis can be made by microscopic identification of anything among any stages of the mites, eggs, and fecal pellets. The skin scraping method using mineral oil is most valuable technique for diagnosis. Mineral oil is

S

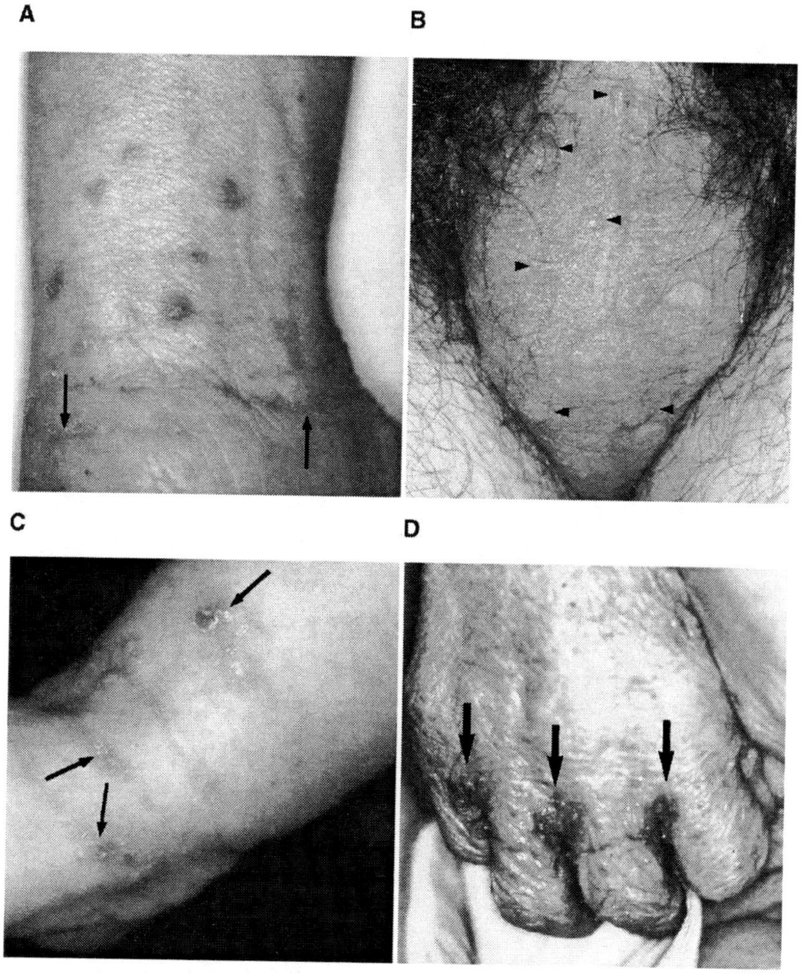

Fig. 2 Clincal findings of scabies: burrows (arrows) and papules on the wrist (**A**), pruritic papules and nodules (arrow heads) on the scrotum (**B**), burrows and erythematous vesicles (arrows) on the ankle of an infant (**C**), and thick crusted lesions (arrows) between the fingers of the old cachexic patient with Norwegian scabies (**D**).

dropped onto a sterile scalpel blade and allowed to flow onto the suspected burrows, which is then vigorously scraped with a blade about six to seven times. The oil and scraped material are then transferred to a glass slide, covered with a cover slip, and examined microscopically (7).

Treatment of Scabies

The standard treatment of scabies is based on topical scabicides. Choice of treating measures may depend on whether the sufferer is an adult, child, or infant and whether the scabies is complicated by secondary infection or eczematization. Two important things must be kept in mind in treating the scabies. One is that without exception everyone living in the patient's house should be treated at the same time. The other is that the prescribed topical drugs should be thoroughly rubbed into all the skin of the patient's body and limbs whether there are any lesions to be seen or not. No part of the skin from the chin downward should be left untreated. The face, head, and neck must also be treated in babies.

Topical treatment

Permethrin cream (5%) is an excellent scabicide with low toxicity. It is the first choice for children and infants older than two months (7). Aqueous emulsion of 25% benzyl benzoate is very effective but should be avoided for infants. The painting should be repeated once, either the next day or one week later. Gamma-benzene hexachloride cream or lotion (Lindane) is also effective with one proper application but is contraindicated in infants, small children, pregnant women, and in those with

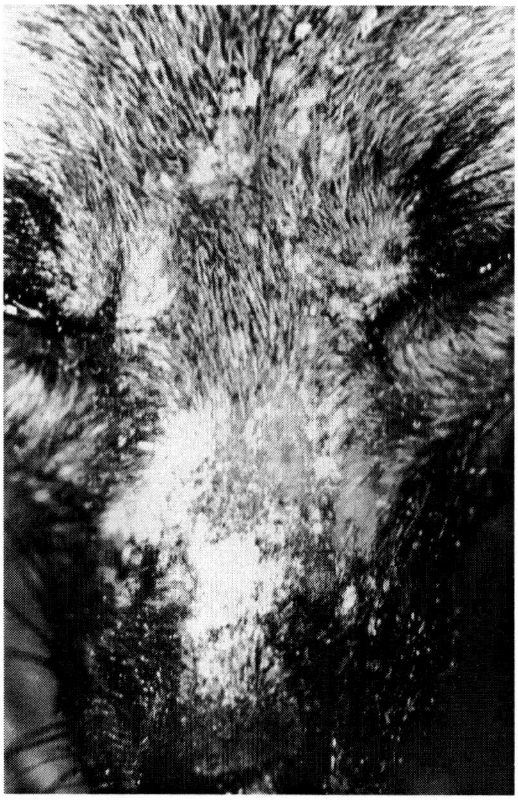

Fig. 3 Canine scabies: hair loss and thick yellowish-brown crusted and scaly lesions on the muzzle and around the eyes after six months of infestation.

extensive inflammatory lesions. Crotamiton cream is a good medication and safe for all patients, but its cure rate is lower than the others. Sulfur ointment, 10% for adults and 2.5% for infants, can be used for any case of scabies (2).

Oral treatment

One to three doses of ivermectin (200 μg/kg) at one- or two-week intervals is very effective for severe scabies, especially for refractory Norwegian scabies (8). The drug has been used extensively for the control of a wide variety of parasites in farm and domestic animals. In humans, ivermectin has been widely used since 1987 to control endemic onchocerciasis. Oral ivermectin is easy, quick, safe, and well tolerated, with maximal patient compliance but it is not yet widely used throughout the world for the treatment of scabies.

REFERENCES

1. Fain, A. Epidemiological problems of scabies. Int. J. Dermatol. **1978**, *17* (1), 20–30.
2. Alexander, J.O. *Arthropods and Human Skin*; Springer-Verlag: Berlin, Germany, 1984; 227–292.
3. Heilesen, B. Studies on *Acarus scabiei* and scabies. Acta Derm. Venereol. **1946**, *26*, 11–315 (Supp. 14).
4. Mellanby, K. *Scabies*; Oxford University Press: London, U.K., 1943.
5. Kim, K.W.; Oh, Y.J.; Cho, B.K.; Houh, W.; Kim, J.A.; Lee, Y.S. Norwegian scabies. Ann. Dermatol. **1990**, *2* (1), 50–54.
6. Hong, J.K.; Jang, I.G.; Cho, B.K.; Lee, W.K. A clinical and histopathological study of experimental canine scabies. Ann. Dermatol. **1998**, *10* (4), 238–246.
7. Orkin, M.; Maibach, H.I. Scabies and Pediculosis. In *Dermatology in General Medicine*; Freedberg, I.M., Eisen, A.Z., Wolff, K., Austen, K.F., Goldsmith, L.A., Katz, S.I., Fitzpatrick, T.B., Eds.; McGraw-Hill: New York, 1999; 2677–2684.
8. Giudice, P.; Marty, P. Ivermectin. Arch. Dermatol. **1999**, *135* (6), 705–706.

SCOUTING

Stephen Lefko
Monsanto Company, DeKalb, Illinois, U.S.A.

S

INTRODUCTION

Despite significant advances in biological, cultural, chemical, and genetic pest management tactics, knowledge of a developing pest problem is still the key to making economical pest management decisions. Scouting is the practice of gathering the knowledge required to make wise pest management decisions; therefore, it is an essential component of the integrated pest management (IPM) philosophy. Often scouting is confused with IPM and a movement toward this synonymy highlights its importance. IPM practitioners rely on scouting programs to decide if implementing a tactic is justifiable in their production system. Making the decision to implement a pest management tactic is becoming easier as the practice of scouting gains attention, new scouting programs are developed, and existing programs are improved.

EVOLUTION OF CROP SCOUTING

Scouting programs have evolved greatly over the past few decades. This is a result of our burgeoning knowledge base of pest biology and population dynamics, crop genetics and crop physiology, and how these factors interact with the environment. Early scouting programs were subjective, primarily carried out by producers, and decisions were based mostly on experience. Today, agriculturists are using scouting programs based on revolutionary science and robust statistical models, and often they are collecting data and resolving problems more efficiently with hand-held digital technology.

These developments, along with social, environmental, and legislative forces have turned scouting into a profitable discipline. The cost of professional crop scouting is directly proportional to the crop value and incidence or quantity of inputs such as pesticides and fertilizer. In 2000, professional firms that scouted corn and soybean charged between \$3 and \$7 per acre for service during one growing season. One full-time scout usually services between 1500 and 5000 acres per year and updates field records weekly. More labor-intensive field crops like potato and sugarbeet command up to \$15 per acre and the cost for orchard crops is even higher. Now, scouting is practiced by personnel in most sectors of agriculture including progressive producers, industry representatives, professional scouting firms, commodity groups, and university extension personnel.

ELEMENTS OF A SCOUTING PROGRAM

Scouting programs are structured similar to contingency plans where the results from one activity are required to justify the next course of action. The elements of a generalized scouting program include: 1) the detection of the pest(s) and confirmation of pest identity, 2) sampling pest density or level of infection, 3) following decision guidelines and, if necessary, 4) prescribing an appropriate management tactic and confirming its efficacy. This process sounds simple, but the challenge is greatly compounded by the diversity of pest species and their unique interactions with crop plants and the environment.

Pest Detection and Identification

Scouting programs ensure producers the information opportunity to make wise economic decisions. The process begins by putting the expertise or mechanism of pest detection in place. Efficient scouts can minimize their operational costs by synchronizing detection efforts with crop development, environmental cues, or historical data on pest occurrence.

In field crops the cues that trigger scouting are often crop stage, accumulation of heat units, or weather conditions outside of average. In the U.S. Corn Belt, a season-long complex of field corn pests forces scouts into the field immediately after planting and this activity continues until after flowering. Seeds and seedlings may be subject to injury by soil-inhabiting insect pests or infection with seedling diseases. Root, stalk, and leaf-feeding insect pests may appear throughout the growing season and their presence can increase the likelihood of secondary pests and or diseases. And finally, pests that may damage the harvestable unit are monitored near the end of the growing season. Particular threats to crop yield may not persist but scouting likely will as the weather changes, the crop develops, and new pest species appear.

Keeping track of weather events is vital to efficient and effective scouting programs. For instance, insect development is directly related to temperature, and degree-day models often trigger scouting activities. Armed with these models, scouts can predict when insects will mature to destructive stages and predict when crops may be vulnerable.

Migratory insect pests of corn, like the black cutworm and corn earworm, are less predictable since they rely on early-summer winds to transport them to more northern locations. In this example the problem of prediction is overcome using a network of pheromone traps. The trap network connects southern overwintering sites with the northern locations that occasionally suffer loss. Trap catches are updated regularly and the progress of migration warns growers to begin scouting.

Experience, historical data, and predictive models can be useful for anticipating when and where pests may be active. However, there is no substitution for the proper identification of pests; this may be the scout's most difficult task of all. Numbers of pest species usually are in the dozens for a particular crop, and this dilemma is compounded because not all pests are equally important. In some instances the method of pest detection is quantitative and a scout can skip to the decision making step of the contingency plan. For most pests their detection triggers the sampling component of a scouting program.

Sampling

Sampling is an attempt at efficiently and accurately estimating the level of infestation or infection by a pest species. This estimate is converted to an expectation of economic loss and weighed against the cost of mitigating the pest problem. In every sampling program the accuracy of the final estimate is directly proportional to the cost of gathering samples. Crop protection researchers struggle to minimize the costs associated with scouting and highlight these two considerations: the correlation between sampling unit and economic loss, and the impact pest dispersion has on the sample number and accuracy of sampling results.

The sampling unit used by scouts is dependent on the crop, the pest, and the stage of development for both of these parameters. The unit must be highly correlated with loss and meaningful when compared against pest management decision guidelines. As a result, sampling plans usually focus on the damaging stage(s) of pests or the stage that immediately precedes the damaging stage. Insects are typically sampled using traps, nets, soil samples, or in situ counts. These techniques relate insect number to plants or plant parts, a unit area of land, length of row, or a period of time. Indices of injury symptoms also are used to quantify the potential for loss. Some common indices are percentage of defoliation, percentage of stand loss, percentage of infected plants, or the amount of injury to the harvestable unit.

Scouts must constantly weight the value of producing accurate recommendations against the cost of collecting additional samples. Dispersion patterns differ among pest species and these patterns usually dictate the sample pattern and sample size necessary to generate accurate estimates. Most insect pests follow an aggregated or clumped dispersion pattern. Therefore, the potential exists for scouts to generate biased results if samples were more often collected inside or outside of the clumps. Scouting program developers study these patterns and usually describe a pattern to follow when sampling specific pests. Alternatively, patterns of infestation or infection sometimes correlate with soil types; areas receiving particular cultural practices; and edge-effects from grass, waterways, or even placement in a greenhouse. Historical data or experience usually make sampling these areas more efficient and result in better pest management decisions.

Making Pest Management Decisions

Wise pest management decisions are based on a firm understanding of how the level of pest pressure impacts productivity. Crop protection scientists work to determine the economic threshold (ET) for pests. An ET is a pest density that justifies a curative management tactic that prevents the pest population from growing to an economic proportion.Thresholds are crop and pest specific, and the unit associated with a threshold is usually identical to the sampling unit prescribed in the sampling program. Prescribing a pest management tactic is justified when these thresholds are exceeded.

Prescribing a Management Tactic

Research is producing new biological, cultural, chemical, and genetic pest management tactics too numerous to describe. Scouts must weigh the economic impact of each tactic with its availability, efficacy, and practicality for implementation. Scouts prescribe tactics knowing the potential for failure is real. Some sources of failure may be environmental, chemical, biological, or related to timing. Therefore, the final step in the contingency plan is to confirm the tactic reduced the pest density below the ET. More sampling may be necessary if this result isn't obvious.

SCOUTING CHALLENGES

Although the science and economics behind scouting programs are sound and objective, adoption of scouting practices is relatively slow. Ideally, producers weigh the investment in time and cost of supplies against the perceived risk of economic loss from pests. This statement; however, has many inherent assumptions that likely slow the rate of adoption. Some of these assumptions are:

- understanding the economic advantage
- correct identification of pests
- understanding pest biology and crop response to pests
- willingness to execute complex scouting programs
- observing results after implementation

Studies investigating the success of scouting programs underscore economics as the factor most responsible for producer participation. Crop yield and pesticide costs are variables that weigh heavy in the mind of the producer. Scouting programs can increase profitability by reducing pesticide use in cropping systems with occasional pests but historically prophylactic pesticide applications. Similarly, scouting sometimes leads to more frequent pesticide applications targeting perennial pests that were previously undetected. In both instances the economics favor the producer even though the pesticide inputs differ greatly between scenarios. These are some of the challenges that crop protection researchers and practitioners face in their pursuit of a recipe for sustainable agriculture.

BIBLIOGRAPHY

Bechinski, E.J. Designing and Delivering In-the-field Scouting Programs. In *Handbook of Sampling Methods for Arthropods in Agriculture*; Pedigo, L.P., Buntin, G.D., Eds.; CRC Press: Boca Raton, FL, 1994; 683–706.

Ferguson, W.; Yee, J. Evaluation of professional scouting programs in cotton production. J. Prod. Agric. **1993**, *6*, 100–103.

Gray, M.E.; Luckmann, W.H. Integrating the Cropping System for Corn Insect Pest Management. In *Introduction to Insect Pest Management*; Metcalf, R.L., Luckmann, W.H., Eds., 3rd Ed.; John Wiley & Sons, Inc.: New York, 1994; 507–542.

Matthews, G.A. The importance of scouting in cotton IPM. Crop Prot. **1996**, *15*, 368–374.

Pedigo, L.P. *Entomology and Pest Management*, 3rd Ed.; Prentice-Hall: Englewood Cliffs, NJ, 1999; 691.

Topliff, L.A.; Schnelle, M.A.; Pinkston, K.N.; Cuperus, G.W.; von Broembsen, S. *Scouting and Monitoring for Pests in the Commercial Greenhouse*; OSU Extension Facts No. 6711, Oklahoma State University Cooperative Extension Service, 1992; 8.

SEED TREATMENTS

Mark M. Stevens
Yanco Agricultural Institute, Yanco, New South Wales, Australia

INTRODUCTION

Chemical, physical, and biological seed treatments are used for many purposes. These include the protection of seeds from pests and pathogens during storage, the protection of crops from seedborne, soilborne and foliar pests and pathogens that cause crop damage after sowing, and the stimulation of seed germination and subsequent plant growth (1–3). This article deals only with seed treatments applied to protect crops from pests and diseases to which they may be exposed after planting.

Seed treatment is an effective and efficient method of protecting plants from pests and diseases during the critical and highly vulnerable crop establishment period. While it is widely used in some crops, it is underutilized in others, and considerable scope exists for developing seed treatments for pests and diseases affecting a broader range of plant species (2). The advantages of seed treatments include their ease of use, the precise placement of pesticides in the root zone of the growing plant (2–4), and the immediacy of their effect after sowing. By replacing broad-area sprays with seed treatments spray drift can be avoided and chemical application rates lowered on a unit/area basis (3, 4). To achieve these benefits several practical constraints need to be overcome. Seed treatments must be nonphytotoxic at field rates (1, 4), and must be uniformly applied, often necessitating dedicated application equipment. Since seed treatments are applied in anticipation of pest or disease activity, the target organism must exceed economic threshold levels on a regular basis. While seed treatments are an extremely valuable form of pest management, they may need to be followed by additional crop protection measures unless pest and disease activity is restricted to the early stages of crop growth.

CROP PROTECTION USES

Seed treatments can be divided into three main categories: physical, biological, and chemical. Physical seed treatments, such as steam, heat, or irradiation have a limited role in protecting crops from pests and diseases, since they can act only on pathogens or insect pests that may be present on or in the seed prior to sowing—for example, fungal spores may be dormant on seeds during storage, but only infect the plant and cause disease after sowing and seed germination. Physical treatments do not confer protection against soilborne pests and diseases that may damage the seed after planting (1).

Biological treatment involves treating seeds with organisms that have a detrimental effect on pests or pathogens. In recent years there has been increased interest in the development of biological seed treatments for crop protection, however biological treatments are currently used primarily for promoting crop growth. Typical of this use pattern is the inoculation of legumes with symbiotic nitrogen-fixing bacteria (1, 3, 5).

Chemical treatments are the most widely used form of seed treatment applied against pests and pathogens that affect crops in the period immediately after sowing. The use of botanical treatments to protect seeds either in storage or after planting dates back to at least Roman times (6). The early development of seed treatments was heavily biased toward fungicides, starting with the use of organic salts (particularly copper sulphate), formalin, and subsequently organomercury compounds in the early part of the twentieth century (2, 5, 6). Since the middle of the twentieth century a broad range of new fungicides has become available, many with systemic activity, and these compounds have increased the opportunities for broadening the usage of seed treatments. The development of insecticidal seed treatments has lagged behind that of fungicides, and has only become significant since the 1960s when methiocarb and carbofuran were developed. The development of new systemic insecticides (such as imidacloprid) may well stimulate the increased use of insecticidal seed treatments (2). Although seed treatment seldom involves the use of herbicides, some limited research has been conducted on using seed as a herbicide delivery system. This research has largely been restricted to broad-

S

Table 1 World seed treatment market by product segment (1997 estimates)

Category	% of world market
Fungicides (F)	60
Insecticides (I)	25
Mixtures (F + I), and others	15
Total	100

(From Novartis Crop Protection AG.)

cast or aerially sown lowland rice, a unique situation in which herbicides can freely move from the seed and redistribute in standing water.

Benefits of Seed Treatment

Seed treatments, when used in appropriate situations, have many advantages over foliar or soil-applied chemicals. The first and most obvious of these is that the biologically active agent is localized in the area where its effects are needed—the root zone of the developing seedling (2–4). In comparison to alternative means of treatment, there is minimal risk of pesticides entering off-target areas. Spray drift is avoided, and potential losses through volatilization or surface runoff are dramatically reduced. A corollary of this benefit is that chemical application rates per unit area can be reduced (3, 4) leading to both economic and environmental benefits. Economic benefits can also be achieved through avoiding the labor and fuel costs associated with applying soil treatments and foliar sprays after sowing. The other major benefit of seed treatment is the immediacy of effect. Once the seed is sown, protection is immediately available to the seed and, in the case of systemic compounds, uptake by the plant will begin as soon as germination is underway. This can be particularly advantageous when dealing with soil-dwelling pests and pathogens that can cause substantial crop damage before their presence is recognized.

Limitations of Seed Treatment

Although seed treatments allow reductions to be made in overall pesticide inputs, they often involve the seed and young seedling coming into direct contact with high pesticide concentrations. Compounds that show no phytotoxicity as foliar sprays may have adverse effects on germination and plant growth when applied to the seed at concentrations necessary for effective pest and disease control. Phytotoxicity problems with seed treatments can usually be detected in the early stages of experimentation, either in the laboratory or field, and their resolution can often be achieved through changes in pesticide formulation or application technology (e.g., the use of chemical "safeners," or the substitution of wet slurry application with film coating or seed pelleting). In some cases alternative active ingredients may need to be found. The risk of phytotoxicity is complicated by the variable response of different crop cultivars to particular treatments (7), and sometimes by the variable response of individual cultivars to different stereoisomers of the same compound (8).

Biological seed treatments containing beneficial fungi or bacteria have to meet a complex set of requirements. The organism itself needs to be efficacious, easy to produce in commercial quantities, and may need to be resistant to desiccation, depending on the application techno logy. Delivery systems that give bioprotectants a competitive advantage over other microorganisms are often required, and may involve inclusion of a food base in the seed treatment, along with pH modifiers and coprotectants (3, 9).

Developing appropriate application technology often represents a significant obstacle in the commercialization of seed treatments. In most cases, the efficient use of seed treatments requires an even loading of pesticide to be applied to individual seeds, and this can be difficult to achieve (5), at least on a commercial scale. In recent years considerable improvements have been made in seed treatment technology, both in the design of machines used for established seed treatment techniques (e.g., wet slurry application), and in the development of new technologies, such as film coating and seed pelleting (1, 2, 3, 5). While these processes have improved the uniformity of seed loading, their use may substantially increase production costs (3), and growers will adopt new technologies only if they are economically viable alternatives to existing pest and disease management practices. Pelleted

Table 2 Fungicide and insecticide seed treatment market by crop (% of total): 1996 data

Crop	Fungicides	Insecticides
Cereals	63	22
Canola	—	15
Potatoes	8	—
Sugar beet	—	14
Rice	8	—
Cotton	6	12
Maize	5	34
Vegetables	5	—
Other	5	3
Total	100	100

(From Ref. 10.)

Table 3 World seed treatment market by region: 1997 estimates

Region	% of world market
Western Europe	38
NAFTA[a]	25
Eastern Europe	19
Latin America	12
Asia/Pacific	3
Africa, Near/Middle East	3
Total	100

[a]NAFTA = North American Free Trade Agreement Countries.
(From Novartis Crop Protection AG.)

seed suitable for use with precision sowing machinery can now be formulated containing a combination of germination enhancers, trace nutrients, and crop protection compounds, however this sort of treatment is currently economically viable in only limited situations. An example is the use of hybrid seed in floriculture, where the untreated seed is expensive and germination can be unreliable, however the potential economic return from the crop is very high. In such a situation, obtaining the maximum return from every seed planted justifies the treatment cost.

Seed treatments are preemptive strategies for crop protection. If the target pest or pathogen does not occur at economically significant levels after planting the treatment has effectively been wasted. Unwarranted seed treatments result in economic losses to growers, the unnecessary introduction of pesticides into the environment, and may contribute to the development of pesticide resistance. In the case of seed-borne pathogens, the likelihood of disease outbreaks can largely be predicted by testing the seed before sowing (5). Maintaining records of pest and disease incidence on individual farms can help determine whether the frequency of outbreaks justifies the routine use of seed treatments for their control.

Seed treatments are most effective when employed against pests and diseases associated with early crop establishment. Although the development of systemic fungicides and insecticides has extended the period of crop protection that can be achieved by seed treatment, this ongoing activity comes at a cost—the longer the period of systemic activity, the greater the risk that pesticide residues may be detectable in the crop at harvest (2).

CURRENT USAGE PATTERNS AND FUTURE OPPORTUNITIES

Seed treatments constitute approximately 3% of the world market for crop protection products (*Source*: T.J. Martin, personal communication). Fungicides dominate the seed treatment market, accounting for an estimated 60% of total sales (Table 1). Cereal crops account for 63% of fungicidal treatments and 22% of insecticidal treatments (10). More than one-third of insecticidal seed treatments worldwide are applied to maize alone (Table 2). Geographically, seed treatment usage is concentrated in Western Europe and the NAFTA countries, which together account for an estimated 63% of the world market (Table 3). In other areas, such as the Asia/Pacific region, seed treatments have been substantially underutilized.

The development of biological seed treatments offers important opportunities for nonchemical pest and disease management, particularly in environmentally sensitive areas. New application technologies have increased the uniformity of pesticide loading on seeds, thereby improving treatment efficacy, while the development of new systemic fungicides and insecticides has made seed treatment a viable option for the control of foliar pests and diseases during the early stages of crop growth. The potential benefits of these advances have not yet been fully realized in many crops. Significant opportunities exist to expand the use of seed treatments to cover more pest management situations.

REFERENCES

1. Desai, B.B.; Kotecha, P.M.; Salunkhe, D.K. *Seeds Handbook. Biology, Production, Processing and Storage*; Ch. 17, Marcel Dekker, Inc.: New York, 1997; 503–519.
2. Schwinn, F.J. Seed Treatment—A Panacea for Plant Protection. In *Seed Treatment: Progress and Prospects*; Martin, T.J., Ed.; British Crop Protection Council: Farnham, UK, 1994; 3–14, BCPC Monograph No. 57.
3. Taylor, A.G.; Harman, G.E. Concepts and technologies of selected seed treatments. Ann. Rev. Phytopathol. **1990**, *28*, 321–339.
4. Hewett, P.D.; Griffiths, D.C. Biology of Seed Treatment. In *Seed Treatment,* 2nd Ed.; Jeffs, K.A., Ed.; BCPC Publications: Thornton Heath: UK, 1986; 7–15.
5. Maude, R.B. *Seedborne Diseases and Their Control. Principles and Practice*; Ch. 7, CAB International: Wallingford, UK, 1980; 114–178.
6. Jeffs, K.A. A Brief History of Seed Treatment. In *Seed Treatment,* 2nd Ed.; Jeffs, K.A. Ed.; BCPC Publications, Thornton Heath: UK, 1986; 1–5.
7. Stevens, M.M.; Fox, K.M.; Coombes, N.E.; Lewin, L.A. Effect of fipronil seed treatments on the germination and early growth of rice. J. Pestic. Sci. **1999**, *55* (5), 517–523.
8. Ito, A.; Saishoji, T.; Kumazawa, S. Synthesis of stereoisomers of ipconazole and their fungicidal and plant growth inhibitory activities. J. Pestic. Sci. **1997**, *22* (2), 119–125.
9. Harman, G.E. Seed treatments for biological control of plant disease. Crop. Prot. **1991**, *10* (3), 166–171.
10. Agrow Reports. *Seed Treatments: Products and Markets*; PJB Publications, Richmond: UK, 1999.

SEEDING RATE

Randy Anderson
USDA-ARS, Brookings, South Dakota, U.S.A.

S

INTRODUCTION

Producers are seeking alternative pest management strategies to supplement pesticides in order to increase profitability, manage pest resistance, and protect the environment. One strategy is including cultural practices in the production system that favors the crop over pests. An example is increasing the crop's seeding rate to improve weed control; weed growth can be reduced 5–15% with this strategy. Furthermore, producers can enhance the effect of higher seeding rates on weed growth by combining this practice with other cultural practices, such as nitrogen fertilizer placement or planting the crop in narrower rows. Combining several cultural practices into a cultural system increases the crop's competitiveness with weeds as well as improves its tolerance to weed interference, thus reducing crop yield loss.

ADVANTAGES FOR PEST MANAGEMENT

Producers can improve the competitive advantage of crops over pests with cultural strategies such as increasing a crop's seeding rate (1). For example, weed biomass can be reduced 5–15% in small grain crops such as winter wheat (*Triticum aestivum* L.) and row crops such as corn (*Zea mays* L.) or sunflower (*Helianthus annuus* L.). Producers also can favor crops over insects with this strategy; increasing seeding rate of canola (*Brassica* spp.) reduces the impact of flea beetles (*Phyllotreta* spp.) and root maggot (*Delia* spp.) feeding injury (2, 3).

Producers can further enhance the impact of increased seeding rate by combining it with other cultural practices; this approach leads to synergism among cultural practices, especially in suppressing weeds. For example, jointed goatgrass (*Aegilops cylindrica* Host.) and feral rye (*Secale* spp.) are common weeds infesting winter wheat. Because herbicides are not available to control these weeds in winter wheat, producers rely on cultural practices to reduce their seed production, with the goal of reducing future infestations (4). Producers in semiarid regions generally plant semidwarf cultivars of winter wheat at a seeding rate of 40–45 kg/ha with N fertilizer applied broadcast before planting. Increasing the seeding rate to 73 kg/ha improved competitiveness of winter wheat such that seed production of jointed goatgrass and feral rye was reduced 5–8% (Fig. 1). However, planting a taller cultivar at 73 kg/ha and banding N fertilizer by the crop row reduced seed production of these weeds 43–45%. A cultural system composed of increased seeding rate plus two other cultural practices increased the crop's competitiveness sixfold compared to a single cultural practice.

Increased seeding rate combined with other cultural practices also improves a crop's tolerance to weeds. This benefit was demonstrated with corn (*Zea mays* L.), where yield loss due to summer annual grass (*Setaria* spp.) interference was 33% when corn was planted at 37,060 plants/ha in rows 76 cm wide with N fertilizer applied broadcast (conventional system in Fig. 2). If seeding rate was increased from 37,060 to 47,000 plants/ha, yield loss declined to 29%, whereas combining the higher seeding rate with N fertilizer applied in a band next to the crop row reduced yield loss to 23%. However, when increased seeding rate and N banding were combined with a row spacing of 38 cm, yield loss was only 8%; a fourfold difference in yield loss compared to the impact of a single cultural practice. These results show that seeding rate can play an integral part in a cultural systems approach to strengthen the natural competitiveness of the crop and minimize weed impact on yield.

Ancillary Benefits of High Seeding Rates

Producers of small grain crops, such as wheat, accrue additional benefits by increasing seeding rates. Wheat planted at higher rates produces more crop residue, which reduces weed seedling emergence in following crops. For example, weed seedling densities are 20–30% less in corn

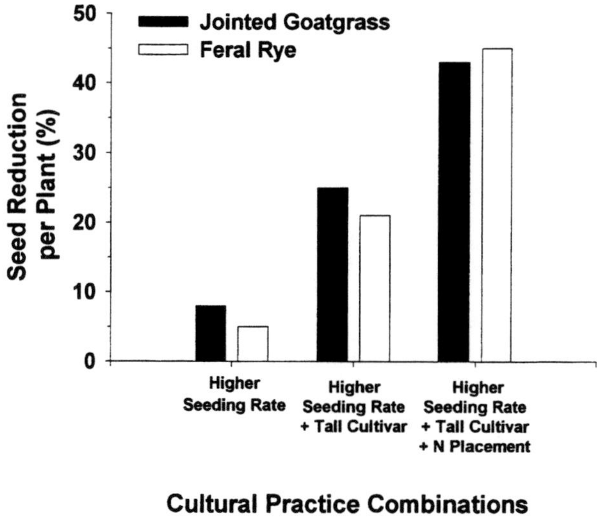

Fig. 1 Effect of combining cultural practices on seed production of jointed goatgrass and feral rye growing in winter wheat, in a semiarid climate at Akron, Colorado, USA. (Adapted from Ref. 4.)

or sunflower if crop residue quantities from the preceding wheat crop are increased 30%. A further benefit of higher seeding rates is that more crop residue after harvest improves soil protection from erosion.

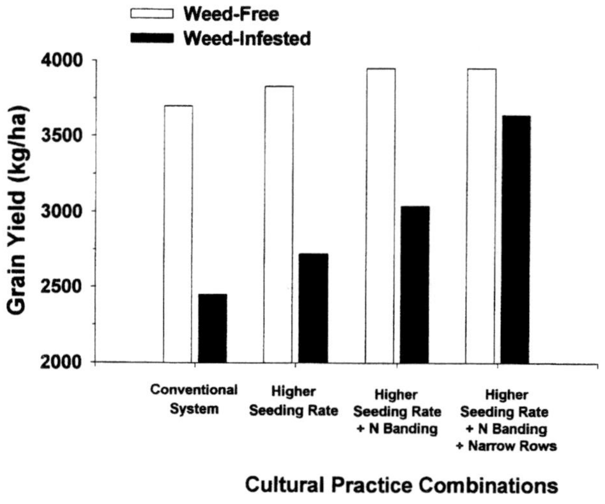

Fig. 2 Grain yield of corn in weed-free and weed-infested conditions as affected by cultural practice combinations, in a semiarid climate at Akron, Colorado, USA.—Conventional system was 37,060 plants/ha planted in 76-cm wide rows, with N fertilizer broadcast at planting. (Adapted from Anderson, R.L. Improving Weed Control in Corn and Sunflowers with Narrow Rows. In Proceedings, 11th Annual Meeting; Colorado Conservation Tillage Association: Akron, CO, 1999; 88–97.)

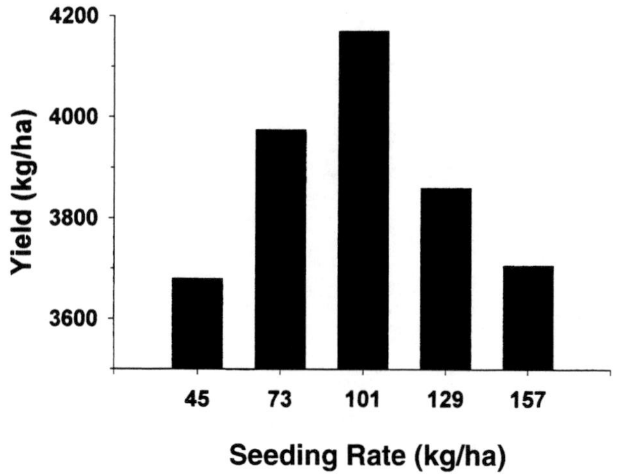

Fig. 3 Yield of winter wheat at various seeding rates, in a semiarid climate at Akron, Colorado, USA. (Adapted from research in progress, USDA-ARS Central Great Plains Research Station, Akron, CO.)

DISADVANTAGES OF HIGH SEEDING RATES

All crops have an optimum seeding rate range. If this range is exceeded, crop yield can be reduced because the crop begins competing with itself. For example, winter wheat grown in a semiarid climate yielded the most when planted at 101 kg/ha (Fig. 3). When seeded at higher rates such as 129 or 157 kg/ha, yield declined 5–10%. Additionally, input costs increase when more crop seed is planted. Therefore, if producers do not accrue a weed management or crop yield benefit, net economic return for the crop will be less. Another possible disadvantage is that higher seeding rates lead to denser crop canopies, which, in some weather conditions, can result in more crop diseases. Also, in more humid regions, excess crop residue can be detrimental to following crops because soil temperatures remain cooler during the early growing season and delay early crop growth.

FUTURE IMPACTS

An alarming trend with current production systems is the development of weed resistance to herbicides. To counter this trend, producers will need integrated weed management systems, where multiple control tactics are used for weed management in contrast to present systems that rely heavily on herbicides. Increased seeding rate, used in conjunction with other cultural practices, can be a key component in integrated weed management systems. This approach may reduce the need for herbicides and also delay development of weed resistance. Furthermore, as planting equipment improves in its capability to arrange plants in more equally spaced geometry, seeding rates will be critical to optimize crop planting patterns.

REFERENCES

1. Holtzer, T.O.; Anderson, R.L.; McMullen, M.P.; Peairs, F.B. Integrated pest management of insects, plant pathogens, and weeds in dryland cropping systems of the great plains. J. Prod. Agric. **1996**, *9*, 200–208.
2. Dosdall, L.M.; Herbut, M.J.; Cowle, N.T.; Micklich, T.M. The effect of seeding date and plant density on infestations of root maggots, *Delia* spp. (Diptera: Anthomyiidae) in canola. Can. J. Plant Sci. **1996**, *76*, 169–177.
3. Dosdall, L.M.; Dolinski, M.G.; Cowle, N.T.; Conway, P.M. The effect of tillage regime, row spacing, and seeding rate on feeding damage by flea beetles, *Phyllotreta* spp. (Coleoptera: Chrysomelidae), in canola in Central Alberta, Canada. Crop Protection **1999**, *18*, 217–224.
4. Anderson, R.L. Cultural systems can reduce reproductive potential of winter annual grasses. Weed Technol. **1997**, *11*, 608–613.

BIBLIOGRAPHY

Barton, D.L.; Thill, D.C.; Shafii, B. Integrated wild oat (*Avena fatua*) management affects spring barley (*Hordeum vulgare*) yield and economics. Weed Technol. **1992**, *6*, 129–135.

Fischer, R.A.; Miles, R.E. The role of spatial pattern in the competition between crop plants and weeds: a theoretical analysis. Math. Biosci. **1973**, *18*, 335–350.

Lewis, W.J.; van Lenteren, J.C.; Phatak, S.C.; Tumlinson, J.H. A total system approach to sustainable pest management. Proc. Natl. Acad. Sci. **1997**, *94*, 12243–12248.

Regnier, E.E.; Bakelana, K.B. Crop planting pattern effects on early growth and canopy shape of cultivated and wild oats (*Avena fatua*). Weed Sci. **1995**, *43*, 88–94.

Swanton, C.J.; Murphy, S.D. Weed science beyond the weeds: the role of integrated weed management (IWM) in agroecosystem health. Weed Sci. **1996**, *44*, 437–445.

SHEEP PEST MANAGEMENT (ARTHROPODS)

John E. Lloyd
University of Wyoming, Laramie, Wyoming, U.S.A.

INTRODUCTION

Several arthropod pests affect the health, value, and productivity of sheep. In the United States, where sheep numbers have declined for 60 years from more than 50×10^6 to fewer than 8×10^6, there has been relatively little research on the control or the economics of control of arthropod pests of sheep. Although certain management practices can suppress pest arthropod levels, effective, noninsecticide control measures, for the most part, are lacking. Often sheep are raised under conditions of minimum management, and arthropod pests, which are not usually life-threatening, are to be endured. Small farm flocks are not usually a major source of income and are often untreated. Larger sheep flocks numbering in the thousands, however, usually are treated to control these pests. In other parts of the world where sheep are a much more important commodity, research on management of arthropod pests is ongoing.

SHEEP KED

The sheep ked, *Melophagus ovinus* (L.), is a permanent blood-feeding ectoparasite of sheep. It is worldwide in distribution. Often mistaken for a tick, sheep ked is a wingless fly that lives in the fleece of the host near the skin surface. Transmission of sheep ked is normally by direct contact between animals.

In response to irritation caused by keds, sheep may roll on the ground and become stranded on their backs. Death loss of sheep that die in this position is termed ''back loss'' by sheep producers. Sheep keds negatively affect weight gains of lambs. Studies have shown, depending on diet, a 1.4–3.6 kg difference in live weight between ked-infested and uninfested feeder lambs (1) and a 1.1 kg per lamb carcass difference at slaughter (2). Ked-free ewes may produce 11% more wool (1) and lambs may produce an additional 0.7 kg of wool (2) per animal. Sheep keds also negatively affect the quality grade of the wool and cause a blemish defect of the pelt known as ''cockle''(3).

If keds are present on inspection following spring shearing then all animals in the flock should be treated topically with an approved insecticide. The shearing process results in a 70% reduction of sheep ked populations. All animals in the flock should be treated and subsequent exposure to untreated animals should be avoided. New additions to the flock should be treated and held in isolation prior to introduction into a flock.

Topical applications of liquid insecticide formulations may be applied in dilute form as a whole body spray, dip or plunge vat treatment, or a shower. In more concentrated form, certain insecticides may be applied as a low-volume spray or pour-on. In these concentrated forms the insecticide migrates through the fleece from the site of application.

SHEEP LICE

Sheep lice, like the sheep ked, are permanent ectoparasites. They are much smaller, however, and may be difficult to see. Sheep may be infested with both chewing (biting) and sucking lice. The sheep biting louse, *Bovicola ovis* (Schrank), has mandibulate mouthparts and feeds on the skin surface, sloughed epidermal tissue, and sebaceous secretions. The sucking body louse, *Linognathus ovillus* (Neumann), the African blue louse, *Linognathus africanus* (Kellogg and Paine), and the sucking foot louse, *Linognathus pedalis* (Osborne), have sucking mouthparts and are blood feeders.

Lice eggs are cemented individually to the wool fibers or hairs close to the skin. Nymphal lice and adults live permanently within the fleece and transfer primarily by direct contact with other sheep.

Lice are irritating, causing considerable rubbing and scratching, which results in abrasions of the skin and tearing and pulling of wool. In severe cases sucking lice may cause anemia. In an Australian study (4) clean wool production of Merino sheep was reduced by 0.3–0.8 kg per sheep by the sheep biting louse, and infested sheep yielded a lower-quality fleece. Live weight of sheep was not affected, however. In a New Zealand study (5) there were no significant differences in host body weight or clean or greasy wool production of Romney-cross sheep lightly infested with sheep biting louse.

The producer must be careful not to introduce lice-infested sheep into a flock. If animals become infested, a number of insecticide treatments are available. Most of the topical treatments for sheep ked control may also be used for lice control. Two treatments 2–3 weeks apart may be necessary because the egg stage may be resistant to insecticides. Every animal in a flock must be treated, and they should be treated at approximately the same time. Outside the United States, macrocyclic lactone-based insecticides and bacterial insecticides are under study for lice control.

SHEEP BOT FLY

The sheep bot fly, *Oestrus ovis*, occurs worldwide. In hot, dry climates it may be active year round. In temperate regions there may be one or two generations per year. The adult sheep bot fly has a bee-like appearance, but does not sting or bite. The female deposits tiny larvae directly in the nostrils of the host. Sheep react by running or walking with their noses close to the ground, or by huddling in groups. The larval sheep bot fly lives on mucous surfaces of the nasal passages and sinuses of sheep and goats. The full grown maggot eventually leaves the sinuses and is discharged by sneezing or coughing. An external sign of infestation is the appearance of a runny nose, which increases in severity as the infestation progresses. Control of *O. ovis* infestations from sheep reduces nasal discharge (6) and may increase live weight gains by 4% (7).

Sheep bot fly can be controlled by treatment of the infected host with a systemic parasiticide. Sheep bot may be effectively controlled if all sheep within an area are treated annually (8).

WOOL MAGGOTS OR FLEECE WORMS

The larvae of several fly species, though normally carrion feeders, are able to adapt to a parasitic existence by infesting the wounds of living animals. Larvae of certain of the blow flies may also infest the fleece of living sheep and feed on wool yolk, skin exudate, and the skin itself. These are the wool maggots or fleece worms. In North America, depending on region and time of year, four species commonly infest sheep: northern black blow fly, *Protophormia terranovae* (Robineau-Desvoidy); black blow fly, *Phormia regina* (Meigen); green bottle fly, *Phaenicia sericata* (Meigen); secondary screwworm fly, *Cochliomyia macellaria* (Fabricius).

Establishment of blow fly larvae on the host is referred to as ''strike.'' An existing infestation attracts additional gravid females, and animals may become heavily infested. Larvae of several fly species may be present at the same time. The maggots are irritating and the sheep become restless. In Australia, breech strike by the Australian sheep blow fly, *Phaenicia cuprina* (Wiedemann), reduced body weights by 2.4 kg and fleece weights by 0.15 kg (9). In extreme cases, the skin may be destroyed with a considerable loss of blood and serum. Bacterial infection and toxemia may occur with death in a few days.

Odors and dampness in the wool, attractive to the ovipositing fly, are enhanced by urine, feces, sweat, or other body fluids. Sheep with a wrinkled skin and fine wool tend to be more susceptible. The ''Mules'' operation, developed for Merino-fine-wooled sheep in Australia, is a surgical operation to remove wooled skin from each buttock. Other procedures to reduce contaminated fleece include crutching (clipping the breech area) and tail docking. Prompt burial of dead animals may help reduce fly abundance. Infested flocks may be clipped and/or treaed topically with an insecticide. Individual animals may be spot-treated with sprays, salves, and smears. Outside the United States insect growth regulators are used extensively as topical treatments and vaccines are under consideration.

REFERENCES

1. Nelson, W.A.; Slen, S.B. Weight gains and wool growth in sheep infested with the sheep ked *Melophagus ovinus*. Expt. Parasitol. **1968**, *22*, 223–226.
2. Everett, A.L.; Roberts, I.H.; Naghski, J. Reduction in leather value and yields of meat and wool from sheep infested with keds. J. Am. Leather Chemists Assoc. **1971**, *66*, 118–130.
3. Everett, A.L.; Roberts, I.H.; Willard, H.J.; Apodaca, S.A.; Bitcover, E.H.; Naghski, J. The cause of cockle, a seasonal sheepskin defect, identified by infesting a test flock with keds (*Melophagus ovinus*). J. Am. Leather Chemists Assoc. **1969**, *64*, 460–476.

4. Wilkinson, F.C.; De Chaneet, G.C.; Beetson, B.R. Growth of populations of lice, *Damalinia ovis*, on sheep and their effects on production and processing performance of wool. Vet Parasitol. **1982**, *9*, 243–252.
5. Kettle, P.R.; Lukies, J.M. Long-term effects of sheep body lice (*Damalinia ovis*) on body weight and wool production. N. Z. J. Agr. Res. **1982**, *25*, 531–534.
6. Horak, G.; Snijders, A.J. The effect of *Oestrus ovis* infestation on merino lambs. Vet Record **1974**, *94*, 12–16.
7. Drummond, R.O.; Lambert, G.; Smalley, H.E., Jr.; Terrill, C. E. Estimated Losses of Livestock to Pests. In *Handbook of Pest Management in Agriculture*; Pimentel, D., Ed.; CRC Press: Boca Raton, FL, 1981; 1, 111–127.
8. Meleney, W.P.; Cobbett, N.G.; Peterson, H.O. Control of *Oestrus ovis* in sheep on an isolated range. J. Amer. Vet. Med. Assoc. **1963**, *143*, 986–989.
9. Gill, D.A.; Graham, N.P.H. Studies on fly-strike in merino sheep. No. 3—The influence of fly-strike and conformation on body-weight and fleece-weight of merino sheep at "Dungalear," New South Wales. J. Counc. Sci. Ind. Res. **1939**, *12* (4), 319–329.

SHIFTING CULTIVATION

S

Joseph D. Cornell
Charles A.S. Hall
State University of New York, Syracuse, New York, U.S.A.

INTRODUCTION

Shifting cultivation (also known as swidden or slash-and-burn agriculture) is a set of agricultural activities that integrates food production into more natural ecosystems by alternatively cultivating and then abandoning fields (1). Fields are abandoned after only a short period of cultivation in response to the loss of productivity, due in large part to the spread of pest and weed species (2). Shifting cultivation has been practiced by humans for millennia (3), although today it is practiced primarily by people living in tropical forest environments (4). Shifting cultivation is currently practiced by 200–500 million people worldwide (5) and is the most extensively practiced type of agriculture throughout the humid tropics (6). Shifting cultivation has played a pivotal role in the development of human agricultural systems serving as a probable link between early human societies based on hunting/gathering activities and later societies based on more intensive forms of agriculture (7).

THE PROCESS

Shifting cultivation has changed little over time and usually follows the same sequence of events (8). First, near the beginning of the driest part of the year, people cut down the trees and other vegetation in what will become a new field. The cut vegetation, or "slash" is left on the site and allowed to dry. Some valuable timber species such as cedar and mahogany, as well as useful tree species that provide shade, fruit, fiber, or forage, may be left standing (9). In addition, some trees may not be cut down entirely, but trimmed so that they will not grow back until about the time when the field will be abandoned (10, 11).

Near the end of the dry season the dried vegetation is burned, returning the nutrients in the vegetation to the soil in the form of ash (12, 13). Around the beginning of the rainy or wet season farmers plant a suite of various crops directly into the ash-covered soil. The main food crops used throughout the tropics include maize (*Zea mays*), rice (*Oryza sativa*), beans (*Phaeseola* spp.), sweet potatoes (*Ipomea batatas*), yams (*Dioscorea* spp.), squash (*Cucurbita* spp.), taro (*Colocasia esculenta*), and manioc (*Manhiot esculenta*), as well as tree crops such as plantains and bananas (*Musa* spp.), mango (*Mangifera* spp.), avocado (*Persea* spp.), pandanus nut (*Pandanus* spp.), breadfruit (*Artocarpus altilis*), and peach palm (*Bactris gasipaes*).

The choice of crops that are planted depends upon cultural, local, and even personal preferences. In the Amazon region, where shifting cultivation is still practiced by several dozen different ethnic groups, the main staple crops are manioc, plantains, bananas, and, to a lesser extent, maize (4, 14). In Papua New Guinea, manioc, taro, sweet potatoes, and yams are the main crops planted among groves of pandanus trees (15, 16). In Central America, farmers plant maize intercropped with beans and squash (10). In the Philippines, Conklin (17) reported that the Hanunóo planted fields primarily devoted to root crops (manioc) as well as fields devoted to grain (maize and upland rice).

Crops may be planted as monocultures, as mixed crops (polycultures), or in other arrangements such as concentric rings of one crop surrounding another (4). Regardless of how the main crops are arranged, shifting cultivators almost invariably cultivate dozens, if not hundreds of additional plants for food, fiber, medicine, spices, dyes, and other materials. For example, Lacandon Maya farmers in Mexico may cultivate up to 75 plant species within a single hectare (18).

For the first few years, the farmer will weed each plot one or more times during the growing season. But, after time, productivity declines as soil nutrients are depleted and as pest and weed species proliferate (2). As productivity declines, each farmer shifts the production of annual crops to newer fields. In the absence of weeding, more natural vegetation is allowed to take over the plot, which quickly becomes what is called "secondary" forest, as distinguished from undisturbed or "primary" forest.

The seeds for the new forest vegetation come from nearby secondary and primary forest and from seeds remaining in the soil that have survived burning and cultivation (13). Regrowth begins quickly, often before a field is actually abandoned (2). While the forest vegetation regrows, the field is said to be ''fallow,'' and may remain so for as little as 4 years, or as long as 20 years or more (19).

Invariably, forest fallows are managed to provide a variety of products including food, fiber, medicinal plants, and game (9, 20). This extends the productive life of fields and significantly increases the economic return (21). But, because of the disturbance caused by cultivation and the selective favoring of useful species, secondary forest fallows often contain higher densities of these selected species than surrounding primary forest (22). As a result, the structure and species composition of even very old secondary forest fallows (200–400 years old) can be noticeably different than that of true primary forest (23, 24).

Nutrients lost during cultivation are slowly regained, primarily by the biomass growing on the previously cultivated area, and can approach predisturbance levels (25), but only after sufficiently long fallow periods; usually 10–20 years (26). Saldarriaga and Uhl (27) estimated that recovery of forest biomass may take as long as 140–200 years for some Amazon floodplain ecosystems, but they noted that this was 5–7 times longer than for most other areas in South America. The recovery of nutrients occurs more quickly however because of the higher concentration of mineral nutrients in rapidly growing leaves and stems as compared to the concentration of nutrients in the trunks of trees, which grow much more slowly, but will eventually account for the bulk of biomass (25). Once the forest vegetation has regrown and nutrient levels recovered sufficiently, the forest may be cut again and the cycle of cultivation begun once more. Given long enough fallow periods, forest fallows may be used again and again over long periods of time with little loss of productivity (28).

KEYS TO SUCCESS

One key to the success of shifting cultivation is the ways in which it mimics the structure and dynamics of the surrounding forest (1, 9). By planting a variety of crops, through intercropping, and by retaining or planting useful trees, farmers unintentionally recreate some of the diversity and structure of the natural forest (29), making shifting cultivation a form of agroforestry (30, 31).

Unlike other forms of agroforestry or agriculture however, shifting cultivation fields tend to be small, irregularly shaped, and dispersed within a mosaic of primary and secondary forests; all of which aids in the regeneration of forest vegetation (3). In addition, fields are cultivated for only a short period of time before being allowed to revert back to forest. The distribution and timing of shifting cultivation therefore is similar to the natural patch-phase dynamics of natural ecosystems (32) making it especially suited to the tropics (28).

Another key to success is the agricultural and ecological information shifting cultivators have gained through many years of experience (33). Shifting cultivators often have detailed knowledge of local vegetation and soils, and their management (34, 35). For example, by carefully timing the preparation of new fields and planting, farmers reduce the amount of time that soils are exposed to rain during the wet season, thereby reducing soil erosion (17, 36). Shifting cultivators also reduce erosion by ''carpeting'' their fields with crops, by leaving weed debris as mulch, and otherwise by reducing the exposure of bare soil to sun and rain (20). But, perhaps the most striking way in which shifting cultivators employ their knowledge is in their attempts to control the weeds and pests that affect productivity.

By carefully burning new fields farmers temporarily eliminate or reduce pests and weeds without unduly harming soil nutrients, structure, or mycorhhizae (13). Burning is particularly important for controlling weed species, which would otherwise compete with crops for space and nutrients. Although farmers will weed new fields once or twice a year, hand weeding becomes increasingly difficult due to the speed with which weeds grow, and also due to the fact that many weed species are armed with thorns, stinging hairs, and caustic saps. As much as half of the labor used in tropical agriculture is spent in weeding (6). Even so, eventually weeds will choke out most annual crops, and may represent the most important reason for the loss of productivity (37). The only way to bring the weeds under control again, short of using commercial herbicides, is to wait until the forest vegetation has regrown so that the field can be burned once more.

Burning also eliminates pest populations and reservoirs of potential plant pathogens. The initial removal of pests from a new field is particularly important because, while fields may be weeded periodically, outbreaks of pests, particularly insect pests, are much harder to control. Instead, more sophisticated shifting cultivators use their detailed knowledge of ecological relationships to control pest populations. One way farmers protect their crops from pests is to take advantage of natural competitors and predators. For example, the Kayapo of Brazil bring the nests of *Azteca* ants into their gardens in order to deter leaf cutter ants and termites (38, 39). Farmers in the Amazon leave dead trees standing in their fields to attract hunting birds that prey on the rodents and birds that damage crops (40).

Another way farmers throughout the tropics protect their crops is by taking advantage of the natural protection afforded by secondary plant compounds. Many cultivated plants, such as black pepper (*Piper nigrum*), and passion fruit (*Passiflora edulis*) contain high levels of secondary compounds that provide natural protection from pests. When planted in polycultures, or when shredded and spread as mulch, these plants can provide protection for other crops (41). Likewise, when planted in concentric circles, some crops may serve as a barrier to pests and offer some protection for more vulnerable crops in the center (4). Shifting cultivators also extract mild pesticides from locally available plants and plant ashes (9, 29).

Shifting cultivators also control insect pests, perhaps unintentionally, by planting a wide variety of crops (42). Shifting cultivators typically plant dozens of different crops, and different varieties of each crop, to increase the variety in their diet and to obtain necessities such as medicines, dyes, fibers, spices, and building materials. Planting a variety of crops may also help to reduce the risks of crop failure by reducing the incidence and severity of insect pest outbreaks. Root (43) suggested two theories to account for this effect. One theory suggests that crop diversity encourages the presence of more competitors, predators, and parasites, which help to control pest species. Another theory suggests that crop diversity may interfere with the growth and reproductive success of specialized insect pests by denying them an abundant and concentrated source of food. Both theories however are based upon the high spatial diversity of polycultures, which recreates to some extent the ''patchiness'' of the natural tropical environment where diversity is often very high, but the frequency of any given species is low. Likewise, the use of small fields dispersed throughout the forest also increases the ''patchiness'' of shifting cultivation and may help to reduce the threat of insect pest outbreaks as well (29, 44). In turn, reducing the incidence of insect pests probably also helps reduce the spread of crop diseases, which are often spread by insect vectors (45).

SUSTAINABILITY

When practiced by indigenous peoples over long periods of time, shifting cultivation can be a highly productive and ecologically sensitive form of agriculture (3). As long as sufficient amounts of land exist, and as long as subsistence needs are met, shifting cultivators can produce food in the same area for long periods of time with little or no losses in productivity (28, 33). Increasingly however, an abbreviated form of shifting cultivation is being blamed for causing deforestation throughout the tropics (3, 5). Immigrants and settlers are using ''slash-and-burn'' techniques to clear tropical forests. Often, settlers and migrants can claim only land that they clear of forest; an incentive which can result in high rates of deforestation. This is not true of shifting cultivation, however, as these people rarely complete even one cycle of cultivation and fallow. Instead, the people often sell the land they clear, which often becomes pasture for cattle (25). The farmers then move on to clear new lands deeper in the forest (46).

This new pattern of land use is reinforced by economic incentives and politics. Whereas most traditional people rely on shifting cultivation for subsistence needs, settlers and new migrants into tropical areas often are driven by the need to generate income and by the desire for profit. This reliance on economic returns is a double-edged sword. If clearing and cultivating their land is profitable, settlers may be encouraged to sell their land and repeat their success elsewhere. If, however, they cannot support themselves on their land, new settlers may be forced to sell the land they have cleared to pay their debts and to move on to clear new lands. The net result is increased land clearing throughout large regions of the tropics, often encouraged and subsidized by governments eager to alleviate the effects of poverty, landlessness, and population growth (3).

Even traditional shifting cultivation, however, can become unsustainable under economic and population pressures (32). As a general rule, shifting cultivation requires at least 10 to 20 times as much land to be held in forest fallows as is currently under cultivation for it to be sustainable. This large amount of fallow, relative to cultivated land, is necessary to ensure that the regenerative capacity of the forest is not exceeded. In practice an even greater total amount of land is required to be left in primary and secondary forest in order to protect the integrity of ecological services, to protect watersheds, to prevent soil erosion, and to serve as a source of plant germplasm and game animals vital to shifting cultivators. Under increased pressure from population growth, migration, or a shift in economics, shifting cultivators are forced to increase the total area in cultivation, to cultivate marginal lands, to shorten fallow cycles, to decrease the total area in fallow (32), and to otherwise exceed the natural regenerative capacity of the forest (47). The result is the loss of productivity, an increase in the rate of deforestation, the loss of local biodiversity, and the loss of sustainability (8, 33).

REFERENCES

1. Geertz, C. *Agricultural Involution: The Processes of Ecological Change in Indonesia*; Association of Asian

Studies Monographs and Papers; University of California Press: Berkeley, CA, 1970; 11, 176.
2. Janzen, D.H. Tropical agroecosystems. Science **1973**, *182*, 1212–1219.
3. Sponsel, L.E.; Bailey, R.C.; Headland, T.H. *Tropical Deforestation: The Human Dimension*; Columbia University Press: New York, 1996; 365.
4. Beckerman, S. Swidden in Amazonia and the Amazon Rim. In *Comparative Farming Systems*; Turner, B.L., Brush, S.B., Eds.; Guilford Press: New York, 1987; 428.
5. Myers, N. *The Primary Source: Tropical Forests and Our Future*; W.W. Norton: New York, 1992.
6. Tivy, J. *Agricultural Ecology*; Longman Scientific & Technical/John Wiley & Sons: New York, 1990; 288.
7. Boserup, E. *Population and Technology*; Basil Blackwell: Oxford, UK, 1981.
8. Jacobs, M. *The Tropical Rain Forest: A First Encounter*; Springer-Verlag: New York, 1981; 295.
9. Denevan, W.M.; Treacy, J.M.; Alcorn, J.B. Indigenous Agroforestry in the Peruvian Amazon: The Example of the Bora Utilization of Swidden Fallows. In *Change in the Amazon Basin*; Hemming, J., Ed.; University of Manchester Press: Manchester, UK, 1984.
10. Ewell, P.T.; Merril-Sands, D. Milpa in Yucatan: A Long-fallow Maize System and Its Alternatives in the Mayan Peasant Economy. In *Comparative Farming Systems*; Turner, B.L., Brush, S.B., Eds.; Guilford Press: New York, 1987; 428.
11. Gomez-Pompa, A. On Maya silviculture. Mexican Studies **1987**, *3* (1), 1–17.
12. Nye, P.H.; Greenland, D.J. Changes in the soil after clearing tropical forest. Plant and Soil **1964**, *21* (1), 101–112.
13. Ewel, J.; Berish, C.; Brown, B.; Price, N.; Raich, J. Slash and burn impacts on a Costa Rican wet forest site. Ecology **1981**, *62* (3), 816–829.
14. Hecht, S.B.; Posey, D.A. Preliminary results on soil management techniques of the Kayapo Indians. Adv. in Econ. Bot. **1989**, *7*, 174–178.
15. Dwyer, P.D. *The Pigs That Ate the Garden: A Human Ecology from Papua New Guinea*; University of Michigan Press: Ann Arbor, MI, 1990; 241.
16. Cook, C.D. The Divided Island of New Guinea: People, Development and Deforestation. In *Tropical Deforestation: The Human Dimension*; Sponsel, L.E., Bailey, R.C., Headland, T.H., Eds.; Columbia University Press: New York, 1996; 365.
17. Conklin, H.C. Hanunóo Agriculture in the Phillipines. In *FAO Forestry Development Paper No. 12*; FAO: Rome, 1957; 209.
18. Nations, J.D.; Komer, D.I. Central America's tropical rainforests: positive steps for survival. Ambio. **1983**, *12* (5), 232–238.
19. Turner, B.L., II; Brush, S.B. *Comparative Farming Systems*; Guilford Press: New York, 1987; 428.
20. Alcorn, J. *Huastec Mayan Ethnobotany*; University of Texas Press: Austin, TX, 1984; 982.
21. Montagnini, F.; Mendelsohn, R.R.O. Managing forest fallows: improving the economics of swidden agriculture. Ambio. **1997**, *26* (2), 118–123.
22. Chazdon, R.L.; Coe, F.G. Ethnobotany of woody species in second-growth, old-growth, and selectively logged forests of Northeastern Costa Rica. Conserv. Biol. **1999**, *13*, 1312–1322.
23. Gomez-Pompa, A.; Flores, J.S.; Sosa, V. The ''Pet-Kot'': a man-made tropical forest of the Maya. Interciencia **1987**, *12* (1), 10–15.
24. Unruh, J.D. Canopy structure in natural and agroforest successions in Amazonia. Trop. Ecol. **1991**, *32* (2), 168–181.
25. Whitmore, T.C. *An Introduction to Tropical Rain Forests*; Clarendon Press: Oxford, 1990; 226.
26. Ramakrishnan, P.S.; Toky, O.P. Soil nutrient status of hill agro-ecosystems and recovery patterns after slash-and-burn agriculture (jhum) in North-Eastern India. Plant and Soil **1981**, *60*, 41–64.
27. Saldarriaga, J.G.; Uhl, C. Recovery of Forest Vegetation Following Slash-and-Burn Agriculture in the Upper Rio Negro. In *Rain Forest Regeneration and Management*; Gomez-Pompa, A., Whitmore, T.C., Hadley, M., Eds.; Man and the Biosphere Series; Parthenon Publishing: New York, 1991; 457.
28. National Research Council. *Sustainable Agriculture and the Environment in the Humid Tropics*; National Academy Press: Washington, DC, 1993; 702.
29. Clay, J.W. *Indigenous Peoples and Tropical Forests*; Cultural Survival, Inc.: Cambridge, MA, 1988; 116.
30. Denevan, W.M.; Padoch, C. Swidden-fallow agroforestry in the Peruvian Amazon. Adv. in Econ. Bot. **1987**, *5*, 1–8.
31. MacDicken, K.G. Agroforestry Management in the Humid Tropics. In *Agroforestry Classification and Management*; MacDicken, K.G., Vergara, N.T., Eds.; Wiley-Interscience: New York, 1990; 382.
32. Coomes, O.T.; Grimard, F.; Burt, G.J. Tropical forests and shifting cultivation: secondary forest fallow dynamics among traditional farmers of the Peruvian Amazon. Ecol. Econ. **2000**, *32* (1), 109–124.
33. Park, C. *Tropical Rainforests*; Routledge: London, 1992; 188.
34. Salick, J. Ecological basis of Amuesha agriculture, Peruvian upper Amazon. Adv. in Econ. Bot. **1989**, *7*, 189–207.
35. Gomez-Pompa, A. Learning from Traditional Ecological Knowledge: Insights from Mayan Silviculture. In *Rain Forest Regeneration and Management*; Gomez-Pompa, A., Whitmore, T.C., Hadley, M., Eds.; Man and the Biosphere Series; Parthenon Publishing: New York, 1991; 457.
36. Posey, D.A.; Frechione, J.; Eddins, J.; da Silva, L.F.;

Myers, D.; Case, D.; MacBeath, P. Ethnoecology as applied anthropology in Amazonian development. Human Org. **1984**, *43*, 95–107.

37. Sanchez, P.A.; Benites, J.R. Low-input cropping for acid soils of the humid tropics. Science **1987**, *238*, 1521–1527.
38. Anderson, A.B.; Posey, D.A. Management of a tropical scrub savanna by the Gorotire Kayapo of Brazil. Adv. in Econ. Bot. **1989**, *7*, 159–173.
39. Posey, D.A. Indigenous management of tropical forest ecosystems: The case of the Kayapo Indians of the Brazilian Amazon. Agroforestry Syst. **1985**, *3*, 139–158.
40. Frechione, J.; Posey, D.A.; da Silva, L.F. The perception of ecological zones and natural resources in the Brazilian Amazon: An ethnoecology of Lake Coari. Adv. in Econ. Bot. **1989**, *7*, 260–281.
41. Gradwohl, J.; Greenburg, R. *Saving the Tropical Forests*; Island Press: Washington, DC, 1988; 214.
42. Letourneau, D.K. Two Examples of Natural Enemy Augmentation: A Consequence of Crop Diversification. In *Agroecology. Ecological Studies 78*; Gliessman, S.R., Ed.; Springer-Verlag: New York, 1990; 380.
43. Root, R.B. Organization of plant-arthropod association in simple and diverse habitats: the fauna of collards (*Brassica oleraceae*). Ecol. Monogr. **1973**, *43*, 95–124.
44. Pimentel, D.; Levin, S.A.; Olsen, D. Coevolution and the stability of exploiter-victim. Am. Naturalist **1978**, *112* (983), 119–125.
45. Powers, A. Cropping Systems, Insect Movement, and the Spread of Insect-Transmitted Diseases in Crops. In *Agroecology. Ecological Studies 78*; Gliessman, S.R., Ed.; Springer-Verlag: New York, 1990; 380.
46. Rudel, T.; Horowitz, B. *Tropical Deforestation: Small Farmers and Land Clearing in the Ecuadorian Amazon. Methods and Cases in Conservation Science*; Columbia University Press: New York, 1993; 234.
47. Longman, K.A.; Jeník, J. *Tropical Forest and its Environment*, 2nd Ed.; Longman Scientific & Technical/ John Wiley & Sons: New York, 1987; 347.

S

SHORT SEASON CROPS

Chris Sansone
Texas Agricultural Extension Service, San Angelo, Texas, U.S.A.

INTRODUCTION

Integrated pest management (IPM) is a system that controls pests and contributes to long-term sustainability by combining judicious use of biological, cultural, physical, and chemical tools in a way that minimizes the risks of pesticides to human health and the environment (1). One of the cultural tools is the use of short season crops and short season production strategies. Short season production is a form of pseudoresistance, a term applied to apparent resistance that results from ephemeral characteristics in potentially susceptible host plants (2). Short season production includes both varietal selection and manipulation of inputs. Profitability from the use of short season crops is a result of normal yields with reduced inputs or if yield reductions occur losses are offset by the savings from reduced inputs. Although other crops have adopted short season strategies, cotton is the model system for this cultural management tool.

IPM programs should emphasize systems or strategies that result in managing pest populations. Although early IPM programs emphasized crop monitoring and pesticide timing, recent emphasis has been placed on designing fully integrated programs (3). One of the important components of an IPM program is the use of cultural tools. Pests can be managed by varying water, fertilizer, and other crop production inputs. Plant or varietal selection can also impact pest density. Host plant resistance has been developed for a wide variety of diseases and insects in a wide variety of crops. However, crops can be managed to escape damage from pests with characteristics best described as pseudoresistance. Three types of pseudoresistance can be distinguished: 1) Host Evasion. The plant passes through the most susceptible stage quickly or when pest numbers are reduced. 2) Induced Resistance. This is temporarily increased resistance resulting from some condition of the plant or environment. 3) Escape. This refers to a lack of infestation of or injury to the crop (2). Short season crops and production strategies take advantage of host evasion and escape to produce a crop while minimizing inputs. Host plants that allow the changing of planting dates, shortening of key development periods, or shortening of the crop season are examples of this strategy (4). The adoption of the strategy depends on producers making a normal yield with reduced inputs, or if yields are adversely affected the cost savings associated with the reduction of additional inputs are greater than the loss of income associated with the yield loss.

COTTON AS THE MODEL

The use of cotton as a textile is documented thousands of years before the birth of Christ (5). Early cotton production relied on varieties that were slow to fruit (180 days) and on hand harvesting that extended well into winter and early spring. Occasional pest outbreaks would occur, especially with bollworms (*Helicoverpa zea* [Boddie]) and tobacco budworm (*Heliothis virescens* [F.]), however insects were not an important economic consideration until 1892, when the boll weevil (*Anthonomus grandis* [Boheman]) entered Texas. The slow fruiting cottons and extended harvest exacerbated the boll weevil situation (6). Recommendations in 1899 included planting early maturing cottons and destroying all cotton stalks promptly after harvest (6). Cotton production survived by the gradual shifting of acreage to the west to escape the boll weevil and the adoption of faster fruiting cotton (130 days). During World War II, cotton production expanded with mechanization and the development of synthetic pesticides. Shortening the production season no longer seemed important, as producers were able to increase yields with added irrigations, nitrogen, and a shift to longer season varieties by controlling insects with the new insecticides. Producers were soon making multiple applications for the bollworm/tobacco budworm complex and resistance to these insecticides developed in the 1960s (7). Producers shifted to the organophosphates but these products had short residual action, required multiple applications, were less efficacious, and the tobacco budworm developed resistance to these chemicals. The short season production scheme developed soon after resistance to the organophosphates occurred. The cornerstone of the system was the development of fast fruiting varieties. These varieties completed the majority of their flowering by day 30 of bloom (8). These varieties are also ready for harvest at 130–140 days compared to 160–180 days for later maturing varieties (9). Success of the short season strategy for cotton grown outside the rolling plains of Texas also depends on early planting. Cotton planted 30–45 days after optimum planting dates will still be exposed to high populations of boll weevils and the bollworm/tobacco budworm complex (7). The final component of the short season strategy for cotton is the proper management of water and nitrogen inputs (10, 11). Although the system was originally developed to manage boll weevils and the bollworm/tobacco budworm complex, the system has also been used with great success to manage pink bollworm [*Pectinophora gossypiella* (Saunders)] (12). The success of the system can be measured by the reduction in insecticide use between 1964 and 1976. In 1964 more than 19 million pounds of insecticide were used in Texas. By 1976, insecticide use had declined to 2 million pounds. By comparison, states east of Texas treated the 1976 crop with more than 53 million pounds—on less acreage.

OTHER CROPS

The short season production scheme has been adapted to other crops with varying results. The melon industry has used the concept with wide success. Producers use short season production strategies to meet demand early in the marketing season, to avoid the whitefly complex, and to avoid the aphid transmitted viruses. Key components of this system are the use of transplants, black plastic mulches, and drip irrigation (13, 14). Short season soybean production can be used in the mid-South to lessen risk by increasing yields with planting dates in April and early May and planting soybean varieties in maturity group III, IV, and V (15). This system results in higher yields by avoiding low rainfall periods during critical water usage periods (bloom and pod-fill), by avoiding foliage-feeding insects and stink bugs, and by avoiding extensive rain and wind associated with the tropical systems in September and October that disrupt harvest (15). The short season production scheme has not been as successful in corn production. Yields are typically greater in late season varieties compared to early season varieties and cost savings resulting from reduced inputs usually do not cover the loss in yield (16, 17).

CONCLUSION

Short season crops can be an important cultural practice to avoid pests. The adoption of short season crops has usually developed due to the inability to economically control pests although some short season crops are grown to take advantage of more favorable weather conditions. The greatest success of this strategy has been in cotton and melon production. Most cotton producers are now growing varieties that are ready for harvest in 120–130 days. Melon producers have been able to take advantage of higher prices early in the season and reduce or avoid insecticide applications by escaping whiteflies and virus transmitting aphids. Producers should understand that incorporating short season crops includes not only early maturing varieties but also varying and managing other production inputs such as planting date, irrigation, and fertility.

REFERENCES

1. Sorenson, A.A. *Integrated Pest Management and Sustainable Agriculture: Complementary Production Strategies for the 21st Century*; Center for Agriculture in the Environment, American Farmland Trust: DeKalb, IL, 1995.
2. Horber, E. Types and Classification of Resistance. In *Breeding Plants Resistant to Insects*; Maxwell, F.G., Jennings, P.R., Eds.; John Wiley & Sons: New York, 1980; 15–22.
3. Benbrook, C.M. *Pest Management at the Crossroads*; Consumers Union: Yonkers, NY, 1996; 272.
4. Harris, M.K. Arthropod-Plant Interactions Related to Agriculture, Emphasizing Host Plant Resistance. In *Biology and Breeding for Resistance to Arthropods and Pathogens in Agricultural Plants*; Harris, M.K., Ed.; Texas Agricultural Experiment Station Publication MP-1451, 1979; 23–51.
5. Cohn, D.L. *The Life and Times of King Cotton*; Oxford University Press: New York, 1956; 286.
6. Parker, R.D.; Walker, J.K.; Niles, G.A.; Mulkey, J.R. *The "Short-season Effect" in Cotton and Escape from the Boll Weevil*; Texas Agricultural Experiment Station Publication B-1315, 1980; 45.
7. Walker, J.K.; Frisbie, R.E.; Niles, G.A. A changing perspective: *Heliothis* in short-season cottons in Texas. ESA Bull. **1978**, *24*, 385–391.
8. Walker, J.K.; Niles, G.A. *Population Dynamics of the Boll Weevil and Modified Cotton Types*; Texas Agricultural Experiment Station Publication B-1109, 1971; 14.
9. Heilman, M.D.; Namken, L.N.; Norman, J.W. Lukefahr. Evaluation of an integrated short-season management production system for cotton. J. Econ. Entomol. **1979**, *72*, 896–900.
10. Boman, R.K.; Raun, W.R.; Westerman, R.L.; Banks, J.C.

Long-term nitrogen fertilization in short-season cotton: interpretation of agronomic characteristics using stability analysis. J. Prod. Agric. **1997**, *10*, 580–585.
11. Steger, A.J.; Silvertooth, J.C.; Brown, P.W. Upland cotton growth and yield response to timing the initial post-plant irrigation. Agron. J. **1998**, *90*, 455–461.
12. Chu, C.C.; Henneberry, T.J.; Weddle, R.C.; Natwick, E.T.; Carson, J.R.; Valenzuela, C.; Birdsall, S.L.; Staten, R.T. Reduction of pink bollworm (Lepidoptera: Gelechiidae) populations in the Imperial Valley, California, following mandatory short-season cotton management systems. J. Econ. Entomol. **1996**, *89*, 175–182.
13. Hartz, T.K.; Mayberry, K.S. Valencia. In *Cantaloupe Production in California; Division of Agriculture and Natural Resources Publication 7218*; University of California, 1996; 3.
14. *Vegetable Growers' Handbook*, 2nd Ed.; Cotner, S.D., Dainello, F.J., Eds.; Texas Agricultural Extension Service: College Station, TX, 1998; 223.
15. Boquet, D. Yield and risk utilizing short-season soybean production in the mid-southern USA. Crop. Sci. **1998**, *38*, 1004–1011.
16. Howell, T.A.; Tolk, J.A.; Schneider, A.D.; Evett, S.R. Evapotranspiration, yield and water use efficiency of corn hybrids differing in maturity. Agron. J. **1998**, *90*, 3–9.
17. Jarvis, J.L.; Guthrie, W.D.; Robbins, J.C. Yield losses from second-generation European corn borers (Lepidoptera: Pyralidae) in long-season maize hybrids planted early compared with short-season hybrids planted late. J. Econ. Entomol. **1986**, *79*, 243–246.

SITE-SPECIFIC FARMING/MANAGEMENT (PRECISION FARMING)

S

Adrian Johnston
Potash and Phosphate Institute of Canada, Saskatoon, Saskatchewan, Canada

INTRODUCTION

Precision agriculture, or site-specific management, refers to technology that allows the improved management of small production areas within a single field. It involves the application of a specific management practice to a specific field location and in a single field operation. The goal of precision agriculture is to manage a field by regions rather than using uniform application of inputs and management, thereby optimizing the crop response based on the production potential and constraints of the specific region of the field. There are a number of terms used to define precision agriculture, including precision farming (PF), farming by the foot, target farming, and site-specific farming. In general, all of these terms have the same meaning.

Associated terminology and technology used with precision agriculture includes yield monitors, variable rate technology (VRT), global positioning systems (GPS), and geographic information systems (GIS). Yield monitors are either flow or load cell sensors used to monitor yield during the harvesting operation. Examples include grain flow sensors on a wheat combine, or weight sensors on the conveyor of a potato digger. The term VRT refers to the application of crop production inputs at variable rates, using controllers on the application equipment, such as sprayers and fertilizer applicators. An example of VRT is the variation in fertilizer N application across a field that would be recommended from the collection of a number of soil samples coming from the various parts of the field. Global positioning system refers to a satellite-based locating system, which provides us with an earth-based position identified using longitude and latitude. Some GPS units are also capable of providing elevation. When linked with a yield monitor, the harvested yield can be identified with a specific location in the field, using GPS, resulting in a yield map showing crop yield variation as a function of field position. Similarly, if the goal is to vary input application with VRT, an application map can be developed and rates changed "on-the-go" using GPS to verify location within a field. Geographic information systems are data management systems that allow the layering of a number of data sets related to information collected from a specific field. For example, a series of soil samples collected from specific locations in a field can be arranged with a crop yield map, providing an overlay of these two sets of information and the ability to evaluate relationships that may exist. The use of GIS also allows for multiple yield maps collected from a field over a period of years to be layered as a means of identifying areas of the field that require individualized management focus.

APPLICATION OF THE TECHNOLOGY TO AGRICULTURE

The current focus of precision agriculture technology is based on optimum management of inputs in a mechanized agricultural system. However, it more likely resembles the situation used in nonmechanized subsistence agricultural production systems, where manual labor allows for the focus of inputs to specific parts of small fields in an attempt to optimize production response and resource use. Under expansive agricultural systems, which generally use uniform application of inputs such as fertilizer and pesticides across a field, producers are evaluating precision agriculture technology as a means of reducing the cost of production. The variability in pest infestations, such as weeds and plant diseases, would allow for pesticides to be applied only to those areas where they are required, or using some range of rates that would optimize the return on the input cost. Where at one time a uniform application of an herbicide to an entire field was used to control a weed that occurred in patches, application only to those areas where the weed patches exist in the field can reduce total pesticide use and cost to the producer. With increased precision in GPS comes the development of guidance for farm machinery, with the current focus on precision application of pesticides and seeding equipment. The ability to avoid double application as a result of overlapping with equipment can have a significant impact on input use.

FUTURE STATUS

Precision agriculture is a new and exciting technology being developed as a tool to improve our understanding of how crop yields vary across the landscape, and allow the farmer to vary the application of production inputs. The ability to vary inputs should prove an effective means of improving both input use efficiency and profitability, while at the same time minimizing the total amount of pesticides applied. Ongoing research and development will focus on the agronomic application of the data collected. It is likely that those farmers with highly variable soil and landscape conditions will be the ones to derive the greatest benefit from variable rate technology. However, a yield monitor with GPS will prove to be an effective tool on any farm in gathering crop response data to a variety of new technologies being tested. This will ultimately allow an individual farmer to make his own on-farm assessment of the suitability and profitability of new technology, inputs, and farming practices.

BIBLIOGRAPHY

Carr, P.M.; Carlson, G.R.; Jacobson, J.S.; Nielsen, G.A.; Skogley, E.O. Farming soils, not fields: a strategy for increasing fertilizer profitability. J. Prod. Agric. **1991**, *4*, 57–61.

Pennock, D.J.; Zebarth, B.J.; de Jong, E. *Landform Classification and Soil Distribution in Hummocky Terrain*; Saskatchewan, Canada, 1987; 40, 279–315, Geoderma.

Robert, P.C.; Rust, R.H.; Larson, W.E. *Precision Agriculture*; ASA, CSSA, SSSA: Minneapolis, MN & Madison, WI, 1996; 1222.

Robert, P.C.; Rust, R.H.; Larson, W.E. *Site Specific Crop Management. Research Issues*; ASA, CSSA, SSSA: Minneapolis, MN & Madison; WI, 1993; 395.

Sawyer, J.E. Concepts of variable rate technology with considerations for fertilizer application. J. Prod. Agric. **1994**, *7*, 195–201.

SLEEPING SICKNESS

Jamie Stevens
University of Exeter, Exeter, United Kingdom

S

INTRODUCTION

Trypanosomes are responsible for a range of debilitating and often fatal diseases of humans and livestock across Africa (1, 2). In humans *Trypanosoma brucei* causes sleeping sickness, which is endemic in 36 sub-Saharan countries between latitudes 14°N and 29°S, where the parasite is transmitted by the bite of tsetse flies of the genus *Glossina*. The disease, which if untreated is invariably fatal, manifests as a wasting condition with increasingly severe degradation of physical condition and impaired mental state, including sleep disorders. The two human infective forms of the parasite, *T.b. gambiense* and *T.b. rhodesiense*, are responsible for chronic and acute forms of the disease in West\Central and East Africa, respectively. World Health Organization (WHO) estimates for 1998 (3) suggest 60 million people are at risk, with an estimated 300,000 new cases of sleeping sickness occurring annually; however, less than 4 million people are under surveillance and less than 30,000 new cases are actually diagnosed and treated annually.

THE DISEASE

Sleeping sickness, or human African trypanosomiasis, occurs following inoculation of trypanosomes from the bite of an infective tsetse fly. The two forms of the disease occur in discontinuous foci. Infection with *T.b. rhodesiense* (East Africa) is more acute, with overt clinical signs and symptoms developing within days of infection; infection with *T.b. gambiense* (West/Central Africa) is generally more chronic, with months and sometimes years elapsing prior to the appearance of often initially less pronounced symptoms (4, 5). In both forms, the disease progresses in two distinct stages: the hematolymphatic phase (stage I) and the meningoencephalitic phase (stage II). Both stages are characterized by a very broad range of signs and symptoms including:

Stage I—canker (the primary lesion), enlargement of the posterior cervical lymph nodes (Winterbottom's sign), and (also in stage II) endocrinological disorders ("moon face") and intercurrent infections, for example, lung infections.

Stage II—sleep disturbances and disruption of the circadian rhythmicity of the sleep-wake cycle, from where originates the name "sleeping sickness," neurological disorders, and deterioration of consciousness leading to coma and death.

THE PARASITES

Trypanosoma brucei infects both humans and a range of animals that can act as reservoirs of the disease. *T.b. rhodesiense* represents a true zoonosis; *T.b. gambiense* is maintained in humans, with pigs implicated among the more important animal reservoirs (6–8). The life cycle of *T. brucei* is shown in Fig. 1.

THE VECTORS

The distribution of sleeping sickness corresponds to the geographical distribution of the tsetse fly across sub-Saharan Africa (Fig. 1). Within this region West/Central Africa and East Africa are characterized by particular vegetation types and combinations of tsetse fly species. Local variations in terrain provide a patchwork of epidemiologically significant microhabitats and climates, for example, riverine or gallery forest in semiarid scrubland. The two groups of tsetse fly involved in sleeping sickness transmission are *Glossina morsitans*-group flies, which inhabit savannah scrublands grazed by livestock and endemic animals in East Africa, and *Glossina palpalis*-group flies, which occupy primarily riverine habitats and humid forests in West/Central Africa, where they frequently feed on humans in areas of increased human-fly contact, for example, waterholes.

CONTROL

Population Screening

Clinical signs are generally unspecific and early diagnosis, particularly of *T.b. gambiense*, is difficult. A variety of

Encyclopedia of Pest Management

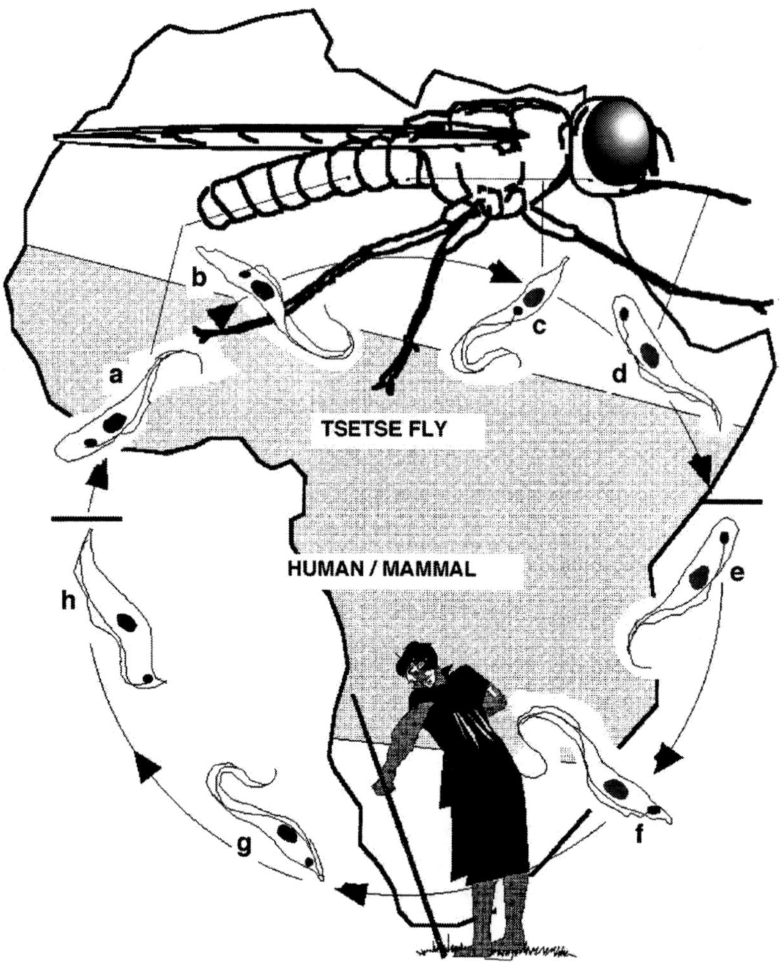

Fig. 1 The life cycle of *Trypanosoma brucei*. Stages a–d occur within the tsetse fly and stages e–h occur in humans and other mammalian hosts; a: procyclic form, which divides and multiplies in the tsetse midgut; b: proventricular form, which leaves the midgut to infect the proventricular region via the peritrophic membrane before migrating to the salivary glands; c: epimastigote stage, which undergoes a further round of extensive division while attached to the salivary gland epithelium; d: infective metacyclic form, which detaches from the microvilli and develops a surface coat like that of bloodstream forms before migrating to the proboscis; e: metacyclic form inoculated into the human\mammalian host by the bite of a tsetse fly; f: the highly motile, long slender, rapidly dividing form and that most commonly observed when screening infected patients; g: the intermediate form, which heralds the end of rapid division; h: the nondividing short stumpy form which, together with the intermediate form, is taken up by the tsetse fly when it feeds on a mammalian host to continue the cycle. The lines from trypanosome life stages a, c, and d indicate the site of development in the tsetse fly of each parasite form: a: procyclic\midgut;c: epimastigote\salivary glands (which extend into the abdomen); d:infective metacyclic\proboscis. The shaded portion of the map denotes the area in which sleeping sickness foci occur. (From Ref. 5.)

direct methods (e.g., slide preparations and centrifugation) can be employed to visualize parasites in blood, lymph node aspirate, and cerebrospinal fluid; these work best when large numbers of parasites are circulating in patients' blood/lymphatic systems, which is generally the case for *T.b. rhodesiense*. However, only serological tests realistically can be used in large-scale field-based population surveys, for example, the antibody-based card agglutination test for trypanosomes (CATT) and the recently developed card indirect antigen test for trypanosomiasis (CIATT).

Antiparasitic Drugs

The arsenical compound melarsoprol (Mel B) remains perhaps the best available drug, especially for the treatment of late stage infections and despite its side effects that cause serious encephalopathy in 5–10% of patients, of which half may die. Given the lack of a lucrative commercial market for sleeping sickness drugs, there is little research into new treatments; all currently available pharmachemicals, except eflornithine (registered in the U.S. in 1990, after originally being developed as an anticancer treatment), were developed before 1950. Accordingly, the need to maximize the effectiveness of available drugs and, in particular, to optimize combination therapy, is paramount. Drugs currently available for the treatment of sleeping sickness are summarized in Table 1.

Cost/Benefit: Treatment of Sleeping Sickness

WHO estimates from 1995 indicate sleeping sickness treatment costs per patient as (stage I) pentamidine $107, suramin $114; (stage II) melarsoprol $253, eflornithine $675. These figures include the costs of hospitalization, drug transport, and administering the drug, but do not include the cost of any associated screening program. The higher costs of treating stage II infections reflects the increased complexity of treating patients with cerebrospinal infections, for example, melarsoprol treatment requires a longer period of hospitalization, while the benefits of using eflornithine (to treat infections refractory to melarsoprol) must be weighed against the fact that it is extremely expensive and requires a long and more complex drug administration regime. Overall costs of control break down into the costs of population screening, detection and treatment. The relative costs of these activities change according to sampling strategy and disease prevalence. For all control\treatment strategies, the total cost of control per patient is relatively greater at lower levels of disease prevalence.

Vector Control

Some of the first attempts to control human sleeping sickness employed so-called‘‘ecological’’ methods. These involved the clearance of vegetation to destroy fly habitats and the shooting of wild animals. The approach, which was popular up until World War II, was abandoned on ethical and environmental grounds (9).

From the early 1950s ground spraying, and latterly aerial application, became the main method of tsetse control (initially using dieldrin and DDT, more recently using pyrethroids). Today environmental concerns (pollution and damage to nontarget organisms) and the comparatively high costs of spraying (particularly aerial application) have caused a decline in use of the technique.

The use of tsetse traps and screens (targets) has become ever more popular since the late 1970s. Traps and screens are highly selective, have an immediate effect in reducing fly density, provide a barrier to reinvasion, and their effectiveness can be enhanced using chemical attractants. Traps kill flies by retaining them until they die from dehydration and\or can be impregnated with insecticide and screens (simple rectangles of tsetse-attracting blue and black cloth suspended from a frame) are always impregnated with insecticide. Both cause minimal environmental damage.

Cost/Benefit: Vector Control

Geographical and other site-related factors can significantly affect the costs of different tsetse eradication techniques.Nevertheless, an overall pattern is apparent: aerial spraying is generally the most expensive method of control (and the effects are immediate and significant) and traps and screens are cheaper than or at least as cost effective as ground spraying. Even allowing for reapplication of insecticide, simplicity of design and associated increased

Table 1 Drugs currently available for the treatment of sleeping sickness

Drug	Chemical\Trade name	*T.b. gambiense* Stage I	*T.b. gambiense* Stage II	*T.b. rhodesiense* Stage I	*T.b. rhodesiense* Stage II
Suramin	Bayer 205, Germanin®	a,b	Ineffective	a	c
Pentamidine	Pentamidine isethionate, Pentacarinat®	a	d	Unreliable	Unreliable
Melarsoprol	Mel B, Arsobal®	e	a	e	a
Nifurtimox	Bayer 2505, Lampit®		a,f	Unknown	Unknown
Eflornithine	DFMO, Ornidyl®	g	a,f,g	Unreliable	Unreliable

[a]Drug of choice.
[b]In combination with pentamidine.
[c]In combination with melarsoprol.
[d]Prior to commencing melarsoprol.
[e]Not generally used in stage I infections due to high risk of adverse effects.
[f]Including treatment of melarsoprol refractory patients.
[g]Highly effective, but costly: ~$1070 per course of treatment.
(From Van Nieuwenhove, S. Present Strategies in the Treatment of Human African Trypanosomiasis. In *Progress in Human African Trypanosomiasis, Sleeping Sickness*; Dumas, M., Bouteille, B., Buguet, A., (Eds.); Springer-Verlag: Paris, 1999; 253–80.)

field life results in screens being generally more economical than more complex trap designs.

OUTLOOK

In the long term, disease control through control of the vector is technically possible. However, trapping effort needs to be maintained even after the threat of sleeping sickness has apparently disappeared; in the past an absence of tsetse flies in traps coupled with no new cases of sleeping sickness has led to the discontinuation of control programes. Across sub-Saharan Africa sleeping sickness is reemerging, often in conjunction with social and ecological upheaval, which both disrupt normal patterns of human-fly contact and degrade the health and resistance of affected populations (10). In southern Sudan the prevalence of infection has jumped to nearly 20% [based on screening ~1400 people in 16 villages (11)]. Increased levels of disease are reported from the Central African Republic and the Democratic Republic of Congo (formerly Zaire), where disease prevalence levels are at their highest in 60 years. In several countries where African trypanosomiasis is endemic, political turmoil and economic crisis have eroded the public health infrastructure so severely that it is questionable whether the undertaking of effective control programes are within their budget. However, with no new commercial drugs available, continued surveillance and screening coupled with the training of local health workers appears to offer the best hope for the future.

REFERENCES

1. World Health Organization, African Trypanosomiasis (Sleeping Sickness) Control Programme, Geneva. www.who.org/emc/diseases/tryp/index.html (accessed March 2000).
2. Food and Agriculture Organization of the United Nations, Programme Against African Trypanosomiasis (PAAT), Rome. www.fao.org/waicent/faoinfo/agricult/aga/agah/pd/vector.htm (accessed March 2000).
3. World Health Organization. *Control and Surveillance of African Trypanosomiasis—Report of a WHO Expert Committee*; Technical Report Series No. 881, WHO Publications: Geneva, 1998; 113.
4. *Progress in Human African Trypanosomiasis, Sleeping Sickness*; Bouteille, B., Buguet, A., Dumas, M., Eds.; Springer-Verlag: Paris, 1999; 344.
5. Seed, J.R.; Hall, J.E. Trypanosomes Causing Disease in Man in Africa. In *Parasitic Protozoa*, 2nd Ed.; Kreier, J.P., Baker, J.R., Eds.; Academic Press Limited: London, 1992; 2, 85–155.
6. Hoare, C.A. *The Trypanosomes of Mammals. A Zoological Monograph*; Blackwell Scientific Publications: Oxford, 1972; 749.
7. Stevens, J.R.; Noyes, H.A.; Dover, G.A.; Gibson, W.C. The ancient and divergent origins of the human pathogenic trypanosomes, *Trypanosoma brucei* and *T. cruzi*. Parasitology **1999**, *118*, 107–116.
8. Stevens, J.R.; Gibson, W. The molecular evolution of trypansomes. Parasitol. Today **1999**, *15* (11), 432–37.
9. Barrett, J.C. *Economic Issues in Trypanosomiasis Control*; NRI Bulletin 75, Natural Resources Institute: Chatham, UK, 1997; 183.
10. Lyons, M. *The Colonial Disease: A Social History of Sleeping Sickness in Northern Zaire, 1900–1940*; Cambridge University Press: Cambridge, UK, 1992; 335.
11. National Center for Infectious Diseases/Centers for Disease Control and Prevention. In *West African Trypanosomiasis Resurging in Southern Sudan*; Focus, Vol. 6, No. 5, Office of Health Communication, NCID: Atlanta, 1997; 2–3. www.cdc.gov/ncidod/focus/vol6no5/trypan.htm (accessed May 2001).

SOIL EROSION AND PESTICIDE TRANSLOCATION CONTROL

S

Monika Frielinghaus
Institute of Soil Landscape Research, ZALF Müncheberg, Germany

INTRODUCTION

Soil Erosion—A Global Problem

One-third of the world's agricultural soils are reported to be in a degraded state. Water and wind erosion contribute to approximately 84% of the observed damage. The worldwide soil loss from soil erosion is estimated at about 75 million tons per year (1). Soil is picked up in Florida and Brazil each year and blown across the Atlantic Ocean to Africa. Also, each year when the Chinese till their soil in the spring, Chinese soil is detected in Hawaii (2). Most soil loss can be traced to inappropriate land management practices. Viewed broadly, this scenario creates an agricultural dilemma of global proportions. Pesticides, used in nearly all agricultural systems, are translocated by wind and water erosion and also cause environmental damages (Fig. 1).

SOIL EROSION PROCESSES

Water erosion is composed of two processes: detachment and transport. Raindrops hit the ground with velocities between 10 and 20 miles per hour. Water flow concentrates in wheels, rills, and gullies with an increased runoff velocity and serious rates of soil detachment result. The potential water erosion rate is a factor of a high rainfall amount or intensity, soil erodibility (instability), slope grade and distance, and type of land use. The real water erosion risk depends on the crop type or the crop residue cover rate, as determined by the method of crop production management.

The development of an erosion network results in a high soil loss per area due to the transport of sediment in rills or gullies. The first phase in the development of rills is the concentration of the sheet-wash runoff flow water into linear paths. The rill formation may be influenced by a variety of different factors like soil crusting, trafficked wheels, and reduced infiltration by soil compaction. Rills and gullies often develop close to one another. The second phase is the heightened concentration of water in morphological deep lines. The hydrological power of transport water as well as the offsite risk for chemical transport into the environment is increased with the development of erosion flow patterns.

Wind erosion results from wind moving across a dry soil surface and dislodging soil particles by pressure and lifting forces. The process is self-perpetuating; blowing sediment disturbs additional particles, which are then lifted into the air stream.

The potential wind erosion rate is influenced by high wind velocity soil erodibility (size/weight of particles), the degree to which the landscape is wind-exposed, and land uses. The real wind erosion risk depends on the actual soil moisture and the real soil cover rate.

The determination of wind erosion transport distances is complex and insufficient research for conclusive statements is available at the present time.

SOIL TRANSPORT AND PESTICIDE MOVEMENT DUE TO EROSION

A significant amount of data have been published on pesticide losses in surface runoff water and sediment (3). To minimize the potential of pesticide pollution, the basic physical, chemical, biological, and hydrological processes influencing the transport of the pesticides from the application location to adjacent ecosystems have to be determined.

The soil and matter transport process is dynamic. In addition to the total pesticide losses in surface runoff, peak pesticide concentrations can also have a significant impact on adjacent ecosystems. Processes involved in pesticide surface runoff and erosion include: 1) water flow on a nonhomogenous soil surface characterized by rills and inter-rills, 2) the displacement of pesticides from soil to the surface runoff stream, and 3) the addition of eroded particles. The relative contributions of each process depend on the soil characteristics and the types and binding configurations of the chemicals involved. Pesticides are categorized based on their association with solid and liquid phases. Atrazine, simazine, diuron, isoproturon, and alachlor, for example, are mainly transported in surface run-

Encyclopedia of Pest Management

A

B

Fig. 1 **(A)** Translocated sediment by water erosion close to a water source. **(B)** Translocated sediment by wind erosion close to a water source.

off water. In contrast, 90% of applied trifluralin and aclonifen was adsorbed onto eroded particles in pesticide studies. Lindane has an intermediate status, with a 37% adsorption and erodible level.

Rainfall simulation studies on small field plots provide information about worst cases for pesticide translocation (rainfall intensity of about 70 mm h^{-1}, 100 mm accumulated rainfall) (4). The pesticides of the first group with water solubility concentrations of 65–700 mg L^{-1}, respectively, (e.g., isoproturon and dichlorprop-p) and of the second group with solubility concentrations <1 mg L^{-1} (e.g., bifenox) were tested under two different soil cover conditions (Table 1).

The results demonstrate that the influence of soil cover characteristics on concentrations and total runoff losses of pesticides for the high soluble group is restricted to the start of the rain event after pesticide application and the runoff rate. The sediment concentration of pesticides in the fairly soluble group were comparable to those of the first runoff event independent of the time lapse between the application and rainfall. The fairly soluble pesticides are thus exiting the field sites only adsorbed to eroded sediment. The cumulative soil loss is the most reliable indicator explaining the decrease in the sediment and pesticide concentrations during each rainstorm as described by a power function.

Pesticide transport by wind may occur by isolated pesticide displacement or as a sediment-bounded form. At the present, only initial findings of the quantity of translocated pesticides are available. It is assumed that the horizontal transport of particulate bounded pesticides is more than 50% of the total loss resulting from extreme wind erosion events. The most intensive investigations to date have been carried out by Larney (5). After 13 wind erosion events, he determined the herbicide loss at about 1.5% of total amount that was applied and integrated into the upper humus horizon on steppe soils in Canada. Losses of herbicide from a nonintegrated application on soil surfaces were approximately 4.5%. The amount of pesticides that remain a suspended part of the translocated sediment is unknown. Applying pesticides on an uncovered soil surface by preemergent management may result in high environmental risks dependent on the application date, the method of application, the adsorption, and persistence properties as well as the property for vertical translocation. A classification system dependents on the binding configurations of the chemicals involved, as available for water erosion, has not yet been developed.

SOIL EROSION AND PESTICIDE TRANSLOCATION CONTROL

The most effective system for prevention of soil erosion is soil conservation tillage (nonplowing till systems) (6). A closed vegetation cover protects the soil surface from the initial soil detachment action caused by raindrop splash forces or wind power and results in a lower volume of runoff. Plant roots help retain soil particles and reduce the soil transport load. The extent of soil cover material, green plant mass, or crop residues can be influenced by the farmer's management and soil tillage practices.

The temporal and spatial distribution of the plant or residue soil cover are decisive for protection of the soil surface. About 2 tons per hectare dry matter, or 30%–50% plant surface cover, is effective for initial soil protection efforts (7) (Table 2).

S

Table 1 Average pesticide losses in relation to amount of applied pesticide (% of applied amount)

Name of pesticide	Pesticide transporting medium	Time lapse between pesticide application and rainfall event				
		2 h	1 d	3 d	5–7 d	14 d
Application on bare soil surface sites (preemergence application):						
Isoproturon	Total	4.5–12.4	4.0–17.2	5.4–13.3	2.1–11.2	n.e.
	In runoff water	3.8–9.8	2.9–13.9	4.0–9.8	1.4–9.0	n.e.
	In eroded sediment	0.7–2.6	1.7–3.3	1.4–3.5	0.7–2.2	n.e.
Dichlorprop-p	Total	2.1–10.4	2.2–16.7	5.9–10.4	1.2–8.3	n.e.
	In runoff water	1.9–9.8	1.8–15.3	5.3–9.3	0.9–7.0	n.e.
	In eroded sediment	0.2–0.6	0.4–1.4	0.6–1.1	0.3–1.3	n.e.
Bifenox	Total	15.6–19.0	19.3–21.6	14.3–17.2	9.3–13.8	n.e.
	In runoff water	<0.1–0.9	<0.1–0.3	<0.1–0.3	<0.1–0.2	n.e.
	In eroded sediment	14.7–19.0	19.0–21.6	14.0–17.2	9.3–13.6	n.e.
Application on small covered soil surface (barely with 3–5 leaves):						
Isoproturon	Total	2.8–13.3	1.8–16.4	4.1–11.2	1.5–6.0	1.6
	In runoff water	2.2–11.4	1.5–13.5	3.3–8.6	1.0–4.6	1.1
	In eroded sediment	0.4–1.9	0.3–2.9	0.8–2.6	0.5–1.4	0.5
Dichlorprop-p	Total	0.8–10.0	0.9–9.2	1.0–8.3	0.7–4.1	0.8
	In runoff water	0.7–9.4	0.8–8.4	0.8–7.5	0.6–3.7	0.7
	In eroded sediment	0.1–0.6	0.1–0.8	0.2–0.8	0.1–0.4	0.1
Bifenox	Total	11.2–15.0	7.8–15.5	7.8–15.9	3.4–9.8	5.2
	In runoff water	<0.1–0.9	<0.1–0.2	<0.1–0.4	<0.1–0.3	<0.1
	In eroded sediment	11.2–14.1	7.8–15.5	7.4–15.9	3.4–9.5	5.2

(Adapted from Ref. 4.)

The rate of soil cover is a highly effective indicator for assessing the risk of water and wind erosion. This indicator addresses the questions: 1) how much soil cover is necessary to reduce the threat of erosion and pesticide translocation for a high-risk area; 2) how much cover can be realized dependent on crop type, crop rotation, tillage, and management practices in different global regions. Based on this analysis, appropriate preventive management practices can be required.

Table 2 Correlation between cover rate, runoff, and soil loss, as determined from long-term experiments

Soil cover (%)	Plant residues, (t ha^{-1}) dry matter	Runoff (% of rain)	Relative soil loss (%)
0	0	45	100
20	0.5	40	25
30	1	35	8
Approx. 50	2	<30	3
Approx. 70	4	<30	<2
>90	8	<30	<1

The greatest erosion risk occurs after seedbed preparation, which is characterized by the lowest degree of soil surface roughness and bare soil exposure. Crop selection, improved crop rotations, and a change in soil tillage practices are established methods for increasing the soil cover.

Other more basic management methods such as contour tillage or strip cropping, which are practiced in some parts of the world, are not as effective as conservation tillage in minimizing soil erosion (8).

While conservation systems significantly reduce the rate of soil loss, the volume of surface water runoff is only about 30% less than the runoff volume from uncovered slopes. As a result, the risk for soil erosion and pesticide transport should be addressed by an appropriate selection and a carefully timed application of pesticides. Products with low water-soluble active substances should be preferred for slopes with a water erosion risk. It is advisable to replace products with isoproturon or isopropylamin with products containing pendimethalin. For purposes of crop protection, water-soluble products which infiltrate into the upper soil layer should be used on areas at risk to wind erosion. Preemergence management on bare soil surface should be avoided on areas with a high erosion risk.

REFERENCES

1. Oldeman, LR. *Global Extent of Soil Degradation*; ISRIC Bi-annual Report 1991-1992, International Soil Reference and Information Centre: Wageningen, 1992.
2. Simonns, M. Winds toss Africa's soil, feeding lands far away. New York Times **1992**, A1–A16, October 29.
3. Lundbergh, I.; Kreuger, J.; Johnson, A. Pesticides in Surface Waters. In *A Review of Pesticide Residues in Nordic Countries, Germany, and the Netherlands and Problems Related to Pesticide Contamination*; Council of Europe Press, 1995; 54.
4. Klöppel, H.; Haider, J.; Kördel, W. Herbicides in surface runoff: a rainfall simulation study on small plots in the field. Chemosphere **1994**, *28* (4).
5. Larney, F.J.; Cessna, A.J.; Bullock, M.S. Herbicide transport on wind-eroded sediment. J. Enviro. Qual. **1999**, *28*, 1412–1421.
6. Hurni, H. Soil Conservation Policies and Sustainable Land Management: A Global Overview. In *Soil and Water Conservation Policies and Programs*; Napier, T.L., Napier, S.M., Tvrdon, J., Eds.; CRC Press: New York, 2000; 19–30.
7. Frielinghaus, M.; Winnige, B. In *The Use of an Indicator System for Crop Residue Management and Soil Erosion Control*, Proceedings of the 15th Conference of the ISTRO, Fort Worth, Texas, USA, 2000; CD.
8. Schwertmann, U.; Rickson, R.J.; Auerswald, K. *Soil Erosion Protection Measures in Europe*; Soil Technology Series 1, Catena-Verlag Cremlingen, Germany, 1989; 216.

SOLARIZATION

James J. Stapleton
University of California, Parlier, California, U.S.A.

INTRODUCTION

Solarization is a natural, hydrothermal process of disinfesting soil of plant pests that is accomplished through passive solar heating. Plastic mulch is placed over soil during hot weather, creating a greenhouse-like environment that can raise soil temperature from levels slightly above ambient to more than 70°C (158°F). Although varying widely, the heat damage threshold for mesophilic pest organisms tested during solarization is generally in the range of 39–42°C (102–108°F) (1). Solarization occurs through a combination of physical, chemical, and biological effects, and is compatible with other disinfestation methods, such as organic amendments, biological control organisms, or pesticides (2, 3). Commercially, it is used on a relatively small scale worldwide as a substitute for synthetic chemical toxicants, but its use is expected to increase as methyl bromide, the major chemical fumigant, is phased out due to its ozone-depleting properties. On the other hand, solarization is an important and widespread practice for home gardeners (4). Solarization, as any other soil disinfestation method, has both benefits and drawbacks. While it is simple, safe, and effective to use within its limitations, and can be readily combined with biological and chemical control measures, solarization is dependent on high air temperatures, is most effective near the soil surface, does not consistently control certain heat-tolerant pests, should be done during the hottest part of the year (possibly interfering with planting schedules), and requires disposal of plastic film (4).

HISTORY AND USAGE

Various forms of heating, including crop stubble burning, solar soil roasting, hot water drenching, and steam injection, have been used—some since ancient times—to rid soil of agricultural pests. The modern technique of soil solarization was developed in Israel in the 1970s, when agriculturists noticed that soil covered by plastic mulch during the summer became quite hot (2). This phenomenon was subsequently exploited to control soil-borne pests, and researchers around the world were soon investigating the biocidal potential of solarization. Commercially, the principal use of solarization on a treated area basis is probably in conjunction with greenhouse grown crops. For example, soil in more than 4500 ha (11,120 ac) of greenhouses was reported to be undergoing routine commercial solarization in Japan in 1991 (5). Greenhouse production is an ideal system for using solarization, since greenhouses are not used during the summer in many parts of the world because of excessive heat. Therefore, during the summer off-season, greenhouses can be closed to maximize heating, and soil can be effectively solarized. Greenhouse solarization is now being commonly used in other Mediterranean and Near-Eastern locations (6, 7).

Another application for which solarization has come into common use, particularly in developing countries, is for disinfestation of seedbeds, containerized planting media, and cold-frames. As for use in greenhouses, these are ideal niches for solarization, since individual areas to be treated are small, soil temperature can be greatly increased, the cost of application is low, the value of the plants produced is high, and the production of disease-free planting stock is critical for producing healthy crops (4).

Around the world, solarization for disinfesting soil in open fields is being implemented at a relatively slow but increasing rate. It has been mainly used on a commercial basis in areas where air temperatures are very high during the summer and much of the cropland is rotated out of production due to excessive heat, such as the central and southern desert valleys of California and the Near-East. Most growers in California who are now using solarization in production fields are those that have some aversion to the use of methyl bromide or other chemical soil disinfestants, either because of their close proximity to urban or residential areas, personal preference, or because they are growing for organic markets (1). Implementation of production field solarization in other areas, with suitable but more tropical climates appears to be progressing at a similar rate (4, 8).

Encyclopedia of Pest Management

In addition to commercial use, the importance of solarization in home gardening and subsistence production is widely recognized. Although most of these users do not use chemical soil disinfestants under any circumstances, solarization has been widely embraced and mainstreamed by gardeners, and contributes to improved plant health and production in these settings (1, 9).

Most transparent polyethylene films are suitable for conducting solarization. However, use of lower quality films may be undesirable due to premature degradation of the plastic, leaving numerous fragments which are difficult to retrieve. Higher quality film more resistant to weakening by ultraviolet light is preferable. The thickness, or gauge, of the film is relatively unimportant. Certain plastics manufacturers produce films specially designed for solarization (2, 3, 10). Most farm supply outlets and many nurseries stock or can order suitable films. The practical value of soil solarization for the end user, as with any pest management strategy, must be assessed by several factors, including pesticidal efficacy, effect on crop growth and yield, economic cost/benefit, and user acceptance. Its routine use as a viable alternative to chemical fumigants in several areas of the world indicates that solarization has already achieved limited user acceptance (1, 4).

THE PRINCIPLES OF SOIL SOLARIZATION

The principal mechanism of solarization is usually the physical inactivation of soil-borne pathogens and pests by heat. The ''heat dosage'' of solarization consists of soil temperature × time, and is affected by numerous factors, including diurnal air temperature (the hotter the better, day and night), solar radiation intensity (the higher the better), wind speed and duration (less wind allows greater heat retention), precipitation (cloudy sky and water on the film surface lower soil temperature), soil texture (soils with high clay content tend to retain more heat), soil color (darker soils absorb more heat), soil moisture content (moist soils allow better heat transfer), and characteristics of the plastic mulch used (color, transparency, permeability). Soil temperature during solarization can range from ambient (under unsuitable conditions or deep in soil) to more than 70°C (158°F) (1–4).

Another critical treatment component is the thermal sensitivity of the target pest(s), which varies widely among species. Most mesophilic pest organisms (fungi, nematodes, weed seed) that have been tested began to show reduced viability (over long treatment periods) at 39–42°C (102–108°F). Soil that is moist rather than dry prior to solarization will stimulate microorganisms to break dormancy from their survival structures and commence active metabolism, thus becoming more susceptible to the biocidal effects of treatment. In many cases, it is not necessary to kill pest organisms; they may be weakened by ''sub-lethal'' heat [in general, soil temperatures below 39–42°C (102–108°F)] to the extent that they are unable to cause damage to plants and/or are more susceptible to chemical toxicants or to attack by antagonists (11). On the other hand, solarization of containerized soil may produce very high temperatures [more than 70°C (158°F)] which are lethal to most soil-borne pests within hours, and approach the heat levels produced during soil disinfestation using aerated steam (1, 9).

Mathematical models for predicting treatment duration and efficacy (i.e., when a solarization treatment is ''done'') by soil temperature alone, generally, have not been successfully implemented as agricultural production tools because of the passive and complex mode of action of the process, over a broad range of target organisms. Nevertheless, because of the potential utility of such predictive models, they continue to be a focus of development in many areas of the world (1).

In addition to physical heating factors, complex biological changes occur when soils are solarized. These changes have been shown to play important roles in the overall disinfestation treatment. The effects of solarization tend to be more pronounced on soil-borne pests, which are often more stringently dependent upon their host(s) for survival rather than other, more competitive soil microflora, many of which are antagonists of weakened pest structures (2, 3). The antagonists tend to tolerate solarization, or rapidly recolonize the soil once the treatment has ended. For example, there have been several reports of the rapid proliferation of fluorescent pseudomonads in the soil and plant rhizosphere, following solarization. Also, *Bacillus* spp., many of which are antibiotic-producing antagonists, can survive solarization due to their heat tolerant characteristics, and more extensively colonize the rhizosphere of subsequently planted crops (1–4).

Apart from the physical and biological effects of solarization, important chemical changes also occur in heated soils that often result in increase of soluble mineral nutrients, following treatment. These chemical changes can be another important factor in the ''increased plant growth response'' (IGR) phenomenon often observed in conjunction with solarization and other methods of soil disinfestation. The increase in available mineral nutrient concentrations, particularly nitrogen, in solarized soil is often equivalent to that of recommended preplant fertilization for crops (2, 3).

There is now a lengthy roster of soil-borne pests which are controlled or partially controlled by solarization, including in excess of 40 fungal plant pathogens, more than 25 species of nematodes, numerous weeds, and a few bacterial pathogens. In addition to these major pathogens that are reduced by solarization, a number of minor pathogens also are controlled. This is one of the reasons that IGR is often observed after solarization, similar to that commonly found after chemical fumigation. Solarization has frequently stimulated IGR in plants even when no major pathogens can be isolated, and reductions in the overall number of soil microorganisms have been significantly correlated with increased plant growth, following treatment (1, 4).

STRATEGIES FOR INCREASING THE BIOCIDAL EFFECTS OF SOIL SOLARIZATION

Solarization can be an effective soil disinfestant in numerous geographical areas for certain agricultural and horticultural applications. Nevertheless, there are inherent limitations, and situations are presented where it may be desirable to increase the efficacy and/or predictability of solarization through combination with other methods of soil disinfestation. Since solarization is a passive process with biocidal activity dependent to a great extent upon local climate and weather, there are occasions when even during optimal periods of the year, cool air temperatures, extensive cloud cover, frequent or persistent precipitation events, or other factors may not permit effective soil treatment. In these cases, integration of solarization with other disinfestation methods may be essential to increase treatment efficacy and predictability (1, 4). Studies have shown that solarization may be beneficially combined with other chemical and biological control methods. Recently, considerable interest has been generated regarding the use of organic amendments, including certain crop residues, with biocidal properties, in combination with solarization to achieve "biofumigation." One promising combination of organic amendments with solarization uses residues of cruciferous plants, which release a number of biotoxic volatile compounds into soil during the decomposition process (12).

Perhaps, the solarization combination most likely to be widely implemented, is that employing chemical pesticides. As methyl bromide is phased out, many current users will turn to other pesticides for soil disinfestation. Combining these pesticides (perhaps at lower dosages) with solarization (perhaps for a shorter treatment period) may prove to be the best option for users who wish to continue using chemical soil disinfestants (1, 4, 8, 13).

S

REFERENCES

1. Stapleton, J.J. Soil solarization in various agricultural production systems. Crop Prot. **2000**, *19*, 837–841.
2. Katan, J. Soil Solarization. In *Innovative Approaches to Plant Disease Control*; Chet, I., Ed.; John Wiley & Sons: New York, 1987; 77–105.
3. Stapleton, J.J.; DeVay, J.E. Soil Solarization: A Natural Mechanism of Pest Management. In *Novel Approaches to Integrated Pest Management*; Reuveni, R., Ed.; Lewis Publishers: Boca Raton, FL, 1995; 309–322.
4. Stapleton, J.J. Solarization: An Implementable Alternative for Soil Disinfestation. In *Biological and Cultural Control Tests for the Control of Plant Diseases*; Canaday, C., Ed.; APS Press: St. Paul, MN, 1997; 12, 1–6.
5. Horiuchi, S. Solarization for Greenhouse Crops in Japan. In *Soil Solarization*; DeVay, J.E., Stapleton, J.J., Elmore, C.L., Eds.; Plant Production and Protection Paper 109, FAO: Rome, 1991; 16–27.
6. Abu-Gharbieh, W. Pre- and Post-plant Soil Solarization. In *Soil Solarization and Integrated Management of Soil-borne Pests*; Stapleton, J.J., DeVay, J.E., Elmore, C.L., Eds.; Plant Production and Protection Paper 147, FAO: Rome, 1998; 15–34.
7. Grinstein, A.; Ausher, R. Soil Solarization in Israel. In *Soil Solarization*; Katan, J., DeVay, J.E., Eds.; CRC Press: Boca Raton, FL, 1991; 193–204.
8. Chellemi, D.O.; Olsen, S.M.; Mitchell, D.J. Effects of soil solarization and fumigation on survival of soil-borne pathogens of tomato in Northern Florida. Plant Dis. **1994**, *78*, 1167–1172.
9. Stapleton, J.J.; Elmore, C.L.; Devay, J.E. Solarization and biofumigation help disinfest soil. Calif. Agric. **2000**, *54* (6), 42–45.
10. Elmore, C.L.; Stapleton, J.J.; Bell, C.E.; DeVay, J.E. *Soil Solarization—A Nonpesticidal Method for Controlling Diseases, Nematodes, and Weeds*; University of California DANR Publication 21377: Oakland, CA, 1997; 1–13.
11. Tjamos, E.C.; Fravel, D.R. Detrimental effects of sublethal heating and *Talaromyces flavus* on microsclerotia of *Verticillium dahliae*. Phytopathology **1995**, *85*, 388–392.
12. Gamliel, A.; Stapleton, J.J. Improvement of soil solarization with volatile compounds generated from organic amendments. Phytoparasitica **1997**, *25* (Suppl.), 31S–38S.
13. Eshel, D.; Gamliel, A.; Grinstein, A.; Di Primo, P.; Katan, J. Combined soil treatments and sequence of application in improving the control of soil-borne pathogens. Phytopathology **2000**, *90*, 751–757.

SPRAY DRIFT

Franklin R. Hall
Ohio State University, Wooster, Ohio, U.S.A.

INTRODUCTION

Most pesticides (ca. 60% + are herbicides) are applied via liquid applications that involve atomization of liquids. Pesticide drift as defined by the National Coalition on Drift Minimization is "the physical movement of pesticide through the air at the time of pesticide application or soon thereafter from the target site to any nontarget site." Drift is undesirable because it results in waste of scarce resources and represents an inefficient use of application equipment and user time. Drift can result in underapplication of valuable actives, and can injure nontarget plants and adjoining waters and organisms. It is not a new problem nor, short of a ban on pesticides, can it be totally eliminated (1).

Two types of drift can be observed: vapor and droplet. Vapor drift, which is the airborne movement of evaporated chemical (highly volatile materials), can occur even after the droplet is deposited on a leaf surface. The second and more prominent, droplet drift, is the movement of spray droplets in liquid form. The larger the droplet, the faster it will fall to the ground and the less time it has to be reacted on by air currents (1, 2). Current spray nozzles produce a wide range of droplet sizes. It is generally accepted that a reduction of fines less than 150 μm in a spray cloud will significantly reduce the problems of drift hazards. The greater the proportion in these small categories, the higher the risks of significant movement off target. The lifetime of a droplet decreases with increasing temperatures and decreasing humidity due to evaporation. This results in an increasing number of droplets that are susceptible to drift. The recent introduction of more powerful chemistries increases the risks of even small amounts of movement resulting in significant impact to nontarget vegetation, etc. Off-target deposits, which decrease asymptotically with increased distance from the point of application, are essentially determined by the equipment, operator skills, and weather (3).

DRIFT TRENDS

Trends affecting drift and its increasing importance include costs of crop protection chemistries, that is, waste is costly; the environmental concerns; the rural/urban interface interactions caused by changing population demographics; increasing regulatory constraints, use of buffers, windbreaks, etc., and need for buffer areas to protect/isolate transgenic or organic crop plantings; and lastly, the new actives, albeit using less materials, that are more powerful and more vulnerable to environmental degradation. The need for improved profitability has led to larger farm sizes, assessment of input reduction scenarios, and greater crop diversity, all of which increase risk. Pesticide drift adds to these increasing farmer concerns. Regulatory authorities may limit use of certain products to certain buffer zones, which means users have to keep specific distances from bodies of water when using these products. Recently, the NL instituted guidelines for tree crops requiring the use windbreaks to mitigate the movement of pesticides onto adjoining waters or other vulnerable sites. The new technologies of precision agriculture promise much in improving targeted delivery and reduced applications in general, thus reducing drift impacts on nontargets (3).

It is generally considered that aerial application produces more drift then conventional ground (boom spraying) while tree crop sprayers utilizing air blast technologies are more prone to drift. The ratios of drift (related to field sprayers) occurring with current sprayers are rated as field crops/vineyards/fruit crops/hops as 1/6/15/25 (4). Spray drift can differ with crop growth, architecture, release heights above the crop, method of delivery (air vs. ground boom), weather, and user skills (2). Tree crops offer significantly higher off-site risks than spraying field crops. Prebloom delivery in tree crops can allow ca. 40–50% off-site movement versus ca. 10–15% at full leaf. Overtree tunnel sprayers plus recycling equipment can

Table 1 Emerging technologies and strategies affecting placement efficiency (drift) of pesticides

Technology/strategy	Impact
General	
Regulatory structure	FQPA, risk reduction, labeling specifics, application restrictions
Population demographics	Increasing conflicts at rural/urban interface; communication/proactive stance essential
Pesticide policy	IPM, risk reduction to aid targeting; farm policy increases crop diversity, decision making, and drift concerns
Ag economics	Global competition, prices pressure input reductions; ag business integration will affect user strategies, capabilities, and options
Knowledge	Decrease in applications via thresholds, prediction models, and CV resistance/tolerance to pests all aid drift reduction goals
Delivery technologies	
Seed coatings	Increased seed value and early plant health protection requirements suggest new opportunities for target-specific controlled release AI's/fertilizers
Herbicide weed wipers	Declining use—speed of travel and appropriate AI plus weed complexity issues
Sensors	Tree crop sensors (on/off), sprayer match to foliar target requirements, weed ID technologies with GPS/GIS
Nozzles	Drift and VRT nozzles, plus drift classification schemes, standardization, calibration/output clinics, closer crop/nozzle distances
Shrouds	Shrouds/shields, and air assistance useful in expanding weather-limited operating conditions plus reducing off–target percentages
User strategies	
Strategy integration	Coarse nozzles, reduce sprayer to target distances, reduce pressures, use drift mgt. adjuvants, ID and avoid vulnerable sites, and adjust strategy for high-risk sites, and changes in weather constraints
Restricted delivery	Real-time drift predictors, drift models aid precision timing and delivery
Buffers/windbreaks	Space and vegetation mitigate off-target movement; some regulatory and label requirements
Cropping strategies	More plants/unit ground, less off target losses, increased high-density plantings
Education	Mgt./training of new skills, more pro-active users; high-risk site ID with appropriate drift minimization tactics
Genetic engineering	Crop protection from within reduces need for external use of pesticides

reduce target losses to less than 1% but are limited by availability of dwarfing stock. The use of 110-degree nozzles in Europe allows reduced release heights (above the crop) but slower travel speeds. In the United States, the predominant use of 80-degree nozzles requires higher release heights (more potential drift), but allows greater travel speeds (2).

The major factors affecting drift include the critical spray characteristics (affected by nozzle type and design and operating parameters, i.e., orientation, release height, pressure, etc.), the equipment and application technologies, the chemistry (evaporation potential), the weather (temperature, relative humidity atmospheric stability, etc.) and operator care and skill (see Table 1 for a summary). The nozzle/equipment/operator factors contribute to ca. 64% of the drift problem. While all these factors are important, spray droplet size is the most critical with smaller sizes most prone to movement by local air currents (3). Biological data also infer that smaller sizes yield the highest mortality (hence crop protection) and thus we have the dilemma of greater biological effect versus environmental risks. Other delivery strategies such as banding versus broadcast treatments reduce use rates (and drift) by 40–50%. Drop nozzles increase coverage by 20–40% but require lower ground speeds. Seed coatings virtually eliminate drift; and while chemistry/uses have been limited to date, biotechnology technologies that increase seed and early plant health values suggest significant new opportunities.

EQUIPMENT CHANGES

Significant changes in nozzle design have taken place for the explicit purpose of reducing spray drift (4). This has been achieved by taking the flat fan design and changing it from a straight bore to one that incorporates a preorifice and an expansion chamber. The result is to reduce liquid velocity and pressure at the outlet orifice creating larger droplets and potentially less drift. Examples include Raindrop, DriftGuard, Turbo TeeJet, Turbo Flood, and CP nozzles and more recently the air-induction nozzles such as TurboDrop and AI TEE Jet. Other nozzle designs include the twin-fluids, AIRTEC and AirJet and recently the Synchro system that allows some on-the-go changes in spray volume without changing spray cloud characteristics (4).

Two main types of sprayers that reduce drift include the air-assisted and electrostatic sprayers. One group of air sprayers has atomizers located outside but directed into the air stream. The second has atomizers mounted within the air stream (often referred to as an air shear sprayer). Electrostactic charging of sprays has long been touted as an effective means of improving deposition on targets (and thus reducing off-target movement) but is still limited in field applications. The use of shields/shrouds to decrease spray drift has shown positive results on spray drift reduction. However, the problems of wetting and dripping, speed of travel, and proper use strategies (and costs) have thus far limited use in the United States versus wider use in Western Canada.

CHEMICAL CHANGES

Chemistry technologies have resulted in the development of certain adjuvants to minimize the production of small driftable droplets during the atomization process (5). These are mostly polymeric and act mostly as spray thickeners. Most have the disadvantage of losing their thickening properties when subjected to the shearing action of many pump/recirculating systems commonly used in pesticide applications. A new guar gum based product has an advantage of pump shear stability. The use of adjuvants to reduce drift while also offering crop performance benefits do have difficulties in that spray angles and pattern uniformities are changed and require user knowledge and adjustment to achieve desired effects. Some chemistries are more greatly affected than others. Uncertainities also arise in that coarse nozzles and the resulting impacts on coverage/efficacy have not been well defined nor have the interactions between spray thickeners and drift reduction nozzles.

Buffer zones can be useful in mitigating the amount of spray drift arriving at unintended sites. However, the amount of buffer space can be very costly to smaller growers and buffer effectiveness can be easily reduced with negligence on good application practices and drift reduction techniques such as temporal changes, proper calibration of sprayers, use of shields, proper adjustment of heights, use of drift nozzles, adjuvants, etc. Legislation on use of windbreaks for tree crops (NL) may be utilized in other countries. Buffer zone surface roughness (type of vegetation and soil conditions) can reduce off target droplet trajectory as it affects both boundary layer thickness and shear stress profiles (3).

New technology with use of GPS/GIS, modeling, etc., present new opportunities to agricultural applications in that precision farming implies less off-target movement because of more precise placement (6–9). Use of models, AgDrift, etc., to predict likely movement of spray away from intended target areas under different conditions leads to an improved decision making (3). Temporal adjustments, with the use of real time drift predictors, could aid the process of improved targeting and reduce drift. Education remains a primary need although the Coalition on Spray Drift Minimization recently completed a new education thrust emphasizing user training and drift minimization strategies (Hewitt, pers. com. and Ref. 3).

Mitigation strategies for minimizing drift and its impact has and will take on a larger role in developing improved strategies for agriculturists (Table 1). The major thrusts include new equipment and nozzles (a nozzle selection standardization effort by the United Kingdom coalition including drift ratings as well as nozzle pattern/atomization parameters (10). This effort should improve decisions about nozzle choices and conditions of use. Shrouds and air-assist technologies while successful, can be costly and also overutilized in more adverse weather conditions. Training of users needs further work as calibration clinics continue to show hands-on calibration techniques result in improved attention to details by users. Adjuvants, buffers, windbreaks, crop matching tools, and real-time drift prediction all add to the reduction of off target movement and thus a minimization of unintended pesticide impact. *If strategies are well integrated and introduced to users, the delivery efficiencies will be greatly improved and off-target concerns greatly minimized.* This will take a greater integration of program efforts aimed at user education of precision tools and techniques with the new crop protection chemistries of the new millenium. A lack of attention to this will surely result in the designed reduction in use profiles of these crop protection tools now seen as useful for a robust and competitive U.S. agriculture.

REFERENCES

1. Proceedings of the North American Conference on Pesticide Spray Drift Management, Portland, Maine, March–April 29–1, 1998; Buckley, D., Ed.; 285.

2. Hall, F.; Fox, R. The Reduction of Pesticide Drift. In *Pesticide Formulation and Adjuvant Technology*; Foy, C., Pritchard, D., Eds.; CRC Press: Boca Raton, FL, 1996; 209–239.
3. Hewitt, A.J. Spray drift modelling, labelling and management in the U.S. Aspects of Biology **2000**, *57*, 11–20.
4. Ganzlelmeir, H. In *Plant Protection—Current State of Technique and Innovations*, Proceedings of the 9th IUPAC Conference, London, UK, 1999; Brooks, G., Roberts, T., Eds.; Royal Society of Chemistry: Cambridge, UK, 100–119.
5. Proceedings of the Fifth International Conference on Adjuvants for Agrochemicals, Memphis, TN, 1998; McMullan, K.P., Ed.; Organizing Comm. Adjuvants Conf. 495.
6. Morgan, M.; Ess, D. *The Precision-Farming Guide for Agriculturists*; John Deere Publishing: Moline, IL, 1997; 117.
7. National Research Council. *Precision Agriculture in the 21st Century*; Academic Press: Washington, DC, 1997; 490.
8. Symposium. In *The State of Site-Specific Management for Agriculture*; Sadler, E., Pierce, F., Eds.; Soil Science Society of America and American, Agronomy Society: Madison, WI, 1997; 430.
9. *National Conference on Pesticide Application Technology*, Ridgetown, Aug. 10–11, 1995; College of Agriculture: Ridgetown, Ontario, 203.
10. Southcombe, E.; Miller, P.; Ganzelmeir, J.; Van de Zande, J.; Miralles, A.; Hewitt, A. In *The International (BCPC) Spray Classification System Including a Drift Potential Factor*, Proceedings of the British Crop Protection Conference—Weeds 3, 1997; Farnham, Surrey, UK, 371–380.

S

THE STERILE INSECT TECHNIQUE

E.S. Krafsur
Iowa State University, Ames, Iowa, U.S.A.

INTRODUCTION

The principle underlying the Sterile Insect Technique (SIT) is a simple one: female insects inseminated by sterile males lay eggs that do not hatch, thereby reducing the reproductive potential of a population in proportion to the rate of sterile matings. This ''birth control'' is continuous when sterile insects are released over large areas and multiple generations. As a target population declines, sterile mating rates may increase until the population disappears. Where immigration rates are substantial, cost-effective, and environmentally friendly suppression can be achieved. For the successful application of SIT, it is irrelevant how many times the female mates, because subsequent inseminations of sterile- or fertile-mated females can be by either sterile or fertile males. Success requires only that sterile mating rates are sufficient to cause a decline in population density.

There are several ways that insects can be sterilized. Hybrid sterility and exposure to ionizing radiation are two sterility-inducing methods that are environmentally harmless. Certain alkylating agents termed ''chemosterilants'' can be applied but their use causes environmental contamination as they accumulate in the food chain. The advantages of successful SIT practice include the suppression or total elimination of the target species from large areas and the reduction or elimination of insecticide usage. As a result, secondary pest problems decrease and biological control and other integrated pest management approaches become viable in the absence of insecticide treatments against the key pests. Aerially distributed sterile insects provide cheaper and more thorough coverage of treated areas than is possible by applying conventional ground based control methods. SIT should always be considered when evaluating the feasibility of areawide control programs (1).

PRINCIPLE

Knipling (2) developed a model that presents the essentials of SIT (Table 1). Its assumptions include a constant per capita reproduction rate per generation, that matings among target females are directly proportional to the relative frequency of sterile and fertile males, and a target population closed to immigration and without age structure. Reproduction rates and sterile fly release rates can be varied easily in this simple ''spreadsheet'' model to show how effective SIT can be. The model shows that when fertility is greater than the reciprocal of the target population's net reproductive rate R_o, then the size of a target population is unaffected by SIT. Thus, sterile mating rates must exceed $1 - R_o^{-1}$ to achieve a downward trend in population density. Over the long term, the *average* value of R_o for most species approximates to 1, but where density dependent regulation occurs, populations will show much greater transient reproduction rates and these must be taken into account. It is also efficient to exploit any natural periods of low population density to apply SIT.

Knipling's model indicates a sharp threshold between sterile male releases achieving a continuous decline in target population density or having no effect at all (Table 2). In terms of the number of generations required to achieve eradication, there is little advantage in greatly increasing sterile fly release rates once the critical level of sterile matings is exceeded. Small changes in sterile mating rates about the equilibrium point will have large effects. In natural populations there will be a lag time of ca. one generation between the first sterile male releases and the exposure of all females to sterile males. This is because natural populations have age structure and the females of many species become inseminated only once, early in their adult lives. Thus when planning a SIT program, a knowledge of target population age structure is most helpful and it is necessary to realize that releases must be made continuously for up to 10 or even more generations before resources can be switched to new target populations.

HISTORY AND METHOD

Both Potts and Vanderplank suggested that the sterile male progeny of interspecific matings between the tsetse

S

Table 1 Knipling's model (2) showing expected insect population sizes after releases of sterile insects (A net per capita reproduction rate of 5 × per generation is assumed)

Natural population N_t	No. of sterile insects released	Ratio, sterile: fertile	No. of insects reproducing N_{t+1}	Natural population
1,000,000	9,000,000	9:1	100,000	500,000
500,000	9,000,000	18:1	26,316	131,580
131,580	9,000,000	68:1	1,907	9,535
9,535	9,000,000	942:1	10	50
50	9,000,000	180,000:1	≈ 0	≈ 0

flies *Glossina morsitans* and *G. swynnertoni* might be used to control *G. m. centralis*. They carried out small field trials in Tanzania beginning in 1942 (3). Using the same principle, Davidson and colleagues released sterile male hybrids from the cross of male malaria mosquitoes *Anopheles arabiensis* x female *An. melas* in an effort to suppress a natural *An. gambiae* population in Upper Volta in 1969. It has long been known that when two related species or subspecies are crossed it is the heterogametic sex (males in most insects, females in Lepidoptera) that is sterile. Hybrid sterility is an attractive option for tsetse control because there are no damaging effects of ionizing radiation, but only males can be released and they must be sexually compatible with females in the target population.

Table 2 The number of generations elapsed from beginning sterile male releases to achieving hypothetical eradication

Per capita reproductive rate	% Sterile matings	No. of generations to eradication
2X	50.0	∞
	50.001	30
	50.01	24
	50.1	18
	50.5	14
	51.0	13
	60.0	7
	90.0	4
	99.0	3
5X	80.0	∞
	80.001	21
	80.01	17
	80.1	13
	80.5	10
	81.0	9
	90.0	5
	99.0	3
10X	90.0	∞
	90.001	19
	90.01	15
	90.1	11
	90.5	9
	91.0	8
	99.0	3

Runner in 1916, using the cigarette beetle, was the first to demonstrate that X-rays induced sterility. In 1937, Knipling raised the possibility that the New World screwworm, *Cochliomyia hominivorax*, might be controlled if the males could somehow be sterilized. A safe and effective means of inducing sterility is to treat insects with ionizing radiation, which induces dominant lethal mutations in their gametes. It is nearly 50 years since the first crucial experiment of releasing radiosterilized sterile screwworms on Sanibel Island (near Tampa, Florida) was carried out. That experiment nearly eliminated the population and was followed by a larger trial in which the species was eradicated from Curacao in the Netherlands Antilles. The Curacao experiment convincingly demonstrated the feasibility of SIT. The chief technical achievement, however, was to develop means of mass rearing screwworms for sterilization, packaging, and aerial distribution (4).

The early success of SIT initiated the development of new and flourishing lines of research. Investigations into details of insect reproduction from developmental, physiological, genetic, cytological, behavioral, evolutionary, and productivity standpoints is now considered essential to the development of biorational control measures. No less important is the bioengineering necessary for the efficient production and distribution of sterile insects—so called methods development.

PRACTICAL APPLICATIONS

Two programs stand out, both directed against highly fecund, colonizing species. SIT eradicated the New World Screwworm from North and Central America and from the Libyan Arab Republic (5, 6). In the United States, screwworms were confined in winter to the southernmost regions from which they spread northward each summer. Most of Mexico was infested throughout the year. Large scale SIT operations eradicated screwworm from the United States in the 1960s and 70s, from Mexico in the 1980s, and most of Central America in the mid-1990s (4). At this writing, efforts are being made to eradicate screwworms from Panama where a permanent barrier zone will be maintained in the Darien Gap by continuous sterile fly releases. And Jamaica is now being treated to sterile fly releases with eradication anticipated to occur in the year 2001 (7).

Screwworms are obligate parasites of mammals and can kill even large animals. Adult females become inseminated, undergo vitellogenesis, and oviposit within a week where mean temperatures exceed 21°C. They continue to mature successive batches of 250–400 eggs at 3- to 4-day intervals. Thus, explosive screwworm outbreaks can occur but *average* adult densities tend to be low compared with house flies or mosquitoes. Research has shown, however, that screwworm densities show highly patchy, or clumped, distributions, which means that local densities can be very much greater than the average density (5).

The medfly, *Ceratitis capitata* Wiedemann is a boom and bust, colonizing species that originated in Africa and has spread through much of the world's tropical and subtropical zones (8). Medfly larvae enjoy a wide host range of more than 350 species including apples, citrus, and many other high-value fruits and vegetables. Adult females may live more than 30 days and generate up to 1000 progeny in that time. Thus the medfly, like the screwworm, is capable of explosive outbreaks. It is frequently introduced into new territories where it establishes colonies that must be eliminated. Using SIT, eradication of medfly from Chile was achieved in 1996. Mexico and the United States are free of medfly despite numerous introductions of the pest. The melon fly was eradicated from Japan and the Queensland fruit fly from Western Australia. There are other examples of medfly and fruit fly eradication but space is too limited to identify them here.

The medfly is one of the world's most important economic pests of fruit and vegetables. Commercial fruit must be sprayed with insecticides throughout the growing seasons or losses may reach 100% of production. Furthermore, as a result of medfly presence, stringent quarantine requirements are imposed on the movement of agricultural products from infested regions, closing access to major export markets. Fruit from countries where medfly becomes established is embargoed by the United States, Chile, New Zealand, Australia, Japan, Hong Kong, Taiwan, Korea, and countries in Southeast Asia. The cost to producers and shippers of treating potentially infested fruit with post-harvest treatments such as methyl bromide or cold temperatures is enormous. In the United States, state and federal regulatory agencies have established protocols to prevent, detect, and treat medfly and other tephritid fruit fly infestations.

The use of SIT against medflies and other fruit flies is increasing as international commerce increases. This is because released sterile males are inherently more effective at finding their wild cousins than are human inspectors and because SIT is most effective in eliminating small, scattered populations (4).

Under the auspices of the Joint FAO/IAEA (International Atomic Energy Agency) Division, the technology of producing sterile medflies is highly advanced (7). Male-only strains, based on genetic sexing that permits the elimination of females early in development, are now mass produced and aerially released. There are great advantages in using such genetic sexing strains, including much cheaper production, packaging, and distribution costs, vastly greater sterile mating rates, and more accurate monitoring of female target population densities. Sophisticated mass rearing methods have been developed that allow the rapid substitution of a laboratory adapted strain by a new strain. As a result, the production cost has now decreased to ca. $250 per million sterile male pupae. There are more than 10 fruit fly mass rearing facilities in various regions of the world, producing ca. 2 billion flies weekly. Now, new factories are needed to meet the increasing international demand for sterile males.

LIMITATIONS OF SIT

To be susceptible to SIT, target populations must be sexually reproducing. Because SIT effectiveness in inversely density dependent, it is most effective against scattered populations of low average density. Widespread, dense populations are considered less susceptible because of the enormous numbers of sterile flies that must be produced, packaged, and released. In an areawide, integrated approach, however, it is feasible to suppress target populations by conventional means in anticipation of applying SIT. Before initiating an eradication program, it is important to assess the reinvasion potential. Released sterile insects should not cause damage, for example, by blood-feeding or by oviposition punctures.

A successful SIT program first must have a good measure of political and financial support from the commodity groups affected and the community at large. Even though benefits of pest eradication are substantial, the cost of SIT is high, and strong financial backing is essential to complete an eradication program or maintain a suppression program. It is also necessary to have the general public well informed, for example in medfly eradication programs in California and Florida where urban populations are subject to aerial sterile fly releases.

PROGNOSIS

In concept and application, SIT is a powerful method of eliminating a pest species from large geographic areas. The method is particularly effective in eliminating scattered clusters of pests that otherwise would escape detection, but it is also effective in sterile fly barrier programs (e.g., the Mexico–Guatemala and Chile–Peru borders) where releases prevent an exotic pest from becoming reestablished. The increase in trade of agricultural products requires the establishment of pest-free areas to meet quarantine requirements and SIT is an effective method to achieve this. In designing areawide programs SIT should always be considered and a simulation model developed to explore its economic feasibility over a long time frame. Some economic models suggest that SIT can be used as a suppression approach to replace continuous insecticide applications with sterile insect releases. Indeed, the economic advantages of both areawide suppression and eradication via SIT are compelling medfly programs in Israel, Jordan, and Palestine (9). The recent success of SIT against tsetse flies on Zanzibar (10) has encouraged a rapidly developing suite of technologies, led by the IAEA, that may soon allow the treatment of significant areas of Africa. Improvements continue in the application of SIT against the pink bollworm, codling moth, and other Lepidoptera. Indeed, there are a large number of insect pests that may prove to be highly susceptible to SIT (7).

ACKNOWLEDGMENTS

Thanks to Jorge Hendrichs, Don Lindquist, and Alan Robinson for helpful comments on the manuscript. This is Journal Paper no. J-18441 of the Iowa Agriculture and Home Economics Experiment Station, Ames, Iowa Project Nos. 3447 and 3457.

REFERENCES

1. Klassen, W. Eradication of introduced arthropod pests: theory and historical practice. Misc. Publ. Entomol. Soc. Am. **1989**, *73*, 1–29.
2. Knipling, E.F. The Basic Principles of Insect Population Suppression and Management. In *U.S. Dept. Agric. Handbook No. 512*; USDA: Washington, DC, 1979; 659.
3. Vanderplank, F.L. Experiments on the hybridization of tsetse-flies (*Glossina*, Diptera) and the possibility of a new method of control. Trans. Roy. Entomol. Soc. London **1947**, *98*, 1–18.
4. Krafsur, E.S. Sterile insect technique for suppressing and eradicating insect populations: 55 years and counting. J. Agric. Entomol. **1998**, *15* (4), 303–317.
5. Krafsur, E.S.; Whitten, C.J.; Novy, J.E. Screwworm eradication in North and Central America. Parasitol. Today **1987**, *3* (5), 131–137.
6. Lindquist, D.A.; Abusowa, M.; Hall, M.J.R. The new world screwworm fly in Libya: a review of its introduction and eradication. Med. Vet. Entomol. **1992**, *6* (1), 2–8.
7. In *Area-Wide Control of Fruit Flies and Other Insect Pests*, Joint Proceedings of the International Conference on Area-Wide Control of Insect Pests, May 28–June 2, 1998, and the Fifth International Symposium of Fruit Flies of Economic Importance, Penang, Malaysia, June 1–5, 1998; Tan, K.-H., Ed.; Penerbit Universiti Sains Malaysia, Malaysia, 2000; 782.
8. Fruit Flies, Their Biology, Natural Enemies and Control. In *World Crop Pests*; Robinson, A.S., Hooper, G., Eds.; Elsevier Science Publishers B.V.: Amsterdam, 1989; 1, 372, Vol. 2, p. 390.
9. Enkerlin, W.; Mumford, J. Economic evaluation of three alternative methods for control of the Mediterranean fruit fly (Diptera: Tephritidae) in Israel, Palestinian territories, and Jordan. J. Econ. Entomol. **1997**, *90* (5), 1066–1072.
10. Vreyson, M.J.B.; Saleh, K.M.; Ali, M.Y.; Abdulla, A.M.; Zhu, Z.-R.; Juma, K.G.; Dyuck, V.A.; Msangi, A.R.; Mkonyi, P.A.; Feldmann, H.U. *Glossina austeni* (Diptera: Glossinidae) eradicated on the island of Unguja, Zanzibar, using the sterile insect technique. J. Econ. Entomol. **2000**, *93* (1), 123–135.

STERILITY CAUSED BY PESTICIDES

William W. Au
The University of Texas, Galveston, Texas, U.S.A.

INTRODUCTION

Pesticides belong to a unique group of synthetic chemicals that have high biological activities and that are released legally and extensively into our environment. Therefore, there has been continued concern about their potential hazard to living organisms. A major concern is whether prolonged exposure to pesticides can cause reproductive problems leading to the extinction of species. A variety of studies have been conducted to elucidate the reproductive hazards of pesticides in native species, experimental animals, and human beings.

MECHANISMS OF ACTION OF PESTICIDES

Pesticides can be subdivided into several major categories: organophosphates, carbamates, organochlorines, synthetic pyrethroids, and others. Their principal mechanisms of action include inhibition of cholinesterase, perturbation of microsomal enzyme production, and damage to nervous systems (1). Therefore, it is possible that excessive exposure to pesticides can interfere with gametogenesis, sexual activities, and reproduction leading to the expression of sterility.

OBSERVED EFFECTS IN NATIVE ANIMALS AND IN EXPERIMENTAL SYSTEMS

Among the pesticides, organochlorines are characterized by their persistence in the environment, and the potential for both bioaccumulation and transfer of the pesticides up the food chain. Therefore, the widespread use of organochlorines in the past has been documented to cause contamination of wildlife and reduction of their populations. For example, the exposure is associated with a significant reduction of fish populations such as trout and salmon (2). Subsequently, the populations of predatory birds were significantly reduced (2). The devastating effects in migratory birds and in other wildlife were also demonstrated (3, 4). In these cases, failure to reproduce appropriately has been shown to be a major cause for the decline of the populations.

In studies using experimental animals under controlled exposure conditions to pesticides, organochlorine pesticides such as methoxychlor have been reported to cause reduction of fertility and litter size (5, 6), and kepone to cause anovulation (7, 8). Organophosphate pesticides have been shown to induce premature ovulation and to perturb oocyte development (9).

OBSERVED EFFECTS IN HUMANS

Very few pesticides have been shown systematically and consistently to cause sterility in humans. An exception is the exposure to a nematocide, 1,2-dibromo-3-chloropropane (DBCP). In the 1970s, workers in several pesticide manufacturing plants were reported to have fertility problems. From a systematic investigation, the infertility based on reduced sperm counts was shown to be associated with testicular function alteration and with exposure to DBCP rather than to other pesticides (10). Subsequently, the same group of scientists found that the reduction of sperm count was associated with an occupational exposure as short as 3 months (11) and with an employment duration-dependent effect (Table 1A). As shown in the table, as many as 76.5% of the workers who had been exposed to DBCP for more than 42 months were oligospermic. Among all the affected workers, many were azospermic or sterile.

Long-term follow-up studies of DBCP production workers showed that some of the affected workers did regain fertility and testicular function, and many of them were able to have children. Their offspring appeared normal and healthy. However, these workers predominantly had female offspring (Table 1B), ranging from 58.6 to 84.6% for the recovery duration from 5 to 17 years (12). As shown in the table, the highest female to male offspring ratio was found among workers within 5 years of recovery (13, 14). It appears that the recovery is a slow process and

Table 1 Reproductive problems from exposure to dibromochloropropane (DBCP)

A. Oligospermia	
Months of exposure to DBCP[a]	% Workers with oligospermia
0	2.9
1–6	8.3
7–24	28.6
25–42	66.7
>42	76.5
B. Offspring	
Years after recovery from oligospermia[b]	% Females in offspring
5	84.6
8	78.9
17	58.6

[a]Data derived from 10.
[b]Data derived from 13–15.

complete recovery with respect to the sex ratio was achieved only after 17 years (15). The observation confirms the previous recommendation of using altered sex ratios as an indication of reproductive hazards associated with pesticides (16). In another study, males infertile due to poor sperm quality were more likely than expected to be in the agricultural occupations with exposure to pesticides (17). Papaya fumigant workers with exposure to ethylene dibromide were reported to have significantly reduced sperm count per ejaculate (the percentage of viable and motile sperm) and increases in the proportion of sperm with abnormalities (18). Abnormal pregnancy outcomes (miscarriages and preterm deliveries) were associated with exposure to a variety of chemicals in combination with pesticides (atrazine, glyphosate, organophaosphates, 4-[2,4-dichlorophenoxy]butyric acid) in males (19). On the other hand, fertility in traditional male farmers, compared with organic farmers (who do not use pesticides), was not influenced by exposure to pesticides, based on the time taken to have the youngest child (20).

Among females, exposure to DBCP in pesticide manufacturing plants appears to have no effects on their fertility based on a limited survey (21). A study was conducted to investigate the relationship between the plasma level of organochlorine pesticides and the diagnosis of endometriosis, and no association was found (22). On the other hand, among women with medically confirmed infertility, exposure to pesticides was shown to be a significant contributing factor (23). Furthermore, the mechanism appears to be due to abnormal ovulation.

FUTURE CONSIDERATIONS

Based on the mechanisms of action of pesticides (1, 24) and on observations in animals, it is highly likely that overexposure and/or prolonged exposure to pesticides can cause reproductive problems in human. However, adverse reproductive effects have not been demonstrated unequivocally with modern pesticides. One reason is that the human population is usually exposed to much lower doses of pesticides, except in accidental exposure conditions, than those used in animals that have been shown to cause sterility. Under this condition, any adverse effects in human would be very small. Therefore, investigations using inappropriate study protocols may have generated inconsistent observations. Future studies should be conducted by using large enough populations and by minimizing multiple confounding factors. At this stage of our knowledge, it is fair to state that the potential impact of modern pesticides on sterility in humans has not been clearly demonstrated yet. However, based on the known biological activities of pesticides, they should be considered hazardous chemicals and should be handled with extreme caution.

REFERENCES

1. Kaloyanova, F.P.; el Batawi, M.A. *Human Toxicity to Pesticides*; CRC Press: Boca Raton, FL, 1991.
2. Pimentel, D. *Ecological Effects of Pesticides on Non-target Species*; U.S. Government Printing Office: Washington, DC, 1971.
3. Gard, N.; Hooper, M. An Assessment of Potential Hazards of Pesticides and Environmental Contaminants. In *Ecology and Management of Neotropical Migratory Birds*; Martin, T., Finch, D., Eds.; Oxford University Press: Oxford, 1995; 294–310.
4. Stinson, E.; Bromely, P. *Pesticides and Wildlife: A Guide to Reducing Impacts on Animals and Their Habitat*; Publication No. 420-004, Virginia Department of Game and Inland Fisheries, Virginia, 1991.
5. Gray, L.E.; Ostby, J.S.; Ferrell, J.M.; Sigman, E.R.; Goldman, J.M. Methoxychlor induces estrogen-like alterations of behavior and the reproductive tract in the female rat and hamster: effects on sex behavior, running wheel activity, and uterine morphology. Toxicol. Appl. Pharmacol. **1988**, *96*, 525–540.
6. Gray, L.E.; Ostby, J.S.; Ferrel, J.M.; Rehnberg, G.; Linder, R.; Cooper, R.; et al. A dose-response analysis of methoxychlor-induced alterations of reproductive development and function in the rat. Fundam. Appl. Toxicol. **1989**, *12*, 92–108.
7. Eroschenko, V.P. Estrogenic activity of the insecticide chlordecone in the reproductive tract of birds and mammals. J. Toxicol. Environ. Health **1981**, *8*, 731–742.

8. Guzelian, P.S. Comparative toxicology of chlordecone (kepone) in humans and experimental animals. Annu. Rev. Pharmacol. Toxicol. **1982**, *22*, 89–113.
9. Rattner, B.A.; Michael, S.D. Organophosphorous insecticide induced decrease in plasma luteinizing hormone concentration in white-footed mice. Toxicol. Lett. **1985**, *24*, 65–69.
10. Whorton, D.; Milby, T.H.; Krauss, R.M. Testicular function in DBCP exposed pesticide workers. J. Occup. Med. **1979**, *21*, 161–66.
11. Whorton, D.; Krauss, R.M.; Marshall, S. Infertility in male pesticide workers. Lancet **1977**, *ii*, 1259–1261.
12. Goldsmith, J.R. Dibromocholorpropane: epidemiological findings and current questions. Ann. of the New York Academy of Sci. **1997**, *831*, 300–306.
13. Potashnik, G.; Goldsmith, J.; Insler, V. Dibromochloropropane-induced reduction of the sex-ratio in man. Andrologia **1984**, *16*, 213–218.
14. Potashnik, G.; Yanai-Inbar, H. Dibromochloropropane: an eight-year re-evaluation of testicular function and reproductive performance. Fertil. Steril. **1987**, *47*, 317–322.
15. Potashnik, G.; Porath, A. Dibromochloropropane: a 17-year reassessment of testicular function and reproductive performance. J. Occup. Environ. Med. **1995**, *37*, 1287–1292.
16. James, W.H. Offspring sex ratio as an indicator of reproductive hazards associated with pesticides. Occup. Environ. Med. **1996**, *52*, 429–430.
17. Strohmer, H.; Boldizsar, A.; Plockinger, B.; Feldner-Busztin, M.; Feichtinger, W. Agricultural work and male infertility. Am. J. Ind. Med. **1993**, *24*, 587–592.
18. Ratcliff, J.M.; Schrader, S.M.; Steenland, K. Semen quality in papaya workers with long term exposure to ethylene dibromide. Br. J. Ind. Med. **1987**, *44*, 317–326.
19. Savitz, D.A.; Arbuckle, T.; Kaczor, D.; Curtis, K.M. Male pesticide exposure and pregnancy outcome. Am. J. Epidemiol. **1997**, *146*, 1025–1036.
20. Larsen, S.B.; Joffe, M.; Bonde, J.P. The asclepiod study group. Time to pregnancy and exposure to pesticides in Danish farmers. Occup. Environ. Med. **1998**, *55*, 278–283.
21. Marshall, S.; Whorton, D.; Krauss, R.M.; Palmer, W.S. Effect of pesticides on testicular function. Urology **1978**, *11*, 257–259.
22. Lebel, G.; Dodin, S.; Ayotte, P.; Marcoux, S.; Ferron, L.A.; Dewailly, E. Organochlorine exposure and the risk of endometriosis. Fertility and Sterility **1998**, *69*, 221–228.
23. Smith, E.M.; Hammonds-Ehlers, M.; Clark, M.K.; Kirchner, H.L.; Fuortes, L. Occupational exposures and risk of female infertility. J. Occup. Environ. Med. **1997**, *39*, 138–147.
24. Sharara, F.I.; Seifer, D.B.; Flaws, J.A. Environmental toxicants and female reproduction. Fertility and Sterility **1998**, *70*, 613–622.

STICKY TRAPS

Mary L. Cornelius
Southern Regional Research Center, New Orleans, Louisiana, U.S.A.

S

INTRODUCTION

Sticky traps are used for the monitoring and control of agricultural and urban pests. For sticky traps, such as panels and spheres, the exposed surface of the trap is coated with an adhesive substance that does not dissolve in water so that traps can be left outside for several weeks or months. Pests become immobilized when they contact the sticky surface of the trap. Other traps are designed to draw pests inside the trap structure. Once pests enter traps, they land on sticky inserts or bottom liners located inside of traps. Sticky traps are generally used for control of pests in homes, gardens, or greenhouses. In large-scale agricultural or urban settings, sticky traps are generally used for detection and monitoring of pest populations, rather than for control. Sticky traps generally use visual and/or olfactory cues to attract pests and their effectiveness depends upon the attractiveness of these cues to the target pest. Also, trap location is often a very important factor in determining the effectiveness of traps for intercepting pests.

VISUAL CUES

Yellow sticky traps have been used to monitor or reduce populations of flying pest insects, such as aphids, whiteflies, and rust flies. Many herbivorous insects are attracted to yellow. The attraction to yellow is considered to be a natural response to plant foliage (1). For example, yellow sticky traps have been effective for controlling leafminer populations in greenhouses. Yellow sticky traps are also used to monitor pest dispersal or migration. In Texas, a whitefly monitoring program used yellow sticky cards to determine that the activity of migrating adult whiteflies correlated with the maturation and defoliation of the cotton crop (2). White sticky traps have also been used successfully to monitor insect pests. For example, white sticky cards are used for monitoring the tarnished plant bug and the European apple sawfly in New England orchards (3). Studies using a range of colors determined that a light shade of blue was the most attractive color to adult thrips. Blue sticky cards, either hung above plants or placed on stakes in individual plant pots, can be used to detect thrip infestations. Thrip populations can be monitored in large greenhouse crops by hanging thin blue sticky strips above plants.

Fruit fly pests are attracted to spheres that resemble host fruit. Extensive studies on the response of fruit fly pests to sticky traps of different shapes, sizes, and colors determined that spheres were generally more attractive than cubes, cylinders, or rectangles of equivalent surface area. Studies also determined that red spheres were the most effective for capturing the apple maggot fly, *Rhagoletis pomonella* (Walsh), and green spheres were most effective for capturing the walnut husk fly, *R. completa* (Cresson) (4). Sticky sphere traps are currently used to monitor fruit fly pests in commercial orchards.

Sticky traps can be used in conjunction with light to attract insects that fly at night. For example, winged reproductive forms of certain termite species swarm at night in order to find mates and establish new colonies, and are strongly attracted to light. In a research program to control Formosan subterranean termites, *Coptotermes formosanus* (Shiraki), in the French Quarter of New Orleans, Louisiana, sticky traps are used to monitor termite populations in treated and untreated areas. Sticky white cards are hung from street lamps. During swarming season, large numbers of winged termites are captured on these sticky cards, and provide an estimate of the termite population in the vicinity.

COMBINATION OF VISUAL AND OLFACTORY CUES

Sticky traps often use a combination of visual and olfactory cues to capture pests. A synthetic apple odor, butyl hexanoate, has been used successfully in combination with red sticky spheres to increase captures of apple maggot flies. Plastic vials filled with the synthetic fruit odor are hung in trees along with red sticky spheres (5).

Sticky traps have also been baited with protein odors to capture fruit fly pests, such as the Pherocon AM trap which consists of a yellow sticky board baited with a vial filled with a protein bait. Another trap that uses both visual and olfactory cues is an open-bottom, cylindrical trap made from opaque green plastic with openings around the midline of the trap. This trap has been designed to monitor fruit fly pests, including the Mediterranean fruit fly, *Ceratitis capitata* (Wiedemann), and the Mexican fruit fly, *Anastrepha ludens* (Loew). An insert made from fluorescent green adhesive paper is hung in the center of the trap to capture flies that enter the trap. Flies are attracted into the trap with odors emitted from protein lures, such as a newly developed three-component hydrolyzed protein lure, composed of ammonium acetate, putrescine, and trimethylamine (6).

TRAP LOCATION

Sticky traps need to be placed in a location that will maximize the likelihood that the trap will intercept the target pest. Information about the biology of the target pest is necessary in order to determine how the position of the trap in the field affects trap captures. For example, factors such as height above ground, proximity to fruit and foliage, and distance from the outside edge of the tree canopy influence the response of apple maggot flies to red sticky spheres (7). Also, captures of the blueberry maggot fly, *Rhagoletis mendax* (Curran), on yellow sticky board traps are greater when the boards are positioned in a "V" orientation with the sticky side facing downward above the canopy of blueberry bushes, than when traps are hung in a vertical orientation because of the upward flight of flies within the canopy (8). Blue sticky strips, used for detecting thrips, are hung vertically so that the top of the strip is 2 ft. above plants, and the bottom of the strip is an inch or two above the plant canopy. Strips are positioned in this way because most adult thrips fly within a 2-ft. zone above the plants (9).

BARRIERS

Sticky traps can also be used as barriers to protect plants against pests. For example, certain ant species protect aphids, scales, and mealybugs from their natural enemies because the ants feed on a sugary substance, known as honeydew, excreted by these insects. The presence of ants tending these pests greatly increases pest infestations on plants. Sticky ant barriers can be applied to trunks or stalks of woody plants to keep ants away, and allow natural enemies to reduce pest populations.

BENEFITS AND COSTS

Sticky traps are nontoxic, relatively inexpensive, and easy to use. Sticky traps have been used successfully to control pests in homes, gardens, and greenhouses. For example, sticky ant barriers are commonly used to prevent ants from tending aphids on rose bushes and sticky yellow cards are frequently hung or staked among greenhouse plants to reduce whitefly populations.

Sticky traps have also been used successfully to detect and monitor pest populations as part of an integrated pest management program in large-scale agricultural and urban settings. Sticky cards have been used to capture small insect pests, such as fungus gnats, that are difficult to sample by other techniques. Certain sticky traps, such as red spheres baited with butyl hexanoate, attract only the target pests and have minimal impact on beneficial insects. In contrast, yellow and white sticky boards are attractive to many insects and may capture beneficial insects, as well as pests.

Sticky traps are also used to evaluate the effectiveness of different control strategies. For example, the parasitic mite, *Varroa jacobsoni* (Oudemans) is a serious pest of the honey bee, *Apis mellifera* L. Sticky-board collection devices that allow mites to fall through, but that protect bees from contacting the sticky surface, have been used to estimate the number of mites in honeybee colonies killed by different acaricide treatments (10). Effectiveness of large-scale termite control is being monitored, in part, by sticky traps to assess populations of swarming termites.

Using sticky traps for control of pests is generally not cost-effective for large-scale commercial operations because of the labor involved in maintaining traps. Sticky traps often become inundated with insects quickly, and have to be collected and changed frequently in order to maintain their effectiveness. For example, during the peak of the swarming season, sticky cards used to monitor winged termites must be changed every few days. Handling sticky traps can be a messy, time-consuming job. Traps that use sticky inserts or linings inside of traps are less labor-intensive and easier to maintain. However, trap lures may need to be changed regularly in order to maintain their effectiveness. Also, the visual/olfactory cues associated with traps usually attract pests from relatively short distances. Therefore, it is necessary to use large numbers of sticky traps to control pests. However, new lures that are extremely attractive to target pests and new trap designs that combine visual and olfactory cues and that are easier to maintain are currently being developed. These new traps have the potential to increase trap captures substantially and may provide cost-effective, nontoxic methods to control pests through mass trapping.

S

See *Trapping Pest Populations*, pages 855–858; and *Pest Population Monitoring*, pages 587–589.

REFERENCES

1. Prokopy, R.J.; Owens, E.D. Visual detection of plants by herbivorous insects. Ann. Rev. Entomol. **1983**, *28*, 337–364.
2. USDA Whitefly Knowledgebase. www.ifas.ufl.edu (accessed Nov. 7, 2000).
3. National IPM Network. Integrated Pest Management in the United States. www.reeusda.gov/agsys/nipmn/index.htm (accessed Oct. 28,1999).
4. Katsoyannos, B.I. Responses to Shape, Size, and Color. In *World Crop Pests: Fruit Flies: Their Biology, Natural Enemies and Control*; Robinson, A., Hooper, G.H., Eds.; Elsevier Science Publishers: Amsterdam, 1989; 3A, 307–324.
5. Reissig, W.H.; Stanley, B.H.; Roelofs, W.L.; Schwarz, M.R. Tests of synthetic apple volatiles in traps as attractants for apple maggot flies (Diptera: Tephritidae) in commercial apple orchards. Environ. Entomol. **1985**, *14*, 55–59.
6. Epsky, N.D.; Heath, R.R. Exploiting the interactions of chemical and visual cues in behavioral control measures for pest tephritid fruit flies. Florida Entomologist **1998**, *81* (3), 273–282.
7. Drummond, F.; Groden, E.; Prokopy, R.J. Comparative efficacy and optimal positioning of traps for monitoring apple maggot flies (Diptera: Tephritidae). Environ. Entomol. **1984**, *13*, 232–235.
8. Prokopy, R.J.; Coli, W.M. Selective traps for monitoring *Rhagoletis mendax* flies. Prot. Ecol. **1978**, *1*, 45–53.
9. Olkowski, W.; Daar, S.; Olkowski, H. *Common-Sense Pest Control: Least-Toxic Solutions for Your Home, Garden, Pets, and Community*; Taunton Press: Newton, CT, 1991; 715.
10. Calderone, N.W. Evaluating subsampling methods for estimating numbers of *Varroa jacobsoni* mites (Acari: Varroidae) collected on sticky-boards. J. Econ. Entomol. **1999**, *92* (5), 1057–1061.

STORED FOOD PEST MANAGEMENT

Judy Johnson
Horticultural Crops Research Laboratory, Fresno, California, U.S.A.

INTRODUCTION

Worldwide losses of stored foods caused by insects, mites, vertebrates, and microorganisms are estimated at 9–20%, with developing countries most severely affected (1). Economic losses include actual consumption of product by pests; product quality degradation through biochemical changes, product contamination with toxins, webbing, cast skins, hairs or feces, and the cost of control. Minimizing these losses requires implementation of a thorough pest management program that should include proper facility design, careful sanitation, maintenance of suitable storage environments, and judicious use of pesticides.

INTEGRATED PEST MANAGEMENT STRATEGY FOR STORAGE ENVIRONMENTS

Because foods are susceptible to attack along the entire processing and marketing chain, pests must be controlled in diverse storage environments, including on-farm product storage, warehouses, food processing facilities, packinghouses, and retail outlets. Although the wide range of products and facilities under consideration requires that the specific tactics employed will vary, the strategy for an effective pest management program remains constant.

1. Identify and understand the nature and scope of the pest problem.
2. Design facilities, storage, and equipment to exclude pests and reduce their harborages as much as possible.
3. Maintain a good sanitation program.
4. Monitor pests through trapping, product sampling, and inspection.
5. Maintain a proper storage environment to minimize pest populations.
6. Select and use treatment methods with care and only when necessary.

COMMON STORED FOOD PESTS

The first step in designing any pest management program is to understand the pest problem involved. A brief listing of stored food pests is found in Table 1.

Microorganisms

Most foods are susceptible to spoilage caused by fungi and bacteria, unless protected in some manner. These organisms destroy foods through rot, or may reduce food quality by causing softening, odors or off flavors; changing the texture or color; reducing the nutritional value; or contaminating the food with toxins (1–3). Generally, fungi and yeasts are more important decay organisms in unprocessed cereal grains and fruits, while bacteria are more important in vegetables, meat, fish, and dairy products. The amount of available water (water activity, or A_w) largely determines which organisms are capable of growth in a particular foodstuff. Bacteria normally require foods of higher A_w than fungi; foods with an A_w of 0.6 or less do not support microbial growth.

Insects and Mites

Numerous insect and mite species may be found in or near food stores (1, 4, 5). Species that reproduce in bulk foods and scavengers in food storage facilities may become serious pests, consuming large quantities of product as well as contaminating product with their bodies, feces, webbing, and exuvia. Infestations at the consumer level result in returned product, loss of consumer confidence, and may affect international trade through phytosanitary or quarantine restrictions.

Vertebrate Pests

Vertebrate pests (rodents, bats, and birds) consume or contaminate product in food storage facilities (1–3). Their presence constitutes a health hazard as they carry ectoparasites and pathogens that may be transmitted to hu-

S

Table 1 Common pests of stored foods

Pest	Examples	Type of damage
Microorganisms (molds, yeasts and bacteria)	*Aspergillus, Mucor, Penicillium, Botrytis, Streptococcus, Lactobacillus, Saccharomyces*	Spoilage, quality loss through biochemical changes, toxin production
Grain weevils	*Sitophilus* spp.	Larvae feed within whole grains
Grain borers	*Rhizopertha dominica, Prostephanus truncatus*	Larvae feed within whole grains, also attack processed foods
Grain beetles	*Oryzaephilus* spp., *Cryptolestes* spp., *Tenebroides mauritanicus*	Adults and larvae feed on variety of foods, many feed on molds
Flour beetles	*Tribolium* spp., various Tenebrionidae	Adults and larvae feed on wide variety of raw and processed foods
Mealworms	*Tenebrio* spp.	Adults and larvae feed on grain and animal products
Dermestid beetles	*Attagenus unicolor, Dermestes* spp., *Trogoderma* spp.	Larvae feed on wide variety of raw and processed foods
Spider beetles	*Ptinus* spp.	Scavengers within storage facilities
Miscellaneous beetles	*Lasioderma serricorne, Stegobium paniceum, Carpophilus* spp. *Typhaea stercorea, Blapstinus* spp.	Adults and larvae feed on wide variety of raw and processed foods
Grain moths	*Sitotroga cerealella, Corcyra cephalonica*	Larvae feed within whole grains
Flour moths	*Plodia interpunctella, Ephestia kuehniella, Cadra cautella, Pyralis farinalis*	Larvae feed on a variety of foods, reduce quality through production of webbing and feces
Booklice	*Liposcelis* spp.	Develop in moist, poorly stored foods
Ants	*Iridomyrmex humilis, Monomorium* spp., *Camponotus* spp.	Scavengers within storage facilities
Cockroaches	*Blattella* spp., *Blatta orientalis, Periplaneta* spp.	Scavengers within storage facilities
Flies	*Musca* spp., *Fannia* spp., *Drosophila* spp.	Breed in a variety of organic matter, causes food contamination
Mites	*Acarus* spp., *Tyrophagus* spp.	Develop in moist, poorly stored foods
Birds	Starling, grackle, pigeon, sparrow	Scavengers within storage facilities, causes food contamination
Rodents	*Rattus* spp., *Mus musculus*	Scavengers within storage facilities, causes food contamination

mans. Additionally, their nests may provide harborage for storage insects and mites.

STRATEGIES AND TACTICS

Facilities and Equipment Design

Buildings and storage structures should be designed to prevent pest invasion (1). Doors, windows, vents, and other openings should fit tightly and be well screened. Outside ledges that may provide roosts for birds should be avoided. Areas that provide harborage for insects, both outside and within the facility, should be identified and, if not eliminated, made easily accessible for cleaning. Outside lighting should be designed to avoid attracting insects into work and storage areas. As a rule, mercury vapor lamps should not be used near entrances, as they are more attractive to insects. Properly designed, pest resistant packaging can prevent infestations throughout the marketing chain. Food storage in developing countries, particularly in tropical regions, offers unique problems, due both to increased humidity and temperature and to a lack of resources. In some areas, traditional underground storage structures offer good insulation and protection against insects, and may be improved by adding inexpensive plastic liners. Large, temporary, airtight, grain storage structures have been constructed inexpensively using conical butyl-rubber bags (6). Product may be protected on the farm or in warehouses by storage in smaller, insect-tight containers or bags.

Sanitation

The importance of a good sanitation program cannot be overemphasized. The accumulation of any waste material attracts scavengers and serves as reservoirs for many storage insects and microorganisms. A thorough program includes regular cleaning of floors, walls, and equipment, prompt clean-up of spills; and daily removal of waste. Facility and equipment design is crucial to the success of any sanitation program (1). Raw product should be stored in structures that are easily cleaned. Surfaces that accumulate processing dust should be avoided or made easily accessible for cleaning. Conveyor belts should be designed to minimize spillage of product and be easily accessible for cleaning. Regular inspection of the facilities is also essential.

Manipulation of the Storage Environment

Growth of pest populations can be reduced or prevented by manipulation of the storage environment (1, 2, 7). Storing foods at the lowest practical temperature and moisture content generally minimizes growth of decay organisms and the development of insect populations. The practice of aerating bulk-stored grain is an effective management technique as it may reduce both grain moisture and temperature. The utility of aeration may be enhanced with the addition of chilled air. Freeze-outs may be used where winter temperatures drop below zero to control insects in flour mills and other food processing facilities simply by turning off the heat. Heat has also been used to disinfest entire facilities.

Modification of the storage atmosphere to reduce oxygen or increase CO_2 levels can reduce or eliminate pest populations. Modified atmospheres may be obtained by adding CO_2 or N_2 supplied from compressed gas cylinders, or with exothermic generators or membrane gas separation systems. Reduced oxygen atmospheres can also be obtained within airtight storage containers by the respiration of microorganisms and insects. This last method may be useful in some developing countries, where traditional underground storage structures are often airtight or nearly so.

Monitoring of Pest Populations

Critical to an effective management program is knowledge of the severity and extent of pest problems. Regular facility inspections can supply some information, particularly concerning vertebrate infestations (1). Rodent infestations can also be monitored with traps, glueboards, and tracking boards. Inspection of product samples for insect infestations is useful in bulk stored grain and may be required by regulatory agencies (5), but population levels in other commodities may be too low for direct sampling to be effective. Particularly for these products and for processed foods, insect populations are better monitored with pheromone or food bait traps. Several designs of aerial, surface, and probe traps are available commercially for monitoring stored food insects (5). Automated monitoring systems for bulk product using probe traps or acoustical detection methods are currently in development.

Unlike many pest management programs for field crops, stored food pest management programs rarely use well-defined action levels in the form of insect counts from samples or traps to make treatment decisions. However, useful models are being developed for management of bulk stored grain pests (7). The utility of monitoring data for most food storage facilities lies in locating and determining the extent of infestations and evaluating the success of management programs. Monitoring data require careful interpretation. Treatment decisions normally are made based on changes in trap or sample numbers, rather than an absolute action threshold.

Trapping

Traps for both insects and vertebrates may be useful for reducing pest infestations. Trapping for insects on the outside perimeter of a facility may intercept invading insects before they can infest product, but are not effective at low trap densities. A carefully designed and maintained trapping program for rodents may reduce or eliminate infestations, and is useful in situations where other control measures are not appropriate (1). Rodent control through trapping is labor intensive, is rarely feasible in large facilities, and should always be accompanied by exclusion methods and a good sanitation program.

Biological Control

Stored food insect pests have numerous natural enemies capable of exerting significant levels of control (1, 7). Although there are documented examples of control by naturally occurring parasitoids or predators, more reliable control is obtained by inundating the contained and protected stored food environment with the release of commercially available natural enemies. This tactic is appropriate for bulk stored grain, where the beneficial insects are removed before final processing, or where product is already packaged, but may be considered a potential product contaminate when applied to finished, unpackaged commodities. At present, the U.S. EPA has granted an exemption from tolerance for beneficial insects only in stored raw whole grains and packaged food in warehouses.

Microbial agents have been shown to persist for long periods in storage environments, where they are protected from heat and ultraviolet light. The insect pathogen *Bacillus thuringiensis* (*Bt*) has been used as a grain protectant, effective against externally feeding moths. However, resistance to *Bt* has been documented in the Indian meal moth. A granulosis virus has been shown to provide effective control of the Indian meal moth in grains, dried fruits, and nuts, but is not yet registered.

Chemical Control

Chemical pesticides are an important part of most stored food pest management programs. Because of the lack of clear action levels, pesticides are often used on a prophylactic basis within stored food systems. There are four broad categories of pesticides used in food storage systems: residual pesticides, space treatments, baits, and fumigants (1). The exact combination and use of pesticides will depend upon the particular pests and facility. In addition, traditional storage methods in developing countries often include the addition of native plants that contain compounds that kill or repel pest insects.

One of the goals of any well-planned pest management program is the reduction of chemical pesticide use. This goal is difficult to obtain for stored foods, because the costs associated with control failures are high (regulatory action, loss of consumer confidence, lawsuits, and endangering public health), the tolerance for infestations is low, and pesticides are cost effective. Pesticide cost increases, development of resistance, recent regulatory actions concerning fumigants, and public concerns about pesticide residues encourage stored food pest management programs that reduce reliance on pesticides, particularly fumigants. Alternative product disinfestation methods under consideration to replace fumigation are heat treatments, cold treatments, modified atmospheres, and ionizing radiation.

S

REFERENCES

1. *Ecology and Management of Food-Industry Pests*; Gorham, J., Ed.; FDA Technical Bull. 4, Association of Official Analytical Chemists: Arlington, VA, 1991; 595.
2. *Storage of Cereal Grains and Their Products*; Christensen, C., Ed.; American Association of Cereal Chemists: St. Paul, MN, 1982; 544.
3. *Stored-Grain Ecosystems*; Jayas, D., White, N., Muir, W., Eds.; Marcel Dekker, Inc.: New York, 1995; 757.
4. *Insect and Mite Pests in Food: An Illustrated Key*; Gorham, J., Ed.; Agriculture Handbook No. 655, U.S. Department of Agriculture: Washington, DC, 1991; 767.
5. *Stored Product Management*; Krischik, W., Cuperus, G., Galliart, D., Eds.; Circular No. E-912, Oklahoma Cooperative Extension Service: Stillwater, OK, 1995; 242.
6. Navarro, S. In *Distribution and Abundance of Insects in Butyl-rubber/EPDM Silos Containing Wheat*, Proceedings of the XV International Congress of Entomology, Washington, DC, August, 1976; Entomological Society of America: College Park, MD, 1976; 680–687.
7. *Integrated Management of Insects in Stored Products*; Subramanyam, B., Hagstrum, D., Eds.; Marcel Dekker, Inc.: New York, 1996; 426.

STRATEGY

Ron B. Hammond
Ohio State University, Wooster, Ohio, U.S.A.

INTRODUCTION

Important to the concept of pest management is the development of an overall plan to exclude, alleviate, or eliminate a real or perceived pest problem—the pest management strategy (1). The plan, or strategy, depends on the particular pest and the crop and can be pest exclusion or the modification of the pest, the crop, or a combination of both (1–3). The overall aim is to reduce pest status, which depends on the economics involved and the pest and crop characteristics (4). Once a strategy is developed, the methods of implementation need to be chosen. Methods, known as tactics, are numerous not only in number but also preferably in use. There are usually a number of tactics available for a particular pest problem. Tactics include regulatory control, biological control, host plant resistance through classical breeding and biogene technology, cultural control, mechanical control, and finally, pesticides. The use of multiple tactics to reduce pest status is a basic tenet of pest management.

Four basic pest management strategies have been developed:

1. Do nothing.
2. Reduce pest population density.
3. Reduce crop susceptibility to pest injury.
4. Combine reduced population numbers with reduced crop susceptibility.

DO-NOTHING STRATEGY

In a pest management program where pest population density is assessed with appropriate surveillance procedures and found to be below the economic threshold, the decision to NOT take any therapeutic action is the most appropriate decision a grower should make. The most frequent situation where the do-nothing strategy is used is with pests that cause indirect injury to the crop. In many instances, slight plant injury does not cause economic injury and, thus, can be tolerated. Often times, the injury is aesthetic in nature, causing only perceived damage. Implementing costly tactics, which is usually a pesticide application, will often cause more money to be spent on management than is gained in preventing the injury. The do-nothing strategy is often the ultimate strategy to use in a pest management program, but requires reliable knowledge of the crop, pest, and economics involved.

REDUCE PEST POPULATION STRATEGY

The most widely used strategy in excluding or alleviating real pest problems is reducing the population density of the pest. With certain pests, regulatory measures are often taken to prohibit a pest's entrance into an area where it could become established. For an existing pest with a history of economic problems, tactics are often used a priori, that is, in a preventive manner. This is often done with various cultural, biological, mechanical, and host plant resistance tactics, and as a last resort, with pesticides. When situations allow the use of economic thresholds and surveillance, and the management of a pest is required, tactics, often pesticides, are employed in a therapeutic manner.

Two slightly different objectives are desirable in reducing pest population densities. If the long-term density of a pest, the general equilibrium position (GEP), is low compared to the economic threshold, the strategy should be to dampen population peaks (Fig. 1A). This prevents economic damage from occurring during outbreaks.

If the GEP is close or above the economic threshold, the best approach is to attempt to lower the GEP so that the economic threshold is not reached. There are two ways to achieve this: 1) reduce the environmental carrying capacity, which is the maximum pest density a given

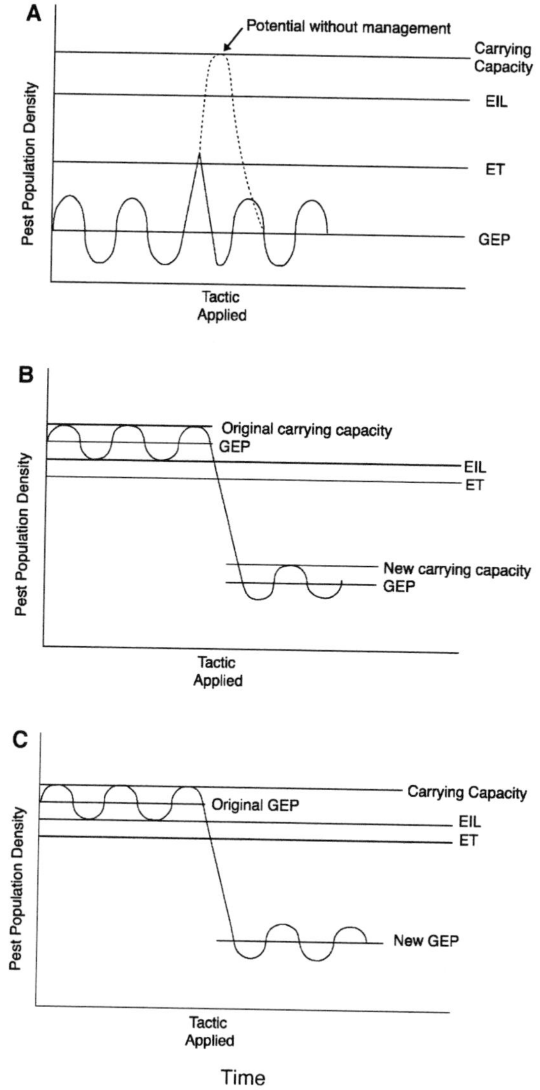

Fig. 1 Graphs showing the management strategies of (**A**) dampening population peaks and lowering the general equilibrium potential by (**B**) reducing the environmental carrying capacity or (**C**) reducing the inherited reproductive and/or survival potential of the pest. EIL, economic injury level; ET, economic threshold; and GEP, general equilibrium position. (Adapted from Ref. 1.)

environment will support (Fig. 1B), or 2) reduce the inherited reproductive and/or survival potential of the pest (Fig. 1C). The former is done by reducing the inherent favorableness of the pest's environment, such as rotating to a nonhost crop, using tillage to reduce insect density or plant disease inoculum, or using pest resistant crops that impart a population density reducing factor. The second approach is accomplished by altering the specific ability of the pest to reproduce or survive. Sterile male insect releases and release of mating disruption chemicals are examples of this approach.

REDUCE CROP SUSCEPTIBILITY STRATEGY

Rather than attempting to reduce the pest population density, an alternative strategy is to reduce the susceptibility of the crop. There is no direct impact on the pest; rather, the pest is tolerated. The are two primary tactics in a reduced crop susceptibility strategy, the first being host plant resistance through a resistance mechanism known as tolerance. Plants are able to tolerate a pest population density without economic losses that normally would be of economic concern. Tolerance is under genetic control and, thus, can be a goal in a plant-breeding program. The second tactic has been known as pseudoresistance, or more recently as ecological resistance, and is not under genetic control. The susceptibility of a crop is lowered through its interaction with its environment. Lowering the susceptibility can be achieved by improved fertilization and irrigation, or by upsetting the synchrony between a pest and its susceptible stage with crop planting dates or crop varieties with a different maturity. The reduced crop susceptibility strategy is one of the most effective and environmentally desirable strategies available.

COMBINATION STRATEGY

The desired approach for any pest management program is the combination of all strategies in a coordinated, integrated fashion. Often attempts can be made to first prohibit a pest from establishing in an area. If the pest is already present, a grower should use the do-nothing strategy, if appropriate. If possible, there should be an attempt to reduce a crop's susceptibility to a pest. Growers then should use nonchemical means to reduce pest population densities in a preventive manner, whether through biological, mechanical, or cultural control, or host plant resistance. In those situations with a history of significant economic pest problems, a pesticide can be used in a preventive manner (often done with herbicides and fungicides) (2, 3). As a last resort, when surveillance indicates that a pest population density has reached the economic threshold, the strategy would call for application of a pesticide in a therapeutic manner.

History has shown that the use of a single strategy and/or tactic ultimately is subject to failure and that a multifaceted approach is more desirable for sustainability and environmental compatibility. Indeed, the use of multiple strategies and tactics, combined with pest surveillance and use of thresholds, are the basic tenets in pest management programs.

See also *Cure (Therapeutics),* page 179; *Decision Making,* pages 184–186; *Integration of Tactics,* pages 416–419; *Pest Concept,* page 575; *Pest Status,* pages 590–592; *Prevention,* pages 664–665; *Principles of Pest Management with Emphasis on Plant Pathogens,* pages 666–669; *Sampling,* pages 745–746; *Systems Management,* pages 826–828; and *Tactics,* pages 829–830.

REFERENCES

1. Pedigo, L.P. Pest Management Theory and Practice. In *Entomology and Pest Management*, 2nd Ed.; Prentice-Hall: Upper Saddle River, NJ, 1996; 271–292.
2. Ashton, F.M.; Monaco, T.J. Weed Management Practices. In *Weed Science: Principles and Practices*, 3rd Ed.; John Wiley & Sons: New York, 1991; 34–67.
3. Maloy, O.C. Principles and Concepts of Control. In *Plant Disease Control: Principles and Practice*; John Wiley & Sons: New York, 1993; 13–54.
4. Funderburk, J.; Higley, L.; Buntin, G.D. Concepts and directions in arthropod pest management. Adv. Agron. **1993**, *51*, 125–172.

STRUCTURAL PESTS

H. Robinson William
Urban Pest Control Research and Consulting, Christiansburg, Virginia, U.S.A.

S

INTRODUCTION

Wood is the most common construction material available, and is used for residential and commercial buildings from the tropics to temperate zones. However, this cellulose material also serves as a primary food or nest site for a large number of insects in these geographic regions. Major wood-infesting pests include termites, beetles, and carpenter bees and ants. Their pest status is based on the long-term presence of colonies or feeding of individual larvae in walls and floors, and the weakening or destruction of the load-bearing capacity of foundation and support timbers.

Termites are the most important pest of wood. Colonies of these insects feed year round and cause millions of dollars of damage to residential, commercial, and agricultural buildings. They contribute to ecological damage through the need for chemical control methods and replacement timbers. Wood-boring beetles feed on hardwood and softwood timbers and cause millions of dollars in damage to structural timber and furniture. The pest status of carpenter ants and bees is based on the presence of colonies and tunneling in exposed wood. Although carpenter ants nest in wood and do not eat it, the nests may be large and weaken infested wood. Carpenter bee nests can result in wood damage.

TERMITES (ISOPTERA)

Termites are among the few insects that utilize cellulose as food. Cellulose is a major constituent of wood, and the majority of hardwood and softwoods are susceptible to termite damage. Under natural conditions subterranean and aboveground nesting termites feed on the roots and stems of living and dead plants and trees, and play an important ecological role. In the household environment they damage or destroy paper and wood timbers.

Termites are social insects that can live in large colonies numbering more than a million individuals. A mature colony consists of three major castes: workers, soldiers, and reproductives. Workers constitute the great majority of the colony, and they actively feed on wood and also feed other members of the colony. They are usually pale colored, with rounded heads and soft bodies. Soldiers are larger than workers, with well-developed heads and mandibles. Their primary role is defense of the colony; they are fed by workers. Reproductives include the queen, who is responsible for egg laying, and secondary queens that also produce eggs. Sexually reproductive males and females are produced in large, mature colonies. They are usually winged and fly from the colony in large numbers (swarms) at certain times of the year.

Subterranean Termites

Subterranean termites nest in soil or wood in contact with the ground. They enter wood or cellulose material above ground through earthen tubes extended from foraging tunnels in the soil. These termites require moisture and almost always have the primary portion of the colony in the soil. In cases of excess moisture above ground a portion of the colony, including secondary queens, may be moved above ground. Colonies may persist for many years, and once mature produce winged reproductives yearly. Common pest species are in the genera *Reticulitermes, Coptotermes,* and *Mastotermes.* Several species may be actively causing structural damage in one geographic region.

Drywood Termites

Drywood termites are aboveground nesting species. Colonies are generally small and characterized by the absence of mud tubes or tunnels, and fecal "pellets" in the nest. These termites can infest wood with a low (2–4%) moisture content, and may be found in wall or attic timbers. Although colonies may contain only a few hundred to a few thousand individuals, damage can be extensive if uncontrolled. Common pest species are in the genera *In-*

cisitermes and *Cryptotermes,* and they are in all major geographic regions.

Pest Management

Termite pest management includes both nonchemical and chemical methods. Reducing exposure and contact of structural wood to soil, maintaining dry conditions, and frequent inspections can limit subterranean termite damage. Physical barriers include a bed of basalt or granite particles, or stainless steel mesh over the foundation and soil. Extreme heat, cold, and microwaves have been used to control small, aboveground colonies. Biological agents include soil nematodes and pathogenic fungi. A few species of trees (such as *Eucalyptus* spp.) are resistant to termite attack. Remedial control is based on establishing a barrier in the form of insecticide-treated soil around the outside and inside perimeter of the structure. Toxic baits combined with soil monitoring can be used for long-term prevention.

BEETLES (COLEOPTERA)

Wood-boring beetle pests spend some stage of their development in wood, deriving food or harborage. Typically, the larval stage tunnels and feeds in the wood, and the damage done by adults is the emergence hole at the surface of the wood they are exiting. After mating, the female may return to lay eggs at the original site or fly off to find new sites. Adults usually live for only a few weeks, but the larval stages may live for years.

Some wood borers lay their eggs beneath the surface of wood while others oviposit in the cracks of structural timber. Wood-decay fungi are sometimes introduced during oviposition. Larval feeding behavior is generally random and is linked to seasonal changes in wood moisture content and ambient temperature. Adults typically emerge in the spring and lay eggs when wood moisture is increasing. Some species attack only hardwoods, and select wood with sufficient nutritional value for larval development; others preferentially attack softwoods. The wood type infested and the damage caused by larval feeding or adult emergence is characteristic and can be used to diagnose the infesting species.

Hardwoods

Powderpost borers (*Lyctus* spp., *Trogoxylon* spp.) in the family Lyctidae are common beetles attacking structural and ornamental hardwoods. The susceptibility of hardwoods to attack depends on the starch content and the diameter of the pores. Attack is initiated by females that oviposit into specific size pores. Eggs hatch in 7–14 days and larvae tunnel parallel with the wood grain. Larvae feed only in the sapwood portion of the wood. Late-stage larvae feed randomly and galleries become backed with fine-textured frass. Full-grown larvae tunnel close to the surface and create a pupal chamber. Adults develop in about 30 days and exit through 1–3 mm diameter holes they cut in the wood. Powdery frass is ejected through the emergence holes and may collect in small piles on underlying surfaces.

Softwoods

Borers in families Anobiidae (*Euvrilletta* sp., *Anobium* sp.) and Cerambycidae (*Hylotrupes* sp.) are the most common beetles attacking structural softwoods. Larvae feed in the sapwood portion of seasoned softwoods, and development is usually not limited by the age or wood moisture content. Attack is initiated by females that oviposit on the surface or in narrow surface cracks. Eggs hatch in 7–14 days and larvae tunnel randomly in the wood. The frass in galleries is characterized as gritty; it is composed of pellets and dry powder. Full-grown cerambycid larvae tunnel and create an emergence hole, then retreat to a pupal chamber beneath the surface. Anobiids form a pupal chamber below the surface and the adult cuts the emergence hole.

Pest management strategies include chemical and nonchemical methods. There are few parasites and predators associated with beetles infesting structural timber. Extreme heat and cold have been used as a control strategy for small infestations. Chemical control includes gas fumigation and the application of liquid insecticide to the surface or injected below the surface of infested wood. Chemical control is based on establishing a lethal concenration of a residual insecticide below the surface of the wood.

BEES AND ANTS (HYMENOPTERA)

Several species of bees and ants build their nests in sound or moisture-damaged wood. The adult stage of these insects create galleries in wood to rear their larvae or as as a nest site for the colony. They feed on other insects, honeydew, and plant materials such as seeds, pollen, and nectar. Immature stages do not feed on wood. Nest sites may persist for several years, and occasionally result in structural damage. The pest status of carpenter bees and carpenter ants is based on their presence as a nuisance, and less on the physical damage done to infested wood.

Carpenter Ants (*Camponotus* spp., *Crematogaster* spp.)

These large, black or black and red ants typically build their nests in moisture-damaged wood. In natural areas they may be in dead or down trees or in the soil around buried limbs; in the house they are often in window sills, door frames, rafters, and wall framing. After 2–3 years colonies can contain several thousand individuals, and the galleries may be expanded into sound wood. Galleries may be extensive and some areas contain large amounts of coarse, fiberous frass. Colony activity is seasonal; the immatures are reared in the spring and summer, and flights of winged male and female reproductives occurring from spring to fall.

Carpenter Bees (*Xylocopa* spp.)

These large, yellow and black bees burrow into trees and exposed wood to make tubular galleries. The females establish a new nest site or excavate previous galleries. Although solitary bees, several females may sometimes share one entrance hole that leads to separate galleries. Wood removed from the galleries usually accumulates below the round entrance hole. Larval rearing galleries are partitioned into chambers, which are provisioned with food and a single egg. The larva may complete development in a few weeks, but the adult usually does not emerge until the following spring.

Control

Control strategies include a limited number of nonchemical and chemical methods. Limiting exposure of structural wood to water and removing dead trees close to structures can reduce the threat of carpenter ants. Painting exposed wood outdoors can help discourage carpenter bees. Few wood species are resistant to attack from these insects. Chemical control for carpenter ants includes the use of liquid sprays and toxic baits. Baits placed near foraging sites can be effective, along with treating the nest site with a liquid insecticide.

S

BIBLIOGRAPHY

Creffield, J.W. *Wood Destroying Insects, Wood Borers and Termites*; CSIRO Australia: Victoria, 1991; 44.

Eaton, R.A.; Hale, M.D.C. *Wood*; Chapman Hall: London, 1993; 546.

Krishna, K.; Weesner, F.M. *Biology of Termites*; Academic Press: New York, 1969, 1970; 1 & 2, 346.

Robinson, W.H. Structural wood protection. Recent Dev. Entomol. **1998**, *2*, 9–17.

Robinson, W.H. *Urban Entomology, Insects and Mite Pests in the Human Environment*; Chapman & Hall: London, 1996; 430.

Robinson, W.H. Wood Boring, Book Boring and Related Insects. In *Handbook of Pest Control*; Mallis, A., Story, K., Eds.; Franzak & Foster Co.: Cleveland, OH, 1990; 283–311.

SURFACE WATERS: PESTICIDE POLLUTION

Victor de Vlaming
Water Resources Control Board, Sacramento, California, U.S.A.

INTRODUCTION

In numerous aquatic ecosystems across the United States losses of biodiversity and significant population declines of multiple species have been documented over the past 40 years. These aquatic ecosystems are imperiled due to human activities such as habitat destruction, damming and diversion of waters, introduction of exotic species, and release of toxic concentrations of chemicals, including pesticides, into surface waters.

The discipline of ecology has demonstrated that on smaller scales and on the scale of the Earth, all things are interrelated and interdependent. Thus, all life, including human existence, on Earth is dependent upon healthy and interdependent ecosystems. The focus of this article is on the health of surface waters (i.e., aquatic ecosystems), particularly their constituent biota, yet the information presented is reflective of overall environmental health.

PESTICIDE USE

Quantities of pesticides used in the United States have increased approximately 50-fold since the 1960s such that more than a billion pounds of pesticide active ingredients are used each year (1). Hundreds of different pesticides have been developed for application in agricultural and other settings to control weeds, insects, fungus, and other pests. The characteristics of pesticides have evolved over the past 30 years from chemicals such as the organochlorines (OC) that are very persistent in the environment but tend to have relatively low toxicity, to chemicals such as organophosphorus (OP) and carbamate insecticides that are toxic at very low concentrations (e.g., parts per trillion) but generally are not very persistent. The majority of pesticide use is agricultural (70–80% of total), but there is also use in urban areas for gardens, lawns, homes, and buildings, as well as in forestry, along roads and railways, and in various industrial and commercial situations.

PESTICIDE TRANSPORT AND PATTERNS OF OCCURRENCE

Unfortunately pesticides do not always remain where they are applied. Off-site movement occurs irrespective of applications made according to label instructions. Off-site movement can occur by aerial drift of sprays, by evaporation, in storm and irrigation water runoff, and by seepage. Surface waters are vulnerable to pesticide contamination because most agricultural and urban areas drain into streams and rivers. Since pesticides are designed to be lethal to organisms, they pose a significant risk to biota when they enter aquatic ecosystems. Many of the currently used pesticides are so lethal that runoff of less than 1% of the quantities applied in a watershed into surface waters can have profound effects on aquatic biota.

The most extensive program for monitoring surface waters for chemical contaminants in the United States is the National Water-Quality Assessment (NAWQA) Program of the U.S. Geological Survey. Initiated in 1991, the focus in NAWQA has been on major watersheds distributed throughout the United States, encompassing 60–70% of national water use. Pesticides have been one of NAWQA's top priorities. Much of the information summarized below was collected in the NAWQA program (1–3).

Pesticides have been detected in every region of the United States where surface waters were analyzed for these chemicals. Although no individual study analyzed for every pesticide, a wide variety of pesticides including insecticides, herbicides, and fungicides have been identified in surface waters throughout the United States. The distribution of pesticides in surface waters generally follows geographic and seasonal agricultural patterns and also the influence of urban areas. That is, frequency of detections and highest concentrations are recorded in streams and rivers where agriculture is a major land use and pesticide use is intense. In most agricultural areas, the highest concentrations of pesticides occur as seasonal pulses, with duration of a few weeks to several months.

Frequency of occurrence and highest concentrations also are associated with seasonal application patterns, being greatest coincident with or after applications. In urban-dominated streams, seasonal patterns are less obvious, pesticide concentrations being elevated for longer periods. In these urban streams, insecticides are detected at higher frequencies and concentrations than in streams draining agricultural areas. The largest areas where high quantities of pesticides are applied to crops occur in California, Florida, the Midwest, the lower Mississippi River Valley, and the coastal areas of the Southeast.

Herbicides are detected more frequently, and at the highest concentrations, than other pesticides in surface waters. Considering data collected from across the United States, herbicides occurring most frequently and at the highest concentrations are atrazine, simazine, metolaclor, prometon, DEA, alachlor, and cyanozine. The insecticides measured most frequently and at the highest concentrations are diazinon, chlorpyrifos, carbaryl, carbofuran, and malathion.

Pesticides are encountered most often and at elevated concentrations in streams and rivers draining agricultural areas just prior to and during the growing season. Also, a greater number of pesticides occur in surface waters coincident with these periods. These mixtures of pesticides have a potential for additive and synergistic adverse impacts on aquatic ecosystem health. An exposure pattern that is developing is one of long-term exposure to relatively low concentrations of pesticide mixtures punctuated with seasonal pulses of high concentrations, the effects of which are not currently known.

Off-site movement is a critical issue with regard to pesticide contamination of aquatic ecosystems. The relationship between quantities of pesticides applied (amount per unit area and total area of application) and detection frequency, as well as concentrations in surface waters was stated above. While this is a general principle, it does not apply to all pesticides and situations. Some pesticides, which are used rather extensively in agriculture, are seldom detected in surface waters. This low detection rate in surface waters relates to physical/chemical and/or degradation properties. Some such properties can result in pesticides adsorbing to particles (organic or soil) that may reduce off-site movement while different properties of other pesticides favor rapid degradation.

One might conclude that use of pesticides with a lower potential for off-site movement would reduce risk of impacting aquatic ecosystem health. While there is some truth in this idea, such physical/chemical properties often render pesticides more persistent. Furthermore, organic and soil particles to which pesticides are adsorbed can be transported by erosion (associated with rainfall, irrigation, or wind) into streams and rivers. Pesticides adsorbed to particles can settle into surface water sediments and have deleterious effects on bottom-dwelling organisms. Such physical/chemical properties also tend to result in bioaccumulation by aquatic organisms. These pesticides can bioaccumulate in aquatic species to levels that are detrimental and/or biomagnified to adverse levels in the food chain as contaminated organisms are eaten.

Examination of fish and bivalve tissues reveal that they are being exposed to a variety of bioaccumulable pesticides in both agricultural- and urban-dominated waterways. Residues of some pesticides in fish tissues can be such that they are harmful to human health. Organochlorine (e.g., DDT, chlordane), pyrethroid, and other hydrophobic insecticides are examples of pesticides that adsorb to organic materials, are persistent, bioaccumulate, and biomagnify. Unfortunately, there is a paucity of information on what pesticide tissue residue levels are deleterious to organisms or to other species that eat them.

EFFECTS OF PESTICIDES

The occurrence of pesticides in surface waters of the United States is widespread. What is the significance of this phenomenon? Adverse effects of chemicals, including pesticides, are determined by concentration, as well as by duration and frequency of exposure. The federal Clean Water Act (CWA) was enacted to protect human and aquatic organism health, requiring that no chemical can occur in surface waters at toxic concentrations. CWA requirements are implemented through enforceable water quality standards for specific chemicals and toxicity. Water quality standards and criteria have been established by various agencies for only a few pesticides. These standards and criteria are an estimate of a chemical concentration in water below which detrimental effects are not expected to occur. Comparing concentrations of a chemical in surface water to such standards and criteria provides an indication of potential antagonistic impacts on aquatic biota.

Standards for human health apply to treated drinking water supplied by community agencies. Therefore, the standards do not apply directly to most surface waters. While these standards do not pertain to concentrations of pesticides in surface waters, they do afford a benchmark to which measured pesticide concentrations can be compared. Chemical analyses of surface water samples do not include all pesticides; however, few pesticides included in analyses were detected at concentrations exceeding any drinking water standards. The pesticides most often exceeding standards are the triazine and acetanilide herbicides, atrazine, alachlor, cyanazine, and simizine (1).

Some, but not all, treatments of water to be used for drinking destroy or remove these herbicides.

For the more than 120 pesticides detected in surface waters there exist only 13 U.S. Environmental Protection Agency criteria developed for the protection of aquatic ecosystem health. For most currently used insecticides and for all herbicides there are no criteria for the protection of aquatic life. Canada has a larger number of aquatic life criteria, which are more stringent than U.S. criteria. U.S. pesticide aquatic life criteria are commonly exceeded in streams and rivers collecting from agricultural lands and/or urban areas. Aquatic life criteria of four OP insecticides, azinphos-methyl, chlorpyrifos, diazinon, and malathion are the most frequently exceeded by concentrations in surface waters. Data collected from across the United States indicate that azinphos-methyl and chlorpyrifos exceed aquatic life criteria for more days per year than other insecticides monitored (1, 3). Major concerns regarding OP insecticides are that they are toxic to aquatic species at very low concentrations, different OPs repeatedly co-occur in surface waters, and their toxicity is additive.

In some regions of the country, where their use is high, carbamate insecticides are threats to aquatic biota. More than 20 years after being banned, OC insecticides continue to be detected in surface waters and sediments at concentrations that exceed aquatic life criteria. Especially in streams and rivers draining agricultural areas, but also in urban-dominated streams, two or more pesticides often cooccur at concentrations that exceed their respective aquatic life criteria. Several studies from across the country reported high occurrences of diseased, deformed, and highly parasitized fish, as well as fish with a high incidence of tumors in surface waters where pesticide concentrations exceed aquatic life criteria and/or are elevated.

Measuring pesticide concentrations in surface waters does not furnish direct information of bioavailability (the percentage of the analytically measured amount of a chemical that produces toxic effects) or toxicity to aquatic biota. Toxicity testing of surface waters provides a direct measure of capacity of these waters to support healthy aquatic organisms. Standardized toxicity tests with aquatic species are available that measure lethal and sublethal (e.g., inhibition of reproduction, growth, etc.) effects. As a diagnostic tool for assessing water quality, toxicity testing has several merits. Toxicity tests afford an integrative measure of adverse effects of chemicals on organism health and viability, as well as the bioavailabity of chemicals (4). Also, results of these toxicity tests have been reliable predictors of impacts on aquatic ecosystem biota. Chemical analyses of water cannot provide such information.

Statewide monitoring programs in California have disclosed that, on a seasonal basis, agricultural- and urban-influenced streams and rivers are lethal to test species (4, 5). Complex toxicological, chemical, and physical procedures (Toxicity Identification Evaluation–TIE) that specifically identify the chemical(s) causing toxicity demonstrated that the pesticides most commonly responsible for the surface water toxicity are diazinon and chlorpyrifos. Carbofuran, malathion, carbaryl, methyl-parathion, thiobencarb, diuron, and molinate also have been shown to becauses of toxicity in California's surface waters. These surface water toxicity testing programs are not common in other regions of the United States. Such surface water toxicity is likely to occur in most urban streams and in surface waters collecting runoff from areas where agriculture is the predominant land use and where pesticides are used intensively.

GAPS IN KNOWLEDGE

Pesticide adverse impacts on aquatic ecosystem health throughout the United States are most likely underestimated for several reasons. 1) The number of U.S. streams and rivers thoroughly monitored is very limited so that the distribution and extent of pesticide contamination is unknown. 2) Most investigations have been incomplete for one or more reasons, including too few sampling sites, sample collection was in frequent, and study duration was abbreviated. 3) None of these monitoring investigations included the complete range of pesticides that could impact aquatic biota. 4) Analytical detection limits for pesticides have been a problem in assessing detrimental effects on aquatic ecosystem health. In most monitoring projects analytical detection limits for many pesticides were higher than known toxic effects on aquatic species, as well as above aquatic life criteria. 5) Sublethal and delayed effects of pesticides, including bioaccumulation and biomagnification responses, generally are not evaluated. For example, several pesticides, including alachlor, atrazine, 2,4-D, metribuzin, trifluralin, aldicarb, carbaryl, parathion, some pyrethroids, benomyl, mancozeb, maneb, zineb, and ziram, commonly used in U.S. agriculture, have been shown to disrupt endocrine systems in some aquatic species (6, 7). Existing aquatic species toxicity screening procedures are inadequate for assessing endocrine disruption. 6) Seldom are indirect effects considered. For example, direct adverse effects on zooplankton during a period when they are critical food for larval fish could indirectly impact fish populations. 7) As stated above, aquatic organisms are exposed to pesticide mixtures in many watersheds across the United States. Assessments of impacts infrequently involve analysis of exposures to multiple chemicals.

Pesticides, especially insecticides, are having widespread impacts on surface water quality and aquatic ecosystem health throughout the United States. To reduce risks of pesticide impacts on aquatic ecosystem health, measures should be identified, developed, and implemented to eliminate or reduce off-site movement of these chemicals. More extensive and thorough monitoring of pesticides and of toxicity caused by these chemicals is advisable to assess the extent of pesticide-caused water quality degradation, as well as the effectiveness of remediation projects.

REFERENCES

1. Larson, S.J.; Capel, P.D.; Majewski, M.S. *Pesticides in Surface Waters: Distribution, Trends, and Governing Factors*; Ann Arbor Press, Inc.: Chelsea, MI, 1997; 373.
2. Larson, S.J.; Gilliom, R.J.; Capel, P.D. Pesticides in Streams of the United States—Initial Results from National Water-Quality Assessment Program. *Water Resources Investigations Report 98-4222*; U.S. Geological Survey: Sacramento, CA, 1999; 92.
3. Gilliom, R.J.; Barbash, J.E.; Kolpin, D.W.; Larson, S.J. Testing water quality for pesticide pollution. Environ. Sci. Technol. **1999**, *33*, 164A–169A.
4. de Vlaming, V.; Connor, V.; DiGiorgio, C.; Bailey, H.C.; Deanovic, L.A.; Hinton, D.E. Application of whole effluent toxicity test procedures to ambient water quality assessment. Environ. Toxicol. Chem. **2000**, *19*, 42–62.
5. Kegley, S.; Neumeister, L. *Disrupting the Balance: Ecological Impacts of Pesticides in California*, Californians for Pesticide Reform, San Francisco, CA, 1999; 99.
6. *Environmental Endocrine Disruptors: A Handbook of Property Data*; Lawrence, H.K., Ed.; John Wiley & Sons, Inc.: New York, 1997; 1232.
7. Moore, A.; Waring, C.P. Sublethal effects of the pesticide diazinon on olfactory function in mature male atlantic salmon Parr. J. Fish Biol. **1996**, *48*, 758–775.

SUSTAINABLE AGRICULTURAL PRACTICES

Clive A. Edwards
The Ohio State University, Columbus, Ohio, U.S.A.

INTRODUCTION

Agriculture productivity worldwide has increased dramatically since the 1950s, by more than three times, particularly in the developed countries. The main bases of these greatly increased yields have been the introduction of new seed varieties, greatly increased use of inorganic fertilizers, extensive use of synthetic pesticides, and increased irrigation. However, these impressive yield increases were not sustained and the annual rates of yield increase began to decline after the 1980s due to intensive soil cultivations, continuous monoculture or biculture cropping, soil erosion, overuse of fertilizers that increased pest incidence, and overuse of pesticides that killed natural enemies and lessened biological control mechanisms.

Since the 1980s in Europe and North America there has been growing support for more sustainable agriculture that depends less on synthetic inputs based on fossil fuels and uses more integrated biological and cultural inputs to supply nutrients and control pests. Integrated pest management is a major component of such sustainable agricultural systems described by Edwards (1, 2).

THE CONCEPT OF SUSTAINABLE AGRICULTURE

There are many definitions of sustainable agriculture, but most incorporate the following characteristics: long-term maintenance of natural resources and agricultural productivity, minimal environmental impacts, adequate economic returns to farmers, optimal production with minimal chemical inputs, and provision for the social needs of farm families and communities. All definitions explicitly promote environmental, economic, and social goals and the need for an interdisciplinary system approach to agriculture.

A definition that my colleagues and I have developed (3) that receives broad acceptance isSustainable agriculture involves integrated systems of agricultural production with minimum dependence upon high inputs of energy in the form of synthetic chemicals and cultivation, which substitute cultural and biological techniques for these inputs. They should maintain, or only slightly decrease, overall productivity and maintain or increase the net income for the farmer on the sustainable basis. They must protect the environment in terms of soil and food contamination, maintain ecological diversity and the long-term structure, fertility, and productivity of soils. Finally, they must meet the social needs of farmers and their families and strengthen rural communities in a sustainable manner.

Many farmers and agricultural scientists view the various practices they use as completely independent of each other. They rarely consider how the amounts of fertilizer they use affect pests, diseases, or weeds. Neither is the impact of cultivations on pest disease and weed problems a factor in deciding the type of tillage a farmer uses. Even with integrated pest management systems, it is rare for any account to be taken of the impact of herbicides on pests and diseases, of insecticides on diseases, or of fungicides on pests.

Farming practices include five important interactive components each of which effect each other (4). The four main inputs are **cultivations, nutrient supply, pesticides,** and **cropping patterns**. Central to this pattern and affected by all is **farm economics**, which encompasses inputs such as land, labor, buildings, machines, chemicals, and seed balanced against profits from yields. A farming system is not just a simple sum of all of its inputs and components, but rather is a complex system with intricate interactions.

In conventional ''higher-input'' farming, large yields often can be obtained without any appreciable attention to interactions between inputs. For instance, if heavy fertilizer use renders a crop much more susceptible to pests and diseases through production of lush, soft growth, this can be compensated for by increased pesticide usage. The decrease in natural pest and disease control caused by herbicides, through loss of foliar and habitat diversity, can be compensated for by increased use of insecticides and fungicides. Any effect of pesticides on earthworms and other soil organisms that promote organic-matter turnover, nutrient cycling, and soil fertility is covered by using more inorganic fertilizers. When chemical inputs are lowered in sustainable agriculture we need to understand how the inputs impact upon each other in much more detail so that the beneficial effects of interactions can be maximized. Consider the major inputs into farming systems in more detail.

Nutrient Inputs

At lower input levels, small increases in the amounts of inorganic fertilizers can produce dramatic increases in crop yields. However, as applications are increased, so the response of the crop diminishes exponentially and eventually levels off. At a certain point, the cost of the fertilizer equals the value of the increase in yield of the crop and it is important to use amounts of inorganic fertilizers considerably less than this. Reductions in the amounts of fertilizers used can be compensated for by use of rotations, particularly those involving legumes as a source of nitrogen and other nutrients, and by the use of organic amendments. Other practices that can minimize fertilizer use include the use of regular soil analyses to assess actual fertilizer needs, growing crop varieties with lower nutrient demands, and accurate placement of inorganic fertilizers in the crop row where they have maximum effect on the crop and do not contribute to weed growth between rows. Large applications of nutrients can increase the susceptibility of crops to pests and diseases.

Pest Management

Pesticides are often used on a recommended or insurance basis regardless of the amount of pest infestation, and such applications may be unnecessary and/or economically unsound. The amounts of pesticides used could be decreased greatly if alternative techniques are used. Insecticide use can be lessened if integrated pest management techniques involving criteria such as economic thresholds, pest forecasting, resistant varieties, rotations, and biological and cultural pest control are used (5). All of these must be integrated into farming systems for pest and disease management as is described later in the section (4).

Cultivations

Traditionally, land in developed countries was cultivated annually to a depth of 7.5–30.0 cm, with the soil completely inverted. This involves a high consumption of energy, particularly in difficult and compacted soils. For the past 30–40 years, there has been a progressive trend toward a range of lesser cultivations with corresponding decreases in energy inputs.

Conservation techniques that lessen the amount of cultivation required, compared with deep plowing, include i) shallow plowing, ii) chisel plowing, iii) deep subsoiling, iv) harrowing, v) shallow-tine soil loosening, and vi) no till (direct drilling). All of these techniques tend to decrease soil erosion, create a much more natural soil structure that improves both drainage and water retention, and favor biological and natural mechanisms of pest and disease control.

Rotations and Cropping Patterns

There has been a trend in farming over the past 40–50 years in developed countries toward monoculture or cropping with only two annually alternating crops. When chemical inputs in terms of fertilizers and pesticides are decreased, it usually becomes essential to increase biodiversity through crop rotations, which provide nutrients and lessen pest and disease attack. More sustainable cropping systems that maximize biological inputs demand a range of different crops in either space or time. This can also be achieved through rotations, intercropping two crops, or through agroforestry, a mixture of crops and trees (usually legumes).

PEST MANAGEMENT AS A MAJOR COMPONENT OF SUSTAINABLE AGRICULTURE

The U.S. National Research Council (5) reported on sustainable agriculture and identified the two most important components of sustainable agricultural systems as integrated pest management (2) and integrated nutrition management (6). *Integrated pest management* seeks to control pre and postharvest weeds, arthropod and vertebrate pests, and pathogens using biological and cultural techniques together with minimal levels of synthetic pesticides. *Integrated nutrient management* seeks to provide plant nutrients through the optimal use of on-farm biological resources (including manures, plant rotations, cropping patterns, and legumes) and minimal inorganic fertilizers. Both integrated pest and nutrient management depend on conserving biological diversity and soil organic matter and, thus, on a sound understanding of biological processes and ecological interactions (5).

In integrated pest management, the minimal use of insecticides can be supplemented by other techniques such as i) minimal use of insecticides; ii) better insecticide placement and formulations; iii) more crop rotations; iv) appropriate cultivations; v) timing of crop sowing; vi) use of controlled weed growth to encourage natural enemies of pests or provide alternative food sources for pests; vii) use of biological insecticides based on insect pathogens; viii) use of nematodes that attack insects; ix) release of parasites and predators of pests; x) use of pheromones and repellents; xi) release of sterile male insects where appropriate; xii) use of crop varieties resistant to pests; xiii) use of crop varieties with toxins implanted by genetic engineering; xiv) encouragement of natural predators by maintenance of biological diversity; xv) use of trap crops; and xvi) innovative cultural techniques such as strip cropping and intercropping, which increase diversity of plants and animals and enhance natural controls (1–7).

Fungicide use can be minimized by i) disease forecasting, ii) use of rotations, iii) better application techniques for fungicides, iv) timing of crop sowing, v) use of disease antagonists, and vi) use of tolerant or resistant crop varieties.

Herbicide use can be minimized by i) use of mechanical weed control, ii) use of rotations, iii) strip-cropping, iv) use of live mulches, v) use of mycoherbicides, and vi) release of pests and pathogens of weeds (1–6).

CONCLUSIONS

From these discussions it is clear that not only is integrated pest management one of two key components of sustainable agricultural systems, but also that many of the cultural and biological techniques and practices involved, can be incorporated only by a whole-farm systems approach to the control of pests and supply of nutrients. Clearly, as agriculture moves from its current chemical dependency to a longer-term, more sustainable mode, biological and cultural inputs to nutrient supply and pest control will become increasingly important.

REFERENCES

1. Edwards, C.A.; Stinner, B.R. The Use of Innovative Agricultural Practices in a Farm Systems Context for Pest Control in the 1990s. In *Pests and Diseases*; Brighton Crop Prot. Conference, 1990; 7C-3, 679–84.
2. Edwards, C.A. In *The Role of Integrated Pest Management in Sustainable Agricultural Systems*, Proceedings of the Philadelphia Society For Promoting Agriculture 1990–1991, Philadelphia Society For Promoting Agriculture, Somasundaran, P., Ed.; Philadelphia, 1991; 30–39.
3. Edwards, C.A. The concept of integrated systems in lower input sustainable agriculture. Amer. J. of Sust. Agric. **1987**, *II* (A), 148–152.
4. Edwards, C.A. The importance of integration in sustainable agricultural systems. Agric. Ecosys. Env. **1989**, *27*, 25–35.
5. A Plan for Collaborative Research on Agriculture and Natural Resource Management. In *Toward Sustainability*; National Research Council: Washington, DC, 1991; 145.
6. Edwards, C.A.; Grove, T. Integrated Nutrient Management for Crop Production. In *Toward Sustainability*; National Research Council: Washington, DC, 1991; 105–108.
7. Edwards, C.A.; Lal, R.; Madden, P.; Miller, R.H.; House, G. *Sustainable Agricultural Systems*; Soil and Water Conservation Society: Ankeny, IA, 1990; 696.

SWINE PEST MANAGEMENT (ARTHROPODS)

D.W. Watson
North Carolina State University, Raleigh, North Carolina, U.S.A.

S

INTRODUCTION

Arthropods of importance to swine production include the itch mite, hog louse, house fly, stable fly, and cockroach. Their management reduces the risk of disease and production losses resulting from poor growth, development, and feed conversion. Biting and nonbiting insects occasionally associated with swine include mosquitoes (Culicidae), black flies (Simuliidae), biting midges (Ceratopogonidae), little house flies (*Fannia canicularis*), dump flies (*Hydrotea aenescens*), fruit flies (*Drosophila repleta*), and moth flies (Psychodidae) (1).

UNITED STATES SWINE PRODUCTION

Pork production has suffered an economic downturn in the past decade. According to the National Agriculture Statistics Service, the total U.S. swine inventory was estimated at 61.2 million as of December 1, 1997. The value of this inventory was estimated at U.S. $4,985,532,000, an average of $82/head. The monetary return on swine in 1997 averaged U.S. $52.90/hundred-weight. Swine inventories increased to 62.2 million in 1998. However, the return on slaughter pigs was poor bringing only U.S. $34.40/hundred-weight, a 28% reduction. Nationally the number of farms housing swine declined an average 2.2% per year from 1988–1992.

COMMON PESTS OF SWINE

''Mange'' is an inflammatory skin condition caused by the itch mite, *Sarcoptes scabiei*, an obligate parasite that burrows in the epidermis of pigs (1). Mature mites mate on the skin surface and the newly fertilized females construct new tunnels in which they lay eggs. Larval and nymphal mites continue to expand the tunnels before developing into adult mites (Table 1).

Infested animals scratch vigorously and rub against objects for relief. Infested animals develop inflamed, thickened skin and encrusted lesions generally beginning on the inner side of the ear and spreading over the head along the neck, and across the body. Mites can be found on pigs throughout the year but are most noticeable and readily spread during the winter. The primary means of transmission is direct contact with infested pigs. People, equipment, and rodents may transiently acquire and transmit mites.

Traditional sampling for mites relies on the extraction and microscopic examination of mites from skin scrapings from the inside surface of the ear. Recently indirect Enzyme Linked Immunosorbent Assay (ELISA) tests have been developed for detection of *Sarcoptes scabiei*, providing certain and prompt diagnosis of an existing or prior infestation.

Itch mites are highly prevalent among swine. Most economic impact studies report significant improvements in animal performance in the absence of mite infestations. In growing pigs, average weight gain reductions ranging from 3.3% to 9.0% and reduced feed efficiencies of 3.1% and 9.0% have been reported (1). Treated sows consumed 4.3 lb. less feed and produced litter weights 9.1 lb. heavier than untreated sows. Piglets from treated sows had an 0.11 lb advantage average daily weight gain and were 12.8 lb. heavier at slaughter.

The hog louse, *Haematopinus suis*, is an obligate blood-feeding parasite (1). Distributed in clusters, louse infestations generally start around the ears and expand to other body regions. Predilection sites include the ears, neck, skin folds, and inner legs. Although lice do survive several days in warm bedding, the primary method of transmission is direct contact with infested hogs.

The hog louse is common in the United States. Although the hog louse has been incriminated in the transmission of eperythrozoonosis, swine pox virus, and other diseases, basic economic data concerning effects on growth rate and feed efficiency have not been developed (1).

Hog louse population estimates are based on whole body or limited area counts, and ranking the infestation levels (i.e., light, moderate, and heavy). These methodologies are subject to considerable variability arising from

Table 1 Approximate developmental time (in days) for arthropod pests of importance to swine production

Pest species	Egg	Larva	Life stage Nymph	Pupa	Adult	Egg to adult
Itch mite	3–5	2–4	4–6	None[a]	?	10–14 days
Hog louse	10–14	None	10–14	None	28	14–21 days
House fly	1–2	4–6	None	3–4	14	10–14 days
Stable fly	2–4	10–14	None	6–8	14	20–30 days
Cockroach						
Oriental	42–81	None	360–365	None	?	>365 days
German	28	None	60–90	None	200	120 days

[a] Life stage not present
From Refs. 2 and 3.

animal size and color, and to accumulations of dirt and debris that may adhere to the hair coat.

Strict biosecurity, through restricted movement of people and equipment, and rodent management, prevents the establishment of itch mites and hog louse. The isolation and prophylactic treatment of all new stock for a minimum of 30 days ensures the introduction of pest-free animals to the herd. Injectable endectocides, and pour-on or spray formulations of ectoparasiticides are highly effective for the management of itch mites and lice on swine. Routine treatment is recommended for large commercial swine operations. Biological control, host resistance, and insect growth regulators are not available for the control of mites and lice on swine.

The house fly (*Musca domestica*) and stable fly (*Stomoxys calcitrans*) are relatively minor pests in liquid waste management systems (1). Largely unsuitable for fly development, lagoons provide a relatively inexpensive, simple means of treating swine waste, reducing nutrient content and biological activity prior to application to land. Fly management in solid waste systems is problematic, particularly when pigs are kept on straw bedding. Manure and urine, mixed with bedding, promote fly growth and development.

Female house and stable flies deposit their eggs in decaying organic matter, such as wet bedding, waste or wet spilled feed, or manure mixed with straw or other bedding materials where the larvae grow and develop (Table 1). Prepupal larvae seek a drier environment for pupation and the subsequent eclosion of adult flies. Adult house flies feed on animal secretions, manure, and other organic matter commonly found in the swine facility. Biting stable flies feed several times a day.

Numerous sampling methods have been used to monitor adult house fly populations including sticky traps, light traps, baited traps, sticky cards, spot cards, and scudder grids. Emergence traps have been used to sample house flies from larval breeding sites. Spot and sticky cards have the most utility. Standardized techniques for sampling stable fly populations have not been applied to swine production. Animal bite counts routinely used to determine stable fly populations on cattle would probably be suitable for swine. Other sampling methods used to monitor adult stable fly populations include alsynite traps, sticky traps, and sticky cards.

Direct economic losses in pork production resulting from house and stable fly infestations have not been documented. The house fly has been incriminated in the mechanical transmission of more than 100 pathogenic organisms, most of which are bacteria and viruses including poliomyelitis, foot and mouth disease, *Salmonella* and *Escherichia coli*, gastroenteritis, and others. The house fly is a nuisance and public health concern, and failure to properly manage house flies often results in litigation. Stable flies caused no measurable effects on weight gains, feed consumption, and feed conversion with 3 to 7 stable flies per pig under environmentally controlled conditions. However, accumulated stress factors, especially heat, may contribute to the economic impact of stable flies in swine. Other economic losses lie in the potential of stable flies to serve as vectors for disease, for example, eperythrozoonosis.

The management and control of house flies and stable flies include the application of conventional insecticides as space sprays, residual surface sprays, feed additives, larvicide sprays, or baits. Recent emphasis has been placed on the use of quick knockdown space sprays in an effort to reduce the onset of insecticide resistance and conserve biological control agents. Larvicides may be applied to manure and other larval development sites as a spot treatment, however total reliance on this method encourages the development of resistance. Baits may also help reduce house fly populations.

Sanitation reduces habitats that promote fly growth and development. In systems designed to handle manure in a solid or semisolid form, the manure should be mechanically removed weekly or more frequently and applied to fields. If land application is not available, the manure can be composted or stored in a covered stockpile (4). Fly development is significantly reduced in liquid manure systems. Frequent inspection and maintenance of feed, water, and waste management systems is critical to manure management.

The use of biological control agents to control house and stable flies in livestock has relied on the use of parasitoids (Pteromalidae). Augmentation of natural parasitoid populations may be included in the overall fly management plan (1). Entomopathogenic fungi show promise in the laboratory but have not been developed.

The cockroach has become an increasingly important pest in swine production in recent years (5). The female cockroach deposits eggs in a protective case called an ootheca. The oriental cockroach, *Blatta orientalis*, deposits the ootheca within a few days. The German cockroach, *Blattella germanica*, however carries the ootheca for several weeks, limiting the potential for predator or parasitoid attack. Gregarious by nature, cockroaches feed on a variety of foods, including animal feeds and feces.

Cockroach populations are usually sampled at night when the insects are most active. Examining wall voids, cracks, and crevices for hiding cockroaches and numerous oothecae is indicative of population size. Pyrethroids are often used to flush cockroaches from hiding sites during the day. Baits and sticky cards may also be used.

The economic importance of cockroach infestations in swine production has not been documented. German and oriental cockroaches are recognized as mechanical disease vectors, and hypersensitivity to cockroaches is particularly common among people (5). Swine suffer from a number of mechanically transmitted disease agents, including mycotoxins, bacteria, and viruses. Although biosecurity may be in force, cockroaches readily move between barns and nurseries, weakening the biosecurity effort.

Control and management of oriental and German cockroaches include the application of conventional insecticides as residual surface sprays and baits. Pyrethroid and organophosphate insecticides are commonly used in swine production. Although insecticide resistance has developed in the urban setting it has not been documented in swine production. To reduce pesticide exposures, inorganic insecticides such as boric acids and diatomaceous earth have also been used.

The entomopathogenic fungus, *Metarhizium anisopliae*, has been used to help manage cockroaches. Parasitoids, targeting the oothecae, have been cultured and released to help reduce cockroach populations. Parasitoids are least effective against the German cockroach that carries the ootheca for several weeks.

FUTURE CONCERNS

Waste management continues to be a serious concern in swine production. Although liquid waste systems are relatively pest free systems, environmental concerns are directing the industry to reevaluate solid waste systems (4). Essential to the survival of the industry is the development of alternative manure handling systems that alleviate environmental concerns, reduce manure solids, recycle nutrients, yet provide a value added product such as feed supplements, compost, or vermiculture.

Basic studies of dispersal, population dynamics, and behavior under diverse production systems and waste handling systems are lacking for flies and cockroaches. Critical to developing integrated pest management (IPM) strategies is the understanding the biology and ecology of naturally occurring biological control agents. Nuisance and public health hazards must be studied, including the role of cockroaches and flies in the transmission of disease and food-borne pathogens. Alternatives for conventional insecticide applications are needed, as is the development of safe, effective, and environmentally compatible integrated pest management strategies.

REFERENCES

1. Holscher, K.H.; Williams, R.E.; Hoelscher, C.E.; Strother, G.R.; Bay, D.E.; Riner, J.L.; Lyon, W.F.; Moon, R.D.; Hagsten, I. Research and Extension Needs for Swine IPM. In *Integrated Management Programs for Livestock and Poultry*; Hogsette, J.A., Geden, C.J., Eds.; Research and Extension Needs for Res., Agric. Res. Div. Institutes of Agric. and Nat. Resources, Univ. Nebraska: Lincoln, NE, *in press*.
2. Harwood, R.F.; James, M.T. *Entomology in Human and Animal Health*, 7th Ed.; Macmillian Press: New York, 1979; 548.
3. Williams, R.E.R.D.; Hall, A.B. Broce; Scholl, P.J. *Livestock Entomology*; John Wiley & Sons: New York, 1985; 335.
4. Mikkelsen, R.L. Agricultural and Environmental Issues in the Management of Swine Waste. In *Use of By-products and Wastes in Agriculture*; Rechcigl, J.E., MacKinnon, H. C., Eds.; Spec. Publ. 668, Amer. Chem. Soc.: Washington, DC, 1997; 110–119.
5. Waldvogel, M.G.; Moore, C.B.; Nalyanya, G.W.; Stringham, S.M.; Watson, D.W.; Schal, C. In *Integrated Cockroach (Dyctioptera: Blattellidae) Management in Confined Swine Production*, The 3rd International Conference on Urban Pests, Grafické závody Hronov Publishing: Prague, Czech Republic, 1999.

SWINE PEST MANAGEMENT (VERTEBRATES)

Robert M. Corrigan
RMC Pest Management Consulting, Richmond, Indiana, U.S.A.

INTRODUCTION

Vertebrate pests can have serious economic impact on commercial swine production facilities in the United States. The most significant pests include the domestic rodents, a few bird species, and sometimes coyotes and feral dogs. The impact of a particular vertebrate pest group on swine production depends on the type of operation and/or the severity of the infestation. Managing vertebrate pests must be done with an integrated pest management (IPM) approach to achieve long-term, cost-effective results.

ECONOMIC SIGNIFICANCE OF VERTEBRATES

Rodents

Perhaps more than any other pest group that affects swine production—insects included—rodents are among the most economically significant. The rodent species of most importance are the house mouse (*Mus musculus*) and the Norway rat (*Rattus norvegicus*). In the Southern and coastal states, the roof rat (*Rattus rattus*) may also be problem. Rodents impact on almost all swine production factors (Table 1).

In confined swine facilities, rodent damage to insulation and structures can amount to thousands of dollars in a matter of months. And, long-term energy losses and the indirect environmental impact to the overall health of a swine herd as a result of such energy losses magnifies the expense. Pork producers who have sustained damage to their operations from rodents have reported repair expenses ranging from $15,000 to $35,000 (year 2000 estimates).

Rodents also destroy doors, floors, walls, ceilings, and various types of wooden equipment from their constant gnawing activities. Electrical and mechanical malfunctions often result when rats and mice gnaw on the many different types of equipment and utility wires within swine facilities such as ventilation systems, computers, and various power lines. Explosions and fires result when rodents gnaw into flexible gas lines used on space heaters in confinement buildings. Rodents also gnaw into water and feed lines causing annoying and costly spillages. Structural damage to swine buildings also occurs from the burrowing activities of rodents. Rats burrow extensively beneath concrete slabs of bins and building foundations causing severe cracking and/or collapse.

Rodents may also play a significant role in the maintenance and transmission of swine diseases such as leptospirosis, trichinosis, erysipelas, swine dysentery, and others. However, the actual occurrence of swine diseases in rodents, and the degree to which they contribute to disease problems on swine farms is poorly documented, and thus poorly understood. Still, an effective biosecurity system cannot be assured if rodents are present around swine operations.

When rodent infestations are allowed to become severe, the contamination of hog feed from rodent excreta can be considerable. Rats void between 12 and 16 ml of urine and up to 50 droppings daily. Mice can void up to 100 droppings and urinate hundreds of times in tiny amounts, directly onto hog feed. Feed contaminated with rat or mouse urine may be rejected by pigs. Moreover, rodent urine and excreta may contain disease organisms that can be directly transmitted to swine during their feeding.

Feed consumption by rodents is primarily a problem with rat infestations. The average rat eats about 6–8 oz of grain per week, or approximately 20 lb of grain per year. In cases of severe rat infestations (e.g., 200 rats) grain loss can be as high as 4000–5000 lb of grain per year. Additional quantities of grain are lost when rodents gnaw holes in feed sacks and wooden feed bins.

And finally, in addition to all the costs, damage, and labor mentioned above, there is also the cost of the labor associated with rodent extermination campaigns.

Pest Birds

The most common bird pests affecting swine production include the European starling (*Sturnus vulgaris*), the English sparrow (*Passer domesticus*), the feral pigeon (*Columba livia*), and occasionally a variety of blackbirds.

Table 1 Cost factors affected by rodent infestations within confined swine facilities

Rodent	Attacks on pigs	Disease transmission	Facility maintenance, utilities repairs	Feed losses	Pest management labor
Norway rat/roof rat	Minor	Low–moderate	High	Moderate	Moderate
House mouse	None	Low–moderate	High	Minor	Moderate–major

Although it is well known that birds can be serious pests of swine production, little quantitative data exist that summarizes the losses. Pest birds consume feed, contaminate feed with droppings, may transmit diseases, destroy building insulation, and can be general annoyances to the swine producer. The significance of each pest bird on these factors is presented in Table 2.

In areas of abundant starling populations, it is not uncommon for several thousand starlings to descend on a swine feedlot each morning during the winter months. These massive starling flocks may travel up to 30 miles daily visiting different swine operations, and thus act as potential disease conduits between infected and uninfected herds. Sparrow infestations often cause extensive damage to rigid foam insulation inside swine buildings and shelters via their pecking activity. Pigeon flocks create damage from their droppings, nesting materials, feathers, and carcasses, generally contributing toward unsanitary conditions in a swine operation.

Like rodents, bird infestations may also contribute in the transmission of various diseases to swine herds. These include transmissible gastroenteritis (TGE), salmonella, avian tuberculosis, and others. The transmission of TGE from starling flocks and sparrows among and within swine herds has been documented in several different studies. But overall, the role of the pest bird species in transmitting swine diseases remains sketchy and is likely to be herd-specific. In operations where good management practices are in place, the disease transmission threat from birds (and rodents) is likely to be low. Nevertheless, as was mentioned for rodents, to maintain effective biosecurity programs around swine facilities, pest bird infestations should not be tolerated.

Predators

The coyote (*Canis latrans*) represents the most significant predator threat for rangeland and pasture swine operations. Occasionally, feral dogs are also reported attacking some swine.

For the most part, coyotes are opportunists around outdoor swine operations, taking prey that is the easiest to capture. Often times, these are the piglets—especially unprotected newborns or sick piglets. Coyotes are also drawn to those operations that leave swine carcasses or any type of carrion on the grounds for prolonged periods. Once rewarded with carcasses, coyotes may develop a habit of visiting the same operation repeatedly.

VERTEBRATE PEST MANAGEMENT PROGRAMS

Managing any type of pest infestation around swine facilities involves employing an integrated pest management approach (IPM). This is particularly true of vertebrate pests, as the restriction of food and/or harborage will have a dramatic effect on rodent, bird, and predator populations.

Good swine management practices such as pest exclusion and sanitation practices, *integrated* with the use of nonchemical and chemical control approaches constitutes an overall IPM program. To the busy swine producer, it sometimes seems easier to quickly apply some type of

Table 2 Potential for conflict associated with bird infestations and swine facilities

Conflict	Starlings	Sparrows	Pigeons	Blackbirds
Feed consumption	High	Low–moderate	Moderate–high	Moderate
Contamination with droppings	High	Moderate	Moderate–high	Low
Disease transmission	Moderate	Low–moderate	Low–moderate	Low
Insulation destruction and losses	None	High	None	None
Management costs of severe infestations	Moderate–high	Moderate	Moderate	Low–none

(Adapted from Ref. 1.)

poison bait, or just set a trap or two. But such shallow efforts are rarely cost-effective long term, and the producer in these cases, merely harvests some of the pests—only to be quickly replaced from the established pest population. It can't be emphasized enough that regardless of the vertebrate pest involved, IPM programs are truly the wisest and most cost-effective and time-efficient approaches.

Rodent IPM

Managing and eliminating rats and mice around swine facilities involves a committed, scheduled program. For rats, it is both attainable and desirable to eliminate all the rats on the premises, and then to monitor each month for any new immigrants. Achieving zero mice—especially inside confined facilities—may not be practical, or even possible. But it is entirely possible, even in large complex facilities, to keep mouse populations to low numbers (e.g., less than a dozen mice).

Rodent IPM programs begin with keeping swine facilities as clean and organized as possible. Eliminating unnecessary clutter, junk piles, overgrown weeds, feed spills, and carrion, and preventing their recurrence has dramatic consequences to rat and mouse populations. Moreover, practical sanitation programs force rodents to explore and interact with traps and baits.

Where feasible, denying rodents entry to swine buildings (i.e., confined swine facilities) via rodent proofing provides long-term protection. Several excellent publications present ''how-to'' approaches for rodent proofing agricultural buildings.

For minor infestations of rats and mice, strategically positioned traps (based on previous ''night watches'') can be effective. But in serious infestations, trapping programs are too labor intensive. In the majority of rodent infestations, a poison baiting (rodenticide) campaign is usually the primary control approach. Recent research has shown that the most effective use of rodenticides depends heavily upon careful and strategic location of the baits. The layperson's approach to putting out baits in corners of barns and buildings, or simply stuffing bait packets down rat burrows, has little long-term effect on reducing an actual rodent population, regardless of the bait brand used.

For effectively controlling mice inside confined swine operations, for example, research has shown baits often have to be placed directly in the pathways of the local mouse colony's primary feeding areas. This must be done both on floor areas as well as along the various off-floor areas (e.g., along wall ledges, pen dividers, off-floor utility lines, etc.). Such strategic baiting efforts are crucial to baiting program success because a significant number of mice in these facilities never travel along the floors, and thus they will not be subject to programs that utilize floor baiting programs only.

It must be stressed again, however, that the use of rodenticide baits alone will rarely be cost effective on a long-term basis if the conditions allowing rodents to enter the premises, hide, and feed at will are not addressed via sound IPM programs.

Despite ongoing advertisements claiming ''new technological breakthroughs,'' there are currently no electronic machines utilizing ultrasonic or electromagnetic means that have any scientific data to prove they actually control rodents. These and similar electronic ''gizmos'' should be avoided.

Occasionally, swine producers attempt to utilize cats and/or dogs in attempts to control rodents. But dogs or cats killing a few rodents on a weekly basis can never keep up with a quickly multiplying rodent infestation.

Bird IPM Programs

When large numbers of birds are present within and around swine facilities, sound IPM programs, similar in concept to those described for rodents, are essential. Additionally, conducting a situations analysis as it pertains to each bird pest and each swine operation is also important. Thus, prevention efforts utilizing good production practices that render facilities less optimal for the proliferation of birds will significantly reduce the severity of a particular bird infestation.

Swine feeds, for example, are a particularly attractive food source in the winter when snow cover and frozen ground impedes feeding in other areas. And, birds require a steady food supply to sustain their high metabolic demands. Typically, large bird flocks at swine units are associated with large quantities of palatable feed accessible to birds throughout the day. Thus, one IPM approach is to avoid providing constant daily exposure of feed to birds during the daylight hours. Quickly eliminating any large grain spills that occur during filling and loading operations is also important.

Bird exclusion programs are appropriate for confined swine and partial confinement operations. During the past decade, new bird exclusion technology has exploded and many bird denial products and mechanical repellents offer economical and effective options for commercial swine producers.

Direct control measures such as the use of avicides (i.e., toxic bird baits), and traps to manage birds vary in effectiveness with the species, their numbers, and the nature of the problem. No single bird toxicant, trap, or repellent exists that works on all the different bird pests.

For low to moderate bird infestations, outstanding "how-to" literature has been produced for managing and preventing bird infestations around swine facilities. But for severe or complex bird infestations, bird management specialists associated with the USDA Wildlife Services program, or local Cooperative Extension agency should be consulted. Such specialists can help assess the specific program needed for the particular problem and help design the most cost-effective IPM program.

Predator Management

Managing the damage from coyotes and other predators begins with good swine management programs of keeping newborns, piglets, or ill pigs protected. Pig carcasses must be removed or properly discarded as soon as possible, so as not to attract, feed, and/or train coyotes or any other scavenging mammal to the swine production site.

In areas of heavy predator pressure, exclusion programs utilizing net wires, and/or electric fences properly constructed and maintained can significantly aid in reducing predation losses. Trapping and/or hunting programs conducted periodically can also reduce the number of coyotes in a local area.

REFERENCE

1. Glahn, J.F.; Johnson, R.J.; Germer, L.E. Bird Management at Swine Facilities. In *Pork Industry Handbook*; PIH 134, Purdue University, Cooperative Extension Service: W. Lafayette, IN, 1995; 6.

BIBLIOGRAPHY

Baker, R.O.; Boadman, G.R.; Timm, R.M. Rodent Proof Construction and Exclusion Methods. In *Prevention and Control of Wildlife Damage*; Hygnstrom, S.E., Timm, R.M., Larson, G.E., Eds.; University of Nebraska, Cooperative Extension: Lincoln, NE, 1994; B137–B150.

Center For Wildlife Damage Control. *Biology, Behavior, Management Strategies, Tools, Supplies, and Expert Contacts*, 1998. http://www.ianr.unl.edu/wildlife/

Corrigan, R.M.; Towell, C.A.; Williams, R.E. In *Development of Rodent Control Technology for Confined Swine Facilities*, Proceedings of the 15th Vertebrate Pest Conference; Borrecco, I.E., Marsh, R.E., Eds.; University of California: Davis, CA, 1992; 280–285.

Green, J.S.; Henderson, F.R.; Collinge, M.D. Coyotes. In *Prevention and Control of Wildlife Damage*; Hygnstrom, S.E., Timm, R.M., Larson, G.E., Eds.; University of Nebraska, Cooperative Extension: Lincoln, NE, 1994; C51–C56.

Hygnstrom, S.E.; Timm, R.M.; Larson, G.E. *Prevention and Control of Wildlife Damage*; University of Nebraska, Cooperative Extension: Lincoln, NE, 1994; 683.

National Pork Producers Council. *Updates in the Management of Rodents for the Swine Industry*; National Pork Producers Council: Des Moines, IA, 2000. http://www.nppc.org

Timm, R.M.; Fisher, D.D. In *An Economic Threshold Model for House Mouse Damage to Insulation*, Proceedings of the 12th Vertebrate Pest Conference; Salmon, T.P., Ed.; University of California: Davis, CA, 1986; 237–241.

Timm, R.M.; Marsh, R.E.; Hygnstrom, S.; Corrigan, R.M. Controlling Rats and Mice in Swine Facilities. In *Pork Industry Handbook*; PIH 107, Purdue University, Cooperative Extension Service: W. Lafayette, IN, 1996; 6.

SYNCHRONY AND ASYNCHRONY PLANTINGS

Ida Nyoman Oka
Amlapura (Karangasem), Bali, Indonesia

INTRODUCTION

One cultural control of pest infestation is synchronized planting—sewing crops over a wide area and rotating with other food crop species (maize, peanut, soybean, and vegtables). The aims of synchronized planting is to prevent certain rice pest species from continuously building up on rice plantings. No serious pest outbreaks have been reported on the old rice varieties in a synchronized system.

Short duration rice varieties (110–120 days) released by IRRI and those by the national rice programs rice can be planted twice a year. Fig. 1 types (A) and (B) illustrate cropping systems of technically well-irrigated areas in West, Central, and East Java, and other rice centers of the country.

Rice in the wet season may be immediately followed by rice in the dry season (Fig. 1, Type A), or rice in the dry season may be immediately followed by rice in the wet season (Fig. 1, Type B) (1). Rice crop plantings only twice per year with short idle periods make the synchronized planting difficult for farmers in a region. However, asynchrony (staggered planting) is typical and provides an ideal environment for certain arthropods and small mammalian rice pests to build up. Under these conditions, food is always in abundance and pest outbreaks take place.

DISCUSSION

A number of rice pests are constraints for rice production. Crop losses vary depending on the pest species, intensity of damage, varietal reaction to the pest, growth stage of the plant, and the types of rice (upland/irrigated rices). For example, epidemic types of insect pests, such as the brown planthopper, *Nilaparvata lugens* (Stal), and the Rice Tungro Virus (RTV) disease, vectored by the green leafhopper, *Nephotettix virescens*, may cause total crop losses.

The brown planthopper [*Nilaparvata lugens* (Stal)], the most feared epidemic rice pest throughout Southeast and South Asia, may totally destroy rice crops. For example, in the 1976–77 rice season in Indonesia, more than 450,000 ha of rice were damaged, reducing the total yield of milled rice to about 364,500 t. The pests suck the sap and kill the plant. The pest also transmits two persistent rice viruses, i.e., the Grassy Stunt Virus (GSV) and the Rugged Stunt Virus (RSV).

The Rice Tungro Virus (RTV) is transmitted by four species of the green leafhopper, among which is *Nephotettix virescens* (Distant). In asynchrony (staggered) planting, the initial inoculum (Xo) of the RTV survives in the diseased rice stubble and ratoons. These plant remains are favored breeding grounds of the green leafhopper populations. Populations carrying the virus are always in abundance in the asynchrony plantings. The most dangerous virus period to the rice plants is when they are in the seedbeds and after transplanting in the fields until the plants are about one month old. When the plants get older they become more tolerant to the disease. The nonpersistence type of the virus and the high mobility of the leafhopper makes the disease highly epidemic, covering thousands of hectares of rice fields within a very short time.

About 10 species of rats attack rice, with *Rattus argentiventer* Rob and Kloss being the most common. Because mature rice is always present in staggered plantings, the rat populations multiply. Within 6 months, a female rat may reproduce four times, with 5 to 10 young each generation. Thus, rat populations can cause major losses in a relatively short period.

In asynchrony (staggered) plantings, chemical control measures become ineffective for insects, diseases, and rats. The farmers tend to increase the frequencies of chemical applications up to 5 or more a season. Baiting the rats with rice grains mixed with rat poison also becomes ineffective, because the rats prefer to feed on the ripening rice grain found in abundance in the region.

On the other hand, synchronized plantings of rice in space and rotated in time with early maturing food crops, is generally able to prevent the continuous build up of most of the important rice pests. Pest population build up is limited because their preferred crops are absent for a

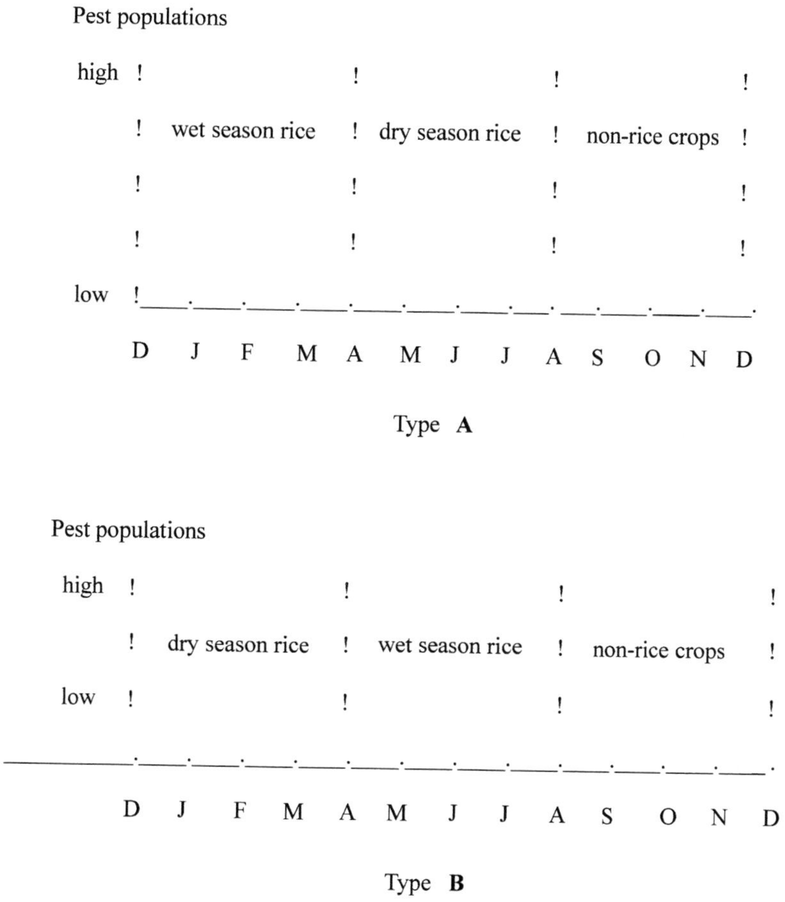

Fig. 1 Cropping systems in well-irrigated areas (staggered planting). Type (**A**): wet season rice—dry season rice—non-rice crops; type (**B**): dry season rice—wet season rice—non-rice crops.

period of time. Fig. 2 type (A) and type (B) suggested synchronized rice cropping systems in some rice centers of West, Central, and East Java and in other provinces of intensive rice production.

Fig. 2 type (A) illustrates the cropping system for one year consisting of wet season rice—dry season rice—early maturing nonrice crops. The wet season rice is planted early in December and harvested in April. During April, the field is being prepared for the dry season rice crop. During soil preparation the remaining rice stubble and ratoons are destroyed by plowing. The grasses grown on the dykes are cut or entirely destroyed. This sanitation destroys the remaining insect pest populations. Also, the initial inoculum (Xo) of the virus diseases (RTV, GSV, and RSV) are drastically reduced. The dry season rice is planted around May and harvested in September, followed by short maturity nonrice crops. Soil preparation for the next wet season rice is carried out in late November.

Wet season rice is planted in December and harvested in April and followed by the planting of short maturity food crops, which is harvested sometime in late June to give time to prepare the soil for the next dry season rice crop to be planted in early July. The dry season rice crop is harvested around November. Soil preparation for the next wet season rice crop is carried out late in November after harvest of the dry season rice crop. Sanitation and rat control programs are carried out as in the type (A) cropping system.

These synchronized cropping systems allow the rat population to have only two peaks of reproduction. Field studies carried out in Northwest Java demonstrated that during harvest time in August and September in type (A) synchronized cropping systems, the percentages of pregnant rats ranged from 71% to 93%. In April the following year—harvest time for the wet season rice—the percentage of pregnant rat populations reached 95%. In May the number of pregnant rats was only

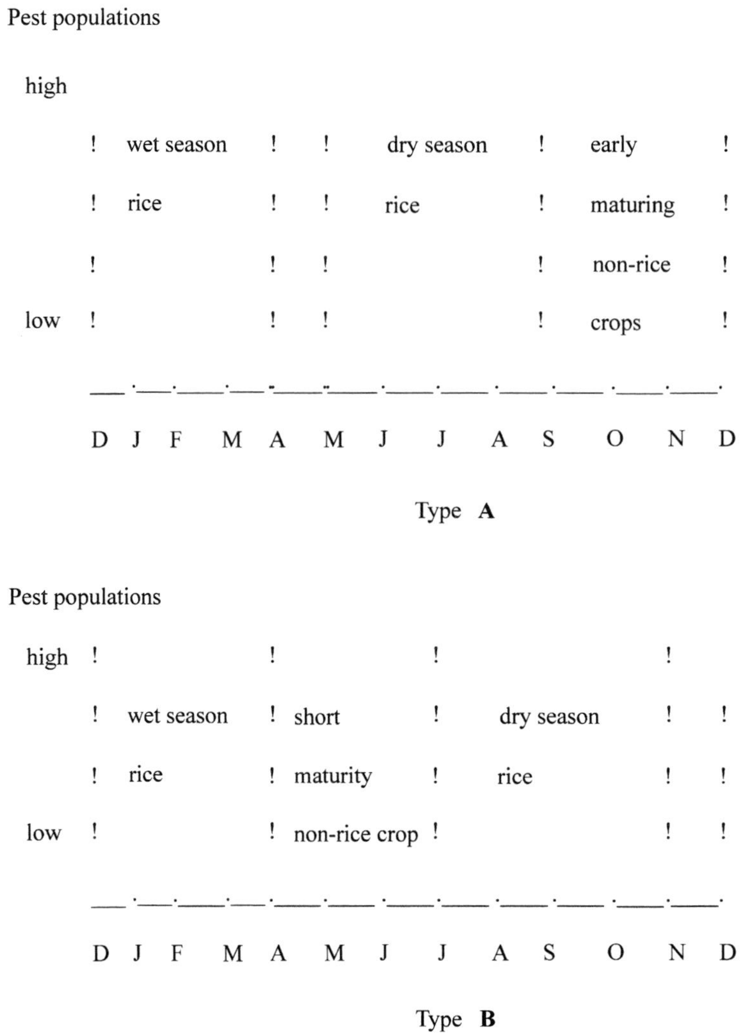

Fig. 2 Suggested synchronized cropping systems in technically well-irrigated rice centers. Type (**A**): wet season rice—dry season rice—early-maturing non-rice crops; type (**B**) synchronized planting system: wet season rice—short maturity non-rice crops—dry season rice.

about 68%. In the asynchronous (staggered) planting systems, however, the generations are overlapping. The distinct two peaks of the rat populations in the synchronized systems make control measures more efficient and easier as follows:

During soil preparation live rat holes found in the dykes and river-beds are closed with mud. Only a few holes may be left open and the rats killed by sulfur fumes. Escaped individuals are clubbed or killed by trained dogs. These mechanical methods are able to drastically reduce the rat populations.

Some investigators question whether asynchronous plantings promote pest problems, because they give the rice pests a constant source of food. This analysis, however, is based solely on the plant–herbivore model and does not consider the role of natural enemies and alternative prey (2). Settle et al., examined the trophic level differences between fields from Northwest Java (synchronized over wide areas) and Central Java, which is less synchronous (the areas of synchrony are smaller), and suggested that predators are late coming into the fields in Northwest Java compared with Central Java. This is because the fields in Northwest Java area are barren and dry during the preceding long-fallow period. It takes about 65 days after transplanting for predator populations in Northwest Java to reach high densities that are reached in Central Java after only 11 days.

The rice region in Northwest Java, which is planted synchronously, covers between 5000–7000 ha (Fig. 2 Type A), with much less plant diversity as compared to

S

those areas of Central Java. However, if a long drought exists, the dry season rice area has to be limited or the fields have to be left fallow for long periods before the next wet season rice is planted. This situation is detrimental to the survival of the natural enemies (predators) of the brown planthopper (BPH). This may not be the case if the wet season rice can be immediately followed with dry season rice and thereafter planted with short duration secondary food crops, such as in Central Java, North Sumatra, and Bali.

CONCLUSION

Synchronized planting systems integrated with resistant rice varieties (Ir 26, 36, 42, 56, 64, 70, etc.) drastically reduce both intensity and severity of rice pests. Serious outbreaks of the RTV disease in Central Java in 1995–96 was successfully overcome after the farmers returned to synchronous planting integrated with varieties resistant to the green leafhopper. There is no need to apply insecticides, because most of the frequently applied insecticides cause extensive outbreaks of the brown planthopper. The insecticides select pests for resistance to the insecticide while destroying the natural enemies. This results in pest outbreaks. Without insecticides, which conserve the natural enemies of the pests, there are no outbreaks. So integration of cultural controls, synchronized planting, sanitation, resistant rice varieties, and abundant natural enemies help stabilize the pest populations at low levels. Production costs also are reduced.

This integrated rice pest management system covering extensive areas is only possible with strong farmer organizations that enable the farmers to act together simultaneously in the production process.

REFERENCES

1. Oka, Ida Nyoman. The Potential for the Integration of Plant Resistance, Agronomic, Biological, Physical/Mechanical Techniques and Pesticides for Pest Control in Farming Systems. In *Chemistry and World Food Supplies, The New Frontiers*; Pergamon Press, 1983; 173–184, Chemrawn II.
2. Settle, William, H.; Hariawan, H.; Astuti, E.T.; Cahyana, W.; Hakim, A.L.; Hindayana, D.; Lestari, A.S. Parjaningsih. Managing tropical rice pests through conservation of generalist natural enemies and alternative prey. Ecology **1996**, *77* (7), 1975–1988.

SYSTEMS MANAGEMENT

Glen C. Rains
University of Georgia, Tifton, Georgia, U.S.A.

Dawn M. Olson
W. J. Lewis
Crop Protection and Management Research Unit, USDA-ARS, Tifton, Georgia, U.S.A.

James H. Tumlinson
Center for Medical, Agricultural, and Veterinary Entomology, USDA-ARS, Gainesville, Florida, U.S.A.

INTRODUCTION

There is a growing consensus that a fundamental shift in pest management strategies that rely on whole-system management and promotion of sustainable principles is needed. Therapeutic-based approaches to pest management require continued use of inputs to keep pests below economic thresholds, and are not sustainable. Farmers that rely on therapeutic inputs for primary pest management create other pest control problems such as pest resistance, increased secondary pest populations, toxic residues from chemicals, and pest resurgence. Systems management can provide long-term sustainable pest control with minimal use of therapeutic inputs, while maintaining production and reducing cost.

The foundation for such a shift involves an appreciation of the interactive webs in ecosystems whereby solutions to problems and net benefits of any action taken are determined at the ecosystem level. Such pest management is primarily based on managing the inherent strengths of the system and secondarily on the use of softer interventions (1). Several interdependent strategies are needed to shift current programs to these goals.

SYSTEMS PEST MANAGEMENT USING SUSTAINABLE PRINCIPLES

A systems approach to pest management must apply sustainable principles to enhance overall agro-ecosystem health. An agro-ecosystem can become sustainable through the application of six defining principles (2). If the agro-ecosystem is managed to foster diversity in organisms, interdependent relationship among organisms, efficient use of materials, self-regulating populations, self-renewing processes, and self-sufficiency, it is sustainable. Diversifying the system not only buffers the system against large-scale crop diseases and environmental catastrophes, but also fluctuations in commodity prices. Landscape ecology through refugia management, crop rotations, designing field borders, and inter-cropping, are examples of interdependent multi-tactic strategies that promote plant and soil health, species presence, and positive species interactions. Interdependent multi-tactic strategies, along with a diversification of the agro-ecosystem, helps incorporate the other four principles. For example, self-regulating mechanisms are dependent on a diverse and interdependent array of organisms that help stabilize the system.

SYSTEMS MANAGEMENT STRATEGIES

An understanding and wise use of inherent strengths of the agro-ecosystem is the foundation for systems management strategies. By utilizing several interdependent tactics these inherent strengths can be optimized. Plant defenses, soil quality, biodiversity, natural enemies, and crop attributes are some of the in- herent properties of the system that should be maximized.

A systems management strategy must not only be sustainable in the long-term, it must also be cost-competitive in the short-term. Adoption of strategies that show little or no economic advantage are not readily adopted.

There are three strategic lines to follow when developing a systems-managed agro-ecosystem for long-term sustainability: 1) habitat management, 2) crop attributes and multi-trophic interactions, and 3) therapeutic back-up.

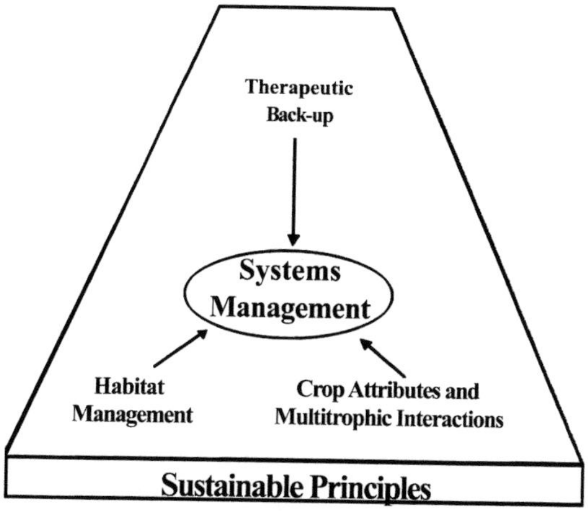

Fig. 1 Three system strategies provide a solid framework from which to build a pest management system. These strategies must also embrace the sustainable principles that will provide long-term pest management.

All three strategic lines must be simultaneously pursued while retaining the sustainable principles that will provide a long-term framework for pest management (Fig. 1).

Habitat and Other Cultural Management

Perennial cropping is a strategy for fostering a balance in populations of beneficial pest organisms. Development of a perennial system provides year-round cover of vegetative biomass on the land, reducing the amount of disruption to the agro-ecosystem that occurs during conventional farming operations. The ecological growth curve (3) illustrates how interactions within systems progress through time. Annual cropping systems mostly operate in the rapidly increasing portion of this curve where the largest fluctuations in species numbers occur. These systems are ecologically unstable because they seldom have time to reach the plateau of the curve where feedback mechanisms stabilize species interactions. The perennial system should consist of diverse cropping tactics that operate near the top of the ecological growth curve year-round. At this level, species interactions are stabilized and mediation of pest populations are optimized. For example, winter cover crops of crimson clover, used in conjunction with strip-till planting, have been shown to harbor beneficial organisms that remain through the summer and help protect cotton from pest damage.

Another extremely important habitat management strategy is the utilization of refugia around and within fields that help boost populations of beneficial organisms, or attract pests away from the primary crop. Certain crops and perennials provide better refugia than others. The particular selection will depend on the particular pests that need controlling. The proper combinations of crops, border refugia, and cover crops can provide shelter and food for beneficial organisms, reducing the time for population ramp-up to counter pest outbreaks. Field intercropping can provide refugia for beneficial organisms or draw pests away from the primary crop. In one example, black-eyed peas were used between and around a pecan orchard as a trap crop to draw stink bugs away from the pecans, reducing damage from 12 to 13% to less than 1% (4). Another practice is to install buffer strips around the field, consisting of the proper plant mix to harbor populations of beneficial organisms. Soil quality can improve the sustainability of beneficial organisms residing in or on the soil, reducing the levels of soil-borne insect and pathogenic activity on plants. Increase of soil complexity can also improve the nutrient availability to the plant, increasing its vigor and health, and thus improve its resistance to opportunistic organisms. Soil quality and biodiversity are generally increased by reducing tillage operations, leaving crop residue on the soil after harvest, and by adding natural soil amendments, such as compost and manure.

Proper planting dates and harvest dates can be critical to the proper mediation of populations and dynamics of weed, insect, and pathogenic pests. Certain pest and beneficial organism populations can be devastated by conducting field operations at particular times during the life cycles of these organisms. One example is the recommended multi-tactic cultural practices for managing peanut (*Arachis hypogaea*) production to reduce the incidence of Tomato Spotted Wilt Virus (TSWV). The virus is vectored by several species of thrips (*Thysanoptera*) and planting dates, tillage practices, peanut variety, and row pattern have been shown to affect the incidence of the virus (5).

Crop Attributes and Multitrophic Interactions

Plants play an active role in managing pest populations and dynamics by their physical, chemical, and biological characteristics. The defensive mechanism depends on the particular plant, pest, and natural enemy interaction involved. Plants usually have a combination of direct physical and chemical defenses against pests, as well as indirect strategies such as extra-floral nectaries, and other resources to attract natural enemies. Most varieties of cotton (*Gossypium hirsutum*) have extra-floral nectaries on the underside of the leaf that attract beneficial insects.

Recent findings demonstrate the importance of chemical signaling in the defensive strategies of the plant. Of particular interest are certain chemically-mediated interactions between the plant, pest, and beneficial organism. Herbivore feeding can induce biosynthetic pathways within the plant that release volatile chemicals that act as a signal that attracts parasitoids of the pest to the plant (6). Induced chemical signals indirectly attract natural enemies of the pest, while constitutive defenses (physical and chemical) repel and slow pest feeding.

A systems management strategy must take into account these multi-trophic interactions and optimize genetic capacity and phenotypic expression. If interactions are ignored, plant defenses are not utilized efficiently, thereby violating principles for a sustainable system. The interplay between plant production and plant protection traits must also be considered. Current reductionist paradigms choose strategies to maximize crop production, while developing a separate strategy to prevent pest damage using therapeutic inputs.

Agronomic management plays an important role in the plants ability to express constitutive and induced plant signalling (7). If the ability of plants to emit herbivore-induced signals is reduced significantly due to stresses on the plant caused by insufficient nutrients or water, beneficial organisms cannot find their host/prey as efficiently.

Therapeutic Role

Since an agro-ecosystem is not a naturally occurring system, there are times when interventions are necessary to bring pests under control and help bring self-regulating mechanisms back to a sustainable level. But, the least ecologically disruptive management practice available to reduce the disruption to the agro-ecosystem should be practised. Disruptions can directly reduce populations of beneficial organisms, as well as indirectly reduce the capacity of plant signalling to attract natural enemies. Research has shown that the flight response, longevity, and fecundity of a parasitoid is dramatically reduced when a systemic chemical pesticide is applied (8). In the case of a selected chemical application over a small area, precision farming techniques may be an efficient means to apply a therapeutic in the right amount and only in the location where needed, reducing the potential disruption to the system.

FUTURE INVESTIGATIONS

To fully develop systems management strategies utilizing proper habitat and plant attributes, a full understanding of the levels of interactions and diversity that make an agro-ecosystem sustainable is required. Strengthening and developing inherent properties should be focussed on by conducting on-farm systems research, where interactions and diversity are carefully monitored and the collective net benefits of plant protection as well as plant production are considered.

Although new technologies can be useful tools to pest management, development and use of these technologies must be redirected to foster a systems approach to sustainable pest management.

Current research developments in biotechnology and precision agriculture have concentrated on improving the therapeutic approach to pest management. Bio-engineered crops, global positioning systems (GPS), crop sensors, and geographical information systems (GIS) must be developed and utilized to enhance the long-term inherent properties of the system and not rely on the ''quick-fix'' short-term solution. The central question should be, why is the pest a pest, and what steps are necessary to manage that pest using a systems approach based on sustainable principles.

REFERENCES

1. Lewis, W.J.; van Lenteren, J.C.; Phatak, S.C.; Tumlinson, J.H. In *A Total System Approach to Sustainable Pest Management*, Proceedings of the Natlional Academy of Sciences USA, 1997; 9412,243–12,248.
2. Lewis, W.J.; Jay, M.M. *Ecologically–Based Communities—Putting it All Together at the Local Level*; Kerr Center for Sustainable Agriculture: Poteau, OK, 2000, 49.
3. Odum, E.P. *Fundamentals of Ecology*; W.B. Saunders Co.: Philadelphia, PA, 1969.
4. Fritz, M.; Mudd, D.; Berton, V.; Kirschenmann, F.; Landis, D.; Lewis, J.; Liebman, M.; Luna, J.; Magdoff, F.; Neher, D.; Phatak, S.; Prokapy, R.; Rajotte, E.; Teasdale, J. A whole-farm approach to managing pests. Sustainable Agric. Network Bull., 20. http://www.sare.org/farmpest/index.htm (accessed April 2001).
5. Brown, S.; Culbreath, A.; Baldwin, J.; Beasley, J.; Pappu, H. Tomato spotted wilt of peanut—identifying and avoiding high risk situations. Univ. Ga. Coop. Ext. Serv. Bull. **2000**, *1165*, 11.
6. Paré, P.W.; Tumlinson, J.H. De novo biosynthesis of volatiles induced by insect herbivory in cotton plants. Plant Physiol. **1997**, *114*, 1161–1167.
7. Cortesero, A.M.; Stapel, J.O.; Lewis, W.J. Understanding and manipulating plant attributes to enhance biological control. Biological Control **2000**, *17*, 35–49.
8. Stapel, J.O.; Cortesero, A.M.; Lewis, W.J. Disruptive sublethal effects of insecticides on biological control: altered foraging ability and life span of a parasitoid after feeding on extrafloral nectar of cotton treated with systemic insecticides. Biological Control **2000**, *17*, 243–249.

TACTICS

Vincent P. Jones
Washington State University, Wenatchee, Washington, U.S.A.

T

INTRODUCTION

Pest management tactics are the methods used to reduce economic loss caused by pest activities. The methods vary widely and depend on the product to be protected, its value, and the specific pest and natural enemy complex. The different types of tactics can be broken down into three major groups (1), but any given tactic may fall into several groups. The groups are:

1. Those that change the favorability or carrying capacity of the environment (=ecological management);
2. Those that decrease the reproductive capacity of the pest;
3. Those that decrease the survival of the pest within the environment.

Tactics from these groups are integrated as necessary to reduce damage to an acceptable level while providing the maximum benefits in terms of cost and ecosystem stability.

ECOLOGICAL MANAGEMENT

The focus in ecological management is to reduce pest problems by modifying the environment at critical points in the pest's life cycle (1). For these techniques to be successful, a thorough understanding of the relationship between the pest and all potential hosts must be known. Example tactics include elimination of alternate hosts, removal or destruction of crop residues that are breeding or overwintering sites, tilling the soil to expose or bury a vulnerable stage of the pest, or use of trap crops to divert the pest away from the main commodity. In post-harvest situations and some urban situations, temperature extremes alone or in combination with high carbon dioxide or nitrogen, or low oxygen atmospheres are used to eliminate pests.

Ecological management also seeks to break up the spatial and chronological synchrony of the pest with its host (1). Reducing spatial continuity can be applied at both the single field level, where plant spacing or plant diversity is modified, or at the area-wide level, where the mosaic of crops planted within a large area is designed to prevent movement of pests from one favorable crop to another.

MANAGEMENT BY ALTERING REPRODUCTION

Chronological synchrony of the pest and its host can be broken up through the practices of crop rotation, crop fallowing, early harvest, or early or late planting (1). Crop rotation is most efficient when the pest reproduces before the crop is planted, if the pest has a narrow host range, and if the damaging stage is relatively immobile. When these conditions are met, rotation of the host crop with a non-host crop (or over a period of several years, several non-host crops) can effectively minimize pest damage. These techniques can also be used for veterinary pests by rotation of pastures or rotation of the type of animal kept in a particular pasture. For crop fallowing to be a good management practice, the fields need to be kept free of crop residue and weeds that might support the pest population. Fallowing may also require several years to reduce pest population levels, which limits its usefulness in many systems. Early harvest and early or late planting disrupt the synchrony of the pest with the susceptible stages of the host and can drastically reduce damage.

Methods to reduce the reproductive rate of the pest are becoming more common as our knowledge of pest biology and behavior increases. Examples include the sterile male technique, sex pheromone-based mating disruption to reduce or delay mating (2), attraction of one sex to a lure combined with pesticide (=male annihilation, attracticide), or the use of sub-lethal pathogens that reduce pest reproductive capacity (such as *Nosema locustae* for locust control). While each of these tactics has been used successfully in different systems, their applicability is generally restricted to a few systems. This is partly a result of their specificity, but also because these tactics are generally more effective on a large scale where the dispersal of pests from outside the treated area is minimized.

MANAGEMENT BY ALTERING SURVIVAL

Methods of reducing survivorship are probably the best-known ways to reduce pest populations. These include pesticides, use of natural enemies (=biological control), resistant cultivars (including transgenic insecticidal cultivars) (3), and irradiation for post-harvest situations. From a management perspective, pesticide use requires knowledge of the effect on both the pests and their natural enemies. Without this information, pesticides may cause instability in the production system by eliminating natural enemies, and in some cases, direct reproductive stimulation of the pest species (hormolygosis). Resistant cultivars may simply tolerate damage well enough that the general equilibrium position of the pest is below the economic threshold or they may reduce pest survivorship or reproduction (particularly when using transgenic insecticidal cultivars).

The conservation and augmentation of natural enemies is the core of many management strategies. In pesticide-dominated systems, natural enemies can be conserved by using selective pesticides, or by changing their dose, formulation, application time, or location of application. Conservation can also be achieved by cultivating plants near the fields to provide nectar or alternate hosts during times when the main pest is at low numbers or not present (4). Augmentation is when natural enemy populations are increased by releasing laboratory-reared individuals. In many cases, the augmentation tactic uses the natural enemies as a ''biotic pesticide'' where direct mortality is expected within a short period, but the natural enemy is not expected to persist in the environment. Examples of this include the use of *Bacillus thuringensis* and other non-persistent disease organisms (such as nematodes, viruses, or fungi), or mass release of *Trichogramma* egg parasitoids for caterpillar control.

The final area of natural enemy use is classical biological control. This is generally used when an exotic pest is introduced into an area without natural enemies (5). Natural enemies are imported into the new system to permanently establish the natural enemy and reduce the pest population level to sub-economic status. However, the degree of control can be variable depending on a broad range of factors including the geographic distribution of the pest, the searching ability of the natural enemy, different host plants, and the effect of other natural enemies already present in the system.

The diversity of pests and the systems they occur in have spawned a large number of potential tactics. For these tactics to reduce pest damage to sub-economic levels, in most systems they must be carefully integrated with a thorough understanding of the relationship between the pest and the crop (or animal) system to be protected. Ideally, proper integration results in a sustainable system that is economically viable and that requires few alterations unless new pests are accidentally introduced into the production system.

REFERENCES

1. Pedigo, L.P. *Entomology and Pest Management*, 3rd Ed.; Prentice-Hall: Upper Saddle River, NJ, 1999; 691.
2. Carde, R.T.; Minks, A.K. Control of moth pests by mating disruption: successes and constraints Annu. Rev. Entomol. **1995**, *40*, 559–585.
3. Gould, F. Sustainability of transgenic insecticidal cultivars: integrating pest genetics and ecology. Annu. Rev. Entomol. **1998**, *43*, 701–726.
4. Bottrell, D.G.; Barbosa, P. Manipulating natural enemies by plant variety selection and modification: a realistic strategy? Annu. Rev. Entomol. **1998**, *43*, 347–362.
5. Van Driesche, R.G.; Bellows, T.S., Jr. *Biological Control*; Chapman and Hall: New York, 1996; 539.

THRESHOLDS FOR PESTICIDE TREATMENT

Uri Regev
Ben-Gurion University, Beer-Sheva, Israel

T

INTRODUCTION

Pest management involves three basic and interrelated decisions: deciding whether and when pest control is needed, the choice of control technology, and its level or quantity. These decisions are critical for food production and have been central research subjects since the early days of chemical pesticide use in the 1940s. The complexity of biological and economic models that dealt with these questions called for a simple criterion to enable the farmer to make pest management decisions, a criterion that is even more important when control is carried out by central governments rather than by individual farmers. Early policies of applying a given dose of chemicals every fixed number of days (as a kind of insurance against pests) proved disastrous once resistance to pesticides began to develop and ecological and environmental damages of chemical pesticides became a subject of serious concern. It also became evident that pesticides could cause pest resurgence and secondary pest outbreaks, and hence farmers began to refrain from indiscriminate pesticide use. The goal of economic thresholds (ET) is to give the farmer a simple "rule of thumb," based on pest population levels, for determining when to initiate pesticide use in order to minimize unnecessary applications. As will be shown, formulating such a decision rule does not escape the complexity of the basic decisions, namely, if, when, how much, and what choice and mix of control to apply. This task is further complicated in developing countries, where basic data are not readily available.

DEFINITIONS

The ET determines the necessity for initiating control measures, the timing, and the dose of specific pest control treatment based on the incidence of insect pests, diseases, and/or weed pests. Common definitions for the economic (or the action) threshold include the following:

1. The pest population that is large enough to cause damage;
2. The minimum pest population level for which it is profitable to apply a specified dose;
3. The pest population that produces incremental damage equal to the cost of preventing that damage;
4. The pest population level subsequent to treatment with a preset dosage level of a specified pesticide, selected to maximize profits;
5. The density of pests where action (control/management) must be taken to prevent the population from reaching the economic injury level (a complex concept that encompasses market policy, availability, competition, etc., rather than just yield losses).

Some of these definitions are vague and others confuse marginal and total profits as well as pretreatment and posttreatment pest levels. In the following, an economically consistent definition of the ET is expounded.

Like any other agricultural input, economic considerations for pest management require that pesticides be applied at the dosage (X) that simultaneously satisfies the following three conditions:

1. marginal costs equal marginal benefits,
2. marginal costs are rising with X, and
3. profits are non-negative.

These conditions should be evaluated for each of the available control technologies in order to select the one most profitable, including not using any controls at all. It should be noted that pest management often involves a number of control tactics applied together, so that the dosage X could represent the level of pesticide combinations rather than a single pesticide. Fig. 1 presents the *total* benefits and costs of pesticide use (X) per season for high and low levels of pretreatment pest densities. The benefits of pesticides include reduced pest damage, thus, the benefits function increases with pest densities. The horizontal difference between the benefit and cost functions represents the profit from pesticide use. Fig. 2 presents another view, namely the *marginal* costs and benefits (the slopes of the corresponding functions in Fig. 1) of the above two pest densities. It shows that the

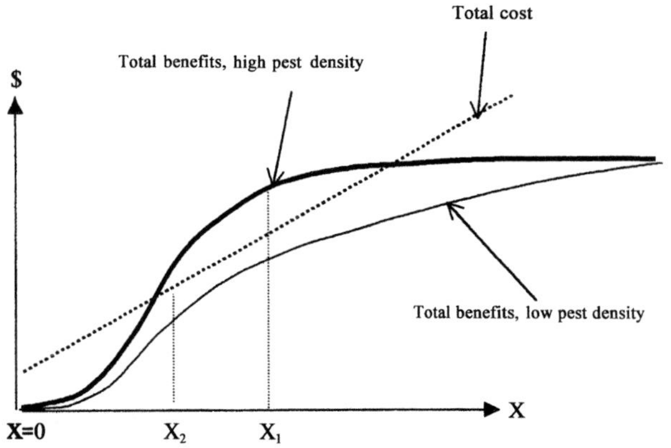

Fig. 1 Total benefits (prevented losses) and costs of pesticide dosage.

high pest density is the ET, since the optimal pesticide level X_1 meets all three conditions. The low pest population is below the ET and therefore no pesticide should be applied because profits are negative. This is observed in Fig. 2 by the difference between the areas under the marginal benefits and marginal costs [the area marked (+) less the area marked (−)].

Based on the preceding discussion,

> *The* ***economic threshold*** *is the minimal pretreatment pest population level that warrants pesticide application when the amount of pesticide used is calculated to maximize profits.*

Bear in mind that any specific economic threshold involves the choice of control tactics and depends on its optimal usage level and timing of the selected technology. When many applications per season are required, pesticide use becomes a dynamic decision process. In such cases, the determination of the ET must take into consideration plant developmental stages and its varying susceptibility to pest damage. In some crops, the ET is likely to be high at the beginning of the season, low in the middle, and high again toward the end. In others, this phenology may be reversed or be quite different.

PROBLEMS IN EVALUATING BENEFITS AND COSTS FOR THE ECONOMIC THRESHOLD

Although the theoretical discussion seems to provide simple and viable solutions for pest management decisions, in reality many issues and problems must be considered for determining ETs. These include disruption of the ecological equilibrium between pests and natural enemies, development of pest resistance to pesticides, air and water pollution by pesticides, pesticide residues in the food chain, and problems of risk and uncertainty. Disregarding these problems can bias the calculation of the ET as well as the choice of the best control tactics.

Pesticide applications may disrupt the ecological equilibrium in the field and the surrounding area, causing pest resurgence and secondary pest outbreaks. For example, many studies have shown that due to crop–pest–natural enemy interactions, a low pest population at an early stage of plant growth may actually contribute to yield increase, so that pesticide applications at that stage may be harmful. This could result from the fact that pests facilitate some physiological function of fruit shedding early in the plant's phenology that leads to lower energy losses and hence higher yield. Disruption of ecological equilibrium greatly complicates the evaluation of the benefits of control for determining the ET, since it requires that indirect benefits and costs of the ecological microenvironment in the field should also be taken into account. Using only

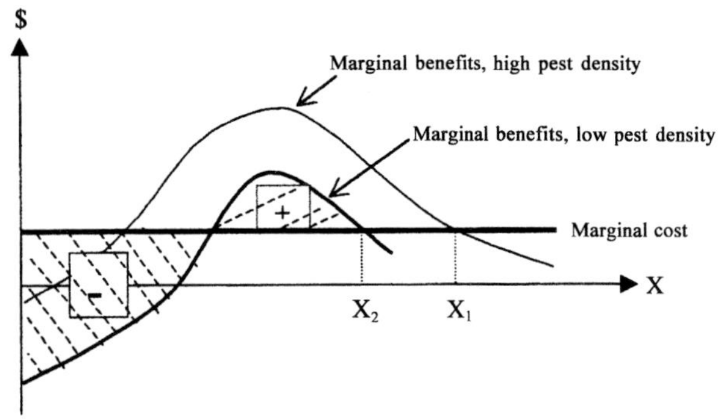

Fig. 2 Marginal benefits (prevented losses) and costs of pesticide dosage.

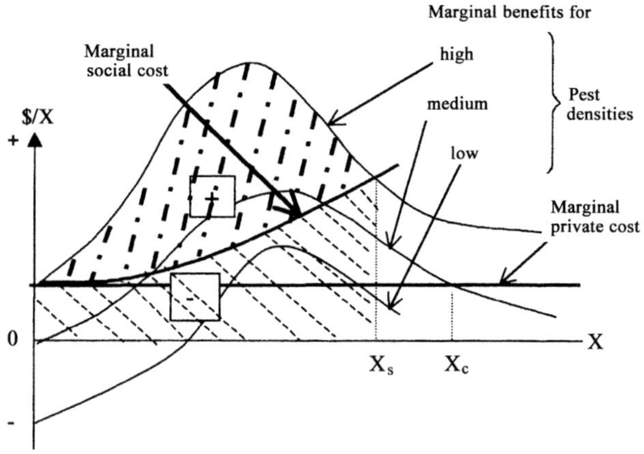

Fig. 3 The effect of private and social marginal costs of pesticides on the economic threshold.

direct benefits of pesticides for the determination of the ET could bias it downward.

Another problem in calculating the ET is the development of pest resistance to pesticides. This is a long-term effect of pest control that reduces the future effectiveness of pesticides and their resulting benefits. This problem is even more complex because due to the mobility of many insect pests resistance development in any specific field is the result of pesticide use in the entire region rather than in a farmer's own field (known as the common property resource problem). For this reason, resistance development will not commonly enter the cost calculations of an individual farmer and is referred to as the external cost of pesticides. The economic costs of pesticides to society are composed of the sum of private cost and external cost. Competitive farmers, however, consider only the private cost for their pest control decisions and this will tend to lower the ET and stimulate overuse of pesticides.

Additional costs arising from external environmental side effects of chemical pesticides include air, soil, and water pollution. These costs are not considered by pesticide users since they are not liable for them in a competitive economy. Furthermore, estimating the cost of these effects is difficult and often unreliable. Chemical residues in the food chain are another source of external cost that is not considered by farmers because consumers are often unaware of residue levels. However, consumers are becoming increasingly alert to this problem, as reflected in market prices, and this trend has begun to affect farmers' control tactics.

The addition of the above external effects (resistance, pollution, and food residues) to the market costs form the social cost of pesticides. Fig. 3 schematically illustrates the impact of social costs on the ET and the level of pest control. In this scenario, the individual farmer considers only the marginal private cost, so that the medium pest density constitutes the ET for him (similar to the situation underlying Fig. 2), and the pesticide will be applied at the level X_C. However, if farmers are forced or encouraged to consider the social cost, Fig. 3 shows that profits become positive only for the high pest density [the area marked with (+) is larger than that marked with (−)]. The latter becomes the social ET and pesticides are applied at an optimal dosage X_S. The conclusion of this discussion is that the consideration of social costs leads to both a higher ET and a lower pesticide dosage.

Another major difficulty in estimating the benefits of pest control is the uncertainty involved in these decisions and the risk attitude of the farmer. Risk aversion may lead farmers to adopt a low ET, thereby overusing a short-sighted "effective" control technology as a form of "insurance" against pest damages.

In summary, the economic threshold might appear to be a simple rule of thumb for pest control decisions, but its calculation involves many difficulties that are embedded in plant–pest–natural enemy–human interrelationships.

BIBLIOGRAPHY

Gutierrez, A.P.; Caltagirone, L.E.; Meikle, W. Evaluation of Results: Economics of Biological Control. In *Handbook of Biological Control*; Bellows, T. S., Ed.; Academic Press: San Francisco, 1999; 243–252, Ch. 10.

Gutierrez, A.P.; Daxel, R.; Quant, G.L.; Falcon, L.A. Estimating economic thresholds for bollworm, *Heliothis zea* Boddie, and Boll Weevil, *Anthonomus grandis* Boh, damage in Nicaraguan cotton, *Gossypium hirsutum* L. Environ. Entomol. **1981**, *10*, 872–879.

Gutierrez, A.P.; Wang, Y.; Regev, U. An optimization model for lygus hesperus (*Heteroptera: Miriadae*) damage in cotton: the economic threshold revisited. Can. Entomol.; **1979**; *111*, 41–54.

Headley, J.C. Defining the Economic Threshold. In *Pest Control Strategies for the Future*; Metcalf, R.L., Ed.; National Academy of Science: Washington, DC, 1972; 100–108.

Hueth, D.; Regev, U. Optimal agricultural pest management with increasing pest resistance. Am. J. Agric. Economics **1974**, *56*, 543–551.

Moffitt, I.J.; Hall, D.C.; Osteen, C.D. Economic thresholds under uncertainty with application to corn nematode management. So. J. Agric. Economics **1984**, *16*, 151–157.

Plant, R.E. Uncertainty and the economic threshold. Economic Entomol. **1986**, *79*, 1–6.

Regev, U.; Shalit, H.; Gutierrez, A.P. On the optimal allocation of pesticides: the case of pesticide resistance. J. Environ. Economics Manage. **1983**, *10*, 86–100.

Talpaz, H.; Frisbie, R. An advanced method for economic threshold determination: a positive approach. So. J. Agric. Economics **1975**, *7*, 19–25.

Weersink, A. Defining and measuring economic threshold levels. Can. J. Agricu. Economics **1991**, *39* (4), 619–625.

TICKS

Daniel E. Sonenshine
Old Dominion University, Norfolk, Virginia, U.S.A.

INTRODUCTION

Worldwide, tick parasitism and tick-borne diseases cost the livestock industry more than $7 billion (1), mostly due to losses to cattle. The most serious losses are due to two tick-borne diseases caused by protozoan parasites of the genera *Babesia* and *Theilieria*. Many domestic animals die as a result of tick paralysis and tick toxicoses. In addition, heavy tick infestations of livestock, poultry, and even wildlife (e.g., deer) often lead to death or injury from anemia or exsanguination. For example, Hereford cattle in a drylot infested with Gulf Coast ticks, *Amblyomma maculatum* Koch, gained 27 kg less than uninfested animals during a 7-week period. Additional economic losses result from damage to hides and lost milk production as a result of tick infestations (2).

Control of ticks and tick-borne diseases is accomplished primarily by acaricides, especially organophosphorus compounds, carbamates, pyrethroids, formamidines, and avermectins. The pyrethroids are among the safest and most effective pesticides and are now widely used for tick control (2).

The most common methods used to deliver acaricides include dipping, spraying, pour-ons, spot-ons, and dusts. For cattle, dipping is done by driving the animals through a fenced enclosure toward a concrete bath filled with dip containing a water emulsion of the acaricide. The animals are forced through the bath on the entry side and the acaricide-impregnated liquid is allowed to pour off back into the dip as they emerge on the exit side. Despite their total immersion, some ticks hidden in sheltered locations (e.g., between the toes or in the ears) often are missed and survive to lay eggs and reestablish the pest population, for example, the Gulf Coast ticks (*Amblyomma maculatum*), which survive in the ears of cattle. Although dipping is efficient, the animals are stressed and dehydrated by the need for periodic forced round-ups and violent cattle drives in order to immerse them in the dips. However, cattle dips are expensive, use a lot of water (a problem during drought periods) and, being stationary, require rounding up the animals and bringing them to the dips. A popular alternative is the spray technique, using high-pressure sprayers to provide an acaricide-containing mist that can reach every part of the animal's body. Although one person can carry some sprayers, the larger sprayers are mounted on trucks that can be driven to the corral where the cattle or other livestock are enclosed, and then driven to the next corral to repeat the process. This is much less stressful for the animals. Moreover, spray formulations can be adapted to minimize water usage, a major benefit during drought conditions. A more recent innovation is the pour-on (or spot-on), a formulation in which acaricide is mixed with surfactants (i.e., detergents) that spread the liquid over the hair coat by the animal's natural movements. Occasionally, dusts such as the familiar "flea powders" for pets, which are effective against ticks as well as fleas, are also used. In this case, the acaricides are mixed with talc and deposited directly onto the animal's fur. Attached ticks can be encouraged to detach rapidly when the infested host is treated with a detaching agent such as Amitraz. Long-lasting efficacy (e.g., several months) with a single treatment can be achieved by incorporating acaricides into plastic. Acaricide-impregnated plastic ear tags are used for control of the Gulf Coast tick *Amblyomma maculatum*, the brown ear tick, *Rhipicephalus appendiculatus*, and other livestock pests, while the familiar flea and tick collars are widely used for control of ticks on cats and dogs. Rarely, a well-tolerated acaricide such as ivermectin can be administered into the blood of livestock or pet animals, where it functions as a systemic, providing long-lasting and effective tick control. In cattle, ivermectin provided excellent control of selected tick species for 2–3 months (2).

CONTROL OF TICK-BORNE BABESIOSIS IN CATTLE

Babesiosis is a serious, often fatal disease caused by Babesia *species*, especially *B. bigemina* and *B. bovis*. In cattle, these parasites are transmitted by cattle ticks, such as *Boophilus microplus* or *B. annulatus*.The traditional control methods in which acaricides are delivered to control ticks have been only partially successful. In a few cases, eradication may be practicable, for example, the Cattle Tick Fever Eradication Program that led to the eradication of Texas Cattle Fever by eradicating its tick vectors, *B. annulatus* and *B. microplus* from the United States. However, attempts to eradicate *B. microplus* in other areas (U.S. Virgin Islands, Argentina, Uruguay, Australia, and Papua New Guinea) have been unsuccessful, despite reductions of more than 99% of the tick populations in some localities. Vaccines and chemotherapeutic agents have been used to protect against or kill the parasites in cattle blood. Although live attenuated vaccines have been used to protect cattle against *Babesia* parasites in the past, more recent research has concentrated on recombinant vaccines. In Australia, scientists at the CSIRO have purified and cloned three protective antigens derived from crude extracts of *B. bovis* that resulted in 95% reduction in parasitaemias in treated cattle. Two of the proteins expressed, when used in combination, proved almost as effective as the live attenuated vaccine. Vaccines have also been developed against the cattle tick vectors. Most promising are the vaccines based on recombinant proteins from the midgut of cattle ticks. One such formulation, Gavac, showed 55–100% efficacy in control of *Boophilus microplus* in field trials with grazing cattle in Cuba, Brazil, Argentina, and Mexico and a 60% reduction in the number of acaricide treatments (increasing the interval between treatments by 32 days). This resulted in a savings of $23.40/animal/year (3). In Australia, studies are in progress to develop a vaccine for control of the cattle tick *Boophilus microplus*. A recombinant antigen vaccine has been tested and there are plans for its commercial development. Although it is possible that antigen-resistant strains of cattle ticks may ultimately appear, large-scale vaccination of cattle herds with the recombinant vaccine offers a promising alternative to acaricides. Chemotherapeutic agents used against *Babesia* include diminazene diaceturate, imidocarb, and buparvaquone (4).Treatment with aminoguanidine has also proven beneficial, ameliorating the levels of parasitaemias and the febrile response in acute infections (5).

CONTROL OF HEARTWATER

Heartwater is a serious, often fatal disease of cattle and other large ruminants caused by a type of rickettsia known as *Cowdria ruminantium*. These microorganisms live in tiny cluster-like colonies in the cells of their mammalian hosts, especially in the endothelial cells lining the capillaries. Common symptoms are high fever and edema, especially fluid accumulation around the pericardium (hence the name heartwater). In Zimbabwe, a recent (1999) estimate of annual losses due to this disease was US $5.6 million, mostly due to the costs of acaricides. Treatment is often done with antibiotics, especially tetracycline and doxycycline, which are effective when administered in large doses early in the course of the disease. Control has been done primarily by means of tick control. Heartwater is still controlled by large-scale use of acaricides, delivered primarily by dipping and spraying. In the Caribbean, where heartwater is a serious problem, a 5-year program to eradicate the local vector tick, the tropical bont tick (*Amblyomma variegatum*), was initiated in 1994, using pour-on formulations of acaricides. However, despite a budget of more than US $16 million, eradication does not appear to have been achieved. An

Table 1 Control of *Amblyomma hebraeum* ticks on cattle using tail tags impregnated with the attractant-attachment-aggregation pheromone and cyfluthrin, flumethrin or alphacypermethrin during two field trials in Zimbabwe[a]

	Percent Control											
	Week of trial 1						Week of trial 2					
Life stage/ Treatment	2	4	6	8	10	12	2	4	6	8	10	12
Males												
Cyfluthrin	98.4	98.8	97.8	95.8	97.5	98.4	99.6	97.4	99.5	98.0	99.1	99.6
Flumethrin	89.8	94.1	92.1	92.3	91.3	91.4	95.3	87.9	97.2	92.3	91.7	91.8
Alphacy permethrin	–	–	–	–	–	–	67.2	64.8	32.5	64.4	68.8	74.3
Females												
Cyfluthrin	100	100	99.1	100	98.2	100	100	94.3	100	98.5	98.8	100
Flumethrin	94.7	100	98.0	98.7	97.3	100	100	94.3	100	95.1	97.1	93.5
Alphacy-permethrin	–	–	–	–	–	–	6.5	56.4	19.3	71.3	71.3	74.8

[a]From Ref. 6.

innovative alternative to the traditional large-scale acaricide treatment is the use of pheromone-impregnated decoys to attract the bont ticks, *Amblyomma hebraeum* and *A. variegatum*to specific sites on cattle and kill them when they attach nearby. Field trials with the tail tags (6) have demonstrated excellent efficacy for periods of up to three months against *A. hebraeum* in Zimbabwe (Table1). The choice of acaricide is very important.Control of bont ticks with cyfluthrin and flumethrin was 99.3% and 95.1% over the three-month trial period, respectively, whereas control using alpahcypermethrin for the same period was only 79.1%. Efforts are being made to make this tick control device available for sale. If successful, it will be the first commercial application of pheromone-assisted tick control.

FUTURE NEEDS

Traditional tick control by dipping, spraying, pour-ons, or spot-ons are only partially effective due to their failure to eliminate all ticks, the tick's high reproductive rate, and the development of resistance. As a result, the injurious effects of tick infestations and the diseases that they transmit continue to limit the potential for growth in the livestock industry. However, promising new technologies such as antitick vaccines, recombinant DNA vaccines against the tick-borne pathogens, and pheromone-assisted tick control have been developed and tested. Efforts should also be directed to identifying antimicrobial genes that may kill invading parasites in their arthropod vectors, similar to the discoveries made in mosquitoes (7). If commercialized and used worldwide, such products may replace acaricides as the preferred method for reducing or even eliminating tick infestations and tick-borne diseases (8). [See also the articles Public Health; Vectors of Human Diseases; Acaricides in this encyclopedia].

REFERENCES

1. Graham, O.H.; Hourrigan, J.L. Eradication programs for the arthropod parasites of livestock. J. Med. Entomol. **1977**, *13*, 629–658.
2. Drummond, R.O.; George, J.E.; Kunz, S.E. *Control of Arthropod Pests of Livestock: A Review of Technology*; CRC Press: Boca Raton, 1988; 1–245.
3. De la Fuente, J.; Rodriguez, M.; Montero, C.; Redondo, M.; Garcia-Garcia, J.C.; Mendez, L.; Serrano, E.; Valdez, M.; Enriquez, A.; Canales, M.; Ramos, E.; Boue, O.; Machado, H.; Leonart, R. Vaccination against tick (*Boophilus spp.*): the experience with the Bm86-based vaccine gavac. Genet. Anal. **1999**, *15*, 143–148.
4. Bruning, A. Equine Piroplasmosis: An update on diagnosis, treatment and prevention. Br. Vet. J. **1996**, *152*, 139–151.
5. Gale, K.R.; Waltisbuhl, D.J.; Bowden, J.M.; Jorgensen, W.K.; Matheson, J.; East, I.J.; Zakrzewski, H.; Leatch, G. Amelioration of virulent *Babesia bovis* infection in calves by administration of the nitric oxide synthase inhibitor aminoguanidine. Parasite Immunol. **1998**, *20*, 441–445.
6. Norval, R.A.I.; Sonenshine, D.E.; Allan, S.A.; Burridge, M.J. Efficacy of pheromone-acaricide tail-tag decoys for control of bont ticks, *Amblyomma hebraeum*, on cattle in Zimbabwe. Exper. Appl. Acarol. **1996**, *20*, 31–46.
7. Beernsten, B.; James, A.A.; Christensen, B.M. Genetics of mosquito vector competence. Microbial Mol. Biol. Rev. **2000**, *64*, 115–137.
8. Sonenshine, D.E. *Biology of Ticks*; Oxford University Press: New York, *2*, 333–371.

TILLAGE

Karen L. Bailey
Bruce D. Gossen
Agriculture and Agri-Food Canada, Research Centre, Saskatoon, Saskatchewan, Canada

T

INTRODUCTION

Many agricultural systems rely on tillage to bury crop residues; to level, consolidate and warm the seedbed in the spring; to reduce surface compaction; to incorporate pesticides and fertilizers; and to reduce weed, disease, and insect problems. Conventional tillage mixes or inverts the soil surface layer by discing, cultivating, or plowing, usually in the fall after harvest and again in the spring prior to seeding. After two tillage operations, little residue from the previous crop is left on the soil surface. If the land is left fallow, additional tillage is often required during the growing season to control weeds. Specialized machinery has been developed for these processes. For heavy operations such as breaking new land, deep trash burial, deep banding of fertilizer, or some weed control and seeding, a heavy plow such as a chisel plow, moldboard plow, or tandem disk plow, is used. Field cultivators and discers provide shallower tillage for seedbed preparation, chemical incorporation, and most weed control and seeding.

Conventional tillage exposes topsoil to wind and water erosion. It also causes breakdown of organic matter, depletion of soil moisture, and spread of salinity. Therefore, reduced tillage practices (also known as conservation tillage) are being adopted in many areas to combat soil degradation (1). Under zero-till (also known as no-till) management of annual crops, the crop stubble and residue are left untouched from harvest to seeding; also, the new crop is planted with a minimum amount of soil disturbance. Minimum-till usually maintains the stubble after harvest, but a tillage operation is made before seeding to prepare a good seed bed. Reduced tillage is a common term that includes both zero and minimum tillage.

CHANGES IN SOIL PROPERTIES WITH TILLAGE

The physical and biological properties of soil are changed with each tillage operation. Tillage knocks down and then incorporates standing stubble and crop residue, leaving the soil surface exposed to heat and wind. This increases heating and evaporation. In contrast, surface residue acts as an insulator, reflects solar radiation, and conserves soil moisture. Consequently, reduced tillage practices result in soils that warm more slowly in the spring and are cooler during the growing season than soils under conventional tillage. In the top 2 in. of soil, the difference in soil temperature between zero-tillage and conventional tillage is about 0.5–2.0°C; differences decline at lower depths (2).

Moisture retention in the soil is improved under reduced tillage because of reduced evaporation, increased snow trapping, and reduced surface run-off due to better water infiltration. In semiarid, temperate regions like the Canadian prairies, each tillage operation results in a loss of 0.5 in. of soil moisture.

Tillage also increases the rate of organic matter breakdown. Organic matter is essential for healthy soils because it increases the soil's capacity to hold water and nutrients, while reducing its susceptibility to erosion and crusting. It also provides a habitat and food source for microorganisms, insects, arthropods, and other organisms that are needed to recycle nutrients, aerate the soil, and improve soil aggregate structure. Greater availability of oxygen and microbial activity near the soil surface assist in the conversion of surface residues to soil organic matter, especially in the top 2 in. of soil.

Under conventional tillage management, deterioration of soil structure frequently has been observed at three levels: 1) a crust at the soil surface that retards crop emergence and water infiltration, 2) beneath the crust, fine aggregates of soil form clods and compact easily when wet, and 3) a compacted layer develops just below the tillage depth, which creates a barrier to penetration of water and roots. After only 2–3 years of zero-till management, the top 1–2 in. of soil become more crumbly and less subject to crusting due to higher organic matter content, and the lower layers become firmer but less compacted.

IMPACT OF TILLAGE ON PESTS

Changes in soil properties associated with tillage practices have a direct impact on the weed, disease, and insect pests of agroecosystems, and some of these changes benefit crop production. Residue at the soil surface provides both pest and beneficial organisms with more favorable habitats for growth and survival. Wind-disseminated weed species (dandelion, foxtail barley, fleabane, narrow-leaved hawk's beard, etc.) and volunteer crops are commonly associated with reduced tillage systems, while invader species like wild oat and millet that require soil disturbance to germinate and establish, are associated with conventional tillage (3). In contrast, other pests are indifferent to tillage system. Perennial weeds such as quackgrass, Canada thistle, and perennial sowthistle establish and spread in all tillage systems.

Environment

Annual weather conditions and other large-scale environmental factors often have a greater influence on pest populations than tillage management. However, tillage practices can change the environment within a field, which can in turn affect pest populations. For each species of microorganism (and to a lesser extent, insects and weeds), there is an optimum temperature and moisture regime for growth and survival. Small changes in environment can favor one species and inhibit another. For example, higher soil moisture levels under reduced tillage are associated with reduced survival of fungal overwintering structures such as sclerotia of *Sclerotinia sclerotiorum* (the cause of white mold of canola and other crops), because they favor the bacteria that attack these structures. Similarly, populations of one cereal root pathogen (*Bipolaris sorokinina*) decline under reduced tillage, but other pathogens (e.g., *Fusarium* spp., which also cause root rot in cereals) increase (4).

Crop Residue

Crop residues are composed of 15–60% cellulose, 10–30% hemicellulose, 5–30% lignin, 2–15% protein, and 10% soluble compounds. Rapid rates of residue decomposition occur when there is a low proportion of cellulose, hemicellulose, and lignin, a low carbon:nitrogen ratio, high moisture, and an active population of soil microbes. Residue breaks down more quickly when buried than when left at the soil surface, because the extreme fluctuations in moisture and temperature that occur at the soil surface are not conducive for growth of most microorganisms. When residue is buried by tillage, it takes about one year for leaf pieces to decompose, and 2–4 years (or more) for decomposition of tougher stem and root pieces.

The availability of crop residues has an important impact on pest populations (5). Predators of insect pests often rely on nonpest species such as mites, slugs, and insects that are involved in the breakdown of crop residue as alternate food sources when pest populations are low. As a result, insect pest populations are often lower where crop residues are present at the soil surface. Many plant pathogens live in the crop residue, and often can persist and continue to sporulate much longer on residue on the soil surface than in residue buried by tillage. For example, *Mycosphaerella pinodes* survives much longer in pea residue at the soil surface than when buried 2 in. in soil. In contrast, the survival of *Colletotrichum truncatum* is higher on buried lentil residue than at the soil surface. Management of residue (presence/absence, thickness, placement) can be manipulated to affect the survival of many pest species.

Beneficial Organisms

Microorganisms are primarily responsible for breaking down crop residues. The quantity and quality of the residue can influence the size and composition of the microbial community. Tillage distributes the microbial biomass through the soil profile to a depth of up to 6 in., whereas with zero-till, microorganisms are concentrated in the top inch. Tillage decreases the total microbial biomass and the activity of the remaining population by exposing the organisms to unfavorable environmental conditions, such as drying, ultraviolet light, and heat. Under reduced tillage, fungal populations in the soil become twice as large as under conventional tillage. Although a few of these fungi cause damage to crops, most are beneficial. Under zero-till, the number of fungi with antagonistic activity toward plant pathogens increases by as much as 95%.

Vesicular-arbuscular mycorrhizae are beneficial soil fungi that assist plants in taking up nutrients such as phosphorus. The number of mycorrhizal associations in plant roots increase under reduced tillage. Similarly, populations of earthworms and predatory soil-inhabiting arthropods such as ground beetles and spiders increase under reduced tillage (6). There also may be positive interactions occurring among plant and insect species. For example, parasitic wasps of the corn cutworm increase with higher levels of crop residue and higher densities of some weed species that occur predominately in minimum-till fields. Reduced tillage increases the sustainability of the production system by taking advantage of natural processes for pest management.

REFERENCES

1. http://www.ctic.purdue.edu/CTIC/CTIC.html (accessed March 2000).
2. http://paridss.usask.ca/index.html (accessed March 2000).
3. Derksen, D.A.; Lafond, G.P.; Thomas, A.G.; Loeppky, H.A.; Swanton, C.J. Impact of agronomic practices on weed communities: tillage systems. Weed Sci. **1993**, *41*, 409–417.
4. Bailey, K.L. Diseases under conservation tillage systems. Can. J. Plant Sci. **1996**, *76*, 635–639.
5. http://ianrwww.unl.edu/pubs/plantdisease/g804.html (accessed March 2000).
6. Stinner, B.J.; House, G.J. Arthropods and other invertebrates in conservation-tillage agriculture. Ann. Rev. Entomol. **1990**, *35*, 299–318.

T

TOXINS IN PLANTS

David S. Seigler
University of Illinois, Urbana, Illinois, U.S.A.

INTRODUCTION

Most plants have the ability to synthesize and accumulate toxins capable of modifying the fitness or behavior of herbivores or pathogens. Although the presence of either constitutive (always present) or inducible defensive compounds may increase fitness of the plant, there is a metabolic cost associated with production, transport, and accumulation of these toxins. Further, much evidence indicates that there is a complicated balance between this internal cost and various ecological effects, such as release from herbivory, that may be observed by altering defensive compounds or allomones in plants (1).

THE EVOLUTIONARY ROLE OF PLANT TOXINS

Most natural populations of plants, herbivores, or pathogens are variable and are under selective pressure. Resistant variants of these groups may increase in populations, whereas less resistant cohorts may be eliminated. Chemical defenses involving toxins are thought to provide resistance, for plants, to attack by other organisms. Nonetheless, if the total expenditures of making, compartmentalizing, transporting, or detoxifying toxins exceed the benefits derived, those plants may be noncompetitive in regard to undefended plants. The need for defensive compounds may not occur in all individuals of a population every growing season. For this reason, inducible defenses may avoid many of the costs required for either constitutively formed defenses (always available), or preformed chemical defenses (usually smaller amounts of more active defensive compounds are involved and the plant can still respond rapidly). Many kinds of defensive substances are produced in response to stress, such as that brought about by herbivory. Phytoalexins are commonly produced in response to fungal or bacterial attack (1).

Compounds with the ability to reduce fitness of consuming organisms belong to many groups of plant secondary metabolites. The amount and type of each compound present, as well as the composition of secondary metabolite mixtures, may differ within individual plants, among plants of natural populations, and from species to species. Among the most active substances are certain acetylenic derivatives, quinonoid compounds, polyketides, coumarins, stilbenes, some types of flavonoids, tannins, peptides, nonprotein amino acids, cyanogenic glycosides, glucosinolates, a variety of terpenes, and alkaloids. Many fatty acid derivatives, phenylpropanoids, stilbenes and related compounds, isoflavonoids, and terpenes are noted for their effects on fungi (2). The resistant effects are accomplished via a wide variety of biological lesions in the organisms affected.

COSTS AND BENEFITS OF TOXINS IN PLANTS TO RESIST ATTACKS OF HERBIVORES AND PARASITES

Although it has been suggested that toxins provide an adaptive advantage for plants, in practice, it is often difficult to measure either the costs or the benefits associated with defensive substances (1). Costs are expressed as the reduction in fitness, or one of its components, for resistant genotypes relative to susceptible genotypes in the absence of herbivores, where the benefits associated with resistance cannot be expressed (3). For example, there is a direct decrease in the number of seeds produced and the amount of furanocoumarins accumulated in the seeds of *Pastinaca sativa*. This decrease in seeds represents a direct "cost" or reduction in fitness, although the increase in furanocoumarins represents an increase in fitness or ecological "benefit," that offsets the cost (4). When 2% of the leaf material was removed from these plants, there was a short-duration decrease in photosynthesis, an overall 8.6% decrease of biomass after four weeks, and an increase in respiration that almost exactly corresponded to an increase in synthesis and accumulation of furanocoumarins. The physiological cost of making the compounds is offset by the benefit of increased furanocoumarin concentration and a corresponding decrease in herbivory (5).

In a critical examination of studies of such costs, not all of which were chemical, approximately 50% showed a cost, 5% showed a benefit, and 45% showed no difference or nonsignificant differences between resistant and susceptible plants (6). It has been noted, however, that demonstrations of cost often have been successful in systems in which resistance mechanisms are identified, and unsuccessful in systems in which the resistance mechanisms have not yet been characterized (7). When the benefits accrued from toxin production and accumulation exceed the costs for a particular set of conditions, the presence of new defensive compounds in plants should be favored. However, few studies that have attempted to measure the benefits of induced responses, when herbivores are present, are convincing (1). The plant may be protected from some enemies, but this gain may be offset by concomitant kairomonal or attractive roles of the modified complement of secondary metabolites, resulting in increased damage from other pests. Autotoxicity may be increased when a compound is introduced into a plant that normally does not have it because the plant lacks appropriate degradative enzymes, or the ability to transport and accumulate the compound properly (1).

ROLE THAT USE OF TOXINS HAS PLAYED IN AGRICULTURAL CROPS

The presence of toxic secondary metabolites has been linked to resistance of many crop plants to herbivores, bacterial and fungal attack, and to crop yields. In general, the presence of defensive compounds, or allomones, would be predicted to reduce the deleterious effects of herbivores and pathogenic bacteria and fungi, to increase plant fitness and, optimistically, to increase yields of plant-derived farm products.

Many important crop plants contain toxins. These include barley (alkaloids), brassicaceous plants (cabbage, rapeseed or canola, mustard, etc.) (glucosinolates), cassava (cyanogenic glycosides), carrots (acetylenic derivatives), celery, parsley, and parsnips (furanocoumarins), citrus crops (monoterpenes, furanocoumarins, alkaloids), cotton (gossypol), cucurbits (squashes, melons, gourds, and similar fruits) (cucurbitacins), flax (cyanogenic glycosides and lignans), grapes (stilbenes and tannins), legume fruits (beans, peanuts, peas, soybeans) (isoflavonoids), lima beans (cyanogenic glycosides), onions (sulfur-containing amino acids), potatoes and tomatoes (steroidal alkaloids), sorghum (cyanogenic glycosides), sugar cane (stilbenes), sunflowers (diterpenes), sweet potatoes (sesquiterpenes), and wheat (DIMBOA) (2).

Resistance has been enhanced in many crop plants by either traditional plant breeding techniques in which specific toxins have not been identified, or more recently by new molecular techniques (1). Increases in amount or introduction of toxins appears to be an important element of new crop plants for resistance to herbivory or to pathogenic fungi or bacteria. If resistance is costly, some of the costs may be overcome by agricultural practices such as fertilization (1). However, breeding for changes in secondary metabolite chemistry of some crop plants has produced effects other than those desired. For example, gossypol, a dimeric sesquiterpene produced in glands on the cotton boll, is associated with resistance to bollworm larvae, rabbits, rodents, and other herbivores. Because this compound is also responsible for the toxicity of cottonseed meal and other cotton products for humans and livestock, lines selected for low gossypol content have been selected. These lines are more susceptible to insect attack (8). In another example, the presence of furanocoumarin phytoalexins in celery infected with the fungus *Sclerotinia sclerotiorum* caused workers to develop severe skin irritation. Lines that contained primarily xanthotoxin and 4,5′, 8-trimethylpsoralene in quantities 10–15 times higher than usual had been selected for disease and insect resistance (9). Although damage to the plants was reduced, the increased presence of the plant toxins also reduced the value of the final product for human or domestic animal consumption. As an additional example, in cabbage (*Brassica oleracea*), there is a correlation between the content of glucosinolates and disease resistance. Wild populations of *Brassica* species usually have higher levels of glucosinolates and are more disease resistant than cultivated plants. Selective breeding for milder flavored individuals favored by human consumers has contributed to a lack of resistance to powdery mildew and other pests in cultivated lines (10). One final example involving cassava (*Manihot esculenta*) illustrates the balance of beneficial and harmful effects of plant toxins appropriately. This important crop plant is well known for the cyanogenic glycosides found in the tuberous-roots, but is also the major starchy food of millions. These toxins are linked to reduction in herbivory and fungal attacks on the plant, but also produce numerous human victims. Recently, however, the biosynthetic pathway leading to the two cyanide-releasing compounds, linamarin and lotaustralin, has been studied in detail and the genes responsible for selected steps have been cloned. It is within our grasp to produce cassava plants truly free of these toxins for the first time, decreasing or eliminating human poisoning by the plant, yet maintaining taste properties considerably desirable by many consumers (11). Whether cyanide-free plants will be sufficiently resistant to herbivores or pathogens remains to be seen.

FUTURE PROSPECTS

Clearly, selection of cultivars rich in toxins or incorporation of toxins into crop plants is promising, but not without limits. As indicated above, these processes may lead to either increased or decreased, but possibly more predictable or sustainable yields. In other instances, increased amounts of toxins may limit the usefulness of the crop. Recent use of corn (*Zea mays*) containing the peptide Bt-toxin suggests, however, that we are creating crop plants with limited variation and introducing them into environments with herbivores that are highly variable in many characters including the ability to deal with the introduced toxin. Plants may incur indirect evolutionary costs if they deploy defenses constantly. As noted by Karban and Baldwin (1), because the herbivores (or pathogens) are variable, selection for resistance will typically occur, either with synthetic pesticides or with introduced plant toxins. To avoid resistance problems, these authors suggest that plant breeders and molecular biologists should consider changing the emphasis from traits that can be expressed inducibly to traits that make plants more variable.

REFERENCES

1. Karban, R.; Baldwin, I.T. *Induced Responses to Herbivory*; University of Chicago Press: Chicago, 1997; 1–319.
2. Seigler, D.S. *Plant Secondary Metabolism*; Kluwer Academic Publisher: Boston, 1998; 1–759.
3. Elle, E.; van Dam, N.M.; Hare, J.D. Cost of glandular trichomes, a ''resistance'' character in *Datura wrightii* regel (Solanaceae). Evolution **1999**, *53*, 22–35.
4. Zangerl, A.R.; Berenbaum, M.R. Cost of chemically defending seeds: furanocoumarins and *Pastinaca sativa*. Am. Nat. **1997**, *150*, 491–504.
5. Zangerl, A.R.; Arntz, A.M.; Berenbaum, M.R. Physiological price of an induced chemical defense: photosynthesis, respiration, biosynthesis, and growth. Oecologia **1997**, *109*, 433–441.
6. Bergelson, J.; Purrington, C.B. Surveying patterns in the cost of resistance in plants. Am. Nat. **1996**, *148*, 536–558.
7. Berenbaum, M.R. Evolutionary Ecology of Plant/Herbivore Interactions. In *Evolutionary Ecology: Perspectives and Synthesis*; Fox, C.W., Roff, D.A., Fairbairn, J., Eds.; Oxford University Press: New York, *in press*.
8. Gershenzon, J.; Croteau, R. Terpenoids. In *Herbivores: Their Interactions with Secondary Plant Metabolites*; Rosenthal, G.A., Berenbaum, M.R., Eds.; Academic Press: San Diego, 1991; *1*, 165–219.
9. Beier, R.C.; Nigg, H.N. Natural Toxicants in Foods. In *Phytochemical Resources for Medicine and Agriculture*; Nigg, H.N., Seigler, D.S., Eds.; Plenum: New York, 1992; 247–367.
10. Greenhalgh, J.R.; Mitchell, N.D. The involvement of flavor volatiles in the resistance to downy mildew of wild and cultivated forms of *Brassica oleracea*. New Phytol. **1976**, *77*, 391–398.
11. Moeller, B.L.; Seigler, D.S. Biosynthesis of Cyanogenic Glycosides, Cyanolipids, and Related Compounds. In *Plant Amino Acids*; Singh, B.J., Ed.; Marcel Dekker, Inc.: New York, 1999; 563–609.

TRADITIONAL CROPPING SYSTEMS

H. David Thurston
Cornell University, Ithaca, New York

T

INTRODUCTION

The latest key word in agriculture is sustainable. Everyone is for sustainability; environmentalists, chemical companies, molecular biologists, and even the public and politicians. Can we learn something about sustainable agriculture from the 10,000 years of experience in traditional agriculture? I believe we can. Some ancient farmers developed sustainable agriculture systems and practices that allowed them to produce food and fiber and manage pests for thousands of years with few outside inputs. Many of the successful cropping systems of ancient farmers have been forgotten, abandoned, or ignored in developed countries, but are still used by traditional farmers in a wide variety of environments in developing countries. There is considerable evidence showing that traditional farmers experiment and innovate, but most of the useful traditional systems, practices, and materials used today in agriculture were probably developed empirically through millennia of trial and error, selection by farmers, and careful observation. Most pest management practices used by traditional farmers consisted of cultural controls, yet little information is available in an easily accessible or understandable form on these cultural practices. The disease and insect resistance found in traditional landraces has also had profound effects on "modern agriculture," as most of our present cultivars evolved from these ancient plant materials that traditional farmers have been selecting for millennia. Landraces are usually genetically diverse and are usually adapted to their environment and endemic pests. Although they not necessarily high yielding, they generally are dependable and are stable in yielding some harvest under all but the poorest of conditions. Today, relatively few pesticides are used by traditional farmers because of their high cost, but pesticide misuse is common. Unfortunately, farmer expectations for pesticides are often unrealistically high. For example, Rosado and Garcia (1) interviewed 59 traditional farmers in Tabasco, Mexico, regarding their management methods for web blight of beans (*Thanatephorus cucumeris*). Disappointingly, although they used several cultural methods of management, all of the farmers interviewed said they were expecting a chemical solution to the problem.

TRADITIONAL FARMER KNOWLEDGE

The knowledge of traditional farmers regarding their cropping systems is often impressively broad and comprehensive. An example illustrates this point. Conklin (2) described the agricultural knowledge of the Hanunóo, a mountain tribe of Mindoro in the Philippines as amazingly wide, accurate, and practical. They distinguished 10 basic and 30 derivative soil and mineral categories and understood the suitability of each for various crops, as well as the effects of erosion, exposure, and overfarming. Their more than 1500 useful plant types included 430 cultigens, and they distinguished minute differences in vegetative structure.

TRADITIONAL PRACTICES

Practices for pest management used by traditional farmers in their cropping systems include the following: altering of plant and crop architecture, biological control, burning, adjusting crop density, depth or time of planting, planting diverse crops, fallowing, flooding, mulching, multiple cropping, planting without tillage, using organic amendments, planting in raised beds, rotation, sanitation, manipulating shade, and tillage. Most, but not all, of the systems using these practices were sustainable in the long term. A few of the above practices require high organic inputs and some practices have high labor requirements. Some practices have multiple benefits. For example, the use of mulches prevents erosion, improves soil quality, manages weeds, lowers soil temperatures, conserves moisture, and may aid in the management of soilborne diseases. Mulches also reduce rain splashing, an important means of dissemination for numerous bacterial and fungal pathogens. Thus, today the use of mulches is increasingly recommended in agricultural development efforts.

TRADITIONAL CROPPING SYSTEMS

Innumerable traditional systems are recorded and many are still in use by indigenous farmers. For example, although the cropping system including maize, beans, and squash was basic in the highlands of Ecuador, Kirkby et al. (3) found more than 100 distinct crop associations within a 30,000 hectare area near Riobamba.

Janzen (4) suggests that tropical agroecosystems are misunderstood in the temperate zones and mismanaged in the tropics. My observations span more than 40 years in the tropics, and I would agree that most tropical agroecosystems are poorly understood by those living in the world's temperate areas. For example, slash-and-burn agriculture is far more important than most people realize. Estimates suggest that today there are from 200 to 500 million slash-and-burn cultivators. The slash-and-burn system is most commonly used today in tropical areas, but has extended in the past into subtropical and temperate zones. Such systems using fire have been in existence since the Neolithic era. Plots are partially cleared from the forest growth, the cut vegetation dries and is burned, and crops are planted in the ashes. Plots can be used for several years and then gradually abandoned to natural vegetation for fallow periods of up to 20 years or more. As crops are harvested from the slash-and-burn plots long after natural vegetation begins to return, fallow is often incomplete in the slash-and-burn system. Slash-and-burn agriculture may be one of the few really sustainable agriculture systems from the perspective that no or very little fossil energy is needed for continuous crop production.

Lesser known systems are the slash/mulch agricultural systems that can be defined as those in which vegetation is slashed or cut in situ to produce a mulch that is subsequently used for a crop (5). Slash/mulch systems are often overlooked or mistaken for slash-and-burn systems. In comparison to slash-and-burn systems little information can be found regarding slash/mulch systems, but such systems are probably far more important (especially in the hot, humid tropics) than most authorities realize. There are many areas in the tropics where it is so wet that it is impossible to burn, and thus over time slash/mulch systems were developed by indigenous groups of the Americas that utilized decomposing mulch as a nutrient source. Traditional farmers in many areas of Costa Rica grow beans (*Phaseolus vulgaris*) using a slash/mulch system called in Spanish "frijol tapado," which in English means "covered beans." The procedure consists of broadcasting bean seeds into carefully selected weeds, then cutting and chopping the weeds with a machete so the broadcasted bean seeds are covered with a mulch of weeds. A semideterminate type of bean, between a bush and a climbing bean, is planted. The beans grow through the mulch and eventually cover it. This combination of mulch and bean plants effectively prevents weed growth and appears to conserve soil moisture. In addition, the mulch prevents soil splashing, which was found in a Costa Rican study by Galindo et al. (6) to be the most important source of inoculum of *Thanatephorus cucumeris*, a fungus which causes a severe bean disease called web blight. Although yields were low, traditional farmers used the system because of its low risk, its small investment in labor (primarily to cut weeds), and because there was always some yield even when prolonged periods of rain produce conditions that allowed disease to destroy beans under clean cultivation systems.

Innumerable cropping systems are found within tropical agroforestry systems (7). Perhaps the most ubiquitous are the gardens called home gardens or backyard gardens that are found around the homes of many traditional farmers in the tropics. Gardens are usually multistoried systems dominated by tall trees, and in some tropical locations they closely mimic the natural forest. An extensive literature on home gardens is available (8).

The rice paddy system may be our oldest sustainable agricultural system (8). Farmers have flooded their rice paddies for millennia. For centuries rice grown in this system has annually produced a ton or two of rice per hectare without high inputs of fertilizer or pesticides. Although weed control is probably the major benefit of the rice paddy system, flooding rice fields also manages some pests. Flooding reduces the number of fungal propagules, insects, and nematodes in the soil and, by controlling weeds that may harbor rice pathogens and insects, reduces disease and insect damage. Kelman and Cook (9) wrote: "The practice of flooding fields for paddy rice and the use of organic material as fertilizers are apparently key factors in the general absence of soilborne diseases in China."

CONCERNS

Far too many giant development projects in developing countries have failed dismally, often with serious ecological consequences, because sufficient understanding of traditional agriculture was lacking. It is impossible in the space available to do more than mention the innumerable cropping systems found worldwide. Traditional agricultural practices should be understood and conserved before they are lost with the rapid advance of modern agriculture in developing countries.

REFERENCES

1. Rosado-May, F.J.; Garcia-Espinosa, R. Estrategias empiricas para el control de la mustia hilachoza (Thanatephorus cucumeris Frank Donk) de Frijol Comun en la Chontalpa, Tabasco. Rev. Mex. Fitopatologia **1986**, *4*, 109–113.
2. Conklin, H.C. An ethnoecological approach to shifting agriculture. Trans. N.Y Acad. Sci. **1954**, *7*, 133–142, Series 2.
3. Kirkby, R.; Gallegos, P.; Cornick, T. *Metodologia para el Desarrollo de Technologia Agricola Apropiada para Pequeños Productores, Experiencias del Proyecto Quimiag-Penipe*, Ecuador. Ministerio de Agric. y Ganaderia, Ecuador; 38, 1980.
4. Janzen, D.H. Tropical agroecosystems. Science **1973**, *182*, 1212–1219.
5. Thurston, H.D. *Slash/Mulch Systems: Sustainable Methods for Tropical Agriculture*; Westview Press: Boulder, CO, 1997; 196.
6. Galindo, J.J.; Abawi, G.S.; Thurston, H.D.; Galvez, G. Source of inoculum and development of bean web blight Costa Rica. Plant Dis. **1983**, *67*, 1016–1021.
7. *Agroforestry in Sustainable Agricultural Systems*; Buckles, D., Lassoie, J.P., Fernandes, E.C.M., Eds.; CRC Press: Boca Raton, FL, 1999; 416.
8. Thurston, H.D. *Sustainable Practices for Plant Disease Management in Traditional Farming Systems*; Westview Press: Boulder, CO, 1992; 279.
9. Kelman, A.; Cook, R.J. Plant pathology in the people's Republic of China. Annu. Rev. Phytopathol. **1977**, *15*, 409–429.

FURTHER READING

Agroecology and Small Farm Development; Altieri, M., Hecht, S., Eds.; CRC Press: Boca Raton, FL, 1990; 262.

Agroforestry in Sustainable Agricultural Systems; Buck, L.E., Lassoie, J.P., Fernandes, E.C.M., Eds.; CRC Press: Boca Raton, FL, 1999; 416.

Multiple Cropping Systems; Francis, C.A., Ed.; Macmillan: New York, 1986; 383.

Huxley, P.A. *Plant Research and Agroforestry*; ICRAF: Nairobi, 1983; 617.

King, F.H. *Farmers of Forty Centuries*; Harcourt Brace: New York, 1911; 379.

Traditional Agriculture in Southeast Asia; Marten, G.G., Ed.; Westview Press: Boulder, CO, 1986; 358.

Okigbo, B.N. *Development of Sustainable Agricultural Production Systems in Africa*; IITA: Ibadan, Nigeria, 1989; 66.

Wilken, G.C. Good Farmers. In *Traditional Agricultural Resource Management in Mexico and Central America*; Univ. of Calif. Press: Berkeley, CA, 1987; 302.

TRANSGENIC CROPS (ANNUALS)

Donald Boulter
Vaughan A. Hilder
University of Durham, Durham, United Kingdom

INTRODUCTION

The development of molecular genetics has made it possible to isolate genes from any class of organism and to incorporate one or more genes stably into the genomes of most (potentially all) crop plants using either the *Agrobacterium* method or the gene gun. Such plants are called transgenics or genetically modified organisms (GMOs); any gene may be transferred, but initially in this field genes encoding insecticidal proteins are used. Insect-tolerant transgenics are the main alternative to synthetic organic pesticides for crop protection against pests. They have some cost and technical advantages over chemical pesticides and from virtually zero acreage in 1996 their use has soared and is predicted to continue to increase as partial substitution products for chemical pesticides. So far only a few genes have been used (1), but developments in genomics, functional genomics, and bioinformatics should identify many more in the near future. New technologies have associated potential risks and the public expects the policy makers to keep these to acceptable levels. All are agreed, therefore, that developments must be effectively regulated; existing regulations are based on scientific evidence, although public perception of the technology involves additional factors.

IMPLEMENTATION AND GROWING USE WORLDWIDE

Commercial, insect-tolerant, transgenic crops use the δ-endotoxin encoding gene derived from the bacterium Bacillus thuringiensis (Bt), modified to enhance the levels of expression in transgenic plants (2) (Table 1). Bt toxins are specific for different orders of insects (Lepidoptera, Coleoptera, and Diptera) although the susceptibility of species within a "susceptible" order varies enormously. Toxicity results from δ-endotoxin binding to gut glycoprotein receptors of susceptible insects leading to cytolysis and death. Different Bt toxins have the same mode of action but may recognize different receptors; thus several varieties of Bt-corn are based on the *cry*1A gene whereas others are based on *cry*9C, both targeted at European corn borers. It is claimed that the latter would still be effective against insect populations that had developed resistance to the former. As a spray, formulations based on Bt have been in limited field use for the control of the larvae of lepidopteran pests for more than 40 years.

Most commercial Bt crops are grown in North America and to a lesser extent in Argentina. China also grows a substantial acreage, e.g., ca. 250,000 acres of Bt cotton in 1998. In the United States, out of a total of ca. 80 million acres of corn nearly 17 million were Bt corn, and of a total of ca. 12 million acres of cotton, more than 6 million were Bt cotton (1998). Other countries have plans to grow Bt crops in the future (3).

Whereas many different crops have been commercialized for herbicide tolerance (the trait accounts for 71% of all transgenic crops), only three Bt crops—corn, cotton, and potato—are currently commercialized. Many more annual crops, including alfalfa, aubergine, oilseed rape, rice, soya bean, tobacco, and tomato, are successful at advanced stages of field trialing (4).

The Bt-crop acreage is predicted to increase dramatically, principally in America. However, contrary to this optimistic scenario, it has been predicted that the development of Bt-resistant insect populations will negate the early gains within 3–10 years, depending on factors such as acreage, presence of other Bt crops, alternative hosts, etc. Currently, regulations (Bt corn and cotton) or voluntary guidelines (Bt potato) require the planting of refugia (up to 20% recommended) of conventional varieties. This, together with the use of resistance management, for example, high dose/refugia, is claimed to ensure the technology can keep ahead of resistance build up, since there are some 200 known Bt genes (although some of these have been shown to be cross-resistant). The efficacy of different management strategies, however, is controversial.

Table 1 Commercial Bt crops

Name	Company	Crop	Target[a]	''Parent'' Bt
YieldGard	Monsanto	Corn	ECB, SWCB, FA	*Cry*1Ab
Bt11	Novartis	Corn	ECB	*Cry*1Ab
NatureGard	Mycogen	Corn	ECB	*Cry*1Ab
Knockout	Novartis	Corn	ECB	*Cry*1Ab
Bt-Xtra	Monsanto	Corn	ECB	*Cry*1Ac
Starlink	AgrEvo	Corn	ECB	*Cry*9Ca
Bollgard	Monsanto	Cotton	BLW, BDW	*Cry*1Ac
NewLeaf	Monsanto	Potato	CPB	*Cry*3Aa

[a]ECB: European corn borer, SWCB: southwestern corn borer, FA: fall armyworm, BLW: bollworms, BDW: tobacco budworm, CPB: Colorado potato beetle.

ADVANTAGES

Transgenics, where appropriate, can be equally effective and less costly to use than chemicals. For example, the economics of growing transgenic cotton in the midsouth and southeast of the United States in 1997 were better than those for conventional cotton by greater productivity of $24.43 per acre and when insect control costs are included (one rather than four to five pesticide applications) the increase in return rises to $39.86 per acre. This example illustrates the main cost factors whereby transgenics benefit the biotechnology farmer. Similar benefits have been demonstrated with other transgenic Bt crops. Farmers also benefit from transgenics conferring all season protection so simplifying the decision when to spray (often difficult if the insects are inconspicuous and considerable damage may have occurred before the farmer suspects that anything is amiss). Time and fuel savings are also possible and other collateral advantages may arise for example, Bt corn is associated with fewer storage molds.

A technical advantage of Bt transgenics versus chemicals follows from the fact that most sprayed chemical molecules do not reach their target and so pollute the environment whereas with transgenics only crop-eating insects are exposed. Furthermore, Bt is a targeted pesticide; for example, Bt-corn kills only lepidopterans (the main pests in many areas) and is nontoxic to other types of insects, all animals, and humans.

RISKS

As well as benefits, new technologies have potential risks. Bt crops might cause environmental damage, for example, escape of insect resistance genes creating ''super-weeds''; contamination of organically grown crops; health risks, for example, allergens. A minority claim they tamper unethically with nature, a view usually associated with antagonism to high-tech agriculture in general. Another objection comes from organic farmers (3% of U.S. agriculture) on the grounds that Bt transgenics but not Bt sprays (used by them) will lead to insect resistance, so obviating the use of both; they also fear that transgenic pollen will contaminate that of organic crops.

Scientific risk assessment is based on the real or estimated size of a hazard times the likelihood of it occurring. The general approach (the substantial equivalence view) has been to compare the characteristics and production of GMO crops and conventional breeding (see Table 2). Based on the large body of information on possible environmental and health risks of the latter and a decade of data from thousands of small scale field trials of GMO crops, scientific societies, committees, and governmental regulatory bodies in several countries have concluded that, with the statutory regulations and oversights required for GMO, but not conventional crops, the former (including Bt crops) are likely to be at least as safe as the latter. Nevertheless, there remain some possible long-term impacts on the environment, for example, reduction in biodiversity and health, for example, use of antibiotic markers such as ampicillin, which need addressing. It has been suggested that long term, scientifically based monitoring of farm grown crops is the way forward. Undoubtedly, there will be a need for sophisticated agronomic management of Bt crops.

Today's trend is for regulation to be open and involve the general public in the decision making process. The public perception of Bt crops involves factors (outrage factors) that may have a greater influence on their acceptability than scientific evidence. The impact of these factors varies both with time and place. For example, while Bt foods are now accepted in the United States, sufficient numbers of the British public have expressed

Table 2 Substantial equivalence convention

S. No.	Conventional breeding	Transgenics
1[a]	Homologously recombines many ''resident'' genes (ca. 1000) between the parents, usually by sexual crossing.	Small number of transgenes of foreign origin (i.e., genes not available to conventional breeding) introduced technologically at random in the genome.
2	Usually no tissue culture step, although sometimes used to provide somaclonal variation.	Involves tissue culture step. Any unwanted somaclonal variation is selected out.
3	Many genes transferred with the gene of interest. The other genes are not characterized but usually are deleterious resulting in the need to select from very large populations. Typically 10 to 15 year breeding and trialing program for a new variety.	The introduced genes are characterized in detail (sequence; gene product; expression level, which can be affected by position effects; copy number; and inheritance). In addition to the genes of interest, usually a selectable marker (antibiotic resistance, herbicide tolerance) and some DNA from the transforming vector is introduced. Few, rather than many, recombined genes that can be individually monitored means a much shorter breeding program.

[a]Genes, whether resident or transgenes, can be unstable; products may interact with those of other genes; outcrossing may occur; and genes may display position effects and gene/environment effects. Single additions of genes such as the modified Bt toxin gene are unlikely to behave very differently in these respects from resident genes.

concerns that food retailers, especially the supermarkets, have replaced many products containing Bt corn with nonengineered alternatives and, where unable to do so, wish to label them if suitable tests become available. This was the outcome of the influence of outrage factors and established scientific evidence played little part. Factors included lack of choice (nonvoluntary); historical analogies for example, eugenics (memorability); lack of trust in science and scientists due to their public disagreement and the aftermath of the bovine spongiform encephalopathy (BSE) crisis (nonknowability); the need for Bt crops not established, nor who would benefit, customer or multinational (equivalent to nonvoluntary risk); concerns about damage to the environment (unnatural, technological); and loss of biodiversity (unethical). The history of world-changing technologies shows that their introduction disturbs the *status quo* and generates controversy. Conflicts arise within business and between it and regulators, special interest groups, and the general public. Mistakes occur occasioning more controversy. This is becoming apparent with GMOs, but in the end the benefits of improved living standards and utility prevail; the public have an enduring belief in science's ability to improve our circumstances. It is a major challenge, therefore, to establish a publicly trusted regulatory infrastructure that will allow the undoubted advantages of GM crops to be developed as rapidly as is compatible with social values.

THE FUTURE

Insect tolerant transgenics will significantly alter the practice of agriculture. The demand for food worldwide will continue to increase; it is conservatively estimated that the world population in 2020 will be 7.7 billion. Producing the additional food (80% of which must be produced in developing countries, mainly on existing arable land) will necessitate large increases in crop productivity; that is, the development of an intensive but sustainable agriculture requiring new mechanisms of technology transfer, economic incentives, and institutions. A major factor for success will be provision of adequate protection against pests. Chemicals will play a major role but insects are already resistant to many chemicals (approaching 1000) and the outlook for new chemical pesticides is very limited because of production costs (greater than transgenics) and for technical reasons (>90% of chemical pesticides attack one of only three common insect target sites). Unlike transgenics, chemicals are a nonrenewable resource and ultimately unsustainable. Transgenics will become a more important alternative with lower costs, technical advantages, and potential for attacking a diversity of insect target sites. Thus insect peptide hormones; wasp, spider, and scorpion venoms; antibodies; lectins; protease inhibitors; vegetative insecticidal proteins; ribosome inactivating proteins; various enzymes; and nonprotein insecticides (e.g., limonene-enhanced corn) have insecticidal activity although it is unlikely that many will be commercialized, having one or more disadvantages for example, chronic action, general toxicity, unpatentability, etc. The first commercial crops using Bt have shown significant pest control benefits, but like chemicals, especially if deployed on a large scale, will lead to resistant insect populations developing. Furthermore, many important crop pests are not susceptible to known Bts—for example, the corn rootworms, *Diabrotica* spp. So alternatives will be needed and non-Bt transgenics are nearing commercialization. Unlike these first generation, single gene transgenics, a second generation will involve the use of multigene domains and protein genetic engineering to hit specific targets (5). Understanding how multigene systems interact at the cellular level combined with the use of highly specific time/place promoters will require sophisticated agronomic management systems.

REFERENCES

1. Hilder, V.A.; Boulter, D. Genetic engineering of crop plants for insect resistance—A critical review. Crop Prot. **1999**, *18*, 177–191.
2. ISB/NBIAP. Special Issue on Bt, 1995. shttp://gophisb.biochem.vt.edu/news/1995/news95.dec.html#dec9511 (accessed Aug. 1999).
3. UCS. *Gene Exchange: Transgenic Crops Around the World*, 1998. http://www.ucsusa.org/Gene/w98.world.html (accessed Aug. 1999).
4. APHIS. Current Status of Petitions, 1999. http://www.aphis.usda.gov/bitech/petday.html (accessed Aug. 1999).
5. Boulter, D. Insect pest control by copying nature using genetically engineered crops. Phytochemistry **1993**, *34*, 1453–1466.

T

TRANSGENIC CROPS (PERENNIALS)

Matthew Escobar
Abhaya M. Dandekar
University of California, Davis, California, U.S.A.

INTRODUCTION

The ecological consequences of global pesticide use have magnified the importance of alternative pest control technologies. The advent of plant genetic engineering in the 1980s has provided a biological alternative to chemical pesticides that, if managed correctly, can have minimal undesired environmental impact. Nearly 70 million acres of pest-resistant and herbicide-resistant transgenic annual crops are cultivated worldwide (1). The success of transgenic corn and cotton may provide a model for application of this technology to perennial plants. Perennials constitute a considerable component of world agriculture including a $7 billion annual crop in California alone (2). Increases in land prices and cost of labor may further shift agricultural emphasis to perennials and high-value horticultural crops. Because of their large size, extended juvenility, and high investment/plant ratio, perennial crops present a unique set of pest management problems. These problems, and the potential solutions provided by genetic engineering, will be discussed below.

PATHOGEN CONTROL

Because each mature perennial represents a large monetary investment, potentially lethal plant pathogens such as *Phytophthora cinnamomi*, plum pox potyvirus, and *Erwinia amylovora*, represent a severe threat to growers. Genetic engineering has allowed an expansion of classical breeding approaches to disease resistance and has also provided several novel plant protection strategies. Classical receptor kinase-type plant disease resistance genes such as *Pto* from tomato have been transformed into other plant species and confer resistance in these heterologous systems (3). In addition, the ectopic expression of phytoalexins and defense proteins such as chitinases can provide further alternatives to chemical control of fungal pathogens in perennials (4). The greatest commercial success has been achieved in viral control, where pathogen derived resistance strategies have led to several commercial virus-resistant cultivars. Most notably, expression of a viral coat protein in papaya (cv. Sunset) conferred resistance to papaya ringspot virus, a major pest of the Hawaiian papaya industry. The resultant ''Rainbow'' papaya was the first commercially released transgenic perennial and is currently cultivated on more than 1000 acres in Hawaii (5). Thus, perennial crops have benefited and will continue to benefit from the virtually unlimited pool of germplasm provided by genetic engineering.

WEED CONTROL

Herbicide-resistant crops currently represent the largest share of the transgenic crop market, and herbicide resistance genes could be easily transferred to a variety of perennial crops (6). However, for many perennial crop systems weed control is not a significant problem. Wide plant spacing in orchards allows weed control by mowing, discing, or flaming. Further, in most mature orchards the canopy shades the ground, limiting weed growth. Herbicide resistance could be a beneficial trait for forestry and the nursery trade, but would likely be unutilized in mature orchards.

INSECT CONTROL

The major focus of genetic engineering for insect resistance over the past decade has been the *Bacillus thuringiensis* (Bt) insecticidal crystal proteins (ICPs). Bt has been exploited commercially for the past 40 years as a highly specific and rapidly biodegradable component of integrated pest management systems. Bt ICPs disrupt insect midgut membranes and are toxic in the ppb range against a variety of Coleoptera and Lepidoptera, including important orchard pests like codling moth and navel orange worm that are currently controlled almost exclusively by chemical insecticides (7). The chemical resynthesis of several ICP-encoding genes allowed high-level expression in plants, creating the first commercially via-

Table 1 Engineered insect resistance in perennial food crops

Crop/transgene	Insects targeted
Apple/*Cry*IAc	Lepidoptera
Apple/Cowpea trypsin inhibitor	Lepidoptera, Coleoptera
Cranberry/*Cry*IAa	Lepidoptera
Grapevine/*Cry*IAc	Lepidoptera
Grapevine/GNA lectin	Homoptera, Lepidoptera
Pear/unspecified *Cry*	Lepidoptera, Coleoptera?
Persimmon/*Cry*IAc	Lepidoptera
Sugarcane/unspecified *Cry*	Lepidoptera, Coleoptera?
Sugarcane/GNA lectin	Homoptera, Lepidoptera
Walnut/*Cry*IAc	Lepidoptera

ble transgenic insecticidal plants (Bt corn, Bt cotton, and Bt potato). Bt ICPs have also been expressed in several perennial crops including apple, persimmon, and walnut (see Table 1) (7).

Several other proteins that interfere with insect digestion, such as α-amylases, lectins, and proteinase inhibitors, have been expressed in transgenic plants. Unfortunately, these proteins possess an acute toxicity several orders of magnitude lower than Bt ICPs, limiting their commercial applicability. Recently, novel microbial insecticidal proteins with toxicity comparable to ICPs have been identified. These proteins, such as cholesterol oxidase, Vip3A, and Tca may provide the basis for the next generation of insect-resistant transgenic plants (8, 9).

THE FUTURE

Maintaining the efficacy of pest-resistant crops requires strategies for resistance management. Currently, growers of Bt crops are required to cultivate nontransgenic refugia in order to maintain a significant population of insects with a Bt-susceptible genetic background. Presumably these susceptible populations will slow the development of individuals homozygous for an ICP resistance allele. Alternately, stacking multiple pest-resistance genes in a single crop cultivar could also diminish the chances of fixation of resistance alleles.

Perhaps the most concerning aspect of the future of transgenic perennial plants is the economics of commercialization. The enabling technologies of plant transformation are under the patent control of a small number of plant biotechnology companies, and to assure a rapid return on development costs these companies have priced transgenic seed based upon their "value added" compared to conventional seed. For perennial crops, the value added for three to four decades of decreased orchard management and pesticide use could be hundreds of dollars per tree. A price increase of this magnitude would not be acceptable to growers, especially considering the recent volatility in consumer acceptance of transgenic foods. In addition, most perennials are sold through small owner-operated nurseries, so market changes tend to be slow and fragmented when compared to the response of large multinational annual seed companies. Thus, significant nonscientific challenges may hinder the broad commercial cultivation of transgenic perennials in the near future.

REFERENCES

1. Abelson, P.H.; Hines, P.J. The plant revolution. Science **1999**, *285*, 367–368.
2. Williams, M.L. California Fruit and Nut Crops. In *1998 California Agricultural Resource Directory*; California Department of Food and Agriculture: Sacramento, CA, 1998; 19–39.
3. Hammond-Kosack, K.E.; Jones, J.D.G. Plant disease resistance genes. Ann. Rev. Pla. Physl. Mol. Biol. **1997**, *48*, 575–607.
4. Bolar, J.P.; Norelli, J.L.; Wong, K.; Hayes, C.K.; Harman, G.E.; Aldwinckle, H.S. Expression of endochitinase from *Trichoderma harzianum*, transgenic apple increases resistance to apple scab and reduces vigor. Phytopathology **2000**, *90*, 72–77.
5. Suwenza, L.; Manshardt, R.M.; Fitch, M.M.M.; Slightom, J.L.; Sanford, J.C.; Gonsalves, D. Pathogen-derived resistance provides papaya with effective protection against papaya ringspot virus. Mol. Breed. **1997**, *3*, 161–168.
6. Botterman, J.; Leemans, J. Engineering herbicide resistance in plants. Trends Genet. **1998**, *4* (8), 219–222.
7. Dandekar, A.M.; McGranahan, G.H.; Vail, P.V.; Uratsu, S.L.; Leslie, C.A.; Tebbets, J.S. High level expression of full length cryIA(c) gene from *Bacillus thuringiensis* in transgenic walnut. Pla. Sci. **1998**, *96*, 151–162.
8. Schuler, T.H.; Poppy, G.M.; Kerry, B.R.; Denholm, I. Insect-resistant transgenic plants. Trends Biotechnol. **1998**, *16*, 168–175.
9. Escobar, M.; Dandekar, A.M. Development of Insect Resistance in Fruit and Nut Tree Crops. In *Molecular Biology of Woody Plants*; Jain, S.M., Minocha, S.C., Eds.; Kluwer Publications: Dordrecht, The Netherlands, 2000; 2, 395–417.

T

TRAP CROPS

Fred A. Gray
David W. Koch
University of Wyoming, Laramie, Wyoming, U.S.A.

INTRODUCTION

The use of trap crops is a unique cultural method of pest control that has been utilized for many years. Other terms used for trap crops include trap plants, decoy crops, and catch crops. Trap crops usually refer to susceptible plants that are used to reduce or lure pests away from their primary host plant or crop, thereby lowering damage. They primarily have been used to control insect pests, insect vectors, and plant parasitic nematodes. Their use to control other disease pathogens, such as fungi and bacteria, has been limited. Trap crops have been used when plant resistance to pests has not been available in the primary crop and when chemicals were either not available or too costly and when crops are organically produced. Trap crops are an ideal addition to an integrated pest management program, in that they target specific pests, while enhancing natural control and beneficial organisms. Pesticides, on the other hand, tend to be broader spectrum.

USE OF SUSCEPTIBLE TRAP CROPS

Insect Control

Trap crops have been used worldwide for management of insect pests in several crops. One of the best examples of trap cropping to reduce insect damage is that used to control the cotton boll weevil. Narrow strips of cotton plants are established along the outer edge of fields several weeks ahead of the main cotton crop. Boll weevils are attracted to these early planted strips where they concentrate and are killed by one of several methods. One method consists of applying granules of a systemic insecticide at planting and again as a side dressing several weeks later. Emerging boll weevils concentrate on the trap cotton because of its more advanced growth stage and are killed by the insecticide during or following feeding. Another method consists of applying multiple sprays of a toxic insecticide spaced several days apart. Both methods have been effective in controlling this cotton pest (1). Another classic example is the use of alfalfa as a trap plant for managing lygus bugs (*Lygus* spp.) in cotton fields (2, 3). It was shown that lygus bugs preferred and concentrated in alfalfa strips that were interplanted within the cotton fields. Many growers in California harvest alfalfa forage fields in modified strip-cutting by dividing fields in half. Each half is cut at alternate two-week intervals. This keeps a constant habitat of green, succulent alfalfa for lygus bug feeding. Since the lygus bugs prefer alfalfa, damage to the cotton crop is reduced.

Similar methods have been used to control crop disease caused by viruses that are transmitted by insect vectors (3). Several rows of corn, rye, or other high-stature crops are grown along the periphery of fields of susceptible vegetable crops such as peppers, squash, or beans. Invading virus-borne aphids are intercepted and thus trapped in these tall plants where they feed and transmit their viruses. Since most of the aphid-borne viruses are nonpersistent in the aphid, viruses will not be infective when they arrive on the susceptible vegetable crop. In this way, trap crops reduce the number of virus-borne aphids reaching the primary crop, thus reducing crop loss (3).

Use of trap crops for insect control of major crops was recently reviewed (4). Adoption of trap cropping at the farm level has been slow. This has been primarily due to the fact that trap cropping is usually more specific in its control (usually one insect in a single crop) as compared to most insecticides. However, potential use of trap crops for insect control in sustaining agroecosystems worldwide is tremendous.

Nematode Control

One of the earliest examples of trap crop use in the United States was the early destruction of potato plants to control the potato eelworm (potato cyst nematode) *Globodera rostochiensis* (5). This practice was further studied for its potential to reduce soil populations of *G. rostochiensis* (6). The destruction of potato plants before nematode reproduction occurs may reduce soil nematode densities by as much as 80%. Its effectiveness relates to the fact that most cyst nematodes, including the potato cyst nematode, are stimulated to hatch by root exudates released from their host plant. Following hatching, juveniles move toward and penetrate small feeder roots where they become entrapped. The trap crop is plowed down before reproduction occurs. Recent work in controlling the tobacco cyst nematode, *Globodera tabacum tabacum*, has shown both tobacco, tomato, and night shade (*Solanum ptycanthum*) to be effective in its control (6). Soil population of *G. tabacum tabacum* was reduced 80% by seeding and destroying nightshade or susceptible tomato plants. Nematode populations were reduced as much as 96% by destroying tomato or resistant tobacco plants. It was found that these trap crops must be destroyed four weeks after planting to prevent nematode reproduction. Of these plant species, tomato, which is resistant to *G. tabacum tabacum*, would be the most economical choice.

If susceptible trap crop roots are not destroyed prior to nematode reproduction, the soil nematode population will increase and result in damage and loss in the succeeding primary susceptible crop (7). Therefore, there is a real danger of reducing yields when using susceptible trap crops and if used, they should be closely monitored to avoid crop damage.

USE OF RESISTANT TRAP CROPS FOR NEMATODE CONTROL

Specialized resistant trap crops have been developed in Europe to control the sugar beet cyst nematode (SBCN), *Heterodera schachtii* (8). Cultivars of oil radish, *Raphanus sativus* spp. *oleifera*, and white mustard, *Sinapis alba*, which are known hosts of the SBCN, have been bred to prevent reproduction following penetration of roots (9, 10). The SBCN is a sedentary endoparasite, which after penetration in roots loses its ability to move, thus becoming entrapped. After root penetration of these resistant cultivars, most juveniles develop into males rather than reproductive females, thus further reducing potential damage in the primary crop (9).

Unlike susceptible cultivars that have been used for many years as nematode trap crops, SBCN-resistant varieties do not require timely destruction to be effective. Several European companies now have varieties of radish and mustard trap crops, which are widely used in Germany and increasingly in other European countries (8). In Germany, these crops are sown following small grain harvest in late summer and allowed to grow into the late fall. When soil moisture is available and soil temperature in the root zone is favorable, soil populations of *H. schachtii* are effectively reduced. The reduction is greater than that of a fallow period or with a nonhost crop. However, because resistance in these varieties is not absolute, they will not reduce nematode populations below a certain level, estimated at 0.5–3.0 eggs/g of dry soil. However, when soil populations are much higher, use of these specialty crops can be very effective.

Recent research in the western U.S. has shown that German-bred trap crop varieties of Adagio radish are effective when timely planted following malt barley harvest (11–13). Although trap crops are usually not utilized as animal forage in Europe, they have been shown to provide highly nutritious forage for lambs and cattle in Wyoming (14), thus making this control method very economical (15). Other attributes in these trap crops include residual nitrogen removal, increased soil organic matter, and increased general soil health. When used properly (16), these trap crop varieties are comparable or superior to a systemic nematicide and provide a low risk control method with better profit potential than soil fumigation (17). Seed of these previously tested trap crop varieties are now being planted in sugar beet growing areas of Wyoming, Montana, Nebraska, Colorado, and Idaho for SBCN control. Growers can now purchase seed of Adagio radish, as well as several other European-developed trap crop radish and mustard varieties from seed companies in the United States.

Trap crops will most likely play an important role as a nonpesticide alternative in sustainable integrated pest management systems, particularly for plants and crops grown for certification under state or federal organic programs.

REFERENCES

1. Pfadt, R.E. Insect Pests of Cotton. In *Fundamentals of Applied Entomology*, 4th Ed.; Macmillan Pub. Co.: New York, 1985; 357.
2. Stern, V.M.; Mueller, A.; Sevacherian, V.; Leigh, T.F. Strip cutting alfalfa for lygus bug control. Calif. Agric. **1964**, *18* (4), 4–6.
3. Sevacherian, V.; Stern, V.M. Host plant preferences of lygus bugs in alfalfa-interplanted cotton fields. Environ. Entomol. **1974**, *3*, 761–766.
4. Agrios, G.N. Control of Plant Diseases. In *Plant Pathology*, 4th Ed.; Academic Press: New York, 1997; 187–188.
5. Javaid, I.; Joshi, J.M. Trap cropping in insect pest management. J. Sustainable Agric. **1995**, *5*, 117–136.
6. Carroll, J.; McMahon, E. Experiments on trap cropping with potatoes as a control measure against potato eelworm (*Heterodera schachtii*). J. Helminthology **1939**, *17*, 101–112.
7. LaMondia, J.A. Trap crops and population management of *Globodera tabacum tabacum*. J. Nematology **1996**, *28*, 238–243.

8. Jenkins, W.R.; Taylor, D.P. Biological Control of Nematodes. In *Plant Nematology*; Reinhold Publishing Corporation: New York, 1967; 234–235.
9. Cooke, D. Europe goes green to control beet cyst nematode. B. Sugar **1991**, *59*, 44–47.
10. Müller, J.; Strudel, W. Der einfluβ der kulturdauer verschiedener zwischenfrüchte auf die abundanzdynamik von *Heterodera schachtii* Schmidt. Nachrichtenbl. dt. Pflanzenschutzd. (Braunschweig) **1983**, *35*, 103.
11. Gardner, J.; Caswell-Chen, E.P. Penetration, development, and reproduction of *Heterodera schachtii* on *Fagopyrum esculentum, Phacelia tanacetifolia, Raphanus sativus, Sinapis alba*, and *Brassica oleracea*. J. Nematology **1993**, *25*, 695–702.
12. Hafez, S.L.; Hara, K. *Heterodera schachtii* populations can be reduced by planting a trap crop or applying low rates of temik to rotation crops. J. Sugar Beet Res. **1989**, *26* (1), A10.
13. Koch, D.W.; Gray, F.A. Nematode-resistant oil radish for control of *Heterodera schachtii.* I. Sugar beet–barley rotations. J. Sugar Beet Res. **1997**, *34*, 31–43.
14. Koch, D.W.; Gray, F.A. Trap crops: a promising alternative for sugar beet nematode control. Univ. of Wyo. Coop. Ext. Serv. Bull. **1999**, B-1029R.
15. Yun, L.; Koch, D.W.; Gray, F.A.; Sanson, D.W.; Means, W.J. Potential of trap crop radish for fall lamb grazing. J. Prod. Agric. **1999**, *12*, 559–563.
16. Jennings, J.W.; Held, L.J.; Koch, D.W.; Gray, F.A. Economics of growing trap crop radish and grazing lambs with a sugar beet and malt barley rotation. Univ. Wyo. Coop. Ext. Ser. Bull. **1999**, B-1077.
17. Koch, D.W.; Gray, F.A.; Yun, L.; Jones, R.; Gill, J.R.; Schwope, M. Trap crop radish use in sugar beet–malt barley rotations of the Big Horn Basin. Univ. Wyo. Coop. Ext. Ser. Bull. **1999**, B-1068.

TRAPPING PEST POPULATIONS

Crawford McNair
Simon Fraser University, Burnaby, British Columbia, Canada

INTRODUCTION

The use of semiochemicals (message-bearing chemicals) to lure pest insects into traps dates back at least to 1893. In that year, mass trapping of gypsy moth males using the female-produced sex pheromone (a semiochemical that carries a message within a species) was tested with live female moths as bait. In the early 1970s, rapid growth in the identification and synthesis of insect pheromones produced euphoric claims for the potential of pheromones as weapons against insect pests. These expectations were frequently disappointed, and the pessimistic overreaction to these failures has retarded the development of mass trapping. Fewer than 1% of the approximately 10,000 species of insects and mites that damage agricultural crops are currently controlled using traps baited with semiochemicals. However, the development of new techniques and innovative combinations of attractants and tactics have laid the groundwork for a wider use of semiochemical-based mass trapping.

SUCCESSES

Several of the most successful applications of semiochemical-based mass trapping have been developed against beetles. The attractants commonly have been aggregation pheromones used by many beetle species to attract both sexes to a host: the ability to trap both males and females greatly enhances the likelihood of success. For 20 years, barrier trapping around lumber-storing areas in British Columbia, Canada, has protected stored lumber from attack by ambrosia beetles (1), with a high benefit-to-cost ratio. As a major part of an integrated pest management (IPM) approach to a massive infestation in Norway by the bark beetle, *Ips typographicus*, 600,000 traps baited with a synthetic aggregation pheromone mixture captured 3 billion beetles in 1979, more than 99% *Ips* species (2). However, it was difficult to evaluate the contribution of trapping to success because of the lack of an adequate study control and ignorance of the size of the natural *Ips* population. Recently the rare Torrey pine, threatened by an outbreak of another *Ips* species, has been successfully protected by a barrier line of aggregation pheromone-baited traps between dead, previously infested trees and the trees' intact ecosystem (3).

Factors conducive to success in mass trapping are listed in Table 1. Each is then examined in more detail.

INSECT INCIDENCE AND BIOLOGY

Thorough study of the target insect is a common factor among successful implementations. Borden and colleagues carefully surveyed the spatial and temporal distribution of the main ambrosia beetle species prior to their effective program (1). Development of a pheromone-based trapping system for *Rhynchophorus palmarum*, a weevil pest of coconut and oil palm, followed years of study by North American and Costa Rican scientists (4). Detailed upwind-flight studies of long-range ovipositional host-finding in the navel orangeworm explored the potential for a novel means of control by incorporating a toxicant into a bait attractive to gravid females (5). Basic traits of a species biology such as clumped distribution of boll weevils within cotton fields may complicate their control. It is important to know the number of females that a male insect is capable of mating and the distance gravid females will travel to lay their eggs. Trapping usually succeeds best against low populations; with "eruptive" pests, it is better suited for periods of low population density than against vast outbreak populations.

RESPONSE TO SEMIOCHEMICALS

The ability of the semiochemical source to outcompete the signal of potential mates or hosts is basic to success. Among polygamous species, lures that attract females will reduce the population more effectively than those that attract only males. Geographically separated populations of pest species sometimes differ in preferred semiochemical blend or ratio, so trapping decisions should be based on data derived from the target population.

Encyclopedia of Pest Management

Table 1 Prerequisites for mass trapping success

1. The insect's incidence and biology (especially its reproductive biology) have been thoroughly studied.
2. The insect responds effectively to an attractive semiochemical or combination of semiochemicals. Success is more likely when both male and female insects are attracted.
3. The semiochemical(s) has been fully identified, and a synthetic version is commercially available and appropriately dispensed.
4. An economically feasible mass trapping technology has been developed, and the traps exclude predators and parasites of the pest insect.
5. The insect is the main pest, and not part of a complex that needs insecticidal treatment.
6. Trapping is a component within an IPM strategy.
7. Pheromones and other attractants are combined.

SEMIOCHEMICAL IDENTIFICATION, SYNTHESIS, AND RELEASE

Optimal synthetic semiochemical lures are usually identical with the natural structures and bouquets (6). A moth can perceive a pheromone below nanogram level ($<10^{-9}$ g). For some species, a few parts of a percent of impurities in a synthetic blend may decrease or obliterate the attraction of males: thus, it is important to use pure synthetic compounds. Early identifications (1960s and '70s) often mistook major components for the full attractive blend. Advances in identification and synthesis of minor components present in miniscule quantities have resulted in more effective synthetic blends, while technical advances have produced lures (e.g., high-tech biopolymers and paraffin) with a constant, adequate release rate over the full period of pest susceptibility.

TRAP TECHNOLOGY

Rate of capture depends upon the numbers of insects attracted to the trap's vicinity, the numbers entering, and the numbers retained—attraction, entry, and retention (7). Visual and olfactory cues interact: for example, many bark beetles are attracted to black traps whose vertical profile mimics tree trunks (Fig. 1); adult corn rootworms are best captured in white traps (similar to preferred white flowers); and multicolored traps are optimal for the beet armyworm. Color attraction may even differ seasonally, between males and females, or between virgin and mated females, opening the way to sophisticated manipulations of pest populations, such as, for Mediterranean fruit fly, the use of orange traps (more attractive to females) in conjunction with the release of sterile males (8). Structural features affect behaviors used to enter and exit traps, and a discrete, well-structured semiochemical plume facilitates entry. Sticky traps generally retain insects better than container traps; but when used against high pest populations, they rapidly lose their retention because they fill up with captured insects. Creative approaches to retention include a system of funnels and light manipulation to prevent escape of adult sheep blowflies (Fig. 1), and a baffled trap lid to retain the palmetto weevil *Rhynchophorus cruentatus* (Fig. 1). Trap design must be tested for individual species: sticky wing traps (Fig. 1) maximize captures of many moth species, but are ineffective against the European corn borer.

Trap placement and density are also important. Two main tactics exist for placement: barrier trapping and in-field trapping. Barrier traps surround the resource to be protected or are interposed between it and the source of pest insects. In-field traps, spaced throughout the resource, aim for the minimum number of traps that will attract pest insects throughout the resource (the number needed may increase with pest population). With either tactic, traps may increase attacks on closely adjacent trees or plants. In the long struggle to protect elm trees against elm bark beetles (the main vector of Dutch elm disease), barrier trapping has generally proved more effective than placement on elm trees (6). Optimal trap height can vary from ground level to tree crown, but around 1–2 m above ground level is effective for many species.

Traps that "filter out" predators or parasitoids of the target insect through careful entry-hole sizing or screen filters (9), or that provide an escape route, facilitate combining biological control with trapping.

Development of traps that are cheap, readily available, convenient, and low-maintenance such as traps from 1-L oil cans for pink bollworm (Fig. 1) is crucial to wide-scale practical implementation, especially in developing countries.

DEALING WITH INSECT COMPLEXES

Switching from insecticide-based control to mass trapping frequently "releases" other pest species formerly controlled by insecticide; but it may prove possible to trap for more than one cooccurring species, as with the two vine-

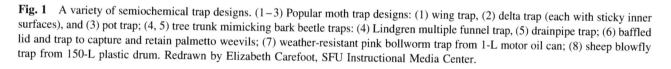

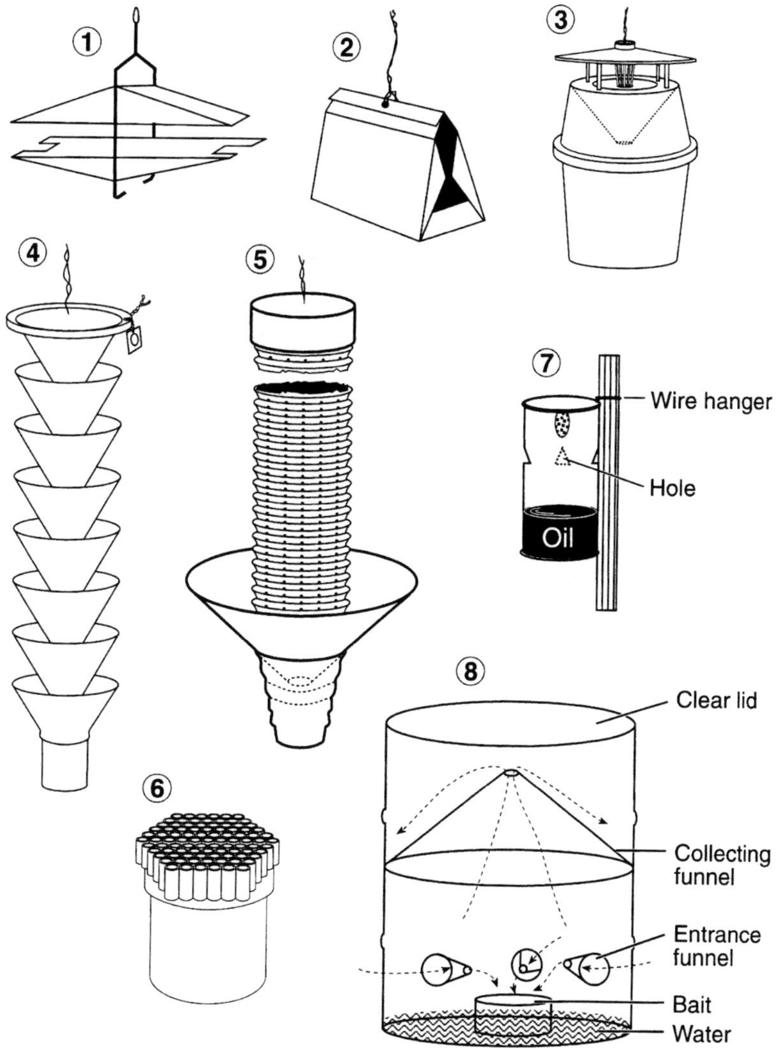

Fig. 1 A variety of semiochemical trap designs. (1–3) Popular moth trap designs: (1) wing trap, (2) delta trap (each with sticky inner surfaces), and (3) pot trap; (4, 5) tree trunk mimicking bark beetle traps: (4) Lindgren multiple funnel trap, (5) drainpipe trap; (6) baffled lid and trap to capture and retain palmetto weevils; (7) weather-resistant pink bollworm trap from 1-L motor oil can; (8) sheep blowfly trap from 150-L plastic drum. Redrawn by Elizabeth Carefoot, SFU Instructional Media Center.

yard pests grape berry moth and redbanded leafroller. Trapping with volatiles emitted by common foods, or with shared components of sex pheromones, may trap several closely associated species, but reduced captures of at least one species usually result from such combinations, and spacing between traps becomes vital.

TRAPPING AS A COMPONENT OF IPM

While trapping sometimes gives stand-alone control, it is usually best viewed as one component in an IPM program tailored to the pest. Most commonly, trapping programs rely on sanitation (elimination of food, breeding sites, and harborage) or on insecticides to reduce populations. Trapping programs can be tailored to mesh with other tactics, as when traps that preferentially capture females are combined with the release of sterile male insects. In future, several semiochemical-based tactics may be combined. Miller and Cowles (10) have advocated ''stimulo-deterrent diversion'' (or ''push-pull''), a bipolar strategy that deters pests from attacking or ovipositing on a valued resource, while providing them with an attractive natural or synthetic alternative. Traps baited with a high pheromone dose may improve the level of control from pheromone-based mass trapping by capturing some of the ''least confused'' males that may otherwise find and mate with females.

COMBINING SEX PHEROMONES WITH OTHER ATTRACTANTS

Combining pheromones with other olfactory attractants probably holds the greatest potential for developing more effective trapping tactics—especially for pest moths, where undue reliance has been placed on female-produced sex pheromones that attract only males. The breakthrough 1980s finding that crude almond oil attracted more gravid female navel orangeworm moths than the ''natural'' oviposition target, mummy almonds (5), stimulated research; and volatiles from host plants, nectaring sites, and larval frass have recently been found to increase captures for a number of moth species, including European corn borer, cabbage looper, peachtree borer, and two clothes moth species. The superiority of combined lures has been shown for chrysomelid and nitidulid beetles, for palm and palmetto weevils, and in Japan for several scarab beetle species. Combined attractants have recently been tested against flies that vector diseases of livestock (tsetse fly and sheep blowfly), or that infest crops [olive fruit fly, apple maggot, melon fly, Mexican fruit fly, and—using food, pheromone, visual, and acoustic cues—Carib fly (8)].

REFERENCES

1. Borden, J.H. Strategies and Tactics for the Use of Semiochemicals against Forest Insect Pests in North America. In *Pest Management: Biologically Based Technologies*; Lumsden, R.D., Vaughn, J.L., Eds.; American Chemical Society: Washington, DC, 1993; 254–279.
2. Bakke, A.; Lie, R. Mass Trapping. In *Insect Pheromones in Plant Protection*; Jutsum, A.R., Gordon, R.F.S., Eds.; John Wiley & Sons: Chichester, U.K., 1989; 67–87.
3. Shea, P.J. Use of Insect Pheromones to Manage Forest Insects. In *Biorational Pest Control Agents: Formulation and Delivery*; Hall, F.R., Barry, J.W., Eds.; American Chemical Society: Washington, DC, 1995; 272–283.
4. Oehlschlager, A.C.; Chinchilla, C.M.; Gonzalez, L.M.; Jiron, L.F.; Mexzon, R.; Morgan, B. Development of a pheromone-based trapping system for *Rhynchophorus palmarum* (Coleoptera: Curculionidae). J. Econ. Entomol. **1993**, *86* (5), 1381–1392.
5. Phelan, P.L.; Baker, T.C. An attracticide for control of *Amyelois transitella* (Lepidoptera: Pyralidae) in almonds. J. Econ. Entomol. **1987**, *80* (4), 779–783.
6. Lanier, G.N. Principles of Attraction-Annihilation: Mass Trapping and Other Means. In *Behavior-Modifying Chemicals for Insect Management*; Ridgway, R. L., Silverstein, R.M., Inscoe, M.N., Eds.; Marcel Dekker: New York, 1990; 25–45.
7. Sanders, C.J. The Further Understanding of Pheromones: Biological and Chemical Research for the Future. In *Insect Pheromones in Plant Protection*; Jutsum, A.R., Gordon, R.F.S., Eds.; John Wiley & Sons: Chichester, UK, 1989; 323–351.
8. Epsky, N.D.; Heath, R.R. Exploring the interactions of chemical and visual cues in behavioral control measures for pest tephritid fruit flies. Florida Entomologist **1998**, *81* (3), 273–282.
9. Ross, D.W.; Daterman, G.E. Pheromone-baited traps for *Dendroctonus pseudotsugae* (Coleoptera: Scolytidae): influence of selected release rates and trap design. J. Econ. Entomol. **1998**, *91* (2), 500–506.
10. Miller, J.R.; Cowles, R.S. Stimulo-deterrent diversion: a concept and its possible application to onion maggot control. J. Chem. Ecol. **1990**, *16* (11), 3197–3212.

ULV APPLICATIONS

G.A. Matthews
Imperial College of Science, Technology, and Medicine, Berkshire, United Kingdom

INTRODUCTION

Ultra-low volume spraying has been defined as the application of the minimum volume that will achieve economic control. This is generally less than 5 litres per hectare for field crops and uses smaller droplets than low volume spraying, with oil or other relatively low volatility liquids used as a carrier instead of water. In practice, ULV ''waterless'' spraying is often at 0.5–1.0 litre per hectare from aircraft and up to 3 litres per hectare with ground equipment.

TECHNIQUES

ULV spraying has distinct logistic advantages with aerial spraying and has been recommended especially for control of pests such as mosquitoes and locusts, e.g., technical malathion has been applied as a ULV spray without further formulation. The spores of a fungus *Metarhizium anisopliae* var *acridum* have been suspended in an oil-based formulation and applied successfully as a myco-insecticide against locusts with as little as 0.5 L/ha from aircraft (1). ULV sprays are also used on cotton, bananas, forests, and other crops.

Formulation of the pesticide is critical for ULV sprays. As the spray volume applied is so small, it is essential to use smaller droplets for good coverage, and to ensure that the droplets, once formed, do not shrink due to the evaporation or volatility of the carrier liquid. A vegetable or mineral oil carrier may be an effective diluent, but it may be too viscous and affect the flow rate to the nozzles and atomization of the spray. Furthermore the amount of pesticide carried by an oil is limited, so the active ingredient is usually dissolved in a solvent before it is combined with an oil. Some organic solvents may not be suitable due to the risk of phytotoxicity, their volatility, or their high cost, so a mixture of solvents may be used (2).

ULV sprays need to be applied with a narrow droplet spectrum to optimize delivery of the pesticide. Controlled droplet application (CDA) devices are therefore generally used, but some applications have been made with low output hydraulic nozzles, or by reducing the flow rate to air shear nozzles on mist blowers. On aircraft the Micronair rotary atomizer was developed initially in the 1950s for ULV sprays against locusts, but is now widely used for crop spraying (Fig. 1). This atomizer consists of a corrosion-resistant, monel metal wire gauze cylinder rotating around a fixed hollow spindle mounted on a boom behind the trailing edge of the aircraft wing. Speed of rotation is controlled by adjustment of the pitch of a series of balanced blades, clamped in a hub, that form a fan. Droplet size spectra for the Micronair AU5000 have been reported (3).

Vehicle-mounted spinning disc sprayers (4) have replaced the exhaust-gas nozzle sprayer (5) developed initially for locust control (Fig. 2).

For locust control a droplet size of 60–120 μm VMD is recommended. These small droplets will be distributed downwind, but unlike water-based sprays subject to evaporation in hot, dry conditions, relatively involatile ULV droplets do not get smaller in flight and will be deposited over a wide swath. Hooper and French (6) compared measured deposits with those predicted by the Forest Service Cramer, Barry, and Grimm (FSCBG) model (7). The swath width will depend on the release height and wind speed, but in control operations a fixed track spacing is recommended. Using aircraft a track spacing of 100 m is possible for locust control where there is sparse vegetation, but this is reduced to 25 m for vehicle-mounted equipment and 10 m for hand-held spinning disc sprayers (see the article on ''Controlled Droplet Application''). Narrower track spacings are used for crop spraying where droplets are filtered effectively by foliage, and turbulent mixing occurs.

When spraying a forest, ULV insecticide sprays are collected more efficiently on foliage with less lost to the ground. In an experiment using an aircraft with Micronair equipment, 1 litre of an insecticide in butyl dioxytol was compared with a water-based formulation of the same

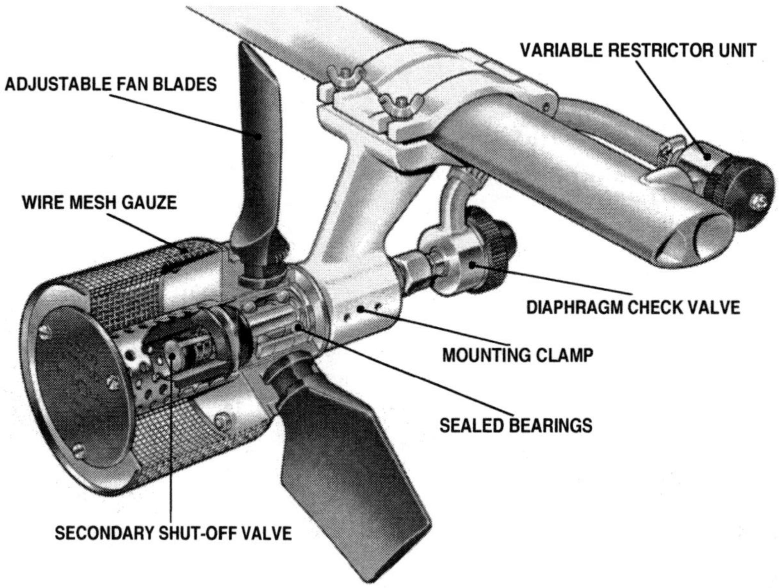

Fig. 1 The Micronair AU5000.

insecticide applied at 20 litres per hectare. Samples indicated that 94.5% of the ULV spray was collected by the foliage and insects in the tree canopy. In contrast only 41.7% of the spray was in the trees with the 20 L/ha treatment, with more lost to the ground (38%) and outside the treated block (20%) (8). Nevertheless, with aerial application, a wider no-spray buffer zone (9) is normally required by legislators to protect areas downwind, especially in noncrop areas or with arable crops that filter less spray than trees.

Small-scale farmers in semi-arid areas have applied ULV sprays with hand-carried CDA spinning-disc sprayers, especially on cotton at up to 3 litres per hectare.

One important advantage of UL formulations is that the spray deposits on foliage tend to be more resistant to removal by rain.

Fig. 2 Vehicle-mounted spinning disc sprayer.

REFERENCES

1. Langewald, J.; Ouambama, Z.; Mamadou, A.; Peveling, R.; Stolz, I.; Bateman, R.; Attigon, S.; Blanford, S.; Arthurs, S.; Lomer, C. Comparison of an organophosphate insecticide with a mycoinsecticide for the control of Oedaleus Senegalensis (Orthoptera: Acrididae) and other Sahelian grasshoppers at an operational scale. Biocontrol Sci. and Technol. **1999**, *9*, 199–214.
2. Maas, W. *ULV Application and Formulation Techniques*; NV Philips: Gloeilampenfabrieken, Eindhoven, 1971.
3. Hooper, G.H.S.; Spurgin, P.A. Droplet size spectra produced by the atomization of a ULV formulation of fenitrothion with a Micronair AU5000 rotary atomizer. Crop Prot. **1995**, *14*, 27–30.
4. Hewitt, A.J. Droplet size spectra produced by the X15 stacked spinning-disc atomizer of the Ulvamast Mark II sprayer. Crop Prot. **1992**, *11*, 221–224.
5. Sayer, H.J. An ultra low volume spraying technique for the control of the desert locust Schistocerca Gregaria. Bull. Ent. Res. **1959**, *50*, 371–386.
6. Hooper, G.H.S.; French, H. Comparison of measured fenitrothion deposits from ULV aerial locust control applications with those predicted by the FSCBG aerial spray model. Crop Prot. **1998**, *17*, 515–520.
7. Tseke, M.E.; Bowers, J.F.; Rafferty, J.E.; Barry, J.W. FSCBG: An aerial spray dispersion model for predicting the fate of released material behind aircraft. Environ. Toxicol. and Chem. **1993**, *12*, 453–464.
8. *Aerial Application of Insecticide Against Pine Beauty Moth*; Holden, A.V., Bevan, D., Eds.; Forestry Commission: Edinburgh, 1981.
9. Payne, N.J. Developments in aerial pesticide application methods for forestry. Crop Prot. **1998**, *17*, 171–180.

UN DEVELOPMENT PROGRAM/UNIDO EXPERIENCE TOWARD RISK REDUCTION IN PESTICIDE PRODUCTION AND USAGE IN ASIA AND THE PACIFIC REGION

S.P. Dhua
Regional Network on Safe Pesticide Production and Information for Asia and the Pacific (RENPAP) UNDP/UNIDO, New Delhi, India

INTRODUCTION

The United Nations Industrial Development Organization (UNIDO) has been actively involved in the field of pesticide development in the countries of the Asia Pacific Region for almost 20 years. While during the '80s UNIDO mainly was involved in preparing feasibility studies for the establishment of pesticide plants with emphasis on producing quality formulations, in the '90s the emphasis has been shifted toward risk reduction through promoting cleaner production and environment management.

PESTICIDE REDUCTION IN THE ASIA PACIFIC REGION

The Regional Network on Pesticide for Asia and the Pacific (RENPAP) made a humble beginning in the early 1980s supported by United Nations Development Program (UNDP) and executed by UNIDO. Initially, nine Asian countries formed a network to discuss and understand common problems and exchange views and experience. Taking note of the gains of participating in the Network program activities with regard to the development and production of user and environmentally friendly varieties of pesticides and formulations, the number of participating countries gradually grew to 15 covering almost half of the world's population and one fifth of the world's surface. This makes RENPAP one of the largest networks on pesticides, and the participating countries are Afghanistan, Bangladesh, People's Republic of China, India, Indonesia, Islamic Republic of Iran, Republic of Korea, Malaysia, Myanmar, Nepal, Pakistan, Philippines, Sri Lanka, Thailand, and Vietnam. The program is implemented in collaboration with the Food and Agricultural Organization of the United Nations (FAO), the World Health Organization (WHO), the Economic and Social Commission for Asia and the Pacific (ESCAP), the International Cooperation Centre in Agronomic Research for Development (CIRAD), Danish Agency for Development Assistance (DANIDA), and other UN international agencies. The main objectives of the program in its new phase of implementation are:

> Promotion of cleaner production and environmentally sound management of pesticides, promoting safety, health, and environmental protection encompassing establishment of mechanisms for waste management, pollution prevention, monitoring of residues in the ecosystem, safe disposal of obsolete pesticides, and establishment of an information exchange system aiming at reduction of risks in the production and use of toxic crop protection chemicals, thereby increasing agricultural production and ensuring safety to the farmers/workers and the environment.

The five immediate objectives of the program are to

1. Promote development and production of the latest varieties of user and environment friendly water-based pesticide formulations; promote development and production of Bt-based biopesticides; upgrade production technology and promote large-scale production, quality assurance, and use of neem-based pesticides; strengthen the quality assurance capabilities of the member countries.
2. Encourage ecological risk assessment needed for protecting the environment; upgrade facilities for monitoring the pesticide load on the environment such as air, water, and soil in the ecosystem.
3. Upgrade standards for the safety of the workers engaged in the production, transport, and use of pesticides; ensure safety of farmers through the adoption of safe application technologies with the introduction of improved applicators manufactured to suit the exact needs of the field.

4. Adopt improved technologies for the treatment and disposal of effluents and obsolete pesticides through recycling/treatment essential for ensuring sound management of chemicals.
5. Strengthen information technology systems for pesticides to provide the member states with appropriate alternatives to hazardous materials needed to reduce the risk of toxic crop protection chemicals.

The program development process is based on a bottom up approach with the major activities being performed at the level of National Coordinating Unit in the respective member countries of the Network (1). The basic strategy of the program is to pool all available resources of the participating member countries and through well-coordinated networking, upgrade these to international standards through state of the art equipmentation and training and attend to the identified pressing problems of risk management on a selective basis.

Experience gained so far shows that the success of regional projects primarily depends on the comprehensive involvement of the participating countries of the network and particularly the nodal technical officers. The best way to involve people in the implementation of such programs appears to hinge on the decentralization of the key activities based on the availability of resources in terms of technical backup facilities and trained personnel in the identified member countries. The success of the RENPAP program has been on account of adoption of such a decentralized approach by setting up of eight technical coordinator units in the member countries of the network based on their capabilities and interest. The following eight technical coordinator units (TCUs) in the selected countries have been established and strengthened to carry out the various activities of the program to bring in effective waste production in the toxic pesticides production and use:

TCU on User and Environment Friendly Pesticide Formulation Technology India, is pursuing research and development work for the elimination of hazardous and environmentally polluting pesticides and replacing these with user and environment friendly products.

TCUs on Biopesticides and Botanical Pesticide Development, China and Thailand, for developing and promoting biopesticides and botanical pesticides as an effective means to reduce environmental pollution, as these are basically natural products and are biodegradable too.

TCU on Pesticide Application Technology, Malaysia, for promoting improved application technology for effective control of targeted pests, weeds, fungi, etcetera and thereby avoiding, to a great extent, environment pollution.

TCU on Ecotoxicology, Pakistan, for bringing in awareness for preserving the environment through the protection of fauna and flora and developing models for ecological risk assessment for protecting the environment.

TCU on Industrial Hygiene and Occupational Health Safety, Philippines, for providing information on industrial hygiene and occupational health safety for ensuring protection of the workers health in the chemical production facilities.

TCU on Monitoring of Pesticide Residue in Air, Water, and Soil, Republic of Korea, for strengthening infrastructural facilities and expertise for environmental monitoring to reduce the presence of pesticide residues in the ecosystem.

TCU on Industrial Safety, Environment Protection, and Effluent Treatment and Disposal, Indonesia. Through this TCU, RENPAP has adopted a consensus for disposal of discarded obsolete pesticides and hazardous wastes.

TCU on Development and Use of Computer Software for Pesticide Market Data Input, Storage, Retrieval, and Dissemination, India and Thailand, has established a database network for information exchange pertaining to risk reduction in the production and use of pesticides.

RENPAP through these TCUs has been serving the needs of the various member countries of the network that are at different levels of technological development. Through this network, countries that are more advanced and have better facilities share their expertise, knowledge, and experience with their fellow members who are less developed. Since the needs and expectations of various countries are different, RENPAP renders its assistance through its TCUs in the following areas: 1) organization of regional workshops for providing hands-on training, 2) training of trainers, 3) technology exchange 4) consultancy, 5) trouble shooting, and 6) assisting member countries in strengthening their capabilities in specialized areas of need.

RENPAP so far during the current phase conducted 10 workshops training programs of 5–10 days duration each in the field of pesticide formulations and quality control (2, 3), application technology (4), biobotanical pesticides (5), production and quality control of biopesticides (6), pesticide residue monitoring (7), effluent and disposal of wastes (8), industrial hygiene and occupational health and safety (9), pesticide application technology (10), ecotoxicology (11), and data collection systems (12), and trained as many as 162 expert nominees of the member countries.

These workshops and training programs are highly rated by the member governments, and the trained experts in turn have been organizing follow-up workshops and training programs to spread the knowledge and expertise far and wide for the overall benefit of pesticide production, usage, and control. There is active exchange of technologies between the member countries that has helped this region in developing and producing water-based environmentally friendly safer formulations; botanical pesticides, namely neem; and biopesticides, namely Bt; and the countries offering such technologies are India, People's Republic of China, and Thailand.

RENPAP Pesticide Database

A dependable pesticide database is invaluable for the pesticide industry and agriculture in the member countries. Therefore, one of the major activities of RENPAP includes pesticide data collection and dissemination. The program has been providing to the member countries better and easier access to an ensemble of reliable pesticide product information and economic data on pesticide production and consumption. This activity has been strengthened with the financial assistance of the government of France and in collaboration with the Center for International Cooperation in Agronomic Research for Development (CIRAD), France. Two TCUs have been set up, one each, in India and Thailand. Regional database centers in Delhi and Bangkok and all the member countries have been fully equipped with computer hardware and software to maintain and update the pesticide database. Data collection experts from the member countries have been specially trained on the data collection methodologies and installation and use of computer software for data input, storage, retrieval, and dissemination (12). The database is centrally coordinated from Delhi.

The RENPAP database is the only one of its kind in the Asia Pacific region providing reliable data on all aspects of pesticides including production, import, export, tariff, etcetera (13–17). The database created by RENPAP is assisting the member countries in phasing out the toxic persistent and hazardous pesticides and replacing these with user and environment friendly ones. The economic database is being utilized by the member countries to chalk out their production strategies. The pesticide industry is benefited from the database with regard to production planning of technical as well as pesticide formulations. UNIDO has plans to include in its database data on the Pollution Release and Transfer Register (PRTR) in the Asia Pacific region and also data on pesticide poisoning data in collaboration with International Program on Chemical Safety (IPCS)/WHO.

As spin-offs of the successful implementation of the RENPAP program, the following UNIDO executed country programs emerged with strong support of the member governments, the United Nations Development Program and bilateral agencies:

1. Sustainable Soil Fertility and Pest Control Program—CPR/91/120 in the People's Republic of China. The main objectives of this program have been to produce effective and environment friendly pesticide formulation products, to serve as a catalyst for efficient production of nitrogenous fertilizers, and to ensure balanced application of fertilizers for increasing the use efficiency of these inputs. The program developed was based on the success of a country project in India using the technology transfer between the two country projects in India and the People's Republic of China.
2. The Ecotoxicology Center in Pakistan. This program was developed as a country program in Pakistan with the financial assistance of DANIDA and the Government of Pakistan to provide the member countries guidance and training in the field of assessment of ecotoxicological impact of pesticide pollutants in the environment.
3. Strengthening of the Pesticide Development Center in India. This country program was launched with the main objective of developing user and environment friendly varieties of pesticide formulations. These new formulations were developed, scaled up, and transferred to the sponsoring industries for large-scale production and usage, much needed for protection of the environment on the one hand and providing safe products to the farming community on the other.
4. Technical Support for Development and Cleaner Production of Neem Products as Environmentally Friendly Pesticides—IND/97/958 in India. This project aims to promote the production and usage of neem-based ecofriendly biodegradable pesticides throughout the country.

Of the six program areas identified in Chapter 19 of Agenda 21 of the United Nations Conference on Environment and Development (UNCED), RENPAP clearly addresses the following four areas pertinent to environmentally sound management of toxic chemicals:

1. Establishment of Risk Reduction Programs. RENPAP promotes Bt-based pesticides and neem-

based botanical pesticides to reduce the risk through its TCU in China for Bt development and production, and another TCU in Thailand for the development and production of neem-based pesticides. RENPAP has made concerted efforts for risk reduction of toxic chemicals through the following measures:

1.1. Assisting the countries to develop water-based pesticide formulations to replace organic solvents and petroleum-based formulations that are unsafe to the farming community and the environment.
1.2. Promoting toxico-vigilance and prevention of pesticide poisoning by establishing a TCU on industrial hygiene and occupational health and safety in the Philippines.
1.3. Prohibiting the use of toxic chemicals—banning of BHC by the government of India and banning of a dozen toxic chemicals in the Philippines.
1.4. Life cycle approaches to chemical management activities through promotion of industrial safety and end of the pipe treatment with the support of the Indonesian government.

2. Strengthening of national capabilities and capacities for management of chemicals. The TCU on ecotoxicology hosted by the government of Pakistan and the TCU on monitoring of air, water, and soil pollution by pesticides hosted by the Republic of Korea (equipped with state-of-the-art equipment and training facilities) is strengthening national capabilities for the management of chemicals.
3. Information exchange on toxic chemicals and chemical risks. The pesticide data collection network hosted by TCUs in India and Thailand satisfies the need for exchange of information that is being used by the member countries for reducing the use of toxic pesticide chemicals in agriculture.
4. Expanding and accelerating international assessment of chemical risks. UNIDOs RENPAP is a participant in the International Program on Chemical Safety of the WHO/UNEP as well as the London Guidelines and Basel Convention dealing with transboundary movements of banned/restricted pesticides and hazardous chemical wastes.

IMPACT OF THE PROGRAM

RENPAP has been able to create a significant impact in all spheres of its activities aimed at cleaner production and regeneration of the environment and providing safety to the farmers and the workers at the production centers as well as at the field level in its 15 member countries. The RENPAP has made significant progress in the following areas:

1. Bringing an awareness of the problems regarding pesticide use, including safety to human health and the environment.
2. Intensifying efforts for development and promotion of safer pesticide formulations and enhancing analytical capabilities for pesticide analysis and quality assurance.
3. Advising and assisting member countries in banning toxic pesticides.
4. Strengthening the capabilities in residue analysis of pesticides and also in the analysis of the pesticide residues in air, water, and soil.
5. Providing information on technologies for pesticide application and pesticide disposal.
6. Providing information on industrial hygiene to protect the workers in pesticide production facilities.
7. Developing a database on regional pesticide markets/economic data.
8. Highlighting the need for ecological risk assessment to protect fish and wildlife from pesticide exposure.
9. Assisting in the safe use of pesticides including packaging to prevent pollution and poisoning at the user end.

REFERENCES

1. RENPAP Secretariat. *RENPAP Booklet: Regional Network on Safe Pesticide Production and Information for Asia and the Pacific*; UNDP: New Delhi, India.
2. Garg, P.K. *Technical Report: UNIDO/RENPAP—Workshop on Development and Production of User and Environment Friendly Pesticide Formulations and Quality Control*, New Delhi, India, Feb. 20–25, 1995; UNIDO, DP/ID/SER.A/1729.
3. Malkanthi, R.P.R. *Technical Report: UNIDO/RENPAP Workshop on Production of User and Environment Friendly Pesticide Formulations and Quality Control*, New Delhi, India, April 21–26, 1997; UNIDO, DP/ID/SER.A.
4. Thornhill, E.W. *Technical Report: UNIDO/RENPAP Workshop on Application Technology*, Serdang, Selangor, Malaysia, Sept. 20–25, 1993; UNIDO, DP/ID/SER.A/1696.
5. Prasertphon, S. *Technical Report: UNIDO/RENPAP Expert Group Meeting on Policy Issues in the Region for*

Bio and Neem (Azadirachta indica) Based Pesticides Development, Bangkok, Thailand, Sept 1–3, 1994; UNIDO, DP/ID/SER.A./1735.

6. Tianjian, X. *Technical Report: UNIDO/RENPAP Workshop on Production and Quality Control of Bio-pesticides (Bacillus thuringiensis)*, Wuhan, People's Republic of China, Oct.–Nov., 31–9, 1995; UNIDO, DP/ID/SER.A/1765.
7. Baloch, U.K. *Technical Report: UNIDO/RENPAP—Workshop on Pesticide Monitoring in Soil, Air and Water*, Sueweon, Republic of Korea, May 13–18, 1996; UNIDO, DP/ID/SER.A.
8. Manadhar, P. *Technical Report: UNIDO/RENPAP Workshop on Preservation of Environment through Proper Control of Effluents and Disposal of Wastes from the Pesticide Production Units*, Indonesia, July 15–19, 1996; UNIDO, DP/1D/SER.A.
9. RENPAP Secretariat. *Technical Report: UNIDO/RENPAP Workshop on Industrial Hygiene and Occupational Health and Safety*, Davao City, Philippines, Dec. 5–9, 1994; UNIDO, DP/ID/SER.A/1730.
10. Mamat, M.J. *Technical Report: UNIDO/RENPAP Workshop on Safe Application of Pesticides: Pesticide Application Technology*, Malaysia, Sept. 23–28, 1996; UNIDO, DP /ID/SER.A.
11. UNIDO-PARC. *Technical Report: UNIDO/RENPAP Workshop on Eco-Toxicology*, Islamabad, Pakistan, March 27–31, 1994; UNIDO, DP/ID/SER.A/1708.
12. Ramdev, Y.P. *Technical Report: UNIDO/RENPAP Workshop on Data Collection System*, Bangkok, Thailand, Sept. 9–10, 1994; UNIDO, New Delhi, DP/ID/SER.A.
13. RENPAP. *Gazette—Special Issue: Supply of Pesticides in Nine Countries*; UNIDO/RENPAP: New Delhi, 1985; 37.
14. RENPAP. *Gazette—Special Issue: RENPAP Pesticides Data Collection System*; UNIDO/RENPAP: New Delhi, India, 1988; 1–54.
15. RENPAP. *Gazette—Special Issue: RENPAP Pesticides Data Collection System*; UNIDO/RENPAP: New Delhi, India, 1989; 1–42.
16. RENPAP. *Gazette—Spotlight—Pesticides Data Collection System*, 3rd Ed.; UNIDO/RENPAP: New Delhi, India, 1991; 1, 1–51.
17. RENPAP. *Gazette—Spotlight—Pesticide Data Collection System India, Sri Lanka, Thailand*; UNIDO/RENPAP: New Delhi, India, 1996; 1–39.

USE OF PREPLANT AND POSTHARVEST IMMERSION FOR DISEASE MANAGEMENT IN HORTICULTURAL CROPS

Robert J. McGovern
University of Florida, Bradenton, Florida, U.S.A.

INTRODUCTION

Propagation of many horticultural crops is asexual, involving cutting or division. These propagation techniques may allow transmission, buildup, and dissemination of pathogens especially in geophytic plants (those grown from bulbs, corms, rhizomes, tubers, etc.) by creation of infection sites through mechanical injury. Plant losses from infection of propagative material have exceeded 50% in some crops. Dipping, rinsing, or soaking plant material in aqueous solutions, dispersions, or emulsions of fungicides, bactericides, nematicides, biological controls, or other materials prior to planting may help to increase plant establishment by decreasing pathogen populations already present and preventing subsequent infection in the field.

Losses from postharvest diseases in fruits and vegetables can also be substantial due to their high moisture content, averaging 10–30% in many crops. Spoilage in specific crops, including accumulation of harmful mycotoxins, often exceeds these estimates especially in developing countries. Therefore, fungicide and bactericide immersion following harvest is routinely used to prevent or reduce postharvest losses, especially in those crops bound for the fresh market.

PREPLANT TREATMENT

Specific geophytic crops that have benefited from preplant immersion in pesticides or other materials to reduce pathogen buildup include caladium, daffodil, garlic, gladiolus, iris, onion, and potato. Preplant root dips in pesticides or biocontrols have also reduced subsequent disease in onion seedlings, roses, and strawberry transplants. Examples of preplant immersion for disease reduction are presented in Table 1.

The success of preplant pesticide dips is affected by population level and location of target pathogens, presence of multiple pathogens, concentration, penetration, stability, and phytotoxic potential of the pesticide, duration of treatment, water temperature and pH, and plant dormancy. Pathogens located on plant surfaces such as the sclerotial form of *Rhizoctonia solani* (black scurf of potato) are more easily eradicated by immersion than pathogens such as *Erwinia carotovora* (soft rot and black leg of potato) that can penetrate more deeply into plant tissue. *Fusarium* spp. represent another difficult-to-eliminate group because of their extensive penetration, often through vascular tissue. Pesticides with systemic activity have an advantage over those that work by contact in treating deep-seated infections. Presoaks to eliminate air pockets and activate pathogens before treatment, and use of surfactants or penetration enhancers also may be beneficial.

Preplant fungicide dips have been used synergistically with hot water against pathogens that have shown sensitivity to thermal inactivation such as *Fusarium oxysporum* f. sp. *gladioli* (corm rot and wilt of gladiolus). Obviously, the thermal stability of fungicides must match the temperatures used during these soaks. Adsorption of pesticides on soil particles and plant material, biodegradation through microbial activity, and adverse solution pH may negatively impact the effectiveness of pesticides used in immersion treatments (e.g., a solution pH of 6.5–8.5 ensures the optimal antimicrobial activity of sodium hypochlorite).

As in other areas of agriculture, increasing attention is being focused on low-toxicity and sustainable approaches to preplant treatment of propagative material for disease management. Ammonium bicarbonate, a constituent in many baked goods, applied as a soak combined with routine hot water treatment reduced the severity of *Fusarium solani* in caladium seed tubers. Dips in a spore suspension of a nonpathogenic strain of *Fusarium moniliforme* significantly reduced subsequent infection and bulb rot in gladiolus caused by *F. oxysporum*. A similar approach used nonpathogenic *F. oxysporum* to protect strawberry against pathogenic strains. A classic preplant use of biocontrol is the suppression of crown gall in roses (*Agrobacterium tumefasciens*) by root dips in solutions containing *A. radiobacter*.

Table 1 Examples of preplant immersion of horticultural crops for disease management

Crop	Plant organ treated	Pathogen(s)	Pesticide(s) and biocontrols	Comments
Allium sativa (garlic)	Bulbs (seed cloves)	*Ditylenchus dipsaci*	Abamectin,[a] formaldehyde, sodium hypochlorite[a]	Used with or following hot water dips
Allium cepa (onion)	Sets	*Fusarium oxysporum*	Benomyl	
A. cepa	Roots	*Sclerotium cepivorum*	Procymidone[a]	
Ananas comosus (pineapple)	Crowns	*Phytophthora parasitica; P. cinnamomi*	Fosetyl Al,[a] metalaxyl,[a] phosphoric acid[a]	
Asparagus officianalis	Root crowns	*Fusarium* spp.	Benomyl	
Caladium x hortulanum	Tubers	*Fusarium solani*	Ammonium bicarbonate,[a] azoxystrobin,[a] thiophanate methyl	Used with or following hot water treatment for *Meloidogyne* spp.
Dieffenbachia maculata	Cuttings	*Erwinia* spp.	Streptomycin	
Fragaria x ananassa (strawberry)	Roots	*Fusarium oxysporum*	Nonpathogenic isolates of *F. oxysporum*[a]	
Gladiolus x hortulanus	Corms, cormels	*Fusarium oxysporum*	Thiophanate methyl,[a] nonpathogenic *F. moniliforme*[a]	Used with hot water treatment of cormels[a]
Iris sp. (Dutch iris)	Bulbs	*Ditylenchus dipsaci*	Formaldehyde	Used with hot water
Lilium longiflorum (Easter lily)	Bulbs	*Fusarium oxysporum*	Thiophanate methyl	
Narcissus pseudonarcissus (daffodil)	Bulbs	*Ditylenchus dipsaci, Fusarium oxysporum, Mucor plumbeus, Penicillium* spp.	Formaldehyde + hot water	
Rosa hybrida (rose)	Roots	*Agrobacterium tumefasciens*	Streptomycin, *Agrobacterium radiobacter*	
Solanum tuberosum (potato)	Tubers	*Rhizoctonia solani*	Benomyl, iprodione, formaldehyde, thiabendazole	

[a]The efficacy of these materials has been demonstrated only experimentally.

POSTHARVEST TREATMENT

Postharvest immersion of fruits, vegetables, and some ornamentals in pesticides or other substances is often essential to prevent potentially catastrophic losses. Practiced for many decades, disease management after harvest addresses infection related both to mechanical injury while handling, and latent infections from the field. Dips, rinses, or soaks to eliminate or prevent infection are often integrated with other postharvest activities such as washing or hydrocooling. However, where technology for environment modification is not available, as in some developing countries, postharvest fungicide immersion provides the only technique to prolong shelf life and reduce losses. Examples of postharvest immersion for disease reduction are presented in Table 2.

Most of the same factors that influence the efficacy of pre-harvest immersions are also important in postharvest systems. Other considerations for postharvest treatment of edible crops include precise environmental control to preserve freshness and retard disease, low damage thresholds because of esthetic considerations, and consumer safety issues raised in recent regulatory measures (i.e., the U.S. Food Quality Protection Act). In addition, the rapid development of fungicide resistance in genera such as *Botrytis* and *Penicillium*, particularly to benzimidazoles (benomyl, thiabendazole), necessitates careful fungicide selection and rotation.

Table 2 Examples of postharvest immersion of horticultural crops for disease management

Crop	Plant organ treated	Pathogen	Pesticide(s)	Comments
Carica papaya (papaya)	Fruit	*Colletotricum gloeosporioides*	Benomyl, thiabendazole (TBZ)	
Citrus spp. (orange, grapefruit, lemon, etc.)	Fruit	*Penicillium* spp.	Benomyl, borax, calcium chloride,[a] imazalil, sodium carbonate, sodium bicarbonate,[b] sodium o-phenylphenate (SOPP), (TBZ), *Candida saitoana* plus 2-deoxy-*D*-glucose[c]	Also enhanced biocontrol activity of the yeast Pichia guilliermondii.[a] Effectiveness increased by combining with sodium hypochlorite[b]
		Geotrichum candidum	SOPP, guazatine	
Cyphomandra betacea (tamarillo)	Fruit	*Colletotrichum acutatum, C. gloeosporioides, Glomerella cingulata, Diaporthe phaseolarum, Phoma exigua*	Imazalil,[a,c] imazalil + prochloraz [c]	Most effective when combined with hot water treatment[a]
Daucus carota (carrot)	Roots	*Chalara elegans*	Calcium propionate,[c] potassium sorbate,[c] sodium hypochlorite	
		Erwinia spp., *Pseudomonas* spp.	Sodium hypochlorite	
		Rhizoctonia carotae	SOPP	
Gladiolus x hortulanus	Corms	*Fusarium oxysporum*	Benomyl, TBZ	
Litchi chinensis (lychee)	Fruit	*Penicillium* spp.	Benomyl combined with hot water	
Lycopersicon esculentum (tomato)	Fruit	*Erwinia* spp., *Geotrichum candidum, Rhizopus stolonifera, Pseudomonas* spp.	Bromine and chlorine compounds	
Malus domestica (apple)	Fruit	*Gloeodes pomigena, Zygophiala jamaicensis*	Sodium hypochlorite, calcium hypochlorite	
		Penicillium expansum	Calcium chloride,[a] Pseudomonas syringae,[a] Candida saitoana + 2-deoxy-*D*-glucose[c]	Combining the two treatments was synergistic[a]
		Nectria galligena	Benomyl	
Mangifera indica *(mango)*	Fruit	*Colletotrichum gloeosporioides*	Benomyl + hot water, prochloraz[c]	
		Alternaria alternata	Prochloraz[c]	
Musa spp. (banana)	Fruit	*C. gloeosporioides*	Benomyl, imazalil, TBZ	
Persea americana (avocado)	Fruit	*C. gloeosporioides*	Benomyl, BHA,[c] BHA + prochloraz (reduced rate),[c] TBZ	
Prunus spp. (cherry, nectarine, peach, etc.)	Fruit	*Botrytis cinerea*	*Candida oleophila, Pseudomonas syringae*	

(Continued)

Table 2 Examples of postharvest immersion of horticultural crops for disease management *(Continued)*

Crop	Plant organ treated	Pathogen	Pesticide(s)	Comments
		Monilinia fructicola	Benomyl, *Bacillus subtilis,*[c] *Pseudomonas cepacia,*[c] *P. corrugata*[c]	
		Penicillium expansum	Dicloran, *Candida oleophila, Cryptococcus infirmo-miniatus,*[a,c] *Pseudomonas syringae*	Best results were obtained with a preharvest iprodione spray and modified atmosphere packaging[a]
Rosa hybrida	Flowers	*Botrytis cinerea*	Polyoxin B + hot water[c]	
Solanum tuberosum (potato)	Tubers	*Erwinia carotovora*	Calcium nitrate[a]	Vacuum infiltration used
		F. solani	Benomyl, TBZ	
		Helminthosporium solani	Potassium sorbate,[c] sodium carbonate[c]	
		Rhizoctonia solani	Boric acid[c]	

[c]The efficacy of these materials has been demonstrated only experimentally.

Increasing public and governmental concern about possible nontarget pesticide effects on the environment and human health has motivated much recent research on the development of low-impact and reduced-toxicity alternatives for postharvest disease management. Approaches investigated include use of plant nutrition to enhance resistance, safer control substances, and biological control.

Penicillium rot in apples was reduced by immersion in calcium chloride (presumably by strengthening cell walls and membranes against fungal attack, and/or direct pathogen inhibition), but depending on fruit maturity was sometimes phytotoxic. However, combining pressure infiltration of calcium chloride at reduced levels with application of the biocontrol *Pseudomonas syringae* synergistically reduced Penicillium rot, without apple damage. Calcium chloride dips also reduced Penicillium rot in grapefruit, and enhanced the biocontrol activity of the yeast *Pichia guilliermondii.*

The food additive BHA decreased anthracnose (*Colletotrichum gloeosporioides*) of avocado apparently by enhancing the natural resistance of fruit. This additive also performed well with a reduced rate of a conventional fungicide, prochloraz. The food preservative potassium sorbate was effective against black root rot (*Chalara elegans*) in carrots, and silver scurf (*Helminthosporium solani*) in potato, while another innocuous substance, sodium bicarbonate, decreased Pencillium rot in citrus.

Research on biocontrol for postharvest disease management has primarily focused on fruit from temperate regions. Brown rot (*Monilinia fructicola*) was suppressed in a number of crops (cherries, nectarines, and peaches) through use of bacterial biocontrols including *Bacillus subtilis, Pseudomonas cepacia,* and *P. corrugata.* The yeast *Candida oleophila* decreased postharvest infection of nectarines by *Penicillium expansum,* alone and in combination with the fungicide dicloran, while *Candida saitoana* plus 2-deoxy-D-glucose controlled *Botrytis cinerea* and *Penicillium spp.* in apple and citrus. Penicillium rot in cherry was suppressed by combining preharvest sprays of the fungicide iprodione with postharvest dips in the yeast *Crytococcus infirmo-miniatus,* and modified atmosphere packaging.

OTHER CONSIDERATIONS AND FUTURE PROSPECTS

Preplant and postharvest immersion for pathogen reduction cannot be viewed in a vacuum; to be effective this techniques *must be combined* with other control strategies. Components of such an integrated approach include the use of certified pathogen-free propagative material, natural or enhanced resistance, soil sterilization, crop rotation, application of fungicides and bactericides during production, management of insects and other pests, optimal horticultural practices (irrigation and fertilization), proper handling after harvest (avoiding wounds and, correct transport and storage environment), and sanitation (disinfestation of containers and structures).

A deficit continues to exist in managing postharvest diseases in many tropical crops, especially in important yet lesser-studied staple commodities (aroid root crops, cassava, yams, etc.), and in the control of bacterial soft rot of many crops. Management of fungicide resistance will continue to be challenging, but this may be aided by the identification of effective and environmentally sound adjuncts and alternatives.

ACKNOWLEDGMENT

This work was supported by the Florida Agricultural Experiment Station, and approved for publication as Journal Series No. N-01854.

BIBLIOGRAPHY

Conway, W.S.; Sams, C.E.; Kelman, A. Enhancing the natural resistance of plant tissues to post-harvest diseases through calcium applications. HortScience **1994**, *29*, 751–754.

Eckert, J.W.; Ogawa, J.M. The chemical control of post-harvest diseases: subtropical and tropical fruits. Ann. Rev. Phytopathol. **1985**, *23*, 421–454.

Hulse, J.H. Food science and nutrition: the gulf between the rich and poor. Science **1982**, *216*, 1291–1294.

Wilson, C.L.; Wisniewski, M.E. Biological control of postharvest diseases of fruits and vegetables: an emerging technology. Ann. Rev. Phytopathol. **1989**, *27*, 425–441.

VECTORS OF HUMAN DISEASE

Christopher F. Curtis
London School of Hygiene & Tropical Medicine, London, United Kingdom

V

INTRODUCTION

The pathogens that cause many human diseases make use of insect, tick, or mite vectors to have themselves carried to their human hosts (1, 2). As shown in Table 1 the pathogens include viruses, rickettsiae, bacteria, protozoa, and nematode worms. The vectors may pick them up while biting humans or animals, or by making contact with infected feces or body secretions. There may be an elaborate process, which takes several days, of development and multiplication in the vector including (in the case of *Plasmodium* malaria parasites) sexual reproduction and eventual infestation of the salivary glands with a form (the sporozoite) with which can be carried into a new host with the mosquito's saliva while it is blood feeding. Such elaborate processes can only proceed in compatible vectors—in the case of the species of *Plasmodium* that cause human malaria, these are most species of the human-biting *Anopheles* mosquitoes (but, even among these, there are a few cases of incompatibility). In the case of nematodes there is development in the vector but no multiplication—sexual reproduction occurs in the human host. In the case of viruses, such as the four serotypes of dengue, and bacteria, such as the causative agent of the bubonic plague, *Yersinia pestis*, there is multiplication in the vector and this can only transmit to a new host when this process is complete. Some of the bacteria causing diarrhea (especially *Shigella*) and trachoma (*Chlamydia*) have recently been proved, via carefully monitored fly control trials, to be transmitted to a significant extent by domestic flies (*Musca domestica* and *M.sorbens*) (3, 4). In these cases transmission does not involve prolonged development and multiplication inside the insect—transmission is mechanical on the outside of the fly or via swallowing and vomiting up of the pathogen. The virus that causes Rift Valley Fever is similarly carried mechanically by *Culex* mosquitoes that act as "flying syringe." These cases of mechanical transmission involve active flight to bite a human, to walk on a child's face, to imbibe eye secretions, or to feed on human food and to vomit pathogens on to it. It seems appropriate to consider such flies or mosquitoes as "vectors." On the other hand the term "intermediate host" is more appropriate for the snails and *Cyclops* in which, respectively, *Schistosoma* species and guinea worms (*Dracunculus medinensis*) pass though important stages of their life cycles.

THE IMPACT ON HUMANS

Table 1 summarizes available information on the human impact of the vector-borne diseases. Malaria is now by far the most important of them. In its heartland in tropical Africa its impact in terms of loss of disability adjusted life years (DALYs—a measure that attempts to combine the killing and disabling effects of diseases (5)) is comparable in importance to other major killers such as diarrheal diseases, acute respiratory infections, and tuberculosis but AIDS deaths are now overtaking those from all of these other diseases.

Lymphatic filariasis does not kill people (except by stimulating suicides) but the gross disfigurements known as elephantiasis (which it causes) make it second, after mental illness, as a cause of disablement. Onchocerciasis ("river blindness") is another form of filariasis but its symptoms are quite different—severe dermatitis causing chronic itching and, in the worst cases, the filarial larvae invade the eye and cause blindness.

Until recently, diarrheal diseases and trachoma would not have appeared on a list of diseases that are known to be vector borne, but the above-mentioned proofs of reductions of incidence following fly control show that they should be on the list. And, even though they also have non-insect routes of transmission, they should have a high position on the vector-borne list because they are such important diseases. Cockroaches may well have a role similar to that of flies as mechanical vectors of pathogens but this has not been proved. As indicated in Table 1, in the past

Table 1 Summary information about vector-borne diseases of humans, arranged approximately in order of worldwide importance

Disease (pathogen genus and class)	Vectors (generic and common name)	Route of infection of vector	Affected areas of world	Human impact
Malaria, (four spp. of *Plasmodium*, protozoa)	*Anopheles* mosquitoes	By biting humans	Tropical Africa, south Asia, tropical America, Melanesia, formerly in temperate zone	c. 300×10^6 clinical cases & 10^6 deaths/year
Lymphatic filariasis, (*Wuchereria, Brugia,* nematodes)	*Culex quinquefasciatus, Anopheles, Mansonia, Aedes,* mosquitoes	By biting humans, (rarely monkeys)	Tropical Africa, south Asia, Polynesia, N.E. Brazil	Elephantiasis is second largest cause of disablement
Diarrheal diseases (*Shigella* etc., bacteria)	*Musca domestica,* house flies (as well as nonvector routes)	By contact with feces	All warm countries	3×10^6 deaths/year
Trachoma (*Chlamydia*, bacterium)	*Musca sorbens*, (as well as contagion)	By contact with eye secretions	Tropics (formerly temperate zone also)	Largest single cause of preventable blindness
Dengue (virus)	*Aedes*, mosquitoes	By biting humans	S.E.Asia, trop. Amer., Polynesia	50×10^6 infections/year; 250,000 severe cases/year
Leishmaniasis (several spp. of *Leishmania*, protozoa)	*Phlebotomus* (Old World), *Lutzomyia* (New World), sand flies	By biting rodents, dogs or humans	Warm countries	1.5×10^6 cutaneous and 0.5×10^6 visceral cases/year
Onchocerciasis (*Onchocerca*, nematode)	*Simulium* black flies	By biting humans	Tropical Africa and Americas	Millions blinded or with severe itching
Sleeping sickness (*Trypanosoma brucei* group, protozoa)	*Glossina*, tsetse flies	By biting humans or animals	Tropical Africa	Disastrous epidemics in early 20th centuary and recent resurgence
Chagas disease (*Trypanosoma cruzi*, protozoon)	*Triatoma, Rhodnius,* Triatomine bugs	By biting humans or animals	Tropical America	Incidence of this disease much reduced due to successful spraying program
Typhus (*Rickettsia prowazeki*, rickettsia)	*Pediculus humanus* body louse	By biting humans	Tropics, formerly temperate zone also	2 million died of typhus in Russian Civil War (1917–22)
Bubonic plague (*Yersinia*, bacteria)	*Xenopsylla*, rat flea (also contagion)	By biting rodents	Foci in tropics and temperate zone	One third of population of Europe killed in 14th cent.
Japanese encepalitis (virus)	*Culex tritaeniorhynchus* mosquitoes	By biting pigs and birds	East Asia	40,000 cases/yr, 11,000 deaths/yr
Yellow fever (virus)	*Aedes, Haemogogus,* mosquitoes	By biting monkeys	Tropical Africa, tropical America (not Asia)	Disastrous epidemics in 18th cent. in Caribbean
Lyme disease, (*Borrelia, burgdorferi*, spirochete)	*Ixodes*, hard ticks	By biting field mice	Temperate zone	Now the only important vector-borne human disease in USA
Relapsing fever (*Borrelia duttoni & B. recurrentis*, spirochete)	*Ornithodoros*, soft ticks & *Pediculus*, body lice	Trans-ovarial or by biting humans	Foci in Africa and Middle East	
Loiasis (*Loa*, nematode)	*Chrysops*, deer flies	By biting humans	West and Central Africa	ca 10^6 people affected
Scrub typhus *Rickettsia tsutsugamushi*, rickettsia	*Leptotrombidium*, chigger mites	Trans-ovarial or by biting humans	Western Pacific	

there have been catastrophic epidemics of typhus, bubonic plague, and yellow fever (6), but at present these diseases are limited to a few foci.

Table 1 does not exhaust the list of medically important arthropods. Limitation of space prevents inclusion of all the many arthropod-borne viruses (arboviruses), most of which are only of any importance in localized areas. Also excluded are arthropods that are not vectors but are harmful via their stings (scorpions, wasps, etc.), via their allergenic feces that trigger asthma (house dust mites) or as irritating nuisances (head lice, biting mosquitoes, and *Culicoides* midges).

Vector-borne diseases may be controlled in a few cases by vaccines (yellow fever, Japanese and tick-borne encephalitis, and Lyme Disease and, perhaps within the next decade, malaria and dengue). For leishmaniasis, elimination of the animal reservoir of the parasites has been attempted with varying success. For all except the viruses there are drug treatments for human patients, but these may be too expensive for the medical systems of developing countries to afford, they may cause dangerous side effects (African and American trypanosomiasis), and, in the case of *P. falciparum* malaria, the drugs are now facing serious problems of drug resistance. Therefore, vector control should have an important role to play in controlling most of the vector-borne diseases.

CONTROL OF VECTORS

Vectors may be attacked at the larval or adult stages (7–9). Where the adults live in or enter houses, spraying the inside surfaces of walls and ceilings with residual insecticides such as DDT (10), pyrethroids, or organophosphates may be a highly cost-effective control method. This has been the main method of control of vectors of malaria and of the house-entering species of sand fly. Spraying with deltamethrin is currently the main method being used by the Southern Cone Project in South America against *Triatoma infestans*, the main vector of Chagas disease (American trypanosomiasis). The success of this program has greatly reduced the importance of this disease in recent years.

For malaria vector control there is increasing effort being put into pyrethroid treated bed nets because these are a more targeted way of using expensive insecticides and because most biting by *Anopheles* mosquitoes occurs in the middle of the night when most people are in bed. Significant reductions in rates of all-cause child mortality have been registered in several African trials (11).

Insecticides may also be used against adult insects in the form of ULV space sprays. These have been used with varying success against *Aedes* vectors of dengue, for aerial spraying against tsetse, and for controlling domestic flies in the trials that demonstrated significant control of diarrhea and trachoma. However, it is recognized that this method is unlikely to be sustainable in the long term. For both tsetse (7) and house flies, much effort has been devoted to the development of traps that can be made locally and serviced and used on a sustainable basis.

Control of larval mosquitoes is the main method of attempting to control dengue. It should have a supplementary role in drug-based programs now aimed at eliminating filariasis as a public health problem (12) and, in appropriate circumstances, can have a role in control of vectors of malaria and Japanese encephalitis. *Aedes* vectors breed mainly in containers and can be attacked by encouraging or enforcing clearing of water-holding garbage (discarded tires etc.), screening water containers, and/or treating them with appropriate larvicides such as temephos or *Bacillus thuringiensis israelensis* (Bti) (9). *Culex quinquefasciatus* breeds in highly polluted water such as in open drains, cess-pits and pit latrines. Organophosphate resistance is now widespread in this species and the most successful recent control programs have concentrated on clearing open drains so that they flow freely, applying floating layers of styrofoam beads to pit latrines, or using *Bacillus sphaericus* (12). *Anopheles* and *Culex tritaeniorhynchus* breed in clean water out of doors, including irrigation water. In some cases, but not all, it is feasible to control the breeding of these mosquitoes by careful management of irrigation in coordination with farmers' interests and/or to stock potential breeding sites with larvivorous fish.

Simulium breeds in fast-flowing water. In the successful West African Onchocerciasis Control Program, these sites have been targeted for weekly aerial spraying with temephos or, where resistance to this insecticide was detected, Bti or a range of other environmentally acceptable insecticides (13).

The breeding of domestic flies should be manageable by appropriate disposal of garbage and animal and human feces.

In the remaining foci of bubonic plague, rat fleas should be controlled by insecticidal treatment of rat-infested areas before use of rodenticides—if this sequence is not observed fleas may be driven from dying rats on to humans.

Body lice live mainly in clothing and should be controlled by regular changes of clothing. They can be disinfested by heating to <60°C. Low doses of pyrethroids are effective against lice and one of the welcome side-effects of programs of pyrethroid treated bed nets against malaria vectors has been elimination of lice.

REFERENCES

1. Lane, R.P.; Crosskey, R.W. *Medical Insects and Arachnids*; Chapman & Hall: London, 1993.
2. Peters, W. *A Colour Atlas of Arthropods in Clinical Medicine*; Wolfe: London, 1992.
3. Chavasse, D.; Shire, R.P.; Murphy, O.A.; Huttley, S.R.A.; Cousens, S.N. Impact of fly control on childhood diarrhea in Pakistan: a community randomised trial. Lancet **1999**, *353*, 22–25.
4. Emerson, P.; Lindsay, S.W.; Walraven, G.E.J.; Faal, H.; Bogh, B. Reduction of trachoma and diarrhea by fly control. Lancet **1999**, *353*, 1401–1403.
5. World Bank. In *World Development Report 1993*: Investing in Health; Oxford University Press: New York, 1993.
6. Busvine, J.R. *Insects, Hygiene and History*; Athlone Press: London, 1976.
7. *Appropriate Technology in Vector Control*; Curtis, C.F., Ed.; CRC Press: Boca Raton, FL, 1990.
8. *Integrated Mosquito Control Methodologies*; Laird, M., Miles, J.W., Eds.; Academic Press: London and New York, 1983; 1 and 2.
9. Rozendaal, J. *Vector Control: Methods for Use by Individuals and Communities*; WHO: Geneva, 1997.
10. Curtis, C.F.; Lines, J.D. Should DDT be banned by International Treaty. Parasit. Today **2000**, *16*, 119–121.
11. Lengeler, C. Insecticide Treated Bednets and Curtains for Malaria Control (Cochrane Review). In *The Cochrane Library*; Issue no. 3, Update Software: Oxford, 1998.
12. Maxwell, C.A.; Mohammed, K.; Kisumku, U.; Curtis, C.F. Can vector control play a useful supplementary role against bancroftian filariasis. Bull. World Health Organ. **1999**, *77*, 138–143.
13. Hougard, J.-M.; Poudiougo, P.; Guillet, P.; Back, C.; Akpoboua, L.K.B.; Quillévéré, D. Criteria for the selection of larvicides by the onchocerciasis control programme in West Africa. Ann. Trop. Med. Parasitol. **1993**, *87*, 435–442.

VEGETABLE CROP PEST MANAGEMENT (PLANT PATHOGENS)

V

Mary Ruth McDonald
University of Guelph, Guelph, Ontario, Canada

INTRODUCTION

The goal of any pest management program is to maintain crop yield and quality using the most effective and efficient means available. This can be accomplished in many ways. Cultural controls and resistant varieties should be used whenever possible. Where fungicide use is called for, disease forecasting systems can increase its effectiveness. While all approaches are available to commercial growers and home gardeners, disease forecasting generally focuses on commercial vegetable production because of the time and cost of weather instruments and scouting.

PEST MANAGEMENT

Many disease management options are available for vegetables. These include site selection, sanitation, cultural controls including crop rotation, the use of resistant varieties, and for diseases, the application of fungicides, biological controls, or materials that induce systemic resistance. Systemic fungicides have expanded disease control options and the effectiveness of certain forecasting systems. Recently, the introduction of new reduced risk materials and biological controls is further reducing the environmental impact of disease control, while products that induce systemic resistance in plants are extending disease control options beyond conventional fungicides, and also targeting bacterial diseases.

DISEASE FORECASTING

Advances in the knowledge of the biology of plant pathogens, and the environmental and crop conditions favorable to disease development, have led to several disease forecasting systems that can improve disease control, reduce the number of fungicide sprays needed, or both. These have been most successful on foliar diseases of vegetable crops. Forecasting or predictive systems for soilborne diseases are still in the developmental stages.

Systems that have been developed for forecasting vegetable diseases generally involve one or more of the following approaches:

(a) general weather patterns that are favorable for disease development,
(b) crop phenology or plant growth stage,
(c) monitoring the crop or the environment,
(d) integrated approaches that focus on combining several techniques to deal with a single disease or allow for the management of several diseases in a single system.

Weather Patterns

The first methods of disease forecasting were based on general weather patterns associated with disease development. In 1954, a system was developed to predict the appearance of late blight of potato, caused by *Phytophthora infestans* (1). The initial appearance of late blight was forecast 7–14 days after the first occurrence of ten blight-favorable days. A day was blight-favorable if the five-day average temperature was below 25.5°C and total rainfall for the last 10-day period was $\geq$3.0 cm. The first forecast for botrytis leaf blight of onions (*Botrytis squamosa*) also depended on weather patterns. Simard (2) reported a correlation between outbreaks of leaf blight and rainfall. If June rainfall was close to the 31-year average, then leaf blight would develop.

Crop Phenology

The age, size, or growth stage of a plant can be an important criterion for decision making in disease management. Susceptibility may change with age or growth stage, and plant size can affect the microclimate, and subsequently disease development. Crop growth stage is the most important factor in the control of sclerotinia drop of lettuce, caused by *Sclerotinia sclerotiorum* and *S. minor*. For effective control, fungicide sprays must be applied at the correct stage of crop growth (3). One application at the four-to-six leaf stage was effective for disease caused by S. minor, while a second application at the rosette stage was required if *S. sclerotiorum* was also present.

Crop phenology is also used to predict the timing of the first fungicide application for early blight (*Alternaria solani*) on potatoes in Colorado (4). Day-degree accumulations above 7.2°C effectively predicted the appearance of the first lesions. The number of accumulated day-degrees varied according to location; 361 for the San Luis Valley and 625 for northeastern Colorado. This system was effective because the appearance of the first early blight lesion is closely related to plant maturity, and plant maturity is related to day degrees. Since late blight is not common in this area, several fungicide sprays could be saved by using this system, which was more efficient than field scouting or spore trapping.

Monitoring the Crop or the Environment

Many forecasting systems use a critical disease level, at which a spray program should begin. This is most useful for foliar diseases that have several cycles during a growing season and can be determined by field scouting or by estimating disease development based on environmental conditions. Temperature, leaf wetness, relative humidity, and rainfall are some of the main factors that are monitored. Disease development is estimated, based on the number of days with weather or microclimate conditions that contribute to sporulation or infection, to arrive at a certain critical disease severity index or critical level of disease. Once the spray program has begun, further modifications in the timing of the sprays can be made according to the estimated risk of disease.

In Ontario, Canada, carrot leaf blight is caused by both *Alternaria dauci* and *Cercospora carotae*. The disease can be managed using a weather-timed spray program that calls for the first fungicide spray when 1–2% of the leaf area is infected. This corresponds to blight on 25% of the leaves in a 50-leaf sample collected from the middle canopy. Once the spray program starts, subsequent sprays can be timed based on forecasted rain and a table outlining the leaf wetness required for infection at different temperatures (5).

Disease forecasts based on predetermined critical levels of disease are effective in many cases, but are not suited for use with diseases that can develop rapidly. These include diseases caused by pathogens in the Peronosporales, such as late blight of potatoes and downy mildew (*Peronospora destructor*) of onions. Onion downy mildew may destroy the foliage completely during the course of four infection cycles. For effective control, fungicide application should begin at the time of the first sporulation-infection period; this can be accomplished using the Downcast program (6). The microclimate within the canopy must be monitored from the time the crop is in the field. Disease prediction is based on conditions that are favorable for sporulation and subsequent infection; which are temperatures between 6 and 22°C, high relative humidity and at least 3 h of leaf wetness that resulted from rapid deposition of dew overnight.

Cumulative disease indices

In Quebec, where carrot leaf blight is caused by *C. carotae*, the spray threshold is 50% incidence on a sample of 50 leaves. Fungicide use was reduced by 40% when field scouting was used to determine the threshold. The timing of the first spray could also be estimated using a cumulative infection equivalence (CIE) method (7). An infection equivalence value of 0, 1, or 2 was assigned for each day, based on temperature, leaf wetness, and relative humidity over 90%. Beginning the spray program at a CIE of 18 resulted in the greatest reduction in fungicide use with no reduction in yield.

The management of botrytis leaf blight of onion depends on initiating a spray program at the critical disease level of one lesion per leaf (8). This can be determined by field scouting, and sequential sampling. However, the initial symptoms may be difficult to identify and field scouting is time consuming. The Botcast program (9) uses temperature and leaf wetness information to predict the development of botrytis leaf blight up to the critical disease level, and to provide a recommendation on the timing of the first spray. Air temperature, leaf wetness, and relative humidity are monitored within the onion canopy to obtain an inoculum value and an infection value. These combined give a disease severity index (DSI) of 0, 1, or 2 for each day. Daily DSIs are added to obtain a running total. An accumulated DSI of 25 indicates that the first spray should be applied before rain is forecasted. The second threshold of 35 indicates that a fungicide should be applied immediately.

Blight-Alert (10) was also developed to forecast botrytis leaf blight of onion. It also incorporates a critical disease level of one lesion per leaf, and allows for timing of subsequent fungicide applications using an inoculum production index (IPI) and National Weather Service forecasts for probability of precipitation. The IPI is based on temperature, relative humidity, and crop development, and forecasts the production of inoculum. The most effective system used the IPI in combination with a probablility of precipitation of greater than 30%.

INTEGRATED DISEASE MANAGEMENT

Integrated management of diseases may involve several methods of disease control with elements of disease forecasting. For example, the best control of late blight of potatoes was achieved when fungicide rates were reduced on cultivars that expressed polygenic resistance to the disease, and also when the application of the fungicide was timed according to Blitecast (11). Integrated management of lettuce drop continues to be a challenge (12). No resistant commercial cultivars or effective biological controls are available; roguing diseased plants is uneconomical. Changing from furrow irrigation to subsurface drip irrigation was effective, but is relatively expensive, so growers continue to rely on fungicide applications timed according to crop growth stage.

Some vegetable production systems allow concurrent forecasting and management of more than one disease. Potato Disease Management was developed at the University of Wisconsin to forecast development and fungicide scheduling for both late blight and early blight of potato. This approach was also effective on potatoes grown in the eastern United States (13).

For processing tomatoes, the FAST system was first developed to forecast the development of early blight (*Alternaria solani*). Later, Tom-Cast added anthracnose (*Colletotricum coccodes*) and Septoria leaf blight (*Septoria lycopersici*) into the forecasts. These systems can reduce fungicide application up to 50% compared to calendar-based sprays. Work is continuing to incorporate cultural practices such as conservation tillage, crop rotation, and the use of resistant cultivars (14).

CONCLUSIONS

Disease management systems are in use for several vegetable crops. These maintain or improve the yield and quality of the crop while allowing for efficient application of fungicides. These systems will continue to evolve to incorporate more diseases and new technologies.

REFERENCES

1. Hyre, R.A. Progress in forecasting late blight of potato and tomato. Plant Disease Reptr. **1954**, *38*, 245–253.
2. Simard, T.; Crete, R.; Tartier, L. Climate and disease development on muck grown vegetables south of Montreal, Quebec, in 1968. Can. Plant Disease Service **1968**, *48*, 124–127.
3. Patterson, C.L.; Grogon, R.G. Differences in epidemiology and control of lettuce drop caused by *Sclerotinia minor* and *S. sclerotiorum*. Plant Diease **1985**, *69*, 766–770.
4. Franc, G.D.; Harrison, M.D.; Lahman, L.K. A simple day-degree model for initiating chemical control of potato early blight in Colorado. Plant Disease **1988**, *72*, 851–854.
5. Sutton, J.C.; Gillespie, T.J. Weather-timed sprays for carrot blight control. Ontario Ministry of Agriculture and Food Factsheet **1979**, *79–035*, 3.
6. Jesperson, G.D.; Sutton, J.C. Evaluation of a forecaster for downy mildew of onion (*Allium cepa L.*). Crop Protection **1987**, *6* (2), 95–103.
7. Abraham, V.; Kushalappa, A.C.; Carisse, O.; Boureois, G.; Auclair, P. Comparison of decision methods to initiate fungicide applications against cercospora blight of carrot. Phytoprotection **1996**, *76* (3), 91–99.
8. Shoemaker, P.B.; Lorbeer, J.W. Timing initial fungicide application to control botrytis leaf blight epidemics on onions. Phytopathology **1977**, *67*, 1267–1272.
9. Sutton, J.C.; James, T.D.W.; Powell, P.M. Botcast: A forecasting system to time the initial fungicide spray for managing botrytis leaf blight of onions. Agric. Ecosyst. Environ. **1986**, *18*, 123–143.
10. Vincelli, P.C.; Lorbeer, J.W. Blight-Alert: A weather-based predictive system for timing fungicide applications on onion before infection periods of *Botrytis squamosa*. Phytopathology **1988**, *79*, 493–498.
11. Fry, W.E. Integrated control of potato late blight—effects of polygenic resistance and techniques of timing fungicide applications. Phytopathology **1997**, *67*, 415–420.
12. Subbaro, V.K. Progress toward integrated management of lettuce drop. Plant Disease **1998**, *82* (10), 1068–1078.
13. Shtienberg, D.; Fry, W.E. Field and computer simulation evaluation of spray-scheduling methods for control of early and late blight of potato. Phytopathology **1990**, *80*, 772–777.
14. Gleason, M.L.; McNab, A.A.; Pitblado, R.E.; Ricker, M.D.; East, D.A.; Latin, R.X. Disease-warning systems for processing tomatoes in Eastern North America: are we there yet? Plant Disease **1995**, *79*, 113–121.

VEGETABLE CROP PEST MANAGEMENT (WEEDS)

Wayne Thomas Lanini
University of California, Davis, California, U.S.A.

INTRODUCTION

Weeds reduce yields of vegetable crops by competing for light, water, and nutrients. Weeds also reduce hand harvest efficiency by hiding the fruit or impeding workers. Weeds can act as hosts of insect and disease pests, which then attack the crop. Yield loss due to competition varies considerably depending on the crop, weed density, time period in which weeds are present, weed species, and resource availability. Slow-growing vegetable crops, such as bell peppers or onions, are very sensitive to weed competition, whereas fast-growing vegetables, such as cucumbers, are less sensitive. Vegetables have the most value when they arrive at the market during a certain time interval and planting time is adjusted accordingly. As a result of market-driven planting times, the ability of a vegetable crop to compete with weeds is often compromised. Management of weeds generally involves a combination of herbicides, mechanical implements, and hand labor. As consumer demand for vegetables free of pesticide residues has increased, growers have responded by exploring strategies for reducing herbicide use.

WEED MANAGEMENT STRATEGIES

The majority of vegetable growers use herbicides for weed control. Preemergence treatments are applied before planting and incorporated mechanically or applied after planting and incorporated by irrigation or rainfall. Preemergence and postemergence herbicides are applied as a broadcast treatment across the entire bed or as a band treatment centered on the seed line (Fig. 1). Postemergence treatments are also applied as directed treatments, avoiding contact with the crop. Due to a limited number of registered herbicides available for vegetable crops and poor control of some weeds by herbicides, cultivation is often used. Generally, one to five cultivation operations are used per vegetable crop. Hand weeding is used by almost all vegetable growers to manage weeds that were not controlled by herbicides or machine cultivation. The high value of the vegetable crop permits the expense of hand weeding, which would not be practical in lower-value crops. In transplanted vegetables, larger plants allow tillage equipment to move more soil into the seed line to bury small weed seedlings. Crop rotation is generally used. Rotation to another crop with a different set of weed control options allows control of problem weeds.

STRATEGIES FOR REDUCING HERBICIDES

Due to high cost of herbicides and a demand by consumers for pesticide-free vegetables, growers have reduced their herbicide inputs. Treating a narrow band centered on the seed line reduces the amount of herbicide and the possibility of carryover into the next crop. In areas where cultivation is possible, preemergence treatments are made on a 15 to 25 cm band centered on the seed line. The area outside the seed line is cultivated to control weeds. Precision planting and cultivation equipment can allow narrower band application.

Mulches block light, suppressing weed germination or growth. Many materials are used as mulches including plastic, wood chips, straw, hay, and newspaper. Black plastic mulches are commonly used for weed control in vegetable crops, as they completely block light. Recently, a clear, infrared transmitting (IRT) plastic has been introduced. The IRT plastic blocks certain wavelengths of light but allows others to pass, which heats the soil for better early season crop growth. Plastic mulches are generally placed on the beds and their edges covered with dirt to prevent their blowing away. Drip irrigation is needed under the plastic mulches to provide the crop with moisture. Certain weeds such as nutsedge are not completely controlled by plastic mulches, as they can penetrate the plastic. Other weeds can grow in the openings provided for crops. Additional problems with plastic mulches include

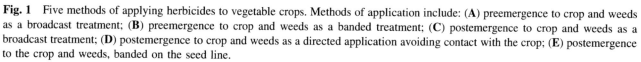

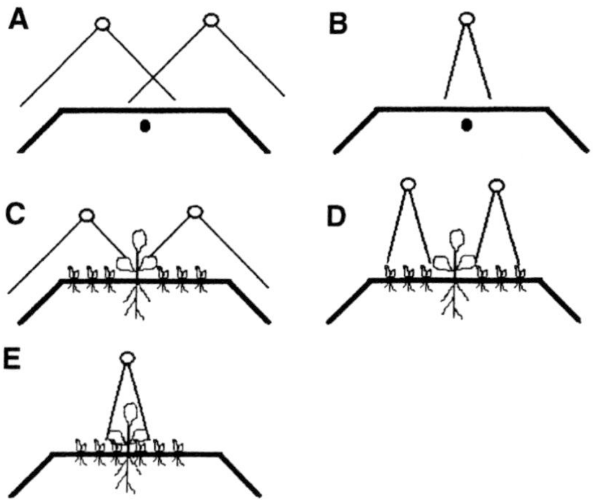

Fig. 1 Five methods of applying herbicides to vegetable crops. Methods of application include: (**A**) preemergence to crop and weeds as a broadcast treatment; (**B**) preemergence to crop and weeds as a banded treatment; (**C**) postemergence to crop and weeds as a broadcast treatment; (**D**) postemergence to crop and weeds as a directed application avoiding contact with the crop; (**E**) postemergence to the crop and weeds, banded on the seed line.

maintaining them in place under windy conditions, disposal after crop harvest, and cost. Organic mulches must be maintained in a layer four or more inches thick to provide adequate light blockage. Since organic mulches are permeable, no special irrigation system is required.

Soil solarization is used to partially sterilize the top layers of soil prior to planting. Soil solarization involves placing a clear plastic mulch over tilled, moist soil to allow the solar energy to heat the soil and kill germinating weed seeds. To be most effective, solarization should be performed during the periods of maximum solar radiation (generally summer), and generally takes four or more weeks. Either the plastic mulch is removed before planting or holes are burned into it for transplanting.

Cultivation is probably the most widely used method of weed control in vegetable systems. The goal of cultivation is to remove weeds as close to the seed row as possible without disturbing the crop. In most cases precision cultivation can take care of weeds on more than 90% of the bed. Mechanical cultivation uproots or buries weeds. Weed burial works best on small weeds, whereas larger weeds are better controlled by destroying the root-shoot connection or by slicing, cutting, or turning the soil to separate the root system from contact with the soil. Cultivation is effective against almost all weeds except certain parasitic forms, such as dodder. Effective cultivation requires good land preparation for precision and accuracy. Listing is often a critical step, as straight rows allow precision planting and close cultivation. Shallow cultivation usually is best, because it brings fewer weed seeds to the surface. Level beds allow greater precision in depth of tillage. Cultivation requires relatively dry soil conditions; subsequent irrigation should be delayed long enough to prevent rerooting of weeds.

Even the best cultivators will not eliminate all weeds; thus hand weeding is often needed. Hand weeding controls weeds in crop rows and weeds missed by other methods of weed control. Hand weeding is most successful when weeds are small. Crop damage can also occur when removing weeds closely associated with the crop. Hand weeding provides no residual control.

Early occurrence of persistent weeds may be more damaging to crop yield than are weeds that establish late. Late-season weeding often disturbs crop root systems or knocks off flowers or fruit and consistently results in reduced yields. Obviously, late-season cultivation to reduce weed seed production must be weighed against the potential for yield loss.

Irrigation can be used to assist weed management. Pregermination of weeds by irrigating (or rainfall) allows them to be killed by light cultivation or flaming, just prior to planting the crop. In areas of limited rainfall, another option exists. After weeds are killed by cultivation, the top 1–3 in. of soil can be allowed to dry to form a dust mulch. At planting the dust mulch is pushed away and large seeded vegetables such as corn or beans can be planted into the zone of soil moisture. The seed of these crops can germinate and grow with no supplemental irrigation that would otherwise germinate another flush of weeds. Additionally, drip irrigation tape buried below the surface of the bed can provide moisture to the crop and minimize the amount of moisture that is available to weeds on the surface. If properly managed, this technique can provide significant weed control during the nonrainy periods of the year.

Crops that grow vigorously can often out-compete weeds. Increasing crop density by increasing the in-row spacing or utilizing closer row spacing improves crop competition. This allows the crop to close the canopy more rapidly and reduces the ability of the weeds to compete. Some crops compete effectively with weeds if given an early competitive advantage (i.e., tomatoes, beans, and sweet corn) while others never establish a competitive canopy (i.e., onions and garlic). Transplanting is used to provide a head start for the vegetable plants, allowing them to be more competitive with the weeds. With help from subsequent cultivation and/or hand weeding operations, the crop can develop a full canopy and out-compete the weeds.

Flamers can be used for weed control in vegetable crops, with propane-fueled models being most common. Heat causes the cell sap of plants to expand, rupturing the cell membrane this process occurs in most plant tissues at about 130°F. Flaming can be used prior to the emergence of the crop in slow-germinating vegetables such as peppers, carrots, and parsley. In addition, flaming can be used over the top of an emerged crop such as garlic or as a directed treatment to the base of tougher crops such as sweet corn when they are 30 cm or more in height. Weeds must have less than two true leaves for greatest flamer efficiency. Grasses are harder to kill by flaming because the growing point is below the ground. Typically, flaming can be done at 3–5 mph through fields, although this depends on the heat output of the unit being used. Best results are obtained under windless conditions, as winds can prevent the heat from reaching the target.

Geese are sometimes used for weed control in vegetable crops. Most types of geese will graze weeds. However, smaller geese are considered best as they generally walk around delicate crop plants, rather than over them. Geese prefer grass species and will eat other weeds and crops only after the grasses are gone and they become hungry. Care must be exercised when using geese to avoid placing them near any grass crops, i.e., corn, sorghum, small grains, etc., as these are their preferred food. Certain other crops of fruit might also be vulnerable, such as tomatoes when they begin to color, thus requiring that geese be removed from tomato fields at certain times.

BIBLIOGRAPHY

Elmore, C.L. Weed Control by Solarization. In *Solarization*; Katan, J., DeVay, J.E., Eds.; CRC Press: Boca Raton, FL, 1991; 61–72.

Masiunas, J.; McGiffen, M.; Wilen, C.; Bell, C.; Lanini, T.; Derr, J.; Kolasani, G. Integrated Weed Management in Horticultural Crops. In *Weed Management in Horticultural Crops*; McGiffen, M., Ed.; ASHS Press: Alexandria, VA, 1997; 1–16.

Zimdahl, R.L.; Burrill, L.C.; Deutsch, A.E. *Weed Crop Competition: A Review*; International Plant Protection Center: Corvallis, OR, 1980.

VERTEBRATE PESTS

Robert M. Timm
University of California, Hopland, California, U.S.A.

INTRODUCTION

Despite efforts to control predation, substantial numbers of cattle, sheep, goats, and poultry are lost to large mammalian predators that include members of the canid family (for example, coyotes, wolves, dogs, jackals, dingoes, and foxes) and the felid family (for example, lions, tigers, leopards, jaguars, and bobcats). Historically, ranchers and landowners that settled in North America were relatively successful in solving predation problems by reducing predator numbers over local areas or regions. In the past few decades, conservation efforts have been aimed at restoring populations of large mammalian predators such as mountain lions and wolves. Coupled with increased regulation and prohibition of traditional predator damage control tools such as toxicants and traps, this has reduced the ability of ranchers and wildlife damage control specialists to effectively prevent or reduce predation on livestock.

ASSESSING PREDATION PROBLEMS

When a livestock manager finds an animal that has recently died, it must first be determined whether the animal was killed by a predator or died of another cause. If predation is the cause, an effective solution to the problem depends on identifying the predator species responsible. Evidence of predation is often present when large animals are injured or killed by predators. An experienced observer who is familiar with the habits and behavior of predators in an area can distinguish predation from scavenging and can often identify the type of predator responsible. Healthy prey, when attacked, show evidence of hemorrhage or bleeding, while carcasses of animals dying from other causes do not. Signs of a struggle or abnormal behavior of livestock in the area can provide additional information that predation has occurred. Predation is easiest to confirm if discovered soon after it has occurred. Scavenging and carcass decomposition are rapid during warm weather and these obscure evidence. Evidence on the prey carcass such as tooth or talon punctures, hemorrhages, claw marks, bruises, or broken bones, may be sufficient to identify the species responsible. Predator tracks, feces, and other signs in the area may also assist in identifying the predator. However, when attacks occur on small prey such as newborn lambs or kid goats, or on poultry, the victims may simply disappear without a trace.

ECONOMIC LOSSES

Worldwide, no accurate estimates are available of the value of livestock lost to predators. In the United States, predators caused losses of cattle and calves totaling an estimated $39.6 million in 1995. Sheep and goats lost to predators were valued at more than $17.7 million in 1994. These losses represent 2.7% of cattle and calf losses and 38.9% of sheep and lamb losses to all causes. The coyote was the most important livestock predator, causing $21.8 million in cattle and calf losses, $11.5 million in sheep and lamb losses, and more than $1.6 million in goat and kid losses. Other significant predators of livestock in the United States include dogs, mountain lions, eagles, bears, foxes, and bobcats. While predation is often the most significant cause of loss to sheep producers, this is rarely the case for cattle producers.

CONTROL MEASURES

Preventing and controlling such conflicts between humans, livestock, and predators requires a degree of understanding of the biology and behavior of the predator species, as well as a concerted effort on the part of the landowner or livestock manager. These predators are highly adaptable, highly mobile, behaviorally complex, and at times unpredictable. Accurate determination of the predator responsible for damage to livestock is an essential first step in preventing further damage. While landowners can take certain actions to prevent or reduce predation, the services and expertise of a professional damage control specialist are often required to solve problems caused by specific offending animals.

Preventive Measures

Livestock management

Selecting pastures and rangelands known to have less history of predator problems can reduce predation. Some predators tend to avoid areas of higher human activity, so keeping young lambs and kids closer to barns, houses, or other areas of high human use, as well as penning them at night, can at times be effective. Herding of livestock also generally reduces predation due to human presence. Changing lambing, kidding, and calving seasons may help reduce predation by coyotes. In some areas, coyote predation is highest during the time coyote pups are being whelped and fed by adults. Shed lambing, kidding, and calving usually reduces predation, at least during the period of confinement. Removal of carrion from pastures may help limit predator populations.

Exclusion

Fences of various designs can be instrumental in protecting livestock from predators. Standard woven-wire livestock fencing can be partially effective in excluding coyotes and dogs, and in directing their movement in more predictable ways so as to make other control tools such as traps and snares more effective. It is considerably more difficult to exclude mountain lions and bears with conventional fence designs, as they are capable of jumping, climbing, or otherwise defeating various types of fences. The cost of designing and building predator-proof fence to protect rangeland is often prohibitive. Recently, the introduction of New Zealand-style high tensile, electric fence design and technology into North America has been successful in excluding most coyotes and dogs and occasionally other predators. Significantly less costly to construct than standard livestock fencing, it can be a viable tool in certain situations. However, its cost of maintenance is generally higher than that of conventional fencing. When the fence becomes inoperative due to short-circuits on accumulated wet vegetation, fallen tree limbs, or from wires slipping out of insulators and grounding to fence posts, it ceases to be an effective barrier to predators.

Frightening and repellents

Lights, sounds, and other stimuli have been used in attempts to dissuade predators from entering pastures or corrals where livestock are kept. Their effectiveness is highly variable and is often of short duration, as predators eventually adapt to such stimuli. No chemical repellents tested have shown significant promise in repelling predators from attacking livestock, after rather intensive research efforts. The killing behavior of coyotes and other predators is often so inherent in their nature that even extremely irritating chemical compounds do not easily dissuade it. Yet, many types of novel stimuli (lights, sounds, and odors) can temporarily interrupt coyotes' killing patterns.

Guard animals

Increasing use of guard animals to protect livestock has shown some success in the United States in recent years. Guard dogs, often European breeds such as Great Pyrenees, Anatolian Shepherd, Akbash, and Komondor, have been bred for their defensive nature when bonded to livestock and can protect flocks of sheep and goats. They are most effective with single flocks that are located in relatively small, open pastures, or when used in concert with full-time herders. Similar efforts to employ guardian llamas, burros, donkeys, and other animals that are aggressive toward coyotes or other predators have shown some promise. The success of guard animals is highly dependent upon the inclination of the individual guard animal to stay near or with livestock. Such efforts may also require considerable time, energy, and expense on the part of the livestock manager. Guard dogs have occasionally become problems themselves by killing wildlife or even young livestock, or by being unable to effectively deal with numerous, aggressive predators. Other types of guardian animals also can create management problems by being aggressive or dominant toward livestock, especially when desirable supplemental feed is provided, or by making it difficult to herd livestock with working dogs.

An estimated 70% of all sheep in the United States in 1994 were protected from predation by use of one or more of the above-mentioned preventive measures. These measures, even when employed in combinations, are not totally effective. When predation occurs, livestock producers often must take timely corrective actions or risk losing further animals.

Corrective Measures

Lethal control

When predation problems cannot be averted by preventive measures, their solution typically requires removal of the offending predators or local reduction of predator populations. In the case of domestic dogs or coyotes, the landowner or livestock manager may accomplish this. However, where large predators such as wolves, bears, or mountain lions are involved, state or federal regulations may require proof of damage as a prerequisite to issuing a depredation permit. In some instances, only government agency personnel may legally remove the responsible predator. Because predators that begin killing livestock are prone to continue this behavior, these individuals are usually killed rather than live-captured and relocated elsewhere.

Trapping and snaring

Traps and snares are primary tools of professional damage control specialists. Recent developments in trap and snare technologies have allowed these tools to be used to more selectively target the offending predator species. The selectivity and effectiveness of traps and snares is dependent upon the expertise of the individual setting the devices. In some states, legislation or voter initiatives have restricted or eliminated the use of these devices in recent years.

Shooting

Shooting is frequently employed to control free-ranging dogs or other predators that are attacking or harassing livestock. Professional predator damage control personnel may employ shooting to control depredating coyotes or other predators by lying in wait at the site of predictable predation, or by calling the predators by mimicking the sound of a wounded rabbit or other prey. In some localities, specialists used trained dogs to lure coyotes into the open, where they can then be shot. To remove offending bears or lions, specialists often employ trained trailing dogs that can follow the scent of the predator from a fresh kill. The specific offending predator can then be located, shot, and removed. However, shooting sometimes cannot be safely or legally done in areas near human habitation. Government agencies have effectively used aerial hunting, using either fixed wing aircraft or helicopters, to solve coyote predation problems in the western United States. This method can only be effectively employed on relatively flat, open terrain.

Denning

Professional damage control specialists in the United States often seek to locate coyote and red fox dens in the spring, and to remove the pups as well as the adults, as a means of preventing further depredation on livestock. In some situations, removal of the pups alone will reduce the adults' food demands sufficiently to stop livestock predation.

Toxicants

During much of the twentieth century, the use of toxicants such as strychnine, thallium sulfate, sodium cyanide, and sodium fluoroacetate (Compound 1080) was a mainstay of efforts to control predation, particularly by canids, in North America. Most predacides used in the United States were banned nationally in 1972. The ban subsequently was relaxed to permit very limited use of a few specific predacide methods to control depredating coyotes and a few other canids. Such uses are limited to agency personnel or others specifically trained and certified in the use of these toxicants.

BIBLIOGRAPHY

Connolly, G.E. Predators and Predator Control. In *Big Game of North America*; Schmidt, J.B., Gilbert, D.L., Eds.; Stackpole: Harrisburg, PA, 1978; 359–394.

Dolbeer, R.A.; Holler, N.R.; Hawthorne, D.W. Identification and Control of Wildlife Damage. In *Research and Management Techniques for Wildlife and Habitats*; Bookhout, T.A., Ed.; The Wildlife Society: Bethesda, MD, 1994; 474–506.

Vertebrates, World Animal Science B2. In *Parasites, Pests and Predators*; Gaafar, S.M., Howard, W.E., Marsh, R.E., Eds.; Elsevier: Amsterdam, 1985; 389–455.

Hygnstrom, S.E.; Timm, R.M.; Larson, G.E. *Prevention and Control of Wildlife Damage*; Cooperative Extension Division, IANR, University of Nebraska-Lincoln, USDA-APHIS-ADC, and Great Plains Agricultural Council: Lincoln, NE, 1994; 848.

National Agricultural Statistics Services (NASS). Cattle Predator Loss, Mt An 2 (5–96); Agricultural Statistics Board, U.S. Department of Agriculture: Washington DC, 1996; 23.

National Agricultural Statistics Service (NASS). Sheep and Goats Predator Loss, Lv Gn 1 (4–95); Agricultural Statistics Board, U.S. Department of Agriculture: Washington DC, 1995; 16.

Roy, L.D.; Dorrance, M.J. *Methods of Investigating Predation of Domestic Livestock*; Alberta Agriculture: Edmonton, Alberta, 1976; 54.

U.S. Dept of Agriculture. Animal Damage Control Program. *Final Environmental Impact Statement*, Animal and Plant Health Inspection Service; Washington DC, April 1994; 1–3.

Wade, D.A. *Impacts, Incidence and Control of Predation on Livestock in the United States, with Particular Reference to Predation by Coyotes*; Special Publication No. 10, Council for Agricultural Science and Technology: Ames, IA, 1982; 20.

Wade, D.A.; Bowns, J.W. *Procedures for Evaluating Predation on Livestock and Wildlife*; B-1429, Texas Agric. Extension Service: College Station, TX, 1982; 42.

VERTEBRATE PESTS OF CATTLE

Michael J. Bodenchuk
U.S. Department of Agriculture, Salt Lake City, Utah, U.S.A.

INTRODUCTION

Domestic cattle can be affected by vertebrate pests, especially by predators of calves or adult cattle during calving periods. While most predation is by larger members of the mammalian order Carnivora, some predation is attributed to avian predators. Additionally, vertebrates serve as vectors for a number of cattle diseases, including rabies. Prevenive measures include increased use of animal husbandry techniques. Selective removal of individual predators through either lethal or nonlethal methods is also useful where economically and ecologically practical.

U.S. CATTLE PREDATORS

Most predation on cattle is to newborn calves, which are especially vulnerable in open range calving situations. Common U.S. predators of calves include coyotes (*Canis latrans*), black bears (*Ursus americanus*) and mountain lions (*Felis concolor*). Where populations are established, gray wolves (*Canis lupus*) also prey on cattle. Although certainly capable of cattle predation, grizzly bear (*Ursus horribilis*) depredations are infrequent, due primarily to the low number of cattle in grizzly bear range. Calves outgrow vulnerability to most predators at about 181 kg (400 lb), although bears, wolves, and mountain lions will kill cattle up to 454 kg (1000 lb). Adult cattle are generally only vulnerable to predation during the calving season when cows are incapacitated in the act of giving birth.

Avian predation is rare, though verified instances occur annually in many states. Ravens (*Corvus corax*) prey on newborn calves before they are able to stand. Black vultures (*Coragyps atratus*) prey on calves less than three weeks old but may also attack cows giving birth. Golden eagles (*Aquila chrysaetos*) have been documented killing newborn calves but attacks on calves up to 141 kg (310 lb) have been documented.

Nationwide, predation affects less than 1% of the calves born annually. Losses are limited by adjusted livestock management practices and an active predation management program in states where losses regularly occur. Calf losses to predators are not distributed evenly and for range-calving beef cattle operations in the western United States, losses in the absence of management may approach 5%. Even with an aggressive predation management program in place, losses will occur, but should be limited to about 1%.

DISEASE THREATS

Vertebrates serve as both reservoirs and vectors for a few diseases that affect cattle. Rabies has been spread to cattle by coyotes, gray foxes (*Urocyon cinereoargenteus*), and raccoons (*Procyon lotor*). In the northeastern United States the raccoon rabies epidemic especially affected dairy cattle, due to confinement of cattle that places the entire herd at risk of exposure to a single rabid raccoon. Rabies is fatal to unvaccinated cattle if the bite is undetected or untreated, and the disease can be spread to humans.

Starlings (*Sturnus vulgaris*), brown-headed cowbirds (*Molothrus ater*), rock doves (*Columbia livia*), and other blackbirds (subfamily Icterinae) frequent dairies and beef cattle feedlots, consuming and contaminating cattle feed with feces. Diseases spread to cattle consuming contaminated feed include coccidiosis and histoplasmosis. Though rarely fatal, these diseases reduce livestock gains or milk production and usually require veterinary treatment.

PREVENTION AND CONTROL OF DAMAGE

Increased use of livestock husbandry during calving periods can prevent or reduce predation threats. Confined birthing areas and regular inspection are the most frequently used methods to reduce predation on cattle, although their application may be limited in dry regions where confinement requires supplemental feeding. Adjusting the timing of calving to where abundance of alternate prey is maximized and predator numbers are minimized can also be effective. Removal of livestock carcasses will prevent attracting cattle predators (especially bears, coyotes, and vultures) to calving pastures. Guard animals have not been extensively tested for cattle operations, although there seems to be some evidence that a guard dog bonded to individual cattle may be effective in small pasture situations. Removal of roost sites may reduce eagle and vulture predation. Scaring devices, such as pyrotechnics, also may work, although they must be used diligently and may require limited lethal control to reinforce their efficacy. Keeping calves in protected areas until they outgrow their vulnerability to predators is a common practice where grazing seasons allow.

Removal of individual predators has been used extensively to reduce predation on cattle. In much of the country, calving coincides with coyote pup rearing, and the combination of increased nutritional requirements for the pack and easy prey in the form of calves causes numerous predation incidents. Removal of the adult, territorial coyotes within calving areas prior to calving can prevent most losses. If depredations have begun, removing coyote pups from the associated den will also cause depredations to cease, even if the adults are not to be removed. For larger, less common predators, selective removal of the offending individual after losses are initiated is a common and effective solution.

Individual predators can be removed through shooting (including aerial shooting), traps, and neck or foot snares. Animals caught in traps or foot snares may be relocated if suitable habitat exists and relocation is judged biologically sound. A sodium cyanide ejector, called the M-44 device, is employed for coyotes where depredations occur. The top of the device is baited with a fetid meat bait and when a coyote pulls at the baited portion a spring loaded plunger injects a small amount of toxicant into its mouth, resulting in rapid death.

Aversive conditioning methods, including electronic shocking collars on individual predators and repellants, are under development but hold little promise for widespread use. Their applications will likely be limited to individuals of endangered predatory species in reintroduction programs. Similarly, the use of reproductive inhibitors is also under development, but holds promise in limited applications.

WORLDWIDE IMPACTS

Throughout the cattle producing regions, predation is fairly common. In Central and South America, jaguar (*Felis onca*) and mountain lion, called puma, are frequent predators of cattle. In sub-Sahara Africa, leopards, jackals, and lions are also cattle predators. Because of increasing human development, cattle predation in these areas can be excessive and abundant predator populations may limit the use of the landscape for cattle production.

Throughout much of Central and South American cattle areas, vampire bats also serve as vectors for rabies. In many areas, cattle serve as the major source of blood meals for vampire bats, and when rabies infects the bat population, cattle are frequently infected. While rabies vaccinations are available, they are not routinely used in remote locations.

FUTURE CONSIDERATIONS

With increasing demand in remote areas of the world for nutrition, increased use of land for cattle production can be expected. However, large predators, such as bears and jaguars, are limited in their range and conflicts between cattle and these species could result in significant declines in predator numbers. Additional research will focus on meeting ecological and human needs in resolving these conflicts.

BIBLIOGRAPHY

Hygnstrom, S.E.; Timm, R.M.; Larson, G.E. *Prevention and Control of Wildlife Damage*; University of Nebraska, Cooperative Extension Service: Lincoln, NE, 1994; 820.

Lowney, M.S. Damage by black and turkey vultures in Virginia. Wildl. Soc. Bull. **1990–1996**, *27* (3), 715–719.

Phillips, R.L.; Cummings, J.L.; Notah, G.; Mullis, C. Golden eagle predation on domestic calves. Wildl. Soc. Bull., **1996**, *24* (3), 468–470.

Smith, B.P. *Large Animal Internal Medicine*, 2nd Ed.; C.V. Mosby Press Co.: St. Louis, MO, 1996; 2112.

VIRULENCE IN PLANT PATHOGENS

Etienne Duveiller
CIMMYT, Kathmandu, Nepal

INTRODUCTION

Every plant pathogen has a host–range determined by host–pathogen interactions. If a pathogenic microorganism never attacks a plant species, the latter is a nonhost, and there is immunity (1). Disease resistance of field crops varies with cultivars and is generally under the control of one to many resistance genes. Virulence is the specific ability of a pathogen to overcome a host gene for resistance (2). Environmental conditions may greatly affect resistance, virulence, and disease reaction types and, hence, the development of epidemics (1). The term aggressiveness describes differences between pathogenic strains based on the amount of disease they cause; it is also a measure of the rate at which a virulent strain produces a given amount of disease. However, the usage of the terms is sometimes unclear (3, 4).

DISEASE RESISTANCE

Breeding for disease resistance is the most economic and environmentally sound strategy to reduce crop losses due to biotic stresses, particularly in developing countries, where fungicides are not easily available. Since plant pathogens evolve along with host plants, resistant varieties may become susceptible due to pathogen adaptation. Susceptible cultivars need to be replaced with new resistant genotypes. A gene-for-gene relationship has been demonstrated to operate in many pathosystems (1, 5). Based on this concept, a resistant reaction occurs when an allele for resistance in the host interacts with an allele for avirulence in the pathogen (1).

Race-Specific and Race-Nonspecific Resistance

In a pathogen species, a race is defined by a group of strains that is virulent on a set of differential host genotypes (1, 2). Race-specific resistance in a host plant is only effective against some pathogen strains. It is usually characterized by strong effects, and is controlled by a single dominant gene also called major gene. Easily detectable on seedling plants, it is generally nondurable because new races may arise from recombination or mutation (2, 6, 7). It thus has a significant interaction or specificity. In contrast, race-nonspecific resistance is equally effective against all strains evaluated and does not show an interaction. It is generally characterized by partial or incomplete adult plant resistance, and controlled by several minor genes whose small additive effects need to be identified from experimental field variation. This partial resistance reduces the disease development rate and is expected to be more durable (6).

Race-specific resistance has been largely used in many crops. In wheat rusts, major genes have been identified over years based on the detection of low infection type in genotype testing at the seedling stage against multiple pathotypes (2). Surveys, race analyses, and monitoring nurseries have contributed to determining the combination of virulence genes prevailing in pathogen populations (2). This resistance may appear durable but is expected to break down sooner or later. It is easily identified in segregating populations. Gene pyramiding is used to increase genetic complexity and delay resistance breakdown. However, due to the risks associated with resistance based on major genes, germ plasm improvement programs are now focusing on durable resistance in order to avoid "boom-and-bust cycles" (1). If race-nonspecific resistance is researched in a pathosystem where race-specific resistance also occurs, advances will be made by selecting plants showing intermediate resistance, after discarding susceptible and very clean plants (6).

Strategies of Resistance Management

Cultivar diversification with different resistance genes and durable type resistance are critical for avoiding epidemics. Reducing crop uniformity and, hence, the probability of pathogen adaptation may significantly increase the durability of resistance based on major genes. Rotation and intercropping are measures used by farmers to reduce crop uniformity and pathogen levels; this decreases disease pressure and possible changes in virulence genes (1). Gene deployment using cultivars carrying different resistant genes in different fields of a same farm is another strategy (6, 8). At the regional level, allowing only the cultivation of recommended cultivars with different genes in different parts of a geographic area is expected to reduce the spread of pathogens with new virulence (1, 6, 9). Sowing cultivar mixtures is an approach to increase the genetic complexity at the field level: the average disease level is generally lower in each component of the mixture than in pure cultivars (6).

FUTURE CONCERNS

Reducing crop vulnerability through the release of different cultivars with different resistance genes is an important component of integrated pest management. But complex races may also arise from extensive use of major gene combinations (1, 2). Cultivar replacement is slow in many developing countries. When a high-yielding cultivar is grown over a wide area, the risk of pathogen adaptation increases. In advanced countries, the selection pressure resulting from pesticide use may induce pathogen mutation for tolerance.

Screening for disease resistance is currently based on phenotypic observation using the virulence gene spectrum available in a given area. The identification of molecular markers for genes conferring disease resistance is anticipated to increase the screening accuracy of resistant parental lines (1). Similarly, the use of genetically modified disease resistant genotypes is expected to increase in the coming years though not without legal considerations regarding free access to germ plasm and concerns over environmental issues (1).

V

REFERENCES

1. Lukas, J. *Plant Pathology and Plant Pathogens*, 3rd Ed.; Blackwell Science: Oxford, 1998; 274.
2. Roelfs, A.; Singh, R.; Saari, E. Rust Diseases of Wheat. In *Concepts and Methods of Disease Management*; CIMMYT: Mexico, 1992; 81.
3. Bos, L.; Parlevliet, J. Concepts and terminology on plant/pest relationships: toward a consensus in plant pathology and crop protection. Annu. Rev. Phytopathol. **1995**, *33*, 69–102.
4. Shaner, G.; Stromberg, E.; Lacy, G.; Barker, K.; Pirone, T. Nomenclature and concepts of pathogenicity and virulence. Annu. Rev. Phytopathol. **1992**, *30*, 47–66.
5. Leach, J.; White, F. Bacterial avirulence genes. Annu. Rev. Phytopathol. **1996**, *34*, 153–179.
6. Parlevliet, J. What Is Durable Resistance? A General Outline. In *Durability of Disease Resistance*; Jacobs, T., Parlevliet, J., Eds.; Kluwer Academic Publishers: Dordrecht, The Netherlands, 1993; 23–39; Proceedings of a Symposium on Durable Resistance, Wageningen, Feb. 22–24, 1992.
7. McIntosh, R. The Role of Specific Genes in Breeding for Durable Stem Rust Resistance in Wheat and Triticale. In *Breeding Strategies for Resistance to Rusts of Wheat*; Simmonds, N., Rajaram, S., Eds.; CIMMYT: Mexico, 1988; 1–10.
8. Ahmed, H.; Finckh, M.; Alfonso, R.; Mundt, C. Epidemiological effect of gene deployment strategies on bacterial blight of rice. Phytopathology **1997**, *87*, 66–70.
9. Poehlman, J. *Breeding Field Crops*, 3rd Ed.; AVI Publishing Company: Westport, CT, 1987; 724.

VIRUS CONTROLS OF PESTS (INSECTS AND MITES)

Brian A. Federici
University of California, Riverside, California, U.S.A.

INTRODUCTION

Viruses are microscopic pathogens composed of a nucleic acid core, either RNA or DNA, and an outer shell consisting of either protein or protein and a lipid envelope. All viruses are obligate intracellular parasites, meaning they cannot reproduce virus particles (virions) outside a living cell. Viruses cause disease by taking over a cell and forcing it to make thousands of copies of the virion. This process kills the cell and can either make the host ill, as do the viruses that cause the common cold, or kill the host, as in case of virulent viruses such as the rabies virus. All forms of life are attacked by viruses, but fortunately individual viruses typically are very specific and attack only one life form. Bacterial viruses only attack bacteria, plant viruses only attack plants, and insect viruses only attack insects. Thousands of viruses have been identified in insects. As these are safe for humans and other vertebrates, the most virulent, the baculoviruses, have been developed as biological insecticides for key insect pests, mainly caterpillars, that attack vegetables, horticultural and field crops, and forests.

VIRUSES THAT ATTACK INSECTS

Insects and mites have been evolving for at least 400 million years, and during this period many different types of viruses have evolved with them. The most common types are the iridescent viruses, cytoplasmic polyhedrosis viruses, entomopoxviruses, and baculoviruses, but only members of the latter group have been developed widely as insect control agents. Though most commonly reported from lepidopterous insects, baculoviruses also have been reported from mosquitoes (order Diptera), caddisflies (order Trichoptera), and sawflies (order Hymenoptera) as well as from noninsect hosts including mites and shrimp.

There are two types of baculoviruses, the nuclear polyhedrosis viruses (NPVs) and the granulosis viruses (GVs). Both produce bacilliform virions consisting of a rod-shaped protein/DNA core surrounded by a lipid-containing envelope. Virions of most baculoviruses are occluded in crystalline protein occlusion bodies referred to as polyhedra (Fig. 1). In NPVs, many virions are occluded in each occlusion body, whereas in GVs only one virion is occluded per crystal. As these are granular, they are referred to as granules.

NPVs and GVs are transmitted by feeding. After being eaten by a caterpillar, the occlusion bodies dissolve in the midgut lumen, and the virions invade midgut cells where they reproduce in the cell nucleus. This colonizing phase is followed by passage of the virus to other tissues where a second phase of reproduction occurs that results in billions of polyhedra or granules produced in each caterpillar followed by death.

Baculoviruses as Viral Insecticides

Use of baculovirus insecticides depends on several factors including the importance of the target pest, the amount of virus required for control, the value of the crop, and the cost and availability of alternative control measures. Baculoviruses are good candidates for use where a single lepidopteran species is the major pest for most of the growing season on a crop with a high cash value, and where other available pest control methods are not cost effective. Thus, the NPVs that have been developed are those effective against caterpillars on such crops as cotton, corn, sorghum, soybeans, tomatoes, strawberries, and chrysanthemums. The amount of virus required is assessed in terms of larval equivalents (LEs)—the number of caterpillars that must be used to produce the virus—necessary to achieve effective control (Table 1). In the United States, this can range from 150 LEs/hectare/treatment using the cotton bollworm NPV to control the bollworm and budworm on cotton to 500 LEs of the beet armyworm NPV for control of this pest on lettuce and chrysanthemums.

The largest documented program using a virus to control an insect is the use of the velvetbean caterpillar NPV against this pest on soybeans in Brazil, where the hectares

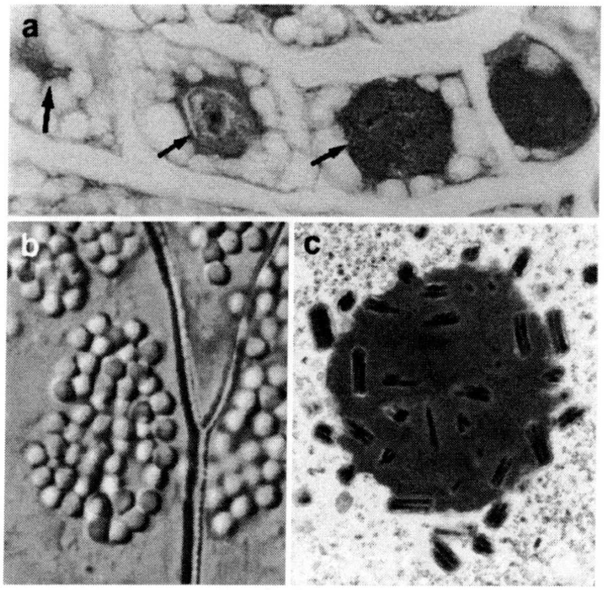

Fig. 1 Nuclear polyhedrosis virus (NPV) in cells of an infected insect. (a) A stained tissue section through the fat body of a caterpillar in an advanced stage of disease. The arrows show stages that represent progressive stages of nuclear infection from early (left) to highly advanced (right). The dense round circles are nuclei filled with viral polyhedra. (b) Phase contrast micrograph illustrating polyhedra of a NPV in the cell nuclei of the tracheal matrix, a tissue responsible in insects for gaseous exchange. (c) Electron micrograph through a polyhedron such as those illustrated in (a) and (b). Note the virions occluded in the protein matrix of the polyhedron.

treated exceed 1 million (2.2 million acres) annually. This single program probably exceeds the total area treated worldwide with all other NPVs and GVs. The velvetbean caterpillar NPV program and a few others are described briefly to provide examples of viruses used in control programs.

Control of the velvetbean caterpillar with its NPV was initiated in the early 1980s and grew by the mid-1990s to a program of 1 million hectares (2.2 million acres) of soybeans. The virus is applied at a rate of 50 LEs/H when caterpillars are young. One application is made per growing season. The cost is about $1.50/H, which is cheaper than chemical insecticides, and the level of control is comparable. The program is coordinated by Embrapa, a state agency that awards contracts for the production of the virus to five private companies. Most virus is produced in caterpillars in the field, where annual yields are 35 metric tons of infected larvae. This program is successful because the virus has to be applied only once per season, soybeans tolerate significant defoliation without loss of yield, and labor costs for virus production are low.

In the United States, the cotton bollworm NPV was the first virus commercialized, being sold originally as Elcar. It was developed for bollworm and tobacco budworm control in cotton, but was not a commercial success due to the advent of the cheaper and highly effective synthetic pyrethroids (chemical insecticides). This virus is used in many countries including China and India to control related caterpillar species on cotton, corn, and sorghum. Total usage worldwide is about 200,000–300,000 hectares, with rates of application in the range of 100–200 LEs/H. Several applications usually are required per season, with the rate and frequency depending on the crop. A new product based on this virus is on the market in the United States under the product name Gemstar.

The armyworm complex includes major pests of field and vegetable crops. The most important pests include the fall armyworm, beet armyworm, and cotton leafroller. Each is attacked by a NPV that is specific, and these have been developed in different countries for armyworm control. For vegetable crops, the beet armyworm NPV is used to control this pest on vegetable and ornamental crops, and is marketed under the tradename Spod-X in the United State, Thailand, and the Netherlands. Application rates range from 100 to 300 LEs/H depending on the crop and insect pressure. Total area treated is about 10,000 H worldwide. The fall armyworm NPV is used to control this pest on maize in Latin America, where 20,000 H are treated annually. The cotton leafworm NPV controls this pest on cotton in the Middle East and Asia, but accurate figures on usage are unavailable.

The granulosis viruses of codling moth, the potato tuberworm, and the cabbageworm are used to varying degrees. The codling moth GV is used against codling moth larvae on apples in Canada, the United States, and several European countries including Switzerland and Germany. The virus is applied at 300 LEs/H, at a cost of $50–80 per application. In cool areas at high altitudes such as in Switzerland, the moth has only one or two generations per year, and thus the virus is cost-effective. In California, however, populations may have to be treated every week, thereby decreasing virus cost-effectiveness. Commercial preparations include Carpovirusine (France), Madex (Switzerland), and Granusal (Germany), with sales worldwide scheduled to reach 60,000 treated hectares worldwide in 1999. The potato tubermoth GV is used in many developing countries to protect potatoes stored above ground. Under the direction of Center International for Potato Research (CIP, Lima, Peru), the virus is produced by ''cottage industries'' in countries such as Peru, Morroco, Libya, China, and Indonesia, and is typically applied to the potatoes as a dust. The cabbageworm GV has been used for many years in China to control cab-

Table 1 Major baculoviruses registered for control of insect pests

Target pest	Virus	Crop	Product name	Producer	Country
Caterpillars					
Anticarsia gemmatalis (Velvetbean caterpillar)	AgNPV	Soybeans	VPN	Agricola El Sol	Brazil
Adoxophyses orana	AoGV	Fruit orchards	Capex	Andermatt Biocontrol	Switzerland
Cydia pomonella (Codling moth)	CpGV	Apples, walnuts	Madex	Andermatt Biocontrol	Switzerland
			Granupom	AgrEvo	Europe
			Carpovirusine	NPP (Calliope)	France
Helicoverpa zea (Cotton bollworm)	HzNPV	Cotton, vegetables	Gemstar	Thermo-Trilogy	Columbia, MD
Heliothis virescens (Tobacco budworm)					
Lymantria dispar (Gypsy moth)	LdNPV	Deciduous forests	Gypcheck	Thermo-Trilogy	Columbia, MD
			Disparvirus	Canadian Forest Service	Canada
Mamestra brassicae	MbNPV	Vegetables	Mamestrin	NPP (Calliope)	France
Orgyia psuedotsugata (Tussock moth)	OpNPV	Douglas-fir forests	TM Biocontrol-1	Thermo-Trilogy	Columbia, MD
Spodoptera exigua (Beet armyworm)	SeNPV	Vegetables, flowers	Spodex	Thermo-Triology	Columbia, MD
Spodoptera littoralis (Cotton leafworm)	SlNPV	Cotton	Spodopterin	NPP (Calliope)	France
Sawfly larvae					
Neodiprion sertifer (Pine sawfly)	NsNPV	Pine forests	Neocheck-S	Canadian Forest Service	Canada
			Sentifervirus		
Neodiprion lecontei	NlNPV	Pine forests	Lecontivrus	Canadian Forest Service	Canada

bageworms on vegetables. The virus, used in either dusts or wettable powders, is produced in larvae in field plots of cabbage and lettuce, and harvested as the infected larvae die.

LIMITATIONS

Virus insecticides are not widely used in industrialized countries where they are of limited availability. Moreover, effective and competitively priced alternatives are available. Among these are new and more specific chemical insecticides, and insecticides based on *Bacillus thuringiensis*, including both bacterial insecticides and transgenic plants. Furthermore, viruses have what are considered key limitations. These include slow speed of kill, narrow target spectrum (host range), little residual activity, and lack of cost-effective cell culture systems for production. Alternatively, these limitations have not prevented virus use in developing countries where NPVs and a few GVs are used, especially on field and vegetable crops, in China, India, and Brazil, as well as in many smaller countries in Latin America, Africa, and southeast Asia. The reasons for this are that chemical insecticides are expensive, their heavy use has resulted in resistance, labor costs for virus production in caterpillars are low, production technology is simple, and registration for use of viruses is either not required or is easily obtained.

IMPROVING VIRAL INSECTICIDES BY GENETIC ENGINEERING

The most significant limitations of viral insecticides are slow speed of kill and narrow host range. When a virus is used against caterpillars, the population typically is a mixture of developmental stages, and may contain more than one pest species. A virulent NPV will kill the early and middle stage caterpillars within 2–4 days. But the more advanced ones may live a week or more and cause extensive crop damage. Thus, approaches to improving viral efficacy aim at developing broad-spectrum viruses that halt larval feeding within 24–48 h of infection, either by death or paralysis. This is done by deleting genes from viruses that delay death, and engineering viruses to produce insect-specific peptide neurotoxins that paralyze or kill the caterpillar directly. The genes for these neurotoxins are obtained from insect-parasitic mites, scorpions, and spiders. These toxins and viruses are tested extensively to make sure they kill only insects. Because NPVs engineered to produce neurotoxins can only reproduce effectively in insects, genetically engineered viruses are safe for humans and other vertebrates.

FUTURE POTENTIAL

Insect viruses, especially NPVs and GVs, will continue to be used and developed as control agents for use in integrated control programs for caterpillar pests. However, increasing their use substantially will require that they become more competitive with other available control agents including products based on the insecticidal bacterium, *Bacillus thuringiensis*, insecticidal transgenic plants, and chemical insecticides. This will require more cost-effective viruses and improved methods for their mass production on a commercial scale using either caterpillar hosts or cell culture technology.

See also *Biological Control of Insects and Mites,* pages 57–60; *Genetic Improvement of Biocontrol Agents,* pages 329–332; *Transgenic Crops,* pages 846–849, pages 850–851.

V

BIBLIOGRAPHY

Federici, B. Naturally Occurring Baculoviruses for Insect Pest Control. In *Biopesticides, Use and Delivery*; Hall, F., Menn, J., Eds.; Humana Press: Totowa, NJ, 1998; 301–320.

Federici, B. A Perspective on Pathogens as Biological Control Agents for Insect Pests. In *Handbook of Biological Control*; Fisher, T., Bellows, T., Caltagirone, L., Dahlsten, D., Huffaker, C., Gordh, G., Eds.; Academic Press: San Diego, CA, 1999; 517–550.

The Biology of Baculoviruses, Practical Application for Insect Control; Granados, R., Federici, B., Eds.; CRC Press: Boca Raton, FL, 1986; II, 276.

Insect Viruses and Pest Management; Hunter-Fujita, F., Entwistle, P., Evans, H., Crook, N., Eds.; John Wiley & Sons: Chichester, U.K., 1998; 620.

The Baculoviruses; Miller, L.K., Ed.; Plenum Press: New York, 1997; 447.

Moscardi, F. Assessment of the Application of Baculoviruses for Control of Lepidoptera. In *Annual Review of Entomology 44*; 1999; 257–289.

Treacy, M.F. Recombinant Baculoviruses. In *Biopesticides, Use and Delivery*; Hall, F., Menn, J., Eds.; Humana Press: Totowa, NJ, 1998; 321–340.

WEED DISPERSAL

W

Anne Légère
Agriculture and Agri-Food Canada, Sainte-Foy, Quebec City, Canada

INTRODUCTION

Weeds of agricultural systems are mainly annuals, biennials, or perennials that produce seeds, vegetative propagules, or both. Production and dispersal of numerous propagules allow weed species to maintain populations where they are already established, colonize new areas, and expand their geographical range. Unlike the parent plant that produced them, seeds are free to roam the world in search of an ideal microsite in which to establish, grow, and reproduce in their turn. However, the seed has no true means of locomotion and no control of its whereabouts. It must rely on external agents to be dispersed beyond the immediate vicinity of its parent. Humans and their activities are primary dispersal agents of weeds in agricultural and other nonrural habitats throughout the world.

DISPERSAL IN SPACE: THE ESSENTIAL ROLE OF DISPERSAL AGENTS

From an anthropomorphic point of view, the strategy of the parent weed plant is to produce as many offspring as possible, while providing each offspring with the minimum resources needed for establishment. A single plant of *Chenopodium album* (lambsquarters) can produce between 30,000 and 350,000 seeds; one population of *C. album* in a corn field was estimated to have produced more than a million seeds per m^2. Most weed seeds are light: a single seed of *C. album* weighs approximately 0.7 mg; seed weight of *Setaria glauca* (yellow foxtail) varies between 1.3 and 4.2 mg. Weed seeds come in many shapes, some with appendages (hooks, barbs, awns), others with sticky mucilage, and surprisingly, many with no special feature. Some of the most noxious weeds also produce vegetative structures such as rhizomes, corms, bulbs, stolons, tap roots, tubers, etc., some of which can be dispersed much like seeds.

Most weed seeds fall and remain near the parent plant thus explaining the gregarious nature and patchy distribution of weed populations in agricultural fields. Assuming a unit plant diameter equal to plant height, most weed seeds disperse generally within 10–15 plant diameters. The fall to the ground is generally passive (passive autochory) but some species such as *Erodium* spp. (filaree) forcefully expel seeds from their fruits (active autochory). Once on the ground, seeds can be dispersed further by agents or forces that are generally unpredictable. Their intervention does not always result in a positive outcome for the seed, which may be destroyed in the process or may end up in an unsuitable site.

Dispersal agents are both abiotic and biotic. Wind (anemochory) will disperse seeds adorned with a plume-like structure or pappus such as found on many species of the Asteraceae family, *Taraxacum officinale* (dandelion) being a well-known example. Most seeds will be carried some distance by water (hydrochory) although some will have better buoyancy characteristics than others. Seeds of *Setaria viridis* (green foxtail) can float for as long as 10 days. Seeds of some 30–50 species have been found in irrigation canals. Birds, mammals, and invertebrates that feed on weed seeds (endozoochory) may also transport seeds over various distances. Earthworms move seeds downward in the soil profile but also return as many as 60–100 seeds per m^2 to the surface in worm casts. Grazing animals, both wild and domestic, can disperse barbed or hooked seeds attached to their fur (ectozoochory). As many as 50–200 *Cynoglossum officinale* (hound's-tongue) burrs have been observed on the faces of cattle. All of these dispersal vectors can contribute to the creation of new foci of infection in spite of the low odds of the seed ending up in an appropriate microsite for germination and growth.

Humans and their activities are the most important dispersal agents of weeds, not only in agricultural systems but in urban and wilderness areas alike. From the earliest times, weed species have crossed great geographical barriers such as oceans and continents by traveling with crops being introduced to new areas. Approximately 50% of the weed flora of the United States is of Eurasian origin, whereas that of Australia partly originates from Eurasia (33%) and the Americas (28%). The success of introduced weed species in new areas is due to the absence of specific enemies such as diseases and insects, but is mostly explained by the fact that human activities recreate the conditions to which weeds are adapted. Soil husbandry contributes to weed dispersal on local and regional scales. Tillage and cultivation disturb the soil and, in doing so, allow weed seeds to move down and up the soil profile. Weed seeds stuck with soil to tillage implements, and vegetative structures such as rhizomes produced by *Elymus repens* (quackgrass) can be carried within and across fields if implements are not cleaned. Weed seeds can be harvested with the crop, redistributed in the field with the chaff, and carried to other fields if the combine is not cleaned before moving on to a new location.

DISPERSAL IN TIME: SHATTERING AND DORMANCY

Weed species produce seeds over a relatively long period of time during the growth season. Pods generally shatter as they mature such that some seeds fall to the soil and become part of the seed bank throughout the season, whereas others may be harvested and removed with the crop. The fate of a weed seed may thus be linked to the time at which it is shed from the parent plant, and also to the timing of crop management operations. In temperate regions of the world, most weed seeds will not germinate as they are shed. This is most fortunate, given that such late germination would generally not allow sufficient time for a plant to complete its life cycle before the onset of killing frosts. When they are shed from the plant, many weed seeds are in a quiescent state and will not germinate in spite of environmental conditions that could be conducive to germination. Dormancy can be attributed to various morphological characteristics and physiological processes. These may include the presence of a hard seed coat, endogenous chemical inhibitors, and seeds with immature embryos requiring a period of after-ripening before germination can occur. Other weed seeds are in a germinable state when they are shed but become dormant when exposed to unfavorable environmental conditions such as low soil moisture or high temperatures. In some weed species, a single plant will produce different types of seeds that vary in terms of dormancy characteristics. Weed seeds may remain dormant for decades, even centuries, awaiting appropriate conditions for germination. The soil seed banks of North West Europe include at least 100 species for which maximum seed longevity is greater than 20 years. Dormant seeds from the seed bank will be induced to germinate when exposed either to environmental conditions or agronomic practices that provide the triggers needed to end dormancy, such as a flash of light, fluctuating temperatures, scarification, leaching of a chemical inhibitor, etc. Dormancy is an insurance policy against ill-timed germination, ensuring weed species survival and dispersal over time.

WANTED: STRATEGIES TO MANAGE WEED DISPERSAL

Dispersal of weeds has long been recognized as a major threat to agriculture. Nevertheless, weeds have continued to disperse to this day in spite of seed acts and noxious weed seed laws introduced in North America as early as the mid-1850s. Sanitation has been advocated as an essential tool in any weed management strategy. However, recent surveys of drill boxes suggest that weed seeds are still being sown with crops. Current agronomic practices continue to foster environments conducive to weed dispersal and establishment, in that most cropping systems still provide empty niches for weed colonization. Dispersal management would require that cropping systems be designed to provide less niche opportunities for weeds, on both temporal and spatial scales. In addition to legislation, sanitation, and improved cropping systems, an apparently simple way to contain dispersal would be to prevent all weed seed production. In general, weed control measures are applied when gains justify costs according to short-term economics. As a consequence, small weed populations may go unchecked and produce many seeds. Even when weed control measures are applied, residual weed populations composed of escapes and late emerging plants may also contribute to seed production. Means of preventing these small weed populations from going to seed are currently few and rarely justifiable economically. However, precision farming and related techniques allowing localized weed control measures may eventually become part of a feasible solution to weed dispersal in agricultural systems, by providing efficient and economical tools for the control of these potential foci of weed infestations.

FUTURE CONCERNS

In spite of laws and quarantine efforts, weed seeds continue to travel with packaging material, soil, ornamentals, nursery stock, animals, vehicles, etc., at a frequency and speed unparalleled in history. Current climatic fluctuations are allowing species to expand their geographical range and become weedy in new areas. Changes in crop husbandry can result in similar outcomes. Clearly, dispersal issues must be addressed from a broad perspective that includes but also reaches far beyond a single field or species. The threat of introducing new species with the capacity to invade and displace native species was recognized in 1999 with the signature by the President of the United States on the Executive Order on Invasive Species, mandating American institutions to interact and direct their efforts in curtailing the introduction and spread of invasive species. This highlights the strategic importance of dispersal and hopefully will provide opportunities to address issues related to weed dispersal and invasive species in a successful and responsible way.

BIBLIOGRAPHY

Baskin, C.C.; Baskin, J.M. *Seeds, Ecology, Biogeography, and Evolution of Dormancy and Germination*; Academic Press: San Diego, CA, 1998; 666.

Buhler, D.D.; Hoffman, M.L. *Andersen's Guide to Practical Methods of Propagating Weeds & Other Plants*; Weed Science Society of America: Lawrence, KS, 1999; 237.

Species Dispersal in Agricultural Habitats; Bunce, R.G.H., Howard, D.C., Eds.; Bellhaven Press: London, UK, 1990; 288.

Plant Invasions: Studies from North America and Europe; Brock, J.H., Wade, M., Pyek, P., Green, D., Eds.; Backhuys Publishers: Leiden, NL, 1997; 223.

Cousens, R.; Mortimer, M. *Dynamics of Weed Populations*; Cambridge University Press: Cambridge, UK, 1995; 332.

Ghersa, C.M.; Roush, M.L. Searching for solutions to weed problems: do we study competition or dispersion? BioScience **1993**, *43*, 104–109.

Harper, J.L. *Population Biology of Plants*; Academic Press: London, UK, 1977; 892.

Thompson, K.; Bakker, J.P.; Bekker, R.M. *The Soil Seed Banks of North West Europe: Methodology, Density and Longevity*; Cambridge University Press: Cambridge, UK, 1997; 276.

WEED ELECTROCUTION

Clément Vigneault
Agriculture and Agri-Food Canada, Saint-Jean-sur-Richelieu, Québec, Ontario, Canada

INTRODUCTION

The concept of weed electrocution was developed in the late 1800s. Several systems were patented in the United States since 1890. They involve the use of electrical energy to kill weeds. When electrical current passes through a plant, heat is generated due to electrical resistance of the plant material (1). The electrical energy increases the temperature of the material and vaporizes water or other liquids contained within the plant (2). Vaporization of liquids increases the pressure inside plant cells, which causes the cell membranes to rupture and consequently kill the plant (3). Compared to chemical or mechanical weed control, this method does not introduce any chemicals into the food chain and does not disturb the soil surface preventing erosion (4). However, electrocution is a time- and energy-intensive method of controlling weeds under field conditions (5, 6). During the 1980s, it was mostly used as an alternative method to destroy persistent weeds in row crops following conventional chemical treatment (6–9). In 1990, theoretical analysis identified the key parameters for the design and the use of weed electrocution equipment and the main problems associated with its application (10). Thereafter, research, development, and the use of weed electrocution were discontinued around the world.

DESCRIPTION OF A WEED ELECTROCUTION SYSTEM

A weed electrocution system mainly consists of a generator, a transformer, an insulator-mounted electrode, and rolling coulters (see Fig. 1). The AC generator driven by the tractor engine through the power take-off shaft generates low voltage electricity (see Fig. 2). The step-up transformer then increases the voltage to 5000–10,000 V. A high voltage wire connects the transformer to an electrode that makes contact with weeds. When the high-voltage electrode touches a plant, the current passes through the plant and into the soil. The current then returns back to the generator via rolling coulters that serve as grounding devices.

THEORETICAL ANALYSIS OF THE POWER REQUIREMENT (10)

The amount of power (P) needed from the generator to electrocute weeds is a function of the voltage (V) and the load resistance (R_1). Neglecting the transformer resistance and assuming an ideal case where all weeds have the same resistance, the load resistance becomes a function of the plant resistance (R_p), the number of weeds in contact with the electrode (n_c), and the soil resistance (R_s).

$$P = \frac{V^2}{R_1} = \frac{V^2}{\left(\frac{R_p}{n_c} + R_s\right)}$$

The number of weeds in contact with the electrode is proportional to the weed population density. As the weed density increases, the number of weeds touching the electrode increases and consequently the power requirement also increases according to Eq. 1. In addition, as the weed density increases, the plant resistance becomes negligible compared to the soil resistance as indicated by Eq. 1. Therefore, a very high weed density increases the power requirement, but most of the energy is absorbed by the soil.

An efficient system should provide enough application time and energy to electrocute the weeds. Longer application time and excess energy, such as slowing down the tractor or increasing the applied voltage, waste resources without increasing weed mortality. A minimum contact time of about 0.2 s between the plant and the electrode is required to ensure the death of young weeds (3). The threshold energy needed for weed electrocution varies largely among species, morphology, age, and size of the

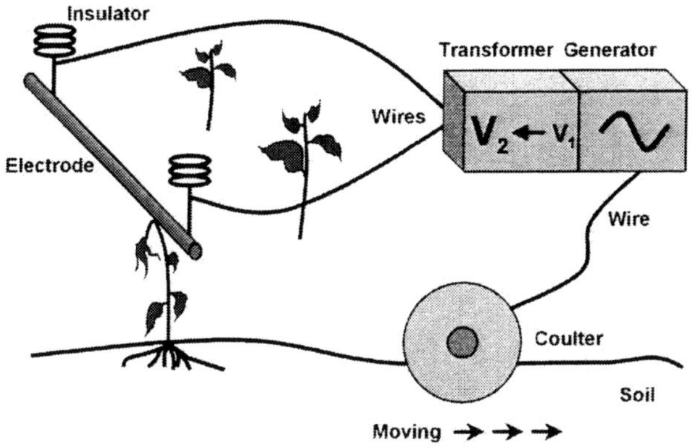

Fig. 1 The main components of a weed electrocution system.

plants (10). These factors alter the plant resistance and its susceptibility to electrocution. In general, the energy requirement increases with the maturity of the plant. The root arrangement also affects the effectiveness of the electrocution (10). For example, plants with large or specialized underground organs normally suffer little root damage. The composition and moisture content of the soil also affect the efficiency of the operation as they alter the soil resistance to current flow. More damage was found in roots under dry soil than in moist soil. The threshold energy required for electrocution can be evaluated in greenhouse conditions. However, the energy requirement for field-grown plants is often several times greater than for greenhouse plants. This is due to additional resistances found in nonconsistent field weeds and nonuniform soil conditions.

ENERGY INPUT COMPARISON WITH CONVENTIONAL CONTROL METHODS

A typical field without prior weed control has a weed density ranging from 50 to 2000 plants/m^2. A weed control study on a field having a weed density of 200 plants/m^2 has shown that electrocution requires approximately 20 times more energy and 50 times longer to apply than the spraying of herbicide (10). For fields with a weed density of 15 plants/m^2, electrocution still requires twice the energy and takes five times longer than the chemical weed control. Mechanical weed control generally requires less energy input than chemical weed control. The proportion of energy required for electrocution compared to mechanical treatment is greater. Furthermore, similar to mechanical weed control, electrocution requires two to three treatments to achieve the desired effect when compared to a single application of herbicide.

CONCLUSIONS

Electrocution is a method that uses electrical energy to control weeds. Interactions between the mechanical and biological parameters involved in the process make it difficult to be applied efficiently and economically under field conditions. The method uses fossil fuels as the source of energy, which create as much pollution as chemical weed control. The use of electrocution may be justified on high-value crops such as herbs or other specialty crops. Also, it may be used in situations associated with crop

Fig. 2 A weed electrocution system ''Lasco Lightning Weeder'' manufactured in the 1980s in Vicksburg, Mississippi, USA.

rotation such as sunflowers growing in soybean fields, where a substantial height difference exists between the undesirable plant and the main crop. Electrocution cannot be used as a primary weed control method, because it becomes cost-effective only in areas with low weed density. With the high cost of application, it can be used in areas where chemical treatment is not acceptable, or in areas where there is a high risk of soil erosion.

REFERENCES

1. Chandler, J.M. In *Crops and Weed Response to Electrical Discharge*, Proceedings of the Southern Weed Science Society, 1978; 31, 63.
2. Diprose, M.F.; Benson, F.A.; Hackam, R. Electrothermal control of weed beet and bolting sugar beet. Weed Res. **1980**, *20*, 311–322.
3. Dykes, W.G. Principles and practices of electrical weed control in row crops. Am. Soc. Agric. Eng. **1980**, 80–1007.
4. Sanwald, E.; Koch, W. In *Physical Methods of Weed Control*, Proceedings of the Brighton Crop Protection Conference on Weeds, 1978; 3, 977–986.
5. Kaufman, K.R.; Schaffner, L.W. Energy and economics of electrical weed control. Trans. Am. Soc. Agric. Eng. **1982**, *25*, 297–300.
6. Diprose, M.F.; Fletcher, R.; Longden, P.C.; Champion, J. Use of electricity to control bolters in sugar beet (*Beta vulgaris* L): a comparison of the electrothermal with chemical and mechanical cutting methods. Weed Res. **1985**, *25*, 53–60.
7. Vigoureux, A. Results of trials carried out in Belgium in 1980 about killing weed beets by electric discharge. Meded. Fac. Landbouww. Ryksuniv. Gent. **1981**, *46*, 163–172.
8. Lutman, P.J.W. A Review of Techniques that Utilize Height Differences Between Crops and Weeds to Achieve Selectivity. In *Spraying Systems for the 1980s*; Walker, J.O., Ed.; Crop Prot. Counc. Monogr. 1980; 24, 291–317.
9. Diprose, M.F.; Benson, F.A. Electrical methods of killing plants. J. Agric. Eng. Res. **1984**, *30*, 197–209.
10. Vigneault, C.; Benoît, D.L.; McLaughlin, N.B. Energy aspect of weed electrocution. Rev. Weed Sci. **1990**, *5*, 15–26.

WILDLIFE KILLS

Pierre Mineau
Canadian Wildlife Service, Ottawa, Ontario, Canada

INTRODUCTION

More than 30 pesticides registered in North America or Europe have been known to result in kills of wild terrestrial vertebrates even when used according to the relatively stringent regulations in force in those countries. Among the species affected, birds figure prominently in the kill record and this, for several reasons. Birds are ubiquitous and visible. In North America, as in several other countries, most species are federally protected from unlicensed taking or kill. Birds are extremely mobile, and cannot be excluded from areas that have been treated with pesticides. Clinics and rehabilitation centers for birds can become a valuable source of information and samples for selected groups of species, most frequently hawks, eagles, and owls. Some bird species are attracted to agricultural fields, and many are economically important to the control of agricultural pests, notably insects. Finally, birds, as a group, are particularly sensitive to some of the more toxic classes of pesticides such as the organophosphorus and carbamate insecticides, and their reproduction has been found to be vulnerable to a wide range of pesticides. New pesticides developed in part for their relative safety to humans have been found to be especially toxic to birds.

Unfortunately, this focus on birds has resulted in virtual ignorance of the effects of pesticides on most other groups of terrestrial wildlife. Small mammal, reptile, and amphibian carcasses are found regularly in the course of pesticide investigations but seldom reported outside of such directed research. Because of the poor state of amphibian diversity worldwide, every possible attempt should be made to factor in those species.

INCIDENT MONITORING

Incident monitoring refers to the capacity of competent authorities to investigate reported kills or conduct spot checks of use conditions. Even if a pesticide has been studied extensively under controlled conditions, unforeseen problems and situations often arise following commercialization of the product. An absence of incident reports does not necessarily mean there are no problems but, conversely, well-investigated incidents and kills can reveal unforeseen aspects of a pesticide or reinforce a suspicion that arose in the course of laboratory or field testing. An incident monitoring scheme will require a network of individuals trained in carrying out pesticide investigations and in proper handling of carcasses and tissue samples, as well as access to a laboratory equipped to perform the required chemical and biochemical analyses (1).

Even where relatively efficient incident monitoring systems are in place, only a very small proportion of kills are ever uncovered. There are several reasons for this: affected wildlife are often dispersed and at relatively low density in farm fields, they often leave the treated area to die, they are likely to seek cover and hide when overcome by the pesticide, they are often cryptic and hard to see, and their carcasses are scavenged rapidly after death. Typical rates of carcass removal by scavengers are 40–90% in 24 h. Farm fields are large; the mechanization and sheer size of modern agricultural machinery often removes the farmer from any "close contact" with the land. The increasing size of farms also means that, when kills occur, often in the few days that follow a pesticide application, the farmer is busy elsewhere, treating another part of the farm. Pesticide intoxication may be a causal factor in a kill visibly caused by something else, for example, intoxicated birds hitting fences, utility wires, cars, or buildings—and not being recognized as a pesticide kill. Also, there is a large difference between casual searching of fields and a well-organized intensive search effort. An intensive search effort consisting of several trained individuals, transects, and repeated, well-timed searches have produced between 10- and 500-fold improvements in carcass detection rates over field inspections carried out once or a few times only by single individuals. Equally important is the motivation and training of the search teams.

Even when incidents are uncovered, often they are not reported. If the kills involve only one or a few individuals, not much importance is attached to the incident even though, for reasons just outlined, these few carcasses likely represent the "tip of the iceberg." Even if the kills are reported, the information often is not centralized and made available to national pesticide regulatory bodies. It is important to understand and recognize biases inherent in any incident reporting system. Some of those biases will depend on how the incident monitoring system is set up and which persons/organizations are responsible. Some biases can be reduced over time, but others are unavoidable. Common biases relate to body size and color of the casualties, numbers and density of the species in any given area, "status" of the species and individual and institutional interests and sensitivities. We expect most kills to be of small-bodied birds widely dispersed in field margins. Yet, such kills are seldom reported.

Despite these limitations, it is important for countries to investigate wildlife kills and make the information available (2). Registration decisions are made on the basis of very limited information. There are large differences in toxicological and ecological vulnerability among species. The ways in which wildlife species are exposed to pesticides are varied and sometimes difficult to predict or study. The behavior of pesticides depends on local conditions although pesticides are often tested under standardized conditions only. The outcome of exposure is also much more variable in the wild. Pesticide exposure can interact with weather, the condition or health of the animal, etc. Therefore, whether or not pesticides are routinely field tested to look for environmental impacts, it is essential to have a good incident monitoring system in place. An incident monitoring system can also be useful to warn manufacturers if their products are abused or used incorrectly.

The usefulness of an incident monitoring scheme will grow as the quality, reliability, and coverage of kill reports increase. Incident monitoring data can be used to

- verify whether registration decisions were appropriate: that is, confirm a risk predicted from lab data, or identify a risk not predicted from lab data
- trigger more systematic field studies
- improve label directions
- allow recommendations on the "best" product to use under some circumstances
- trigger a regulatory review
- ensure that products are being used correctly
- provide data for potential legal action

PESTICIDES RESPONSIBLE FOR MOST WILDLIFE KILLS

The acute oral toxicity (LD_{50}) of a pesticide as well as the extent of its use are good predictors of wildlife kills (3). The dietary toxicity test currently carried out on young birds (dietary LC_{50}) can seriously mislead however. As a rule, insecticides and vertebrate control agents are much more likely than herbicides or fungicides to give rise to wildlife kills. Since the 1970s, poisoning cases involving cholinesterase inhibitors (organophosphorus and carbamate pesticides) have been particularly frequent (4). Fortunately, the measurement of brain cholinesterase levels has proven to be a useful (although certainly not foolproof) diagnostic tool for such incidents. The test has the advantage of being economical and relatively easy to carry out. It is noteworthy that cholinesterase measurements in carcasses do not correlate very well with quantities of the pesticide recovered from the same carcasses. Wildlife kills resulting from newly developed insecticide groups will be harder to elucidate in the absence of such a convenient biomarker. Diagnosis will hinge on sophisticated and costly residue analyses without the benefit of the "smoking gun" that reduced cholinesterase titers represented.

ROUTES OF PESTICIDE EXPOSURE RESPONSIBLE FOR KILLS

Although the use of a toxic insecticide per se does not guarantee an impact, some pesticides are so acutely toxic to birds and other wildlife that it is difficult to use them under any circumstance without causing some wildlife mortality (5). Birds ingest pesticides through their food or through preening or grooming. Despite being feathered, they absorb pesticides through their skin, encountering droplets directly or by rubbing against foliage and other contaminated surfaces. Birds are also exposed through their feet. Finally, they have a very high ventilation rate and inhale vapor and fine droplets. The degree to which each of these routes of exposure contributes to the total dose depends on the crop being sprayed, the chemical, the species exposed, and environmental factors. The ecology of the species (i.e., feeding preferences and behavior) along with the characteristics of the chemical (i.e., its persistence, tendency to bioaccumulate, and pharmacokinetics) and the intended use go a long way in determining the nature and the scope of the impacts on wildlife. Climatic factors as well as condition factors, such as nutritional status, disease, and parasite load also exert an influence, directly on the toxicity of the pesticides to the organism, and indirectly through their influence on food consumption. These are yet additional reasons why the kill record is not entirely consistent between chemicals and between applications of the same chemical.

Several specific routes of exposure are featured in the wildlife kill record.

Abuse and Misuse

Pesticide abuse is the deliberate use of a pesticide in a nonauthorized fashion, usually to poison wildlife species considered to be pests. In the United Kingdom, as well as in several European countries, officials estimate that deliberate bird kills due to pesticide abuse outnumber cases where label instructions were strictly followed. Between 1978 and 1986, officials in the United Kingdom estimate that, on average, 71% of incidents were the result of abuse. For birds of prey alone, more than 90% of cases recorded between 1985 and 1994 in the United Kingdom were abuse cases. On the other hand, for raptors in the United States during the same period, kills involving labeled uses of pesticides were almost as frequent as abuse cases. This difference appears to be wholly attributable to the high toxicity of insecticides used in the United States. Abuse generally involves baits of some kind, the only limit being the imagination of the perpetrator. Typically, liquid insecticides are poured or injected and applied to seed, bread, meat, etcetera. Granules are sprinkled or mixed into a paste. Because of the high concentration of pesticide involved in abuse cases, carcasses are usually found in close proximity to the site of baiting thus biasing the kill record through a higher recovery of carcasses. The choice of chemicals used in abuse cases reflects availability and toxicity. Pesticides typically used in deliberate poisoning attempts include carbofuran, aldicarb, monocrotophos, parathion, mevinphos, diazinon, and fenthion, chemicals that are all recognized as being inherently very toxic to vertebrates in general and birds in particular. The main problem of course is that the baits are often indiscriminate in the species that they kill. Secondary poisoning is also frequent when predators or scavengers take dead or debilitated prey with a highly concentrated bait in their gut.

Pesticide misuse refers to a pesticide application that is not exactly as specified by the label. This may be an application at a rate that is higher than specified, or to a crop or pest other than those listed. Pesticide misuse is difficult to establish, especially after the fact. Also, in some cases, it becomes very difficult to distinguish a misuse from a normal agronomic use when the label contains instructions that are vague, difficult, or impossible to follow. What constitutes a misuse in one jurisdiction may indeed be an approved use elsewhere.

Persistent Bioaccumulating Substances

The use of most persistent organochlorine pesticides (POPs) such as DDT, aldrin, dieldrin, heptachlor, chlordane, HCB, HCH, mirex, and toxaphene has been canceled or significantly curtailed in most countries. However, some POPs continue to be a concern because of persisting local contamination in areas of high historical use, and because of continuing uses in some parts of the globe. High residue levels are found in birds that migrate to areas where these products are still used. These birds in turn pass the residues along the food chain, where they hamper the full recovery of populations of species such as the peregrine falcon (*Falco peregrinus*) in North America. Kills resulting from persistent organochlorine pesticides often involve individuals under food stress and in which brain residues achieved lethal levels following a remobilization of body lipid stores.

Granular Formulations and Treated Seed

Granular insecticides and treated seeds are frequent routes of exposure and intoxication in wildlife. Granular insecticides were designed for convenience, safety to applicators, and time release of the chemical; yet for birds, granular formulations of the more toxic insecticides such as aldicarb, parathion, carbofuran, fensulfothion, phorate, terbufos, fonofos, disulfoton, diazinon, and bendiocab are repeatedly associated with bird mortality.Several bird and small mammal species have a fatal attraction to granular formulations, mistaking them either as dietary grit or as a food source. The most attractive granules are those made of sand (silica) or an organic base such as dried corn (maize) cobs. Somewhat less attractive are clay, gypsum, and coal granules. Exposure can also occur via invertebrates, especially earthworms to which granules easily adhere. Secondary toxicity is likely in predators and scavengers that eat their prey whole or ingest their gastrointestinal tract contents. In Canada and the United States, there have been cases of poisoning of waterfowl foraging in puddles in fields more than six months after applications of granular insecticides because of specific soil conditions that may retard breakdown (6). Granules that are friable and disintegrate quickly when exposed to moisture are best for birds, but they are the products least convenient to farmers. Regardless of the type of carrier, a pesticide granule is likely to be a problem if a lethal dose can be obtained in a few granules only.

To date *no* agricultural machinery or application technique can achieve complete incorporation of granular insecticides below the soil horizon. Birds have also been known to probe the soil for granules or to pull up germinating seeds with granules attached. The worst applications are those made above the soil surface and "banded" or "side dressed" over or to the side of the seed furrow. In carefully controlled engineering trials, between 6 and 40% of applied granules were left on the soil surface. The same equipment can achieve radically different soil incorporation when used by different individuals under different conditions.

Treated seeds present a similar engineering problem. As with granules, more seeds are left on the surface wherever the seeders have to turn or negotiate obstacles. Small spills are part of normal farming practice and can occur anywhere depending on topography and soil conditions but more often at field edge. Historically, seed dressings were one of the main sources of bird exposure to organochlorine and mercurial compounds. Poisoning incidents with seed dressings are still relatively frequent because several bird species make heavy use of waste (or even planted) grain in fields. The size and type of seed dictate which bird species are at risk. Since use of organochlorines and organomercurials has declined, kills have been recorded with cholinesterase-inhibiting insecticides such as carbophenothion, chlorfenvinphos, isofenphos, bendiocarb, disulfoton, furathiocarb, and fonofos. Some kills have been recorded with newer insecticides as well, for example, imidacloprid although it is not yet known how serious or frequent a problem this will become.

Liquid Formulations on Vegetation: The Grazing Problem

Grazing birds are particularly vulnerable to foliar applications of pesticides. Kills have been recorded with several cholinesterase-inhibiting pesticides such as parathion, diazinon, carbofuran, isofenphos, dimethoate, and triazophos. Grazers typicallyinclude geese, ducks, and coots (families *Anatidae* and *Rallidae*). These birds eat large quantifies of foliage because they do not digest cellulose. Fertilized areas are particularly attractive to grazing species that can detect the high nitrogen levels. Golf courses attract grazers because the turf is cut frequently, watered, and fertilized, and courses often have other attractions such as ponds and drainage streams. More than 100 cases of waterfowl mortality were recorded due to the use of diazinon on turf (7) before the pesticide was withdrawn from golf courses and sod farms in the United States. Other well-documented problems are kills of ducks and geese in alfalfa fields treated with carbofuran and of sage grouse (*Centrocercus urophasianus*) feeding on alfalfa crops treated with dimethoate or on potato foliage and weeds in potato fields sprayed with methamidophos.

Liquid Formulations on Insect Prey: The Gorging Problem

Bird species that feed on agricultural pests such as grasshoppers, leatherjackets (larvae of the crane fly), grubs, and cutworms are at high risk of poisoning. Kills of these species are all the more tragic because they are beneficial to agriculture.Some species are particularly vulnerable because they specialize in insect outbreaks. These birds take advantage of pest control operations that result in insects becoming either debilitated or more visible following treatment. In a recent case in Argentina, approximately 20,000 Swainson's hawks were poisoned within the span of a few weeks after feeding on grasshoppers sprayed with monocrotophos (8). As with carbofuran, the extreme toxicity of this product means that it is difficult to find use patterns that do not result in bird kills.

Vertebrate Control Agents: Unintended Victims

Rodenticides as a rule are not specific to their intended targets and cause direct impacts to nontarget species. Only a detailed knowledge of the habits of the target species and use of specific baiting locations or specialized bait holders can reduce kills of nontarget species. More problematic is secondary poisoning. Unfortunately, the trend has been for the more recent, more efficacious ''single feed'' anticoagulants to present a greater hazard to predators than the older products. Compounds such as difenacoum, brodifacoum, bromadialone, difethialone, flocoumafen, and other similar ''super'' coumarin-type products present a problem where the target species are likely to be predated or scavenged (9). The use of thallium and endrin to control rodents has also been shown to have disastrous consequences on raptors.

Fenthion, an organophosphorus ''insecticide'' used to control pest birds in Africa (e.g., *Quelea quelea*) and in North America (e.g., by means of the Rid-a-Bird™ perch system), has given rise to frequent secondary poisoning (10). Secondary poisoning is also very likely following the use of toxic organophosphorous or carbamate products for the control of parakeets, doves, and other seed eaters. The use of organophosphorus pesticides such as famphur and fenthion for the treatment of parasites in livestock frequently lead to wildlife kills. Famphur, which is one of the leading causes of eagle poisonings in the American Southwest, persists on the hair of cattle up to 100 days after treatment. Magpies are poisoned when they eat the hair, and eagles when they scavenge the magpies. Medicated feed at livestock feed yards is another high exposure situation. Sparrows, starlings, and other birds pick up the feed and subsequently are scavenged by hawks and eagles.

Forestry Insecticides

Forestry uses of toxic insecticides deserve special consideration because the terrain and method of application result in kills being difficult or impossible to detect. In a forestry situation, critical wildlife habitat is sprayed directly, and a large number of individuals of many species are exposed to the chemical. In Canada, the forestry insecticides phosphamidon and fenitrothion were canceled after impacts on birds were judged unacceptable. Although fenitrothion is not as acutely toxic as a number of other anticholinesterase insecticides used in agriculture, its use in forestry led to severe and widespread inhibition of brain acetylcholinesterase in a number of songbird species as well as some reports of kills.

See *Bird Impacts*, pages 101–103.

REFERENCES

1. ASTM Standard Guide for Fish and Wildlife Incident Monitoring and Reporting. In *Standard E, 1849–96*; American Society for Testing and Materials: Philadelphia, 1997.
2. Greig-Smith, P.W. Understanding the Impact of Pesticides on Wild Birds by Monitoring Incidents of Poisoning. In *Wildlife Toxicology and Population Modeling*; Kendall, R. J., Lacher, T.E., Eds.; SETAC Special Publication Series, CRC Press, Inc.: Boca Raton, FL, 1994; 301–319.
3. Mineau, P.; Baril, A.; Collins, B.T.; Duffe, J.; Joermann, G.; Luttik, R. Reference values for comparing the acute toxicity of pesticides to birds. Rev. Environ. Contamination Toxicol. **2001**, *170*, 13–74.
4. Mineau, P.; Fletcher, M.R.; Glazer, L.C.; Thomas, N.J.; Brassard, C.; Wilson, L.K.; Elliott, J.E.; Lyon, L.A.; Henny, C.J.; Bollinger, T.; Porter, S.L. Poisoning of raptors with organophosphorous pesticides with emphasis on Canada, U.S. and U.K. J. Raptor Res. **1999**, *33* (1), 1–37.
5. Mineau, P. *The Hazard of Carbofuran to Birds and Other Vertebrate Wildlife*; Canadian Wildlife Service, Technical Report Series No. 177, Environment Canada: Ottawa, 1993; 1–96.
6. Elliott, J.E.; Wilson, L.K.; Langelier, K.M.; Mineau, P.; Sinclair, P. Secondary poisoning of birds of prey by the organophosphorus insecticide, phorate. Ecotoxicology **1996**, *5*, 1–13.
7. Frank, R.; Mineau, P.; Braun, H.E.; Barker, I.K.; Kennedy, S.W.; Trudeau, S. Deaths of Canada geese following spraying of turf with diazinon. Bull. Environ. Contam. Toxicol. **1991**, *46*, 852–858.
8. Hooper, M.J.; Mineau, P.; Zaccagnini, M.E.; Winegrad, G.W.; Woodbridge, B. Monocrotophos and the Swainson's hawk. Pestic. Outlook **1999**, *10* (3), 97–102.
9. Stone, W.B.; Okoniewski, J.C.; Stedelin, J.R. Poisoning of wildlife with anticoagulant rodenticides in New York. J. Wildl. Dis. **1999**, *35* (2), 187–193.
10. Hunt, K.A.; Bird, D.M.; Mineau, P.; Shutt, L. Secondary poisoning hazard of fenthion to American kestrels. Arch. Environ. Contam. Toxicol. **1991**, *21*, 84–90.

Index

(Continued from inside front cover)